About Island Press

Island Press is the only nonprofit organization in the United States whose principal purpose is the publication of books on environmental issues and natural resource management. We provide solutions-oriented information to professionals, public officials, business and community leaders, and concerned citizens who are shaping responses to environmental problems.

In 2002, Island Press celebrates its eighteenth anniversary as the leading provider of timely and practical books that take a multidisciplinary approach to critical environmental concerns. Our growing list of titles reflects our commitment to bringing the best of an expanding body of literature to the environmental community throughout North America and the world.

Support for Island Press is provided by The Bullitt Foundation, The Mary Flagler Cary Charitable Trust, The Nathan Cummings Foundation, Geraldine R. Dodge Foundation, Doris Duke Charitable Foundation, The Charles Engelhard Foundation, The Ford Foundation, The George Gund Foundation, The Vira I. Heinz Endowment, The William and Flora Hewlett Foundation, W. Alton Jones Foundation, The John D. and Catherine T. MacArthur Foundation, The Andrew W. Mellon Foundation, The Charles Stewart Mott Foundation, The Curtis and Edith Munson Foundation, National Fish and Wildlife Foundation, The New-Land Foundation, Oak Foundation, The Overbrook Foundation, The David and Lucile Packard Foundation, The Pew Charitable Trusts, Rockefeller Brothers Fund, The Winslow Foundation, and other generous donors.

Predicting Species Occurrences

We dedicate this book to two individuals, friends and colleagues, who, although they worked in very different disciplines, contributed greatly to our ability to model species occurrences.

Dr. John Estes, geographer, visionary mapper of the planet's surface, and builder of bridges among disciplines,
and
Dr. Jared Verner, wildlife biologist, who pioneered the use of wildlife habitat models in land use planning.

Predicting Species Occurrences

Issues of
Accuracy
and Scale

Edited by

J. Michael Scott
Patricia J. Heglund
Michael L. Morrison
Jonathan B. Haufler
Martin G. Raphael
William A. Wall
Fred B. Samson

ISLAND PRESS

Washington • Covelo • London

Library of Congress Cataloging-in-Publication Data

Predicting species occurrences : issues of accuracy and scale edited by J. Michael Scott ... [et al.].
        p. cm.
Includes bibliographical references (p.)
    ISBN 1-55963-787-0 (cloth : alk. paper)
  1. Zoogeography. 2. Animal populations. 3. Habitat (Ecology) 4. Spatial ecology. I. Scott, J. Michael.
    QL101 .P64 2001
    578'.09—dc21

British Cataloguing-in-Publication Data available.

Printed on recycled, acid-free paper

Manufactured in the United States of America
10 9 8 7 6 5 4 3 2 1

Contents

PART 3—Modeling Tools and Accuracy Assessment

PART 4—Predicting Species Presence and Abundance

Foreword

Peter H. Raven

One of the central problems in ecology is how pattern and scale influence the distribution and abundance of organisms that we see. We seek to understand and predict the ecological causes and consequences of global climate change and of human-caused changes in the environment. There exists a great deal of literature, both basic and applied, that attempts to explain the patterns that we observe in nature and yet we continue to struggle with this important problem. Part of the difficulty lies in the fact that there is no single fundamental scale at which ecological phenomena should be studied (Levin 1992). Ecosystems and their associated populations vary over a range of spatial, temporal, and organizational scales, and the mechanisms driving the patterns we see may be operating at very different levels. Every ecosystem exhibits variability and patchiness on a variety of levels, and each interacts with other systems to promote or inhibit the phenomena of persistence or coexistence among species (Chesson 1986; Levin 1992). To address questions regarding population phenomena, we need to find ways to quantify these patterns in time and space, to understand how pattern may change with scale, and to understand the causes and consequences of the patterns we see (Wiens 1989b).

The use of descriptive statistics is a starting point for understanding pattern, but correlations are not a substitute for mechanistic understanding (Lehman 1986). We now recognize there is no single correct scale or level at which to describe a system, and also that not all scales serve equally well for describing patterns (Levin 1992). In the absence of a complete inventory of the distribution and abundance of the world's species (Raven and Wilson 1992), a problem that will not soon be resolved, models are one way to gain and apply information about the patterns of the distribution of biodiversity in space and time. But we must have at hand a variety of models with different levels of complexity that will allow us to determine the appropriate levels of aggregation and simplification for each question (Levin 1992). As practitioners of the science of conservation biology, we recognize that timely policy and management decisions must be made in the face of uncertainty. To accomplish this, we commonly now rely on remote sensing and spatial statistics to quantify patterns at broad scales, but we still need a great deal of experimental evidence to provide information about the mechanisms driving the patterns. Finally, we need to provide measures of the accuracy of our predictions so that we can make these decisions with some level of confidence.

The thrust of the conference, Predicting Species Occurrences: Issues of Scale and Accuracy, from which this book was developed, was to discuss the status of our ability to model species distributions with regard to concerns about pattern and scale. This volume is a compendium of research that defines the current state of our knowledge of modeling species occurrences; it reiterates the theoretical and basic ecological needs remaining as we try to understand and predict species occurrences and distribution. This book is a continuation of the major effort conducted in 1984 that culminated in *Wildlife 2000: Modeling Habitat Relationships of Terrestrial Vertebrates* (Verner et al. 1986b)—an assessment of the state of our knowledge fifteen years ago. This current volume evaluates the progress we have made since the publication of *Wildlife 2000*. In it, the authors reiterate the need for continued support of basic and applied ecological studies of pattern and scale. Beyond that, the authors emphasize the need for measures of model accuracy when policy and management decisions are involved.

I hope that this book will help to build stronger bridges between those scientists who conduct basic ecological studies in the field and their colleagues who use field data to create increasingly accurate and useful model predictions.

Preface

Michael F. Goodchild

In late 1989, I met Brad Parks at an EPA (Environmental Protection Agency)-sponsored workshop near Detroit, Michigan, and we persuaded each other that much could be gained by encouraging greater interaction between the geographic information system (GIS) and environmental modeling communities. GIS technologies were fast gaining recognition among scientists and policy makers as tools for the analysis of problems in a spatial context. Environmental models were also proliferating and winning acceptance as approaches for numerical analyses and prediction. If they could be integrated, these applications offered great promise in bridging some of the gaps between scientific research and policy development. Out of that conversation grew a series of International Conferences/Workshops on Integrating GIS and Environmental Modeling, in Boulder (1991), Breckenridge (1993), Santa Fe (1996), and Banff (2000). (Proceedings are available, respectively, as Goodchild et al. (1993), Goodchild et al. (1996), http://www.ncgia.ucsb.edu/conf/SANTA_FE_CD-ROM/main.html, and http://www.colorado.edu/research/cires/banff/.) It was extremely satisfying to see our vision broadened and enriched at the Predicting Species Occurrences: Issues of Scale and Accuracy conference in Snowbird, Utah, in October 1999.

Modern science and problem solving require extensive collaboration between specialists, and the high costs of data acquisition and model development argue strongly for effective mechanisms for sharing information. Effective communication is particularly difficult across the boundaries of disciplines that may use the same terms in different ways. One important component of community building is the construction of a common language (see particularly the arguments presented by Michael Morrison and Linnea Hall in Chapter 2). Internet technologies have helped enormously to foster a culture of collaboration and sharing among scientists. At the same time, these developments have pointed to some critical weaknesses in our ability to deal effectively with pervasive problems among GIS and environmental modeling processes, including temporal variability, scale, representation, accuracy, visualization, and spatial context (for a comprehensive discussion, see the research themes of the University Consortium for Geographic Information Science, UCGIS [1996] and http://www.ucgis.org). These themes form the core of this new book.

One weakness in our ability to predict species occurrences is rooted in the development of accurate models and is a major topic of this volume. The problem with accuracy stems from the fact that the real geographic world is infinitely complex such that it would require an infinite and therefore impossible amount of information to characterize it completely. Any description of any aspect of the world must therefore be an approximation, generalization, or abstraction that almost certainly omits much of the detailed information that organisms sense and respond to. This missing information constitutes uncertainty, in the sense that a user of a database has a degree of uncertainty about the true conditions existing in the real world (for a review of uncertainty in spatial ecology see Hunsaker et al. 2001). It also follows that an infinite number of potential approximations exist or, in other words, that an infinite number of ways can be found to represent any aspect of the world in digital form. Some of these will do much more damage to our ability to model and predict than others, because their associated uncertainty translates or propagates into uncertainty over model predictions. Finding the metrics that allow us to evaluate a representation's performance in environmental modeling and prediction is a vitally important task. The metric long favored by cartographers and producers of geographic databases—the scale or representative fraction of a paper map—is insufficient. There are still an infinite number of possible digital representations at any given scale depending on how the objects represented in the database are specified. Advances in computing power over the past few decades have made it possible to handle much larger volumes of data. We have the ability to model more accurately, as Dean Stauffer argues in Chapter 3, and as many other chapters in this book demonstrate. However, given that variability in data stemming from unknown sources ultimately limits our ability to reduce the uncertainty in our predictions, authors in this volume demonstrate the importance of providing a measure of accuracy for all our future modeling and mapping efforts.

Geographic information systems have inherited representations of the world developed by cartographers. Today, we have computers capable of capturing not only how the world *looks*, in the form of largely static geographic data, but also how it *works*, in the form of codes that implement environmental process models. We must recognize that the data populating our GIS databases today were largely created to appear on maps, and many of them were obtained by digitizing or scanning those same maps. They use the scales and accuracies that were devised long ago to serve the needs of mapping. Vegetation-cover mapping practice, which emphasizes the delineation of homogenous areas of cover class separated by sharp boundaries, made sense to users of maps, but it may be totally inappropriate for today's GIS uses, including dynamic modeling of ecological processes. The concept of a patch as an ecologically meaningful unit may appear superficially similar to the concept of an area of uniform cover class, but at a more fundamental level they may have little relationship. Similarly, the pixels of a remotely sensed image may appear to be convenient units for modeling ecological process, but they originated in the geometry of an instrument on a satellite and were in no way informed by ecological reality. So bringing GIS and environmental modeling together offered promise at several levels: improved support for environmental modeling through better tools for sharing and managing data and for visualizing and disseminating results, and also improved representations in GIS that were more appropriate for a new generation of requirements.

Carol Johnston (1993) presented an appealing conceptual structure for spatially explicit modeling of ecological populations that was firmly grounded in theory. She postulated that K strategists base their survival and success on local resources and are relatively easy to model in GIS, provided local resources are represented in suitable ways. Because organisms respond to resources over a local neighborhood, in ef-

fect integrating through a spatial convolution function, it is essential that the representation have an appropriate spatial resolution that is no coarser than the convolution, if the convolution is precomputed, and much finer if the convolution must be computed on-the-fly by the model. Vector representations, which lack explicit spatial resolution, are likely to be unsuitable in these models.

On the other hand, r strategists present much more difficult problems. Here, success is determined by the intrinsic rate of natural increase of the population and hence on spatial interaction between organisms. However, spatial interaction presented enormous problems to cartographers, and very little progress has been made in the development of effective representation of interactions in GIS databases. Multiagent models attempt to model interaction at the individual level and clearly do not scale effectively to large populations. Island models deal with interaction by assuming it to be perfect within the island and absent otherwise, and it is tempting to think that landscapes can be similarly fragmented into isolated patches. But simple rules for the identification of patches (e.g., every watershed is a patch, every area of homogeneous land cover class is a patch, every Landsat pixel is a patch) are unlikely to have much grounding in ecological reality. Jason Dunham, Bruce Rieman, and James Peterson explore this theme in relation to modeling fish occurrences in Chapter 26; in Chapter 59, David Theobald and Thompson Hobbs present an alternative approach that explicitly addresses gradients; and in Chapter 63 Thomas Sisk, Barry Noon, and Haydee Hampton create a patch model that explicitly recognizes within-patch heterogeneity.

The growth of the World Wide Web and associated data archives, digital libraries, and clearinghouses have created an environment conducive to data sharing and access—environmental modelers are now blessed with an abundant supply of geographic data. However, this series of technological developments has in turn raised awareness of the weaknesses of much of the data supply. Too often, the rules used to create data are not known, or are too subjective, or data have not been effectively validated, or quality is too uncertain and variable. Too often, the high cost of accurate data forces us to accept low-resolution data without adequate ways of assessing what has been lost (see, for example, the analysis presented by Kathryn Thomas and her coauthors in Chapter 10). Using data that are too coarse can affect not only the goodness of fit of the model, but also its structure, as Catherine Johnson and her coauthors show in Chapter 12. In the best of all possible worlds, questions of structure would be resolved by resorting to theory, but in Chapter 15 Claudine Tobalske shows how alternative theories can often lead in conflicting directions.

New acquisition systems are solving some of these problems—for example, the data becoming available from the Shuttle Radar Topography Mission of 2000 clearly have some advantages over previous digital elevation data, and new satellite sensors offer improved spatial, temporal, and spectral resolution. There has also been substantial improvement in software support for modeling. GISs aimed specifically at dynamic modeling have appeared, such as the University of Utrecht's PCRaster (http://www.geog.uu.nl/pcraster/), and powerful scripting languages have been defined (Van Deursen 1995). Many of the chapters in this volume illustrate the benefits of these changes, and the software environments described in the chapters of Part 4, "Predicting Species Presence and Abundance," are very different from those available ten years ago.

The world has invested vast resources over the centuries in creating representations of geographic knowledge that are in many cases highly inappropriate for ecological modeling. We need a new generation of representations that are specifically designed to make ecological sense and to do minimal damage to the accuracy of ecological models that use them for prediction. Specifically representations could

- Maintain explicit spatial resolution (rasters are preferable in this respect to vectors)
- Support multiple representations (e.g., preserve both raw data and more generalized interpretations based on them) to implement reasoning and modeling across spatial scales, a theme explored by Brian Maurer in Chapter 9 and other chapters in Part 2, "Temporal and Spatial Scales"
- Interrupt continuity at ecologically significant barriers rather than in areas of steep gradient (e.g., bounding areas of approximately homogeneous cover)
- Support the representation of interaction (e.g., through vector fields, metamaps [Takeyama and Couclelis 1997], or attributed relationships)
- Support the measurement and modeling of uncertainty, and its propagation through models to confidence limits on outputs
- Support the observation and modeling of change through representations that are fully spatiotemporal (e.g., define objects as regions in three-dimensional space-time, or as space-time trajectories)

All of these examples are part of a much longer-term transition as GIS evolves away from its cartographic roots and moves to center stage in support of environmental modeling.

They are some of the important changes that will be needed in the future as tools evolve to support the needs of better environmental modeling and species prediction. The chapters in this volume demonstrate the state of our knowledge regarding environmental modeling and the progress being made in linking such models with GIS technologies. They point the way for future investigations and identify problems that will need resolution before more progress can be made. Together they demonstrate our progress toward further developing and integrating sound reasoning and prediction across multiple spatial scales, developing representations that make ecological sense, measuring and reporting model accuracy, incorporating measures of uncertainty, and supporting the observation and modeling of change over time and at multiple spatial scales. We have come a long way in the past decade, but there is much opportunity for further progress.

Introduction

*J. Michael Scott, Patricia J. Heglund,
Michael L. Morrison, Jonathan B. Haufler,
Martin G. Raphael, William A. Wall, and
Fred B. Samson*

The complexity and inherent variation in species and their responses to physical and biological factors at multiple scales as well as the dynamic nature of environments and species ranges (Botkin 1990) make predicting species occurrence, abundance, or viability with high levels of precision and accuracy difficult. Predictive models are by necessity, low-dimensional abstractions of *n*-dimensional forces acting on individuals. Our attempts to predict species occurrences have been hampered by mismatches between the spatial and temporal scales at which we make measurements and the scale at which ecological phenomena influence patterns of species occurrences and an incomplete knowledge of a species' life requirements. In many respects we are still in the age of exploration and discovery that was enjoyed by nineteenth-century naturalists. Consider that one hundred of the breeding bird species of North America are each the subjects of fewer than five publications in the refereed literature (J. T. Ratti and J. M. Scott personal communication). Of those species whose ecology and biology are well documented, there is a clear bias toward game species (Boone and Krohn, Introduction to Part 3 this volume; Karl et al. 1999). This lack of information on natural history of species severely limits our ability to confidently predict species occurrence.

Factors affecting species occurrences, abundance, and population viability in time and space occur along gradients but are most often modeled as discrete variables (Lawton et al. 1994). Conditions in one area of a species range (e.g., massive die-off on wintering grounds, weather conditions during migration) may determine a species occurrence, abundance, or reproductive rates (Beard et al. 1999) at another (Wiens and Rotenberry 1981b). Our ability to predict with confidence a species status in an area is further complicated by their often-nonlinear responses to habitat (Heglund, Chapter 1; Austin, Chapter 5). The size and ecological context of habitat patches influence not only whether a species occurs there but possibly many of its demographic values as well. Additionally, conditions occurring at several spatial scales can influence species occurrence and associated ecological processes (Wiens 1989a). All of these factors as well as previous events (e.g., pollution, drought) influence a species' occurrence. Areas occupied one year may be unoccupied the next with seemingly no detectable differences in the habitat conditions at the site. Nature is indeed more complex than we can envision.

Modeling efforts are often conducted in response to the needs of society. The resultant science-policy interface is a world in which many of us are uncomfortable. Nonetheless, if we are to meet the needs of society we must be able to effectively communicate the results of our research to local, state, and national managers and policy makers. These individuals are required to make day-to-day decisions often under severe budget constraints and deadlines and with incomplete and highly uncertain information. They require answers for questions that range from a fine-scale "What's the impact of different timber harvest regimes on red-cockaded woodpeckers (*Picoides borealis*) in a 40-hectare forest?" to the coarse-scale "Which areas of a sagebrush steppe are most important to the maintenance of regional biodiversity?" Predicting species responses to such scenarios requires researchers to think at a variety of spatial and temporal scales and at different levels of biological organization (demes, populations, metapopulation, subspecies, and species). Models are often developed for single species for specific areas tens of hectares in size, less often for the range of a species, and almost always independently of ecological process or context.

All too often there is a disconnect between researchers and managers. It is not unusual for managers to attempt to apply the results of a model developed to predict a species response across millions of hectares to a 10-hectare site without fully recognizing the limits of the model. To avoid such misuses all parties must become better communicators. Researchers must conduct research at scales that are appropriate to the question being asked and the management challenge at hand. Products must be presented not only in the peer-reviewed literature, where they are subject to the scrutiny of colleagues, but also in formats that are more meaningful to managers (e.g., workshops, one- or two-page research information brochures, interviews, and video tapes). Ideally scientists should work with managers to interpret management implications of the research and clearly state potential applications along with any inherent limitations of the models.

Demographic units above the level of individuals and demes rarely can be found within a single game management unit, nature reserve, or county, while few species spend their lives within a single country. Managers are increasingly responsible for more species of concern and for a wider variety of taxa. Whereas their attention was formerly focused most often on charismatic megafauna or on those species hunted or fished for sport or commercial gain, managers must now address the conservation of a broad representation of taxonomic entities including vascular and nonvascular plants, fungi, invertebrates, and lesser-known vertebrates. The proposed regulations under the National Forest Management Act and the Criteria and Indicators emerging from the Montreal Process meeting held in 1995 (Anonymous, undated, http://www.mpci.org/whatis/criteria_e.html) are two examples of this broader mandate to manage and conserve biological diversity. The Montreal Process identified seven criteria to provide member countries with a common definition of what characterizes sustainable management of temperate and boreal forests. The seven criteria are (1) conservation of biological diversity; (2) maintenance of productive capacity of forest ecosystems; (3) maintenance of forest ecosystem health; (4) conservation and maintenance of soil and water resources; (5) maintenance of forest contribution to global carbon cycles; (6) maintenance and enhancement of long-term multiple socioeconomic benefits to meet the needs of society; and (7) legal, institutional, and economic framework for forest conservation and sustainable management. There are sixty-seven associated indicators for these seven criteria, many of which will re-

quire modeling species occurrences. These comprehensive mandates have brought into sharp focus the debate over the proper balance between species-based (fine filter) and ecosystem-based (coarse filter) approaches to conservation. And in either case, development of reliable models for predicting system responses to environmental change is essential.

Increasingly, management efforts are conducted in partnership across traditional political boundaries in the United States (Cox et al. 1994, Scott et al. 1993, Stein et al. 2000) and elsewhere (Brunckhorst 2000). Management and policy decisions require predictions of species occurrences across the range of species. Our ability to do this requires that we work at multiple levels of physical and temporal resolutions, incorporating ecology, wildlife biology, and biogeography. There is increasing recognition that environmental planning is best done when the full range of a species occurrence is considered (Heglund, Chapter 1; Heglund et al. 1994). Yet most efforts are site specific. Additionally, we most frequently model presence/absence and abundance despite indications that these are likely poor indicators of habitat quality (Van Horne 1983). But, as evidenced by papers presented at this symposium, we are making progress. We now find examples of better approaches to modeling species occurrences and to developing demographic models that go beyond presence/absence.

The international symposium Wildlife 2000: Modeling Habitat Relationships of Terrestrial Vertebrates (Verner et al. 1986b) set the stage in 1984 for periodic review of our progress on modeling species occurrences. With the increased awareness of the shortfalls of current modeling efforts we invited scientists and managers from government, private businesses, and academia to a symposium to assess what has been accomplished and what hurdles to advancement remain in our attempts to understand and predict the distribution and abundance of species.

Fifteen years later, 325 individuals from fourteen countries assembled at Snowbird, Utah, October 18–22, 1999, to discuss advances in our efforts to model species distributions since Wildlife 2000: Modeling Habitat Relationships of Terrestrial Vertebrates (Verner et al. 1986b). These individuals came from a variety of disciplines (geography, ecology, botany, etc.)

and age classes. Approximately 20 percent of the participants were students.

In seeking participants for the symposium Predicting Species Occurrences: Issues of Scale and Accuracy, we made a special effort to ensure participation of individuals with different approaches to modeling. It was our hope that doing so would stimulate discussion, sharing of ideas, and future collaborations. Our goal in hosting this symposium was to examine the theoretical basis for model development, assess the degree to which assumptions were met, discuss the current applications of modeling techniques, and focus on the future of modeling for natural resource conservation and management. We paid particular attention to the various ecological and behavioral characteristics of species related to their distribution and abundance, problems related to scale (extent and grain) of various phenomena (e.g., time, space, habitat), problems with current methods of model validation, and how all these affect model performance and reliability. Our objectives were to review the theory and practice of predicting species distribution, abundance, and viability; to develop our understanding of management and policy frameworks within which models are applied; to identify the strengths and weaknesses in the application of various modeling techniques; and to examine the potential for linking models to one another for large-area management programs. The following chapters are representative of the variety of topics covered at Snowbird. You will recognize that there are areas of emphasis that are underrepresented relative to their importance in predicting species occurrences. These inequalities should provide a direction for future efforts. Because of the large overlap among authors in references cited, we have merged the literature cited section of each chapter into a single Literature Cited section at the back of the book. In addition to saving space, it provides a solid introduction to the literature on modeling species occurrences. You will see much innovation in the work presented herein, and hopefully you will be moved to follow one or more of these pathways to an ever-increasing effort to conserve our natural resources for future generations.

Past efforts to model species occurrences have been biased toward vertebrates—especially birds (Verner et al. 1986b) and traditional game species (Dettmers et al., Chapter 54; Karl et al. 1999—so we sought out those working on other taxa to build diversity into this conference. Despite this effort, vertebrates, especially birds, were the most frequently modeled group, just as they were in Wildlife 2000, and so they are the most frequently discussed group in this book. Happily, however, you will still find numerous discussions of fungi, plants, butterflies, small mammals, and amphibians in the pages that follow.

A recurring theme among these papers is the difficulty of assessing the accuracy of species occurrence models. Sample sizes of several hundred independent observations may be needed, (Karl et al., Chapter 51) a difficult task when dealing with less-common species. It is a task made even more daunting when we realize that most species are uncommon (Preston 1948, 1962a,b). Litigation over land use practices requires that we use defensible methods of data acquisition and measures of the reliability of our predictions of species presence, absence, abundance, and viability. Thus, considerations of accuracy and scale are important if we are to understand the relationships between species and their environment.

It is these issues (scale and accuracy) that we asked the participants of the Snowbird symposium to address. One of the most compelling ideas resulting from this effort was the determination that no silver bullet or single model will suffice. Different species and different ecological settings require different approaches.

The chapters in this book were selected from the eighty-two oral and 125 poster presentations given at the conference. All manuscripts were sent out for review by two or more referees and fifty-nine of those manuscripts may be found in the pages of this book. The manuscripts are grouped into six parts. Within each part, species are placed in phylogenetic order. Each part is preceded by an introduction in which the contents of the chapters are discussed, their limitations and strengths presented, and future directions to improve the accuracy and precision of modeling the patterns and processes associated with species occurrences are noted.

In Part 1, "Conceptual Framework," the history of the theory and practice of modeling is reviewed and a standard terminology is presented. This part is preceded by a detailed discussion of the ecological context in which predictions of species occurrences are made and our struggles to understand their influence. In Part 2, "Temporal and Spatial Scales," one of the most difficult topics dealt with at the symposium is discussed in terms of its influence on patterns and processes of species distributions. A recurring theme throughout this part is "What are the consequences of asking questions at the wrong scale?" There are detailed discussions of state-of-the-art modeling tools and descriptions of methods for assessing model accuracy in Part 3, "Modeling Tools and Accuracy Assessment." Two important discussions in this part are by Alan Fielding, who considers what the appropriate characteristics of an accuracy measure might be, and the disquieting findings by Jason Karl and his colleagues that statistically defensible assessments of accuracy require a very large number of independent observations (Karl et al., Chapter 51). Efforts as diverse as Generalized Linear Modeling, Generalized Additive Modeling, General Algorithm for Rule-set Prediction and Neural Network modeling used to predict presence/absence and abundance of species are discussed in Part 4, "Predicting Species Presence and Abundance." In Part 5, "Predicting Species: Populations and Productivity," several examples of how spatially explicit data on demographics can provide important information for managers are presented. Finally, in the concluding part, "Future Directions," John A. Wiens provides an in-depth review and synthesis of the symposium. Wiens provides guidance for future directions and cautions regarding misuse of models.

You will find two underrepresented topics in this volume. They are vegetation classification systems and the use of spatial statistics in modeling. The selection of an appropriate classification system is of critical importance to predicting species occurrence. Most habitat models depend on a map of biologically meaningful areas delineated across a landscape as the basis for making predictions of species occurrence, abundance, or fitness. Failure to address the effects of specific classification systems and the quality of their associated habitat attribute data can result in major errors in model performance. Whereas several papers in the symposium did address the question of the resolution of the mapped units used in model predictions, the topics of classification system effects and habitat attribute quality and accuracy were not adequately addressed. This, then, represents an area where significantly more work may be needed. Further, failure to adequately address these factors will result in the accuracy of the relationships described in many models being questioned, when these relationships may in fact be valid but are being masked by our inability to describe the actual location of quality habitat within the landscape.

Despite the usefulness of spatial statistics for modeling species distributions, there was little discussion of their use at Snowbird or in the pages of this book. We think that their widespread use by biologists awaits the development of user-friendly programs in widely available statistical packages (e.g., SAS, SPSS). Just as this volume used the results of Wildlife 2000 as its foundation, it is our hope that it will serve as a foundation for the next generation of wildlife habitat relationship models.

Acknowledgments

A symposium the size and complexity of that held at Snowbird, Utah, and this book are the result of the efforts of many individuals. Kathy Merk took charge of the tracking of correspondence registration of participants and supervision of the correspondence with authors. At Snowbird, she and her husband Larry Merk were responsible for supervision of onsite registration preparation of meeting brochures and travel arrangements for many of the participants. Kathy was a key participant in making this book possible from the first meeting of the steering committee to the mailing of the finished manuscripts to the publisher. Her participation and leadership have been critical. Robbyn Abbitt, Jason Karl, Nancy Wright, and Leona Svancara were responsible for all audiovisual support at Snowbird and for shuttling participants to and from the airport in Salt Lake City. We thank Sandy Andelman, Terri Donovan, Thomas Edwards, Michael Goodchild, Edward O. Garton, Jon Haufler, Carolyn Hunsaker, Francis James, Brian Maurer, Michael Morrison, and Stan Temple for serving as session chairs. In addition to their duties as session chairs, Michael Goodchild,

Edward O. Garton, Francis James, and Brain Maurer provided detailed comments on manuscripts in their sessions that significantly improved the quality of the manuscripts. We also thank the many student assistants at Snowbird who helped the presentations flow smoothly. Financial support for the symposium was provided by U.S. Geological Survey, U.S. Fish and Wildlife Service, U.S. Forest Service, Boise Cascade Corporation, Potlatch Corporation, University of Idaho, Idaho Cooperative Research Unit, Bureau of Land Management, American Fisheries Society, The Wildlife Society, and the Idaho Chapter of the Wildlife Society.

We wish to thank Doris A. Baldry, Stefan A. Sommer, Charles R. Peterson, George W. LaBar, Leona K. Svancara, and the Idaho Natural Heritage Center for kindly donating use of their photographs for the cover.

The development of this book was no small task, and we thank all those who reviewed manuscripts and the authors for their responsiveness in returning draft revisions. Barbara Dean and Barbara Youngblood of Island Press walked us gently through the editorial procedural process, providing encouragement and guidance along the way. Judy Scheel labored over format and style issues. Andrea Reese took on the challenge of creating a unified literature cited section. She and Naomi Bargmann also corresponded with authors regarding missing references and spent long hours in the library checking citations against original publications. We greatly appreciate their efforts.

Introductory Essay:
Critical Issues for Improving Predictions

Michael A. Huston

Natural resource management and conservation depend on accurate information about the distribution and response dynamics of natural resources, which include populations of plant and animals that can change dramatically over time and space. Unfortunately, the complex interactions that typify ecological processes have been an impediment to understanding and predicting these shifting patterns. Theories about the effects of a specific environmental condition on some ecological property have been difficult to test because many different environmental conditions are correlated with any specific property. Such covariance also hinders the development of predictive statistical models because of the difficulty of distinguishing the "causal" predictor from other correlated factors that have no causal role. Nonlinear response dynamics can lead ecologists to opposite conclusions about the response of an ecological property to environmental conditions. Finally, even the statistical analyses used to quantify these patterns don't work very well. Statistical methods based on the standard assumptions of linear relationships, normally distributed "errors," and uncorrelated independent variables, can actually prevent the identification of the "functionally significant" relationships necessary to understand ecological processes and predict ecological patterns.

The three primary impediments to developing a coherent conceptual framework for ecological predictions are closely interrelated: (1) mismatches between the spatial and temporal dimensions of ecological measurement and the dimensions at which hypothesized processes operate, (2) misunderstanding of ecological processes, and (3) use of inappropriate statistics to quantify ecological patterns and processes. Each of these problems stems from the failure of ecologists to develop a coherent set of theories that apply at clearly defined spatial and temporal dimensions. Although it is beyond the scope of any single chapter to develop or present such a theoretical framework, it may be useful to look at examples of how new approaches in each of these areas can lead to more powerful insights than those derived from traditional ideas.

After defining the spatial and temporal requirements for ecological sampling, this introductory essay will discuss three examples of theories that are relevant to the distributions of organisms and the structure of ecological communities. Each theory describes how certain processes are expected to produce particular ecological patterns. Each theory can be identified as being relevant to patterns that appear at particular resolutions of measurement, and each theory has implications for the statistical analysis of ecological patterns as well as for the management and conservation of natural resources, including endangered species and biodiversity.

Spatial and Temporal Issues in Sampling

Biological processes occur over time spans ranging from microseconds to millennia and over distances from nanometers to continents. However, most ecological processes occur over a more limited range of spatial areas and temporal durations that depend on the specific processes and organisms involved. As a consequence, understanding the dynamics and regulation of a particular process (or phenomenon such as species distribution) requires measuring the process rate, the results of the process, and the factors that influence the process at appropriate spatial and temporal scales.

The concept of "scale" has long been recognized as an important issue in ecology. Nonetheless, much of the current confusion about the relationships between patterns and processes in ecology (e.g., local versus regional control of species diversity) stems from a failure to collect, analyze, and interpret data at the appropriate scale for the process of interest (cf. Cornell and Lawton 1992; Huston 1999). Although the term "scale" is often used loosely as a general concept related to the size of something, the technical use of the term in ecology has become muddled as ecologists have attempted to address phenomena that occur over large areas (e.g., "landscapes") and at different "levels of organization" (e.g., individual organisms versus populations, cf. Morrison and Hall, Chapter 2).

Landscape ecology has extended the definition of scale to include the total size of the area being considered independent of the resolution at which an area (or time period) is measured or represented (Turner et al. 1989a), thus creating a definition that has two independent and unrelated elements ("grain," defined as resolution, and "extent," defined as total size). The confusion caused by this expanded definition is magnified by the fact that geographer/cartographers have always referred to scale specifically in terms of resolution, so that a map at the scale of 1 foot to 24,000 feet is a larger-scale map (higher resolution) than a map at the scale of 1 foot to 100,000 feet. Ecologists, by incorporating the size of the total area (or "extent") into the concept of "scale," have reversed this definition by referring to the geographer's small-scale map as "large scale" because it covers a larger total area than the high-resolution map.

Historically, the limitations of storage media created an inverse relationship between spatial resolution and the size of the area that could be represented on a single unit of the storage medium (i.e., the piece of paper on which a map is printed). However, with the development of electronic digital storage devices, this inverse correlation has been greatly relaxed to the extent that the resolution and the size of the area (or time period) represented are essentially independent (e.g., Franklin 1995; Scott et al. 1993). Combining two independent properties (resolution and total size) in the single word "scale" has created a term that has no precise definition. Because the common ecological usage of "scale" has no clear meaning, it cannot be used comparatively (e.g., "large scale," "small scale"). It is best treated as a generic term for all of the issues related to the size (area or duration) of samples, phenomena, or processes (Peterson and Parker 1998a; Csillag et al. 2000). Two of these issues are obviously sample resolution (the cartographic definition of scale), and the total size of the set from which samples

are taken (the statistical "population" of space, time, or organisms).

Sample resolution includes two distinct concepts. One is the size of a single sample, which can be an area of space, a period of time, or a group of organisms or objects, and is the minimum unit between which differences can be resolved. The size of the unit sample used in species occurrence modeling varies over several orders of magnitude from field plots or quadrats to pixels of satellite imagery used to characterize conditions over large areas. The size of the unit sample influences the value of parameters that are associated with each unit area. Specifically, information on heterogeneity within the unit sample is lost when a single value is used to characterize a specific property (e.g., vegetation type) over a large area. This inevitably results in the loss of information on rare conditions (Trani, Chapter 11; Henebry and Merchant, Chapter 23) and a consequent degradation of predictive ability (Boone 1997; Thomas et al., Chapter 10; Debinski et al., Chapter 44; Gonzalez-Rebeles et al., Chapter 57; Hunsaker et al., Chapter 61; Young and Hutto, Chapter 8). In general, higher spatial resolution leads to better predictive models, assuming data quality and size of the sample region are not compromised.

The second concept related to sample resolution is measurement sensitivity, or the ability to distinguish differences between the measured values of a property of two different sampling units. Measurement sensitivity can be a function of the detection limits of a chemical analyzer or the precision of a field measurement. Variation in measurement sensitivity produces data resolutions that range from continuous quantitative values, through ordered rankings, to presence/absence information, each of which requires different analytical methods (Guisan, Chapter 25).

In addition to the spatial and temporal resolution, the size of the overall area from which samples are collected (the sample region or sample period) is also important. The sample region (also called "extent," see Morrison and Hall, Chapter 2) is a size issue that is independent of the size of the sampling unit (i.e., the resolution, or "grain"). As with resolution, the size of the total area that should be sampled depends on the processes of interest and the questions being addressed.

The size of the sample region (or sample period) affects two distinct properties that are critical for predicting ecological patterns such as species occurrence. The first is the total amount of whatever is contained in the area. Obviously, a larger area will have more of everything (e.g., forage, prey animals, nesting sites) than a small area with identical properties. This property is most important for issues related to population size and behavior. Understanding the factors that determine how a wolf pack uses its range requires a study area at least as large as the range and fully containing at least one complete range. However, there is no constant size that will always meet this objective, since the quality of "habitat" varies from one location to another and from one year or decade to another. In productive environments, a much smaller range size is needed to support a predator, or a viable population of predators, than in an unproductive environment, although the complexities added by animal territoriality may complicate the expected pattern (see Smallwood, Chapter 6).

The second property of size is the environmental variability contained within a sample region (or encountered over time). Again, all things being equal (which they rarely are), large areas will generally include a greater range of environmental conditions (i.e., greater heterogeneity or variability), than a smaller area. Of course, most environmental variability results from the interaction of geology, topography, and weather, with the obvious consequence that an area in a mountainous region will have a much greater range of conditions than an area of the same size in the plains.

The range of conditions sampled in a study (assuming an appropriate resolution and sample density) will determine the accuracy, precision, and generality of the resulting predictions (see Austin, Chapter 5). Studies that include only a narrow range of environmental conditions (regardless of the total size of the sample region or study area) are likely to have very low precision, because the unpredictable (random) effects of dispersal, disease, or other factors that are not correlated with environmental conditions will be much greater than any predictable responses to environmen-

tal conditions. The responses of an organism to the environment (i.e., whether it is a generalist or specialist) also affect the range of conditions available for prediction. The habitat use pattern (i.e., specialization or "niche width") of a species within the study area has a strong effect on the accuracy properties of the predictive model that can be developed (Fielding, Chapter 21; Johnson et al., Chapter 12; Garrison and Lupo, Chapter 30; Hepinstall et al., Chapter 53; Dettmers et al., Chapter 54).

Increasing the number of samples in a region with low variability will increase the accuracy of the prediction, but the precision will still be low (i.e., a low proportion of variance will be explained). On the other hand, studies that include extremely broad ranges of conditions, most of which are unsuitable for the organism or property of interest, will produce high accuracy and precision over the sample region but will be of little use for predicting variation in population density, or even where the species is most likely to occur, within the range of conditions it can survive (see Elith and Burgman, Chapter 24). Models of this type may be of little or no value outside the study area, where the range of conditions is different or narrower (e.g., in potentially suitable habitat) because they are not related to the processes that determine species abundance and occurrence.

A third property forms the linkage between sampling resolution and the size of the sample region: sample density, or the number of measurements at a specific resolution per unit area. Satellite images, photographs, phonographs, and maps have a sample density of 1.0, which is to say that the entire area (or time period) is sampled at the maximum resolution of the method. However, ecological samples rarely have a density of 1.0, because of the time and effort required to collect quantitative environmental data. Typically, only a few percent of a total area (e.g., 1 hectare) are actually sampled with plots of a given resolution (e.g., one hundred 1-square-meter plots is 1 percent of 1 hectare). Adequate sampling involves not only sampling at the appropriate resolution (time interval or plot size) for the process of interest, but also having a high enough density of samples to resolve the spatial or temporal patterns in sample values over the focal area (or time period) (Austin, Chapter 5; Karl et al., Chapter 51).

The accuracy of predictions of species distributions, or of any ecological phenomenon, depends on whether scale issues are handled properly. The critical issue is not the specific resolution or total size of a study, but whether the resolution and size are appropriate for the phenomena being studied and the hypotheses to be addressed. The single most important issue with regard to accuracy of prediction is whether the data are collected at a resolution and density appropriate to identify and characterize the processes that determine the pattern of interest (e.g., species distribution).

Understanding the relationship between a specific pattern and the processes that produce it is essential for accurate prediction (see Van Horne, Chapter 4; Smallwood, Chapter 6). Perhaps the largest impediment to accurate prediction is the fact that many patterns can be detected at resolutions far coarser than the resolution needed to understand the processes that produce the pattern, which is also the resolution needed to predict the pattern. Presence/absence of a species is a phenomenon that produces patterns that can be represented at very low spatial resolution. However, the processes that result in presence/absence are the population dynamics that influence population growth rates, dispersal and colonization, and ultimately extinction or survival. A species that is present in only a small portion of a large area would be counted as present in the large area. However, the average or median environmental conditions for the large area are likely to be very different than the conditions of the small subset of the area where the species occurs. Thus, predictions of species occurrence based on properties of the large area (i.e., low spatial resolution measurements) are unlikely to produce accurate predictions of where the species actually occurs or to lead to an understanding of the processes that lead to the low-resolution pattern.

For example, mismatches between the size of the area in which species diversity is measured and the size of the area over which ecological processes that influence diversity actually operate has led to an overestimation of the influence of "large-scale processes" on patterns of species diversity (Huston 1999). The relative importance of local processes (e.g., competition) versus regional processes (e.g., evolutionary history) in controlling local species diversity is an issue of both theoretical and practical interest (Ricklefs 1987; Ricklefs and Schluter 1993). The critical issue is whether (1) local diversity "saturates" at some level that is independent of the total number of species in the region around the locality (implying local diversity is controlled by local processes), or (2) local diversity continues to increase as regional diversity increases (implying local diversity is controlled by regional processes).

A recent study that compiled data on the species richness of different types of organisms on different continents over spatial areas ranging from "local" to "regional" in size found that local species richness was linearly correlated with regional species richness (Caley and Schluter 1997). This lack of evidence for local "saturation" of species richness supported the conclusion that local diversity is controlled primarily by regional processes and that local processes, such as competition, are relatively unimportant. However, a reevaluation of these results revealed a mismatch between the resolution of species richness data and the distance over which local ecological processes operate (Huston 1999). The size of the area designated as the "region" was 500×500 kilometers, which is reasonable. However, the size of the "local" area was 1 percent of this area, or an area of 50×50 kilometers, which is vastly larger than the area over which local interactions occur between organisms. Furthermore, the physical heterogeneity included within an area of 2,500 square kilometers is generally large enough that many different types of environments are represented, corresponding to many different "habitat types." Since organisms that occupy different habitat types rarely interact, any sample that includes multiple habitats cannot reflect limitation of species richness by local interactions, since multiple localities with species that do not interact are aggregated (cf. Cornell and Karlson 1996; Karlson and Cornell in press; Huston 1999).

When species richness is evaluated in areas small enough for local interactions (e.g., competition, facilitation, predation) to occur, the saturating effect of local interactions is clearly seen in a specific range of environmental conditions predicted by nonequilibrium theory, but not under all conditions (Huston 1999; Lord and Lee 2001; Karlson and Cornell in press). The effects of specific ecological processes can only be detected if ecological patterns are measured at the appropriate scale of resolution.

Unless ecological patterns, and the environmental conditions that influence the processes that determine the patterns, are quantified appropriately, species occurrences, patterns of species distributions, and variation in population viability cannot be understood or predicted. The issues of sample resolution, sample density, and size (or duration) of a study must be adequately addressed before it is possible to quantify and understand the complex interactions of ecological processes.

Ecological Processes and Theoretical Ecology

Our understanding of ecological processes is expressed in the body of hypotheses that can be called ecological theory. Although much of theoretical ecology seems to have little practical value (cf. Peters 1991; Sarkar 1996; Suter 1981), theory potentially has value far beyond the satisfaction of intellectual curiosity. Hypotheses that are sufficiently robust to avoid definitive falsification not only provide a statement of our understanding of particular ecological processes, but also improve predictions by identifying the critical independent variables and describing their effect on ecological processes and properties. In addition to refining predictions of known phenomena, strong theories can identify phenomena or patterns not previously known to occur. Such patterns may be discovered only because the theory predicted that they should occur. Discovery of such previously unreported phenomena clearly represents strong support for a theory, which would otherwise be considered falsified by the fact that the predicted phenomena were not known to occur.

Although there are many subdisciplines of ecology with distinct sets of theories, such as behavioral ecology or aquatic ecology, the primary concern of this book is that area of ecology that addresses issues of community structure, including species distributions and species diversity. Community ecology is a broad subdiscipline spanning population ecology and ecosystem ecology and including processes operating over a range of dimensions from molecular to continental. Although there are many theories about various aspects of community ecology, it lacks a rigorously tested and widely accepted theoretical framework. Lack of such a framework has been a major impediment to sound resource management and conservation planning.

Just as the individual organism has long been recognized as the fundamental unit of natural selection, the individual organism with its behavior and physiology shaped by natural selection is also a fundamental unit of ecological processes. Over the past decade, there has been increasing recognition that the individual organism can serve as a fundamental unit that allows integration of processes across dimensions ranging from molecular and genetic to ecosystems, landscapes, and even the globe (Huston et al. 1988; DeAngelis and Gross 1992; Huston 1994; Moorcroft et al. in press; Roloff and Haufler, Chapter 60). In most ecological research, observations and measurements of individual organisms are the basic units of data, which may be aggregated statistically to higher organizational levels such as populations, communities, ecosystems, landscapes, and biomes.

From the perspectives of both data analysis and computer modeling, aggregation from individual organisms to populations, communities, and ecosystems is a straightforward process (Huston et al. 1988). Ecological theories and models are now being developed using the individual organism as the basis for understanding higher-level phenomena (e.g., Smith and Huston 1989; Huston 1994; Moorcroft et al. in press; Shapiro et al., Chapter 49; Raphael and Holthausen, Chapter 62; Hartless et al., Chapter 39; Gross and DeAngelis, Chapter 40; Roloff and Haufler, Chapter 60).

The critical requirements for the success of this individual-based approach are (1) a thorough, quantitative understanding of how individuals of particular species respond to the range of environmental conditions they may experience and (2) a quantitative description of the relevant environmental conditions in the specific areas over which we desire to understand

and predict ecological patterns at spatial and temporal resolutions appropriate for the organisms of interest. Our increasing ability to describe the physical environment at high spatial and temporal resolutions using satellite-based sensors and GIS (geographic information system) technology, as well the ever-increasing knowledge base of the physiology and behavior of organisms, make this mechanistic approach to ecology feasible (Scott et al. 1993; Franklin 1995; Trani, Chapter 11). Many of the advances in our ability to predict wildlife habitat quality and species distributions are the direct consequence of higher spatial and functional resolution (e.g., vegetation type and structure, Young and Hutto, Chapter 8; DeAngelis et al. 1998). Explaining and predicting patterns that can be detected at a coarse resolution (e.g., species occurrence) generally requires information on processes and environmental conditions that must be collected at a much finer resolution. The importance of basing predictions on processes rather than on pattern correlations, and of measuring both patterns and processes at sufficiently high levels of spatial resolution, are themes that recur in many of the chapters in this volume.

The following three examples illustrate how different types of theories, all of which have the individual organism as the fundamental unit of process and measurement, can contribute to understanding ecological processes, to identifying the appropriate resolutions at which patterns are predicted to appear, and to suggesting statistical methods for quantifying those patterns. In each example, failure to understand the regulation of ecological processes leads to either (1) underestimation of the significance and miscalculation of the shape and magnitude of a causal relationship between environmental conditions and ecological processes or patterns, or (2) complete failure to detect existing causal relationships between environmental conditions and ecological processes or patterns.

Interactions of Limiting Resources: Liebig's Law of the Minimum

In 1840, J. Liebig published the seminal observation that even if most of the chemical elements needed by plants are abundant, plant growth can still be limited to low levels if a single critical element is in short sup-

ply. Much of the success of modern agriculture is built upon this insight, particularly for crops on the highly weathered soils of the humid tropics and arid Australia, where even micronutrients such as molybdenum or copper can limit plant growth (Jones and Elliott 1944; Hingston et al. 1980; McArthur 1991; Burvill 1979). Agronomists have long recognized that there are other types of interactions between nutrients, such as complementarity, that must be considered in crop management and yield prediction (Heady et al. 1955). Nonetheless, the generality of the law of the minimum has held up for over a century and a half and provides a conceptual framework that is useful for a wide range of ecological processes in addition to plant growth.

Virtually every ecological process is influenced by multiple factors, some of which are resources that can be depleted by organisms and others of which are regulators, such as temperature, that organisms cannot affect (Huston 1994). Although any resource or regulator can potentially occur at levels that limit the rate of a process, it is unlikely that the same single factor will always be limiting. Spatial and temporal variation in resources and regulators results in a continual shifting of limitation from one factor to another. For example, in seasonal climates, water is likely to limit plant growth during the dry season regardless of the levels of soil nitrogen or phosphorus, but during the wet season it is probable that nitrogen, phosphorus, light, or some factor other than water will limit plant growth.

This shifting between limiting factors has unfortunate consequences for understanding and predicting ecological processes, including processes as fundamental as plant growth. When processes are regulated according to the law of the minimum, variation in a nonlimiting factor has little or no effect on the rate of the process of interest. For example, measurements of the growth response of plants will have little correlation with variation in soil nitrogen, one of the most important potentially limiting plant resources, if some other resource, such as water, is currently at limiting levels. This lack of any correlation between plant growth and nitrogen obviously does not mean that nitrogen is unimportant for plant growth or is not a major regulator of plant growth rates but simply that nitrogen is not limiting under some or all of the conditions when the measurements were made. Process rates will be correlated only with variation in the limiting resource, and only under these conditions will regression analysis reflect the potential effect of the limiting resource on the process. Consequently, the "true" or maximum potential response to any specific resource or regulator of a process can only be quantified when all other resources or regulators occur at nonlimiting levels.

Unfortunately, most ecological research under field or experimental conditions measures only a subset of the factors that potentially limit the process under investigation. Ecological studies typically include a mixture of measurements made under both limiting and nonlimiting conditions and thus produce datasets with high variance (and low correlation coefficients) that hinder or prevent detection of the actual response of a process to a specific factor (cf. Huston 1997).

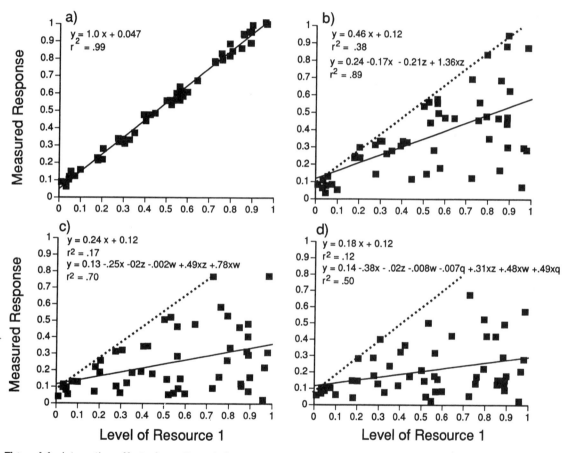

Figure I.1. Interactive effect of one through four limiting resources on an ecological response regulated according to Liebig's Law of the Minimum. (a) Response of a hypothetical measured response to random variation in Resource 1, when Resource 1 is the only limiting factor. The only variance is added random error. (b) Observed response to Resource 1, with random variation in one additional limiting resource. The solid regression line is the predicted response using only Resource 1. The dotted lines in b, c, and d indicate the upper bound of data, which is a close approximation of the "true" response of (a). The lower regression equation is based on precise measurement of both resources (x and z), with multiplicative interaction term. (c) Observed response to Resource 1, with random variation in two additional limiting resources. The solid regression line is the predicted response using only Resource 1. The lower regression equation is based on precise measurement of all three resources (x, z, and w), with multiplicative interaction terms. (d) Observed response to Resource 1, with random variation in three additional limiting resources. The solid regression line is the predicted response using only Resource 1. The lower regression equation is based on precise measurement of all four resources (x, z, w, and q) with multiplicative two-way interaction terms. Note that (1) the departure of the statistical relationship of the response to Resource 1 from the actual response (a) increases with additional limiting factors (b, c, d); (2) the increasing variance of the measured response at increasing levels of Resource 1 with one or more additional limiting factors (i.e., variance amplification); and (3) that the same phenomenon occurs with complex nonlinear relationships as with the linear relationship illustrated here.

It is usually impossible to determine whether the measured factors are actually limiting at the time the measurements are made. If the measured factor is the only limiting condition, there is likely to be a strong correlation between the process rate and the level of the factor, providing a good estimator (e.g., regression) of the effect of the factor on the process (Fig. I.1a). However, if some of the measurements are made at times or locations where the measured factor is not the limiting condition, there should be little or no correlation and no predictive relationship between the measured factor and the process. As additional factors become limiting at the places or times that measurements of the response are made, the correlation between the response and the hypothesized driving variable becomes rapidly weaker (Figs. I.1b,c,d). If all of the periodically limiting factors are identified and measured, more of the response variability can be statistically "explained," but the standard linear (or nonlinear) regression still cannot describe the true relationship between each of the limiting factors and the measured response.

This problem has been receiving increasing attention in the ecological and statistical literature (e.g., Thompson et al. 1996). A number of approaches to quantifying responses under conditions with multiple limiting factors have been developed, involving the use of "upper bound" or "envelope" statistical approaches (e.g., Maller 1990; Kaiser et al. 1994). One recent contribution describes a relatively simple and intuitive approach to quantifying the actual response of an ecological process to a single limiting factor in a situation with multiple limitations (Cade et al. 1999). The "quantile regression" approach provides a standardized method (with computer software available) for identifying and quantifying the upper bound of the cloud of points (i.e., high variance) of a typical ecological response (Fig. I.2).

Thus, the high variance that is often found in correlations between ecological processes and presumed causal factors may not be sampling error or random "noise" but rather the mechanistic consequence of shifts between limiting resources or the effects of other limiting factors such as mortality or dispersal (see O'Connor, introduction to Part 1). Although Liebig's Law of the Minimum was originally proposed in rela-

tion to plant growth, the same phenomenon can occur with any process that is regulated by more than one factor, which includes virtually all ecological processes. For example, the successful establishment and growth of a particular plant species may require a particular minimum level of nitrogen above which plant growth and biomass during succession is positively correlated with nitrogen availability. However, if for some reason seeds of the species never reach a given area, or the seed supply is extremely low, then the abundance of the plant may be very low under favorable nitrogen conditions.

Identifying the true relationship between a species' abundance and environmental conditions is complicated by the "false negatives" that result from the failure of species to colonize all areas where it could potentially thrive (see Fertig and Reiners, Chapter 42; Fleishman et al., Chapter 45). Although many different factors can potentially limit an ecological response, in many cases it is the availability of the organisms themselves that determines the presence or absence of a particular species and such ecological properties as species richness (e.g., Tilman 1997; Hubbell et al. 1999; Fleishman et al., Chapter 45).

This limitation of the dispersal and spread of organisms is an inevitable natural phenomenon that varies predictably in response to local environmental conditions and the life history and dispersal properties of the organisms themselves (Harrison et al. 1992; Nekola and White 1999). Dispersal is a process that is strongly influenced by chance and probability, and it is well known that plants and animals whose propagules (e.g., spores) are born passively on the wind tend to quickly colonize all suitable habitats. Birds obviously tend to reoccupy suitable habitats following local extinction much more rapidly than less-mobile animals, such as turtles or frogs. As a consequence, the probability of a species actually occupying an area of suitable habitat increases with the mobility of the species but decreases with the frequency of climatic (or other) events that kill or displace organisms.

It is in the arena of dispersal that the theory of island biogeography (MacArthur and Wilson 1967) is most relevant, along with such conservation biology issues as corridors, "matrix" properties, and the like. The probabilistic nature of dispersal results in high

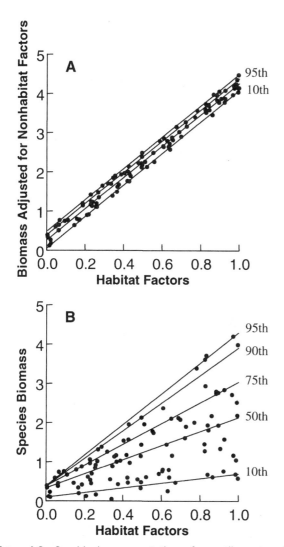

Figure I.2. Graphical representation of quantile regression method illustrating subdivision of data into quantiles to quantify maximum response (presumed response when only the x-axis factor is limiting). (A) Relationship of response (biomass) to independent variable (habitat factors) with effects of additional limiting (nonhabitat) factors controlled by adjusting biomass for nonhabitat factors. (B) Relationship of biomass to habitat factors without controlling for nonhabitat effects (i.e., assuming nonhabitat factors are unknown and/or unquantified). Note that the 95th quantile regression in B closely approximates the "true" relationship shown by the adjusted data in A. (Fig. 2 from Cade et al., 1999).

predictability at coarse resolution over large areas in the distribution of organisms among suitable habitats but unfortunately also leads to low predictability of presence or absence at high resolution (i.e., for specific small localities).

This phenomenon is a critical issue in the prediction of species occurrence and undoubtedly accounts

for a large proportion of errors of commission in which a species is predicted to occur in an area but is not observed there. In these cases, prediction of the suitability of a habitat to support a species (presumably based on the population responses of the species to relevant environmental conditions) may be correct, but the limiting factor of dispersal (or recent mortality, or inadequate time for a population to build to detectable levels following mortality, etc.) prevents the predicted population from being actually observed (see Fleishman et al., Chapter 45). Methods to predict the expected level of errors of commission can potentially be based on estimates of disturbance frequencies and locations (see Guisan et al. 1999) and estimates of dispersal rates based on organismal properties, distance to source populations, and geographical barriers or restrictions on dispersal (e.g., Nathan et al. 2001). Use of an upper quantile relationship (see Cade et al. 1999) in developing predictive models of species occurrence could reduce the confounding effects of dispersal limitations on models of potential species abundance and distributions. Evaluation and comparison of the accuracy of species occurrence models are addressed in many of the chapters in this volume (e.g., Hepinstall et al., Chapter 53; Johnson et al., Chapter 12; Johnson and Sargeant, Chapter 33; Fielding, Chapter 21; Karl et al., Chapter 51; Pearce et al., Chapter 32).

Another important consequence of the interactive effect of multiple limiting resources on a process is the amplification of variance in the measured process in response to increasing levels of the limiting resource that is measured as a driving variable. This phenomenon of increasing variance (the variance amplification hypothesis) has important implications both for the quantification of ecological processes (e.g., prediction of habitat quality) and for the possible destabilization of ecological interactions in response to increasing levels of limiting factors (e.g., temperature, nitrogen, carbon dioxide).

The interaction of multiple limiting factors, including processes that cause mortality, as well as resource availability and dispersal probability, produces a pattern in which unpredictability (i.e., variance) increases as environmental conditions become more "favorable," where favorability is interpreted as higher levels of one or more limiting resources (e.g., more fertile soils, more favorable temperature, optimal moisture conditions). Under such conditions of high levels of all limiting factors, biological responses such as plant growth or the rate of increase of animal populations, can be extremely high. However, when a single limiting factor is reduced to a low level (e.g., a drought), the biological response can suddenly drop to a very low level, leading to the impression that the biological response is unstable and unpredictable.

Although this type of "instability" can be very dramatic, it is a completely predictable consequence of the variance amplification hypothesis, based on the interaction of multiple limiting factors. Such sudden changes in ecological responses that have been interpreted as "chaos" (e.g., Tilman and Downing 1994) or "instability" caused by low species diversity (Tilman 1996) can be simply understood as the variance amplification that results from interacting limiting resources (Huston 1997). In addition to increasing response variance along gradients of increasing resource availability, other complex patterns can appear along gradients of environmental conditions, including complete reversal of response dynamics from one end of a gradient to the other.

Single-factor Response Reversals: The Unimodal Diversity-productivity Curve

Predictive models require quantification of the relationship between the environmental property of interest (e.g., population size, species presence, species richness) and the environmental conditions hypothesized to influence that property. This quantification may be purely descriptive, as in the case of a statistical regression, or may be mechanistic at some level, as with population models or plant succession models. In either case, the predicted relationship is based on or tested against observations (at the appropriate spatial and temporal resolution and density) from experiments or field sampling, ideally over the full range of environmental conditions to which the model will be applied. The confounding effects of multiple limiting factors on the quantification of cause-and-effect relationships (whether linear or nonlinear) were discussed above. Other types of problems can occur as a result of nonlinear relationships, which are common in ecological phenomena.

A potentially confusing type of nonlinear relationship is the unimodal or "humpbacked" response curve in which the response variable first increases at low levels of an environmental variable and then decreases at higher levels. With patterns of this type, data collected at high-enough resolution along different portions of an environmental gradient can lead to statistically significant, but opposing, conclusions about the relationship between an ecological response and environmental conditions, depending on where the data were collected. This response reversal is an inherent feature of any distribution that has a central maximum with declining values toward higher and lower levels, including both the Gaussian or "normal" curve, and the more complex, skewed curves typical of species distributions along environmental gradients (e.g., Mueller-Dombois and Ellenberg 1974; Austin 1980, 1999b; Austin et al. 1990; Austin and Smith 1989). This phenomenon underlies one of the longest-running debates in ecology: the relationship between productivity and species diversity.

Numerous studies have found that species diversity *increases* with increasing productivity, measured as net primary productivity of plants, secondary productivity of animals, or some factor presumed to be correlated with productivity, such as nutrients, temperature, or water availability (see recent reviews in Rosenzweig and Abramsky 1993; Huston 1994; Waide et al. 1999). Examples are found in plants (Grime 1979; Huston 1979, 1980), fish (Dodson et al. 2000), and mammals (Abramsky and Rosenzweig 1983; Owen 1988). However, numerous other studies have found that species diversity *decreases* with increasing productivity, particularly in plants, both aquatic (e.g., phytoplankton) and terrestrial (Al-Mufti et al. 1977; Huston 1979, 1980, 1994; Keddy 1989; Keddy et al. 1997; Tilman 1987, 1993; Abramsky and Rosenzweig 1983). There are numerous theoretical reasons, supported by a fair amount of data, to believe that the maximum diversity of plants and animals are likely to occur under different environmental conditions (Huston 1994). This is more likely to be true for generalist species at the herbivore trophic level and above and less likely to be true for specialists on plants, whose diversity is likely to be correlated with plant diversity (Huston 1994; Huston and Gilbert 1996).

Focusing on plants alone, it is evident that the increases of diversity with productivity tend to occur under very unproductive conditions, where any improvement in conditions allows more species to survive. Most of the decreases in diversity are found at the more productive end of an environmental gradient, where competitive interactions among rapidly growing plants lead to the elimination of less-competitive (generally smaller or more slowly growing) species (Grime 1973a, 1979; Huston 1979, 1980; Bakker 1989; Reader and Best 1989; Berendse 1994). Together these patterns make a unimodal response curve, which was first described by Grime (1973a,b, 1979) and later documented in marine, aquatic, and terrestrial ecosystems around the world (Huston 1979, 1980, 1985, 1994; Keddy 1989; Dodson et al. 2000; Rosenzweig and Abramsky 1993; Grace 1999; Waide et al. 1999). Although productivity (plant growth rate) influences the rate at which competitive differences are expressed in the absence of mortality-causing disturbances, productivity itself is a complex response to a variety of environmental conditions (including temperature, light, water, and mineral nutrients) each of which can vary independently over environmental gradients (as well as over time) and each of which can have independent effects on the growth and survival of plants.

Ecological studies that cover a large portion of an appropriate environmental gradient (e.g., plant productivity or some factor such as a soil nutrient that is strongly correlated with productivity) are likely to reveal unimodal abundance distributions of individual species, as well as a unimodal diversity pattern (e.g., Guo and Berry 1998; Grace 1999). However, the three different conclusions about diversity responses that can be drawn from statistically significant (or insignificant) correlations along different portions of a productivity gradient are wrong when applied to the entire gradient: (1) diversity increases monotonically with productivity, (2) diversity is uncorrelated with productivity, and (3) diversity is negatively correlated with productivity.

Further confusion about the relationship between species diversity and productivity (or other environmental factors) results from failure to measure diversity at resolutions appropriate to ecological processes and environmental variability (Cornell and Lawton 1992; Huston 1999), and from failure to consider the limits imposed by regional species pools on the range in diversity that is potentially detectable (e.g., with only three species in a region, the range in diversity is 0–3, which is less likely to reveal statistically significant variation in diversity than a situation in which the potential range in diversity is 0–20 [cf. Huston 1999; Lord and Lee 2001]).

Much of the confusion about the unimodal diversity-productivity relationship has resulted from incomplete sampling of the productivity gradient and sampling at incompatible resolutions for linking patterns to processes and environmental conditions (e.g., Waide et al. 1999). However, in addition to species diversity, unimodal responses of other ecological properties are quite common and can lead to the same types of confusion and erroneous conclusions about functional relationships that occur with diversity and productivity. Examples include the abundance of species along an environmental gradient. Along any gradient associated with the same processes that produce a unimodal pattern of species richness, it is inevitable that most species will increase in abundance at low values of the environmental factor and then decrease in abundance (relative and probably absolute) at high values of the factor.

This raises the question, "why is a species rare or absent in what is apparently optimal habitat within dispersal distance of known populations?" There are a number of potential explanations for this phenomenon, all of them linked to limiting factors other than the physical environment. A decline in the abundance of many species under favorable abiotic conditions is most likely to result from an increasing frequency and intensity of biotic interactions. In the case of plants, increased intensity of competition for light under productive stable conditions often eliminates smaller, shade-intolerant species and leads to a reduction in species diversity (e.g., Guisan et al. 1998; Grace 1999; Reader and Best 1989; Berendse 1994).

Other interactions may also negatively affect smaller or less-abundant plant species, including increased densities of pathogens or herbivores (Connell 1978; Hubbell 1980; Augspurger 1983; Clark and

Clark 1984; Nathan et al. 1999). Similar interactions may lead to the reduction or elimination of animal species, including direct competition (such as aggression or interference competition, Brown 1973) as well as higher predation pressure caused by the higher predator densities supported by other prey species (i.e., "indirect interactions," Holt 1984). These phenomena emphasize the importance of distinguishing between the physiological (potential) niche and the ecological (actual) niche of organisms. This refinement of the "niche concept" makes the distinction between the physiological limitations to species distributions and the biological limitations, and it demonstrates the complex distribution patterns that can result from these interactions (Ellenberg 1956; Mueller-Dombois and Ellenberg 1974; Walter 1964/1968, 1970; Austin and Smith 1989; Austin, Chapter 5).

Even an appropriate statistical description of a unimodal distribution of species abundance along an environmental gradient may not lead to better predictions of the spatial distribution of the species. The problem is that the observed abundance distribution of a species represents its "ecological niche, "which may not include the full range of conditions under which the species could potentially be found, and, in particular, may not include the physical conditions where the species actually grows best (in the absence of negative biotic interactions such as competition). The observed distribution of a species may differ from its potential distribution (or "physiological niche"), if the species is excluded from the conditions where it grows best by competition or other negative effects of other species. Depending on the pattern of overlap of the physiological responses of potential competitors, the realized niche of a species that is a poor competitor under optimal environmental conditions may be displaced into a skewed or even a bimodal distribution along the environmental gradient. These alternative (i.e., non-Gaussian) patterns of species responses along environmental gradients were first described by Ellenberg (1956) and Walter (1964/1968) and conceptually developed in the context of continuum theory by Austin (1980; Austin et al. 1990) and Austin and Smith (1989).

Two strikingly different patterns of fundamental niches (physiological responses) can be identified. Along resource gradients, such as soil nitrogen or other consumable resources, most species tend to have overlapping physiological optima at the high end of the gradient. However, for other environmental factors that are not affected or used by organisms, such as temperature, no single level is superior and species tend to have dissimilar physiological optima distributed along the gradient. In the case of resource gradients, the actual conditions in which species are found (the "ecological niche") is often displaced toward lower resource levels as a result of the dominance of better competitors under the most favorable resource conditions. For regulator (e.g., temperature) gradients, the ecological niches of poorer competitors can be displaced in either direction or may be split into a bimodal distribution (Austin 1980; Austin and Smith 1989). There are very few cases in which symmetrical bell-shaped species distributions are found in nature (Austin 1980, 1999b, Chapter 5).

Although the complex patterns of species distributions in relation to environmental conditions (ecological niches) that are found in nature seem to make the quantitative prediction of species occurrence even more difficult, a focus on the fundamental (or physiological) response of species helps resolve this difficulty. Understanding the physiological optima of a rare species provides critical information about the environmental conditions (e.g., soil nutrients and moisture, temperature, vegetation structure, etc.), where it could potentially occur, even if it is rarely found under those conditions. This information is critical for restoration, conservation, and management of rare species because these are the conditions in which, with appropriate management, the species actually does best. Management, such as control of predators or restoration of a different (and possibly more natural) disturbance regime may allow some species to thrive in areas where they are rarely found under present conditions (Abbott 2000; Risbey et al. 1999).

Differences in mobility and resource specialization among organisms have the consequence that distribution patterns of fundamental and realized responses along resource gradients tend to be very different for plants versus animals. Most plants have their greatest survival, growth, and fecundity under similar favorable conditions of nutrients, water, and light. However, animals have a much greater range of opportunities for specialization in resource use, and thus even potential competitors are likely to have different optima in their fundamental responses (Huston 1994).

The complexity of ecological responses along "single-factor" gradients results from the fact that different ecological processes predominate on different parts of the gradient. This shift in processes is most conspicuous along resource gradients and is unlikely to occur as consistently along regulator gradients. At very low resource levels, most species grow poorly, if they can survive at all. With increasing levels of the limiting resource, more species can survive, and all species grow better. Further increases in the limiting resource potentially allow all species to grow even better, except that a few of the species that grow best typically dominate and eliminate other species that are poorer competitors for some other resource (often light in the case of plant competition). Thus, the dominant process along the gradient shifts from physiological tolerance to low resource levels at the low end of a particular resource gradient, to competition at the high end (Grime 1979). Addition of a second factor along an environmental gradient adds further complexity to ecological responses as illustrated below by the interaction between disturbance and productivity.

Multifactor Response Reversals: Complex Effects of Mortality on Diversity

The effect of periodic mortality of organisms on ecological processes and patterns has long been recognized (Andrewartha and Birch 1954; Loucks 1970; Connell 1978; White 1979). However, the use of mortality-causing disturbances, such as clearcuts or fires, as environmental management tools has only recently become common practice (Wright and Bailey 1982; Romme and Despain 1989; White et al. 1991) and remains politically controversial in many regions.

The intermediate disturbance hypothesis, attributed to Connell (1978; see also Fox 1979), summarizes the general observation (Paine 1966; Loucks 1970; Lubchenco 1978; Sousa 1979; White 1979) that high levels of species diversity are often found in situations with a moderate amount of mortality, while both low and high levels (intensity and/or frequency) of mortality tend to be associated with lower levels of diversity.

The processes that produce this unimodal diversity pattern are analogous to those that produce the unimodal diversity-productivity relationship but are reversed along the primary axis. Thus, high levels of mortality create conditions in which few species are able to survive, analogous to the effects of extremely low productivity, while low levels of mortality allow competitive exclusion to occur and reduce species diversity, analogous to the effects of high productivity.

In spite of the logical soundness and convincing anecdotes that support the intermediate disturbance hypothesis, many systematic efforts to test it have found that it does not work very well. Two recent reviews (Feminella and Hawkins 1995; Steinman 1996) on the effects of grazing (a type of mortality-causing disturbance) in aquatic systems concluded that the effects of grazing on species diversity of algae were not predictable. Some studies found that grazing increased species diversity while other studies found that it decreased diversity, but there was no consistent pattern. This situation provides an example of a fundamentally important ecological pattern that was not detected until it was predicted by a theory and the data were reanalyzed appropriately.

The interactive effects of mortality and productivity on species diversity were first predicted by the dynamic equilibrium model (Huston 1979) and summarized as a three-dimensional response surface of diversity in relation to orthogonal axes of mortality frequency versus the rate of population growth and competitive displacement (both of which are often correlated with productivity and the levels of limiting resources). This model predicted that the single-factor conditions (i.e., either mortality level or population growth rate) under which highest diversity should occur would shift, depending on the level of the other factor (i.e., population growth rate or mortality level). The consequence of this interaction was that the same quantitative increase in disturbance level could either increase species diversity (under high-productivity conditions) or decrease species diversity (under low-productivity conditions), with analogous responses to a similar change in productivity under different disturbance regimes (Fig. I.3).

Unfortunately, the predictions of the dynamic equilibrium model have never been systematically tested in field experiments, largely because of the size and complexity of experiments involving the factorial interaction of enough treatment levels to detect nonlinear (and potentially unimodal) responses. A number of small experiments and studies have produced results consistent with the predictions of the model (Huston 1994). In spite of the implications of these predictions for conservation and resource management, there was no systematic effort to evaluate the model until a recent effort involving "meta-analysis" (Osenberg et al. 1999) of published experiments on the effect of grazing on species diversity (Proulx et al. 1996; Proulx and Mazumder 1998).

Grazing fits within the definition of a mortality-causing disturbance because it results in the mortality of all (in the case of grazing of phytoplankton) or part (in the case of most terrestrial grazers) of an organism and thus affects both survival and competitive interactions. Lawnmowers act as mortality-causing disturbances for herbaceous plants, killing a variable proportion of each plant and affecting both survival and competitive interactions, as do logging, ice storms, windstorms, fires, and other natural and anthropogenic disturbances.

The relative ease of experimental manipulation of plant grazers in aquatic and terrestrial systems has led to a large number of published grazing experiments. These studies were reviewed by Proulx and Mazumder (1998) for their meta-analysis, which clearly showed the reversal of grazing effects on species diversity (plant species richness) between productive and unproductive environments predicted by the dynamic equilibrium model (Fig. I.3). This analysis found that most of the published experiments in systems that could be classified as productive (based on measured nutrient levels, precipitation, or qualitative assessments such as oligotrophic versus mesotrophic) showed an *increase* in plant diversity in response to grazing. In contrast, all of the studies conducted in nutrient-poor, unproductive systems found that plant diversity *decreased* with increased intensity of grazing.

The consistency of grazing effects on plant diversity in terrestrial, aquatic, and marine systems suggests that the predictions of the dynamic equilibrium model (Huston 1979) are applicable to a wide range of

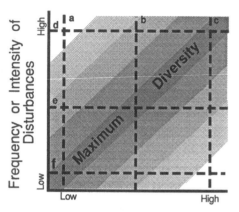

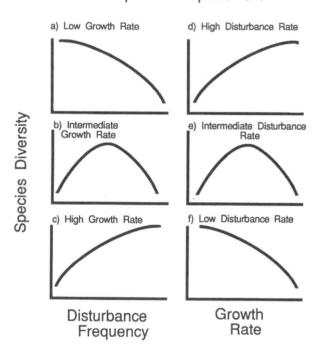

Figure I.3. Predictions of the dynamic equilibrium model of species diversity. Both disturbance type and frequency, and growth rates (plant productivity, population growth rates) vary across landscapes in response to geology, topography, and climate. The effect on species diversity of a change in either disturbance frequency or population growth rates can be reversed from one region to another, depending on local conditions of growth rates and disturbance frequency, respectively. Note that the response of diversity to a given change in disturbance frequency is predicted to reverse between environments with low growth rates (a) and environments with high growth rates (c), and that the response of diversity to a given change in productivity (growth rate) is predicted to reverse between environments with a high disturbance rate (d) and environments with a low disturbance rate (f) (from Huston 1979, 1994).

ecosystems with different types of mortality as well as different controls on productivity (e.g., nitrogen, phosphorus, water). Similar patterns of response to disturbance can be produced by computer simulation models of plant competition (Doyle 1981; Huston 1994).

The effects of disturbances on species diversity represent the summation of disturbance effects on the distribution and occurrence of the many species that are potentially present in any particular environment. The same basic framework of interacting productivity and disturbance can also be used to predict when and where particular species are likely to be found. The dynamics of plant succession produce a shifting pattern of species distribution across landscapes, particularly in situations where productivity is high enough that competitive exclusion reduces diversity in late succession (Smith and Huston 1989; Smith and Urban 1988; Huston 1994). It is important to recognize that the reduction in diversity that potentially occurs as a result of competitive exclusion can only be found in groups of organisms that can compete with one another. In contrast, the diversity of broad groups of dissimilar organisms with little potential interspecific competition is more likely to increase monotonically with decreasing mortality and increasing productivity (Huston 1994).

In spite of the complexity of many ecological patterns, it is possible to make accurate predictions about ecological phenomena that vary in response to environmental conditions, as shown by the recent demonstration of the reversal of the effect of grazing between productive and unproductive environments (Proulx and Mazumder 1998). Both a sound conceptual framework based on an understanding of ecological processes and a sampling design with the appropriate resolution, density, and total area are essential for making accurate predictions of ecological responses.

Conclusions

Our efforts to understand and predict variation in the abundance and distributions of species, including the diversity of species in any particular area, have been hampered by inadequate theories, mismatches be-

tween processes and sampling scales, and inappropriate statistical methods. A stronger theoretical framework that addresses multiple interacting processes and limiting factors can help resolve the patterns underlying the complexity of nature and also contribute to identification of appropriate sampling scales and the development of statistical methods more appropriate for quantifying causal relationships than those that have been traditionally used.

Elements of a conceptual framework for predicting species occurrence include (1) recognition that the interactive effects of multiple limiting factors require new statistical approaches for quantifying ecological processes; (2) matching the spatial and temporal dimensions of measurements of ecological patterns to those of ecological processes for hypothesis testing and model development; (3) planning of sampling designs and model development based on the probability that ecological responses may reverse and processes change in relative importance along the environmental gradients found on all landscapes; and (4) recognition that the interaction of population dynamics and competitive processes with mortality-causing disturbances and other factors that affect population dynamics can produce complex responses along either disturbance or productivity gradients that completely reverse between different environments. The complexity of ecological processes should be seen as a stimulating challenge, not an insurmountable barrier, to improving our predictions of species occurrence.

Acknowledgments

This work was supported by the U.S. Environmental Protection Agency's STAR Grant Program, and the Program for Ecosystem Research, Environmental Sciences Division, Office of Health and Environmental Research, U.S. Department of Energy. Oak Ridge National Laboratory is managed by UT-Battelle, LLC for the U.S. Department of Energy, under contract DE-AC05-00OR22725. This is publication number 5038 of the Environmental Sciences Division, Oak Ridge National Laboratory.

PART 1

Conceptual Framework

The Conceptual Basis of Species Distribution Modeling: Time for a Paradigm Shift?

Raymond J. O'Connor

In setting out to review the conceptual basis of species distribution modeling in the light of the contributions to this volume, I am mindful that any mature discipline carries within it what might best be characterized as intellectual baggage. The discipline has its own set of jargon to provide shorthand to its key concepts and their components (Morrison and Hall, Chapter 2). Its concepts typically carry within themselves a set of assumptions and logical treatments. A long history in a discipline then honors this logical framework as tradition, leaving the fundamentals all too often unscrutinized for long periods. Finally, a mature discipline typically passes through stages of being overwhelmed by an accumulated mass of contradictions between its conventional wisdom and emerging data, leading to a Kuhnian paradigm shift. The contributions to the present book reflect all of these elements, setting the scene, I believe, for the volume as a whole to be both the last hurrah of the old way of conceptualizing species distribution modeling as a process of determining and modeling habitat correlates under equilibrium conditions and the foundation for a new paradigm characterizing species distributions as nonequilibrium spatial dynamics under habitat constraints. Here I want to portray chapters in this section as collectively delineating the structure of a discipline struggling under the weight of its accumulated contradictions and as offering signposts to the new paradigm.

The Lessons from History

An old adage holds that those who fail to study history are doomed to repeat its mistakes. In Chapter 3, Dean F. Stauffer shows this to have been true of species modeling as the "dawning of the quantitative ecology era" replaced the ever-growing volume of natural history studies that filled the first half of the twentieth century. One of the problems with quantitative ecology is that everyone feels obliged to conform to its forms, irrespective of the ability of their data and their training to support such activity. A continued supply of top-quality natural history studies would define the factual corpus of ecology and provide the raw material for rigorous analysis by technically competent modelers and theoreticians, a goal endorsed by Van Horne (Chapter 4) and Wilcove and Eisner (2000). Instead, though, one is expected to have a hypothesis guiding every study, leaving no room for high-quality research into the natural history of a previously unstudied species or system. As a result, the literature has seen a plethora of supposedly quantitative studies that cloud more than they illuminate: one cannot make progress in science if models are built on unspecified or unchecked assumptions and if generalizations are attempted across model results supposedly comparable but actually disparate (Murray 2000). There is, alas, no room in current ecology for rigorous determination of ecological facts: despite the majority of graduate ecologists being destined for careers in dealing with facts, too often the required formalism is met with toy hypotheses exuding statistical jargon. The most telling phrase in Stauffer's historical review should give us pause in this respect: "MacArthur (1958) published a rigorous and quantitative, *though not highly statistical* [my italics], analysis of niche relationships of five warbler species." How often can one say that of a modern paper in which all too often the high statistical content has crowded out all semblance of rigorous logic. Michael Austin (Chapter 5) reiterates that a knowledge of the distribution of species is central to the conservation of biodiversity, yet, he states, no amount of statistical modeling can compensate for a poorly defined problem. Continued development of rigorous statistical approaches to analyzing habitat data, assisted by the spread of easy computation in the form of computing power and of packaged statistical analysis, has been unaccompanied, even to this day, by corresponding development of rigorous logic.

Stauffer's conclusion is that despite our powerful analytical tools there are limits to the precision of the models we develop because of the noise inherent in the systems with which we work, and that the simplest analyses that meet our objectives will likely be best. I emphasize, however, the distinction between "simplest" and "simplistic." The use of powerful tools may tempt us into the equivalent of modeling the orbits of the planets with epicycles, getting ever-closer fits to the data but hiding from us the need to develop the equivalent of Newton's Laws.

Failure in the Practice of Distribution Modeling

One of the repeating elements in Stauffer's history of habitat modeling is how individual researchers periodically rose to challenge the prevailing bandwagons of near-dogma. Not all were successful. I find it curious that, despite their initial impact on the field, Alldredge and Ratti's (1986) urging that models be tested against artificial data with known structure is still largely ignored fifteen years later and has to be reiterated both by Stauffer and by Heglund. A handful of researchers have attempted to assess model validity and assump-

tions by examining a crosssection of published models, but this goes only so far in protecting the scientific community from spurious models. Models always involve some methodology and its associated assumptions. With an established methodology, the responsibility of the modeler is to check the validity of the underlying assumptions *for the particular organism being modeled*. With any new modeling technique, however, a modeler should recognize his or her extended responsibility to understand the functioning of the model *well enough to ensure that it does behave as expected*. (Van Horne, in Chapter 4, extends this idea to the checking of boundary conditions for models.) This is possible only through testing with data (artificial if necessary) of known structure and by demonstrating that the modeling technique correctly recovers that structure. After all, most models are constructed because the context of their application is too complex to be represented by a simple procedure: by that same fact one should anticipate that the model is too difficult to validate by inspection alone. In many respects such testing parallels the "order-of-magnitude" calculations of engineering and physics: one re-calculates one's work with conveniently rounded numbers (allowing simple tracking of the computations) to ensure that in working through the full specification one didn't do something as stupid as shifting a decimal place. The approximate calculation may yield an answer that is in error by 10, 20, or 30 percent (and a further approximate calculation often yields an estimate of how big this error is likely to be), but an approximate answer that is different by a factor of two or three indicates the need to revisit the original calculation. The recovery of known structure from datasets with known (often simple) structure provides equivalent confidence that the modeling procedure is not in major error. (One might note here, too, that checking the assumptions underlying even a standard modeling technique requires a level of care greater than the current norm among ecologists. Contrary to common belief, acknowledging the presence of the problems of non-normal data or of sample sizes limited by logistical resources doesn't justify ignoring the problem thereafter.)

Chapter 8, by Jock S. Young and Richard L. Hutto, is another of these periodic challenges to current assumptions. In it Young and Hutto set out to describe the use of large-scale exploratory studies in determining bird-habitat relationships by exploring the implications of the various phenomena they encounter with their focal species, the Swainson's thrush *Catharus ustulatus*. The real merits of this chapter are its insights into the process of modeling birds in relation to habitat, particularly in its general analysis of the issues involved in using spatially extensive, data-collection programs as the basis of studying birds and habitat. Considerable thought is given by Young and Hutto to just how variables enter the models under development, and they clearly have little trust in the structures created by multivariate analysis. First, they argue—correctly in my view—that the major value of the models is in identifying those variables that are either biologically linked to the bird's abundance or are surrogates whose investigation may yet reveal variables with biological significance. Second, their logic leads inexorably to their conclusion that most statistical models are likely to perform poorly when tested by management manipulation of the critical variables in the model: the manipulation is likely to disrupt the very correlation structure that leads to the model in the first place. Much of the field of species distribution modeling has historically been crippled by a lack of understanding of these two points. Indeed, one can make a case for this lack of understanding being *the* factor most responsible for the dichotomy (emphasized by Stauffer in Chapter 3 and by Van Horne in Chapter 4) between managers and researchers. As is inevitable with inductive studies, one cannot extend their arguments to conclude that all such studies will be compromised in this way but, perhaps more importantly, in the light of their study one cannot in future blithely assume that correlation structures will persist under management manipulation.

Young and Hutto's signposts to improved modeling fly in the face of the current fad for hypothetico-deductive research. Myopic emphasis on hypothetico-deduction analysis acknowledges only those process variables causally traceable to the abundance of the species and asserts their experimental study to be the only way to do science. Such extreme reductionism allows no value to the broad sweep of extensive statistical analysis delineating both the candidate variables for process models and the surrogate correlates likely to stimulate new paradigms as to process. Since adherents of these views are often the reviewers of science-funding applications, broad exploratory research is hard to fund as science. Conversely, such work is more easily funded as management-oriented research where solutions are needed urgently and in the extreme the resulting empiricism rejects the need for process-based variables on the grounds of pragmatism, hoping that a statistical model will provide the desired predictive power.

Young and Hutto's notion of the destruction by management experiment of the very correlations that originally generated the predictive models challenges this converse and readily explains why so many predictive models have proved disappointing in practice. Granted this, one has to call for a more realistic view of the role of hypothetico-deduction in ecology. Just as the demise of natural history studies noted by Stauffer (and cogently lamented by Wilcove and Eisner [2000]) has, to a degree, reduced ecology to a collection of case studies, the lack of a large-enough corpus of statistical studies identifying which classes and combinations of variables are most frequently implicated in wildlife-habitat relationships has resulted in a scattering of species studies too diverse in scope and methodology to generate the broad sweep of pattern that stimulates truly general hypotheses. Reductionist science may well be the apex of the pyramid of science, but it is founded on an ever-widening base of more factual studies as one descends the pyramid. Hypothetico-deductive science functions best in fields with a large body of factual information, and in the case of wildlife habitat modeling this fact needs to be recognized as much by science funding as by management funding. Little is gained by the creation of toy hypotheses to justify as hypothetico-deductive those studies that have intrinsic value as providers of factual information. Young and Hutto's message regarding the breakdown of statistical correlations under management manipulation is both fine evidence about the need for experimental hypothetico-deductive testing of causal models and a strong case for supporting exploratory correlative research to inform such experiments. Young and Hutto's chapter does much to clarify the strengths and weaknesses of complementary hypothesis-based and empirical research in this field.

Logic and Reasoning in Distribution Modeling

Chapter 7, by Kristina E. Hill and Michael W. Binford, offers a sweeping critique of the quality of logical reasoning in contemporary habitat modeling that complements the concerns of Hutto and Young and of Stauffer. Although parts of their discussion are framed in terms of the need to meet legal constraints when land use planning based on habitat models enters the courts, their emphasis on examining the fundamental assumptions of the models rather than on further theoretical proliferation is salutary, reinforcing the same arguments touched on in this section by Stauffer, by Young and Hutto, by Van Horne, by Smallwood, by Austin, and by Huston in the volume's Introductory Essay. Their logic itself results in distinguishing between two classes of models: (1) *forecasting models* based on linking known occurrences to predictor variables that may then estimate the probability of finding the species present at other locations, and (2) *exploratory models* that in effect model habitat potential on the basis of environmental similarity of areas of unknown use to conditions in areas of known use. This latter class of models help clarify what variables might be involved in causal linking of habitat to species occurrence, paralleling the similar role specified by Young and Hutto for multivariate exploration in their context. Hill and Binford (Chapter 7) also anticipate Young and Hutto's (Chapter 8) conclusion about unstable correlation matrices in their excoriation of ad hoc techniques for modeling. They do so, however, with a qualification rare in ecology, that there exist theories of measurement that can guide the formulation of practical models with limited input of ecological theory. This is an area in which most ecologists are totally untrained, resulting in a world-view that sees variability among organisms or plots as paramount and measurement issues as negligible or unimportant.

One other element in Hill and Binford's treatise—raised there only in passing—deserves mention here. Although the logical and practical difficulties of using inferential correlation and regression statistics as the basis for ecological modeling are widely (though not universally) appreciated, there are also significant ethical issues in the use of these methods in testing models intended for use in planning and management. As any experienced reviewer can testify, all too often the only guiding ethic in the writing of ecological papers appears to be the need to conform to the formalism of a "scientific" hypothesis within the paper. How often does the introduction to a habitat modeling paper imply great potential in the approach being presented as a tool to solve hypotheses about all sorts of management or conservation problems, only to show in the discussion section little evidence of any concern as to whether the model presented is actually suitable for its stated purpose? To be sure, caveats are expressed, reservations entered, and caution urged, but these are fig leaves protecting authors from being crucified for failing to discuss limitations: the mind-set appears to be, "If the manuscript gets past the reviewers it must be all right." It is worth considering, especially in the light of Hill and Binford's comment, whether there may be serious ethical questions to be addressed as to how we typically represent the utility of our modeling work.

Statistical Tools and Techniques

Several chapters in this section, including those by Patricia J. Heglund (Chapter 1), and Beatrice Van Horne (Chapter 4), and also the volume's Introductory Essay by Michael A. Huston, address the underlying assumptions made in technical approaches to species distribution modeling. The statistical techniques typically used in such modeling fall into a small number of categories: a group of related regression and correlation techniques, including simple and multiple linear regression and various multivariate methods; logistic regression; GAP (Gap Analysis Program) and habitat suitability index (HSI) methods; and various so-called modern regression methods. Each technique carries its own set of assumptions about the data it models. These chapters and Huston's essay each focus on the major shortcomings of current understanding of these assumptions.

Heglund and Van Horne emphasize that too much of extant modeling effort is based on relatively uncritical application of statistical techniques to the distribution modeling in hand, at the expense of the sound knowledge of environmental processes and an understanding of the nature of species response functions that are at the heart of robust models of such distributions. A glance at the standard works on wildlife modeling shows that the dominant model in use is one of linear regression. However, anyone using linear regression as the basis of distribution modeling is essentially making a particular set of ecological assertions. If the independent variable in the regression is X (say), these assertions include (1) the species is limited by the amount of habitat of type X, (2) doubling X will induce a doubling of the species abundance (subject to the modification implied by any intercept), and (3) the species distribution is in equilibrium with X. Note that these assertions are ecological assumptions, not statistical assumptions. Huston's chapter elaborates how spatial variation in limiting factors introduces variance that obscures, and even biases, the outcome of regression analysis. The assumption of equilibrium is, however, even more likely to be problematic: for most species population levels are likely to be below equilibrium levels as a result of the vagaries of year-to-year impacts of climate and weather on overwintering survival and on reproductive output, and this alters the outcome of any statistical analysis (Heglund, Chapter 1; Van Horne, Chapter 4). The principles identified by Huston as to spatial variation in limiting factors are also as relevant to nonequilibrium population analysis. Similarly, the assumption of linear response to change in X carries implicit assumptions about scale of per capita resources, an issue illustrated in more detail in one of K. Shawn Smallwood's examples in Chapter 6. What is extraordinary is how poorly understood the limiting natures of the assumptions of this model are, despite the widespread reliance on regression.

Logistic regression has become popular recently as its use in published papers has proliferated. As Van Horne cynically but correctly notes in Chapter 4, one form of habitat modeling becoming more popular than another always deserves to be questioned to determine whether the replacement constitutes progress. Given that logistic regression can model the binary variable of presence or absence, its use really consti-

tutes no more than the extension of regression methods to data at a lower level of measurement. In addition, assumptions about a binary response need to be examined in the light of Kunin and Gaston's (1997) useful formulation of the "area of occupancy" concept: if there are "holes" within a range boundary, should one model the species being potentially anywhere within the range or as being precluded from occurring within the holes? This question also raises the issue of nested correlates: if range is set by long-wavelength variables and occupancy by shorter wavelength variables, the two cannot be simultaneously captured as a binary response. Finally, the issue of equilibrium raised above continues to be critical, and the statistical assumptions inherent in an underlying linear function are now replaced with the functional assumptions of the logistic. Thus the use of logistic regression still requires the analyst to confirm the validity of the assumptions made, a requirement largely ignored in the studies published to date. Indeed, logistic regression is in danger of being viewed as "the silver bullet which will solve all model building problems" all too often sought in wildlife research (L. McDonald personal communication).

GAP and HSI models have been widely used to characterize species distributions. What they strictly predict, however, is an envelope of environmental and vegetation requirements within which the species *may* occur. These models have little utility for tracking changes in distribution. Both GAP and HSI map components of the environment have been shown to be linked to the presence or abundance of the species of interest at a certain time or in a certain region, but their methodology does not ensure that the species-component relationship will persist over time or across regions: mapping the amount of hedgerow on farmland may be well correlated with the abundance of a small songbird species when a GAP assessment is prepared, but only the hedges, and not the birds, persist if the farm is sprayed with a pesticide toxic to birds. Obvious as this may seem, the assumption is regularly overlooked whenever workshop discussion turns to tracking distribution over time. That three chapters here—those of Young and Hutto, Heglund, and Van Horne (Chapter 4)—find it necessary to point this out is a serious indictment of the quality of current modeling.

Heglund portrays the advantages of generalized linear models (GLM) and generalized additive models (GAM) as regression procedures suited to distribution modeling. Even so, the potential of GLM and GAM approaches are limited by issues associated with multicollinearity and with limitations of stepwise regression. The application of these methods to distribution data introduces a new element, the need to take into account the spatial autocorrelation intrinsic to distribution. Spatial coherence typically inflates the significance of the GLM/GAM predictors and may also bias them. For now a critical safeguard with these models is to plot the residuals from the models so as to reveal spatial patterning.

Heglund also emphasizes that the environmental predictors used in distribution modeling must be ecologically meaningful. Even then, the long wavelengths of environmental data mean that these approaches will achieve relatively high predictive success over coarse resolution but are still unable to successfully predict the local occurrence of the species within the general envelope thus identified because local issues are dominant locally. As I noted above, this is a significant and all-too-often-overlooked problem for logistic models.

Techniques are available to discriminate among the performances of the potential models listed above. As John Wiens points out in the concluding chapter of this volume, the use of Akaike's information criterion (AIC) has considerable potential in evaluating the relative merits of competing models. Burnham and Anderson (1998) and Anderson et al. (2000) summarize many of the major problems associated with simple-minded hypothesis testing of models and provide an excellent introduction to the use of AIC in the simultaneous evaluation of multiple models.

The Need for a Paradigm Shift

The above sweep through the major arguments concerning the conceptual basis of our discipline shows that limitations of current concepts are marked and aggravated by generally poor implementation of the available approaches despite a long tradition of in-

formed commentary by the more percipient practitioners. So where do we go in the face of a history of wise advice about this litany of modeling problems that has failed to resolve the problems? Huston's introductory essay and Smallwood's chapter develop thoughtful accounts of how ecologists repeatedly fail to understand ecological patterns and instead wind up applying inappropriate statistics to quantify them. One issue is that the scale (extent and grain) at which analysis is conducted is often wrong for the scale of the processes of interest, for sparsely distributed sample points may characterize gross distribution but not cast light on local processes. Smallwood argues that if the laws of thermodynamics primarily influence spatial distributions of animals, then assuming a random distribution as the null pattern is inappropriate and reduces an analyst's capacity to recognize meaningful agreggations. He further challenges us, as modelers, to make certain we understand issues of scale and their influence on demographic organization. Far more significant in my view is Huston's emphasis on factors as limiting agents to species abundance. Huston's argument contains two key assumptions. The first is that most species *can* be limited by any of a variety of factors. The second is that the influence of any given ecologically relevant factor is not *additive* to the influence of any other factor; instead, typically only one is limiting in any particular situation. Huston's essay shows how misleading correlative analyses can be under these assumptions, even to the point of reversing the direction of the effect on abundance attributed to any given factor. It is my belief that Huston's review, augmented by the points raised by the other authors I cited above, is actually the recognition of the accumulation of a critical mass of contradictions within the prevailing paradigm that immediately precedes its overthrow (Kuhn 1970). I want now to use these ideas to provide the basis for a paradigm shift in how we model species distributions.

Suppose we view any data distribution with respect to any factor of interest (a habitat or environmental factor) not as a function of that factor as a correlate but as the simultaneous outcome of multiple factors *only one of which is limiting in any particular location*. Figure P1.1 presents some artificial data to illustrate how allowing the limiting factor to change from place to place translates into real world datasets. For this illustration, we assume that there are three potential limiting factors influencing the abundance of a species (implied by the height of the curve) and labeled X_1, X_2, and X_3. The figure plots abundance data from seventeen locations against the value of variable X_3 on the abscissa and shows a unimodal response envelope to the maxima. Unimodal responses are regularly modeled—for example, Jongman et al. (1987)—but as a function rather than as a bounding envelope of the type discussed by Huston and regularly seen in habitat data (e.g., O'Connor and Shrubb 1986 p. 137; O'Connor 1987a p. 40). For fourteen of the locations the abundance of the species is limited by variable X_3, the least severe of the three factors. Point A is a location in which all factors are favorable, and the species is in optimal conditions with respect to X_3, so abundance is greater there than anywhere else. At location B abundance is only half that of location A because the (suboptimal) value of X_3 is even more limiting than at A (and neither X_1 nor X_2 are limiting). At location C, however, factor X_3 is optimal but abundance is lower than at site A because factor X_2 is limiting. At location E factor X_3 is suboptimal but would have allowed abundance to reach the same value as in location B were X_2 not also suboptimal. We would see this limitation by factor X_2 manifest in the data points for locations C and E lying at the envelope edge in a plot of abundance against variable X_2 (as sketched in Fig. P1.1) if we knew enough to exclude the fourteen locations not limited by X_2. Finally the most severely limiting variable X_1 acts only in a further subset of locations (here of just location D), preventing abundance rising to the level of site A even though conditions in X_3 and X_2 are optimal.

The model of Figure P1.1 formalizes the thinking in Huston's introductory essay in respect of data patterns, and a few further comments are in order. First, the envelopes for factor X_2 and X_1 lie along axes orthogonal to that of X_3. Second, and more critically, their envelopes, when rotated, will not form the nested curves quite as illustrated. All fourteen locations along the X_3 envelope will then lie in a column centered on the optimal value of X_2 but at abundance levels dictated by the value of X_3 they experience. The same is true also, in turn, in respect of X_1. This is difficult to

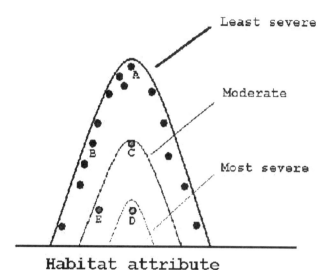

Figure P1.1. Artificial data illustration showing how allowing the limiting factor to change from place to place translates into real world data sets.

portray visually while retaining display of the limitations imposed at locations C, D, and E. One should also note that although this portrayal resembles Brown's (1995) idea of maximum abundance occurring where multiple factors are simultaneously favorable, the present world view of the maximum being where all limiting factors are simultaneously *permissive* has very different implications from their being *favorable*. Brown's prediction that densities should in general be unimodal within a species range—a pattern that is certainly not commonplace (Maurer 1990)—now gives way to a more topographically diverse surface generated by Huston's pattern of shifting limiting factors.

How might one implement the analysis in practice? Plot abundance against each variable of interest and find the data distribution most similar to that of Figure P1.1 for variable X_3. Determine the values of abundance at the envelope edge using one of the edge identification techniques discussed by Huston (though I would opt for fitting a unimodal curve rather than a quantile linear regression) and express abundance at all locations as a percentage of the envelope maximum at that X_3 value. (For example, locations A and B would score 100, location C about 50, location E about 40, and location D about 20). Each score will have some uncertainty associated with the precision of the edge determination algorithm. Repeat this independently for the other candidate limiting variables. Considering the plot against X_2, since all the points limited by X_3 are located in a column at the optimal (least limiting) value of X_2 but at levels determined by the X_3 constraint, point A will still earn a score of 100 but point B will now earn only about 50 and so on for the rest of the fourteen locations. All other points (here only three are illustrated) will score values determined both by their edge or interior position and by the scaling imposed by location A's abundance serving as curve maximum, with points such as E yielding scores of 100. Thus, if enough locations limited by X_2 are in the sample, they generate a group of scores of 100. This logic prevails recursively. With a uniform density of points across a plot such as Figure P1.1, histograms (one for each factor) of these scores against abundance will then yield an ordered series, with the least-limiting factor uniquely yielding scores of 100 at the highest abundances (here, from the seven points above the B-C level). A block of additional scores of 100 then enters in the histogram for the second least limiting factor (though X_3 still generates 100s at these abundances because of points such as B and the three below it and to its left), and so on. Recursively removing points limited (scores near 100) by the outer variables will reduce successive overlaps in the original, facilitating identifying the sequence of limiting variables.

Do we have any evidence that Huston's arguments can be used to synthesize such a dramatically different way of accounting for species distributions? I believe that we do. My colleagues and I, among others, have recently modeled a variety of species and species richness distributions by use of classification and regression tree (CART) analysis. What is valuable about CART analysis in the present context is that the data are recursively partitioned into subsets on the basis of *constraints* rather than correlates. The occurrence or abundance of a species at each binary split into subsets is greater or smaller according to whether a variable exceeds a stated value; those subsets are again examined for further predictive rules, and so on, eventually yielding (on the basis of statistical optimization rules) a series of subsets—in our data typically regional in distribution—of sites in which abundance is predicted conditional on one or more

constraint rules. Such rules are of the type that July temperatures exceed 25 degrees Celsius, that elevation be below 500 meters, that cropland forms no more than 15 percent of the local land cover, or that average forest patch size exceeds 2,500 hectares. Good examples of such models are available for plants (Iverson and Prasad 1998), lizards (Hollander et al. 1994), birds (O'Connor et al. 1996, 1999; Hahn and O'Connor, Chapter 17), amphibians (Guerry 2000), and butterflies (Lawler, Bartlett, and O'Connor in preparation). In addition, the use of GAM models described by Heglund in Chapter 1 can be regarded as defining species distributions in terms of constraint envelopes. Since these CART and GAM models are data hungry, they cannot readily be implemented with the relatively small datasets typically found in species habitat studies. The approach sketched out from Figure P1.1 seems to have potential with these smaller datasets, though much remains to be filled in to achieve a robust analytical implementation of that thinking. In particular, the treatment of variance associated with census uncertainties and sampling, and the treatment of nonequilibrium populations in the analysis require detailed consideration. It is also important to recognize that implementation of my constraints model involves statistical and analytical assumptions that will be wrong in particular cases. Time, and experience with the application of the concept to particular cases, will inevitably yield examples where constraint models fail. Meanwhile, though, the concept of constraints models provides a unified framework within which to think about many of the issues raised in this book— the nonequilibrium situations emphasized by Wiens, the hierarchical action of effects with different spatial wavelengths and boundary conditions (Van Horne, Austin, Chapters 4 and 5), understanding demographic organization in space (Smallwood, Chapter 6), the relative utility and limitations of forecasting and exploratory models (Young and Hutto, Hill and Binford Chapters 8 and 7), the regular (but generally partial) success of very disparate classes of models such as HSI and GAP models and the various linear and nonlinear regression models (multiple authors), and the need to reconcile complexity of data patterns with simpler causal processes (Stauffer, Chapter 3). Perhaps most important of all is that it raises new questions that do not emerge naturally from the older concepts of habitat correlates: as just one example, can Brown and Maurer's (1987) ideas of animal abundance peaking where multiple environmental factors coincide in being favorable be reconciled with Kolasa and Pickett's (1989) theory of hierarchical community organization solely by nesting Brown and Maurer's factors by wavelength (i.e., with smaller wavelength factors nested within long wavelength ones)?

In summary, therefore, it is time to acknowledge the need for a scathing indictment of the poor practices that have accumulated in species distribution modeling over the decades. The chapters I discuss above collectively reiterate the well-known problems with such modeling, but these are problems that have chronically been ignored by practitioners. I believe that this situation results from the underlying models being incor-

rect. Most of the authors I cite above identify problems with the underlying assumptions we make in our analysis of distributions but, absent an alternative paradigm, such work in essence asks us to be more cautious about our use of results. It is not until one takes up Huston's alternative that one has in place the ingredients for an alternative world view. The arguments presented from Figure P1.1 assert that analysis of constraints rather than of correlates is the only paradigm that accommodates the logic of local carrying capacity being set by the most severe of the multiple alternative limits. The logical evidence for this view is thoroughly developed by Huston; the empirical evidence is evident in the regression trees and GAM results to hand to date; and the only puzzle is why it has taken so long to acknowledge that a habitat constraint paradigm must replace that of habitat correlates.

Foundations of Species-Environment Relations

Patricia J. Heglund

Ecology is a science of contingent generalizations, where future trends depend . . . on past history and on the environmental and biological setting.

—Robert May

Historically, humans have used a combination of intuition and observation to locate resources for food and shelter. Over time we have sought to quantify the patterns that we see and to pass this information along to others. Our ability to detect habitat use patterns and to determine their cause is strongly affected by the scale of space or time at which they are studied (Wiens 1989c). Today more than ever, effective conservation actions will often require specific, predictive, accurate models of relations between species and their environment (O'Neil and Carey 1986; Stauffer, Chapter 3). In 1984, the international symposium Wildlife 2000: Modeling Habitat Relationships of Terrestrial Vertebrates was held at Fallen Leaf Lake, California, to discuss the development and application of models used in predicting the response of wildlife to habitat changes (Verner et al. 1986b). The focus of the Predicting Species Occurrences: Issues of Scale and Accuracy symposium, fifteen years later, was to provide a forum for discussing the current state of our knowledge and to identify future directions with regard to our ability to develop and assess predictive species-environment models. In this chapter, I revisit some of the ecological foundations upon which species-environment modeling efforts have been built as a means of setting the stage for the chapters that follow. To do this, I briefly highlight the following foundational concepts: (1) that space and time will influence the patterns we see; (2) that relations between species and their habitats are essentially nonlinear and depend on the scale of our investigations; and (3) that the dynamical nature of species distributions will influence the generality and accuracy of models over space and time. I then briefly discuss which foundations I consider weak and require additional work.

Overview of Species-Environment Models

Species-environment studies examine the general environmental characteristics associated with the distribution of a given species. Models based on measures of these conditions are used to predict patterns in species diversity as well as to make inferences about the attributes and adaptations of the species for which they have been measured. Underlying species-environment models is the premise that predictable relations exist between the occurrence of a species and certain features of its environment (the *niche* as defined by Grinnell 1917; the *niche-gestalt* of James 1971) and that the distributions of species have adaptive significance (Hildén 1965; Rosenzweig 1981).

Models may either address single species or more complex multispecies assemblages when identifying relations between occurrence and associated environmental features at a variety of scales (local to biogeographic). Examples of species-environment models include prediction of species occurrence, distribution and abundance using habitat suitability, habitat capability, pattern recognition, and wildlife-habitat relations models (Morrison et al. 1998). These models encompass an overwhelming array of statistical and nonstatistical methods employed in the assessment of species in relation to their environment. Modeling techniques include, but are not limited to, expert opinion, correlation, ordination, gradient analysis, reciprocal averaging, multidimensional scaling, linear and nonlinear regression, as well as numerous other multivariate methods (Gauch and Chase 1974; Verner et al. 1986b; Morrison et al. 1998; Austin, Chapter 5; Jones et al., Chapter 35).

Geographic information system technology allows ecologists to extend species-environment models into a spatial dimension (Goodchild, preface this volume; Scott et al. 1993; Knick and Rotenberry 1998). Again, a wide variety of statistical techniques used in modeling and mapping occurrence, diversity, or probability of use exist, including more recent innovations such as gap analysis, genetic algorithm for rule-set production (GARP), artificial neural networks, and fuzzy set theory (Scott et al. 1993; Knick and Rotenberry 1998; Hill and Binford, Chapter 7; Rotenberry et al., Chapter 22; Lusk et al., Chapter 28). Many of these techniques hold much promise for extending the predictive capability of models from simple presence/absence and abundance to modeling population vital rates in response to fluctuating environmental conditions (Landstrom et al. 1998; Spitz and Lek 1999).

Temporal and Spatial Patterns

A guiding principle of resource selection is that species use among environments should coevolve with the qualities of those environments (Hildén 1965; Rosenzweig 1981). Many species-environment studies focus on habitat relations, but clearly there is more to the occurrence of a species than habitat alone. At the most basic level, a knowledge of biogeography and evolu-

tion are essential to creating good predictive models (Morrison et al. 1998). The distribution and abundance of organisms we see today were set in motion millions of years ago, influenced by climate, plate tectonics, and competition. Thus, a major goal in ecology is to develop our understanding of the factors that determine the patterns of distribution that we see (MacArthur 1972b).

Time Considerations

Our ability to detect resource selection patterns and to determine their cause is strongly affected by the temporal scale or time frame within which they are studied. Time plays an especially important role in how we interpret patterns of resource use. The amount of time available to an organism for habitat selection is highly variable. Constraints on an organism's time may be related to the short-term nature of an important food resource, or, in the case of high latitudes, to the amount of time available for reproduction (Orians and Wittenberger 1991). Time constraints may also be placed on an organism by social pressures wherein better habitats are occupied earlier than poorer ones (Brown 1969a; Fretwell and Lucas 1969). In addition, cues available at the time of selection may not reliably forecast the availability of future resources (Orians and Wittenberger 1991). Populations in variable environments periodically face unpredictable conditions ("Ecological crunches" after Wiens 1977) that affect their reproductive success in a given area. These ecological crunches may strongly influence detectability and patterns that we see within a given time frame. This limits the usefulness of short-term studies and may result in misleading information (Wiens 1981c).

Spatial Patterns

The assessment and prediction of a species occurrence depends on the resolution of environmental patchiness in relation to the scale of exploitation by that species (Levins 1968). Dispersion patterns change as the size of the area analyzed is changed such that over large areas organisms may appear aggregated whereas over smaller areas territoriality may lead to a uniform distribution (Whittaker 1975). These reorganizations reflect both intraspecific and interspecific social interactions and the scale of habitat patchiness or other resources on which species depend. Differences in scale ultimately influence the questions that can be addressed, the sampling procedures followed, the observations obtained, and how the results are interpreted (Wiens 1989b).

Models developed at one scale and applied at a different scale may lead to misleading results (Wiens 1989b; Orians and Wittenberger 1991; Rotenberry et al., Chapter 22). For example, assume we make a simple examination of a species occurrence in relation to a given local habitat feature, and we only sample a portion of the response function that characterizes that gradient. The "true" relationship between the two variables may be quite different from the relationship identified for any portion of the gradient (Wiens and Rotenberry 1981a; Wiens 1989b). Detecting factors used by individuals to select habitats depends on the existence of enough variability among habitats such that they could affect selection. If inappropriate scales of sampling and analysis are used, key factors involved in species-environment relations may not be detected (Orians and Wittenberger 1991). Our investigations may need to encompass a more widely distributed sample of sites that characterize the full range of an environmental variable to better understand a species response to that variable. Thus, it may be as important to sample areas where a species does not occur as it is to sample where a species does occur. In addition, patterns evident at a biogeographic scale may be a consequence of events at a local scale and hence an understanding of local events is necessary to interpret coarse-grain patterns correctly (Wiens 1989b). Alternatively, patterns evident at a fairly fine grain may reflect widespread disturbances or the actual distribution of critical resources (Orians and Wittenberger 1991; Rotenberry 1981).

The Importance of Scale in Modeling

Distributional modeling has historically been conducted at four scales. At the continental, or biome scale, biogeographers frequently define the range of a species based on presence/absence or they portray patterns of species richness. The foundational concept for these coarse-resolution models resides in resource selection theory (Rosenzweig 1981). According to cur-

rent theory, habitat selection occurs on two levels: ultimate and local. (Hildén 1965; Cody 1985b). At the ultimate or evolutionary level, selection for specific habitats is reflected in differential reproductive rates (e.g., species fitness, Levins 1968). The ability of species to occupy a given location is ultimately restricted by aspects of its physiology, ecology, morphology, and behavior (Wiens 1989b). Subsequently, selection is influenced by proximal factors that work at increasingly finer or more local scales (Hildén 1965).

Regional Scale

At a regional scale presence/absence is the most common response variable resulting in the modeling of a species occurrence within its range and is delineated by the distribution of particular habitats, which are often coarsely specified (e.g. Kirtland's warbler [*Dendroica kirtlandii*] and jack pine [*Pinus banksiana*] forests). A few studies conducted at a regional level have resulted in the prediction of species abundance values or species richness, but the value and reliability of these predictions are uncertain without validation (see Part 4, and Chapter 65). Even fewer studies have attempted to predict productivity measures of a species or various changes in their population vital statistics in relation to environmental characteristics or perturbations (see Part 5).

Local Scale: The Fundamental Niche

On a proximate level, habitat selection involves immediately operative factors. A species distribution may further be constrained by geophysical events, resulting in temporary stochasticity in resource levels. Whether or not a species actually occurs in a given location may be constrained by a variety of biological interactions, including competition and predation (Andrewartha and Birch 1954; Hildén 1965; Connell 1980; Wiens 1989b). Thus, the ability of a species to occupy a given place is ultimately restricted by species-specific features.

At the local level, species-environment models are based on the assumption that an individual selects a general location according to certain landscape or topographic features. Because food resources, predator populations, and climatic conditions vary in time and space, organisms may be unable to directly assess critical resources at a given location but rather must rely on certain habitat features that are indirect but stable measures of these resources (Smith and Shugart 1987). Local habitat features may serve as cues that provide organisms with a means of predicting future environmental conditions or evaluating current habitat suitability (Orians and Wittenberger 1991). Such assessment processes can be seen in local, site-specific models that describe the occurrence of a species within a plant association; they are typically a function of microhabitat features. The development of these models, as with all species-environment models regardless of scale, is founded on the concept of the niche (Grinnell 1917; Elton 1930; Hutchinson 1957; Peters 1991; Cao 1995). Many ecologists, as a matter of convenience, view the frequency of distributions of resource use along one or more resource axes as characterizing the niche (Cao 1995).

Grinnell emphasized the environmental requirements of a species and considered the niche a fundamental distributional unit of species (Grinnell 1917). Elton (1930) later defined the niche as the "role" of the species in the community, which is a behavior-based concept. This definition highlighted the role other species play in shaping the expressed niche of an organism. Both definitions are considered conceptually vague and years later a quantitative concept of the niche was proposed by Hutchinson (1957). Based on this concept, the niche is best described by the coordinates of a species with n-dimensional resource axes and combines both the behavioral and the distributional concepts of Elton and Grinnell (Cao 1995; Morrison et al. 1998). Thus, we generally collect data on a multitude of variables within the environment, select from among the measures most strongly related to the occurrence of the species, and devise models that generally describe the location of that species in just a few dimensions (MacArthur 1968; Krebs 1972). Morrison et al. (1998) suggested we focus species-environment studies on what they call a resource axis of the niche because animals select *habitat* only in the broadest geographic sense, but they select *resources* within habitats at a very fine resolution. Herein lies an important distinction: resources are important to the survival and successful reproduction of an individual and the predictive strength of our models lies in our ability to identify and measure resources relevant to the species at hand.

Local Scale: Characterizing the Realized Niche

In reality, the fundamental niche is unlikely to be seen in the real world because the presence of competing individuals necessarily restricts a given species to a narrower range of conditions—its "realized" niche. The foundation of our current modeling efforts lies in the characterization of a species' realized niche rather than simply determining habitat relations. Once patterns in resource use have been recognized for a species or group of species, ecologists then attempt to identify specific elements of the environment associated with the occurrence of an individual or species. The numerous niche-gestalt versions of species-environment modeling (James 1971) generally relate species response in a binomial fashion (presence/absence) with very detailed independent variables. Theoretically, most species should exhibit a unimodal distribution approximating a Gaussian response curve, with a maximum response at some point along an environmental gradient (Gauch and Chase 1974). However, a number of factors (competition, predation, disturbance, etc.) place pressure on an organism and the curve narrows to its realized niche. It is important to note that the realized niche and its optimum may differ from the fundamental niche in both location along a gradient and in the shape of the response (Austin and Meyers 1996). Austin and Meyers (1996) noted that the realized niche could take on a variety of shapes from skewed to bimodal. They cautioned that failure to recognize the various shapes of response curves may result in inefficient or incorrect predictive models. Under niche theory we would expect a species response to any environmental variable to be curvilinear at a minimum. It is therefore somewhat surprising that the occurrence and number of species has been linearly related to a variety of habitat features (MacArthur 1964; Wiens 1989a; James and McCulloch 1990). Linear models may not adequately explain species-environment relations without including quadratic or more complex terms (Meents et al. 1983).

The Dynamical Nature of Populations and Environments

The approaches described above each provide a fairly static picture of populations when in reality species-environment relations are dynamic. Many species-environment models are based on cross-sectional studies of habitat that may be replicated in space, but not in time (O'Connor 1986). Populations fluctuate in abundance between years in response to a number of factors similarly dynamic, including weather, food conditions, habitat, predator abundance, and parasite loads (Wiens 1989b). These factors may affect the overwinter survival and the reproductive output of organisms, thus holding breeding densities below carrying capacity (Van Horne 1983). Variations in populations may be cyclic, allowing for the development of predictions, or episodic over time and space, restricting our ability to make reliable predictions about species-environment relations (Wiens 1989a). Equally important, populations may not respond to environmental changes in the short term (e.g. Wiens and Rotenberry 1985) or even in the long term, leading to what Knick and Rotenberry (2000) have called the "ghosts of habitats past."

Environmental Variation

Population levels are not static from year to year given the variations in environmental conditions that occur. Depending on the species, there may be fairly close tracking of changes in resource levels (e.g., food) or environmental conditions (e.g. precipitation, Wiens 1989a). However, when environments become episodic or fluctuate radically, close tracking of resources is difficult. Alternatively, conditions may fluctuate in a cyclical fashion within or beyond the lifespan of individual organisms (e.g., El Niño) making prediction more feasible (Wiens 1989b).

Populations and intra- and interspecific dynamics vary from location to location. Numerous hypotheses have been proposed to account for spatial variation in the number of species (Begon et al. 1990), including evolutionary history of an area, climate, disturbance, seasonality, energy, competition, and predation. Predicting species occurrence and abundance may be difficult when populations and their requisite resources

or competitors and predators vary spatially and temporally (Wiens 1989b). The environmental patchiness or local environmental heterogeneity influences the spatial variability of any resource or feature relevant to the existence of an organism in that location (Roth 1976). Such predictive models will require, at least, knowledge about individual species-resource requirements and about the production of resources in an area (Wright 1983). Detailed information about resources available to a particular species and how that translates into an estimate of abundance for that species would ideally include some consideration of how much was unavailable due to physical environmental conditions, competitors, and predators, and about the resource requirements of individuals of the species (Wright 1983). Knowledge of the temporal and spatial dynamics of populations in relation to variations within and among their environments is important for several reasons. Although annual changes in populations may be apparent between sampled areas within a single location or among several locations within a given habitat, when samples or locations are combined and annual changes averaged over a broader scale, substantially less variation may be observed (Wiens 1989b).

Intraspecific Dynamics

If a goal of species-environment studies is to identify patterns that reveal underlying ecological processes, the interpretation of species responses to environmental features may be difficult without a knowledge of the distribution of key environmental features and the density of populations in relation to them (Orians and Wittenberger 1991). A broader-scale approach may provide an understanding of certain aspects of population dynamics. Brown (1969a) suggested there might be limits between a species density and habitat characteristics because as a population increases the number of territorial individuals would eventually reach a carrying capacity set by the size of the habitat patch and a minimum territory size. Fretwell and Lucas (1969) expanded this model to describe a hierarchy of habitat preferences, with the highest-quality (as measured by evolutionary fitness) habitats colonized first; that habitat quality might be density-dependent such that quality decreases with increasing density. In this way, an individual breeding in suboptimal but uncrowded habitat might have the same fitness (following the ideal free distribution) as an individual breeding in an optimal but crowded habitat (Fretwell and Lucas 1969).

These dynamic concepts are relevant to species-environment modeling (O'Connor 1986). Fretwell and Lucas' (1969) hierarchy of habitat preference predicts that a greater variety of habitats should be used as population density increases and that individuals will occupy the best habitats when density is low. Additionally, reproductive success should decline as population density increases as more individuals are forced to breed in suboptimal habitats (O'Connor 1986). In this way, we may have an altered view of significance of certain habitat features in relation to the occurrence of a species, and species-environment correlations are likely valid only at low densities. In addition, nonlinearities in species-environment relations are more likely under this scenario. Thus O'Connor (1986) cautions us to remain aware of the dynamical aspects of populations and habitat use and to consider the possible impacts of species distributions that may be influenced by different processes in the development of species-environment models.

Species-Energy Relations

Strong relations have been demonstrated between species richness and temperature, solar radiation, and precipitation at the regional scale (Wright 1983). Currie (1991) has shown that energy, as measured by actual evapotranspiration or potential evapotranspiration, within a system is an underlying mechanism driving species richness. Latitudinal gradients in species diversity (Pianka 1966) and in various metrics related to the ecology of species, including clutch size, have been recognized (Lack 1947). Ashmole (1963) is credited with suggesting that latitudinal gradients in clutch size might well be related to energy as measured as a ratio of summer measures of actual evapotranspiration to winter measures of actual evapotranspiration (Ricklefs 1980). Thus, species richness in any given area is ultimately limited by physical constraints such as energy availability (Currie 1991) and seasonality (Ashmole's hypothesis after Ricklefs 1980). The con-

cepts embodied in species-energy theory and equilibrium biogeography can potentially be used not only to predict patterns of species number but also to address in detail the abundances of individual species and their probabilities of being present or absent (Currie 1991).

Modeling in the Absence of Adequate Data

Although a variety of strong relations have been identified between species and their environments, the extent of a study in time or space will have an influence over the relations observed. As modelers, we attempt to—and perhaps even believe that we can—perceive the environment in the same way as our species of interest (Best and Stauffer 1986; Morrison et al. 1998). Ecologists still do not fully understand how ecological and historical phenomena interact to determine diversity or the exact distribution of a species (Schluter and Ricklefs 1993). Currently, we make assumptions regarding the nature of what we consider important environmental metrics, whether they be causal or simply serve as a surrogate measure with some indirect and often unknown relation to a causal factor.

The life-history information necessary to develop reliable ecological models for any local population is limited or lacking (Laymon and Barrett 1986). Despite the large number of wildlife-related studies conducted annually, only a small number report on species-environment relationships (Karl et al. 1999). Much of the available natural history information largely involves a few well-studied taxa such as game species and other politically important species (e.g., threatened and endangered species, or species of special concern, Karl et al. 1999). Large area mapping and modeling efforts typically do not include collection of field data. For mapping purposes, modelers rely on information from the literature. Much of the ecological literature is either at a finer scale (grain) than can be mapped with today's technology or it resides in often hard-to-find state and government reports or student theses, or it simply does not exist (Laymon and Barrett 1986; Karl et al. 1999). What we end up with are an excess of measured variables and a lack of information regarding which ones are important. A thorough knowledge of the natural history of species and their life requisites is essential for understanding the relations between species and their habitat and for determining which particular habitat components are relevant to the survival and productivity of each species (Rosenzweig 1981; Morrison et al. 1998).

Weakness in the Foundations

To date, most of the advances in the study of species-environment relations have involved refinements of analytical techniques (Morrison et al. 1998). Unfortunately, linear models have frequently been used to describe relations between an organism and its environment that are essentially nonlinear (Meents et al. 1983; James and McCulloch 1990). Although strong linear relations have been demonstrated among species assemblages and a variety of habitat features, including vegetation structure and composition (MacArthur 1964), the distribution, or response, of an organism in regard to a given environmental variable is generally considered nonlinear and should be modeled as such (Whittaker 1975; Gauch and Chase 1974; Austin 1976; Heglund et al. 1994).

A common criticism of species-environment models is that most are based on correlations and provide little insight pertaining to the mechanisms underlying species-environment relations (Capen 1981; Capen et al. 1986). A better knowledge of the ecological and evolutionary causes of the species environment relations we see will lead us to a rationale for selecting the appropriate variables to measure. It is critically important to understand the relations underlying the observed patterns before we can devise or implement effective management plans and to do so at the correct scale. Morrison et al. (1998) caution that if our fundamental approaches to modeling are flawed, the analytical tools we use to develop models are of little importance. Additionally, we recognize that if correlative factors are weakly to moderately related to causal factors, then their predictive value is likely poor, but we seldom test the accuracy of our models.

This is, in part, due to insufficient field data that are most often drawn from a restricted range of observations from a small area due to time and cost constraints. These data are then used to develop models that are often applied over a wider area representing a novel array of environmental conditions (Johnson 1981b; Green 1979). Many models are developed specifically for use at one scale or another, and it has long been recognized that extrapolation of results to multiple scales is limited by the nature of the data. Ideally, models should be developed and tested using independent data sets derived from field studies that draw samples from the range of populations densities and habitats used by each species of interest (Johnson 1981b).

Finally, we should keep in mind that the statistical distribution of a species may not adequately represent the actual distribution of a species. The dynamical nature of populations and habitats can lead us to develop predictive models that give an altered view of the significance of certain habitat features in relation to the occurrence of a species. In future modeling efforts, we should try to consider all of the possible influences on species distributions discussed herein and throughout this book and focus more effort on determining underlying causes of these relationships.

Acknowledgments

It is with great humility that I recognize the multitude of authors who put forth the foundations and original ideas summarized but not necessarily cited in this chapter but who are included in the Literature Cited section of this book. I thank R. J. O'Connor for providing candid evaluations of earlier drafts, and I greatly appreciated additional reviews by J. Pearce, J. T. Rotenberry, and J. M. Scott. My work on this project was graciously supported by Dr. J. Olson and Mr. D. Pritchard of Potlatch Corporation, Lewiston, Idaho, and the U.S. Fish and Wildlife Service, Division of Refuges, Anchorage, Alaska.

Standard Terminology: Toward a Common Language to Advance Ecological Understanding and Application

Michael L. Morrison and Linnea S. Hall

Standardized, operational definitions are essential if different workers are to make similar measurements of similar entities. Operationalization is the practical specification, or measurement, of the range of phenomena that a concept or term represents. Because no definition is so precise that all uncertainties about its meaning are removed, initially unnoticed ambiguities in the definition will likely lead to misapplication of the concept (Peters 1991:77). Thus, authors who provide new or modified definitions of old terms, no matter how precise those definitions may be, are undoubtedly adding to a conflagration of terms and further confusion. Standardization and operationalization of terms is also important for the users of the results of scientific research.

Growth of ecological understanding depends in large part on the exploration of new concepts. As noted by Peters (1991:78), it is normal practice for good scientists to entertain seemingly irrational, vaguely formed, and poorly defined concepts and their associated terms, because these concepts and terms are often the building blocks that lead to strong theory. However, because a concept is a general notion of how a process behaves, it is difficult to operationalize it. And, unfortunately, because concepts are not open to testing until they are recast as testable hypotheses, the early products of conceptual development often multiply and then are difficult to extirpate (Peters

1991:79). As developed by Goldstein (1999), when conservation planning and management are based on novel constructs that are imprecisely defined, then the strategy itself is founded in a vacuum.

Peters (1991:20) thought that although nothing is gained by arguing definitions, differences in the meanings behind words could be important. If a clear, operational definition is not developed, different users of a term may develop independent, often inconsistent definitions. In this way the original concept diversifies in meaning until any single meaning of the concept appears restrictive and inappropriate. A concept soon carries so many meanings that one can never be sure which is intended. Likewise, Fauth (1997) noted that definitions are not simply fodder for semantic arguments. Rather, definitions arise from the operational requirements and structure of theory; the challenge is to decide which definition best promotes further theoretical development.

As noted by Peters (1991:81–82), clarity can only be achieved if a term is defined at each use, but this proliferation of definitions usually only confounds the problem. As summarized by Peters (1991:82), Hurlbert (1981) found twenty-seven definitions for "niche," MacFadyen (1957) found seven for "community," and Hawkins and MacMahon (1989) found three for "guild." Recently, Hall *et al.* (1997b) reviewed fifty papers from prominent journals and books in the wildlife and ecology fields between 1980 and 1994 for uses of the term "habitat." They concluded that habitat terminology was used vaguely and imprecisely in 82 percent of the articles reviewed and that few papers provided definitions. For example, they found that "habitat type" was used—without definition—to describe both a simple vegetation association and the habitat used by one or more animal species (see our definitions below).

Thus, there is both theoretical and practical utility in establishing precise and consistent definitions of terms. This is not to say that definitions cannot be changed over time. But, we argue that any substantive changes must be well developed, justified, and critically peer reviewed before use. Science advances more by testing theory than it does through arguing over the definition of terms. However, the dramatic creation and misuse of terms makes it difficult for anyone to apply the results of different studies. And, there is no justification for being sloppy with the use of terms, nor should we accept vague and self-serving justifications for failing to adhere to rigorous standards.

In this chapter, we first develop a hierarchical context within which to place ecological studies, and then we give definitions following from this context. Definitions must be developed in this manner; that is, they must follow *from* a theory or concept. There is no heuristic value in developing ecological terms in isolation from a conceptual framework.

The terms and definitions provided below are meant as a guide for all authors whose works appear in this volume, as well as for other ecologists. For consistency, authors have agreed to follow our terminology or to provide a rationale for using alternative definitions. This will assure readers that terms will be defined and adhered to throughout this volume so that meanings are not equivocal. Additional terms are defined in the Appendix.

Hierarchical System Organization: Level and Scale

The terms *level* and *scale* are not synonymous (King 1997:198), and unfortunately, ecologists and geographers have used different definitions of "large" and "small" scale. Scale refers to the resolution at which patterns are measured, perceived, or represented. Scale

can be broken into several components, including grain and extent. *Grain* is the smallest resolvable unit of study, for example a 1×1-meter quadrat. Grain generally determines the lower limit of what can be studied. *Extent* is the area over which observations are made and the duration of those observations; for example, the boundaries of a study area, a species range, or duration of study (Milne 1997). Thus, many characteristics of wildlife will vary with scale, such as habitat, animal density, patch geometry, or resource availability (*habitat* and *resource* are defined below). Different patterns we observe in these characteristics with changes in scale reflect transitions between the controlling influence of one environmental factor over another (Milne 1997).

Thus, grain and extent should define the scale of each study. Because four combinations of grain and extent are possible, the term *scale* should not be used without first specifying which combination is intended: small-scale studies should be of small grain and either small or large extent; large-scale studies should be of large grain and either small or large extent. To define these terms otherwise only leads to confusion among workers and makes it difficult to place study results in the context of overall theoretical considerations (e.g., foraging, movement patterns, population control).

To further complicate matters, "scale" for a geographer or cartographer refers to the relationship between distances on a map and distances on the ground. Thus, 1:24,000 is a larger scale than 1:100,000. Technical limitations (e.g., amount of time and money, computer storage capabilities) limit the application of large mapping scales to large areas. Thus, the small-grained study of an ecologist can be diagrammed or mapped at a relatively large *mapping scale* if the spatial extent of the study area is not too large. If the extent is too large, then the study elements can only be depicted at a small mapping scale. But, the mapping scale is not directly related to the grain of the ecological study being pursued.

Workers should be careful not to confuse mapping scales with the ecological scale necessary to investigate a concept of interest. Jenerette and Wu (2000) discussed these definitions of scale in detail and urged ecologists to be explicit when defining their use of the term.

Level is a relative ordering of system organization (after King 1997). Invoking the term *system* implies that a set of connected parts form a whole or work together in some fashion. The *levels* of organization in a hierarchically organized system are not isolated or independent. The elements at one level emerge as a consequence of the interactions and relationships among elements of the next lower level. These interactions and relationships can be quantified by differences in rate structure, frequency of behavior, frequency of interaction, and interaction strength. Thus, the higher levels provide the context for the lower levels.

King (1997) concluded that levels of organization should always be extracted from data and should not be imposed by a priori assumptions about what levels of explanation should be. For example, a landscape is often considered a hierarchical "level." But, do the patches within the landscape *interact* to form the landscape? It cannot be presumed that they do. Thus, it is better not to refer to a landscape as the "landscape level" but rather as just "landscape" until interactions are demonstrated. "Level" is used to indicate "the level of organization revealed by observation at the scale (grain and extent) under consideration."

Definitions

The following definitions were offered as a common vocabulary to authors, and it is hoped they will also provide guidance for readers.

Habitat

The term *habitat* is a concept and, as such, cannot be tested per se. Habitat is understood, even by the lay public, to mean a place where an animal resides. The traditional definition of habitat is useless as a predictive tool; it is a "concept cluster" (sensu Peet 1974) with numerous similar, but not identical, definitions. There is no "theory of habitat." Rather, habitat is a concept that serves as an umbrella under which specific relationships between an animal and its surroundings are stated as testable hypotheses.

Habitat has a spatial extent that is determined during a stated time period. Thus, the physical area occupied by an animal can be described (by the observer) by both extent and grain. The various factors we commonly recognize as components of habitat—cover, food, water, and such—are contained within this area. The direct or indirect *use* of these factors can be measured and quantified; numerous authors (review in Hall et al. 1997b) have recognized the term "habitat use." A functional relationship between resources (defined below) and animal performance is assumed; the observer often does not define a specific area, or he or she might produce a user-defined area that is perhaps based on animal activity. Thus, habitat is a convenient boundary for measurement of vegetation, various other resources, and the environment. A priori, user-defined boundaries are convenient but are likely artificial, because the resources contained within those spatial areas will, of course, vary over time both in response to abiotic factors and as a result of use of resources by animals. The spatial extent of the habitat and the grain at which measurements are taken should be defined to increase communication among workers. Descriptions of habitat should consider the dynamic nature of the components. Terms such as *microhabitat*, *mesohabitat*, and *macrohabitat* refer to grain size and are usually used to characterize the continuous nature of the factors that can be measured within an animal's habitat. Here again, technical limitations prevent microhabitat from being described and measured over a large extent at large mapping scales. Macrohabitat usually includes measures such as canopy cover and tree density, whereas microhabitat will include shrub-stem density and pebble cover.

Quantifying habitat quality requires discovering which relationships determine individual fecundity/survivorship at the appropriate scale (grain and extent). The strength and frequency of interactions between the individual and its environment define the performance of the animal (e.g., survival, fecundity) and are considered niche relationships (discussed below). Thus, we can draw boundaries around where the animal performs activities and interacts with the biotic and abiotic characteristics, which can then be called the spatial extent of the habitat. Within the spatial extent we can then define the spatial and temporal resolution of our observations—the habitat grain.

Habitat can certainly be used to develop general

descriptors of the distribution of animals. However, we fail repeatedly to find commonalties in "habitat" for most populations across space because we are usually missing the underlying mechanisms (e.g., size distribution of prey, forage nutrients, competitive factors) that determine occupancy, survival, and fecundity. Habitat per se can only provide a limited explanation of the ecology of an animal. Other concepts, including *niche*, must be invoked to more fully understand the mechanisms responsible for animal survival and fitness. As we have seen in many wildlife studies (e.g., Collins 1983; Mosher et al. 1986; review by Morrison et al. 1998), "high-quality habitat" varies in physical attributes for a species across its range because it is not measuring mechanisms. Rather, our statistical models of habitat are, at best, analyzing surrogates of these mechanisms. Habitat is a useful concept for describing the physical area used by an animal and should probably retain its simplicity for ease of communication among scientists, managers, and the public.

Niche

Wiens (1989b:146) called the *niche* one of the most variably defined terms in ecology. Two primary meanings have been ascribed to it:

The *Grinnellian niche* of a species is the set or range of environmental features that enable individuals to survive and reproduce. Grinnell's (1917) focus was on factors determining the distribution and abundance of species.

In contrast, the *Eltonian niche* pertains to a species' functional role in the community, especially with regard to trophic interactions (Elton 1927). Hutchinson (1957) reinforced and expanded this concept of the niche by mathematically describing a large number of environmental dimensions, each representing some resource (see below), or other important factor, on which different species exhibit frequency distributions of performance, response, or resource utilization (Wiens 1989b:146). Collectively, the dimensions define an *n*-dimensional space and the frequency distributions for a species define an *n*-dimensional hypervolume within this space—its niche. And, according to Peters (1991:91), the hypervolume is an infinitely large set of properties that cannot be operationalized.

Subsets of properties can be quantified, however, which makes the hypervolume useful conceptually but precludes testable hypotheses.

Grinnell's niche was thus primarily autecological, whereas the Hutchinson niche was synecological and was concerned with the relative position of a species within a community of species. Thus, the view taken results in a different emphasis of study: individualistic studies under the Grinnellian view and studies of communities under the Hutchinson view. As summarized by Wiens (1989a:147), there is nothing inherent in the Hutchinson concept that excludes incorporation of Grinnell's concepts. The factors emphasized by Grinnell may also be considered as dimensions within the Hutchsonian niche.

Note that under the Hutchsonian niche multiple species are assumed to interact within a community; the community is assumed to exist. In contrast, Grinnell's objective was to understand population regulation in terms of resources that limit the distribution and abundance of a single species; this understanding must precede analysis of communities (Wiens 1989b:Table 6.1). Hutchinson's view thus can be interpreted to suggest that a *level of organization* (the community level) *exists*. As summarized by Wiens (1989b:178–179), patterns that emerge from studies of groups of species (e.g., "guilds," "assemblages") may be consequences of imposing an arbitrary arrangement that is actually structured in some other way or not at all.

Hutchinson's view of the niche thus violates our adoption of rigorous criteria for identifying organizational levels. We agree with Wiens (1989b:176–177) that much insightful work has resulted from studies of "niches" (as well as of "communities" and "guilds"; see below). However, we reiterate that no organizational structure should be preimposed before it is determined from measured observations.

Arthur (1987) recommended that we follow MacArthur's (1968) quantification of the niche, which plots utilization against some quantifiable resource variable—the resource utilization function (RUF). Arthur argued that it was better to build complexity as needed, as with RUFs, rather than to dissect it as when using a hypervolume concept. RUFs describe the choice of resources by animals; these choices can be constrained by predators, competitors, and other factors. We think that this type of approach is preferable because it makes far fewer assumptions about organizational structure, and it can be tuned to fit specific questions.

It is not our intent to recommend study designs and methods here (see Morrison et al. 1998 for recommendations on habitat, niche, and resource studies). However, we recommend that the following standards should be followed when discussing the niche:

1. Clearly describe the niche concept being assumed (e.g., see Wiens 1989b:Table 6.1 for guidance). For example, what ecological hierarchy, if any, is being assumed? Refer to the above discussion of biological *levels* for guidance.
2. Describe the range of resources thought to influence the species' distribution, abundance, or interactions, and the specific subset being studied. This description is, of course, closely related to the description under (1), above.
3. Describe the specific relationship(s) being examined or tested. That is, describe the RUFs (or other defined terminology).
4. Clearly separate resources (e.g., food, space, minerals, nest sites) from constraints on the use of those resources (e.g., predation, competition, activity time).

Guilds

Wiens (1989b:156–159) presented a brief but insightful review of the guild concept. Root (1967) defines guilds as "a group of species that exploit the same class of environmental resources in a similar way." Wiens (1989a:156) identified three key elements of a guild: (1) species are syntopic (co-occurrence in the same habitat); (2) similarity among species is determined by their use of resources (niche requirements) rather than by their taxonomy; and (3) competition among species is especially important. Most applications of the guild concept have concentrated on subsets of species within the same family. Jaksic (1981) and MacMahon et al. (1981) argued that this misinterprets Root's (1967) intentions. A resource (e.g., seeds of a certain size) is used by a host of species, such as ants, lizards, rodents, and birds. Thus, restrict-

ing guild membership to only one taxonomic category captures only a segment of the potential interactions.

Jaksic (1981) identified two types of guilds: a *community guild* included all species that were known to use a specific resource in a similar way; an *assemblage guild* included species within a specific taxonomic category. He made these distinctions because he thought it was not possible to include all species using similar resources in a single study. As summarized by Wiens (1989b:159), the Root (1967) concept represents clusterings of species overlapping in resource use, whereas the Jaksic-MacMahon concept examines how several consumers use a resource and how different resources share consumers.

However, Jaksic's (1981) community guild should be avoided because it mixes two related and overlapping terms. Furthermore, a community can conceptually be subdivided into guilds (Wiens 1989b:159); thus, a level of organization is implied without the benefit of data. Likewise, assemblage guild should be avoided because when following Jaksic's definition it gives the false impression that a guild is being analyzed (when, in fact, only taxonomically related species are under study).

Thus, we adopt the term *species assemblage* when one is simply studying some group of species for any number of interesting reasons. This is the same terminology that can be used to identify a "community," because in both cases we are simply choosing a group of species for which we have no knowledge of the organizational structure (i.e., level). As noted by Wiens (1989b:159), the primary value of the guild concept is to focus attention on sets of species sharing positions in niche space, or influencing resource dynamics in similar ways, so that the consequences of these similarities may be determined. We emphasize the latter part of Wiens' comments: it is the *consequences* of the interactions between species and their resources, as reflected in survival and fecundity, that are of primary importance in understanding ecological relationships. We should avoid artificially assigning a level of organization (i.e., calling our work a study of "guilds") when, in fact, no such level has been quantified. As summarized by Wiens (1989b:178–179), just because an ecological *concept* such as the guild may be a handy way to group species, it is not necessarily biologically valid. This is because of the arbitrary nature of most guild classifications: "patterns" that arise from such studies may be the result of imposing an arbitrary arrangement on a group of species.

Community

A survey of the definitions of "community" (Wiens 1989b:3–5) shows that the co-occurrence of individuals of several species in time and space is common to most. Additionally, most also stress the importance of interdependencies among the populations under study. Wiens (1989c:257–258) stated that "regardless of how one defines 'community,' the community being investigated and the criteria used to determine its membership should be described explicitly." He went on to add that we "should not become overly concerned with the semantics of communities or with whether or not communities are 'real' or possess holistic properties." He concluded that multispecies assemblages do occur in nature, and thus we should concentrate on identifying interactions among these species. Wiens chose to "accept the operational utility of talking about bird communities as assemblages of individuals of several species that occur together." We understand this viewpoint but suggest that Wiens' conclusion does not advance the explicit study of communities. Rather, it furthers the arbitrary nature of our definitions and the haphazard way in which we conduct our "community-level studies." Likewise, we disagree with Fauth et al. (1996) that a community can be described simply by placing boundaries around a study site. We suggest that when the level of the community cannot be identified from data, the term "species assemblage" be used instead of "community." And, like Wiens, we urge that the composition of this assemblage and its boundaries—no matter how arbitrary—be explicitly stated. This is similar to the recommendations of Jaksic (1981) regarding terminology applied to guilds. Our arguments given for "guild" and "niche" apply here, and we will not reiterate them.

Population

All studies must identify the population of inference. *Population* has usually been defined as a collection of individuals. However, few studies explicitly identify the *boundaries* of the population under study. Thus, the applicability of research results beyond the individual animals actually being studied is unknown. Because most species are composed of numerous ecotypes and often numerous subspecies, extrapolating beyond the individuals under study is seldom warranted and can lead to misuse of results.

Individual animals interact with resources. Two or more animals interact individually and collectively with each other (e.g., through copulation, predation, and competition) and with the environment, forming populations (or subpopulations or metapopulations). Thus, the population has an organizational level. However, our understanding of this level will largely be based on the scale—extent and grain—chosen for study.

Peak (1997:70–72) noted that several major problems arise when discussing populations: (1) the number of sampled populations may not represent the number of relevant populations, and (2) the level of aggregation associated with the sample is likely not the appropriate level to quantify dynamics. Cooke (1997:188–189) noted that the foundation of population studies rests on the claim that a population can be characterized from local observations. However, although the traditional principles of population dynamics are the basis of many resource management decisions, treatments are rarely applied to entire populations but rather to demes that are treated as if they were populations, or to individuals as representatives of "the population." Treating demes, which are subsets of a population, may be simpler than treating populations but may create changes in dynamics that are too complex to interpret readily.

The term "population" does define a conceptual unit that has properties that an observer can define. The problem arises when we try to establish the domain of inference from population studies. That is, having studied a "population" (i.e., some assemblage of individuals) and derived values for various measures of that population, the problem is to determine to what do we apply these measures? It is an example of the distinction between statistical description (e.g., a mean and standard deviation) and statistical inference (e.g., how widely can we apply the statistical description?). Our above comments on defining a population apply primarily to the latter, while the former remains a legitimate use of the concept of population (J.A. Wiens, personal communication).

Thus, we recommend that the following standards be used in all studies:

1. Clearly define the group of animals being studied with regard to spatial extent (i.e., define your study area).
2. Explicitly describe how the study "population" was identified. That is, was a convenient area selected (e.g., a research station, study plot, island), or were ecological data (i.e., demographic or genetic) actually used to define the population (or deme or subpopulation) boundaries?
3. Discuss how changes in grain and extent might influence study results.
4. Describe the current taxonomic classification of the species (i.e., species, subspecies) and known or suspected ecological classification (i.e., range of environmental conditions occupied, the ecotype).
5. Describe the potential applicability of your results in light of 1–4, above.

Ecosystem

Fauth (1997) discussed the concept of *ecosystem* in light of recent ecological theory. As is usually done, Fauth concluded that the basic spatial and temporal dimensions of an ecosystem were user defined, rendering an ecosystem an adimensional conceptual unit. He further noted that ecosystem boundaries could be artificial (a square meter of lawn) or natural (a pond), depending upon one's goals. Fauth then categorized the components within this boundary as either biotic or abiotic. He defined all members of the biota that occur within a bounded area as a *community* and all abiotic factors within the same area as the abiotic environment. Interactions within and between these two components form the ecosystem.

In most cases, researchers will have little knowledge of an ecosystem, including its interactions and boundaries. It is probably safer and less confusing to simply refer to the scale (grain and extent) of the study and restrict discussion to the functions and relationships observed in the area. Thus, we recommend that the following standards be used when discussing ecosystems:

1. A level of organization should not be assumed; rather, a statement should be made describing the specific functions and interactions under study or potentially influencing the study.
2. Other potential influential factors and interactions not studied should be mentioned, along with their potential impacts on study results.
3. The spatial and temporal extent of the study should be described (see also "Population"), no matter how arbitrary.
4. The term "ecosystem" should be avoided altogether and instead replaced with the descriptions suggested in 2 and 3 above.

Landscapes

Landscape can be defined as a spatially heterogeneous area used to describe features (e.g., stand type, site, soil) of interest. King (1997:205–206) described a *landscape* primarily by its spatial extent. The term *landscape level* is best understood as *the level of organization revealed by observation at the spatial extent of the landscape*, but only after that level and the associated hierarchical organization have been determined from the data collected (King 1997:205–206).

A serious problem with application of the term "landscape" is that it is usually taken to mean relatively large areas (1–100 square kilometers) that are composed of interacting ecosystems (Forman and Godron 1986; Davis and Stoms 1996). However, the perception of "landscape" to a small animal is likely much different than that perceived by a large animal. As reviewed by King (1997:204), the fundamental themes of landscape ecology are not scale dependent or limited only to spatial extents greater than a few square kilometers. Questions of how spatial heterogeneity influences biotic and abiotic processes—a theme defining landscape ecology—can be addressed at virtually any spatial scale. Thus, by adopting an organism-centered view, a landscape can credibly be described using a microscope *or* a satellite. Thus, we recommend that no area limitation be placed on the landscape. Describing a landscape in whole kilometers will be appropriate for certain applications (e.g., gap analysis), whereas describing it in a few square meters will be appropriate other uses (e.g., salamander niche relationships).

Resources

Wiens (1989c:262) noted that "Although 'resources' are involved in most explanations of community patterns, all too often they have been defined in ad hoc ways, rarely measured directly, or inferred to be limiting on the basis of faith rather than evidence." Many of the items we label as resources (or natural resources) are actually artificial constructs that do not have clear biological definitions or justifications. For example, the "rangelands" resource or "rangeland vegetation" is composed of numerous abiotic and biotic properties and cannot be quantified, per se (Morrison and Marcot 1995). Rather, the actual resources in a rangeland include air, water, minerals, soil, sunlight, flora, and fauna. Thus, to really define a resource, the area of interest must be explicitly identified as to its spatial extent and broken down into its measurable component elements.

Little attention has been given to the identification and measurement of resources. As summarized by Wiens (1989b:321), almost any environmental factor that correlates with the distribution and abundance of a species has been called a resource. Most studies of "resource partitioning" employ circular logic: variables that are used differentially by species are termed resources, and the coexistence of species is then explained by the partitioning of those resources. Furthermore, Wiens (1989b:323) pointed out that without a precise definition of the resources present, it is not possible to derive accurate patterns of resource use or niche relationships (see Tilman's [1982] work on resource competition in plants for a possible exception).

Wiens (1989b:321) provided two critical features for the definition of a resource: (1) it must be used by the organism, and (2) it must be at least potentially limiting to individual fitness and/or population dynamics. He included the limitation criterion because his focus was on the potential for competition. We think, however, that the criterion of limitation is unnecessary for our application. Requiring a resource to be potentially limiting invokes competition as an assumed driving force in species interactions; a level of organization is thus required before an item—even if it is consumed and is essential to survival—be considered a resource. Thus, we instead define a resource as any biotic or abiotic factor that is directly used by an organism. Resources that *are* limiting to an organism could then be referred to as "limiting resources."

Wiens (1989b:321–323) also noted that it is critical that the differences between resource abundance, availability, and use be distinguished to be certain which one is actually being measured. *Resource abundance* is the absolute amount (or size or volume) of an item in an explicitly defined area. For example, the number of food items in a 1-hectare area. *Resource availability* is the amount of a resource actually available to the animal (i.e., the amount exploitable). For example, the number of food items in a 1-hectare area that an ungulate can reach. Finally, *resource use* is a measure of the amount of the resource directly taken (e.g., consumed, removed) from an explicitly defined area. For example, the number of food items in a 1-hectare area that an animal consumed in a six-hour sampling period.

Thus, we recommend that all "resources" invoked by an investigator as an ecological explanation for observed patterns should be clearly identified and discussed with regard to their *use* by the organism(s) under study, and careful distinctions should be drawn between the abundance, availability, and use of the resources by the organisms.

Concluding Remarks

Definitions have little heuristic value if created outside an explicit theoretical framework. Definitions made in the absence of such a framework are, in essence, "definitional orphans." Thus, for the ecological disciplines, we advocate the use of standardized terms based as far as possible on operationalized concepts. In this chapter we have suggested standardized definitions for many critical and commonplace terms in ecology. Authors may disagree with a definition but if so should provide their own justified and operationalized definition so that further ambiguity in the use of terminology is avoided.

Acknowledgments

We thank J. Michael Scott for suggesting the topic and providing suggestions. Anthony W. King, John A. Wiens, Pat Heglund, Jason Karl, and K. Shawn Smallwood provided many insightful comments on drafts of the paper.

Appendix

Provided here are definitions of standard terms used in studies of the distribution and abundance of animals. Terms followed by "see text" are discussed in detail in the main text of this chapter.

Abundance The number of individuals (Lancia et al. 1994); contrast with Density.

Accuracy The nearness of a measurement to the actual value of the variable being measured; not synonymous with Precision (Zar 1984:4).

Assemblage A group of species under study; see discussion of "community" in text.

Census A complete enumeration of an entity (modified from Lancia et al. 1994).

Community The co-occurrence of individuals of several species during a specified time and space that are interacting and show some degree of interdependencies; see text.

Complexity Relative comparisons of grains separated by a given distance.

Density The number of individuals per unit area (Lancia et al. 1994).

Distribution The spread or scatter of an entity within its range.

Ecosystem The specific functions and interactions under study or potentially influencing the study (within an explicitly defined area); see text.

Extent The area over which observations are made and the duration of those observations; see text.

Grain The spatial and temporal resolution of observations; the smallest resolvable unit of study; see text.

Guild A group of species that exploit the same class of environmental resources in a similar way (Root 1967); see text.

Habitat The physical space within which the animal lives, and the abiotic and biotic entities (e.g., resources) in that space (see also Hall et al. 1997b); see text.

Habitat Availability The accessibility and procurability of physical and biological components in a habitat (Hall et al. 1997).

Habitat Avoidance An oxymoron that should not be used; wherever an animal occurs defines its habitat.

Microhabitat, Mesohabitat Relative terms that refer to the grain size of the area over which habitat is being measured (see also Hall et al. 1997b).

Habitat Preference Used to describe the relative use of different locations (habitats) by an individual or species.

Habitat Quality The ability of the area to provide conditions appropriate for individual and population persistence (Hall et al. 1997b).

Habitat Selection A hierarchical process involving a series of innate and learned behavioral decisions made by an animal about what habitat it would use at different scales of the environment (Hutto 1985:458, Hall et al. 1997b).

Habitat Use The way an animal uses (or "consumes," in a generic sense) a collection of physical and biological entities in a habitat (Hall et al. 1997b).

Home Range The area traversed by an animal during its activities during a specified period of time.

Landscape A spatially heterogeneous area used to describe features (e.g., stand type, site, soil) of interest; see text.

Landscape Feature Widespread or characteristic features within the landscape (e.g., stand type, site, soil, patch); see text.

Level The level of organization revealed by observation at the scale under study (King 1997); see text.

Metapopulation Strictly, a system of populations of a given species in a landscape linked by balanced rates of extinction and colonization. More loosely, the term is used for groups of populations of a species, some of which go extinct while others are established, but the entire system may not be in equilibrium (Pickett and Rogers 1997).

Model Any formal representation of the real world. A model may be conceptual, diagrammatic, mathematical, or computational.

Model Calibration The estimation of model parameters from data.

Model Parameterization The process of specifying a model structure (see Conroy and Moore, Chapter 16).

Model Validation Comparison of a model's predictions to some user-chosen standard to assess if the model is suitable for its intended purpose.

Model Verification The demonstration that a model is formally correct.

Niche The strength and frequency of interactions between the individual and entities (e.g., resources, other animals) in its habitat; see text.

Patch A recognizable area on the surface of the earth that contrasts with adjacent areas and has definable boundaries (Pickett and Rogers 1997).

Population Classically, a collection of individuals; see text.

Precision The closeness to each other of repeated measurements of the same quantity; not synonymous with accuracy (Zar 1984:4).

Range The limits within which an entity operates or can be found.

Resolution The smallest spatial scale at which we portray discontinuities in biotic and abiotic factors in map form (Hargis et al. 1997).

Resource Any biotic and abiotic factor directly used by an organism; see text (see also Manly et al. 1993).

Resource Abundance The absolute amount (or size or volume) of an item in an explicitly defined area.

Resource Availability A measure of the amount of a resource actually available to the animal (i.e., the amount exploitable).

Resource Preference The likelihood that a resource will be used if offered on an equal basis with others (Manly et al. 1993).

Resource Selection The process by which an animal chooses a resource.

Resource Use A measure of the amount of resource taken directly (e.g., consumed, removed) from an explicitly defined area.

Scale The resolution at which patterns are measured, perceived, or represented. Scale can be broken into several components, including grain and extent; see text.

Scale of Observation The spatial and temporal scales at which observations are made. Scale of observation has two parts: extent and grain.

Sensitivity Analysis A process in which model parameters or other factors are varied in a controlled fashion (see Conroy and Moore, Chapter 16).

Sink Populations In a landscape, a population or site that attracts colonists, while not supplying migrants to other sites or populations (Pickett and Rogers 1997).

Site An area of uniform physical and biological properties and management status (contrast with Study Area).

Source Population In a landscape, a population or a site that supplies colonists to other patches (Pickett and Rogers 1997).

Study Area An arbitrary spatial extent chosen by the investigator within which to conduct a study (contrast with Site and Scale).

Territory The spatial area defended (actively or passively) by an animal or group of animals.

Viability Strictly, the ability to live or grow. In conservation biology, the probability of survival of a population for an extended period of time.

Linking Populations and Habitats: Where Have We Been? Where Are We Going?

Dean F. Stauffer

As we consider the state-of-the-art of modeling animal-habitat relationships, it is useful to consider the work of those who have come before. Understanding the historical context of our current approaches to modeling and analysis can help to provide a sense of "place" in the discipline, and also, if we are attentive, may help us to avoid errors and unproductive research avenues identified by those who have explored the limits of the field in the past.

As a discipline, the arena of quantitative ecology and, specifically, modeling wildlife-habitat relationships, is relatively young, yet much has been accomplished in the past five decades. My goal here is to provide an overview of how our approach to analyzing wildlife-habitat relationships has evolved over the past fifty years. I will consider changes in the use of computers, statistics, and philosophical trends over this period. Clearly, those topics I address reflect my experiences and biases and may not cover all topics that others may consider important. The space available does not allow complete coverage of all relevant and important literature; thorough historical reviews have been provided by Karr (1980) and Block and Brennan (1993).

Several general stages have been identified in the development of natural history disciplines. The cataloging stage came first, perhaps beginning with Aristotle (Block and Brennan 1993). This stage evolved into a natural history era during the mid-1800s. Through the first part of this century, we can note a dramatic increase in natural history work on a variety of species. The mid-1950s might be considered the dawning of the quantitative ecology era. It is the period from the 1950s to present that I wish to address.

For the most part, this overview considers the development of computing and statistical methods and how they have influenced the analysis of wildlife-habitat relationships. I conducted, with the assistance of undergraduate students at Virginia Polytechnic Institute and State University (Virginia Tech), a review of 415 habitat papers published in the *Journal of Wildlife Management* (JWM) for the period 1965–1996. For each paper, we noted which statistical methods were used to analyze wildlife habitat. I summarized results for decades and half-decades: 1965–1969, 1970–1979, 1980–1989, and 1990–1996 to provide an overview of how use of statistical procedures has changed over time. Clearly, this is not a complete coverage of the field (N = 1, 0 df) but I assume that the trends in JWM reflect general trends in analytic approaches, even though there will be some variation among specific disciplines. For the ensuing discussion, I consider development of approaches during four general eras. Within each, I will consider the state of computer resources, statistical methods used, and major analytic and philosophical trends during the period.

Where Have We Been?

The Qualitative-Quantitative Transition: The 1950s and 1960s

In the mid-1950s, we can see the beginning of the move from qualitative natural history research to work that was more quantitatively focused. During this time, Hutchinson (1957) presented the notion of an animal's niche being an *n*-dimensional hypervolume. This concept of species being arrayed along multiple environmental gradients influenced how many in the future would look at wildlife-habitat relationships.

At the same time, MacArthur (1958) published a rigorous and quantitative, though not highly statistical, analysis of niche relationships of five warbler species. Not long after, MacArthur and MacArthur (1961) presented their initial work relating bird species diversity (BSD) to foliage height diversity (FHD) and introduced the use of H′ as a measure of diversity. His regression equation of BSD on FHD might be considered one of the first wildlife-habitat models.

Even as some researchers were becoming more quantitative in their approach to habitat analysis, others were wishing for a consistent way to evaluate habitat. Hanson and Miller (1961:75) stated, "The work of game managers would be aided if they could readily identify some attribute of cover that permits rapid estimation of carrying capacity for bobwhite." They clearly were seeking a wildlife-habitat model that could be applied in a management context.

Although we see the stirrings of quantitative approaches in the 1960s, of the thirty-four JWM papers reviewed for the latter part of this period, only nineteen used statistics. During this time, much habitat analysis was qualitative and descriptive, relying on terms such as "very dense, dense, open, very open" (Bendell and Elliot 1966) with little accompanying data to tell us just how dense "dense" was. When statistical analyses were used, the dominant methods applied were *t*-test, ANOVA, Chi-square, correlation, and simple linear regression (Fig. 3.1). Access to computers was limited and the majority of calculations were carried out by hand or with rotary calculators and slide rules; as a result, relatively simple statistics were typically used, and use of computers warranted a special note. For example, Klebenow (1969) reported the use of an IBM 1620 and noted that his analysis was limited to twenty variables because of computing limitations.

As the 1960s ended, we saw that a theoretical framework for habitat and niche analysis was evolving and that statistics were beginning to be used, although their use was limited by computational capabilities.

The Era of Multivariate Muddles: The 1970s

The 1970s represent a decade during which there was tremendous growth in computer availability and power, with an accompanying increase in the use of computationally intensive statistical procedures. During this time, there was a substantial increase in the use of statistical analysis of habitat in JWM papers (Fig. 3.1). Most noticeable is the greater number of papers using regression approaches, particularly multiple regression after 1975. I attribute this to the availability of statistical computer software that would determine the inverse of the X′X matrix, which is an unpleasant undertaking, at the least, with a rotary calculator, or that previously would have required custom computer programs. During this time, we also saw an increase in the application of nonparametric statistics. However, in JWM the use of multivariate statistics was relatively low compared to its published use in other journals, such as *Ecology*.

This growth in statistical applications parallels the continued growth of computing power. Whereas at the beginning of the decade many researchers were analyzing their data by hand, the majority of analysis was done with computers by 1979. During this time, we saw the development of comprehensive statistical packages such as BMDP, SAS, and SPSS. We were no longer dependent on custom programs developed at individual computing centers for analysis of data. Also during this time came the ability to store data on 9-track tapes or card decks, making it more portable, but it also meant we were developing the means to make mistakes more quickly than ever before.

The most dominant analytical feature of the 1970s was the growth in application of multivariate methods such as discriminant function analysis, principal com-

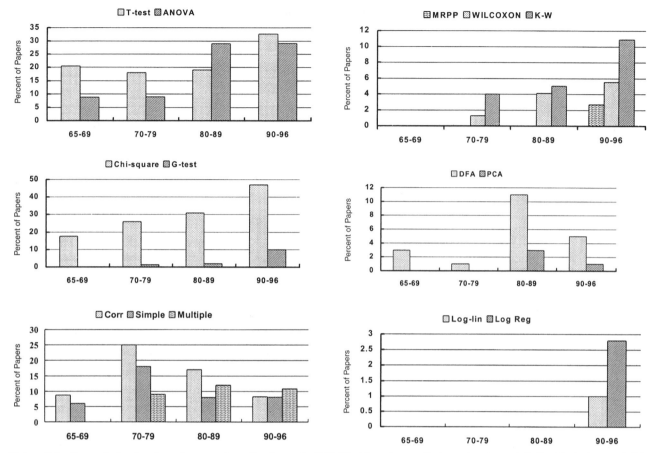

Figure 3.1. Percentage of reviewed papers from the *Journal of Wildlife Management* that used various statistical techniques at different time periods. Sample sizes for each time period were 1965–1969, 34; 1970–1979, 76; 1980–1989, 195; and 1990–1996, 110. Statistical methods summarized are t-test, analysis of variance (ANOVA), Chi-square, G-test, simple correlation (Corr), simple linear regression (Simple), multiple linear regression (Multiple), multiple response permutation procedures (MRPP), Wilcoxon rank-sum test (WILCOXON), Kruskal-Wallis test (K-W), principal components analysis (PCA), discriminant function analysis (DFA), log-linear analysis (Log-lin), and logistic regression (Log Reg).

ponents analysis, and factor analysis. The first published applications of multivariate statistics to wildlife habitat I am aware of actually come from the late 1960s: Cody's (1968) analysis of grassland bird habitat and Klebenow's (1969) work on greater sage-grouse (*Centrocercus urophasianus*). However, I believe the seminal work was that of James (1971). She introduced to us the idea of *niche-gestalt* as a description of a bird's habitat, and operationalized the multidimensional niche of Hutchinson (1957) through her analysis. This paper had a great influence on me as a graduate student and how I looked at habitat, and I suspect many others also were energized and motivated by her work. Subsequently, throughout the 1970s multivariate analyses were applied to a variety of taxa in a diversity of habitats (e.g., Martinka 1972;

Anderson and Shugart 1974; Dueser and Shugart 1978).

A potential problem with the application of multivariate statistics, however, was that anyone capable of entering their data into a matrix that could be processed by a statistical program could conduct a multivariate analysis. It was not uncommon during this time to see applications of new and unfamiliar multivariate procedures that failed to address assumptions concerning the use of such techniques. As a result, many of the analyses carried out were of limited value.

Another aspect of wildlife habitat research that was addressed during this time was the question of when a resource, such as a cover type, is preferred. Neu et al. (1974) presented an approach using chi-square to assess habitat preference in animals. We terrestrial workers had lagged somewhat behind our aquatic colleagues in addressing this question (Ivlev 1961). This approach provided a template for quantitative analysis of habitat-use-availability data and meshed nicely with the increasing number of radiotelemetry studies. However, as with the multivariate analyses, assumptions about the method being used were often not considered.

As researchers were forging ahead with sophisticated statistical techniques and powerful (or so it seemed at the time) computers to seek the truth underlying how animals were relating to their habitats, managers were seeking ways in which to use such information to address their needs. The National Environmental Policy Act of 1969 (U.S. Laws, Statutes, etc., Public Law 91-190) required that the impacts of activities using federal funds be described prior to action. This included impacts on wildlife habitat. As a result, approaches were devised to allow the assessment of these impacts. Prevalent among the methods proposed were the Habitat Evaluation Procedures (HEP) developed by the U.S. Fish and Wildlife Service (USFWS 1980, 1981a) that made use of habitat suitability index (HSI) models and the wildlife-fish habitat relationships models of the U.S. Forest Service (Thomas 1982). Pattern recognition (PATREC) models, based on Bayesian statistics, also were adapted during this time to address natural resource problems (Williams et al. 1978). As these approaches were developed, the groundwork was likewise laid for potential conflicts between researchers, who were seeking "truth," and the managers, who were seeking tools to help them meet their resource management needs.

During this era, we see relatively little consideration of the precision or accuracy of the models that were being developed. A statistically significant (e.g., $P < 0.05$) result for a regression or multivariate analysis usually was taken to mean that the model was "good," and the models were accepted somewhat uncritically, perhaps because many researchers did not fully understand the assumptions underlying the procedures. Classification accuracy of models was typically assessed by how well the model predicted the data used to build the model, which, as we would expect, indicated the models predicted well.

Thus, in the arena of habitat analysis, the decade closed with many on the multivariate bandwagon, convinced that with adequate computer power and sophisticated statistics there was no problem that could not be overcome. However, even as we forged ahead, there were some who reevaluated what we had done and offered cautions. For example, Roth (1976) found that in some habitats, the BSD-FHD link of MacArthur (MacArthur and MacArthur 1961) didn't apply in all cases and that horizontal diversity within a habitat may be as important, or more so, than vertical diversity in influencing bird diversity. At the close of the decade, Karr (1980) cautioned care in the application of multivariate statistics, noting that we should be aware of biologically—and not necessarily statistically—significant relationships in our data.

Manifest Destiny and Gadflies: The 1980s

As we came into the 1980s, the enthusiasm of the previous decade continued, and it seemed there was nothing we couldn't accomplish with more computing and statistical power. This period saw the advent of the personal computer for data analysis and, by the end of the decade, only $5,000 would purchase a 386-20 megahertz machine with at least a 200-megabyte hard drive and 4 megabytes of RAM. In JWM, we saw during this time a continued increase in the use of various statistical methods to analyze habitat relationships (Fig. 3.1). There was an increase and then a decline in multivariate methods during the decade and a general decline in the use of regressions. However, there also was an increase in the use of nonparametric statistics, which likely represents an increasing awareness of the assumptions associated with parametric methods and the desire to not violate them (but see Johnson 1995). For the first time, the use of log-linear methods, which are computer intensive, also appears (Fig. 3.1). Multivariate methods continued to see substantial use in this decade, and a symposium dedicated to multivariate analysis was held in 1980 (Capen 1981).

Progress was made in the arena of preference assessment and Johnson (1980) provided a new technique, based on ranks, to complement and at times replace the chi-square approach of Neu et al. (1974). A little later, Alldredge and Ratti (1986) presented an evaluation of the behavior of several preference assessment approaches applied to a variety of data sets with known attributes. They demonstrated the low power of tests done with few animals and that the number of animals, observations per animal, and number of habitat types all will affect interpretation of preference. At the end of the decade, Thomas and Taylor (1990) presented a very clear description of three basic study designs that lead to preference assessment analyses. Their paper has helped to remove confusion that some researchers may have had concerning what their exact sampling unit was.

As the habitat analysis bandwagon rolled on through the 1980s, a notable addition was the analysis of landscapes. This period saw the emergence of landscape ecology as a discipline that came into its own during this time (Forman and Godron 1986). Analysis of large-area data was greatly facilitated by the increased availability of remotely sensed data (e.g., Palmeirim 1988), geographic information system (GIS) software, and powerful computers capable of handling large data sets. These technologies allowed us to measure habitat efficiently at new scales and to relate animal occurrences to these large-scale patterns (e.g., Lyon et al. 1987). As this new discipline developed, a new set of habitat metrics was generated as we sought to understand the relationship between patches, edges, and corridors. Unfortunately, it appears to be easier to invent a "new" measure of a landscape than it is to clearly explain exactly what that measure means. Although numerous new metrics have been suggested, many are redundant and difficult to understand or to relate in a meaningful way to animal populations (e.g., as the fractal dimension of the landscape changes, what does it really mean?).

Perhaps the distilling event of the decade was the convening in 1984 of researchers working on modeling wildlife-habitat relationships at the Wildlife 2000 conference. This resulted in the publication of *Wildlife 2000: Modeling Habitat Relationships of Terrestrial Vertebrates* (Verner et al. 1986b) and serves as a relatively complete summary of modeling approaches, issues, and concerns as they were at that time. A particularly useful aspect of the published proceedings was the presentation of managers' and researchers' views of various issues within modeling. Through these essays, it was recognized that these two groups have different needs and agendas that should be considered, and their sometimes-conflicting views gained recognition.

Up to this time, a large number of models of various types had been developed, but seldom were they actually tested against independent data sets to evaluate their accuracy. The 1980s saw many tests conducted to assess the accuracy of models representing wildlife-habitat relationships. A number of the papers presented at the Wildlife 2000 conference addressed model accuracy (Verner et al. 1986b). Sweeney and Dijak (1985) tested a PATREC model for ovenbirds (*Seiurus aurocapillus*) and found it predicted presence/absence on independent sites with 94 percent accuracy. Other representative model tests for this period include Morrison et al.'s (1987) evaluation of regression models and Lancia and Adams' (1985) tests of habitat suitability models. Morrison et al. (1987) concluded that their regression models were useful for predicting presence/absence but lacked the precision to adequately project bird abundance, whereas Lancia and Adams (1985) concluded that HSI models predicted adequately but pointed out that scale is an important consideration when testing models. Also during this time, O'Neill et al. (1988a) provided information on how to adjust HSI models to adapt to local conditions and in doing so helped to bridge the researcher-manager gap.

As researchers forged ahead with their efforts to conduct studies on wildlife habitat using the myriad of new techniques available, several works were published that considered the status of the field. Romesburg (1981) issued a challenge to the wildlife community to "shape up." He provided a thoughtful and, perhaps to some, annoying essay that pointed out that a substantial amount of the research done in wildlife amounts to a whole lot of nothing much, based on what he termed "general purpose data." He encouraged the use of the hypothetico-deductive (HD) method whereby clear hypotheses based on the best understanding of the system being studied would be postulated. Once the hypotheses were established, then a study could be designed to test the hypotheses. I believe a solid contribution of his efforts was to remind us that it is important to think before we go into a study; we can't simply assume that with enough data, the computer will sort it out. Romesburg's

(1981) paper led to a particular emphasis for using the HD method in the mid- to late-1980s for papers in JWM. However, this may have led to some high-quality work not being published. When we have a base of information to work from then the HD method may work well, but in many cases we still need to establish the pattern (i.e., descriptive research) for a system before we can develop a hypothesis to explain the pattern (e.g., Wiens 1989b).

Much of the modeling done (and still being done today) has used some measure of population density or abundance as the response variable that is related to some suite of habitat characteristics. Van Horne (1983) brought to our attention that at times density may be a misleading indication of habitat quality and that demographic information on survival and fecundity will aid greatly in establishing the quality of a particular habitat. She describes cases when density is a poor indicator, and her work helped many to think more clearly about what they were doing when modeling wildlife habitat. One other influential work in the early 1980s was that by Hurlbert (1984). He clearly pointed out the importance of understanding the concept of the sampling or experimental unit and how pseudoreplication can invalidate many of our conclusions drawn from statistical analyses. I suspect that most workers have been guilty of pseudoreplication at some time in their career. For example, I was able to miraculously generate 1,349 data points for a regression analysis from twenty-eight study sites (Stauffer and Best 1980, 1986), yet this point passed by two editors and numerous referees. Pseudoreplication still remains a problem today and is something we all need to guard against.

Near the end of the decade, Rexstad et al. (1988) took multivariate statistics to task. They analyzed a data set of random data composed of numbers from sources such as phone book numbers, liquor prices, produce prices, and mean temperatures. They were able to demonstrate "significance" within the analyses and provided a useful warning concerning the apparently arbitrary nature of interpretation of some multivariate techniques. Bunnell (1989) presented an enjoyable discourse on the need to bring together researchers (cerebral anarchists in his terminology) and managers (alchemists) when modeling so that

there is clear communication concerning goals and objectives, and he provided suggestions on how this might be done.

As we approached the next decade, we saw that the tremendous growth in quantitative approaches continued to address issues at different scales and took advantage of exponentially increasing computing power. However, during this time we were also treated to discourses that reflected on our work and provided guidance to ensure we correctly approached wildlife habitat analysis.

Refinement and Moderation: The 1990s

The theme and patterns of habitat analysis established in the 1980s continued into the 1990s. Computer power has continued to increase, and we now have on our desks computers more powerful than the mainframe computers some of us used for our graduate work. All major statistical analysis packages are now available for personal computers, and the concept of the "Computer Center" is foreign to most of our graduate students.

In this decade, we can see continued refinement of statistical approaches. Particularly noticeable is the advent of logistic regression and permutation procedures (Mielke 1986). It seems that during the last ten years, researchers have paid more attention to assumptions associated with various statistical procedures, and, as a result, application of various tests and analyses are generally appropriate. One trend that carried over from the 1980s is the use of power analysis. Toft and Shea (1983) recommended the use of power analysis to evaluate the probability of nonsignificant tests actually detecting a difference. As we entered the 1990s, we began to see power analysis show up in some papers. However, Steidl et al. (1997) credibly pointed out the fallacies associated with post hoc power analyses. They emphasized clearly that the appropriate use of power analysis is for study design prior to data collection; it should not be used as a tool to justify or discuss nonsignificant results.

The 1990s also saw considerable attention given to hypothesis testing. As we left the 1980s, the HD method was the desired approach, and we encouraged our graduate students to establish clear, testable hypotheses prior to conducting their study. As this statistical rigor was being emphasized for inferential statistics, others were suggesting alternatives. Reckhow (1990), among others, encouraged the use of Bayesian statistics to evaluate research results rather than the more traditional parametric statistics. Johnson (1999a) developed a case for the Bayesian approach and pointed out the common mistakes and assumptions that are made in traditional hypothesis testing. He encourages the use of evaluating data with point estimates and confidence intervals. Johnson (1995) has additionally pointed out that in many cases nonparametric statistics are inappropriately applied and that in most cases, parametric tests, such as a *t*-test, are robust and perform well. James and McCulloch (1990) provided a thorough overview of the use of multivariate statistics in ecology and found the applications by researchers wanting in many cases. They indicated clearly the situations for each method is appropriate. These works and others have helped to refocus attention on biological, rather than statistical, results.

We also saw during the 1990s further refinement of methods used at the landscape level, which is closely tied to technology development. At the landscape scale, a major new approach is the development of spatially explicit models (e.g., Pulliam et al. 1992, Rickers et al. 1995) that tie population dynamics to specific locations or patches on the ground. Turner et al. (1995) have pointed out how such models can help to meet many of the needs of natural resource managers who have responsibility to manage large tracts of land. Our ability to generate a plethora of landscape metrics increased during this decade (McGarigal and Marks 1995), but our ability to understand the nature of these measures still lags behind our capacity to create new indices. Distributional properties of these measures are generally not well known, are usually nonlinear, and are sometimes multimodal as landscapes change (Trani and Giles 1999). Hence, as we use such measures as predictor variables in models, it may not be clear just what the resulting model represents.

Determining habitat preference assessment has continued to be an important component of many studies, and a relatively new approach, compositional analysis, has come to be the method of choice to evaluate resource selection (Aebischer et al. 1993). This approach addresses perceived problems with approaches such as Neu et al. (1974) and fits well within the framework of study types outlined by Thomas and Taylor (1990). Additionally, a wide variety of approaches was synthesized and presented by Manly et al. (1993).

We continued to develop models from a diversity of habitat analyses during this period, but the tendency to develop, rather than to test, models continued (Morrison et al. 1992). During this period, however, assessing accuracy of a variety of models was undertaken (e.g., Timothy and Stauffer 1991; Flather and King 1992; Block et al. 1994; Heppell et al. 1994; Adamus 1995; Nadeau et al. 1995). Generally, these and other studies indicated we can achieve moderate to good accuracy in predicting presence/absence, but abundance is more difficult to precisely predict.

Methods to model habitat quality in a management context continued. Particularly notable was the development of the hydrogeomorphic method (HGM) (Smith et al. 1995), which is designed to evaluate wetland quality and complements HEP. Roloff and Kernohan (1999) summarized seventeen studies that tested fifty-eight HSI models; upon finding that the majority of the tests were deficient, they provided alternative guidelines for effective and valid model tests. Van Horne and Wiens (1991) reviewed forest bird HSI models and assessed ways to develop multispecies HSI models. They provided suggestions for incorporating large-area variables into these models and emphasized the need for testing model components and the final product.

Even as we developed more management models, the apparent gulf between researchers and managers had not narrowed appreciably (Turner et al. 1995) but more efforts were being made to develop models that were based on data typically available to resource managers (e.g., Dettmers and Bart 1999; Penhollow and Stauffer 2000). Starfield (1997) very nicely discussed the difference between models designed to discover "truth" and those to be used as a problem-solving tool. He developed a strong argument for better communication between decision-makers and scientists.

Where Are We Going?

Clearly, we've come a long way since the 1950s with our approaches to analyzing and modeling wildlife-habitat relationships. The development of technological aids has greatly influenced how we have been able to analyze data at multiple scales of extent and grain. As computer power continues to increase along with additional development of analytical techniques, we will be able to develop increasingly complex models. I anticipate that we will continue to see greater integration of different scales (extent and grain) of habitat information with a greater reliance on GIS and remotely sensed data. Some promising new approaches include fuzzy logic and neural networks (Lusk et al., Chapter 28), spatially explicit models (Raphael and Holthausen, Chapter 62), and CART regression (Clark et al. 1999); how well these will work remains to be seen. Increasing computing power should also allow continued use of statistical methods such as permutation tests and Bayesian approaches as appropriate.

The pattern of the past seems to be that as a new method or technique appears, it generates a great deal of excitement, researchers quickly apply it to their work, and, for a period of time, it is discussed frequently in the literature and at professional meetings. It is only later, as we come to better understand the limitations of and assumptions behind the new method or technique, that it is applied less frequently, but perhaps more appropriately. If the method proves to be truly useful for some situations, even as it falls from grace as the *method du jour*, it will continue to be used in those situations where it works well. I suspect that, as new approaches continue to be developed, we will see this pattern repeated.

As we try to bridge the gap between the "real world" needs of managers and the more esoteric domain of researchers at universities and other institutions, I am optimistic that progress will be made. As our natural resources continue to be impacted, it is critical that we have the means at our disposal for wise stewardship. One aspect of this will include the use of tools such as models that clearly allow the evaluation of potential impacts and management alternatives. Resource managers will continue to need accurate models that can be readily applied over large areas to help them address management challenges. I believe that much of the past distrust managers had for researchers resulted from the fact that many of the models developed were for small areas and required inputs of habitat data not typically present in management databases. As researchers continue to develop models at landscape scales that are more appropriate to management needs, I am optimistic that the gap between managers and researchers will narrow. We must communicate better so that researchers can develop the methods managers need to meet their immediate needs; in an adaptive management context, doing so may prove fruitful (e.g., Conroy and Moore, Chapter 16), but it will require effort on both sides of the fence.

A concern for the present and future is the information glut. New journals continue to be published, many now in electronic format, and it has become difficult to keep up with the relevant work within any subject area. The result might be a duplication of effort and repetition of research rather than research built upon what has come before. For example, two recent papers (Hepinstall and Sader 1997; Tucker et al. 1997) presented the use of Bayesian statistics to model suitability of landscapes for birds. This would appear to be an innovative approach to such a task. However, a quarter of a century ago, Williams et al. (1978) presented the development of PATREC models, based on Bayes' theorem, that can be used to evaluate habitat quality. Reference to the work on PATREC models in the 1970s does not appear in these recent papers. Here is a case where the approach has been reinvented. (I do not intend to detract from the solid quality and useful results of these papers; rather, I use this to point out how difficult it is to keep up with everything that has been done.) Symposia and works such as this volume contribute greatly to the sharing of information and to allowing all to stay up with current work. Such meetings should be encouraged in the future.

As our computing power increases, our models can (but shouldn't necessarily) become more complex. A model may be considered to be a hypothesis based on the data used for its development. It is important that models (i.e., hypotheses) continue to be tested and evaluated. I fear at times that model complexity is equated to model quality and that we may have exceedingly high expectations for our model outputs. The data used to develop models (both response variables and predictors) are measured with error (Maurer, Chapter 9), and it is not likely that we will ever create models that predict with great precision (e.g., number of animals per unit area within 0.1 individuals per hectare) with any degree of reliability. A variety of factors influence the detectability of species in field studies (Ralph and Scott 1981; Thompson et al. 1998), yet most modeling efforts have usually treated the presence/absence or abundance of the species being modeled as a known entity. Doing so is likely to hinder model accuracy. We must recognize that we are often predicting the *probability of detecting* a species, given a set of habitat conditions, and not necessarily its true abundance or presence. Recently, species detectability has received consideration in modeling efforts (Conroy and Noon 1996, Boone and Krohn 1999, Schaefer and Krohn, Chapter 36); improving the accuracy of our models should prove to be a fruitful area of research in the future.

We will continue to improve our ability to more precisely and accurately measure habitat features at large and small scales and to estimate species abundance or presence with greater accuracy. I believe, however, our ability to develop models that adequately predict species abundance or density will be limited because of the inherent noise in the systems within which we work. In many (or most) cases, we may well be satisfied with being able to accurately predict presence or absence, or perhaps general categories such as none, low, medium, and high, with some degree of consistency. For example, Jorgensen and Demarais (1999) found it difficult to develop models that predicted exact species richness of small mammals, but they were able to reasonably predict habitats with high and low species richness, which may be adequate to meet management needs. The simplest model that meets a stated goal is likely to be the best one. Although such models may not provide the "truth" that assures us another publication in a rigorous journal, they may provide the problem-solving capability needed by a manager (Starfield 1997).

The more-sophisticated statistics and faster computers now readily available allow us to make mistakes more quickly than ever before. A firm foundation in the basics of experimental and survey design is necessary; we should always be aware of what the sampling/experimental unit is for the problem at hand (Hurlbert 1984). It is critical that all researchers have a solid understanding of their objectives and of the tests or procedures appropriate for their data. Given the ease with which most of the current analytical programs can be applied, it is quite simple to conduct nonsensical analyses and, unless one is paying close attention, to believe in the results. No amount of statistical wizardry can compensate for a poor study design or inadequate data. Thus, it is incumbent upon those who have responsibility for training the upcoming generations of researchers and managers to ensure they are well grounded in these basics. I believe it is particularly important that we teach our students to think conceptually. Serious reflection on the what, why, and how of data analyses, before touching a computer, should be required of all students (and established researchers). Doing so should help provide students with a sense of how their work adds to previous research and help ensure appropriate application of methods.

I am optimistic about the future of modeling the relationships between animals and their habitats. We have learned much over the years and have new technologies and methods available that allow us to address needs of single species and communities at multiple scales. To the extent that we can learn from our mistakes, build upon our successes, and temper our expectations with pragmatism, we should be able to continue to develop the ability to analyze and model animal responses to habitat in contexts valuable to both researchers and managers.

Acknowledgments

I appreciate the assistance of C. Darby with the review of JWM papers. This manuscript was improved by the review and suggestions of R. Oderwald, M. Reynolds, J. Berkson, D. Whitaker, T. Fearer, K. Mattson, J. M. Scott, and an anonymous reviewer; all errors remain my own.

Approaches to Habitat Modeling: The Tensions between Pattern and Process and between Specificity and Generality

Beatrice Van Horne

In ecology, as in other sciences, theoretical models summarize important processes and relationships without claiming to predict occurrences in specific situations. Applied habitat models, on the other hand, make quantitative predictions that can be evaluated with empirical data. Because models make verifiable predictions, applied habitat modeling is an activity most ecologists approach with some trepidation. Most agree that quantitative models provide an essential opportunity to make ideas about habitat relationships of wildlife species explicit, quantifiable, and testable. Such models foster communication among researchers and managers by minimizing misunderstandings associated with more qualitative descriptions of wildlife-habitat relationships. Of course, because any model is a simplification of a complex biological system, it cannot be perfectly predictive. Hence, a model may be thought of as a hypothesis that we know is false for any given biological system, but we may choose to keep because it is useful (Box 1976). The fear ecologists feel in constructing models results from knowing that the model is open to criticism because it is ultimately false. Any modeling implementation will include unrealistic biological assumptions, not meet statistical assumptions, omit causal relationships, and/or fail to meet modeling objectives. We handle our models with skepticism because we are aware of these foibles and know that the modeling approach we have used is likely to be heavily criticized and replaced with new approaches in the future. Unfortunately, misunderstanding the power of a given approach for making certain types of predictions or answering some types of questions may lead to misuse of models or misplaced criticism. Sometimes such criticisms leads to the premature abandonment of promising approaches.

In this brief discussion of habitat models, my objective is to clarify the strengths, weaknesses, and appropriate uses of different types of models. It is my hope that, if these are understood more clearly, we may be less likely to be blown by the winds of fashion in choosing modeling approaches and more likely to make real progress in constructing appropriate and useful models. Block and Brennan (1993) have discussed historical approaches to studying and modeling bird-habitat relationships, and Rosenzweig (1989) has summarized experimental approaches to developing ecological models of habitat selection by small mammals. In my discussion, I will assume Block and Brennan's (1993) definition of habitat as the subset of physical environmental factors that a species requires for its survival and reproduction. It follows from this definition that individuals in higher-quality habitat will have greater survival and reproduction, and hence fitness, than individuals in poorer-quality habitat.

Modeling techniques are closely tied to modeling objectives. Just as there is no perfect personal automobile, there is no single "best" modeling approach.

When choosing a personal automobile, customers incorporate tradeoffs between size and gas mileage, clearance and stability, or smoothness of the ride and precision of handling according to their larger objectives such as minimizing cost, maximizing off-road capabilities, or impressing the neighbors. Similarly, any modeling approach involves a trade-off. Adopting an appropriate modeling strategy involves matching management objectives with model capabilities. Researchers need to assess the generality they wish to achieve with their modeling effort and balance that against the power to use the model to make specific predictions (Levins 1966). Many modeling efforts represent an attempt to quantify imperfect knowledge. Models may be useful in determining the relative value of management alternatives but not be able to provide accurate predictions of the effects of these alternatives.

Generality Versus Specificity in Modeling

Models that make specific predictions are likely to be limited in their application because of variability in both time (Best and Stauffer 1986) and space. Because even the most complete model cannot include all biologically important causal relationships for a given location, one cannot expect accurate predictions over time. Such models may describe occurrences in the past accurately but are not useful for forecasting (Morrison et al. 1992). A model that works well for a given location, such as one that predicts the number of burrowing owls (*Athene cunicularia*) based on the number of prairie dog burrows available, may not be useful in another location, such as an area where prairie dogs are absent and burrowing owls nest in sites other than prairie dog burrows (Haug et al. 1993).

Usually, general models are simpler than models that aim to make specific predictions. Using fewer terms enhances the clarity of such models and increases their applicability to a broad range of systems. General models may either describe broad patterns (general empirical models) or be more theoretical than specific models (Fig. 4.1). General empirical models lose the ability to make specific predictions because they are less likely to be directly related to processes than are either the specific models (Bissonette 1997) or

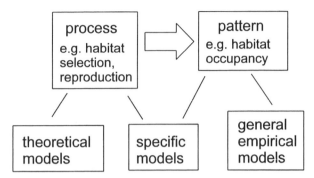

Figure 4.1. In reality, the influence of habitat on a species is determined by the pattern-process relationship. Theoretical, specific, and general empirical models incorporate varying amounts of pattern and process.

the theoretical models. Researchers construct theoretical models to understand the important relationships that govern population numbers. For instance, much recent work has been framed in terms of "bottom-up" or "top-down" population regulation (see Power 1992). A theoretical model would describe when and how each type of regulation is primary across a wide range of populations of a given species or across a broad range of species. This could have important implications for managing a target species. A specific model would document the nature of population regulation for an intensively studied population but would have little explanatory power across the species range or among species.

Model Validation and Boundary Conditions

Because general theoretical or empirical models do not make precise predictions, and because specific models make predictions only for a target population at a certain time (or under certain conditions), habitat models cannot be tested with new data sets in the same way that a hypothesis can be tested. Thus, the emphasis on external validation of models is often misguided. One can, however, test the importance of processes thought to be important in driving observed habitat relationships and test for the boundary conditions under which these processes may become relatively unimportant. Boundary conditions describe the range of a given habitat variable or variables within which the model is applicable, or the model "domain." For example, at the northern end of their range in winter, belted kingfishers (*Ceryle alcyon*) feeding on freshwater minnows may be limited by the availability of ice-free rivers (Kelly and Van Horne 1997), whereas farther south other factors may determine habitat occupancy. Hence, a model describing habitat occupancy in terms of ice-free rivers would apply only when and where lotic waters were subject to freezing. Similarly, availability of old-growth forest may determine numbers of northern spotted owls (*Strix occidentalis caurina*) (Forsman et al. 1996a,b), whereas availability of a mixture of forest types influences Mexican spotted owls (*Strix occidentalis lucida*) (Ganey and Dick 1995; Gutiérrez et al. 1995). Rather than attempting to construct a single model that can either be applied across the entire range of a species or restricting application of the model to the place in which data were collected, it would be more useful to broaden sampling or experimental sites to identify boundary conditions in forest cover, temperature, moisture, and similar factors within which the model is applicable. The use of gradients in sampling design has been more common among botanists than zoologists but is essential for identifying boundary conditions and thresholds. Austin (Chapter 5) discusses the design of gradient-based sampling for modeling relationships between fauna and vegetation.

Such boundary conditions for continuously distributed variables can only be identified when gradients are investigated as part of the study design. Animal ecologists do not usually test boundary conditions because most work is done within or among discrete habitat types. They may array experimental blocks along a gradient of interest, but such sampling may be too discontinuous to identify boundary conditions. This contrasts with the approach of plant ecologists for whom sampling along gradients (and hence, determining boundary conditions) is more common (Austin 1999b). These differences in study design may in part result from the different statistical and/or experimental approaches that are used by these two groups. Animal ecologists often begin their investigations with ANOVA approaches and block experimental designs, whereas plant ecologists often begin with ordination.

It appears that models cannot be universally validated because they are too general to make precise and testable predictions, apply only within tight and often unknown boundary conditions, or have failed to incorporate process and therefore rely on correlations that may be spurious. Adopting an appropriate modeling strategy involves matching management objectives with a realistic assessment of model capabilities. In particular, researchers need to assess the generality they wish to achieve with their modeling effort and balance that against the power to use the model to make specific predictions. The use of alternative models in an adaptive framework (Conroy and Moore, Chapter 16) can greatly enhance the predictive power of models if carried out over a long-term period.

Too often, modeling approaches are chosen on the basis of currently accepted "fashions" rather than a realistic assessment of objectives and capabilities. When a new form of habitat modeling becomes more popular than an existing technique, it is useful to consider whether this replacement represents real progress. The answer to this query is not simple, because there is no good standard for measuring "progress." Changes in approaches to modeling take place for several reasons. Often, they are prompted by criticism of existing approaches. Usually these criticisms are targeted at the assumptions of the models and therefore driven by a misunderstanding of the necessary simplification. Because the factors involved in evaluating models are complex, it is sometimes difficult to discern valid and useful criticism combined with the development of useful new approaches from criticism based on misunderstanding of the assumptions and objectives of existing models. Sometimes changes are made possible by the development of new statistical techniques and/or advances in computer hardware and software.

Matching Goals with Techniques

The goals of habitat modeling fall into two broad categories. The first is a general inventory to establish links between habitat conditions or distribution and biodiversity (Cablk et al. Chapter 37; Scott et al. 1993). From this assessment, species that are rare, have restricted ranges, or are unexpectedly absent may be identified. Such an inventory may serve as the basis for management decisions based on the notion of maximizing biodiversity. The second goal is to establish links between habitat condition and the viability of one or several species. Assessing viability requires knowledge of the processes influencing changes in density, which may allow managers or researchers to predict the effects of ongoing or expected habitat change on the target species.

Assessing Biodiversity: Predicting Species Occurrence in a Landscape

Organized inventories of species in which presence/absence is associated with habitat categories may be used to identify areas of high biodiversity to develop protection schemes that maximize the total number of protected species or to identify a loss of species associated with anthropogenic habitat changes, including those resulting from fragmentation. Multispecies inventory is an ambitious undertaking, and a relatively high level of error is expected and tolerated so that an expansive view can be developed (Noss et al. 1997). Lack of attention to processes driving the observed patterns of habitat occupancy make these approaches easy both to criticize and to misuse, but such approaches are an essential first step in identifying conservation concerns.

Methods for systematically organizing information about multiple species and the habitats they occupy fall broadly under the WHR (wildlife habitat relationships) rubric (Nelson and Salwasser 1982; Patton 1978; Thomas 1979; Verner and Boss 1980; Hoover and Wills 1984). Such methods may help to organize information during the inventory phase but do not include or predict information about densities, population trend, or habitat patch size and configuration.

The Gap Analysis Program (GAP) of the USGS Biological Resources Division (Scott et al. 1991a, 1991b, 1993) is a systematic means of inventorying vegetation using GIS (geographic information system) mapping procedures, predicting the occurrence of wildlife species in different habitats, and using this information to identify priority areas for protection. Used appropriately, gap analysis can encourage consideration of biodiversity patterns at a broad scale (among habitats) rather than focusing on local, within-habitat biodiversity. Because it is conducted over large areas with coarse-grained resolution, gap analysis is less useful

for evaluating management techniques (e.g., burning or logging), for predicting the effects of landscape pattern (e.g., importance of ecotones or patch sizes), or for understanding the processes governing the presence and absence of species.

One danger of biodiversity models is that the output of the models can allow managers to follow prescriptions thoughtlessly. For instance, the model may suggest that protecting a series of habitat types in certain proportions will maximize biodiversity. Such a prescription, however, ignores individual habitat requirements of species (e.g., a minimum area or configuration to protect a sustainable population), species interrelationships (e.g., loss of a keystone predator [Paine 1966] may have ramifications that loss of an herbivore would not), and species valuations. Species valuations may be based on species' popular "charisma," trophic level, whether or not a species is native and/or endemic to a particular area, or rarity outside the focal area of study. Some species may have value because they attract popular attention (e.g., raptors, grizzly bears, apes, porpoises, penguins). Species at a higher trophic level (e.g., wolves, spotted owls) may be more valued for the role they play in the community or because their habitat requirements may provide an "umbrella" for prey species. Defining native species may present difficulties where species' ranges have expanded as a result of anthropogenic change. For instance, riparian-associated species, such as indigo buntings (*Passerina cyanea*) and yellow-bellied sapsuckers (*Sphyrapicus varius*), have expanded their ranges westward in North America along the Platte River because of its now-artificial flow regime, enhancing the biodiversity of riparian areas. Rarity outside the focal area is tied to the notion of regional- or gamma-diversity and can present difficulties in species inventory approaches (Samson and Knopf 1982). Species contributing to higher diversity along the Platte River are more likely to be represented elsewhere in North America than are members of the relatively depauperate species community of the surrounding prairie (Knopf 1985, 1986).

How fragmentation effects are incorporated into biodiversity models is another area of difficulty. Because individual species differ in home-range size, minimum population sizes, effects of habitat edges, and other factors, the effects of a given level or form of fragmentation of habitat will vary among species (Bierregaard et al. 1992). Such variation is incorporated implicitly into species/area-based models, such as those based on the Theory of Island Biogeography (MacArthur and Wilson 1967), but such a level of generalization does not allow for predictions about effects on individual species. Use of species/area curves to model biodiversity may mask important relationships between the invasion of exotics and species diversity, as when invasion rates increase with biodiversity (Stohlgren et al. 1999).

Assessing Species Viability: Predicting the Effects of Habitat Change

A species may be highly valued or of special concern either in its own right or because it serves as an indicator or umbrella species. In such cases, focal studies may be used to predict the effects of anthropogenic change or management actions on the viability of the species.

Species-habitat Correlations

Analyses of the relationship between the co-occurrence of individuals of a species and habitat variables have often been used to identify high-quality or essential habitat for a species. Conceptually, this approach follows from the *n*-dimensional niche described by Hutchinson (1957), in which the niche of an organism was described by *n* axes, each representing an environmental variable. It follows that, if we correlate the occurrence of an organism with a series of environmental variables in a given study area, we may be able to describe the niche of the organism and, hence, predict its pattern of habitat occupancy in other areas. The process of gathering data for such analyses is straightforward. Individuals are surveyed and their presence or absence is associated with habitat variables collected nearby. The underlying philosophy of such an approach is that a higher level of objectivity is achieved by collecting the data and massaging them to life so that they may speak to us about the organism's view of its environment, rather than imposing our own intuitions or natural history-based knowledge to structure the analysis.

Problems with such a correlative approach are found in at least four areas. The first of these is in reducing the number of variables. All analytic approaches using these sorts of data must deal with the statistical problem that one can measure many more variables that may influence species' occurrence than there are observations or samples in most of these data sets. During the 1970s, the use of multivariate approaches based on either principal components or discriminant analysis was common (Capen 1981). Because such multivariate approaches use the correlation structure within the habitat variables to produce a few synthetic axes that can later be interpreted, it was thought that these provided an objective means of discerning habitat relationships. This idea was bolstered by the view that organisms viewed their environment in a synthetic or "gestalt" fashion (James 1971). Multivariate approaches faltered, however, when it was pointed out that assumptions of these analyses were seldom met, so that the significance values assigned to habitat relationships were questionable (James and McCullough 1990). The problem of reducing the number of variables still exists, however. Although we can all agree that it is important to take the point of view of the organism in choosing the variables to measure (Morrison et al. 1992, 1998), this is exceedingly difficult. Consequently, the variables chosen for correlation are generally those that we can measure easily.

The second problem in using correlations arises from the shape of the variable-response relationship. Hutchinson's concept of the niche (Hutchinson 1957), and explorations of the nature of niche breadth and overlap that followed, envisioned a normal, Gaussian response by the organism to a gradient in the variable or resource. Where such a curve exists, we may record an increasing, nonexistent, or decreasing response depending on the range of the variable that is measured (Fig. 4.2). Yet most correlative approaches assume that if no linear response is detected, a given habitat variable is not an important determinant of species occupancy (Austin 1999a). Statistical significance (still usually at $P < 0.05$) has become the measure of "significance" of habitat variables. Logistic regression allows for one form of nonlinearity in response (Hassler et al. 1986), but there are many nonlinear forms

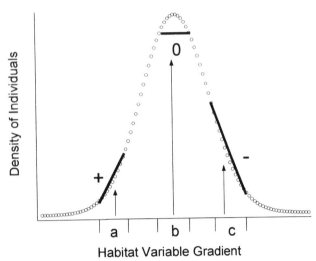

Figure 4.2. Limited sampling along a gradient in a habitat variable may produce a positive (a), nonexistent (b), or negative (c) correlation with a species response variable such as density if the species response is a variant of a normal curve. It may then be useful to describe the boundary conditions for wildlife-habitat correlations. These boundary conditions may be set by the habitat variable in question or by other habitat variables.

possible in ecological data. Hypothesizing the best model to fit the data a priori is often difficult, and trial and error exercises of model fitting weaken final inferences.

The third problem, related to the second, is that the nature of the organism response will vary with scale. Wiens (1989a) pointed out that the nature of perceived species-habitat relationships changed with the scale (extent and grain) at which they were measured. As we scale up or down, the domain investigated, the observation set, and the language used will often change (Bissonette 1997a). In other words, types of variables and the scale the measurements encompass change. High density of a species within a 1-hectare area may be qualitatively different and have a different effect on population change than high density of the same species over a 100-hectare area. Because habitat selection may itself take place at multiple scales (Bissonette et al. 1997a; Hildén 1965; Hutto 1985; Wiens et al. 1987), it may be necessary to incorporate measurements of variables at these multiple scales into predictive models of habitat use. *It is therefore important to carefully match the question to the scale and to ask questions at different scales of extent and levels of resolution.*

The fourth problem is the classic conundrum that

correlation does not signify causation or process. Because causal factors may change without concomitant changes in the values of the habitat variables measured, correlations may break down in space and time. A special case of this results form a source-sink habitat structure (Pulliam 1988) that can cause densities to build up in low-quality habitats (Van Horne 1983) as a result of social structuring (Fretwell 1972). Similarly, loss of habitat may cause increased density in remaining habitat. There may be little effect on population processes where habitat is not limiting, as in some situations with migratory birds (Goss-Custard et al. 1994). In such situations, correlations of habitat variables with species density are likely to overestimate the effect of habitat loss because they don't incorporate long-term increases in density in the remaining habitat, nor do they include information about the effect of such increased density on population processes (Goss-Custard et al. 1994).

Because ecological systems are highly variable, an experimental approach is necessary to define the cause-effect relationship. I caution, however, against viewing experimental approaches as the solution to all dilemmas in understanding relationships between wildlife and their habitats. Experiments are limited in scope and can only test for effects of one or a few variables in isolation. In addition, not all experiments elucidate cause-and-effect processes. For example, an experiment determining habitat association will not necessarily expose the cause of that habitat association. Often, the most profitable procedure is to mix experimental and descriptive approaches. Descriptive approaches might define patterns of habitat occupancy for a squirrel across its range, while experiments might be used to distinguish the roles of habitat structure and mast availability in determining habitat use or patterns of survival and reproduction at a particular site.

Knowledge-based Models

Apparent instability of habitat correlations has resulted in part because such relationships mask the processes that drive population change and, in part, because the relative importance of different processes may change with time, space, and scale. Frustration with such instability led to the development of the

HEP (habitat evaluation procedures; USFWS 1980, 1981a) and HSI (habitat suitability index; Schamberger et al. 1982) approaches. HEP is a systematic means of assessing habitat conditions and management alternatives using habitat units of assigned qualities, while HSI is a means of assessing habitat quality. Although HSI models may make use of correlations, they differ fundamentally from the correlative approaches described above in that the key causal relationships are not expected to emerge from the analysis. Rather, they rely on knowledge of natural history and processes that is derived from researchers familiar with the target species. Development of an HSI model often involves assembling such "experts" and forcing them to draw relationships between habitat quality and key life-history attributes (even in the absence of $P < 0.05$ data). Thus, a synthetic variable uses logical relationships (either/or/and) to derive a single measure of habitat suitability. Used properly, such models represent a valuable approach to structuring our knowledge of animal-habitat relationships and identifying areas in which further research is needed (Van Horne and Wiens 1991). There is the opportunity to include nonlinear relationships in these models where we have sufficient knowledge of the shape of these relationships. As currently constructed, however, we cannot expect these simplistic models to make precise predictions about the nature of habitat occupancy. Hence, it is difficult to "validate" such models or even to understand what might constitute validation. Further, they don't tell us much about population change or viability, nor do they do a good job of handling fragmentation effects or other spatially mediated processes.

Andrewartha and Birch (1984) developed the envirogram approach, a diagrammatic means of conceptualizing the factors influencing species abundance. In their scheme, factors that influence the species directly (what I would call key processes) fall into four categories they refer to as resources, mates, malentities (e.g., weather, competitors), and predators. Outside of this centrum, a web of biotic and abiotic factors influences the species indirectly; the research objective is to determine which of these pathways are important. Thus, the indirect effects must be explicitly tracked through the factors producing the direct effects, forc-

ing researchers to conceptualize the cause-effect linkage between habitat variables and species occurrence.

Burnham and Anderson (1998) describe the use of Akaike's Information Criterion (AIC; Akaike 1978a,b; 1981a,b) to select the most appropriate model from a set of possible models or hypotheses determined a priori. When information about an animal's natural history is used to establish the set of possible models, this approach combines knowledge-based and correlative approaches while avoiding the artificial standards (e.g., $P < 0.05$) for selecting an appropriate model. This approach may be particularly useful for identifying variables representing real effects on the species prior to experimental manipulation.

Population Models

Viability models are classic population or metapopulation models linked to habitat or environmental conditions to predict changes in population size that may result from different management policies. These are used increasingly to address problems with endangered or threatened species (Noss et al. 1997). These models may be useful to encapsulate our best predictions of population-habitat associations to allow for relative comparisons of management policies. Because of the difficulties of understanding causes of population change and modeling stochastic events, they are not likely to predict accurately the patterns of population increase, stasis, or decrease.

Where factors influencing survival and reproduction are documented, a life-table approach may provide important insight into the effects of habitat change. For instance, if the life table is more sensitive to small variation in adult survival than to variation in reproductive rates or juvenile survival, factors influencing the quality of wintering habitat may be more critical than those influencing breeding habitat. Observed life histories are an integration of genetic or phylogenetic and phenotypic factors. Differences among habitats in life histories may be important (Van Horne 1983). To an extent, life history can buffer habitat effects, as, for example, where the organisms can respond to decreased adult survival with increased investment in current reproduction. Where this is the case, models using demographic constants, such as per capita birth or death rates, may be erroneous. Recent advances permit demographic parameter estimation from mark-recapture models and ecological covariates (White and Burnham 1999). This approach has been used to rank habitats according to their relative effects on life-history traits of northern spotted owls (Franklin 1997).

Individual-based Models

Models that track individual responses to habitat and project them to population responses are more likely to include causal processes than those relying strictly on correlations or population-level responses. Such models, however, are highly specific and tend to be very complex. Goss-Custard et al. (1994), for example, have built individual-based models that predict the proportion of Eurasian oystercatchers (*Haematopus ostralegus*) emigrating or dying because of habitat loss or change. These predictions are then used to project population numbers at larger scales.

Mangel and Clark (1988) have translated dynamic programming approaches used in engineering for the ecological sciences. Their approach to predicting individual behavior allows the organism to make choices based on fixed or stochastic values of risk and reward associated with these choices as they influence survival and/or reproduction, the components of fitness. Choices among habitat types can be modeled in terms of causal factors, such as predation probability and food availability, associated with each habitat patch. Physiological states that are correlated with survival and reproduction, such as the probability of maintaining a threshold level of body fat, can then be used as the currency to evaluate the choices. For example, Farmer and Wiens (1999) have used such an approach to gain insight into migratory movements between stopovers and reproductive patterns in pectoral sandpipers (*Calidris melanotos*).

Models in biology are often based on new tools that become available. Radiotracking (telemetry) represents one such tool. Manly et al. (1993) describe resource selection functions that can be used to interpret results independent of the number and types of habitat identified. Early modeling of habitat preference based on comparison of used and unused habitat as determined by radiotracking, however, was roundly and justifiably criticized (Hobbs and Hanley 1990), be-

cause the investigators' choice of habitat to be included in the available but unused portion greatly influences the outcome of the analysis.

Modeling the Effects of Animals on Their Habitats

Most approaches discussed thus far have considered habitat as a factor that influences animal populations and have ignored influences in the opposite direction. The concept used is one of a habitat (usually defined by vegetation composition or structure) that is taken as a given, which the organism may then choose or not choose to occupy. Over time and/or as densities increase, however, animals may modify their biotic and abiotic habitats, indirectly influencing their own density as well as that of other species by activities such as grazing, predation, digging, or scavenging. In extreme examples, such as the American beaver's (*Castor canadensis*) modification of its habitat through dam construction, the term "ecosystem engineers" is used (Lawton and Jones 1995). Community or ecosystem models are more likely to address such effects because the former models include feedbacks between species, while the latter emphasize dynamic flows of energy and nutrients that essentially change the habitat available over time.

The Use of GIS Information in Habitat Models

Geographic information systems (GIS)-based information has vastly expanded our ability to construct spatially explicit models of animals in actual or projected habitat configurations. Because this tool encourages us to use actual habitat maps, it may impart a feeling of power and omniscience on the part of the habitat modeler. The emergence of new tools in ecology is often followed by a period of euphoria and rampant collection of new types of data (e.g., genetic techniques, radiotracking) before methods of analyzing and using such data have been critically evaluated. GIS opens many new possibilities, but I would like to post some cautionary notes before we leap wholesale into this new, spatially explicit world. The first of these is that problems in variable selection are increased by orders of magnitude in spatially explicit landscapes. The number of variables that become available in landscapes is staggering (cf. Gustafson 1998); these range from geometric variables, such as patch size, perimeter, shape, density, connectivity, and fractal dimension, to continuous variables sampled at points using trend surfaces, correlograms, semivariograms, autocorrelation indices, or interpolation such as kriging, to synthetic variables such as viscosity. With so many variables possible, allowing the important variables to emerge from an "objective" analysis (that is, allowing the data to speak for themselves) is virtually impossible, and the process of a priori variable selection supersedes later analytic processes in importance. It seems evident that an informed choice of variables, based on knowledge of processes that drive species occurrence, is essential to the success of such analyses.

A second difficulty with GIS approaches is the scale imposed by pixel size and/or by analytic limitations that require pixels of a minimum size. Appropriate pixel size depends on whether the analysis is individual- or population-based. In the former case, it should be based on the minimum size of habitat unit that influences habitat choice, while in the latter case it should be based on the size of a single within-season home range or the minimum habitat patch that can support an individual.

A third difficulty is that the variables that define a given pixel may not be highly correlated with those that influence the organism directly. If this is the case, it is impossible to develop a predictive habitat model based on GIS procedures alone.

A fourth difficulty is that spatially explicit analyses tend to be highly empirical and have relatively little theoretical foundation. Hence, the ability to generalize from a single analysis is severely limited.

Fifth, GIS analyses are usually pattern- rather than process-based (but see Goodchild et al. 1993, 1996). That is, it may be simplest to match animal occurrence with pixel type; this matching may give some clues about possible processes driving populations, but it is well removed from these processes. It does not, therefore, follow that if we manage to increase habitat features associated with a certain pixel type in which a target species is commonly found, we will necessarily increase the numbers of individuals of that species. For instance, it may be that the species is found in "wetland" pixels on our map, and all such pixels on our map are near a river. Proximity to the river may

be critical to the species, but this would not be evident from our analysis. Any subsequent attempt to increase wetlands away from the river would then fail to increase the target species. Knowing how the animals use their habitat would be critical to interpreting our initial GIS analysis.

Finally, I would like to highlight a related problem. It is expensive, difficult, and time-consuming to put people into the field. It is much simpler to work directly with maps and computers. It is easy to succumb to the temptation to believe, for instance, that maps of vegetation types based on GIS are accurate and to shortcut the field verification of vegetation composition. Similarly, researchers may believe that they know which animals are associated with each vegetation type, without having to verify that the animals are actually there. It is easy to think that we know which variables to include in our analyses without actually observing the annual pattern of habitat use of the target species. It is easy to ignore variation in survival and reproduction across the landscape that may be obscured by dispersal into low-quality habitats. The use of GIS encourages us to cut corners and to believe results without field verification. Usually, error associated with measurement, scaling, and extrapolation is ignored in describing the variances associated with model output. We need to make sure the standard of acceptance of GIS analyses does not encourage laziness and corner cutting.

Conclusions

Habitat models that are very specific will not be useful for making predictions of management effects across time and space unless enormous resources are devoted to their development, as occurred for the spotted owl (Forsman et al. 1996a). They may, however, elucidate processes that are important in driving population change; such processes can be incorporated into more-generalized models that do have predictive power. This process of generalizing requires that boundary conditions for a given habitat model be defined. Such definition may require the use of gradient-based rather than strictly categorical or ANOVA-based approaches. Models that achieve generality by collecting data at a very broad scale may be useful in the definition of problems across habitat types, in initial inventories, and in identification of large tracts of land for preservation. Because they are so distantly related to the processes that drive population change, however, they will not be useful in identifying the effects of process-oriented management policies regarding timber harvest, stream flow, grazing, hunting pressure, and the like. If we pay close attention to matching model objectives to the model type selected, rather than thoughtlessly embracing the latest habitat analysis technique, we are likely to make better progress toward meeting management objectives. Certainly, we should match the questions we ask to the scale at which we gather information and model. We should also de-emphasize model validation and accept the position that habitat models are a means of quantitatively assembling our best knowledge of animal-habitat relationships to make the most informed decisions possible and to identify research needs, rather than expecting the models to be predictive with $P < 0.05$. Finally, we should take advantage of GIS approaches to enhance our understanding of spatially explicit processes driving population change rather than using the tools and maps as an excuse to become lazy about the field work necessary to understand these processes.

Case Studies of the Use of Environmental Gradients in Vegetation and Fauna Modeling: Theory and Practice in Australia and New Zealand

Michael P. Austin

Knowledge of the distribution of species is central to the conservation of biodiversity. Predicting the distribution of species whether plant or animal demands a high degree of knowledge in ecology, statistical modeling, geographic information systems (GIS), and remote sensing to be cost-effective. Wildlife scientists (Verner et al. 1986a; Scott et al. 1993) and ecologists (Margules and Austin 1991) have recognized the need to predict distribution but have developed their own paradigms (Margules and Redhead 1995; Scott et al. 1996). A degree of convergence between paradigms with respect to spatial modeling is now emerging (Lindenmayer et al. 1990a, 1991b; Pereira and Itami 1991; Mladenoff et al. 1995; Pausas et al. 1995), as evidenced by contributions in this book (e.g., Young and Hutto, Chapter 8; Fertig and Reiners, Chapter 42). Any study of species distribution has several components. A certain ecological theory is assumed, either explicitly or implicitly. Environmental variables are selected as important for estimating species distribution. Certain sources of data or methods of survey design are accepted as suitable. Decisions are made about the use of GIS or remote sensing. Habitat scoring systems are devised or particular statistical modeling methods adopted. These decisions will reflect the research paradigm of the researcher but governing all this will be the pragmatic decisions necessitated by available data, skills, time, and resources. In this chapter, attention is focused on the role of environmental variables in ecological theory, survey design, and statistical modeling as used to predict species distributions, particularly in Australia.

Theory

The common and pragmatic assumption of homogeneous vegetation communities for conservation planning does not have a sound theoretical base. The continuum concept where species composition varies continuously along environmental gradients (Austin and Smith 1989) is more consistent with early quantitative studies (Whittaker 1956; Curtis 1959) and recent statistical modeling studies (Austin et al. 1990; Austin et al. 1994; Leathwick and Mitchell 1992; Leathwick 1995). Each species shows individualistic distribution patterns in relation to environmental variables, particularly climatic variables such as temperature and solar radiation, and local variables such as topographic position. Homogeneous communities of co-evolved species with discrete boundaries along environmental gradients do not exist. However, the most frequent combinations of environmental conditions in a landscape will have characteristic combinations of species that can be recognized as communities for management purposes (Austin and Smith 1989). Shifting combinations of environmental conditions in the landscape and climatic gradients across regions will restrict such community types to local regions only (Austin 1991).

Whittaker (1956, 1960) and Curtis (1959) pioneered the continuum concept for plants and demonstrated the individualistic behavior of species, but it is not often considered for vertebrate fauna (Verner et al. 1986b; Austin 1999a). Yet, the distribution of a species in an environmental space, the niche hypervolume of Hutchinson (1957), is a basic tenet for developing a predictive model of a species distribution. For successful modeling, however, it is necessary to recognize the nature of the environmental space being used. Three idealized types of environmental gradient can be distinguished, though all kinds of intermediates may occur (Austin and Smith 1989; Huston 1994; Austin 1999b).

Indirect gradients (complex gradients, Whittaker 1978) have no direct physiological effect on growth. Altitude is an example; it is correlated with variables that do have an effect on growth, such as temperature and rainfall. The correlation of altitude with both temperature and rainfall will be location-specific and will confound any interpretation. Analysis of species environmental niche using indirect gradients will give predictive models that are only valid locally and will lack robustness.

Direct gradients (regulator gradients; Huston 1994) are those that have a direct effect on growth but are not consumed (e.g., temperature). Temperature may have numerous direct effects. There is only limited theory to guide what the expected shape of a species response to a temperature variable might be, beyond the probability that it will be unimodal (Austin 1992). Predictions may be robust if a suitable expression of the temperature gradient is found.

Resource gradients are those where the resource (e.g., soil nitrogen) is consumed by a plant. The proximal causal resource will often be unknown (e.g., phosphate concentration at the surface of the root hair). Surrogate distal variables such as available soil phosphate are usually used. The species response function is likely to be hyperbolic or perhaps unimodal. The more proximal the variable the more robust will be the predictions.

These gradients are idealized contrasting types. Soil moisture may be a resource gradient at low levels but an indirect gradient at high levels. The waterlogged soil at high levels will be anaerobic and toxic to many terrestrial plants; the direct variable may be lack of oxygen or presence of sulfide ions. Knowledge of environmental processes and the likely nature of species response functions is the key to developing robust models of species distributions.

Models for fauna differ from those of plants with variables for food supply, nesting sites, protection from predators, and size of territories more important than environmental conditions. Choice of appropriate predictors depends on having a suitable theoretical framework, knowledge of relevant environmental processes, and ecological expertise on the organisms being studied. No amount of statistical modeling can compensate for poor definition of the problem.

Data

There exist a wide variety of data types arising from biological sources and GIS and remote sensing developments. Large conservation studies have had to depend on the data available to them rather than the data most suitable for prediction (Scott et al. 1996). Frequently the only data available are herbarium records or museum specimens. Such presence records without equivalent records of absence can only be used for statistical models that are conditional on the presence of the species. For example, models of species abundance can be developed based on the premise that the species is present. Methods exist for predicting the potential occurrence of species based on presence data and climate records (Busby 1991; Austin 1998). These methods predict the climatic envelope within which a species may occur, but they can say nothing about where it will be absent within climatic limits of the envelope.

A minimum data set suitable for statistical modeling is the presence/absence records for a geocoded plot of specified size. If the plot's location has been recorded, then environmental predictors can be captured via remote sensing and GIS layers. Numerous surveys exist that can provide such data for plants. Data for animals is less readily available and is usually conditioned on collecting effort. Collation of existing vegetation surveys in southeastern New South Wales

(NSW) has provided a data set of over nine thousand plots (Austin et al. 1994). The problem with collated data sets of this kind is that they may represent a biased sample from the region (Stockwell and Peterson, Chapter 48).

Survey Design

Environmental gradients can play a significant role in survey design where unbiased species data from geocoded locations is unavailable, and cost-effective surveys are needed (Margules and Austin 1991). To ensure a representative sample of the range of vegetation composition, a survey should sample the environmental space defined by the major influential environmental gradients in the region. Choice of the gradients constitutes a hypothesis stating the presumed principal environmental controls in the region.

Austin and Heyligers (1989, 1991) describe a procedure called gradsect sampling that is designed to provide a representative sample of the environmental space of a region. Transects are selected to traverse the major environmental gradients of a region in such a way as to minimize travel times and access problems. This gives a biased sample; areas not on the chosen gradsects cannot be selected for sampling. The aim of such a survey is to sample the potential range of variation in vegetation composition, not to obtain an unbiased estimate of some mean value for the region. The survey design approach has been termed the *SR³* strategy, where S = Stratification, R = Representation, R = Replication, and R = Randomization (Austin 1998). Where access is less of a problem, a modified version of the design can be adopted.

A recent vegetation survey of the Mid-Lachlan Valley in central New South Wales (Austin et al. 2000) provides an example (Fig. 5.1) for an area of 22,500 square kilometers. Available GIS layers for the study area consisted of a digital elevation model (DEM) and polygon-based soil landscape maps (Kovac et al. 1990 unpublished maps). Climate layers for mean annual temperature and mean annual rainfall were derived from the DEM using ANUCLIM (Hutchinson et al. undated). Four environmental factors were assumed to be important in the area: temperature, rainfall, soil type, and topographic position. The first two were

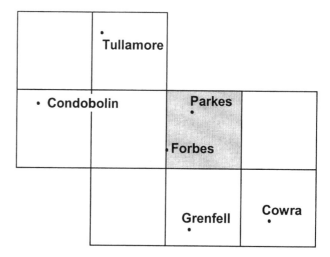

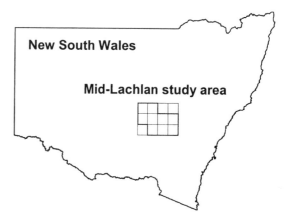

Figure 5.1. Location map for Mid-Lachlan survey in New South Wales, Australia.

considered to be direct gradients. Soil type is a surrogate categorical variable for soil nutrients and soil moisture. Topographic position is an ordered categorical variable representing an indirect gradient, the soil catena along which numerous soil variables are confounded. Mean annual temperature (9.1 to 17.0 degrees Celsius) and mean annual rainfall (350 to 1,330 millimeters) were divided into seven and eight classes, respectively. There were 153 soil landscape types representing extensive depositional landscapes separated by low north-south mountain ranges with shallow skeletal soils. When these landscape types were overlaid with the climatic layers in the GIS, they formed 628 environmental stratification units (ESUs) for the area, due to the correlation between GIS layers.

The SR[3] strategy was achieved by

1. Geographical stratification (S): The primary stratum was the nine 1:100,000 mapsheets of the area. The ESUs on each mapsheet were then sampled independently.
2. Environmental representation (R): A representative sample of the environmental space is obtained by sampling each ESU occurring on the mapsheet (see Fig. 5.2 in color section).
3. Replication (R): This was achieved by taking samples from each ESU in proportion to the area of each ESU on the mapsheet following rules similar to those used by Austin and Heyligers (1989).
4. Randomization (R): This is only strictly feasible for those ESUs that have large scattered areas on a single mapsheet. However, the repeated sampling of the same ESU on different mapsheets will result in a systematic random sampling of the geographical range of the ESU.

A two-stage sampling procedure is introduced to incorporate the fourth potential predictor, topographic position, because it operates at a different geographical scale from the other variables. At each sampling site (approx. 1 square kilometer), three plots from different topographic positions were measured for vegetation composition (Fig. 5.3). There were two pragmatic decisions made for logistical reasons. Only presence/absence of trees and shrubs taller than 1.5 meters were recorded for each plot, and plots wherever possible were located on roadsides and public lands to avoid the time-consuming effort of seeking permission to sample on private land.

This survey design is explicit, consistent, and repeatable. Data from plots provide the evidence for interpretive vegetation maps, and that evidence is only as good as the survey design (Table 5.1). The design depends on extensive knowledge of the variables influencing the composition of eucalypt forests and woodlands (Austin et al. 1997). There is one assumption among the many of this approach, which is not always made explicit: the vegetation is in equilibrium with the environment as defined. There is growing circumstantial evidence from New Zealand that this assumption may not be true. There is evidence that species distributions are not yet in equilibrium after a volcanic

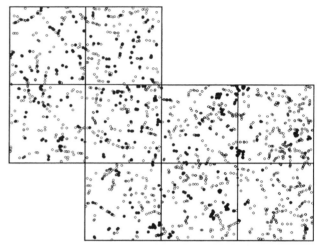

Figure 5.3. Plot locations sampled based on the SR[3] survey strategy in Mid-Lachlan study area after taking account of existing survey plots.

eruption eighteen hundred years ago (Leathwick and Mitchell 1992) or after the last glacial period due to the slow rates of re-invasion of the dominant species from the genus *Nothofagus* (Leathwick 1995, 1998; Leathwick et al. 1996). Provided the primary geographical stratum—the map sheet—is used in the design of surveys, this need not be an issue. It will however require explicit consideration in any statistical modeling that is done.

Statistical Modeling

The use of statistical models for prediction depends on the availability of suitable environmental predictors.

TABLE 5.1.

Distribution of plots sampled in relation to the climatic gradients used to stratify the Mid-Lachlan survey. Nonexistent combinations are indicated with a dash (—).

Rainfall	Temperature						
	> 17	16–17	15–16	14–15	13–14	12–13	<12
< 400	32	3	—	—	—	—	—
401–475	64	503	65	—	—	—	—
476–550	46	444	76	—	—	—	—
551–625	—	95	305	164	—	—	—
626–700	—	—	29	197	67	0	—
701–775	—	—	—	29	71	0	—
776–850	—	—	—	—	46	25	—
> 850	—	—	—	—	—	23	13

These must be ecologically meaningful and exist as a GIS layer for the entire region for which predictions are required. This reduces the number of possible predictors, although the increasing use of remotely sensed variables may compensate for this (Aspinall and Veitch 1993). The recent availability of new methods of regression modeling has allowed the development of models more consistent with ecological theory (Austin et al. 1994; Franklin 1995). Both generalized linear modeling (GLM; McCullagh and Nelder 1989; Nicholls 1989, 1991) and generalized additive modeling (GAM; Hastie and Tibshirani 1990; Yee and Mitchell 1991) have been used in vegetation studies in Australia (Austin et al. 1994; Austin and Meyers 1996) and New Zealand (Leathwick 1995, 1998). Boyce and McDonald (1999) provide a recent review discussing some faunal examples; see also Özesmi and Mitsch (1997). GLM greatly expands modeling opportunities compared with classical linear regression, in particular making possible the easy analysis of presence/absence data with both continuous variables and factors (Nicholls 1989; Austin et al. 1990). GAM allows the fitting of nonparametric functions to the data using a smoothing algorithm while maintaining significance testing (Yee and Mitchell 1991). The great advantage is that the exact shape of a species response to an environmental predictor does not have to be specified prior to fitting the model. Given that theory suggests that a great variety of shapes are possible (Austin and Smith 1989; Austin 1999a), this is an important advantage. There are numerous other statistical methods that deserve attention; decision trees (Breiman et al. 1984; Walker 1990: Stockwell et al. 1990; Lees and Ritman 1991), LOWESS (Locally Weighted Sums of Squares, Cleveland 1979; Currie 1991), and neural networks (Caudill 1990; Fitzgerald and Lees 1992). How to evaluate the relative performance of these methods is a continuing issue (Austin 1994; Austin et al. 1995; Walker and Aspinall 1997).

Environmental predictors for eucalypt species modeling at a regional scale in Australia have generally been climatic variables with temperature being the most important (Austin et al. 1997). Other useful predictors have been lithology, topographic position, solar radiation corrected for aspect and slope, and soil fertility index. Some early work at a more local scale

indicated that moisture stress based on soil water balance models and interspecific competition could contribute to statistical models but could not be used for GIS-based prediction in their current form (Austin et al. 1997). Leathwick (1995, 1998; Leathwick et al. 1996) has successfully used soil moisture models together with climate variables and interaction terms for forest models in New Zealand. Interestingly, solar radiation is a more important predictor than rainfall in New Zealand as compared with Australia.

None of the array of statistical modeling techniques currently available to ecologists is without problems. The regression-based techniques of GLM and GAM are both limited by multicollinearity and the inadequacies of stepwise regression procedures (James and McCulloch 1990). No consensus yet exists on the choice of spatial modeling method. Comparative evaluations tend to champion one method. Two recent papers have recommended neural networks over multiple regression (Lek et al. 1996c; Guégan et al. 1998). In one case (Guégan et al. 1998), only linear terms were used in the regression, and, in the other case (Lek et al. 1996c), use of best current practice with GAM and appropriate analysis of residuals might have yielded entirely different conclusions.

Comparison of the performance of different techniques using real data is fraught with difficulties. Three techniques may give one answer, while another three give a different answer but with similar explanatory power. Two possible conclusions might be that (1) all six techniques provide an inadequate answer or (2) each group of techniques finds half the answer, and no technique gives the full answer. Alternatively, model specification of predictors was inadequate, and better ecological and statistical skills would have given a much better model independent of the particular technique used. Comparisons of statistical modeling techniques are only possible where "truth" is known. To generate realistic "true data" requires theoretical knowledge of how species respond to environmental gradients and knowledge of the environmental processes, which generate the multicollinearity between environmental predictors.

Austin et al. (1995) have attempted to investigate this difficult problem. They constructed artificial data based on explicit ecological theories about how plant

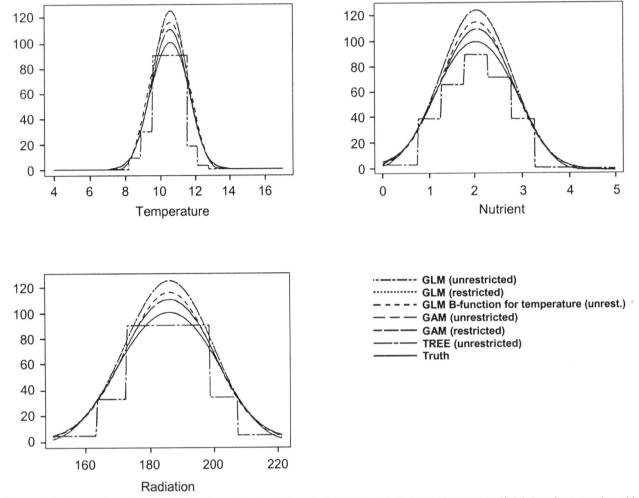

Figure 5.4. Comparison of response functions fitted by different statistical modeling techniques to artificial data for a species with a Gaussian response function to the proximal "causal environmental gradients." See text for description of terms.

species respond to environmental gradients using the computer package COMPAS (Minchin 1987). Two types of species response were generated based on different theories, Gaussian (bell-shaped) response functions (Gauch and Whittaker 1972), and skewed β-functions (Austin et al. 1994). The assumption was that the best technique should be robust to changes in response function. These responses were then linked to two sets of environmental predictors. The first set comprised three constructed proximal causal variables (temperature, radiation, and soil fertility). The second set was derived from the direct gradients via plausible environmental process models. For example, the indirect gradient variables latitude, longitude, and altitude were derived from temperature using a real data set and a thin-plate spline surface fitted to climate station

data (Hutchinson 1997). Aspect and slope were derived from radiation using a real data set and a complex trigonometric function. In addition, a random variable unrelated to the species was added to the environmental data. The difficulties of evaluation are not discussed here (see Austin et al. 1995).

Figure 5.4 shows an example of the type of results obtained in the simplest case using the predictor. A β-function is a particular parametric function suitable for fitting certain skewed responses (Austin et al. 1994). Data restriction is applied when the range of observations along the environmental gradient clearly exceeds the species environmental niche. This can reduce prediction errors at the edge of the species niche (Austin and Meyers 1996). The true Gaussian curves are recovered by all the techniques but with varying degrees of over- or underestimation. Use of the indirect predictors results in much more complex failures of model-fitting (see Austin et al. 1995). Four conclusions from the study are relevant here:

1. A random variable may be selected as a highly significant predictor by GLM and GAM unless special attention is paid to error functions and the response curves obtained.
2. Of the methods tested, GAM is the most successful at recovering the true relationships but is not without problems.
3. The correct mix of ecological and statistical expertise is more important than the particular technique used.
4. It is much more difficult to obtain satisfactory unbiased predictive models with indirect gradients as predictors.

The use of environmental gradients in predictive models of species distribution is a necessary but insufficient condition for successful modeling. Species models often show spatial autocorrelation of the residuals implying that the significance estimates for the predictors are biased and inflated. Most species distributions are spatially autocorrelated as are many environmental variables (e.g., rainfall and temperature). Use of such predictors will remove some of the autocorrelation. Any remaining autocorrelation may be due to an unknown environmental variable having spatial autocorrelation or to species not being in equilibrium with their environment. This lack of equilibrium can arise from historical factors related to dispersal rates of species or local extinction from suitable habitat due to, for example, plant collecting or hunting.

Smith (1994) used a proximity variable, the occurrence of the species within a local neighborhood of the observation after fitting an environmental model. Leathwick (Leathwick et al. 1996; Leathwick 1998) has used this approach to analyze forest distribution patterns in New Zealand. The importance of the proximity variable for southern beech *Nothofagus* species was ascribed to the poor dispersal and invasive capabilities of these forest dominants and helped to explain the existence of "Beech Gaps" (Leathwick 1998). Augustin et al. (1996) have approached this problem from a different aspect studying red deer (*Cervus elaphus*)

distribution in Scotland (see also Boyce and McDonald 1999). Testing for spatial patterns in the residuals from environmental models needs to become standard procedure in any species distribution modeling.

Case Studies of Fauna Modeling in Australia

The understanding and modeling of the distribution of forest fauna in southeastern Australia has increased greatly in the last fifteen years (Braithwaite et al. 1989; Cork and Catling 1996; Woinarski et al. 1997; Landsberg and Cork 1997; Lindenmayer et al. 1999). Braithwaite and colleagues published a series of papers on the distribution of arboreal marsupials (Braithwaite 1983; Braithwaite et al. 1983, 1984). They found that 52 percent of the forested area on the far south coast of New South Wales had no arboreal marsupials while 63 percent of the arboreals were found on 9 percent of the area. This difference is related to differences in foliage inorganic nutrient concentrations of potassium, nitrogen, and phosphorus as arboreal marsupials are predominantly folivores. Foliage nutrients are correlated with the occurrence of certain eucalypt forest communities and hence to variations in lithology (Braithwaite et al. 1984). A threshold in foliage nutrient levels is apparent below which arboreals are almost absent and above which other factors appear to operate (Fig. 5.5, Braithwaite 1984).

Stockwell et al. (1990) modeled the distribution of the arboreal marsupial, the greater glider (*Petauroides volans*), in a nearby area using regression and decision trees with stand condition, slope, site quality, and foliage nutrients as predictors. Pausas et al. (1995) used GLM to model the original data of Braithwaite et al. (1984) on species richness of arboreal marsupials. Two models were developed: one for low foliage nutrient concentrations and the other for high nutrient concentrations. The model at low nutrients contained the predictors foliage nutrient concentration, soil nutrient index, topographic position, hole index, and bark index. The first three and the fifth relate to food quality. Hole index is a measure of nesting sites and depends on tree age and tree species. Bark index is a measure of the amount of decorticating bark at the site, an indicator of insect availability. Above the

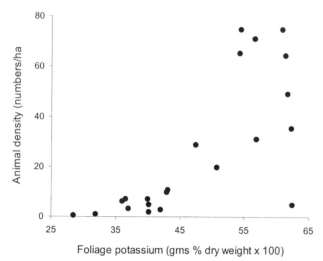

Figure 5.5. Numbers of arboreal marsupials in relation to foliage potassium concentration for each eucalypt community. Redrawn from Braithwaite et al. (1984).

foliage nutrient threshold, only forest structural features were important. The model consisted of the predictors: number of trees with diameter greater than 60 centimeters, an indicator of potential nesting hollows, and the proportion and basal area of small trees. This model was later used in a forest dynamic simulator for predicting the habitat quality for arboreal marsupials under different logging regimes (Pausas et al. 1997; Pausas and Austin 1998).

Lindenmayer and colleagues have studied arboreal marsupials in Victoria 400 kilometers from the previous studies (Smith and Lindenmayer 1988; Lindenmayer et al. 1990a,b). The eucalypt forests of the Central Highlands of Victoria are markedly different from those of the south coast of New South Wales. They tend to form monospecific even-aged stands of either *Eucalyptus regnans* or *E. delegatensis* (Lindenmayer et al. 1990a) as compared with multispecies eucalypt communities (Braithwaite et al. 1984). Two arboreal species, mountain brushtail possum (*Trichosurus caninus*) and the greater glider, were modeled using Poisson regression (Lindenmayer et al. 1990a). The model for the possum indicated that the preferred areas (3 hectares) had high numbers of hollow-bearing trees, high basal areas for *Acacia* species, and occurred in gullies with few shrubs. The model for the greater glider contained only the number of hollow-bearing

trees and stand age, the species preferring older stands originating before 1900.

These models were subsequently tested on an independent data set and shown to be robust (Lindenmayer et al. 1994). Another study (Lindenmayer et al. 1993) with the same species compared their occurrence in intact forests and in forested wildlife corridors. They concluded that the probability of occurrence for the two species did not differ between the two habitats. The model for the greater glider was combined with a GIS to predict the species spatial distribution (Lindenmayer et al. 1995). A kernel-smoothing procedure was used to allow for the spatial dependence present in the data. The focus of these studies and others (Lindenmayer et al. 1991a,b,c) has been on the existence and characteristics of hollow-bearing trees as the primary habitat limitation for these species in even-aged regrowth forests. A recent attempt to model arboreal marsupial species distribution at a landscape scale was unsuccessful except for one species (Lindenmayer et al. 1999). The environmental predictors were estimated for 20- and 80-hectare circles surrounding the 3-hectare survey sites. The only species, the yellow-bellied glider (*Petaurus australis*) for which any of the predictors were significant is also the only species with a home range of equivalent size (40–60 hectares) to the sampling units used. The other three arboreal species have ranges less than 6 hectares in size (Lindenmayer et al. 1999). The size of plot used for modeling must relate to the home range of the species. Foliage nutrient concentration does not appear to have been examined in this part of Victoria.

Catling and Burt (1994, 1995a,b) studied ground mammals in an area on the south coast of New South Wales, north of the area studied by Braithwaite et al. (1984). They found a response to foliage nutrient concentration different from that for arboreals. Abundance of native small mammals (less than 200 grams) was negatively correlated with nitrogen, phosphorus, and potassium in the foliage but positively related to foliage magnesium (Catling and Burt 1995a). Habitat complexity (Newsome and Catling 1979) had the highest positive influence on ground mammal abundance. Catling and Burt (1995b) examined the distribution and abundance of ground-dwelling mammals in relation to different types of environmental gradients, including direct gradients (temperature and rainfall) and indirect gradients (aspect, slope, and topographic position). Although species abundances were sensitive to many of the predictors, none were as important as habitat complexity (e.g., height and cover of shrubs and abundance of logs) reflecting disturbances such as grazing, fire, and logging. Catling et al. (1998) fitted GLM models using both environmental gradients and habitat variables. The general conclusion was that variables related to habitat complexity, eucalypt community, and a nutrient factor based on lithology were important determinants of species distribution.

The introduced predators, European red fox (*Vulpes vulpes*) and the domestic cat (*Felis catus*), were abundant in southern NSW and medium-sized mammals were absent or in low abundance. In northern NSW, Catling and Burt (1997) found increased abundance of medium-sized native mammals (200–6,000 grams) and greater diversity of small mammals. The fox was absent from large parts of the region and in low abundance elsewhere, suggesting that predation is a controlling factor.

The importance of different environmental gradients and habitat variables, such as tree hollows, varies markedly between regions for arboreal marsupials. The south coast of NSW is characterized by a diversity of mixed eucalypt communities on a wide variety of predominantly low-nutrient lithologies (Braithwaite et al. 1984). The Central Highlands of Victoria has mainly monospecific tree communities on relatively nutrient rich lithologies. The dominant species *E. regnans* and *E. delegatensis* are readily killed by fire and develop even-aged stands with overstories of tall dead trees. It is not surprising, therefore, that food quality is more important in southern NSW and the presence and nature of tree hollows more important in Victoria. The ground mammals on the south coast of NSW behave differently to arboreal mammals in relation to foliage nutrients though ground mammal abundance and bird species richness behave similarly to foliage magnesium (Braithwaite et al. 1989; Catling and Burt 1995b). Habitat complexity and the risk of predation from introduced foxes are the main determinants. The relevant environmental gradients or habitat variables for inclusion in predictive models of fauna distribution vary even for the same species from region to region. Statistical models and their predictor variables are highly contingent on the precise combinations of environmental and biotic conditions in a specific region.

Discussion

The ecological theory that determines the success of predictive species modeling differs radically between plant and animal ecology, yet the central concept of a species niche is common to both. The physical environment in terms of climate and soils is clearly more important for plants. The extent to which biotic predictors are more important than environmental predictors will depend on the exact circumstances. The existence and nature of source and sink areas within the range of an animal species is ultimately contingent on environmental differences between the areas. Morton (1990), Braithwaite and Muller (1997), and Soderquist and MacNally (2000) provide examples of the importance of this effect from different parts of the Australian continent. Difficulties in incorporating these effects in modeling arise when species populations are not in equilibrium with the environment and invasive species are displacing others. The recent results of Leathwick (1998) in New Zealand suggest that plant ecologists will need to consider species dispersal abilities and historical disturbance patterns in order to develop models that are more robust. Spatial autocorrelation can no longer be ignored in species modeling. Statistical models of fauna in Australia (Catling et al. 1998; Lindenmayer et al. 1999) have yet to demonstrate the importance of climate predictors. However, BIOCLIM models of fauna presence records reviewed in Burgman and Lindenmayer (1998; see also Nix [1986]) indicate that climate may have considerable importance for many faunal groups (cf. Currie 1991). The relative importance of environmental and habitat variables for fauna groups needs to be determined but will depend on the scale of the study. Large areas or areas with steep environmental gradients are likely to have climatic gradients to which animals are sensitive. Where steep environmental gradients exist, stratified surveys based on the gradients are likely to be necessary if only to reduce the bias in existing records.

The choice of statistical modeling technique remains uncertain; the best practice may in fact be to combine several techniques. The strengths and weaknesses of different techniques will need to be evaluated with "true data." The construction of realistic ecological data has yet to be fully solved. The transfer of techniques from one research area to another without testing whether the technique is compatible with the new theoretical framework is a common problem (Austin 1999a,b). Statistical modeling cannot substitute for ecological insight, appropriate environmental gradients, and knowledge of the processes linking environment with biota.

Acknowledgments

Thanks are due to D. J. Grice for help with the manuscript and to E. M. Cawsey, D. J. Grice, S. V. Briggs, B. L. Baker, and M. M. Yialeloglou for their collaboration on the Lachlan survey and the Department of Land and Water Conservation, NSW, for access to unpublished soil information and maps.

Habitat Models Based on Numerical Comparisons

K. Shawn Smallwood

Habitat Model Types and Analysis

Habitat analysis typically involves either a comparison of used versus not used environmental elements or a test for disproportionate *use* of environmental elements from a measured set of *available* environmental elements. This chapter will focus on the latter approach (Table 6.1). The measure of use can include numerical spatial patterns (distributions) based on an individual's locations or on collections of individuals (Fig. 6.1). Using estimates of density or the number of individuals to perform habitat analysis also involves the use of pattern analysis, because density and spatial distribution are interlinked (Taylor 1961; Taylor et al. 1978). If organisms were patterned uniformly or randomly across a landscape or region, then habitat analysis would not be informative. The species would be considered as simply ubiquitous. However, species do not distribute themselves this way. Numbers vary spatially, and implicit in most habitat models is the assumption that numerical peaks coincide with locations where habitat is of the highest quality (but see Van Horne 1983 for warnings against this assumption).

Using densities or numbers of individuals for habitat analysis, the measurement units for the response side of the analysis can range in demographic organization from subpopulations and populations to megapopulations (sensu Garshelis and Visser 1997)

and metapopulations (sensu Hanski and Gilpin 1997). These response units are measured on geographic areas that encompass at least several home ranges to regions (Fig. 6.1). Many habitat models include the term N to represent number of individuals, D to represent density, or some other term representing number or spatial intensity in the environment. The terms N and D are explicitly represented in some habitat models (e.g., Table 6.1, Equations 6.1, 6.5, 6.11, 6.12, 6.13) and are needed for frequency counts and ratios used in the other models. Therefore, habitat analysis is scale-dependent because N and D are scale-dependent (Verner 1981; Blackburn and Gaston 1996; Smallwood and Schonewald 1996). Furthermore, these models do not specify which demographic unit is being represented by N or D, nor the contexts of demography, season, interannual variability, or condition of the landscape, all of which can affect N or D (Cyr 1997; Smallwood in press).

Habitat models using numerical comparisons have been reviewed already (e.g., Morrison et al. 1998). Van Horne (1983) pointed out that density is a misleading indicator of habitat quality, mainly because density estimates are frequently too narrow in their representation of a population's success. According to Van Horne (1983), habitat studies are often spatially and temporally inadequate for including areas occupied during all seasons, nor do they include the full

TABLE 6.1.

Habitat models based on numbers or densities.

Equation Number and Model Name	Model Structure	Explanation
Measures of effect, no P-value		
6.1. Edge index (Helle and Jarvinen 1986)	$E = \dfrac{D_i}{D_j} * 100\%$	D_i and D_j are densities at sites i and j
6.2. Index of selection (Paloheimo 1979)	$S = \dfrac{r_i}{p_i}$	r_i is the percentage of the ith element used, and p_i is the percentage available in the measured set
6.3. Index of electivity (Ivlev 1961; Jacobs 1974; Gordon 1989)	$E = \dfrac{r_i - p_i}{r_i + p_i}$	
6.4. Usage-availability rank difference (Johnson 1980)	$t_{ij} = rank\ r_i - rank\ p_i$	
Hypothesis tests with P-values		
6.5. Isodar Theory of habitat selection (Morris 1990)	$D_1 = \dfrac{a_1 - a_2}{b_1} + \dfrac{b_2}{b_1} * D_2$	in the ith habitat, a_i = maximum fitness, b_i = per capita reduction in fitness
6.6. χ^2 test with Bonferoni Z test (Pearson 1900; Neu et al. 1974)	$\chi^2 = \sum \dfrac{(O - E)^2}{E}$	O = observed, and $E = p_i * N$ = expected number of observations
6.7. Log-likelihood ratio criterion (Fisher 1924, 1950; Everitt 1977)	$\chi^2 = 2 \sum O * \log_e \dfrac{O}{E}$	
6.8. Measure of aggregation (David and Moore 1954; Pielou 1977)	$\chi^2 = \left(\dfrac{1}{n\bar{x}}\right) \sum_{i=1}^{n} (x_i - \bar{x})^2$	x_i = number of occurrences in ith element; x = mean of occurrences among n elements
6.9. Ecological order (Smallwood 1993)	$M_i = \dfrac{O}{E}$	M_i = the number observed as a multiple of the number expected in the ith element
6.10. Negentropy (Fisher 1924; Shannon and Weaver 1949; Rothstein 1951; Phipps 1981)	$H' = \sum_{i=1}^{s} O \log_e \dfrac{O}{E}$	Use of information; use of the measured set of choices; deviation from uniform association indicates negative entropy
6.11. Surface regression model (Rotenberry 1986)	$D = b_0 + \sum_{i}^{h} b_i X_i + \sum_{i}^{h} b_{ii} X_i^2$ $+ \sum_{i}^{h} \sum_{j}^{h} b_{ij} X_i X_j$	X_i = the ith habitat, h is the number of habitat variables, and bs are fitted coefficients
6.12. Correlation (Wiens and Rotenberry 1981)	$d = ah_i$	In a strip transect, h_i is the ith habitat attribute expressed as a continuous variable, e.g., percent cover
6.13. Patch shape/Type (Otis 1998)	$\ln\{E(D_{ij})\} = \beta_0 + \ln(A_{ij})$ $+ \beta_1(S_{ij}) + \sum \beta_i P_{ijk}$	A_{ij} = area of the jth patch in the ith environmental element; S = shape, which is a function of the patch perimeter and its area; P = patch perimeter

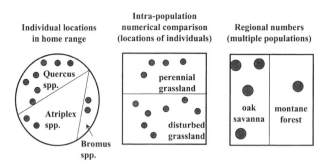

Figure 6.1. Numerical data in habitat models can include individual locations within a home range or any number of individuals with or without reference to a demographic unit. The grain of the measured set of environmental elements typically grades from micro- to macro-habitat representation as the numerical data grades from individual locations to regional in scope.

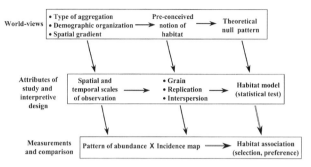

Figure 6.2. The effectiveness of numerically based habitat models and the measurements of use and availability ultimately depend on the attributes of the field studies and the world-views of the analyst, which have primacy over the actual measurements of use and availability.

range of numerical responses to environmental changes, nor do they address the social interactions leading to occupation of both source and sink areas. Vickery et al. (1992) found density to be a poor predictor of nest success among three species of passerine bird species, although they did not account for the effect their variable plot sizes had on density (Smallwood 1999). Nevertheless, unburdened by any possible effect of variable plot sizes on density, Morris (1989) found mean litter size of the white-footed mouse (*Peromyscus leucopus*) to decline with increasing density. Alldredge et al. (1998) discussed faulty assumptions underlying various habitat models, such as the assumption that the sample of animals is random (rather than belonging or not belonging to a population), that each observation is independent of the next, that each animal's habitat selection was independent of others in the sample, that the availability of environmental elements is known rather than estimated, that availability is constant over the period of the study (rather than changing seasonally or for other reasons), and that detectability of animals is constant among the environmental elements sampled. Smallwood (1993) began a discussion on the theoretical foundations of these models, focusing on the role of thermodynamics in resource selection models.

In this chapter, I will frame habitat analysis within the larger context of world-views on how and why animals distribute themselves and how and why analysts apply experimental design principles and statistical tests to habitat studies based on relative number or

density. Figure 6.2 depicts a hierarchical framework for discussing habitat analysis, first considering world-views (paradigms), then attributes of study and interpretive design, and finally the measurements and comparisons, the results and interpretation of which depend on world-views, and study and interpretive design. I intend to demonstrate that the available habitat models are inadequate by themselves for characterizing habitat of animal species; these models are no substitute for long-term research experience, although their predictions can be used as testable hypotheses. Just as habitat analysts need operational terms with which to work (Morrison and Hall, Chapter 2), they also need a larger framework with which to interpret applications of the habitat models based on numerical comparisons.

The Theoretical Null Pattern

The significance of a measured numerical pattern (spatial distribution) to habitat use is decided largely by its comparison to a theoretical null pattern. Measured numerical patterns can be uniform, random, aggregated, or regular, and null patterns can be uniform, random, or both (Fig. 6.3). Deviation of a measured pattern from the null pattern is assumed to be caused by a relationship between the species and the energy or material resources composing the environmental elements (Smallwood 1993). The theoretical foundations of the null pattern are therefore critical for interpreting the association. The null pattern in thermodynamics is uniformity, which connotes the equilibrium cold state, or

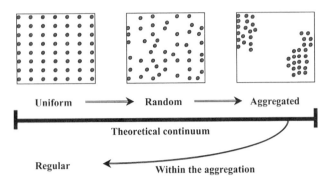

Figure 6.3. Random patterns occur somewhere along a gradient ranging from uniformity (regularity) to aggregation, thereby reducing the distance that can be measured between an observed aggregated pattern and the null pattern of randomness.

lack of energy relationships (Hutchinson 1953). The null pattern in information theory is also uniformity, which connotes no use of the available information (Shannon and Weaver 1949; Kullback 1959; Phipps 1981). Uniformity is the state of maximum entropy in both thermodynamics and information theory. The null pattern in ecology is randomness, which connotes lack of influence from the locations of other individuals (Fisher 1950; Taylor 1961). The null pattern chosen by the habitat analyst identifies the world-view of the analyst regarding how and why nature is organized, and it determines the set of alternative results.

Habitat analysts usually explicitly or implicitly assume that spatial distributions are determined by some form of energy relationship(s) between the organisms and their environment (Hall et al. 1997b; Morrison et al. 1998). Although rarely discussed explicitly, many ecologists hold the world-view that biological processes are governed ultimately by the laws of thermodynamics (Hutchinson 1953). The measures of effect used in Equations 6.6–6.10 are derived from these laws, and all the other models in Table 6.1 (except Equation. 6.5) are theoretically and mathematically related to Equations 6.6–6.10 (Smallwood 1993). Many aspects of habitat are assumed to be energy-related, and the goals of wildlife are usually assumed to be resource acquisition and total fitness (Southwood 1977; Rosenzweig 1985; Wiens 1989b,c), which are energy-related and energy-dependent, respectively. Habitat elements other than energy are recognized as important, such as water, nutrients, refugia, travel corridors, nest sites, and so on. However,

even the response variable, density, has its conceptual origins in the physical sciences and is energy-dependent. High-quality habitat is said to have greater levels of energy available (Hall et al. 1997b).

However, if the laws of thermodynamics largely govern the spatial distributions of animals, then assuming randomness as the null pattern is not only inappropriate on theoretical grounds but also reduces the analyst's capacity to recognize meaningful aggregations and selection of environmental elements (Smallwood 1993). Random patterns occur somewhere along a gradient that ranges from uniformity (regularity) to aggregation (Fig. 6.3). This gradient is represented by probability values of the χ^2 distribution, the lower-tail values of which correspond with uniformity and increasingly upper-tail values grading through randomness and eventually aggregated patterns. The distance that can be measured between an observed aggregated pattern and the null pattern of randomness is less than between the aggregated pattern and the null pattern of uniformity. Therefore, assuming the null pattern is random instead of uniform washes out some of the potential significance that can be attributed to an observed aggregated pattern. The world-view of the analyst is at least implicitly expressed by the types of habitat model and statistical test used, because these models and tests are structured either on the assumption that the null pattern is uniform (e.g., χ^2 tests) or random (e.g., ANOVA tests).

Only aggregated patterns of animals can lead to inferences about habitat use. Purely uniform or random patterns of use will fail to reveal any selection of environmental elements from a measured set because no disproportional patterns will be evident (Fig. 6.4). A regular pattern of distribution might indicate that the study area included territories of individuals numbering fewer than occurred in the larger population (Fig. 6.5). In other words, the study area was smaller than the area occupied by the population, which often arranges itself into regularly spaced home ranges or territories held by its members. Similarly, random patterns have been observed when progeny (of insects) dispersed passively in air or water (Taylor et al. 1978), or when occurrence was scarce or at the edge of a larger aggregation (Fig. 6.5).

Recognizing whether the spatial pattern is uniform, random, aggregated, or regular depends on both the

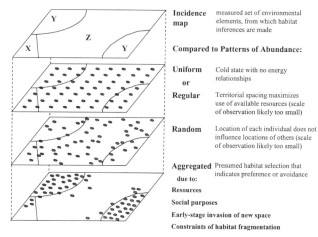

Figure 6.4. To interpret meaningful associations, patterns of distribution are compared to a map of the measured incidence of environmental elements, composing a measured set. The meaning of an association ultimately depends on the world-view of the analyst and on the theoretical foundations of the statistical test or measure of effect being used. It also depends on the type of aggregations that are being compared to the measured set of available environmental elements.

spatial scale of observation and knowledge of the species' demographic organization. Examining the spatial relations of individuals from a portion of a population can mislead an analyst into thinking the pattern of distribution is random or uniform, when it is more likely regular at the scale appropriate for measuring the population and aggregated at the regional scale. For example, prairie dog (*Cynomys* spp.) burrows occur fairly regularly within colonies (Tileston and Lechleitner 1966), which are themselves aggregations in the region (Koford 1958). Of course, information about the spatial distribution is lost when transforming the locations of individuals into numerical estimates and then to densities. Comparing N or D for habitat analysis cannot lead to conclusions about demographic organization unless the spatial areas are included in the comparison (Smallwood 1999).

Types of Aggregation

Aggregations can form for several reasons (Fig. 6.6), and each of these reasons poses a unique implication for interpreting the habitat association. These reasons bear on the choice of the theoretical null pattern and the habitat model. It will be important for wildlife biologists to determine the relative occurrence frequen-

cies of each of the following four types of aggregation: resource, demographic, early-stage, and constrained.

Individuals may aggregate around a centralized or patchy resource, forming a *resource aggregation*. Competition for a limited resource can force some animals to live on the fringe of the resource patch, or even to spill over into ecological sinks. Such an aggregation can extend beyond the boundary of the resource patch just because off-patch individuals get sufficient access to the resource or because those exploiting the resource generate progeny that disperse

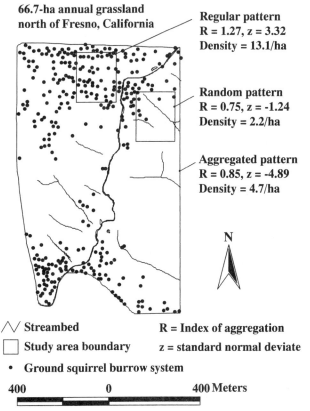

Figure 6.5. The size and position of study areas inform of different spatial patterns among burrow systems of California ground squirrels (*Spermophilus beecheyi*). Using the pacing method for mapping burrow systems of pocket gophers (Smallwood and Erickson 1995), I mapped the approximate centers of ground squirrel burrow systems on 66.7 hectares of annual grassland in the low-elevation foothills north of Fresno, California, during April and May 2000. A nearest-neighbor-distance method (Clark and Evans 1954) indicated that ground squirrel burrow systems were regularly distributed within a larger aggregation and randomly distributed at the edge of the aggregation. The larger aggregation was only recognizable within the boundary of the largest (66.6 hectares) study area. The density of burrow systems was also greater within the aggregation compared to the edge.

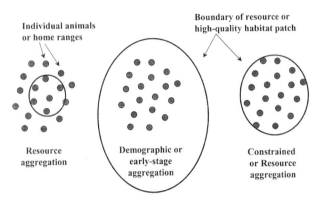

Figure 6.6. The types of aggregations observed largely determine the pattern of distribution and the association with the measured set of environmental elements. Demographic, resource patchiness, habitat fragmentation, and early-stage filling of ecological space can affect our perception of use and availability.

outward, beyond the patch. In either case, and in the case of other possible explanations, the number of individuals associated with the resource-based aggregation need not be bound within the resource patch.

Individuals also aggregate with conspecifics to improve fitness through reproduction, rearing of young, and group cooperation such as foraging and predator avoidance. They may choose places to live based on the momentum of the congregation rather than the energetic or nutrient accessibility of the habitat patch (Alatalo et al. 1985). *Demographic aggregations* may have spatial limits that are imposed by behaviors or by uncertainties of habitat quality outside the aggregation (Stamps 1991). Demographic aggregations conceivably could be confined to space that is smaller in extent than the resource patch (Taylor and Taylor 1979). Social constraints could therefore limit our observed use of an environmental element (resource) to less than that expected based on its availability. Dasgupta and Alldredge (1998) devised a behavioral dependency parameter for use with χ^2 tests, involving situations where multiple individuals are observed together. However, aggregations can occur without individuals being observed together, per se (e.g., *Puma concolor*, Smallwood 1997), but which may have formed to serve social and demographic needs nevertheless (Lloyd 1967).

An early-stage aggregation can also form while immigrants invade previously unoccupied habitat, which could have been cleared of former occupants, or

which was recently discovered or became available to the species. An invasion of pocket gophers (*Thomomys* spp.) into new stands of alfalfa (Smallwood and Geng 1997) is a good example. Early-stage aggregations are combined resource and demographic aggregations but temporarily give different results when analyzing habitat selection based on use and availability methods. Such aggregations provide the best opportunity to observe preference of the available environmental elements because they are relatively free of demographic constraints (e.g., competition and territoriality). However, as numbers increase due to continued immigration, social mechanisms such as territoriality force later arrivals into less-preferred locations, thereby affecting our perception of use (Fig. 6.7). An example of early-stage aggregations can be found in "Intrapopulation Numbers," later in this chapter.

Constrained aggregations result from habitat fragmentation (sensu Wilcox and Murphy 1985) or the division of previously contiguous habitat patches (Addicott et al. 1987). They may also result from intolerable environmental conditions occurring naturally outside the aggregation, including the occurrence of competitors or predators (Hutchinson 1953; Koford 1958). Patches of low-quality habitat are occupied by constrained aggregations because the species is left with no other place to go (du Toit 1995). Low-quality habitat can appear to be high-quality habitat when loss of habitat constrains individuals to fragmented habitat patches or to peripheral habitat areas. Habitat analysis using constrained aggregations can be misleading due to packing of individuals into the habitat fragment or the inclusion of the intolerable areas in the measured set of environmental elements. Unfortunately, many special-status rare species are found only in fragmented habitat, or in habitat at the boundary of their tolerable conditions (e.g., Scott et al. 1986), so their constrained aggregations should not be used alone to interpret habitat associations. For example, giant garter snakes (*Thamnophis gigas*, federally endangered) occur in marshes and adjacent irrigation canals where marshes in the Central Valley of California have not been converted entirely to agricultural fields and houses. The study of historical data and taxonomically and functionally similar species might help investigators interpret habitat associations for special-

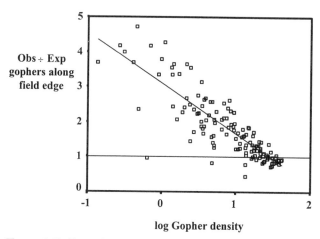

Figure 6.7. The relationship between selection for the field edge and pocket gopher (*Thomomys bottae*) density across alfalfa fields (*Medicago sativa*). The observed value divided by the expected value of 1.0 indicates the edge was used in proportion to its availability.

status rare species. For example, habitat descriptions of *T. sirtalis*, *T. couchi*, and *T. elegans* might provide insight into the historical habitat of *T. gigas*.

The prevailing view among those using the models in Table 6.1 appears to be that resource aggregations result in our measurements of disproportionate use of environmental elements. For example, Rotenberry et al. (Chapter 22) present a habitat selection model for which they assume that most wildlife observations are made where these animals want to be—where they derive some benefit from resources. However, the models in Table 6.1, and that of Rotenberry et al., can yield the same measures of disproportionate use when applied to aggregations that were forced by habitat fragmentation or the need to congregate. Applied to *T. gigas*, these models will indicate that *T. gigas* selects irrigation canals, probably because the landscape matrix is annual field crops where the snake cannot live. The models in Table 6.1 will not discriminate among aggregations influenced by various disparate factors, so they cannot be relied upon to measure habitat quality.

Demographic Organization

The demographic units represented by the numbers compared also bear on habitat analyses (Van Horne 1983). Whether or not demographic organization can be related to use of environmental elements, it affects the pattern of distribution observed. I recently discov-

ered a range in size of study areas in which the number of individuals might be given some meaning in terms of demographic organization (Smallwood 1999). As I had also found for mammalian carnivora (Smallwood 1999), nearly half the numerical estimates of northern goshawk (*Accipiter gentilis*) increased proportionally with increases in the sizes of their corresponding study areas (scale domain A, Fig. 6.8). Given that species of carnivora and northern goshawk typically space themselves at fairly even distances due to territory maintenance, I interpreted the aforementioned pattern as an indication that these estimates were made at areas smaller than those occupied by the "population." Each species in Smallwood (1999) appeared to have what I termed a *threshold area*, at which the number of individuals ceased to increase proportionally with increasing study area size (as in Fig. 6.8). The threshold area was the low end of a spatial scale domain (B), in which the number of individuals varied considerably but did not regress on study area size with a slope significantly different from zero. I interpreted the numerical estimates within scale domain B to represent populations because no other demographic units have been defined for the clusters of twenty-five to sixty adults that are typical of this scale domain (Smallwood 1999).

Regardless of whether the estimates in scale domain B represented distinct populations, the theoretical foundation of animal ecology includes the population as a key demographic unit with functional, goal-directed significance. Odum (1959) and Dasmann (1981) defined a population as some collection of organisms of the same species occupying a particular space and sharing a suite of attributes representing a unique organizational structure. Yet our use of N (number of individuals) and D (density) in habitat and other analyses is usually given no meaning with respect to the population concept. Usually, no demographic unit is attributed to N or D in habitat models. In conducting habitat analysis using a measured set of environmental elements, what does it mean to compare three individuals in element X to fifty in element Y and to four hundred in element Z, when the three individuals are a small portion of a population, the fifty compose an entire population, and the four hundred are from six populations? The three in element X

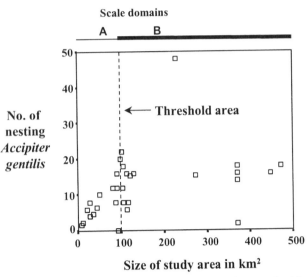

Figure 6.8. Based on the published estimates (summarized in Smallwood 1998), the number of nesting pairs of northern goshawks (*Accipiter gentilis*) increases proportionally with increasing size of study area until the study area is at least 100 square kilometers (scale domain A). Study areas larger than 100 square kilometers include aggregations that no longer increase with increasing study area size (scale domain B).

may occur in an ecological sink after having been forced there by the fifty, which themselves occur in element Y out of proportion to Y's availability but well within the space needed to support a population.

That the number of individuals in spatial scale domain B did not correlate with study area size also revealed a mathematical artifact in measuring animal density (Smallwood 1999). The number of individuals in domain B is relatively similar in magnitude to those in larger-area domains (see Smallwood 1999). Regressing density on its corresponding study area size will force a slope of −1 when density was calculated by dividing a constant number by a variable area. Similarly, regressing density on its corresponding study area size will force a slope of 0 when density was calculated using the numbers that increased proportionally with study area size, such as those in scale domain A and those forming the transitions between scale domains B and larger. Therefore, density appears to be a continuous variable only in the absence of demographic organization, and its discontinuity can confound habitat analyses based on the models in Table 6.1.

Spatial scale (extent) domains of distribution pro-

vide habitat analysts with a means to attribute social meaning to the frequency counts used in the habitat models. These domains revealed a possible demographic context against which frequency data can be compared. I lack the directed field research evidence needed to conclude that the numerical estimates in spatial scale domain B were distinct populations. However, the population concept can still be used to discuss the intent of habitat analysts and the meaning of their results.

Home Range

Habitat analysis within the individual's home range fails to reveal the significance of the location of the home range. Differential use of areas within the individual's home range is interesting at the microhabitat level, but the location of the home range itself can be influenced by the location of the population and the social and demographic status of the individual (Van Horne 1983). An individual's home range can encompass low-quality habitat simply because it was forced to live there in order to participate as a member of the population (Stamps 1991). Leger et al. (1983) found that California ground squirrels (*Spermophilus beecheyi*) adjust their behaviors to access microhabitats that would otherwise pose increased risks of predation. Animals do not always go to a place because it provides the highest-quality habitat, but once at the place, they make the best of it. Reports of habitat studies based on telemetry locations of individuals often provide no information of the demographic status of the individual or of the individual's spatial relationship to the population.

Intrapopulation Numbers

Comparing numerical estimates made at the subpopulation level (scale domain A) can confound the habitat analysis with the proportional relationship between the estimated number of individuals and study area size (Smallwood 1999), unless study sites (plots) were chosen randomly or systematically from a region, used and unused sites were compared, or long-term observations were used. Habitat analysis at this numerical level risks observing proportional patterns of use simply because individuals or their home ranges are regularly distributed among microhabitat elements due to territoriality or other social interactions that partitioned the resources at this level. Habitat analysis at this level of organization may or may not contribute useful inference regarding habitat, as social interactions are driving the spatial distribution within a larger area of habitat. Unless home range size can be related to habitat quality, habitat identified from a home range or from within the bounds of a population was pseudoreplicated (Hurlbert 1984; Aebischer et al. 1993).

As an illustration of apparent habitat use based on intra-population numbers, I compared a measure of resource selection (Equation 6.9 in Table 6.1) to Botta's pocket gopher (*Thomomys bottae*) density to test whether measured selection for the edge of alfalfa (*Medicago sativa* L.) fields was density-dependent:

$$\frac{Observed\ gophers\ at\ edge}{Expected\ gophers\ at\ edge} = \frac{gophers\ along\ edge}{gophers\ in\ field \cdot \left(\dfrac{ha\ along\ edge}{ha\ of\ field}\right)}$$

The density estimates were from 134 counts of burrows across thirty-seven alfalfa fields in Yolo County, California, during 1992 through 1994 (Smallwood and Geng 1997).

At low density, gophers occurred along the field edge at nearly five times the number expected by chance (Fig. 6.7). However, increasing density reduced my ability to quantitatively represent gopher selection for the field edge ($r^2 = 0.71$, Root MSE = 0.49, P < 0.001):

$$\frac{Observed\ gophers\ at\ edge}{Expected\ gophers\ at\ edge} = 3.09 - 1.43\ \log\ Density$$

The obvious preference for the edge grew increasingly hidden as gophers invaded the interior of the field and approached a regular distribution due to saturation of territorial space, represented by a ratio of 1.0 between the observed and expected gophers at the field edge. I do not think that gophers favored the edge any less as density increased, but rather their preference for the edge grew increasingly less recognizable as the overall density in the field increased (Fig. 6.7). As gophers saturated the field interior, the clustering along the edge became hidden, just as

Hansen and Remmenga (1961) found a clustered gopher distribution to vanish as the increasing density shifted the gopher population to a regular distribution. This density-dependence of distribution was recognized before (Taylor et al. 1978), but the density-dependence of measuring selection preference or avoidance will require habitat analysts to reconsider current reliance on measures and statistical tests based on use and availability of resources. Measured selection of environmental elements will be density- and therefore area- and time-dependent (also see Smallwood 1995), just as information theoretical measures (e.g., H', the composite index of species richness, $\log_2 S$ and evenness, J) are dependent on sample size (Cousins 1977) or on the spatial scale at which these measurements are derived.

This density- or scale-dependence also bears on the timing of habitat analysis. Measures of resource use will vary depending on the stage of ecological succession of a site or the stage at which a site is being colonized. Preference can be clearly observed during the early stages of colonization, but during the later stages the preferred resources can be hidden by overflow of individuals into relatively less-preferred conditions.

Interpopulation Numbers

Comparing numbers (densities) between populations poses two potential obstacles to recognizing true patterns of habitat association. The spatial shifting of aggregations summarized by Taylor and Taylor (1979) involves clustering at a subset of the available high-quality habitat patches. Taylor and Taylor (1979) proposed four hypotheses to explain the frequently observed spatial shifting of aggregations: (1) populations must move once they deplete their most limiting resources; (2) population members shift locations innately so as to prevent the exhaustion of resources; (3) dispersal and territory establishment of the next generation also establishes the location of the next aggregation, while the previous aggregation senesces; and, (4) a combination of hypotheses 1–3. Occupied habitat patches are either not different from many of those that are unoccupied at the time of the analysis or they are ephemeral in their quality. In the latter case, the results of habitat analysis are ephemeral. In the former case, habitat analysis might measure use as less than availability, and a strict adherence to the numerical comparison might mislead the investigator to conclude that the occupied habitat patches are less preferred.

Today's absence at a site can be tomorrow's presence, as the spatial distribution is unique for each generation (Taylor and Taylor 1979). Habitat measured at time t_1 can represent the species' habitat at times t_2 and t_3 only if the same generation and the same environmental conditions span times t_1 through t_3. If the species shifted locations between generations, as is common according to Taylor and Taylor (1979) and den Boer (1981), or if environmental conditions changed between times t_1, t_2, and t_3, then the analysis at time t_1 will be inadequate and misleading if the goal is to describe the species' habitat; it will contribute to a belief that the species relies on a narrower range of environmental conditions than it actually does. The exception will be constrained aggregations, which will be at the same locations at all times, so long as none of the aggregations go extinct. However, habitat analyses using constrained aggregations are predisposed to mislead simply due to the constraints on availability of suitable environmental elements.

An additional obstacle is the meaning we attach to numerical variations of each population. Population number can vary greatly, with increases and decreases lagging behind changes in conditions of resources. Interannual variability in N typically cannot be characterized adequately until monitoring has been conducted spanning multiple generations (Cyr 1997). Therefore, when comparing N or D from population X to that of population Y, and N or D differs, can we really conclude that habitat quality differs between the sites occupied by populations X and Y?

Presumably, sociality and energy and material resources maintain a carrying capacity just below which exists an optimal density. Within the spatial bounds of a population, density can vary somewhat but must be constrained by territoriality or the habitat element in most limited supply. Comparing locations occupied by populations, density might be nearly the same within the bounds of each population (Smallwood 1999). Comparing density within the spatial bounds of a population or between high-density populations is un-

likely to reveal meaningful differences and is therefore relatively uninteresting. In other words, density needs to vary considerably to reveal possible differences in habitat quality, but intrapopulation regularity of distribution and mutually high-density populations will not vary sufficiently in N or D to provide inference regarding habitat selection.

The terms N and D in habitat models can be representative of whole populations, and the occurrence frequency or spatial areas occupied by these populations compared to the measured set of environmental elements within a region or the species' range. The N or D representing a whole population can simply be replaced by a 1 to denote presence or a 0 to denote absence, and little information would be lost so long as the populations are discrete and easily bounded. Representing whole populations, presence or absence might be just as informative as N or D. If this step were taken, then the boundary of the area that is occupied by the population would need to be identified. The area occupied by the population might be more reliable as a measure of use than either N or D.

Megapopulation or Regional Numbers

Demographic organization is again implicitly ignored by habitat models that compare densities to measure use and availability of elements between regions. Regions are usually large-enough areas to include multiple populations. Garshelis and Visser (1997) termed the collective abundance from regions as megapopulations because they did not know what else to call it. Metapopulations, on the other hand, theoretically organize within regions (Hanski and Gilpin 1997), but empirical evidence is lacking for knowing the bounds of a metapopulation or how many populations might compose a metapopulation. Regional patterns of distribution are still largely theoretical, and the term megapopulation indicates that this theory is weak.

Comparing densities or use of regions poses the additional problem of comparing the availability of disparate measured sets of environmental elements (Wiens and Rotenberry 1981b). For example, Smallwood and Fitzhugh (1995) compared the number of puma track sets to represent use of available vegetation complexes and topographic categories across all the sampled areas of California, which included areas spanning the full north-south and east-west extents of the state within the species' range. The topographic categories were common across the state, consisting of ridges, mountain peaks, basins, canyons, and so on. These macrohabitat categories were fairly comparable across California, much like Anderson's (1981) use of vegetation structural elements across the United States. However, sage-juniper forests are limited to north-central and eastern California, and chaparrals occur more to the west; they are not interspersed. Comparing use to availability of vegetation categories seemed uninformative on a statewide level, especially when puma populations occurred in both sage-juniper and chaparral. Knowing the range of environments used by pumas in California is useful, and knowing that pumas are more rare in the Mojave Desert than in the Klamath Mountains is also useful, but habitat analysis involving use and availability comparisons is probably best conducted within regions, within which key elements are more likely to be naturally interspersed.

Gradient of Abundance

Animal density is often represented with contours, which illustrate relatively smooth gradations in density from high to low across landscapes or regions (e.g., Taylor and Taylor 1977; Wiens and Rotenberry 1981b; Cody 1985a,b; Scott et al. 1986; Root 1988a,b,c; Price et al. 1995; Morrison et al. 1998). These gradients are mathematically derived using averaging and interpolation, whereas the actual spatial distribution may not be smoothly graded. This transformation of field observations into density contours expresses a world-view of distribution that acknowledges aggregation as the norm for animal species but also facilitates the potentially erroneous idea that animal density corresponds to habitat quality. Habitat quality is assumed to be energy-related, as discussed previously, so it is related to the world-view that the laws of thermodynamics govern the spatial patterning of animal species, also previously discussed. However, this world-view, that density grades smoothly across the landscape, can be supported by few examples, just as sharply bounded aggregations can be supported by few examples (Morrison et al. 1998). It may be that spatial distributions of

animals can be better represented by categories, including population occurrence, trace activity (i.e., a few individuals in a large area), and absence. Lidicker (1995) also summarized categorizations of habitat quality, which were thought to bear directly on densities. Habitat analysts need to test whether animal aggregations are discrete or graded, whether population boundaries can be identified, and whether densities are categorical in correspondence with habitat quality. Empirical evidence should be the foundation of theory, and theory that bears on gradients of abundance also bears on habitat analysis.

It is well documented that individuals can occur in what are regarded as ecological sinks for the species. In fact, numerical estimates in ecological sinks can always be greater than zero just because dispersing animals are forced into the sink where the rates of recruitment outpace mortality. Habitat models based on numerical comparison will thus give ecological sinks some positive habitat value in such cases, whereas the true value should be negative (less than zero) with respect to the functionality of the population. Density estimates alone cannot inform of sink conditions (Lidicker 1975, Van Horne 1983), because density estimates cannot be negative—they range from zero to some presumed carrying capacity. Learning of sink conditions requires intensive study, making use of more information than numerical estimates (e.g., Morris 1989).

Preconception of Habitat

The map of available environmental elements is made by the habitat analyst. Some a priori notion of habitat inevitably goes into construction of the incidence map and will affect the typology of the map and its spatial grain (Austin, Chapter 5). For example, Smallwood and Fitzhugh (1995) used typical home range size of puma to decide on the quadrat size and transect lengths for counting track sets in California. The volunteer biologists who selected the exact locations of the transect segments believed that roads along ridgetops would produce more tracks. We later learned that roads along streams were most productive. If we were to start a new sampling program for puma track sets, it would differ markedly from the

1985 design, and the results would likely differ as well.

Preconceived notions of habitat influence the study's location, spatial extent, grain of mapping, and ultimately the replication and interspersion of the environmental elements in the map. Habitat analysts often collect the use and availability data from locations where the species was known to occur, often in abundance. By siting habitat studies this way, analysts may often force results that conform to preconceived notions of habitat. Experimental design principles should be applied to mensurative as well as to manipulative studies (Smallwood 1993). The spatial extent and grain of the study should be appropriate to measuring the species' differential use of the environment. Replication and interspersion of treatments (i.e., environmental elements considered to be available to the species) also must be incorporated into mensurative studies (also see Otis 1998; Austin, Chapter 5). Along a transect or within a defined study area, replication of available elements is achieved through multiple occurrence of each environmental element in the measured set of elements (e.g., vegetation complexes). Those elements occurring once are pseudoreplicated (sensu Hurlbert 1984) and association of the species with that element should be considered dubious pending further research (e.g., environmental elements X and Z in Fig. 6.4 are pseudoreplicated, while Y is measured as two replicates). Interspersion is achieved by each element's occurrence between other elements in the measured set, such that gradient effects do not cause spurious relationships between measured use and availability. Thus, replication and interspersion of various environmental elements can be achieved when the program of observation includes two or more patches of each environmental element within the study area or along the transect.

Given that the analyst is constrained by observing extant environmental conditions, the observed use of the environment by an individual, population, or larger social unit need not reflect the species' perception of what constitutes habitat or high-quality habitat (Morrison et al. 1998). Much of what analysts may perceive as a numerical response to extant environmental conditions actually may be responses to the combination of both relic and current habitat and de-

mographic conditions. Habitat models implicitly assume that deviations from the theoretical null patterns of distribution are immediate responses of individuals to resources that are available in the measured space. However, demographic organization can represent a long-term response to energy and material resources in the environment experienced by the species, and it can pose much of the information useful to the individuals. If a suite of environmental and demographic conditions typified the success of the species over long time periods, then the species likely developed perceptions of habitat and demographic organization that are relatively reliable evolutionarily. Such stability could be achieved by building the responsible neurons early in ontogeny (Coss and Goldthwaite 1995). How such perceptions would be manifested in the changed landscapes of modern times can affect our interpretations of habitat.

Habitat analysis is bound to include plenty of noise caused by conserved use of information from relic environments, which today may be missing from the studied environment. For example, after Hutto (1990) experimentally reduced food supplies from under the bark of specific trees, he found no difference in visits per tree or time spent at the tree by birds that depend on this food supply during the winter months in the boreal forest. In another example, the carrying capacity of pocket gophers in alfalfa fields is determined not by the food source, which is plentiful, but by the home range size. The home range size was established by natural selection in past environments where food supplies were typically more limited. Pocket gophers in alfalfa are so constrained by evolutionarily designed perceptions of space that they cannot fully exploit alfalfa stands, which are perhaps the most abundant food source these animals have ever encountered. Management or policy decisions should acknowledge a reasonable level of uncertainty in habitat analysis or assessment and should be very conservative.

Conclusions

Comparing numerical terms for use and availability cannot reliably characterize habitat without considering the environmental and demographic contexts of the numbers. The measurements and resulting associations are determined largely by the attributes of the field study design, as well as those of the analytical, interpretive design. These attributes are themselves determined largely by the analyst's preconception of habitat, demographic organization of the species, how the species is patterned spatially and temporally, and the reasons for animals to congregate or associate with environmental elements out of proportion to the element's availability (Fig. 6.2). At our current state of knowledge, world-views may bear more heavily on habitat analysis than does the procedure for measuring occurrence frequencies in environmental elements. Some of these world-views can be tested as hypotheses by directed field research. For example, how often do animals aggregate for reasons that do not bear directly on the availability of a food resource? Are all aggregations truly populations? How sharply bounded are animal populations? Answers to these and other related questions can provide habitat analysts with meaning for numerical comparisons that the models do not provide.

The effectiveness of habitat analysis based on numerical comparisons is largely dependent on knowing how and why animals distribute themselves. The models in Table 6.1 will be prone to inappropriate application and erroneous interpretation until analysts largely agree on the extent to which animal species respond numerically to energy availability, information presented by the past and current environments, or to predators and competitors. Each model has a theoretical root or history, and we need to decide whether the theoretical foundation is consistent with what we are attempting to measure and how we should interpret the resulting patterns.

Van Horne's (1983) argument that density is a misleading indicator of habitat quality remains valid, and based on recent research on density, it is all the more clear that numerical comparisons are currently of dubious utility to habitat analysis. Density estimates are sensitive to whether they are derived from population "isolates" or from sampling of the statistical universe (Preston 1962a), and they are sensitive to the size of the study area examined (Smallwood and Schonewald 1996). Until much more basic research on animal distribution has been conducted, the models in Table 6.1 should be used cautiously because their use can translate into inappropriate management decisions, sometimes with possible dire consequences for the species.

Acknowledgments

I thank Michael Morrison and Linnea Hall for helpful discussions on habitat. I thank Michael Morrison, Leona Svancara, Martin Raphael, and J. Michael Scott for critical review comments.

The Role of Category Definition in Habitat Models: Practical and Logical Limitations of Using Boolean, Indexed, Probabilistic, and Fuzzy Categories

Kristina E. Hill and Michael W. Binford

The use of categories is a fundamental part of human reasoning (Rosch 1975; Lakoff 1987). Conceptual categories are essential for the articulation of biological theories. The biological species itself is a category with an interesting history of conflicting definitions, as are the notion of habitat and theories of the niche (Mayr 1982). Set theory is the branch of mathematics that deals primarily with numerical descriptions of membership in categories, which allows us to represent categories in logical statements (Ayyub and McCuen 1987; Burrough 1989, 1992; Ross 1995). Our argument is that categories are an unavoidable component of model construction and use and are either implicitly or explicitly represented in a model. Even models that are composed only of mathematical statements rely on underlying conceptual categories, such as "habitat" or "suitability." Managers and planners, in turn, often use models as the basis for defining (and defending) categories like "suitable habitat" (Steinitz 1969). Models are frequently used in applied settings to determine whether a certain set of environmental conditions is "better" or "worse" for a threatened species, terms which themselves represent categories. Models are also sometimes used to make decisions about whether certain geographic areas will be altered by human land use, perhaps permanently. Our concern is that habitat models are sometimes used in ways that are inappropriate, given the logical assumptions that may be inherent in such models. These logical assumptions derive in part from the types of categories that are implicitly or explicitly used in model construction. We are also concerned that the ad hoc use of categories in habitat model construction limits the contribution that these models can make to theory.

A Taxonomy of Category Types for Habitat Models

Quantitative modeling must be informed by theories of measurement and systems of logic if it is to be useful in building bodies of theory (Stevens 1946) and if it is to be used as defensible support for strategic decisions (Steinitz 1979; Lemons 1996). This imperative emerges from guidelines for theoretical rigor as well as from standards of common sense. In 1946, facing a lack of guidance for researchers who wished to measure human responses to environmental stimuli, Stevens proposed a classification of measurement scales (nominal, ordinal, ratio, and interval). He included a table of appropriate mathematical transformations that could reasonably be applied to each type of measurement. This classification was a significant aid to consistency in subsequent modeling activities. We argue that a similar kind of taxonomy is needed today with regard to categories, particularly in the use of geographical data for habitat models (Robinson

TABLE 7.1.

Summary of differences between discretely defined categories and ambiguous categories.

	Discretely defined categories	Ambiguous categories
Logical basis	Classical set theory	Fuzzy set theory
Membership Concept	Discrete, with absolute thresholds	Uses a gradient of similarity to a clearly defined best example
Membership Function	Boolean (yes or no; 0 or 1)	Graded (0 to 1 scale)
Means of Representing Uncertainty	Probability	Possibility
Source of Uncertainty	Stochasticity	Ambiguity
Operator for Finding Intersection	Multiplication	Minimum function

and Frank 1985; Rotenberry 1986; Goodchild and Gopal 1989; Stoms et al 1992; Conroy and Noon 1996; Hunsaker 1996; Flather et al. 1997; Clark and Shutler 1999).

In contrast to Stevens, our focus is on the types of categories that can be used to construct models rather than on the scales of measurement that may be used to represent phenomena. We are concerned that researchers and managers do not currently share a common framework for identifying the kinds of uncertainty that can be represented by different category types. More broadly, we believe there is a need for an approach to modeling that identifies the kinds of uncertainty that can be accounted for when an investigator constructs a habitat model and when a manager subsequently applies that model.

Our proposed approach is based on the simple observation that there are two fundamentally different types of categories: (1) discretely defined (or "discrete") categories, and (2) ambiguous categories. Discrete categories are based in the logic of classical set theory, while ambiguous categories have their basis in fuzzy set theory. Table 7.1 provides a summary of the differences represented by these two category types.

Discretely defined categories can be defined as categories that use discrete thresholds to define limits for a variable, regardless of whether the variable itself is numerically continuous or discrete. Membership in such a category is absolute. The logic of membership in discrete categories is based on Aristotle's Law of the Excluded Middle, which states that either an element belongs to a category or it does not. The concept of discrete categories was fundamental to Aristotle's approach to categorization and to the development of logical constructs that subsequently affected mathematical philosophy and the development of classical set theory (Apostle 1980). Moreover, the assumptions of probability theory require that one must be able to define a discrete state or event in order to assign a probability to its occurrence (Chernoff and Moses 1959; Ross 1995).

Ambiguous categories, also referred to as vague or fuzzy categories, are categories that have been defined using the concept of similarity to an ideal state or event. Membership in these categories is frequently defined using a gradient rather than a discrete threshold. If at least one limit for a category is defined using a gradient, then this category must be considered ambiguous—even if a discrete threshold is used for the definition of membership at the other limit. The theory that underlies ambiguous categories has been described as a generalization of classical set theory that relaxes the Law of the Excluded Middle (Ross 1995). This is now widely referred to as fuzzy set theory.

As a generalization of classical set theory, fuzzy set theory allows membership in a category to be described using either a gradient or a discrete threshold, or some combination of both. It is important to note that both discrete and fuzzy categories allow for the representation and consideration of a particular kind of uncertainty. When used in combination with probabilistic reasoning, discrete categories allow a model to represent uncertainties that result from randomness. Fuzzy categories, on the other hand, represent uncertainties that are non-random. Gradients of membership in these categories are based on similarity, not stochasticity. The different degrees of membership that may be assigned to an element represent the degree of similarity between that element and some ideal state or event. It is impossible to refer to the probability of this similarity, since probability theory requires the identification of a discrete state or event. Instead, fuzzy set theory provides the term "possibility" to refer to the likelihood of similarity (Dubois and Prade 1980).

Model Objectives, Underlying Theory, and the Logic of Category Definition

When building a model, an investigator must judge whether the concepts that underlie the model are inherently discrete, inherently ambiguous, or both. This determination provides the model with an explicit foundation in current theoretical understandings of both logic and biological phenomena such as the spatial distribution of a biological species or the responses of individual organisms to environmental stimuli. Without this grounding in theory, models run the risk of misinterpretation and a lack of defensibility in an applied context (Greenhouse 1997). The development and testing of models will also be unlikely to contribute to future theoretical understanding if the models do not explicitly incorporate testable hypotheses that originate in a body of theory.

Habitat models frequently address (1) the likelihood of the occurrence of a species at a given location, (2) the likely abundance of a species at a given location, or (3) the potential of a given location to serve as habitat for a particular species. Each of these three modeling objectives implies a different set of logical and theoretical assumptions, and each allows model users to make different interpretations of model output in an applied context. The fundamental difference we see among these models derives from the logical basis each uses for making predictions. A probabilistic model uses a pattern of discretely defined states or events to predict the likelihood of a discretely defined state or event. A fuzzy model, on the other hand, uses

the concept of similarity to predict the possibility of a state or event that is ambiguously defined. The problem for the model builder, and for the user, is to choose the type of model that is appropriate to a given question.

Probability theory may be used as the basis for prediction if the occurrence of an individual organism at a specific sampling location is defined as a discrete event and the likelihood of occurrence is determined using a record of observations at that location over time. Predictions of occurrence can then be made using methods derived from the limiting frequency interpretation of probability theory. This interpretation holds that over a large number of instances, a discrete event can be predicted to occur with a known frequency (Hacking 1975). Limiting frequency approaches to probability are the basis for most inferential statistics and are familiar to most biological researchers. However, this approach is not logical when a model is constructed using anything other than discretely defined states or events. The use of probability theory is not appropriate for models that describe habitat suitability using the concept of similarity to some expected optimal conditions.

The concept of suitability may contain both discrete and ambiguous dimensions. Some defining criteria can be discrete, such as whether or not a particular tree is wide enough to accommodate the body size of a pair of cavity-nesting birds. There is likely to be a discrete lower limit along this dimension of suitability. But the upper limit of this category can be ambiguous. There may be no trunk diameter that is too wide for a nesting pair of these same birds. Other criteria that define suitability may contain ambiguity in both their upper and lower limits. Canopy density, for example, may be used in some habitat models as a proxy measurement of food availability for an insectivorous bird species. There may be no discrete upper or lower limits for this dimension of suitability for the species of interest, except when there is a total absence of tree canopy.

If a habitat model describes the potential suitability of a given location to serve as habitat for a particular species, then its logic must be able to account for categories that allow a gradient of membership. The desire to define "suitable sites," whether for nesting, foraging, or other activities, requires an investigator to answer the question of whether upper and lower limits of environmental variation that affects the responses of individual birds can be discretely defined. Answering this question requires a theoretical framework that offers explanations of animal behavior and perception (Fig. 7.1).

Theories of the niche, for example, provide a theoretical concept that can inform the definition of habitat suitability. Using these theories, suitability would be defined using the concept of similarity to an optimum condition. The appropriate logic for representing similarity to an ideal state or event is fuzzy set theory. Habitat models that address suitability using theories of the niche necessarily rely on an underlying logic of ambiguous categories. The only way to avoid this is for an investigator to employ an underlying theoretical concept that does *not* rely on the idea that an organism has different levels of response to environmental gradients and that an optimal level exists for that organism.

The niche concept is described using similarity to an optimal state. Most habitat suitability models rely on this logic as well, since they are explicitly or implicitly based on theories of the niche. The familiar concept of a "suitability index" represents, in logical terms, a fuzzy set (Hill 1997)—as does the biological concept of the niche (mathematical representations of the niche as a fuzzy set are presented by Cao 1995 and by Bock and Salski 1998). An investigator who uses similarity-based concepts to construct a habitat model should approach model construction and testing using fuzzy logic as the basis for reasoning with and about such a model. Otherwise, the interpretation of the model's results will be inadequately grounded in both theory and logic.

Of course, it is also possible to define "suitability" and "potential habitat" using the limiting frequency approach of counting occurrences or recording abundance over time at a given location. But unlike similarity-based reasoning, this method alone does not provide a logical or theoretical basis for making predictions in new locations where sampling has not been conducted over time, unless the expected variation in habitat selection by an organism is assumed to occur only as a result of stochastic processes. Predictions

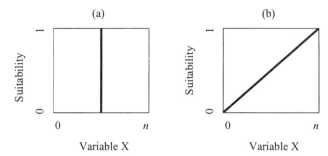

Figure 7.1. Use of two different functions to define habitat suitability along a range of values for the environmental variable *X*: (a) a Boolean function with a discrete threshold, and (b) a continuous linear function.

that are based on non-random variations in an organism's process of habitat selection require an underlying theoretical concept that describes that organism's responses to its environment. Theories of stochasticity alone do not provide such a description.

In common practice, correlational studies that observe the presence or abundance of a bird species and one or more environmental conditions over a set of sampling locations are used as a basis for habitat suitability models (USFWS 1981a,b). Probability theory is typically used to determine whether the variation in abundance is greater than would be expected by chance. In our view, there is no conflict between the use of these methods and the use of a similarity-based logic in constructing model categories. The results of correlational studies can be the basis for the construction of ambiguous categories as well as discrete categories. The difference in practice that we do recommend, however, affects both (1) the choice of

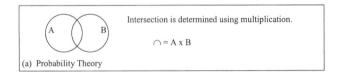

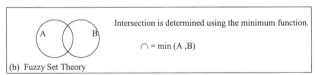

Figure 7.2. The operations used to find the intersection of sets in probability theory (a) and in fuzzy set theory (b). Similar operations are used in models based on Boolean logic and in models constructed using indices, but the choice of an operation is frequently treated as if it were an ad hoc decision.

mathematical operations used to combine variables within a model and (2) the testing, use, and interpretation of the model in an applied context. We discuss each of these issues below and provide an example from a case study by Hill (1997).

Choosing Mathematical Operators for Combining Categories in a Model

Stevens pointed out that each scale of measurement should be subjected only to those mathematical operations that are appropriate to that type of scale (Stevens 1946; Hopkins 1977). Similarly, the underlying logic of the two category types we identify (discrete and ambiguous) restricts the range of mathematical operations that can be applied in a model (Hill 1997) (Fig. 7.2).

Our argument requires us to first characterize the logical operations used to combine habitat variables in a model. If we wish to know where individuals of a species will be able to find both food and a nest site in the same place, then the appropriate logical operator is "and." We are looking for places where the statement that "Condition A (food availability) exists *and* Condition B (nest site availability) exists" is true. The corresponding term for this logical operator in mathematical set theory is "intersection" (Tomlin 1990; Burrough and McDonnell 1998).

If the categories defined in a model are all discrete, then the intersection of two or more categories may be found using the "and" statement provided by Boolean logic, in which only two states can exist (true and false, or yes and no, or 0 and 1). The mathematical operator required to produce an answer of "true" (i.e., represented by the number 1) when both Condition A and Condition B are also "true" (i.e., represented by the number 1) is multiplication. We could use a similar logic to treat Condition A or Condition B as probabilities. Probability theory finds the intersection of two probabilities (stated as the likelihood that Condition A *and* Condition B are true) using multiplication. Thus the intersection of discrete categories, whether these are treated as a Boolean "truth" or as a probabilistic likelihood, is found using multiplication.

However, if any of the categories defined in a model use gradients to define membership, then Boolean logic and probability theory are not useful. The inter-section of two categories where at least one category limit is ambiguous may be found using operations defined by fuzzy logic (Zadeh 1965). The mathematical operator that corresponds to a fuzzy "and" statement is the minimum function (or a weighted minimum function, if the investigator believes there is reason to weight the variables that were used to define the categories) (Dubois and Prade 1980; Ross 1995). This function is, in a sense, more conservative than the multiplication operator, because it returns the lesser of the two values it combines. Fuzzy logic uses this more conservative function because it recognizes that the gradations in membership used in two different categories represent "apples and oranges"—in other words, the two categories refer to different things, perhaps using different scales, and therefore cannot be manipulated via multiplication as if they were ratio- or interval-scale measurements.

An Example: Building a Habitat Model for the Black-capped Chickadee

The black-capped chickadee (*Poecile atricapilla*) is a common species in several regions of the United States. A habitat suitability index (HSI) model was developed and tested for this species by the U.S. Fish and Wildlife Service (USFWS) using species abundance and vegetation data from floodplain forests of the South Platte River in Colorado (Fig. 7.3) (Schroeder 1983a, 1990).

Hill (1997) was interested in the effect that category definition might have on the model's ability to predict observed abundance values and its usefulness for defining a geographic area that would contain all of the sampling locations where chickadees had been recorded. These two model performance issues are particularly important to planners and managers since they must frequently use habitat models to establish the geographic boundaries of management zones. Planners and managers often have insufficient resources to collect field data to develop and calibrate habitat models specific to their own management areas, particularly when they must respond rapidly to emerging land-use conflicts. This need accounts for the use of the USFWS HSI model in this case study as if it were a generic suitability model that could be applied anywhere.

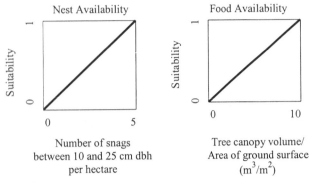

Figure 7.3. U.S. Fish and Wildlife Service habitat suitability index model for the black-capped chickadee (*Poecile atricapilla*)(Schroeder 1983a, 1990). The final habitat suitability value is obtained by taking the minimum value of these two variables.

Hill (1997) created several versions of the USFWS black-capped chickadee HSI model, varying only the number and type of categories used in the model (Figs. 7.4, 7.5, and 7.6). Bird abundance data were obtained for model validation from an area of approximately 375 hectares at the Harvard Forest Long-Term Ecological Research site at Petersham, Mass., by R. Lent. Sixty-seven randomly established sample sites were used to census the area during the months of May and June 1993 (R. Lent unpublished). Figure 7.4 shows the distribution of chickadee abundance in relation to canopy volume. Canopy volume was estimated from a measure of basal area using a relationship presented by the author of the USFWS HSI model (Schroeder 1990). The second variable, number of dead snags per hectare, was not influential in this example since the number of snags in the sampling locations consistently exceeded the threshold value defined in the HSI model as suitable habitat (ten snags per hectare).

The first alternative version of the HSI model prepared by Hill (1997) was Boolean and contained only two categories, "suitable habitat" and "unsuitable habitat" (Fig. 7.5). Bayer and Porter (1988) demonstrated better predictive performance when they simplified HSI models using Boolean categories created by setting the 0.5 suitability index level as a threshold value. Hill (1997) used the same threshold value to discretize the continuous habitat suitability output of the chickadee model. This discretized version can be thought of as a probabilistic model of suitability. But the biological basis for this construction of the model

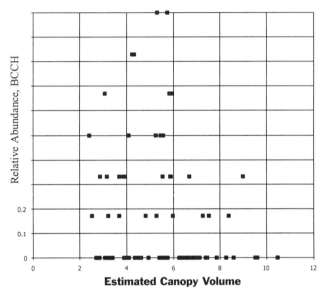

Figure 7.4. Relative abundance of black-capped chickadees (*Poecile atricapilla*) versus estimated canopy volume in the Harvard Forest study area.

is not clear. It represents the hypothesis that the black-capped chickadee uses only forested areas with estimated canopy volumes of greater than 7.5 cubic meters per square meter. We are not aware of any theoretical basis for that discrete hypothesis. If indeed there is none, then discretizing the continuous output of a habitat suitability model is simply an ad hoc attempt to improve model performance. It does not represent a hypothesis based in current theory and is unlikely to contribute to a theoretical understanding of species occurrences.

The USFWS HSI model can also be constructed using ambiguous categories. Hill (1997) assembled these using an approach common in fuzzy systems engineering (Kosko 1992; Ross 1995). Three symmetrical and overlapping fuzzy categories were used to describe habitat suitability along a gradient of canopy volume: high suitability, moderate suitability, and low suitability. Fuzzy set theory allows sets to overlap where elements belong to more than one set. In practical terms, this means that a canopy volume estimate of 7.5 cubic meters per square meter would belong to the set of "high suitability" sites to a degree of 0.5 and to the set of "moderate suitability" sites to a degree of 0.5. A membership value of 1.0 indicates an area of

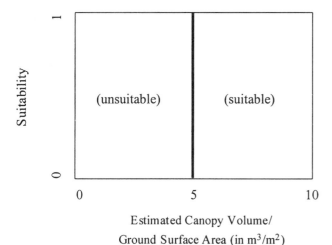

Figure 7.5. A two-category Boolean version of the habitat suitability index model for the black-capped chickadee (*Poecile atricapilla*). A threshold of 0.5 was used to discretize the original, continuous function published by Schroeder (1990).

the graph that has no overlap between categories (Fig. 7.6).

The use of three overlapping categories instead of a single fuzzy set (i.e., an index) has become well-established in the design of engineering control systems, where the three overlapping categories might program a microprocessor to adjust the response of a motor to a measured environmental condition such as ambient air temperature (Kosko 1992). The overlap among categories helps to prevent the phenomenon of "overshoot" in mechanical systems that respond to environmental inputs.

Hill (1997) suggested that using overlapping categories can help biologists and land managers represent and perhaps test the concept of a realized niche (Hutchinson 1957), while most single-category suitability indices attempt to represent only the fundamental niche. Her argument was that phenomena such as interspecific competition or the temporal order of patch colonization by different species may be indirectly visible in the relationship between the observed abundance or presence of a species and the suitability ratings provided by each of three overlapping fuzzy categories. For example, if the category "moderately suitable" best predicts the abundance or occurrence of the species, that might be interpreted as evidence that the species is not occupying its optimal habitat rather than as evidence that the suitability model itself is wrong. Bird abundance data from the Harvard Forest,

for instance, showed that red-breasted nuthatches (*Sitta canadensis*) were most abundant at sample sites with the greatest canopy volume. The black-capped chickadee might be expected to occupy that same optimal habitat but instead was most abundant at sites with moderate canopy volume.

Abundance is not an ideal indicator of habitat suitability (Van Horne 1983), and the suboptimal distribution of chickadees found in this case may have occurred for various historical or biological reasons. But this suboptimal distribution cannot even be perceived if a planner or manager simply applies a single-category model that rates suitability using expectations of optimal conditions. The typical suitability index model may simply appear to fail in its predictions without allowing a researcher or manager to form new hypotheses about the biological environment in which the species of interest has become established. On the other hand, application of a three-category model in which the category "moderate suitability" is the best predictor of abundance raises the question of why this category predicted best.

Testing the Accuracy of Habitat Models

Tests used to establish the predictive ability of habitat models require that the data used in these models

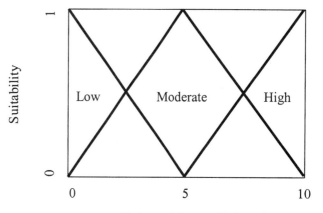

Figure 7.6. A three-category fuzzy set version of the habitat suitability index model for the black-capped chickadee (*Poecile atricapilla*). The triangular functions are used in a symmetrical pattern to represent an even distribution of the categories along the variable's range, since no more specific hypothesis has been stated.

must meet appropriate statistical assumptions. Most contemporary tests of scientific models have been applied with the assumption that the data should meet the requirements of inferential statistics and probability theory. For instance, a correlation coefficient has often been used to test predictions of suitability against actual abundance. The suitability ratings produced by the original USFWS HSI model for the black-capped chickadee (Schroeder 1990) did not show a significant correlation with that species' abundance in the Harvard Forest case study described above (Pearson's $r = -0.18$, $P > 0.10$). Similarly, a Boolean version of this model did not perform well in a *t*-test that compared the mean suitability rating of sites with chickadees to the mean suitability rating of sites without chickadees ($p > 0.10$).

Perhaps our first question should be whether the concept of suitability meets the basic requirement that we must begin by defining a discrete state or event in order to use probability theory. We argue that it does not, except in the unusual case where the biological theories that underlie the suitability model identify *only* discrete thresholds of biological response. This requirement excludes habitat models that are based on theories of the niche, since this theory predicts gradients of species response. If we wish to maintain consistency between theory and method in habitat suitability analysis, we must look for both models and tests that do not require the definition of discrete states or events. Those tests should be able to enhance the defensibility of habitat models in applied decision-making contexts as well as establish the conditional usefulness of a model for theory building.

We find that the issue of defensibility is often not well understood. From an applied perspective, there are two ways a habitat suitability model can fail: (1) it can fail to predict the abundance of a species where it is indeed abundant, and (2) it can overpredict, rating many or all locations suitable although the species has not been detected there. The first case represents an error of omission in which the model used by a manager to predict habitat use omits areas where the organism is abundant. The second case represents an error of commission. Both types of errors undermine the defensibility of a model in an applied setting. Yet, the presumption in natural resource management

must be that errors of omission are worse than errors of commission if we wish to implement a precautionary principle for biological conservation (Shrader-Frechette and McCoy 1993).

In order to avoid an inappropriate reliance on probability theory via the use of inferential statistics, Hill (1997) developed two simple tests for the sake of argument. These tests are pragmatic and are driven by (1) the planner or manager's ethical interest in avoiding irreversible harm to a species of conservation concern, and (2) a need to enhance the defensibility of habitat models. Her first test was designed to identify and reject models that produce numerous errors of omission. It was intended for application with models that try to predict the abundance of a species using an estimate of habitat potential. The test plots habitat suitability (i.e., habitat potential) against relative abundance and treats the number of times when relative abundance exceeds the predicted habitat potential as instances of model failure (Fig. 7.7).

This test allows models to be accepted or rejected according to the degree to which they underestimate habitat potential but does not measure the tendency of a model toward overestimation. A second test was in-

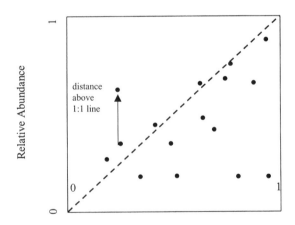

Figure 7.7. A simple test for habitat models intended to assess potential habitat rather than to forecast the actual abundance or occurrence of a species. This test counts only instances of abundance exceeding the expected habitat potential as failures of the model. Both maximum distance of points above the 1:1 line and total distance of points above the 1:1 line may be useful measures to compare the performance of different models.

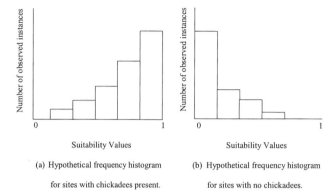

(a) Hypothetical frequency histogram

for sites with chickadees present.

(b) Hypothetical frequency histogram

for sites with no chickadees.

Figure 7.8. A simple test to assess the degree to which a model of potential habitat overestimates the potential for a species to occupy a geographical area. The measured skewness of a frequency histogram should be a higher number in a histogram of suitability values for sites with no chickadees present than in a histogram of sites with one or more chickadees present. Ideally, the skewness should be positive where no individuals of the species are observed and negative where individuals are observed.

tended to account for this by measuring the degree of overestimation, specifically, in models that predict potential habitat suitability. It simply compares the degree of skewness in a frequency distribution of habitat suitability ratings at sample points where the species is present to the skewness of a distribution of suitability ratings at sample points where no individuals of that species were detected (Fig. 7.8). Skewness should be greater where no individuals of that species were observed (and ideally should be highly positive, although even the best model may not produce this result). In the case of a model that has more than two categories, the tests described above would be applied to each category individually.

Implications for the Use of Habitat Models in Planning and Management

When habitat models are applied to land management and land-use decisions, we must be clear about precisely what is being modeled. Models that are based on limiting-frequency definitions of the probability of occurrence for a species should not be used as if they were models of potential habitat, and models based on similarity should not be misinterpreted as predicting the probability of a species' occurrence. Each type of model relies on a different underlying logic, which we

argue should be made explicit in each modeling effort. We have recommended specifically that a taxonomy of category types should be used to determine when it is appropriate to use probability theory or fuzzy set theory in constructing habitat models. This also allows a model to explicitly represent two different kinds of uncertainty: (1) uncertainty that results from random variation, and (2) uncertainty that results from gradients of species response to environmental conditions. Discrete categories can be used along with probability theory to represent the first kind of uncertainty, while fuzzy categories represent the second.

Serious errors can be introduced by the application of habitat models when researchers or planners and managers ignore or misinterpret the fundamental logic of the model used. The most significant sources of error are (1) the use of discrete thresholds for convenience when these are not supported by either theory or evidence, and (2) the use of an inappropriate mathematical operation to produce output values from a habitat model. These errors are particularly important in determining the appropriate geographic area needed to conserve a given species' habitat.

In situations where discrete thresholds are used inappropriately, geographic boundaries are introduced in the mapping of management zones that are actually just data artifacts, with no clear connection to a theoretical framework. Frequently, these reified boundaries are misinterpreted as having genuine biological meaning—particularly when many map layers have been combined in an analysis and the source layers where these data artifacts originated are no longer individually visible. The spatial consequences can be significant in the final map of areas to be conserved if a significant proportion of the area has values close to the discrete threshold value used in one or more category definitions. These areas represent "gray zones" that almost but don't quite meet the model conditions (or conversely, there may be extensive zones that were just barely included by the model). A legal suit could be mounted, either by conservation or development interest groups, contesting the designation of this intermediate zone. Such a suit could significantly impact conservation efforts. Sensitivity analyses should be conducted before deciding on the functions and thresholds that will be used for category definitions to reveal the risk of this type of error.

When inappropriate mathematical operations are used to find the intersection of two or more categories in a model, the spatial errors created when these models are used to plan habitat reserves can be even more difficult to detect. A final map of habitat suitability generated by multiplying two suitability indices (an operation appropriate only for probability estimates or Boolean categories) will differ from a final map generated by taking the minimum value (as is appropriate for fuzzy sets). In general, the greatest difference between taking the minimum value and using multiplication is that the final suitability values will be highest when all constituent variables have values in the midrange of rated suitability. In other words, if one suitability variable is rated with a suitability of 0.5, and a second is also rated 0.5, the combined value under multiplication (0.25) is much lower than the value that would be obtained using the minimum function (0.5). Depending on the threshold value used to draw a boundary around suitable habitat areas, a planner or manager could be confronted with maps that show two quite different geographic habitat areas that have been designated using models that are otherwise identical.

Recommendations for the Development and Use of Habitat Models

In summary, researchers and managers alike need habitat models that can explicitly represent their assumptions about uncertainty. This uncertainty may arise from ambiguities in the definition of suitability, or from stochastic processes that affect species distributions. Our conclusion is that models that seek to predict the potential abundance or presence of a species using the conceptual framework provided by theories of the niche should represent the inherent ambiguity of suitability using a similarity-based logic. Models that seek to predict the likelihood of species

occurrence or abundance using probability theory must first define a discrete state or event. Historical occurrences or abundance levels may be that event. But it is not clear to us that there is a general theoretical basis that provides a definition of discrete states for habitat suitability.

The theory of fuzzy sets is being developed rapidly. This theoretical work is likely to be directly applicable to the work of ecologists and managers who seek to describe habitat suitability using models that are both conservative (in the sense that they help to avoid irreversible loss of species) and defensible in an applied setting. We recommend that managers and planners who wish to use previously published habitat models should reconstruct these models using three overlapping fuzzy sets. This simply requires that three symmetrical, triangular functions be superimposed on the range of variability in each constituent variable of the habitat model and that the minimum function be used to provide a final suitability rating. The model categories can be tested for errors of omission and commission using simple tests like the ones we presented above. Model calibration may indicate that the moderate suitability category (or the low suitability category) is the better predictor according to these simple tests. In that case, we would suggest that managers might look for limitations on habitat use that could be caused by factors not represented in the suitability model, such as competition or predation.

Although this recommendation may seem simplistic to researchers, we are sympathetic to managers' needs to use approximate reasoning to come to defensible conclusions in situations of uncertain knowledge. We believe that the alternative approach we describe offers opportunities to reason approximately, develop new hypotheses, and make decisions that implement a precautionary principle. On the research side, we believe that the issues we have raised regarding category definition point to a need for closer integration of models with underlying biological theories, as attention to the defensibility of models in applied settings increases.

Use of Regional-scale Exploratory Studies to Determine Bird-habitat Relationships

Jock S. Young and Richard L. Hutto

The wide geographic extent of regional bird monitoring programs usually makes them nonexperimental and exploratory in nature. In such studies, variables are often chosen by expedience, and sources of variation are not controlled. Even so, there is great potential to learn something meaningful about bird-habitat relationships when bird distribution or abundance information is linked with additional information about vegetation characteristics at the sites (Wiens and Rotenberry 1981a; Ralph et al. 1995a).

One goal of bird-habitat relationship studies is to identify environmental conditions that presumably control the distribution and abundance of a bird species. Knowledge of the biologically important variables would help us make more informed management decisions and more accurate predictions of bird occurrence in new, unsurveyed sites. Because we cannot measure all biologically important variables, however, the resulting models are heavily influenced by the choice of variables and the methods used for exploratory analysis. Consequently, there is a critical need to discuss the best approaches for getting the most out of such observational data sets.

In this chapter, we present analyses of data from the U.S. Forest Service (USFS) Northern Region Landbird Monitoring Program (Hutto and Young 1999). Full-scale monitoring began under this program in 1994. One aim was to conduct long-term monitoring,

so a series of permanently marked transects were located randomly within a geographic stratification scheme. This is not the most efficient design for discerning habitat associations, because a preponderance of points lie in common cover types (Austin and Meyers 1996; Heglund, Chapter 1; Austin, Chapter 5). Nonetheless, we collected data on local vegetation characteristics in the area surrounding each point to obtain as much basic habitat-relationship information as possible to supplement the population monitoring data and to help explain the possible reasons for any population declines that may be detected later.

The distributions of various landbird species across cover types in the northern Rocky Mountains have already been published (Hutto and Young 1999). However, the eighteen cover types used in those analyses pooled together a diverse assemblage of vegetation structures. No bird species was detected at all points within any given cover type. Some of those absences may have been due to sampling error, population fluctuations, or chance, but we assume that there were also absences due to finer-resolution variation in habitat characteristics among the points *within* a single cover type. We need to know what these features are if we are to manage bird species successfully.

In this chapter, we discuss methods for this second step: building regression models to expose finer-resolution patterns of occurrence due to the continuously variable nature of vegetation features. We apply our proposed model-building methods to a single example species, the Swainson's thrush (*Catharus ustulatus*). We then examine several subsets of the data to determine the consistency and accuracy of our results under different conditions. Even though this is an exploratory exercise, and even though we make use of uncontrolled data not designed primarily to expose habitat relationships, the resulting descriptive models should at least suggest ecological relationships and help focus future research.

Methods

Sample points were distributed across twelve national forests of the USFS Northern Region, in northern Idaho and western Montana. Transects were geographically stratified by 7.5-minute topographic quadrangle maps throughout nonwilderness Forest Service lands and part of the Potlatch Corporation lands in central Idaho (Fig. 8.1). Potential transect start points were located by positioning a random point within each quarter of the quadrangle maps and then finding the nearest point on an unpaved, secondary or tertiary, open or closed road or trail (Hutto and Young 1999). Transects were selected randomly from this potential set, subject to logistical constraints. The ten points along each transect were placed at least 250 meters (straight-line distance) apart. Each point was sampled once during the breeding season in each of three consecutive years (1994 to 1996). There were some changes in transect locations between years. In this chapter, we used the 428 transects that were visited in all three years, although point selection criteria discussed below resulted in only 292 transects being represented in the data set used for analyses.

Bird Survey Methods

Our field technique followed recommendations discussed by Ralph et al. (1995a) and methods described by Hutto et al. (1986). All observers participated in a one-week training session. Points were visited once each breeding season between mid-May and mid-July. All birds seen or heard during a ten-minute count period were recorded, and the distance to each was estimated (Hutto and Young 1999). Field observers generally began counts about 15 minutes after sunrise

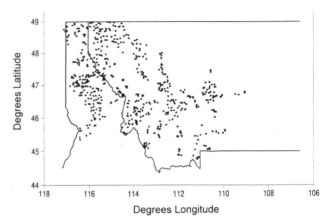

Figure 8.1. The distribution of permanent landbird monitoring transects in northern Idaho and western Montana.

(after the predawn chorus), and generally completed counts within four hours. Counts were not conducted on days with continuous rain or strong winds.

Vegetation Survey Methods

Field observers first determined whether a 100-meter-radius circle around a survey point could be considered a homogeneous cover type. If so, further vegetation measurements were then taken on relatively continuous variables representing vegetation physiognomy and floristics (Hutto and Young 1999).

The selection of variables to measure was based largely on our understanding of avian ecology from the literature and personal observation. Vegetation variables involved structural characteristics of the vegetation at different layers, as well as tree species compositions. Emphasis was placed on variables that were of potential biological importance for one or more of the bird species analyzed and could be collected quickly in the field by trained workers.

We estimated the tree species composition of the canopy layer because of the evidence that floristics may be important in habitat relationships (e.g., Mac Nally 1990b). Different tree species have different architecture and different invertebrate assemblages (e.g., Recher et al. 1991). Bird species forage non-randomly among plant species (Airola and Barrett 1985; Rotenberry 1985), and nests are often placed in some tree species preferentially over others (e.g., Martin 1992). General surveys of cover types (e.g., Hutto and Young 1999) have shown that many bird species are nonrandomly distributed across stands of different tree species.

Because it had been determined qualitatively that the vegetation cover was homogeneous out to 100 meters, we assumed that quantification of vegetation variables within a 30-meter-radius circle (excluding the road corridor) was sufficient to represent the entire area. Therefore, all vegetation variables used in this report were estimated to 30 meters, except for counts of large-dbh (diameter at breast height more than 40 centimeters) trees (LGTREE), which were based on an 11.3-meter-radius circle (excluding the road corridor). Ocular estimates of the following variables were conducted within the 30-meter-radius circle (Hutto and Hoffland 1996; Hutto and Young 1999): HEIGHT—the typical height of the tree canopy layer; CANOPY—the percent cover of canopy trees (larger than saplings); SAPLING—the percent cover of sapling trees (between 5 and 10 centimeters dbh); SHRUB—the percent cover of tall shrubs (multistemmed woody plants greater than 1 meter tall); BUSH—the percent cover of low shrubs (less than 1 meter tall); and GROUND—the percent cover of grasses and forbs.

For tree species composition, we estimated the percentage of the total canopy cover made up by each of the tree species indicated below (some associated species were lumped together): PIPO—percentage of canopy made up by ponderosa pine (*Pinus ponderosa*); PSME—percentage of canopy made up by Douglas-fir (*Pseudotsuga menziesii*); LAOC—percentage of canopy made up by western larch (*Larix occidentalis*); PICO—percentage of canopy made up by lodgepole pine (*Pinus contorta*); SPRFIR—percentage of canopy made up by spruce/fir (*Picea engelmannii, Abies lasiocarpa*); MESIC—percentage of total canopy cover made up by western red cedar (*Thuja plicata*), western hemlock (*Tsuga heterophylla*), and grand fir (*Abies grandis*); and DECID—percentage of total canopy cover made up by deciduous trees (*Betula papyrifera, Populus tremuloides,* and *P. trichocarpa*).

Each year, new field observers independently estimated the values of vegetation variables in conjunction with the collection of bird data (although tree species composition was estimated in 1994 only). Except when noted, the data from the three separate years were averaged for each point.

Analysis Methods

Points with more than one cover type within 100 meters were excluded from the analyses, so the estimates of vegetation structure could reasonably be expected to represent an average of a relatively homogeneous area around the point. The bird data were also limited to 100 meters so that both the bird and vegetation data represented samples of the 3.14-hectare (100-meter-radius) area surrounding the point. Some authors recommend the use of a 50-meter radius for bird data, but the effect of restricting count radius is not uniform across species (Wolf et al. 1995). Wide-ranging species with loud calls, such as the common raven (*Corvus corax*) and pileated woodpecker (*Dryocopus pileatus*), were detected within 50 meters only 5–15 percent of the time. Small birds with soft or high-pitched songs, such as the brown creeper (*Certhia americana*) and golden-crowned kinglet (*Regulus satrapa*), were detected within 50 meters 85–90 percent of the time. Some other species whose songs are unmistakable, such as the olive-sided flycatcher (*Contopus cooperi*) and varied thrush (*Ixoreus naevius*), were also identified at greater distances, perhaps due to observer confidence. Thus, it may be best to vary the cutoff radius for different species. Because between-species comparisons are not recommended for point counts (Wolf et al. 1995), the different radii should not be a concern. In the case of the Swainson's thrush, only 35 percent of detections were within 50 meters, whereas 80 percent were within 100 meters. The song carries well and is easily identified, so we used all detections within 100 meters. To test the effect of this decision, however, we also constructed a model based on a 50-meter-radius plot.

To reduce the confounding effects of very different cover types, some of which we know this species would not occur in, we modeled the habitat associations of the Swainson's thrush within the subset of conifer forest cover types. We included all points with conifer trees ranging from 5 to 35 meters tall and from 1 to 80 percent canopy coverage. By restricting the data set, we changed the question from the distribution of a bird species across a wide array of cover types to a more refined distribution within a subset of cover types.

Although different points on a transect can be in

different cover types and can be argued to be independent choices in habitat selection made by bird species with territory sizes of a couple hectares or less, the relative health of the local population, local meteorological conditions, and so forth, will always produce some dependence in the data within each transect. Nonetheless, we used individual points as sample units because (1) combining data from all points on a transect would create meaningless sample units with respect to vegetation variables, given that transects run through a series of different cover types; (2) given a mixture of cover types on each transect, and the elimination of points near edges, we included, on average, only 3.8 points per transect; and (3) our emphasis on the relative importance of variables rather than strict rules for inclusion of variables in the model made the sample size a less pressing issue, although it was still important relative to the ability of the data to support a model with many parameters. We also present models based on one point per transect.

For the main model of our example species (Swainson's thrush), we pooled the three years of data by averaging the vegetation estimates over the three years and by counting the presence of the species in any of the three years as a presence for that point. This method allowed inclusion of sites where the species was simply missed in some of the three years. Alternative approaches are addressed below.

The statistical importance of a variable in any modeling procedure depends on how close the mathematical form of the model is to the form of the true relationship. The simplicity of linear regression has led to its adoption as the typical method in wildlife habitat studies (Young 1996). However, a linear model assumes that a unit-unit relationship holds true for the entire range of an environmental attribute (Meents et al. 1983) so that if more is better, then a lot more must be much better (Johnson 1981b). However, simple niche theory assumes that organisms respond to most important resource gradients in a unimodal fashion. This has been standard procedure among plant ecologists for decades (e.g., Whittaker 1967; Austin 1976; ter Braak and Prentice 1988), but animal ecologists have been much slower to embrace it (Young 1996; Heglund, Chapter 1; but see Meents et al. 1983; Heglund et al. 1994, etc.). There is no particular rea-

son why the relationship must be Gaussian or even symmetrical (Austin 1976); the mode may not even be the optimum (Austin 1980). We do not really know what the shape is likely to be in any particular case, so we used the simplest method possible to pick up at least some of any unimodal signal, while adding only one parameter, which was the addition of a quadratic term (ter Braak and Prentice 1988).

A significant quadratic term can result from nonlinear but monotonic relationships, such as asymptotes, as well as from unimodal relationships. Such response curves would be expected if there were a threshold in the response or if we did not have the complete gradient for the variable. In such cases, however, the linear regression would also show a strong relationship, so the quadratic term may not be necessary to determine the importance of a variable (although it may help improve predictions). The inclusion of the quadratic term is even more important in cases where linear regression indicates no relationship.

If a bird species is associated with one tree species, it is likely to show a quantitative response to others because the proportions are interdependent. One way around this might be an ordination procedure, although this may result in gradients due to overall productivity and/or forest structure rather than direct effects of tree species composition. We opted to use the direct variables to more easily interpret direct effects of tree species. We did not consider quadratic terms for the tree species variables because sparse data (many zeros) would make the additional parameter less supportable, and because such relationships would have little logical interpretation.

We visited point counts only once per year. Because single visits do not commonly produce multiple detections of any one bird species, we used logistic regression (Hosmer and Lemeshow 1989) to analyze the effects of vegetation variables on the presence or absence of each bird species at the points. In addition, biases due to detectability and observer variability should be less pronounced in presence/absence data than in abundance data.

To begin the model-building process, we discarded variables that had exceptionally high p-values in simple regressions and variables that made no biological sense for the particular species (in this case the Swainson's thrush). We then used the Akaike Information Criterion (AIC) for model selection (Akaike 1974; McQuarrie and Tsai 1998). AIC incorporates the tradeoff between bias and variance as variables are added to a model, and it provides a straightforward comparison between models that does not depend on a hypothesis-testing framework (Burnham and Anderson 1998). It moves the emphasis away from *p*-values and arbitrary cutoff criteria (Johnson 1999a), and extracts more information from the data regarding the relative strength of evidence for each variable (and model).

When choosing among models using statistical inference, it is best to work with only a few select models chosen a priori on biological grounds (Burnham and Anderson 1998). However, our data set was both correlative and unstructured, and we knew little about the expected relationships for many species, so we considered this an exploratory analysis. The best method of variable selection for modeling such a data set has been the subject of much discussion. Stepwise selection methods do not necessarily identify the "best" model even from a statistical perspective (James and McCulloch 1990). On the other hand, all-possible-subsets procedures will inevitably lead to overfitting of the data, because the model thus chosen will be highly specific to the data at hand (Burnham and Anderson 1998). In fact, no method can produce the "true" biological model from correlative data, and some overfitting is perhaps inevitable. We chose the most influential variables by forward selection, with AIC as the selection criterion for each step. As inclusion of variables became more uncertain, we modified the procedures to more closely resemble an all-possible-subsets methodology. This allowed the comparison of many likely models and embraced the idea of alternative models and model-selection uncertainty. Although we report the model thus chosen, the goal was not to produce a single final model but rather to determine the strength of evidence for the inclusion of each variable (Burnham and Anderson 1998: 202).

Accuracy Issues

We did not have independent data for testing the accuracy of our models, but we have analyzed the data in several different ways to get an indication of the robustness of the models in terms of the variables included. We performed a cross-validation procedure by splitting the full database in half. We sorted transects by latitude and longitude and selected every other transect for each subset of the data. We then built a logistic regression model for each subset and compared the classification accuracy of each model when predicting the observed data for the other (test) subset relative to the training set. For a classification accuracy assessment that was independent of the cut-point threshold, we used receiver operating characteristic (ROC) plots (Swets 1988; Fielding and Bell 1997; Pearce et al., Chapter 32).

We also checked the consistency of results by building a model for each of the three years separately. Examining each year separately is not an independent validation, of course, because we sampled the same points and, in some cases, the same individual birds returning the next year (or their philopatric offspring). However, the results from three consecutive years should be consistent if we expect the models to perform well on independent data. We used the same methods as above to build multiple logistic regression models for 1994, 1995, and 1996 separately, and compared the classification accuracy to that of the original three-year model using ROC plots. We used the same data set for the vegetation variables, with averages across all three years, so that any year-to-year variability was due to the bird data only, whether from sampling error or from actual changes in occupation of sites.

Sample Unit Considerations

We tested the sensitivity of our results to pseudoreplication issues by redoing the analyses with one randomly selected point from each transect. We had the luxury of doing this because of the large sample size in our regional program. We selected two subsamples, each with one randomly selected point from each transect. The second subsample was selected without making the points from the first set available for selection (i.e., sampling without replacement). The fifty-eight transects with only one available point were randomly divided between the two subsamples. This produced two subsamples of 263 points, each with only one point per transect and with no points in common. Multiple logistic regression models were built for these data sets using the same methods as above, and the classification accuracy was compared to the original model using ROC plots.

Poisson Regression

When we pooled the bird data from three years at each point, we obtained considerable variation in abundance for some bird species, which was lost when we converted the data to presence or absence for logistic regression (Mac Nally 1990a). A point where a Swainson's thrush was detected in only one year (or was mistakenly identified) was given the same importance value as a point with many territorial thrushes singing every year. To better differentiate the relative use of the sites by this species, we reanalyzed the three-year data set using the summed abundances and Poisson regression (Jones et al., Chapter 35). Count data are more likely to follow a Poisson distribution than any other readily available distribution (but see White and Bennetts [1996] for a recommendation of the negative binomial distribution), and the method is fairly robust, requiring only that the variance in the data be proportional to the mean (McCullagh and Nelder 1989).

Regional-scale Considerations

In any study of habitat use, the set of "available" locations must be carefully chosen (Johnson 1980). If the species is not present in some areas for any reason other than the variables we have measured in the study, then it would be misleading to dilute the data with such absences, or "naughty noughts" (Austin and Meyers 1996), in potentially suitable habitat that is not occupied for other reasons (e.g., climate or landscape-scale factors, or current or historical dispersal barriers). If some measured vegetation variables also change across the same gradient, then it may look like those measured variables are controlling the distribution rather than the unmeasured factors that are truly limiting.

More than one scale is involved here. If data cover a large region, the geographic range of a species may not extend throughout the entire area. But even within a species' range, there may be suitable habitat that is not occupied due to landscape-scale factors. In a study of local-scale factors, both of these problems might be addressed by using only those transects along which a particular bird species was detected. Those occupied

transects are ones in which the range, landscape, and season are apparently appropriate for the bird species' presence. If landscape-scale factors are to be included in the models, then we would want to analyze all transects within the occupied range.

Choosing the best approach to this problem may be a subjective exercise. For example, we detected the Swainson's thrush only rarely in south-central Montana (Fig. 8.2). However, because this area was well within the geographic range of the species (Montana Bird Distribution Committee 1996), we felt that it still could have been present in appropriate habitat. It therefore would be reasonable to use all occupied transects as our method to control for landscape for this species. However, because this abundant species was found on about 85 percent of the transects, this alternative procedure was not likely to produce different results. This method may thus be more useful for less-common species. We decided to restrict the data based on geographic area. Because we were interested in the effect of the geographic distribution of western larch on the importance of that variable in the habitat models, we restricted the data to the geographic range of the larch (all forests west of the Continental Divide except for the Bitterroot National Forest), which was also the area where Swainson's thrushes were most

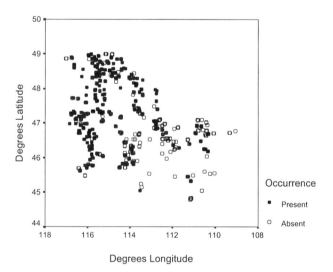

Figure 8.2. The geographic distribution of the Swainson's thrush (*Catharus ustulatus*) across all 1,102 points used in the analyses. Closed circles indicate a presence of the Swainson's thrush in any of the three years and open circles indicate absences in all years. Points within transects are nearly congruent in this depiction.

common. We built a new logistic regression model using this subset of the data.

Another regional-scale consideration is a potential change in habitat use in different parts of a species' range. We did not pursue this question because, with this kind of correlative data, it would be very difficult to show that the inevitable differences between the models for two areas were due to actual biological differences in habitat selection rather than different competitive environments or simply sampling error.

Results

The final data set included a total of 1,102 points on transects visited in each of three years. We considered the area around a point to be relatively homogeneous (no edges) if only one (559) or none (543) of the three observers thought otherwise.

Almost any data set involving multiple vegetation variables will include a number of intercorrelations among the predictor variables. The highest correlations among the predictor variables in our data were among canopy cover, canopy height, and number of large trees, especially the latter two measures of tree size (Table 8.1). The proportion of mesic conifer species (western red cedar, western hemlock, and grand fir) in the canopy was also highly correlated with those three variables, especially canopy cover. Sites with more ponderosa pine had the lowest average canopy cover. The greatest understory development was under canopies of larch or, secondarily, spruce/fir. The proportions of ponderosa pine and Douglas-fir were negatively related to understory. Because we had already combined the most important species associations (spruce/fir and cedar/hemlock/grand fir), most of the correlations among conifer species variables were negative. The largest correlation coefficient (*r*) was less than 0.5 (Table 8.1), so with our sample size it should be possible to tease apart the effects of all variables, at least to some extent.

Swainson's thrushes were detected in at least one year at 555 of the 1,102 points. Therefore, the categories of presence and absence were nearly equal for the main analysis of the three-year data set using logistic regression. The most important variables in this main model (logistic regression of three-year averages;

Table 8.2) appeared to be understory cover consisting of tall shrubs and conifer saplings, positive associations with larch and mesic tree species and, to a lesser degree, canopy cover.

The classification accuracy (at a cut-point of 0.5) of the main model was about 72 percent. When only the strongest variables were used (CANOPY, SAPLING, SHRUB, SHRUB^2, LAOC, and MESIC), then the classification accuracy was still 71 percent.

When the data set was split in half for cross-validation, each half had an internal classification accuracy of about 73 percent. When each of the resulting models was used to predict presence in the other half of the data, the ROC plots (Fig. 8.3) indicated a classification accuracy for this test set that was nearly as high as for the training set. In fact, the internal classification accuracy of the models for each half were similar to that for the main model (Fig. 8.4a).

When each year was analyzed separately, Swainson's thrushes were detected on 353 of the 1,102 points in 1994, 294 in 1995, and 281 in 1996. There were some differences in the apparent importance of the vegetation variables in models for the three different years (Table 8.2), most notably for canopy cover and tree species composition. The internal classification accuracy for these models was not quite as good as that for the main model (Fig. 8.4b).

To determine the sensitivity of our results to the use of points as sample units, we also randomly selected two subsets of the data that consisted of single points from each of 263 transects. Swainson's thrushes were detected on 128 and 134 of the 263 points in these two separate subsets. The models based on these two data sets were quite different from one another (Table 8.2), with more variables being included in the second model (including tree size). This second model was also the only model that did not include western larch as a tree species associate (although it would have if only positive tree associations were allowed). Although the internal classification accuracy for these models was better than that for the main model (Fig. 8.4c), validation of each model using the other subset as testing data gave relatively poor accuracy (Fig. 8.5).

The restriction of data to detections within a 50-meter radius did not change the core variables of the model (Table 8.2). The data supported fewer minor

TABLE 8.1.

Nonparametric correlation coefficients (Kendall's tau-b) for bivariate comparisons of predictor variables.

	SAPLING	SHRUB	BUSH	GROUND	HEIGHT	LGTREE	PIPO	PSME	LAOC	PICO	SPRFIR	MESIC
CANOPY	+0.09	−0.04	−0.03	−0.12	+0.35	+0.31	−0.12	+0.02	−0.01	−0.04	−0.06	+0.24
SAPLING		+0.06	+0.02	−0.03	−0.11	−0.13	−0.18	−0.15	+0.15	+0.04	+0.10	+0.16
SHRUB			+0.47	+0.03	<0.01	<0.01	−0.02	+0.03	+0.25	−0.16	+0.08	+0.10
BUSH				+0.10	+0.01	−0.02	<0.01	<0.01	+0.21	−0.07	+0.09	+0.04
GROUND					−0.03	−0.03	<0.01	+0.03	+0.08	<0.01	<0.01	−0.04
HEIGHT						+0.37	+0.04	+0.06	−0.04	−0.15	−0.03	+0.23
LGTREE							+0.10	+0.09	−0.04	−0.24	+0.02	+0.16
PIPO								+0.09	−0.13	−0.24	−0.22	−0.15
PSME									0.05	0.24	0.31	0.23
LAOC										+0.03	+0.07	−0.03
PICO											+0.07	−0.27
SPRFIR												−0.12

TABLE 8.2.

Order of selection for variables included in multiple regression models of the habitat relationships of the Swainson's thrush (*Catharus ustulatus*), using AIC (Akaike information criterion).

		Models[a]								
Variable		Poisson N = 1102	Logit 100m	Logit 50 m	1994 N = 1102	1995 N = 1102	1996 N = 1102	Tr A N = 263	Tr B N = 263	West N = 749
CANOPY	+		7c	4c	6c	3c		3c	4c	4c
	b∓		8c	9c		5c			4c	
SAPLING	+	2c	2c	2c	2c	2c	3c	2c	1c	1c
	b±						5c	2c		
SHRUB	+	1c	1c	1c	1c	1c	1c	1c	2c	3c
	b±		6c	3c	4c	6c	4c			7c
BUSH	+			7c	7	11	7		5c	
	b±			8c	8	12	8			
GROUND										
HEIGHT	+		9		9	8c			3c	
	b∓		11		10	9c			3c	
LGTREE	−		12c		11				8	6c
PSME	−									
LAOC	+	3c	3c	5c	5c	4c	2c	4c		2c
PIPO	−		10c			10	6		7c	
PICO	−								6c	
SPRFIR	+		13			7c				
MESIC	+	4c	4c	6c	3c					5c
DECID										

[a]All models used logistic regression except for the Poisson model.

All models used the same vegetation data, with the variables averaged over three years.

The first three models used accumulated data for abundance or presence of Swainson's thrush (*Catharus ustulatus*), over all three years.

The models designated by dates used presence data from each of the three years separately.

The Tr A and B models are based on two separate sets with one randomly chosen point per transect, and the last column is based on the subset of transects from west of the Continental Divide.

[b]Indicates quadratic term for indicated variables; signs are for linear and quadratic term, respectively.

[c] Indicates variables included under traditional hypothesis-testing methods.

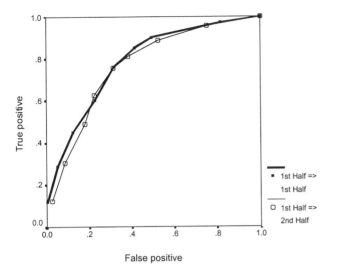

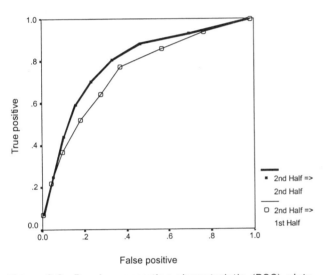

Figure 8.3. Receiver operating characteristic (ROC) plots showing the classification accuracy of the models based on each half of the data set in predicting the data for the same half (training data) and the other half (test data).

although it was less likely to indicate nonlinear relationships and it did not include tree-size variables.

We detected the Swainson's thrush on 493 of 749 points in the northwestern part of the region. The resulting logistic regression model was similar to that obtained for the full data set of 1,102 points (Table 8.2), although without canopy height, SPRFIR, and the quadratic term for canopy cover.

The AIC method usually indicated additional variables beyond those included by traditional hypothesis testing with alpha equal to 0.05. Because of the many models we examined in this exploratory analysis, it is likely that there was some overfitting of the data when the best model was chosen according to AIC.

Discussion

Two main goals of building regression models in habitat relationships are to identify biologically important variables and to predict the occurrence of bird species at previously unsampled sites. The first goal, identifying environmental conditions that a bird species needs to be present and successful, is of obvious scientific interest. In addition, it is only through understanding the true biological processes involved in a species' distribution that we can hope to reach meaningful management recommendations. Determining the important variables can be difficult for a number of reasons. We know we have not measured many potentially important biological variables, such as food resource availability (Hutto 1990) or specific nest sites (Martin 1992). In addition, the biological importance of the measured variables cannot be directly confirmed from a correlative analysis. It is necessary to assume that the importance of the larger biological effects will be revealed by the statistical model, especially if several subsets of the data are examined, but there is no way to know how much of the observed effect is due to actual biological processes or to sampling error. This inherent model uncertainty should encourage us to emphasize the strength of evidence for each variable rather than try to decide which quantitative model is "correct." The observed evidence of relative importance must then be used to form hypotheses for further investigation.

variables, with a shift to understory variables rather than tree size. The internal classification accuracy for this model was not quite as good as that for the main (100-meter) model (Fig. 8.4d).

The abundance of Swainson's thrushes at the 555 points varied between one and eleven (sum of three visits; some high numbers brought the accuracy of the abundance data into question). More than one individual was detected at 348 points. Therefore, there was considerable variation in counts for use in a Poisson regression analysis. The resulting model was similar to that obtained by logistic regression (Table 8.2),

In this study, the various regression models of

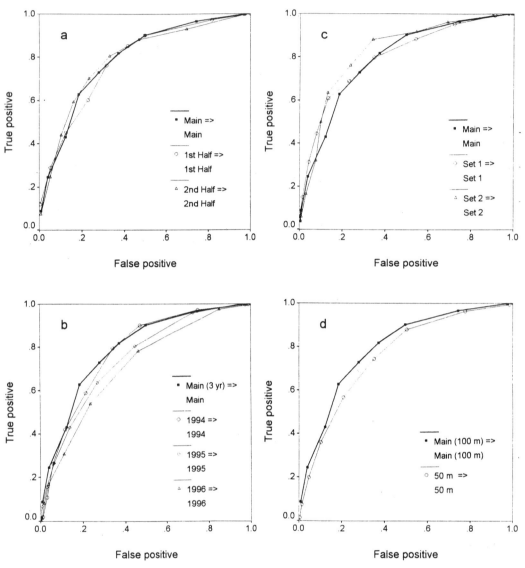

Figure 8.4. Receiver operating characteristic (ROC) plots comparing the internal classification accuracy (resubstitution) of different models to the main model (see text). (a) Accuracy of models for each half of the data compared with main model; (b) accuracy of models for each year of data compared with main (three-year) model; (c) accuracy of models for each of two subsets with one point per transect compared with main model; and (d) accuracy of model based on 50-meter radius compared with main (100-meter) model.

Swainson's thrush habitat relationships were fairly consistent in that they all included the same set of strongly influential variables (Table 8.2). The variables that appeared to be the weakest predictors of Swainson's thrush occurrence in any one model were the same variables that were less consistently included in the other models. All of the variables chosen by AIC but not by hypothesis testing methods were in this category. In fact, a model with only the strongest and most consistent variables had nearly the same predictive ability as the full model. This suggests that the weaker variables were either biologically unimportant or inconsistently correlated with the true controlling variables. The increased resolution necessary to understand the possible effects of these variables would require a much larger sample size or a more intensive, controlled study.

This is a first attempt at getting a list of variables, more or less in order of statistical (but not necessarily biological) importance, *within this data set*. We can never be sure if the model reflects true biological relationships without confirming the results with inde-

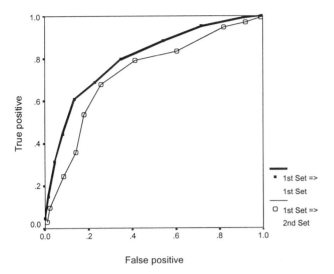

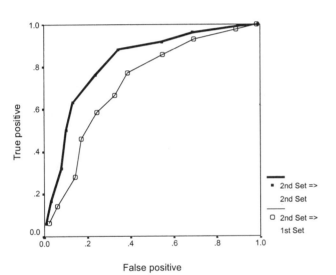

Figure 8.5. Receiver operating characteristic (ROC) plots showing the classification accuracy of the models based on each of two subsets of data (with one point per transect) in predicting the data for the same subset (training data) and the other subset (test data).

pendent or experimental data, but managers can still benefit from such a model because it helps focus future studies and provides a first approximation of the important controlling variables, which can aid in management decisions.

A first step for managers would be to look closely at the variables that were most consistently included in the models. Clearly, the understory is critical for the Swainson's thrush. Both tall shrub and conifer sapling cover were included in every model, usually with the first or second strongest associations. Because shrub

cover and sapling cover were only weakly correlated (r = +0.06; Table 8.1), it seems that conifer saplings may provide an adequate substitute for this bird species as understory structure. Also, there appeared to be a threshold (asymptote) in the relationship of bird occurrence and understory cover (Fig. 8.6), indicating that 20–30 percent understory cover provided maximum benefit, as might be expected for a shrub-nesting species that forages more generally (Ehrlich et al. 1988). Because management practices tend to increase the amount of land with this level of understory cover, this species is not likely to be of management concern.

We do not know of any particular biological reason for larch to be important to the Swainson's thrush, and this demonstrates the ambiguity of exploratory analyses. Larch cover was correlated with shrub cover (r = 0.24), but both variables were strongly significant in most multivariate models, so it is difficult to know whether this was a true biological relationship or an artifact of confounding variables. The fact that larch is restricted to west of the Continental Divide was our main reason for limiting the bird data to this western region for one model. In this way, we discovered that larch was still an important variable for the thrush within the tree's geographic range, so the apparent association between the bird and tree species was not an artifact of geography. Further study would be necessary to determine if the retention of larch in the landscape is as important for this species as it is for many cavity-nesting birds (McClelland 1977), but this may be a good example of a relationship that was not apparent using simple cover type distributions (Hutto and Young 1999).

The negative association of Swainson's thrush occurrence with ponderosa pine could be due simply to a positive association with other tree species, or perhaps the thrush does not do well in that type of tree architecture. Alternatively, it may have more to do with ponderosa pine stands typically having low canopy cover or minimal understory. Although the multivariate analyses may have been able to tease these apart, some residual effect probably remained.

We did not have an independent data set with which to test our models. However, we used a number of internal validation and classification accuracy

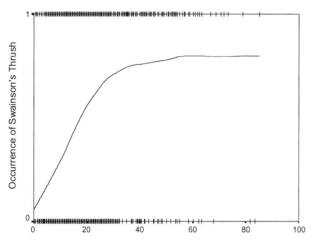

Figure 8.6. Occurrence of the Swainson's thrush along an environmental gradient representing percent coverage of tall understory vegetation (sum of tall shrub and conifer sapling variables). Absence = 0; Presence = 1; curve generated by LOWESS smoothing.

procedures to explore the robustness of our results. We assumed that the most consistently included variables were more likely to have some biological foundation. This is not only of scientific interest, but should also increase the usefulness of the results for predicting the presence of the Swainson's thrush at new sites, and for estimating its probable response to management decisions.

The cross-validation procedure seemed to show that we have created relatively robust and useful models. Models based on each of the two halves of the data set not only had classification accuracies nearly as large as the full model (Fig. 8.4a), but also the consistency of the models in predicting the other half of the data was encouraging. These results also suggest that doubling the sample size did not result in a greatly improved model.

The combination of three years of data appeared to improve the predictive power of the habitat model relative to the models based on single years. This is understandable given the sampling methods. A point count survey provides an incomplete sample of the birds in a given area. It is probably common for a species to be present but not detected. There is also true year-to-year variation in bird occupancy. It is important to design surveys to sample a representative

cross-section of this variation (i.e., multiple years, places, etc.). In this respect, the Northern Region Landbird Monitoring Program may be unique. Samples were large in comparison with other controlled studies of habitat use, and data collection was repeated over several years. The results of these analyses suggest that improving the accuracy of data at each point may be more critical than increasing the total number of points. We are beginning to test these ideas by examining the accuracy of models for other bird species.

Vegetation variables are subject to measurement error and observer variability. There was considerable variation in the estimates by the three different observers at each point over the years. When the separate years were analyzed using the separate estimates of vegetation rather than the three-year average, there was a much greater difference between years than that shown in Table 8.2. This suggests that observer variability can be a serious problem, especially if vegetation is measured quickly by crews primarily trained to identify birds, or if only one year is available. We took the average of all three years for this reason, and it has prompted us to subsequently collect more vegetation data at the points, using experienced forestry crews and additional plots.

A plot radius of 50 meters is often recommended for comparison of bird abundance between different cover types because vegetation density can affect the detectability of individual birds. In addition, if both the bird and vegetation data were accurate, we would expect the 50-meter-radius model to have greater classification accuracy because the bird data would be more tightly associated with the vegetation near the point. In this study, however, the 50-meter-radius model was slightly less accurate then the main model using a 100-meter radius. This suggests that a 50-meter radius may have been insufficient to accurately represent occupancy in the stand. This is even more likely to be the case for less-common species. We conclude that the 100-meter-radius cutoff not only resulted in an adequate model, but also it may be preferable in studies with only one or a few visits to a point, where we are most likely to have an incomplete inventory.

Most bird-habitat relationship models explain only a small proportion of the variance in bird presence or abundance (Maurer 1986; Morrison et al. 1987). This is due to a variety of factors that have often been mentioned (e.g., Rotenberry 1986; Wiens 1989b,c), and most of these are probably exacerbated by the nature of large exploratory studies. It is important for us to both realize the limitations of the method and to design surveys and analyses to decrease the effects of these problems as much as possible.

In spite of the numerous reasons that regional-scale monitoring data might not be conducive to rigorous habitat analyses, we determined a suite of vegetation characteristics that were strongly correlated with the presence of Swainson's thrush in forest stands (Table 8.2). We think it is very important that such data are used as fully as possible, as long as the results are not overinterpreted. Managers must be made aware of model uncertainties so that potential problems are not overlooked when final decisions are made (Conroy and Moore Chapter 16).

We may also wish to use these bird-habitat relationship models for the second main goal of building regression models—predicting the likelihood of a particular bird species being present at new, unsurveyed sites. Such predictions would be more robust and less location-specific if the predictor variables were more relevant to biological processes (Austin and Meyers 1996), but this is not absolutely necessary for a useful model as long as the new sites requiring prediction have the same correlational linkages between the measured surrogate variables and the true variables that influence bird occurrence. In any case, the expense of measuring predictor variables over wide regions may be prohibitive unless remotely sensed data can be used. There is little reason to develop models for region-wide prediction until we know what variables are likely to be available to managers over all target areas. We can then determine if models based on those variables are adequately robust for management needs.

PART 2

Temporal and Spatial Scales

Role of Temporal and Spatial Scale

Michael L. Morrison

I will briefly summarize the key findings of each chapter within this section and conclude with recommendations for future directions in issues of habitat scale. Issues of scale were brought to the forefront of habitat analysis in the 1980s, as indicated by the attention given this topic in *Wildlife 2000: Modeling Habitat Relationships of Terrestrial Vertebrates* (Verner et al. 1986b). In comparing conclusions drawn in Verner et al. (1986b) with those in this section overview and the literature in general (e.g., see Morrison et al. 1998 for a review), I think that (1) more scientists are aware of spatiotemporal influences on habitat use, and (2) more researchers are incorporating issues of scale into their studies. In essence, *Wildlife 2000* raised the issue, and this volume indicates how far we have followed through with our studies in the intervening years. But, as I conclude below and elsewhere (Morrison et al. 1998), the studies that hold the most promise for advancing our knowledge of wildlife-habitat relationships are rarely conducted.

Recommendations

The major recommendations of authors in this chapter can be summarized as follows. I list only the authors who provide in-depth discussion and data about a particular subject or point.

- Iterative testing of alternative models is needed (Conroy and Moore, Chapter 16; Maurer, Chapter 9; Zabel et al., Chapter 19).
- Habitat evaluation is influenced strongly by spatial scale (Cogan, Chapter 18; Hahn and O'Connor, Chapter 17; Johnson et al., Chapter 12; Trani, Chapter 11).
- There is no "correct" scale; the appropriate scale is goal dependent (Johnson et al., Chapter 12; Tobalske, Chapter 15; Trani, Chapter 11).
- Validation and sensitivity analysis is a necessary part of model evaluation (Thomas et al., Chapter 10; Tobalske, Chapter 15; Trani, Chapter 11).
- Evaluation of temporal influence on habitat models is needed (Greco et al., Chapter 14; Wright and Fielding, Chapter 20).
- The influence of population density on models is necessary (Johnson and Krohn, Chapter 13).
- Reliable field data on animal distribution and abundance is needed (Johnson and Krohn, Chapter 13; Tobalske, Chapter 15).

The theme that emerges from these recommendations is that we are still doing a poor job of thoroughly evaluating our habitat models—hence the emphasis on iterative testing, validation, and sensitivity analysis. Further, there appears to be confusion among many researchers regarding the "best" scale at which to operate. There seems to be an unstated assumption in the literature—especially that of conservation biology—that broad "landscape" is the most appropriate scale for modeling. Two points are worth making here. First, the scale one works at is dependent on the question being asked and how varying scale influences results. Second, "landscape" to a human is likely very different from "landscape" to a small animal. As developed by Morrison and Hall in Chapter 2, the ecological concept of landscape need not be a large area. Last, the interrelated issues of temporal scale and population density, and their influence on habitat use, have received little attention. Johnson and Krohn in Chapter 13 have done a good job of raising these issues.

Maurer critiqued Morrison and Hall and others to set the stage for promoting what he described as an alternative view of defining level and scale. Maurer (Chapter 9) suggested that the Morrison and Hall approach requires observers to define levels in a system. However, Morrison and Hall promote using *data* to define levels of organization rather than a priori setting them (as is done in most studies). Maurer expands on this view by using knowledge to test various models that make predictions about the relationship between a level and scale. Morrison and Hall and others did not develop this approach but certainly did not exclude it. In fact, Maurer's argument coincides with the spirit and intent of Morrison and Hall, who contend that an observer-based decision on a specific relationship is not appropriate.

Conclusions

After reviewing the habitat literature (e.g., Morrison et al. 1998), including the chapters in this section, I think we should give serious consideration to the following points:

• Terminology still needs to be better standardized within this volume by adhering more closely to the definitions provided by Morrison and Hall or by providing specific alternative definitions. Failure to clearly define terms, and relationships between them (e.g., habitat versus niche), can cause confusion when attempting to compare studies.

• The easy studies have been done. Ultimately, for many applications we need to quantify those habitat factors that determine recruitment into the adult breeding population. Other models are surrogates of habitat quality (which are often appropriate for a specific application). Maurer (Chapter 9) noted that although mechanistic models are often preferable, the appropriate data are seldom available.

• Recognizing that scale is goal-dependent is, of course, critical but can also direct us toward potentially weak and misleading results because of our tendency to a priori set the scale. The results are likely to be artifacts of such arbitrary decisions.

• Little attention is being given to temporal changes in habitat use. Differences between seasons, subtle changes within a season, and variations across years receive little work.

• The size of the area sampled and its relation to population dynamics is seldom studied, nor is the location of the study in relation to the range of the species (i.e., whether it is on the edge of the range or near the center). The failure to consider population dynamics when developing habitat models is probably the major weakness in the study of wildlife-habitat relationships.

We have certainly advanced in our studies of wildlife-habitat relationships, as shown by the data presented in this volume and the questions raised regarding our approaches. However, several central areas of research are not being adequately explored, and we remain sloppy in our use of terms and explanations of the associated concepts.

Predicting Distribution and Abundance: Thinking within and between Scales

Brian A. Maurer

Modern conservation biology and wildlife management both require ever-increasing sophistication from models intended to either describe or predict how many individuals of a particular species exist and where in a particular landscape they are found. Once data are available, decisions must be made based on information that is either incomplete, or of unknown reliability. How can biologists and managers handle the complexity and uncertainty inherent in the systems with which they deal? An emerging paradigm intended to address complexity and uncertainty is based on the idea that processes in systems like wildlife populations in human-dominated landscapes occur at different temporal and spatial scales (e.g., Bissonette 1997a,b).

Part 2 is about how information from different temporal and spatial scales can be used in models to strengthen biological inferences and conservation policies. Generally, the chapters in this section examine wildlife-habitat systems or decision-making problems from a multiscale perspective. That is, each system considered is assumed to be influenced by a complex set of causes whose effects on system behavior occur at more than one scale. Although they do not explicitly consider such multiscale processes, Conroy and Moore's (Chapter 16) emphasis on comparison of alternative models in an adaptive management framework is broad enough to be used with models that explicitly incorporate processes at different scales.

Here, I consider two related issues dealing with temporal and spatial scales. The first is the definition and meaning of the term "scale." A standard definition has been offered by Morrison and Hall (Chapter 2), but having such a definition to build on doesn't mean that we understand the usefulness and limitations of the concept that generated the definition. Thus, the second issue is determining how scale as a concept can be applied to analyses of the behavior of complex wildlife-habitat systems or conservation decision systems. I will consider the chapters in this section in light of these two issues. Assuming that these chapters represent the state of the art, it is important to understand how close we are to being able to apply appropriate concepts of temporal and spatial scale to assessing the reliability of the complicated models we are capable of developing with modern computer and remote sensing technologies.

Definitions and the Meaning of Scale

Morrison and Hall (Chapter 2) suggest a standard definition of scale as "the resolution at which patterns are measured, perceived, or represented." They draw their definition of scale from King (1997), who defines scale as "the physical dimensions of a thing or event." Physical dimensions, King (1997) explains, imply that measurements are taken by an observer. That is, a scale does not exist without an observer. A fact yet to become well established in ecology is that an observer takes measurements in *units* (e.g., meters, joules, etc.), and manipulations of those measurements ought to include the units rather than exclude them (Schneider 1994). Thus, scales do not exist without observer-defined units of measurement. The fact that so many statistical calculations are done without reference to those units, especially in multiscale analyses, can lead to confusion regarding the meaning of those calculations.

Following King (1997), Morrison and Hall point out that the term "scale" does not mean the same thing as the term "level." They offer in the appendix to Chapter 2 the confusing definition of level as "the level of organization revealed by observation at the scale under study." This is clarified in both King (1997) and by Morrison and Hall in Chapter 2. Level refers to a rank within a hierarchically organized system. Basically, hierarchies are formed as aggregations of systems into larger systems (although there are many variations on this theme). Any collection of aggregations defines a level in the system. For wildlife biologists, this means that a level refers to a hypothesized aggregation (e.g., individuals within populations, patches within landscapes). Confusion often arises because the scale an observer uses need not correspond to any particular level (Schneider 1994; King 1997).

What is needed to reduce such confusion is a distinction between the two types of models represented by levels and scales. A level is a theoretical construct used to induce conceptual order to the thinking of scientists studying complex systems. A scale is an empirical construct used to organize data collected by scientists measuring complex systems. Relating a range of scales to a set of levels can be viewed as a model-fitting procedure. Given that there may be more than one set of levels that might describe a complex system like a wildlife population, a scientist might have several different empirical statements incorporating different ranges of scales that might be evaluated according to some "goodness-of-fit" criterion to choose the "best" alternative. So, in studies of complex systems, which comes first, the level or the scale?

Some authors advocate that recognition of levels should be extracted from data rather than assumed to exist a priori (King 1997). This approach draws upon treatments of ecology such as Peters' (1991) *A Cri-*

tique for Ecology, which advocates that induction is the only reliable way to understand ecology. The inductionist view, however, limits the progress that can be made in a field of science (Pickett et al. 1994; Maurer 1999). A complementary approach is the deductionist's view. According to this view, rather than making assumptions about the existence of levels, scientists define levels in a manner such that they may be evaluated empirically. That is, levels are defined a priori and used to construct a set of alternative models of the process being examined. The models are used to make predictions about measurements taken at different scales. The model defines a relationship between a level (e.g., population) and a scale-dependent process (e.g., population dynamics). Data collected at appropriate scales are then used to construct tests of the predictions made by each model using a well-defined goodness-of-fit criterion. As Conroy and Moore (Chapter 16) suggest, in a decision-making context, this may be an iterative procedure, where what is learned at one iteration informs the choices of models constructed and scales examined for the next iteration.

Modeling Variable Populations in Space and Time

Given the problems inherent in using the concept of scale in a scientifically useful manner, what is missing in many current applications of scale to wildlife-related problems? I suggest that most often, what is lacking is a clear a priori definition of levels of interest and the implied processes that result from those definitions. In the first part of this chapter, I describe a relatively general definition of levels that might apply to a wide variety of wildlife-habitat systems. This model implies that spatial and temporal patterns of populations within landscapes are complex because the processes that cause them are thought to operate at many spatial and temporal scales (Villard et al. 1998). Furthermore, these different processes are often stochastic in some sense, so that the effects they produce on a population are not consistent over space and time.

In the face of such complexity, it is unlikely that any single modeling approach will be successful for all purposes. In the second part of this chapter, I describe the strengths and limitations of different modeling approaches within the context of the modeling goals of a study or management activity.

Mechanistic and Phenomenological Models of Populations

Because populations are complex and are regulated by processes at multiple scales, it is likely that no single modeling technique will capture every important spatial or temporal aspect of a population. Different kinds of models have different uses and limitations. To understand these limitations, I outline a framework of the major pathways of causation assumed to underlie patterns of population dynamics and dispersion within a landscape.

Consider the following conceptual model as defining a set of levels that describe the complexity of a wildlife-habitat relationship. At the lowest level, the availability and quality of appropriate habitat directly controls the ability of individual organisms within a population to undergo their life histories (Hildén 1965; Rotenberry 1981). Individuals with access to sufficient resources are able to survive and accrue enough energy to reproduce. Availability of habitat will also influence the likelihood that individuals will disperse in and out of the population (Newton 1998). The summed patterns of reproduction, survival, and dispersal across all individuals in a population determine the rates of change that the population will experience. When these rates are played out in space and time, patterns of population dynamics and dispersion within a landscape emerge (Fig. 9.1). Note that this model of causation explicitly recognizes that different processes operate at different levels. There are three levels: the individual level, the population level, and the landscape level.

Two general types of models have been used to describe this general view of how populations change in space and time. The first type of model can be termed "mechanistic." The intent of such models is to describe the detailed relationship between relevant features of the habitat and their effects on the life histories of individuals within the population (Fig. 9.1). Typically, the behavior of each individual in the population is modeled in relationship to patterns of re-

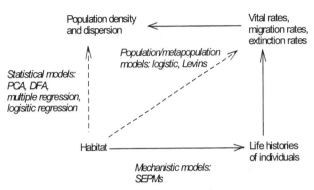

Figure 9.1. Solid lines represent a simplified schematic of pathways of causation relevant to predicting abundance and distribution. Mechanistic models attempt to model directly the link between habitat characteristics and life histories of individuals. Population/metapopulation models are phenomenological because they attempt to model the relationship between habitat characteristics and population vital rates or metapopulation colonization/extinction rates without reference to individual life histories. Statistical models attempt to relate the outcome of population and individual level processes to habitat indirectly, without explicit reference to spatial or temporal dynamics.

source distribution within a specific landscape (Pulliam et al. 1992; McKelvey et al. 1992; Judson 1994; Raphael et al. 1998; Villard et al. 1998). Individual behaviors are often modeled as stochastic processes, and the results of the model generally attempt to describe in statistical terms the relationship between habitat attributes and the dynamics and dispersion of the population. Such models are often used to describe, at least qualitatively, the ability of the population to persist given different manipulations or changes that might be applied to the habitat.

The second type of model encountered can be considered "phenomenological," that is, such models attempt to describe the relationship between habitat and population dynamics indirectly, without incorporating data on the actual life history mechanisms that are ultimately responsible for population change. There are two types of phenomenological models. Statistical models seek only to uncover a correlation between population and habitat patterns. Models of this sort have been widely used to assess which particular aspects of the habitat are related to some population attribute such as density or dispersion (James and Shugart 1970; James 1971; Smith 1977; Gauch 1982; Pielou 1984). Correlations between populations and habitat characteristics uncovered by such techniques have no essential cause-and-effect relationship, and hence such statistical models have been criticized as being of limited scientific value (Karr and Martin 1981; Rexstad et al. 1988).

The other class of phenomenological models attempts to relate habitat to population parameters, such as birth and death rates, or to metapopulation processes of extinction and colonization (Fig. 9.1). Population parameters such as birth and metapopulation parameters such as extinction rates are not mechanistic in the sense that they describe statistically a large number of individual events. In population models, rates are calculated on a per capita basis, which is essentially the same as averaging attributes of individual organisms, such as clutch size, nesting success, etc., across a large number of organisms. Typically, individual rate parameters are used to estimate the trajectory the population is expected to follow over time and can be expanded to examine patterns of temporal dynamics in space (e.g., Lele et al. 1998). In metapopulation models, rates of colonization and extinction are calculated on a per patch basis, again, essentially an averaging of many events across a large number of patches. Colonization and extinction rates are used to examine patterns of persistence of a species within a collection of natural or human-created habitat patches (McCullough 1996).

Multiscale Models and Reliability Assessment

Given the three types of models described above, what are the strengths and limitations of each type, and how is this likely to affect their use in management of complex wildlife-habitat systems? It should be clear after briefly reflecting on the structure of Figure 9.1, that each kind of model is explicitly tied into a different scale of measurement and so presumably describes different levels of organization. Mechanistic models attempt to describe wildlife-habitat systems at a scale corresponding to individual organisms. They can be used to make statements about higher levels by aggregating or averaging their output over large collections of individuals (e.g., Noon and McKelvey 1996a). Population/metapopulation models do not consider the detail of individual life histories but instead assume a scale of measurement that implies a spatial aggregation of individuals. Metapopulation models differ

from population models primarily by making assumptions about the dispersal abilities of individuals. If the spatial structure of the habitat is discontinuous, and the size of the discontinuities are larger than the dispersal ranges of individuals, then a metapopulation model is assumed to be superior to a population model for describing the spatial pattern of the system. Statistical models forego any detail about causation in favor of different types of statistical models that correlate attributes of populations at the landscape scale to habitat features.

The strength of a mechanistic model is that, if properly validated, it provides a wealth of detailed information about how habitat affects individual organisms. Thus, any descriptions obtained from aggregating the results of such models can be defended on scientific grounds as a valid description of the causes and effects that generate a spatial pattern in a landscape. But the wealth of detail of such models carries with it a limitation. Because the ecosystems in which individual organisms live are constantly changing in space and time, parameter estimates obtained at one location or time may not be valid at other times and places. Moreover, the structure of a mechanistic model must match the details of the natural history of individual species and thus must be created anew for each species. Conceivably, model structure could even be different for the same species if it lives in different habitats at different geographic locations.

Population/metapopulation models at least partially overcome the limitations in time and space of mechanistic models because they average events across space (and conceivably over short periods of time such as a single breeding season). However, averaging across individuals means that variability among individuals must be considered as "process error." For example, many attempts to estimate extinction times are based on stochastic population models that incorporate "environmental" and "demographic" variability (Leigh 1981; Goodman 1987). From the perspective of a mechanistic model, this variability would be implicit in the different conditions experienced by each individual. A population or metapopulation model may be more general, and one type of model may be applicable to a wide variety of species or locations.

Statistical models are the most general of all and

have no restrictions on the kind of species or habitats that can be modeled. They require only that data meet specific statistical criteria (such as normality, independence, etc.). Often, these criteria are easily met by transformation, initial exploratory data analyses, and diagnostic statistics. The price paid for such flexibility is that such models say little about causation. Only very general assumptions about how habitat variables affect population dispersion and dynamics need be made, such as "population density increases when a particular habitat variable increases." Statistical models make statements exclusively about pattern at the landscape level. Paradoxically, this is true regardless of the scale at which habitat is measured. For example, one common way of developing a statistical model is to measure habitat variables within the territory or home range of individual organisms. Correlation is established using techniques such as principal components analysis, logistic regression, and so forth. However, since these statistical techniques consider neither population dynamics nor the behavior or life history of the individuals within a territory, they are conceptually linked to the landscape level rather than to the level of the individual organism. There are no specific cause-effect assumptions needed in order to construct the model.

The strengths and limitations of the different types of models described previously have important implications for how they are used. There are two general uses for these models: statistical description and statistical prediction. The differences are best recognized by examining the goodness-of-fit criterion that is used to evaluate the model's reliability.

Statistical description is accomplished by examining how well a model describes the data that are used to estimate its parameters. This is what is often referred to as "calibration" of a model. Typical goodness-of-fit criteria are "proportion of variance explained statistics" such as R^2 values for regression-type models, chi-square statistics based on "observed minus expected" calculations, and statistics based on likelihood ratios. The emphasis of statistical description is to produce a model that faithfully represents the statistically important aspects of a data set. What these aspects are, of course, depends on the kind of model being used.

Statistical prediction is accomplished by examining how well a model describes new data that were not involved in the estimation of its parameters. Typically, a data set is divided into an "estimation" data set used to obtain parameter estimates and a "validation" or "test" data set for which values are predicted using the model generated from the estimation data set. Goodness-of-fit criteria include cross-validation statistics and likelihood ratios, where likelihoods are calculated separately for the estimation and prediction data sets. Statistical predictions made by mechanistic or population/metapopulation models tend to be more reliable than those made by statistical models. This is probably because statistical models tend to have relatively little cause-effect assumptions built into them.

Temporal and Spatial Scales: An Assessment

The preceding discussions recognized three important aspects of the modeling process when constructing multiscale models of complex wildlife-habitat systems. The first is the philosophical approach used to assess the relationship between scales and levels: inductive if levels are inferred from the data, deductive if data are used to test predictions about processes inferred logically from a theoretical model describing levels. The second is model type: mechanistic models that describe the details of organismal life histories, population/metapopulation models that describe vital rates or colonization/extinction rates, and statistical models that are used to establish correlations between habitat features and population patterns. The third is the modeling goal: statistical description if the model is evaluated with data used to estimate its parameters, statistical prediction if the model is evaluated with data not used to estimate its parameters.

Decisions regarding each aspect of the modeling process are most appropriately made based on the context in which the model is constructed and the ultimate uses that will be made of the model. The choice of philosophical approach is not arbitrary. Inductive inference of levels is appropriate where there is a general lack of information about the nature of the system being studied. The goal of using an inductive approach is to establish to whatever degree of reliability the data allow what natural levels might exist in the system being studied. When there is some prior knowledge about how the system might be expected to behave, then this information can be used to construct deductive predictions, which can then be assessed using appropriate statistical techniques.

The type of model chosen often reflects constraints on how much time or effort can be expended collecting data relative to the size of the system being considered. Because mechanistic models are data intensive, they are most likely only applicable when logistical constraints allow collection of large amounts of data. For the same amount of effort, population and metapopulation models generally allow for the collection of data across larger spatial and temporal scales than mechanistic models. The tradeoff is between intensive information about relatively few locations versus larger amounts of less-intensive information. Statistical models can be constructed using vast amounts of information (e.g., remotely sensed images) and across the largest spatial and longest temporal scales. The amount of information about the processes of interest is relatively low in statistical models, but this low information content allows for broad, extensive coverage.

If a model is to be used only to infer something about the particular system from which data were collected, or to manage that system over a time period when conditions aren't expected to change, then statistical description should suffice as a method to assess the reliability of the model. If a model is to be generalized to other systems or used to guide management decisions about other systems, then it is important to assess its reliability at making statistical predictions.

The chapters in this part illustrate a variety of combinations of the aspects of multiscale modeling I have described above (Table 9.1). Three chapters described studies of decision-making systems that incorporated wildlife-habitat models; the others described the development of wildlife-habitat models. Two important points emerge from Table 9.1. First, there is a preponderance of statistical models in these chapters. A cursory scan of titles of chapters in other sections suggests similar patterns throughout this volume. The paucity of other types of models, I think, reflects the reality of conservation decision making with limited resources. Time, money, and person-power are too limited to be expended on intensive studies except in the most exceptional cases. Metapopulation models, viability analyses, and individual-based models are most often associated with species of economic (e.g., game species) or legal (e.g., endangered species) importance. Most decision making in conservation biology and wildlife management, it seems, will continue to be based on statistical models, implicitly limiting in-

TABLE 9.1.

Classification of chapters in Part 2 by the different aspects of multi-scaled modeling they incorporate.

Chapter Author	Study type	Philosophical approach	Model type	Reliability assessment
Cogan	Decision making	Inductive	Statistical	Description
Zabel et al.	Decision making	Inductive	Statistical	Prediction
Conroy and Moore	Decision making	Deductive	Metapopulation	Prediction
Trani	Model development	Inductive	Statistical	Description
Tobalske	Model development	Inductive	Statistical	Description
Hahn and O'Connor	Model development	Deductive	Statistical	Description
Thomas et al.	Model development	Inductive	Statistical	Description
Greco et al.	Model development	Deductive	Statistical	Prediction
Johnson and Krohn	Model development	Deductive	Statistical	Prediction
Johnson et al.	Model development	Deductive	Statistical	Description
Wright and Fielding	Model development	Deductive	Statistical	Prediction

ferences about processes to those that can be measured at landscape scales. If most multiscale models are in fact statistical models, this means that most multiscale models do not attempt to identify levels by varying the scale at which data are collected. Rather, they attempt to infer properties of lower levels (e.g., habitat suitability, population persistence) from models restricted to the landscape scale. The models are multiscale because data are collected at more than one scale. The scales at which data are collected, however, do not necessarily correspond to any particular hierarchical level. It is not impossible to make inferences in such circumstances. It simply means that what inferences are made are likely to contain much more uncertainty than models where measurement scales correspond closely with process scales.

The second point that emerges from Table 9.1 is the need for reliability assessments to be based on statistical prediction rather than statistical description when a model is to be used to make predictive statements. Every chapter in this section clearly indicates that their models are intended to be used as predictive tools, that is, to make statements about systems other than the one in which the models were generated. Yet, four of the chapters provided reliability assessments based on measures of statistical description, not statistical prediction. It is incumbent upon the users of multiscale models to make sure they understand the degree to which the reliability of the model they are using has been assessed and act accordingly. Less credibility in the decision-making process should be assigned to models evaluated by statistical description.

Recommendations

Given that, most of the time, decision makers will be limited to statistical models that are conceptually linked to landscape scales, it is important that a large number of alternative models be evaluated whenever possible. Zabel et al., Chapter 19, for example, initially examined nearly one hundred models in the first phases of their decision-making process. A smaller number were selected using statistical description to be examined more rigorously using a validation data set. This kind of iterative use of models in the decision-making process is crucial when individual models

contain a large degree of uncertainty. Conroy and Moore (Chapter 16) describe a rigorous method for such iterative decision making and illustrate their method using a set of simple metapopulation models for two species with conflicting habitat requirements. The message to managers should be clear: when decisions are to be based on models containing relatively large degrees of uncertainty, be expansive in evaluating models that are to be used in the decision-making process.

If large numbers of models are to be evaluated, more sophisticated reliability assessments are needed (Maurer 1998). This is especially true when models vary in complexity. With statistical models, it is often possible to transform variables to allow the use of likelihood based model evaluation criteria, such as Akaike's Information Criterion (AIC; see Hilborn and Mangel 1997 for a discussion of this statistic). AIC can be used in mechanistic and population/metapopulation models when assumptions are made about the statistical distributions of model parameters (e.g., colonization and extinction rates).

Although it might be preferable to have detailed mechanistic models to base conservation decisions on, this will rarely be possible. Thus, it is very important that when we have detailed life history information about individuals, attempts are made to understand how this information can be used to understand patterns that emerge at larger spatial scales and longer temporal scales. For example, Stith et al. (1996) use information on dispersal of individual Florida scrub-jays (*Aphelocoma coerulescens*) to define degrees of connectedness among local subpopulations and to infer the spatial structure of the entire geographic population of the species in Florida. Although not directly applicable to other species, such studies can provide a way to construct hypotheses about spatial structure of populations that can be tested using data collected at landscape scales. Greco et al. (Chapter 14) describe an example of this kind of inference using data on the yellow-billed cuckoo (*Coccyzus americanus*) along the Sacramento River in California.

Explicit consideration of spatial and temporal scales in modeling wildlife-habitat systems provides a paradigm upon which defensible conservation decisions can be made. Although such models will continue to have large degrees of uncertainty associated with them, when integrated into an adaptive management framework, decisions can be made and re-evaluated as necessary in relatively objective manner. This type of adaptive decision making is preferable to abandoning conservation policy decisions to purely political processes.

A Comparison of Fine- and Coarse-resolution Environmental Variables Toward Predicting Vegetation Distribution in the Mojave Desert

Kathryn Thomas, Todd Keeler-Wolf, and Janet Franklin

A major constraint to mapping vegetation in arid areas is the cost of obtaining suitable imagery for direct detection of the vegetation types being mapped. Vegetation is usually sparse and of low structure. Direct detection requires very high-resolution imagery, which is usually cost prohibitive. Indirect methods, such as predictive modeling, can be employed in the mapping procedure to augment the use of lower-resolution imagery. In this analysis, we examine whether environmental variables derived from digital data can substitute for field-derived observations, and we examine the effect of resolution differences between field-collected variables (fine resolution) and map-derived variables (coarse resolution) in estimating vegetation distribution.

In the Mojave Desert of California, vegetation types were mapped at a 5–10-hectare resolution using a combination of (1) interpretation of 1:32,000 true-color aerial photography, (2) delineation on remotely sensed imagery (SPOT panchromatic satellite imagery with 10-meter resolution obtained from California Department of Fish and Game (copyright CNES/SPOT Image Corp. 1994), and (3) predictive modeling in a geographic information system (GIS) environment. Distributions of vegetation types were predicted by applying decision tree analysis (Michaelsen et al. 1987; Franklin 1995) to over two thousand vegetation plot samples. The independent variables used for prediction were obtained from coarse-resolution maps and include macro topography derived from a digital elevation model (DEM) (elevation, slope, and aspect), terrain variables (landforms and rock/sediment composition), and regional climate variables (precipitation and temperature).

The data set used to develop predictive vegetation models included plot samples that were obtained in the field using a two-stage random stratified sampling design (Franklin et al. in press) and those obtained from existing vegetation studies. Elevation, slope, and aspect were directly measured in the field and terrain variables were described.

The predictive modeling approach is based on the assumption that the distribution of vegetation is correlated with environmental factors that can be measured in the field and from digital maps. However, environmental variables measured in the field usually describe a variable, such as slope angle or landform type, at a different resolution than the value derived from a coarse-resolution map of that variable. In order to examine this issue we asked these questions:

1. What is the strength of relationship between environmental variables derived from fine-resolution field observations with the same environmental variables derived from coarser-resolution digital maps?
2. What is the relationship of the environmental variables, both fine and coarse resolution, to vegetation types?

The study area for this project is a 5-million-hectare area in the eastern Mojave Desert of California. This area includes all of Death Valley National Park, the Mojave National Preserve, Fort Irwin Military Reservation, China Lake Naval Weapons Center, Marine Corps Air Ground Combat Center, public land managed by the Bureau of Land Management (BLM), and some private land. The Department of Defense Legacy Program and the Strategic Environmental Research and Development Program (SERDP) funded the project. It was conducted by a team consisting of federal (U.S. Geological Survey), state (California Department of Fish and Game), and university (San Diego State University) researchers. The final products, including a map of actual vegetation types for the central section of the Mojave Ecoregion, can be found on the Mohave Desert Ecosystem Program web site at http://www.mojavedata.gov.

Methods

Measurements of plant species composition and associated environmental variables were made during the fall of 1997, winter and spring of 1998, and spring of 1999 on 1,000-square-meter releves. During the 1997 and 1998 field-sampling season, the releves were placed using a random stratified sample based on representative sampling of environmental types at a 1-kilometer resolution. Environmental types were characterized by four climate variables (average winter and summer precipitation and average January minimum and July maximum temperature), geologic substrate (based on a digitized 1:750,000 geological map of California (California Department of Conservation, Division of Mines and Geology, originally compiled by Charles W. Jennings, 1977), and topographic position (Franklin et al. in press). During the spring of 1999, releves were placed nonrandomly in order to increase the sample size for certain rare or undersampled vegetation types. The coordinate position of each releve was determined using a global positioning system (GPS) with at least 5-meter accuracy in most cases.

Classification of Vegetation Types

Vegetation types as defined by alliances were assigned to each plot sample in a two-step process. Grossman et al. (1998) defines alliances by their floristic composition within the National Vegetation Classification System hierarchy (FGDC 1997) and notes they occur at the two lower levels of the standardized classification system. Species data for each plot was standardized to a common nomenclature using The Plants Database (NRCS 1999). A combination of classification algorithms (Twinspan, indicator species analysis, and cluster analysis) (McCune and Mefford 1997) was applied independently to existing vegetation plot data and the data collected in our 1997–1999 surveys. T. Keeler-Wolf and K. Thomas (unpublished data) developed concordance rules between the existing and new data to define vegetation alliances and applied these rules to the total data set in order to identify consistent species groupings.

Fine-resolution Environmental Variables

The fine-resolution variables used in the analysis were derived from five field-collected variables: elevation (meters), slope (degrees), aspect (degrees), landform, and rock/sediment composition. Elevation was determined using a global positioning system or, in a few cases, a 1:24,000 topographic map. Aspect was determined by aligning a compass to the direction that water would be expected to flow from the plot and measured as the degrees from north. Slope was measured in that direction with a clinometer. Aspect was converted to a "southwestness" index using the transformation, $((\cos (\text{aspect}-255) + 1)* 100)$. This southwestness index varied from 0 to 200, with a value of 300 assigned to flat terrain. Aspects and slope measurements were made over a slope distance of approximately 90 meters.

The field crew, working in pairs, visually determined landform and geological substrate categories. They used a 38-category classification of types defined using a preliminary classification developed for a parallel landform and rock/sediment composition-mapping project sponsored by the Legacy Program at Louisiana State University (R. Dokka, Louisiana State University, pers. com.). The categories were aggregated into fewer types. The seven recoded landform categories were (1) rocky highland, (2) arroyo, (3) upland alluvial deposits, (4) wash, (5) fluvial floodplain, (6) playa, and (7) dunes and sand sheets. The five composition categories were: (1) igneous, (2) metamorphic, (3) calcareous carbonate, (4) evaporite, and (5) sedimentary. The six-person field crew received orientation to recognizing landform and composition categories, but they were not specifically trained in geomorphology or geology.

Coarse-resolution Environmental Variables

Coarse-resolution variables were derived from digital maps in a GIS environment. The UTM location of each field sample was associated with digital maps of environmental variables in order to obtain the value at that location for the variable. Elevation (meters) and slope (degrees) were derived from USGS 30-meter digital elevation models (DEMs) for the study area. Aspect was also derived from the DEM and transformed into a southwestness index using the same transformation as was used for the fine-resolution environmental variables. Landform and rock/sediment composition were derived from the digital maps developed at Louisiana State University for each of these features. The nominal resolution for the landform and rock/sediment composition mapping is a 10-hectare minimum mapping unit. The coarse-resolution landform and rock/sediment compositions were recoded into the same aggregated categories as the fine-resolution landform and rock/sediment composition.

Relationship of Fine- and Coarse-resolution Environmental Variables Pairs

A Shapiro-Wilk W test for normal distribution was conducted with the plot sample values for elevation, slope, aspect, and southwestness. None of the sample distributions were normal. Accordingly, a nonparametric measure of association was conducted for each fine/coarse-resolution pair of environmental variables. Significance of each comparison was tested with the Spearman's Rho and Kendall's tau-b test. Correlation between the fine/coarse resolution pairs of nominal variables was determined with two-way contingency table analysis. The Kappa statistic was used to measure the degree of agreement (0–1) between the pairs of nominal variables where they have the same set of values. All statistical analysis was performed using JMP IN software, version 3.2.6 (Sall and Lehman 1996).

Relationship of Fine- and Coarse-resolution Environmental Variables to Vegetation Types

One thousand sixty-four (1,064) plot samples were used in this analysis. Samples deleted included those for which complete field measurements were not obtained. The first three eigenvector axis scores were determined for each of the samples. Scores were obtained using detrended correspondence analysis (DCA) (Hill 1979) in PC-ORD version 3.14 (McCune and Mefford 1997) The scores were initially determined for all samples pooled ($n = 1064$) without any stratification for alliance type or downweighting of rare species. A second set of scores were determined for a dataset ($n = 1,039$) with "playa"-related plots removed (samples identified as *Allenrolfea occidentalis* Shrubland Alliance, *Suaeda moquinii* Intermittently Flooded Shrubland Alliance, *Prosopis glandulosa* Woodland Alliance, and those Sparsely Vegetated Alliance plots dominated by *Distichlis spicata* or *Pluchea sericea*. Correlations between each pair of numerical environmental factors, the fine resolution and the coarse resolution, and each axis score were determined using Spearman's Rho and Kendall's tau-b test for continuous variables (elevation, aspect, southwestness). One-way analysis of variance was used to calculate the significance of the relationship between the nominal variables (landform and rock/sediment composition) and the ordination axis scores.

Results

Fine-resolution and coarse-resolution elevation measures are highly correlated with each other (tau-b 0.97, $P < .0000$). Fine- and coarse-resolution slope, aspect, and southwestness measures are more moderately correlated with each other (slope tau-b 0.64, aspect tau-b 0.53, and southwestness tau-b 0.56, all $P < .0000$).

Landform categories are likewise moderately correlated between the fine- and coarse-resolution observations (Kappa = 0.52). A contingency table with cross tabulations (Table 10.1) shows the specific pairs of variables. If the coarse-resolution data are assumed to be the "true" assignment of landform types, the field crew's assessment for Playa is the most accurate (12/13, 92 percent), followed by Upland Alluvial Deposits (75 percent), Rocky Highland, and Dunes and

TABLE 10.1.

Comparison of field measurement of landform (fine resolution) to measurement from digital maps (coarse resolution).

| Coarse resolution | Fine resolution landform | | | | | | | | |
	Rocky highland	Arroyo	Upland alluvial deposits	Wash	Fluvial floodplain	Playa	Dunes and sand sheets	Row totals	Row %
Rocky highland	465	36	97	36	9	3	0	646	72
Arroyo	2	2	5	11	2	0	0	22	9
Upland alluvial deposits	20	6	216	38	2	5	1	288	75
Wash	0	0	10	3	0	0	0	13	23
Fluvial floodplain	0	0	6	2	4	2	1	15	27
Playa	0	0	0	0	1	12	0	13	92
Dunes and sand sheets	2	0	3	2	1	3	28	39	72
Column totals	489	44	337	92	19	25	30	1036	
Column %	95	5	64	3	21	48	93		100

Sand Sheets (72 percent each). Fluvial Floodplain (27 percent), Wash (23 percent), and Arroyo (9 percent) have the lowest correspondence.

Pictures taken at the plot locations by the field crew were examined to understand why seemingly obvious landform features (arroyos, washes, dunes, and sand sheets) were often mis-assigned. Mismatches in thematic interpretation were noted. The field crew often called arroyos either washes or arroyos. The field crew sometimes labeled sand sheets on slopes as part of the larger landform (e.g., Upland Alluvial Deposits) even though the substrate was sandy. Mismatches in the resolution of interpretation also occurred. For instance, changes in slope direction (which may occasionally support drainage) were labeled as Rocky Highland in the coarse dataset and Arroyo in the fine dataset. Other resolution mismatches were noted, for example where a plot appeared to include both Upland Alluvial Deposit and Wash, the label applied by the field crew varied. Interpretation errors in the coarse dataset also caused mismatches. In several cases where the field crew called a plot Upland Allu-

TABLE 10.2.

Comparison of field measurement (fine) of rock/substrate to measurement from digital maps (coarse).

| Course resolution | Fine resolution Rock/substrate composition | | | | | | |
	Igneous volcanic	Metamorphic	Calcareous carbonate	Evaporite	Sedimentary	Row totals	Row %
Igneous	394	0	7	0	14	415	95
Metamorphic	74	0	4	0	2	80	0
Calcareous carbonate	37	2	38	0	15	92	41
Evaporite	29	0	0	7	15	51	14
Sedimentary	176	5	33	1	20	235	9
Column totals	710	7	82	8	66	873	
Column %	55	0	46	88	30		100

vial Deposit and the coarse label identified the feature as Wash, it was noted that features occurred in or near the plot that may have been misinterpreted as wash in the coarse data (i.e., a dirt road, a mined area, desert pavement). Variability in interpretation among field-crew members could not be determined because they worked in rotating pairs and each landform identification was a team report.

Rock/sediment composition has poor correlation (Kappa = 0.18) between the GIS-derived variables and the field-observed variables. The contingency table (Table 10.2) suggests that the field crew could reasonably recognize igneous substrate (394/415, 95 percent). However, the crew attempted classification in only 82 percent of the plots (873/1,064) and 81 percent of the time (*n* = 710) they determined the plot substrate was igneous. It seems that the crew only felt comfortable recognizing igneous substrate.

T. Keeler-Wolf (unpublished data) described forty-two vegetation types alliances and each plot (observation) in the modeling dataset was assigned to one of these types. The alliances were each represented by varying numbers of samples, ranging from 2 to 311 (Table 10.3). Each plot also received scores on the DCA axes based on the results of the ordination analyses. Inspection of the individual plot scores determined that certain vegetation types, in particular playa types, were largely influencing the ordination scores for the 1,064 samples. Therefore, results are presented for the ordination based on 1,039 observations (with playa types removed). With these plots removed, elevation was the variable most strongly correlated with DCA axes 1 and 2 (Table 10.4). Slope and southwestness were significantly but weakly correlated with DCA axis 1. In all cases correlation of coarse variables to axis scores was equal to or slightly higher than those derived from fine variables (Table 10.4).

Both the fine- and coarse-resolution-derived values for landform are most strongly related to the second axis (Table 10.5). The coarse-resolution-derived values for rock/sediment composition also shows significant relationship to the second axis. The differences between the fine/coarse pairs for each axis score were greater than those for continuous variables. For the

TABLE 10.3.

Preliminary alliance types described by Keeler-Wolf (unpublished data).

Preliminary alliances	No. samples
Acacia greggii Shrubland Alliance	16
Allenrolfea occidentalis Shrubland Alliance	8
Ambrosia dumosa Dwarf-shrubland Alliance	34
Artemesia nova/Mortenia utahensis Dwarf-shrubland Alliance	5
Artemesia tridentata Shrubland Alliance	10
Artemesia tridentata-Ephedra viridis Shrubland Alliance	4
Atriplex canescens Shrubland Alliance	3
Atriplex confertifolia Shrubland Alliance	44
Atriplex hymenolytra Shrubland Alliance	29
Atriplex polycarpa Shrubland Alliance	3
Coleogyne ramosissima Shrubland Alliance	39
Encelia farinosa Shrubland Alliance	13
Ephedra nevadensis Shrubland Alliance	10
Ephedra viridis Shrubland Alliance	3
Ericameria nauseousus Shrubland Alliance	7
Eriogonum fasciculata Shrubland Alliance	10
Grayia spinosa Shrubland Alliance	15
Hymenoclea salsola Shrubland Alliance	20
Juniper spp. Wooded Alliance	14
Larrea tridentata Shrubland Alliance	126
Larrea tridentata/Ambrosia dumosa Shrubland Alliance	311
Larrea tridentata/Encelia farinosa Shrubland Alliance	43
Lycium andersonii Shrubland Alliance	7
Menodora spinosa Shrubland Alliance	10
Pinus monophylla Wooded Shrubland Alliance	9
Pinus monophylla Woodland Alliance	5
Pinus monophylla/Juniperus spp. Wooded Shrubland	28
Pleuraphis rigida or *P. jamesii* Herbaceous Alliance	16
Prosopis glandulosa Shrubland Alliance	10
Prunus fasciculata Shrubland Alliance	13
Psorothamnus spinosa Wooded Alliance	5
Salizaria mexicana Shrubland Alliance	14
Senna armata Shrubland Alliance	2
Sparsely Vegetated Type	56
Suaeda moquinii Intermittently Flooded Shrubland Alliance	5
Viqueria parishii Shrubland Alliance	6
Viqueria reticulata Shrubland Alliance	6
Yucca brevifolia Wooded Shrubland Alliance	41
Yucca brevifolia/Coleogyne ramosissima Wooded Shrubland Alliance	14
Yucca brevifolia/Juniperus spp. Wooded Shrubland Alliance	10
Yucca brevifolia/Pleuraphis spp. Wooded Herbaceous Alliance	4
Yucca schidigera Shrubland Alliance	44

TABLE 10.4.

Correlation of environmental variables with ordination scores for alliances.

	DCA Axis 1		DCA Axis 2		DCA Axis 3	
	Fine	Coarse	Fine	Coarse	Fine	Coarse
Elevation	0.62	0.63	0.31	0.31	0.03	0.03
Slope	0.07	0.11	0.02	0.04	−0.19	−0.2
Aspect	−0.02	−0.04	−0.02	−0.03	0	0.03
Southwestness	−0.07	−0.08	0	0.01	−0.03	−0.03

Note: Shaded cells indicate $P < .001$.

TABLE 10.5.

Landform and rock/sediment composition relationship with ordination axis scores for alliances.

	DCA Axis					
	DCA Axis 1		DCA Axis 2		DCA Axis 3	
	Fine	Coarse	Fine	Coarse	Fine	Coarse
Landform	8	14	31	21	1	16
Rock/Sediment Composition	13	8	7	10	27	1

Note: All cells $P < .05$
Cell value = F statistic; n = 1,039 sites
Critical Value (95% CI) for all fine landform = 3.7
Critical Value (95% CI) for all coarse landform = 1.8
Critical Value (95% CI) for all fine rock/sediment composition = 5.6
Critical Value (95% CI) for all coarse rock/sediment composition = 1.7

first and third axis, the fine-resolution variables have a stronger relationship to rock/sediment composition.

Discussion

This paper investigates and tests the assumption that the use of independently derived digital data (DEM, landform, and geology) is useful for vegetation prediction and mapping. If digitized or interpolated environmental data are reliable for the analysis of ecological determinants of vegetation patterns across gradients, some effort could be eliminated during large-resolution botanical surveys. Plant ecologists can then estimate the composition, structure, and cover of vegetation in samples and rely on independently derived environmental data for the investigation of vegetation-environment patterns.

The values for environmental variables that we compared are reasonably well correlated between field (fine) and GIS (coarse) data sources except for the rock/sediment composition. The highest correlation is for elevation. This is not surprising as both the fine- and coarse-resolution sources of data are derived from accurate methods, global positioning system (accuracy within meters) and 30-meter digital elevation models (DEMs). The digital elevation model is very acceptable as a substitute for field-derived elevation readings.

Slope, aspect, and southwestness were only moderately correlated between the fine (field) and coarse (GIS) variables. Slope, aspect, and southwestness were integrated over a 90×90-meter area that may accentuate systematic and nonsystematic errors within the DEMs. In the field, one would expect that some subjectivity is involved in determining the aspect and slope, especially in uneven terrain. Other researchers have found correlations between field-measured and DEM-derived slope and aspect of 0.38 to 0.48, or root mean square errors ranging from 9 to 56 (for example Davis and Goetz 1990; Franklin 1998; Wise 1998). Despite this, the correlation for fine- and coarse-resolution values with the ordination axis is similar. At the resolution of this project, 5 hectares, use of DEMs to calculate slope, aspect, and southwestness appear to be a reasonable substitute for field measurement.

The lack of correlation between the fine and coarse description of landform is significant. Differences in definition of the landform categories as applied in the field versus the GIS mapping, scale of interpretation, and mistakes in interpretation can explain differences between the descriptions. The fine-resolution assignment of landform type by the field crew appears to be more correlated with vegetation composition than with the coarse-resolution map (Table 10.5). The aggregation of landforms may have masked important influence of some landform categories.

The poor correlation between the fine- and coarse-resolution rock/sediment composition measures can be attributed to field-crew error. However, the F statistic is low but significant (Table 10.5) for both the fine- and coarse-resolution measures. It appears that in the Mojave Desert few geological composition classes are significant determinants of plant distributions. Alternately, it may be that the salient substrate features are not captured by the classification used regardless of the resolution. For example, it is known that some species respond to caliche layers (McAuliffe 1994) and others to salinity gradients (Hunt 1966; T. Keeler-Wolf unpublished data; Wallace et al. 1982; West 1983).

The results reported can be used by land managers and researchers to set guidelines for collection of field data and use of independently derived digital data. Digital elevation models and derived variables are suitable surrogates for medium-resolution interpretation (5 hectares). The usefulness of landform data may vary by landform category. It is evident that resolution can influence interpretation, particularly for drainages. Although it seems that a desert wash is an easy feature to define, the study demonstrates our observers placed drainages in a variety of categories. Not only are standardized definitions of landform features needed but also calibration in interpretation is recommended. Although the field crew received at least two training sessions in substrate identification, more substantial training in geology was needed to eliminate uncertainty.

Summary

In summary, field-collected environmental variables and site/sample-specific environmental variables derived from a digital elevation model are equivalent for the purpose of modeling medium-resolution Mojave vegetation patterns. Substrate data derived from digital sources are not as readily interchangeable with field observations of substrate. However, some of these differences, at least for landforms, may be the result of different definitions for the same category. A different aggregation of the categories may have yielded a higher correlation between the variables and less difference in their correlation with ordination scores.

This study supports the use of independently derived digital data but emphasizes the need to consider the resolution, underlying purpose, and classification of the physical data before it is used as a surrogate to field-collected environmental data. The accuracy of the predictions based upon digital surrogates to field-collected data can be adversely affected if categorized data is casually used without consideration of the resolution of the field-collected training data. We recommend that elevation from DEMs and derived products such as slope, aspect, and southwestness can be readily used in predicting meso-scale vegetation patterns. Interpreted data such as landforms and rock/substrate can also be used; however, more caution is advised.

It is recommended that users of a vegetation map developed by prediction with coarse-resolution model parameters be informed of the methodology of map development and the resulting limitations and/or cautions that may be resolution related (Maurer, Chapter 9). For example, land managers seek information on vegetation distribution for management planning yet are often skeptical of the outcome (K. Thomas U.S. Geological Survey personal observation). These users may wish to use vegetation maps at a scale parallel to that from which a field crew observes data. This chapter shows how differences in the fine view and the coarse view become more apparent for variables that have been derived from interpretation in the field rather than measurement (e.g., landform, rock/substrate composition). Land managers must be cognizant of these limitations and be advised by the map developer of the influence of resolution on map development and application. As pointed out elsewhere in this book (Henebry and Merchant, Chapter 23), "we are still learning to use the 'macroscope'," both the users and developers of predictive mapping.

The Influence of Spatial Scale on Landscape Pattern Description and Wildlife Habitat Assessment

Margaret Katherine Trani (Griep)

A fundamental theme of landscape ecology centers on the influence of spatial pattern upon the abundance and dynamics of species (Levin 1992). Landscape ecology includes the study of the patterns in communities and ecosystems, and the processes that affect those patterns. The structure and dynamics of communities are strongly influenced by the variation in patterns over large regions.

Landscape pattern description is influenced by the scale of observation and can alter the description of species distributions (Meentemeyer and Box 1987; Holling 1992; Levin 1992). If spatial scale influences landscape analysis, then the habitat assessment resulting from such analyses may be affected. The analysis of landscape pattern occurs at several scales (grain and extent); landscapes described at one scale are unlikely to be the same as those described at another. How to integrate landscape measurements made at disparate scales (Musick and Grover 1991) and how to extrapolate information from one spatial scale to another remains a problem. There is a need to consider how the scale of examination may limit or add to observable relationships (Allen and Starr 1982). Quantitative analysis of these relationships may clarify these concerns.

Landscape area (extent) and the resolution of observation (grain) characterize spatial scale. Spatial resolution, as used here, refers to the level of detail inherent in spatial data (i.e., the smallest discernible spatial

unit). The objective of this chapter is to examine how changes in spatial scale influence landscape pattern analysis. First, changes in pattern metrics as a function of spatial scale are presented. Second, semivariogram analysis is used to assess the variability of metric behavior. Finally, the implications for wildlife habitat evaluation are discussed.

Methods

Fourteen pattern metrics that express aspects of spatial heterogeneity, fragmentation, and edge characteristics were selected for analysis (Table 11.1). Selection was based on their potential relevance to wildlife resources (Trani 1996; Trani and Giles 1999). These metrics are described briefly below.

The spatial heterogeneity metrics express the complexity and variability among the land classes occurring on a landscape. The Simpson and binary comparison matrix indices are based upon the number and proportions of land classes. Evenness refers to how abundance is distributed among the classes present on the landscape, while interspersion reflects the arrangement of those classes.

Fragmentation metrics describe the amount of forest cover on a landscape. This includes patch metrics that reflect the count (number of forest patches), size (mean patch size), or degree of isolation (interpatch distance) on a landscape. Fragmentation Index I

TABLE 11.1.

Landscape pattern metrics selected for spatial scale modeling.

Landscape metric	Selected reference
Spatial Heterogeneity	
Simpson Index	Pielou 1977
Landscape Evenness	Romme 1982
Interspersion	Eastman 1997
Binary Comparison Matrix	Murphy 1985
Fragmentation	
Fragmentation Index I	Monmonier 1982
Fragmentation Index II	Ripple et al. 1991
Percent Interior Forest	Dunn et al. 1991
Number of Forest Patches	Trani 1996
Mean Patch Size	Dunn et al. 1991
Interpatch Distance	Urban and Shugart 1986
Percent Forest Cover	Lauga and Joachim 1992
Contiguity Index	LaGro 1991
Edge Characteristics	
Total Forest Edge	Ranney et al. 1981
Convexity Index	Berry 1991

reflects the number of distinct landscape regions on a map relative to the total number of map pixels. Fragmentation Index II (the average distance to non-forested areas) and percent interior forest (the amount of forest area remaining after buffer removal from the edge of each forested tract) describe the distribution and amount of forest cover. Forest contiguity expresses the spatial connectedness or the unbroken adjacency of a landscape.

The edge metrics characterize areas where two different land classes come together. Total forest edge refers to the length of edge that exists at the interface between forest and other land classes, while the convexity index is a perimeter-to-area ratio that describes the amount of edge per unit area of forest.

Twenty-two forested landscapes (scale 1:24,000) that represent a broad array of landscape conditions were selected from the George Washington and Jefferson National Forests, Virginia. USDA Forest Service resource specialists visually regrouped the landscapes; the criteria used to assign membership centered on the spatial arrangement and the number of vegetation communities and land classes. The maps were placed

into one of two categories: simple or complex landscape characteristics. Maps represented the simple landscape group with relatively few polygons per landscape area, such as those with large, continuous forest blocks with few land classes. The complex landscape group contained numerous polygons per landscape area, characterized by numerous forest patches, irregular forest boundaries, and diverse arrangements of land-use classes.

These maps were used as initial conditions for the cartographic modeling process. Digitized using a 30-meter pixel (0.09 hectare) resolution, the maps served as a baseline reference for making comparisons with alternative maps of coarse resolution (pixel size greater than 30 meters). Producing the base map at the finest resolution available made it possible to take measurements over an increasing range of spatial scales. Changes in spatial scale were modeled using Idrisi (Eastman 1997). The baseline map was repeatedly generalized at a sequence of pixel sizes (30, 60, . . . 420 meters) using 30-meter increments; each map was reduced by the number of rows and columns while pixel size was enlarged simultaneously. Each succeeding image was derived from the 30-meter baseline reference map and was independent of each preceding image. This procedure provided a consistent, repeatable means of map generalization. The suite of landscape metrics was tabulated at each level of spatial resolution.

Regression analysis based on the sum of least squares was used to generate trend lines illustrating the relationship between each metric and spatial resolution. Trend lines were used solely to provide a useful summary of the data and have no statistical significance. The plotted lines having the best fit were selected on a visual basis. Semivariogram analysis was used to examine metric variability as a function of spatial scale. The semivariogram summarizes spatial variation in magnitude and general form (Oliver and Webster 1986), relating semivariance to the spatial distance between measurements (Curran 1988). The semivariance, y_h, is defined as

$$y_h = \sum_{i=1}^{n-h} (Z_i - Z_{i+h})^2 / 2n$$

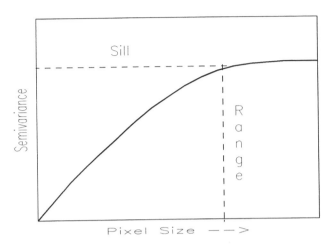

Figure 11.1. General form of the semivariogram. The sill represents the maximum level of semivariance observed. The range indicates the place on the x-axis where semivariance reaches 95 percent of the sill.

where h is the distance over which y_h is measured, Z_i is the value of an metric taken at resolution level i (e.g., 30 meters), Z_{i+h} is another measurement taken h levels away (e.g., 60 meters), and n is the number of observations used in the estimate of y_h. Semivariance is one-half of the mean squared differences between metric values separated by a level h apart. The semivariogram becomes a plot of y_h as a function of h.

At 30-meter resolution, the value of a metric is compared to itself and the semivariance is zero. At 60 meters and beyond, semivariance rises when the comparisons are increasingly different from those obtained at 30 meters. This increase continues until the values are no longer related to each other and their squared difference becomes equal to the average variance of all samples. The graphical line of y_h levels off and becomes flat (the *sill*), representing the maximum level of semivariance observed (Fig. 11.1). The point on the x-axis where semivariance reaches approximately 95 percent of the sill is the *range* and serves as an estimate of area similarity (Yost et al. 1982). At distances closer than the range, the values are considered scale-dependent (i.e., they become more alike with decreasing distance between them). Scale-dependence describes the relationship between the magnitude and variability of a process and the scale of measurement. Beyond the range, metric values are considered scale-independent (i.e., the values do not reflect the scale of

measurement). The general form of the semivariogram relies on pixel area, the spacing of those pixels, and the pattern metric computed based on landscape characteristics.

Results

Spatial scale (grain) influenced pattern metrics in a variety of ways, and several factors changed metric values repeatedly. The first factors related to the underlying pattern of the landscape included the number, size, shape, and distribution of land classes. These are described with the specific pattern metrics. The second group of factors reflected the sampling procedure that was used in the spatial scaling process; these included sampling intensity and sample unit size.

Sampling intensity. During the modeling process, each succeeding map had a reduction in the number of pixels representing the pattern of the original landscape. Each landscape image contained 25 percent of the pixels contained in the previous map. The reduction caused problems inherent to small sample sizes (e.g., the smaller the sample, the less representative it becomes of the sampled landscape). Accuracy of several metrics was closely linked to sample size. At each new resolution, a subset of pixels was selected to represent the landscape; as each subset became smaller, it no longer accurately reflected the original landscape.

The variance associated with the estimates of metric means appeared inversely proportional to sample size (O'Neill et al. 1991). Variance increased as sampling intensity decreased with the increase of pixel size. The degrees of freedom for each variance estimate were also reduced at each level; the reliability of semivariance estimates decreased as the resolution was reduced. This occurred because the number of observations decreased with increasing distance. Fluctuations in pattern descriptions for the new maps were dependent upon those pixels retained and upon those pixels selected for removal. In a very small area, the selection of the initial pixel had a profound effect on the semivariogram.

Sample unit size. Sampling intensity was reduced as spatial scale and pixel size increased. The interplay between pixel size and landscape area influenced several aspects of landscape description. When the pixels were small in relation to patch size, the landscape was sampled with pixels small enough to lie wholly within those patches. This resulted in little change in the distribution of patches; landscape pattern description remained similar to the preceding level. Small pixels increased the likelihood that a patch would be retained into the next resolution level. As pixel size increased but remained small enough that there was little chance that the pixel would cross patch or land class boundaries, a shift in patch distribution occurred with minor changes in land class proportions.

As pixel size reached and exceeded patch size, retention of patches and land classes became unpredictable; there was either a loss of patches or a coalescence of like-patches. This loss (or gain) contributed to the variance observed at the coarse-resolution levels. Spatial relationships among the land classes were also altered with large pixels. A marked loss in landscape detail occurred and continued as pixel size exceeded the known median patch and land class areas.

Pixel size set a minimum threshold for land class features. An increase in this threshold at each resolution level influenced the heterogeneity and the patch metrics through the loss of landscape features. Pixel size set the minimum edge-to-edge distance between patches; the interpatch distance could not be less than the current pixel size. Figure 11.2 presents the relationship between spatial resolution and sampling intensity.

The progression of the evenness index across the range of resolution levels is depicted in Figure 11.3. (The Simpson index, not shown, exhibited a similar trend.) The continuity of both metrics during changes in scale was unique among the other pattern metrics. Values remained fairly stable until a 270-meter pixel size was reached. At that point, both indices became quite variable with each resolution reduction, dependent upon the conditions encountered. Large pixels either omitted some land classes or resulted in dramatic proportional changes within the remaining classes. The vertical axes of the trend graphs (and subsequent figures where noted) were scaled by dividing metric values at each resolution level by the values obtained at the 30-meter resolution level (EXP_{Nm} / EXP_{30m}). Scaling the axes to 1.0 allowed direct comparison (i.e., making them unitless) across several different metrics.

The influence of spatial resolution on metric behav-

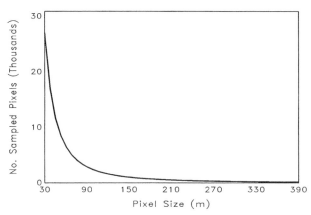

Figure 11.2. The relationship between spatial resolution (pixel size) and sampling intensity (number of sampled pixels).

ior was influenced by landscape complexity. For simple landscapes, the evenness metric becomes unpredictable as pixel size increases: one large pixel contributes to proportional evenness, the next pixel may reduce the index value. In contrast, there was a consistent trend for those values computed from the complex landscapes. The loss of one land class from a complex landscape having forty-five classes represents a 2 percent loss, while the loss of one land class from a simple landscape with nine land classes represents an 11 percent loss.

The horizontal line between 30 and 210 meters on the semivariogram indicates intervals where the spatial pattern is stable and not scale dependent. The abrupt semivariance rise following 270 meters indicates that values measured beyond this level are becoming different. The marked reversal of slope after reaching a maximum is uncommon (Oliver and Webster 1986), suggesting that further fluctuation in evenness values may occur beyond 420 meters.

Reducing spatial resolution resulted in striking increases for both the interspersion and binary comparison matrix metrics (Fig. 11.3). Logarithmic regression curves were computed for the relationship between metric behavior and pixel resolution. Interspersion increased at a constant rate (Interspersion = −8.32 + 2.54 LN Resolution; r^2 = 0.95) and binary comparison matrix behaved similarly (binary comparison matrix = −6.79 + 2.14 LN Resolution; r^2 = 0.95). Values obtained at 420-meter levels were six to eight times greater than baseline values, a magnitude far surpass-

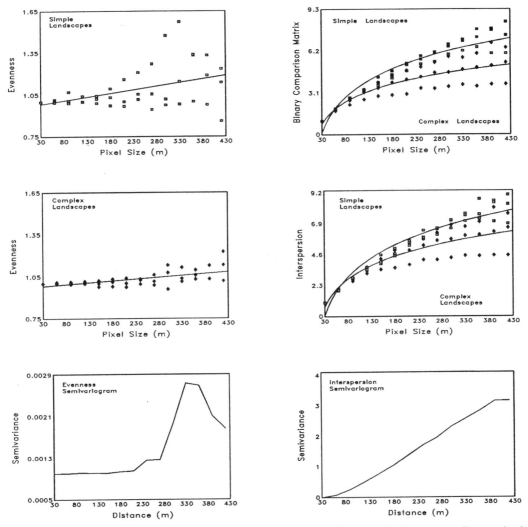

Figure 11.3. The influence of spatial scale on the heterogeneity indices with their corresponding semivariograms. Plot markers identify simple (square) and complex (diamond) landscapes. Vertical axes are scaled to 1.0 for direct comparison. Trend lines are derived from regression analysis.

ing that observed for the Simpson and evenness indices. Interspersion and binary comparison matrix are based on a three-by-three roving window that covers proportionally more area with each resolution level. Landscape complexity also influenced metric behavior; values routinely overestimated the spatial heterogeneity of simple landscapes.

The interspersion semivariogram depicts the classic semivariance curve. (Binary comparison matrix, not shown, emulated interspersion). Spatial dependency exists at each resolution until 390 meters (a sill) is reached. Each successive landscape map has a pattern directly dependent on the resolution at which it is

measured. The change in form beyond 390 meters approximates the range.

The reduction in spatial resolution resulted in a marked decline in convexity and total forest edge length (Fig. 11.4). Edge detail was suppressed at each successive resolution, reflected by the exponential regression trend lines for forest edge (LN Forest Edge = 0.039 − 0.001 Resolution; r^2 = 0.973) and the logarithmic trend for convexity (Convexity = 1.60 − 0.156 LN Resolution; r^2 = 0.971). The edge length and convexity measured at the coarsest resolution was 50 percent less than baseline. Each change reflected the loss of edge detail as patch shapes were repeatedly

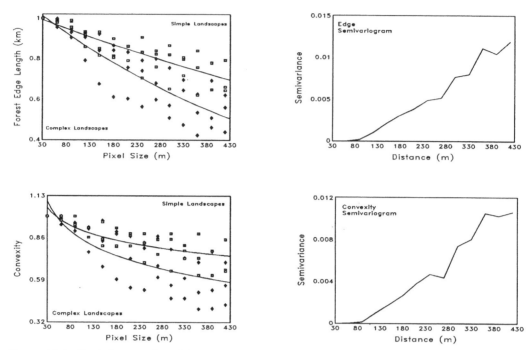

Figure 11.4. The influence of spatial scale on the edge metrics with their corresponding semivariograms. Plot markers identify simple (square) and complex (diamond) landscapes. Vertical axes are scaled to 1.0 for direct comparison. Trend lines are derived from regression analysis.

simplified. The more complex the landscape pattern, the greater the potential for the loss of edge detail.

The semivariograms for edge and convexity show a tendency toward a linear trend in semivariance with small-scale noise, suggesting weak spatial dependence. Semivariance change is almost imperceptible between 30 and 90 meters. The semivariogram for forest edge length exhibits unbounded variance (Curran 1988). Convexity semivariance beyond 360 meters fluctuates near 0.01 (a possible sill). These values are considered spatially independent if this fluctuation continues.

The influence of changing spatial scale on forest contiguity and the fragmentation indices are depicted in Figure 11.5. Fragmentation Index I values at the coarsest resolution were 170 times greater than baseline values, the magnitude of which was unsurpassed by any other landscape metric (LN Fragmentation Index I = −6.31 + 1.89 LN Resolution; r^2 = 0.973). This index expresses the ratio between polygon number and total map pixels. With each resolution reduction, the denominator decreases fourfold. In contrast, changes in polygon number (the numerator) are gradual in comparison. This resulted in an exponential in-

crease for the index over all landscapes. The semivariogram illustrates nonstationary processes (Oliver and Webster 1986); the range for finite variance is not reached.

Dramatic changes were also observed for the Fragmentation Index II (the mean distance to nonforest pixels). The repetitive loss of forest boundary detail and small forest clearings resulted in an increase in metric values with each resolution change (LN Fragmentation Index II = −0.116 + 0.004 Resolution; r^2 = 0.884). Application to complex landscapes resulted in a fivefold overestimation, while values computed from simple landscapes were three times baseline values. Values rose slightly approaching 180 meters and increased quickly with succeeding changes in resolution. Semivariance does not reflect the scale of measurement within this interval. Peak height and spacing mirror the different patterns emerging during scale changes; distance to nonforest was quite variable within these periods.

Changing spatial scale resulted in a steady decline in contiguity values for all landscapes (LN Contiguity = 0.206 − 0.057 LN Resolution; r^2 = 0.914). The

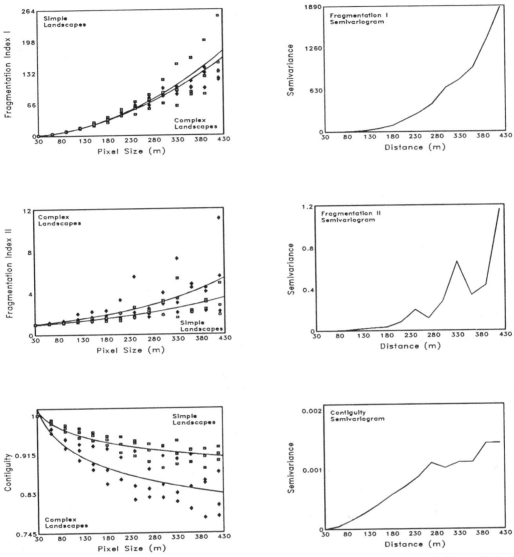

Figure 11.5. The influence of spatial scale on forest contiguity and the fragmentation indices with their corresponding semivariograms. Plot markers identify simple (square) and complex (diamond) landscapes. Vertical axes are scaled to 1.0 for direct comparison. Trend lines are derived from regression analysis.

mean rate of information loss for simple landscapes was 9 percent, while that for complex landscapes was 17 percent. Contiguity values were increasingly variable with each succeeding resolution, dependent upon the distribution of remaining forest pixels. Spatial dependency existed until 270 meters; semivariance then changes direction until a range is reached at 390 meters.

Figure 11.6 illustrates the relationship between spatial resolution and percent forest cover (Forest Cover = 0.991 + 0.0003 Resolution; r^2 = 0.827). Metric values remained relatively constant through 90 meters,

increasing until the coarsest resolution was reached. The distribution of forest and nonforest within a landscape influenced the variability observed for forest cover estimation. Forest cover was overestimated on continuous forest landscapes, particularly forest areas compact in shape. Elongated or irregularly shaped nonforest areas were lost quickly during the scaling process, resulting in the overestimation of forest cover. In contrast, landscapes with compact areas of nonforest (e.g., agricultural fields) often retained these areas during the resolution changes.

The variability of forest cover estimation is evident

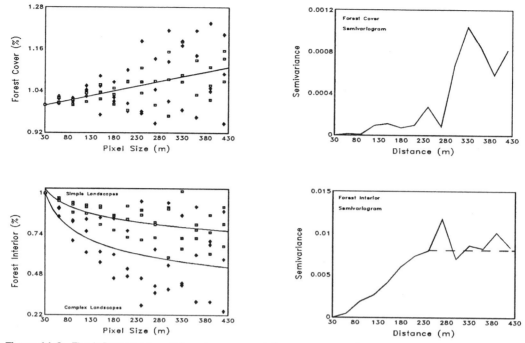

Figure 11.6. The influence of spatial scale on percent forest cover and forest interior with their corresponding semivariograms. Plot markers identify simple (square) and complex (diamond) landscapes. Vertical axes are scaled to 1.0 for direct comparison. Trend lines are derived from regression analysis; the line for forest cover represents both simple and complex landscapes. The dashed line on the forest interior semivariogram indicates a possible sill.

by the slope reversals in the semivariogram. The variable period height reflects the fluctuating distribution of forest pixels emerging from each change in resolution. At fine resolution levels, semivariance does not reflect the scale of measurement; as resolution becomes coarser, semivariance rises.

Figure 11.6 also depicts the relationship between spatial scale and percent forest interior. Forest interior was quite sensitive to changes in resolution; the net change in values was negative (LN Forest Interior = 0.588 − 0.170 LN Resolution; r^2 = 0.744). Metric values fell at 60 meters and continued to drop throughout each level. This steady reduction was a function of the loss of edge detail as forested areas were repeatedly simplified, the shrinkage of forested areas, and the relationship between forest buffer and pixel size. There was a greater likelihood for the loss of forest interior on complex landscapes. At the coarsest resolution, values were reduced by 50 percent compared to a 25 percent reduction measured on simple landscapes. The loss of interior forest ac-

celerated where several nonforested areas were dispersed throughout a landscape.

The forest interior semivariogram depicts scale dependence between 30 and 230 meters. Semivariance values beyond 230 meters fluctuate slightly above and below 0.0075 (the sill); values beyond this distance are considered spatially independent (Palmer 1988).

Figure 11.7 illustrates the association between spatial resolution and the number of forest patches. (Mean patch size and interpatch distance, not shown, depict a similar trend.) The variability observed for the patch metrics was higher than that for any other metric group. Since each stage of the scaling process was independent of the last, the opportunity for disappearance or coalescence of forest patches varied with each resolution.

The influence on forest patch number was negative; 45 percent of the patches were lost by the coarsest resolution level (LN Forest Patches = 0.088 − 0.002 Resolution; r^2 = 0.858). Patch detection was suppressed as pixel size exceeded patch size; the rate of patch loss was greater on landscapes comprising several small

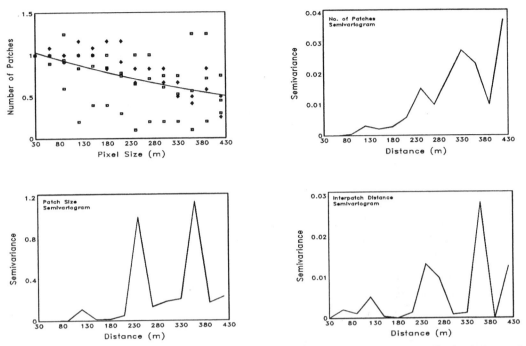

Figure 11.7. The influence of spatial scale on forest patch metrics. Plot markers identify simple (square) and complex (diamond) landscapes. Vertical axes are scaled to 1.0 for direct comparison. Trend lines are derived from regression analysis; the line for number of patches represents both simple and complex landscapes.

patches than on landscapes composed of a few large patches. Landscapes with clumped patch arrangements lost those patches at a slower rate than those landscapes with uniform patch distributions. The associated semivariogram highlights the reversal of variability that occurs at several levels; a finite semivariance level was not reached.

The disappearance of forest patches during changes in spatial resolution also influenced mean patch size and interpatch distance. When pixel size was less than patch size, the likelihood was high that the patch would remain during the next resolution. As pixel size reached and exceeded patch size, retention of those patches became unpredictable, resulting in an r^2 of 0.441 for mean patch size and an r^2 of 0.184 for interpatch distance. As the resolution decreased, mean patch size could increase (e.g., the selected patch area is magnified by the new pixel size) or decrease (e.g., the nonselected patch area is replaced by nonforest pixels). Patch distribution and size influenced interpatch distance. If a patch was removed during the scaling process, mean interpatch distance declined. However, if a patch proximal to another patch is re-

moved, mean distance rises. Each patch metric poses predictability problems at different spatial scales.

The semivariograms for mean patch size and interpatch distance show prominent semivariance fluctuation, suggesting that the pattern described varies continuously with scale. The stability of the semivariance observed between 30 and 90 meters suggests that patch size exceeded pixel size. Semivariance for interpatch distance increased immediately following the 30-meter baseline, reflecting the influence of patch number and distribution.

Discussion

Landscape pattern reflects the number of distinctive classes, their sizes, shapes, and the distances between them. While observing how pattern analysis changes as a function of spatial resolution, particular features appear to repeatedly influence patterns observed at coarse resolution levels. A consideration of these factors may aid in determining the adequacy of resolution levels for describing landscapes (Wehde 1982).

The size and shape of land classes influenced their

persistence during changes in spatial scale. Compact-shaped classes had a higher probability of being retained during resolution reductions than those land classes occurring as elongate features. When forest patch size is larger than the current pixel, there is relatively little change in the distribution or proportion of land classes. When pixel size exceeds patch size, patch retention became unpredictable and the spatial relationships between land classes altered.

The spatial arrangement of land classes and forest patches also influenced their retention. Rare land classes distributed in patchy arrangements disappeared more rapidly than did contiguous classes. Classes that were clumped disappeared slowly with increasing pixel size and classes that existed as large, continuous units were retained at coarse levels. In contrast, the loss of pattern information was pronounced at coarse levels when one or more land classes were scattered as small patches or distributed in a noncontinuous manner.

The changes observed during spatial scaling were influenced by landscape complexity. Dale et al. (1989) found that the rate of information loss was greater for complex landscapes; however, the loss of spatial heterogeneity was more striking on simple landscapes. The addition or omission of a single land class on a simple landscape resulted in substantial changes in both the Simpson and evenness indices. The interspersion and binary comparison matrix metrics regularly overestimated the heterogeneity of simple landscapes, suggesting that scale problems occur in spatially homogeneous landscapes.

There were pattern metrics that were not influenced by landscape complexity. The rate of pattern loss for the fragmentation indices, percent forest cover, and the patch metrics was influenced by factors other than complexity (e.g., the distribution of land classes and the pixel-patch size relationships).

There were instances where pattern loss was greater on complex landscapes (e.g., forest-edge length, convexity, forest interior, and forest contiguity). These landscapes quickly lost edge and shape detail with the continual smoothing of land class boundaries that occurred during changes in scale. These metrics were also influenced by the shifting distribution of forest accompanying boundary changes.

General Trends in Landscape Pattern Change

Modeling spatial scale is a homogenizing process from which emerge several trends. A pattern metric may either increase or decrease depending upon the characteristics of a landscape image at a particular resolution. Spatial scaling systematically favors land classes for retention or exclusion based on their area and location. The sampling process also excludes land classes based on their characteristic size and shape. If a forest patch or other land class area drops below the resolution of the landscape image, the resulting pattern description is modified dramatically.

Landscape features that do remain after the scaling process are methodically altered in size and shape. The subsequent analysis of these generalized images results in land class omission or the exaggeration of areas in relation to their true areas on the ground.

Much of the change observed with transformation of spatial scale centers on the loss of discernable detail and includes the following: removal of small inclusions and patches results in reassignment to another land class; enlargement of these features results in area overestimation. Spatial heterogeneity is averaged out, masking localized variability. Boundary smoothing results in the loss of edge and shape detail; the shifting of boundaries distorts land class proportions as edge pixels are transformed from one land class to another. Pixel redistribution influences the connectivity of forest and other land classes. The spatial arrangement of land classes is altered, while the relationships between like pixels are obscured (Chou 1991).

The changes observed may vary with the generalization algorithm. However, information will be lost when using taxonomic, spatial, or sampling generalization. With reduction in resolution, error is introduced by discarding different features of the original pattern (Hole and Campbell 1985); each change cannot ensure that landscape features will be represented at coarse resolution levels.

Variability of Pattern Metrics and the Semivariogram

Modeling variance may illuminate the influence of spatial resolution on landscape pattern, while the use of mean estimates for pattern metrics may mask landscape variation. The timing and magnitude of variability with respect to spatial resolution provides useful information for landscape analysis. Several approaches are based on the observation that variance increases as transitions are approached in hierarchical systems (O'Neill et al. 1986). If variance is measured as a function of spatial resolution, increased variance may indicate the approach of a change in pattern (Turner et al. 1989b).

Metric variability associated with changing grain size may define the limits within which meaningful extrapolation is possible. Extrapolation error will differ among the metrics selected based on their behavior. The semivariogram assesses metric constancy; the sill and the range together serve to identify scales where extrapolation is possible and intervals producing comparable landscape descriptions.

Semivariograms have been used for describing the variation in vegetation structure (Palmer 1988) and digital imagery (Weishampel et al. 1992); for detecting patterns in canopy structure (Cohen et al. 1990); for optimal resolution determination (Atkinson et al. 1990); and for mining applications (Curran 1988). In this study, the semivariogram has proven to be a useful tool for identifying how spatial resolution influences landscape pattern description.

The semivariogram parameter, y_h, describes the magnitude of a metric's variation, while the form of the curve provides insight into the nature of that variation. It is useful for identifying metrics that change direction predictably and those that are unpredictable. In addition to identifying the inherent variation within spatial scaling, semivariogram analysis may suggest the sampling resolution required for characterizing landscape pattern.

Several pattern metrics may be candidates for extrapolation to other scales. The predictable behavior displayed by the binary comparison matrix, interspersion, Simpson, and evenness metrics suggests their potential for extrapolation. Other metrics showing consistent rates of change with bounded variance over partial intervals include forest contiguity, convexity, and forest edge length. It is possible that these metrics could be extrapolated between 30 and 240 meters.

Semivariogram analysis also identified intervals where the semivariance was stable as grain size changed. These intervals suggest the potential for extrapolation within that interval, presenting the limits of tolerance for choosing resolution level. For example, landscape description was equivalent between 30 and 90 meters for forest cover and the number of patches.

The use of semivariograms identifies metrics that exhibit dramatic variability during the spatial scaling process. For metrics demonstrating unbounded variance (e.g., fragmentation indices I and II), landscape description cannot be made at one spatial resolution based on another. The variability observed for percent forest interior and forest cover values at different levels and spatial resolution indicates that the choice of resolution needs to be carefully considered prior to their use for landscape analysis. Patch metric variability indicates that pattern misrepresentation can occur at coarse resolution levels; the semivariograms imply the potential error associated with extrapolation may be great across spatial scales. Examining the semivariograms of landscape metrics demonstrates the importance of using variance to discriminate metric usefulness at different spatial scales.

Spatial Scale and Wildlife Habitat Assessment

The landscape metrics examined here are valuable descriptors of wildlife habitat. Their behavior at different spatial scales highlights the importance of understanding how pattern influences species occurrence. Only at the proper spatial scale do metrics have potential meaning for resource managers. A metric that overestimates a desirable habitat component at coarse resolution levels concludes in an inflated assessment of habitat potential. Similarly, a metric that minimizes undesirable habitat components results in an overrating of habitat condition.

One objective of habitat relationship research is to identify biologically important variables that have the ability to predict species occurrence (Young and Hutto, Chapter 8). Habitat quality for many species contains a spatial component related to the arrangement and amount of habitat elements across the landscape. If the spatial scale of measurement alters the values of these variables, this influences our ability for predicting species occurrence and for assessing habitat conditions.

TABLE 11.2.

The influence of spatial scale on landscape pattern description and wildlife habitat evaluation.

Species	Common name	Key reference[a]	Landscape metric	Influence on — Metric[b]	Influence on — Evaluation
Canis latrans	Coyote	Thomas 1979	Simpson Index	Increase	Overestimation
Urocyon cinereoargenteus	Gray fox	Fritzell 1987	Binary Comp. Matrix	Increase	Overestimation
Phasianus colchicus	Pheasant	Yahner 1988	Interspersion	Increase	Overestimation
Accipiter nisus	Sparrowhawk	Hunter 1990a	Forest Edge Length	Decrease	Underestimation
Catharus fuscescens	Veery	Hamel 1992	Forest Edge Length	Decrease	Overestimation
Martes pennati	Fisher	Rosenberg and Raphael 1986	Fragmentation Index I	Increase	Underestimation
Meleagris gallopavo	Eastern wild turkey	Gustafson et al. 1994	Distance to Nonforest	Increase	Underestimation
Martes americana	Marten	Bissonette et al. 1989	Distance to Nonforest	Increase	Overestimation
Seiurus aurocapillus	Ovenbird	Sweeney and Dijak 1985	Percent Forest Cover	Variable	Unpredictable
Wilsonia citrina	Hooded warbler	Whitcomb et al. 1981	Percent Forest Interior	Decrease	Underestimation
Dryocopus pileatus	Pileated woodpecker	Schroeder 1983b	Mean Patch Size	Variable	Unpredictable
Streptopelia turtur	European turtle dove	Van Dorp and Opdam 1987	Interpatch Distance	Variable	Unpredictable
Ursus americanus	American black bear	Rudis and Tansey 1995	Forest Contiguity	Decrease	Underestimation

[a]Source of observations citing the importance of landscape pattern.

[b]Influence of spatial scale when measured from fine to coarse levels of resolution.

Table 11.2 suggests how landscape description at different spatial scales may influence wildlife habitat evaluation. The examples indicate how failure to consider spatial scale may result in misleading habitat assessments. For instance, coyotes (Canis latrans) and gray fox (Urocyon cinereoargenteus) use a diversity of habitats. Both species hunt in open shrub habitat for small prey and den in downed logs in old-growth stands. At coarse resolution levels, habitat suitability is overestimated using spatial diversity metrics such as the binary comparison matrix and Simpson indices.

The arrangement and distribution of vegetative communities (interspersion) is important to several species. Pheasants forage in agricultural fields, preferring fields juxtaposed with hay crops for nesting and roosting cover. Coarse resolution levels overestimate the value of this landscape characteristic.

Forest edge length reflects the transition between different community types. Edge characteristics are also important descriptors for raccoon (Procyon lotor)

habitat. This species uses edge habitat for foraging and travel (Pedlar et al. 1997). The European sparrowhawk (Accipiter nisus) and the Eurasian badger (Meles meles) are also frequently associated with edge. Habitat suitability is underestimated for each of these species when landscapes are analyzed using relatively large pixels.

Fragmentation occurs when a tract of forest or other habitat is broken into patches. This process influences a number of species. Fisher (Martes pennanti), gray fox, and ringtail cat (Bassariscus astutus) are sensitive to fragmentation (Rosenberg and Raphael 1986). Large pixels result in significant increase in the distance-to-nonforest and other fragmentation indices, introducing error into habitat assessment for these species. The hooded warbler, the worm-eating warbler (Helmitheros vermivorus), and the black-and-white warbler (Mniotilta varia) are also intolerant of fragmentation. The habitat for these forest interior specialists is underestimated at coarse resolution levels.

The persistence of many species may be linked to the number, size, and degree of isolation of forest patches. Patch isolation is often a significant predictor of relative abundance for many bird species. However, habitat suitability assessment is unpredictable using patch metrics derived from large pixels. Habitat connectivity facilitates dispersal and enhances habitat quality by connecting patches of critical habitat. Forest contiguity is an important explanatory variable for predicting habitat quality for black bear (*Ursus americanus*). Landscape analysis at coarse resolution levels reduces contiguity estimates, underestimating habitat suitability for this species.

Several studies have examined the influence of spatial scale on habitat assessment. Applying a marten (*Martes americana*) habitat model across several resolution levels, Schultz and Joyce (1992) documented changes in home range suitability prediction. Increasing pixel size resulted in simplified landscapes and the loss of small forest patches; the ability to recognize pattern detail was lost. The mean area of suitable home range was greater at fine resolution levels. Schultz and Joyce (1992) found that changes in forest patch size and distribution modified habitat assessment.

Laymon and Reid (1986) investigated the influence of resolution level on a spotted owl (*Strix occidentalis*) habitat suitability model. Using 4-hectare pixels resulted in accurate predictions of home range use when compared to known owl locations. In contrast, 16-hectare pixels masked small habitat pockets of important value. The usefulness of model prediction declined appreciatively with the coarseness of spatial scale. In another study on home range estimation, Hansteen et al. (1997) assessed the effect of spatial resolution (0.06–100-meter pixels) on estimates of tundra vole (*Microtus oeconomus*) occurrence. Polygon estimates of home range size increased substantially as a function of resolution; large pixel sizes produced estimates of home range area in locations where voles were absent.

Johnson et al. (Chapter 12) reported that spatial scale influenced variable approximation relative to amphibian occurrence. Both local habitat parameters and landscape structure (fragmentation and edge contiguity) were important for treefrog prediction. The relationship between species occurrence and explanatory variables was a function of the scale of analysis.

Stoms (1992) examined the effects of resolution change on species richness prediction in Sierra Nevada landscapes. As pixel size increased, there was a decline in the number of mapped habitat types and the number of predicted species. When landscape elements dropped below the resolution of the coverage, richness dropped dramatically. The use of 100-hectare pixels resulted in a substantially different pattern of species richness than that observed for smaller pixels.

MacNally and Quinn (1998) found that monitoring striated thornbill (*Acanthiza lineata*) at different spatial resolution levels (15–45-hectare pixels) caused dramatic fluctuations in density prediction. In a similar example, Haufler et al. (1999a) discuss a study that compared mapping resolution using 1,000-hectare and 9.5-hectare pixels. Using fine resolution levels, sixty-nine community types were detected that supported ninety-eight bird species. In contrast, twelve community types were identified (supporting sixty-seven species) using 100-hectare pixels.

The choice of spatial scale has a profound influence on both the strength and the nature of wildlife-habitat relationships. Analyzing landscapes at an inappropriate scale may not accurately describe existing pattern but instead reflect artifacts of the scale of measurement. Whether the scale at which landscape characteristics are measured matches the scale at which those attributes influence wildlife species remains an important question. The measurement of spatial pattern in turn has implications for the prediction of species occurrence.

Selecting the Appropriate Spatial Scale

Landscape pattern description is but one source of uncertainty for predicting species occurrence (Maurer, Chapter 9). Measurement error in landscape predictor variables can bias metric values leading to unreliable predictions of species occurrence. Pattern analysis is most useful when the scale of analysis matches the scale at which species use the landscape. For example, Cogan (Chapter 18) discusses critical biodiversity sites that are detectable only when using fine resolution levels. These remnant areas have high habitat value and are important for predicting rare plant populations.

It is unlikely that a single scale of spatial resolution can be identified for all species in a given landscape (MacNally and Quinn 1998). Consideration should be given to species needs (e.g., habitat preferences and the distribution of those habitats). For small mammals, the arrangement and patchiness of herbaceous understory may be essential, while for a gray fox, the interspersion of vegetation communities on a landscape may be important. Determining the appropriate scale for describing pattern and for predicting species occurrence should consider the differences in the home range sizes, geographic distributions, and habitat requirements. Laymon and Reid (1986) reported that a pixel size comprising 1–2 percent of home range was useful for predicting spotted owl habitat. For other species, the need for travel corridors or a diversity of forest types may suggest another scale of analysis. The heterogeneity of the landscape is yet another consideration. Young and Hutto (Chapter 8) detected fine resolution patterns of occurrence within cover types for Swainson's thrush (*Catharus ustulatus*). Differences in site occupancy reflected fine-scale variation in habitat characteristics. This knowledge could be used to guide managers toward an appropriate spatial scale for this species.

It is evident that species habitat requirements (e.g., minimum area limitations) will affect the choice of spatial scale. Species exploit resources within a different range of scales; large species have home ranges encompassing hundreds of kilometers, while small species use modest ranges. The choice of spatial scale (grain and extent) for wildlife habitat evaluation should reflect species range, seasonal area use, and the landscape characteristics influencing its lifestyle.

There is no single correct scale for describing a system (Levin 1992). The appropriate scale will depend on management objectives, the species involved, the underlying landscape pattern, and the processes believed to be important. If the assessment focuses upon interior species habitat, the loss of boundary detail accompanying coarse scales may yield unacceptable results. The conservation of rare habitats dictates fine resolution analyses whereas large-area analyses may

suffice for monitoring wetland loss. Patterns of habitat association for many species may be dependent on the scale of investigation. Different processes will be important at different spatial scales.

Management Implications

Landscape analysis is a compromise between spatial scale and the accuracy of pattern description. Often the resolution of mapping data is based upon available funds, storage, or data availability. However, it is important to consider the level of effort and the quality of the resulting analysis. Landsat imagery (30-meter pixels) may adequately differentiate pattern for many species but may be too coarse for small mammals or too detailed for black bear habitat assessment.

The ability to assess environmental conditions will reflect the resolution of the data used for analysis. The changes that occur as resolution is reduced may not produce acceptable results; analyzing patterns at coarse resolution levels accompanies the loss of localized variability and boundary detail. A manager should recognize how much detail will be retained during landscape analysis and whether this detail is sufficient for the species of concern. The loss of detail is related to the number of sampled pixels used for describing a landscape. Selecting resolution presents many of the same problems observed with selecting sample size. Coarse resolution levels result in small samples and associated problems of induction.

This chapter proposes that the relationship between pattern and species occurrence may be altered when applied at different spatial scales. This suggests a number of steps for the resource manager. First, geostatistical techniques such as semivariogram analysis can identify transitions between spatial scales. Second, a sensitivity analysis can identify whether the same pattern emerges at different levels of resolution. If pattern changes dramatically at a particular resolution level, the decision is made to lose the previous level of detail or to reverse course and select a finer resolution level. Spatial analysis, as outlined above, can be useful for selecting the appropriate scale for landscape analysis.

Management objectives are related to ecosystem processes. If an objective is to mimic natural disturbance regimes using silvicultural treatment, fine resolution analyses may be required. Conversely, if a management objective is to describe the pattern of regional defoliation, coarse resolution analyses may be adequate. In either case, spatial resolution is guided by the phenomena of interest and the management objective.

Economic constraints will continue to dictate the nature of landscape analyses; the objective is to maximize the information obtained at an acceptable level of confidence. Spatial scale should be the first consideration in landscape analyses, since scale directly influences the nature of the final results (Meentemeyer and Box 1987). Selecting a spatial scale for providing specific information should accompany consideration of the pattern information excluded from that scale. Understanding those limitations will guide the selection of appropriate scales to meet management needs.

There is a growing demand for landscape analysis, driven by the premise that ecological processes can be predicted by pattern (Gustafson 1998). The relationship between landscape pattern analysis and spatial scale has important implications for both resource management and wildlife habitat evaluation. Selecting the appropriate scale (extent and resolution) may be critical for successful landscape analyses on forested landscapes and for the prediction of species occurrence.

Acknowledgments

Special thanks are extended to Drs. Dean Stauffer, James Campbell, and Timothy Gregoire for their statistical and spatial analysis expertise. Acknowledgment goes to Drs. Robert Giles, James Smith, and Paul Angermeier for their counsel during the initial research. Reviews of an earlier draft by R. H. Giles, J. M. Scott, and two anonymous reviewers are appreciated. The USDA Forest Service, Southern Region, supported this research.

Predicting the Occurrence of Amphibians: An Assessment of Multiple-scale Models

Catherine M. Johnson, Lucinda B. Johnson, Carl Richards, and Val Beasley

Widespread reports of local and regional amphibian population declines recently have focused attention on factors potentially influencing the distribution and abundance of amphibians (see Blaustein and Wake 1990; Pechmann et al. 1991; Wake 1991; Blaustein et al. 1994; Green 1997). These declines are generally attributed to habitat loss and fragmentation; yet, in many cases it is unclear whether declines are human induced or simply represent natural population fluctuations (Pechmann and Wilbur 1994; Blaustein et al. 1994). Furthermore, the status of a given species may be dependent on the spatial scale of interest, with some species appearing to be in decline in a given geographic area while remaining relatively constant at other scales of spatial occurrence (Pechmann and Wilbur 1994; Hecnar and M'Closkey 1997).

The distribution of local amphibian populations may be greatly influenced by site-specific habitat factors (e.g., wetland water chemistry, vegetative composition, and interspecific interactions); however, a broader perspective is required to understand factors affecting regional populations (Hecnar and M'Closkey 1996). Landscape-level land-use practices can have both direct and indirect effects on wetland habitats and amphibian populations (Green 1997; Lehtinen et al. 1999). In regions dominated by intensive agricultural practices and urbanization, wetlands have been destroyed at an alarming rate, with some states losing more than 80 percent of their original wetland acreage since 1780, primarily as a result of agricultural activities (Dahl 1990). Extant wetlands in such stressed landscapes also may be modified to the extent that they become unsuitable for many species as a result of pollution, introduced species, vegetative composition changes, altered hydrologic regimes, or other anthropogenically induced changes.

Most regional amphibian populations could be described as metapopulations (Levins 1969, 1970; Hanski and Gilpin 1991) whose stability is dependent on a balance of population extinction and colonization rates. As such, the extinction of many individual populations due to local or broad-ranging habitat changes, with no concurrent colonization of nearby sites, could have a long-term negative effect on regional populations. The loss and fragmentation of nearby upland habitats, such as forest and grassland, also may render a wetland unsuitable for many amphibian species, especially those that spend most of their life cycle in those habitats. Furthermore, changes in broad land-use patterns that affect the ability of individuals to move between local amphibian populations could ultimately result in the loss of a given species over a broad geographic area. Such movements appear to be facilitated by certain landscape features, such as stream corridors, and obstructed by others, such as roads and highly fragmented habitats

(Seburn et al. 1997; deMaynadier and Hunter 1998; Gibbs 1998; Lehtinen et al. 1999).

Since both local habitat parameters and landscape structure could contribute to declines in amphibian populations, a multiscale perspective is needed to obtain a holistic view of amphibian species status. Studies conducted at several spatial scales can provide a better resolution of the scale(s) at which these species are responding to environmental heterogeneity and the interrelationships among scales (Wiens 1989a). However, few studies have examined amphibian populations at both site-specific and large-area scales simultaneously, and assessments of the effects of landscape structure are even more rare. Such an understanding is critical if we hope to predict the potential effects of future land-use changes on regional and global populations.

To assess the relative influence of local habitat variables, landscape structure, and anthropogenic influences on anurans, we analyzed the relationship between species occurrence and habitat parameters at three spatial scales. Scale, in the context of this manuscript, refers to the extent over which observations are made. Three scales of explanatory variables were included in this study: site-specific (i.e., physical habitat parameters), local landscape (within 2 kilometers of a site), and broad landscape (within 10 kilometers). The objectives of our analyses were to: (1) assess the importance of various explanatory variables in predicting species occurrences over a broad geographic area, and (2) compare the predictive abilities of models derived using site-specific data versus landscape data.

Study Area

The study extends from central Minnesota through Wisconsin and northeastern Illinois and is entirely located within the Eastern Broadleaf Forest Eco-Province (Bailey 1983; Fig. 12.1). Current land cover within this region ranges from forest to intensive agricultural or urban land uses.

Site Selection

Thirteen regional watersheds were randomly selected from a list of candidate fourth-order watersheds in the study region. Candidate wetlands (predominantly palustrine emergent, open water, and aquatic bed types) were identified on National Wetland Inventory (NWI) and Wisconsin Wetland Inventory maps. Field reconnaissance was conducted to select four to eight wetlands in each watershed that represented a range in wetland quality (based on visual assessments) and a gradient of surrounding land-use types. Most sites were rejected because they no longer existed, we were unable to acquire permission to access them, or they were threatened by development (e.g., close proximity to new development, or impending sale of the land).

Methods

We conducted night-time anuran calling surveys at each wetland three times during the 1998 and 1999 field seasons (in March–April, May–June, and late July) and once in 2000 (early April) to determine species richness and relative abundance levels at each site. These surveys were conducted in accordance with the North American Amphibian Monitoring Program (NAAMP) protocol (Lannoo 1998a) to ensure consistency of anuran species richness and abundance data. Each set of calling surveys was initiated in the southern part of the study region and progressed northward, and all wetlands in a watershed were surveyed during the same evening. During each survey, air and water temperature and ambient weather conditions were recorded in addition to the presence and abundance ranking for each frog species.

Daytime visits were made to each site, in conjunction with evening calling surveys, to gather data regarding the physical, chemical, and biological characteristics of the wetlands. Wetlands were characterized in terms of their size, general morphology, hydrologic regime, dominant wetland class, and vegetative composition. Aerial photographs and aerial compliance slides (U.S. Dept. of Agriculture, County Soil Conservation Service Offices) were examined to assess juxtaposition to other wetlands and potential anthropogenic stressors and to describe surrounding land-use types.

Water-quality measurements conducted during each of the surveys included pH, alkalinity, conductivity, ammonia, nitrate, temperature, and dissolved oxygen. These measurements were made immediately upon ar-

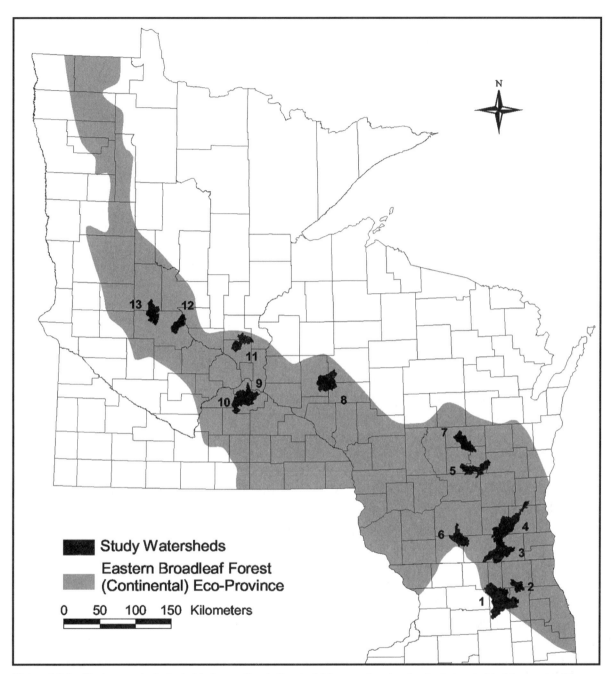

Figure 12.1. Study area, including sixty-two wetland sites in thirteen regional watersheds located in Minnesota, Wisconsin, and northern Illinois.

rival at a given site using a water-quality probe (YSI 600XL sonde with a 610 dm data logger, Yellow Springs Instruments, Yellow Springs, Ohio). Maximum water depth and change in water level since the previous visit also were measured during each site visit.

During 1998, macroinvertebrates were collected using time- and area-constrained dipnet surveys (mod-

ified from Lenat 1988); samples were sorted and identified in the field to the lowest-possible taxon. During the third survey, voucher specimens were collected from each site, preserved in a 10 percent ethyl alcohol solution, and identified to family in the laboratory. Landowner survey data were used in conjunction with dipnet surveys to determine whether fish were present in each wetland.

Amphibian larvae and eggmasses encountered during dipnetting surveys were identified whenever possible. Visual encounter surveys (Crump and Scott 1994) also were conducted for anurans in conjunction with daytime site visits. During June–July 1998 and July 1999, an attempt was made to capture metamorphosing frogs at all sites. These species data were combined with the calling survey, dipnet, and visual encounter survey information to develop a list of frog species present at each site.

Landscape Data

Local and broad-scale spatial data, including land use, land cover, hydrography, wetlands, roads, Quaternary geology, and watershed boundaries were incorporated into a geographic information system (GIS) database. Land cover data were obtained from the Illinois Natural History Survey (Illinois Department of Natural Resources 1996), Upper Midwest Gap Analysis Program (Lillesand et al. 1998), and Minnesota Land Management Information Center. Wetland data were acquired from the National Wetlands Inventory (NWI) and the Wisconsin Wetland Inventory.

The U.S. Census Bureau's TIGER (1995) data were used to describe roads and hydrography. Other regional spatial data were collected from the USGS 7.5-minute digital elevation models (DEMs) and the Quaternary Geologic Atlas of the Conterminous United States. Ancillary landscape data (e.g., finergrain land-use and land-cover data in the immediate vicinity of study sites) were collected during site visits and incorporated into the overall spatial database. Landscape structure, including fragmentation patterns, patch density, connectivity, nearest-neighbor distances, and other landscape metrics, was quantified using FRAGSTATS (McGarigal and Marks 1995).

Analysis

We used Systat 7.0.1 (SPSS 1997) for all statistical analyses. Explanatory variables (i.e., wetland characteristics and landscape variables) were screened using Principle Components Analysis (PCA) and Pearson's correlation coefficients to remove colinearity and reduce the total number of variables used in subsequent analyses. The remaining explanatory variables were checked for normality and appropriate transforma-

tions were applied where necessary. Variables were then separated into three groups: (1) site-specific wetland characteristics, (2) local landscape variables, and (3) large area landscape variables. Site-specific data included both physical and biotic measurements taken during on-site field visits. Because of the diurnal and seasonal variability in water-quality parameters, and because we were interested in relative differences between sites rather than absolute values, we used only ranks of the maximum or minimum values for these parameters in our analyses. Local-scale landscape variables included spatial data within a 1–2-kilometer buffer around each site, collected via aerial-photo interpretation and available state and national digital data sets. Broad-scale landscape variables included similar spatial data within a buffer of 10 kilometers.

To assess whether any explanatory variable was consistently associated with presence or absence of individual anuran species, simple relationships were examined for all species using Spearman rank correlations. Sites were grouped by presence or absence of each species. The Mann-Whitney U-Test was used to test for significant differences in explanatory variables between groups (occupied versus unoccupied sites) for eight of the ten species encountered in this study: the chorus frog (*Pseudacris triseriata*), American toad (*Bufo americanus*), Cope's gray treefrog (*Hyla chrysoscelis*), eastern gray treefrog (*Hyla versicolor*), spring peeper (*Pseudacris crucifer*), wood frog (*Rana sylvatica*), leopard frog (*Rana pipiens*), and green frog (*Rana clamitans*). Analyses of wood frog occurrences were confined to data from eight watersheds located in northern Wisconsin and Minnesota because of this species' limited geographic range in the study area (Casper 1996; Harding 1997). The ranges of the bullfrog (*Rana catesbeiana*) and mink frog (*Rana septentrionalis*) each incorporated only three to five of the watersheds in the study area, and so these species were eliminated from all analyses.

To determine whether the habitat relationships observed at the eco-province scale (i.e., our entire study area) also occurred over smaller areas, we reexamined the species occurrence data within two geographic regions, hereafter referred to as bioregions, based on amphibian community faunal regions as described by Brodman (1998). The southern grouping included

TABLE 12.1.

Species occurrences (presence or absence within each site) by watershed, from April 1998 to April 2000.

Watershed (n[a])	Bufame[b]	Hylchr	Hylver	Psucru	Psutri	Rancat	Rancla	Ranpip	Ransep	Ransyl
1 (5)	5	1	0	1	5	3	3	1	0	0
2 (4)	3	1	0	0	3	3	4	3	0	0
3 (4)	4	4	2	2	3	0	4	4	0	0
4 (8)	6	7	8	6	7	2	8	6	0	0
5 (4)	3	2	3	3	3	0	4	2	0	1
6 (4)	1	3	0	4	3	0	4	4	0	0
7 (3)	2	1	1	3	3	0	2	2	0	3
8 (5)	3	1	4	4	3	0	4	4	1	3
9 (5)	4	4	2	1	5	0	4	5	0	1
10 (4)	2	1	0	1	3	0	1	4	0	1
11 (7)	3	6	5	7	7	0	5	6	0	4
12 (4)	2	2	2	3	2	0	1	4	1	2
13 (5)	3	5	3	2	3	0	2	5	2	5
All Sites (62)	41	38	30	37	50	8	46	50	4	20

[a]n refers to the number of sites in each watershed.

[b]See note in table 12.2 for Latin names.

thirty-two sites in seven watersheds located in Illinois and southern and eastern portions of Wisconsin. The northern group consisted of thirty sites in six watersheds, including all those in Minnesota and the most northwestern watershed in Wisconsin.

In addition to examining correlations among species occurrences and explanatory variables, habitat occupancy models were developed for the spring peeper, gray treefrog, and wood frog using logistic regression. These species were selected for model development because the other species encountered occurred at either very few (20 percent or fewer) or most sites (80 percent or more), or were habitat generalists. Since we wanted to develop parsimonious models that could easily be interpreted, we considered only main effects in the model. Beginning with a model that included all explanatory variables remaining following exploratory analyses, we manually removed and added variables to achieve a best-fit model, similar to an automated stepwise regression procedure. Significance of variables included in the final model for each scale was assessed at an alpha level of 0.05 (Hosmer and Lemeshow 1989). Goodness-of-fit of each logistic model was reflected by both the percentage of sites correctly classified (the cor-

rect classification rate or CCR) and the McFadden's Rho-squared value (ρ^2). Three scale-specific models were developed for each species using each of the three groups of explanatory variables (i.e., site-specific, local-landscape, and broad-landscape scales), as well as one that incorporated variables from all three scales (Allvars). The models were compared for each species to determine the ability of data from each scale, and from all scales combined, to predict occurrence. Final models were tested against a reserve data set of sixteen sites added to the study in 1999.

Results

Ten anuran species were observed in the sixty-two wetlands included in this study. The number of sites these species occurred in ranged from four (mink frog) to fifty (leopard and chorus frogs; Table 12.1). Over two hundred explanatory variables were assembled from field and spatial data collection and spatial data analyses. After removal of collinear variables, the number of explanatory variables was reduced to sixty-two, including eighteen site-specific, twenty-six local-landscape, and eighteen broad-landscape variables (Table 12.2).

TABLE 12.2.

Individual explanatory variables showing a significant difference between sites with or without each of eight anuran species.

Scale	Variable	Bufame	Psutri	Rancla	Ranpip	Hylchr	Hylver	Psucru	Ransyl
Site-specific	Natural				*			**	
	WetlandType								**
	Topography					***			
	Rowcrop50							*	
	Woods50						**	*	***
	MaxdepthDecr								*
	MinWaterTemp			**					
	MinpH			***				*	**
	MinConductivity						**	*	
	MaxAmmn			*					*
	MinDOsurf				**				
	TotalInsect				*				
	Fish								**
2 Km	AWMeanShapeIndex				***	*			
	SimpsonEvenIndex				*		*	*	
	ContagionIndex				**		*		
	PatchLandSimilIndex	*							
	PatchEdgeContrastIndex						*	*	*
	EmergentWet1k	*			**				
	PalustrineWet2k						*	**	
	EmergentWet2k	**							
	PalustrineWetCount1k			**					
	PatchDensityAg								
	EdgeContrastIndexAg							**	
	PatchFractalDimAg					*			
	Intersp/juxtaposAg						***	*	
	PercentUrban				*				
	PatchDensityUrban				**				
	Intersp/juxtaposUrban						**	**	
	PercentForest						**	**	
	CoreAreaCVForest			**					
	Intersp/juxtaposForest			*					
	PercentWetland	**					**	**	
	PatchDensityWetland						**	**	
	PatchFractalDimWetland			**					
	PatchFractalDimOW				**			**	
	CoreAreaCVOW								
10 Km	No. Patches					**	**		
	ContrastWtEdgeDensity				*				
	MeanShapeIndex				**	**			
	NearestNeighborCV			**			**		
	Intersp/juxtaposLandsc					*	***	*	
	PalustrineWet10k					*	***	***	
	Freeway	*		*		*			
	LocalRoads					***	***		
	PatchFractalDimHDUrban					**			

TABLE 12.2. (Continued)

Individual explanatory variables showing a significant difference between sites with or without each of eight anuran species.

Scale	Variable	Bufame	Psutri	Rancla	Ranpip	Hylchr	Hylver	Psucru	Ransyl
10 Km	Intersp/juxtaposHDUrban		*	**	**		*		
	PatchDensityAg								*
	CoreAreaCVGrass					*		**	
	PercentForest						**	***	
	PatchDensityForest			***	**				
	Intersp/juxtaposForest						***		*
	PercentOW					*	***	*	
	PatchDensityOW						***	***	*
	EdgeContrastIndexOW			***					

Note: Mann-Whitney *p*-values: $P \le .05 = *$, $P \le .01 = **$, and $P \le .001 = ***$. Scale refers to the spatial extent from which the variables were derived: site-specific, local landscape (2 kilometers) or broad landscape (10 kilometers); descriptions of each variable are given in the Appendix.
Bufame = American toad, *Bufo americanus*
Psutri = chorus frog, *Pseudacris triseriata*
Rancla = green frog, *Rana clamitans*
Ranpip = leopard frog, *Rana pipiens*
Hylchr = Cope's gray treefrog, *Hyla chrysoscelis*
Hylver = eastern gray treefrog, *Hyla versicolor*
Psucru = spring peeper, *Pseudacris crucifer*
Ransyl = wood frog, *Rana sylvatica*

Individual Species-Habitat Relationships

Spearman rank correlations between each species and explanatory variable are presented in Table 12.1. Although no two species exhibited the same pattern of relationship with all variables, five species did show similar patterns for many variables. The treefrogs, spring peeper, wood frog, and leopard frog had similar relationships (i.e., either negative or positive) with 47 percent of the variables assessed. Green-frog occurrences were inversely related (i.e., had an opposite sign) for the majority of the variables that the aforementioned species had in common.

Results of the Mann-Whitney U-test indicated that the occurrence of six species was significantly associated with variables in all three spatial scales (Table 12.2). The exceptions were the American toad, which showed no strong correlation with any of the local variables measured and with only one of the 10-kilometer variables, and the chorus frog, which exhibited a significant difference in occurrence associated with only one explanatory variable. The spring peeper and gray treefrog exhibited many similarities in the pattern of relationships with explanatory variables, especially with the landscape metrics.

When the study area was split into northern and southern bioregions for analysis, no consistent relationship between a given species occurrence and the explanatory variables was observed for the two regions at either the site-specific or the local-landscape scale. Significant relationships between explanatory variables at these scales were almost completely different for the two regions examined for both the spring peeper and green frog (Table 12.3). Only the broad-scale landscape variables (i.e., 10 kilometers) showed some similar relationships for the two groupings.

Site Occupancy Models

The best-fit logistic regression models for individual species had McFadden's Rho-squared (ρ^2) values ranging from only 0.09 to 0.64 and correct classification rates (CCRs) ranging from 56 to 85 percent (Tables 12.4–6). The spring peeper CCR values for the site-specific, local landscape, and broad-landscape models were 57, 79, and 68 percent, respectively (Table 12.4).

TABLE 12.3.

Explanatory variables with significant Mann-Whitney associated with the occurrence of spring peepers (*Pseudacris crucifer*) or green frogs (*Rana clamitans*) in two ecoregions.

Scale	Spring Peeper			Green Frog		
	Variables	NW	SE	Variables	NW	SE
Site-specific	Natural	**		WetlandType		**
	Woods50		*	MinWaterTemp	**	
	MinpH	*		MinpH	*	
	MinConductivity		*	MaxAmmn	*	
	MaxNitrate	*		MinDOsurf	*	
	TotalInsect	*				
2 Km	AWMeanShapeIndex		**	EmergentWet1k		*
	PatchEdgeContrastIndex	**		EmergentWet2k	*	
	PalustrineWet2k	**		PalustrineWetCount1k	*	
	EdgeContrastIndexAg		**	EdgeContrastIndexAg	*	
	Intersp/juxtaposAg		*	MeanPatchFractalDimAg		*
	Intersp/juxtaposUrban	*	*	Intersp/juxtaposForest		*
	PercentForest		*			
	PercentWetland	**				
	PatchDensityWetland	*				
10 Km	MeanShapeIndex		*	ContrastWtEdgeDensity		*
	NearestNeighborCV		*	NearestNeighborCV	*	*
	Intersp/juxtaposLandsc	*	*	Intersp/juxtaposLandsc		*
	PalustrineWet10k		**	Intersp/juxtaposHDUrban		*
	PatchfractlDimHDUrban		*	PatchDensityForest	*	
	CoreAreaCVGrass		**	EdgeContrastIndexOW	*	
	PercentForest	**	**			
	PatchDensityForest		*			
	PatchDensityOW	**	***			

Note: *NW* refers to sites in Minnesota and northwestern Wisconsin (*n* = 30); those in the southeast include SE Wisconsin and Illinois (*n* = 32). *Scale* refers to the spatial scale from which the explanatory variables were derived; descriptions of variables are given in the Appendix. *P*-values (*P* ≤ .05 = *, *P* ≤ .01 = **, and *P* ≤ .001 = ***).

However, when explanatory variables for all three scales were incorporated into model development, the CCR increased to 85 percent; the general fit of the model (as described by the ρ^2 value) also increased. This model also correctly classified 87.5 percent of sites in the reserve data set (Table 12.7). This final model (Allvars) included both site-specific variables (artificial versus natural wetland and distance to the nearest wetland) and landscape variables (number of wetlands within 1 kilometer, wetland patch and agricultural edge contrasts, landscape diversity, and urban patch interspersion). In comparing the different scale models, the local landscape (1–2 kilometer) model was the best predictor of spring peeper presence, correctly classifying 21 percent more of the sites than did the local model.

The site-specific scale model for the gray treefrog, like that for the spring peeper, was the poorest predictor of species site occupancy (56.4 percent; Table 12.5). However, while the local landscape model was the best individual-scale predictor for the spring peeper, the broad landscape variables provided a better model for predicting gray treefrog occurrence (71.7 versus 60.6 percent). The Allvars model for this

TABLE 12.4.

Results of four habitat occupancy models for spring peepers (*Pseudacris crucifer*), developed using variables from three discrete spatial scales, and a combined set of variables.

Scale	n	ρ^{2a}	CCR[b]	Variable[c]	Sign	P
Site-specific	62	.087	57.3	MinConductivity	−	**
2 Km	62	.509	78.8	SimpsonEvenIndex	−	***
				Intersp/juxtaposUrban	+	**
				PalustrineWetCount1k	−	**
				EdgeContrastIndexAg	+	**
				EdgeContrastIndex	−	**
				PalustrineWet2k	−	*
				PercentWetland	+	*
10 Km	62	.266	67.7	PatchDensityOW	+	***
				CoreAreaCVGrass	+	***
Allvars	62	.644	85.1	EdgeContrastIndexAg	+	**
				PalustrineWetCount1k	−	**
				SimpsonEvenIndex	−	*
				PatchEdgeContrastIndex	−	*
				Intersp/juxtaposUrban	+	*
				WetlandDist	+	*
				Natural	+	*

Note: Explanatory variables and their individual *p*-values are presented for each model.

[a]McFadden's rho-squared (ρ^2) is an estimate of the fit of the overall model.

[b]CCR represents the percentage of correctly classified sites.

[c]Descriptions of variables are given in the Appendix.

TABLE 12.5.

Results of four habitat occupancy models for gray treefrogs (*Hyla versicolor*), developed using variables from three discrete spatial scales, and a combined set of variables.

Scale	n	ρ^{2a}	CCR[b]	Variable[c]	Sign	P
Site-specific	62	.294	56.4	Woods50	+	**
2 Km	62	.176	60.6	Intersp/juxtaposAg	+	***
10 Km	62	.387	71.7	Intersp/juxtaposLandsc	+	**
				PercentOW	+	*
Allvars	62	.433	74.6	Intersp/juxtaposLandsc	+	***
				Woods50	+	**

Note: Explanatory variables and their individual *p*-values are presented for each model.

[a]McFadden's rho-squared (ρ^2) is an estimate of the fit of the overall model.

[b]CCR represents the percentage of correctly classified sites.

[c]Descriptions of variables are given in the Appendix.

species included only two variables, one from the site-specific scale (presence of woods within 50 meters of the wetland) and one from the broad-landscape scale (landscape patch interspersion); this multiscale model again provided the best ρ^2 and CCR values (0.43 and 74.6 percent). Similar results (CCR = 75 percent) were obtained when tested against the reserve data set (Table 12.7).

The wood-frog models showed a different pattern than those of the spring peeper and gray treefrog. The

TABLE 12.6.

Results of four habitat occupancy models for wood frogs (*Rana sylvatica*) developed using variables from three discrete spatial scales and a combined set of variables.

Scale	n	ρ^{2a}	CCR[b]	Variable[c]	Sign	P
Site-specific	38	.230	64.8	Woods50	+	**
2 Km	38	.087	56.4	PatchEdgeContrastIndex	−	**
10 Km	38	.136	58.7	PatchDensityOW	+	**
				Intersp/juxtaposForest	−	*
Allvars	38	.528	79.3	WetlandType_2	+	*
				Woods50	+	**
				PercentUrban2k	−	*

Note: Because of the wood frog's geographic range, only sites in watersheds 5 and 7–13 were used for this analysis. Explanatory variables and their individual p-values are presented for each model.

[a]McFadden's rho-squared (ρ^2) is an estimate of the fit of the overall model.

[b]CCR represents the percentage of correctly classified sites.

[c]Descriptions of variables are given in the Appendix.

percentage of correctly classified sites was higher for the site-specific model (64.8 percent) than for either the local-scale (56.4 percent) or broad-scale landscape model (58.7 percent; Table 12.6), though none of the individual-scale models performed well. All three of these models were relatively simple; the site-specific and local-landscape models each included only one explanatory variable (woods within 50 meters of the wetland, and patch edge contrast, respectively), and the broad-landscape model included two (patch density of open water, and the forest interspersion and juxtaposition index). The combined model resulted in a much higher correct classification rate (79.3 percent) than any of the individual scale models and incorporated both site-specific and local-landscape scale variables (wetland type, presence of woods within 50 meters, and percentage of urban land within 2 kilome-

ters). This model also correctly predicted wood frog presence on 75 percent of the sites added in 1999 (Table 12.7).

Discussion

The eight frog species considered in our analyses exhibited a wide range of associations with habitat variables at all three spatial scales examined (site-specific, local-scale, and broadscale landscapes). The diversity of these relationships indicates that no single environmental factor or set of factors from those examined has an overriding influence on all or even most anuran species in this region. Although it is not surprising that each species is associated with a different suite of site-specific wetland characteristics, the variation in response to landscape variables, especially those related

TABLE 12.7.

Percentage of sites correctly classified (CCR) using the final (Allvars) habitat occupancy models for spring peepers (*Pseudacris crucifer*), gray treefrogs (*Hyla versicolor*), and wood frogs (*Rana sylvatica*).

Common name	Species code	Model data		Test data[a]	
		n	CCR	n	CCR
Spring peeper	psucru	62	85.1	16	87.5
Gray treefrog	hylver	62	74.6	16	75.0
Wood frog	ransyl	38	79.3	16	75.0

[a]Test data refer to those sites added to the study in 1999.

to anthropogenic disturbances (e.g., indices of habitat fragmentation and road densities), was somewhat unexpected.

Increased habitat fragmentation resulting from habitat destruction is considered one of the most important factors causing amphibian declines in industrial regions (Blaustein et al. 1994). However, measures of habitat fragmentation should be viewed cautiously in terms of their relationship to species distributions. The landscape included in our study area is dominated by agricultural land (Fig. 12.2), with wetlands, forested areas, and other "natural" habitats generally distributed as smaller patches throughout an agricultural matrix. The juxtaposition of many, relatively small nonagricultural patches interspersed within the agricultural matrix (rather than just a few larger "natural" patches) may benefit the long-term survival of local populations by decreasing the distance frogs must travel across a hostile environment (e.g., agricultural fields) to reach other suitable habitats. Viewed in this light, the strong positive relationship between landscape patch interspersion and the occurrence of species such as the spring peeper and gray treefrog is consistent with what is known about the biology of these species. Although these anurans appeared to favor a more highly fragmented landscape in this region, their presence also was associated with a low degree of edge contrast (i.e., the land immediately surrounding the wetland was not agricultural or urban). Both species also exhibited a strong tie to forest cover at all three spatial scales.

The results of these analyses indicate that use of wetlands as breeding habitat by many anuran species in the midwestern United States is dependent on explanatory factors operating at various spatial scales. In addition, the significance of the relationship between these explanatory factors and species occupancy is strongly influenced by the geographic regions analyzed. For example, the relationships between spring peeper occurrences and explanatory variables shifted as the spatial scale of analyses changed from the entire study area to the two bioregions assessed (Table 12.3). Although the presence of woods within 50 meters of a site, the percentage of forest cover within 2 kilometers, and the density of forest patches within 10 kilometers all were significantly associated

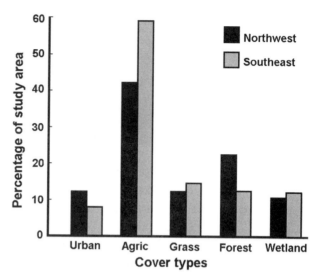

Figure 12.2. Percentages of different land cover types in the study area. The northwest and southeast bioregions are based on amphibian community faunal regions as described by Brodman (1998).

with the presence of spring peepers in the southeastern bioregion, those relationships were not mirrored in the northwestern region. This may be due to the smaller overall percentage of forest cover in the southeastern portion of the study area (Fig. 12.2). Lehtinen et al. (1999) also found differences in species-habitat relationships for the American toad at different geographical scales of analysis. Although they observed a significant relationship between species occurrence and forest cover over the entire study area (twenty-one wetlands in central and southwestern Minnesota), that relationship was not detected when the study area was subdivided into prairie versus forest ecoregions.

The spatial scale of these analyses also may account, in part, for the relatively low correct site classification rates observed in the three final logistic models. Because the relationships between species site occupancy and the explanatory variables differed greatly when viewed at a smaller geographic scale (i.e., southeastern and northwestern bioregions versus the overall study area), it seems likely that different pressures are acting on the amphibian communities in each region. The significance of variables that have a strong influence on species occurrence at a finer scale (e.g., the northwestern sites only), could be lost when sites in a different landscape matrix (e.g., southeastern sites) are included in the analysis.

Conversely, analyses that are limited to more-restricted geographic areas may miss broad-scale landscape patterns that influence the ability of individuals to disperse between wetland habitats and maintain regional populations.

The predictive ability of our models probably could be improved with additional data, since our models were based on only two years of survey data and one year of environmental data. In a three-year study, Hecnar and M'Closkey (1997) found high turnover rates for green frogs in individual wetlands, with some sites occupied only one in three years. Similarly, Skelly et al. (1999) found that changes in the distribution of fourteen amphibian species were surprisingly common between surveys conducted in 1967–1974 and 1988–1992. The physical and chemical properties of wetlands also may vary dramatically between years as a result of stochastic events, such that assessments based on one year of data are not necessarily characteristic of the general condition of the wetland. Finally, we may have failed to incorporate, or used an inappropriate scale of measurement for, variables that are critical to site selection by these species. Given the importance of interpopulation proximity to the persistence of amphibian metapopulations (Sjögren-Gulve 1994), data regarding the distance to other occupied wetlands probably would increase the predictive ability of our models, as might inclusion of occurrence or abundance data for other species, especially potential predators or competitors. In addition, more detailed habitat data (e.g., dominant vegetation types, soil chemistry data, or continuous water chemistry measurements) could prove important in discerning differences between wetland habitats.

Since there is no single correct scale at which models should be developed, we must determine the appropriate scale of spatial analysis based on the problem or question at hand (Wiens et al. 1986; Wiens 1989a; Levin 1992). Many of the anuran species included in this study have both site-specific wetland requirements (e.g., wetland hydroperiod or water-quality tolerances) and surrounding landscape requirements (e.g., wooded habitat within dispersal distance); therefore, a multiscale approach is necessary to understand the mechanisms controlling the abundance and distribution of these species' populations (Goodwin and Fahrig 1998). The results of the logistic models developed in this study support such an approach, indicating that anuran species occurrence in wetlands can best be predicted with a combination of variables based on different spatial scales. Our analyses also indicate the importance of considering variables associated with landscape structure (e.g., fragmentation and edge contrast) in addition to more traditional landscape measures (e.g., percent cover). The interpretation of such variables should be made with caution, however, and should include careful consideration of the consistency of the landscape matrix across the region studied.

Given the broad scale of our analyses, these models are of limited use to managers attempting to predict species occurrences at individual wetlands. As has been stressed repeatedly in this book (e.g., by Van Horne, Maurer, and Wiens [Chapters 4, 9, and 65, respectively]), the results of our study point to the limitations of predictive models in their spatial application. Relationships between species and individual explanatory variables varied when assessed at different spatial scales (extent) and when compared across different geographic regions at similar spatial scales. However, informative patterns emerge in the relationships between species occurrence and explanatory variables, and comparisons between geographic regions may be useful. For example, variables that consistently appear as significant predictors at varying spatial scales or geographic regions could be considered important when developing broad-scale conservation plans. Managers concerned about localized population declines, on the other hand, may want to focus on variables whose negative associations with species occurrence are limited to their geographic area of concern.

Acknowledgments

Funding for this study was provided by the U.S. Environmental Protection Agency as part of the STAR grant program (Grant no. GAD R825867). We gratefully acknowledge the contributions of Joseph Murphy, Anna Schotthoefer, and Anton Treml, who collected and processed much of the field data used in this study.

Thanks also are due to Tom Hollenhorst and Connie Host for the acquisition and compilation of the spatial (landscape) data used herein, and to Malcolm Jones and three anonymous reviewers for helpful comments on a previous version of this manuscript. Finally, we thank the many landowners who allowed us access to wetlands on their land. This is contribution number 271 from the Center for Water and the Environment of the Natural Resources Research Institute, University of Minnesota, Duluth.

APPENDIX.

Explanatory variables used in model development, including the abbreviation used and a brief description of each. Refer to McGarigal and Marks (1995) for more detailed explanation of these landscape metrics.

Scale	Variable	Description
Site-specific	Natural	natural vs artificial wetland
	Wetlandtype	dominated by one of three wetland types (1 = wet meadow; 2 = marsh; 3 = shallow open water)
	Size	wetland size
	WetlandDist	distance to nearest wetland
	Topography	subjective ranking of topographic relief around wetland—from 1 (relatively flat) to 4 (steep)
	Rowcrop50	presence/absence of row crops within 50m of wetland
	Woods50	presence/absence of woodland within 50m of wetland
	MaxDepthdecr	max decrease in wetland water depth over the field season
	MinWaterTemp	min wetland water temperature (water temp measured at each of three surveys)
	MinpH	min wetland pH (measured at each of three surveys)
	MinConductivity	min wetland conductivity (measured at each of three surveys)
	MaxAmmn	max ammonium concentration (measured during first two surveys)
	MaxNitrate	max nitrate (measured during first two surveys)
	MinDOsurf	dissolved oxygen just below water surface (measured during last two surveys)
	TotalInsect	number of insect families observed in wetland
	InsectOrder	number of insect orders observed in wetland
	Gastropoda	presence/absence of gastropods in wetland
	Fish	presence/absence of fish in wetland
2 Km	AWMeanShapeIndex	area-weighted mean shape index for the landscape within 2 km of wetland study sites
	SimpsonEvenIndex	Simpson's evenness index for the landscape within 2 km
	ContagionIndex	landscape contagion index for areas within 2 km
	PatchLandSimilIndex	patch landscape similarity index
	PatchEdgeContrastIndex	patch edge contrast index
	Roads1k	weighted index of road density within 1 km
	EmergentWet1k	area of emergent wetland within 1 km
	PalustrineWet2km	area of palustrine wetland within 2 km
	EmergentWet2k	area of emergent wetland within 2 km
	PalustrineWetCount1k	count of palustrine wetlands within 1 km
	PercentAg	percent agricultural land within 2 km
	PatchDensityAg	patch density of agricultural land within 2 km
	EdgeContrastIndexAg	area-weighted mean edge contrast index for agricultural areas within 2 km
	PatchFractalDimAg	area-weighted mean patch fractal dimension for agricultural areas within 2 km
	Intersp/juxtaposAg	index of juxtaposition and interspersion for agricultural areas within 2 km

APPENDIX. (Continued)

Explanatory variables used in model development, including the abbreviation used and a brief description of each. Refer to McGarigal and Marks (1995) for more detailed explanation of these landscape metrics.

Scale	Variable	Description
	PercentUrban	percent urban land within 2 km
	PatchDensityUrban	patch density of urban land within 2 km
	Intersp/juxtaposUrban	index of juxtaposition and interspersion for urban areas within 2 km
	PercentForest	percent forested land within 2 km
	CoreAreaCVForest	patch core area coefficient of variation (variation in size of forested core areas)
	Intersp/juxtaposForest	index of juxtaposition and interspersion for forested areas within 2 km
	PercentWetland	percent of emergent wetland habitat within 2 km
	PatchDensityWetland	patch density of emergent wetlands within 2 km
	PatchFractalDimWetland	mean patch fractal dimension of wetlands within 2 km (complexity of patch perimeters)
	PatchFractalDimOW	mean patch fractal dimension of open water areas (ponds) within 2 km
	CoreAreaCVOW	patch core area coefficient of variation (variation in size of pond core areas)
10 Km	#Patches	number of patches in the landscape within 10 km of wetland study sites
	ContrastWtEdgeDensity	contrast-weighted edge density of landscape within 10 km
	MeanShapeIndex	mean shape index of landscape within 10 km
	NearestNeighborCV	nearest-neighbor coefficient of variation of landscape within 10 km
	Intersp/juxtaposLandsc	index of landscape juxtaposition and interspersion within 10 km
	PalustrineWet10k	area of palustrine wetland within 10 km
	Freeway	measure of highway density within 10 km
	Localroads	measure of local road density within 10 km
	PatchFractDimHDUrban	area-weighted mean patch fractal dimension for high density urban areas within 10 km
	Intersp/juxtaposHDUrban	index of juxtaposition and interspersion for high density urban areas within 10 km
	PatchDensityAg	patch density of agricultural land within 10 km
	CoreAreaCVGrass	patch core area coefficient of variation (variation in size of grassland core areas)
	PercentForest	percent forested land within 10 km
	PatchDensityForest	patch density of forested areas within 10 km
	Intersp/juxtaposForest	index of juxtaposition and interspersion for forested areas within 10 km
	PercentOW	percent of open water habitat within 10 km
	PatchDensityOW	patch density of ponds within 10 km
	EdgeContrastIndexOW	area-weighted mean edge contrast index for ponds within 10 km

Notes: Scale refers to the spatial extent from which the explanatory variables were derived (site-specific: physical, chemical, and biological wetland data derived from field visits; 2 km: spatial data for areas within 1–2 km of each wetland study site; and 10 km: spatial data for areas within 10 km of study sites.

Dynamic Patterns of Association between Environmental Factors and Island Use by Breeding Seabirds

Catherine M. Johnson and William B. Krohn

Habitat requirements for the establishment of seabird colonies include suitable breeding sites, adequate food supplies within foraging range, and safety from terrestrial predators (Birkhead and Furness 1985; Cairns 1992). The presence of a seabird colony on an island indicates that the area meets these requirements and provides "suitable" habitat, whether of high or low quality. However, not all suitable sites may be occupied by a species at a particular time, and unoccupied sites may have substantial conservation value (Wiens 1996b). Since entire seabird colonies may temporarily or permanently abandon islands (Mendall 1936; Drury 1973; Buckley and Downer 1992), the identification of suitable, unoccupied habitats can be important for long-term conservation planning. Habitat occupancy models can be used to identify characteristics associated with suitable habitats. They also may be effective tools for predicting which sites may be used in the future and for examining the potential effects of specific habitat management plans on a given species. However, the predictive ability of such habitat occupancy models may be limited to relatively large-area analyses (Rotenberry 1986).

Since both physical habitat structure and social factors affecting populations may change over time, the results of habitat models must be interpreted carefully with regard to temporal effects (O'Connor 1987b; Wiens 1989a; Krohn 1996; Van Horne, Chapter 4).

The ability of an island and the surrounding aquatic environment to provide suitable habitat for a particular species may change over time as the local environment changes. Although more-plastic species may be able to cope with dramatic changes in the physical characteristics of an area, populations of species that are more closely tied to a specific food resource or habitat type may not be able to adapt to changing resource levels. Habitat models generally are tied to population and resource levels and spatial distributions that exist at a particular time. Though a habitat model may be highly robust for the time period in which it was developed, its ability to predict long-term habitat use is dependent on both changes in the environment and changes in a species' population level over time (Krohn 1996). Thus, habitat suitability or quality assessments made at any one point in time may be misleading when used for long-term conservation planning, especially in the case of highly stochastic environments such as the insular habitats found along the North Atlantic coast.

The number and types of habitats occupied by a species may shift over time, with more varied habitats used as population densities increase (Svardson 1949; O'Connor 1986; Rosenzweig 1991). It is generally assumed that individuals will select the highest-quality available habitats first (Brown 1969b; Fretwell and Lucas 1969). At higher densities, individuals are likely to be found in suboptimal and sink habitats (Van

Horne 1983; O'Connor 1987a; Pulliam 1988; Danielson 1992). As such, analyses of populations at peak densities could mask the existence of habitat preferences, and assessment of habitat quality may be informative only at lower densities (O'Connor 1986). However, since most usable sites are occupied at high densities, habitat occupancy models could be used to identify the characteristics that distinguish suitable sites, of either high or low quality, from unsuitable sites.

Coastal Maine has over three thousand islands and ledges, providing nesting habitat for a variety of seabirds and other waterbirds. The diversity of insular and aquatic habitats in this area and the availability of historic seabird census information make coastal Maine a good location for the development and testing of temporal seabird habitat models. The number of seabirds breeding on Maine's coastal islands has increased steadily since their near extirpation at the beginning of the twentieth century, though populations of some species have declined since the mid-1900s (Drury 1973; Krohn et al. 1992; Johnson and Krohn 1998). The number of islands used by nesting seabirds also has increased; however, seabird colonies still occur on only about 10 percent of the islands along the Maine coast (Conkling 1995). In addition, the availability of specific aquatic food resources (e.g., fish and shellfish) in the Gulf of Maine has changed drastically since the late 1800s (Ojeda and Dearborn 1990; NOAA 1991). The Atlantic cod (*Gadus morhua*), haddock (*Melanogrammus aeglefinus*), halibut and smaller groundfish that supplied a superabundant food resource for cormorants at the beginning of the twentieth century were decimated by the 1970s, and the Atlantic herring (*Clupea harengus*) and capelin (*Mallotus villosus*) populations that provided the primary food supply for the large alcid colonies along the Newfoundland coast were reduced to residual levels by the early 1980s (Mowat 1984).

In a previous study, habitat (island) occupancy and colony size models were developed for five seabird species breeding in coastal Maine using 1977 survey data (Johnson 1998). The resultant models appeared to be robust when tested against reserve data from that time period. However, we wondered how accurate these models would be at predicting future habitat use for long-term conservation planning. In order to test the predictive ability of such models, we developed temporally distinct island occupancy models for three seabird species and colony size models for one species based on survey data collected from several time periods. The data used to develop these models came from surveys conducted along the Maine coast at various times between 1941 and 1997. The variables included in each temporal model for a given species were compared to determine the utility of these variables and the overall habitat models as long-term indicators of island occupancy. Models also were tested using data from the later surveys to determine their long-term predictive ability.

Study Area

The study area is located in mid-coastal Maine and extends from Pemaquid Point eastward to Schoodic Point (Fig. 13.1). This area encompasses over sixteen hundred islands and ledges, ranging in size from a few square meters to thousands of hectares. Approximately 10 percent of these islands currently are known to support at least one seabird colony, defined as a group of at least five nesting pairs of a given species. Three of the twelve seabird species breeding in Maine were considered in this study: the double-crested cormorant (*Phalacrocorax auritus*), common eider (*Somateria mollissima*), and great black-backed gull (*Larus marinus*). These species were chosen because of the availability of comprehensive, historical survey data for the study area.

Selection Of Occupied and Unoccupied Islands

All occupied islands in the study area are larger than 0.01 hectare and smaller than 500 hectares in size. Larger islands (more than 500 hectares) generally are inhabited by people or mammalian predators, and the smallest islands (less than 100 square meters) provide little suitable habitat for seabird colony establishment. Thus, islands selected for use in this study were limited to the size range of occupied islands; this limitation allowed us to focus on other, less-obvious, habitat-selection factors. The islands that met this size criteria (approximately 1,130) were divided into occu-

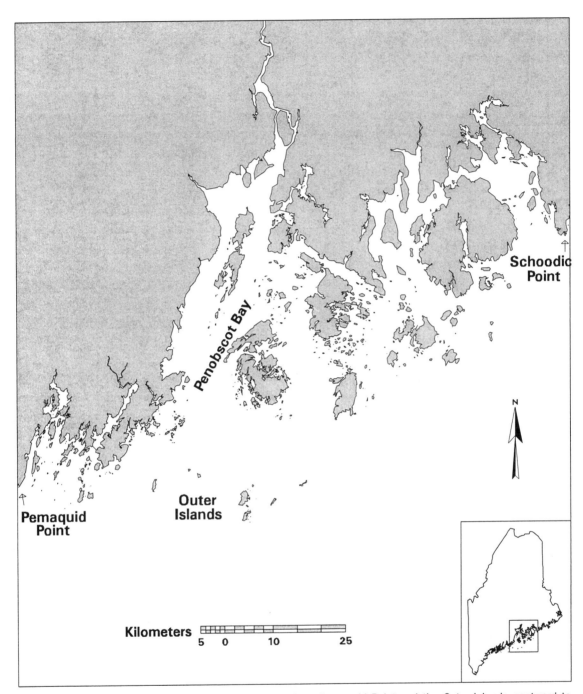

Figure 13.1. Mid-coastal Maine study area, extending from Pemaquid Point and the Outer Islands eastward to Schoodic Point.

pied (by any of the species considered in this study) and unoccupied groups, based on 1976–1977 survey data (Korschgen 1979). Five hundred islands were randomly selected from within each group for analysis at a ratio of 1:3 (i.e., 125 islands from the occupied group and 375 from the pool of unoccupied islands). Aerial photographs (9×9, 1:12,000-scale color photos)

were reviewed to eliminate islands that were connected to the mainland or populated islands (with roads) at high tide. After dropping islands for which aerial photographs were unavailable, a total of 306 islands were included in subsequent analyses. For individual species island occupancy models, all islands supporting a colony of that species were included, as were an equal number of randomly selected, unoccupied (by that species) islands.

Model Development

Island incidence data and nest counts were used as response variables for island occupancy models and colony size models, respectively. Double-crested cormorant incidence and nest count data were from surveys conducted during 1943–1944 (Gross 1944a), 1976–1977 (Korschgen 1979), and 1994–1997 (Johnson 1998). For common eiders, we used occurrence data collected during 1941–1943 (Gross 1944b), 1976–1977 (Korschgen 1979) and 1984 (Maine Department of Inland Fisheries and Wildlife [MDIFW] unpublished data). Great black-backed gull occurrences were based on surveys conducted during 1944 (Gross 1945), 1976–1977 (Korschgen 1979), and 1994–1996 (MDIFW unpublished data; Johnson 1998). Nest count data used for this study were limited to those made using ground nest counts for 1944 and 1977 and aerial counts for 1997. Although these methodologies have different biases, restricting each time period to a consistent methodology ensured that all estimates within the same survey period were comparable.

We used a geographic information system (GIS) (ArcInfo, Version 7.03, ESRI) to develop a spatial database of environmental and anthropogenic factors potentially associated with the use of islands in this area by nesting seabirds (Johnson 1998). Spatial variables were chosen for inclusion in this database based on the general habitat requirements for each species considered and the practicality of obtaining the data. These variables fell into three broad categories: physical island characteristics, foraging habitat, and human disturbance and predation. Refer to Johnson (1998) for an explanation of the methodology used to develop explanatory variables used in these habitat models.

Logistic regression (Systat 7.0; SPSS Inc.) was used to develop species island occupancy models and linear regression for colony size models. All explanatory variables with non-normal distributions were transformed to approximate a normal distribution. For each species and survey period, all occupied islands and an equal number of randomly selected unoccu-

pied islands were selected for model development. Explanatory variables were screened using Principle Components Analysis (PCA) and Pearson's correlation coefficient to remove colinearity and reduce the total number of variables used in the creation of habitat models. Variables associated with land use also were removed for the analysis of historical (pre-1976) island occupancy, due to a lack of information regarding these factors.

Because we wanted to develop parsimonious models that could be easily interpreted, we considered only main effects in the model. Beginning with a model that included all explanatory variables remaining following exploratory analyses, we manually removed and added variables to achieve a best-fit model, similar to an automated stepwise regression procedure. Goodness-of-fit of each logistic model was reflected by both the percentage of correctly classified sites and the McFadden's Rho-squared value (ρ^2). Since the primary objective of these models was to predict the probability of occupancy of an island, the percentage of correctly classified sites, or the correct classification rate (CCR), represents the most meaningful measure of model quality (Ryan 1997). However, when comparing models, it also is useful to use McFadden's ρ^2 values as a supplementary goodness-of-fit statistic (Hosmer and Lemeshow 1989). The ρ^2 value ranges from 0 to 1, with higher values reflecting more significant results, similar to the R^2 of a linear regression; however, ρ^2 values tend to be much lower than R^2, with values over 0.20 considered to represent satisfactory models.

After assessing the fit of each model for the time period for which it was developed, we tested that model against survey data from a later time period. For example, the model developed for cormorants based on survey data from 1944 was used to predict habitat occupancy during 1977 and 1997. The percentage of correctly classified sites was determined for each time period and the predictive ability of the model across different time periods was assessed.

Colony size models were developed for cormorants using multiple linear regression (Systat 7.0; SPSS). In addition to explanatory variables, nest count data were transformed to approximate normal distributions. A manual stepwise regression process also was

TABLE 13.1.

Explanatory variables considered in developing habitat models for double-crested cormorants (*Phalarcrocorax auritus*), common eiders (*Somateria mollissima*), and great black-backed gulls (*Larus marinus*).

Variable	Type[a]	Description
Area	C (log)	Island size (m^2)
Perimeter	C (log-base 10)	Island perimeter
Maxel	C (log(log))	Maximum elevation (m)
Ledge	C (log)	Coverage (m^2) of ledge or boulders
Cobble	C (log)	Coverage (m^2) of cobble or gravel
Rock	C (log)	Sum of Ledge and Cobble
Sand	C (log)	Coverage (m^2) of sand
Lowcov	C (log)	Sum of Herb, Shrub, and Open coverages
Forest	C (log)	Coverage (m^2) of forest or woodland areas
Disturb	C (log)	Coverage (m^2) of residential or agricultural areas
House_0	B	Absence of house, on or contiguous to island
Land_0	B	Absence of suitable boat landing areas
Mainmin	C (square root)	Distance to the mainland
Roadmin	C (cube root)	Distance to islands with maintained roads
Bigidist	C (cube root)	Distance to the nearest island greater than 50 ha in size
Minlf	C (cube root)	Distance to the nearest town landfill
Boatmin	C (cube root)	Distance to the nearest boat launch
Total2	C (none)	Total area of aquatic habitat within 2 km of island
Lev4km2	C (square root)	Total area of water 21–50 m in depth within 2 km of island
Total4	C (none)	Total area of aquatic habitat within 4 km of island
Lev2km4	C (none)	Total area of water 11–20 m in depth within 4 km of island
Lev5km4	C (log)	Total area of water 51–100 m in depth within 4 km of island
Area8km	C (cube root)	Total area of water less than 10 m in depth within 8 km of island
Aqua2km	C (square root)	Amount of marine aquatic bed habitat within 2 km of island
Estu4km	C (log)	Amount of estuarine habitat within 4 km of island
Estu8km	C (log)	Amount of estuarine habitat within 8 km of island
Anadmin	C (log)	Distance to the nearest anadromous
Salmmin	C (log)	Distance to the nearest Atlantic salmon stream

[a]Type refers to continuous (C), rank (R), or binary (B) variables, with transformations used for analysis noted in parentheses.

used for these models. The models were assessed using the adjusted R^2 values and explanatory variables using their individual P values from the overall model.

Results

Data for a total of sixty-four explanatory variables were assembled for the islands used in this study. As a result of the exploratory data analyses, the number of explanatory variables was reduced to between nineteen and twenty-two for inclusion in development of each species' habitat models; selection of final variables was based on a review of Pearson's correlation coefficients between individual species' survey data and explanatory variables. Table 13.1 includes all variables used in at least one species model development.

Double-crested Cormorant

Predictive models were developed for the occupancy of islands by double-crested cormorants during three survey periods: 1944 (1942–1944), 1977 (1976–1977), and 1997 (1994–1997). In the earliest period examined (DCCO44), occupancy was negatively associated with the extent of aquatic bed habitat within 2 kilometers of an island (AQUA2KM); this was the only

TABLE 13.2.

Results of habitat occupancy models for double-crested cormorants (*Phalarcrocorax auritus*), common eiders (*Somateria mollissima*), and great black-backed gulls (*Larus marinus*) over discrete time periods.

Species	Year	N	ρ^{2a}	CCR[b]	Variable	Estimate	P
			Overall model		**Explanatory variables[c]**		
Double-crested	1944	30	0.328	83.3	aqua2km	−0.258	.007
cormorant	1977	102	0.627	84.3	forest	−1.836	.955
					aqua2km	−0.172	.021
					lowcov	0.215	.039
					total2km	0.003	.020
	1997	106	0.635	89.6	lowcov	0.204	.048
					forest	−1.847	.965
					house_0	7.973	.964
					maxel	4.125	.023
					aqua2km	−0.225	.002
Common eider	1943	33	0.490	84.8	lowcov	0.535	.011
					aqua2km	−0.163	.055
	1977	158	0.582	86.1	lowcov	0.504	<.001
					bigisle	0.646	<.001
					house_0	1.449	.001
	1984	159	0.651	86.2	lowcov	0.439	<.001
					house_0	2.566	<.001
					total2km	0.003	.001
					aqua2km	−0.140	.001
Great black-	1944	38	0.796	94.7	lowcov	0.868	.014
backed gull					aqua2km	0.286	.065
	1977	128	0.703	90.6	lowcov	0.549	<.001
					forest	−0.421	<.001
					bigisle	0.686	.001
					aqua2km	−0.151	.005
	1995	134	0.664	92.5	lowcov	0.540	<.001
					forest	−0.478	<.001
					total2km	0.002	.004

[a]McFadden's rho-squared (ρ^2) is an estimate of the fit of the overall model.
[b]CCR represents the percentage of correctly classified sites.
[c]Explanatory variables are presented along with their coefficients.

meters of an island (AQUA2KM); this was the only variable included in the model. Eighty-three percent of the islands used in model development were correctly classified; however, the ρ^2 value (0.33) indicated only a satisfactory model fit (Table 13.2).

The 1977 cormorant model ($n = 102$) showed a positive relationship between site occupancy and both the extent of herbaceous and low shrub cover (LOW-COV) and the total amount of water in the immediate

vicinity (2 kilometers) of an island (TOTAL2KM). A negative relationship was apparent with both forest cover (FOREST) and AQUA2KM. This model was relatively robust, with a ρ^2 value of 0.63 and CCR of 84 percent. The 1997 model again indicated that the presence of nesting cormorants was positively associated with the amount of low cover and negatively associated with FOREST and AQUA2KM. However, the remaining explanatory variables differed, includ-

TABLE 13.3.

Comparison of the percentage of correctly classified sites (CCR) determined for island occupancy by colonies of double-crested cormorants (*Phalarcrocorax auritus*), common eiders (*Somateria mollissima*), and great black-backed gulls (*Larus marinus*) during different survey periods.

Species	Model year[a]	Test year[b]	N	Correctly classified sites (%)		
				Occupied	Unoccupied	Overall
Double-crested cormorant	1944	1944	30	71.4	93.7	83.3
		1977	102	56.9	82.4	69.6
		1997	106	53.8	91.7	72.6
	1977	1977	102	78.4	90.2	84.3
		1997	106	80.8	88.9	84.9
	1997	1997	106	92.3	87.0	89.6
Common eider	1943	1943	33	93.8	76.5	84.8
		1977	158	79.7	84.8	82.2
		1984	159	80.8	92.6	86.8
	1977	1977	158	91.1	81.0	86.1
		1984	159	91.0	77.8	84.3
	1984	1984	159	89.7	81.5	86.2
Great black-backed gull	1944	1944	38	94.7	94.7	94.7
		1977	128	81.5	82.5	82.0
		1995	134	80.6	80.6	80.6
	1977	1977	128	95.4	85.7	90.6
		1995	134	94.0	76.1	85.1
	1995	1995	134	91.0	94.0	92.5

[a]Model year refers to the survey data used to develop that model.
[b]Test year refers to the survey data used to test the model.

(MAXEL) and the absence of houses on the island or contiguous islands (HOUSE_0).

The presence of aquatic bed vegetation within 2 kilometers of an island was the most significant variable affecting nesting cormorant use of islands during all time periods examined. Despite this consistency, however, the ability of the 1944 model to correctly predict site occupancy in 1977 and 1997 was 10–14 percent lower than that for 1944 (Table 13.3), and the omission error rate was more than 40 percent for both periods. In addition to AQUA2KM, the extent of low cover and forest were important variables in both the 1977 and 1997 models. Though these models were not identical, the 1977 model was able to correctly classify occupancy in 84–85 percent of sites in both 1977 and 1997 (Table 13.3). In fact, the omission error rate for the 1977 model (i.e., the percentage of occupied islands classified as unoccupied) was lower when tested against 1997 data (Table 13.3).

Colony Size Analysis

The relationships between full surveys (i.e., all colonies surveyed in a given year) and explanatory variables indicated little temporal change in the relationship between the number of nesting pairs on an island and explanatory variables over the different survey periods (Table 13.4). The 1944 and 1997 models both included LOWCOV and LEV2KM4 (the extent of coastal water 10–20 meters deep within 4 kilometers of an island), while the 1997 model was limited to LOWCOV. Although the first model explained a significant amount of variability in nest counts ($R^2 = 0.66$), neither the 1977 nor 1997 model was particularly good $R^2 = 0.21$ and 0.22, respectively). Since each model includes data for all islands with cormorants during that survey period, however, it is difficult to distinguish potential differences between islands with long-established colonies and recently colonized islands.

TABLE 13.4.

Comparison of the colony size models developed for double-crested cormorants (*Phalarcrocorax auritus*) using survey data from different time periods.

Survey year	N	R^2	Explanatory variables[a]	P
1944	13	.662	lev2km4	< .005
			lowcov	< .01
1977	51	.207	lev2km4	< .01
			lowcov	< .05
1997	50	.216	lowcov	< .05

[a]Variables defined in table 13.1.

To elucidate the differences between older and more recently established colonies, and the changes that take place within a colony over time, we divided the data into three distinct island groups, based on the age of the colony, for further analysis. Specifically, DCCO44 was composed of those islands colonized by cormorants prior to 1945; DCCO77 included islands colonized after 1944 but prior to 1978, and DCCO95 was limited to those islands first colonized after 1977. In each case, the size of recently established colonies appears to be most strongly influenced by the aquatic (foraging) environment, specifically LEV2KM4 (Table 13.5). The size of established colonies (twenty years or more), however,

is more closely tied to island size or the amount of low cover on an island. In the case of colonies established prior to 1945, island size alone was able to explain 54 percent and 33 percent of the variability in colony sizes for 1977 and 1997, respectively, but only 2 percent of the variability in the size of colonies during 1944.

Common Eider

Two variables were included in the 1943 model for common eiders. The occurrence of nesting eiders was positively related to the amount of low cover and negatively associated with AQUA2KM (Table 13.2). The overall model fit ($\rho^2 = 0.49$) was satisfactory and 84.8 percent of sites were correctly classified. The presence of eiders in 1977 was positively correlated with increasing distance from islands more than 50 hectares in size (BIGISLE), the amount of low cover, and the absence of houses on or adjacent to an island (HOUSE_0). The overall fit for this model ($\rho^2 = 0.58$) was fairly good and was supported by a CCR of 86 percent of the 158 sites included. The CCR for the 1984 model also was 86 percent, with a ρ^2 of 0.65. Like the 1977 model, both LOWCOV and HOUSE_0 were positively associated with the presence of eiders. However, distance to islands was not included in this model, being replaced by

TABLE 13.5.

Comparison of the colony size models developed for double-crested cormorants (*Phalarcrocorax auritus*) using survey data from different time periods.

Year of colonization[a]	Survey year	N	R^2	Explanatory variables[b]	P
Prior to 1945	1944	13	.662	lev2km4	< .005
				lowcov	< .01
	1977	13	.764	area	< .001
				lev5km4	< .005
	1997	12	.636	area	< .05
				lev2km4	< .05
1945–1977	1977	38	.302	lev2km4	< .001
				roadmin	< .05
	1997	38	.431	lowcov	< .001
1978–1997	1997	11	.860	roadmin	< .001
				total2km	< .05
				estu4km	< .05

[a]Year of colonization refers to the first survey, conducted after 1900, for which a cormorant colony was noted on an island.

[b]Variables defined in Table 13.1.

a negative association with AQUA2KM and a positive association with TOTAL2KM.

Variables included in all three eider models were similar, with LOWCOV the most important predictor in each. In addition, negative associations with AQUA2KM or the presence of houses on or adjacent to an island, each occurred in two of the three models. The predictive ability of these models also differed little across the time periods examined. The model developed from 1943 data correctly predicted between 82 and 87 percent of occupied islands for all three time periods assessed (Table 13.3).

Great Black-backed Gull

Like the earliest eider model, the 1944 model for black-backed gulls included two explanatory variables, LOWCOV and AQUA2KM (Table 13.2). However, this simple model had a very high ρ^2 value (0.80) and correctly classified 95 percent of sites. The 1977 model also included LOWCOV and AQUA2KM as variables but was more complex, including two other explanatory variables. Island occupancy for this time period was positively associated with increasing distance from the mainland or islands larger than 50 hectares and negatively associated with the amount of forest cover. This model also had a high ρ^2 value (0.70) and a CCR of 91 percent. The 1995 model again included LOWCOV and FOREST; however, the third explanatory variable was a positive association with the total amount of coastal water within 2 kilometers. This model had a ρ^2 of 0.66 and a correct classification rate of 92.5 percent.

As with the previous two species, a single variable, LOWCOV in this case, exerted the most influence on models across all three time periods. The ability of early models to predict future occupancy was relatively good (80–85 percent), but did exhibit a noticeable decrease across time (13–14 percent from 1944 to 1977 or 1997; Table 13.3).

Discussion

Island Occupancy

The extent of low cover on an island was the most significant explanatory variable for all eider and gull models and also was significant in the latter two models for cormorants. Along the Maine coast, both eiders and black-backed gulls use dense herbaceous or shrub cover for nesting sites (Blumton et al. 1988; C. Johnson personal observation). However, cormorants prefer to nest on bare ground or in trees (Mendall 1936; Cowger 1976; Nettleship and Birkhead 1985). Since cormorant nests usually are not located in herbaceous or shrub vegetation, the importance of low cover may be associated with the protection it affords young from predators, especially gulls, when adults are absent from the colony. Eiders nesting in dense cover have been shown to have lower predation and higher success rates than those nesting in less-concealed areas (Choate 1967). In addition, islands with dense herbaceous or shrub cover may provide nesting birds with more protection from winds and other harsh environmental conditions than those with little or no vegetative cover.

A negative association with extensive areas of aquatic bed vegetation in the immediate vicinity of an island (i.e., within 2 kilometers) also was an important variable in the earliest model for each species. Previous analyses (Johnson 1998) indicated that the presence of aquatic bed habitat could be acting as a surrogate for intertidal areas in the model. As such, this negative relationship may be associated with avoidance of predators. Larger islands are more likely to support populations of predatory mammals, and intertidal (i.e., low tide) connections to these islands could allow predators to move more easily onto smaller islands.

The earliest models, when fewer than forty islands in the study area were occupied by each species, were quite simple, limited to one or both of the abovementioned variables, AQUA2KM and LOWCOV. These variables appear to be related to two of the three primary requirements for colony establishment: the presence of suitable breeding sites and safety from terrestrial predators.

The ability of the 1943 eider model to predict occupancy forty years in the future was quite good (87 percent). However, the predictive abilities of the earliest cormorant and gull models were less satisfactory, especially in the case of cormorants. The omission error rate for the 1944 cormorant model was between 40 and 50 percent, when tested on data from 1977 and 1997. The difference in predictive ability of the eider model versus the cormorant and gull models could be related to the ability of these species to exploit varied habitats and food resources. The more stenotypic eider nests almost exclusively in dense, low herbaceous and shrub cover and relies on motile or sessile food resources, such as urchins, mussels, and snails (Choate 1967; Cantin et al. 1974). Cormorants and gulls, on the other hand, are able to adapt to different nesting conditions as well as to changing food-resource levels and locations. Cormorants are opportunistic feeders that may readily switch between prey when a particular food item becomes scarce (Mendall 1936; Ross 1974; Blackwell et al. 1995). The diet of great black-backed gulls is extremely flexible, including fish and shellfish, seabirds and their eggs, other vertebrates, insects, and refuse (Buckley 1990). Thus, as seabird populations expanded and available resource levels shifted along the Maine coast from the early 1900s to the 1970s and 1990s, it is likely that cormorants and gulls were better able to exploit a wider variety of habitats and food resources.

Populations of each species increased greatly from the mid-1940s to 1977 (Korschgen 1979; Johnson and Krohn 1998). The 1977 models all had a corresponding increase in complexity, incorporating additional variables for a best-fit model. Although there was little change in population size or the number of occupied islands in any of the species considered between 1977 and 1995, the explanatory variables included in the models did change over this time period. These changes could simply reflect a difference in the random selection of unoccupied islands (since each only included about half the available pool of islands). However, a review of the islands occupied in 1977 and the 1990s indicates a change in occupied islands as well, at least for cormorants and gulls. Although most seabirds exhibit strong site fidelity (Cairns 1992), it is not uncommon for entire colonies of some species to shift breeding locations (islands) between years (Mendall 1936; Buckley and Downer 1992). In the case of double-crested cormorants, only 60 percent of the islands in the study area used for nesting at some time

from 1994 to 1997 were occupied during all four years (Johnson 1998).

Despite the change in model composition, however, the ability of 1977 models to predict island occupancy seven to twenty years in the future was good for all species (84–85 percent; Table 13.3). The robustness of the 1977 models in predicting future island occupancy could be indicative of habitat saturation by all three species in the area. Since an increasing number of habitats are used as populations increase (Svardson 1949; O'Connor 1987b; Rosenzweig 1991), species at saturated densities could be expected to be occupying most usable habitat types in an area. Thus, if an island occupancy model is developed using data from a population at a very high density, it should, theoretically, be able to identify all potentially usable habitat. Populations of both double-crested cormorants and common eiders were considered to be near saturated densities for the region in the late 1970s and early 1980s (Mendall 1976; Krohn et al. 1992; Krohn et al. 1995; Johnson and Krohn 1998).

We did not explicitly test for spatial autocorrelation in our logistic models because of the clumped spatial pattern of habitat (i.e., islands) in the coastal system studied. Although several methods are available to estimate the degree of spatial autocorrelation that often is exhibited by ecological data (Sokal and Oden 1978; Cliff and Ord 1981; Legendre 1993), most of these methods (e.g., Moran's I, Mantel test, semivariogram) were designed primarily for data collected from a regular lattice or measured on a continuous scale (Cressie 1993; Legendre and Legendre 1998). Little attention has been given to developing similar techniques for binary data or data collected on an irregular lattice, such as that presented in this chapter (Cressie 1993; but see Fingleton 1983). At this time, we are not aware of any software package that includes methods for directly estimating the effect of spatial autocorrelation for inclusion in our logistic regression models (but see Smith 1994 and Klute et al., Chapter 27). Since we did not explicitly account for potential spatial dependence in our model, the results of our significance test could be somewhat biased (i.e., the Type I error rate could be greater than the declared alpha level); however, such effects are unlikely to change the direction of the observed relationships.

Colony Size

As mentioned previously, there was little difference in temporal habitat associations between the full survey colony size models for double-crested cormorants. However, interesting changes in habitat relationships emerged when colonies of different ages (i.e., islands that were initially colonized at different times since 1900) were analyzed separately. In a crude sense, this allowed us to follow individual sites or colonies through time. As the colony aged, there was a shift in emphasis from aquatic and disturbance-related factors to physical island characteristics for explaining most of the variability in colony size, though aquatic variables were still important. This distinction was lost when looking at models developed for all colonies surveyed during a given period (i.e., the full-survey models).

The size of recently established colonies can be misleading and models developed during that time period may not be reflective of factors that will ultimately regulate colony size. Although many of the islands along the Maine coast exhibited a rapid increase in the number of nesting cormorants after initial colonization, over a long period (e.g., twenty to fifty years), colony size generally decreased to a relatively consistent, much lower level (Johnson 1998). Krohn et al. (1992) noted a similar overall decline in the number of nesting pairs of eiders per island and suggested that a steady decline could be indicative of a stabilizing population. The importance of considering colony age in interpreting models based on colony size provides further support for the need for caution against using density alone as an indicator of habitat quality (Van Horne 1983; Vickery et al. 1992).

Management Recommendations

The results of this study indicate that seabird occupancy models may be good predictors of future island colonization if species either are habitat specialists or are at or near saturation density in a region. However, caution should be exercised when using the results of models developed for less-specialized species to predict future island use, especially when at low densities in a region. The use of colony size as an indicator of habitat quality also should be considered cautiously, since the size of a recently established colony often is misleading. In addition, the location of occupied versus unoccupied, and high- versus low-quality sites, may shift over long time periods as the result of broad habitat changes resulting from environmental or human-induced changes (Helle and Jarvinen 1986; Greco et al., Chapter 14). Interspecific interactions (e.g., predation) also can result in displacement of seabird colonies from formerly high-quality sites to new areas (Buckley and Buckley 1984; Moors and Atkinson 1984). Coastal environments are especially prone to such fluctuations as a result of both environmental stochasticity and anthropogenic influences.

Acknowledgments

Funding for this study was provided by the Cooperative Unit Program, Biological Resources Division of the U.S. Geological Survey and by the Department of Wildlife Ecology, University of Maine. We thank Brad Allen, Maine Department of Inland Fisheries and Wildlife, for his interest in this project and for providing historical seabird data. We also recognize the contribution of the many individuals, mostly state and federal employees, who collected seabird data over the decades. We thank the Maine Department of Marine Resources and the Maine Geological Survey for providing access to aerial photographs and bathymetric data used in this study. We are grateful to M. T. Jones and three anonymous reviewers for helpful comments on an earlier version of this manuscript. This chapter is a contribution of the Maine Cooperative Fish and Wildlife Research Unit (BRD/USGS, MDIFW, UM, and the Wildlife Management Institute, cooperating) and is publication number 2414 of the Maine Agriculture and Forest Experiment Station.

Geographic Modeling of Temporal Variability in Habitat Quality of the Yellow-billed Cuckoo on the Sacramento River, Miles 196–219, California

Steven E. Greco, Richard E. Plant, and Reginald H. Barrett

The western subspecies of the yellow-billed cuckoo (*Coccyzus americanus occidentalis*) is a riparian-obligate forest-interior species associated with large blocks (more than 41 hectares) of riparian forest vegetation (Laymon and Halterman 1987a; Laymon and Halterman 1989; Laymon et al. 1997) within the major alluvial meandering system of the Sacramento River. Because of the precipitous decline in the yellow-billed cuckoo's population and the scarcity of riparian habitat resources remaining in California (Bay Institute 1998), the yellow-billed cuckoo is presently a state-listed endangered species (Steinhart 1990).

The portion of the riparian ecosystem occupied by the yellow-billed cuckoo is a highly dynamic zone of the floodplain where forest structure and floristic composition can change significantly in less than a decade. The changes result in a shifting mosaic controlled by three major agents on the Sacramento River: (1) natural river channel migration and floodplain dynamics, such as erosion, deposition, scour, and flooding (physical processes); (2) succession and high growth rates of riparian vegetation (biological processes); and (3) cultural modification of the landscape for flood control and agricultural land development (e.g., regulated flows from dams, levee construction, and land clearing for crops).

Based on these three agents of change the main hypothesis for this investigation was: As a consequence of the extensive, rapid changes to the structure of the riparian landscape, the suitability or quality of habitat with respect to its potential value for nesting and foraging for the yellow-billed cuckoo can change rapidly through time, forming a shifting mosaic of habitat patches that vary in area. The hypothesis was tested by modeling the habitat relationships of the yellow-billed cuckoo within a 23-river-mile study reach on the Sacramento River using spatial data from 1997 and then applying the habitat model to five historical land-cover data sets. The historical land-cover data, spanning the years 1938 to 1987, is used for comparative purposes to monitor the probable historical development (formation and extinction) of suitable habitat patches of the yellow-billed cuckoo through time. Knowledge of habitat patch dynamics can help managers develop more effective recovery plans for the yellow-billed cuckoo on the Sacramento River.

Background

The western subspecies of the yellow-billed cuckoo is a Neotropical migrant that winters in South and Central America and breeds in western North American riparian forests and floodplains during the summer months (Gaines 1970; Halterman 1991; Ehrlich et al. 1992). The yellow-billed cuckoo once bred abundantly throughout the western Pacific states, but today the northern range is limited to the middle reaches of the Sacramento River (Laymon and Halterman 1987a; Zeiner et al. 1990). The decline in abundance of the yellow-billed cuckoo is mainly attributed to conversion of essential riparian floodplain habitats to agricultural and flood control land uses and also may have been augmented by the widespread application of pesticides (Gaines and Laymon 1984; Laymon and Halterman 1987a; Steinhart 1990). Additionally, the migration pathways of the cuckoo have been fragmented over the past 150 years of settlement.

Historically, the yellow-billed cuckoo was described as common in the Central Valley of California (Belding 1890); however, by the 1940s the population's decline was noted by Grinnell and Miller (1944). Population censuses intermittently collected between 1972 and 1990 indicated the Sacramento River population of the yellow-billed cuckoo had declined from an estimated 96 pairs in 1973 (Gaines 1974), to approximately 60 pairs in 1977 (Gaines and Laymon 1984), and a four-year survey by Halterman (1991) found the total number of pairs between 1987 and 1990 to vary between twenty-three and thirty-five pairs. The yellow-billed cuckoo was listed by the state of California as threatened in 1971 and then as endangered in 1988 (California Department of Fish and Game 1998). For a description of the subspecies status of the yellow-billed cuckoo, see Ridgway (1887); Bent (1940); Laymon and Halterman (1987a); Banks (1988); Franzreb and Laymon (1993). The American Ornithologists' Union listed the western subspecies in its checklist from 1895 to 1957 (AOU 1895, 1910, 1931, 1957).

Also known as the California cuckoo or the western yellow-billed cuckoo (Ridgway 1887; Martin et al. 1951; Ehrlich et al. 1992), the bird is medium-sized, averaging 27–30 centimeters in length and 60 grams in weight. The yellow-billed cuckoo arrives in California from South America between late June and early July and departs California with its young by mid-August to September. Past field studies of the yellow-billed cuckoo's nesting and foraging habitat requirements (Gaines 1970; Gaines 1974; Laymon and Halterman 1987a; Halterman 1991; Laymon et al. 1997) show it is a riparian-obligate forest-interior species that has a home range between 17 and 41 hectares (about 40–100 acres) (Laymon 1980; Laymon and Halterman 1987a; Laymon and Halterman 1989; Laymon et al. 1997). The preferred habitat is a mosaic of riparian forest vegetation consisting of willow (*Salix* spp.) and cottonwood (*Populus fremontii*) forests in combination with open-water habitats such as an oxbow lake or backwater channel. Dense vegetation less than 20 meters in height is especially important for nesting while both low and high vegetation are used for foraging (Laymon et al. 1997). Preferred prey include green caterpillars, hornworms, katydids, tree frogs, grasshoppers, and cicadas. The average clutch is two to four eggs and the young are fledged at an age of one month. In years of high food abundance, double and triple brooding has been observed (Halterman 1991; Laymon et al. 1997).

Wildlife Habitat Relationship Modeling

It is widely recognized that models of species-specific habitat relationships and population demographics can be useful in gaining greater insight into development of a conservation strategy for recovery of a species of special management concern (USFWS 1981a; USFWS 1981b; Verner et al. 1986b; National Research Council 1995; Bissonette 1997b; Noss et al. 1997). The objective of our study was to develop a habitat suitability model to predict presence or absence of yellow-billed cuckoo. This was done using a modified version of the California wildlife habitat relationships (CWHR) system land-cover classification scheme (Mayer and Laudenslayer 1988) to model the habitat requirements of the yellow-billed cuckoo.

It is important to note that the predicted habitats of the CWHR system represent only *potential* habitat, since for example nonhabitat factors that affect abundance, such as competition, predation, disease and stochastic processes, are not considered (Airola 1988; Garrison 1993). For a discussion of gap analysis and application of species-distribution maps (e.g., using wildlife habitat relationship models) see Scott et al. (1987), Burley (1988), Davis et al. (1990), Lopez (1998), and Savitsky et al. (1998). For more information on wildlife habitat relationship modeling, see Verner et al. (1986b) and Morrison et al. (1992).

Materials and Methods

Study reach description. Throughout this study, specific locations and reaches of the river were spatially referenced using the river-mile marker system established by the U.S. Army Corps of Engineers (e.g., US-ACOE 1991). The study reach for this investigation was located between river-mile 196 at Pine Creek bend and river-mile 219, near Woodson Bridge State Recreation Area (Fig. 14.1a, b). This extent of the river was selected because (1) yellow-billed cuckoo surveys were available, (2) there were few levees to constrain flood flows, and thus geomorphic processes such as active channel migration and bend cutoff had occurred over the past thirty years resulting in extensive riparian forests, (3) historical aerial photographs were available, and (4) several publicly owned lands were accessible for research purposes (Fig. 14.1b). The publicly accessible areas include the Pine Creek Wildlife Area and Wilson Landing (both owned and managed by the California Department of Fish and Game) as well as the River Vista Unit at Merrills Landing, and Foster Island (owned and managed by the U.S. Fish and Wildlife Service).

The six available surveys for presence of the yellow-billed cuckoos in the study reach were dated intermittently from 1972 to 1990 (Gaines 1974; Gaines and Laymon 1984; Laymon and Halterman 1987b; Halterman 1991). All of these surveys had found the Pine Creek Wildlife Area, located at the southern end of the study reach, to consistently be one of the highest-density breeding areas of the yellow-billed cuckoo on the river.

Many historical aerial photographs of the Sacramento River were available in stereo-pair format for the years spanning 1937 to 1997. The six time periods selected and mapped for this study were 1938, 1952, 1966, 1978, 1987, and 1997. These dates were chosen to assess landscape changes over approximately decades (the mean paired age difference was 11.8 years).

Land-cover mapping methods. For this study a modified version of the CWHR land-cover classification system was used. Three primary habitat variables were interpreted within the extent of the study reach: (1) land-cover type, including riparian, valley oak woodland, annual grassland, freshwater emergent wetland, riverine, lacustrine, gravel bar, orchard, cropland, pasture, and urban/developed; (2) tree size in three height categories: low (less than 6 meters), medium (6–20 meters), and high (greater than 20 meters); and (3) canopy cover, in four classes: sparse (10–24 percent), open (25–39 percent), moderate (40–59 percent), and dense (60–100 percent). For comparison (a category crosswalk) of the land-cover types used to map riparian vegetation in this study and the equivalent CWHR system land-cover types, tree size classes and canopy cover classes, see Greco (1999).

The modified mapping system was tested using 1997 aerial photography within the extent of the study boundaries defined by the flight lines as depicted in Fig. 14.1b. The land-cover and woody vegetation structural classes were delineated and field verified in 1997 before applying the mapping methods to the historical photographic sets. Each aerial photographic image was manually interpreted under a stereoscope by marking geographic control points and stratifying the riparian landscape into homogeneous land-cover variables delineated as polygons. The delineation process employed a grain size, or minimum mapping unit, of 20 meters. Forest stand height classes were visually estimated for each polygon using the stereoscope and canopy cover was estimated using a standard photo interpretation tree stocking density scale (Aldrich et al. 1984).

The interpreted mapping was then converted to digital format and imported into the geographic information system software application (GIS), ArcInfo Version 7.1 (ESRI 1997) and converted to polygon coverages. The polygon coverages were then attributed and transformed to geographic coordinates using control points derived from corresponding (matched) coordinate pair locations from each historical photographic set to a 1978 (black and white) orthophotographic set of quadrangle maps from the U.S. Geological Survey at a mapping scale of 1:24,000.

Validation of land-cover mapping. The 1997 land-cover mapping was field verified in late summer 1997 at four sampling locations that were publicly accessible from either land or water. Only the riparian land-cover types were selected for field verification; the

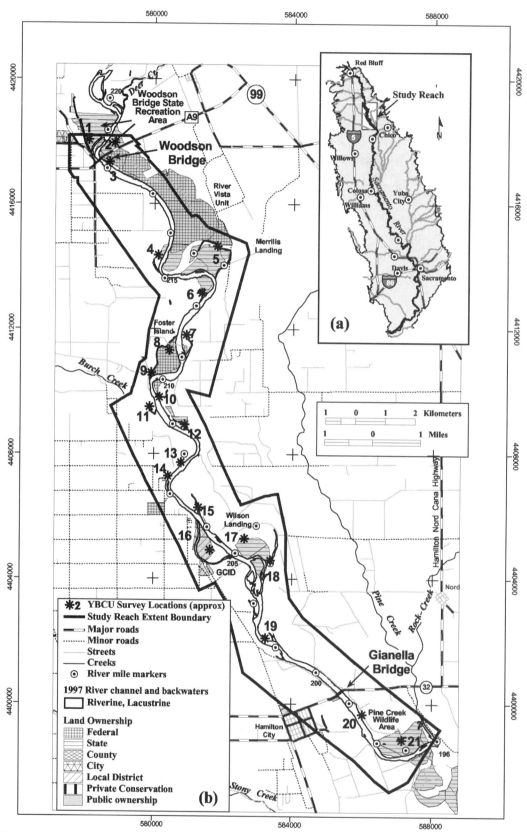

Figure 14.1. (a) Study reach location within the Sacramento Valley, (b) the study reach public lands, river-mile markers, approximate locations of yellow-billed cuckoo (*Coccyzus americanus*) (YBCU) detection surveys, and the extent of the analysis boundary common to each temporal data set from river-miles 196–219 (grid tics in UTM, Zone 10).

agricultural land-cover types were not verified. A complete list of polygons was generated and sorted by location, height class, and canopy cover class. Sample size was determined for the "stand-height" variable using Stein's two-stage method (Steele and Torrie 1960). From this sample-size assessment, a stratified random set of sixteen polygons was selected from the polygon list using a random number table.

The height of each sample forest-stand polygon was measured from an observation point using a survey laser instrument (Criterion model 400; Laser Technology 1996) at a mean distance of approximately 50 meters. At each forest stand, twenty measurements were collected from restricted random compass bearings using a random number table. The forest stand canopy cover was measured using a vertical densitometer (sighting tube) (Geographic Resource Solutions 1997), by taking one hundred measurements over a randomly positioned 100-meter transect with perpendicular cross lengths each 20 meters long (Ganey and Block 1994). See Greco (1999) for more information regarding the field methods and results.

Based on positive results from field verification of the 1997 land-cover mapping of the study reach, the same mapping process was then applied to the historical photographs from 1938, 1952, 1966, 1978, and 1987. An error check was performed on each mapping year's final land-cover data set to verify for database attribute accuracy and polygon boundary errors and omissions.

Habitat Modeling of the Yellow-billed Cuckoo

The geographic habitat model developed in this investigation to assess the quality of yellow-billed cuckoo habitat on the Sacramento River was implemented using ArcView GIS version 3.1 (ESRI 1996a), Spatial Analyst extension version 2.0 (ESRI 1996b), and the native scripting language Avenue (ESRI 1996c). For a description of the GIS procedures and Avenue code developed to implement the model, see Greco (1999).

The purpose of the model was to identify potential suitable or optimal habitat patches within the extent of the study reach (river-miles 196 to 219) by assessing attributes of the land-cover data based on the known nesting, feeding, and cover habitat requirements of the yellow-billed cuckoo. The model generates a habitat index score (scaled from 0 to 1) for a particular patch that reflects the relative strength or magnitude of the suitability of the patch to support a breeding pair of yellow-billed cuckoos. The model does not predict carrying capacity (population density per unit area). A "patch" in this investigation was defined as the geographic area of the following contiguous land-cover categories: riparian, freshwater emergent wetland, or lacustrine. A patch is also referred to as a forest stand, polygon, zone, or "zonal area" in the following descriptions involving spatial data processing.

HSI model parameters. A habitat suitability index (HSI) model for the yellow-billed cuckoo was developed from a wildlife habitat relationship model published by Laymon and Halterman (1989) in combination with an additional forest structure variable derived from Laymon (1980) and Laymon et al. (1997) (Table 14.1). The HSI model variables used for this study were (1) patch area, (2) patch width, (3) patch distance-to-water, (4) within-patch ratio of high vegetation to low and medium vegetation, and (5) within-patch ratio of young floodplain to old floodplain. Two versions of the HSI model were tested; a "reduced model" that used variables 1–3 (listed previously) and a "full model" that used all five variables. Each of the yellow-billed cuckoo HSI variables is depicted in graphical format in Fig. 14.2 (a–e). The habitat variables identified by Laymon and Halterman (1989) were patch vegetation species, patch area, patch width, and patch distance-to-water (see Table 14.1).

The variables "patch area" and "patch width" were measured using the Spatial Analyst (ESRI 1996b) zonal functions. The patch "distance-to-water" variable was measured by generating a distance grid from lacustrine and riverine land-cover types within the study reach. The variable "patch vegetation species" was specified by Laymon and Halterman (1989) as "willow-cottonwood," which was a subset of the mapped land-cover category "riparian" in the 1997 land-cover mapping. To identify the species association within riparian forest patches the surrogate variable "floodplain age" was used as an indicator for all

TABLE 14.1.

Wildlife habitat relationships model variables for the western yellow-billed cuckoo (*Coccyzus americanus*) on the Sacramento River, California (adapted and modified from Laymon and Halterman 1989; Laymon et al. 1997).

Habitat suitability (index score)	Land cover / vegetation type	Patch area (ha)	Patch width (m)	Patch distance-to-ratio water (m)	Zonal area sum of riparian tree height classes (H:L+M)	Zonal area sum ratio of floodplain age (yrs.) (young:old) (<60:>60)
Optimum (1.0)	Riparian/ Willow-Cottonwood	>80	>600	<100	0.8–1.249	2.1–4.0
Suitable (0.66)	Riparian/ Willow-Cottonwood	41–80	200–600	<100	0.25–0.799/ 1.25–2.0	1.1–2.0 4.1–7.0
Marginal (0.33)	Riparian/ Willow-Cottonwood	17–40	100–199	<100	<0.249	0.6–1.0 >7.1
Unsuitable (0.0)	Riparian/ Willow-Cottonwood	<17	<100	>100	>2.1	<0.5

willow-cottonwood associations (see Greco 1999). A zonal sum ratio of young floodplain to old within each patch using the floodplain age variable was employed as a means to distinguish between those patches most likely to contain the willow-cottonwood association from those containing predominantly older upper-floodplain tree species such as valley oak (*Quercus lobata*) and California black walnut (*Juglans hindsii*). The zonal sum ratio was defined as the sum total area of all of the cells of young floodplain (less than sixty years) divided by the sum total area of all of the cells of old floodplain (more than sixty years) within a patch. To calibrate this variable an estimated monotonic distribution was applied (see Fig. 14.2e). A patch analysis of the Pine Creek Wildlife Area suggested a zonal area sum ratio of 3:1 (young floodplain to old floodplain) is an optimal proportion for the yellow-billed cuckoo.

The forest structural composition variable added to the Laymon and Halterman (1989) model is based on two yellow-billed cuckoo nesting behavior and habitat structure studies from research conducted on the Sacramento River by Halterman (1991) and the Kern River Preserve by Laymon et al. (1997). These authors reported that nesting was most frequent in vegetation below 20 meters in height. Therefore, the model variable specifies that each patch must posses some fraction in the height size class low or medium. To cali-

brate the forest structure variable, a home range analysis of four breeding pairs in 1979 at the Pine Creek Wildlife Area (Laymon 1980) was used to derive optimal (birds breeding) and unsuitable (no birds) proportions of vegetation height. The analysis for tree height proportions used the 1978 land-cover data set and suggested a zonal area sum ratio of 1:1 (high to medium plus low vegetation) was an optimal proportion and an upper limit of 2:1 was estimated to be unsuitable. A monotonic distribution was approximated for the calibration of this relationship (Fig. 14.2d). The assumption behind the calibration of this variable was that forest structural characteristics are a controlling variable influencing the cuckoo's presence in these forest patches. However, other habitat factors may be limiting their presence. Habitat variables were combined to generate a score using a geometric mean equation with each variable receiving equal weight. The geometric mean approach was selected because it limits habitat suitability for a particular patch to those patches that contain all the necessary attributes for suitable habitat (Cooperrider 1986). Implementation of the model was accomplished with grid-based processing (at 0.04-hectare cell size) to perform the digital cartographic overlay mapping of the land-cover habitat variables.

HSI model validation. The HSI model output for the study reach was validated using the pooled

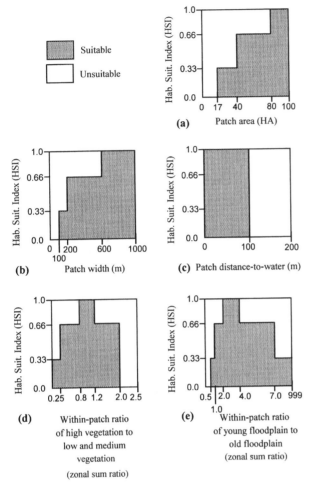

Figure 14.2. **(a–e)** The five habitat suitability index variables used to identify habitat for the yellow-billed cuckoo (*Coccyzus americanus*) on the Sacramento River.

observations from previous census surveys of the yellow-billed cuckoo (Gaines 1974; Gaines and Laymon 1984; Laymon and Halterman 1987b; Halterman 1991) plus a survey conducted by this investigation in July and August 1998 for a total of eleven survey years between 1972 and 1998. The pooled observations were grouped into twenty-one common observation locations within the study reach (Fig. 14.1b). The twenty-one observation locations were then converted to presence-absence values and compared to the output from the HSI model. The results of the model were tested by performing two contingency table analyses using a chi-square test statistic (a Likelihood Ratio, SAS Institute 1994) and Cohen's Kappa (*K*) (Cohen 1960). Kappa is related to chi-square but it de-emphasizes the omission and commission error

terms by placing more weight on the results the model predicts correctly.

Yellow-billed cuckoo surveys were conducted by this investigation over land in late July 1998 and by boat within the main channel and backwaters of the Sacramento River in early August 1998. The surveys were conducted within the study reach south of the Woodson Bridge State Recreation Area to Pine Creek Wildlife Area at river-mile 196. These surveys followed the sampling protocol described by Laymon et al. (1997).

HSI modeling with the historical land-cover data sets. The final stage in this investigation was the application of the geographic HSI model to the five historical land-cover data sets. This was accomplished by adapting the HSI Avenue script developed for the 1997 land-cover mapping to the data sets from 1938, 1952, 1966, 1978, and 1987.

Results

Land-cover Mapping and Validation

The results from the 1997 land-cover mapping (Fig. 14.3) and the field verification study indicated that 87.5 percent of the polygons were correctly classified for the variables "stand height" and "vegetation canopy cover" (Tables 14.2 and 14.3), which is viewed as being within an acceptable target range for land-cover mapping accuracy (Morgan and Savitsky 1998:171). The forest stands in the tall (high) height

TABLE 14.2.

Error matrix for polygon forest stand height classification.

		Photo-interpreted stand height class			
		Low	Medium	High	Total *N*
Field-measured stand height class	Low	6	0	0	6
	Medium	2	4	0	6
	High	0	0	4	4
	Total *N*	8	4	4	16

Note: Proportion correct (accuracy): 14/16 × 100 = 87.5%
Errors denoted in gray box.

TABLE 14.3.

Error matrix for polygon canopy cover classification.

		Photo-interpreted canopy cover class				
		Sparse	Open	Medium	Dense	Total N
Field-measured	Sparse	2	0	0	0	2
canopy cover	Open	1	1	0	0	2
class	Medium	0	1	3	0	4
	Dense	0	0	0	8	8
	Total N	3	2	3	8	16

Note: Proportion correct (accuracy): 14/16 × 100 = 87.5%
Errors denoted in gray boxes.

class were interpreted with greater accuracy and precision than the medium or low height classes due to the greater variability in stand heights in the low and medium stands (Table 14.4).

Habitat Modeling of the Yellow-billed Cuckoo

HSI modeling results. The output of two versions of the HSI model applied to the 1997 land-cover data set are shown in Figure 14.4 (see color section). The two versions were a reduced model using three variables, which did not provide satisfactory results, and a full model using all five variables (Greco 1999). It was evident that the full model was more discriminating in predicting suitable and optimal patches than the reduced model. No patches scored in the marginal category.

Validation of HSI modeling results. Five yellow-billed cuckoos were detected at two of the thirteen

sample locations in 1998. The two detection locations were Pine Creek Wildlife Area (PCWA) at river-mile 197 and the River Vista Unit at Merrills Landing near river-mile 214. Based on the behavior of the detected birds, two of the four cuckoos detected at the PCWA, were presumed to be part of a mated pair, and the other two were presumed to be unmated males. Only one cuckoo was detected at the River Vista Unit and was presumed to be an unmated male.

A pooling of all the yellow-billed cuckoo census data for twenty-one patch locations for presence or absence is presented in Table 14.5 along with a comparison of the resultant scores generated by each HSI model for each location. The overall accuracy (the proportion of locations correctly predicted) of the full model was 81 percent and for the reduced model 48 percent. A presence-absence error matrix was generated for each respective HSI model (Tables 14.6 and 14.7). A chi-square test (Likelihood Ratio, SAS Institute 1994) applied to the results of the HSI models showed the full model, which used all five variables, was highly significant ($G^2 = 12.63$, df = 1, P = 0.0004), whereas the results of the reduced model were only marginally significant ($G^2 = 3.18$, df = 1, P = 0.0746). The Kappa (K) statistical test, which is a measure of agreement between the observed and predicted data, indicated that the full model was moderately significant (K = 0.63, se = 0.154); however, the reduced model was not significant (K = 0.19, se = 0.108) (see Fielding, Chapter 21, for further discussion of the Kappa statistic).

From the standpoint of wildlife habitat relationship modeling, an omission error can have a greater

TABLE 14.4.

Summary of forest stand polygon field sampling results.

Stand height class code	Stand height class	Stand height range (m)	Stand height mean (m)	Stand height sd (m)	Precision (%) (d)	Significance level (α)	Stein's two-stage sample size (N)	Polygons field sampled (N)
H	High	>20	22.7	1.6	10	0.05	3	4
M	Medium	6–20	12.7	3.2	20	0.10	4	4
L	Low	<6	5.8	1.9	20	0.10	6	8
L + M	Low and Medium combined	<20	8.1	4.1	20	0.10	12	12

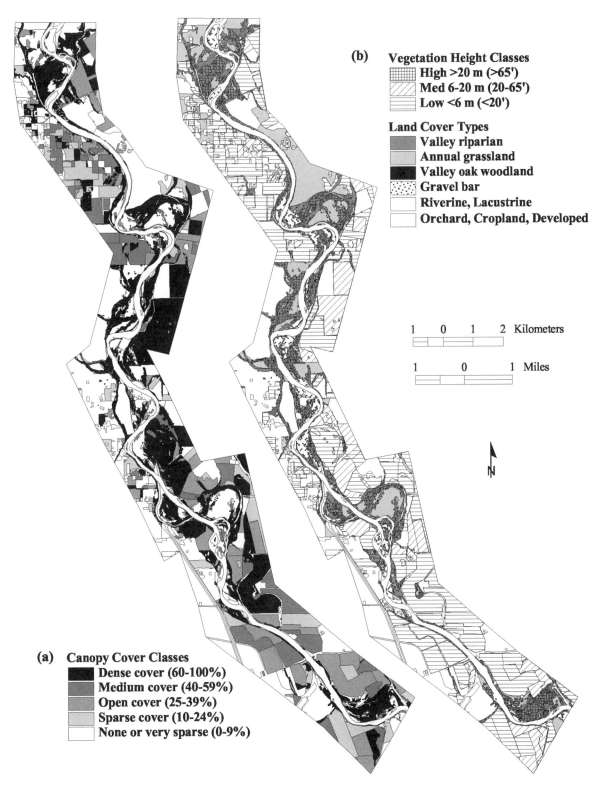

(b) Vegetation Height Classes
High >20 m (>65')
Med 6-20 m (20-65')
Low <6 m (<20')

Land Cover Types
Valley riparian
Annual grassland
Valley oak woodland
Gravel bar
Riverine, Lacustrine
Orchard, Cropland, Developed

1 0 1 2 Kilometers

1 0 1 Miles

N

(a) Canopy Cover Classes
Dense cover (60-100%)
Medium cover (40-59%)
Open cover (25-39%)
Sparse cover (10-24%)
None or very sparse (0-9%)

Figure 14.3. Results from the 1997 aerial photograph land-cover interpretation of the variables (a) canopy cover, and (b) vegetation height class and land-cover habitat type. Habitat types are depicted in various patterns while the size classes (forest stand height) are depicted in various hatching patterns.

TABLE 14.5.

Summary of pooled yellow-billed cuckoo (*Coccyzus americanus*) field observations and habitat suitability index (HSI) model predictions within the study reach.

Patch location number	River mile	Year observed[a]	YBCU birds obs. (total no.)	Reduced model HSI score	Reduced model HSI prediction (±)	Full model HSI score	Full model HSI prediction (±)
1	219.0	1972, 1977, 1987	16	0.87	(+)	0.67	(+)
2	218.5	1972	0	0.47	(−)	0	(+)
3	218.0	1972	0	0	(+)	0	(+)
4	215.2	1972, 1977	2	0.47	(+)	0.54	(+)
5	213.5	1987, 1998	2	0.87	(+)	0.77	(+)
6	212.2	1998	0	0.47	(−)	0.47	(−)
7	211.5	1998	0	0.75	(−)	0.77	(−)
8	211.0	1987–1990, 1998	0	0.87	(−)	0	(+)
9	210.0	1998	0	0.87	(−)	0	(+)
10	209.8	1998	0	0.47	(−)	0	(+)
11	209.5	1972	4	0.60	(+)	0.54	(+)
12	209.0	1998	0	0	(+)	0	(+)
13	207.6	1998	0	0.47	(−)	0.59	(−)
14	207.2	1987–1990	0	0.60	(−)	0	(+)
15	206.5	1998	0	0.60	(−)	0.54	(−)
16	205.5	1987–1990, 1998	0	0.75	(−)	0	(+)
17	204.5	1987, 1989	8	0.87	(+)	0.84	(+)
18	203.0	1987–1990	1	0.87	(+)	0.84	(+)
19	201.5	1977	1	0.47	(+)	0.47	(+)
20	199.0	1972	0	0	(+)	0	(+)
21	197.0	1972, 1977, 1987–1990, 1998	135	0.87	(+)	0.92	(+)

Note: YBCU = yellow-billed cuckoo; [a] = see text for census study references; (+) = correctly predicted; (−) = incorrectly predicted.

negative biological consequence than a commission modeling error (Morrison et al. 1992; Garrison 1993). For example, in this study, an omission error would be an instance where the HSI model failed to predict the presence of potential yellow-billed cuckoo habitat where cuckoos have been observed in field surveys. A commission error, on the other hand, would be an instance where the model predicted the cuckoo's presence in a patch where no yellow-billed cuckoos have been detected by field surveys. "Optimal or suitable"

TABLE 14.6.

Error matrix for the yellow-billed cuckoo (*Coccyzus americanus*) reduced habitat suitability index (HSI) model (three variables).

N = 21		Field observation Present	Field observation Absent
HSI Reduced Model Prediction	Present	8	10
	Absent	0	3

TABLE 14.7.

Error matrix for the yellow-billed cuckoo (*Coccyzus americanus*) full habitat suitability index (HSI) model (five variables).

N = 21		Field observation Present	Field observation Absent
HSI Full Model Prediction	Present	8	4
	Absent	0	9

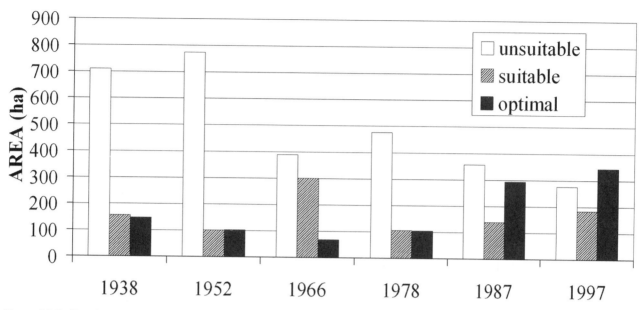

Figure 14.6. Trends in the total area and quality of yellow-billed cuckoo (*Coccyzus americanus*) habitat from 1938 to 1997 on the Sacramento River, river-miles 196–219.

habitats are not always 100 percent occupied (Van Horne 1983; Morrison et al. 1992) and may be occupied in one year and not in the next. Therefore, commission error is a more "preferable" type of ecological error than omission error (Garrison 1993). Both of the spatial HSI models in this study had no omission error whereas the commission error was 19 percent for the full model and 48 percent for the reduced model. Since the yellow-billed cuckoo is presently a relatively rare species (i.e., endangered) and a habitat specialist on the Sacramento River, the commission error rates may have been reduced with larger sample sizes for detection (i.e., a "rarity effect;" see Karl et al., Chapter 51.

Results of HSI model application to historical land-cover data sets. The final step in this investigation was to apply the full HSI model to the historical land-cover data sets interpreted from 1938, 1952, 1966, 1978, and 1987 (see Fig. 14.5a–e in color section). The trend in the quantity of optimal habitat appears to have decreased from 1938 to 1966 and increased from 1966 to 1997 (Fig. 14.6). The spatial distribution of suitable and optimal habitat for the yellow-billed cuckoo shifted considerably from 1938 to 1997 (Fig. 14.5a–f). Some general patterns of the rates of habitat patch formation and extinction are evident.

Perhaps the most revealing example of habitat for-

mation is the Pine Creek Wildlife Area (PCWA) at river-mile 196 (at the southern extreme of the study reach). This location has had consistently high occupancy by yellow-billed cuckoos since 1977. In the extensive census work done by Halterman (1991) on the Sacramento River (from Red Bluff to Colusa), the PCWA location had the highest occupancy (total number of birds) of yellow-billed cuckoos as compared to any other sampled location on the river during the 1987–1990 time period. Results of the modeling indicate that the riparian forests at the Pine Creek bend began to form during the late 1960s and steadily progressed into the 1970s to become optimal habitat in 1978 (Fig. 14.5d). Prior to the 1970s (Fig. 14.5a–c), however, optimal habitat patches were apparently located 9–20 kilometers to the north of Pine Creek bend near the Glen-Colusa Irrigation District (GCID) canal pumping facility and Snaden Island (near river-mile 206).

The formation of riparian forests and oxbow lakes at the River Vista Unit (Merrills Landing at river-mile 214) between 1978 and 1987 suggests that the optimal conditions (as defined in the WHR model in Table 14.1) for yellow-billed cuckoo habitat formation can occur within a nine-year time span. Some locations may require longer time; for example, the Pine Creek Wildlife Area took twelve years to form. In terms of persistence, the southern tip of the riparian forests at Kopta Slough (near river-mile 219) appears to have lasted as suitable habitat for thirty-one years while the location near GCID at river-mile 205.5 evidently lasted twenty-eight years (1938–1966). Between 1938 and 1997, the mean percent of size (height) classes found *not* to change between time periods was 17 percent for low vegetation, 33 percent for medium-sized vegetation, and 62 percent for tall vegetation (more than 20 meters). Maintenance of low- and medium-sized forests appears to be primarily a function of substrates made available from channel migration and meander bend cutoff processes (Greco 1999).

Several extinction patterns are also evident from the modeling results, though we can only speculate as to the cause due to the extensive forest clearing conducted by farmers throughout the study reach. The largest patch to lose habitat value for the yellow-billed cuckoo was located at river-mile 215 (right bank) depicted in 1938 (Fig. 14.5a). This result from the model suggests that habitat values can be altered in less than fourteen years from what appears to have been a river-meander-induced change (from optimal to unsuitable habitat conditions). Hence, the erosive forces of the channel eliminated essential habitat conditions, including low willow (*Salix* spp.) forests and a backwater channel. The GCID location lost its suitable yellow-billed cuckoo habitat between 1966 and 1978. Examination of the GCID location suggests that vegetation growth rates and shifts in vegetation species composition (i.e., succession) were responsible for the change in habitat value within a period of less than twelve years.

Discussion

The concepts and theories of shifting mosaics within the patch dynamics approach to studying heterogeneous landscapes (i.e., Bormann and Likens 1979; Pickett and White 1985; Malanson 1993; Pickett and Rogers 1997) provide a framework within which to evaluate yellow-billed cuckoo core habitat dynamics. The concept of a "minimum dynamic area" as described by Pickett and Thompson (1978) is an approach applicable for anticipating changes to the shifting habitat mosaic of an endangered species. The

proposed Sacramento River Conservation Zone (SRCZ) as detailed in the Sacramento River Conservation Area Handbook (California Resources Agency 1998) is a good starting point for defining this type of a management strategy.

Pickett and Rogers (1997) argued that the shifting mosaic approach is robust enough to unify population ecology with community and landscape ecology because spatial gradients can have great influence on metapopulation dynamics and ecosystem functions. The spatial arrangement of habitat patches and metapopulation distributions on the landscape are important factors for estimating the potential value of a landscape to a species of concern (Hanski and Gilpin 1991; Harrison 1991; Harrison and Fahrig 1995). Predicting how a mosaic of suitable habitat will change with respect to time (Dunn et al. 1991) is a serious management challenge when habitat resources are limited, as is the case on the Sacramento River.

Understanding the turnover rates of habitat features and the sustainability of suitable habitat through time (habitat demography) is clearly critical to long-range recovery planning of endangered species. This investigation shows there has been replacement and an increasing trend of suitable and optimal habitat for the yellow-billed cuckoo since 1978 within the extent of the study reach. This replacement and enhancement of habitat was enabled by the processes of river meander dynamics that create complex geomorphological floodplain surfaces and oxbow lakes (through channel cutoff events) upon which riparian forests colonize and evolve (see Malanson 1993). Essential forest types for the yellow-billed cuckoo are low and medium-sized willow-cottonwood associations (Halterman 1991; Laymon et al. 1997). The low- and medium-height forest stands have the highest rates of turnover in the riparian landscape as shown by Greco (1999). Management should anticipate these shifts by periodically reassessing reserve area functionality and designating new reserve locations in areas undergoing geomorphic changes.

Conclusion

The decline of the yellow-billed cuckoo in the western United States is a prime example of the cumulative effects of the widespread removal of floodplain habitats (Gaines 1977) and extensive alterations to the fluvial geomorphological processes that create and maintain habitat for endangered species (e.g., Scott et al. 1997). A critical point raised by Pickett and Rogers (1997:121–122) is that "sustainability must be judged in part on the maintenance of community function" and "[l]ack of information on the function of patch mosaics is currently the largest limit to ecological knowledge needed to manage patch dynamics effectively." Our results support the validity of this statement in terms of management of riparian forested floodplains on the Sacramento River, especially for the management of endangered species such as the yellow-billed cuckoo.

To effectively plan a conservation strategy for the yellow-billed cuckoo on the Sacramento River, the dynamics of the riparian floodplain landscape must be taken into consideration. A goal of a recovery plan for the yellow-billed cuckoo along the whole river should emphasize the need to restore ecosystem processes that maintain the forest types that constitute suitable and optimal habitat for the cuckoo. A sustainable reserve system for the cuckoo should anticipate shifts in the habitat mosaic. Restoration of the riparian ecosystem on the middle and lower reaches of the Sacramento River (above and below the study reach) should focus on maintenance of channel hydrodynamic processes, such as channel migration and bend cutoff, that give rise to complex riparian floodplain forest mosaics. Distance between levees should be widened in several locations both above and below the town of Colusa (near river-mile 145) to allow for more-extensive riparian forest and floodplain formation. As a long-term conservation strategy, anticipatory and adaptive management should identify currently active meander lands and soon-to-be-active lands (including acquisition of both sides of the river channel) as high priority reserve areas to allow for essential habitats to form in the future to maintain viable breeding populations of yellow-billed cuckoos on the Sacramento River.

Acknowledgments

We would like to thank the California Department of Water Resources, Northern District Office, Red Bluff, California, for funding provided under Agreement No. B-81714, and the U.S. Fish and Wildlife Service, Ecological Services, under Agreement No. FWS 14-48-11300-97-J-146, Sacramento, California. Special thanks to the John Muir Institute of the Environment, Center for Integrated Watershed Science and Management, the Geology Department, the Department of Agronomy and Range Science, and the Department of Environmental Design, University of California at Davis. A grateful thanks to Garrett V. Lee for his critical field and lab assistance and review of this manuscript. Further background and detail about this chapter are presented in a dissertation by Greco (1999) from the Graduate Group in Ecology, University of California, Davis.

Effects of Spatial Scale on the Predictive Ability of Habitat Models for the Green Woodpecker in Switzerland

Claudine Tobalske

Scale influences every aspect of ecological research (Wiens 1989a; Levin 1992), and models relating the distribution of wildlife species to characteristics of their environment are no exception. Three different spatial scales can characterize wildlife habitat relationships (WHR) models: (1) the grain of the species' distribution data; (2) the grain of the habitat variables; and (3) the extent of the study area. Changes in any of these are likely to affect the predictive ability of models. Increasing the grain size of the species' distribution data may reveal patterns hidden by individual variability, as with the American redstart (*Setophaga ruticilla*), a species that is more selective at the territory than at the nest-site level (Sodhi et al. 1999). Alternatively, it may lower the classification success of models developed for species that base habitat selection on microhabitat characteristics (such as the dusky flycatcher, *Empidonax oberholseri*; Kelly 1993). The value taken by habitat variables is also likely to be influenced by the grain at which they are collected. For example, the characteristics of a landscape—the relative proportion of land cover types and their arrangement in patches—are grain dependent: increasing the minimum mapping unit (MMU) of a raster map will affect landscape composition and configuration through the loss of certain cover types (Turner et al. 1989a,b). These changes will, in turn, affect modeling output. Predictive accuracy of the model may increase,

because the "noise" that obscured patterns is eliminated; or it may decrease, because small but important habitat patches associated with the species' presence are gone.

These issues are especially important when the modeler has little control over the scale of the data used to develop the model. The current proliferation of breeding-bird distribution atlases offers a wealth of distribution data that can be used to derive WHR models (Gates et al. 1994; Tobalske and Tobalske 1999). These data, however, are traditionally presented in a grid format at a single, fixed scale that may not be biologically meaningful to the species being modeled and may not maximize classification success. Because habitat selection may be a hierarchical process occurring at several scales (Hutto 1985), patterns of bird-habitat relationships are likely to vary with the scale of investigation (Wiens 1985; Wiens et al. 1987). Indeed, several studies have shown the value of viewing habitat selection at more than one scale (e.g., Virkkala 1991; Bergin 1992; VanderWerf 1993). If this is not possible, then the scale of analysis should at least be compatible with the goals of the study. For example, Fielding and Haworth (1995) elected to work with 1-square-kilometer atlas grid cells because this grain was appropriate for a wide-ranging study aimed at identifying suitable nesting habitat rather than specific nest sites.

In this study, I use atlas distribution maps of the

green woodpecker (*Picus viridis*) from two Swiss breeding-bird atlases to assess the effect of changing the grain of distribution data (atlas cell size) and the grain of habitat variables (MMU) on model classification results.

Methods

The Orbe Valley and Geneva Canton are both situated in the western part of Switzerland, although a portion of the Orbe Valley extends into eastern France (Fig. 15.1). Similar in extent, they present very different landscapes (see Table 15.1, see Fig. 15.2 in color section). The Orbe Valley is characteristic of high-elevation valleys of the Jura mountain range (PNRHJ 1988): the valley floor is open pasture surrounded by dense, unbroken mixed forests dominated by Norway spruce (*Picea abies*) and European beech (*Fagus sylvatica*). Urban development is minimal and scattered. Elevation ranges from 972 to 1,669 meters. Forestry, dairy farming, and tourism are the principal economic activities. By contrast, Geneva Canton is a highly developed agricultural landscape dominated by crops and fields, with important urban and aquatic components (the city of Geneva and the Lake of Geneva). Forests, mostly deciduous, occur as small patches embedded in the agricultural matrix, and elevation ranges from 328 to 563 meters.

Digital Database

Both atlases used in this study present green woodpecker breeding distribution data in the form of 1-square-kilometer (100-hectare) grid cells (Géroudet et al. 1983; Glayre and Magnenat 1984). Breeding green

woodpeckers were censused in 49 of the 273 cells (17.9 percent) of the Orbe Valley and in 203 of the 306 cells (66.3 percent) of Geneva Canton. I used the geographic information system (GIS) software ArcInfo version 7.0.3. (ESRI 1995) on a Unix workstation to digitally recreate the atlas grids.

The models were built from a pool of seven variables: five land-cover classes, mean elevation, and edge density. These variables were selected based on their potential importance to the green woodpecker (Tobalske 1998) and because they were available for both study sites. Land cover was extracted for each atlas cell from a 1988 classified Landsat TM image with a 25-meter pixel resolution (0.0625-hectare MMU). The supervised classification was labeled by intensive groundtruthing of a topographically diverse 15-by-15–square-kilometer area, and by comparison with existing, fine-scale land-cover maps (Vuillod 1994), and resulted in five land-cover classes: oak-hornbeam-beech (Decid), pure beech (Beech), beech-fir-spruce (Conif), conifer plantations (Planted), and not forested (Open). I computed edge density (Edgeden) between forest and nonforest patches for each cell, after grouping the first four classes into one, and vectorizing the resulting file. I obtained mean elevation (Meanelev) for each cell by averaging elevation values from a 50-meter digital elevation model purchased from the French Institut Géographique National.

Modeling Procedures

I used multiple logistic regression (LR) to create models to classify the presence and absence of the green

TABLE 15.1.

Land cover composition of the Orbe Valley and Geneva Canton, Switzerland, as obtained from a classified Landsat TM satellite image.

Land cover class	Orbe Valley		Geneva Canton	
	Area (ha)	Percent	Area (ha)	Percent
Oak-hornbeam-beech forest	0.81	0.00	3,055.13	8.19
Beech forest	3,409.19	10.59	1,994.69	5.35
Beech-fir-spruce forest	16,071.44	49.94	594.44	1.59
Conifer plantation	861.31	2.68	275.94	0.74
Not forested	11,841.17	36.79	31,377.44	84.13
Total	32,183.92	100.00	37,297.64	100.00

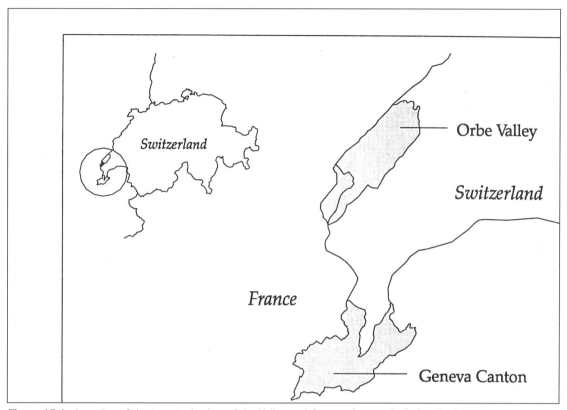

Figure 15.1. Location of the two study sites, Orbe Valley and Geneva Canton, in Switzerland.

woodpecker. To prevent multicollinearity, I computed Pearson product-moment coefficient (r) between all pairs of variables and eliminated one variable from pairs with r greater than 0.7 (Green 1979). The decision about which variable to eliminate was based on the results of univariate LR (log likelihood and Wald statistic; SPSS 1990). Parsimonious models were developed from the remaining pool of variables using both forward and backward stepwise selection procedures. Stepwise procedures were employed because the pool of variables from which the models were built was already small and because these procedures provided an objective, repeatable approach to model building. When the two procedures resulted in different models, I retained the better of the two models based on log likelihood, Wald statistic of the predictor variables, and improvement of the model over chance classification as estimated from Cohen's Kappa (K) (Titus et al. 1984). Addition of variables in the forward procedure was based on the Wald statistic (P-of-entry = 0.05); removal of variables in the backward procedure was also based on the Wald statistic (P-of-removal = 0.1). Because the output of LR is probabilistic, allocation of cases to predicted groups (presence or absence) required that a cutoff be defined; I retained the midpoint between the mean probabilities for the presence and absence cells (Fielding and Haworth 1995). Even though this rule may not maximize Kappa, it was adopted because of its objectivity and consistency.

To assess the influence of atlas cell size on classification results, I created new, scaled-up distribution maps by grouping four adjacent 100-hectare atlas cells. If at least one of the four cells was coded as presence, the new, 400-hectare cell was coded as presence (Fig. 15.3). Because this coding depended on which cells were aggregated, four maps were created to cover all the possible allocations of 100-hectare cells, and a model developed with each. Aggregates that had only two or three cells (along the edge of the study areas) were dropped from analysis. The high proportion of presence cells in Geneva Canton resulted in only three aggregated cells being coded as absence (Fig. 15.3), so

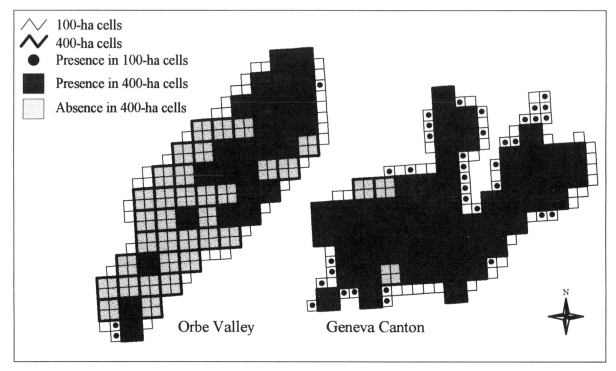

Figure 15.3. Effect of aggregating four cells of the census grid on the distribution of presences and absences for the green woodpecker (*Picus viridis*) in two study sites in Switzerland.

large-cell models could only be built for the Orbe Valley.

The influence of the grain of habitat variables on predictive ability was assessed for both study sites by resampling the classified Landsat image from a 0.0625-hectare MMU to a 1-hectare MMU using a rule-based algorithm (Ma 1995). Previous manipulations showed that increasing the MMU from 1 hectare to 2 or 4 hectares resulted in little additional changes (Tobalske 1998). The LR modeling approach described above was used to derive the following models: (1) 100-hectare cells, 0.0625-hectare MMU (both study sites); (2) 400-hectare cells, 0.0625-hectare MMU (Orbe Valley only); (3) 100-hectare cells, 1-hectare MMU (both study sites); and (4) 400-hectare cells, 1-hectare MMU (Orbe Valley only). Because high rates of presences and absences correctly classified can be obtained by chance alone (Morrison 1969; Capen et al. 1986), I used improvement over chance classification (Kappa) to compare the models' predictive accuracy. After comparing the performances of several confusion matrix-based measures, Fielding and Bell (1997) concluded that Kappa is one of the

most suitable because it makes use of all available information in the confusion matrix.

Results

Geographic location, atlas cell size, and MMU all influenced the composition of the regression equations and the classification accuracy of the models predicting green woodpecker presence and absence (Table 15.2). Elevation was the only consistent variable and entered all the models negatively. The main difference between models for the Orbe Valley and models for Geneva Canton was the sign reversal of the variable Conif. Classification accuracy, as measured by Cohen's Kappa, varied from 0.22 (Orbe Valley, 0.0625-hectare MMU, 100-hectare cell size) to 0.53 (Orbe Valley, 1-hectare MMU, 400-hectare cell size).

Increasing the MMU from 0.0625 hectare to 1 hectare had a strong influence on edge density values (Table 15.3) and reversed the significance of Edgeden in univariate LR (from nonsignificant to significant in the Orbe Valley and from significant to nonsignificant in Geneva Canton, even using a *P*-value as high as

TABLE 15.2.

Regression equations and classification results (percent presence and absence correctly classified, and Kappa value) for logistic regression models developed for the green woodpecker (*Picus viridis*) in two study sites in Switzerland.

Site	MMU (ha)[a]	Cell size (ha)[b]	Regression equation	P (%)	A (%)	Kappa
Orbe Valley	0.0625	100	2.0957 − 0.0029Meanelev − 0.0284Conif +			
			0.3865Planted	61.2	69.2	0.22
		400	0.39 − 0.0678Conif + 1.1775Planted	80.0	70.8	0.51
	1	100	4.7992 − 0.0061Meanelev + 0.02Edgeden			
			+ 1.0318Planted	61.2	72.3	0.25
		400	8.9399 − 0.0094Meanelev + 0.0497Edgeden	77.8	75.0	0.53
Geneva Canton	0.0625	100	8.7277 − 0.0218Meanelev + 0.0158Edgeden	68.5	68.9	0.35
	1	100	12.0752 − 0.0291Meanelev + 0.292Conif	83.7	36.9	0.22

[a]Models were created from a classified Landsat TM scene with two different minimum mapping units (MMU).
[b]Distribution data were extracted for two different cell sizes.

0.2). It did not affect the composition of the two landscapes as much as it affected their configuration: mean patch size increased dramatically (Table 15.3) as small patches of less than 16 pixels (1 hectare) were eliminated. In Geneva Canton, the 1-hectare MMU model resulted in fewer misclassifications among the presences (83.7 percent correctly classified; Table 15.2), but fewer absences were correctly classified than with the 0.0625-hectare MMU model (36.9 percent versus 68.9 percent; Table 15.2), so overall model performance as measured by Cohen's Kappa was lower at 1-hectare MMU (0.223 versus 0.347; Table 15.2). In the Orbe Valley model, the variable Conif was replaced by Edgeden at 1-hectare MMU, but this had little effect on classification results ($K = 0.218$ at

0.0625-hectare MMU versus $K = 0.250$ at 1-hectare MMU; Table 15.2).

Spearman rank correlations, univariate LR results, and regression equations for the four models created in the Orbe Valley from the 400-hectare atlas cells were similar within MMUs, so I kept only the best model at each MMU for comparison. At 0.0625-hectare MMU, the presence/absence ratio changed from 0.22 (49/224) with 100-hectare cells, to 1.25 (30/24) with 400-hectare cells. At 1-hectare MMU, this ratio changed from 0.22 to 0.84 (27/32). At both MMUs, increasing atlas cell size from 100 to 400 hectares more than doubled Kappa (Table 15.2). This increase resulted from a better prediction of green woodpecker presences for the 400-hectare cell models

TABLE 15.3.

Characteristics of two study sites at two different minimum mapping units (0.0625 hectare and 1 hectare): composition (percentage of five Landsat TM classes), mean patch size (hectares, in parenthesis), and edge density (meters per hectare^{-1}; mean value computed from 100-hectare cells).

	Orbe Valley		Geneva Canton	
	0.0625 ha	**1 ha**	**0.0625 ha**	**1 ha**
Decid	0	0	8.2 (0.3)	6.4 (7.3)
Beech	10.6 (0.2)	4.9 (4.0)	5.4 (0.2)	2.0 (4.4)
Conif	49.9 (3.1)	58.6 (79.5)	1.6 (0.1)	0.5 (4.9)
Planted	2.7 (0.1)	0.0 (2.00)	0.7 (0.1)	0.2 (2.9)
Open	36.8 (1.3)	36.5 (41.3)	84.1 (9.3)	90.9 (721.4)
EdgeDen ± SD	106.9 ± 49.3	47.5 ± 31.7	83.9 ± 48.7	17.5 ± 23.1

compared to the 100-hectare cell models (from 61.2 percent to 80.0 percent at 0.0625-hectare MMU and from 61.2 percent to 77.8 percent at 1-hectare MMU; Table 15.2). Classification rates of absences remained fairly constant, around 70 percent at both MMUs (Table 15.2). Differences in K values between 100- and 400-hectare models therefore resulted from a decrease in omission errors (presences predicted as absences), not in commission errors (absences predicted as presences).

Discussion

The goal of this study was to assess how changes in grain of the habitat and distribution data affected the classification results of models developed for the green woodpecker in two Swiss areas. The results suggest that model performance was a function of the landscape characteristics of the study sites, the MMU of the habitat map, and the size of the atlas distribution grid cells.

The models developed for the Orbe Valley and Geneva Canton comprised different variables, and increasing the MMU of the land-cover map had different consequences on the predictive ability of the models in each site (Table 15.2). Although species-habitat associations may vary geographically (e.g. Collins 1983; Shy 1984), the differences observed between the two study sites are more likely an artifact caused by the scale of the study and by the variables used to derive the models. In the Orbe Valley, univariate LR analyses showed that the presence of the green woodpecker was negatively associated with the variable Conif and, conversely, positively associated with the variable Open, and that there was no significant correlation with edge. The exact opposite was found in Geneva Canton: positive correlation with Conif, negative correlation with Open, and strong significance of the variable Edgeden ($P < 0.0005$). This apparent contradiction disappears when the structure and composition of the entire landscapes are considered instead of composition within individual atlas cells. Indeed, forest patch characteristics differ between the two sites (Fig. 15.2). In the Orbe Valley, they tend to be large and unbroken. The green woodpecker is known to avoid closed, dense coniferous forests, favoring in-

stead open or broken deciduous or mixed forests with grassy fringes or clearings (Cramp 1985; Spitznagel 1990; Hågvar et al. 1990; Angelstam and Mikusinski 1994); hence, the negative correlation with Conif. By contrast, in Geneva Canton, forest patches are smaller and scattered in the agricultural matrix. Although considered more an arboreal than a forest species (Cramp 1985), the green woodpecker still requires forest patches for nesting. Hence, the positive correlation between the species' presence and forest classes (and the correlated variable Edgeden) in Geneva Canton. Because the models did not incorporate patch configuration attributes such as patch size, fundamental differences between the two sites could not be taken into account during the modeling phase.

Scale has been defined as the interaction of grain and extent, where grain relates to the level of resolution (i.e., MMU) and extent relates to the largest entities that can be detected in the data (size of the study area or duration of time under consideration; Allen and Hoekstra 1991; Turner et al. 1989a,b; 1993). Using this definition, the Orbe Valley and Geneva Canton study sites were at similar scales; however, because of the presence of larger forest patches, the Orbe valley can be considered a "coarse-grained" landscape whereas Geneva Canton may be considered a "fine-grained" landscape (Forman and Godron 1986). Spatially explicit models that incorporate information about patch size and arrangement (Van Horne 1991) are likely to have higher predictive capabilities than composition-based models, because landscape patterns exert a strong influence on species' distribution (Hansen and Urban 1992; Gustafson et al. 1994; Lescourret and Genard 1994; Farina 1997). Woodpeckers, because of their large territories, are likely to be affected by the spatial patterning of the landscape (Angelstam 1990). Unfortunately, gridded data are poorly suited to extracting configuration variables such as patch size (Tobalske and Tobalske 1999), so I could not test whether the inclusion of spatial variables in the models resulted in higher classification results.

The choice of the grid cell size for breeding-bird atlases (and other distribution atlases) is a compromise between the level of detail sought and the human resources available to conduct censuses. A 100-hectare cell size was retained for both areas, but for larger sites, even this coarse level of sampling may not be possible (Jovéniaux 1993). In the Orbe Valley, increasing the cell size to 400 hectares almost doubled classification success (K), possibly by clarifying bird-habitat association patterns. Heikkinen (1998) suggested that distribution patterns of rare plant species richness in a Finnish reserve may have been more obscured at the 1-kilometer grid scale he used for his models than at either finer or broader scales. The number of 400-hectare absence cells in Geneva Canton was too small (Fig. 15.3) to allow models to be developed, so it was not possible to assess whether the classification improvement observed in the Orbe Valley was site-specific or a more general pattern for the green woodpecker. This illustrates the influence of scale in data-collection procedures: using 400-hectare cells, virtually all of Geneva Canton appeared suitable for the nesting green woodpecker, but this was not the case with 100-hectare cells (Fig. 15.3). The proportion of cells in which the green woodpecker was predicted to be present also increased with increasing cell size in the Orbe Valley, where the ratio of presences over absences reversed from 0.22 at the 100-hectare scale to 1.25 at the 400-hectare scale (at the 0.0625-hectare MMU). The loss of information resulting from aggregating distribution squares could have been lessened by using an index of abundance, in other words the number of 100-hectare cells in each 400-hectare cell in which the species was recorded, as input to the LR procedure (Gates et al. 1994).

It is also important to note that, despite being a better measure of accuracy than percent presence and absence correctly classified, Kappa also has limitations; in particular, it is likely to be sensitive to sample size (Fielding and Bell 1997; Fielding, Chapter 21). Aggregating the 100-hectare cells into 400-hectare cells led to a drop in sample size (from 273 to 54 cells at the 0.0625-hectare MMU and from 273 to 59 cells at the 1-hectare MMU) and an increase in K values. However, if sample size exerted a strong influence on Kappa, opposite results (higher K values for the 100-hectare cell models) would have been expected, because commission error rates have been shown to decline with increasing number of observations (Karl et al., Chapter 51). This was not the case in this study:

commission error rates were similar, and omission rates decreased, between 100-hectare cell models and 400-hectare cell models.

In general, models developed from atlas (gridded) data may be more prone to omission errors than other types of models (e.g., those developed from point data) because the position of the grid cells has no relation to the spatial distribution of land-cover types. If a species is present in a cell dominated by unsuitable habitat, as would be the case if the cell happened to encompass only the edge of a highly suitable habitat patch, then the model will fail to predict a presence. This was observed for bird-habitat models in northwest Scotland, where grid cells of coastal nests were dominated by sea (Fielding and Haworth 1995), and for black woodpecker (*Dryocopus martius*) habitat models in the Jura, France, for cells bordering large forested patches but composed mostly of open habitat (Tobalske and Tobalske 1999).

Finally, validation should be an integral part of model development (Morrison et al. 1992), and whenever possible it should be conducted using an independent data set (Fielding and Bell 1997; Fielding, Chapter 21). I did conduct such an external validation by applying the models developed in the Orbe Valley to data in Geneva Canton and vice versa. In general, the models performed poorly when applied to the other area. In only one instance (100-hectare cells, 0.0625-hectare MMU model from the Orbe Valley applied to Geneva Canton) was improvement over chance classification statistically significant ($P \leq 0.05$). Increasing cell size or MMU did not improve predictive accuracy, and models developed with data from Geneva Canton performed poorly when applied to the Orbe Valley (see Tobalske 1998 for full results). Similarly, models developed for woodpecker species in the adjacent French Jura using a local atlas failed to correctly predict species distribution in the Orbe Valley and Geneva Canton (Tobalske 1998). Differences between the atlases (different grid cell size, surveys conducted by different people, etc.) probably contributed to these results.

Although limited to one species, two study sites, two MMUs, and two atlas cell sizes, the present study demonstrates the overwhelming influence of scale on model composition and predictive accuracy, and the site-specificity of this influence. In one site, classification results were little affected by increasing the MMU of the land-cover data but were sensitive to the grain of the species distribution data. In the other, classification success was higher at the finer MMU than at the larger one. Extending the study to incorporate additional species and scales may further elucidate the relationships between model predictive accuracy and scale, but the results are likely to be site- and species-specific.

Before using atlas data for model development, several issues should be addressed, especially if management is the intended purpose of the models: (1) Is the scale of the atlas grid compatible with the goal of the study? Clearly, a large cell size (and few, if any, atlases use cell sizes smaller than 100 hectares) will not be appropriate if habitat management is to focus on individual nest sites. (2) Which independent variables should enter the model, at what scale, and what is their availability? Atlases typically cover relatively large areas; the development of remote-sensing technology now provides land-cover data for such areas, but other habitat data may only be available patchily (e.g., snag density). (3) How reliable are the distribution maps of the atlas? Atlas quality may vary, especially if the area covered is large, and false-negatives (failure to report a species in a given cell) are likely to occur for rare, secretive, or highly mobile species (Johnson and Sargeant, Chapter 33). If reliability is questionable, data manipulation may be required before proceeding with the modeling phase; Johnson and Sargeant (Chapter 33) present several methods for im-

proving the quality of atlas data. Conversely, models developed from atlas data can also be used to improve the atlas itself: errors of commission may indicate cells in which a species did occur but went undetected (Tobalske and Tobalske 1999). Finally, (4) modelers should consider not only the availability of data for model development, but also for external model validation. The cost associated with collecting independent data may preclude accuracy assessment over large areas. If data from another atlas are to be used for testing the accuracy of the models, it is essential to make sure that they are compatible with data used in model development (such as similar cell size and amount of effort applied during the census).

Acknowledgments

I am grateful to P. Géroudet, D. Glayre, and the other authors of the two breeding bird atlases for permitting use of their distribution data. I also wish to thank R. Redmond of the Wildlife Spatial Analysis Lab (WSAL; University of Montana, Missoula, USA) for providing access to a GIS station. The Laboratoire Environnement et Paysages of the Geography Department of the University of Besançon, France, provided the raw Landsat TM image required to run the MERGE program. C. Winne and T. Taddy (WSAL) helped me download the image and run the MERGE program; I am thankful to all the members of WSAL for sharing their technical knowledge and for providing a wonderful work environment. J. M. Scott provided useful suggestions on an earlier version of the manuscript.

16

Wildlife Habitat Modeling in an Adaptive Framework: The Role of Alternative Models

Michael J. Conroy and Clinton T. Moore

Habitat relationship models (e.g., Verner et al. 1986b; henceforward habitat models) purport to establish a quantitative relationship between measures of the physical and vegetation characteristics of a *habitat* (Morrison and Hall, Chapter 2), including vegetation composition, structure, and spatial arrangement of surrounding habitats, and the presence or absence, abundance, or persistence of one or more species in a *landscape* (Morrison and Hall, Chapter 2). With the rapid development of geographic information systems (GIS) and associated computing algorithms, it is now possible to encode mathematical rules describing presumed habitat-population relationships and to rapidly perform complex analyses of the predicted impacts of various arrangements of land cover and vegetation characteristics. For instance, presumed habitat-species occurrence relationships are a crucial part of gap analysis (Scott et al. 1993), as well as forest-planning tools such as FORPLAN (Johnson et al. 1980).

In this chapter, we review some approaches used to evaluate the accuracy and predictive ability of habitat models. We suggest that standard model validation approaches are ambiguous and that assessment of the reliability of habitat models is most meaningful when models are a part of formal optimization procedures in which management actions are selected so as to achieve a specific, quantitative objective. Decision the-

oretical methods allow for the incorporation of sources of uncertainty in this process, one of which is model reliability. Finally, we think that most conservation decisions are based on a relatively small number of assumptions about ecological pattern and process, and that formal consideration of models based on alternative assumptions is needed. Habitat models can be thought of as tools for translating alternative assumptions into testable predictions, and management can be thought of as the means of providing the experiment under which model predictions can be tested and models and decisions adaptively improved.

Assessing Model Reliability

It is not our intent to provide an exhaustive review either of model assessment in general or of habitat models in particular. Nonetheless, several commonly agreed-upon principles will be relevant to the ensuing discussion. Model *parameterization*, *verification*, and *calibration* (Morrison and Hall, Chapter 2) are all critical parts of model development (Conroy et al. 1995). In practice, each of these steps, even if taken, may be inadequate to assure a model's reliability. For example, model verification demonstrates neither the truth nor the usefulness of a model, only the model's internal consistency; thus, a verified model may nonetheless be inadequate for management if it is based upon faulty assumptions or logic. Likewise, statistical estimates of model parameters often cannot be obtained, especially when key model states or parameters simply cannot be observed, as is frequently the case for highly parameterized models (e.g., spatially explicit population models; Conroy et al. 1995; Dunning et al. 1995; Pulliam et al. 1992). One approach is to use values based on general knowledge or assumptions about the animal's life history, which are then adjusted so as to provide overall model agreement with observations. However, the resulting parameter values are not bona fide statistical estimates, are likely not unique, and may not have biological meaning. A more serious concern is the likelihood that prediction beyond the range of data used to calibrate the model, frequently necessary in management, may prove unreliable.

Regardless of the method of model calibration, there remains the issue of whether the model will in fact be useful for making management decisions (Van Horne, Chapter 4). *Validation* (Morrison and Hall, Chapter 2) probes beyond whether the model appears reasonable, and fits data; it also examines how well the model might perform under conditions different from those in which the model was constructed. However, validation can be difficult in practice, for several reasons. First, statistical uncertainty in the data, imprecision in model predictions, or both may result in low power of statistical validation tests to discriminate between observations and predictions (Mayer and Butler 1993). Thus, failure to reject the null hypothesis that the model and data agree is weak support in favor of the model; in fact, it might simply be an artifact of insufficient sampling effort. Second, field measurements must be of appropriate resolution to validate a habitat model. Consider an artificial example of a species existing in a landscape containing three habitat types, each with different predicted values (under a habitat model) for the species, depending on the *habitat quality* (Morrison and Hall, Chapter 2): high (predicted number = 2/10 hectares), marginal (predicted number = 1/10 hectares), and low (predicted number = 0/10 hectares) (Fig. 16.1a). Suppose we are capable of exactly enumerating the population in each 10-hectare block, and we observe one animal in each block. Under this scenario, we would have obtained a poor correspondence between the model pre-

dictions and observations, with the numbers agreeing in only 44 percent (six of sixteen) of the comparisons. We would probably conclude that this model was "invalid," or in other words was a poor representation of the relationship between habitat and abundance. Suppose instead that we were only capable of counting the total number of animals on each 40-hectare block (but could do so without error) and were incapable of assigning these counts to habitats other than the array of habitats occurring in each 40-hectare block (Fig. 16.1b). Under this scenario, we would have 100-percent agreement between the model predictions and observations and might be inclined to consider this a valid model.

The above artificial example, while highly contrived, illustrates the point that the selection of the spatial scale is a subjective matter (Trani, Chapter 11) but one that may strongly influence the outcome of model validation (Laymon and Reid 1986; Elith and Burgman, Chapter 24). Reliance upon presence/absence statistics in lieu of counts is another form of coarsening of the data that may result in apparent concurrence between model and data when finer resolution of the latter may have resulted in model rejection. At an even finer resolution, predictions based on abundance or density alone will be inadequate for validating source-sink or other models in which habitat quality cannot be inferred from density, regardless of spatial scale (Pulliam 1988; Van Horne 1983; Conroy and Noon 1996; Maurer, Chapter 9).

Sensitivity analysis (Morrison and Hall, Chapter 2) is often advocated as a practical alternative to true validation. However, these arguments are frequently not convincing, particularly if a model is to be used in decision making. Rather than assuring decision makers of the robustness of the model and of GIS, insensitivity to input errors should be a warning that the model also may be insensitive with respect to making predictions. For some applications, it may be sufficient that the model is capable of ordering alternative management actions, with respect to their relative impact on the resource objective (Hamilton and Moller 1995). However, we are less sanguine than Hamilton and Moller (1995) about even this utility for models and instead propose that unverified assumptions and unreliable parameter values may render as unreliable

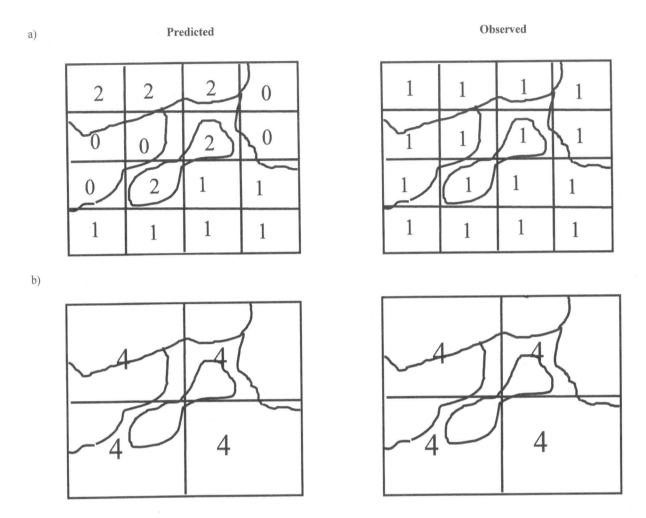

Figure 16.1. Hypothetical comparison between predictions of a simple landscape model and observations under a null model of no habitat affinity; cell values represent predicted and observed counts (a) by 10-hectare block, (b) by 40-hectare block.

even ordinal statements about the relative impacts of various management alternatives; an example of how this might occur follows.

The Role of Alternative Models

As seen in the previous section, validation of habitat models presents serious methodological challenges. However, validation alone cannot resolve whether the model under consideration is superior to a plausible alternative model, in particular, a model that may imply a different course of management. We address this issue in a more appropriate decision-making context. Assume that we have a model (*M*) and an alternative

model (*M′*), and that both models are plausible, that is, at least some theoretical or empirical support exists for each (e.g., Pascual et al. 1997). It may be that we have performed model validation tests and both models are valid (i.e., neither model is rejected in comparison to the data available). A natural question for a decision maker is: what difference will it make to my decision if I place full faith in model *M* versus if I place full faith in model *M′*?

A simple illustration can be used to make this point by returning to the artificial example in Figure 16.1. Suppose that *M* corresponds to the model predicting that the species has specific affinities, as predicted by the map in Figure 16.1a ("*Predicted*") and that model

M' corresponds to the situation in Figure 16.1a ("*Observed*") in which the species is distributed perfectly evenly among the sixteen 10-hectare blocks. Clearly, belief in one or the other of these models will make a difference in how habitats should be managed. Under the *M* scenario, management presumably would be directed toward certain habitats—in other words those that are preferred by the species, assuming a goal of conserving this species. Under the *M'* scenario, management favoring these habitats would appear to be unwarranted, particularly given that such management would no doubt have inherent costs (e.g., tradeoffs with other objectives). Clearly, from the standpoint of decision making, injudicious choice of the spatial scale for model prediction might potentially result in a critical loss of information.

Note that, depending on the scale at which observations are made, the models are both valid and therefore equally plausible (Fig. 16.1b), or one model appears to have more empirical support than the other (Fig. 16.1a). Thus, the usual approach for validation—of comparing predictions with independent observations—may be indeterminate, depending on the scale chosen. Also, note that sensitivity analysis contributes little to the resolution of model uncertainty. The fact that either model, or both, depending on the spatial scale, is relatively more or less sensitive to changes in parameter values sheds no light upon the question of which (if either) model will better inform decision making.

Decision Making under Uncertainty

If habitat or other models are to be useful to managers, they must be capable of making predictions about the consequences of management decisions that are better than the educated guesses that managers would make on their own in the absence of models. The fact that mathematical algorithms can join together hundreds or thousands of habitat models and rapidly display the results using GIS should be small comfort if critical model components are poorly substantiated by evidence (Van Horne, Chapter 4). Even in those cases where models seem to do a reasonable job of prediction, our earlier discussion should convince readers of the risks of blind application of valid models to decision-making problems.

On the other hand, we recognize that decisions must be made and that imperfect models, validated at inappropriate scales of resolution, or perhaps not at all, may be all that are available. Even under ideal circumstances, assumptions about biological mechanisms will not be perfectly understood, and thus it will not be possible to make unambiguous predictions about the impacts of management decisions whether or not models are used to make these predictions. Obviously, biological systems, even if well understood, are subject to intrinsic variability, but of special concern here is what we term *structural uncertainty*. That is, more than one mechanism (or model) might plausibly explain and predict the potential response of the system to management, and we are uncertain as to which is better *for a given management goal*. Formally, we are faced with making a decision or action, *a*, from a set of possible or feasible decisions $a \in \{a_1, a_2, \ldots, a_n\}$. Any decision we make will result in an outcome that will have a value to us, which we will denote as $u(a)$. This value may be in terms of species conservation, economic gain, or perhaps a tradeoff between one or more goals (e.g., species conservation versus economic gain). Assuming that such an objective value can be ascertained or agreed upon, a rational decision maker (Lindley 1985) will seek to select that decision that will result in the greatest value for the objective. However, uncertainty exists as to what the actual objective value or utility will be for any decision. First, consider uncertainty induced by environmental or demographic variability. Let $E[u(a|\theta)]$ represent the average or expected value or utility of decision *a*, assuming that a particular model or parameter value (represented by θ) is known to be true. This value is obtained by averaging over the statistical distribution of uncertain outcomes *x* resulting from each possible decision *a*

$$E[u(a|\theta)] = \int_x u(a|x; \theta)\, f(x|\theta)\,dx$$
$$\text{for } a \in \{a_1, a_2, \ldots, a_n\}$$

where $u(a|x; \theta)$ is the value of decision *a* given outcome *x* under model θ and $f(x|\theta)$ is a statistical distribution of these outcomes, where θ is assumed known.

Until this point, we have assumed that the model (as expressed by $f(x|\theta)$) is correctly specified, and that any deviations of model predictions from outcomes must be due to environmental or demographic factors.

Here we switch our focus to structural (model) uncertainty. Let $p(\theta_i)$ represent a probability distribution reflecting our uncertainty in θ, which takes on values θ_i, $i = 1, \ldots, m$ under each of m alternative models. This uncertainty may include statistical error, but more generally includes bias due to incorrect model assumptions. The average value of the decision is now

$$E[u(a)] = \sum_{i=1}^{m} \int_x u(a|x;\theta_i)\, p(\theta_i) f(x|\theta_i) dx \qquad (16.1)$$
$$\text{for } a \in \{a_1,\ a_2,\ \ldots,a_n\}$$

$\sum_{i=1}^{m} p(\theta_i) = 1$; when θ is continuous, the summation operator changes to integration over θ. By definition, the optimal decision satisfies

$$\max_a E[u(a)]$$

and must be found by averaging over the uncertain environmental and demographic conditions (i.e., values of x) and structural uncertainty (i.e., values of θ). Ignoring either source of uncertainty will result in suboptimal decision making. Conversely, reduction of either source of uncertainty will improve decision making. Obviously, there is little that can be done about environmental and demographic uncertainty, beyond including components of each in the decision-making model.

On the other hand, structural uncertainty *can* be reduced, theoretically to zero, if additional information (data) can be obtained that places higher probability on certain model structures (values of θ) than on others. We describe below how this source of uncertainty can be reduced via adaptive management. For now, we focus on the impact of structural uncertainty on decision making. Consider a case where an optimal decision a is sought, and consider only the average response across environmental and demographic conditions, assuming that a given model of ecological processes is true. Suppose that there are two alternative models of this process, which we shall label θ_1 and θ_2, and that our degree of belief in each model is $p(\theta_1)$ and $p(\theta_2) = 1 - p(\theta_1)$, respectively. The expected value of any candidate decision a, taking into account only structural uncertainty, is

$$E[u(a)] = u(a|\theta_1)p(\theta_1) + u(a|\theta_2)p(\theta_2).$$

Clearly, structural uncertainty exists any time that $0 < p(\theta_1) < 1$. However, notice that this uncertainty is only important in the decision-making process to the extent that the values of the resulting decisions would be different, or in other words

$$u(a|\theta_1) \neq u(a|\theta_2).$$

Conversely, if the models predict the same outcome for any given decision, or if that outcome is equally valued to the decision maker, then uncertainty about the ecological process is not relevant to the decision process. This can be illustrated by a simple numerical example. Suppose that for each of the above two model structures we obtain values for decision a of $u(a|\theta_1) = 4$ and $u(a|\theta_2) = 7$. Suppose that there is a competing decision, a', for which the corresponding values are $u(a'|\theta_1) = 6$ and $u(a'|\theta_2) = 4$. If there is complete uncertainty about which model correctly describes the process, then $p(\theta_1) = p(\theta_2) = 0.5$ and the values for each decision are given by

$$E[u(a)] = 4(0.5) + 7(0.5) = 5.5$$

and

$$E[u(a')] = 6(0.5) + 4(0.5) = 5.$$

Therefore, the optimal decision is a. Suppose however that additional knowledge accumulates (e.g., from a monitoring program carried out on the managed system) so as to place more faith in model 1, such that $p(\theta_1) = 0.8$. Now the decision values are

$$E[u(a)] = 4(0.8) + 7(0.2) = 4.6$$

and

$$E[u(a')] = 6(0.8) + 4(0.2) = 5.6$$

and the optimal decision is now a'. This approach thus places the issue of model reliability (and its resolution) squarely in the context of optimal decision making. That is, we are no longer comparing a single model to an arbitrary measure of accuracy but instead are asking which decision should we make given two or more plausible models and an assessment of relative belief in each model. In some cases (e.g., where theory or data provide justification), we may be justified in giving one model more weight; in others (e.g., either model is theoretically justifiable, and both seem valid given current data), we may not. In either instance, we have an objective means for making a decision, taking into account model uncertainty.

Adaptive Management

As shown above, model uncertainty must be considered, along with other sources of uncertainty, in making optimal conservation decisions. Because our knowledge of systems will always be imperfect, and parameters will always be estimated with error, model uncertainty can never be eliminated. However, model uncertainty can and should be reduced. One method to reduce model uncertainty is adaptive optimization, as incorporated as a part of adaptive resource management (Walters 1986). The basic steps of adaptive optimization are

1. Define a resource objective (e.g., species conservation, as above)
2. Delineate a set of feasible management alternatives $\{a_1, a_2, \ldots, a_n\}$
3. Develop models $\theta \in \{\theta_1, \theta_2, \theta_3, \ldots, \theta_m\}$ that predict the impact of the decision on the objective
4. Identify and quantify the relevant sources of uncertainty in (3)
5. Implement the decision that appears to be optimal given (4)
6. Compare predictions under each model to data ($\underline{x}$) collected following management
7. Compute a likelihood $L(\underline{x}|\theta)$ for each model given these data; these likelihoods reflect the relative agreement of the observed data to the predictions of each model
8. Update the model probabilities from Bayes' Theorem

$$p(\theta_i|\underline{x}) = \frac{p(\theta_i)L(\underline{x}|\theta_i)}{\sum_{j=1}^{m} p(\theta_j)L(\underline{x}|\theta_j)}; \theta_i \in \{\theta_1, \theta_2, \ldots, \theta_m\}, \quad (16.2)$$

where $p(\theta_i \mid \underline{x})$ is the posterior probability of θ_i, conditioned on having observed the data $\underline{x}$
9. Incorporate these new model weights in prediction and decision making at the next decision opportunity

Thus, adaptive resource management provides a mechanism for feedback of information following management, which in turn reduces model uncertainty and promotes further understanding of system processes. Because of the long-term nature of many conservation problems, that feedback may be slow or may not occur at all at a given location (e.g., once a reserve is built, there will likely be little interest in revisiting the decision). However, knowledge gained through monitoring one system should inform future decision making in similar systems. In other conservation problems, for instance those involving forest cutting practices, decisions may be regularly revisited and the information gained from one decision cycle will provide direct feedback for future decision cycles.

Case Study: Habitat Management for Population Persistence under Uncertainty

We illustrate the above principles with an example of landscape management in which the objective is the maintenance of populations of two forest species. The two species have resource needs that pose a potential conflict for management in the sense that provision of resources for one species may remove resources for the other. This example, although hypothetical, is similar to a problem we are currently investigating involving forest management in the Piedmont National Wildlife Refuge (PNWR) in central Georgia. At PNWR, a primary management emphasis is that of maintaining viable populations of the endangered red-cockaded woodpecker (*Picoides borealis*, henceforth woodpeckers), with the long-term goal of tripling the 1998 refuge population (Richardson et al. 1998). However, concern exists that aggressive management favoring woodpeckers, including maintenance of low densities of understory and midstory vegetation via prescribed burning, may adversely affect species of birds and other organisms that depend on these vegetative strata for shelter, foraging, or nesting. Previous research (Powell et al. 2000) has addressed the specific concern that woodpecker habitat management reduces fitness of the wood thrush (*Hylocichla mustelina*, henceforth thrushes) as measured by adult and juvenile survival and reproduction rates. Results to date suggest that woodpecker management, at least as currently practiced at PNWR, has minimal if any impact on thrush fitness. However, this system exhibits great spatial and temporal variability in demographic parameters, which together with estimation error induces uncertainty in these conclusions. Further, there is no assurance that results observed by Powell et al. (2000) would extend to a more aggressive management regime than that which occurred during the study. Current understanding of the effects of woodpecker management (e.g., Richardson et al. 1998) may be inadequate to accurately predict whether such management would enhance the long-term viability of woodpeckers—the primary goal of its management. These factors, taken together, suggest that forest management at PNWR and similar systems, in addition to being influenced by system uncertainty and statistical error, may be relatively sensitive to structural assumptions in models used to predict the impact of management decisions on objective values that include both a woodpecker component and a component reflecting other resource goals.

We describe a simplified, artificial system that nonetheless captures some of the essential elements of management at PNWR. Here we reduce the resource management objective to a tradeoff between a species favored by understory vegetation reduction (represented by woodpeckers) and another favored by its retention (represented by thrushes). However, the statements that "woodpeckers are favored by burning" and "thrushes are favored by exclusion of burning" result from assumptions in the models underpinning the decision analysis. We thus formulated explicit alternatives to these assumptions to mimic extremes in the relationship between population response and management that might be consistent with real field data. Specifically, we considered alternatives that propose that populations do not respond to management actions. We incorporated these different hypotheses in eight alternative models, as described below.

System Features and General Assumptions

The landscape was represented as a 10×10 square grid. Any cell in the grid could be occupied by a woodpecker, a thrush, or by both species. From an initial distribution of woodpeckers within the landscape, models predict a resulting distribution of woodpeckers following a single 10-year time step. These models alternatively suggest that woodpecker population growth is, or is not, dependent on distance to nearest-

neighbor source sites, and is, or is not, dependent on woodpecker response to habitat management through controlled burning. In contrast, we model thrush occupancy only in a habitat-suitability context and do not consider an initial distribution of thrush. That is, following management a thrush occurs in a cell with probability that does, or does not, depend on the burning status of that cell. Generally, we want to maximize population growth of woodpeckers and density of thrushes through appropriate selection of one of a few decision alternatives. Our aim is to look at every alternative for each combination of models and for certain initial distributions of woodpeckers.

Habitat-occupancy Models

For the woodpecker, we modeled single-time-step cell occupation probabilities conditional on current cell occupation status, habitat treatment, and distance to nearest occupied cell. That is, we built expressions for the conditional probabilities

$$\Pr\{X_i(1) = x_1 | X_i(0) = x_0, d_i(0) = d_0, H_i = h_i\},$$

where $X_i(t)$ is a random variable indicating occupation status of landscape cell i at time t, $x_t = 0$ (unoccupied) or 1 (occupied), $d_i(0)$ is the decision variable for cell i, d_0 is the decision value (1 = burned, 0 = not burned), H_i is a random variable, and h_i is a distance value.

Given that landscape cell i is currently occupied by a woodpecker (i.e., $X_i(0) = 1$), we used the following expression as the model of cell occupancy probability at time 1:

$$\Pr\{X_i(1) = 1 \mid X_i(0) = 1, d_i(0) = d_0\} = \begin{cases} p_0, & d_0 = 1 \\ p_0', & d_0 = 0 \end{cases} \quad (16.3)$$

where p_0 and p_0' are user-selected probabilities. Because $x_t = 0$ or 1, $\Pr\{X_i(1) = 0 \mid X_i(0) = 1, d_i(0) = d_0\} = 1 - \Pr\{X_i(1) = 1 \mid X_i(0) = 1, d_i(0) = d_0\}$. Thus, the probability of woodpecker persistence is sensitive to the management decision, where the degree of sensitivity is reflected in the difference $p_0 - p_0'$.

Given that landscape cell i is not currently occupied by a woodpecker, we expressed the probability of cell i being colonized at time 1 as a function of distance h_i to the nearest occupied cell and burning status for cell i:

$$\Pr\{X_i(1) = 1 \mid X_i(0) = 0, d_i(0) = d_0, H_i = h_i\}$$
$$= \begin{cases} e^{-h_i/\beta}, & d_0 = 1 \\ e^{-h_i/\alpha\beta}, & d_0 = 0 \end{cases} \qquad (16.4)$$

where α and β are user-controlled parameters. Thus, woodpecker colonization probability is partially dependent on the spatial distribution of woodpeckers. We proposed an alternative model in which colonization probability was not sensitive to h_i:

$$\Pr\{X_i(1) = 1 \mid X_i(0) = 0, d_i(0) = d_0\}$$
$$= \begin{cases} \beta(1 - e^{-a/\beta})/a, & d_0 = 1 \\ \alpha\beta(1 - e^{-a/\alpha\beta})/a, & d_0 = 0 \end{cases} \qquad (16.5)$$

where a is a user-controlled parameter. We derived this model by integrating the functions in equation 16.4 over the interval 0 to a and then dividing the result by the length of the interval to obtain a uniform probability mass over 0 to a.

For both the woodpecker persistence and the colonization models, we considered forms in which occupation probabilities were not dependent on the habitat decision. For the persistence model, we used the expression

$$\Pr\{X_i(1) = 1 \mid X_i(0) = 1\} = (p_0 + p_0')/2. \qquad (16.6)$$

For the spatially dependent colonization model, we used the expression

$$\Pr\{X_i(1) = 1 \mid X_i(0) = 0, H_i = h_i\} = (e^{-h_i/\beta} + e^{-h_i/\alpha\beta})/2 \quad (16.7)$$

and for the non-spatially dependent model, we used

$$\Pr\{X_i(1) = 1 \mid X_i(0) = 0\}$$
$$= [\beta(1 - e^{-a/\beta}) + \alpha\beta(1 - e^{-a/\alpha\beta})]/2a. \qquad (16.8)$$

Combinations of these model structures provided four alternative models for the woodpecker response to management and woodpecker spatial distribution:

1. model W_{DS}—decision-sensitive and spatially sensitive; equations 16.3 and 16.4.
2. model W_D.—decision-sensitive and spatially insensitive; equations 16.3 and 16.5.
3. model $W._S$—decision-insensitive and spatially sensitive; equations 16.6 and 16.7.
4. model $W..$—decision-insensitive and spatially insensitive; equations 16.6 and 16.8.

Unlike probability in the woodpecker models, the thrush occupation probability of cell i at time 1 was considered to be solely dependent on habitat treatment in one model alternative (T_D)

$$\Pr\{Y_i(1) = 1 \mid d_i(0) = d_0\} = \begin{cases} q_0, & d_0 = 1 \\ q_0', & d_0 = 0 \end{cases} \quad (16.9)$$

where Y_i is the thrush occupation status (either 0 or 1) of cell i at time 1, and q_0 and q_0' are probabilities set by the user. An alternative model (T.) to reflect decision-insensitivity for thrushes is

$$\Pr\{Y_i(1) = 1\} = (q_0 + q_0')/2 .$$

Thus, the four woodpecker model alternatives in combination with each of the two thrush model alternatives yielded eight alternative system models.

Landscape Simulation

We simulated effects of decisions under each of the species models over a range of initial woodpecker conditions. We considered four types of initial condition: (1) low woodpecker occurrence ($n = 5$ cells occupied), highly clumped; (2) low occurrence, highly dispersed; (3) high occurrence ($n = 20$), highly clumped; and (4) high occurrence, highly dispersed. We used a rejection procedure to generate clumped and dispersed distributions. We calculated an index of clumping K (Krishna Iyer 1949; Pielou 1977) for each randomly generated candidate distribution of n occupied cells, and we assumed that the index followed a normal distribution under random mingling of cells. We accepted the distribution as a clumped distribution if $K \geq 1.282$ (normal critical value at 90th percentile) and as a dispersed distribution if $K \leq -1.282$. We continued this process until we had generated one hundred distributions of each type on the landscape grid.

For each initial distribution of woodpecker occupancy, we simulated a set of management decisions under each of the alternative models. The burning status for cell i, $d_i(0)$, was a random outcome of the decision variables $d^{(1)}$, the proportion of woodpecker-vacant habitat burned, and $d^{(2)}$, the proportion of woodpecker-occupied habitat burned. For a fixed selection of $d^{(1)}$ and $d^{(2)}$, $(100 - n)d^{(1)}$ cells were randomly chosen for burning from the set of woodpecker-vacant cells, and $nd^{(2)}$ cells were chosen at random from the set of woodpecker-occupied cells. We considered four settings of $d^{(1)}$ and $d^{(2)}$:

1. $\{d^{(1)}, d^{(2)}\} = \{0.2, 0.2\}$
2. $\{d^{(1)}, d^{(2)}\} = \{0.2, 0.8\}$
3. $\{d^{(1)}, d^{(2)}\} = \{0.8, 0.2\}$
4. $\{d^{(1)}, d^{(2)}\} = \{0.8, 0.8\}$

Given an initial distribution of woodpecker and values of $d^{(1)}$ and $d^{(2)}$, we drew one hundred random arrangements of the $d_i(0)$. Thus, each of the sixteen combinations of initial conditions and decision variables provided ten thousand random distributions of woodpecker occupancy and burning activity.

All simulations were conducted over a single ten-year time step. Values 0.904 and 0.665 for p_0 and p_0', respectively, correspond to annual persistence rates of 0.99 and 0.96; in other words, annual risk of extirpation is four times as likely for an unburned cell than for a burned cell. We chose values of 0.8 and 0.25 for β and α, respectively, which render colonization unlikely in any burned cell not adjacent to an occupied cell. For unburned cells, colonization is extremely unlikely for any nearest-neighbor distance. We chose values of 0.1 and 0.6 for the thrush occupation probabilities q_0 and q_0', respectively.

For each of the decision simulations, we recorded the woodpecker population growth as $\lambda = \Sigma X_i(1)/n$, and we calculated $w = \Sigma Y_i(1)/100$, the proportion of habitat occupied by thrushes. We combined these quantities in the objective function

$$J = \{\max(0, \lambda - 1)\}^u \, w^v,$$

where u and v were set to the values 1.0 and 0.2, respectively. These values imply that woodpecker population growth is rewarded approximately linearly as long as thrushes occupy a minimum threshold (about 20 percent) of the landscape. Rewards are minimal if the decision grows one species at the expense of the other. We obtained means and variances of ten thousand objective function evaluations for each of 128 initial condition × model alternative × decision alternative combinations.

Simulation Results

Because initial conditions, the decision action, and population responses were all realizations of stochastic processes, values of the objective function were also stochastic. Therefore, for any given population model, each decision was superior to the others in least one simulation simply by chance (Tables 16.1, 16.2). However, the large number of simulations clearly indicated that certain decisions provided the greatest expected value of the objective function and that others were consistently inferior.

The optimal decision depended on accurate identification of the underlying management response model (Tables 16.1, 16.2; Fig. 16.2). For example, given that initial woodpecker population size is 20, then the decision to burn 20 percent of both woodpecker-vacant and woodpecker-occupied landscape cells is the best decision only if one correctly presumes that thrushes respond negatively to fire and that woodpeckers do not respond at all (Fig. 16.2d–f). However, this same decision is the worst that can be made if, in fact, woodpeckers respond positively to fire (Fig. 16.2d–f). The four decisions were equally adequate only in the special case in which neither species responded to fire management.

The parameter values that we chose for the objective function heavily rewarded management directed toward woodpeckers, and this was reflected in how the decision patterns varied among management response models. For the four model types in which woodpeckers were not assumed to respond to fire management (models W..T., W..T$_D$, W.$_S$T., W.$_S$T$_D$), models W..T. and W.$_S$T. provided no trend in mean objective value as extent of burning increased in the landscape, whereas the thrush response models (W..T$_D$ and W.$_S$T$_D$) provided a negative trend as more of the landscape was burned (Fig. 16.2). However, all woodpecker response models provided a positive trend in mean objective value, though the rate of increase was slower when the thrush response was considered (Fig. 16.2, models W$_D$.T$_D$ and W$_{DS}$T$_D$) than when it was not (Fig. 16.2, models W$_D$.T. and W$_{DS}$T.). As objective value parameters are altered to bring management desires for the two species into greater conflict, we would expect the trend in objective value over the decisions under the

TABLE 16.1.

Mean and approximate 99% confidence interval for objective value (J) and frequency of optimality (n_{Opt}) for four decisions under eight alternative system models and two types of spatial arrangements of woodpeckers (dispersed versus clumped), given an initial population of five woodpeckers.

| | Decision[a] | | | | | | | | | | | |
| | $d^{(1)} = 0.2, d^{(2)} = 0.2$ | | | $d^{(1)} = 0.2, d^{(2)} = 0.8$ | | | $d^{(1)} = 0.8, d^{(2)} = 0.2$ | | | $d^{(1)} = 0.8, d^{(2)} = 0.8$ | | |
Model[b]	J	99% CI	n_{Opt}	J	99% CI	n_{Opt}	J	99% CI	n_{Opt}	J	99% CI	n_{Opt}
Initial woodpecker population size = 5, highly dispersed												
$W_{\cdot\cdot}T_{\cdot}$	0.384	(0.376–0.392)	2440	0.385	(0.378–0.393)	2400	0.386	(0.378–0.394)	2537	0.388	(0.381–0.396)	2469
$W_{\cdot\cdot}T_D$	0.412	(0.404–0.420)	2941	0.414	(0.406–0.422)	2868	0.347	(0.340–0.354)	2095	0.345	(0.338–0.352)	2005
$W_{D\cdot}T_{\cdot}$	0.175	(0.169–0.180)	438	0.254	(0.248–0.260)	809	0.521	(0.512–0.530)	3518	0.631	(0.622–0.640)	5124
$W_{D\cdot}T_D$	0.187	(0.182–0.193)	713	0.267	(0.261–0.274)	1154	0.471	(0.462–0.479)	3395	0.559	(0.551–0.567)	4664
$W_{\cdot S}T_{\cdot}$	0.632	(0.622–0.641)	2529	0.626	(0.616–0.635)	2423	0.631	(0.621–0.640)	2436	0.633	(0.623–0.642)	2501
$W_{\cdot S}T_D$	0.674	(0.664–0.684)	3059	0.674	(0.664–0.684)	3056	0.576	(0.567–0.584)	1994	0.557	(0.548–0.565)	1825
$W_{DS}T_{\cdot}$	0.161	(0.156–0.167)	25	0.237	(0.231–0.242)	45	1.045	(1.033–1.056)	4254	1.158	(1.147–1.169)	5622
$W_{DS}T_D$	0.174	(0.168–0.179)	59	0.256	(0.250–0.262)	107	0.939	(0.929–0.950)	4269	1.031	(1.020–1.041)	5488
Initial woodpecker population size = 5, highly clumped												
$W_{\cdot\cdot}T_{\cdot}$	0.385	(0.377–0.393)	2447	0.386	(0.378–0.393)	2483	0.388	(0.380–0.396)	2441	0.388	(0.380–0.396)	2487
$W_{\cdot\cdot}T_D$	0.414	(0.406–0.423)	2972	0.409	(0.401–0.418)	2881	0.346	(0.339–0.353)	2060	0.341	(0.335–0.348)	2013
$W_{D\cdot}T_{\cdot}$	0.178	(0.173–0.184)	512	0.252	(0.246–0.258)	812	0.515	(0.506–0.524)	3500	0.625	(0.616–0.634)	5050
$W_{D\cdot}T_D$	0.189	(0.183–0.195)	728	0.270	(0.263–0.276)	1141	0.471	(0.462–0.479)	3392	0.559	(0.551–0.567)	4670
$W_{\cdot S}T_{\cdot}$	0.474	(0.465–0.482)	2468	0.475	(0.466–0.483)	2447	0.475	(0.466–0.483)	2467	0.476	(0.468–0.485)	2494
$W_{\cdot S}T_D$	0.506	(0.497–0.515)	2971	0.505	(0.496–0.514)	2934	0.433	(0.426–0.441)	2046	0.428	(0.420–0.436)	1977
$W_{DS}T_{\cdot}$	0.115	(0.111–0.119)	56	0.182	(0.177–0.187)	107	0.784	(0.773–0.794)	4000	0.903	(0.892–0.913)	5761
$W_{DS}T_D$	0.122	(0.117–0.127)	97	0.192	(0.187–0.198)	191	0.707	(0.698–0.717)	4080	0.799	(0.790–0.808)	5548

[a]Expressed as proportion of ninety-five woodpecker-vacant cells ($d^{(1)}$) and proportion of five woodpecker-occupied cells ($d^{(2)}$) burned. Each decision was simulated one hundred times under one hundred random woodpecker occupancy distributions.

[b]Model expressed as a character triplet $W_{ij}T_k$, where i indicates woodpecker colonization and persistence probabilities are (i = D) or are not (i = .) sensitive to burning, j indicates woodpecker colonization probability is (j = S) or is not (j = .) sensitive to distance from a nearest-neighbor source cell, and k indicates thrush occurrence probability is (k = D) or is not (k = .) sensitive to burning.

TABLE 16.2.

Mean and approximate 99% confidence interval for objective value (J) and frequency of optimality (n_{Opt}) for four decisions under eight alternative system models and two types of spatial arrangements of woodpeckers (dispersed versus clumped), given an initial population of twenty woodpeckers.

| | Decision[a] | | | | | | | | | | | |
| | $d^{(1)} = 0.2, d^{(2)} = 0.2$ | | | $d^{(1)} = 0.2, d^{(2)} = 0.8$ | | | $d^{(1)} = 0.8, d^{(2)} = 0.2$ | | | $d^{(1)} = 0.8, d^{(2)} = 0.8$ | | |
Model[b]	J	99% CI	n_{Opt}	J	99% CI	n_{Opt}	J	99% CI	n_{Opt}	J	99% CI	n_{Opt}
Initial woodpecker population size = 20, highly dispersed												
$W_{..}T_{.}$	0.016	(0.015–0.017)	1441	0.016	(0.015–0.017)	1517	0.016	(0.015–0.017)	1499	0.015	(0.014–0.016)	1429
$W_{..}T_D$	0.018	(0.017–0.019)	1614	0.017	(0.016–0.018)	1554	0.016	(0.015–0.017)	1417	0.015	(0.014–0.015)	1365
$W_D.T_{.}$	0.002	(0.001–0.002)	124	0.015	(0.014–0.016)	1179	0.016	(0.015–0.017)	1151	0.060	(0.058–0.061)	4837
$W_D.T_D$	0.002	(0.002–0.002)	175	0.015	(0.014–0.015)	1285	0.015	(0.014–0.016)	1169	0.056	(0.054–0.057)	4725
$W_{.}sT_{.}$	0.208	(0.205–0.212)	2557	0.205	(0.202–0.209)	2409	0.207	(0.203–0.210)	2525	0.205	(0.202–0.209)	2430
$W_{.}sT_D$	0.224	(0.220–0.227)	3146	0.218	(0.215–0.222)	2867	0.194	(0.191–0.197)	2122	0.185	(0.182–0.188)	1845
$W_{DS}T_{.}$	0.015	(0.014–0.016)	1	0.061	(0.059–0.063)	8	0.363	(0.359–0.367)	2964	0.478	(0.475–0.482)	6993
$W_{DS}T_D$	0.017	(0.016–0.018)	3	0.066	(0.064–0.068)	19	0.344	(0.340–0.348)	3374	0.427	(0.424–0.430)	6576
Initial woodpecker population size = 20, highly clumped												
$W_{..}T_{.}$	0.016	(0.015–0.017)	1478	0.017	(0.016–0.018)	1515	0.016	(0.015–0.017)	1498	0.017	(0.016–0.018)	1463
$W_{..}T_D$	0.018	(0.017–0.019)	1595	0.018	(0.017–0.019)	1631	0.016	(0.015–0.017)	1441	0.015	(0.014–0.016)	1392
$W_D.T_{.}$	0.002	(0.002–0.002)	137	0.015	(0.014–0.016)	1156	0.016	(0.015–0.017)	1086	0.061	(0.059–0.063)	4872
$W_D.T_D$	0.002	(0.002–0.002)	169	0.015	(0.014–0.016)	1299	0.015	(0.014–0.016)	1173	0.055	(0.053–0.056)	4702
$W_{.}sT_{.}$	0.149	(0.146–0.152)	2473	0.151	(0.148–0.154)	2505	0.150	(0.147–0.153)	2502	0.148	(0.145–0.151)	2426
$W_{.}sT_D$	0.160	(0.157–0.164)	2970	0.157	(0.154–0.160)	2765	0.141	(0.138–0.144)	2261	0.133	(0.130–0.136)	1964
$W_{DS}T_{.}$	0.009	(0.008–0.010)	3	0.045	(0.043–0.046)	37	0.270	(0.266–0.274)	2844	0.383	(0.379–0.386)	7071
$W_{DS}T_D$	0.010	(0.009–0.010)	6	0.046	(0.045–0.048)	56	0.251	(0.248–0.255)	3070	0.342	(0.339–0.345)	6836

[a]Expressed as proportion of eighty woodpecker-vacant cells ($d^{(1)}$) and proportion of twenty woodpecker-occupied cells ($d^{(2)}$) burned. Each decision was simulated one hundred times under one hundred random woodpecker occupancy distributions.

[b]Model expressed as a character triplet $W_{ij}T_k$, where i indicates woodpecker colonization and persistence probabilities are (i = D) or are not (i = .) sensitive to burning, j indicates woodpecker colonization probability is (j = S) or is not (j = .) sensitive to distance from a nearest-neighbor source cell, and k indicates thrush occurrence probability is (k = D) or is not (k = .) sensitive to burning.

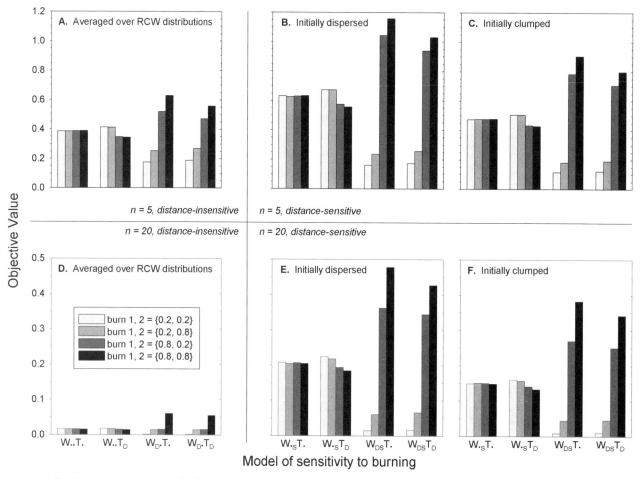

Figure 16.2. Mean objective values for four landscape burning decisions under alternative models of red-cockaded woodpecker (*Picoides borealis*) and wood thrush (*Hylocichla mustelina*) population dynamics. Decisions are expressed in the form {$d^{(1)}$, $d^{(2)}$}, where $d^{(1)}$ represents proportion of woodpecker-vacant habitat burned, and $d^{(2)}$ represents proportion of woodpecker-occupied habitat burned. Shading of decision bars increases from light to dark as extent of landscape burning increases. In each plot, decision results are provided for four models of species response to burning: no response by either woodpeckers or thrushes (model W.$_i$T.), response by thrushes only (model W.$_i$T$_D$), response by woodpecker only (model W$_{Di}$T.), and response by both woodpecker and thrushes (model W$_{Di}$T$_D$), where subscript i indicates whether woodpecker colonization probability is (i = S, plots *b, c, e, f*) or is not (i = .., plots *a, d*) sensitive to distance to nearest source neighbor. Decision outcomes vary according to two initial states of the system: initial population size (plots *a, b, c* for n = 5 and plots *d, e, f* for n = 20), and initial distribution of woodpeckers (plots *b* and *e* for dispersed distributions and plots *c* and *f* for clumped distributions).

joint woodpecker-thrush response model to become quite flat, almost resembling that for the null response model.

We found that making the optimal decision was not dependent on correctly identifying the appropriate distance-sensitivity mechanism for woodpecker colonization (contrast Fig. 16.2a with Fig. 16.2b–c and Fig. 16.2d with Fig. 16.2e–f), the initial abundance of woodpeckers (contrast Fig. 16.2a–c with Fig. 16.2d–f), or the initial distribution of woodpeckers (contrast Fig. 16.2b with Fig. 16.2c and Fig. 16.2e

with Fig. 16.2f). With regard to this latter result, however, we point out that our simulated decisions were carried out by selecting habitat cells completely at random with regard to woodpecker location. Had our decision set also included the selection of cells under some alternative sampling scheme (e.g., probability of selection of woodpecker-vacant cells inversely proportional to distance from woodpecker-occupied cells), we would then expect to find that the optimal decision does depend on correct identification of the initial distribution of woodpeckers.

Discussion

This hypothetical example demonstrates the importance of correct model identification in decision making, or at least the importance of considering a set of reasonable model alternatives. Furthermore, the relative performance of decisions across the model set will vary according to form and parameterization of the objective function.

In our example, we were omniscient observers of the system and could easily understand the implications of each decision under each version of nature. In real systems, however, we are uncertain about the true version of nature, and our observations are incomplete and imprecise, yet we still are faced with making an optimal decision for a management objective. Our real need, therefore, is twofold: to find the decision that maximizes some physical attribute of the system, and to apply the results of the decision action toward the reduction of uncertainty and toward better decision making in the future.

Suppose that we are managing a system that is described by one of the eight models above, but that we are completely uncertain about which model is correct. We will also assume that the initial population of woodpeckers is five and that woodpeckers occur in a dispersed pattern. Then we may apply equation 16.1 to find the optimal decision under uncertainty, using values of J (Table 16.1) for the $\int_x u(a|x;\theta_i)f(x|\theta_i)dx$ and $p(\theta_i) = 0.125$. The maximum value of $E(a)$ is 0.663, which occurs for decision $\{d^{(1)}, d^{(2)}\} = \{0.8, 0.8\}$.

Following the decision action, the system may be observed for a number of years until the time of the next management action. We assume that data are collected according to some design that yields observations at temporal, spatial, and demographic resolutions that are consistent with model predictions. Suppose that data from the field, collected ten years following the decision action, provided a set of values $L(\underline{x}|\theta)$, the statistical measures of agreement between the data and each model θ (e.g., a sum of squared differences between observations and model predictions scaled by a variance measure). Furthermore, suppose that these values redistributed (through equation 16.2) model weight from the equal allocation of 0.125 for each model to the allocation of 0.79 to the model

$W_{.S}T_D$ (woodpecker management-insensitivity, woodpecker distance-sensitivity, thrush management-sensitivity) and 0.03 to the other seven models. Now, if we are again required to choose a management action for the next ten years, and again starting from an initial condition of five dispersed woodpeckers, then reapplication of equation 16.1 under these new weights results in a maximum value of 0.606 for $E(a)$, which occurs for decision $\{d^{(1)}, d^{(2)}\} = \{0.2, 0.8\}$. Thus, at both decision periods we not only made optimal decisions under system uncertainty, we also exploited our decision action and our monitoring data to reduce uncertainty between decision episodes. Note also in this approach that statistical measures of model-to-data agreement are not used to make dichotomous, absolute assignments of model validity or invalidity based on arbitrary criteria. Rather, they are used in a way that allocates more or less credibility to a model over time without ever completely dismissing a contender from the model set.

Summary

We make several observations regarding the use of habitat models in a conservation decision context. First, the assessment of model accuracy (model validation) must be based on observable phenomena so that model predictions can be directly compared to observations. "Suitability," meaning the *potential* of a habitat to provide a portion of the needs of a population, cannot itself be objectively measured, and models dependent on "suitability" as the output cannot be validated (but see Hill and Binford [Chapter 7] for another perspective). Second, even when presence, absence, or numerical abundance can be observed and appear to conform to model predictions, this agreement may constitute weak validation, that is, not exclude competing explanatory or even null habitat models. Weak validation occurs for several reasons, including (1) possible existence of source-sink and other demographic phenomena tending to obscure functional relationships between habitat and populations; (2) weak evidence based on qualitative (e.g., present or absent) versus quantitative comparisons; (3) lack of statistical power; and (4) injudicious choice of spatial scale. Third, arbitrary conventions of accuracy (map or attribute) or precision are irrelevant to decision making and tend to distract from the proper consideration of uncertainty in decision making, which will always be made under uncertainty. The key is to provide tenable alternative models that make different predictions about the relationship between management actions and the objective. Finally, optimal decisions can be made based on current information about the tenability of alternative models (expressed as model weights). Adaptation occurs when uncertainty is reduced (changing model weights) by information feedback obtained in the course of management and monitoring.

Contrasting Determinants of Abundance in Ancestral and Colonized Ranges of an Invasive Brood Parasite

D. Caldwell Hahn and Raymond J. O'Connor

A generalist parasitic species functions on a different ecological scale than any of its individual hosts. Characterizing the niche of a generalist parasite requires sifting through the complex set of environmental variables underlying the distributions of its multiple hosts, then using an analytical technique that can distinguish between the relative influence of the environmental factors and the presence of the hosts themselves. The brown-headed cowbird (*Molothrus ater*) (henceforth "cowbird") is an obligate parasite that never builds its own nest, and it is an extreme host-generalist that parasitizes over two hundred species of North American passerines (Ortega 1998). The cowbird switches among multiple host species in different geographic areas of its range, and it parasitizes hosts with broad geographic ranges as well as hosts with ranges limited regionally or by habitat. The cowbird has a broad geographical range that covers most of the continental United States (see Fig. 17.1 in color section), an extent that few North American songbirds can match (Price et al. 1995). However, the range also has two distinct areas: an ancestral range and a more recently colonized area. The ancestral range lies in the plains and prairies of the central Great Plains, where cowbirds associated with migratory buffalo. The invaded range is distributed both east and west of the central United States and stretches to the Atlantic and Pacific coasts (Rothstein 1994).

The cowbird's range expansion occurred in association with European colonization of North America, so its occupation of the eastern United States is approximately 350 years old and its occupation of the western United States may be as recent as 150 years. The cowbird coexists successfully with domestic livestock and agriculture and also exploits suburban lawns and bird feeders (Ortega 1998).

We hypothesized first that the distribution of the parasitic cowbird would be less influenced by climate and weather factors than are the distributions of other songbirds, and, second, that different niche attributes would characterize the cowbird's ancestral and colonized ranges.

Methods

Our analysis was based on mapping abundance and environmental variables to a spatial grid, followed by statistical analysis to relate cowbird abundance at each location to the environmental conditions and host densities there. Our spatial grid was the hexagonal grid developed for the U.S. Environmental Protection Agency (EPA) for use in the Environmental Monitoring and Assessment Program (EMAP) (White et al. 1992). Each hexagon was approximately 640 square kilometers in area and approximately 12,600 hexagons cover the conterminous United States. A hexagonal grid, unlike a square grid, has a constant center-to-center distance between adjacent grid cells (here 27 kilometers).

For predictor variables we used the land cover class and environmental data compiled by O'Connor et al. (1996). Loveland et al. (1991) used Advanced Very High Resolution Radiometry (AVHRR) meteorological satellite images to derive a prototype land cover classification for the conterminous United States at 1.1-square-kilometer sensor resolution. O'Connor et al. (1996) added an urban class from the Digital Chart of the World (Danko 1992), summarized the representation of each of the 160 cover classes for each of the 12,600 EMAP hexagons, and computed landscape metrics such as patch size distributions, shape complexity, contagion and dominance, fractal dimension, types and frequencies of habitat edges, road abundance, and total length of riparian systems for each hexagon (Hunsaker et al. 1994; O'Connor et al. 1996). Several climate variables—annual precipitation, mean January and mean July temperatures, and annual temperature variation (seasonality)—in the form of long-term climate averages from the Historical Climatology Network (Quinlan et al. 1987; HCN 1996) were also incorporated. The data were modeled with 1-kilometer resolution (except that precipitation was modeled to 10 kilometers and then resampled to 1 kilometer) and were then summarized within each hexagon as average, minimum, and maximum values. Other variables included in the environmental data set were ownership (federal or nonfederal), road density (separately for major and minor roads), and stream density. All were expressed as within-hexagon averages and corresponding extrema (O'Connor et al. 1996).

The bird data analyzed came from the national Breeding Bird Survey (BBS) (Sauer et al. 1997). The BBS comprises bird surveys of a stratified random sample of 25-mile (40-kilometer) lengths of secondary roadside; for each of fifty stops spaced at half-mile (0.8-kilometer) intervals, an observer records all birds registered in a three-minute count. Criteria concerning timing, weather conditions, and so on must be met for the route to be judged of acceptable quality for inclusion in the survey. In the present analysis, some 1,223 routes within the conterminous United States with at least seven high-quality surveys between 1981 and 1990 were used (following O'Connor et al. 1996). For each species of interest (below), the proportion of surveys in which the species was recorded at the site was computed, to provide a measure of *incidence*; this measure is typically correlated with absolute density and is a relatively robust measure of abundance. To obtain an overall index of cowbird host abundance the incidence values for the different species in two lists of host species were summed. One list was of the fifty most frequently parasitized host species and the other was of geographically widespread hosts (see Appendix). In addition, the numbers of these hosts present at each location were determined and used as a predictor variable in analyses below.

Statistical Analysis and Modeling

We used classification and regression tree (CART) modeling (Sonquist et al. 1973; Breiman et al. 1984) to identify the nonlinear relationships between our response variables and the land-use, pattern metric, and climate covariates. Traditional linear regression and correlation techniques assume that independent variables entering the regression model have common effects across the entire sample, an assumption unlikely to be the case here. Moreover, these techniques require explicit specification of terms for interactions. We used the S-Plus (MathSoft. 1995, Seattle, Wash.) implementation of CART (Clark and Pregibon 1992; Venables and Ripley 1994) to partition our response variables recursively with respect to a set of selected covariates. At each node, the independent variable that best discriminated the response variable was used in the tree as the splitting variable for that node. Discrimination was maximized by trying all possible splitting thresholds for all possible prediction variables and choosing the variable and threshold to maximize the differences in the response variable (maximum between-group diversity) before splitting the dataset into two subsets. The process was then repeated independently and recursively on each increasingly homogenous subgroup until a stopping criterion was satisfied. This tree was then pruned back using tenfold cross-validation (Clark and Pregibon 1992). This strategy reduced the propensity of CART models to over-fit the data. Since cross-validation is currently the subject of debate among statisticians (e.g., Miller 1994b) we used the criteria developed by Sifneos and her colleagues (J. Sifneos, D. White, and N. S. Urquhart personal communication) on the basis of extensive experiments in optimizing cross-validation recovery of known data structures. We also perturbed the response variable by 5 percent, and re-ran the model to check for overall consistency in tree structure. We controlled for collinearity problems by randomly perturbing each independent variable in the pruned model by up to 5 percent and re-running the analysis to check for inclusion or omission of the variable in the tree. Variables stable in the face of such perturbation could not be markedly collinear with any other variable in the data set. The models presented here passed all these checks.

Geographical Delineation of Regions

We operationally defined the ancestral zone by overlaying the areas of highest cowbird abundance in the present-day distribution (Fig. 17.1) on a map of Omernik ecoregions (Omernik 1987) and defining the range as those ecoregions with cowbirds. The delineated area ranged from North Dakota south to Texas and east to Indiana, Kentucky, and western Tennessee and Mississippi.

Results

Our findings provide insight into the relative importance of physical and biotic variables in predicting the occurrence of cowbirds.

Environmental Factors Determining Cowbird Distribution

Breeding Distribution. Our first analysis examined the relative importance of different predictors of brown-headed cowbird incidence at a national scale. The model explained a significant percentage (50.9 percent) of the variance in cowbird abundance and identified five major predictors: crops, Conservation Reserve Program (CRP) lands, geography (region), weather, and climate (Table 17.1). The single best predictor of cowbird incidence was crops occurrence, which explained 15.7 percent of the variance in cow-

TABLE 17.1.

Major biophysical predictive factors influencing brown-headed cowbird (*Molothrus ater*) distribution.

Factors	Summer distribution (%)	Winter distribution (%)
Climate	5.3	21.7
Weather	9.0	4.8
Geography/region	9.6	5.9
Conservation reserve program	11.3	—
Crops	15.7	4.0
Total R^2	50.9	36.4

bird incidence. Specifically, soybeans, maize, sunflowers, and sorghum (in that order) each accounted for 3.8 to 2.1 percent of variability in cowbird presence; other crops were combined in a fifth category. The second-best predictor (11.3 percent) was location of CRP lands, which are formerly farmed areas that have been allowed to go fallow for ten years. The other three major predictors identified were geography/region (9.6 percent), weather (9.0 percent), and climate (5.3 percent). The geography/region variable comprised longitude (7.6 percent) and latitude (2.0 percent) components.

Winter Distribution. We compared the major predictors of cowbird incidence on wintering grounds to those on breeding grounds (Table 17.1). Four of the five variables proved to be major predictors for both distributions, although in winter the relative importance shifted away from crops and toward climate. The CRP dropped out in this winter analysis.

Influence of Host Abundance on Cowbird Distribution

When we repeated our analysis of the cowbird's summer distribution with the host variables included, the list of major predictors shifted dramatically from crops to host abundance (Table 17.2, column A). Crops and CRP lands no longer emerged as major predictors, although these factors had accounted for 27 percent of the variance in the previous analysis of cowbird's distribution. Host Abundance Index rose to the top of the list of major predictors (18.9 percent) and accounted for more of the variance in cowbird incidence than had any single variable in the previous

TABLE 17.2.

Major biophysical and avian predictive factors determining distribution of the brown-headed cowbird (*Molothrus ater*).

Predictive factors	A Cowbird range with hosts (national) (%)	B Ancestral cowbird range (regional) (%)	C Colonized cowbird range (regional) (%)	D Grassland passerines' range[a] (%)
Crops	7.1[b]	—	—	33
Climate and weather	16.2	12.8	6.1	18.9
Geography	—	9.5	4.6	7.7
Host abundance	18.9	5.5	26.6	NA
Overall species richness	4.9	—	8.3	NA
Total R^2	47.1	27.8	45.6	59.6

[a]O'Connor et al. (1999:51): obligate grassland passerines of North America: horned lark (*Eremophila alpestris*), vesper sparrow (*Pooecetes gramineus*), lark bunting (*Calamospiza melanocorys*), savannah sparrow (*Passerculus sandwichensis*), grasshopper sparrow (*Ammodramus savannarum*), Baird's sparrow (*Ammodramus bairdii*), Henslow's sparrow (*Ammodramus henslowii*), McCown's longspur (*Calcarius mccownii*), dickcissel (*Spiza americana*), bobolink (*Dolichonyx oryzivorus*), eastern meadowlark (*Sturnella magna*), western meadowlark (*Sturnella neglecta*).
[b]All involved patch size of cropland, pure or mixed.

analysis. A related biological factor, overall species richness, also emerged (4.9 percent).

We present the details of this analysis as they appear in the CART tree in order to explain the sequence and interrelations of the principal factors (Fig. 17.2). The overriding prominence of host abundance in predicting cowbird presence was reflected in its index being the splitting factor at the first tier of the CART analysis. As reflected in the right branch of the tree, when host abundance was high (i.e., a Host Abundance Index value for the route greater than 5.4), cowbirds were recorded in 93 percent of the surveys there. Host abundance was more important than overall avian abundance or diversity, though total species richness did appear as a splitting factor in the second tier of the left branch (segregating the locations in end node A from those in nodes B through E).

The predictive values second and third in importance in this analysis were climate (16.2 percent) and patch variables (7.1 percent). Among the climate variables, seasonality—differences between mean January and mean July temperature—was the major contributor. Areas of high seasonality are those that experience a strong seasonal flush of productivity and that attract a rich assemblage of breeding species that take advantage of the abundant food resources (Ashmole 1963). Figure 17.2 shows that on routes associated with the higher seasonality index (in nodes C, D, and E on the left branch of the tree) 77 percent of the surveys

recorded cowbirds versus a 35 percent incidence for less-seasonal areas (node B). On the right branch, in colder areas associated with lower maximum January temperatures (nodes F, G, H, and I) 94 percent of the surveys recorded cowbirds against only 5 percent of the surveys in the warmer areas of node J. Thus, the northern United States has many more cowbirds than the southern United States but in association with different predictors in different regions.

The specific land cover classes for which patch variables were predictors were dominated respectively by crop/grassland mixtures (node C versus nodes D and E), small forests of maple-birch-beech in corn-soybean areas (nodes F and G), and row crops (nodes H and I). In the case of the row crops variable, it was the size of the patches of row crops relative to the national average for patches of this type that had predictive power: hexagons in which blocks of row crops were relatively small— less than about 5 percent of the national average— were more likely to hold cowbirds than were hexagons with larger expanses of row crops. Patch variables can point to habitat fragmentation effects, and their emergence here indicates their strong predictive value in determining cowbird abundance. All three patch-size variables identified are involved in area sensitivity, with cowbirds more abundant in hexagons with small rather than large patch sizes (note the higher value in the left-hand nodes in each

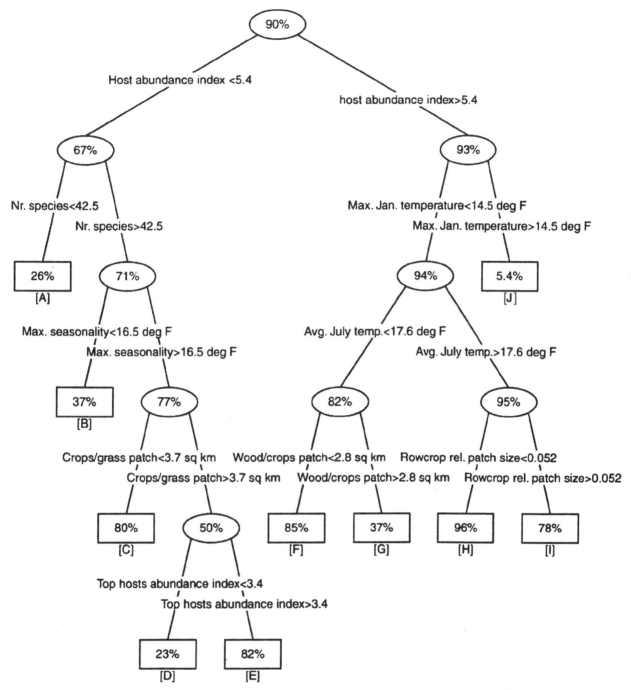

Figure 17.2. Classification and regression tree (CART) model for brown-headed cowbird abundance across the conterminous United States. Numbers inside the oval (intermediate nodes) or rectangles (end nodes) are the percentage of Breeding Bird Survey (BBS) routes, rounded to whole numbers, on which the brown-headed cowbird (*Molothrus ater*) was detected. The splitting variable and its threshold value are shown at each node. The end nodes are labeled from A through J and represent a set of hexagons with the characteristics of the unique set of splitting variables preceding it.

TABLE 17.3.

Predictive model for the probability of occurrence of brown-headed cowbirds (*Molothrus ater*).[a]

Alternative		Host abundance index	Total species richness index	Seasonality index		Probability that cowbirds are present (% routes)
1.	If	> 5.4 spp / route,	[any value}	[any value]	then	93
2.	If	≤ 5.4 spp / route, and				
a.			> 42 spp, and	> 16.5 deg. F,	then	76.7
b.			> 42 spp, **but**	< 16.5 deg. F,	then	36.9
3.	If	≤ 5.4 spp / route, **but**	< 42 spp,	[any value]	then	26.2

[a]Other predictors (e.g., climate, land use, and habitat patchiness) modify these rules locally.

of the sibling nodes involving patch variables: C versus the DE parent, F versus G, and H versus I).

Predictive Model for the Probability of Occurrence of Cowbirds

The information depicted in Figure 17.2 can be recast as a series of explicit predictive rules (Table 17.3). This hierarchy of rules specifies (1) if Host Abundance Index in a hexagon or on a route is greater than 5.4 species/route, then the probability of cowbirds being present is 93.5 percent; (2) if the Host Abundance Index is less than or equal to 5.4 species/route, then two additional factors must be assessed to make a prediction, yielding (a) if the number of species present (total of hosts and nonhosts) is greater than forty-two species and the seasonality index is greater than 16.5 degrees Fahrenheit, then the probability of cowbirds being present is still high (76.7 percent); otherwise (b) in hexagons where Host Abundance Index is less than or equal to 5.4 species and the total number of avian species present is greater than 42, but the seasonality index is less than 16.5 degrees Fahrenheit, the probability of cowbirds being present is only 36.9 percent; and finally (3) if the Host Abundance Index is less than or equal to 5.4 species and the total number of avian species present is less than 42 species, then the probability of cowbirds being present—regardless of seasonality index—drops to 26.2 percent.

Role of Host Species' Habitat Preferences

Shared habitat correlates between host and parasite might explain why host abundance figured so promi-

nently for cowbirds in Table 17.2. Our perturbation tests (see "Methods") precluded simple confounding of variables but more complex commonality in habitat requirements would not necessarily be excluded. We therefore analyzed host abundance data with respect to land cover and climate variables and compared the spectrum of predictors against that for cowbirds. Precipitation was the strongest predictive factor (44.2 percent): host abundance was low where precipitation was low; intermediate in wetter areas where total species richness was low and also in areas where species richness was high but seasonality was weak; and highest where seasonality was high. The second and third major predictive variables were species richness (14.2 percent) followed by seasonality (8.8 percent). Since the variables that predict host abundance are quite different from those identified for the brown-headed cowbird (Table 17.1), it is unlikely that the cowbird-host abundance association above was due to shared habitat requirements.

Regional Analysis: Ancestral Versus Colonized Range

Host abundance was a major predictor in both ancestral and colonized regions, although its influence relative to biophysical factors was markedly different in the two regions (Table 17.2, columns B and C). In the ancestral range, biophysical factors (topography and elevation) carried over four times as much influence (24 percent) on cowbird incidence as did biological ones (5.5 percent). In this model, the principal host abundance factor was presence of host species with geographically broad distributions, and the level was similar to that found for overall species richness (4.9 percent) on the previous national-scale analysis (Table 17.2). The major host abundance indicator that had appeared on the national scale analysis, or in other words the extent of the presence of the fifty most frequently parasitized hosts (18.9 percent), did not emerge as a major predictor within the ancestral range.

In contrast, within the colonized range the relative importance of biological versus biophysical factors was reversed, and host abundance indices carried five times as much influence on cowbird incidence as did biophysical factors (Table 17.2, column C). Specifically, host abundance indices accounted for 34.9 percent of the variability in cowbird incidence, the highest value for a predictor in any of our models of cowbird incidence. Within the general category of host abundance, the specific factor of abundance was an index of the fifty most frequently parasitized cowbird hosts (26.6 percent).

Among the biophysical factors that emerged as predictors in both the ancestral and the colonized range, climate appeared in the cowbird's ancestral range at a level (12.8 percent) similar to that in the national analysis (16.2 percent) (Table 17.2) and with the same components, specifically high average July temperature (8.6 percent) and the degree of seasonality (4.2 percent). Topography, as the importance of lack of elevation, accounted for 9.5 percent of the variation in the model but the crops and patch variable predictors of the continental analyses were not important. The total variance accounted for in this CART model for the ancestral cowbird range was 27.8 percent, a level approximately half that in previous models at the national scale. In the region of colonized range, the physical factors that emerged were topography (4.6 percent) and climate (6.1 percent) (Table 17.2). The total variability accounted for by the CART model for the colonized range was 45.6 percent, again similar to the levels explained by CART in both the national-scale analyses (Table 17.2).

Discussion

The brown-headed cowbird is recognized as a textbook example of a species that must be studied at different scales to answer different questions (Robinson 1999; Morrison and Hahn, in press). Before 1990, most studies of the species were local-scale field studies conducted at sites in the cowbird's ancestral range, and they typically looked at parasite behavioral strategies such as nest-searching or mating system and at the rates and effects of parasitism on different host species (see Ortega 1998). Attempts to extract general principles from local studies often yielded contradictory patterns that reflected the large number of host species and habitats exploited by the brood parasite.

Since 1990, much work has been done at larger areas (extent and resolution) and at sites in the colonized range of the brown-headed cowbird (see Morrison et al. 1999; Smith et al. 2000). The design and location of these studies reflect recent interest in investigating which ecological factors are driving the cowbird's expansion into new habitats and new hosts, particularly in the forest interior. Facilitated by advances in radiotelemetry and GIS (geographic information system) techniques, many of these studies were conducted across landscapes, with a few at the regional scale (Robinson et al. 1995a; Morrison et al. 1999; Smith et al. 2000).

From these studies, generalizations emerged that apply within landscapes but questions remained as to how these patterns might change at larger scales and what other patterns might appear at larger areas. Robinson (1999) summarized the core conclusions from landscape and regional studies: cowbird abundance and parasitism rates are much lower in forested landscapes where foraging opportunities are limited; cowbird abundance decreases with distance from rich feeding areas; and cowbird abundance is correlated with host abundance in landscapes with unlimited foraging habitat. The only national-scale analyses of cowbird abundance are three studies of population trends based on BBS data, which concluded that cowbird numbers are stable nationally, with slight regional increases or decreases in some regions that are not linked to declines or increases in host populations (Maurer 1993; Peterjohn et al. 2000; Wiedenfeld 2000).

Our study provides the first national perspective on the principal factors that underlie the abundance of the parasitic cowbird. Our findings unambiguously identified host abundance as the fundamental predictor of the cowbird's distribution at the national scale. Although avian species distributions are typically constrained by spatially extensive variables such as climate, habitat, spatial patchiness, and microhabitat attributes, we had hypothesized that the distribution of a brood parasite depends as strongly on host distribution patterns as on biophysical factors. Our findings suggest that the distribution of hosts does indeed take precedence over habitat attributes in shaping the cowbird's distribution at a national scale, within an envelope of constraint set by biophysical factors. The importance of hosts can be missed, because an analysis of the predictive values of biophysical factors alone (Table 17.1) yields the result that crops and CRP lands are dominant predictors, a result that fits the profile of the brown-headed cowbird as a bird of the central prairies that associates with agriculture.

Many studies have weighed the relative importance of host availability versus food ability. Our results suggest that while host availability is predominant for the national distribution of cowbirds, food availability (associated with the three patch variables that include foraging habitats) is a major factor in particular habitats. Three different patch types were detected in widely separate nodes in our CART model for cowbird abundance reflecting the importance of this landscape pattern. Robinson (1999) concluded that host abundance was the most influential environmental variable, with the caveat that this was true only when food is sufficient. Several studies have found food availability a more fundamental determinant of cowbird incidence, particularly in habitats such as those where forested areas are extensive (Morrison et al. 1999). The influence of food availability on the winter distribution of cowbirds is a separate question, also much debated, and our analysis showed a sharp repositioning of the major predictors from summer to winter distribution, dropping crops to 4.0 percent and increasing climate nearly fourfold to 21.7 percent (Table 17.1). These results suggest that either food availability is a significant constraint on cowbird populations (Robinson 1999) or energetic constraints limit the cowbird's ability to exploit cold areas (Root 1988c).

Our CART analysis of summer distribution also distinguished the influential size of patch variables and found that cowbirds are more abundant in hexagons with small rather than large patch size. This result confirms the observations of many local studies that larger forest stand size limits cowbird incidence (Morrison et al. 1999). Since the reverse pattern of area sensitivity characterizes Neotropical migrants (i.e. they are more abundant in hexagons with large patch sizes), this result indicates that separate niches still exist between cowbirds and forest-interior nesting birds.

Our analysis identified climate and weather as the second most important predictors. Local- and landscape-scale studies have rarely addressed climate and weather, which illustrates how an overlooked variable can emerge when an analysis shifts from a local to a continental scale. The areas of high seasonality in the north-central region of the United States experience a strong seasonal flush of productivity that attracts a rich assemblage of breeding species that take advantage of the abundant food resources (O'Connor et al. 1996). These are the areas of greatest cowbird abundance. Seasonality is very highly correlated with the portion of Neotropical migratory songbirds in an area, a pattern first hypothesized by Ashmole (1963) and since supported by Wilson (1974), Herrera (1978), and Ricklefs (1980). Thus, cowbirds can exploit both the flush of productivity and the abundance of breeding hosts.

The importance of host abundance to cowbird distribution is further put into perspective by the regional analysis. Distinguishing the cowbird's ancestral range from its colonized range revealed a strong geographical bias in the influence exerted by host species. In the colonized range, host abundance indices carried five times as much influence as biophysical factors. This result may reflect the fact that there is greater variance in incidence of cowbirds in the colonized range and that cowbirds colonize new areas only where conditions are good. In the East, although cowbird hosts are more abundant, much of their breeding habitat is in large forests, where they are inaccessible. In the West, host species are concentrated in riparian areas, which draws cowbirds to those sites. We plan more detailed analyses of the colonized range in which we look at the eastern and western ranges separately, since both the habitat types and host abundance levels are distinctly different (O'Connor et al. 1996).

Host abundance is not the dominant predictor of cowbird incidence in its ancestral range, although it does emerge as a lesser predictor. Here topography and elevation carried over four times as much influence as biological factors (24 versus 5.5 percent). These results may reflect conditions that are relatively uniformly good for cowbirds and consequently the variance in cowbird incidence is lower. In future analyses, we plan to subdivide the area we designated as ancestral into the core area where cowbird abundance is greater than thirty birds per route (i.e., the Dakotas, eastern Nebraska and Kansas, western Missouri, and Iowa) and the surrounding zones where cowbird abundance is eleven to thirty birds per route (see Fig. 17.1).

The analyses in this study accurately distinguished the cowbird's ecological niche as a parasite. The CART results identified a different set of predictors for cowbirds and for their most frequently parasitized hosts, illustrating that the cowbird distribution is not simply the result of shared habitat preferences. Moreover, the CART analysis distinguished a different set of predictors for the parasitic cowbird and the guild of obligate grassland passerines that are its ancestral hosts (O'Connor et al 1999). An important reservation about the findings here is that the CART models we used, despite their sophistication, return only estimates of correlation. Therefore, our conclusions are subject to the normal caveats of correlation analysis, in particular that correlation does not ensure causation. Our emphasis on host availability and patch size arrives at the same conclusions as those of earlier investigators, and since ours are based on analyses with very different biases than those in site-specific studies, this lends strength to all the studies. We distinguished the strong influence of climate and weather, largely overlooked in landscape-scales studies, and we described differences between the cowbird's ancestral and colonized range in the role of host abundance. The broad spatial extent of our analyses provides a robust overview of the correlates of the distribution of the principal North American brood parasite that has not previously been available.

Acknowledgments

We wish to thank Richard L. Jachowski (USGS Patuxent Wildlife Research Center) for long-term encouragement and support of research on brood parasitism and our Biodiversity Research Consortium collaborators B. Jackson and S. Timmons (Oak Ridge National Laboratory) and Carolyn Hunsaker (USDA Forest Service) for provision of landscape metrics; R. Neilsen, D. Marks, J. Chaney, C. Daly, and G. Koerper (EPA Environmental Research Laboratory, Corvallis, Ore.) for assistance in computing climate data; Tom Loveland (EROS Data Center, USGS) for land cover class data; and Denis White (Geosciences, Oregon State University) for assistance with spatial analysis.

We acknowledge financial support for this work, interagency agreements DW12935631 between EPA and USDA Forest Service, and EPA Cooperative Agreements CR818843–01–0 and CR823806–01–0 and USDA Forest Service Cooperative Agreement PNW93–0462 with University of Maine (Raymond J. O'Connor, Principal Investigator) and Award 9711623 from the National Science Foundation to Raymond J. O'Connor and Deirdre M. Mageean. Two anonymous reviewers gave helpful reviews of the manuscript.

Appendix

The fifty most frequently parasitized host species as identified by Friedmann (1963) by a review of published studies. No comparable reassessment of frequency of parasitism has been done, but these fifty hosts still appear representative (Rothstein, pers. comm.; DCH pers. obs). Friedmann designated a first (primary) group of seventeen hosts ("1," more than one hundred records of parasitism), a second group of seventeen hosts ("2," more than fifty records of parasitism), and a third group of sixteen hosts ("3," twenty-five to fifty records of parasitism). Twelve host species that Friedmann designated common hosts of great geographic availability are indicated with a "g." Three host species indicated by an asterisk (*) are obligate grassland species included in Table 17.2.

Acadian flycatcher (*Empidonax virescens*): 3
American goldfinch (*Carduelis tristis*): 2, g
American redstart (*Setophaga ruticilla*): 1
Bell's vireo (*Vireo bellii*): 1
Black-and-white warbler (*Mniotilta varia*): 3

Blue grosbeak (*Guiraca caerulea*): 3
Blue-gray gnatcatcher (*Polioptila caerulea*): 3
Blue-winged warbler (*Vermivora pinus*): 3
Brown thrasher (*Toxostoma rufum*): 3
Chestnut-sided warbler (*Dendroica pensylvanica*): 2
Chipping sparrow (*Spizella passerina*): 1, g
Clay-colored sparrow (*Spizella pallida*): 2
Common yellowthroat (*Geothlypis trichas*): 1, g
Dickcissel (*Spiza americana*): 2, *
Eastern bluebird (*Sialia sialis*): 3
Eastern phoebe (*Sayornis phoebe*): 1
Eastern towhee (*Pipilo erythrophthalmus*): 1, g
Eastern wood-pewee (*Contopus virens*): 2
Field sparrow (*Spizella pusilla*): 1
Gray catbird (*Dumetella carolinensis*): 3
Hermit thrush (*Catharus guttatus*): 3
Indigo bunting (*Passerina cyanea*): 1
Kentucky warbler (*Oporornis formosus*): 1
Kirtland's warbler (*Dendroica kirtlandii*): 2
Lark sparrow (*Chondestes grammacus*): 3, g
Louisiana waterthrush (*Seiurus motacilla*): 2
Yellow-rumped warbler (*Dendroica coronata*): 2
Northern cardinal (*Cardinalis cardinalis*): 2

Ovenbird (*Seiurus aurocapillus*): 1
Painted bunting (*Passerina ciris*): 2
Prairie warbler (*Dendroica discolor*): 3
Prothonotary warbler (*Protonotaria citrea*): 2
Red-eyed vireo (*Vireo olivaceus*): 1, g
Red-winged blackbird (*Agelaius phoeniceus*): 1, g
Rose-breasted grosbeak (*Pheucticus ludovicianus*): 3
Savannah sparrow (*Passerculus sandwichensis*): 3, *
Scarlet tanager (*Piranga olivacea*): 2
Song sparrow (*Melospiza melodia*): 1, g
Swamp sparrow (*Melospiza georgiana*): 2
Willow flycatcher (*Empidonax traillii*): 1
Veery (*Catharus fuscescens*): 2
Vesper sparrow (*Pooecetes gramineus*): 2, *
Warbling vireo (*Vireo gilvus*): 2, g
White-eyed vireo (*Vireo griseus*): 3
White-throated sparrow (*Zonotrichia albicollis*): 3
Wood thrush (*Hylocichla mustelina*): 2
Worm-eating warbler (*Helmitheros vermivorus*): 3
Yellow warbler (*Dendroica petechia*): 1, g
Yellow-breasted chat (*Icteria virens*): 1, g
Yellow-throated vireo (*Vireo flavifrons*): 1, g

Biodiversity Conflict Analysis at Multiple Spatial Scales

Christopher B. Cogan

As society places increasing demands on the environment, methods for predicting and minimizing environmental effects of population growth and land-use change are urgently needed. Much effort is now focused on developing spatial decision support systems (SDSS) for regional planning and policy analysis (Jankowski et al. 1997). These efforts draw from both the social and natural sciences, and a key challenge is reconciling and integrating the disparate paradigms, semantics, measurements, and scales of analysis from these different traditions (Machlis 1992). In this chapter, I present a case study of Santa Cruz County, California, that looks at the integration of ecological and urban planning perspectives as a way to help local and county land-use planners evaluate the potential threats of land-use change to native biodiversity. To date, this critical aspect of the planning process is usually not attempted (Press et al. 1996; Crist et al. 2000). I appraised local ecological resources by considering regional distribution patterns, rarity, uniqueness, and restoration potential. Predicted patterns of future development from an urban growth model were compared to appraised biodiversity value to map relative risk of biodiversity loss and fragmentation. This approach also served to identify areas where conflicts between goals of urban expansion and biodiversity conservation were likely to be greatest.

Biodiversity is a documented metric for ecosystem health (Callicott and Mumford 1997) functioning at particular spatial scales (extent and grain) and organizational levels (Soulé 1991; Caughley 1994; Chapin et al. 1997; Soulé and Mills 1998). Despite this multiscale attribute of biodiversity, there are few examples of analysis that bridge landscapes and habitats (e.g., 500 hectares) to ecoregions or other large planning areas (e.g. 3 million hectares). Noteworthy publications include Hansen et al. (1993), White et al. (1997a), and Smallwood et al. (1998), as well as more theoretical discussions presented by Noss (1987), Noss et al. (1995), Norton and Ulanowicz (1992), and Heywood (1994). In this study, I focused on spatial patterns of species, wildlife habitat types, and areas of special ecological interest resolved at 1–100 hectares over an ecoregional extent. I also considered patterns of habitat fragmentation and suitability of areas for restoration. Biodiversity has several meanings, though, and for this study I focused on an analysis of ecosystem, species, and genetic diversity, which incorporated compositional and structural elements (Noss 1990a) within a county. My comparison of the county biodiversity through time was the temporal equivalent of delta diversity (Whittaker 1977). The use of multiple criteria runs the risk of compounding uncertainty because of incomplete and inaccurate data, but this is an acceptable trade-off for increasing relevance to the biodiversity analysis (Costanza 1992). I emphasize that the final products from this model are not intended to stand alone. Rather, this SDSS represents one component of biodiversity that should be supported by finer, and in some cases coarser, scales of species and habitat analysis.

Habitat loss or fragmentation due to urban development is only one of many anthropogenic impacts on biodiversity, but it now ranks among the principal causes of species endangerment in the United States (Dobson et al. 1997; Vitousek et al. 1997; Wilcove et al. 1998). A variety of methods are available for predicting urban growth—I chose a relatively simple, general model that could be applied over counties or larger areas using 10–50-year planning horizons and generally available data. Because many land-use decisions are made at the county level, this county analysis promotes conservation opportunities (Press et al. 1996). This study also used ecoregional boundaries to ecologically define the extent of the study area. Although these spatial and temporal scales will not apply to all situations, they will be useful in many contexts of biodiversity protection and will provide a starting point for the method I now present.

This Santa Cruz County, California, case study demonstrates how biodiversity value can be assessed as a spatially explicit property and seeks to predict where these values will decrease in the future.

Development of Biodiversity Analysis Submodels

The biodiversity value of a site was based on four criteria, or submodels. *Ecoregional analysis* described the county wildlife habitat types which contributed disproportionately to the regional distribution; *species habitat models* predicted composition of terrestrial vertebrate species within landscapes; *restoration opportunities* were identified; and *special features* indicated presence of special natural landscapes.

Threat or future conflict were assessed based on two additional submodels: habitat-specific conversion to urban land use over the county subregion and landscape pattern analysis based on changes in patch cohesion (Schumaker 1996).

By maintaining each submodel independently, I promoted flexibility in the model as a whole, enabled open dialog between model users and stakeholders, and clarified results. After completion of each submodel, my results were best interpreted individually, though under some circumstances they could also have been additively combined as a weighted linear combination (Eastman et al. 1995) or in an analytic hierarchy (Saaty 1980; Anselin et al. 1989; Saaty and Vargas 1994).

Whereas the relative contribution of each submodel to final decision making was context dependent and adjustable, it was important to include each in the analysis. These four biologic indicators and the landscape pattern analysis were intended to represent a collection of biodiversity "principal components," each representing its own orthogonal metric to the overall representation. This array of components can best be visualized using a process flow chart (Fig. 18.1) to provide an overview of the submodels and their interactions in biodiversity analysis.

Ecoregional Analysis Submodel

The first part of the biologic assessment, ecoregional analysis, was designed to evaluate coarse-scale processes that function across large area extents (millions of hectares) and decadal periods. Using ecoregions defined by physiographic and biologic consistency (Hickman 1993), I combined information on land management, land cover, and habitat area into a single quantitative assessment. The land-cover data used the wildlife habitat relationships (WHR) vegetation classification system (Airola 1988) from the California Gap Analysis Project (GAP), though the same types of data are common in other areas. This submodel was based on five explicit assumptions:

1. Habitat types with more area are more advantageous to species compared to the same habitat type with less area.
2. Land areas under private ownership are more likely to undergo development or habitat loss compared to certain publicly owned parcels such as wilderness areas, wildlife refuges, and parks.
3. Areas of a given habitat type may be rare in a county, but if that same type is common in the ecoregion, these areas may not be as critical for conservation as suggested by an analysis based on county data alone.

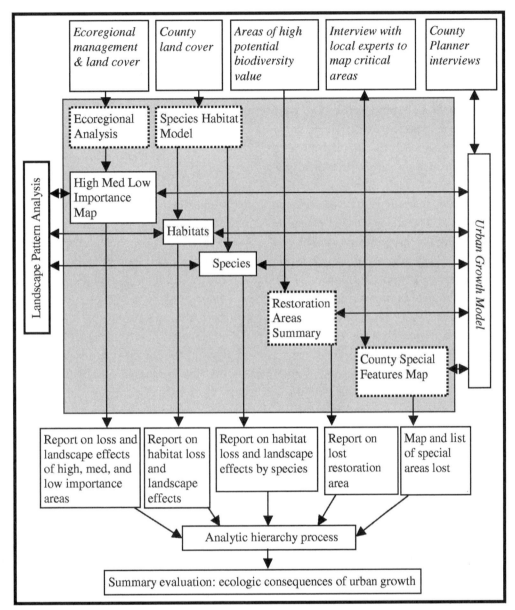

Figure 18.1. Components of the biodiversity model and their interactions. Boxes containing italic text represent input data; central boxes containing dashed lines represent the four biodiversity submodels. Output data are depicted at the bottom of the figure. Double-headed arrows represent possible feedback mechanisms. Components of the model functioning at relatively coarse spatial scales are grouped to the left side and finer scale components are located to the right. The analytic hierarchy process and summary are optional.

4. Areas of a given habitat may be common in a county, but if that same type is regionally rare, these areas may be of more concern for conservation than county data alone suggest.

5. Habitats and their associated species represent a unique community assemblage within an ecoregion. This complex of biotic and abiotic interactions is only partially documented and cannot be considered equivalent to similar assemblages in other ecoregions. See Walker (1992) for further discussions on inter-ecoregional community turnover.

Based on these five assumptions, I assessed the habitat types found in the study area. I began by calculating the ratio of land areas between the county and the ecoregion, summarized by habitat type. Privately held land areas were down-weighted compared to most classes of public lands. This down-weighting was accomplished by counting only a user-selectable fraction (e.g., 75 percent) of each privately held habitat patch. Expected values were derived from the ratio of county size to ecoregion size, and habitat types that were within 20 percent of their expected areas were neutrally weighted. Habitats that exceeded the expected area by more than 20 percent (those concentrated in the county) were categorized as high ecoregional importance and those with less than 20 percent of expected area were categorized as low ecoregional importance. This approach works well if the ecoregion is relatively intact; other areas with extensive land-use alteration could benefit from historic ecoregion descriptors for the county comparison. The quantitative output from this submodel was a numerical score for each habitat type in the county. These scores were then aggregated to depict low-, neutral-, and high-importance areas, which could be graphically represented in a three-color map of the county (see Fig. 18.2 in color section). The results of the ecoregional analysis were not intended to be absolute. Adjustable parameters such as optional weightings for land ownership allowed the model user to fine-tune the model for a given region or particular application. Each biodiversity submodel was designed to capture specific information, leaving other details to submodels that were more appropriate. For example, the ecoregional analysis would not identify a habitat that had been decimated throughout the ecoregion; however, in this situation, the special-features submodel (described below) came into play.

Example of *Ecoregional Analysis*: Redwood forest habitat in Santa Cruz County. Most of the redwoods (*Sequoia sempervirens*) in the central west ecoregion are contained within Santa Cruz County. Consequently, this habitat type has a greater importance in the county than would otherwise be considered using county data in isolation. The high importance areas identified by the ecoregional analysis (Fig. 18.2) were strongly influenced by the restricted distribution of redwood forest.

Species Habitat Submodel

The second part of the biologic assessment utilized species habitat submodels. Using California GAP data

models, I identified habitat areas as landscape features and associated each with a list of vertebrate species (Scott et al. 1993; Hollander et al. 1994; Davis et al. 1998). The GAP models were designed to provide assessments of conservation status for native vertebrate species and natural land cover types (Scott and Jennings 1997). Areas of nonurban, nonagricultural land cover were considered important wildlife habitats in this analysis. For the species habitat submodels, I assume habitat types within an ecoregion are ecologically equivalent. The model output is a species list for each landscape unit in the county.

Example of *Species-Habitat* Relationships: Black salamander. Using a combination of species range and habitat suitability, the black salamander (*Aneides flavipunctatus*) is predicted to occur in two regions within California (California GAP data, Davis et al. 1998). One of these areas is in Santa Cruz County. By analyzing the specific landscape units within the county that are highly suitable for the species, species as well as spatially explicit landscape units could be used to analyze biodiversity.

Restoration Opportunities Submodel

Restoration opportunities, the third component of this four-part biodiversity model, defines areas of potential restoration as areas capable of increased biodiversity value with feasible levels of management change. In many cases, urban areas have low biodiversity value, and it is often not practical to improve them. Sites composed of invasive species monocultures, disrupted wetlands, and timber harvest areas are all candidates for restoration (Dobson et al. 1997). Farmlands may also have low value, though land-use practices can often be modified to improve habitat with little or no financial impact to the farmer (Ratti and Scott 1991). Farmlands are likely to have an increasing role in conservation efforts (Pienkowski et al. 1996), and the role of agriculture in this biodiversity analysis was intentionally included to avoid interpreting biodiversity as a phenomenon limited to nature reserves. As a preliminary approach, this submodel identified all agricultural areas in the county as restoration opportunities, with final output in a map format.

Example of *Restoration Opportunities:* Agriculture. Minor modifications can be made to agricultural areas that can potentially increase biodiversity value (Pienkowski et al. 1996). Examples include larger hedgerows between fields (EIP-Associates 1997) and changes in pesticide application techniques.

County Special Features Submodel

The fourth biodiversity submodel identified and mapped special features found within county. Examples of well-known special features in Santa Cruz County include unique soils and old-growth redwood forests. These data were obtained through interviews with local experts, who often have remarkable insights and information about such features. Complementing these data is a geographic information system (GIS) database, maintained by the California Department of Fish and Game, Natural Heritage Division, on rare plant and animal habitats (Hoshovsky 1988). This portion of the analysis typically functioned at a fine spatial resolution using information about present-day known critical areas. Although these types of data will usually be somewhat biased and incomplete, any knowledge of current problem areas is an extremely valuable component of biodiversity analysis. The purpose of this submodel was to identify and locate these important components of biodiversity. Each special area could then be evaluated for existing or proposed protection status, directly modifying the results of the urban growth model. Without protection status, special areas are subject to the same threats as other areas. Traditionally, the various levels of protection status for a species or habitat have been an important component of conservation strategy. Although the methods described here took advantage of this information, the unprotected special areas and currently unidentified species of concern were likely to be equally important in the fifty-year outlook being considered. Output from this submodel was a GIS database identifying special features areas for protective-status consideration.

Example of *County Special Features*: The Sandhills soil type in Santa Cruz County. Sandhills soils support several local species of plants that are currently threatened (Marangio and Morgan 1987). This soil type is an indicator for important biodiversity areas that are currently (and not just predicted to be) stressed by urbanization. This submodel flagged these areas as "special" without attempting to rank them by size, shape, or other landscape-pattern metrics.

Landscape Pattern Analysis

Landscape pattern analysis is a submodel that complements the four biodiversity submodels for assessing one measure of habitat stress or degradation following land-use change. As patches of habitat change size, shape, and adjacency, habitat quality can also change. Landscape factors that affect habitat quality can be assessed at three stages within the biodiversity analysis: (1) assessment of ecoregional importance, (2) assessment of habitats impacted by urban growth, and (3) habitat evaluation for individual vertebrate species (Fig. 18.1). For this case study, the second stage, evaluating predicted urban growth to assess habitat alteration, was used. Although direct landscape measurements such as fragmentation, contagion, and cohesion are possible, the resulting values are difficult to interpret ecologically. My solution was to apply a habitat cohesion calculation twice—before and after the urban-growth predictions were calculated. In this way, those areas shown to undergo the most radical landscape changes could be flagged for further analysis at the species or habitat-specific level. Submodel output was the ratio of current and predicted landscape patch cohesion scores associated with each habitat type.

Example of *Landscape Pattern Analysis*: Patch cohesion. By comparing this area-weighted metric of each landscape unit before and after applying the urban growth model, habitat types with the greatest change could be flagged. I used patch cohesion (Schumaker 1996) modified as a vector index of habitat pattern calculated as:

$$PC \equiv \frac{\max(\xi) - \xi}{\max(\xi) - \min(\xi)}$$

where

$$\xi = \frac{\Sigma\,P}{\Sigma\,A} \div \left[\left(\frac{1}{2\pi} \right) \frac{\Sigma\left(P\sqrt{A}\right)}{\Sigma\,A} \right]$$

where:

$\xi \propto$ (perimeter : area) ÷ shape index

max (ξ) case of many small patches

min (ξ) case of single large patch

P = perimeter of habitat patch

A = area of habitat patch

PC = patch cohesion, range (0,1)

where patch cohesion (*PC*) was calculated for each habitat type, and comparisons of pre-growth to post-growth models were calculated as $PC_{pre\text{-}growth} / PC_{post\text{-}growth}$. The limit values of ξ were calculated assuming two extreme cases. At maximum ξ, the habitat type consisted of many small circular patches with individual patch area constrained by the minimum map unit (MMU). At minimum ξ, the index was calculated assuming a single large circle of habitat. In both cases, total area of each habitat was set equal to the actual area of that habitat type in the dataset. This submodel allowed ranking of habitat types for degree of cohesion change, identifying imperiled habitats that might be overlooked by area calculations alone. Typically, this analysis would identify vegetation areas that had become fragmented or otherwise eroded into linear shapes with reduced area-to-perimeter ratios. Once identified, heavily altered habitats were prioritized for site-specific analysis.

Development of Urban Growth Predictive Models

The biodiversity and landscape pattern submodels presented an indicator of environmental condition over a countywide area. This information was based on current conditions and provided spatially explicit ecological data, which was used as a starting point for predictive modeling. To examine the effects of increasing urbanization over the next forty years, an urban growth model was applied to my present-day biodiversity analysis. Urban models have recently been developed (Makse et al. 1995; Couclelis 1997; White et al. 1997b; Batty 1998; Landis and Zhang 1998; Makse et al. 1998), which incorporate a broad range of variables and spatial dependencies; however, by following the approach taken by Clarke and Gaydos (1998), I arrived at a model that was transportable and largely independent of spatial grain or extent.

A cellular automata approach to modeling future growth was used, incorporating likelihood of urban sprawl, likelihood of entirely new communities being developed, and likelihood of urban areas expanding along transportation corridors. Model inputs included spatial data on urban areas, roads, land ownership, and topography. Spatial data were resampled to 100-meter grids, though the model was flexible enough to function at other levels of resolution. I also incorporated user-adjustable variables for growth rate, growth type, and time period. Figure 18.3 (see color section) represents output from the urban model showing areas likely to become urbanized by the user-specified target year. These areas of likely future urbanization are a key component in the assessment of future biodiversity change.

Submodel Combination for Composite Valuation

Combining the results of the submodels tended to mask the relationships between different aspects of biodiversity and environmental threats. In some cases, however, it may be necessary to combine data, for instance, when comparing multiple scenarios, performing sensitivity analysis of the submodels, or creating public-presentation summaries. Using the landscape unit of analysis, the combined data can represent a composite valuation of future biodiversity risk for the county area. A simple way to integrate the submodel results was to keep these landscape units as a single common currency, maintaining the option of biodiversity evaluation by species using the WHR models when needed. This approach had the advantage of simplicity, which in some cases might outweigh the disadvantages of data generalization and the uncertainties of the WHR models.

When combining submodel results, several types of weightings are possible, achieved by adjusting the relative input of each submodel. Since the relative weighting of each component will involve some degree of uncertainty and will often involve multiple decision makers, it may be useful to use the Analytic Hierarchy Process (AHP) as a decision analysis tool (Saaty 1994). AHP assists multiple users in building a consensus of component weights and maintains explicit weightings both within and between submodels. The weightings for each submodel are intended to be user-adjustable. For example, some model users may wish to de-emphasize restoration potential whereas other users may prefer to make it a major factor in the analysis. AHP is included in the process flow chart (Fig. 18.1) but is optional.

An Analysis of
Post-urbanization Biodiversity

My analysis of post-urbanization biodiversity is presented as a series of discrete measures. The results are reported in terms of ecoregionally important habitats, compromised within-county habitats, impacted vertebrate species, loss of high-restoration-potential areas, ecologically critical areas of risk, and landscape-pattern changes. These results are spatially explicit and are best presented with the aid of interpretive maps.

Ecoregional analysis addressed only the coarse-grain perspective of biodiversity. In this assessment three of the twelve dominant WHR habitat types in Santa Cruz County were more important than a county perspective alone would show. These habitat types are redwood, montane hardwood conifer, and Douglas-fir (*Pseudotsuga menziesii*) forest. These three habitat types represented the high-priority conservation targets identified by this submodel, without considering the effects of predicted urbanization.

Habitat loss from urban growth is a second discrete measure of biodiversity impact. Here, my cellular automata urban growth model predicted an 11 percent increase in urbanized land cover in Santa Cruz County over the next fifty years. Much of the new growth was predicted to be in the form of contiguous expansion rather than isolated new settlement (Fig. 18.3). Habitat losses resulting from this growth included a 13 percent reduction from the current extent of redwood forests, a 10 percent reduction of montaine hardwood chaparral, and a 4 percent loss of Douglas-fir forests. Santa Cruz County has a long history of human-induced landscape alterations. Because these transitions are largely undocumented, historic reductions to date cannot be calculated. My results from this submodel were therefore based on habitat loss calculated from 1990 levels, which was an arbitrary but logical reference point. Habitats that were already seriously compromised before 1990 should be identified using either the special features submodel or the restoration opportunities submodel.

Species impacts are a third measure of biodiversity loss. Using the California GAP species habitat models, vertebrate species most strongly affected by my future urban-growth scenario included the hermit warbler (*Dendroica occidentalis*), the golden-crowned kinglet (*Regulus satrapa*), and the American shrew-mole (*Neurotrichus gibbsii*). Sixteen avian, two mammal, three amphibian, and two reptile species were predicted to lose at least 10 percent of their present-day habitat in the Jepson Central West ecoregion (Table 18.1).

Areas of potential restoration were evaluated for losses following urban growth. In my model, 30 percent of the new growth forecast was predicted to displace cropland areas, equivalent to a 17 percent loss of county croplands by the year 2040. Other possible restoration areas such as urban riparian zones, impacted coastal zones, and timber harvest areas were not included in this case study.

The county special features analysis identified several types of fine-grain special features in Santa Cruz County. Areas of special concern included unique vegetation complexes associated with sandhills soils, old growth redwood forest, and regions identified as Significant Natural Areas by the California Department of Fish and Game. In this preliminary analysis, I did not assign protective status to special features areas but instead permitted the growth model to forecast urbanization in these regions. In many instances, the special features areas I identified were not at the time protected by state or federal regulation. Allowing a first-iteration growth model to select from these habitats is useful for identifying locations of possible future conflict.

By calculating the patch cohesion landscape metric of each habitat type before and after predicted growth, I was able to rank order the habitats by magnitude of landscape alteration (Table 18.2). Although absolute cohesion value is not directly meaningful (see discussion), it was useful to note the greatest landscape pattern effects in patches of coastal scrub, coastal oak woodland, and redwoods.

Discussion

By analyzing the major components of biodiversity, I sought to clarify the assumptions, strengths, and limitations inherent in such a complex analysis. This deconstruction also permitted me to target the most

TABLE 18.1.

Species most affected by future urban growth in Santa Cruz County, California. Of 260 vertebrate species with habitat in the Jepson central west ecoregion, thirty-nine species are predicted to lose more than 5 percent of their habitat, and the twenty-three species listed here are predicted to lose more than 10 percent of their habitat. Area impact refers to the percentage of the ecoregional habitat predicted for urban conversion. Listed status of "CA SSC" refers to California Species of Special Concern, as described by the California Department of Fish and Game (January 1999).

Common name	Scientific name	Area impact (%)	Listed status
hermit warbler	*Dendroica occidentalis*	40	none
golden-crowned kinglet	*Regulus satrapa*	39	none
shrew-mole	*Neurotrichus gibbsii*	38	none
Vaux's swift	*Chaetura vauxi*	38	CA SSC
pine siskin	*Carduelis pinus*	38	none
MacGillivray's warbler	*Oporornis tolmiei*	38	none
olive-sided flycatcher	*Contopus cooperi*	37	none
Trowbridge's shrew	*Sorex trowbridgii*	35	none
black salamander	*Aneides flavipunctatus*	35	none
Pacific giant salamander	*Dicamptodon ensatus*	34	none
northern alligator lizard	*Elgaria coerulea*	28	none
hermit thrush	*Catharus guttatus*	27	none
sharp-shinned hawk	*Accipiter striatus*	24	CA SSC
rubber boa	*Charina bottae*	24	none
pileated woodpecker	*Dryocopus pileatus*	19	none
Cassin's vireo	*Vireo cassinii*	18	none
white-crowned sparrow	*Zonotrichia leucophrys*	17	none
winter wren	*Troglodytes troglodytes*	15	none
chestnut-backed chickadee	*Poecile rufescens*	12	none
American goldfinch	*Carduelis tristis*	11	none
brown creeper	*Certhia americana*	11	none
flammulated owl	*Otus flammeolus*	11	none
California slender salamander	*Batrachoseps attenuatus*	11	CA threat

objective data sets and clarify where scale dependencies and assumptions regarding the input data were present. The scale dependencies in this biodiversity analysis were revealed when one or more of the scale-specific submodels were judged by the model user to be critical in the analysis. As an example, some sites or species may have had critical biodiversity value due largely to local (small area) effects. In this case study, redwood habitats were of critical importance in every submodel, whereas areas of coastal oak woodland and coastal scrub were emphasized solely through the landscape pattern analysis.

In the ecoregional submodel, data inputs were spatially coarse grained as is appropriate in studies of larger areas. Thematically, I also conducted this analysis at a coarse-grain scale using a generalized habitat classification to describe the county in an ecological context. The ecoregional analysis included an intentional bias for land management, incorporating the judgment of the model user to set the urban conversion probability of public versus private lands. This was done by counting only a user-selectable fraction of habitats where they were privately held. A more explicit approach would have extended the urban growth model from the county to the entire ecoregion; however, the benefits gained from this refinement may not have justified the added complexity in the model.

The county special features data represented elements that were detectable only when using a fine-grain approach to biodiversity analysis. These areas

TABLE 18.2.

Landscape analysis results. Major habitat types in Santa Cruz County, California, listed in order of patch cohesion alteration. Cropland and urban area effects are also included.

Habitat type	Rank order
coastal scrub	most altered
coastal oak woodland	
redwood	
montane hardwood	
chamise-redshank chaparral	
annual grassland	
cropland	
urban	least altered

were the current habitats of concern in the county, often representing fragile remnants of once-larger or more-numerous landscape features. This resolution of analysis and associated small plots may be particularly important for preserving small populations of plants (Lesica and Allendorf 1992). The special features submodel also provided data concerning what were typically the most studied and understood environmental issues in an area and were valuable examples of biodiversity loss in the area. This portion of the data, when combined with larger-extent biodiversity metrics, allowed detailed study results about single species to be leveraged or combined with other data for a more complete understanding of regional biodiversity. In some instances, special areas in the county will be adequately protected by existing state or federal law. At other times, a special area may still be vulnerable, even with state or federal protection, or it may not possess such protection. Because of this political and legal uncertainty, realistic assessment of impacts on special areas is best deferred until county planners can modify the initial iteration of the urban-growth scenario.

The species habitat submodels can function at a variety of spatial resolutions, though the California GAP vertebrate models are largely based upon the WHR models developed for a statewide area. WHR has not been immune to criticism (Block et al. 1994), though validation of these types of models is a challenging problem in itself (Marcot et al. 1983; Karl et al., Chapter 51). The California GAP models offer the re-

finement of updated and improved habitat descriptions; however, this did not permit the models to replace field surveys of local interest areas. It may have been reasonable to fine-tune the vertebrate models using local landcover data; however, this would still have yielded only a first-draft list of species needing verification by field survey.

The landscape pattern analysis submodel addressed the issue that patches of habitat are not simple summaries of their areas, notably that patch shape is an important element in any biodiversity stressor model. Landscape pattern metrics are commonplace in the research literature (Pearson et al. 1995; Schumaker 1996; O'Neill et al. 1997; Wickham et al. 1999) even if they're little used in the planning process. Although there is not universal agreement on what the many landscape metrics mean, it is useful to identify the extreme changes in area/perimeter relationships following habitat modification. The identified habitat types can then be assessed individually to determine if the forecast change will be biologically important. By calculating patch cohesion, which is a variant on the perimeter-to-area ratio, I compared habitat patches before and after the application of the urban growth model. Where projected urbanization most radically altered habitat, as in the case of coastal sage scrub, this landscape analysis was useful to flag specific habitat types for possible future conflicts. Future improvements in habitat map grain and WHR reliability will improve this submodel. However, as fine-grain habitat data becomes available, issues of species perception of habitat edges will also become more critical.

I used an urban growth model to address the issue of stressors to the environment. Buildout plans or other spatially explicit land-use scenarios could also function with my biodiversity model. The approach I have adopted for modeling urban growth is best used as an iterative model, letting the growth-model results be assessed by knowledgeable county planners who guide the tuning of model parameters in a series of model runs. By restricting input data to spatially explicit physical features (slope, roads, protected areas, etc.) and avoiding the less-tenable political and socioeconomic dependencies, I optimized the growth model for a long-term (20–50-year) forecast. Other models incorporating data on politically guided development patterns and population growth may be more appropriate for short-term (5–20-year) forecasts. For the purposes of this analysis, I have used the urban growth scenario generated by my model without the benefit of iterative tuning by county-planning personal. This additional step would implicitly incorporate some of the near-term socioeconomic trends and could have improved the utility of my results. Comparison of multiple urbanization scenarios and sensitivity analysis is also a logical approach to explore.

Starting with the California GAP vertebrate models, the habitat sites encroached upon by forecast urban growth were considered "compromised" regardless of the actual percentage of the habitat patch converted to urban land use. This is a conservative approach, providing a list of species worthy of further analysis to determine if continued habitat pressures could indeed disrupt the species. Few of the species I identified were particularly threatened at the time of the study, but as such, they were also not always considered in the planning process. Many of the species I identified for priority treatment during the next forty years were actually species that marginally included the Jepson Central West ecoregion in the southern fringe of their habitat. Species such as the flammulated owl (*Otus flammeolus*) may be doing quite well in the Sierra Nevada, but it may not be reasonable to assume that the species functions in exactly the same ecological niche across different ecoregions. Other species such as the California giant salamander (*D. tenebrosus*) do not have continuous distributions connecting to another ecoregion, suggesting possible differentiation within the species.

The loss of farmlands to urban development is an issue of general importance, but I limited my discussion here to farmlands as they relate to biodiversity. From this perspective, farmlands are areas that do not currently support a diverse assemblage of native species; however, especially when compared to other more-destructive types of land-use practices, farmlands are resource areas that could be adapted for conservation management. In my model, the forecast 17 percent loss of agricultural areas is largely due to adjacency of existing urbanized areas, ready access to road transportation, and flat, buildable terrain. Other areas could also have been appropriately considered to have restoration value, such as degraded wetlands or any area disrupted to the extent that biodiversity function is compromised.

Conclusion

My analysis has quantified potential future biodiversity loss by reporting the habitat types, vertebrate species, and restoration areas most likely to be adversely impacted in coming decades. The results predict potential biodiversity degradation in specific areas of coastal scrub, redwood forest, coastal oak woodlands, montane hardwoods, and Douglas-fir. I also described potential conflicts in terms of vertebrate species, and raised additional issues of agricultural losses. Each submodel reported spatially explicit results, which when presented as map offers an easily comprehendible form of data presentation (e.g., Figs. 18.2 and 18.3). In several cases—notably for redwood habitats—the submodels provided converging evidence of future biodiversity stress.

Ideally, urban planners, land managers, and biologists will analyze urban-growth scenarios for predicted impacts on biodiversity and use these models for iterative reevaluation of land-use planning and reserve design. Iterations are powerful techniques in biodiversity analysis, allowing the model user to change assumptions and parameters such as ecoregional land ownership, new road placement, and urban growth type. With each iteration, biodiversity losses can be minimized and a more thorough understanding of county ecosystem health can be achieved. My case study of Santa Cruz County, California, was designed to be a general model, allowing it to be applied in other areas.

Each of the submodels function with data that were gathered at particular spatial and temporal scales. Since biodiversity is itself a multiscale phenomenon, I have strived to design an analysis approach that captures this variation. In some cases, portions of the model will not reveal threats to biodiversity, though this may be difficult to determine before compiling the data. By analyzing the fundamental components of biodiversity and reporting the results as discrete submodels, I can determine the scale at which the dominant stressors are functioning and ensure that they are not overlooked. At times, these multiscale processes may be nonintuitive. As an example, a study looking at local-scale dynamics may not reveal background effects from regional-scale stressors. Altering the biodiversity submodels can also reveal hidden scale dependencies. For example, scale sensitivity can be determined by generalizing the ecoregional habitat map. In other cases, the urban growth models can be run with a larger grid size, and the area of analysis can be enlarged or reduced to look for grain- and extent-dependent effects. Regardless of the choice of map scales, some of the biodiversity indicators are inherently scale dependent. Wildlife habitat models function best at larger grain sizes, urban growth models are not appropriate at the submeter level, and areas of special concern are not subject to area or landscape metrics. Because of these multiscale properties, I do not attempt to sum the submodels in terms of land area or species counts but instead consider levels of impact summarized over landscape units as the common theme.

In each submodel, the resolution of data is also related to accuracy of data (see Fielding, Chapter 21). In some cases, actual accuracy is not extremely important, whereas in others it is a critical issue. The urban growth model is predictive, and measures of "hindcasting" will not yield an accuracy assessment of future model performance. More importantly, such models provide county planners with a warning about probable future conflict areas and are designed for iterative tuning and multiple-scenario comparisons. Ideally, actual future development patterns will be proactively guided away from predicted scenarios. In other submodels, such as the species habitat models, the spatial resolution of the data is important for model accuracy. GAP WHR models are being evaluated for accuracy (Edwards et al. 1996; Boone and Krohn 1999); see also Karl et al. and Hepinstall et al. (Chapters 51 and 53), and Garrison and Lupo (Chapter 30). Initial assessment results show encouraging success,

although much room remains for improvement. This issue goes beyond the need for finer-scale habitat maps and corresponding WHR models, though these measures would certainly benefit the analysis presented here. A more difficult task will be to analyze biodiversity in a manner that includes a more sophisticated species-specific habitat analysis, incorporating more of the environmental needs of each species into each submodel.

My case study includes several types of ecological analysis, and it complements other approaches. Species populations can be studied using techniques such as population viability analysis (PVA), which is based upon measures of critical values like environmental and demographic stochasticity (Boyce 1992). Habitats can also be assessed by their component parts, and as shown by this study, ecosystem biodiversity and ecosystem health can be similarly assessed. As population level models are improved and refined, we will be able to incorporate PVA-type data with habitat viability analysis (HVA) and ecosystem viability analysis (EVA) to gain a more complete and consistent understanding of ecosystem health across multiple scales.

Acknowledgments

I thank Frank Davis, David Stoms, Jim Wiley, J. Michael Scott, Mary Anne Van Zuyle, and Uta Passow for reviewing earlier versions of this chapter and providing constructive comments. Their insights are greatly appreciated. I also thank Keith Clarke at the University of California, Santa Barbara, and John Landis at the University of California, Berkeley, for their assistance with urban growth models. I am especially indebted to Tom Bulgerin for his C++ wizardry, which vastly improved the power and flexibility of my urban growth analysis. Peter Stine has generously provided input throughout the research project, and my early collaboration with Steve Minta, Michael Soulé, and Bruce Goldstein was invaluable.

A Collaborative Approach in Adaptive Management at a Large-landscape Scale

Cynthia J. Zabel, Lynn M. Roberts, Barry S. Mulder,
Howard B. Stauffer, Jeffrey R. Dunk, Kelly Wolcott,
David Solis, Mike Gertsch, Brian Woodbridge,
Adrienne Wright, Greg Goldsmith, and Chirre Keckler

Resource managers are increasingly confronted with the problem of how to make informed decisions that rely on new information or tools. This is especially true as we attempt to shift land management practices to a regional or ecosystem perspective. Federal agencies and others have variously defined ecosystem management, which has led to many debates over the concept (Haeuber 1996). Recently, increasing conflicts over resource management have resulted in a number of area ecosystem planning efforts (e.g., Great Lakes–St. Lawrence River Basin, Interior Columbia Basin Ecosystem, Everglades–South Florida, Sierra Nevada Ecosystem Project, and Southern California Natural Communities [Johnson et al. 1999]). There have also been many well-publicized efforts to develop comprehensive landscape plans for managing individual species, focusing mainly on those that are federally listed under the Endangered Species Act (ESA) such as the northern spotted owl (*Strix occidentalis caurina*; USDA/USDI 1994a,b), California spotted owl (*Strix occidentalis occidentalis*; Verner et al. 1992), and grizzly bear (*Ursus arctos*; Burroughs and Clark 1995). For a variety of reasons, not all of these planning efforts have been successful, but all were costly.

Adaptive management has been emerging as a central theme in the management of natural resources on federal lands in the United States, particularly as it applies to the concept of ecosystem management (Wal-

ters and Holling 1990). A true adaptive process involves a rigorous scientific and repeatable approach to resource planning (Holling 1978; Walters 1986). There is a distinction between active and passive adaptive management where in the former there is an active pursuit of information as an objective of the decision-making process (Nichols et al. 1995). Walters (1997) stated that "adaptive management should begin with a concerted effort to integrate existing interdisciplinary experience and scientific information into dynamic models that attempt to make predictions about the impacts of alternative policies." He emphasized that this serves three functions: (1) problem clarification and enhanced communication among scientists, managers, and other stakeholders; (2) policy screening to eliminate options that are least likely to succeed; and (3) identification of key knowledge gaps.

Adaptive management is complex and conceptual, and the methods are ambiguous and rarely or only partially applied (Lee 1993; Gunderson et al. 1995; Walters 1997; Carpenter 1998; Rogers 1998). Failures of traditional management that did not use an adaptive approach have occurred most obviously with problems in large complex ecosystems (Johnson 1999b). Many efforts to implement large-area management plans (including early attempts on the northern spotted owl, hereafter spotted owl, or owl) failed for a variety of reasons. Managers are often limited by one or more of the following: a lack of data, inadequate knowledge or understanding of available data, lack of useful methods to analyze and interpret data in a meaningful way, and/or lack of effective communication with researchers (Arnett and Sallabanks 1998). Modeling has seldom been a part of natural resource management efforts where model predictions could be tested and used to enhance knowledge and improve management (Conroy 1993). In some cases, there may be considerable information, but it may not be useful for informing decision makers. Often there are insufficient resources to use the data, especially in a timely manner. Other problems such as political pressures or resistance to change within federal agencies have also contributed. A few efforts, such as those on North American waterfowl (Nichols et al. 1995), are succeeding because the primary constituents agreed that a problem existed, that specific information was needed to address the issue (and was sought), and how the resulting data would be used to modify management plans (Johnson and Williams 1999).

Many recent books and papers discuss problems and lessons learned from attempts at large-scale resource management, including those of manager-scientist interactions (e.g., Marzluff and Sallabanks 1998; Bormann et al. 1999; Carey et al. 1999; Concannon et al. 1999), but few described specific steps and results that led to successful implementation of a management plan. Although some tools and particularly some of the lessons that were learned from exercises such as those cited above were pertinent to parts of our work, comprehensive guidance or consistent methods had not emerged. This is not unusual given the evolving state of large-landscape-scale assessments under an adaptive management construct.

We were confronted with many of these problems in our roles as managers and regulators of the spotted owl on federal lands in northern California. The questions we sought to address led to a unique collaborative approach in adaptive management that would support informed decision making. We did not begin this effort in a formal structured way. Instead, we went through a process of trial and error until we eventually realized the importance of following a sequential and integrated adaptive approach to addressing species issues at a large-landscape scale. This was

not an easy process, particularly at a spatial scale covering four national forests (more than 2.2 million hectares). Although we experienced lessons similar to those reported elsewhere, we believe our eventual process was unique because we put the main concepts of adaptive management into practice using a genuinely collaborative process. Resource managers and specialists developed hypotheses to test in collaboration with the scientists. We then developed predictive models to apply on a large-landscape basis that addressed a suite of ecological factors. This was a much more direct method than operating separately, as had traditionally been done, and it developed trust, open communication, and understanding among team members. Although the process was time consuming, it turned out not to be difficult. As a result, managers (with scientists' support) will be using the information that was generated for guiding future land-management efforts for the owl. Because we were successful in applying a structured adaptive process, we believe the description of our approach offers a significant learning opportunity that has wider application to future resource planning.

Background

The northern spotted owl has captured the attention and interest of research biologists, land managers, regulators, politicians, lawyers, and the public for over twenty-five years. It has been the focus of numerous management plans by federal, state, and private groups (e.g., USFWS 1990; Thomas et al. 1990; Simpson Timber Company 1992; FEMAT 1993; USDA/USDI 1994a,b; The Pacific Lumber Company 1999).

The Northwest Forest Plan (hereafter Forest Plan) established a system of late-successional reserves (LSRs or reserves) covering over 24 million acres on eighteen national forests and seven Bureau of Land Management districts, including the four national forests analyzed in this effort. Over a one-hundred-year planning period these reserves should provide habitat for multiple late-successional-associated species, including the spotted owl (see Fig. 19.1 in color section). Given this assumption, there was an implicit expectation that further analyses to test and adapt management approaches would occur. These analyses would provide the information necessary to address future changes to species and forest management throughout this period.

Northern spotted owls are among the most-studied and well-known owls in the world (Gutiérrez et al. 1995) and the best-known owl in northern California (e.g., Solis and Gutiérrez 1990; Blakesley et al. 1992; Hunter et al. 1995; Zabel et al. 1995; Gutiérrez et al. 1998; LaHaye and Gutiérrez 1999; Thome et al. 1999; Franklin et al. 2000). However, major gaps in our understanding of spotted owl habitat selection remain, particularly from the large-landscape perspective. Survey results were inadequate for large-scale analyses because of biases in selection of sites (e.g., centered around timber sales), inadequate descriptions of survey boundaries, and to a lesser extent variation in survey protocol. Lastly, data were not always available or in a useable format. Data that were available only partially represented the full range of habitat conditions found within this ecologically heterogeneous area of northern California.

Current Situation

Most planning and regulatory evaluations for owls continue to apply the traditional project-by-project or site-by-site approach. Interagency efforts to plan projects, such as timber harvests, that meet the National Forest Management Act (NFMA), ESA, and Forest Plan requirements are also hampered by organizational and logistical factors. These include differences in terminology, inconsistent habitat descriptions, varying quality of owl-habitat databases, difficulty of evaluating proposed management activities beyond the project or site level to a larger landscape scale, varying opinions of individuals involved, and a lack of methods to adequately assess cumulative impacts of proposed management activities in time and space. As a result, resource specialists mostly rely on their professional judgment to evaluate impacts. In some cases, each national forest, U.S. Department of Interior (USDI) Fish and Wildlife Service (hereafter Fish and Wildlife Service) field station, and U.S. Department of Agriculture (USDA) Forest Service (hereafter Forest Service) ranger district uses its own unique description(s) of suitable habitat or evaluation methods. In

addition, owl surveys are no longer conducted in most areas, resulting in a greater dependence on analyses of habitat rather than on evaluation of known owl nest locations. Consequently, there remains a strong emphasis on evaluating and planning around individual owl sites instead of at larger spatial scales. These issues have resulted in disagreements between the regulatory and land management agencies on owl management, even among personnel from different agencies with similar objectives.

A Collaborative Process in Adaptive Management

In 1995, the Fish and Wildlife Service and Forest Service in northern California began an informal effort to improve the ability of managers to address questions about owl management under the Forest Plan. Early efforts focused on updating the spotted owl habitat database on national forest lands in the California portion of the Klamath Province. Although this informal approach is common to everyday application of resource assessments under the ESA, NFMA, and other laws, these early informal efforts to address large-landscape issues had little success. Finally, in 1997, managers formally directed a team of biologists from the four national forests and the Fish and Wildlife Service's three northern California field offices to improve the basis for resource planning and decision making under the Forest Plan. This project eventually represented a three-way collaboration among resource managers, specialists, and scientists, with each team member bringing their own unique expertise. Although the primary group responsible for this effort consisted of wildlife biologists, we were supported throughout by a variety of specialists, including resource planners, foresters, forest ecologists, silviculturists, geographic information system (GIS) specialists, fire/disturbance modelers, and ingrowth modelers. The term "resource specialist" refers to this larger group. The four major tasks the team undertook were to

1. Update and improve the quality of the forest vegetation databases for owl habitat.
2. Identify and apply more applicable tools to analyze and interpret the data at multiple scales.
3. Determine how to provide the results to decision

makers in a form that would be useful for owl management at larger landscape and longer temporal scales.

4. Create and implement an adaptive approach to owl management on Forest Service lands in northern California.

Because of the importance of these steps to large-area planning, herein we describe the approach, outcome, and implications of our efforts and products.

Collaboration among Resource Specialists

The initial basis for successful resource planning, whether for single or multiple species, is development of a credible up-to-date habitat database and map. The recent improvement and general availability of GISs has greatly increased our ability to develop these products for use at larger spatial scales and with greater spatial consistency. Although existing forest vegetation and spotted owl habitat databases in northern California were about twenty years old, each of the four national forests had recently made efforts to update their timber-attributed databases to support resource planning. These databases set the limits of our efforts to develop habitat descriptions and a new map that reflected spotted owl habitat use in northern California (i.e., we were unable to include some habitat features that we felt were relevant to owls when those features did not exist in the GIS databases).

Map Development, Quality, and Accuracy

We used published information on owl habitat use within the province and expanded the description of owl habitat based on limited analyses of known owl sites and the vegetation types in which they occurred, and the professional judgment of resource specialists knowledgeable about owls in the Klamath Province. The draft descriptions were evaluated and corrected using a modified Delphi approach (Coughlan and Armour 1992) until specialists were comfortable with the quality of the results. To improve our understanding of future habitat conditions and trends, we also used this approach to describe criteria to identify vegetation that would be capable of becoming owl habitat in the future. The resulting map was consistent with our understanding of owl habitat use and was

more amenable to evaluating ecosystems, as required by the Forest Plan.

Development of an acceptable map was more difficult and time-consuming (nearly three years) than anyone on the team expected. The quality and accuracy of the forest vegetation databases and our interpretation of owl habitat relative to those databases were significant issues that had to be addressed. Our efforts were hampered by the fact that the existing GIS vegetation databases among northern California forests were not always compatible (not an uncommon situation among resource agencies and administrative units). For example, each database originated from different mapping efforts, and coding or labeling was not consistent for the same attributes. In addition, resource specialists and managers had rarely questioned the quality of the information contained on old maps, which made it difficult to ensure map accuracy. This resulted in numerous false starts as errors were found and the maps had to be recreated. Eventually a single seamless map of suitable and capable owl habitat across the four forests was completed. Based on our best professional judgment, we assumed this map offered a better basis for analyzing management actions on owls under the Forest Plan.

Collaboration among Scientists

In response to questions raised about the use and quality of the updated vegetation database and habitat descriptions, scientists from the USDA Pacific Southwest Research Station undertook an effort to quantitatively evaluate the effectiveness of these habitat descriptions at predicting owl presence-absence. They also recommended that formally applying a probabilistic approach to modeling the landscape for owl occurrence would significantly enhance the quality of the map. This step represented a significant departure from management and regulatory agencies' traditional approach to using available data and maps. Involvement of scientists required integrating their goals with those of management. Consequently, the following specific goals were agreed upon:

1. Develop habitat models for predicting owl presence-absence using both the old and new habitat descriptions.

2. Determine the optimal spatial scale to apply the models.

3. Compare and rank the various models using objective criteria.

4. Test the highest-ranked models on independent data sets.

5. Evaluate various methods to apply the best model(s) for management needs.

Because of the significance of these products (from an ecological and economic perspective), this science-based modeling approach became the major focus of our effort and laid the foundation of our adaptive process (for a thorough treatment of the modeling effort, see Zabel et al. in review).

Model Development

Developing habitat-based models to predict the presence-absence of wildlife species is relatively straightforward. First, several attributes hypothesized to be important to the species in areas that are occupied and unoccupied (though apparently available to the species) are measured. Then, sites with and without the species are compared to determine which attribute(s) are most closely associated with presence-absence. Alternative models developed in this manner are evaluated and compared, and the best model is selected (e.g., Johnson et al., Chapter 12; Young and Hutto, Chapter 8).

To develop habitat models for this project, we used data from sites that had been randomly selected and surveyed for spotted owls on national forests in northern California. These sites had been surveyed according to a standardized protocol for two consecutive years (1988 and 1989) so that both occupied and unoccupied sites were determined. To facilitate our understanding of owl-habitat associations, we developed models that discriminated between sites with and without owls at three spatial scales using concentric circles that approximated different aspects of an owl's home range size: 200 hectares, 550 hectares, and 900 hectares. Models were developed by placing concentric circles over the vegetation polygons using ARC/INFO software (ESRI 1998) and then calculating the quantity of each covariate within those circles. Three habitat covariates were evaluated: (1) the total

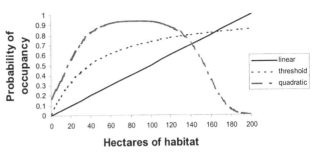

Figure 19.2. Example of linear, quadratic, and threshold forms of the relationship between habitat quantity and probability of owl occupancy.

area of nesting, roosting, and foraging habitat; (2) the total length of linear edge between habitat and non-habitat; (3) and the amount of core area within each polygon (defined by buffering each polygon by 100 meters and determining the interior area). Linear, quadratic, and threshold forms of the relationship between the probability of owl occupancy and the three habitat covariates were then evaluated using logistic regression (sensu Franklin 1997) (Fig. 19.2). Six habitat descriptions were also compared that allowed us to take into account different quantities and forms of relationships between the covariates and probability of owl occupancy.

Ranking and Selecting Models

We developed approximately one hundred models at each of the three spatial scales (200, 550, and 900 hectares). The bias-corrected Akaike's Information Criterion (AIC; see Burnham and Anderson 1998) was used to determine the most parsimonious model(s) that discriminated between occupied and unoccupied owl sites. The two models with the lowest AIC within each of six habitat descriptions, and at each of three spatial scales, were selected for further comparison and testing on independent data (i.e., a total of thirty-six models).

After critically evaluating the merits of both AIC and percentage correct classification, we decided that AIC and the percentage of owl-occupied sites correctly classified would be used to select the best models. Under percentage correct classification, predicted probabilities of occupancy are considered correct (assigned a value of 1) if they exceed some predetermined cutoff point, and incorrect (assigned a value of 0) if they fall below that cutoff point. Although most statistical software packages use 0.5 as the arbitrary cutoff point, there are many instances when 0.5 may be inadequate. Choice of a probability cutoff point is analogous to decisions regarding Type I and II errors. As Nichols et al. (1995) noted, in science there is a strong bias against Type I errors in which a null hypothesis is mistakenly rejected. Therefore, scientists typically assign a low probability (e.g., 0.05) for Type I errors despite the fact that lower probabilities of Type I errors produce higher probabilities for Type II errors (failure to reject false hypotheses) and, hence, to detect real differences. As a result, this places the burden of proof with resource managers. Because of ESA requirements to protect known individuals of a species, the Fish and Wildlife Service was more concerned with errors of omission (i.e., predicting absence when owls were present) than errors of commission (i.e., predicting presence when owls are absent). Therefore, we decided it was more important to correctly predict owl presence than it was to correctly predict their absence. We separately determined an optimal cutoff point for each model based on the following criteria: (1) percentage correct classification of owl-occupied sites was greater than 75 percent, and (2) any loss in percentage correct classification of owl-occupied sites was more than compensated for by a gain in percentage correct classification of unoccupied sites. Final model rankings were based on the average of the AIC rank plus ranks of percentage correct classification of owl-occupied sites. As recommended by Nichols et al. (1995) and Burnham and Anderson (1998), an empirical Bayesian approach was also used to rank the models and compare results.

Model Testing

Models should not be used as the basis for management decisions without testing (Conroy 1993). Testing should be conducted on truly independent data (Fielding, Chapter 21). Therefore, we selected the best two models within each habitat description at each spatial scale and then tested them using eight independent data sets. Each independent study area had been completely censused for owls. Thus, both presence and absence were documented, most over periods of longer than two years. The study areas were well distributed throughout the Klamath Province and provided a representative test of our best models for this region. Again, we used both AIC and the percentage of owl sites correctly classified to evaluate the performance of each model on the independent data sets. This allowed us to compare the accuracy of the twelve best models at each spatial scale. Model ranks for the best models were fairly consistent for both the development data set and the test data sets. In addition, the percentage correct classification of owl-occupied sites was greater than 90 percent for our best models. The approach we developed ([AIC rank + percentage correct classification rank] / 2) and the Bayesian approach gave very similar results for the top models. This further strengthened our confidence in our choice of the best models; they fit all of the independent study areas with a high degree of accuracy. This exercise produced a best habitat model and two potentially competing models. We used the best model to evaluate the quality of owl habitat across the landscape.

Owing to the collaboration of researchers and managers, this phase of our process differed from what would have been done had this been a pure research project. First, we would not have selected the top two models within each habitat description for subsequent testing. Instead, we would have chosen a subset of the top-ranking models. Our decision to keep the best two models within each description was management driven. For example, the top-ranking model (habitat description) currently used by the management and regulatory agencies ranked fifty-fifth using the developmental data set, nowhere near the top twelve models. However, since it was the habitat description being used, it seemed important to give it a "fair chance" in both the model development and testing phases because our results could ultimately lead to a change in that habitat description.

A Framework for Future Collaboration

Although it may seem obvious to some, it is critically important for managers and resource specialists to understand (at least conceptually) the analytical techniques that will be used by those who develop wildlife habitat models. For example, once the resource specialists and managers on our team understood what AIC was, we collectively made purposeful choices regarding the weight we gave it relative to errors of commission. If this had been a pure research project, different criteria may have been chosen and the results may have been less understandable or useful to resource managers. Equally important, the scientists had to understand the needs of managers and specialists. This was an example of the collaborative interaction between scientists and managers that embodies the principles of conservation biology and adaptive management: the application of the best information to management, even in the absence of complete information, where the results of the application will provide new information.

To accommodate the application of habitat models in resource management, we suggest the following hierarchical (adaptive) approach to model development, testing, and application:

1. Resource scientists, managers, and specialists should work together closely from the beginning phases of any planning effort to ensure the usefulness of resulting models.
2. Models should be developed using data from large-enough areas to warrant their application to a variety of conditions.
3. Models should be tested on independent data to evaluate their accuracy.
4. Models should be tested in a manner consistent with how they are intended to be used.

Although incredibly useful for elucidating features of the biology of organisms of interest, many wildlife-habitat models have fallen short of being applied practically. To be fair, many times model application has not been the goal of the scientists, although we suspect that most expect their work will be of practical use. Regarding the concerns of scientists versus managers about models, Salwasser (1986) noted that "determining accuracy is the purview of the scientist; practicality, that of the manager." Therefore, ultimate decisions on the performance of habitat models must be made with or by the managers who will be using them.

Collaboration for Successful Adaptive Management

The final collaborative step in this exercise was for the managers, resource specialists, and scientists to establish a basis for interpretation of owl habitat quality so

TABLE 19.1.

Questions used to identify and prioritize Late Successional Reserves for managing northern spotted owls (*Strix occidentalis caurina*) in the Klamath Province in northwest California.

Question 1	What is the quality of owl habitat within and between reserves and groups of reserves?
Question 2	Does an opportunity exist (and where) to improve owl habitat through silvicultural treatments?
Question 3	Is there a need (and where) to manage for fuel hazard and risk?

that management recommendations pertinent to the scale of the Klamath Province could be developed. Our thoughts on approaches and methods for analysis evolved as we refined our questions about owl habitat relationships through the process of map development and model testing. We eventually realized that many of our traditional ideas about data analysis and methods at the site scale were not appropriate at larger scales. Consequently, we strove to complete our efforts with a more rigorous and collaborative approach to developing management recommendations within an adaptive management framework.

The first step was to jointly refine the questions of management interest in northern California. Table 19.1 identifies the three primary questions that we agreed were the most significant to both regulatory evaluation and owl management under the Forest Plan in the Klamath Province. These questions helped focus our efforts to select appropriate landscape features, evaluate the available data, and use the results to rate habitat quality for spotted owls at different scales.

Application of the Model to the Map

The primary task in the interpretive process was to assess the current habitat quality of individual reserves and their potential quality. Thus, we evaluated how best (or whether it was reasonable) to apply the habitat model at the scale of a reserve or a group of interacting reserves. The models generated spatially explicit predictions within a large landscape, but the absolute results (i.e., the quality of habitat within each reserve) were the values of interest. Using the best model, we applied a hexagonal grid that covered the

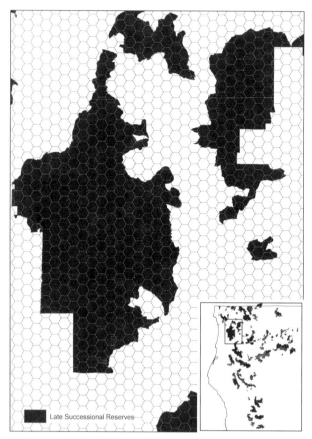

Figure 19.3. Example of hexagon grid applied to a group of Late Successional Reserves in the Klamath Province in northwestern California. Each hexagon has the probability of northern spotted owl (*Strix occidentalis*) occupancy attached to it.

landscape using the scale of the model that performed best: 200 hectares. The spatially explicit predictions were only applicable to this scale. We chose to use hexagons rather than circles to apply our model. When linked in a grid network, hexagons fully cover the landscape, unlike circles, and their shape closely resembles a circle (Fig. 19.3; Noon and McKelvey 1996b). The product of this exercise was a map in which each hexagon contained a probability of owl occupancy that we assumed to represent habitat quality. This approach allowed us to evaluate habitat quality at different spatial scales, from small to large reserves, to groups of reserves and intervening forested land.

Knowledge of landscape patterns and the factors that affect them are important to fully understand resource issues (Concannon et al. 1999). For example, Perry (1995) discussed the importance of considering the role and scale of disturbance in managing land-

scapes. To ensure an adequate suite of factors were included, we identified a set of other qualitative and quantitative factors that were important to evaluate the reserves to complement the modeling results. These factors included the probability and estimated intensity of wildfire, estimates of reserve connectivity based on published spotted owl dispersal distances, and projections of areas that were capable of becoming suitable or higher quality in the future. Because it is also important to know the scale at which these factors interact, we determined the spatial and temporal nature of each and whether they lent themselves to qualitative or quantitative analysis. For example, fire data were provided as probabilities of future occurrence across relatively large areas, while distance between reserves was used to assess connectivity. Data representing these factors were compiled and tabulated from other planning documents or databases developed by the national forests. Although there were concerns about the accuracy and currency of some of these data, there were neither useful methods nor other data to address them. These factors were modeled, reported as percentages, or qualitatively summarized in tabular form to make further comparisons of the reserves.

Application of the Results by Management

To provide the basis for interpretation of the compiled data, a spreadsheet was created that was linked to the updated GIS database. Within this spreadsheet, we divided the more than 2.2 million hectares of national forest lands in northern California into different landscape categories associated with the Forest Plan (reserves, non-reserved or matrix lands, and other administratively reserved areas such as wilderness). This spreadsheet allowed us to easily evaluate and compare results among reserves for a suite of factors pertinent to federal owl management at a large-landscape scale.

The probability results from the hexagon model and the summary data from each of the selected factors were evaluated and numerical ratings or condition indices were generated for each. The resulting table of indices provided the basis for a qualitative cumulative assessment of habitat quality or condition for both current and expected future conditions within each reserve. The indices and base data for

TABLE 19.2.

Summary results of Late Successional Reserve (LSR) analyses for each management question about northern spotted owls in northern California.

Ecological zone	Management questions			
	Number of LSRs analyzed	Number of LSRs with low habitat quality	Number of LSRs with high priority for silvicultural treatment	Number of LSRs with high priority for fuels reduction treatment
Western Klamath	18	0	2	15
Eastern Klamath	5	3	1	2
Western Cascades	2	2	2	0
Modoc	1	1	1	0
Interior Coast	7	3	4	0

each factor were ranked and displayed as a frequency distribution. This frequency distribution was used to determine whether obvious thresholds or cutoff points existed that would relate to levels of management interests or needs. Finally, a new table was created that rearranged the reserves into threshold categories for each factor. This new list was used to rank the reserves for each of the three management questions noted earlier (see Table 19.1). Three separate lists were generated that prioritized reserves (Table 19.2) according to (1) quality of habitat (question 1), (2) need for silvicultural treatment to improve quality of owl habitat (question 2), and (3) need for fuel or wild-fire reduction treatment (question 3). An array of recommendations for management and regulatory use was then collaboratively determined for each reserve based on the cumulative assessment and priority of management need. Recommendations included prescribed burning, thinning, timber harvest, and other activities that would contribute to management of spotted owl habitat at different landscape scales and were specific to the needs of individual reserves. The range of recommendations we developed offers flexibility for resource management at different spatial and temporal scales that endeavors to meet owl conservation needs. As further conditions or data change, the assessment can easily be revised and recommendations adjusted accordingly.

Future Application of Model Results

Adaptive management, in which science is a substantial part of planning, evaluating, and modifying man-

agement strategies, can improve interactions between scientists and managers thereby increasing the effectiveness of planning, allocation, and management of resources (NAS 1997). To ensure the results of our exercise have lasting utility to both the regulatory and management agencies, we developed four products to support future planning and decision making: (1) a comprehensive database and seamless map (and associated metadata), (2) a table that ranks and lists recommendations for each reserve and larger area, (3) guidelines for using the information and tools, and (4) a procedure for incorporating new information and adjusting recommendations. We expect these products to be used by managers to draw reasonable and supportable conclusions about owls and owl habitat at scales much larger than an individual owl site, allowing for more-efficient land management planning and fulfillment of regulatory requirements under the ESA and NFMA. For example, the model can be used as a planning tool to help regulators evaluate potential effects of management activities and to identify areas where projects (management activities) such as timber harvests are most likely to improve owl habitat quality or to minimize the reduction of habitat quality. However, we realize this cannot be accomplished without educating staff and managers to use the products and process we have developed.

We envision an active approach to continuing application of our efforts, as described by Nichols et al. (1995). Adaptive management treats management as an experiment and evaluates whether the desired and hypothesized outcome emerges after some period of time (Gunderson 1999). In our case, the experiment was evaluating a large number of competing habitat models. Our treatment is implementing the new map/model in resource management where implementation has a hypothesized outcome, in particular, stability of owl populations in the long term and a reduction in the rate of population decline in the short term (5–10 years). Although the specific steps to do this have not yet been implemented, we have identified a number of key activities that will allow us to continue to test, improve, and revise both the products and their implementation over time.

To ensure that our initial attempt at applying an adaptive approach succeeds, several critical tasks remain. There is a need to continue to collect and test data because of the complexity of owl-habitat relations, the stochastic nature of forest dynamics, and the recognition that our efforts reflect only what we currently know. This will provide the data to allow our predictions to be tested (Walters 1997). A major step is to integrate this exercise into the owl-monitoring program so that results from studies of spotted owl demographics can be used to link demographic performance with habitat quality and management actions. This will allow us to evaluate associations among management activities (actions), probability of owl occupancy, and demographic parameters. Related to this is the need to test the results of our "Delphi" approach to refine the different habitat descriptions used in the five ecological zones and in particular to investigate the effects of variation in vegetation structure within these different habitat types. We made other assumptions that can and should be tested concurrently with this process. These include evaluating whether the owl is an indicator or umbrella species for other late-successional species, investigating the utility of this process to addressing other species/habitat conflicts, and understanding how forest management and manipulation of habitat quantity and distribution affects spotted owl and barred owl (*Strix varia*) interactions. Future success, however, is predicated upon a continuing effort to improve and maintain GIS vegetation databases, using new remotely sensed and ground-plot data, and ensuring that the databases accurately reflect continuing changes in the forests due to fires and other disturbances. A formalized cyclic approach needs to be un-

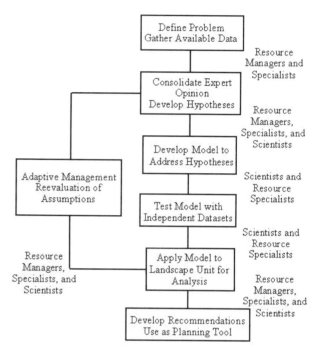

Figure 19.4. Conceptual model of the collaboration process among managers, resource specialists, and scientists.

dertaken in which use and revision of the above products are linked with research and monitoring programs, both at the regional and national-forest scale.

We hope that the application of our best model to the landscape and our focused concern on reserves that are not currently providing well for owls will result in a reduction in the rate of population decrease in northern California in the short term (ten years). However, because of the new approach described here, we must view these products as a first generation that will be improved and revised as we learn more about analyses, management, and owl habitat use and population dynamics at this scale. As a result, we will need to continue to work to integrate this process into future planning efforts so that our respective field units can repeat it indefinitely. This will enable us to routinely test and revise our results and associated management recommendations as new information becomes available—in other words, to practice adaptive management (Fig. 19.4).

Lessons Learned

Although we did not begin this effort under an adaptive management framework, the learning process itself became an adaptive process by default. Over the four years of this effort, we learned much about the steps involved and the problems encountered in using large-scale resource information for management in a structured and adaptive setting. In particular, maintaining contacts among disciplines and between scientists and managers has become more critical than in the past. However, there is an increasing workload associated with resource management, primarily due to the complexity and general lack of understanding about ecosystem management at any scale. Therefore, although the following is by no means an exhaustive list, we offer a synopsis of the key lessons learned that may be helpful to others attempting to undertake similar efforts.

Collaboration

Close collaboration between managers and scientists has not been the traditional approach to resource management. Large-scale resource planning, especially from a landscape perspective, requires an interdisciplinary approach, specialized knowledge, open communication, team compatibility, and integrated thinking. Although it has been recommended that scientists and managers should have a "translator" to foster communication (Schonewald-Cox 1994), we realized it was more critical for our team to have each group do the translating. Through this interaction, we eventually realized that scientists, managers, and resource specialists need to work closely from the beginning phases of any planning effort and should maintain their collaboration into implementation. By applying an adaptive and collaborative approach from the start, research could more easily be directed to support management needs, and management could more efficiently take research findings into consideration, thus reducing uncertainty and improving resource management. Although we did not include representatives from special interest groups on our team, we gave several public presentations to such groups. Based on their comments, all seemed supportive of our process and conclusions.

Changing Paradigms

Dealing with change, particularly change brought about by applying new concepts, was critical in our

endeavor to adaptively manage. However, there continues to be resistance among people and institutions to change. For example, even six years after the Forest Plan mandated the change from project-scale to large-area planning, this shift had not occurred. Seeking, analyzing, and applying new information and methods involves taking levels of risk that make some people and institutions uncomfortable. We believe that people will be more open to change if they are included from the beginning phases of a project rather than having new systems imposed on them from higher levels of government. Adaptive management is a structured and formal approach that requires focused and collaborative efforts to successfully integrate it into everyday operations. That had not been our experience as agency resource specialists or scientists, where it was often treated as an additional task or sometimes a constraint, if it was applied at all. Adaptive management offers a potential solution to dilemmas encountered when managing natural resources, such as uncertainty, conflicting information, and how to evaluate whether management is successful, and if not, why not (Lancia et al. 1996). We need to take a proactive approach to acquire the information necessary to avoid reacting to a problem after it has occurred. This is particularly important given the assumptions that underlie management policies, especially over these large areas.

Temporal and Spatial Scales

Temporal and spatial scale (extent and grain) issues are poorly understood in resource management, particularly when evaluating larger landscape units. We often found that data we had used to make management decisions prior to this project were applicable only at the site or local level and often had little relevance to questions that were pertinent to resource management at larger scales such as reserves or ecological zones. Although not usually considered, we should recognize that landscape goals dictate the level of analysis. By analyzing the context of an action within the larger landscape, we felt that we were better able to understand not just the effects of an action, but also the significance of those effects important to the scale of the Forest Plan. Managers and scientists must continue to ask whether a species needs to be managed at a coarse or fine scale, and to identify questions and apply techniques that are appropriate to that scale.

Data Quality and Availability

There is an increasing need for more resource data that address critical management questions, and particularly for data and maps of known quality and accuracy to carry out large-area assessments such as this one. The quantity of new and existing information about spotted owls is immense (Fig. 19.5). Even with all of the previous work completed on the spotted owl, we were continually surprised at how poorly data were maintained, how inconsistently they were reported, or how few data were accessible or even useful. This problem alone caused the most frequent and longest delays in our effort. This, coupled with the rapid rate of emerging ideas on habitat analyses, resource selection, mathematical and statistical models, and issues of scale, makes it extremely difficult for agency managers and specialists to remain current. As agencies attempt to improve their efforts toward ecosystem management, major emphasis needs to be placed on maintaining and spatially linking data, keeping data accessible, and using long-term data sets, all in collaboration with scientists.

Methods and Tools

There is a general lack of applicable and easily used methods or models for resource specialists and man-

agers to apply when addressing large-landscape-level questions. The lack of supported methods and inconsistent terminology continually hampered our project. In addition, the way in which we used tools such as GIS add a level of complexity, cost, and time that managers and resource specialists are reluctant to fund. Because of skepticism regarding conceptual or theoretical approaches, testing or piloting new methods in real situations with actual data is critical and should be a normal part of the process (Ringold et al. 1999). This is particularly important in testing assumptions and addressing the relationships in species/habitat interactions. We agree with the suggestion of considering multiple models (rather than a single most-probable model) in developing management strategies and then assessing their relative credibility by comparing competing predictions with subsequent observations (Conroy 1993). We also agree with the perspective of Young and Varland (1998) regarding "meaningful" research in a management environment—in other words, research that can be used to help make management decisions.

Conclusions

Our process is amenable to new information (e.g., dispersal data, habitat relationships in additional areas, etc.) that may emerge in the future. It is specifically set up to be an adaptive (repeatable) process that will further the progress we have made through this collaborative effort. Using an adaptive approach should not only change the way we work but also should make our work more efficient and proactive. It is our hope that this analytic process will serve as an effective model during future efforts to develop a comprehensive owl conservation plan for all public and private lands in northern California.

Close collaboration increased our appreciation and understanding of each other's perspectives and priorities. This effort was not easy and was at times frustrating (see Hejl and Granillo 1998 for additional insights). Had we not learned to work together, however, we would not have gained the ability to shift our way of viewing, and thus managing, the landscape from a deterministic to a probabilistic manner. Project impacts would have continued to be evaluated indi-

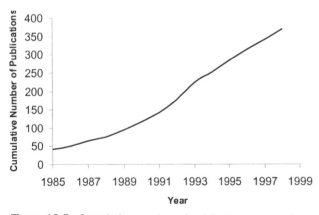

Figure 19.5. Cumulative number of publications on northern spotted owls (*Strix occidentalis caurina*) from 1985 through 1998.

vidually without looking at cumulative effects in time and space. We would have had no quantitative way to guide project planning into the future as our knowledge improved and changed. We strongly encourage this model of collaboration among scientists, resource specialists, and managers because we found it was fundamental to successful adaptive management.

Acknowledgments

We could not have succeeded without the assistance and ideas from a large number of managers, specialists, and scientists. A multitude of meetings, workshops, and individual discussions were held throughout the four-year period, so we apologize for names omitted from this list. Team members: Laura Finley, George Mayfield, (Fish and Wildlife Service); Jesse Plumage (Forest Service). In addition to our other team members, we thank our earlier collaborators for their contributions: Jean Elder, John Hamilton, and Laurie Simons (Fish and Wildlife Service); Steve Clausen, John Larson, Colleen Pelles-Madrid, Greg Schmidt, and Linda Tatum (Forest Service). We especially want to thank Alan Franklin, Kevin McKelvey, Barry Noon, and Darrin Thome, whose ideas benefited this endeavor, and lastly the managers from all three agencies who believed in and supported this effort. J. R. Waters and two anonymous reviewers provided helpful comments on drafts of the manuscript.

Modeling Wildlife Distribution within Urbanized Environments: An Example of the Eurasian Badger *Meles meles* L. in Britain

Amanda Wright and Alan H. Fielding

The impact of urbanization has rarely been addressed within wildlife-habitat models, and yet the rate of urban development is an ongoing issue in wildlife conservation. Most wildlife-habitat models are developed within relatively natural, undisturbed environments. However, their applicability to populations within urbanized locations have rarely been addressed (Boal and Mannan 1998). One such species that is influenced within part of its range by urban development is the Eurasian badger (*Meles meles*). The badger is a protected species within the United Kingdom through the Protection of Badgers Act, 1992, which also covers any sett (the burrow system occupied by badgers) with indications of badger use. Badgers live in social groups, usually occupying one main sett within a defended range. A main sett is distinguished from other setts within a territory by its continuous use by the social group (Kruuk 1978). Sett losses rather than mortality of individual badgers are the most significant population threat (Harris et al. 1995). Consequently, species protection requires knowledge of the distribution of main setts. In urban areas, particularly at the urban fringe, the landscape is changed by rapid land development (Trietz et al. 1992; Bolger et al. 1997). The impact of developments such as housing and road-building schemes, and the need for mitigation in some cases for the translocation of badger social groups, requires knowledge of the distribution and habitat requirements of badgers within urbanized environments. Previous studies on rural badgers (e.g., Wiertz and Vink 1986; Thornton 1988; Cresswell et al. 1990) have shown that setts tend to be more common in well-wooded, hilly areas at an altitude between 100 and 200 meters with a soil that is relatively easy to dig. However, it is not clear that these can be applied to badgers living in an urbanized landscape.

We investigated whether habitat factors considered important for determining badger distribution elsewhere within its range were applicable to main-sett distribution across a heavily urbanized area, with main setts used as a surrogate for social-group distribution. Our main objective was to determine whether areas used for main sett construction could be discriminated from unused areas within the urbanized landscape, but we also wanted to determine whether a model produced by discriminating used from unused sites would be influenced by land-cover change in terms of its stability and applicability.

Study Area

The urbanized landscape used within the study was the county of Greater Manchester, UK (53°29′ N, 2°15′ W), which covers an area of 1,286 square kilometers. It is a heavily urbanized conurbation with a polycentric structure. The human population is just over 2.7 million (Office of Population Census 1991). Although described as an urban county, approximately 50 percent of its area is occupied by farmland, woodland, and upland moorland, which is peripheral to a polycentric urban core. The county therefore has a highly varied landscape enhanced by variation in elevation and geology. See Wright (1997) for a more detailed description.

Methods

Four data layers (elevation, slope, soil, and land cover), which are related to known badger habitat requirements within rural areas (e.g., Neal 1972), were imported into a geographic information system (GIS). Elevation data were acquired from the Ordnance Survey in the form of a digital terrain model (DTM) with a resolution of 50×50 meters. The DTM was reclassified to produce discrete interval scale of 50 meters elevation data. In addition, the DTM was processed within the GIS to obtain slope data. The slope image was reclassified to produce a discrete interval scale of slope in 5-degree divisions. A 100×100-meter pixel resolution image of soils within Greater Manchester was made available to us under license from The Soil Survey and Land Research Centre, Cranfield University. Greater Manchester contains twenty-two soil associations that were reclassified into eight major soil groups as described by Avery (1980). The final data layers were land-cover images of Greater Manchester produced by supervised classification of Landsat thematic mapper (TM) satellite images captured May 1988 and June 1997.

Species data were acquired by surveying random 1-square-kilometer areas of the county for badger setts over June–September 1992–1996. The survey covered approximately 40 percent of the county. Accounts of sightings or records of sett presence from alternative sources (general public, countryside rangers, etc.) were also used to assist the survey. See Wright et al. (2000) for details of the sampling methods. Over the five-year study period sixty-five main setts were located within the county of which thirteen became disused. For model construction, forty-seven used main setts and the thirteen disused setts were used, with an additional five active main setts retained for model validation.

A control group, randomly selected from the population of all available sites avoided by the badgers, against which the data for suitable habitat locations could be compared, was established. The random non-sett sites were obtained using randomized grid coordinates, which did not contain main setts within a 1-square-kilometer area of the grid reference. Locations that had not been surveyed previously, along with those where there was uncertainty about badger activity, were omitted. A data set of six hundred random 1-square-kilometer non-sett squares was created. More absent locations were used than locations of badger presence, since the former were expected to show more variability (Pereira and Itami 1991; Fielding and Haworth 1995).

Statistical Analysis

For both data sets (setts and non-sett squares), a 1-square-kilometer area was extracted from each data layer. The extraction of the habitat data from the four data layers revealed a complex data set of forty-six variables, some of which were highly correlated. The dimensionality of the habitat data was reduced using principal components analysis (PCA) prior to discrimination. This produced fewer dimensions that were internally correlated but that were unique with regard to other derived dimensions (Krzanowski 1988). Only those components having an eigenvalue greater than one were retained for further analysis. A direct discrimination method (James and McCulloch 1990) was employed. The correctly classified rate was tested against that obtained by chance using a normalized mutual information (NMI) measure (Forbes 1995; Fielding and Bell 1997). From the NMI and its variance, a one-tailed z test was calculated to determine statistical significance of the observed accuracy of the classifier. Since the classification produced an accuracy higher than expected by chance, a statistical model was constructed. The model was built using the unstandardized canonical discriminant function coefficients derived within the discriminant analysis. Unstandardized coefficients can be used to calculate the discriminant score for a new case (Huberty 1994), which could be compared with the mean scores calculated for the two groups (sett squares and non-sett squares). The discriminant scores for each case ($N =$

660) were saved and imported into a GIS. These scores were interpolated using an inverse distance weighting method. The resulting surface was a weighted linear combination of the principal component axes. This discriminant score surface was a visual representation of the predictive discriminant model, and hence the surface predicted the presence/absence of main setts. It should be noted that the model only identifies "potential" habitat (based solely on habitat variables included in the analysis), which does not imply that the species is actually present at a given location. This model will be referred to as model one. The whole process was repeated using land-cover data derived from a supervised Landsat TM image captured June 1997 in order to determine the performance of the model after land-cover change within the county. This model will be referred to as model two.

Variable ordering was performed on both models, using the leave-one-out method, omitting one PCA variable in turn, and obtaining the total group hit from which a $Z_{(i)}$ value can be calculated (Huberty 1994). The best predictor was the one associated with the lowest $Z_{(i)}$ value. Ecological modeling has little merit if the predictions cannot be, or are not, tested using independent data (Verbyla and Litvaitis 1989; Fielding and Bell 1997; Beutel et al. 1999). Therefore, three approaches were used to validate the models: (1) training and testing data using cross-validation (Efron 1982), (2) overlaying the locations of setts not used within the model construction onto the models within a GIS and extracting discriminant scores for those locations, and (3) field surveys of unsurveyed areas highlighted as suitable sett locations by the models.

Results

The PCA for model one produced twelve normally distributed axes that explained 72.9 percent of the variation of the original data set. Following input into the discriminant analysis, these axes separated the data into two groups ($\chi^2 = 141.473$, df = 12, $P < 0.0001$). An overall classification accuracy of 81.7 percent was calculated, with 72 percent of sett squares and 82 percent of non-sett squares correctly classified (Table 20.1). An NMI of 0.187 was calculated for the confusion matrix (Table 20.1), where $z = 4.14$, $P < 0.0001$,

indicating a classification significantly better than expected by chance. The main discriminating variables are shown in Table 20.2. The sett squares were discriminated from the random non-sett squares by positive scores for PC 1 (a measure of slope), a positive score for PC 6 (a gradient from woodland and grassland to suburban and urban land cover), and by a negative score for PC 2 (an upland axis). Figure 20.1a highlights areas suitable for sett locations as those primarily found to the east and north of the county, or in other words those areas having a discriminant score of –1 to –3 (group mean discriminant score for sett squares –1.555). However, unsuitable areas border these on both sides, with low-lying urbanized land to the center of the county and upland moorland areas further north and east.

The PCA for model two produced fifteen normally distributed axes that explained 73.8 percent of the variation of the original data set. Following input into the discriminant analysis, these axes separated the data into two groups ($\chi^2 = 176.883$, df = 15, $P < 0.00001$) with an overall classification accuracy of 83.31 percent (Table 20.1). An NMI of 0.113 was calculated for the confusion matrix (Table 20.1), where $z = 4.365$, $P < 0.0001$, indicating a classification significantly better than expected by chance. The main discriminating variables are shown in Table 20.2. The sett squares were discriminated from the random non-sett squares by a positive score for PC 1 (a measure of slope), a positive score for PC 11 (a measure of woodland

TABLE 20.1.

Confusion matrices for the two models showing the predictive discriminant analysis classification results for sett[a] squares and random, non-sett squares.

Actual group	No. samples	Predicted group	
Model one		**1**	**2**
1 Used sett squares	60	43 (71.7%)	17 (28.3%)
2 Non-sett squares	600	108 (18.0%)	492 (82%)
Model two			
1 Used sett squares	60	41 (68.3%)	19 (31.7%)
2 Non-sett squares	600	91 (15.2%)	509 (84.8%)

[a]A sett is the burrow system occupied by the Eurasian badger (*Meles meles*).

TABLE 20.2.

Descriptions of the main discriminating principal components and their constituent variables in predicting the occurrence of Eurasian badger (*Meles meles*) setts.

Model one			Model two		
PCA axes	Variable	Rotated factor loading	PCA axes	Variable	Rotated factor loading
1	Slope 25–30°	0.9184	1	Slope 20–25°	0.9216
	Slope 30–35°	0.8982		Slope25–30°	0.9145
	Slope 20–25°	0.8743		Slope15–20°	0.8685
	Slope 15–20°	0.7894		Slope10–15°	0.7737
	Slope 35–40°	0.7068		Slope30–35°	0.7601
	Slope 10–15°	0.6724		Altitude 251–300 m	0.6127
	Altitude 251–300 m	0.4752		Altitude 300–351 m	0.6126
6	Urban	–0.7732	11	Coniferous wood	0.7240
	Improved grass	0.6646		Deciduous wood	0.5543
	Deciduous wood	0.6382			
	Semi-improved grass	0.4872	5	Improved grass	0.7351
	Suburban	–0.3456		Urban	–0.6342
	Mineral workings	–0.6150			
2	Bracken	0.8525		Semi-improved grass	0.5550
	Altitude 351–400 m	0.7685			
	Rough grassland	0.7615			
	Altitude 301–350 m	0.7466			
	Moorland grassland	0.6812			

Note: PCA 2 for Model one has a negative relationship with the discriminant group mean score for that model. All other axes (for both models) are positively related to the group mean score, or, in other words, the standardized canonical discriminant function coefficient is positive.

cover), and a positive score for PC 5 (a gradient from grassland to urban land cover). Figure 20.1b highlights areas that are potentially suitable for sett locations, which follow a similar spatial arrangement as model one. The group mean discriminant score for sett squares was +1.765 and –0.17681 for non-sett squares for model two.

Model Validation

Cross-validated classifications of the data for model one and two produced an overall classification accuracy of 81.2 and 82.6 percent respectively (Table 20.3). The classification accuracies were significantly higher than expected by chance, with for model one an NMI of 0.182 ($z = 4.09$, $P < 0.0001$) and for model two an NMI of 0.178 ($z = 4.01$, $P < 0.0001$).

A point vector file of the geographical locations of the five setts used for testing the model was overlaid onto the two discriminant score surface images (Fig. 20.1) within a GIS. A query of the discriminant score at those points gave discriminant scores between 0 and –3 for the five locations on model one (sett square group mean score = –1.555). For model two the scores ranged from 0 to +2 (sett square group mean score = +1.765). Therefore, the scores for the five setts used for testing the model indicated that both models could identify suitable sett sites, although model two was more conservative in its predictions than model one.

Two areas highlighted as potentially suitable for sett presence by the models were ground-truthed for sett presence/absence in order to provide additional evidence for the performance of the models. In area 1, an occupied sett was located. Its status as a main sett was not determined due to time restrictions. Within area 2, although an active sett was not located, evi-

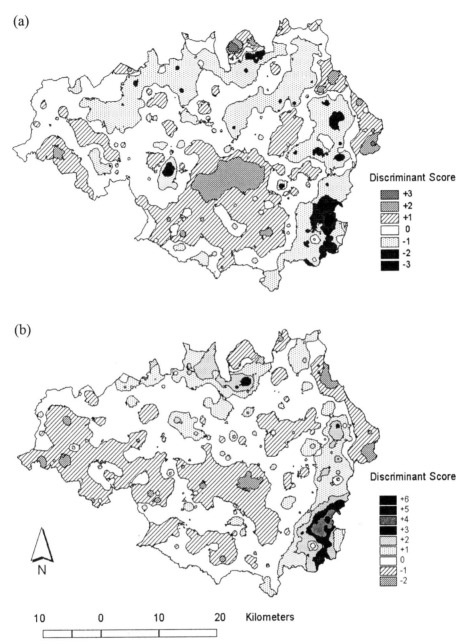

Figure 20.1. Discriminant score surfaces for (a) model one, and (b) model two. The group mean scores for sett squares and non-sett squares were different for the two models, resulting in a change of sign for the same geographical location from model one (sett squares group mean score = –1.555) to model two (sett square group mean score = +1.765). A sett is the burrow system occupied by Eurasian badgers (*Meles meles*).

dence of occupation was found through the presence of hair and an active latrine.

Misclassifications

Both models misclassified some data points, with the misclassifications lying in approximately the same di-

rection. That is, relatively similar proportions of sett and non-sett squares were placed in the incorrect group by the classifier. However, model one misclassified more non-sett squares and model two misclassified more sett squares (Table 20.1). Of the sixty sett squares, seventeen were misclassified by model one

TABLE 20.3.

Confusion matrices for the two models showing the cross-validated discriminant analysis results for prediction of Eurasian badger (*Meles meles*) setts.

Actual group	No. samples	Predicted group	
		1	**2**
Model one			
1 Used sett squares	60	43 (71.6%)	17 (28.3%)
2 Non-sett squares	600	108 (18.0%)	492 (82.0%)
Model two			
1 Used sett squares	60	40 (67.0%)	20 (33.0%)
2 Non-sett squares	600	95 (16.0%)	505 (84.0%)

and nineteen were misclassified by model two (Table 20.1), with fifteen setts common to both models. Of the sett squares misclassified by both models (*N* = 15), twelve contained large areas of suburban and/or urban areas within their sample square. For example, one was adjacent to a large airport. Two were situated on low-lying flat farmland prone to periodic flooding. The final misclassified sett was situated in rough grassland adjacent to open moorland.

Discussion

We have demonstrated that areas selected for use can be discriminated from those avoided across an urbanized landscape (after Morrison and Hall, Chapter 2). Both models successfully classified at a rate considered appropriate for ideal management models (Mosher et al. 1986). The models were validated and again successfully discriminated sett squares from non-sett squares. There was evidence of a reduction in the ability to correctly discriminate sett squares from model one to model two (Table 20.3). This may indicate that the change in land cover from 1988 to 1997 is in a direction that could be considered unsuitable for badger occupation, such as an increase in suburban land cover at the expense of grassland and woodland surrounding a sett site.

Habitat selection procedures are dominated by different variables at different scales (Manly et al. 1993). For example, the habitat selection procedure of the Mt. Graham red squirrel (*Tamiasciurus hudsonicus grahamensis*) is dominated by terrain vari-

ables at the overall landscape, while vegetation characteristics were more important at a finer resolution (Pereira and Itami 1991). We consider this the case within the present study where terrain variables were prominent in discriminating used and unused areas. The main discriminating variables also included land cover as well as terrain descriptors. For example, deciduous woodland, and semi-improved and improved grassland were positive discriminators for sett presence. However, this was to be expected as land cover is correlated with terrain. The generalized habitat associations found within the present study agree with those shown in previous studies (e.g. Wiertz and Vink 1986; Thornton 1988; Cresswell et al. 1990) in that badgers were most common in well-wooded, hilly areas. The habitat associations that emerged probably represent a response to a basic configuration of the environment (James 1971). It appeared that the coarse habitat association, evident for the distribution of badger setts within a heavily urbanized area, was similar to associations found elsewhere within other parts of its range.

The same setts were misclassified by both discriminant analyses (Table 20.1). These were setts situated in nontypical locations (as identified by previous authors: e.g., Thornton 1988; Cresswell et al. 1990; Macdonald et al. 1996), such as on low-lying flatland, or in upland areas without woodland. Those setts located in the more traditional badger habitat of deciduous woodland, with semi-improved or improved grassland, on sloping land were correctly classified as sett squares. The increase in the number of setts misclassified by model two relative to model one may be indicative of the change of land cover within the county over the nine-year period. This can be seen in an increase in land cover considered to be indicative of absences, for example, an increase in suburban land cover (from 25 percent cover within the county in 1988 to 31 percent cover in 1997 as calculated from the classified land cover images).

The misclassifications may have been caused by a variety of factors. First, the cause could be due to the scale of the habitat data: mixed pixels commonly occur in remotely sensed images, especially those with a coarse spatial resolution (for example 30×30-meter pixel resolution for the land cover images). Since image classification routines assume homogeneous pixels, the presence of mixed pixels will degrade classification accuracy (Foody and Cox 1994). This may have led to some inaccuracy in allocating cover types due to the coarse spatial resolution. Image classification may also be a cause of some misclassifications. The creation of training areas is a relatively subjective procedure, with inaccuracies arising from variations in pixel clarity within training sites. With two separate land cover images used within the present study, captured from different years, the training sites used in classifying the 1988 satellite image could not be used to classify the 1997 satellite image. Therefore, variation within the purity of the polygons used for training may have led to some inaccurate labeling of land cover within the sample squares. A second cause of misclassifications may again be due to the land-cover image. Within urban areas, particularly at the urban-rural fringe, temporal change in land use and cover is very rapid (Trietz et al. 1992). Therefore, the land cover surrounding a sett at the time of image capture may not be representative of the land cover within that location when the sett was constructed; in other words, the sett may now be surrounded by less-suitable habitat. Many of the misclassified setts contained large areas of suburban land cover, such as recently constructed housing estates, within their sample squares. This location of setts within suboptimal habitats suggests setts are a valuable resource for badgers (Roper 1993). There is evidence of strong fidelity between a social group and a main sett, with badgers reluctant to leave a well-established sett. A third cause of misclassifications include those setts constructed within apparently atypical habitats. For example, two of the misclassified setts were located on flat land at an elevation of less than 25 meters in an area prone to periodic river flooding. However, at a finer scale the setts were situated on small sloping banks thereby avoiding any potential for flooding. Therefore, it appears that for many misclassified cases, the spatial resolution of the input data contributed to their misclassification.

Habitat selection may have been influenced by temporal changes (Orians and Wittenberger 1991; Morrison et al. 1992; Greco et al., Chapter 14; Johnson and Krohn, Chapter 13) or perhaps by factors not easily quantified, such as anthropogenic disturbance. Skin-

ner et al. (1991) found that the loss of main setts within a county in England over a twenty-year period was attributable to a variety of factors, in particular, agricultural activities and increased urbanization and industrialization. These temporal changes may also account for the present distribution of main setts within Greater Manchester and therefore the difference between used and unused areas. Aaris-Sørensen (1987) found that sett disturbance by the public, and in particular the dog-walking public, was one of the major factors contributing to a 33 percent decline of the badgers in the Copenhagen area over a ten-year period. Although such factors may be influential in distinguishing between used and unused areas, they would be difficult to obtain and quantify across a landscape.

The combination of the habitat data provided a high ability (81.2 percent for model one and 82.3 percent for model two) to correctly classify the sample squares over a heavily urbanized area. However, classification accuracy for sett squares was less than the overall classification accuracy, especially for those setts situated within more urbanized locations. This raises the issue of the cost of the misclassification and the quality of misclassified setts (Fielding and Bell 1997). In terms of the performance of the model, a Type II error could be considered to be more expensive than a Type I error. If the model fails to identify a suitable environment for sett construction by labeling it as an absence site, then its use as a planning aid may be limited. Some setts within Greater Manchester are at risk through development programs. One social group was translocated within the study period due to the construction of an industrial park, and other setts were likewise affected by similar large-scale engineering projects, such as road-bypass schemes. However, with misclassifications inevitable within predictive models (Fielding and Haworth 1995) it seems appropriate to support the call for caution in using the predictions of computer models to make real-life decisions (Corsi et al. 1999), especially the use of models whose predictions have not been tested in the field using independent data.

The value of the models lies within their applicability at the landscape level for the identification of suitable sett sites, although they may be limited by the resolution of the input data. The models could enable threats to individual setts to be placed in a broader context, for example, when assessing planning applications. The usefulness of the models is in their construction and the techniques used within their development, with the results of model two indicating stability within the methodology used. Within a heavily urbanized county with a small badger population, the models were able to identify topographical and land-cover characteristics associated with main sett locations. For setts situated in favorable or typical locations, the models provide a high degree of accuracy and highlight areas of potential sett locations. For those in less favorable locations, the model may provide a means of identifying changes or restrictions in appropriate resources and point toward the need for conservation measures.

Acknowledgments

We would like to thank Phil Wheater for his contribution to the work. We are grateful to Jim Petch and Jo Cheesman for the assistance with the image processing. We would like to acknowledge the many people who provided information on badger distribution throughout the county, especially the various ranger services. Thanks also for the helpful comments of the anonymous referee.

Modeling Tools and Accuracy Assessment

PART 3

Modeling Tools and Accuracy Assessment

Randall B. Boone and William B. Krohn

As the number of ecological and management questions with spatial and temporal components has increased, so has the number of modeling tools used to predict species occurrences and abundances across space and time. In turn, the methods used to assess the accuracy of predictions of species occurrence have become more complex. The chapters of this section reflect that complexity. Authors explore methods of determining sample sizes, use a variety of modeling methods (e.g., classification and regression trees, artificial neural networks, genetic algorithms, Poisson regression, autologistic regression, Pearson's planes of closest fit, proportional odds models, and identification based upon habitat use categories), and employ statistical and graphical reports of model accuracy.

Complexity in models that predict species occurrence and the assessment of these models is warranted; the systems affecting species occurrence are often complex (Huston, O'Connor, Wiens, Introductory Essay, introduction to Part 1, and Chapter 65, respectively). Modelers must concern themselves with how species perceive the world (their habitats) and if humans can perceive the world in similar ways (Morrison et al. 1992). Further, these habitat components must be mapped to be used in spatial modeling. Habitats change over time and space due to disturbance (e.g., fire, hurricane, human development), succession (e.g., Litvaitis 1993), and climate change.

Habitat associations of a given species can vary over time (Morrison et al. 1992; Johnson and Krohn, Chapter 13), across regions (Collins 1983; Krohn 1996; Smith and Catanzaro 1996), *and* across different population densities (Brown 1969a; Fretwell and Lucas 1969). Species can be very common or very rare (Hanski 1982), and a given species can be abundant or rare in different parts of its range or through time (Maurer 1994). Lastly, range edges are difficult to define (Price et al. 1995) and often change (Hengeveld 1990). Given that population levels can vary markedly over time and space, that species interact with hundreds and perhaps thousands of other species, and that each of these populations may vary uniquely or in concert, the prospect of predicting occurrence can be daunting. Yet, the chapters in this section demonstrate that modelers have created methods that do indeed predict occurrence.

A review of the methods used to model species occurrences and abundances is too broad a topic for this brief introduction and is larger yet when accuracy assessment of results is included. We therefore introduce the chapters in this section, grouped into general themes, without delving into details. In our conclusions, we provide detail on selected topics and highlight areas requiring further research, a list that is by no means exhaustive.

Modeling Tools and Accuracy Assessment: The Symposium

A wide variety of topics was included in this section of the symposium. Most chapters include an assessment of the accuracy of demonstration results, but three chapters (i.e., Fielding, Chapter 21; Pearce et al., Chapter 32; and Schaefer and Krohn, Chapter 36) address assessment directly. Robertsen et al. (Chapter 34) use standard modeling methods, then use the results to explore management questions. The remaining fifteen chapters address modeling tools and methods.

Modeling Tools and Methods

Several authors reported on their efforts to improve modeling methods within the context of methods that are commonly used. Henebry and Merchant (Chapter 23) review the difficulties in modeling the occurrence of species in space and time, and suggest pathways for the future. Elith and Burgman (Chapter 24) compared the performance of four methods of species occurrence modeling (i.e., an envelope method of defining habitat, generalized linear models, generalized additive models, and genetic algorithms) predicting the occurrence of rare plants. They found that each model performed well but that the generalized linear and additive models predicted occurrences in novel data somewhat better. McKenney et al. (Chapter 31) used Monte Carlo methods on simulated warbler occurrences to assess the effects of sample size on logistic regression model results. They found that fewer than thirty samples yielded extremely variable results, and more than one thousand samples yielded very precise results. The results of McKenney et al. (Chapter 31) highlight the need for quantitative assessment—qualitatively, a map created from thirty samples and another from one thousand samples may both appear reasonable. Hartless et al. (Chapter 39) describe a factorial design they recommend for use when conducting individual-based modeling and demonstrate the technique while modeling white-tailed deer (*Odocoileus virginianus*) occurrence in Florida. We believe their methods could be used by those employing many modeling methods. Finally, Dunham et al. (Chapter 26) demonstrate how logistic regression can be applied to watersheds (considered patches in their application), with predictions of presence/absence constrained by water flow, for two fish species.

As methods of spatial modeling of species occurrence mature, modelers are encouraged to move away from methods that either violate underlying assumptions of statistical techniques or lead to a loss of information. Autocorrelation in animal occurrence data or predictor variables (e.g., elevation or climate) violates assumptions of independence of points in common regression techniques. Henebry and Merchant (Chapter 23) review the errors introduced by inattention to autocorrelated data, such as elevated significance estimates. Cablk et al. (Chapter 37) also review the effects of autocorrelation on model results and provide an alternative method, classification and regression trees. Cablk et al. (Chapter 37) use tree regression to correlate vertebrate richness in Oregon using a suite of environmental variables (others with a similar approach include O'Connor et al. 1996; Wickham et al. 1997; Boone and Krohn 2000a) and assess those models with an intriguing use of straightforward spatial statistics. Correlations between the environment and species richness tend to be at broad scales (Cablk et al., Chapter 37), making their uses in conservation different from those of detailed species predictions. However, these analyses can give us early indications of how a suite of vertebrates may respond to broadscale environmental changes, such as global warming. Klute et al. (Chapter 27) encourage the use of an autologistic regression model to model autocorrelated presence/absence data rather than use of the typical logistic regression. They describe the modified model and its autocovariance term.

Guisan (Chapter 25) encourages those using ordinal data (e.g., absent, rare, common, and abundant) in species modeling not to simply collapse the data to presence/absence and use logistic regression. Instead, modelers of such data should use methods that make use of these semiquantitative data (e.g., the proportional odds model). In his demonstration set, Guisan (Chapter 25) showed that the semiquantitative analyses yielded results on par with logistic results but with the added value of semiquantitative estimates of abundance. Violation of assumptions of normality in abundance data is another reason researchers will collapse data to binomial responses and use logistic regression, or use linear regression inappropriately. Abundance data often exhibit highly skewed distributions, with zeros (i.e., not found) occurring most frequently. Jones et al. (Chapter 35) demonstrates the use of Poisson regression, which is appropriate for highly skewed data, and, as with the semiquantitative method introduced by Guisan (Chapter 25), abundances are included in the model results. Further, as occurrence counts increase and the distributions approach normality, Poisson regression remains appropriate, allowing a modeler to use a single model form (Jones et al., Chapter 35). Johnson and Sargeant (Chapter 33) provide a means of using smoothing techniques on relatively sparse presence/absence data associated with atlases to create models showing relative probability of occurrence. Atlas maps were modified using simple smoothing methods, as well as methods using additional habitat information, with some evidence that the smoothed maps yielded better predictions than the original atlas data. Finally, Smith and Jenks (Chapter 38) contrast maps showing presence/absence for selected small mammals of South Dakota using typical species/habitat association methods with maps showing relative abundance, created based upon habitats in which trapped mammals had occurred.

In a review, Fielding (Chapter 21) noted that training sites must be representative of areas in which models will be used (see also Krohn 1996). A similar concern is outlined by Rotenberry et al. (Chapter 22), where their original effort used Mahalanobis distances to model sparrow habitat, a method that assesses habitat quality of a given point by comparing its attributes to the centroid of those attributes for occupied habitats. The method performs well when training and testing sets are similar. However, if habitat vastly improves at a site, for example, Mahalanobis distance methods may show the habitat as lower quality, because habitat attributes have moved away from the centroid of (relatively poorer) habitat that had been occupied. Rotenberry et al. (Chapter 22) recommend an alternative termed Pearson's planes of closest fit (e.g., Collins 1983), which uses linear relationships with habitat variables that are consistently important.

Artificial neural network analyses and genetic algorithms are provided as alternatives to traditional methods. These methods, in addition to evolutionary computation (e.g., Fogel 2000), address optimization problems by mimicking biological systems. Neural networks (Lek and Guégan 1999), which mimic learning processes, have several advantages over methods such as multiple linear regression (e.g., no prior assignment of a model form and no assumed distribution of data) (Lusk et al., Chapter 28). Lusk et al. use artificial neural networks to predict bobwhite quail occurrences in Oklahoma, with some success. Elith and Burgman (Chapter 24) include a genetic algorithm in the suite of models they compared. Genetic algorithms generate potential solutions to optimization problems—such as optimizing agreement between species occurrence and environmental variables—by emulating genetic processes (e.g., mutations, crossovers, combinations). Many solutions are tested, and those most successful spawn related solutions that are in turn tested, while others are discarded, ultimately leading to optimal solutions (Fogel 2000).

Accuracy Assessment

Evidence is now strong that modelers are more adept at predicting occurrences and abundances of species that are common than those that are rare (and associated attributes, e.g., secretive, quiet, using few habitats, or inhabiting remote areas) (Edwards et al. 1996; Boone and Krohn 1999; Garrison and Lupo, Chapter 30; Hepinstall et al., Chapter 53; Karl et al., Chapter 51; Schaefer and Krohn, Chapter 36). The stochastic component associated with whether or not a site is occupied by a rare species is larger than for a common species, leading to predicted distribution maps with larger commission errors. Schaefer and Krohn (Chapter 36) partition this commission error into *actual error* due to mistaken modeling form, and *apparent error* due to incomplete field surveys. The authors show that for the birds that bred within Maine, the apparent error was high for rare species inhabiting smaller test sites, suggesting field surveys were indeed incomplete (for direct evidence of incomplete surveys, see Nichols et al. 1998b). Incomplete surveys can reflect low detectability of species. Recognizing that detectability affects the number of surveys required and the power of statistical analyses, Stauffer et al. (Chapter 29) modified power analysis methods to incorporate detectability explicitly. They also showed that to characterize occurrence of marbled murrelet (*Brachyramphus marmoratus*) within a region, single surveys in many places are more beneficial than repeated surveys at given sites (Link et al. 1994; Maurer 1994). Last, Garrison and Lupo (Chapter 30) provide an example, where their efforts to model species ranges based upon habitat maps and associations were more successful for common species than rare species.

Fielding (Chapter 21) warns against overfitting species occurrence data used in training models and the importance of assessment to detect overfitting. Incidences or abundances in training data can often be modeled well, but modelers may describe not only the ecological relationships underlying patterns, but also individual variations in data points, including noise. An example is provided by Lusk et al. (Chapter 28). We agree with the authors that artificial learning techniques such as neural networks will become more important in future modeling work. However, Lusk et al. (Chapter 28) also show how artificial neural networks can overfit data ($r = 0.96$ in training, describing quail occurrence in Oklahoma, $r = 0.42$ in testing).

There is a consensus that multiple assessment terms should be reported when describing the accuracy of species occurrence or abundance models (Guisan, Chapter 25; reviewed in Henebry and Merchant, Chapter 23). Simple overall errors should be reported (e.g., Johnson and Sargeant, Chapter 33) as well as commission and omission errors (e.g., Schaefer and Krohn, Chapter 36). These measures provide straightforward ways of understanding model performance but are less helpful when comparing the relative success of models of different species (Boone and Krohn 1999). Guisan (Chapter 25) reviews statistics such as Kappa, which may be used to report accuracy in contingency tables. When reporting logistic regression assessments, instead of the traditional reporting of logistic regression probabilities less than 0.5 as absent and greater than or equal to 0.5 as present, receiver operating characteristic (ROC) plots (Fielding and Bell 1997; Elith and Burgman, Chapter 24; Fielding, Chapter 21; Guisan, Chapter 25; Pearce et al., Chapter 32) quantify accuracy for the full range of probabilities. Finally, Pearce et al. (Chapter 32) provide a framework for reporting errors in presence/absence models. They encourage the use of discrimination his-

tograms (which quickly identify how well presences and absences are being modeled), ROC plots, and calibration plots (which reflect whether the relative occurrence of species are being modeled well). Statistical measures are also proposed, which reflect agreement, biases in modeling, and spread (Pearce et al., Chapter 32).

Application and Management

Lastly, we cite the modeling effort by Robertsen et al. (Chapter 34), in which they used traditional logistic regression methods to predict the probability of occurrence of six species of birds in Wisconsin. They then linked the predicted distributions to simulated changes in forest cover due to harvest and fire and loss of habitat from development. Their efforts remind us of the importance of linking spatial modeling of species predicted occurrences to management questions.

Conclusions and the Future

In general, we believe that the methods used to model the occurrence and abundance of species are more developed than those used to assess those predictions. The number of chapters in this part of the volume that describe methods attests to this. Methods are improving rapidly, but limitations and research questions remain. As examples, researchers seek to know how animals perceive their surroundings, defining both the coarse- and fine-scale limits upon the distribution of species. Once researchers understand what animals perceive, they must create metrics and, if possible, maps that reflect that perception. Species/habitat relationships are scale (grain and extent) dependent, so modelers must work at the scale most appropriate for the issue at hand. In some cases, this will include working at more than one scale. Modelers are also limited by the continued lack of natural history information for many species. The habitat associations, for example, of common game species in North America have been described in some detail, but many questions remain; our uncertainty of habitat associations of nongame species is often large (Smith and Catanzaro 1996; Karl et al. 1999). How selection within a species varies by sex, season, and social status are often unknown (Morrison et al. 1992). Lastly, how species relate to even simple landscape metrics modelers calculate (e.g., patch size, patch count, area-to-perimeter ratio) is unknown for most species.

The spatial nonindependence of data used in modeling continue to limit our progress, especially if ignored by modelers. The implications are reviewed in this volume by Henebry and Merchant (Chapter 23) and Cablk et al. (Chapter 37), and so we will avoid repetition and simply add our voices to those saying nonindependence cannot be ignored. Instead, we caution against a reflexive removal of spatial patterning or its assignment to a stand-alone coefficient. In some applications, modelers seek to create an optimum predictive surface or to identify local effects on the response, and here, spatial pattern should be removed. In other applications, the structure of the spatial pattern itself is of interest (Legendre 1993). Klute et al. (Chapter 27) demonstrate that coefficients of predictor variables decline when an autocovariance coefficient is included, but that coefficient can mask interesting spatial patterns. Spatial autocorrelation in species occurrences may be due to the underlying population dynamics of the species in which partitioning variation in an autocovariance coefficient is appropriate. In contrast, spatial autocorrelation may indeed be due to a correlation with a spatially autocorrelated predictor variable, such as elevation. Here, reassigning variation from predictors to an autocovariance coefficient would be misleading. A method of contrasting between these sources of variation has been proposed by Borcard et al. (1992) and demonstrated in richness analyses by Boone and Krohn (2000b).

With today's computers and the ready availability of statistical packages, models predicting the occurrences of species can be generated in great numbers. But which model to use? The authors of the chapters consider a variety of approaches and discuss a number of the issues related to this question. A useful approach not stressed is to make quantitative comparisons between models, comparing relative measures of fit, such as AIC (Akaike's Information Criterion) indices. For a complete treatment of this topic, see Burnham and Anderson (1998) and Anderson et al. (2000).

This brief introduction can only stress that many questions in modeling species occurrence remain. Researchers should strive for more rigorous definitions

of the range limits of species and to gain a general picture of their dynamics through space and time (Krohn 1996). The utility of biogeographic relationships in species modeling across scales (extent and grain) is beginning to be explored (e.g., Root 1988c; Walker 1990; Scott et al. 1993; Venier et al. 1999). The effects of population dynamics upon predictions of species occurrence require more insight. As examples, modelers create models for populations that may be at or near carrying capacity and then apply them to areas that are understocked, or vice versa. Such methods may not yield accurate predictions of species occurrence (Krohn 1996). Also, how well species models apply to both source populations and sink populations in a given species remains to be explored (e.g., Pulliam and Danielson 1991).

Methods of assessment are lagging behind advances in methods of modeling. Simply put, it is both more straightforward and more rewarding to create predicted occurrences than it is to assess their accuracy. However, progress is being made in assessment, as reflected in this volume. A sometimes-neglected component of assessment is a consideration of risk, such as the cost of classifying habitats incorrectly (Green 1979; Krohn 1996; Smith and Catanzaro 1996; Fielding, Chapter 21). We also encourage further research into the assessment of individual layers used in species occurrence modeling, as well as how errors propagate in the modeling process (e.g., Haining and Arbia 1993; Krohn 1996). It remains important to assess species occurrences and abundances using a variety of techniques and to report a variety of results (e.g., resampling efforts, assessment with new data, expert opinion, reporting simple errors and statistics, presenting graphical summaries). Experience and analyses demonstrate that assessments using independently collected data require a large number of samples to yield a useful level of power. In general, methods for assessing individual sites appear to be reasonably robust. Methods for assessing species occurrence across a region appear weaker. Methods of assessing the models for a suite of species across a region are still in early development. Lastly, methods used to compare modeling efforts between regions are poorly developed. When modeling a suite of species, such as in gap analysis (Scott et al. 1993), careless comparisons of errors between sites or between states can be misleading. Sites within a region, even if in close proximity, can have different species compositions due to climate, historic, and other space-specific effects.

We reiterate a point already made by Fielding and Bell (1997): tests of model accuracy and the metrics used to report those tests must be related to the intended use of the model. As an example, commission errors may reflect previously occupied habitat when modeling a species with a shrinking geographic range. When modeling a suite of species, calibration plots (Pearce et al., Chapter 32) may be most useful, identifying whether the relative abundances of the species are being predicted well.

Methods of assessment used in the future must recognize that species are not equally likely to be successfully modeled, depending upon detectability, for example. We used these differences a priori to modeling to essentially form hypotheses regarding our assessments (the Likelihood Of Occurrence Ranks [LOORs] of Boone and Krohn 2000a). Fielding (Chapter 21) promotes the creation of assessment techniques that remove the effects of differences in detectability. One method may prove more useful, but at this stage of knowledge, research on both fronts should continue.

Species that are adapted to produce a few highly developed young (i.e., K-selected; reviewed in Pianka 1970) generally compete well in a specific set of habitats. That habitat specificity allows modelers to predict species occurrences reasonably well (although if the species is rare, commission errors may remain high). In

contrast, species that produce many less-developed young (r-selected species) have populations that can be irruptive. Habitats approaching the species' ideal may not contain the species in a given year, but in the following year, even relatively poor habitats may be occupied (Krohn 1992). Modelers have certainly modeled species on both ends of the continuum successfully. Our contention, however, is that commission errors of models of K-selected species are most often associated with the relative abundance of the species, whereas the commission errors associated with r-selected species are associated with the irruptive populations. In practice, assessments of r-selected species may yield more variable results than those for K-selected species. Clearly research on this issue of modeling K- versus r-selected species is needed.

It appears there is no unifying method that will allow us to successfully model the occurrence of all species. Each species can present its own complexities that must be understood and overcome. However, methods for modeling are improving as are methods of assessing the accuracy of predictions. Increased cooperation between disciplines and the expanding information base bode well for future efforts. Modelers must mesh the information and skills geographers can bring to modeling species occurrence, the analyses of pattern and change that landscape ecologists add, the knowledge of the ecological relationships of individual species that wildlife scientists add, the rigor of analyses that statisticians provide, and the knowledge that managers bring to use our predictions to improve conservation efforts.

What Are the Appropriate Characteristics of an Accuracy Measure?

Alan H. Fielding

Ecologists who predict the distribution of wildlife have been exposed to considerable debate about the relative merits of a variety of statistical methods (e.g., logistic regression versus discriminant analysis). There has been much less debate about the appropriate ways to measure our prediction errors (Fielding and Bell 1997). The problem appears to be that we have focused our discussions on the statistical assumptions of the predictive techniques (fitting the data to the model) and have largely ignored the best ways to assess the prediction accuracy. Usually the default options, offered by the software, are all that are reported. In fact, there are a number of alternative approaches to accuracy assessment, including those commonly used with machine learning (ML) models (Fielding 1999). Because machine-learning methods are not predicated on the presumption of well-defined probability density functions there has been little interest in the development of significance-testing methods. This means that accuracy assessments can use methods that are not constrained by normality or maximum-likelihood considerations. Instead of significance tests, models based on machine-learning methods are frequently validated by testing them on independent data sets.

This discussion is restricted to a consideration of the performance of a generalized predictive technique called a classifier that allocates cases to a small number of predefined classes. This definition excludes unsupervised clustering techniques and methods that predict the value of a continuous variable but includes multivariate techniques such as logistic regression, discriminant analysis, and classification and regression trees (CART). It also includes more novel methods, such as artificial neural networks (see Boddy and Morris [1999] for a review of their ecological applications).

It is useful to begin with two fundamental questions. First, why are we making the predictions? For example, is it to gain some insight into the ecology of an organism, or is it to predict the future or current distribution? Second, what level of accuracy do we expect from our model? Ecologists work in a relatively knowledge-poor environment where stochastic events, such as historical "accidents" and opportunism, can play a central role. Even if the real variables that control the distribution of a species can be identified they probably can't be measured directly. Therefore, we should not be too surprised if predictions lack accuracy; indeed, perhaps we should be suspicious if our accuracy is very high.

Imagine a scenario in which you have been asked to predict the future distribution of a species following some proposed environmental impact. The accuracy of your predictions is questioned at a planning inquiry. How confident are you that you can defend your predictions? Amongst the many questions that may be asked in such circumstances is "what do you understand by accuracy?" One definition is that it is the nearness of the predicted values to their actual values. However, this leads to a follow-up question about the nature of the actual values. Unfortunately, in wildlife distribution studies we rarely have a gold standard against which we can judge our predictions. If we were predicting the status of a banknote (forged versus genuine) or the gender of an individual, there is no ambiguity. In wildlife distribution studies, we have the added complications of scale, time, detectability, and biological variability. We must, therefore, begin by setting very clear goals against which we can judge our predictions.

Suppose that we wished to predict the distribution of the peregrine falcon (*Falco peregrinus*) nest sites in Britain. What is the appropriate reality against which predictions should be judged? Is it the current, historical, future, actual or potential distributions? This is important because Johnson and Krohn (Chapter 13) have demonstrated that we need to understand the population dynamics of the species at the time that the data used to develop the model were collected. Similarly, how should a nest site be categorized if it is only occupied when an adjacent golden eagle (*Aquila chrysaetos*) nest site is not occupied? In addition, what is the appropriate scale for such a study? At one extreme, we may be 10 meters out in our prediction of a nest location; at another, we may fail to predict the presence of a nest in a 4-square-kilometer square. Surely the second of these is the more serious error. Changing the scale of a study will influence our accuracy and move our focus away from the decisions made by individuals (behavioral ecology) to the relationships between the climate and the landscape and the broad scale distribution of a species (biogeography).

The Nature of a Classifier

It is important to understand that the separation and identification achieved by all classifiers is constrained by the application of some algorithm to the data. For example, many statistical methods assume that classes are linearly separable. In a linear discriminant analysis, class membership is assigned following the application of some threshold to a score calculated from $\Sigma b_i x_{ij}$, where x_{ij} is the score for case j on predictor x_i.

TABLE 21.1.

A comparison of statistical and pragmatic issues that should be considered during classifier development.[a]

Statistical issues	Pragmatic issues
Model specification	Model accuracy
Parameter estimation	Generalizability
Diagnostic checks	Model complexity
Model comparisons	Cost

[a]Based on Table 1 in Hosking et al. 1997.

The problem then reduces to one of estimating the values of the coefficients (b_i) using a maximum-likelihood procedure that depends on an assumed probability density function. Because the values of these coefficients are unknown, but presumed to be fixed, the emphasis moves toward the structure of the separation rule rather than the individual cases. Using statistical classifiers, we can usually obtain confidence intervals for our parameter estimates. However, these are conditional on having specified the appropriate statistical model. The classification accuracy of a technique such as logistic regression is largely independent of the goodness of fit (Hosmer and Lemeshow 1989). Part of the problem is that classification accuracy is sensitive to the group sizes (prevalence), and cases are more likely to be assigned to the largest group. Thus, the criteria by which a logistic regression should be judged are those relating to the statistical model rather than the classification accuracy.

Perhaps it is time to rethink our approaches to the application of classifiers to ecological problems. In particular, we could move the emphasis away from statistical concerns, such as the goodness of fit, to more pragmatic issues (Table 21.1), such as accuracy and cost.

Some Causes of Prediction Failure

Using the pragmatic criteria in Table 21.1, it is possible to examine situations under which a classifier is likely to be inadequate.

1. The form of the classifier is too complex and overfits the data. This tends to arise when the parameters:cases ratio exceeds some desirable limit and we

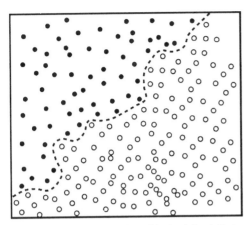

Figure 21.1. A decision boundary (dashed line) that over-fits the data. This boundary is unlikely to separate other samples with the same accuracy.

begin to fit the random noise in the data (Fig. 21.1). This will lead to poor generalization.

2. The form of the classifier is too simple or has an inappropriate structure. For example, classes may not be linearly separable or important predictors have been excluded (Conroy et al. 1995). This will reduce accuracy.

3. The initial class membership is incorrect, or "fuzzy." Most classifiers assume that class membership is known without error. Obviously, if classes are not clearly defined, it becomes more difficult to apply a classification procedure, and any measure of accuracy is likely to be compromised (see Magder and Hughes [1997] and Magder et al. [2000] for examples using logistic regression). Although the assumption of known groups is reasonable for discrete classes such as gender, it is questionable in other situations such as wildlife-distribution models. It is easy to imagine scenarios in which the dichotomization of cases could be compromised by ecological difficulties. For example, we should expect under-recording for rare or cryptic species and those that live in difficult habitats such as the canopy of a tropical rain forest. Boone and Krohn (1999) and Schaefer and Krohn (Chapter 36) have suggested that we can use a priori information on features such as detectability to place the accuracy of our predictions into an appropriate context. If cases are initially misclassified, this is likely to affect the number of prediction

errors. More importantly, misclassified cases will influence the structure of the classifier, leading to bias or classifier degradation; Maurer (Chapter 9) showed how errors in the predictor variables, when combined with errors in the response variable (presence/absence in these examples), can produce a systematic bias. For example, misclassified cases are likely to be outliers on some predictors, but their influence will depend on the classifier. In a discriminant analysis, outliers can have large effects because of their contribution to the covariance matrix. Conversely, in a CART analysis, their influence on the classification rules may be trivial because they are assigned to their own branch (Bell 1999).

4. Training cases (those used to produce the classification rule) may be unrepresentative. If they are, it will lead to bias and poor performance when the classifier is applied to new cases. Careful sampling designs should reduce this problem, but such bias may unavoidable if there is significant regional and temporal variability in the habitat relationships.

Characteristics of an Ideal Accuracy Index

When we make predictions about the distribution of wildlife, we gain little if all that we predict is the distribution of the animals used to produce the predictions (Beutel et al. 1999; Verbyla and Litvaitis 1989). To be most helpful, our predictions must be robust, general, and unbiased. An essential component of any index of prediction accuracy is that it was obtained from data that are independent of those used to generate the prediction rules. In other words, it is important to have some idea about how well the classifier will perform with new data. This is needed because the accuracy achieved with the original data is often much greater than that achieved with new data (Henery 1994). When a classifier is tested, its future error rate is estimated from a confusion matrix (a cross-tabulation of the number of cases correctly and incorrectly assigned to each of the classes) that was obtained, hopefully, from randomly selected members of the population. It can be shown that the expected performance of a classifier is linked to the size of the training set and that larger test sets reduce the variance of the error estimates. Unfortunately, the number of available test sets is frequently small so error rate estimates are normally imprecise.

The two data sets needed to develop and test predictions are known by various synonyms. The terms *training* and *testing* data are used here. The problem now becomes one of finding appropriate training and testing data. Ecologists seem to have paid little attention to the range of available methods or how the choice may influence the estimated error rates. One exception is Verbyla and Litvaitis (1989), who briefly reviewed a range of partitioning methods in their assessment of resampling methods for evaluating classification accuracy.

Resubstitution (reuse of the training data) is the simplest way of testing the performance of a classifier. Unfortunately, this provides a biased assessment of the classifier's future performance, possibly because the form of the classifier has been determined by some model-selection algorithm (e.g., stepwise variable selection). An inevitable consequence of model selection processes is that the final model tends to "overfit" the training data because it has been optimized to deal with the nuances in the training data (Fig. 21.1). This bias may still apply if the same set of "independent" testing data were used to verify the model selection (Chatfield 1995).

The best assessment of a classifier's future value is to test it with some truly independent data—ideally, a sample collected independently of the training data (prospective sampling). Because this is often difficult, a common practice is to split or partition the available data to provide the training and the "independent" testing data. Unfortunately, partitioning the existing data is not a perfect solution since it is less effective than collecting new data. In addition, the inevitable reduction in the size of training set will usually produce a corresponding decrease in the classifier's accuracy. There is, therefore, a trade-off between having a large test set that gives a good assessment of the classifier's performance and a small training set that is likely to result in a poor classifier.

The simplest partitioning method splits the data into two unequally sized groups. The largest partition

is used for training. Huberty (1994) provided a heuristic for determining the ratio of training to testing cases that is based on the work of Schaafsma and van Vark (1979). This heuristic suggests a ratio of $[1 + (p - 1)^{1/2}]^{-1}$, where p is the number of predictors. For example, if $p = 10$, the testing set should be $1/[1 + \sqrt{9}]$, or 25 percent of the complete data set. An increase in the number of predictors is matched by an increase in the proportion of cases needed for training. For example, if there are seventeen predictors, 80 percent of cases are needed for training. In reality, this type of partitioning is a special case of a broader class of computationally more intensive approaches. The first of these, k-fold partitioning, splits the data into k equal sized partitions. Each is used sequentially as a test set, whilst the remaining $k - 1$ sets are used for training. This yields k performance measures. The overall performance is then based on an average over these k test sets.

Leave-One-Out (L-O-O), and the related jackknife procedures, give the best compromise between maximizing the size of the training data set and providing a robust test of the classifier. In both of these methods each case is used sequentially as a single test sample, while the remaining $n - 1$ cases form the training set. Alternatively, a large number of bootstrapped samples (random sampling with replacement) may be used for testing. Efron (1983) and Efron and Tibshirani (1997) developed their *0.632 estimator* to correct the slight bias in error estimates obtained from bootstrapped samples.

Incorporate Costs

The index must take into account the total cost of the prediction errors. Three categories of cost apply to all classifiers. First, there are the predictor costs, which may be complex if some costs are shared between predictors (Turney 1995). For example, most field data will share the overheads associated with transportation to the field. These costs should be considered during the classifier's development (e.g., Schiffers 1997; Turney 1995). Second, there are computational costs, which include the time spent preprocessing the data and learning how to use the classifier and the computer resources that are needed to run the classifier.

TABLE 21.2.

Three examples of confusion matrices[a] and six derived accuracy measures.

		A		B		C	
		+	−	+	−	+	−
	+	95	20	80	5	95	120
	−	5	80	20	95	5	780
CCR[b]		0.875		0.875		0.875	
Sensitivity		0.950		0.800		0.950	
Specificity		0.800		0.950		0.867	
PPP[c]		0.826		0.941		0.442	
Kappa		0.750		0.750		0.540	
NMI		0.480		0.480		0.453	

[a]Columns are actual classes, rows are predicted.
[b]CCR = Correct Classification Rate.
[c]PPP = Positive Predictive Power (true positives/[true positives + false positives]).

Finally, there are the costs associated with the misclassified cases. This section is concerned only with the latter.

In a simple presence/absence classifier, we can make two mistakes (errors of commission and omission) that may have different costs. For example, do the measures shown for the three confusion matrices in Table 21.2 reflect the ecological "value" of the level of predictive accuracy? The answer must be no, because they do not identify all of the important differences between the three matrices. In particular, they do not take account of the misclassification costs.

Measures derived from confusion matrices assume that both error types are equivalent. There are situations, for example in a conservation-based model, where this assumption can be questioned. If a model is used to define protected areas, failure to correctly predict positive locations will be more "costly" (in conservation terms) than would commission errors, or in other words, omission cost (OC) is greater than commission cost (CC). Although these inequalities can be compensated for partly by the choice of error measure and allocation threshold (Fielding and Bell 1997), it is possible to adopt other approaches, such as a cost matrix that weights errors prior to the calculation of model accuracy. For example, we could assign weights by taking into account perceived threats to the species. Although it can be argued that the allocation of costs

must be subjective, unless there are clear economic gains and losses, a failure to explicitly apply costs equates to the implicit application of equal costs that can rarely be justified.

The cost associated with an error depends upon the relationship between the actual and predicted classes. This is because misclassification costs may not be reciprocal; for example, classifying a nest site as a nonnest site may be more costly than the reverse. Lynn et al. (1995) used a matrix of misclassification costs to evaluate the performance of a decision-tree model for the prediction of landscape levels of potential forest vegetation. Their cost structure was based on the amount of compositional similarity between pairs of groups.

If costs are applied, the aim changes from one of minimizing the number of errors to one that minimizes the cost of the errors. Consequently, the imposition of costs complicates how the classifier's performance should be assessed. For example, we now need to take account of the total costs that could be incurred from the imposition of a trivial rule that assigns cases to the least costly class.

Prevalence Independent

The value of the index should be independent of the proportion of cases in each group (prevalence). This is difficult because prevalence can affect the classifier in a number of ways. First, there are ecological implications of low prevalence, possibly arising from the important differences between rare and common species, some of which will influence the model's performance (Boone and Krohn 1999). Karl et al. (Chapter 51) showed, using sampling simulations, how commission errors decreased with increasing sample size, while omission error rates remained constant but with increased precision. Second, there are classifier-specific problems, such as those caused by the use of prior probabilities when assigning cases to groups (Titus et al. 1984; Manel et al. 1999). Finally, some confusion-matrix-derived measures are sensitive to the prevalence (p) of positive cases. For example, even the simple correct classification rate (CCR) is affected by the prevalence since CCR = $p \times$ sensitivity − $(1 − p) \times$ specificity (Ruttiman

1994), where sensitivity is the ratio of correctly predicted positives to the total number of positive cases, and specificity is the ratio of correctly predicted negative cases to the total number of negative cases. Consequently, it is important to avoid these potential pitfalls when a model's performance measures are interpreted in an ecological context (Fielding and Bell 1997).

Threshold Independent

Although there are better measures than the simplistic overall percentage correct (Fielding and Bell 1997), most fail to make use of all of the available information. Most classifiers assign cases to groups following the application of a threshold (cut-point) to some score, for example a discriminant score. Inevitably, this dichotomization of the raw score results in the loss of information; in particular, we do not know how marginal the assignments were. For example, using a 0–1 raw score scale and a 0.5 threshold, cases with scores of 0.499 and 0.501 would be assigned to different groups. In addition, unequal class sizes can influence the allocations, with cases being more likely to be assigned to the larger class. There are several strategies that we can use to deal with this bias. In particular, we can adjust the prior probabilities of class membership or adjust the assignment threshold. Similarly, if we have decided that omission errors are more serious than commission errors, the threshold can be adjusted to decrease the omission rate at the expense of an increased commission error rate. Few ecological studies appear to have addressed this problem (some exceptions are Capen et al. 1986; Fielding and Haworth 1995; Manel et al. 1999; Pereira and Itami 1991). Other thresholds could be justified that are dependent on the intended application of the classifier, for example a "minimum acceptable error" or omission criterion. However, adjusting the threshold does not have consistent effects on confusion-matrix-derived measures. For example, lowering the threshold increases sensitivity but decreases specificity. Consequently, any adjustments to the threshold must be made within the context that the classifier is used.

An alternative to threshold adjustments is to use of

all the information contained within the original raw score and calculate measures that are threshold independent. The receiver operating characteristic (ROC) plot is a threshold-independent measure that was developed as a signal-processing technique. The term refers to the performance (the operating characteristic) of an "observer" (receiver) that assigns cases into dichotomous classes (Deleo 1993; Deleo and Campbell 1990). The technique has been applied widely to clinical problems (Zweig and Campbell 1993) and there has been some recent interest from the ML community (Bradley 1997; Provost and Fawcett 1997). Marsden and Fielding (1999) used ROC plots when assessing the effectiveness of cross-island predictions of the habitat use by parrots on three Wallacean islands, and Manel et al. (1999) used them to overcome the threshold sensitivity of logistic regression when comparing species' distribution models.

A ROC plot is obtained by plotting all sensitivity values (true positive proportion) on the y-axis against their equivalent (1 − specificity) values (commission proportion) on the x-axis (Fig. 21.2). The area under the ROC function (AUC) is usually taken as the index of performance because it provides a single measure of overall accuracy that is independent of any particular threshold (Deleo 1993). It is also invariant to prior class probabilities (Bradley 1997). The AUC has a value between 0.0 and 1.0. If the value is 0.5, the scores for two groups do not differ and the classifier would perform as well as a coin toss. Conversely, an AUC of 1.0 indicates no overlap in the distributions of the group scores and the classifier would never misclassify. An AUC of 0.75 indicates that, on 75 percent of occasions, a random selection from the positive group will have a score greater than a random selection from the negative class (Deleo 1993).

Despite its advantages, the ROC plot does not provide an automatic rule for class allocation. However, there are strategies that can be used to develop decision rules from the ROC plot (Deleo 1993; Zweig and Campbell 1993). Finding the appropriate allocation threshold from a ROC plot depends on having values for the relative costs of commission and omission errors. Assigning values to these costs is complex and subjective, and dependent upon the context within which the classification rule will be used. As a guide-

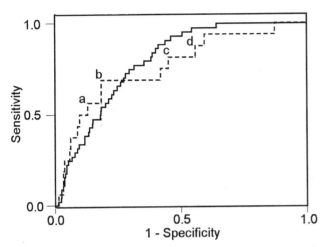

Figure 21.2. An ROC plot for training (solid line, 75 percent of the data) and testing data (dashed line, remaining 25 percent). Four thresholds are marked: (a) 1.13; (b) 0.79; (c) 0.13; (d) –0.16. The data are for the golden eagle (*Aquila chrysaetos*) nest sites described in the example.

line, Zweig and Campbell (1993) suggest that if CC is greater than OC, then the threshold should favor specificity, while sensitivity should be favored if OC is greater than CC. Combining these costs with the prevalence (*p*) of positive cases enables a slope (*m*) to be calculated (Zweig and Campbell 1993) from $m = (CC/OC) \times ((1 - p)/p)$, where *m* describes the slope of a tangent to the ROC plot if it is a smooth and parametric curve. The point at which this tangent touches the curve identifies the particular sensitivity/specificity pair that should be used to select the threshold. If costs are ignored (or assumed to be equal), the tangent is simply the ratio of negative to positive cases. Zweig and Campbell (1993) also describe a related algorithm for stepped nonparametric ROC plots.

Better Than Guessing

Is the level of accuracy better than could be obtained by chance or by the application of some trivial rule (e.g. put all cases in one group)? If one class has a high prevalence, a high CCR is achieved by the simple expedient of assigning all cases to the most common class. For example, if the prevalence of positive cases was 0.01, a CCR of 0.99 is possible if all cases are labeled as negatives. Huberty (1994:105) describes a one-tailed test that can be used to test if a classifier's performance is better than could be obtained by

chance. If *p* is low, the calculation is adjusted to enable a test of improvement over the rate achieved using the trivial rule of assigning all cases to the negative group.

Kappa (*K*), the proportion of specific agreement, is often used to assess improvement over chance. Landis and Koch (1977) suggested that $K < 0.4$ indicates poor agreement, whilst a value above 0.4 is indicative of good agreement. However, *K* is sensitive to the sample size and it is unreliable if one class dominates. The tau coefficient (Ma and Redmond 1995) is a related measure that depends on a priori knowledge of the prevalence rather the a posteriori estimate used by *K*. The more recent Normalized Mutual Information (*NMI*) measure does not suffer from these problems, but it can show non-monotonic behavior (Forbes 1995). It is important to note that all three of these measures do not favor accurate prediction of positive cases (see Table 21.2).

Incorporate the Context of the Predictions

Most error assessments do not take into account the context of any errors. For example, Riordan (1998) noted that if nineteen out of twenty footprints in one track run were assigned to animal A, and one was assigned to animal B, this was an obvious misclassification that could be ignored, since all the prints in one track set must be from the same individual. Similarly, it may be useful to examine the spatial pattern of prediction errors. Buckland and Elston (1993) discussed how patterns in prediction errors could be used to infer spatial patterns of habitat suitability, and Fielding and Bell (1997) described two spatially corrected error rates that are related to the technique developed by Augustin et al. (1996) for incorporating explicit autocorrelation into general linear presence/absence models. The rationale for the spatial weighting is that commission errors adjacent to real positives may be less serious errors than commission errors that are more distant from a real positive. Similarly, it is possible that some prediction errors are a consequence of ecological processes such as interference that have not been incorporated in the classifier, for example, the peregrine falcon–golden eagle example described earlier. In such circumstances, it may be possible to incorporate ecological information about territory size and spacing to weight some of the prediction errors.

An Example

The following example is used to illustrate most of the issues raised in the previous sections. The data are a subset of those used by Fielding and Haworth (1995). The aim is to construct a classifier that is capable of predicting the location of golden eagle nest sites (current, alternative, and historical [since 1960], $n = 60$) on the Island of Mull, Scotland, UK. The potential predictors are the habitat within each of 1,117 one-square-kilometer squares. These consist of the scores from eight principal components that retained 78 percent of the variance from the original twenty-one predictors. Because it is a reasonably well-known technique, discriminant analysis (SPSS for Windows, Release 9.0) was used as the classifier.

The results from a range of analyses are summarized in Table 21.3. Using these data, there is very little difference between the results obtained using resubstitution (equal priors) and those using cross-validation. Unfortunately, we have to be cautious about both analyses because the large number of commission errors has resulted in low Positive Predictive Power (PPP) values. There is only a 12 percent chance that a predicted square actually contains a real nest. The Kappa statistic is quite low, again suggesting that the classifier is not performing well. If the priors are changed to reflect the class sizes—60/1,117 and 1,057/1,117—the CCR rises to an impressive 94.4 percent. Unfortunately, we are no longer able to correctly predict any real nest sites.

In the remainder of the analyses, three partitioning schemes are used. The first applies Schaafsma and van Vark's (1979) rule, which uses the number of predictors to determine the proportion of cases needed for testing. Seventy-five percent of the cases were selected randomly for training. As might be expected, the classifier performed better with the training data. The second scheme was a fivefold partitioning. No attempt was made to retain an equal number of nests within each partition. Although the mean values are

quite similar to the previous results, it is interesting to note the range of values obtained for these five sets. When the second partition was used for testing, the predictions were better than those obtained with the training data. When the fifth partition was used for testing, there was a large discrepancy between the training and testing sensitivity values. These results illustrate clearly that the performance of a classifier is dependent on the composition of the training and testing sets.

The final partitioning scheme attempts to simulate prospective sampling. The island can be split into three geographical regions. The data from each region were used to produce classification rules that were tested on the other two regions. These results illustrate some important points about accuracy assessment. For example, if the CCR is used as an index, it is apparent that the North Mull classifier does not translate well to other regions. However, if sensitivity is used, North Mull is an excellent classifier. This discrepancy arises because almost half of the cases are errors of commission. Although it is not possible to illustrate this here (to protect the nest site locations), the commission errors are clustered around the real nest sites. If a spatial correction is applied (Fielding and Bell 1997), the CCR rises to 67 percent. When the Ross of Mull training data were used for training, a high CCR, combined with good sensitivity, was obtained. However, the classifier performed poorly with the eagle-nest-site test data. Fielding and Haworth (1995) demonstrated that the cross-region predictive success for a range of raptors on the island of Mull and on the adjacent Argyllshire mainland was dependent upon the particular combination of species, training, and testing sets. There was no guarantee of reciprocity. Similar patterns of between-region predictive failure were also found for parrots on Wallacean islands (Marsden and Fielding 1999).

In a final series of tests, the effect of the allocation threshold was investigated. These tests used the classifier trained on 75 percent of the data. The results are summarized in Table 21.4. As expected, changing the threshold altered the classification accuracy. Note that the value of the Kappa statistic rises as the number of

TABLE 21.3.

Results obtained from various permutations of training and testing data sets.

	n[a]	tp[b]	fn[c]	sens[d]	fp[e]	PPP[f]	K[g]	NMI	CCR%[h]
Resubstitution[EP] *	1,117	44	16	0.73	317	0.12	0.13	0.10	70.2
Cross Validated**	1,117	42	18	0.70	320	0.12	0.12	0.08	69.7
Resubstitution[CP]***	1,117	0	60	0.00	3	0.00	-0.01		94.4
75% training	855	33	11	0.75	244	0.12	0.13	0.10	70.2
25% testing	262	11	5	0.69	81	0.12	0.11	0.07	67.2
k fold results, k = 5									
train k = 2, 3, 4, 5	874	34	13	0.72	231	0.13	0.14	0.10	72.1
test k = 1	243	9	4	0.69	79	0.10	0.09	0.06	65.8
train k = 1, 3, 4, 5	895	33	16	0.67	246	0.12	0.12	0.08	70.7
test k = 2	222	9	2	0.82	51	0.15	0.19	0.17	76.1
train k = 1, 2, 4, 5	897	33	13	0.72	240	0.12	0.13	0.10	71.8
test k = 3	220	8	6	0.57	45	0.15	0.15	0.07	76.8
train k = 1, 2, 3, 5	892	36	13	0.73	255	0.12	0.13	0.10	70.0
test k = 4	225	8	3	0.73	73	0.10	0.10	0.07	66.2
train k = 1, 2, 3, 4	910	38	11	0.78	248	0.13	0.15	0.12	71.5
test k = 5	207	6	5	0.54	63	0.09	0.06	0.03	67.1
Training data mean				0.72	244	0.12	0.13	0.10	71.2
Training data std error				0.017	4.04	0.003	0.005	0.008	0.384
Testing data mean				0.67	14	0.12	0.12	0.08	70.4
Testing data std error				0.050	6.410	0.014	0.022	0.025	2.480
Cross-region predictions									
Ben More	583	26	13	0.67	162	0.14	0.13	0.07	70.0
Test set	534	14	7	0.67	96	0.13	0.16	0.12	80.7
Ross of Mull	175	4	1	0.80	27	0.13	0.18	0.21	84.0
Test set	942	28	27	0.51	178	0.14	0.13	0.06	78.2
North Mull	359	11	5	0.69	68	0.14	0.17	0.13	79.7
Test set	758	41	3	0.93	312	0.12	0.12	0.14	58.4

Note: Approximately 75 percent of the data were selected randomly to form the training set, and the remainder formed the test set. These proportions were determined using the rule suggested by Schaafsma and van Vark (1979).

[a] n = number of cases

[b] tp = true positive

[c] fn = false negative

[d] fp = false positive

[e] sens = sensitivity (true positive fraction)

[f] PPP = Positive Predictive Power

[g] K = Kappa

[h] CCR = Correct Classification Rate

*[EP] = equal prior probabilities

**Cross validated = Leave-One-Out testing

***[CP] = class size prior probabilities

TABLE 21.4.

Accuracy assessment after applying different thresholds to a discriminant score.

Threshold[b]	Training[a] AUC = 0.798					Testing[a] AUC = 0.762				
	tp	fn	fp	K	PPP	tp	fn	fp	K	PPP
−0.92	44	0	662	0.023	0.06	15	1	200	*0.018*	0.07
−0.46	43	1	516	0.052	0.08	15	1	171	0.041	0.08
0.00	40	4	353	0.100	0.11	13	3	120	0.073	0.10
0.46	33	11	237	0.134	0.12	11	5	80	0.114	0.12
0.92	38	6	135	0.141	0.22	9	7	41	0.199	0.18
1.38	26	16	49	0.180	0.35	6	10	20	0.227	0.23
1.84	14	30	28	0.161	0.33	4	12	10	0.222	0.29

Note: See Table 21.3 for sample sizes.

[a]tp = true positive, fn = false negative, fp = false positive, K = Kappa (figure in italics have $p > 0.05$).

[b]0.46 is the default threshold.

commission errors declines. Unfortunately, this is at the expense of true positives and is a consequence of the low prevalence of positive locations. Selecting the appropriate threshold for these data is dependent on the intended application. For example, if I wished to be reasonably certain of including 75 percent of nest sites in future samples, the default threshold of 0.46 or below should be applied. Conversely, if I had limited resources and wished to maximize my chances of finding a nest site in an unsurveyed area (large PPP), I would need to apply a threshold of 1.84 or above. If relative costs are assigned to the incorrect predictions, an optimum threshold can be determined from a ROC plot. As the relative cost of the omission errors increases, the slope of the tangent reduces and the threshold declines. For example, using the testing data, a relative omission cost of two suggests a threshold of 1.91, while a relative cost of ten decreases the threshold to 0.79.

Summary

Most ecologists do not undertake the development of predictive models lightly. Considerable effort is put into ensuring that the ecology and the analyses are sensible and appropriate. Unfortunately, this effort is not usually carried through to an appropriate assessment of the accuracy of the predictions. In this chapter, I have attempted to raise the profile of this aspect of distribution modeling. Although it is difficult to be prescriptive, the best method would appear to be the use of the AUC from a ROC plot, because it is threshold independent and provides an indication of how well the two classes are separated (Bradley 1997). I would also emphasize the need for rigorous testing using independent data. Whenever possible, this should involve some truly independent data obtained through prospective sampling. A more detailed discussion of the construction and assessment of classification rules can be found in Hand (1997).

22

A Minimalist Approach to Mapping Species' Habitat: Pearson's Planes of Closest Fit

John T. Rotenberry, Steven T. Knick, and James E. Dunn

Identification of probable use areas by animals is important for land-use planning, identification of conservation regions, and ecological studies of spatial distribution, movements, and resource use. Many statistical methods, often extensions of nonspatial resource selection models (e.g., Manly et al. 1993), increasingly are employed in geographical information systems (GIS) to determine the value of an index of use or likelihood of occurrence of a species at each point or grid cell within a study area based on the multivariate configuration of habitat variables at those points. The resulting maps then depict spatial variation in potential animal use, often at relatively fine resolution over large areas.

An issue that frequently arises in spatial mapping exercises is that whatever model is developed must be applied to points and even landscapes not originally sampled for the target organism. Indeed, the extrapolation of such GIS-based models forms the core of their use in identification of areas of conservation interest. However, spatial mapping extends beyond simply developing and calibrating selection models for prediction; statistical models that have a large R^2 may have little predictive capability (Rotenberry 1986; Gauch 1993; ter Braak and Looman 1995). This problem may be especially pronounced when attempting to predict animal use regions outside the immediate study area or in areas undergoing change, which

may contain landscape configurations not present in developing the original selection model. The solution to this problem is significant not only for the capability to reliably predict animal use areas without additional sampling but also because of the ecological implication that such a model can identify the basic requirements of animals that may be present in any alternative environment.

In this chapter, we discuss one emerging model that has been used in just such a GIS-mapping context, Mahalanobis D^2 (Clark et al. 1993b), briefly describing its biological inferences and how that contributes to its successes and failures (Knick and Rotenberry 1998). We then describe the biological attributes that a more appropriate model should embody. Finally, we provide an alternate statistical model based on the decomposition of D^2 and show how it incorporates those attributes.

Mahalanobis D²

The D^2 technique for mapping the probability of animal use or occupancy of a point is based on the generalized squared distance, $D^2(\mathbf{y})$, as a measure of the dissimilarity between a p-dimensional vector $\mathbf{y}$ and a sample mean vector $\boldsymbol{\mu}$ of the same dimension, in a space standardized by $\boldsymbol{\Sigma}$, the variance-covariance matrix of the sample (Clark et al. 1993b). Formally:

$\mathbf{H}$ = occupied habitat, an $n \times p$ matrix of p variables measured at n points where a species was detected. Frequently, this may be abstracted from a much-larger set of points that were surveyed but some of which did not include the target species.

Eqn. 22.1: Mahalanobis distance D^2:

$$D^2(\mathbf{y}) = (\mathbf{y} - \boldsymbol{\mu})' \, \boldsymbol{\Sigma}^{-1} \, (\mathbf{y} - \boldsymbol{\mu})$$

Where $\boldsymbol{\mu}$ = vector of means based on $\mathbf{H}$ ($p \times 1$) (i.e., the centroid),

$\mathbf{y}$ = vector of measurements on any point ($p \times 1$; may or may not be taken from $\mathbf{H}$); thus $\mathbf{y} - \boldsymbol{\mu}$ is a vector of deviations of a point from a species' mean vector,

$\boldsymbol{\Sigma}$ = variance-covariance matrix based on $\mathbf{H}$ ($p \times p$), and

D^2 is a squared scalar distance, standardized in the $\boldsymbol{\Sigma}$ metric.

Thus, any point may be described by its distance from the centroid of occupied habitat. Presumably, the closer it lies to this centroid, the more it resembles occupied (or occupiable, if we are predicting) habitat. As a hypothetical example, we might measure the size of the contiguous patch of shrubland in which each point occupied by a species lies. This will generate a mean occupied patch size (the univariate equivalent to $\boldsymbol{\mu}$) and its variance (the univariate equivalent to $\boldsymbol{\Sigma}$). Any individual point (occupied or not, from the same or a new sample) will also have a value for patch size (equivalent to $\mathbf{y}$). The difference ("distance") between the point and the mean (scaled by the variance) is a measure of the point's similarity to occupied habitat, and, presumably, more similar is better.

D^2 has been used successfully to link habitat use with a GIS to describe the distribution of species as disparate as black bears (*Ursus americanus*) in Arkansas forests (Clark et al. 1993b) and black-tailed jackrabbits (*Lepus californicus*) in Idaho shrubsteppe (Knick and Dyer 1997). However, an attempt to extrapolate the model developed in the jackrabbit example to shrubsteppe landscapes evolving under either natural or managed patterns of fire and plant community succession was far less successful (Knick and Rotenberry 1998). This latter exercise highlighted the biological inferences underlying the use of D^2 and

pointed out some of its shortcomings, which are discussed below.

One of the advantages of using D^2 is that one need measure only the set of "used" or occupied points. This avoids the ambiguity associated with including "unused" points (points surveyed but at which the species was not detected) in any resource selection model used for mapping. We assume that more often than not a species is at a point at which it is detected because it "wants" to be, that the habitat there satisfies some basic requirement. Points at which the species was not detected are ambiguous in the sense that they could represent habitat that is suitable but currently unoccupied; habitat that is suitable and occupied, but the observer failed to detect the species; or habitat that is unsuitable. Although consideration of "unused" points may be necessary to identify patterns of nonrandom resource use, such patterns will display a strong sampling scale-dependency (Wiens and Rotenberry 1981b; Wiens 1989a), a feature undesirable when extrapolating to new areas (which carries an implicit change in scale). Thus, use of D^2 does not depend on any arbitrary definition of study boundaries.

An additional useful attribute of D^2 is that deviations are scaled by the variance-covariance matrix. This not only standardizes across variables, but also explicitly incorporates their intercorrelations, which otherwise may cause problems in multiple regression-based approaches.

Use of D^2 for mapping explicitly assumes a selection function that is based on an observed multivariate mean ($\boldsymbol{\mu}$) and its covariance matrix ($\boldsymbol{\Sigma}$); the farther away from that mean, in standardized units, of any point $\mathbf{y}$, the less appropriate the habitat. Therefore, *any* deviation from the original habitat mean vector, even if it is in a direction that is actually biologically positive, is translated as less-desirable habitat. Returning to the hypothetical example of shrubland patch size, suppose that the species in question thrives as it finds itself in larger and larger patches, although it can hang on in smaller ones, which are occupied at a low frequency. Further suppose that the originally sampled landscape contained only relatively small patches (hence, the mean occupied patch size would be small). Were we to sample a new landscape with larger patches, mapping D^2 would indicate that points falling in large patches would actually be *less* suitable, because they deviate from the mean occupied patch size in the original sample. A phenomenon similar to this appeared to be responsible for the lack of success in extrapolating jackrabbit distributions in response to vegetation change cited above (Knick and Rotenberry 1998).

A further implication of the assumption that the selection function can be characterized by an observed mean and variance is that the original sample reflects the optimal habitat distribution of the animals in the sampled landscape. As a corollary, it assumes that the selection response has been fully characterized (at least in the vicinity of the mean), or in other words, that μ and Σ fully characterize the species response to habitat. This implies two additional features: the sampled area contains the full range of habitat variation to which the species responds, and we have identified and measured the appropriate variables (i.e., we have not left out any that are important, and we have not included any that are irrelevant). We leave it to the reader to judge the likelihood that these assumptions are met in any particular study, but we think the probability can often be low.

Although D^2 appears to work well for static, well-sampled landscapes where a species' distribution has been fully characterized and one has a strong biological basis for selecting variables to include in modeling, it may not be appropriate for other common situations (Knick and Rotenberry 1998). In particular, it may be prone to fail if applied to areas not included in the original sample or if applied to dynamic landscapes, such as those that are disturbance prone (whether natural, such as fire, or anthropogenic, such as logging), or that are undergoing restoration or succession. Yet, these are the landscapes for which mapping applications may have the most value from a conservation perspective.

An Alternative Model

Mapping techniques based on dissimilarity to an *optimum* configuration may not be ideal for predicting animal use areas in a changing environment because we can never be certain of defining a biological optimum.

We propose, instead, that identification of a *minimum* set of basic habitat requirements of a species is more appropriate for predicting potential animal use in changing environments. We also propose that such requirements are easier to identify correctly from a sample than is an optimal vector.

We assume that there is "habitat selection." For an animal to occur at a point, we assume that the values of some of the variables at the point, either singly or in combination, satisfy some basic requirement of the species. Although not all points that contain appropriate values of the variables are necessarily occupied, we do assume that points that are occupied represent at least some minimally suitable configuration of habitat. Furthermore, some habitat variables, although measured, are irrelevant. Whether correlated with "important" variables or not, we want to ensure that they have only a minimal effect in obscuring the detection of the relevant patterns.

We show below that $D^2(y)$ can be partitioned into p separate components, each representing the independent squared, standardized distance between y and a "plane of closest fit" derived from H (Pearson 1901; Anderson 1958; Collins 1983). We suggest that the plane of closest fit constitutes a first-order, linear approximation to the most basic combination of habitat variables required by a target species.

Partitioning D^2

It is relatively easy to show that D^2 can be partitioned into p separate components.

Eqn. 22.2a: The spectral decomposition of Σ (e.g., Seber 1984) is given by

$$\Sigma = \sum_{j=1}^{p} \lambda_i \alpha_j \alpha_j'$$

Eqn. 22.2b: from which it follows that the spectral decomposition of Σ^{-1} is

$$\Sigma^{-1} = \sum_{j=1}^{p} \lambda_j^{-1} \alpha_j \alpha_j'$$

where $\lambda_1 \leq \ldots \leq \lambda_p$ are the eigenvalues of Σ (for convenience in *reverse order* from the usual principal components analysis presentation), and $\alpha_1 \ldots \alpha_p$ are their associated eigenvectors ($p \times 1$), normalized to length one.

Eqn. 22.2c: Substituting Eqn. 22.2b into Eqn. 22.1 yields

$$D^2(y) = \sum_{j=1}^{p} (y - m)' \boldsymbol{\alpha}_j \boldsymbol{\alpha}_j' (y - m) / \lambda_j$$

To see the partitioning of D^2 more clearly,
Eqn. 22.2d: Let

$$d_j = (y - \boldsymbol{\mu})' \boldsymbol{\alpha}_j$$

Note that d_j will be a scalar, the result of multiplying vectors $(1 \times p)(p \times 1)$. Also note that

$$(y - \boldsymbol{\mu})' \boldsymbol{\alpha}_j = \boldsymbol{\alpha}_j' (y - \boldsymbol{\mu})$$

Eqn. 22.2e: Substituting Eqn. 22.2d into Eqn. 22.2c,

$$D^2(y) = \sum_{j=1}^{p} d_j^2 / \lambda_j$$

And thus D^2 can be partitioned into p scalar components. The task remains to attach meaning to these components, which we do by showing their correspondence to Pearson's planes of closest fit (Pearson 1901).

Ecological Rationale for Pearson's Planes of Closest Fit

We want to identify the *constant relationship* in a species' distribution (i.e., which functions of the variables maintain a *consistent value* where the species occurs). These variables may be thought of as representing basic requisites of the species. Functions that have a relatively high variance (take on many different values) are less likely to be informative since such functions are not restrictive of a species' distribution, at least over the range of variation sampled. This concept has also been presented by Collins (1983) and Knopf et al. (1990), although in substantially different forms.

Imagine that we perform a principal components analysis of **H**. Our interpretation of the resulting components is that they represent a rigid rotation of the original variable axes to a new set of axes (the components) such that the first component accounts for the maximum amount of variation in the original dispersion of data points, the second accounts for the maximum amount of remaining variation and is orthogonal to the first, and so forth. The position of these new axes with respect to the original ones is given by their eigenvectors, and the variances of data projected on these axes are given by their eigenvalues. Conventionally, we focus on the first few components, those with large eigenvalues (i.e., those with large variances). However, given that the interpretation in the preceding paragraph suggests we should be looking for low variances, those components are exactly the wrong ones. The dimensions with low variances are the components at the other end, the ones with the smallest eigenvalues. The idea is that we can partition variation in **H** into components that represent basic habitat requisites versus those that do not. We can then discard from further consideration those components that are not restrictive to species utilization.

Pearson (1901) defined the concept of a plane (or line) of closest fit to a system of points: that plane for which the sums of squares of the perpendiculars from the system of points to the plane is a minimum. Obviously, then, the variance of these projections of points on a vector normal to such a plane will be a minimum as well. This is one of the principal attributes we are seeking; if we can identify such a plane based on measurement of habitat variables, then the distance from any point to that plane carries information about its relative suitability as habitat as related to the basic requirements of a species. Necessarily, each successive plane (1) passes through the centroid of the system, and (2) represents one additional restriction among the variables that characterize the habitat requirements of a species.

Relation between the Partitions of D^2 and Planes of Closest Fit

Eqn. 22.3a: From standard principal components analysis,

$$x_j = y' \boldsymbol{\alpha}_j$$

where x_j is the length of the projection of **y** on an axis defined by $\boldsymbol{\alpha}_j$.

Eqn. 22.3b: Also from principal components analysis,

$$\text{Var}[x_j] = \lambda_j.$$

Eqn. 22.3c: From Pearson (1901), $\boldsymbol{\alpha}_j$ is normal to a $p - 1$ dimensional hyperplane defined by

$$(\mathbf{y} - \boldsymbol{\mu})' \, \boldsymbol{\alpha}_j = 0$$

Therefore, the perpendicular deviation of $(\mathbf{y} - \boldsymbol{\mu})$ from this hyperplane is identical to its projection on the axis defined by $\boldsymbol{\alpha}_j$, so that the variance of these distances is also λ_j.

Thus, the plane of closest fit is $(\mathbf{y} - \boldsymbol{\mu})' \, \boldsymbol{\alpha}_1 = 0$, since the deviation $d_1 = (\mathbf{y} - \boldsymbol{\mu})' \, \boldsymbol{\alpha}_1$ has the smallest variance, namely λ_1. Therefore $d_1^2/\lambda_1 = (d_1/\sqrt{\lambda_1})^2$ represents the square of this deviation in standard measure.

The second-best $p - 1$ dimensional hyperplane, which satisfies $\text{corr}[d_1,d_2] = 0$, is defined by $(\mathbf{y} - \boldsymbol{\mu})' \, \boldsymbol{\alpha}_2 = 0$, with deviation $d_2 = (\mathbf{y} - \boldsymbol{\mu})' \, \boldsymbol{\alpha}_2$ and variance λ_1, and a squared standardized distance of d_2^2/λ_2, and so forth.

Thus $D^2(\mathbf{y})$ represents a sum of squared deviations, in standardized measure, of a particular point with coordinates given by $\mathbf{y}$ from each of p, $p - 1$ dimensional hyperplanes, all of which pass through the centroid of the original p-dimensional sample ($\mathbf{H}$):

Eqn. 22.4a:

$$D^2(\mathbf{y}) = d_1^2/\lambda_1 + \ldots + d_k^2/\lambda_k + \ldots + d_p^2/\lambda_p$$

Our major premise is that not all of the p components of $D^2(\mathbf{y})$ as partitioned above define limiting combinations of habitat variables. Some $p - k$ of these do not define habitat suitability but rather are included in $D^2(\mathbf{y})$ simply because the investigator decided a priori to measure p habitat variables. Certainly the hyperplane corresponding to the first principal component cannot be considered a limitation since the variance of deviations from this hyperplane is λ_p, the maximum possible. Despite the fact that $\boldsymbol{\alpha}_p$ defines the axis of maximum possible variation measured from its normal hyperplane, it continues to contribute to $D^2(\mathbf{y})$. Thus, we propose that habitat suitability for a p-dimensional $\mathbf{y}$ be measured by

Eqn. 22.4b:

$$D^2(\mathbf{y};k) = d_1^2/\lambda_1 + \ldots + d_k^2/\lambda_k$$

for some $1 \le k < p$, where the eigenvalues of $\boldsymbol{\Sigma}$ are ordered $\lambda_1 \le \ldots \le \lambda_p$. Thus, suitability of a particular habitat location $\mathbf{y}$ for a species would be measured in terms of deviations from k basic requirements for that species, to the extent that we are able to know k.

Methods

We compared predictive models for $D^2(\mathbf{y})$ and $D^2(\mathbf{y};k)$ using presence/absence data on sage sparrows (*Amphispiza belli*), a shrubland obligate species breeding in southwestern Idaho (Knick and Rotenberry 1995, 1998; Rotenberry 1998). Much of this region, originally dominated by shrubland communities characterized by sagebrush (*Artemisia tridentata*), winterfat (*Krascheninnikovia lanata*), or shad-scale (*Atriplex confertifolia*), is currently in transition from a shrubland to a grassland-dominated state because exotic annual grasses have changed the size and frequency of wildfires (USDI 1996; Knick and Rotenberry 1997; Rotenberry 1998). Therefore, predicting animal distributions after landscape changes caused either by fires or by restoration efforts is an important management concern.

Our sampling design has been described in detail elsewhere (Knick and Rotenberry 1995, 1999). Briefly, we determined the presence of each species from surveys at 121 sites scattered throughout the Snake River Birds of Prey National Conservation Area (NCA). An additional thirty-nine points located at randomly determined coordinates were surveyed for model verification. We developed $\mathbf{H}$ (and the sample-based equivalents of $\boldsymbol{\mu}$ and $\boldsymbol{\Sigma}$) from the characteristics of each point where the species was detected in at least three of five years. We placed each survey point into our GIS (USACOE 1993) and calculated the total number of shrubland cells and the mean patch size of shrublands for areas within 0.5- and 2-kilometer radii around the point (Baker and Cai 1992). The map depicting the current landscape (see Fig. 22.1 in color section, top) had an 80 percent accuracy in separating shrub- and grass-dominated cells (Knick et al. 1997). We resampled the map to 150-meter cells from an original resolution of 30 meters.

We used a computer simulation to project landscapes in our study area that might result from continuing current trends of extensive fires and subsequent loss of shrublands (Fig. 22.1, middle), or, alternatively, from management for active fire suppression and shrub restoration (Fig. 22.1, bottom) (Knick et al. 1996). Under the first scenario, we expect a decrease in the amount of habitat suitable for sage sparrows, whereas under the second we expect increasing regional suitability based on what we currently know about habitat associations of sage sparrows (e.g., Rotenberry and Wiens 1980; Wiens and Rotenberry 1981a,b; Knick and Rotenberry 1995, 1999; Martin and Carlson 1998; Rotenberry and Knick 1999). We note that we could have specified any arbitrary landscape to examine how its suitability for the target species changed; we simply used the simulation model to generate landscapes that have a good probability of developing from the current configuration.

We measured habitat suitability for the four-dimensional $\mathbf{y}$ represented by any point on the map by $D^2(\mathbf{y};k) = d_j^2/\lambda_j$ (Eqn. 22.4b) for $1 \leq k \leq p$, where the eigenvalues of $\mathbf{\Sigma}$ (or its sample analog) were ordered $\lambda_1 \leq \ldots \leq \lambda_p$. For each landscape (current plus simulated scenarios for burned and restored landscapes), we calculated $D^2(\mathbf{y})$ and $D^2(\mathbf{y};k)$ for each cell.313

Results

The two simulated landscape development scenarios produced fundamentally different landscape patterns (Fig. 22.1). Under continued trends of extensive and repeated fires, the modeled region became dominated by grasslands (Fig. 22.1, middle), whereas fire suppression and restoration activities produced a landscape with extensive coverage of native shrublands

TABLE 22.1.

Mean habitat vector ($\mathbf{\mu}$) for sage sparrows (*Amphispiza belli*).

Scale	Variable	Mean
0.5-km radius	% shrubland cells	73.1
	shrubland patch size	25.1
2-km radius	% shrubland cells	59.2
	shrubland patch size	153.7

Note: 1. $n = 36$ occupied points. 2. Habitat characteristics were determined in a GIS for a landscape classified into shrubland or grassland categories. Percent shrubland cells represents the percentage of 150-meter cells within the sampling radius that were classified into the shrub category. Shrubland patch size represents the mean number of cells for all shrubland patches within the sampling radius.

(Fig. 22.1, bottom), even more so than in the current, fire-dominated landscape (compare to Fig. 22.1, top).

Using the current landscape habitat composition, we developed the sample-based equivalents of $\mathbf{\mu}$ (Table 22.1) and $\mathbf{\Sigma}$ (Table 22.2), then used principal components analysis to generate the associated eigenvalues, λ_j, and eigenvectors, $\mathbf{\alpha}_j$, (Table 22.3) from $\mathbf{H}$, which was defined by $n = 36$ points where we observed sage sparrows in at least three of five years, and $p = 4$ variables.

The map of predicted use areas for the $D^2(\mathbf{y})$ model using four variables for the current landscape (see Fig. 22.2 in color section, top) reflected the distribution of shrublands. As expected, the $D^2(\mathbf{y})$ model predicted use by sage sparrows only in those regions where shrublands remained in the burned landscape (Fig. 22.2, middle). However, the $D^2(\mathbf{y})$ model did not predict use by sage sparrows in the larger shrublands in the restored landscape (Fig. 22.2, bottom). Instead, the $D^2(\mathbf{y})$ model predicted use only on the edges of

TABLE 22.2.

Variance-covariance matrix (**S**) for sage sparrows (*Amphispiza belli*).

		0.5-km radius		2-km radius	
		No. shrubland cells	Shrubland patch size	No. shrubland cells	Shrubland patch size
0.5-km radius	% shrubland cells	0.0585	0.0500	2.4999	26.2929
	Shrubland patch size	0.0500	0.0615	2.2324	39.6685
2-km radius	% shrubland cells	2.4999	2.2324	122.8272	1214.2648
	Shrubland patch size	26.2929	39.6685	1214.2648	38923.342

Note: $n = 36$ occupied points.

TABLE 22.3.

Eigenvectors and eigenvalues of **S**, based on **H** of sage sparrows (*Amphispiza belli*).

	k			
	1	**2**	**3**	**4**
Eigen-value	0.0047789	0.0120878	84.908618	38961.364
Eigen-vector	0.7923471	0.6097501	0.0197589	0.0006765
	–0.610011	0.7923062	0.0116852	0.0010194
	–0.008523	–0.02134	0.9992474	0.0312492
	0.0003523	–0.000554	–0.031266	0.9995109

Note: *n* = 36 occupied points.

large shrubland patches, which were areas that most closely resembled the number of shrub cells and mean patch sizes of the originally sampled landscape.

Eigenvalues associated with each of the four planes represented successively increased relaxation of restrictions on the habitat matrix of sage sparrows (Table 22.3). The frequency distributions of values of d_j^2/λ_j for the combined verification and sample sites where we observed sage sparrows also reflected the decreased deviation from zero (Fig. 22.3). In particular, sage sparrows were much more likely to be observed at points lying close to the planes defined by *k* = 2 and, especially, *k* = 1. There was no discernable pattern of sage sparrow presence with respect to the plane *k* = 3, and a strong negative relationship with plane *k* = 4 (Fig. 22.3).

The pattern of predicted use by sage sparrows in the current and burned landscapes produced by the $D^2(\mathbf{y}; k = 1)$ model resembled that modeled by $D^2(\mathbf{y})$ (compare Fig. 22.4 in color section, top and middle, to Fig. 22.2, top and middle). However, use areas predicted by $D^2(\mathbf{y}; k = 1)$ more closely tracked expected changes in distribution of sage sparrows in the restored landscape than the $D^2(\mathbf{y})$ model. Unlike $D^2(\mathbf{y})$, the $D^2(\mathbf{y}; k = 1)$ model predicted use within the large shrubland patches (Fig. 22.4 bottom).

Discussion

We present an alternative model, $D^2(\mathbf{y}; k)$, that predicts animal use based on a minimum combination of a

species requirements rather than on a habitat's relative similarity to a mean or optimum set of conditions as modeled by $D^2(\mathbf{y})$. The maps of predicted use by sage sparrows, based on a minimum set of habitat conditions and modeled by the value of $D^2(\mathbf{y}; k = 1)$ relative to the first plane, the plane of closest fit, tracked our expected response to changes in landscape configuration. In contrast, the $D^2(\mathbf{y})$ model, although accurate for the landscape configuration in which **H** was obtained (see Knick and Rotenberry 1999 for details), was unable to track changes outside of the original habitat. Our study emphasized that $D^2(\mathbf{y})$ represents the mean habitat vector of species presence only for a specific sampled landscape and does not necessarily represent the mean or optimum set of conditions for the species.

Use of $D^2(\mathbf{y}; k)$ has the potential for coping with an evolving environment in a way that $D^2(\mathbf{y}) = D^2(\mathbf{y}; p)$ does not. For $D^2(\mathbf{y}) = D^2(\mathbf{y}; p)$, habitat suitability is limited in each of the *p* possible dimensions. For $D^2(\mathbf{y}) = D^2(\mathbf{y}; k)$, habitat suitability is limited in only *k* dimensions; unlimited habitat variation may occur in any of the *p* – *k* remaining dimensions without

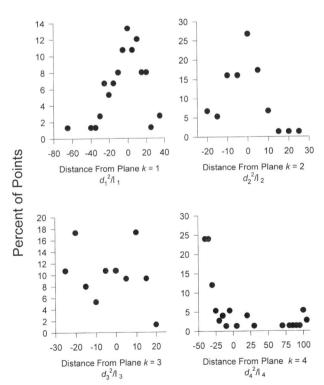

Figure 22.3. Frequency (percent) of points occupied by Sage Sparrows (*Amphispiza belli*) as a function of standardized distance ($\underline{d}_k^2/\lambda_k$) from each of *k* = 4 planes.

affecting the value of $D^2(\mathbf{y};k)$. The only requirement is that habitat variation not proceed in any of the directions parallel to mutually orthogonal axes defined by $\boldsymbol{\alpha}_1, \ldots, \boldsymbol{\alpha}_k$. This is to say that habitat change may occur parallel to any of the axes defined by $\boldsymbol{\alpha}_{k+1}, \ldots, \boldsymbol{\alpha}_p$ without affecting $D^2(\mathbf{y};k)$.

The choice of k is likely to be somewhat qualitative, depending on the magnitudes and relative spacing among the eigenvalues of $\boldsymbol{\Sigma}$, or its sample analog, and the credibility of the GIS results that occur by the use of particular choices of k. Some guidance might be provided by considering a sequence of tests of the sphericity hypothesis based on eigenvalues of the correlation matrix associated with $\boldsymbol{\Sigma}$ (e.g., Morrison 1990).

The question of what constitutes the "best" original variables to measure remains at the discretion of the investigator. In our approach, a good variable is one that takes on certain values that are closely associated with the occurrence of a species. We assume that an investigator has some criteria (e.g., biological intuition, personal experience, previous studies) on which to base initial variable selection. Theoretically, our approach should accommodate the inclusion of less-than-useful variables by shunting them to the less-than-useful components. Whether this extends to variables to which a species responds to in a nonlinear or threshold manner remains to be investigated.

Each of the four planes describing sage sparrow presence represented increasing variance in the combination of variables that described the set of used habitats. In other words, each additional plane represented a less-consistent set of habitat conditions describing the sites used by sage sparrows. To this point, we can offer no more than a geometric interpretation of the plane of closest fit; identification of "important variables" in the original measurement space remains problematic. Although Collins (1983) briefly mentions planes of closest fit in an analysis of geographic variation in avian habitat selection, we do not believe that his interpretation of the technique as involving the variables with the largest correlations (i.e., factor loadings) on the principal components with the lowest eigenvalues is necessarily correct. Correlation assumes a comparison of projections of points on two axes, whereas our problem in-

volves characterizing the association between each measurement axis and a plane.

Future Research

The approach outlined above is clearly only a start, and much remains to be learned of its limitations before it we can confidently substitute it for most previous approaches. For example, our current "verification" of the model consists only of the perception that it produces a distribution of likely sage sparrow habitat that much more closely mirrors our biologically based intuitive expectation than does a competing model. Assessments of accuracy explored in this volume should be applied in a more extensive test of the model. Likewise, our analysis is based on a relatively small number of observations and only four, somewhat artificial, variables. This is not an altogether trivial data set, however. Although the number of observations is low, it is indicative of the likelihood of detecting this species throughout the immediate region. Likewise, these variables were selected because (1) we expected them to differ among our simulated landscapes, and (2) we expected that sage sparrow distribution would be related to these measurements. We are currently working on a larger data set with more species, more species observations (including an independent verification set), and a wider array of variables to probe the model more deeply. Finally, our inclusion of a point within the data set was based only on species presence. If other measures of a species' performance at a point are available (e.g., breeding success), then we may be able to interpret planes of closest fit in a context more directly related to a species' evolutionary fitness.

Three statistical issues remain to be elucidated with further work as well. First, we need to explore differences arising from the use of unstandardized versus standardized data. We employed the former, but it is reasonable to expect that use of the latter may yield different patterns, since eigenvalues and eigenvectors can change with changes in measurement scales. Second, we need to develop a method that permits more than simply a geometric interpretation of a plane of closest fit. Interpreting planes in the context of the habitat variables that were originally measured will assist a manager in developing any habitat management scheme, as well as provide information to a biologist seeking to understand the mechanisms underlying a habitat association. Finally, we need to devise more quantitative methods for selecting k. This is likely to prove analogous to attempts to determine the number of "significant" principal components, a search that is still underway (Jackson 1993).

In conclusion, we believe that the statistical approach we have outlined above contains biologically realistic assumptions, assumptions that are more likely to be met under current sampling regimes than alternative analyses require. Moreover, this technique is more able to cope with evolving environments and is more easily extensible to previously unsampled environments. As such, we believe it will be most useful in identification of areas of conservation interest and their applications to land-use planning.

Acknowledgments

Our attendance at the symposium Predicting Species Occurrences: Issues of Accuracy and Scale was underwritten by the U.S. Geological Survey and the University of Arkansas.

Geospatial Data in Time: Limits and Prospects for Predicting Species Occurrences

Geoffrey M. Henebry and James W. Merchant

Robust integration of time into geospatial (or geographic) information systems (GIS) is a grand challenge but one that must be addressed if we are to monitor, model, and manage biodiversity. In this chapter, we examine some of the deeper conceptual issues that must be confronted in order to predict species distributions in space and time. Conceptual challenges are often resolved through development of particular algorithms and specific technological implementations. It can be difficult for the practitioner or the manager to discern, amidst a choice of GIS tools, the underlying worldview that informs, defines, and limits these tools. We shall touch upon several key challenges to integrating geospatial data effectively into predictive models of habitat. These issues lie deeper than the choice of minimum mapping unit or image resolution, and they highlight current limitations in theory and point toward future directions for research and development. Some of the visceral unease of GIS expressed by Van Horne (Chapter 4) is justified: the tools may appear more robust and more objective than is the case. Yet, as we practice with tools, we learn to use them better, as Stauffer (Chapter 3) relates in his historical review of animal habitat modeling.

There are currently two fundamental approaches to estimating spatial occurrence of species: (1) range maps interpolated and/or extrapolated from sparsely sampled field observations (e.g., museum voucher specimens, county dot maps), and (2) potential range maps developed from modeled habitat availability. In both cases, range maps are typically portrayed as static. We shall focus on issues in modeling habitat within the context of GIS and will highlight some of the current conceptual and technical difficulties encountered when introducing temporal structures into geospatial data. For illustration, we examine the challenges in predicting the spatial distribution of habitat for an anuran species, the Great Plains toad, *Bufo cognatus*. We conclude with a look to the prospects for improving prediction of species occurrences using new types of geospatial data.

The Challenge of Geospatial Data in Time

Species occur in space through time. Each species has a range of tempos and spatial extents and resolutions associated with the resources it requires for persistence. These resources can be available recurrently or episodically in habitat that is distributed across the landscape. Ecological patterns and processes weave their mutual causality through space and time (Turner 1989; Allen and Hoekstra 1992; Brown 1995; Maurer 1999); thus, if species are to be viewed as spatiotemporal entities, then habitat characterizations must also include dynamical portraits of (at least) the limiting resources.

Tracking resource location in space and time

requires time series of geospatial data. Data are geospatial when they are organized with coordinates that relate sampling points to locations on the planet using some two-dimensional or three-dimensional positional reference system. Not all spatial data are geospatial because georeferencing is not always feasible, especially for legacy datasets. Today, GISs are commonly used to facilitate the organization, manipulation, visualization, and analysis of geospatial data (Hunsaker et al. 1993; McCloy 1995). Habitat modeling with geospatial data is, however, fraught with significant methodological challenges (aside from the familiar and chronic problem of data availability). Major issues that need to be addressed include autocorrelation, issues of rescaling, model validation, and data representation, including the representation of time in GIS.

Autocorrelation

The issue of autocorrelation faces any treatment of spatial data. Autocorrelation is a characteristic of data derived from a process that is articulated in one or more dimensions and it describes the error structure of the data (Cliff and Ord 1981; Henebry 1995). Geospatial data frequently exhibit positive spatial autocorrelation, that is, the residual deviations are less variable than random expectation because proximate neighbors are more similar than distant neighbors. This situation complicates model parameter estimation, model error analysis, and identification of significant statistical relationships between variables because it enhances the risk of making a Type I error, or in other words, rejecting the null hypothesis when it should, in fact, be accepted. The naïve regression of one mapped variable against another will fall into this trap of inflated and erroneous significance levels. Spatiotemporal autocorrelation arises naturally from dynamical portrayals of spatially explicit ecological processes (Henebry 1993). Density-dependent processes can readily produce negative spatiotemporal autocorrelation (Henebry 1995). The hazard in this situation arises from residuals that are more variable than random expectation increasing the risk of a Type II error, or in other words, accepting the null hypothesis when it should be rejected. Autocorrelation among

mapped attributes has also been found to be a principal source of error propagation in GIS models based on simple map analyses, such as basic arithmetic operations and superpositioning (Arbia et al. 1998; Griffith et al. 1999).

Autocorrelation can be harnessed, however, as a robust indicator of spatial structure and its temporal development (Henebry 1993; Henebry and Su 1993). Information on the arrangement of habitat involves spatial autocorrelation. Hiebeler (2000) demonstrates how the intensity of positive spatial autocorrelation—via habitat clustering—affects population processes. He argues that much of the ecological research developed on the concept of patch distributions could be more robustly articulated in terms of clustering intensities. Positive spatial autocorrelation is expected to be higher within habitat patches than at the margins (Brown et al. 1995; Maurer, Chapter 9). It is just this positive spatial autocorrelation within habitat patches that may be responsible for the apparent success of Pearson's planes of closest fit (Rotenberry et al., Chapter 22): high positive spatial autocorrelation of a habitat feature minimizes its residual variance and thus the feature appears as a habitat constant.

What are some options for dealing with spatial autocorrelation? There are two general strategies for coping with the negative consequences of spatial autocorrelation on statistical inference: explicit incorporation of local spatial interactions and distribution-free methods. Researchers in the field of spatial econometrics have pioneered the former approach (Anselin 1988; Pace and Barry 1997; Bivand 1998). Autologistic regression is one form of this approach: Klute et al. (Chapter 27) demonstrate how the autologistic model incorporates spatial autocorrelation explicitly and thereby improves predictive performance over standard logistic regression. Distribution-free methods rely on randomization (permutational and combinatorial techniques), resampling (bootstrap and jackknife), and Monte Carlo sampling to assess the spatial, temporal, or spatiotemporal configurations of samples against empirical distributions synthesized using computer-intensive methods (Manly 1997).

Fielding and Bell (1997) in a review of methods for error assessment of presence/absence models touched on how spatial autocorrelation and spatial context affect prediction errors. They recommended some useful strategies for dealing with the spatial dimension, but as presence/absence models typically lack dynamical structures, they did not broach the issue of spatiotemporal autocorrelation and its effects on model predictions and patterns of errors. Fielding (Chapter 21) mentions how differential spatial weighting of prediction errors enables observational context to be introduced explicitly into model assessment.

Rescaling Relationships

Significant differences often exist between the spatial and temporal scales at which data are collected—via remote sensing, GIS, and field methods—and the scales at which ecological and environmental models operate (Goodchild 1997). Moreover, as some processes require analyses at multiple scales, nested and coupled modeling strategies are being used with increasing frequency (Bian 1997). Mismatches between scales of activity, observation, and modeling can lead to inappropriate rescaling of rates and relationships. The general problem of scale dependence has been investigated for spatial data in geography as the Modifiable Areal Unit Problem (MAUP) (Openshaw 1984; Arbia 1989; Jelinski and Wu 1996) and for temporal data in ecology as transmutation (O'Neill 1979; King et al. 1991) and aggregation error (Cale and Odell 1980; Gardner et al. 1982; Rastetter et al. 1992).

The MAUP stems from twin errors. First, there is the problem of assuming implicitly or explicitly that patch-specific data applies to all individuals dwelling in the patch. Geographers have dubbed this imprudent inferential practice the "ecological fallacy." Second, there is the problem of treating spatially aggregated data as individual observations in analysis. There are no natural a priori spatial units; they are all imposed by our observational processes (Allen and Starr 1982; Allen and Hoekstra 1992). Thus, delineations between patches are arbitrary and may be imprecise in location, transitory in duration, and irrelevant to underlying population structure.

The principal undesirable consequence of the MAUP is equivocal statistical analysis: by simply varying either data resolution through aggregation or data allocation through alternative zonations, the entire spectrum of correlations may be extracted from the same dataset (Openshaw 1984; Arbia 1989). By enabling the user to define and redefine areal units, GIS can actually exacerbate the MAUP and promote discovery of spurious correlative relationships (Openshaw and Alvanides 1999).

There are no objective solutions to the MAUP. Rather, its effects can be attenuated by knowing well the system under study in order to be able to specify ecologically meaningful ways to (dis)aggregate data. Huston (1999) argued for the need to distinguish carefully between processes occurring at biogeographic and local scales to predict variation in species assemblages. Trani (Chapter 11) indirectly approached the MAUP through a comparative analysis of the scaling behavior of spatial pattern metrics. The MAUP lurks also in the amphibian habitat modeling effort described by Johnson et al. (Chapter 12). Working with a wealth of more than two hundred explanatory variables and a rich and recent occurrence record, data-reduction procedures were undertaken to attenuate multicollinearity to yield a set of explanatory variables at three spatial extents: site-specific (from field data), local landscape (1–2-kilometer buffer around sites), and broader landscape (10-kilometer buffer). Although they found the variation encountered in species response to landscape habitat metrics unexpected, it is not clear how dependent the results are on the assignment of buffer extents (Johnson et al., Chapter 12).

Transmutation error arises when rates are naïvely extrapolated. For example, rescaling foraging rates observed only during the growing season to an annual basis obscures the temporal clustering of resource capture. Aggregation error arises when multiple state variables with different characteristic rates are lumped into a single rate-limited state variable. For example, rescaling life-stage dependent mortality rates to a single population mortality rate by taking the simple average obscures important demographic dynamics.

The key to handling these kinds of modeling errors is to be aware that they exist and have potential to propagate uncertainty and degrade prediction accuracy. Helpful tutorials illustrating the problems and a range of solutions are available (King 1991; King et al.

1991; Rastetter et al. 1992). In addition, Schneider (1994) provides a useful introduction to dimensional analysis that includes examples of analytical pitfalls commonly encountered when reconciling the scales of observation and prediction.

The development of hierarchy theory in ecology has provided some useful conceptual tools to handle issues of scale deftly (Allen and Starr 1982; O'Neill et al. 1986; Allen and Hoekstra 1992; Johnson 1996). Fractal neutral models (Milne 1991) and allometric power laws (West et al. 1997) offer robust means of reconciling disparate scales (called renormalization by the physicists) as well as identifying changes in scaling behavior. Milne et al. (1992) simulated herbivory across fractally structured landscapes to demonstrate how allometric scaling rules can reconcile the apparent behavioral differences in hypothetical species. Along similar lines, Theobald and Hobbs (Chapter 59) illustrated how the process of habitat fragmentation can be better described in terms of the functional rescaling of species body size and physiology. The common solution to rescaling problems is the judicious application of prior knowledge of how the system works to the modeling process. Modeling should not be conducted in a knowledge vacuum. Geospatial data complement, not substitute for, natural history and field data.

Model Validation

Model validation is never easy. The very spatial extensiveness of geospatial data magnifies the validation problem. For example, methods to determine the accuracy of maps portraying land-cover characteristics derived from coarse-resolution space-borne sensors and mapped over vast areas are simply not well established. Traditional site-specific field-based verification methods, such as advocated by Congalton and Green (1999) and others, would require thousands of samples collected within a narrow time window coincident with sensor overpasses and thereby be prohibitively expensive, if even logistically feasible. Merchant et al. (1994) proposed that, at present, validation of large-area land-cover data sets will need to be a cumulative, multitiered process involving consideration of a number of different types of evidence that, collectively,

will serve to support or refute the accuracy of a given product. Such multiple lines of evidence could include (1) selective site-specific field observations, (2) qualitative assessments of derived products by experts in land-cover characterization, (3) examination of the internal logical consistency among variables within a multidimensional GIS in which land cover is one component layer, and (4) end-users' assessments of the performance of derived habitat data in modeling species occurrences.

The accuracy of maps predicting vertebrate range distributions are even more difficult to assess than maps portraying land cover or land use. Seasonality of habitat use and availability, nocturnal and fossorial habits, and low population densities can lead to low encounter rates. Regional to subcontinental distribution of species, patchily distributed habitat, and interannual climatic variability can further complicate validation efforts.

Design-based inference offers an alternative evaluation framework to assess mapped habitat accuracy in a manner that is not sensitive to spatial autocorrelation. Stehman (2000) offers an accessible entry to this literature. A principal difference between design-based inference and the commonly employed model-based inference is the population targeted by the inference. Model-based inference seeks to discern the process or model that generates the observed sample: from patterns observed in the sample, an underlying process is inferred. Design-based inference takes as the object of interest an actual, well-defined, finite population of observations (such as a map) in order to describe characteristics of the population with some degree of tolerance for uncertainty (Stehman 2000). In sharp contrast to standard assumptions of model-based inference, design-based inference places few constraints on the nature of the observations in order to conduct estimation of population characteristics. Observations are not considered as random variables plucked from a specific probability distribution; rather they are considered fixed values with variability attributed to sampling design. The most important consequence of this perspective is that the confounding effects of spatial autocorrelation on estimation are effectively neutralized, although the precision of estimates is still affected adversely (Stehman 2000). There is, of course, a cost associated with the shift in inference framework and relaxation of assumptions: there is no way to predict the accuracy of unobserved data.

In a simulation study, Karl et al. (Chapter 51) explored how "rarity" affected model accuracy. Estimates of commission error conflate actual failures to observe a species in a predicted area with apparent failures due to low encounter rates (Boone and Krohn 1999). Actual failures point to model lack of fit or, more seriously, model misspecification. Apparent failures, on the other hand, simply indicate inadequate sampling. Karl et al. (Chapter 51) concluded that (1) if species respond to readily observable habitat features, and (2) if knowledge of habitat associations is more accurate for rare species, then habitat-relationship models developed for those rare species ought to perform as accurately as predictive models for common species, despite the high error rates caused by small sample sizes. This provocative conclusion hinges critically on untested assumptions.

Model validation goes beyond assessing a model's predictive performance based on data that are either similar to those used to develop the model (the case of cross validation) or distinctly different (model verification) (Cale et al. 1983; Warwick and Cale 1988). It extends to evaluating model performance under adverse conditions, such as uncertainty in input data and model parameters (Warwick and Cale 1988; Henebry 1995). Computer-intensive model-error analyses using Monte Carlo methods (Kalos and Whitlock 1986; Mowrer 1999) enable wide-ranging explorations of a model's parameter space to estimate model reliability, that is the probability that a calibrated model will correctly predict to within a predetermined level of accuracy (Warwick and Cale 1987).

The advantage of the model reliability approach is that it imposes a decision-making framework on assessment of model performance. The reliability of a model is estimated by the frequency with which Monte Carlo predictions fall within a user-designated accuracy interval at some specified time and/or location. Operations on the empirical distribution of model outputs form the basis for decision statistics. For example, a model may correctly predict species occurrences within a 1-kilometer radius 60 percent of the time, given input data with 15 percent uncertainty,

but that reliability might increase to 75 percent with a reduction of the input uncertainty to 10 percent. For models with multiple variables, the joint reliability is usually not simply the product of the individual reliabilities because variables are typically not independent. For the decision maker, the utility of a model comes from its ease of interpretation and the degree of confidence that can be placed in its predictions. Reliability, therefore, is an attractive decision statistic in that it is easier to grasp than error measures based on sum of squares. Mapping model performance onto a binomial variable leaves no gray areas: either the model performed up to the decision maker's standards or it did not.

For any given model, reliability is not a uniquely determined characteristic. Diverse combinations of accuracy and uncertainty can yield the same reliability. Thus, it is appropriate to construct a database of model performance to aid in model assessment and decision making. This database can then serve for "what if" scenario analysis. For example, to explore the possible consequences of data uncertainty the decision maker specifies what magnitude of deviation is acceptable (accuracy interval) and what proportion of the time the model must provide acceptable answers (reliability). Similarly, to explore the consequences of a stringent prediction requirement (accuracy interval), the decision maker can specify the reliability and the degree of data uncertainty. The decision maker can thus review model performance under a variety of constraint scenarios.

The routine use of Monte Carlo analyses in spatiotemporal modeling of natural resources sciences and management is not yet established, but the need has been recognized (Henebry 1995; Gascoigne and Wadsworth 1999; Mowrer 1999). Likewise, effective techniques for visualization of map error/uncertainty (Fotheringham et al. 1996; Beard and Buttenfield 1999) and understanding of error propagation in GIS are still rudimentary but experience is rapidly increasing (Heuvelink and Goodchild 1998; Griffith et al. 1999). However, it is significant to note that wildlife habitat relationship models constructed primarily through Boolean overlay operators on GIS data (e.g., Dettmers and Bart 1999) are particularly susceptible to complex propagation of errors in attributes and locations. Spatial autocorrelation among both attributes and errors only exacerbates the problem (Griffith et al. 1999). Clearly, there is need to develop new approaches to spatiotemporal model validation. Part of this effort will require educating both producer and user communities about how to grapple intelligently with data and model uncertainty (Mowrer 1999).

Data Representation

The tools and techniques commonly used to represent and manipulate geospatial data carry strong assumptions as to what constitutes the units of analysis. The phrase "units of analysis" refers not to units of specific measurement systems but rather to the conceptual entities that are subject to measurement and analysis. Geospatial entities include axiomatic geometric objects (e.g., points, lines, polygons, polyhedra) that are located within a spatial reference system as well as synthetic geometric objects derived from sensor systems, such as the spatiotemporal trajectory representing telemetered animal movements or an array of pixels portraying hyperspectral radiance upwelling from a landscape. Fisher (1997) urged that the pixel is "a snare and a delusion" because it may represent an unobservable admixture of geospatial entities. Thus, the pixel does not constitute a "proper" geographic object and its ill-defined status often hinders analysis. This warning applies both to imagery per se and to its representation and manipulation within GIS (Cracknell 1998). An issue related to the units problem is the now-tired debate in GIS circles over the relative merits of vector versus raster representations and discrete objects versus continuous fields. Couclelis (1992) offered a clever conceit in her title "People manipulate objects (but cultivate fields)" and argues that human cognition relies on both modes of spatial representation. It is prudent, therefore, to embrace the richer realm of multiple representational modes. This multimodality does then pose a challenge to the data models actually implemented in GIS.

Burrough and Frank (1995), inquiring about the generality of GIS implementations, observed an unresolved and possibly irresolvable tension between the universal data models that computer scientists seek and the ad hoc data models that GIS practitioners use

to address specific problems. They further identified three major groups of GIS users: managers of defined objects (e.g., cadastres, utilities, facilities management); planners and resource managers (e.g., multiattribute evaluation and decision making); and space-time modelers (e.g., environmental scientists broadly construed). What Burrough and Frank (1995) discovered was a profound conceptual disconnect in the GIS community between the units of analysis and the baseline models employed by different disciplinary subgroups. Current GIS implementations are *not* generic and they do not adequately support space-time modeling (Burrough and Frank 1995; Couclelis 1999).

Representing Time

Inclusion of time in GIS is not as straightforward an exercise as might be expected. A major source of difficulty stems from how the increased dimensionality of the data affects what can be assumed about the data. Consider an unordered list, the simplest database structure. It is a collection of zero-dimensional data, database records that lack spatial or temporal relationships with other records. Although this structure is easy to implement and enables efficient querying about the relationships between records, it can permit inferences about relationships that are nonsensical when viewed within the broader context of the data. Classical statistics is built around a central assumption of the zero-dimensionality of data and it is common knowledge that statistically significant but biologically irrelevant relationships can be obtained through the injudicious use of correlation analysis.

Temporal databases introduce an explicit, unidirectional, one-dimensional structure to the data. The "arrow of time" makes temporally oriented queries and logical inferences possible (Snodgrass 1992; Chomicki and Toman 1998). Time-series analysis is the statistical analogy. Spatial databases represent spatial relationships as locations (raster/fields) and/or as entities (vector/objects). Although coordinate systems supply topology, there is no a priori ordering of the directionality of causation in space as there is in time. This has the important consequence of requiring the user to inform the database about the flows of influence among spatially ordered data. The user must specify a model of spatial relationships in order to make meaningful queries. For example, many GIS have a module that introduces the influence of gravity into the database topology in order to analyze drainage patterns. Although topological relationships indicate who is the neighbor of whom, additional information is required to know who are the *effective* neighbors. Different processes can have different effective neighborhoods or corridors at different scales.

The addition of time in a spatial model further complicates the issue of influence and places more responsibility on the user to identify relevant neighborhoods and supply meaningful ordering. Neighborhoods can be discontinuous in space and time due to lagged effects such as the spatiotemporal dispersal in seedbank dynamics and wildfires.

Spatiotemporal Data in GIS

According to Peuquet (1999), there are currently three nonexclusive modes of spatiotemporal data representation of discrete events in GIS: location-based, entity-based, and time-based. Location-based representation uses sequential map layers with each map layer being a "snapshot" of the current state of spatial relationships. Aside from the potential problem of data volume, this approach has the limitation that changes are not explicitly represented but instead must be calculated from the data volume. Further, the timings of changes are not specified other than noting that they occur sometime between successive observations. This temporal imprecision creates problems for data analysis if the time between observations is significantly longer than the tempo of the activity of interest (Ratcliffe and McCullagh 1998). A partial solution is to specify change events using a variable-length list for each spatial object, but the problem of calculating durations persists (Peuquet 1999).

An entity-based representation can treat geographic objects (e.g., points, lines, polygons) as variables in time using "amended" topological vectors (Langran 1992, 1993) or using an object-oriented approach (Raper and Livingstone 1996). With the former approach, the spatiotemporal topology may quickly get unwieldy (Peuquet 1999). With the latter approach, there is the enduring problem of identifying what constitutes well-defined and well-behaved entities with all the attendant subtleties of scale-dependent behaviors

(King et al. 1991; Johnson 1996) and representation of discontinuous "fields" (Couclelis 1992).

Temporal vectors record "events" in time-based representations generating an explicit temporal topology that acts as an adjunct to location or entity representations (Peuquet and Duan 1995; Peuquet 1999). There are three kinds of temporal relationships: (1) ordinal and interval operations along a single temporal distribution of events; (2) Boolean set operations between different event distributions; and (3) temporal rescaling, including extrapolation and generalization (Peuquet 1999). A principal difficulty of this approach is an implicit assumption of "omniscience" associated with the temporal topology. Imprecision in event dating can lead to the loss of event duration as a temporal metric for the event. In this case, an ordinal model can be used, though it is a less-powerful descriptor. In some cases, even temporal ordering may be unknown or unobservable. Fuzziness, or uncertainty about event initiation, duration, and termination, generates a fuzzy temporal topology that limits inferential power. Couclelis (1999) argued for consideration of relative spatiotemporal structures and influence of culture on cognition and event representation.

Distinct from the discrete event approach, spatiotemporal data models for continuous change have been proposed (Erwig et al. 1999; Chomicki and Revesz 1999b). The model of Erwig and colleagues (1999) relies on developing novel abstract data types for spatiotemporal objects, such as moving points and moving regions. The model of Chomicki and Revesz (1999b) is based on parameterized polygons in which the vertices are defined using linear functions of time. Although this approach can model several types of continuous change, including movement, growing, and shrinking, it has other limitations and no current implementations.

Many of the difficulties of incorporating time into geospatial databases stem from the drive to forge universal data models for GIS implementations. Several definitions for spatiotemporal data types have been proposed (Worboys 1994; Peuquet and Duan 1995; Erwig et al. 1998; Chomicki and Revesz 1999a; Erwig et al. 1999). Burrough and Frank (1995) recognized this tension between theory and practice and called for a plurality of approaches rather than a hobbled but generic model. However, there is a significant lag time between academic technical innovations and their implementation in commercial GIS. Ultimately, there is a need to wed GIS with simulation models: "The essential task is to extend a world-history model, consisting of observed events, objects and locations, to a process model that also includes interpretive occurrence rules and patterns expressed as combinations of relationships" (Peuquet 1994, p. 457). GIS then becomes a mechanism to inform simulations by serving geospatial data and then to gather, organize, and visualize the results. Although this approach downplays the generic querying functionality for the resulting spatiotemporal database, it emphasizes the problem-solving and focused-decision-support aspects of modeling within a resource-management context. Progress has been made in this direction: witness the recent proceedings volumes on the topic of GIS and environmental modeling (Goodchild et al. 1993, 1996).

Modeling Geospatial Data through Time

The ecological literature is replete with approaches to predict species occurrences. We shall only touch on general points relevant to GIS. A common approach to spatiotemporal modeling is empirical and relies on opportunistic temporal sampling. Relationships are inferred from one or more slices in time, generalized, and then interpolated (Yang et al. 1998a) or extrapolated, as in Markov transition matrices (Hall et al. 1991; Pastor et al. 1993; Hobbs 1994). Opportunistic sampling can suffer from significant, but largely hidden, observational biases, as shown by research with such spatiotemporal anthropocentric processes as malaria (Schellenberg et al. 1998) and crime (Ratcliffe and McCullagh 1998).

A more robust approach is to represent explicitly significant ecological processes and their interactions: resource availability and capture; organismal reproduction, dispersal, and demise; and habitat heterogeneity and climatic variability. The modeling may emphasize the temporal dynamics of resource availability within the habitat (e.g., Weiss and Weiss 1998), characterize population-level behaviors (Conroy et al. 1995; Turner et al. 1995; Radeloff et al. 1999), or characterize populations of individual-level behaviors

(Shugart et al. 1992; Johnston et al. 1996; DeAngelis et al. 1998; Gross and DeAngelis, Chapter 40).

An example of the prospects for integrated GIS-simulation modeling is the Across Trophic Level System Simulation (ATLSS) that aims to predict responses of several higher trophic level species to various change scenarios in Everglades hydrology (DeAngelis et al. 1998; Gross and DeAngelis, Chapter 40). The modeling approach uses hydrologic models to drive habitat and resource availability models through changes in water depth. Individual-based species models populate and move through the South Florida landscape that is articulated at the spatial resolution of Landsat Thematic Mapper imagery (30 meters). This suite of models interacts hierarchically. The spatiotemporal topology of influence is informed by hydroperiod, which, in conjunction with topography, largely drives habitat suitability for various species (Gross and DeAngelis, Chapter 40).

Practical Limits in Using Current Geospatial Data to Predict *Bufo cognatus* Occurrence

Terrestrial habitats typically lack the a priori structuring of influence offered by flowing waters. Habitat modeling for anuran amphibians is a challenge because of a complex life cycle (Wilbur 1980; Alford 1999) that requires two contrasting habitats: an ephemeral aquatic setting for the larvae and a terrestrial setting for the adults. To illustrate some of the difficulties in predicting geospatial occurrences of species in this taxon, we have chosen *Bufo cognatus*, the Great Plains toad, a grasslands toad found not only across the Great Plains but also in moister areas of the arid southwestern United States and northern Mexico. Nocturnal in habit, it can be diurnally active in the breeding season and can burrow underground and survive several weeks to avoid dry weather (Bragg 1940). *B. cognatus* overwinters in shallow terrestrial burrows but is not tolerant of freezing (Storey and Storey 1986). During the brief breeding season, males congregate only around shallow ephemeral pools, such as in bison wallows and ponding areas in fields, shortly after intense rains and join chorus to attract females (Bragg 1940). The ensuing scramble competition for females is characteristic of explosive-breeding anurans (Wells 1977).

In an insightful paper, Toft (1985) reviewed the current herpetological literature to compare the resource partitioning in habitats across a wide variety of amphibians and reptiles. Her resource categories followed those of Schoener (1974): macrohabitat, microhabitat, food type, food size, diel time, and seasonal time. She discovered that larval amphibians differ significantly from other organisms in their patterns of resource partitioning. Although habitat characteristics are key to determining habitat use for most amphibians and reptiles, seasonal time was the more critical resource for larval amphibians: the duration of time available in an ephemeral pond is limited on one end by the onset of the breeding season and on the other end by increased risk of predation and the pond drying out (Toft 1985). Wilbur (1980) contends that complex life cycles are adaptations to exploit transient opportunities for growth or dispersal. The anuran solution has the advantage of "being able to exploit the rich, but highly transient, aquatic environments that are seasonally available at all latitudes" (Toft 1985, p. 11). A recent review by Alford (1999) stresses the significance of environmental uncertainty on larval growth, development, behavior, and intra- and interspecific interactions.

Anuran habitat modeling is complicated by the fact that what constitutes a resource for the adults—breeding pools—is habitat for the larvae. Given the vagaries of the continental climate of the Great Plains, several years can pass before *B. cognatus* has a robust recruitment year (Bragg 1940). Data from the Nebraska panhandle (Nebraska climate division 1) typify the high interannual variability in spring precipitation these organisms encounter (Fig. 23.1). An approach to species occurrences that relies primarily on the use of a land cover–vegetation alliance class as the primary habitat indicator (e.g., Scott et al. 1993) will tend to overpredict habitat extent for *B. cognatus* because it focuses only on adult habitat. A principal constraint in the species population dynamics is the availability of larval habitat to persist long enough to enable metamorphosis (Wilbur 1980; Toft 1985; Alford 1999). Surveying across genera, Brodman (1998) concluded that physical geography

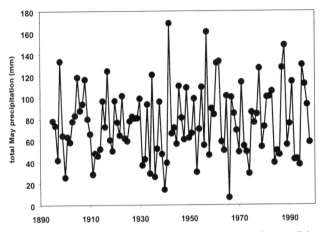

Figure 23.1. Total May precipitation in Nebraska climate division 1: 1895–1998.

was a stronger predictor of amphibian distributions in the U.S. Midwest than either precipitation or vegetation community. Corn and Peterson (1996) argued that changes in disturbance regimes across the Great Plains—in particular, decreased fire frequency and increased extent of grazing lands—may decrease habitat for many amphibians and that the presence of specific habitat is a better indicator of occurrence than vegetation class surrogates.

From a geospatial perspective, the identification of shallow, ephemeral pools is currently a challenge. Such topographic features are too fine to be detected using extant 7.5-minute digital elevation models. The small spatial extent and transitory nature of the pools make them elusive to most space-borne sensors, which have a coarse spatial resolution and long return interval relative to the phenomenon of interest. For example, Landsat 7 (http://landsat7.usgs.gov/), which was launched in April 1999, carries the ETM+ sensor that provides 15-meter (panchromatic) and 30-meter (multispectral) spatial resolution at a return interval of sixteen days (Goward and Williams 1997). Several fine spatial resolution optical sensors with short return intervals are planned for over the next few years (Alpin et al. 1997). For example, a commercial satellite launched at the end of 1999, Ikonos (http://www.spaceimaging.com/), boasts impressive spatial resolutions for both panchromatic (1-meter) and multispectral (4-meter) imagery.

Obscuring cloud cover is always a concern with optical imagery. Imaging radar may prove helpful:

microwaves can penetrate cloud cover and backscattering is sensitive to surface roughness. In the absence of significant wind-driven waves, small bodies of water scatter back few of the illuminating microwaves due to specular reflection off the relatively smooth water surface. Low backscattering is not, however, a unique signature of water: concrete pads, empty parking lots, and radar shadows can masquerade as ponds or ephemeral pools. Moreover, there is a trade-off between the spatial resolution of the image and the intensity of speckle, the high spatial frequency salt-and-pepper noise that is inherent to imaging radar. Principal component analysis of image time series can greatly improve the signal-to-noise ratio by effectively trading time for space (Henebry 1997). Nevertheless, the finest spatial resolution currently available from space-borne imaging radar is 8–9 meters from the fine beam of RADARSAT, and its return interval remains a limitation. Combining multiple sensors may prove a useful, if costly, strategy to monitor such a variable spatiotemporal resource (for adults) or habitat (for larvae).

A secondary constraint on *B. cognatus* distribution may relate to the spatial distribution of soils characteristics. The adults spend much of their life underground in burrows, very shallow during breeding season and deeper during dry spells and during the winter (Bragg 1940). Overwintering adults must burrow below the soil frostline to survive. Depth to soil frostline is variable with soil texture, moisture, thermal conductivity, and intensity and duration of subzero air temperatures. The interaction of the spatial distributions of winter temperature minima, their duration and interannual variability on the one hand, and soil textural characteristics on the other, may restrict occurrence of *B. cognatus* within the northern reaches of its broad latitudinal range, which extends from the southern ends of the Canadian prairie provinces into Mexico (Conant and Collins 1991). High water tables may exclude *B. cognatus* from some areas (Jones et al. 1981) and recently metamorphosed individuals may seek to burrow in the softer earth of cultivated fields rather than in undisturbed, harder surfaces (Smith and Bragg 1949). As the USDA Natural Resource Conservation Service completes the county-level Soil Survey Geographic database (SSURGO) over the next several years, an important gap in geospatial data will be filled and modelers will have access to detailed, field-verified soils inventories at finer spatial resolutions than are currently available from the state-level STATSGO database (State Soil Geographic). For instance, ponding duration is a significant temporal habitat variable listed as available within STATSGO but which in actuality is blank for the Nebraska dataset. (Moreover, the categorical and spatial resolution of the STATSGO data make it ill suited to this particular task.)

A third constraint on *B. cognatus* distribution lies not in specific environmental limits but rather in observational error and bias. A fossorial, nocturnal amphibian, the Great Plains toad is simply difficult to encounter. Bragg (1940) notes that the burrows are difficult to locate once the toad is below the surface: "I have found but few during several years of observation" (p. 332). In nine herpetofaunal surveys in Iowa, Kansas, and Nebraska since 1979, the success of encountering *B. cognatus* is mixed. Ballinger and coworkers (1979) did encounter *B. cognatus* during surveys from 1975 to 1978 but concluded it was the least-abundant toad in western Nebraska. Jones and coworkers (1981) failed to find *B. cognatus* but attributed this to the high water table at the Nebraska site. Lynch (1985) noted that *B. cognatus* were observed only after heavy rains during a multiple-year statewide survey of Nebraska. Christiansen (1981) and Christiansen and Mabry (1985) concluded that *B. cognatus* and other toads in Iowa had maintained their ranges observed in a 1940 statewide survey. Two more recent studies in Iowa (Lannoo et al. 1994; Hemesath 1998) suggest that *B. cognatus* may even be expanding its range eastward; however, a more parsimonious explanation is that *B. cognatus* was merely overlooked in earlier surveys. Heinrich and Kaufman (1985) failed to observe the Great Plains toad at a tallgrass prairie site in the northern Flint Hills of Kansas but surmised that it was likely to occur there. Surveying a different tallgrass prairie site in the northern Flint Hills, Busby and Parmelee (1996) did not find *B. cognatus* in 1993. They noted that it was also not found in surveys from the 1930s but had been observed in the same area prior to 1930. It is possible that the series of droughts during the 1930s caused local extinction of *B. cognatus*.

Blaustein et al. (1994), commenting on the perceived global decline in amphibian populations, argued that amphibian metapopulation dynamics are poorly understood and that recolonization following local extinction events may be hindered by human land-use patterns. Although one-third of these surveys failed to encounter *B. cognatus*, central Oklahoma has hosted a robust, well-studied metapopulation for more than five decades (Bragg 1940; Krupka 1989). Of the 173 voucher specimens of *B. cognatus* in the Nebraska State Museum collected since 1969, 115 were sampled during the 1970s, forty-four during the 1980s, and only fifteen during the 1990s. This apparent decline may be largely attributable to reduced sampling effort: six of the specimens were collected "DOR" (dead on road) and five of those six were collected on the same day at one location. The observational record of *B. cognatus* can be said to exhibit spatiotemporal patchiness, which complicates both model development and validation efforts.

Conclusion

Modeling species occurrences successfully depends ultimately upon blending the spatial perspectives of biogeography and landscape ecology with the temporal emphases of population and community ecology. Thus, there is a deep need to link geospatial data with temporal data in spatiotemporal models of metapopulation dynamics across changing landscapes. There are currently several interrelated limitations toward meeting this goal:

Current Limits

1. Current implementations of GIS are not designed for spatiotemporal modeling.
2. Rescaling of data and relationships is a subtle art because it is largely context-specific.
3. GIS topologies are devoid of a priori causal structures. It is thus critical that domain experts inform these topologies, but this fundamental need is often not recognized as such.
4. Methods to validate models persuasively remain an open question because theory about spatiotemporal processes and patterns, whether from an ecological or statistical perspective, is sparse or immature.

Future Prospects

The prospects for integrating time with geospatial data in models of species occurrence are good given developments on multiple fronts.

GIS Diversity

As more GIS vendors enter the market over the next several years, there will be opportunities to explore new approaches. Object-oriented GIS holds some promise, but there are questions about the robustness of dynamical representations (Raper and Livingstone 1996). As this new technology is largely underexplored, it may require different techniques for validation and error analysis.

Increasing Resolution and Volume of Remote Sensing Data Sets

Earth resources satellites are being designed with increasingly higher spatial, spectral, radiometric, and temporal resolutions. Image sizes are often no longer expressed in megabytes but in gigabytes. In December 1999, NASA began implementing the Earth Observing System (EOS) with the launch of the Terra satellite (http://terra.nasa.gov/), the flagship of the EOS constellation. The data produced by the EOS sensors will be measured in terabytes/day. The Moderate Resolution Imaging Spectroradiometer (MODIS) on Terra, for example, provides coverage of the earth in thirty-six spectral bands at 250 meters to 1 kilometer in resolution (Justice et al. 1998). Together with Landsat 7, Ikonos, and a host of future public and private platforms, including Europe's powerful ENVISAT (http://envisat.esa.int), a veritable flood of observational data will rise during the first decade of the twenty-first century. It is critical that we be able to access, archive, analyze, and share very large data sets and that we be able to integrate EOS data with other remote sensing imagery and with ancillary geospatial data (e.g., digital elevation data, climate data, soils data) via GIS. The volume of data requires a paradigm shift in analytical strategy: from occasional observation to monitoring, from statics to dynamics.

Multitemporal and Integrated Multisensor Analyses

Characterization and mapping of vegetation typically requires at least two or three images acquired at critical times during the growing season to capture essential components of phenology and/or land-management practices (Loveland et al. 1995; Wolter et al. 1995; Congalton et al. 1998). Moreover, there is increasing interest in the remote-sensing community to assess both intra-annual (seasonal) and inter-annual changes in land cover (Maisongrande et al. 1995; Reed and Yang 1997; Egbert et al. 1998; Yang et al. 1998b). Such work can involve acquisition, archiving, and processing of dozens to hundreds of images along with supporting ancillary geospatial data. In addition, it is now well established that there are significant advantages to using data acquired by sensors operating in distinctly different regions of the electromagnetic spectrum. For example, data from microwave sensors provide structural information that complements reflectance data gathered from visible and infrared systems (Imhoff et al. 1997; Chauhan 1997).

As yet more orbital platforms are sent aloft carrying both active and passive sensors with varying combinations of spectral-spatial-temporal resolutions and off-nadir view angles, there will be more opportunities to elicit the spatiotemporal dynamics of habitat at synoptic extents. Similarly, the broader use of the global positioning system (GPS) for tracking individual organisms and populations will generate richer databases from which to study animal movement patterns across the inhabited landscape. Spatiotemporal modeling demands large volumes of data for model development, calibration, validation, and verification. Data and theory move hand in hand. As experience with collecting, handling, and analyzing spatiotemporal data accumulates, we expect to see new concepts and theories emerge. Modeling species occurrence must be an iterative process as long as the landscape changes.

Acknowledgments

This research was partially supported by the Gap Analysis Program, U.S. Geological Survey, Biological Resources Division. G. M. Henebry acknowledges support from NSF DEB 96-96229.

Predictions and Their Validation: Rare Plants in the Central Highlands, Victoria, Australia

Jane Elith and Mark Burgman

Models of the association between species and their environment and the predictive maps based on these associations have been used for a variety of purposes including population estimates and conservation planning (e.g., Corsi et al. 1999; Jarvis and Robertson 1999). They provide a means of applying existing data in a regional context in the many situations where further field surveys are constrained by time or resources (Nicholls 1989). Such models can be tested, and the decision about whether any formal testing is required and about the most appropriate test should be influenced by whether the theoretical content of the model or its operational capability is more important (see Van Horne, Chapter 4; Rykiel 1996). In the context of land-use planning for conservation, validation is important and should be designed to test whether the model, within its domain of applicability, possesses sufficient accuracy to be useful for its intended application (Chatfield 1995; Rykiel 1996).

In this chapter, we present a set of models produced from existing government data for eight plant species in the Central Highlands of Victoria, Australia. Several of the modeling protocols were developed for species models being used by the government in land-use planning, and part of the intention of the study was to investigate success under these protocols. The models were developed through four different modeling methods that utilize either presence-only or presence-absence data. The aim of this chapter is to investigate the accuracy of the models at different extents (see Morrison and Hall, Chapter 2) and to assess the performance of different validation strategies.

Study Area

Land use in Australian forests is currently negotiated through Regional Forest Agreements (RFAs) in regions defined by government. Because we were interested in assessing the adequacy of models that could be created for an RFA, modeling was based on data from one of these regions. The Central Highlands RFA area is a 1.1-million-hectare region northeast of Melbourne, Victoria (see Fig. 24.1 in color section). The landscape is dominated by the deeply dissected landforms of the Great Dividing Range but extends to a flat to undulating coastal plain in the south. The area ranges in elevation from 25 to 1,550 meters, in annual mean temperature from 5.5°C to 14.6°C, and in annual precipitation from 570 to 1,800 millimeters. Floristic communities are diverse and form mosaics throughout the region (Foreman and Walsh 1993). Fifty-six percent of the area is public land, largely covered by native forest, and the remaining private land is mostly cleared and used for agriculture (RFA Steering Committee 1996).

TABLE 24.1.

Details of eight of the modeled plant species.

Species name	Status[a]	Life form	Range (within RFA)[b]	Records in modeling quadrats[c]	Records in modeling subset[c]	Records in independent validation set[c]	Records in independent validation subset[c]
Grevillea barklyana	Rr	shrub	l	15/3507	15/444	8/383	8/100
Oxalis magellanica	r	forb	m	37/3485	28/381	12/379	12/66
Tetratheca stenocarpa	Rr	low shrub	m	54/3468	50/412	17/374	10/101
Wittsteinia vacciniacea	r	low shrub	m	131/3391	129/280	57/334	54/24
Helichrysum scorpioides		forb	(w)	204/3318	195/1459	26/365	17/121
Leptospermum grandifolium		tall shrub	w	179/3343	179/1661	58/333	56/105
Nothofagus cunninghamii		tree	w	509/2843	509/1312	99/292	79/73
Phebalium bilobum		shrub	(w)	119/3403	119/673	63/328	58/35

[a]Conservation status: species that are rare (r); capitals indicate national status and lowercase indicate state classification.
[b]Widespread (w), medium range (m), and localized (l); brackets indicate intermediate classes.
[c]Expressed as number present/number absent.

Species Data

Government flora records for the region are stored digitally in the Flora Information System (FIS) of the Department for Natural Resources and Environment (NRE). From the full set available in September 1996, all quadrat records that were assessed as accurately located (Fiona Cross, pers. comm.) and that provided presence-absence data for all vascular species were extracted. These records were the result of full quadrat searches commissioned since 1976. The searches were often purpose-driven, being part of prelogging surveys, regionwide studies, and targeted sampling of particular habitats, such as rainforests and heathlands. The most common sampling protocols utilized quadrats of 900 square meters (30 × 30 meters, or 15 × 60 meters in riparian vegetation) and searching continued in a sequence of dissimilar microhabitats until no new species had been sighted for ten minutes. The final set of 3,522 quadrats was scattered irregularly over the region (Fig. 24.1), with site location biased toward roadsides and away from ecotones. Sites tended to be clustered in high rainfall areas closer to Melbourne and in certain vegetation types (e.g., in cool temperate rainforest) (Elith et al. 1998). Approximately half of the quadrats were surveyed before 1986. The biases within this data set are similar to those in other sets of data compiled in an ad hoc way. The most likely implications for modeling are that some unique environmental combinations

will not have been sampled and models will not fit well in such regions. In addition, if older records apply to species affected by disturbance or canopy closure the models are likely to be less reliable in these cases. Site location was recorded by latitude and longitude. The absence of original maps for many of the quadrats make cross-checks of the accuracy of site location difficult, but it is likely that 95 percent of the records were located within 250 meters of their true position (Elith et al. 1998).

In the full study, twenty-nine species were selected for modeling; these were all perennial species and represented a range of scarcities, sizes, and life forms. For any modeled species, the full set of 3,522 sites was used to document species presence or absence at each site. We were particularly interested in the comparative success of models for common and rare species, with rarity being expressed either as scarcity in the region or as conservation status. In this chapter, the results for eight of these species are presented; these represent the range of results in the larger set and include four species classified as rare and four more-common species (Table 24.1).

Environmental Data

Models of species distribution are based on the relationship between the presence of species, or presence and absence of species, and environmental data (Table 24.2). Although models can be developed on site-specific data, prediction for reserve planning requires either extensive site-specific data, which is rarely available, or continuous environmental data across the region. The digital data available for this region included climate data estimated from long-term monthly records of temperature, rainfall, and radiation; topographic data derived from a 9-second (about 250 meters) digital elevation model (DEM); and categorical data on rock type and vegetation class (Elith et al. 1998; and see Table 24.3). These produced a total of thirty-five potential predictor variables rasterized to a cell size of 9 seconds in a geographic projection. These variables were sampled at quadrat locations for model development and predictions were made to all cells in the region.

Modeling Methods

Four modeling methods are presented in this chapter: see Table 24.2 for a summary of main references and example applications and software. Since our intention was to assess the adequacy of predictions for regional planning, methods were applied in a way that

could be adapted to many species for the whole region within the restrictive time frames for planning. Rare species were treated in the same way as common ones.

1. ANUCLIM (Nix 1986) is a bioclimatic envelope method in which the climate profile for a species is developed by sampling the estimated climate data (in this case, the first twenty-three variables in Table 24.3) at the sites where the species is known to be present. The potential climatic domain for the species is the multidimensional envelope that encompasses all recorded locations of the species. For prediction, the climatic variables at each cell in the region are used to generate a rank for the cell that reflects its position in the climatic envelope for the species. In our modeling, cells were ranked from 0 to 3 for each species, and rankings were based on pairs of percentiles (0 and 100, 5 and 95, 10 and 90). For instance, a cell outside the range of sampled climates would be given a rank of 0, a cell inside the envelope but only sitting at an extreme percentile such as the 97th percentile would be given a 1, and a cell at the core of the envelope (between the 10th and 90th percentiles) would be given the highest rank.

2. Generalized linear models (GLMs) are a class of

TABLE 24.2.

Modeling methods: key references, examples, and software.

Method	Key references	Example of ecological applications	Software used
Bioclimatic envelopes	ANUCLIM: Nix (1986) and CRES (1998)	ANUCLIM: kauri pine in New Zealand (Mitchell 1992), eucalypts in South Africa (Richardson and McMahon 1992), possums (Lindenmayer et al. 1991b) and myrtle beech (Busby 1986) in Australia	ANUCLIM v1.01 (CRES 1998)
Generalized linear models (GLMs)	McCullagh and Nelder (1989) and Agresti (1996)	Arboreal marsupials in Australia (Lindenmayer et al. 1990a, 1990b, 1991a), eucalypts in Australia (1984, Austin et al. 1983, 1994, 1990), alpine plants in Switzerland (Guisan et al. 1998), and a bird species in America (Akçakaya and Atwood 1997)	S-PLUS v3.3 for Windows (MathSoft 1995)
Generalized additive models (GAMs)	Hastie and Tibshirani (1990)	Trees in New Zealand (Yee and Mitchell 1991), eucalypts in Australia (Austin and Meyers 1996), and wetland plants in The Netherlands (Bio et al. 1998)	S-PLUS v3.3 for Windows (MathSoft 1995)
Genetic algorithms	Mitchell (1996)	GARP algorithm in Australia (Stockwell and Noble 1992) applied to plant species (Elith et al. 1998)	GARP (Stockwell and Peters 1999)

TABLE 24.3.

Details of environmental data offered in modeling.

Data	Type[a]	Source and format	Cell Size or Scale[b]
Annual mean temperature	C		
Mean diurnal range temperature	C		
Maximum temperature of warmest month	C		
Minimum temperature of coldest month	C	All climate grids calculated using	9-second cell size
Temperature annual range	C	(a) ANUCLIM algorithms (Nix	(about 250 m)
Mean temperature wettest quarter	C	1986) constructed from long-term	
Mean temperature driest quarter	C	climate data, and	
Mean temperature warmest quarter	C	(b) the GEODATA National 9-second	
Mean temperature coldest quarter	C	DEM (Australian Surveying and	
Annual precipitation	C	Land Information Group [AUSLIG]).	
Precipitation wettest month	C	Projection was geographic.	
Precipitation driest month	C		
Precipitation wettest quarter	C		
Precipitation driest quarter	C		
Precipitation warmest quarter	C		
Precipitation coldest quarter	C		
Annual mean radiation	C		
Highest monthly radiation	C		
Lowest monthly radiation	C		
Radiation wettest quarter	C		
Radiation driest quarter	C		
Radiation warmest quarter	C		
Radiation coldest quarter	C		
Lithology	F	Derived from Land Systems of Victoria coverage, Dept. Natural Resources and Environment. Data grouped into four classes related to fertility and drainage.	1:250000
Geology	F	Derived from Surface Geology coverage, Dept. Natural Resources and Environment. Data grouped into five classes related to fertility and drainage	1:250000
Elevation	C	9-second DEM (AUSLIG)	9-second cell size
Topographic position	C	Mean difference in elevation between current grid cell and all cells within a 1-km radius	9-second cell size
Wetness index	C	Calculated from flow accumulation, cell width and slope, according to Moore (1993).	9-second cell size
Gully	F	Calculated from a 1:25000 DEM, selecting cells with low topo position at that scale and resampling to 250 m.	9-second cell size
Slope	C	Calculated from DEM.	9-second cell size
Ecological vegetation class (evc)	F	EVC100 coverage, Dept. Natural Resources and Environment. Twenty-nine vegetation classes in this region (Woodgate et al. 1994).	1:100000
Aspect: eastness	C	Calculated from DEM.	9-second cell size

TABLE 24.3. (Continued)

Details of environmental data offered in modeling.

Data	Type[a]	Source and format	Cell Size or Scale[b]
Aspect: southness	C	Aspect (degrees) converted to continuous measures after Pereira and Itami (1991) and Zerger (1995).	9-second cell size
Latitude	C	Created in ArcInfo	9-second cell size
Longitude	C	Created in ArcInfo	9-second cell size

[a]Continuous (C) and factor or categorical (F) variable.
[b]Coverages converted to grids in ArcInfo by transforming to geographic projection, and gridding to 9-second cells aligned with the DEM. Values assigned to grid cells on the basis of the category covering the largest area of the cell.

statistical model that encompasses classical linear regression and analysis of variance and extends to models that can be applied to categorical response data. Logistic regression is the appropriate type of GLM for presence-absence data. It uses the binomial distribution to model the variation in the response, and a logit link (i.e., $\log[\mu/(1 - \mu)]$ where μ is the mean response) to transform the linear predictor to a suitable scale for binomial data. The predictor can comprise one to many variables, and continuous variables can be modeled with a variety of parametric response functions (e.g., linear, quadratic, or cubic). The model can be applied to new data to obtain predictions of the probability of the species presence at each new location. In this study, all available environmental variables were initial candidates as predictors. For each species, the predictor candidate set was reduced to a subset through a univariate logistic analysis, where a variable was retained for the subset if the change in deviance between the model containing the variable and the null model was significant at a p-value of 0.2. In the subset, variables that were correlated were excluded through analysis of variance inflation factors (Sokal and Rohlf 1981; Booth et al. 1994). Models were developed with an automated forward selection procedure in which, at each step, the variable associated with the largest reduction in deviance was retained, and the process was repeated until none of the remaining variables significantly improved the model ($p < 0.05$). If latitude and longitude were included in the subset, they were considered last. Linear, quadratic, and cubic terms were investigated for all continuous variables, but interactions were not considered.

3. Generalized additive models (GAMs) are an extension of GLMs and allow nonparametric, nonlinear data-dependent smooth functions to be fitted to continuous variables in the predictor. This means that a functional form does not have to be specified (cf. GLMs), giving greater flexibility to the modeling process. The final model can include both linear and smoothed terms for continuous variables and categorical variables. Variable selection for GAMs was approached in the same manner as for GLMs. GAMs were developed with four degrees of freedom initially assigned to each continuous variable, with subsequent testing of changes in deviance used to reduce the degrees of freedom where possible. Relationships were smoothed with cubic splines.

4. GARP (Genetic Algorithm for Rule-set Prediction; see Stockwell and Peterson, Chapter 48) is a computer-based modeling approach that uses a genetic algorithm to develop a set of rules that describe the relationship between species data (presence or presence-absence) and environmental variables. The rules are then applied to all other cells of interest for prediction. The genetic analogy is intended to express the idea of an evolving set of rules, from the initial random starting point to one optimally fit for prediction. The resulting predictions express the relative likelihood of the presence of the species at any given site, with sites of presence always assigned the maximum likelihood. Models developed in GARP were constructed with the full set of

Observed

	presence	absence
Predicted presence	a	b
absence	c	d

Figure 24.2. Representation of a confusion matrix.

environmental variables. Since each run of the genetic algorithm provides a unique solution, and maps of the resulting predictions were demonstrably different even with a quick visual assessment, spatial predictions were constructed from the average of ten iterations of the program.

Model Assessment

Measurement of accuracy. Our primary concern was to produce models of species distribution that discriminated well between sites where the species was present and sites where it was absent, so an appropriate measure of model discrimination was required. If presence-absence data are available, the performances of models are often summarized in a confusion or error matrix (Fielding, Chapter 21) that records the counts of observed presence and absence against predicted presence and absence (Fig. 24.2).

Two common measures derived from the confusion matrix are the true positive fraction (TPF) or sensitivity ($a/[a + c]$ in Fig. 24.2) and the false positive fraction (FPF, or [1 – specificity], or in other words $b/[b + d]$). Construction of the confusion matrix requires that a threshold is selected so that the predictions (probability, likelihood, or rank) can be converted into binary data. The selection of such a threshold is not necessarily a straightforward task (Fielding and Bell 1997), particularly for rare species, and excludes some of the information generated by the model.

One threshold-independent measure that is used regularly in clinical medicine is the receiver (or relative) operating characteristic (ROC) curve (Metz 1978; Swets 1988). This plots the TPF against the FPF

over a large range of thresholds (see Fig. 24.2). The area under the curve (AUC) is an important summary statistic for ROC plots. The AUC ranges from 0 to 1, with values less than 0.5 indicating that the model tends to predict presence at sites at which the species is, in fact, absent. A value of 0.5 implies no discrimination (i.e., no consistent difference between predictions from the observed absence group and the observed presence group), which is equivalent to random predictions, and 1.0 indicates perfect discrimination. For binary data, the AUC is identical to the nonparametric two-sample Mann-Whitney statistic (Hanley and McNeil 1982). An AUC of 0.78 can be understood to mean that, if random pairs of predictions were sampled from observed presence and observed absence groups, the prediction for the observed presence would be higher than the prediction for the observed absence 78 percent of the time. Since the predictions are intended for use in ranking the suitability of land for inclusion in reserves, this statistic is an appropriate measure of model usefulness.

In this study, models with AUCs of 0.75 or above will be considered sufficiently discriminatory to be potentially helpful (see Pearce et al., Chapter 32). We calculated nonparametric AUC and DeLong standard errors (DeLong et al. 1988) with routines written for use in the S-Plus statistical program (MathSoft 1995; Elizabeth Atkinson personal communication). All results are presented graphically with confidence intervals based on the standard errors for an individual AUC estimate. However, where modeling methods are tested on a common validation set, the ROC curves are correlated and the unadjusted standard errors are less sensitive than they should be in detecting differences between the AUCs (Hanley and McNeil 1983). In these cases, the results presented in the text are based on tests of significance developed by DeLong et al. (1988), which take into account the correlation of the ROC curves.

Validation data sets. To investigate the effect of the validation set on the conclusions about model discrimination two sets of data have been used:

1. The original data used to develop the model was also used to test discrimination. Although it is well documented that this does not represent true vali-

dation (Fielding and Bell 1997; Chatfield 1995), it is still regularly used and published in government and scientific circles (e.g., Lindenmayer et al. 1991b; Skidmore et al. 1996; NSW NPWS 1998; Franklin 1998). In this chapter, AUCs calculated on this data set will be referred to as AUC_{model}.

2. An independent set of species data was collected in 1997 specifically for validation of the models for these species. Our intention was to visit sites that spanned the range of predictions for each method for each species and to gather a reasonable number of presence records. The selection of the first one hundred cells was based on expert opinion of likely sites for each of the scarcer species. This provided records independent of the models we had created and was intended to supply sites that may include rare species if the models failed. In order to select sites predicted by the models to be very likely to contain the species, the final set was required to include at least three cells for each species/method combination from the top 0.1 percent of the distribution of predictions. If the set did not already contain such cells, they were added through a random selection from the highest predictions. Cells were not visited if a species had previously been found there. In all cases, the lower range of predictions was well represented by sites selected for high likelihood for other species. The final validation set contained 391 sites. These were located by latitude and longitude at the center of the selected cell (all methods predicted presence or absence within 9-second cells). In the field, the locations were converted to Australian Map Grid (AMG) projection and sites located on 1:100,000 topographic maps, or 1:25,000 maps where there was insufficient detail on smaller-scale maps to locate the site reliably. The site was sampled at its given location regardless of whether other mapped data (such as mapped vegetation class) matched the actual situation. Presence-absence data was recorded both within a 900-square-meter quadrat (consistent with the protocol for the modeling data) and within the 9-second (about 250 meters) cell (Elith et al. 1998), and in this chapter, data for the presence-absence of the species within the 9-second cell

are presented. AUCs calculated on this data set will be referred to as AUC_{indep} in the results.

Validation scale. We were concerned that, even though the models were being developed at the regional scale (about 10^6 hectares), they would be applied to decision making within a subcatchment (about 10^4 to 10^5 hectares). A subcatchment is considered a different scale compared with a whole region because it has a smaller extent. The data sets have been used for model assessment in their entirety (i.e., at the regional scale) and as subsets that represent the subcatchment scale for each species. For a subset, the locations of the species in the full modeling set (set 1, above) were taken to indicate the geographic extent of the species in the RFA, and bounds were defined that enclosed most presence locations and extended to the estimated limits of the inhabited subcatchments. The subcatchment limits were defined through a visual interpretation of the region's DEM because we did not have the GIS capacity to do otherwise at the time; subsequently, a more-detailed definition of the subcatchments created through ArcInfo algorithms has indicated that, even though the validation subsets may differ slightly with an automated approach, the conclusions from the study are similar regardless of which limits were used. In summary, the subsets exclude sites that are in subcatchments that have no other record of species presence.

Comparison of Modeling Success: Data Sets

At the regional scale, the modeling data set (set 1) returned AUC_{model} greater than 0.85 for 91 percent of the models and greater than 0.75 for 94 percent of the models. By comparison, observations from the independent validation set (set 2) returned AUC_{indep} greater than 0.85 for 69 percent of the models and greater than 0.75 for 88 percent of them (Fig. 24.3). The general trend was for AUC_{model} to be higher than AUC_{indep}. In most cases, the difference between AUC_{model} and AUC_{indep} was relatively small, with 81 percent of the values for AUC_{indep} being within 10 percent of the values for AUC_{model}. Thirty-one percent of the differences were significant at the 5 percent

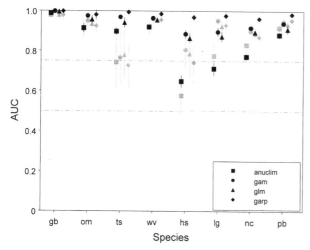

Figure 24.3. Area Under Curve (AUC) summaries of the receiver operating characteristic curves for eight species (indicated by initials of their scientific names), where predictions are tested against the modeling data set (black plots) and the independent data set (gray plots) at the regional scale. Plots show the AUC and its 95 percent confidence intervals for each species/method combination.

level, and in 80 percent of these cases AUC_{model} was higher than AUC_{indep} (Fig. 24.3). These significant differences were restricted to four of the eight species, and the most marked reductions in AUC (AUC_{model} to AUC_{indep}) were centered on two species, (*Tetratheca stenocarpa* and *Helichrysum scorpioides*).

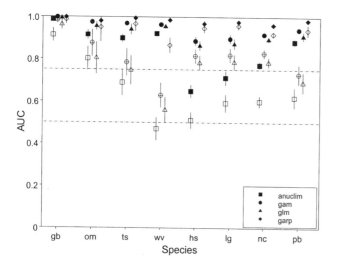

Figure 24.4. Area Under Curve (AUC) summaries of the receiver operating characteristic curves for eight species (indicated by initials of their scientific names), where predictions are tested against the modeling data set at the regional scale (solid black marks) and the subcatchment scale (open black marks). Plots show the AUC and its 95 percent confidence intervals for each species/method combination.

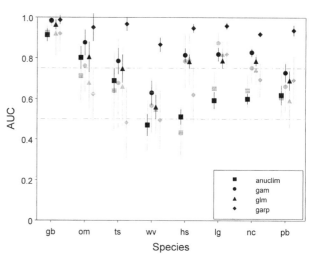

Figure 24.5. Area Under Curve (AUC) summaries of the receiver operating characteristic curves for eight species (indicated by initials of their scientific names), where predictions are tested against the modeling data set (black plots) and the independent data set (gray plots) at the subcatchment scale. Plots show the AUC and its 95 percent confidence intervals for each species/method combination.

AUCs were always lower at the subcatchment extent than at the regional extent (Fig. 24.4). In the subcatchments, 66 percent of the models had AUC_{model} greater than 0.75, and only 34 percent of the models had AUC_{indep} greater than 0.75 (Fig. 24.5). The overwhelming trend was for AUC_{model} to be higher than AUC_{indep}, and the difference between the AUCs tended to be greater than at the regional extent, with only 56 percent of subcatchment values for AUC_{indep} being within 10 percent of the values for AUC_{model} (Fig. 24.5). However, only 22 percent of these differences were significant at the 5 percent level.

Comparison of Modeling Success: Methods and Species

The AUC statistics derived from the independent validation set at the subcatchment scale have been primarily used to study differences in model performance between methods and between species since these data are more likely to represent the true modeling success at a scale appropriate for the intended application of these methods. The results for the eight species reported here are qualitatively the same as for the full set of species considered in the larger study (Elith et al.

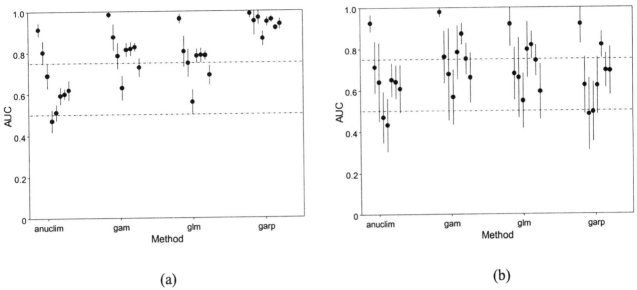

Figure 24.6. Area Under Curve (AUC) summaries of the receiver operating characteristic curves for four modeling methods, where predictions are tested against the modeling data set (a) and the independent data set (b) at the subcatchment scale. Plots show the AUC and its 95 percent confidence intervals, with results for each species grouped by method, and species in the same order as previous figures.

1998). They indicate no consistent significant difference among the discriminatory performance of predictions from the four different methods, although there is an apparent trend toward better discrimination from the GAMs and GLMs. This coincides with a tendency for the methods that use presence-only data (ANUCLIM in this study) to perform less well than the presence-absence models (Fig. 24.6). Significance tests that take into account the correlation of the AUCs indicate that there is no evidence that the AUCs are not equal for six of the species ($p \leq 0.05$). For the remaining two species (*H. scorpioides* and *Leptospermum grandifolium*) there is evidence of inequality, with tests of pairs of methods showing that the GAM for *L. grandifolium* and the GAM and GLM for *H. scorpioides* have significantly better discrimination than models produced with the other methods.

The difference in modeling success between species tended to be more pronounced than the differences between modeling methods (Fig. 24.5). Some species such as *Grevillea barklyana* and *L. grandifolium* appeared to be modeled with sufficient accuracy for the information to be useful in land-use decisions, whereas the models for other species such as *Wittsteinia vacciniacea* and *Phebalium bilobum* were unlikely to be useful. There were no clear associations between modeling success and species' characteristics such as rarity, range within the region, or life-form.

Discussion

It is clear that our perception of whether a model is accurate enough to be useful is affected by the way in which we assess it (see Van Horne, Chapter 4). The results of this study show that use of the modeling data (i.e., the data used to build the model) as a validation set gives an overly optimistic view of modeling success. The result is not surprising and is consistent with the reviews of model uncertainty and its assessment by Fielding and Bell (1997) and Chatfield (1995). In this study, the degree of optimism displayed by the modeling data set varied with the species, the modeling method, and the extent of analysis.

At the regional scale, the most marked overoptimism was focused on two species, *T. stenocarpa* and *H. scorpioides*. A distinctive feature of the presence records for these two species was that they were both found, in the independent surveys, at sites relatively distant (up to 55 kilometers) from any modeling records. These are unlikely to be new occurrences of the species and indicate that either the surveys contributing to the modeling data did not adequately sample the region, missing unique and important environmental combinations, or that the species were present but not correctly identified (this is particularly possible for *T. stenocarpa*, which is difficult to distinguish from *T. ciliata* in some of its forms [D. Frood, personal communication]). The result supports the common-sense notion that models built on inaccurate or unrepresentative records will not predict reliably to new sites. Targeted sampling that filled in "gaps" or deficiencies in data sets (as assessed by geographic or environmental representativeness, by date of survey, or by expert knowledge) would increase model accuracy in these situations.

Some modeling methods appear particularly (and unjustifiably) effective when the modeling data set is used for model assessment. Methods that create a rule or measure an environmental distance (such as GARP in this study) often predict highest likelihood at the original record site. In contrast, methods that fit a prediction surface (such as the GLMs and GAMs) do not necessarily predict high probabilities at presence locations, especially if identical environmental conditions are also associated with absence records. Any method that by default or by consequence of the applied measures predicts highest likelihoods of occurrence at the modeling presence sites will appear to have good discrimination when tested against the modeling data set. This can be seen in the data presented in Fig. 24.6a and 24.6b. GARP appears to perform significantly better than the other three methods when tested against the original data (Fig. 24.6a), but testing with independent data (Fig. 24.6b) indicates no clear distinction between the methods. The only accurate way to test models developed with methods that return highest likelihoods at presence sites is to ensure that the validation data set is independent. If a completely independent data set is not available, data partitioning that excludes some data from the modeling process (see, for example, Fielding, Chapter 21) can achieve a degree of independence that will result in a more realistic picture of model performance.

The same patterns in measurement optimism were apparent at the subcatchment scale; differences tended to be relatively greater but less often signifi-

cant. The significance is affected by the standard errors for these statistics, which increase with decreasing AUC and with increasing imbalance between number of presence records and number of absence records (Hanley and McNeil 1982). Therefore, the confidence limits at this smaller extent tend to be relatively large, especially for rare species. The clearest effect of validation extent is demonstrated in Figure 24.4, which shows that, even for the optimistic tests with the modeling data set, a model may have insufficient discrimination to be useful at a subcatchment scale even though it appears to discriminate well at a regional scale (e.g., see *W. vacciniacea*, *P. bilobum*). The effect of scale will not be the same in every region—it is likely to vary with the homogeneity of the landscape, the match between the environmental patterns and the grain of the mapped environmental variables, and the intensity of sampling. Nevertheless, it is important to consider the scales most relevant to the context and objectives of the modeling and to give attention to assessing the models at a realistic grain and extent.

The differences in model performance between methods tended to be less marked than the differences between species. As assessed with the independent validation set at the subcatchment scale, the trend was for the presence-only method (ANUCLIM), which only uses climatic variables for prediction, to perform less well than the other methods, though the differences were generally not significant. Other studies that have compared methods have generally concluded that the statistical methods (GLMs and GAMs) perform better than ANUCLIM (Ferrier and Watson 1996) and that GAMs may model species responses to habitat more successfully than GLMs (Austin and Meyers 1996; Bio et al. 1998). It is likely that, given a more complete and more accurate modeling data set and more extensive validation data, significant differences between the methods would emerge, on the basis that we would expect the added refinement provided by absence data and additional predictor variables to improve modeling success.

Some species appear to be well modeled at the subcatchment scale, and the success of these models suggests that it is worth putting effort into modeling species distributions, even for rare species (see Karl et al., Chapter 51). The models presented in this study were constructed using approaches that were likely to be applied in the RFAs, but a number of refinements are possible. Filtering the presence-absence data to reduce spatial autocorrelation (Mackey et al. unpublished) or dealing with the autocorrelation within the model (Augustin et al. 1996; Fielding and Bell 1997) may improve the models. The region includes a diverse range of landscapes, and constraining the set of modeling data to exclude absence data outside the environmental range of the species may be effective (Austin and Meyers 1996), although determination of the appropriate range is not straightforward when the data set is not comprehensive. Other refinements could focus on additional sampling to overcome data inadequacies (Neldner et al. 1995), the development of variables more directly related to species distribution (Austin et al. 1984), and alternative approaches to variable selection for any of the methods (e.g., Booth et al. 1994).

Since we cannot identify particular characteristics of species or their data sets that can be used as a guide to the likely success of modeling, there remains a pressing imperative to assess models carefully before using them in decision making. Managers interested in using predictions to discriminate between sites may not have a choice about how the predictions were constructed, but an understanding of the general tendencies of the method will help interpretation of results. For example, envelope methods (such as ANUCLIM) tend to include in the envelope sites of absence, whereas GLMs theoretically predict accurate probabilities of presence. Two issues of practical importance highlighted by our results that can be assessed by a manager are (1) At what scale is the model likely to be effective and will it discriminate between sites at the grain (resolution) required? (2) How has the model been tested? If it has not been tested with independent data at the grain and extent that is relevant to the application, the results need to be assessed with care.

Acknowledgments

We are indebted to many people for help with data collection, analysis, and modeling. In particular, we thank Jennie Pearce, Guy Carpenter, David Stockwell, Graeme Watson, Simon Ferrier, Paul Yates, Michele Arundelle, Fiona Cross, Neville Walsh, Doug Frood, Dale Tonkinson, Andrew Taplin, David Barratt, Brendan Wintle, and four anonymous reviewers. The S-Plus routines written by Elizabeth Atkinson and colleagues were made available by the Mayo Clinic, USA. The research was supported by Project No. FB-NP22 of Environment Australia.

Semiquantitative Response Models for Predicting the Spatial Distribution of Plant Species

Antoine Guisan

According to Zar (1996), four main types of data can be found in biology: (1) data on a ratio, (2) on an interval, (3) on an ordinal (or ordered), and (4) on a nominal scale. The first two data types, which can be further divided into continuous or discrete data, are commonly grouped under the denomination of quantitative data; ordinal data are synonymously called semiquantitative data, and nominal data can also be called qualitative data. To date, statistical model building in ecology has mostly focused on quantitative (e.g., biomass) and qualitative (e.g., vegetation units) responses. However, little attention has been paid to semiquantitative, ordinal responses.

An ordinal scale is defined as an ordering of measurements, with only relative—instead of quantitative—differences between values. For instance, measurements of biological entities in classes that are, for instance, longer, larger, or more abundant than others are clearly ordinal. Ordinal data can also originate from ratio or interval data by slicing a continuous scale, although such procedure inexorably results in a loss of important ecological information. On the other hand, some so-called quantitative data—such as Braun-Blanquet (1964) and other abundance-dominance scales in phytosociology (see Table 25.1)—are best considered ordinal in statistical analyses. Among other ordinal outcomes commonly found in ecology are successive phenological stages of plant-flowering

processes (e.g., Theurillat and Schlüssel 1998) or of insect development (e.g., Candy 1991; Manel and Debouzie 1997), categories of tree height or diameter in forestry (e.g., Schabenberger 1995), or tolerance levels in ecotoxicology (e.g., of an organism to a pollutant; e.g., Ashford 1959). More generally, ecological data might certainly be best analyzed as ordinal where one expects the uncertainty in field measurement to be over an acceptable threshold (e.g., visual abundance classes of animals).

Statistical methods for ordered data have been used since the late fifties (e.g., Ashford 1959). Multiple regression models for an ordinal response were first described in the late sixties (e.g., Walker and Duncan 1967; see McCullagh 1980 or Anderson 1984 for early reviews on this matter). Since then, ordinal regression models have been included in the broader category of generalized linear models (GLMs; see McCullagh and Nelder 1989) and were principally used in epidemic studies, toxicity assessments (bioassays; e.g., Ashford 1959) and social sciences, as these disciplines often have to deal with semiquantitative variables. But few examples of ordinal regression models currently exist in ecology (see, for example, Schabenberger 1995; Guisan et al. 1998), even though in this field, several data are actually on an ordinal scale.

In this chapter, I present one regression model—the Proportional Odds Model—which can be successfully applied to an ordinal response. The use of this model

TABLE 25.1.

The ordinal scale used in the two case studies ranges from 0 to 5. These ordered values correspond to intervals of unequal widths of plant ground cover density. Correspondence is also given with previous scales commonly used in phytosociology (from Guisan 1997).

Classe	Cover (%)	Braun-Blanquet (1964)	Barkman et al. (1964)	van der Maarel (1979)
0	0	—	—	0
1	1–5	1	r, +, 1, 2m	1, 2, 3, 4
2	6–12	2	2a	5
3	13–25		2b	6
4	26–50	3	3	7
5	51–100	4, 5	4, 5	8, 9

is illustrated through case studies of modeling spatial distribution of plant species, together with their ordinal density on the ground, with data from two mountain ranges: a small study area in the Swiss Alps (see Guisan et al. 1998) and the entire range of the Spring mountains in Nevada (see Guisan et al. 1999).

Specific objectives were (1) to set up ordinal regression models for the distribution of plant species' cover density, (2) to implement these models in a geographic information system (GIS), and (3) to assess the adequacy of ordinal model predictions in a general context, in other words, giving no greater weight to any one type of prediction success or of prediction error than another, and to compare them to predictions from (1) Gaussian and Poisson models applied to the ordinal response, and (2) logistic presence-absence models.

Methods

The following examples are taken from two study areas—the Belalp area in Switzerland and the Spring Mountains in Nevada, USA (see Fig. 25.1)—described respectively in Guisan et al. (1998) and Guisan et al. (1999).

Study Areas

The study area of Belalp is a wide, open, north-south–oriented side valley of the Rhone Valley, located in the Aletsch region (Valais, Switzerland; Fig. 25.1a). Elevation ranges between 1,867 and 3,554 meters. Geology is mainly siliceous (gneiss, granite). The climate is subcontinental. Soils are mainly of a podzolic

type. The upper subalpine vegetation is mainly dominated by mesophilous heaths, swards, and fens. The alpine vegetation belt, ranging from 2,300 to 3,000 meters, is dominated by low heaths, swards, and snowbed communities. The landscape has been modified by human activity for centuries through intensive grazing by cattle, sheep, and goats, the main effect being the lowering of the timberline by several hundred meters. At present, grazing is extensive.

The Spring Mountains are located in southern Nevada, 20 kilometers west of Las Vegas, at latitude 36°15′ N and longitude 115°45′ W (Fig. 25.1b). Rising out of the Mojave Desert at 700 meters, the Spring Mountains reach an elevation of 3,600 meters within a distance of 10 kilometers. Plant communities range from Mojave Desert shrub at the base, through Joshua tree woodland, sagebrush, pinion/juniper woodland, and a variety of mixed conifers and aspen at the upper elevations. The highest region supports limber (*Pinus flexilis*) and bristlecone pine (*Pinus longaeva*), with a small area of alpine tundra at the peak. Deep canyons radiate from the highest peaks, creating complex mosaics of vegetation across solar radiation gradients on opposing slopes. Natural and anthropogenic fires and disturbances have added to the complexity of the vegetation mosaic.

The Ordinal Scale

The ordinal scale used in the following case study is derived from the modified Braun-Blanquet (1964) abundance/dominance scale (Barkman et al. 1964; Table 25.1). In phytosociological studies, data are still sampled by attributing to each plant species in a *relevé*

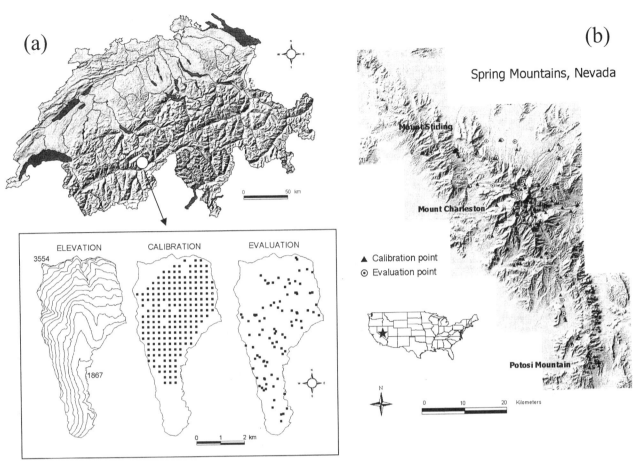

Figure 25.1. Study areas: (a) The Belalp area, near the Aletsch Glacier in the Swiss Alps. (b) The Spring Mountains, near Las Vegas in Nevada.

one of several predefinite classes of percentage ground cover. The Barkman scale is clearly more semiquantitative (i.e., ordinal) than truly quantitative. It takes values 0, r, +, 1, 2m, 2a, 2b, 3, 4b or 5 to characterize vegetation abundance/dominance (Table 25.1), which correspond respectively to no cover, rare species, some individuals present, less than 5 percent cover, less than 5 percent cover but abundant, 6–12 percent, 13–25 percent, 26–50 percent, 51–75 percent, and 76–100 percent cover. In most studies, this scale is transformed into a scale of linear integer values ranging from 0 to 9 (e.g., van der Maarel [1979] scale; Table 25.1) and is then considered a quantitative variable if calculations are to be made (see Jongman et al. 1987), using, for example, a least-square (LS) model. Although the predictions from such models might be acceptable, there are four main limitations to using such a procedure (Guisan and Harrell 2000):

1. On a quantitative-ratio scale, as required by, for instance, an LS model, the differences between successive ordered categories (e.g., between 1 and 2 and between 5 and 6) are assumed to have the same meaning, although actually they have not (see the scale above). If quantitative comparisons are made on such a linearized scale, ratios would be used to calculate estimated coefficients, thus leading to possible misinterpretation, because the scale actually represents intervals on an arbitrary unlinear scale.

2. Linearizing the abundance/dominance scale might be considered a similar process to log-linear transformation, which was used, before the appearance of GLM, when attempting to retrieve a normal error distribution. In that case, the statistical model should be based on a log-normal distribution

rather than on the normal distribution considered in LS models.

3. Modeling an ordered categorical response variable—which is discrete with few values, including *floor* and *ceiling*—using a continuous probability model, was shown by Snell (1964) to cause possible problems.

4. Unless an appropriate link function is used (i.e., other than unity or logarithm, like logit), predictions from such quantitative models may possibly take values much higher or much lower (e.g., negative values) than the maximum or minimum theoretical ordinal class. This is unacceptable from both an ecological and a methodological standpoint.

It would be better to consider each class to be simply higher or lower than the adjacent class, depending on its position along the ordinal scale. This implies fitting a regression model for ordinal response, as proposed by Schabenberger (1995), Manel and Debouzie (1997) or Guisan et al. (1998; see Guisan and Harrell 2000). In the following, I will use the term *cover density* rather than cover abundance to follow the terminology proposed by Morrison and Hall (Chapter 2).

Sampling the Response Variable

In the Belalp study area, the two species data sets used for calibrating and evaluating the model (Fig. 25.1a) are the same as described in Guisan et al. (1998). Calibration points of 4 square meters (N = 205) were sampled following a grid sampling scheme, meaning at all intersection points of a 250×250-meter grid overlaying the whole study area. Sampling on this scale means that autocorrelation is avoided (see Guisan 1997) and ensures that significance tests for selecting predictors remain valid (Palmer and Van der Maarel 1995; Van der Maarel et al. 1995). A set of independent points (N = 92) were later sampled randomly from a 25×25-meter grid overlaying the study area, for the evaluation of model predictions. At each point, localized in the field by means of a Garmin GPS navigator, a map accurate to a scale of 1:10,000 and a Thommen altimeter, ocular estimates of ground cover density were assessed for each species according to the Barkman abundance-dominance scale (Table 25.1).

They were later reclassified into the ordinal scale described in Table 25.1 for fitting the models.

In the Spring Mountains, a data set of 230 plots (generally 20×20 meters), including all higher plant species (upland vegetation only, with associated ocular estimates of ground cover density), was sampled from a 30×30-meter grid overlaying the study area, according to an ad hoc stratified design (see Guisan et al. 1999) by The Nature Conservancy (TNC; Nachlinger and Reese 1996). Each point was localized in the field using a Trimble Geoexplorer GPS navigator with post-processed differential corrections. The original single data set was split into two subsets (Fig. 25.1b), one for calibrating the model (*training data set*) and one for evaluating the model predictions (*evaluation data set*). As for the Belalp data, ocular estimates of ground-cover density were reclassified into the ordinal scale described in Table 25.1.

In both cases, ordinal models were only fitted for those species showing a sufficient variation in ground-cover density over the training data set. As a criterion, species' density was modeled for a species only when at least three classes of ground-cover density could be recorded for it, with each class recorded in a minimum of five plots.

Environmental Predictors

Environmental descriptors—hereafter called *predictors*—used to model species distribution in the Belalp area were obtained from two main sources: (1) a digital elevation model (DEM), obtained from the Swiss Federal Office of Topography (high reliability), or (2) remotely sensed data (black-and-white and color infrared aerial photographs) scanned and rectified at a 1×1-meter grain (resolution). All predictors were calculated (DEM-related predictors) or aggregated (remote sensing data) to the 25×25-meter grain of the DEM. Aggregation was performed using nearest-neighbor assignment. Some of these predictors are described in more detail in Guisan et al. (1998) and can be summarized as follows: (1) annual mean temperature (*amt*, calculated from elevation using a field-calibrated transition formula), (2) slope angle (*slo*), (3) four indices of topographic positions (representing a gradient from ridge top to middle slope to valley, and calculated with different moving windows with respective radii of 125, 250, 500, and 1,000 meters (*tp5*, *tp10*, *tp20*, and *tp40*), (4) two indices of solar radiation (*rad1* and *rad2*), obtained by taking the first two axes of a principal component analysis [PCA; explaining 99 percent of the variance] made on nineteen individual daily sums of solar radiation taken every tenth day from early April to late August), (5) two indices of snow cover (obtained by combining respectively two and four aerial photographs taken at regular intervals during 1996 and 1997: *snowi96* and *snowi97*), (6) the three bands of the color infrared aerial photograph (*cir-1* to *cir-3*), and (7) the potential permafrost (*perm*, modeled with the PERMAKART model; Keller 1992).

However, predictor variables used for explaining species distribution in the Spring Mountains were all terrain related, derived from the 30×30-meter DEM (see Guisan et al. 1999 for more details on their calculation). They were (1) elevation (*elev*), (2) slope angle (*slo*), (3) northness (*nness*) and eastness (*eness*), calculated respectively as the sine and cosine of aspect (in degrees), (4) summer solstice and spring equinox insolation calculated using the SOLARFLUX model (Hetrick et al. 1993; *ssol* and *esol*), and (5) four topographic position indices calculated at the different smoothing levels: 150 meters, 300 meters, 1,000 meters, and 2,000 meters (*tp150* to *tp2000*).

In both studies, predictors were standardized (by removing the mean and dividing by 1 standard deviation), prior to fitting the models, and stored in the GIS.

Statistical Models

In recent years, predictive distribution modeling (see Franklin 1995 or Guisan and Zimmermann 2000 for a review) has been broadly applied worldwide to predict the distribution of plants (e.g., Lehmann 1998; Leathwick 1998; Franklin 1998; Guisan et al. 1998, 1999; Zimmermann and Kienast 1999) and animals (e.g., Aspinall 1992; Augustin et al. 1996; Mladenoff and Baker 1999), species and communities. Despite its intrinsic limitations (pseudoequilibrium assumption, no temporal dimension), this approach is considered a powerful and rapid way to assess the possible impact of a climatic change on the distribution and abundance of plant species (e.g., Lischke et al. 1998; Guisan and Theurillat 2000), to test biogeographic

hypotheses (Mourell and Ezcurra 1996; Leathwick 1998), and for studies in conservation biology (see the many examples in this volume).

Generalized linear models (GLMs; McCullagh and Nelder 1989) are particularly appropriate for such predictive modeling, as attested to by the numerous papers published in this field in recent years (see Guisan and Zimmermann 2000, and see several chapters in this volume, including McKenney et al., Chapter 31; Pearce et al., Chapter 32; Jones et al., Chapter 35; Klute et al., Chapter 27; Fertig and Reiners, Chapter 42; and Vernier et al., Chapter 50). GLMs are an extension of classical linear models. In GLMs, the combination of predictor variables xi (i = 1, . . . , p) produces a *linear predictor* LP, which is related to the expectation $E(Y)$ of the response variable Y through a *link function* g(), such as

$$g(E(Y)) = LP = \alpha + X\beta \qquad (1)$$

where X is a matrix of p column vectors $\{x_1, x_2, \ldots, x_p\}$, is the constant term to be estimated, and $= \{\beta_1, \beta_2, \ldots, \beta_p\}$ is the row vector of p coefficients to estimate for the predictor variables. X denotes the matrix product, so that the ith element of g(E(Y)) is

$$\alpha + \beta_1 x_{i1} + \beta_2 x_{i2} + \ldots + \beta_p x_{ip}$$

Unlike classical linear models, which presuppose a normal distribution and an identity link, the distribution of Y may be any of the exponential family distributions and the link function may be any monotonic differentiable function, like logit (i.e., exp(LP)/ [1 + exp(LP)]) for binomial models, logarithm for Poisson models, or unity for Gaussian models.

All models were fitted in S-Plus (Mathsoft). Ordinal models were fitted by using a Proportional Odds (PO) model (lrm function, Harrell 1999; see Guisan and Harrell 2000 for the mathematical rationale and Guisan et al. 1998 for an ecological application). Poisson, Gaussian, and binomial models were fitted by using the glm function and specifying a correct probability distribution and link function. Final models were fitted and evaluated by using custom S-Plus functions (one for each type of model), which additionally implemented the model in the GIS GRID calculator by automatically generating a custom AML (Arc Macro Language; see Guisan et al. 1999).

For two species in the Belalp study area, (1) a GLM with a Poisson distribution and a logarithmic link, and (2) a GLM with a Gaussian distribution and an identity link, were additionally fitted, with the ordinal abundance as the response, in order to compare the predictions from these two alternative models to those from ordinal models. In both study areas, simple presence-absence GLMs with a binomial distribution and a logistic link (see Nicholls 1989; Guisan et al. 1999) were fitted for the same species, in addition to ordinal models, to evaluate the capacity of ordinal models to predict at least presence-absence correctly.

Evaluating the Predictions

Evaluating a model is a critical task in the overall process of model building (see Fielding, Chapter 21). It is fundamental for assessing the quality of the model and the range of situations in which the model can properly be used (called *model applicability* in Guisan and Zimmermann 2000). However, for the same type of response variable, several measures and procedures are often proposed to evaluate a model, as shown by Fielding and Bell (1997), in the case of a binary presence-absence response. The assessment of model quality and applicability primarily depends, however, upon the choice of a measure that will correctly evaluate the model in the specific context of the study's objectives. The focus of this chapter is, however, on comparing models. Hence, I will discuss the accuracy of model predictions in a general context without assuming different weights for omission and commission errors as would be the case if conservation objectives were the focus (see Fielding, Chapter 21, for more information on cost assessments).

Evaluation of model accuracy was carried out differently for the ordinal and binomial models. Because ordinal predictions are on a continuous scale from 0 to 5 (sum of probabilities), they should be transformed into discrete classes in order to compare them to any observations. For this purpose, I used the threshold calibration method described in Guisan et al. (1998), which consists of choosing the threshold providing the best agreement on the training data set and using it to evaluate the model on the evaluation data set. The calibrated threshold is dependent on the measure of agreement used. Thus, several thresholds

are calibrated if several agreement measures are used. Ordinal predictions were evaluated with Goodman and Kruskal's (1954) γ (Gamma) and Somers' (1962) d_{yx}, two measures of agreement for comparing ordinal variables as discussed in Gonzalez and Nelson (1996) and used by Guisan (1997) and Guisan et al. (1998) in the context of static distribution modeling. Ordinal predictions were also reduced to presence-absence in order to compare them with presence-absence predictions made by logistic binomial models, for the same subset of species.

In the case of presence-absence models, Cohen's (1960) Kappa provides an overall measure of accuracy (i.e., it makes full use of the information in the confusion matrix; Fielding and Bell 1997), although it assimilates omission and commission errors and was sometimes considered to be sensitive to prevalence (i.e., unequal group sizes, in this case between presences and absences), although Manel et al. (in press) did not find any evidence of this. A measure that seems to be less sensitive to prevalence is the normalized mutual index (NMI) introduced by Forbes (1995), although it cannot be calculated if any cell of the confusion matrix is zero, for instance, when one of the two possible error rates (omission or commission) is null (which is exactly what one would expect to obtain in this particular modeling context). Both measures test the model predictions for being different from chance performance, but they are also both dependent upon the choice of a threshold to cut the predicted probabilities (on a scale from 0 to 1) into presence/absence. For these reasons, Fielding and Bell (1997) suggest using the threshold-independent receiver operating characteristic (ROC) method for evaluating model predictions. The ROC method consists in plotting sensitivity (i.e., the true positive rate) against [1 – specificity] (i.e., the false positive rate) and eventually integrating the area under the curve, which can take values between 0.5 (1:1 line; agreement no different from that obtained by chance) and 1.0 (perfect agreement). Here, I give the correct classification rate, the sensitivity and specificity error rates, Kappa, and the ROC plot for comparing ordinal and binomial models (see Fig. 25.5).

Results

Ordinal models were successfully adjusted for several species in both study areas. Their ability to predict ordinal species cover density was compared to that of Poisson and Gaussian GLMs respectively (Table 25.2). All three types of model were fitted, for the same species, with the same subset of predictors. As shown by quantile-quantile plots (QQ-Plots) for the Poisson and Gaussian models and for the same two species, residuals are not distributed on the 1:1 line as would be the case if the postulates regarding the distribution of the response variable were respected (Fig. 25.2). For both species, all predictors explained a significant proportion of the deviance and were thus truly selected in each of the compared models. The proportion of deviance explained, adjusted by the number of degrees of freedom, was similar for ordinal and Poisson models (0.54 and 0.59 respectively) but much lower for the Gaussian model (0.22), although these measures are difficult to compare between GLM fitted with different distributions. For the two species given in the example, ordinal models failed to predict the maximum possible class of five on the training data set but were able to predict the maximum possible class on the evaluation data set. In turn, the Poisson model was able to predict the maximum cover density class on the training data set in the case of *Carex curvula*, although it did not predict classes higher than 3 on the evaluation data set. However, the Poisson model predicted over-high values at other locations in the study area (up to 22 for *C. curvula*; see Fig. 25.3 and 25.4), which correspond to impossible density classes (5 already represents 50–100-percent ground cover). The Gaussian model behaved inadequately in both cases, predicting negative values (down to –6) at many locations and being unable either to fit or to predict the maximum values on both the calibration and evaluation data sets (maximum of three predicted in both cases).

Ordinal Versus Presence-Absence Models

The evaluation of ordinal models was based on the γ and d_{yx} statistics. Their ability to predict at least presence/absence, using the K statistics, was assessed by comparing their predictions, recoded into presence/absence, to those of binomial models.

TABLE 25.2.

Comparison of ordinal, Poisson, and Gaussian GLMs for two alpine species in Belalp.

Model[a]	Ordinal	Gaussian	Poisson
Carex curvula			
amt + amt^2	< .0001	< .0001	< .0001
slo	0.0003	0.0010	< .0001
sc97	0.0080	< .0001	0.0024
D2[b]	0.54	0.22	0.59
max-fitted (5)[c,d]	4	2	4
min-fitted (0)[c,d]	0	–1	0
max-pred (4)[d,e]	4	2	3
min-pred (0)[d,e]	0	–1	0
max-pred-area	5	3	22
min-pred-area	0	–5	0
Trifolium alpinum			
amt + amt^2	< .0001	< .0001	< .0001
slo^2	0.0064	0.0001	0.0355
tp500	0.0002	0.0348	0.0009
cir2^2	0.0001	0.0014	< .0001
cir3^2	0.0012	0.0019	< .0001
D2[b]	0.52	0.25	0.52
max-fitted (5) +>[c,d]	4	2	5
min-fitted (0)[c,d]	0	–2	0
max-pred. (5) [d,e]	5	3	7
min-pred (0)[d,e]	0	–1	0
max-pred-area	5	3	7
min-pred-area	0	–8	0

[a]For each model, the p-value of the deviance reduction chi test is first given for each predictor.

[b]*D2* = overall proportion of deviance explained by each model, adjusted by the number of degrees of freedom used.

[c]*max-fitted* and *min-fitted* = highest and lowest values fitted by the model on the training data set.

[d]The maximum possible value (i.e., observed) is given in brackets.

[e]*max-pred* and *min-pred* = highest and lowest values predicted on the evaluation data set.

Note: *Max-pred-area* and *min-pred-area* are respectively the highest and lowest values predicted over the whole study area. *Min-fitted, min-pred* and *min-pred-area* thus indicate whether negative values were fitted or predicted by the model at any **location** in the study area (usually outside the range of the calibration and evaluation data sets). See Figures 25.2 and 25.3 for visualizing the extent of pixels where negative values are predicted by the Gaussian models.

In Table 25.3, the confusion matrices for *Nardus stricta*, using (a) the calibration and (b) the evaluation data set, show an example of ordinal model evaluation. This model provides a maximum value of 0.82

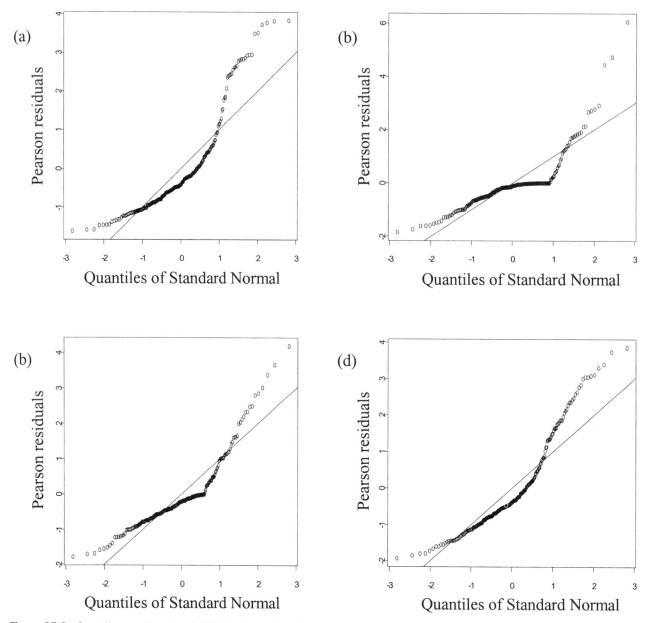

Figure 25.2. Quantile-quantile plots (QQ-Plots) of standard Pearson regression residuals for (a) the Poisson GLM (generalized linear model) of *Carex curvula*, (b) the Gaussian GLM of *Carex curvula*, (c) the Poisson GLM of *Trifolium alpinum*, (d) the Gaussian GLM of *Trifolium alpinum*.

for γ at the calibration (Tables 25.3 and 25.4), which corresponds to cutting the continuous predictions (summed probabilities) at multiple thresholds (pt) of 0.15, 1.15, 2.15, 3.15, and 4.15 (i.e., every successive integer.15, hereafter written [I].15). Recoding the predictions to presence/absence (0 stays 0; [1:5] becomes 1) provides a value of K of 0.61 for the same model. Thresholds at every [I].85 are obtained instead of every [I].15 (pt in Table 25.3) if d_{yx} is optimized in-

stead of γ, corresponding to a maximum value of d_{yx} of 0.691 and an associated K of 0.767 (Table 25.3). Hence, using the threshold providing the best d_{yx} rather than the best γ provided a better prediction of presence/absence for this species, as measured by K, although the reverse was true for other species (Table 25.4).

Overall, the evaluation of ordinal models can be considered fair to good (Table 25.4), as all values of γ

TABLE 25.3.

Comparing predictions to observations in the case of the *Nardus stricta* model. Ordinal confusion matrices and related optimal γ (Gamma) and Somer's d_{yx} agreement measures for ordinal variables. Tables on the left are obtained by using the threshold providing the best γ (see Guisan et al. 1998) and, on the right, the threshold providing the best value for Somer's d_{yx}. Ordinal matrices are further reduced to presence-absence tables, and related K (Kappa) measure of accuracy, to check the ability of ordinal models to predict at least presence-absence correctly.

(a) Calibration (N = 205)

Goodman and Kruskal's γ								Somer's d_{yx}					
	0	*1*	*2*	*3*	*4*	*5*		*0*	*1*	*2*	*3*	*4*	*5*
0	46	1	0	0	1	0	*0*	62	2	1	1	1	0
1	20	2	2	1	0	0	*1*	10	12	3	2	2	0
2	6	12	2	2	4	1	*2*	4	6	3	1	9	3
3	5	5	3	2	8	5	*3*	3	5	6	5	20	19
4	2	5	7	7	30	22	*4*	0	0	1	4	11	9
5	0	0	0	1	0	3	*5*	0	0	0	0	0	0

	0	*1*	pt^a = 0.15						*0*	*1*	pt = 0.85
			$\gamma = 0.819$								$d_{yx} = 0.691$
			$K = 0.611$								$K = 0.767$
0	46	2					*0*	62	5		
1	33	124					*1*	17	121		

(b) Evaluation (N = 92)

Goodman and Kruskal's γ								Somer's d_{yx}					
	0	*1*	*2*	*3*	*4*	*5*		*0*	*1*	*2*	*3*	*4*	*5*
0	13	1	0	0	0	0	*0*	19	2	0	1	1	0
1	7	3	0	1	1	0	*1*	9	6	1	0	3	1
2	9	7	3	0	4	1	*2*	2	5	3	0	1	0
3	1	2	1	3	2	1	*3*	1	3	3	3	10	5
4	2	6	3	0	11	7	*4*	2	4	0	0	3	4
5	1	1	0	0	0	1	*5*	0	0	0	0	0	0

	0	*1*	pt = 0.15						*0*	*1*	pt = 0.85
			$\gamma = 0.644$								$d_{yx} = 0.531$
			$K = 0.432$								$K = 0.544$
0	13	1					*0*	19	4		
1	20	58					*1*	14	55		

^{a}*pt* is the probability threshold.

measured on the independent evaluation data set (held back data) range between 0.64 and 0.94 and between 0.53 and 0.87 for d_{yx} (the latter always provides a value lower than γ when applied on the same confusion matrix; see Gonzalez and Nelson 1996). However, γ and d_{yx} provide different patterns of change when predictions are cut at successive 0.01 probability thresholds (Fig. 25.5). As a result, the maximum value of γ —max(γ)—does not provide the same probability threshold as the one provided by max (d_{yx}). In this respect, results suggest (Table 25.4) that, although optimizing γ at the calibration seems to provide better presence/absence predictions (i.e., K_γ) than does optimizing d_{yx} (i.e., K_D), (for *Cercocarpus*

TABLE 25.4.

Comparison of ordinal and binomial (p/a) models for three species in Belalp (Swiss Alps) and six in the Spring Mountains (Nevada).

Species	Calibration data set[a,b,c]							Evaluation data set[a,b,c]				
	D^2_O	D^2_B	γ	K_γ	d_{yx}	Kd_{yx}	K_B	γ	K_γ	d_{yx}	Kd_{yx}	K_B
Belalp												
Carex curvula	0.542	0.576	0.899	0.592	0.821	0.592	0.769	0.717	0.415	0.561	0.415	0.490
Nardus stricta	0.633	0.575	0.819	0.611	0.691	0.767	0.788	0.644	0.432	0.531	0.544	0.516
Trifolium alinum	0.560	0.437	0.825	0.681	0.686	0.681	0.685	0.716	0.577	0.590	0.504	0.577
Spring Mountains												
Cercocarpus ledifolius	0.542	0.489	0.901	0.649	0.744	0.460	0.658	0.918	0.670	0.707	0.319	0.589
Cleogyne ramosissima	0.694	0.812	0.952	0.889	0.899	0.825	0.903	0.908	0.735	0.790	0.627	0.768
Ephedra viridis	0.658	0.531	0.984	0.699	0.717	0.738	0.739	0.954	0.716	0.701	0.746	0.745
Pinus flexilis	0.535	0.548	0.963	0.684	0.869	0.579	0.699	0.938	0.619	0.851	0.590	0.619
Pinus longaeva	0.727	0.716	0.961	0.788	0.925	0.788	0.877	0.921	0.696	0.864	0.696	0.745
Pinus ponderosa	0.554	0.596	0.944	0.799	0.823	0.496	0.799	0.927	0.640	0.874	0.365	0.703

[a]D^2 = percentage of deviance explained, γ = Gamma, d_{yx} = Somer's d_{yx}, K = Kappa.

[b]For indices : O = ordinal model, B = binomial presence/absence model.

[c]K_γ and Kd_{yx} = Kappa calculated on the ordinal confusion matrix recoded into presence/absence using the threshold providing respectively the optimal γ and the optimal d_{yx}. See main text for details.

ledi-folius, *Coleogyne ramosissima*, *Pinus flexilis*, and *P. ponderosa*) than the opposite situation (for *N. stricta* and *Ephedra viridis* only), an equivalent quality of predictions can also be obtained in a certain number of cases (for *C. curvula*, *T. alpinum*, and *P. longaeva*). At the evaluation, where optimal thresholds are used to assess model accuracy on the independent data set, the latter category includes *C. curvula* and *P. longaeva* but not *T. alpinum* for which a same value of K was actually obtained from two different thresholds (left part of Table 25.4).

Ordinal and binomial models explain, overall, a similar proportion of deviance (D^2) (Table 25.4). However, ordinal models explain a greater proportion of deviance than binomial models in the case of *N. stricta*, *T. alpinum*, *C. ledifolius*, *E. viridis*, and *P. longaeva*, and less in the case of *C. curvula*, *C. ramosissima*, *Pinus flexilis*, and *P. ponderosa*.

The evaluation of the ordinal and binomial models of *C. curvula* was assessed graphically (Fig. 25.5) by displaying the change of six measures of accuracy (*four binary*: correct classification rate, sensitivity, specificity, Kappa, and *two ordinal*: γ and d_{yx}; Figs. 25.5a and 25.5c) obtained by cutting the probability at successive threshold values (using here a 0.01 incre-

ment) between 0 and 1. The ordinal model appears less threshold-sensitive than the binomial model in this case (flatter curves; Fig. 25.5a). Fig. 25.5b and 25.5d show the ROC plots for the ordinal and binomial models respectively. Curves on both plots have similar shapes, although points on the curves for the ordinal model are more concentrated toward higher values of sensitivity and more toward lower values for the binomial model (especially in the case of the evaluation curve). Thus, overall, both models provide similar accuracy when predicting presence/absence (i.e., the area under the curve is visually approximately the same).

Finally, comparing K-values of ordinal models (K_γ and K_D) to K-values of binomial models (K_γ), as well as to the ROC plots, on the evaluation data set shows that (1) ordinal models provide slightly better presence/absence predictions when the threshold is chosen by optimizing γ rather than d_{yx} (K_γ is always higher than K_D, except in the case of *N. stricta* and *E. viridis*), and (2) ordinal models predict presence/absence as well as binomial models (maximum absolute difference between K_γ and K_B of 0.084; see Table 25.4) but are additionally able to predict density cover of plant species satisfactorily.

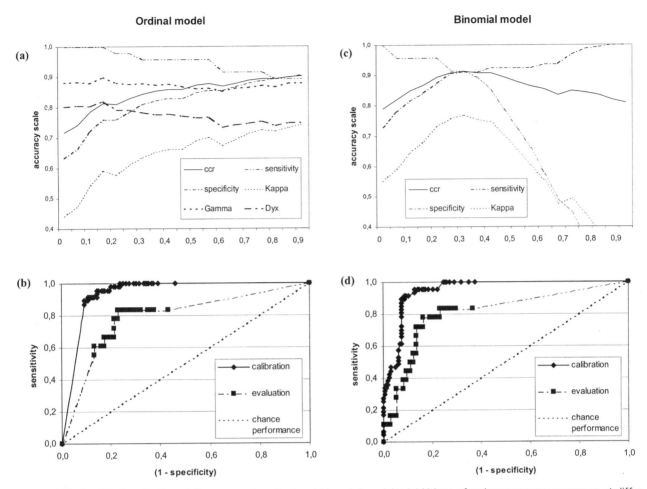

Figure 25.5. Graphical evaluation of *Carex curvula* ordinal and binomial models: (a) Values of various accuracy measures at different probability thresholds (every 0.05 unit between 0 and 1) used for cutting predictions from the training data set into ordinal (γ, d_{yx}) or presence/absence (ccr, sensitivity, specificity, Kappa) values, to compare them to observed values on the same scale; (b) ROC plot for the ordinal model build with ninety-nine thresholds between 0 and 1 (see text); (c) p/a accuracy measures for the binomial model; (d) ROC plot for the binomial model. Prevalence—the proportion of presence compared to the total of observations—is here 0.229. ccr = correct classification rate.

Discussion

My results show that it is worthwhile using appropriate ordinal regression models when the response variable is supposedly ordinal in nature or when a continuous quantitative response variable is sampled in an ordinal fashion (e.g., considering intervals of values). Using a Gaussian model (in our case without log-linearizing the response) proved particularly unsuitable, as this model predicts negative values down to –6, and classes below zero are biologically meaningless. In addition, the Gaussian model was systematically unable to predict the highest class or even the next highest. The Poisson model proved better in this respect as it was able to predict the highest class and cannot predict negative values. The proportion of deviance explained by the Poisson models corresponded closely to that of the ordinal models. However, in a Poisson GLM, the predictions are not constrained to fit the range of possible values for the outcome (although this might perhaps be done), as they are in ordinal (see Guisan et al. 1998; Guisan and Harrell 2000) and binomial models (due to the inverse logit [or probit] link function, which can only predict values in the range 0 to 1; see Guisan et al. 1999). As a result, values of up to 22 were predicted by the Poisson model for *C. curvula*, even though the highest possible ordinal class is 5, corresponding to a ground-cover density of 50 to 100 percent. The procedure for evaluating ordinal models bears the same intrinsic limitations as for evaluating other types of models, as reviewed for presence/absence binary models by Fielding and Bell (1997). Although I used Gamma and Somers' d_{yx} for evaluating ordinal models, other measures are proposed in the literature, such as Kim's d_{xy}, Wilson's d_{yx} or Kendall's tau (see Gonzalez and Nelson 1996 for more details on these measures), which might eventually be more appropriate in other situations. With the exception of Kendall's tau, a threshold needs to be chosen with all these measures of agreement to compare continuous predictions that are sum of probabilities to discrete ordinal observations. As stated by Fielding and Bell (1997), the problem with such threshold-dependent measures of accuracy is their failure to use all the information predicted by the model. Unfortunately, the ROC plot method they recommend for presence/absence models is hardly applicable to ordinal predictions since there are no equivalent measures to *sensitivity* and *specificity* in the case of ordinal confusion matrices.

Hence, I propose using a measure of ordinal agreement (e.g., γ, d_{yx}) conjointly with associated K-values obtained by recoding ordinal data into presence/absence ([0] 0/[1 to 5] 1) at successive probability thresholds and associated ROC plot, because this allows one to check the ability of the ordinal model to predict at least presence/absence satisfactorily, as shown by the models presented in this study. Finally, the implementation of ordinal models into a GIS proved successful and not too complex.

Ordinal response regression models applied to mimic the spatial distribution of plant species can easily be extended to simulate animal distributions when the response is measured on a semiquantitative scale (e.g., visual abundance classes in entomology). Recoding such data into presence/absence, as often seen in literature to fit a logistic regression model, necessarily involves losing ecological information that may be of primary importance in some applications (e.g., conservation studies). In plant ecology, predicting plant assemblages can hardly be made from presence/absence

data alone. Here, simulating the cover density of (at least) each dominant species is necessary to yield realistic predictions of plant communities.

Recommendations for future research in modeling species occurrence include according greater consideration to the probability distribution of response variables and choosing an appropriate statistical model accordingly. This is particularly important if models are to be used in a conservation perspective, for example, for suggesting important areas to protect based on their modeled species pool. GLMs are very effective in this respect, because they allow for consideration of nearly all possible distributions (Guisan and Zimmermann 2000) from qualitative (multinomial) to quantitative (Gaussian, Poisson, negative-binomial, binomial, etc.), with intermediate semiquantitative (i.e., ordinal) and, as a special case (i.e., either quantitative or qualitative), binary distributions (treated as binomial; see Guisan et al. 1999). Finally, more detailed studies on the evaluation of such ordinal models should be encouraged.

Acknowledgments

This study was partly funded by the Swiss National Science Foundation (SNF; ECOCLINE Project), within the framework of the Priority Program for the Environment (PPE), as part of the Climate and Environment in Alpine Region module (CLEAR). My heartfelt thanks go to F. E. Harrell for his constant help on statistical matters and for making his S-Plus functions available to me. Thanks also to R. Gonzalez and L. D. Bacon for providing the S-Plus codes for calculating respectively γ, d_{yx}, and Kappa. A. H. Fielding additionally provided support on model evaluation. S. B. Weiss, A. D. Weiss, and The Nature Conservancy should be acknowledged for making the Spring Mountain and related GIS data available. P. Ehrlich and C. Boggs should also be thanked for providing full CCB support, and J.-P. Theurillat for providing full ECOCLINE project support. The Academic Society of Geneva and the Swiss Center for Faunal Cartography (SCF) generously covered the travel costs to the Symposium.

Patch-based Models to Predict Species Occurrence: Lessons from Salmonid Fishes in Streams

Jason B. Dunham, Bruce E. Rieman, and James T. Peterson

Environmental heterogeneity often produces patchy or discontinuous distributions of organisms. Even broadly distributed species show localized peaks of abundance (Maurer 1999). This is particularly obvious in stream ecosystems, where patch dynamics is a dominant theme (Pringle et al. 1988). Features of the environment that may influence species occurrence in streams are believed to result from a hierarchy of physical processes operating within drainage basins. This idea has formed the basis of several classification schemes for stream habitats (e.g., Frissell et al. 1986; Hawkins et al. 1993; Imhof et al. 1996; Naiman 1998; see Morrison and Hall (Chapter 2) for definition of "habitat"). These classifications provide a useful framework for understanding physical processes that generate stream habitat over areas of varying size and spatial resolution, but they do not explicitly consider how individual species actually perceive or utilize these patchy environments.

To be most useful, patches should be clearly defined by associations between a biological response (e.g., reproduction, migration, feeding) and environmental variability (Addicott et al. 1987; Kotliar and Wiens 1990). Classifications of aquatic habitat based purely on physical characteristics or subdivision of watersheds into arbitrary segments may not adequately describe "patchiness" from an organism's point of view. Lack of attention to realistic scaling of environmental variation and biological responses can produce weak or misleading inferences (Goodwin and Fahrig 1998). Our focus in this chapter is on definition of patches suitable for supporting local breeding populations. This is a key prerequisite for applying ideas from metapopulation and landscape ecology to predicting species occurrence.

Here, we review our attempts to develop patch-based classifications of aquatic habitat and models to predict occurrence of salmonid fishes in streams. We begin with a brief overview of the concept of patchiness. Next, we outline criteria to define the biological response of interest: occurrence of local populations. We then describe models to predict the distribution of local populations within stream basins. These models allow delineation of patches of suitable habitat within watersheds and definition of patch structuring. Patterns of patch structuring and characteristics of individual patches provide the basis for modeling occurrence of local populations. We compare patch-based models of occurrence for two threatened salmonids: bull trout (*Salvelinus confluentus*) and Lahontan cutthroat trout (*Oncorhynchus clarki henshawi*). Finally, we compare our results to alternative approaches to predict occurrence of salmonids and discuss implications of a patch-based approach that should be generally relevant for developing models of species occurrence.

The Concept of Patchiness

The term "patch" has been applied in numerous contexts in ecology (e.g., Pickett and White 1985; McCoy and Bell 1991; Pickett and Rogers 1997; Morrison and Hall, Chapter 2). Our definition of patches parallels the concept of ecological neighborhoods introduced by Addicott et al. (1987). Ecological neighborhoods are defined by a specific biological response and not by an arbitrary temporal or spatial scale or by a perceived boundary or control imposed on the system. A "patch" corresponds to limits or boundaries of environmental conditions that can support a biological response. Patches of environmental conditions potentially suitable to support local populations of a species are often the focus in landscape and metapopulation ecology.

Kotliar and Wiens (1990) provided a general framework for defining patch structure. Given that a biological response is observed to occur within a definable spatial frame, patch structure can be characterized by (1) the degree to which patches can be distinguished from each other and the surrounding environment (patch contrast), and (2) how patches are spatially aggregated. Patch structuring may be characterized by a nested or hierarchical pattern and may vary widely among biological responses.

Patch structuring is not directly synonymous with a specific temporal or spatial scale. For example, patches defined here may vary by an order of magnitude or more in size (patch area). It is definition of common biological responses and environmental criteria for determining patch structure, not spatial or temporal scale per se, that provides a foundation for patch-based models of species occurrence that may be generalized within and among species.

Defining a Biological Response

We were interested in predicting occurrence of fish in patches of habitat suitable for local breeding populations. Patches suitable for local populations should correspond to locations where population growth can be attributed primarily to in situ reproduction, rather than immigration (Addicott et al. 1987). Limited demographic interaction among local populations im-

plies some degree of reproductive (genetic) isolation. Spatial isolation of spawning and rearing habitat for salmonids is reinforced by strong natal homing (Quinn 1993), and patches of suitable habitat may therefore support relatively discrete local populations. Ultimately, it would be desirable to use multiple sources of information to delineate local breeding populations. Several studies have demonstrated the limitations of using limited genetic or demographic information alone to infer population structuring (Ims and Yoccoz 1997; see Utter et al. 1992, 1993 for salmonid examples).

Unfortunately, detailed genetic and demographic data are not available for most systems. For salmonids, we have defined patches of suitable habitat by modeling the distribution limits of smaller, presumably "premigratory" or resident individuals within streams. Larger juvenile and adult salmonids may adopt migratory life histories (Northcote 1997, Fig. 26.1) and range far outside of spawning and rearing areas, but their existence ultimately depends on spawning and rearing habitat. Our delineation of patches for salmonids, then, is based principally on ecological information and an assumption of natal homing. Population genetic analysis (e.g., Kanda

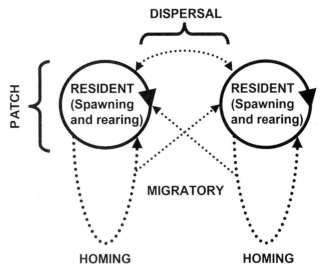

Figure 26.1. Simplified schematic of life-history variation in salmonid fishes. Fish with a "resident" life history spend their entire lives within spawning and rearing areas. Migratory fish use habitats outside of spawning and rearing areas but return faithfully (homing) to breed in natal areas. Some dispersal is possible among both life-history types. Our definition of patches corresponds to the extent of spawning and rearing.

1998; Spruell et al. 1999) for some systems indicates genetic divergence does correspond to juvenile distributions. Our approach to defining patches appears to be a reasonable approximation, but detailed demographic and/or genetic data will be necessary to confirm the structure of any system (Haila et al. 1993; Rieman and Dunham 2000).

Models of Distribution Limits and Patch Delineation

Unlike terrestrial habitats, streams are generally viewed as one-dimensional systems in terms of fish distributions and dispersal. Therefore, boundaries of habitat patches may be delineated in an up- and/or downstream direction. Many factors can potentially limit the distribution of spawning and rearing habitat for salmonids, including natural and artificial dispersal barriers, water temperature, interactions with nonnative salmonids and other fishes, human disturbance, and geomorphic influences. These factors are often not independent. For example, interspecific interactions mediated by water temperature may influence longitudinal distributions of species within streams (De Staso and Rahel 1994; Taniguchi et al. 1998).

In the case of bull trout and cutthroat trout, spawning and early rearing usually occur in upstream or headwater habitats (often fourth-order streams or smaller), so we were particularly interested in factors that determine downstream distribution limits of juveniles. Our two study areas are located at the southern margin of the range for both species, where unsuitably warm summer water temperatures in streams are probably an important factor limiting the amount of suitable habitat (Rieman et al. 1997; J. B. Dunham and B. E. Rieman unpublished data). Local populations of these species in other areas may be delineated by different habitat characteristics, such as availability of high-quality spawning habitat (Baxter et al. 1999; also see Geist and Dauble 1998), barriers (e.g., dams, waterfalls, subsurface flow), and sharp transitions in habitat that occur as tributary streams flow into larger streams or lakes.

There are many recent examples of attempts to classify aquatic habitat for salmonids based on different indicators related to variability in stream tempera-

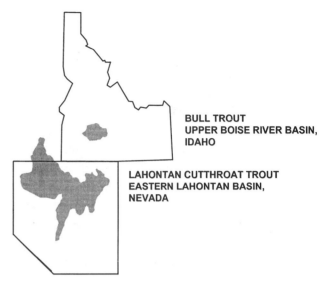

Figure 26.2. Map of study areas: the upper Boise River Basin, Idaho, and the eastern Lahontan Basin, Nevada.

ture. Various researchers have classified thermally suitable habitat from variation in groundwater (Meisner 1990; Nakano et al. 1996), air (Keleher and Rahel 1996), and surface water temperatures (Eaton et al. 1995; Rahel et al. 1996). Our approach is currently based on modeling elevation gradients, which are correlated with temperature (Keleher and Rahel 1996). Our attempts to delineate the amount and distribution of suitable habitat (i.e., patch structure) for salmonids have relied on empirical relationships between downstream distribution limits of juveniles and elevation or geographic gradients (see also Flebbe 1994). Our work has been with Lahontan cutthroat trout in the eastern Lahontan Basin in southeast Oregon and northern Nevada, and bull trout in the upper Boise River Basin in southern Idaho (Fig. 26.2).

Delineation of patches for bull trout in the Boise River Basin relied on information from surveys of juvenile distributions, which suggested a sharp increase in occurrence above an elevation of 1,600 meters (Rieman and McIntyre 1995; Dunham and Rieman 1999). This distribution limit was used to delineate the amount and distribution of suitable habitat patches within the basin. In the case of Lahontan cutthroat trout, a geographic model was necessary to account for changes in the elevation of distribution limits over the eastern Lahontan Basin, which covers a much broader area (Dunham et al. 1999). Geographic location (latitude and longitude) explained over 70 percent of the variation in the elevation of downstream distribution limits for Lahontan cutthroat trout.

Patch delineation involved linking models of downstream distribution limits with a geographic information system (GIS). Predicted downstream distribution limits were used to delineate the size and distribution of watersheds with suitable habitat. We defined patches of suitable habitat as the watershed area upstream of predicted elevations for downstream distribution limits. Defining patches in terms of watershed area is consistent with the view that watershed characteristics have an important influence on stream habitats (Montgomery and Buffington 1998). Local or regional variation in watershed characteristics may have an important influence on the development of stream channels and aquatic habitat (Burt 1992), and patch structuring and patterns of species occurrence may vary accordingly.

An alternative, and perhaps more precise, measure of patch size would be actual length of stream occupied within a watershed. Length of stream occupied requires information on both up- and downstream distribution limits, whereas watershed area requires only information on downstream distribution limits. Stream length might be important where there is strong local variability in climate and geomorphology, or when barriers to fish movement within streams limit upstream distributions. If barriers are important, fish may only be able to occupy a very limited amount of habitat, and patch sizes estimated by stream length and watershed area could differ substantially. Limited evidence suggests the influence of barriers on fish distributions within streams is generally minimal, though important exceptions do exist (e.g., Kruse et al. 1997; Dunham et al. 1999).

Another potentially important localized factor is occurrence of nonnative trout. In the case of Lahontan cutthroat trout, for example, downstream distribution limits were significantly restricted when nonnative trout were present (Dunham et al. 1999). This effect was not consistent or predictable, so we could not simply account for the effect of nonnative trout in defining distribution limits and patch sizes. Earlier models of occurrence of Lahontan cutthroat trout did not detect an effect of nonnative trout (Dunham et al.

1997), but this study did not provide clear definition of patch structure.

Because localized factors within streams (e.g., geomorphic features, nonnative fish) may place constraints on the amount of habitat that can be occupied, patch areas may not reflect the "effective" size of habitat available to fish. To remedy this potential problem, we examine model interaction terms and prediction errors for streams with and without known constraints. The alternative is to directly map local features of stream habitats across large areas to delineate patches, which is often difficult to justify with limited resources.

Modeling Species Occurrence

Delineation of patches and patch structuring within drainage networks provides a template for predicting species occurrence. Our approach is essentially a two-tiered model: (1) define distribution limits in terms of geographic or elevation gradients to delineate suitable habitat patches, (2) predict occurrence of fish within patches. To predict species occurrence, we have focused on influences of the geometry of patches within landscapes, namely the size, isolation, and spatial distribution of patches (see Rieman and Dunham 2000). Here, we focus on patch size. Patch size may be related to fish occurrence because habitats in larger patches may be more complex and resilient to disturbance and should generally support larger populations.

Rieman and McIntyre (1995) used multiple logistic regression to model occurrence of bull trout in the Boise River Basin and found patch area to be the strongest predictor. Other significant factors included patch isolation and road density within patches (Dunham and Rieman 1999). Solar radiation and occurrence of nonnative brook trout were not associated with occurrence, and patterns of occurrence were not spatially aggregated (Dunham and Rieman 1999). Logistic regression of occurrence of Lahontan cutthroat trout in relation to patch area revealed a highly significant ($P < 0.0001$) and positive relationship (J. B. Dunham unpublished data).

Although analyses of data for Lahontan cutthroat trout are preliminary, some interesting common themes are suggested by the results. For both species,

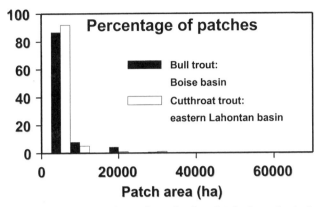

Figure 26.3. Comparison of patch size distributions for bull trout (*Salvelinus confluentus*) in the upper Boise River Basin and Lahontan cutthroat trout (*Oncorhynchus clarki henshawi*) in the eastern Lahontan Basin.

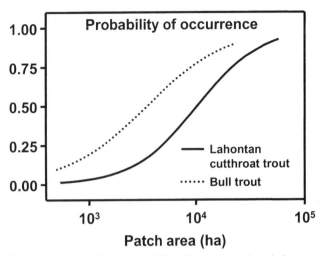

Figure 26.4. Predicted probability of occurrence in relation to patch size (area) for bull trout (*Salvelinus confluentus*) in the upper Boise River Basin and Lahontan cutthroat trout (*Oncorhynchus clarki henshawi*) in the eastern Lahontan Basin.

patch area appears to be a significant correlate of species occurrence. This is a common pattern emerging for many species in both terrestrial and aquatic ecosystems (Bender et al. 1998; Moilanen and Hanski 1998; Magnuson et al. 1998; Hanski 1999) and a general prediction from island biogeography and metapopulation theory (Hanski and Simberloff 1997). More interesting are the details of the relationship between patch size and species occurrence.

First, an examination of patch size distributions reveals that size distributions are remarkably similar for both species and skewed toward very small patches (Fig. 26.3). This means that very few patches are likely to have a high probability of occurrence and that a few large patches may be very important for both species. In terms of total area of potential habitat occupied, bull trout in the Boise River Basin occupy relatively more (46 percent) than Lahontan cutthroat trout in the eastern Lahontan Basin (36 percent). When the actual responses of both species to changes in patch size are compared (Fig. 26.4), it is clear that both species are likely to occur when patch sizes exceed about 10^4 ha in area. Relative to Lahontan cutthroat trout, bull trout are more likely to occur in smaller patches (Fig. 26.4), which may explain why bull trout occupy a larger percentage of suitable habitat overall.

Because a common definition for patches was used for both species, we were able to compare specific responses of each to variability in patch size. In the future, analysis of occurrence in relation to other char-

acteristics of patches may reveal additional insights. Although the biology of these two species differs in important ways (Rieman and Dunham 2000), these results provide general themes to guide efforts to conserve and manage these species, along with important details relevant to particular species or environments they inhabit.

Evaluating Model Prediction

The most relevant measure of a classifier is its expected error rate (EER, Lachenbruch 1975). Among possible EER estimators, leave-one-out cross-validation is a nearly unbiased estimator of out-of-sample model performance (Fukunaga and Kessel 1971) that provides a measure of overall predictive ability without excessive variance (Efron 1983). Leave-one-out cross-validation involved removal of an individual observation from the data set, fitting a model with the remaining observations, and predicting the omitted observation. Model probabilities of occurrence greater than or equal to 0.50 were classified as predicted occurrences. Fits between observed and predicted occurrences were summarized as classification error rates, summarized over all observations and by response.

Because the results for Lahontan cutthroat trout were preliminary, we focused on classification (omission) and prediction (commission) errors for the model of bull trout occurrence reported by Dunham and Rieman (1999). Based on the overall classification error rate (Table 26.1), our logistic regression model was fairly accurate with a 19.7 percent error rate. This suggests a good fit between the data and our logistic regression model, but this simple measure of predictive ability does not reveal insights into potential bias or sources of error. When modeling species occurrences, biological responses are usually approximated assuming some predefined statistical distribution. For example, logistic regression assumes a binomial response distribution and a logit link. Hence, model accuracy is likely a function of how faithfully the distribution approximates the biological response. To illustrate, we fit the bull trout occurrence data to a k-nearest-neighbor (KNN) model, a relatively flexible nonparametric classification technique that does not require distribution assumptions or a strong assumption implicit in specifying a link function (Hand 1982). The overall KNN error rate was 17.3 percent, only slightly lower than the logistic regression model, which suggested a reasonable fit of the data to the assumed binomial distribution. Category-wise (i.e., response-specific) classification and prediction error rates can also be influenced by response-specific sample size (Agresti 1990). In general, error rates are higher for less-frequent responses. For instance, bull trout occurrence had the lowest sample size and the highest prediction and classification error rates for both the logistic regression and KNN models (Table 26.1). Additionally, logit model error rates are influenced by the choice of the baseline category (e.g., modeling presence or absence, Agresti 1990).

Choice of statistical model may be important, but

error in determination of occurrence can also bias model predictions. For example, bull trout in smaller patches may also occur at lower densities, which may affect probability of detecting fish. Another important point is that predictions from the model are not precise. Lower and upper confidence intervals for slope estimates of the patch area effect range from 48 to 61 percent of the point estimate (see Dunham and Rieman 1999). We do not have enough confidence in the precision of our models to believe that model "accuracy" can be reasonably assessed by analysis of errors of omission or commission alone. In other words, there is a need for both statistical and biological "validation" of models. We suspect similar limitations apply to many other models of species occurrence.

Implications for Models of Species Occurrence

The efficacy of a patch-based approach depends on how clearly patches can be defined. Even if the definition of patch structuring is clear, it may not be realistic to treat patches as independent of the landscape in which they are embedded. The degree to which the landscape "matrix" should be considered in models of species occurrence will depend on patch contrast, aggregation, and scale of study (Kotliar and Wiens 1990; Wiens 1996a,b), as well as life history characteristics of the species in question. For example, many salmonids have complex migratory behaviors, and interactions between migratory behavior and habitat outside of spawning and rearing areas (e.g., Fig. 26.1) may affect species occurrence within patches (Rieman and Dunham 2000).

A patch-based approach to modeling patterns of species occurrence has several important advantages. In the simplest sense, a patch-based approach permits stratification of models of species occurrence. In purely statistical terms, stratification may be a useful tactic to increase the precision of model predictions. Lack of consideration or knowledge of patch structure can produce mismatched inferences between patterns of occurrence and habitat characteristics. Many common approaches to subdividing landscapes into pixels, polygons, political boundaries, and so forth

TABLE 26.1.

Summary of leave-one-out cross-validation classification and prediction[a] error rates[b] for bull trout patch occupancy models.

Patch status	n	Logistic regression	K[c]-nearest neighbor
Occupied	29	0.276 (0.276)	0.207 (0.258)
Unoccupied	52	0.154 (0.154)	0.154 (0.120)

[a]Prediction in parenthesis.
[b]Omission and commission errors, respectively.
[c]k = 12 nearest neighbors.

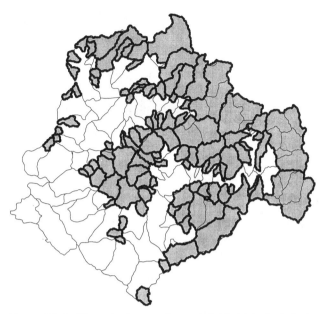

Figure 26.5. Watersheds are often subdivided into hydrologic units (HUs) for classification and analysis of aquatic habitats (Maxwell et al. 1995). This overlay of sixth-field HUs (thin lines) and patches (heavy lines with shading) for bull trout (*Salvelinus confluentus*) in the Boise River Basin shows that patch and HU watershed boundaries can be substantially different.

may not adequately reflect patch structuring (e.g., Fig. 26.5).

Patch Structure and Scale

Patch structure can have important implications for finer-scale models of occurrence. Often, fish-habitat relationships for salmonids are considered within relatively small sites ranging from individual pools and riffles to stream segments (10^0–10^2-meter) (Fausch et al. 1988; Angermeier et al., Chapter 46). At these spatial scales, occurrence of fish among sites is probably not independent, because they are nested within a larger area supporting a population that is influenced by larger-scale environmental variation and fish movement (Schlosser 1995; Gowan and Fausch 1996b).

At finer scales (e.g., stream segments within patches), the existence of suitable but unoccupied habitat is a possibility, especially when patch dynamics are characterized by extinction and/or (re)colonization (Rieman and Dunham 2000). This implies that absence of fish at sites may not be a function of site-

scale habitat quality, but rather a characteristic of a larger patch (e.g., size, isolation), within which sites are nested. As hypothesized by Kotliar and Wiens (1990), larger-scale environmental variation may place important constraints on patterns nested at smaller scales. Nonspatial models of species occurrence implicitly assume organisms are free to select all habitats, but this is not true if external constraints (e.g., spatial isolation, dispersal barriers) or internal constraints (e.g., homing, philopatry) are important (see Rosenberg and McKelvey 1999 for a recent example).

At larger ecological scales, patches supporting local populations may be aggregated within landscapes, perhaps forming metapopulations or "semi-independent networks" (Hanski 1999; Rieman and Dunham 2000). For salmonids in streams, patterns of hydrologic connectivity may produce spatially aggregated clusters or "networks" of patches. Whether defined as metapopulations or otherwise, aggregates of patches supporting local populations can be characterized by a number of larger-scale characteristics, such as number of patches, size distribution, isolation, land form, land type, or climatic associations. Models linking species occurrence to aggregate characteristics may be considered in terms of occurrence of a single or multiple local populations.

Management Applications

One of the strongest motivations for better predictive models of species occurrence is the need for better data to support conservation planning for threatened and endangered species (the two species described here are listed as threatened under the U.S. Endangered Species Act). Key steps in conservation planning include (1) delineation of units for conservation, (2) risk or status assessment of the units, and (3) prioritization of management and species recovery actions. Delineation of biotic units and selection of appropriate biological responses and/or criteria have been the focus of much debate, especially for salmonid fishes (e.g., Nielsen 1995; see also Paetkau 1999). Patch-based models provide a useful context for integrating different elements of biological diversity (e.g., compositional, structural, functional; Franklin 1988) that should be considered in delineation of conservation units at different scales (Noss 1990b). In terms of risk or status assessments, patch-based models provide managers with a useful description of the amount and distribution of suitable habitat (e.g., patch size, isolation). Predictions from models of occurrence can provide important information to address fundamental questions about habitat conservation (e.g., How many habitats? How large? Where to focus?). Other patch attributes (e.g., occurrence of other species, land cover, ownership, climatic data, etc.) can also be added to prioritize species recovery actions.

Patch-based models have provided useful insights for a wide variety of species (see Hanski 1999). Ignoring potential patch structure or applying overly simplistic or unrealistic habitat classifications may result in models with little biological relevance or poor predictive power. The past several years have seen a dramatic increase in our capacity to generate predictor variables (e.g., via GIS, online databases, etc.) and better analytical models to predict species occurrence. As our ability in these important areas increases, however, we should not lose sight of the biological responses we wish to understand *and* predict.

Acknowledgments

Thanks to Debby Myers for GIS support. The Biological Resources Research Center, University of Nevada, Reno, provided additional support. Constructive reviews by Paul Angermeier, Erica Fleishman, Kirk Krueger, and three anonymous reviewers improved earlier versions of the manuscript.

Autologistic Regression Modeling of American Woodcock Habitat Use with Spatially Dependent Data

David S. Klute, Matthew J. Lovallo, and Walter M. Tzilkowski

Autocorrelation is a frequently observed characteristic of spatial variables but is rarely considered in models of wildlife-habitat relationships. Random variables are said to be spatially autocorrelated if neighboring points are more (positive autocorrelation) or less (negative autocorrelation) similar than would be expected for random groups of observations. Thus, the value of a spatially autocorrelated random variable can be in part predicted by values of the variable at neighboring locations (Legendre 1993). The structuring of environmental elements in ecologically functional forms often results in positive spatial autocorrelation, and many ecological principles and theories (e.g., competition, population growth, species-habitat relationships) rely on an assumption of spatial dependencies among interacting elements (Legendre and Fortin 1989). Therefore, explicit consideration of spatial dependencies is fundamental to a complete understanding of ecological processes.

Ecologists frequently use classical statistical techniques (e.g., *t*-tests, linear and logistic regression) that assume independence among data points (Rossi et al. 1992). Spatial autocorrelation violates the assumption of independence and can result in incorrect inference when using classical statistical techniques. Positive spatial autocorrelation results in overestimating the effects of ecological covariates in descriptive or predictive models and in declaring too often that significant

differences exists when in fact they do not (Cliff and Ord 1981; Augustin et al. 1996). Correct inference can be drawn by extracting spatial dependencies and then analyzing residuals or by explicitly modeling spatial autocorrelation (Legendre and Fortin 1989).

Researchers have used a variety of techniques to develop species-habitat relationships models using spatial data. Logistic regression (Pereira and Itami 1991; Mladenoff et al. 1995), Mahalanobis distance (Clark et al. 1993b), log-linear (Homer et al. 1993), and Bayesian techniques (Milne et al. 1989; Aspinall and Veitch 1993) have been used to construct models of species' presence or abundance as a function of ecological covariates. However, none of these studies explicitly examined or accounted for spatial dependencies in response or predictor variables. The extensive use and increased availability of spatial databases and geographic information systems (GIS) warrants increased awareness and application of techniques for dealing with spatial dependencies.

Ecologists are often interested in modeling binary responses (e.g., present or absent, successful or unsuccessful). Logistic regression models the log-odds (logit) of the response variable as a linear function of predictor variables and is well suited to binary response data. However, logistic regression requires independence of observations and is not appropriate for use with spatially autocorrelated data. An appropriate method for modeling binary responses in the presence of spatial dependencies is autologistic regression. The autologistic model incorporates spatial autocorrelation by conditioning the response for a given location on the values of the response at neighboring locations (Gumpertz et al. 1997). As with standard logistic regression, autologistic models can also use relevant covariates as predictors. By explicitly accounting for spatial dependencies, more parsimonious models that provide a better indication of the importance of predictor variables in influencing the distribution of the response variable are expected (Augustin et al. 1996). Autologistic regression is a flexible technique and has been used in ecological applications to model the spatial distribution of bark-beetles (Preisler 1993), disease in bell peppers (Gumpertz et al. 1997), habitat suitability for red deer (*Cervus elaphus*) (Augustin et al. 1996) and the distribution of plant species (Huffer and Wu 1998).

Systematic sampling techniques are often used to assess the presence or abundance of wildlife species. It is expected that response and predictor variables from systematic surveys will frequently be spatially autocorrelated. American woodcock (*Scolopax minor*) are a highly secretive species, although courtship displays by males can be easily observed. However, observation of courtship activity is temporally limited both daily and seasonally. Because of these limitations, systematic roadside surveys (i.e., singing-ground surveys) consisting of ten observation points (i.e., "stops") are used to detect woodcock habitat use and monitor populations (Bruggink 1997). Because of their systematic structure, singing-ground survey results and associated habitat variables are expected to be highly spatially dependent. The spatial distribution of woodcock on singing-ground survey routes may be due to the effects of habitat variables or, alternatively, may be dependent on the response status of neighboring locations. Because singing-ground survey stops within routes are not spatially independent, construction of habitat relationships models must occur at the route level or spatial dependencies must be explicitly modeled at the stop level. Autologistic regression provides a technique that explicitly accounts for the spatial dependencies resulting from route-based survey techniques while allowing investigation of relationships between woodcock presence and associated habitat variables.

We investigated the relative performance of logistic and autologistic regression for modeling habitat relationships of American woodcock based on singing-ground surveys. Our objectives were to (1) quantify the degree of autocorrelation in response and predictor variables from singing-ground surveys and remotely sensed habitat data, (2) determine the effectiveness of logistic regression and autologistic regression for modeling spatial dependencies and habitat relationships, and (3) compare the reclassification performance of logistic regression and autologistic regression models to determine their potential usefulness as predictive models of woodcock habitat suitability. Our hypothesis was models that explicitly incorporated spatial dependencies would provide a more appropriate view of woodcock-habitat relationships and provide better predictive power than could be obtained by using methodologies that did not explicitly incorporate spatial dependencies.

Study Area and Methods

Singing-ground surveys were used to detect presence of singing male woodcock (Tautin et al. 1983). Routes were randomly selected from an existing road database using the ArcInfo GIS software (Environmental Systems Research Institute, Redlands, Calif.). In 1996, sixty-seven singing-ground survey routes were surveyed, primarily in the Ridge and Valley Province of Pennsylvania (Cuff et al. 1989). Sixty-one of these routes were surveyed twice and six were surveyed once. In 1997, twenty-nine different singing-ground survey routes were surveyed. All twenty-nine of these routes were surveyed only once. Starting points of routes were located at easily identifiable points (e.g., road intersection). Ten survey stops were located along each route at 0.64-kilometer intervals.

All singing-ground surveys were conducted by personnel from the Pennsylvania Cooperative Fish and Wildlife Research Unit, Pennsylvania Game Commission, or by experienced volunteer observers between 15 April and 5 May. Surveys began twenty-two minutes after local sunset if the sky was less than or equal to 75 percent overcast and fifteen minutes after local sunset if the sky was more than 75 percent overcast. Surveys were not conducted if the temperature was less than 4.4 degrees Celsius, if wind speeds were greater than 15 kilometers per hour, or if it was raining. Observers listened for two minutes and recorded numbers of singing males heard at each stop. If one or more woodcock was detected, the stop was classified "present"; otherwise, the stop was classified "absent."

Habitat Variables

Habitat variables associated with each singing-ground survey stop were measured within 350-meter radius circular buffers centered on each stop. We chose a buffer of 350 meters because it represents the approximate maximum detection distance for singing male woodcock (Duke 1966). Buffers were constructed using ArcInfo software. We used classified thematic mapper imagery (30×30-meter resolution) developed for the Pennsylvania Gap Analysis project to identify coarse-scale habitat elements associated with the singing-ground survey stops. The imagery was classified to eight habitat types (water, coniferous forest, mixed forest, broadleaf forest, early successional, perennial herbaceous, annual herbaceous, terrestrial unvegetated) using an unsupervised classification method (W. L. Myers, School of Forest Resource, Pennsylvania State University personal communication). We used the FRAGSTATS spatial pattern analysis program to measure five landscape-level and eighteen class-level habitat variables. Landscape-level metrics described the spatial pattern of the entire landscape by considering all habitat types simultaneously. Class-level metrics described the spatial pattern within a landscape of a single habitat type (McGarigal and Marks 1995).

Spatial Autocorrelation and Habitat Modeling

We initially constructed a logistic regression model following the protocol and terminology of Hosmer and Lemeshow (1989). We fit univariate logistic regression models and examined likelihood ratio tests (G tests) for each variable (PROC LOGISTIC; SAS Institute 1985). Variables with P less than 0.25 for G tests were retained for further analysis (Bendel and Afifi 1977; Mickey and Greenland 1989). Variables retained from univariate analyses were subjected to correlation analysis (PROC CORR; SAS Institute 1990a,b) to detect collinearity. Pairs of variables with

r greater than 0.50 were considered for elimination. The decision of which variable to eliminate from each correlated pair was based on significance of the univariate G tests; the more significant variable was retained. After correlation analysis, a multivariate logistic regression model was fit using the remaining variables. Variables with P less than 0.25 (Wald chi-square) were eliminated from the initial multivariate model. All two-way interactions among remaining variables were individually considered. Variables were mean centered before construction of interaction terms to reduce collinearity between main and interaction effects (Neter et al. 1996). Interaction terms with P less than 0.25 from G tests were retained for the final model.

We constructed correlograms of Moran's I to investigate the degree of spatial autocorrelation in response and predictor variables at various scales of influence (Moran 1950; Cliff and Ord 1981). We used the S-Plus spatial statistics module (MathSoft, Seattle, Wash.) to calculate Moran's I for the binary response (present or absent) and main effect terms from the logistic regression model (Kaluzny et al. 1996). Moran's I was calculated in ten distance categories: 0–700, 700–1,400, . . . , 6,300–7,000 meters. We used 700-meter intervals because it was the approximate distance between adjacent stops. We calculated one thousand random permutations of the data to determine if spatial autocorrelation for variables was significantly different from zero (P less than 0.05) in all distance categories.

We used the logistic regression model as a starting point for fitting autologistic models. Logistic regression modeled the logit of probability of presence of woodcock (p_i) as a linear function of predictor variables:

$$\text{logit}(p_i) = \ln\left(\frac{p_i}{1-p_i}\right) = \beta_0 + \beta_1 X_{1i} + \ldots + \beta_n X_{ni}.$$

Where p_i was the probability of presence for American woodcock at the ith stop, β_0 was the model intercept, $\beta_1, \ldots, \beta_n$ were parameter estimates, and $X_{1i}, \ldots, X_{ni}$ were values of predictor variables at the ith stop. Autologistic regression was similar, with an additional term (called the autocovariate) added to account for spatial dependencies in the data by conditioning the probability of presence (p_i) on neighboring stops:

$$\text{logit}(p_i) = \beta_0 + \beta_1 X_{1i} + \ldots + \beta_n X_{ni} + \beta_m auto\,\mathrm{cov}_i$$

where

$$auto\,\mathrm{cov}_i = \frac{\sum_{j=1}^{k_i} w_{ij} y_j}{\sum_{j=1}^{k_i} w_{ij}}$$

was the value of the autocovariate for stop i and β_m was the parameter estimate for the autocovariate. If woodcock were present at stop j then $y_j = 1$, otherwise $y_j = 0$. The autocovariate in our models represented the weighted average of the number of singing-ground survey stops with woodcock present among a set (i.e., clique) of k_i neighbors of stop i. The weight

$$w_{ij} = \frac{1}{h_{ij}}$$

given to stop j was the inverse of the Euclidean distance (h_{ij}) between stops i and j (Augustin et al. 1996). Autologistic models were fit using PROC LOGISTIC (SAS Institute 1990).

Because we could not determine a priori what clique size was most appropriate, we constructed multiple cliques to investigate the effect of size of neighborhood on the autologistic model. We used the Akaike Information Criterion (AIC) to identify the clique size that produced the most parsimonious autologistic model. We defined the first order clique as all neighbors within 700 meters of stop i. Additional cliques were constructed by adding neighbors at 700-meter intervals (second-order clique = all stops within 1,400 meters, third-order clique = all stops within 2,100 meters, etc.) until the most parsimonious model was identified. Cliques were identified using the S-Plus spatial statistics module (Kaluzny et al. 1996).

To investigate remaining spatial autocorrelation after model fitting, we constructed correlograms of Moran's I for Pearson residuals from logistic and autologistic regression models. Autocorrelation analysis of residuals was conducted using the same methods used for response and predictor variables. We assessed differences

in modeling efficiency by examining standard reclassification statistics. We calculated the cutoff probability for reclassification by determining the proportion of the total stops with woodcock present (cutoff probability = stops with woodcock/total stops). We defined sensitivity as the percentage of absent-stops (i.e., no woodcocks observed) correctly classified, specificity as the percentage of present-stops (i.e., one or more woodcocks observed) correctly classified, and overall correct as the percentage of all stops correctly classified. Errors of commission measured the percentage of absent-stops incorrectly classified (100-sensitivity) and errors of omission measured the percentage of present-stop incorrectly classified (100-specificity).

Results

Woodcock were observed at 103 of the 960 total singing-ground survey stops. Woodcock observations tended to be clustered. No woodcock were observed on fifty-four singing-ground survey routes, and all present-stops occurred on the remaining forty-two routes. On routes where woodcock were observed, the average number of present-stops was 2.45.

Preliminary screening of univariate logistic regression models indicated that twelve habitat variables were potentially important predictors of woodcock presence (P less than 0.25, Table 27.1). Through correlation analysis and multivariate model fitting, we converged on a final logistic regression model consisting of four main-effect terms and two 2-way interactions (Table 27.2). All terms in the final logistic regression model were class-level percent cover variables and the model was significant at P less than 0.001.

Moran's I correlograms indicated significant spatial autocorrelation for response and predictor variables from the logistic regression model (Fig. 27.1). Woodcock presence was most strongly autocorrelated at the smallest distance categories (Fig. 27.1A). The percent cover of water was significantly autocorrelated at the four smallest distance categories (Fig. 27.1E); however, other habitat variables were significantly autocorrelated over a larger range of distances (Fig. 27.1B–D).

We fit autologistic regression models with cliques of neighbors less than or equal to 700 meters

TABLE 27.1.

Means, standard errors (SE), and *P*-values for *G*-tests from univariate logistic regression models for habitat variables associated with American woodcock (*Scolopax minor*) singing-ground survey stops in Pennsylvania, 1996–1997.

	Present-stops[a]		Absent-stops[b]		
Habitat variable	Mean	SE	Mean	SE	P
Landscape-level indices					
Number of patches	38.44	1.13	38.73	0.52	0.850
Mean patch size (ha)	1.12	0.04	1.27	0.03	0.066
Edge (km)	8.51	0.20	8.48	0.85	0.918
Shannon diversity	1.27	0.03	1.25	0.01	0.545
Interspersion/juxtaposition	65.13	0.94	63.66	0.39	0.196
Class-level indices					
Percent cover					
Water	3.32	1.08	0.79	0.12	<0.001
Coniferous forest	6.82	1.01	7.55	0.36	0.501
Mixed forest	10.47	1.28	11.00	0.44	0.685
Broadleaf forest	40.71	2.12	39.70	0.81	0.682
Early successional	14.06	1.05	10.16	0.33	<0.001
Perennial herbaceous	15.17	1.44	16.50	0.53	0.405
Annual herbaceous	8.19	1.16	12.27	0.55	0.008
Terrestrial unvegetated	1.27	0.26	1.92	0.14	0.071
Number of patches					
Water	0.72	0.15	0.63	0.06	0.632
Broadleaf forest	6.26	0.34	5.72	0.16	0.127
Early successional	10.50	0.54	9.67	0.21	0.191
Terrestrial unvegetated	1.35	0.20	2.04	0.09	0.008
Edge (km)					
Broadleaf forest	4.91	0.16	4.46	0.07	0.022
Early successional	3.59	0.21	2.83	0.70	<0.001
Mean patch size (ha)					
Water	1.09	0.40	0.21	0.04	<0.001
Broadleaf forest	5.15	0.69	5.73	0.30	0.506
Early successional	0.89	0.07	0.47	0.05	0.447
Terrestrial unvegetated	0.19	0.06	0.19	0.01	0.966

[a]Number of woodcock observed > 0, *n* = 103.

[b]Number of woodcock observed = 0, *n* = 857.

(first order), less than or equal to 1,400 meters (second order), less than or equal to 2,100 meters (third order), and less than or equal to 2,800 meters (fourth order). The third-order autologistic regression model provided the most parsimonious fit based on AIC values. The relative magnitudes of parameter estimates for habitat variables in the autologistic model were consistently less than in the logistic regression model (Table 27.2).

Moran's *I* correlograms indicated significant autocorrelation of Pearson residuals for small-distance categories after fitting the logistic regression model (Fig.

TABLE 27.2.

Parameter estimates from logistic regression and autologistic regression models for predicting American woodcock (*Scolopax minor*) presence from habitat variables associated with singing-ground survey stops in Pennsylvania, 1996–1997.

	Parameter estimates							
Model	Intercept	Water[a]	Early succ.[a]	Ann. herb.[a]	Terr. Unveg.[a]	Water x Early succ.[b]	Ann. herb. x Terr. unveg.[b]	Autocov.[c]
Logistic regression	–2.134	0.043	0.033	–0.022	–0.113	0.006	–0.016	—
Autologistic regression	–3.073	0.003	0.029	–0.014	–0.083	0.005	–0.016	4.679

[a]Main effect terms: Water = Water (%), Early succ. = Early successional (%), Ann. herb. = Annual herbaceous (%), Terr. unveg. = Terrestrial unvegetated (%).

[b]Two-way interactions of mean-centered main effect terms.

[c]Autocovariate term.

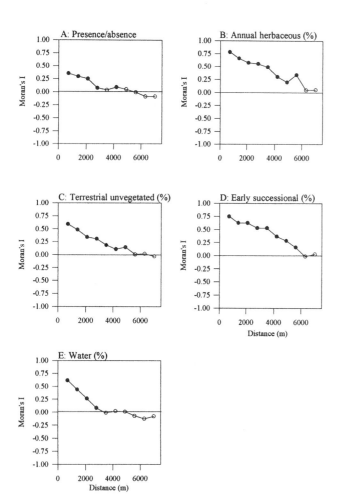

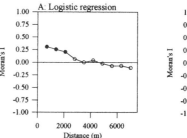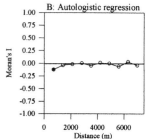

Figure 27.2. Moran's *I* correlograms of Pearson residuals from logistic regression and autologistic regression models for predicting American woodcock (*Scolopax minor*) presence from habitat variables associated with singing-ground survey stops in Pennsylvania, 1996–1997. Filled circles indicate *I* values significantly different from zero ($P < 0.05$).

Figure 27.1. Moran's *I* correlograms of response and predictor habitat variables used in the logistic regression and autologistic regression models for predicting American woodcock (*Scolopax minor*) presence from habitat variables associated with singing-ground survey stops in Pennsylvania, 1996–1997. Filled circles indicate *I* values significantly different from zero ($P < 0.05$).

27.2A). The third-order autologistic regression model effectively accounted for the majority of spatial autocorrelation and completely removed all significant positive spatial autocorrelation (Fig. 27.2B).

Reclassification statistics performed better for autologistic models than for the logistic regression model. Model sensitivity, specificity, and percent correct were 63.1, 56.3, and 62.4 percent, respectively, for the logistic regression model. Errors of commission and omission were 36.9 and 43.7 percent, respectively, for the logistic regression model. Model sensitivity, specificity, and percent correct were 83.4, 66.0, and 81.6 percent, respectively, for the autologistic regression model. Errors of commission and omission were 16.6 and 34.0 percent, respectively, for the autologistic regression model.

Discussion

Models of wildlife-habitat relationships may be affected by spatial autocorrelation in two ways. First, neighboring locations may tend to exhibit similar environmental conditions due to spatial proximity resulting in positive autocorrelation of predictor variables. Second and independent of environmental conditions, the occurrence of the species of interest may not be independent of the occurrence of conspecifics at neighboring locations due to behavioral processes resulting in autocorrelation of the response variable (Augustin et al. 1996). Scale of measurement, species, and behavioral processes will influence whether response variables will exhibit positive autocorrelation (e.g., flocking, lekking) or negative autocorrelation (e.g., territoriality, competition). In both situations, model residuals are expected to exhibit significant spatial autocorrelation if spatial dependencies are not effectively modeled. Therefore, inference based on inappropriate statistical analyses may lead to inaccurate conclusions about species-habitat relationships.

Our analyses indicated significant positive spatial autocorrelation of both response and predictor variables. Due to the systematic nature of singing-ground surveys, these patterns of autocorrelation were expected. If woodcock occurred on a singing-ground survey route, they were commonly detected at more than one stop. Stops on singing-ground survey routes were spaced 0.64 kilometers apart, and thus duplicate counting of individuals at adjacent stops should have been minimal (Tautin et al. 1983). Zero counts, which are common on singing-ground survey routes, and clustering of present-stops within routes produced significant positive spatial autocorrelation. The correlogram of presence/absence indicated significant but relatively weak autocorrelation within small distance categories. Other researchers have noted clumped distributions of woodcock on singing-ground survey routes (Tautin 1982). Clustering of woodcock on singing-ground survey routes likely resulted from neighboring sites with similar habitat conditions. However, clustering of woodcock may have also resulted from behavioral processes related to courtship behavior (i.e., lekking, male-dominance polygyny)

(Hirons and Owen 1982; Oring 1982; Dwyer et al. 1988; Ellingwood et al. 1993).

Habitat covariates in our logistic regression model also exhibited positive spatial autocorrelation. Autocorrelation of water was initially strong but declined quickly. Water elements along singing-ground survey routes were limited primarily to streams; therefore, only short-range positive autocorrelation was expected due to the spatially limited nature of streams. Autocorrelation of the percentage of annual herbaceous, terrestrial unvegetated, and early successional cover persisted across longer distances. We used relatively coarse-grained habitat information (30×30-meter pixels), and this limited our ability to detect fine-scale changes along singing-ground survey routes. Fine-scale habitat characteristics may change significantly within a region although coarse-scale habitat variables remain relatively consistent.

In some situations, ecological covariates alone may be sufficient for eliminating spatial dependencies in model residuals (Gumpertz et al. 1997). However, the Pearson residuals from our logistic regression model exhibited significant positive spatial autocorrelation at small distances indicating logistic regression was ineffective for completely accounting for autocorrelation of response and predictor variables. Use of autologistic regression reduced spatial autocorrelation in model residuals. Examination of the correlogram for the autologistic model showed all positive spatial autocorrelation had been adequately modeled. Small but significant negative spatial autocorrelation persisted for stops 700 meters or less apart. When positive spatial autocorrelation is present at small distances, statistical techniques that do not account for the autocorrelation will result in inflated parameter estimates and declaring too often that results are significant (Cliff and Ord 1981; Legendre and Fortin 1989).

Habitat Relationships

Fine-scaled studies have demonstrated that American woodcock exhibit positive associations with early-successional mesic forests, wetlands, and water, and they exhibit negative associations with agricultural and urbanized lands (Kinsley et al. 1980; Gutzwiller et al. 1983; Hudgins et al. 1985; Straw et al. 1986). Variables in our logistic regression model were consistent with habitat variables known to be associated with woodcock singing-ground use. Variables positively associated with woodcock presence in our logistic regression model were the percent cover of water and early successional habitats and their interaction. Variables negatively associated with woodcock presence were the percent cover of annual herbaceous and terrestrial unvegetated habitats and their interaction. No class-level or landscape-level habitat heterogeneity indices were selected as significant predictors in our logistic regression model. Klute (1999) reported coarse-grained, percent-cover variables were most strongly associated with woodcock presence when measured at small spatial extents (i.e., small buffers). Landscape heterogeneity indices exhibited stronger associations when measured at large spatial extents (i.e., large buffers). The importance of percent-cover variables in our models have been shown to be a direct result of the size of the buffer in which habitat variables were measured (Klute 1999).

The logistic regression model appeared to give a reasonable picture of coarse-scale habitat relationships for American woodcock, based on known fine-scale habitat preferences. However, the logistic regression model was not appropriate for these data as indicated by the spatial autocorrelation in model residuals. The autologistic model effectively accounted for all positive spatial autocorrelation (Fig. 27.2). Furthermore, autologistic regression is expected to be an appropriate model because spatial dependencies may have resulted from both the distribution of habitat variables and processes related to the breeding behavior of male woodcock.

When true spatial dependencies are ignored, parameter estimates in a logistic regression model are expected to be inflated, because part of the effect that is due to the spatial dependence between neighboring locations is attributed to other predictor variables (Augustin et al. 1996; Gumpertz et al. 1997). When interpreting results from our autologistic model, it is important to recognize that parameter estimates represent the effect of habitat variables given the response status of neighboring stops. This effect may be different from the unconditional effect of the habitat variables (Gumpertz et al. 1997). For all habitat variables, introducing the autocovariate resulted in the expected

decrease in the magnitude of the parameter estimates. The parameter estimate for percent cover of water exhibited the largest decrease. The importance of water for predicting American woodcock presence may be much less than expected for coarse-scaled studies, after the spatial dependence on neighboring stops is considered. All other main effects also showed decreased importance when spatial autocorrelation was explicitly modeled. If we had ignored the spatial dependencies in our data and modeled using only logistic regression, we would have overemphasized the importance of all habitat variables for predicting woodcock presence. Such a mistake could lead to inaccurate predictive models, over- or underestimates of suitable woodcock habitat, and poor information for use in regional management decisions.

Model Prediction

Examination of reclassification statistics allowed us to determine the potential usefulness of logistic regression versus autologistic regression as predictive models of woodcock habitat availability. The overall percentage of routes correctly reclassified was higher for autologistic models than for logistic regression models. If our models were to be used for predicting the distribution of woodcock habitat, we expect that the autologistic model would provide more-accurate prediction because it explicitly accounts for spatial dependencies and more accurately models habitat relationships. Model sensitivity (percentage of correctly classified absent stops) increased substantially between the logistic regression and the autologistic regression models. Model specificity (percentage of correctly classified present stops) exhibited smaller increases between the logistic regression and the autologistic regression models. Several factors may have contributed to incorrect classification of both present and absent stops. First, all relevant habitat variables may not have been included. Second, because each survey route was visited only one or two times, woodcock may not have been detected at some stops when they were actually present. This result would have directly contributed to incorrect classification of absent-stops. Undetected woodcock would have also contributed to miscalculation of the autocovariate term in the autologistic models because it would have affected the characteristics of the cliques of some stops. Third, woodcock may not have been detected at some stops with suitable habitat because not all suitable sites are inhabited by woodcock in a given year. Perfect detection and complete use of suitable habitat would be needed to reduce reclassification errors. Regardless of the sources of reclassification errors, an explicit consideration of spatial dependencies improved the predictive capabilities of our models.

Conclusions

Our analyses demonstrated the importance of explicitly considering spatial autocorrelation in the development of models of species-habitat relationships. Systematic sampling often provides a simple, effective, and inexpensive method for determining presence and abundance of wildlife species; however, results from systematic sampling techniques will frequently result in spatial autocorrelation of response and predictor variables. Moreover, spatial autocorrelation is an inherent and necessary component of functional ecological systems, and understanding spatial structure is necessary for a complete understanding of biological populations and habitat selection. Autologistic regression is an appropriate technique for use with a categorical response variable and continuous and categorical predictor variables that exhibit spatial dependence. Furthermore, autologistic regression provides estimates of both the strength of species-habitat relationships and the strength of dependence among spatially neighboring areas, resulting in a more complete description of the factors influencing the distribution of organisms in the environment. This information may be used by land managers to more accurately identify areas of high habitat suitability and to better understand the importance of habitat elements for the species of interest.

Acknowledgments

This research was supported by the Pennsylvania Game Commission, the Webless Migratory Game Bird Research Program (U.S. Fish and Wildlife Service and U.S. Geological Survey, Biological Resources Division), the Pennsylvania Cooperative Fish and Wildlife Research Unit, and the School of Forest Resources at the Pennsylvania State University. We thank G. L. Storm, W. L. Myers, and J. K. Ord for assistance. G. Baumer (Office of Remote Sensing of Earth Resources, Pennsylvania State University) provided access and facilities for geographic analyses. We give special thanks to observers who assisted with singing-ground surveys.

A Neural Network Model for Predicting Northern Bobwhite Abundance in the Rolling Red Plains of Oklahoma

Jeffrey J. Lusk, Fred S. Guthery, and Stephen J. DeMaso

More-accurate predictions of species abundance are necessary for management and conservation to be effectively implemented (Leopold 1933; Peters 1992; Schneider et al. 1992). Such predictions are increasingly important as human impacts on the environment increase. Artificial neural network (ANN) models are extremely powerful and allow the investigation of linear and nonlinear responses. As such, ANN models offer ecologists a powerful new tool for understanding the ecologies of declining species, which can lead to more-effective management (Colasanti 1991; Edwards and Morse 1995; Lek et al. 1996b,c; Lek and Guégan 1999).

Current applications of ANN models include statistical modeling (Smith 1996). In this capacity, ANN models have considerable advantages over traditional statistical models, such as regression. Artificial neural networks are extremely powerful due to their capacity to learn from the data used during training. Another advantage of ANN models over traditional models is that ANNs are inherently nonlinear (Haykin 1999:2). Because most ecological phenomena are nonlinear (Maurer 1999:110), this property of ANN models makes them more useful than standard statistical models that are often limited to linear relationships (Lek et al. 1996b). Even minor nonlinearities in the response of one variable to another can reduce the predictive power of traditional statistical techniques

(Paruelo and Tomasel 1997). Neural networks also do not require any a priori knowledge of the nature of the relationship between predictor and response variables, which makes available nonlinear methods cumbersome (Smith 1996:19–20). ANNs find the form of the response in the data presented to them and, as such, are not constrained to simple curves, as are curvilinear regression techniques (Pedhazur 1982:406; Smith 1996:20). Finally, ANN models are nonparametric (Smith 1996:20). Use of non-normal data for neural model development will not bias the results (Baran et al. 1996).

We developed an artificial neural network model to investigate the influence of weather patterns on the abundance of northern bobwhites (*Colinus virginianus*; bobwhites hereafter) in a semiarid region of western Oklahoma, United States. An understanding of the effects of weather on species abundances is warranted in the light of global climate change (Root 1993; Schneider 1993). We also sought to evaluate the ANN modeling technique. Specifically, we (1) compared ANN model output with that of a traditional multiple regression model, (2) determined which model was better by using a sums of squares criterion (Hilborn and Mangel 1997), and (3) conducted simulation modeling using the ANN and regression models.

Much is known about bobwhite ecology, so it offers an effective means of evaluating the ANN technique and its applicability to management and conservation. Furthermore, an understanding of bobwhite-climate relationships is an important component of management and conservation of bobwhites. Bobwhite abundance has declined over much of their range during the past several decades (Koerth and Guthery 1988; Brennan 1991; Church et al. 1993; Sauer et al. 1997). Bobwhite declines may be accelerated by climate change in some regions of their range (Guthery et al. 2000). Although we cannot manage the weather, we can factor in its effects when making management plans. By working in cooperation with state management agencies, the results of our research can be directly and immediately applied in the field, completing the research-management cycle (Hejl and Granillo 1998; Kochert and Collopy 1998; Young and Varland 1998.

Methods

We modeled bobwhite abundance in the Rolling Red Plains ecoregion of Oklahoma. This ecoregion is in western Oklahoma, excluding the panhandle (Peoples 1991), and occupies 5.7 million hectares. Mean annual precipitation is 58 centimeters (Oklahoma Climatological Survey unpublished data).

Biologists from the Oklahoma Department of Wildlife Conservation counted bobwhites in each county in Oklahoma. Survey routes were established in typical quail habitat (Peoples 1991). Each 32-kilometer route was surveyed twice annually beginning in 1991: once in August and once in October. Surveys were conducted at either sunrise or one hour before sunset. Total number of bobwhites observed per 32-kilometer route was used as an index of bobwhite abundance. Although roadside counts such as these are prone to biases, these surveys are positively related to the fall harvest in Oklahoma ($r > 0.70$, S. DeMaso unpublished data).

Artificial Neural Networks

Artificial neural networks are mathematical algorithms developed to imitate the function of brain cells for the study of human cognition (Hagan et al. 1996; Smith 1996:1; Haykin 1999:6–9). However, early techniques were handicapped by their inability to han-

dle nonlinear relationships (Hagan et al. 1996:1–4; Smith 1996:8). In the 1980s, neural network modeling experienced a renaissance of sorts with the development of a backpropagation algorithm (see below) that is capable of handling nonlinear relationships (Smith 1996:20).

Because of their foundations in cognitive science, many of the terms used to describe aspects of ANNs are derived from neurobiology. What follows is a short explanation of the terminology of neural network modeling and a brief description of how a typical neural model works. A neural network typically consists of three layers: the input nodes, the neurons (also called hidden nodes or processing elements), and the output nodes. However, ANNs with more than one neuron layer are possible. Typically, each node in each layer is connected to each node in the previous layer by synapses (connection weights), and, as such, is termed fully connected (Smith 1996:21). The synapses store the information learned by the model (Haykin 1999) and are analogous to regression coefficients (Heffelfinger et al. 1999). Each input node represents an independent variable. Values of input nodes are scaled so that they range between zero and one (Smith 1996:67). Each neuron processes the input nodes by computing a logistic function from the sum of the inputs:

$$g(u) = \frac{1}{1 + e^{-u}}$$

where u is the weighted sum of the inputs ($w_j x_j$) plus a bias weight (w_b):

$$u = w_b + \sum_{j=1}^{J} w_j x_j$$

(Smith 1996:40). The logistic function above is the most widely used but is not the only function available (Smith 1996:35). The values calculated by the neurons, $g(u)$, are transferred to the output nodes. The output nodes perform a similar calculation and their output is detransformed to obtain a prediction of the independent variable (Smith 1996:22). In backpropagation ANNs, the error between the predicted output and the actual output is calculated and propagated back through the model where it is used to adjust the values of the synaptic weights according to one of a variety of learning rules (Hagan et al. 1996:11–40; Smith 1996:67). The adjustment of the synapses is termed *learning* (Smith 1996:59). This process continues iteratively, with synapses adjusted after each forward pass, and is termed *training*. With each iteration, the ANN learns more about the relationship between inputs and outputs and, therefore, the prediction error decreases. Training is stopped before the model maps the relationship between inputs and outputs exactly. When this occurs, the network is said to be *overtrained* and the model's predictive abilities are diminished when presented with novel data (Hagan et al. 1996:11–22, Smith 1996:113). The use of ANNs in the ecological sciences requires predictability, and there is a trade-off between model generality and accuracy of prediction.

Because ANN models begin training with randomly selected connection weights, the minimum error achieved by a network may not be the global minimum, but only a local minimum (Smith 1996:62). Therefore, an error minimum lower than the one achieved by the network may exist. However, Smith (1996:62) reported that the probability of such local minima existing decreases as more neurons are added to the model. Determining the optimum number of neurons should, therefore, maximize the chances of finding the global minimum in the error surface.

Database Construction

Roadside counts were initiated in Oklahoma in 1991, and, therefore, our database comprised the 1991–1996 bobwhite surveys. We averaged each year's August and October count for our models. The database also included weather and land-use data as independent variables. Weather data were obtained on CD-ROM from EarthInfo (Boulder, Colo.). We extracted mean monthly temperature data for June, July, and August. Seasonal precipitation data were calculated from total monthly precipitation. We divided the year as follows: *winter*—December, January, and February; *spring*—March, April, and May; and *summer*—June, July, and August. Therefore, seasonal precipitation equaled total monthly precipitation averaged for each three-month period. We grouped climate data into these periods because they represent ecologically important phases of the bobwhite's life cycle (breeding, recruitment, and

winter survival). We did not include any time lag for the effects of rainfall on quail abundance because other networks we developed indicated this lag effect was not important to model predictions (J. Lusk unpublished data). We used weather stations closest to each survey route for obtaining weather data. As measures of land use and human impacts, we used cattle density on nonagricultural lands (total head per square kilometer) and the proportion of county area in agricultural crop and hay production (hereafter, agricultural production). We selected these variables because they are likely to have the greatest effect on bobwhite abundance (Murray 1958; Roseberry and Sudkamp 1998). Bobwhite abundance in Florida varied directly with cultivated acreage and inversely with acreage grazed (Murray 1958). These land-use variables were determined at the county level and were extracted from the Oklahoma Department of Agriculture's annual crop statistics for each survey year in the database.

The final variable included in the data set was the number of bobwhites counted during the previous year's survey. The number of bobwhites present in one year is dependent on the number of bobwhites present the previous year. Furthermore, survival and reproduction may be density dependent (Roseberry and Klimstra 1984).

ANN Construction, Training, and Validation

Network architecture. We used a three-layered, back-propagation neural network. The network consisted of a layer of input nodes representing the independent variables, a layer of neurons, and an output node representing the dependent variable. Our model was fully connected (Smith 1996:21). We used a commercial neural-modeling software package (QNet for Windows, ver. 97.02, Vesta Services, Winnetka, Ill.) for ANN development. Including too many neurons in the neuron layer may result in reduced prediction ability and including too few will limit the complexity the network can accurately learn (Smith 1996:120–123). Therefore, we determined the optimal number of neurons experimentally by training models in which the same data set and model parameters were used but the number of neurons was varied. We developed models that contained two to nine neurons. We limited the maximum number of neurons to the number of input variables in the model. We selected the model with best performance gauged as the correlation between the predicted counts obtained from the model and the actual counts in the validation data set.

Training parameters. We used an adaptive learning rule during model training (Smith 1996). In addition, three parameters were adjusted to optimize model performance. These parameters were the number of iterations, the learning rate, and the momentum. The values we selected for the learning rate and momentum were within the range of those found to be most effective in a wide variety of neural network applications (Smith 1996:77–90). The number of iterations controls how long the model has to learn the pattern and relationships among the variables in the model. The larger the number of iterations, the more attempts the network has to minimize prediction errors. We trained our model for ten thousand iterations. We believed that ten thousand iterations would allow the network to find the error minimum and allow us to stop training if the network began to overfit the data. The learning rate controls the magnitude of the corrections of the synaptic weights per iteration based on the direction and magnitude of the change in the prediction error during past iterations (Smith 1996:77). Selection of too small of a learning rate will increase the number of iterations necessary to reach an error minimum. However, selection of too large of a learning rate may make the network unstable, resulting in oscillations in the prediction error (Hagan et al. 1996:5–9). We used a learning rate of 0.05. The final network parameter was momentum. Momentum determines how many past iterations are used in determining synaptic-weight adjustments in the current iteration (Smith 1996: 85–88). Momentum keeps the error corrections moving in the same direction along the error surface (Smith 1996:85). If a large momentum value is used, it will take longer for weight corrections to respond to changes in the prediction error. In other words, synaptic weight adjustments are based on the long-term trend in prediction error, and momentum determines the number of iterations used in determining the long-term trend. We used a momentum of 0.90. This momentum is appropriate for most types of models (Smith 1996:86).

Validation. To assess the predictive ability, accuracy, and reliability of our ANN model, we presented the trained model with data not used in network training. We created a validation data set by extracting 20 percent of the data from the original data set. Data were rank ordered by the number of quail counted, and every fifth record was assigned to the validation data set. There were ninety-eight records in the original database, resulting in twenty records in the validation data set. The systematic removal of the validation data allowed us to gauge the performance of the network over the entire range of the original bobwhite count data. Because the validation data were derived from the original data set and were, therefore, obtained under the same conditions as those used for network training, the network can be considered only validated for this particular ecoregion in Oklahoma (Conroy 1993; Conroy et al. 1995).

In addition to our validation data set, we tested our model with data collected in the same ecoregion but not part of the training or validation data sets. Because this model will eventually be used by managers to predict bobwhite abundance, this test will determine the utility of the model. We presented the trained model with the 1997 data and recorded the accuracy of the predictions.

Regression Analysis

We performed a multiple regression analysis to compare ANN performance with that of this traditional statistical model. We used the same data set used for training and validating the ANN model for the regression analysis. The full-model multiple linear regression included all the independent variables and the dependent variable used in the ANN model. We used the statistical software package Statistix (Analytical Software 1996). We used the Student's t-test for determining which variables were contributing ($P < 0.05$) to the model predictions (Analytical Software 1996). The correlation between each model's predicted and actual bobwhite count was used as an indicator of the relative performance of each model.

Model Comparison

We used the percent contribution of each variable to the ANN model's predictions to identify important variables (Özesmi and Özesmi 1999). The percentage contribution is calculated by dividing the sum-of-squared synaptic weights for the variable of interest by the total sum-of-squared synaptic weights for all variables. We also determined each variable's contribution to the total, unadjusted R^2 using a forward stepwise regression (Wilkinson 1998). We calculated the increase in R^2 after each variable was entered into the model to apportion the amount of variance accounted for to each variable. We then divided each individual R^2 by the total unadjusted R^2 for the model. This gave the percentage contribution of each variable in the regression model to the model's response. This percentage is, therefore, homologous to the percent contribution of the ANN model. Although these percentage contributions are not directly comparable, they allowed us to determine what variables were driving each model.

To determine if the differences in performance were due to the increased power of the ANN modeling technique, or to the increased parameterization of the ANN model, we used a sum-of-squares criterion for model comparison (Hilborn and Mangel 1997:114–117). This technique adjusts the sum of squared deviations (SS) by penalizing parameterization:

$$SS_A = \frac{SS_m}{n - 2m}$$

where SS_m is the sum of squared deviations for the model of interest, n is the sample size used to develop the model, and m is the number of parameters in the model (Hilborn and Mangel 1997:115). This sum-of-squares criterion is similar to Mallow's C_p (Hilborn and Mangel 1997:116). As such, the model with the lowest adjusted sum-of-squares is selected as the best predictor of the dependent variable (Hilborn and Mangel 1997:116). The SS deviations for each model were calculated from the observed and predicted values of the bobwhite counts. We calculated the SS from the training data only, resulting in an n of 78. The ANN model had thirty-four parameters (one for each synapse: nine inputs times three neurons equals twenty-seven, an additional three synapses connecting each of the neurons to the output node, and four bias weights, one for each neuron and output node), and the regression model had ten parameters (regression coefficients, one for each independent variable, and the constant).

Simulation Analyses

Following model training and validation, we used simulations to explore the effects of each independent variable on ANN model predictions (Lek et al. 1996a; Heffelfinger et al. 1999). This allowed us to further evaluate model performance. We constructed simulation data sets in which one independent variable was allowed to vary incrementally between its maximum and minimum value and all other variables were held constant at their mean value. These data sets were then processed through the trained neural network to generate predicted bobwhite counts. Predicted counts were then plotted against the range of the variable allowed to vary to determine the response of network predictions to that particular variable.

Results

We determined that three neurons were optimal for the data set. The ANN model accounted for 78 percent (R^2) of the variation in bobwhite counts in the training data and 32 percent of the variation in bobwhite counts in the validation data (Fig. 28.1). The lower R^2 for the validation data resulted mainly from a single outlier (Fig. 28.1). With this outlier removed, the amount of variation accounted for by the ANN model increased to 52 percent. However, we could find no reason for the large prediction error associated with this data point and so provide both results here. Our test of the network model accounted for 17 percent of the variation in the 1997 data ($R^2 = 0.17$). The full-model regression was not significant and accounted for 6 percent of the variation in bobwhite

TABLE 28.1.

Parsimony analysis of the artificial neural network model and the regression model using the adjusted sum-of-squares (Hilborn and Mangel 1997).

Model	Number of parameters	Sum-of-squares	Adjusted sum-of-squares
Artificial Neural Network	34	2,821.64	282.1
Regression	10	12,950.90	223.3

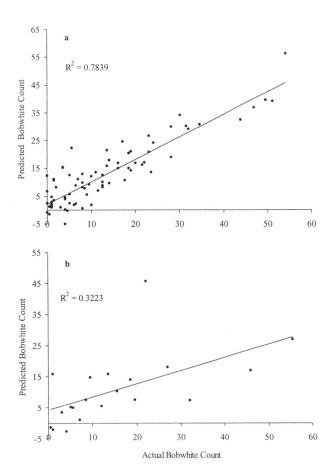

Figure 28.1. Predicted Northern bobwhite (*Colinus virginianus*) counts from the artificial neural network model plotted against the actual values in (a) the training data set and (b) the validation data set for the Rolling Red Plains of western Oklahoma. The trend line represents the linear model regression of predicted bobwhite count on the actual bobwhite count.

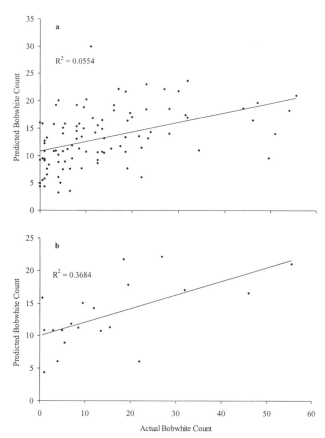

Figure 28.2. Predicted Northern bobwhite (*Colinus virginianus*) counts from the full model regression plotted against the actual values in (a) the training data set and (b) the validation data set for the Rolling Red Plains of western Oklahoma. The trend line represents the linear model regression of predicted bobwhite count on actual bobwhite count.

counts ($F_{9,68}$ = 1.50; P = 0.17; Fig. 28.2). The regression model accounted for 37 percent of the variation in the validation data set (R^2 = 0.37; Fig. 28.2). The sum-of-squares criterion indicated that the regression model (SS_A = 223.3) was the better predictor of bobwhite abundance than the ANN model (SS_A = 282.1; Table 28.1). In other words, the increased predictive power of the ANN model was not enough to warrant increased complexity.

Although it is not possible to determine statistically the significance of the variables in the ANN model, we assume that the importance of independent variables is related to the magnitude of its contribution to predictions. Each of the independent variables contributed some information to the model predictions

(Table 28.2). Mean August temperature and summer precipitation had the highest individual contributions to the network outputs, with a combined contribution of 32 percent (Table 28.2). The remaining variables also contributed to the ANN model's predictions, but to a lesser extent (Table 28.2). There was one variable significant to the regression model: winter precipitation (Table 28.2). Winter precipitation also accounted for 54 percent of the total R^2 of the regression model (Table 28.2). Only spring precipitation and the previous year's bobwhite counts contributed more than 10 percent to the overall R^2 (15 and 11 percent, respectively; Table 28.2). The density of cattle on nonagricultural land contributed nothing to the overall R^2.

The Student's *t*-test we used to determine significant variables in the regression model was limited to linear

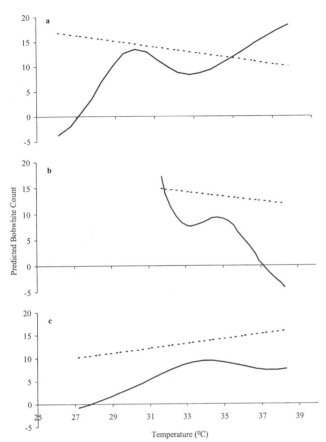

Figure 28.3. Neural network simulation analyses (solid line) and regression predictions (dashed line) of the response of northern bobwhite (*Colinus virginianus*) counts in the Rolling Red Plains of western Oklahoma to mean monthly temperature in (a) June, (b) July, and (c) August. Temperature is reported in degrees Celsius, and the same scale was used for each plot.

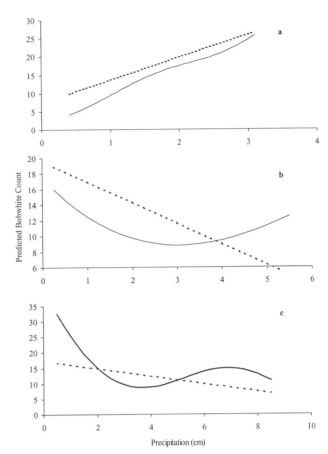

Figure 28.4. Neural network simulation results (solid line) and regression predictions (dashed line) of the response of northern bobwhite (*Colinus virginianus*) counts to seasonal precipitation in the Rolling Red Plains of western Oklahoma. Winter months (a) included December, January, and February; spring months (b) included March, April, and May; and summer months (c) included June, July, and August. Precipitation is reported in centimeters, but each plot has its own scale.

relationships. Such linear relationships did not exist for all variables as indicated by the ANN model. Predicted bobwhite counts increased nonlinearly with increasing June and August mean monthly temperature. Predicted bobwhite counts increased with increasing June temperature until approximately 30 degrees Celsius, after which predicted counts decreased (Fig. 28.3a). Predicted counts also increased with increasing August temperature until 34 degrees Celsius, after which predicted counts also decreased (Fig. 28.3c). The regression model predicted a steadily decreasing count with increasing June temperatures, and a steadily increasing bobwhite count with increasing August temperatures (Figs. 28.3a and 28.3c, respectively). As July temperature increased, the ANN

model predicted that bobwhite counts decreased non-linearly. However, the regression model predicted bobwhite counts would not respond strongly to July temperature, although the regression predictions did decrease with increasing July temperature (Fig. 28.3b).

There was a near-linear relationship between winter precipitation and bobwhite counts as predicted by the ANN model (Fig. 28.4a). The regression model predicted a positive linear relationship (Fig. 28.4a). Increases in winter precipitation were related to increased bobwhite counts, but counts decreased with both spring and summer precipitation (Figs. 28.4b and 28.4c, respectively). These predictions matched those of the regression model, in that they predicted

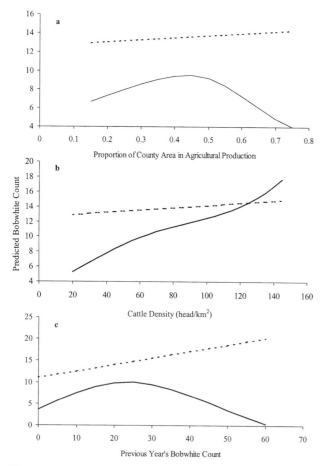

Figure 28.5. Neural network simulation results (solid line) and regression predictions (dashed line) of the response of bobwhite (*Colinus virginianus*) counts in the Rolling Red Plains of western Oklahoma to (a) the proportion of county area in agricultural production, (b) cattle density on nonagricultural lands, and (c) the previous year's northern bobwhite count. Cattle density is reported as total number of head per square kilometer of nonagricultural land.

decreases. However, the ANN model suggested non-linearities in the responses.

Predicted bobwhite counts reached their maximum value at midlevels of the proportion of county area in agricultural production and the number of bobwhites counted during the previous year's survey (Figs 28.5a and 28.5c, respectively). The regression model predicted little response of bobwhite counts to the proportion of county area in agriculture, but there was a positive trend (Fig. 28.5a). The regression model also predicted a linear increase in bobwhite counts with increasing previous year's counts (Fig. 28.5c). Predicted bobwhite counts also increased near-linearly with increasing cattle density, although the regression model

showed little effect of cattle density on bobwhite counts (Fig. 28.5b).

Discussion

The application of ANN modeling techniques to the study of ecological phenomena has great potential for understanding complex, dynamic processes (Colasanti 1991; Edwards and Morse 1995; Lek et al. 1996b). However, to date, little research has made use of this tool. When applied to an ecological research problem, ANN models have consistently outperformed traditional statistical models (Recknagel et al. 1997; Maier et al. 1998). Artificial neural networks have proved highly effective in predicting aboveground biomass in the tallgrass prairie (Olson and Cochran 1998). Compared to regression models, ANNs predicted biomass and described changes in standing biomass with substantially greater accuracy. Heffelfinger et al. (1999) used ANNs to accurately predict call counts and age ratios for Gambel's quail (*Callipepla gambelii*) in Arizona from precipitation and temperature data. Other studies have used ANNs to accurately predict trout (*Salmo trutta*) abundance (Baran et al. 1996; Lek et al. 1996a). Mastrorillo et al. (1997) used a neural model to correctly predict the presence of three small-bodied fish in freshwater streams in more than 80 percent of cases. Özesmi and Özesmi (1999) compared ANNs with logistic regression to classify locations in a GIS database as nest or non-nest sites for red-winged blackbirds (*Agelaius phoeniceus*) and marsh wrens (*Cistothorus palustris*) based on site characteristics. Their ANN models outperformed logistic regressions in all but one case. The better performance of the ANN model resulted because nest-site selection by these marsh-nesting species was a nonlinear process.

For our data set, the regression model performed better than the ANN model based on the adjusted sum-of-squares criterion. Our neural model also performed poorly when presented with 1997 data, but the weather in 1997 was outside the range of conditions used to train the model. We have found that the magnitude of deviations from long-term mean conditions may have a greater effect on bobwhite populations than yearly weather conditions (Lusk et al. unpublished manuscript). This may in part be

TABLE 28.2.

Contribution of each independent variable to the artificial neural network and regression models' predictions of northern bobwhite (*Colinus virginianus*) abundance in the Rolling Red Plains of Oklahoma.

Independent variable	Neural network Percent contribution	Regression model Percent contribution[a]	Regression model *t*	Regression model *P*
Mean June temperature (C)	12.5	2	−0.75	0.4568
Mean July temperature (C)	13.5	1	−0.31	0.7540
Mean August temperature (C)	16.0	5	0.57	0.5702
Winter precipitation (cm)	12.5	54	2.30	0.0245
Spring precipitation (cm)	7.0	15	−1.47	0.1462
Summer precipitation (cm)	16.0	9	−1.06	0.2913
Proportion cropland[b]	7.0	3	0.14	0.8928
Cattle density[c]	7.5	0	0.17	0.8637
Previous year's bobwhite count	8.0	11	1.57	0.2218

[a]Individual R^2 expressed as a percentage of the total R^2 (0.166) accounted for by the model.
[b]Proportion of county area in agricultural production.
[c]Total head per hectare of nonagricultural land.

responsible for the network's poor performance in 1997. However, the additional knowledge gained by using the ANN modeling technique is essential for successful management. Management and conservation decisions based on incomplete or misleading information can only harm the species of concern. Simplicity is only one criterion by which to judge a model's performance. Also important is the ability of the model to approximate the process under investigation (Burnham and Anderson 1998:23). The ANN model provided more biologically meaningful predictions of responses, because the ANN was able to find the nonlinear elements of the responses. We believe that the length of the data set may have limited our results. The six years for which we have data may not have sufficiently captured the response of bobwhites to climate variables. Dynamics in semiarid areas are characterized by episodic events that require long-term data. Model accuracy is a function of sample size (Smith 1996:134). Furthermore, with small sample sizes, such as those used in our study, the effects of noise on the model's performance are amplified, especially if the relationship being modeled is complex (Smith 1996:115). Too small of a sample size can reduce the ability of the ANN model to generalize, but

there are no sample-size restrictions to the application of neural networks (Paruelo and Tomasel 1997).

Using simulations (Lek et al. 1996a, Heffelfinger et al. 1999, Özesmi and Özesmi 1999), ANN models provide information about the effects of the independent variables on bobwhite abundance. This not only provides a better understanding of bobwhite ecology, but also allows us to evaluate the ANN model's explanatory ability. June, July, and August temperatures were important contributors to the model's predictions (Table 28.2); however, August temperature contributed more than June or July temperatures. The higher importance of August temperature may be an artifact of counting quail in the fall. Because climate conditions can affect the daily activity patterns of bobwhites (Roseberry and Klimstra 1984), conditions during the roadside counts may have a larger influence on the network's predictions. This influence is the result of the more-direct effect of the conditions during the count on the count's outcome. Our model predicted that bobwhite abundance would increase with June and August temperature, but only to a certain temperature, after which counts declined. The increase in counts predicted at high June temperatures is probably the result of too few data points in that part of the range, making the predictions susceptible to outliers. Had we limited our simulation data set to within one standard deviation of the mean, the effects of outliers may have been reduced. Predicted bobwhite counts decreased with increasing July temperature. Summer heat decreased California quail (*Callipepla californica*) chick survival in California (Sumner 1935). Quail productivity was negatively associated with summer temperature in northwest Florida (Murray 1958), and July-August temperature was negatively associated with the length of the nesting season and positively associated with nest abandonment in southern Illinois (Klimstra and Roseberry 1975). July temperature decreased the age ratios of Gambel's quail in Arizona (Heffelfinger et al. 1999). Bobwhites in Texas avoided habitat space-time (Guthery 1997) in which the operative temperature was more than 39 degrees Celsius (Forrester et al. 1998).

Our ANN model indicated a near-linear, positive relationship between winter precipitation and predicted bobwhite counts. This near-linearity probably accounts for the significance of this variable in the regression model (Table 28.2). Winter precipitation may indirectly influence bobwhite abundance through increased spring vegetation, seed, and insect production (Swank and Gallizioli 1954; Sowls 1960). Scaled quail (*Callipepla squamata*) abundance in Texas (Giuliano and Lutz 1993) and bobwhite harvest in Illinois (Edwards 1972) were strongly positively correlated with January-March precipitation. Spring and summer precipitation had negative curvilinear relationships with bobwhite abundance. Among gallinaceous birds, young are susceptible to precipitation for the first few days of life (Newton 1998) and increased rain early in the hatching season may lead to increased juvenile mortality (Sumner 1935). Although most studies of the effects of spring precipitation on quail abundance report a nonsignificant relationship (e.g., Campbell 1968; Campbell et al. 1973; Heffelfinger et al. 1999), spring rain might affect breeding behavior adversely, therefore reducing fall abundance.

Similar to the findings of Roseberry and Sudkamp (1998), our model predicted bobwhite abundance to be greatest at intermediate levels of agricultural land use. As agricultural land increases, initially there may not be a net loss of usable space-time for bobwhite.

Bobwhite abundance at low proportions of agricultural use may result from an abundance of mid- to late-successional habitat, less suitable for bobwhites. Similar to the intermediate disturbance hypothesis (Connell 1978), intermediate levels of agriculture may provide bobwhites with more of the habitat components necessary to support large populations than less agriculturally developed lands. Other research has indicated that bobwhites are associated with patchy heterogeneous landscapes with moderate levels of grassland, row crop, and woody edge (Roseberry and Sudkamp 1998). However, as the proportion of agricultural land increases, there is a net loss of usable space-time, any further edge becomes redundant (Guthery and Bingham 1992), and quail abundance declines.

Predicted bobwhite counts increased with increasing cattle density. This is counter to other research that indicates grazing negatively influences quail habitat (Schemnitz 1961). However, Spears et al. (1993) found that site productivity governs the seral stage most important to bobwhites. Early successional stages are favorable for bobwhites on more productive sites, whereas late seral stages are favorable on less productive sites. Because western Oklahoma is semiarid, and therefore less productive, the positive response we found (Fig. 28.5b) is not consistent with expectations.

Predicted bobwhite abundance showed a weak but discernible density-dependent effect in relation to the previous year's bobwhite count. For bobwhite counts higher than about twenty-five, predicted counts for the next fall decreased. The implication of this result is that at current levels of habitat space-time availability, bobwhite abundances above a certain level will adversely affect the population as a whole. In other words, the available habitat space-time can only support a given number of bobwhites, regardless of climate conditions beneficial to bobwhite increase.

Conclusions

We believe ANN modeling techniques offer wildlife managers and conservationists with a valuable and powerful tool for managing species of concern.

Although the ANN model did not outperform the regression model based on the adjusted sum-of-squares criterion, the ANN model did provide a better understanding of how bobwhite abundances in the Rolling Red Plains of Oklahoma respond to climate and land-use variables. Nonlinear relationships, although widespread in nature, are often ignored by researchers (Gates et al. 1994). The ability of the ANN technique to find the nonlinear responses of quail abundance to climate variables makes ANN models preferred to traditional linear and nonlinear techniques that require the specification of the curvilinear response variable. A lack of knowledge of the natural history of many species makes specification of the correct polynomial term a matter of trial and error.

Model validation indicated that the ANN technique was accurate for this region of Oklahoma, but the increase in power was only due to the increased parameterization of the ANN model. However, use of linear modeling techniques may result in a misunderstanding of the factors influencing a particular process. Our regression analysis was only able to identify the linear relationship between winter rain and bobwhite abundance. Any management or conservation plan must take into account climatic factors if it is to be successfully implemented. Furthermore, the ANN model we described can continue to learn as more data become available, and can, therefore, be used as part of an adaptive management plan (Morrison et al. 1998). Our analysis was limited to a six-year data set that may not have represented the entire spectrum of response by bobwhites to climate variables. The predictions of the simulation analyses can be used to generate hypotheses suitable for empirical testing (Recknagel et al. 1997). Simulations also can be used to judge the biological realism of the ANN predictions and increase the understanding of the factors influencing a species' abundance. The use of ANN models also can allow more cost-effective management because the data used to generate the predictions are readily available and cheaply obtained. Our model will be used by the Oklahoma Department of Wildlife Conservation to estimate bobwhite abundances for the management of the fall harvest. A similar modeling effort is underway for Texas Parks and Wildlife Department. We will develop a model that will be used by managers in better managing bobwhites in Texas.

Acknowledgments

We thank the Oklahoma Department of Wildlife Conservation for providing funding for this research. J. Lusk was supported by a Presidential Fellowship for Water, Energy and the Environment from the Environmental Institute at Oklahoma State University, and by a Eugene and Doris Miller Distinguished Graduate Fellowship from the Oklahoma State University Foundation. We thank S. Fuhlendorf, E. Hellgren, H. Wilson, and K. Suedkamp for reviewing the manuscript. Support was also provided by the Department of Forestry at Oklahoma State University, the Oklahoma Agricultural Experiment Station, the Game Bird Research Fund, the Bollenbach Endowment, and the Noble Foundation. This manuscript is approved for publication by the Oklahoma Agricultural Experiment Station.

Incorporating Detection Uncertainty into Presence-Absence Surveys for Marbled Murrelet

Howard B. Stauffer, C. John Ralph, and Sherri L. Miller

There is a long tradition associated with sample surveys for presence or absence of flora or fauna at sites or stations in sampling units where detectability may be an issue. Species may be present at a sampling unit yet fail to be detected. Historically, the problem with detectability has often been ignored. Recently, attention has been given to the development of survey protocols that increase the likelihood of detection. Often, these protocols call for repeated visits to a sampling unit.

It is a key requirement in the design of an increasing number of surveys that the numbers of visits to sampling unit sites ensure a sufficient level of probability, say 95 percent, that species be detected, either in an individual sampling unit, or in the entire survey region, if they are present. In such instances, the specific objective of the survey is to test the null hypothesis H_0 that species are not present, versus the alternative hypothesis H_A that species are present. The 95 percent probability of at least one detection is the *power* of the survey.

When counts are taken to estimate abundance, various strategies have been developed to address the issue of detectability. Capture-recapture methods allow the estimation of recapture probabilities for both closed and open systems (Otis et al. 1978; Pollock et al. 1990). Distance sampling allows the estimation of a detection function to compensate for the loss of detectability at increasing distances away from an observer in line transect and point transect surveys (Buckland et al. 1993). An extensive literature exists describing estimators for these methodologies, both capture-recapture (e.g., Jolly 1965, 1982; Cormack 1968, 1979; Nichols et al. 1981; Pollock 1981; Seber 1982, 1986; White et al. 1982; Burnham et al. 1987) and distance sampling (e.g., Burnham et al. 1980). A modified version of Emlen's method also addresses the issue of detectability for count response (Ramsey and Scott 1981; Scott et al. 1986).

For presence-absence surveys, Azuma et al. (1990) addressed some aspects of this problem using spotted owls (*Strix occidentalis*) as an example. They proposed a fixed number of visits to each sampling unit and a bias adjustment to compensate for false negatives when estimating the proportion of occupied sampling units. Link et al. (1994) found that within-site sampling variability is a significant portion of overall variability in breeding bird surveys, particularly for species with low abundance levels. Pendleton (1995) recommended two strategies for addressing the effects of variation in detectability probabilities in bird point count surveys: standardizing surveys, and obtaining separate estimates of detection rates and adjusting for them. Kendall et al. (1992) commented on the problem with detectability in a power analysis study of grizzly bears (*Ursus arctos*), recommending multiple strata and optimal timing of surveys to enhance the power of the design. Zielinski and Stauffer (1996) were also concerned with detectability probabilities in a power analysis for fisher (*Martes pennanti*) and American marten (*Martes americana*), recommending multiple-station sampling units with repeated visits. Sargent and Johnson (1997) noted the problem of detectability with carnivores due to secretive behavior and low densities.

Predictive accuracy assessment of wildlife habitat relationship models is dependent upon the quality of the response data. Adjustments for the uncertainty of detection with response data must be taken into account. Young and Hutto (Chapter 8) found that problems of detectability can be reduced by using presence-absence responses rather than counts in a survey. They obtained different results in logistic regression and Poisson regression analysis of habitat relationships for Swainson's thrush (*Catharus ustulatus*) over three successive years, partially due to problems with detectability. They collected data on ten-point transects for the three years to mitigate their uncertainty of detection. Karl et al. (Chapter 51) examined the effects of rarity on the predictive accuracy of habitat relationship models. They observed that errors of commission (species predicted but not detected) are either real or apparent. Real errors are caused by species-specific behavior such as the avoidance of humans, cryptic nature, episodic appearance, or temporal and spatial variation. Apparent errors, on the other hand, are caused by inefficient or limited sampling where there is uncertainty of detection. Reed (1996) discussed the influences of detectability in drawing inferences about extinction caused by species density, sampling effort, habitat structure, visibility, observer bias, number of observers, ambient noise, season, and weather. Stauffer (Chapter 3) cautioned against the use of inadequate data in his historical survey of statistical methods applied to wildlife habitat modeling and concluded that simple models, using 0–1 response data, may work best. Authors do not always explicitly address the effect of measurement Type II error caused by problems with detectability (species present but not observed) in their assessments of wildlife habitat relationship model accuracy (e.g., see Conroy and Moore, Chapter 16; Elith and Burgman, Chapter 24; Fielding, Chapter 21; Henebry and Merchant, Chapter 23;

Robertsen et al., Chapter 34; Rotenberry et al., Chapter 22).

The marbled murrelet (*Brachyramphus marmoratus*) is a particularly important case in point. Federally listed as threatened (USFWS 1997) and listed as endangered in California, the murrelet is difficult to detect on land. This species nests in the canopy of trees in mature and old-growth forests. Each pair spends approximately two months of the April-through-September nesting period incubating and feeding one nestling, and the rest of the year is spent at sea (Ralph et al. 1992). Their flight is rapid and often silent. Furthermore, their detectability is often affected by visibility at survey sites (O'Donnell et al. 1995). Estimates of detectability at survey stations with occupied behavior (see discussion) in six different redwood stand types in California have ranged from 29 to 86 percent, with a mean of 59 percent. In individual stands with 25 or more stations surveyed, estimates of detectability have ranged from 12 to 100 percent (H. B. Stauffer personal observations).

Problems with detectability during repeated presence-absence surveys have lacked a statistical model structure to describe the distribution of the possible survey outcomes for sampling units. It is the objective of this chapter to present such a model and describe its practical application. The theory will be illustrated with its application to marbled murrelet surveys in the Pacific coast forests of North America, to an inland survey for murrelets in low-abundance areas within national forests of California.

Methods

Marbled murrelet terrestrial surveys on the Pacific Coast of North America follow a standardized protocol developed by the Pacific Seabird Group (Ralph and Nelson 1992; Ralph et al. 1994). Sampling units, up to 48.6 hectares (120 acres) in size, are surveyed for two-hour visits at dawn. Each sampling unit is surveyed for presence four times each year for two years—a total of eight visits. The station-visits are distributed over the murrelet nesting season. Observers record murrelet activity consisting of visible and audible detections of varying nesting and non-nesting behaviors.

Murrelets are extremely cryptic and individuals are not easily distinguished. Although identifiable as murrelets as they fly into a stand, detections cannot be readily translated into distinct counts of individuals. We have focused our attention on the *presence* of nesting or non-nesting behaviors as an alternative measure of bird activity.

An Inland Survey for Marbled Murrelets in California

We are using data from extensive surveys for murrelets conducted by the United States Department of Agriculture (USDA) Forest Service, Six Rivers National Forest, in low-abundance inland areas identified as Management Zone 2 in California by the Forest Ecosystem Management Assessment Team (FEMAT) (USDA et al. 1993; Hunter et al. 1998). The primary objective of these surveys has been to determine if murrelets are present in specified regions. They used forest type and geographic location to define habitat strata that were surveyed for presence or absence, using 48.6-hectare sampling-unit locations. These sampling units were visited four times per nesting season in each of two consecutive years following the guidelines of the standardized marbled murrelet protocol (above). It was a critical requirement in the design of the survey that sample sizes be sufficient in each stratum to ensure a 95 percent probability of at least one murrelet detection if they were present in 3 percent of the area. Thus, the objective of this survey was to test, for each stratum, the null hypothesis H_0 that murrelets were not present versus the alternative hypothesis H_A that murrelets were present, with a power of 95 percent. They assumed the *confidence* of the survey was 100 percent; in other words, that there would be no significant Type I error, or false positives.

Incorporating Detectability into the Binomial Model

For presence-absence surveys where there is uncertainty of detection, detectability can be incorporated into the binomial model so that options for power and sample size can be selected for the survey design. It can be incorporated using an adjustment to the probability parameter. The binomial distribution $B(X;P,n)$ (Cochran 1977; Särndal et al. 1992; Thompson 1992) is described by the probability distribution

$$B(X = x; P, n) = \binom{n}{x} \cdot P^x \cdot (1 - P)^{n-x}$$

where x is the number of sampling units where the species is present, P is the probability of presence of the species in a sampling unit, and n is the total number of units sampled. Note that x can vary between 0 and n. The model assumes that the total number of sampling units in the sampling frame is large compared to the number sampled, or that the sampling is performed with replacement. Otherwise, the probability P of presence would not remain constant as the sampling proceeds in a draw-sequential scheme (Särndal et al. 1992). The model also assumes complete certainty of detection, if the species is present in a sampling unit. What happens in surveys where complete certainty of detection is not the case? We need to develop an adjusted model that incorporates uncertainty of detection into its assumptions.

An adjusted binomial model $B_d(X;P,n,p,m)$ generalizes the binomial model $B(X;P,n)$ to incorporate detectability, using four parameters: P = the probability of presence; n = the number of units sampled; p = the conditional probability of detection, if present, with one visit to a sampling unit; and m = the number of visits to the units sampled. The model is described by the probability distribution

$$B_d(X = x; P, n, p, m) = \sum_{j \geq x}^{n} \binom{n}{j} \cdot P^j \cdot (1 - P)^{n-j} \cdot \binom{j}{x} \cdot p'^x \cdot (1 - p')^{j-x}$$

where $p' = 1 - (1 - p)^m$ describes the conditional probability of at least one detection, with m visits to a sampling unit, if the species is present. This distribution describes the probability of x, the number of sampling units where the species was present and detected, as the sum of the following probabilities: (1) the probability of sampling x units with the species present, successfully detecting it all x times (of n total sampling units); plus (2) the probability of sampling (x + 1) units with the species present, successfully detecting it x times and failing to detect it once; plus (3) the probability of sampling (x + 2) units with the species present, successfully detecting it x times and failing to detect it twice; . . . ; plus (4) the probability of sampling n units with the species present, successfully detecting it x times and failing to detect it (n–x) times. The binomial coefficients count the number of combinations of such possibilities. Again, note that x can vary between 0 and n.

The B_d model incorporates detectability into the binomial model. It assumes that the sampling units and visits are independent Bernoulli events. It also assumes that the parameters P and p are fixed throughout the population. The B_d model is actually a special case of a compound binomial-binomial distribution (Johnson and Kotz 1969:194, eq. 36). It can be shown directly from basic assumptions of the B_d model, or with algebraic simplification (J. A. Baldwin personal communication), that $B_d(X;P,n,p,m) = B(X;Pp',n)$.

Power for a Sampling Unit: Power_unit

The power of a survey at a single sampling unit, $power_{unit}$, the probability of successfully obtaining at least one detection with repeated visits to a sampling unit, is given by the formula

$$power_{unit} = p' = 1 - (1 - p)^m$$

where p is the conditional probability of detection, with one visit, if the species is present, and m is the number of visits to the sampling unit. We assume the visits are independent and the conditional probability p is constant.

One can calculate $power_{unit}$ by using estimates of detectability p, based upon previous surveys, in the formula. Alternatively, if estimates are not available, one can substitute low values for p and obtain approximate lower bounds on $power_{unit}$.

Conversely, one can calculate the number of visits necessary to ensure desired levels of $power_{unit}$ by substituting the prescribed $power_{unit}$ and lower bounds on p and solving for m in the equation as follows:

$$m = \log(1 - power_{unit})/\log(1 - p).$$

Power for a Regional Survey: Power_region

The power for a target population in an entire survey region, $power_{region}$, can be calculated as the comple-

ment of the probability of zero detections in a survey of the region, $1 - B_d(0;P,n,p,m)$, using the B_d model:

$$power_{region} = 1 - \{1 - P[1 - (1 - p)^m]\}^n = 1 - \{1 - Pp'\}^n.$$

We are referring here to the presence or absence of a target population in a geographical region consisting of multiple sampling units, such as a ranger district, multiple river drainages, or a national forest, typically 10,000 hectares or larger. Conversely, sample size n can be calculated, if the desired $power_{region}$ and number of visits m are specified along with lower bound estimates of P and p:

$$n = \log(1 - power_{region})/\log(1 - P(1 - (1 - p)^m)) =$$
$$\log(1 - power_{region})/\log(1 - Pp').$$

Results

Below, we summarize our results in three parts: (1) incorporating detectability into the binomial model; (2) power for a sampling unit; and (3) power for a regional survey.

Incorporating Detectability into the Binomial Model

Figure 29.1 shows the probabilities for the B_d model $B_d(X;P,n,p,m)$, contrasted with those of the binomial model $B(X;P,n)$, for the case where a small survey of ten sampling units (i.e., n = 10) is conducted in a region where the species is present in 30 percent of the area (P = 30 percent). For the B_d model, we consider the case where the conditional probability of detection with one visit is p = 30 percent, and there are m = 2, 4, and 6 visits to sampling units.

The three pairs of contrasting bar graphs (Fig. 29.1) illustrate that for low values of X, the probabilities that X sampling units, out of the ten sampled, would have detections is greater when detectability is uncertain. For example, note in the figure that the probability of detections at zero of the sampling units (X = 0) for the binomial model (white bar) is approximately 3 percent (i.e., $power_{region} = 97$ percent), whereas the probabilities of X = 0 for the B_d model are approximately 19, 8, and 5 percent (i.e., $power_{region} = 81, 92,$ and 95 percent) with m = 2, 4, and 6 visits, respectively (black bars). With small

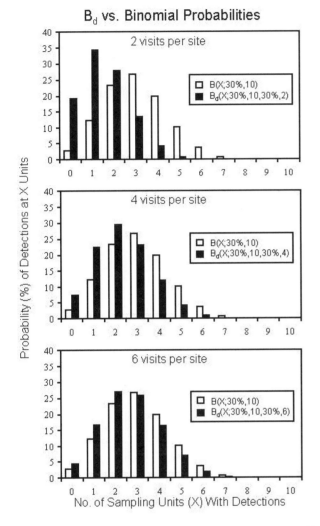

Figure 29.1. Comparison of the B_d model $B_d(X;P,n,p,m)$ with the binomial model $B(X;P,n)$ where X is number of sampling units with detections. We surveyed n = 10 sampling units, each with a probability of presence P = 30%, conditional probability of detection with one visit p = 30%, and m = 2, 4, and 6 numbers of visits to sampling units. The bars show the probability of X sampling units having detections out of ten surveyed. For example, with m = 2 visits (top graph), the probability of X = 0 sampling units with detections with the B_d model is 19 percent (i.e., power$_{region}$ = 81%) (black bar) in contrast to 3 percent (i.e., power$_{region}$ = 97%) for the binomial model with complete certainty of detection (white bar). Additional visits to the sampling units decreases the probability of zero detections to 8 percent and 5 percent (i.e., power$_{region}$ = 92% and 95%), respectively, for m = 4 and 6 visits (black bars) (middle and bottom graphs). The probability of X = 1 sampling units with detections is 34 percent, 22 percent, and 17 percent with m = 2, 4, and 6 visits, respectively, for the B_d model in contrast with 12 percent for the binomial model.

numbers of visits (e.g., m = 2), the B_d model probabilities (black) are much higher for lower X values, in contrast to the binomial model probabilities (white). With fewer numbers of visits, sampling units having the species present will be more likely to have zero detections; the probability of Type II error of false negatives will be greater. As the number of visits increases, the probabilities in the B_d distribution approach those of the binomial model that has complete certainty of detection.

In summary, when detection is uncertain, calculations of power based upon the binomial model will be misleadingly high. This could result in a greater likelihood of false negatives (i.e., undetected presence).

Power for a Sampling Unit: Power$_{unit}$

We calculated the power for a sampling unit, power$_{unit}$, of a detection during at least one visit, with increasing numbers of visits m to a sampling unit (Table 29.1). Table 29.1 presents a range of levels of conditional probability p of detection with one visit: 10, 30, 50, 70, and 90 percent. Fewer numbers of visits are necessary to realize a 95 percent power for a sampling unit as the conditional probability of detection increases. For example, with a 10 percent conditional probability of detecting the birds in one visit, twenty-nine visits are necessary for a 95 percent probability of at least one detection at a sampling unit. With a 90 percent conditional probability of detection, on the other hand, only two visits are required for a 95 percent power of successfully detecting presence.

With a 30 percent conditional probability of detection with one visit, eight visits will ensure an approximate 95 percent power of at least one detection. The current marbled murrelet survey protocol, based upon eight visits to sampling units (Ralph and Nelson 1992; Ralph et al. 1994), ensures an approximate 95 percent power for values of p as low as 30 percent. The Pacific Seabird Group is currently revisiting the protocol to consider revising the number of visits, since some estimates for p, particularly in low-abundance areas, have been falling below the 30 percent threshold. A study is in progress to determine if it will be necessary to revise the protocol, at least for sampling unit locations in some regions.

TABLE 29.1.

Power at a sampling unit (power$_{unit}$) for presence-absence surveys.

p^a (%)	m^b	Power$_{unit}$^{c} (%)
10	2	19.00
10	4	34.39
10	6	46.86
10	8	56.95
10	10	65.13
10	12	71.76
10	14	77.12
10	16	81.47
10	18	84.99
10	20	87.84
10	22	90.15
10	24	92.02
10	26	93.45
10	28	94.77
10	30	95.76
30	2	51.00
30	4	75.99
30	6	88.24
30	8	94.24
30	10	97.18
50	2	75.00
50	4	93.75
50	6	98.44
70	1	70.00
70	2	91.00
70	3	97.30
90	1	90.00
90	2	99.00

^{a}p = conditional probability of detection at the sampling unit, with one visit.

^{b}m = number of visits to the sampling unit.

^{c}power$_{unit}$ = probability of detecting presence at the sampling unit, with m visits (= p′).

Power for a Regional Survey: Power$_{region}$

We calculated sample sizes required to realize specified levels of power$_{region}$ (95, 90, and 80 percent), for varying levels of the probability of presence P (1, 3, 5, and 10 percent), and varying levels of detectability p (10, 25, and 50 percent) (Table 29.2). In these calculations,

TABLE 29.2.

Sample sizes (n) for target populations in a region for presence-absence surveys.

P^a (%)	p^b (%)	m^c	Power$_{region}$^{d} (%)	n
1	10	8	95	525
1	10	8	90	404
1	10	8	80	282
1	25	8	95	332
1	25	8	90	255
1	25	8	80	179
1	50	8	95	300
1	50	8	90	231
1	50	8	80	161
3	10	8	95	174
3	10	8	90	134
3	10	8	80	94
3	25	8	95	110
3	25	8	90	85
3	25	8	80	59
3	50	8	95	99
3	50	8	90	76
3	50	8	80	54
5	10	8	95	104
5	10	8	90	80
5	10	8	80	56
5	25	8	95	66
5	25	8	90	51
5	25	8	80	35
5	50	8	95	59
5	50	8	90	46
5	50	8	80	32
10	10	8	95	52
10	10	8	90	40
10	10	8	80	28
10	25	8	95	32
10	25	8	90	25
10	25	8	80	18
10	50	8	95	29
10	50	8	90	22
10	50	8	80	16

^{a}P = probability of presence.

^{b}p = conditional probability of detection at a sampling unit, with one visit.

^{c}m = number of visits to a sampling unit.

^{d}power$_{region}$ = probability of at least one detection in a region.

we assumed eight visits per sampling unit, corresponding to the marbled murrelet protocol. With P = 1 percent and p = 10 percent, 525 sampling units are necessary to realize a 95 percent power for a target population in an entire region. At the other extreme, with the more optimistic levels P = 10 percent and p = 50 percent, only twenty-nine sampling units are required for 95 percent power.

For the marbled murrelet Zone 2 inland survey in California, lower bound estimates of P = 3 percent and p = 10 percent were assumed and a sample size of 174 was necessary to attain a 95 percent power for each forest habitat stratum (Hunter et al. 1998).

Discussion

In the marbled murrelet protocol, observers note murrelet activity, recording visible and audible detections with behaviors assigned to one of three categories: (1) occupancy: present and exhibiting nesting behavior; (2) presence: present but not exhibiting nesting behavior; and (3) absence. Although we have referred solely to "presence" or "absence" of species, for the murrelet, "occupancy" may be substituted for "presence" for the B_d model, if appropriate to the requirements of a particular survey.

Maximum Likelihood Estimators

In this chapter, we have focused on the assumptions of the B_d model and its probability distribution. We have presented formulas for the calculation of power—for sampling units and for target populations in entire regions—for presence-absence surveys satisfying the assumptions of this model. Such information is useful in the determination of sampling design for such surveys.

For the analysis of data collected from presence-absence surveys, T. A. Max, J. A. Baldwin, and H. T. Schreuder (personal communication) have developed closed-form maximum likelihood estimators for P and p within a probability parameter space, for marbled murrelet and spotted owl survey protocols in the Pacific Northwest. Their owl estimators assume a protocol whereby the number of visits to sampling units is modeled by a negative binomial model: the visits to sampling units are ceased once the behavior (i.e., presence) has been observed, or a specified maximum number of visits has been achieved. Their murrelet estimators alternatively use the assumptions of the B_d model, prescribing a fixed number of visits to each sampling unit. Their estimators assume fixed P and p for a region.

More general maximum likelihood estimators need to be developed, allowing varying P and p for multiple regions, years, and seasons. One approach might use computer optimization routines to approximate maximum likelihood solutions. This context would be analogous to capture-recapture estimators, with capture-recapture heterogeneity corresponding to varying B_d regional P and p, and varying recapture and survival estimators corresponding to varying year and season P and p.

Repeated Visits to Sampling Units—Its Effect on Power for Regional Surveys

Presence-absence surveys have historically focused on locations where species are present at relatively high abundance, to determine behavioral and habitat characteristics. Protocols for such surveys have emphasized repeated visits to sampling units to ensure a high degree of power to detect presence in each sampling unit. Without repeated visits, the conditional probability of detection at specific locations may be low and the probability of not detecting the species unacceptably high.

In surveys, however, where the primary objective is to sample a rare species to determine whether it is present in a region, a more efficient sampling design may be quite different. With this objective, it can be shown that the power of the survey will be effectively increased by sampling additional sampling units rather than by repeatedly revisiting sampling units that have already been sampled, if conditions are reasonably approximated by the assumptions of the B_d model. That is, increasing n is more efficient and cost effective than increasing m. This observation may be surprising at first to surveyors accustomed to existing protocols that have emphasized repeated visits to sampling units.

The reason for this is that the first visit to a sampling unit will provide a maximum amount of "information"—more than a second visit. The amount of information then decreases with each successive visit. Revisiting a sampling unit will indeed increase the probability of detection of the species if it is present, but moving on to new sampling units will increase the probability even more for detection of the species in the entire region. If a sampling unit has been visited once and the species was not detected, a second visit to that sampling unit will have probability $P(1 - p)p$ of detecting the species. A visit to a new sampling unit, however, will have probability Pp of detecting the species. Since $P(1 - p)p \leq Pp$, it is thus the better strategy from a statistical point of view to move on to new sampling units rather than to revisit old ones.

We illustrate this effect with an example. If n = 50 units are sampled in a population with P = 1 percent, an increase in sampling intensity from m = 4 to 8 visits to each sampling unit will raise the power from 37.6 to 39.4 percent. However, if the sample size is raised to n = 100 with m = 4 visits, resulting in an equal number of total sampling unit-visits, the power of the survey will be increased to 61.0 percent. Costs will likely be higher for the latter alternative, to move to new sampling units, but even if n = 75 sampling units are surveyed with m = 4 visits, the power is raised to 50.7 percent. These comparative differences will remain generally true for other cases although the contrasts will be less extreme where the levels of P are higher.

Variation in P and p and Its Effects on Power_{region}

It is disconcerting that in practical application it may not be realistic to make the assumption that P and p are fixed, as in the B_d model. How might variation in the probability of presence P and the conditional probability of detection p affect the B_d model and its power? Species such as the fisher and the American marten (Zielinski and Stauffer 1996) may very well be opportunistic and the probabilities p may increase, or decrease, with time due to the capabilities of the species to adapt their behavior to visiting baited sign detection stations. For murrelets, the effective survey area of a morning's visit to a 48.6-hectare sampling unit is estimated to be approximately 12.2 hectares (30 acres). This reflects an observer's ability to hear and see murrelet behavior that often includes circling in and around the nest area. Therefore, the sampling unit cannot be completely surveyed in a morning's visit and must be surveyed with repeated visits spread over the April–August nesting season and between

years. It is certainly likely in these cases that P and p may vary, geographically, seasonally, and annually.

Feller (1968:230–231) proves the surprising result that the variability of the probability of presence P in the binomial model actually decreases the variance of its estimator. For the conditional probability of detection p, it can be shown, with some elementary probability calculations for the B_d model, that if p varies in a survey at or above a minimal (assumed) fixed value, say p_f, then the power of the survey will be at least as large as that calculated for the fixed p_f. In fact, if p varies symmetrically around a fixed p_f, then the power of the survey can be shown to be at least as large as that calculated for the fixed p_f. These results suggest that the power of the survey will not be reduced by p varying above, or symmetrically around, an assumed fixed average p_f for a survey; in other words, power calculations for regional surveys using the B_d model are robust to those types of variation in p.

Matsumoto (1999) has conducted a sensitivity analysis of power estimates for the murrelet protocol, applied to regions with low species abundance. Her study determined that power estimates are quite robust to varying parameter probabilities P and p within the investigated ranges. She examined varying P and p, assuming low average abundance levels of P = 1, 3, 5, and 10 percent, and average levels of conditional probability of detection p = 10, 25, and 50 percent. Her simulation study examined the effects of varying P and p on estimates of power_{region}, based upon the B_d model assumptions of fixed P and p. She allowed P and p in her simulation to vary, using beta distributions with mean values equal to the assumed fixed values and with varying standard deviations. Her study indicated that power estimates are quite robust to varying parameter probabilities for P and p within those ranges and beta distributed around assumed fixed averages.

Biological and Sampling Components Affecting Presence and Detectability

Errors of commission (species predicted but not observed) in wildlife habitat relationship modeling, both real and apparent, affect the predictive accuracy of wildlife habitat relationship models. We have focused on the statistical aspects of power and sample size selection for presence-absence surveys for a species characterized by an uncertainty of detection. A number of biological components affect both detectability p and presence P. Real errors are caused by species-specific behavior, such as avoidance of humans, cryptic nature, episodic appearance, or temporal and spatial variation. Such behavior occurring globally throughout the survey region affects P, the probability of presence of the species. Apparent errors, on the other hand, are caused by similar behavior occurring dynamically within sampling units. This affects p, the conditional probability of detection of the species, if present. Other influences on apparent error, such as species density, sampling effort, habitat structure, visibility, observer bias, number of observers, ambient noise, season, and weather affect the detectability p. It has been beyond the scope of this study to investigate the contribution of each of these biological and sampling components to P and p. Future investigators are well advised to examine the relative effects of each of these contributors to detectability in their species surveys.

Conclusions

By incorporating uncertainty of detection into survey design and analysis, the predictive accuracy of wildlife habitat relationship models can be improved. The adjusted binomial B_d model provides a method for incorporating uncertainty of detection into presence-absence surveys. The B_d model is useful for both the design and analysis of the survey. For the design, it allows the calculation of the number of visits necessary at sampling units to ensure a prescribed power, or probability of detection, when the species is present. It also allows the calculation of sample sizes and power for regional surveys. For the analysis, it provides a model for estimating the parameters P, the probability of presence, and p, the conditional probability of detection if the species is present, based upon presence-absence data from a survey, using maximum likelihood. Moreover, although its application has been illustrated here for a particularly challenging species, the marbled murrelet, it is sufficiently general to be applicable to presence-absence surveys of other species in sampling units or regions, wherever detectability is of concern.

Acknowledgments

We particularly thank James Baldwin, Timothy Max, and Teresa Ann Matsumoto for their many helpful comments and insights regarding the B_d model. John Hunter, Kristin Schmidt, and Lynn Roberts provided much collaborative support for the application of the B_d model ideas to regional surveys. Barry Noon, Kim Nelson, Tom Hamer, J. Michael Scott, and the anonymous reviewers have also provided helpful critique. The U.S. Forest Service, Six Rivers National Forest, and Humboldt State University provided partial funding and support for this project.

Accuracy of Bird Range Maps Based on Habitat Maps and Habitat Relationship Models

Barrett A. Garrison and Thomas Lupo

Maps illustrating the range or distribution of an organism are fundamental sources of natural history information used in many biological conservation efforts (see Price et al. 1995). In addition, large-area wildlife conservation planning efforts such as gap analysis (Scott et al. 1993; Davis et al. 1998) use range maps as one source of information to predict species occurrences and assess biodiversity. Range maps are developed many ways (Csuti 1996), including the traditional method using manually delineated ranges that often result in a few cohesive polygons that represent general ranges on relatively small-scale maps (e.g., 1:5,000,000). Recently, range maps are being developed by linking habitat maps with habitat relationship models in a geographic information system (GIS). Data on species occurrences (e.g., museum records, checklists, etc.), topography, and climate also can be linked in the GIS as additional data for the maps or to test map accuracy. Habitat maps and habitat relationships models are becoming increasingly available, so model-based maps are becoming the standard method for mapping species' ranges.

Model-based maps have several advantages over manually drawn maps including (1) automation, because maps can be produced by computers; (2) increased precision, because model-based ranges can be more complicated and more biologically based; (3) consistency, because habitat relationship models are

linked with habitat maps so maps of different species are based on the same types of information; and (4) flexibility, because different models, including mechanistic types (Maurer, Chapter 9), and different data can be used to develop several maps that are verifiable.

Increased automation, precision, consistency, and flexibility, however, may not necessarily result in more-accurate range maps because model-based map accuracy is highly dependent on the model and data used to develop and test maps. Map accuracy must be known if conservation efforts are to use these maps, because fiscal and logistic resources could be misused and incorrect conservation actions could be taken based on inaccurate maps. Species distribution and habitat maps combined with habitat relationship models can predict species composition used to identify potential conservation areas (Scott et al. 1993). Overpredicting range (commission error) may lead to misapplication of species-focused conservation efforts, because predicted species may be absent and habitats and locations may be overvalued, if species presence is a criterion for acquisition or management. However, underpredicting range (omission error) also may misapply conservation efforts due to the species being present when predicted absent. Habitats and locations may be undervalued with omission errors as predicted species richness will be lower than actually occurs.

Range map error may be due to errors with habitat

polygons and other spatial data on which ranges are based (e.g., topography, climate) and/or with habitat relationships models and locational data (Edwards et al. 1996; Krohn 1996; Karl et al., Chapter 51). Furthermore, differential accuracy and error patterns may occur due to species-specific ecological attributes, including aggregation patterns, home-range size, niche width, range size, and population abundance, trend, or stability (Krohn 1996; Boone 1997; Hepinstall et al., Chapter 53; Karl et al., Chapter 51). Accuracy is affected by test data that are limited to certain species, locations, time periods, or habitats.

Accuracy of model-based species' range maps for large numbers of species and large geographic areas is rarely determined. Tests of model-based maps usually have focused on single species (Hollander et al. 1994) or on many species from several small areas (Edwards et al. 1996; Krohn 1996; Boone 1997). Tests generally are not performed to determine how species-specific ecological attributes affect map accuracy. Using species checklists from relatively small-sized conservation areas (e.g., wildlife refuges, national parks), Krohn (1996) and Boone (1997) found that map accuracy was affected by some ecological attributes. Because large-area conservation efforts are using model-based maps, testing should be done over large areas (i.e., states and provinces) and should involve many species. In this study, we evaluated how ecological attributes affected accuracy of range maps for one hundred species of birds that breed in California. In this chapter, we attempt to determine whether differential error patterns exist and what their causes might be so that these patterns can be considered when model-based maps are used for wildlife conservation.

Methods

We developed range maps for one hundred species of breeding birds by randomly selecting them from the 184 species with population trends reported for California from the Breeding Bird Survey (BBS) for 1966–1996 by the Biological Resources Division of the U.S. Geological Service (Sauer et al. 1997). Range refers to the mapped area in which the species is predicted to occur when breeding. Our maps functionally defined the extent of each species' breeding distribu-

tion in California using the definition of Morrison and Hall (Chapter 2). As of 1994, three hundred and twenty-five species of birds were known to breed in the California (Small 1994) so BBS data were available for 57 percent of the state's breeding birds. BBS methods and data biases are described by Sauer and Droege (1990) and Price et al. (1995).

The habitat map developed for California's Gap Analysis project (Davis et al. 1998) was the basis for each species' range map. Habitat polygons were identified using the wildlife habitat classification system of the California Wildlife Habitat Relationships (CWHR) system (Mayer and Laudenslayer 1988) such that "habitat" in this context is a distinct association of dominant plant species meeting CWHR classification criteria. This definition of "habitat" differs from that of Morrison and Hall (Chapter 2) because we worked with mapped and classified polygons. Polygons had wetland attributes, which we could not use because these data were lacking for 35 percent of the polygons and for large areas of the state (B. Garrison and T. Lupo unpublished data). The habitat map had a minimum mapping unit of 100 hectares, and the average habitat polygon was 1,930 hectares (Davis et al. 1998). In a GIS, the habitat map was combined with suitability values for breeding habitats predicted by CWHR habitat relationships models (Garrison and Sernka 1997). Habitats were suitable for breeding if the suitability value was rated by CWHR as Low, Medium, or High (see Garrison and Sernka 1997 for definitions of these ratings). The map was further refined by retaining habitat polygons that occurred in counties where the species was known to breed based on county bird checklists provided by the California Bird Records Committee (R. Erickson and M. Patten unpublished data). Polygons were not "clipped" to county boundaries, so there was occasional overlap into counties where the species was not known to occur.

Accuracy Assessment

Six life-history attributes and three measures of population dynamics (hereafter called ecological attributes) were used as independent data to determine if they were responsible for possible error patterns in five measures of map accuracy. Our analysis followed the general approach discussed by Krohn (1996) and Boone (1997) and further refined by Hepinstall et al. (Chapter 53). Size (square kilometers) of the species' breeding range delineated using the habitat map, habitat relationship models, and county checklists was used as the independent variable for range extent. Number of habitats modeled as breeding habitat for each species was used as the independent variable for niche width. Primary habitat association (terrestrial or aquatic) was determined using the major habitat use pattern described by Zeiner et al. (1990). Species' seasonality (summer or yearlong) was categorized using Small (1994). We also determined whether vocalizations (songbird) (Ehrlich et al. 1988; Zeiner et al. 1990) were the primary method of detecting breeding individuals (yes or no) on BBS routes. Population aggregation pattern (territorial or colonial) was categorized using Ehrlich et al. (1988) and Zeiner et al. (1990), and relative abundance, population trend and trend P-values for BBS data from California from 1966 to 1996 (Sauer et al. 1997) were the three measures of population dynamics. The number of BBS routes per species from which population trends were calculated averaged 69.2 plus or minus 46.9 standard deviations (range 14–178).

BBS records from 1977–1996 were used to test map accuracy. Start locations of individual 40-kilometer BBS routes were point locations for the presence or absence of each species. Species were present on a route if detected at least twice (two years) during the twenty-year sample period, otherwise, the species was absent. We chose detections from at least two years per BBS route as the minimum because we felt that detections for one year out of twenty years was too infrequent to represent the species' breeding range and may have an unreasonable likelihood of misidentification.

We used 1:100,000-scale quadrangle maps for California (n = 99) as the test grid to calculate map accuracy. A standard 2×2 error matrix (Congalton 1991) was calculated for each species by determining agreement between quadrangles where the species was present or absent by the range map and present or absent from BBS data. The 1:100,000-scale quadrangle maps were appropriate for our accuracy assessment given the large size of habitat polygons and species' ranges, occurrence of the species anywhere along the 40-

kilometer route, and need to determine error patterns for an area as large as California. Larger-scale quadrangle maps (1:24,000, 1:62,500, etc.) have greater resolution (Stoms 1992) but bird-occurrence data and GIS coverages were not congruent with these scales. Furthermore, we were interested in evaluating how range map accuracy was affected by species' ecological attributes, not in measuring absolute error rates. The nine ecological attributes are not affected by map scale so their effects could be tested with grids of various scales. It would be appropriate, however, to reevaluate our results should data (e.g., each stop along the 40-kilometer route) become available that are more appropriate for analysis using larger-scale quadrangle maps or individual habitat polygons.

Statistical Analysis

Using the 2×2 error matrix for each species, we calculated levels of presence (present by map and BBS), absence (absent by map and BBS), and total (presence plus absence accuracy) accuracies. We also calculated commission (present by map but absent by BBS) and omission (absent by map but present by BBS) errors. Using backward stepwise multiple regression, the three accuracy and two error measures (hereafter called accuracy measures) were dependent variables, while the nine ecological attributes were independent variables. Because of deviations from normality, arcsine (radian degrees) transformations were applied to proportion values of the accuracy measures and BBS trend P-values. Square-root transformations were applied to range size, number of breeding habitats, and BBS trend, and log_{10} transformations were applied to BBS abundances (Zar 1996).

Backward stepwise multiple regression was conducted using a general linear models procedure (SPSS 1998). Habitat association, aggregation, songbird, and seasonality were categorical variables, while range size, number of breeding habitats, BBS trend, BBS relative abundance, and BBS trend P-values were continuous variables. All nine independent variables were initially tested singly with the five accuracy measures, and independent variables with individual F-test results $P < 0.05$ were removed and entered into the model until all remaining variables had $P < 0.05$. Interactions were not tested because of the large

TABLE 30.1.

Values of nine ecological attributes used to test accuracy of breeding range maps for one hundred species of birds in California (see text for attribute definitions).

Measure	Mean	Median	Std. dev.	Min.	Max.
Continuous variables					
Range size (km²)	164,348	151,684	105,173	2,446	404,123
No. breeding habitats	22.6	22.0	12.3	4.0	52.0
BBS trend[a]	3.6	0.1	9.9	−10.2	31.6
BBS trend P-values	0.3	0.2	0.3	0.0	1.0
BBS relative abundance	5.2	2.5	11.1	0.1	89.2
Categorical variables					
Habitat association	Terrestrial: 80 spp., Aquatic: 20 spp.				
Seasonality	Summer: 24 spp., Yearlong: 76 spp.				
Songbird	Yes: 68 spp., No: 32 spp.				
Aggregation	Territorial: 87 spp., Colonial: 13 spp.				

[a]Percent change in Breeding Bird Survey population index between 1966 and 1996.

number of two-way interactions (n = 36) in the initial model, small number (n = 1–3) of independent variables remaining after the stepwise regression, and low levels of multicollinearity between independent variables.

Results

Most species were territorial songbirds that were yearlong residents of terrestrial habitats (Table 30.1). Breeding range size averaged 164,348 square kilometers (median = 151,684 square kilometers); the marsh wren (*Cistothorus palustris*) and golden eagle (*Aquila chrysaetos*) had the minimum and maximum ranges, respectively. Birds were modeled to breed in an average of twenty-three habitats (median = 22), and the American white pelican (*Pelecanus erythrorhynchos*) and western gull (*Larus occidentalis*) bred in the minimum number of habitats, while the American kestrel (*Falco sparverius*) bred in the maximum number of habitats. BBS route indices averaged 3.6 percent change (median = 0.9 percent), and most route indices were not statistically significant (Table 30.1).

Mean total accuracy was 65.7 percent (median = 67.7 percent), and the double-crested cormorant (*Phalacrocorax auritus*) and brown-headed cowbird (*Molothrus ater*) had the minimum and maximum total accuracy levels, respectively. Presence and ab-

sence accuracy averaged 49.1 percent (median = 47.5 percent) and 16.6 percent (median = 10.6 percent), respectively (Table 30.2). The western gull and brown-headed cowbird had the minimum and maximum values, respectively, for presence accuracy. Eight species and the black-billed magpie (*Pica hudsonia*) had the minimum and maximum values, respectively, for absence accuracy. Commission error was the greatest source of error, averaging 33.3 percent (median = 29.8 percent), while omission error was the lowest source of error, averaging 1.0 percent (median = 0.0 percent) (Table 30.2). The brown-headed cowbird and snowy egret (*Egretta thula*) had the minimum and maximum commission errors, respectively. The marsh wren had the greatest omission error, while fifty-three of the one hundred species had no omission errors. Detections averaged 48.6 of the ninety-nine 1:100,000 grid cells (median = 47.0, sd = 22.8, minimum = 11, maximum = 91, n = 100).

Regression equations with the best fit explained 18–63 percent (R^2 = 0.175–0.628) of the variance in the five accuracy measures (Table 30.3). Presence and absence accuracies had 52–63 percent of the variance explained by the regressions, while total accuracy and commission error had 42–44 percent of the variance explained. Omission error was poorly explained by the ecological attributes as 18 percent of the variance was explained only by range size.

TABLE 30.2.

Values of five accuracy measures from testing breeding range maps for one hundred species of birds in California against Breeding Bird Survey (BBS) detections.

Measure	Mean	Median	Std. dev.	Min.	Max.	95% Conf. interval	
						Lower	Upper
Accuracy (%)							
Total	65.7	67.7	16.4	26.3	95.0	62.4	68.9
Presence	49.1	47.5	23.1	11.1	91.9	44.5	53.7
Absence	16.6	10.6	17.3	1.0	73.7	13.2	20.0
Error (%)							
Commission	33.3	29.8	16.6	5.1	71.7	30.0	36.6
Omission	1.0	0.0	1.7	0.0	12.1	0.7	1.3

One to three of the nine ecological attributes were retained (P < 0.02) by the stepwise regressions. BBS abundance was part of the regressions for all accuracy measures but omission error. Total and presence accuracies increased with increasing abundance, while absence accuracy and commission error decreased with decreasing abundance (Fig. 30.1). Presence accuracy increased and absence accuracy and omission error decreased with increasing range size (Fig. 30.2). Seasonality, number of breeding habitats, whether the species was a songbird or not, BBS trend, and P-value of the BBS trend were not retained (P < 0.06) in the general linear models for any accuracy measure (Table 30.3).

Range size and BBS abundance each had the greatest standardized coefficient for two of the four measures with more than one attribute in the equation (range size was the only attribute for omission error), indicating they played the greatest role in determining accuracy (Table 30.3). The standardized coefficient for range size was greater than the coefficient for BBS abundance in the two equations, with both attributes indicating that range size played the greatest role in determining accuracy for those measures (presence and absence accuracy). Habitat association and aggregation were part of the regression equations for three accuracy measures. Total and presence accuracy were greater and commission error was lower if the bird was a territorial species, and total accuracy was greater and absence accuracy and commission error

was lower if the species associated with terrestrial habitats (Table 30.3).

Discussion

Model-based range maps were most accurate for breeding birds with the following attributes: (1) were relatively abundant; (2) had relatively large breeding ranges; (3) were territorial; and (4) were associated with terrestrial habitats. Species that were less abundant, were colonial, associated with aquatic habitats, and had relatively small ranges had model-based maps that were comparatively inaccurate. Colonial species associated with aquatic habitats had particularly high rates of commission and omission errors.

Of the nine ecological attributes, BBS abundance and range size were the most important variables explaining model-based range map accuracy when compared to BBS occurrences. Habitat association and aggregation pattern were the next-most important variables. Moderate to high levels of variance (42–63 percent) explained by most regression models indicated that few ecological attributes are needed to explain map accuracy. Low variance for omission error (18 percent) explained by the regression model is largely due to the narrow range of omission error values.

Prediction error is related to species life history (Flather and King 1992; Krohn 1996; Boone 1997; Hepinstall et al., Chapter 53; Karl et al., Chapter 51).

TABLE 30.3.

Results of backward stepwise multiple linear regression of effects of nine ecological attributes on model-based breeding range map accuracy for one hundred species of birds in California.

Variables	R^2	Steps	Coefficient	SE	Std. conf.	Tolerance	df	F	P-value
Total accuracy	0.442	6							
BBS abundance			0.247	0.047	0.414	0.939	1, 96	27.67	0.000
Habitat association			−0.088	0.023	−0.319	0.838	1, 96	14.65	0.000
Aggregation			−0.078	0.027	−0.239	0.886	1, 96	8.68	0.004
Presence accuracy	0.628	5							
Range size			0.000	0.000	0.575	0.900	1, 96	76.98	0.000
BBS abundance			0.309	0.048	0.405	0.985	1, 96	41.71	0.000
Aggregation			−0.068	0.027	−0.161	0.912	1, 96	6.11	0.015
Absence accuracy	0.522	6							
Range size			0.000	0.000	−0.791	0.780	1, 96	98.06	0.000
Habitat association			−0.107	0.019	−0.469	0.747	1, 96	33.00	0.000
BBS abundance			−0.088	0.036	−0.178	0.946	1, 96	6.03	0.016
Commission error	0.423	6							
BBS abundance			−0.181	0.040	−0.366	0.939	1, 96	20.88	0.000
Habitat association			0.073	0.019	0.321	0.838	1, 96	14.31	0.001
Aggregation			0.074	0.022	0.273	0.886	1, 96	11.00	0.001
Omission error	0.175	8							
Range size			0.000	0.000	−0.418	1.000	1, 98	20.80	0.000

Prediction error is also influenced by model complexity and scale and data resolution (Karl et al. 2000). Species' life history also influences responses to landscape patterns (Hansen and Urban 1992). Boone (1997) categorized species by their likelihood of occurring on range maps, and he found that range size, abundance, and number of habitats used were the most important of ten variables explaining species occurrence. We found that abundance, range size, aggregation, and habitat association were the most important variables explaining map accuracy. We conclude that some population and habitat-use attributes have little effect on map accuracy because seasonality, population trend, and habitat niche were not important in the regressions. Differences between the two studies could be due to the larger ranges of birds in California, more habitats in California, different bird species

and sample sizes, test data differences, and inclusion of different variables in the regressions.

Birds that are abundant, have large breeding ranges, associate with terrestrial habitats, and are territorial are well suited for automated model-based range map development because they are usually relatively well known and widely distributed. In addition, habitat maps with large polygons generally portray terrestrial habitats more accurately than they do aquatic habitats. Furthermore, territorial birds usually vocalize, so they are more easily detected with BBS than birds that are colonial or do not vocalize. Territorial individuals are also more evenly distributed along BBS routes than are breeding colonies, which tend to occur at lower frequencies and are more localized. Moreover, the strong effect of BBS abundance on accuracy indicates that map accuracy may be influenced by the test data because abundant species are

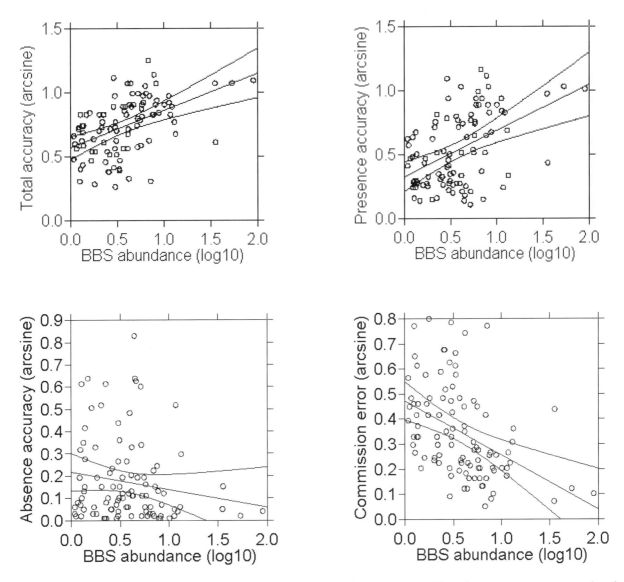

Figure 30.1. Scatterplots of BBS abundance (log$_{10}$ transformed) with four measures of breeding range map accuracy (arcsine transformed) for 100 species of birds in California. Lines represent linear regressions ($P < 0.02$) and 95% confidence ellipses. See Table 30.3 for regression equations.

more effectively detected and counted with BBS methods (see Sauer and Droege 1990).

Habitat maps and habitat relationships models generally have more types of terrestrial habitats than aquatic habitats (see Mayer and Laudenslayer 1988), so terrestrial species have more opportunities to have suitable habitat polygons placed in their ranges. Our results compare well with Boone (1997), who found that range map accuracy increased when compared to occurrence data from larger areas (i.e., larger polygons), and accuracy was lower for species that are

rare, colonial, difficult to observe, and have narrow niches and large body sizes.

Model-based maps overpredicted species' ranges (commission error) by approximately one-third, while ranges had very little underprediction because omission error averaged 1 percent and over half the species had no omission errors. Accuracy may have changed had we used smaller grids from larger-scale maps since overall map error declines with smaller grids (Boone 1997). Increases in size of accuracy assessment areas decreases commission error and does not affect

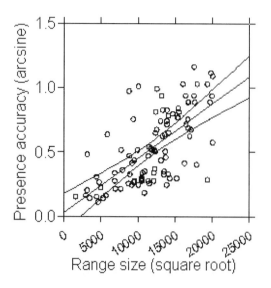

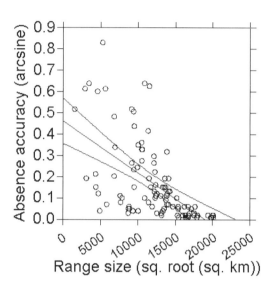

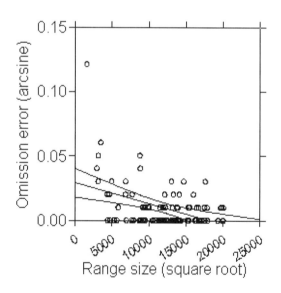

Figure 30.2. Scatterplots of range size (square root transformed) with three measures of breeding range map accuracy (arcsine transformed) for 100 species of birds in California. Lines represent linear regressions (*P* < 0.001) and 95% confidence ellipses. See Table 30.3 for regression equations.

omission error (Garrison et al. 1999). The 1:100,000 grids are large assessment areas (average = 4,936 square kilometers, n = 99) so commission error in our study is likely due to models, BBS test data, and species life history attributes.

Interaction occurs among model accuracy, species abundance, and sample sizes of test data (Karl et al., Chapter 51). Commission error, in particular, decreases as sample size increases (Karl et al., Chapter 51), so our relatively high level of commission error (33 percent) may be due to the small number of species detections. Our average total error of 34 percent (range 5–74 percent) was slightly greater than

that of Boone (1997), who found disagreement (both commission and omission error) <1–38 percent between forty-seven bird-range maps generated from breeding atlases and other published information with maps generated using BBS data.

Based on our results, wildlife managers should be careful when using automated model-based range maps for conservation. Errors of commission and omission increased or decreased depending on several ecological attributes, including range size, aggregation, vocalization, and habitat association. Higher commission error for species with smaller ranges may overemphasize conservation, but species with smaller ranges tend to be more rare (Meffe and Carroll 1994) so overprediction errors may be acceptable to wildlife managers. Species that were colonial and associated with aquatic habitats had more commission and omission errors, respectively, so conservation efforts may overemphasize or underemphasize, respectively, these species if model-based range maps are used. We suspect total and presence accuracy and commission error would have been greater for aquatic species had there been wetland attributes for all polygons. California's

Gap Analysis Project habitat map has large polygons, many terrestrial habitats, and few aquatic habitats, so the map's scale is more appropriate for habitats dominated by trees or shrubs (Davis et al. 1998).

Wildlife managers can rectify model-based map errors that we identified in several ways. Focused mapping efforts could be done for aquatic habitats to more accurately portray the areal extent and distribution of these important wildlife habitats. Habitat maps could be used from smaller areas because they would likely have smaller mapping units, and aquatic habitats might be mapped at a scale commensurate with terrestrial habitats.

Data sources other than BBS might be more appropriate for testing distribution maps for more rare species and colonial species. Species lists from conservation areas may be more appropriate than BBS data for developing and testing maps because these lists are more thorough and less subjective since most if not all species present in an area are typically documented (Edwards et al. 1996; Krohn 1996; Boone 1997). However, conservation areas may not be distributed throughout a species' range, so parts of the range may not be tested.

BBS data are limited to breeding birds that occur in relatively high numbers and can be readily detected using point counts (Sauer and Droege 1990; Price et al. 1995). BBS routes are systematically located throughout the United States along roads such that it is difficult to encounter all species. Yet, BBS routes were more appropriate for our study because they were evenly distributed throughout California compared to conservation areas, which were primarily located in wetland, coastal, and alpine areas. Moreover, BBS routes encompass larger areas than many conservation areas, and hence BBS data are more spatially consistent with California's Gap Analysis Project habitat map. In states and provinces, sample sizes of BBS routes are generally less than that needed for precise estimates of prediction accuracy (Karl et al., Chapter 51). BBS stop data, while not available for our work, would have given us much-greater sample sizes. BBS data are limited to breeding species, so maps for non-breeding species need different data. Lastly, conservation efforts should only involve species with ecological attributes that are appropriate for habitat relationship models, habitat maps, and locational and test data used by wildlife managers. Species with ecological attributes that are not appropriate should not be included in the effort or different models, maps, and data should be used.

Acknowledgments

Richard Erickson and Michael Patten deserve special thanks for providing the county checklists, and we extend our sincere appreciation to the county coordinators who developed the checklists. Kevin Hunting and J. Michael Scott provided constructive comments on earlier versions of the manuscript.

A Monte Carlo Experiment for Species Mapping Problems

Daniel W. McKenney, Lisa A. Venier, Aidan Heerdegen, and Mick A. McCarthy

Much ecological research attempts to describe species distributions and abundance in ways that can be mapped. Managers need maps or spatial predictions to make decisions about activities ranging from timber harvest scheduling to nature reserve design (Hof 1993). But we will never be able to know the entire truth about distributions and abundance because we cannot sample everywhere. This is particularly true in places like Canada and Australia where there remains much unstudied territory and relatively few biologists. Hence, some form of modeling is required to produce inventories or maps. Some researchers have used interpolation techniques such as kriging to describe the patterns of interest (e.g., McKenney et al. 1998; Villard and Maurer 1996). Others use field observations and statistical relationships with environmental gradients to make spatial predictions (e.g., Venier et al. 1998, 1999; Mackey 1994). In the latter case, relationships between species occurrence and biotic and/or abiotic conditions can be mapped using spatial estimates of the independent variables.

Generally a successful derivation of a statistical relationship is related, among other things, to the size of the sample and the strength of the relationships between the dependent and independent variables. Various randomization procedures such as bootstrapping and jackknifing (see for example Manly 1997) have

been developed to help assess the veracity of relationships between dependent and independent variables. Use of tools such as these are increasing and clearly add to the strength of conclusions drawn from ecological research (Pitt and Kreutzweiser 1998). However, applications of these types of tests often focus on parameter estimation and the confidence limits around these using the actual observation data (e.g., see Hilborn and Mangel 1997), not on the resultant maps. The robustness of the maps depends on how well the observation data sample the environmental space in the area being mapped. Our approach is to simulate what we will call a "true" map of a species' distribution based on a statistical model. This map is compared to predictions from new models based on a series of increasing sample sizes. Controlled simulation experiments such as these are not uncommon (e.g., Common and McKenney 1994; Kennedy 1992; see also Virtanen et al. 1998).

In this chapter, we present a Monte Carlo simulation experiment of the reliability of maps derived from logistic regression models as a function of sample size. The experiment is set in the context of bird distribution models in the Great Lakes Region of North America where we have been modeling species' occurrence in relation to climate and vegetation gradients (Venier et al. 1999). Logistic regression models provide probability of occurrence estimates (values between 0 and 1) based on a binary (0,1) response variable. Logistic regression techniques are widely used in ecological re-search where presence/absence data are available (Hosmer and Lemeshow 1989; Collett 1991).

An Overview of the Experimental Design

The overall design of the experiment was to predefine a distribution of probability of occurrence for a particular species (in this case magnolia warbler, *Dendroica magnolia*) as a function of environmental variables using logistic regression modeling for the Great Lakes region of North America. Our selection and presentation of this species is purely for illustrative purposes. We converted the probability estimates to binary values (0,1) and sampled this presence/absence truth using a range of sample sizes. For each sample we recreated the logistic regression model and predicted probabilities of occurrence using the same set of environmental variables. For each new model, we compared the true or baseline probability of occurrence with the predicted distribution of probability of occurrence. The experimental design allows us to examine differences in maps as a function of sample size. To simplify the experiment, we forced the new models to use the same environmental variables as the original model.

Study Area and Background

Our study area includes the Great Lakes Basin of Canada and the United States, an area of approximately 2.3 million square kilometers divided into roughly 100,000 cells for this study. Breeding Bird Atlas data from across the region were compiled (Niemi et al. 1998) and used to generate the initial model. The model used to define the distribution was developed using the approach described in Venier et al. (1999). Numerous single and multivariate logistic regression models were examined that included a range of both climatic and vegetation cover parameters. The probability of occurrence (*p*) was described as a function of four variables, including two climate variables (precipitation during the growing season (PGS), and mean annual precipitation (MAP) and two land-cover variables (amount of deciduous forest, ADF and amount of mixed forest, AMF). The equation is:

$$logit(p) = \ln[(p/(1-p)] = -0.0273 - 0.0221(PGS) + 0.0119(MAP) + 0.0221(ADF) + 0.0492(AMF).$$

The spatial climate data for the region was developed using the ANUSPLIN model of M. Hutchinson at the Australian National University (e.g., Hutchinson 1987, 1995, 1998). Mathematical surfaces were created for 1961–1990 monthly mean maximum and minimum temperature and precipitation (McKenney et al. 1999). Weather stations (n = 2503) throughout the region were used for these surfaces. A number of secondary climate variables such as growing-season length were derived from these primary surfaces. Land-cover variables were derived from AVHRR (Advanced Very High Resolution Radiometer) satellite land-cover classification available from the United States Geological Survey at approximately 1-kilometer resolution. The amount of each land-cover type was summarized in 5×5-kilometer squares to better match the resolution of the Atlas data. Climate estimates were taken from the centers of the squares.

Monte Carlo Protocol

The distribution for the magnolia warbler (see Fig. 31.1 in color section) was the probability of occurrence grid for the region generated from the logistic regression model noted above. The probabilities were converted to binary values (0,1) for the purposes of logistic regression modeling using a uniform random number generator between zero and one. The cell was assigned a value of one (occupied) if the random value was below the probability of occurrence and zero (unoccupied) otherwise. For example, if the true probability of occurrence were 0.80, then the algorithm would generate an occupied cell in 80 percent of the cases and an unoccupied cell in 20 percent of the cases.

The 0,1 grid was then sampled (without replacement) for a variety of sample sizes (Table 31.1). This particular sampling scheme, which shows the pattern of results more clearly, was developed after several trials, including multiple runs at a single sample size. For each sample, the logistic regression model was developed using the same four variables that were used to define the distribution. The logistic regres-

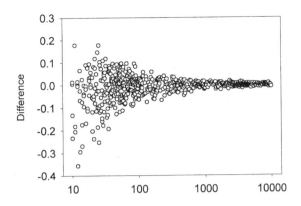

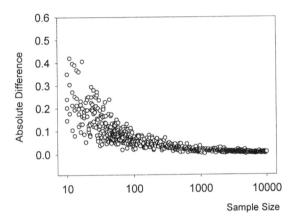

Figure 31.2. Mean difference and mean absolute difference between true and estimated probabilities of occurrence at sample sizes ranging from 10 to 10,000.

TABLE 31.1.

Sampling protocol used in the Monte Carlo experiment.

Sample size	Number of samples	Every N steps
10–59	4	1
60–99	2	1
100–199	1	2
200–499	1	4
500–999	1	10
1,000–4,999	1	25
5,000–10,000	1	100

TABLE 31.2.

Mean absolute differences between the "true" and estimated probabilities for sample sizes ranging from 10 to 10,000 for the magnolia warbler (*Dendroica magnolia*).

Sample size	Magnolia warbler			Species 2	Species 3	Species 4
	Mean	Min	Max	Mean	Mean	Mean
10	0.250	0.08	0.42	0.094	0.175	0.3210
100	0.060	0.02	0.08	0.044	0.054	0.0675
200	0.042	0.02	0.07	0.025	0.031	0.0525
500	0.020	0.01	0.05	0.014	0.0205	0.0305
1,000	0.014	0.00	0.02	0.011	0.015	0.0210
10,000	0.007	0.00	0.01	0.002	0.0055	0.0090

sion equation was then resolved for the region using the gridded estimates of the climate and land-cover variables thus producing a gridded probability of occurrence map. This predicted probability of occurrence was compared to the "true" probability in a cell-by-cell fashion. Mean absolute difference and the mean difference between predicted cell values and true cell values were calculated. Other diagnostics such as mean square errors and standard deviations of absolute differences were calculated but not reported here because they do not substantively add to the interpretation of the results. In addition, the parameter estimates for the environmental variables in each model were stored and examined as a function of sample size. We repeated this procedure for three other models and compared the results for consistency. Note our focus was comparing predicted probabilities rather that comparing the binary values. The latter could lead to comparisons of false

positives and negatives and other diagnostics that typically arise in logistic regression modeling.

Results and Discussion

The mean of the absolute difference summarizes the accuracy of the model over the entire study area (Fig. 31.2). Each point in the figure represents an individual run at the given sample size. The figure shows that at the smallest sample sizes (thirty or less), individual models could be very good (mean absolute differences almost as good as the highest sample) or very poor (mean absolute differences greater than 0.4). Mean absolute differences (the aggregate value for all 100,000 cells) range from 0.08 to 0.42 at small sample sizes and fall below 0.01 after one thousand observations (Table 31.2). In a real-world application, only one sample is available out of an infinite number of possibilities. This suggests that

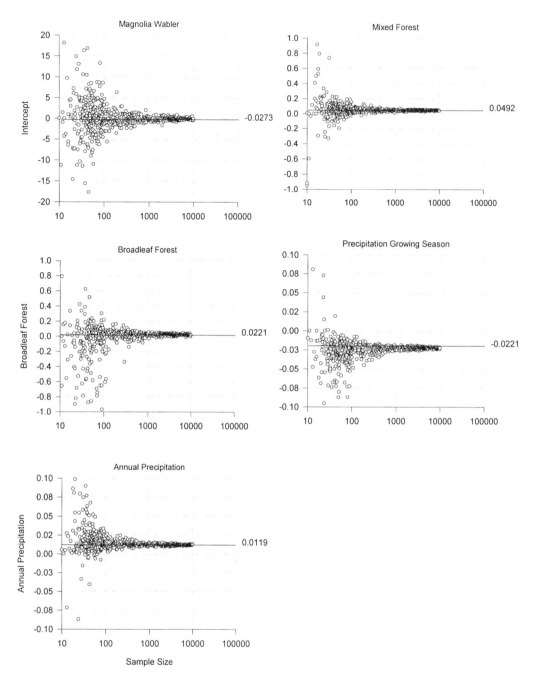

Figure 31.3. Intercept and parameter estimates across sample sizes with the true parameter values shown.

models derived from small samples are often very inaccurate and little confidence should be placed in maps from such models. Nevertheless, it is difficult to assess the biological or management significance of the size of these errors in the absence of a decision-making context.

The mean difference (Fig. 31.2) provides an indication of the direction of error in single runs. At small sample sizes, mean differences were both positive and negative, ranging between 0.2 and –0.4. Sample sizes over one hundred have mean differences between +0.1 and –0.1. Beyond one thousand samples, the mean differences have little variation around zero. The distribution appears roughly symmetrical; however, we have not tested for bias.

Parameter estimates from runs are all centered on

the true parameter value (Fig. 31.3). Again, with small sample sizes there is much variation around the true value. When interpreted in the context of the spatial predictions (i.e., Fig. 31.2), small perturbations in the parameter estimate can lead to large changes in predicted probabilities. The implication of parameter estimate error in an individual run is difficult to assess *a priori*. These results suggest that ecological interpretations of individual parameters could be problematic at smaller sample sizes because their signs range from positive to negative. Clearly, the closer the true value is to zero, the more problematic the interpretation of the sign even with large sample sizes.

Last, we ran the same type of experiment for three other species (i.e., model configurations). Results were similar (Table 31.2). Mean absolute errors declined with increasing sample size and generally did not go below 1 percent until the sample size was greater than one thousand.

Conclusion

We created a Monte Carlo experiment to examine the effect of sample sizes on the predictive accuracy of broadly scaled species distribution models. We note that most researchers would undertake validation tests of statistical models of species occurrence by withholding actual observation data (e.g., Pearce and Ferrier 2000). If sufficient data are available, then this is clearly the most effective way to test the veracity of the statistical models; however, even this approach does not deal explicitly with the mapping problem *per se*. The mapping problem is connected to the representativeness of the observation data and the nature of the environmental space over which the spatial predictions (maps) are being made.

For this particular experimental configuration, we found that map reliability asymptotes around one thousand observations. We reported mean errors and mean absolute errors and also summary statistics on the parameter estimates of the models. Maps based on small sample sizes (less than one hundred) were unstable (Karl et al., Chapter 51; Garrison and Lupo, Chapter 30; and Elith and Burgman, Chapter 24). In fact, parameter estimates varied in their signs even at larger sample sizes. Three other models had similar results. As with all Monte Carlo experiments, the specific results are of course contingent on the formulation, but we anticipate that they are reasonably representative of studies at this scale.

These results are conservative in that we forced new models to use the same set of environmental variables as the original model. Future experiments will include variable selection problems and explore various representations of rarity and nonrandom sampling schemes. The ultimate test of reliability would be to embed the problem in a decision-making framework. Future applications could do so by using the spatial predictions in, for example, reserve design problems. Given the conservative nature of the experimental design, the results would seem to reinforce a sense of caution in our ability to map species occurrences in the absence of large sample sizes. Experiments that include other sources of error are likely to further decrease our confidence in mapping exercises.

Finally, we note that Monte Carlo experiments are valuable tools and could be more widely used for species mapping problems. Species mapping applications are often derived from small samples and inferences made across large geographic regions. Monte Carlo experiments like these are possibly the only approach to gain insights on interactions between dependent and independent variables and sample sizes.

Acknowledgments

We thank Kathy Campbell and Pia Papadopol for assistance in the manuscript preparation.

Measuring Prediction Uncertainty in Models of Species Distribution

Jennie L. Pearce, Lisa A. Venier, Simon Ferrier, and Daniel W. McKenney

Statistical modeling of biological survey data in relation to mapped environmental variables is often used as a surrogate for direct species distributional information. Such modeling can provide cost-effective, definitive, and explicit spatial information for use in regional conservation planning and management. To date, distributional models have generally been derived by modeling species presence/absence data collected at field sites in relation to mapped environmental predictors, using logistic regression or related modeling techniques (Osborne and Tigar 1992; Buckland and Elston 1993; Augustin et al. 1996; Venier et al. 1999). These models are then applied to environmental layers held within a geographic information system (GIS) database to extrapolate predicted likelihood of occurrence across the entire region of interest.

To use models of species distributions effectively in conservation planning, it is important to determine their predictive accuracy (Edwards et al. 1996; Boone and Krohn 1999). Predictions from such models will always contain a level of error resulting from a wide range of factors, including insufficient sample size, measurement error in the biological survey data, measurement error and insufficient spatial resolution in the mapped environmental variables, and failure to incorporate critical habitat variables and other factors (e.g., predation, competition, dispersal) into the modeling. Evaluation of the nature and magnitude of pre-

diction error assists in determining the suitability of models for particular applications and in identifying specific weaknesses requiring correction. Such evaluation also facilitates comparison of modeling techniques and competing models.

Performance of distributional models may be assessed at a number of scales (extent and grain). At the broadest scale, maps of predicted distribution may be examined visually by experts to determine whether the predictions provide a reasonable estimate of the range limits of each species. This is an important step if maps of species distribution are to be used in decision making. It is important for all stakeholders involved in, or affected by, a planning decision to have confidence in the broad-scale accuracy of modeled distributions. Visual examination of such maps does not, however, tell us much about predictive performance at finer spatial resolutions, for example at the scale of a grid cell. Nor does it tell us much about the reliability of predicted probabilities of occurrence or specific sources of error in such predictions. More-detailed quantitative assessment of the magnitude, nature, and potential sources of prediction error requires statistical evaluation of the agreement between predictions from a model and observations within an independent validation data set (i.e., data collected at sites other than those used to develop the model).

Pearce and Ferrier (2000) present a broad framework for evaluating the fine-scaled performance of models using independent data. They describe approaches for addressing two commonly asked questions relating to distributional models: (1) does a given model provide a good rank index of the occurrence of a species across a range of sites for a given application, and (2) does the model provide an accurate estimate of the probability of detecting the species at each site for a given application? Although these two questions are similar, they require very different methods of accuracy assessment and of addressing the needs of different applications. The first question relates to the ability of a model to rank areas within a region in terms of their likelihood of being occupied by the species in question. This information is important if a model is to be used to design a reserve system, to locate potential survey sites, or to explore relationships between habitat characteristics and species occurrence. The methods used to address this question do not, however, tell us much about how we might correct for specific sources of error within a model. The second question concerns the accuracy with which a model predicts the probability of a given species occurring at a site, not just in terms of rank order but also in terms of absolute value. This second type of assessment also provides more detailed information about specific sources of error in a model, thereby providing guidance as to how the performance of a model might be improved.

Our objective in this chapter is to use the evaluation framework of Pearce and Ferrier (2000) to assess the performance of a selection of distributional models developed using logistic regression. These models relate the distribution of bird species in the Great Lakes region to a number of environmental variables.

The Models

We evaluated logistic regression models for ten forest songbird species (Table 32.1). These species were selected because there is a larger overlap of their ranges with the Great Lakes region, they use a variety of forest habitat types, and they include common and rare species. These models were developed using Breeding Bird Atlas data compiled for the Great Lakes Basin as part of a project of the Great Lakes Protection Fund (NRS 795-2467, Forest Bird Biodiversity: Indicators

of Environmental Condition and Change in the Great Lakes Watershed). The models relate the presence/absence of breeding birds to broad climate and satellite-derived land cover using 1,042 selected sites from the study area. Models were developed using logistic regression. Variables considered for entry into each model were chosen to be biologically meaningful to each species. Models presented here are those from the set of possible models that had the greatest explanatory power within the model development data set and that were biologically interpretable. We evaluated the predictive performance of these models using data from 260 sites withheld for this purpose from the model development data set.

The Evaluation

As outlined in the introduction, the evaluation framework of Pearce and Ferrier examines two broad questions concerning the predictive accuracy of models. These two questions form the basis for the current evaluation.

Does a Model Provide a Good Rank Index of Species Occurrence across a Range of Sites?

The discrimination ability of a logistic regression model refers to the capacity of the model to correctly discriminate between occupied and unoccupied sites. This can be examined graphically by plotting the distribution of predicted probabilities associated with the occupied sites and the distribution of predicted values associated with unoccupied sites. If these two distributions are plotted as histograms on the same set of axes, then the amount of overlap of the two distributions provides an indication of the discrimination ability of the model. Discrimination histograms were derived for each of the ten species models (Fig. 32.1) by dividing the predicted probability range (zero to one) into ten evenly spaced classes. The histograms for each species provide two types of information. First, they describe the refinement of the predictions. That is, they show the range of predicted values obtained within the validation sample. A well-refined model should generate predictions that span the entire zero-to-one probability range. Second, the histograms depict the degree of overlap in predicted probabilities

TABLE 32.1.

Logistic regression models describing the distributions of forest songbird species within the Great Lakes Basin.

Common name	Scientific name	Model
American redstart	*Setophaga ruticilla*	$4.89 - 0.26(X1) - 0.05(X14) - 0.04(X3) + 0.01(X4) - 0.04(X11) - 0.02(X11)$
Bay-breasted warbler	*Dendroica castanea*	$-3.05 + 0.01(X7) - 0.02(X10) + 0.01(X8) + 0.015(X6)$
Black-and white-warbler	*Mniotilta varia*	$-41.29 + 3.76(X1) - 0.08(X1)^2 - 0.29(X3) + 0.005(X3)^2 + 0.04(X12) + 0.05(X14)$
Blackburnian warbler	*Dendroica fusca*	$-1.26 + 0.003(X6) - 0.003(X9) - 0.02(X10) + 0.003(X10)^2$
Black-throated blue warbler	*Dendroica caerulescens*	$-70.6 + 5.24(X1) - 0.12(X1)^2 + 0.40(X2) - 0.15(X3) + 0.01(X4) + 0.003(X5) + 0.003(X6) - 0.01(X10)$
Black-throated green warbler	*Dendroica virens*	$5.30 - 0.18(X1) - 0.22(X2) + 0.003(X5) + 0.007(X6) - 0.00002(X6)^2 - 0.016(X10) + 0.00002(X10)$
Chestnut-sided warbler	*Dendroica pensylvanica*	$-78.51 + 6.05(X1) - 0.12(X1)^2 + 0.01(X4) - 0.06(X10) + 0.05(X12) - 0.09(X13) + 0.16(X14) - 0.01(X14)^2$
Magnolia warbler	*Dendroica magnolia*	$6.83 - 0.48(X1) - 0.01(X10) + 0.51(X2) - 0.06(X3) - 0.01(X10) + 0.01(X5)$
Nashville warbler	*Vermivora ruficapilla*	$10.12 - 0.28(X1) - 0.32(X3) + 0.005(X3)^2 + 0.001(X5) + 0.004(X6) + 0.002(X9) - 0.005(X10)$
Tennessee warbler	*Vermivora peregrina*	$-78 + 6.41(X1) - 0.15(X1)^2 + 0.85(X2) - 0.06(X3)$

Note:

X1	Maximum temperature in the hottest quarter	X9	Grasses or Brush (MSS)
X2	Mean diurnal temperature range	X10	Agriculture (MSS)
X3	Precipitation seasonality	X10	Dryland, Cropland, and Pasture (AVHRR)
X4	Precipitation in the hottest quarter	X11	Woodland/Cropland Mosaic (AVHRR)
X5	Hardwood (MSS)	X12	Broadleaf Deciduous Forest (AVHRR)
X6	Conifer-Hardwood Mix (MSS)	X13	Evergreen Coniferous Forest (AVHRR)
X7	Conifer (MSS)	X14	Mixed Forest (AVHRR)
X8	Bare Ground (MSS)		

associated with occupied and unoccupied sites. These two histograms may best be depicted as two distributions by plotting the midpoint of each bar and joining the midpoints by lines. This is the approach taken in Figure 32.1. If a model has good discrimination ability, the predicted values for occupied sites will be higher on average than those for unoccupied sites. In Figure 32.1, the distribution of predicted values for occupied sites (shown by the solid line) lies to the right of that for unoccupied sites (shown by the dotted line) for all ten models, although there is often a considerable degree of overlap between the two distributions.

The discrimination ability of logistic regression models is often quantified by calculating statistics from a 2×2 classification table of predictions and observations (e.g., Edwards et al. 1996; Boone and Krohn 1999; Hepinstall et al., Chapter 53; Karl et al., Chapter 51; Schaefer and Krohn, Chapter 36). A species is predicted to be present or absent at a site based on whether the predicted probability for the site is higher or lower than a specified threshold probability. A problem with this measure of discrimination accuracy is that the measure is sensitive to the location of the threshold probability. Different threshold values will give very different assessments of model accuracy. The choice of an appropriate threshold is difficult, and often arbitrary, although there are some guidelines depending on the intended use of the model (Fielding and Bell 1997).

A more universal accuracy measure should describe the accuracy of the system, not just its performance in a given scenario (i.e., for a given threshold value). One such measure is the area under the relative operating characteristic (ROC) curve. An ROC curve is a plot of the sensitivity and false positive values obtained by considering a large number of threshold probability values. For a given threshold, sensitivity is the proportion of occupied sites correctly classified by the model

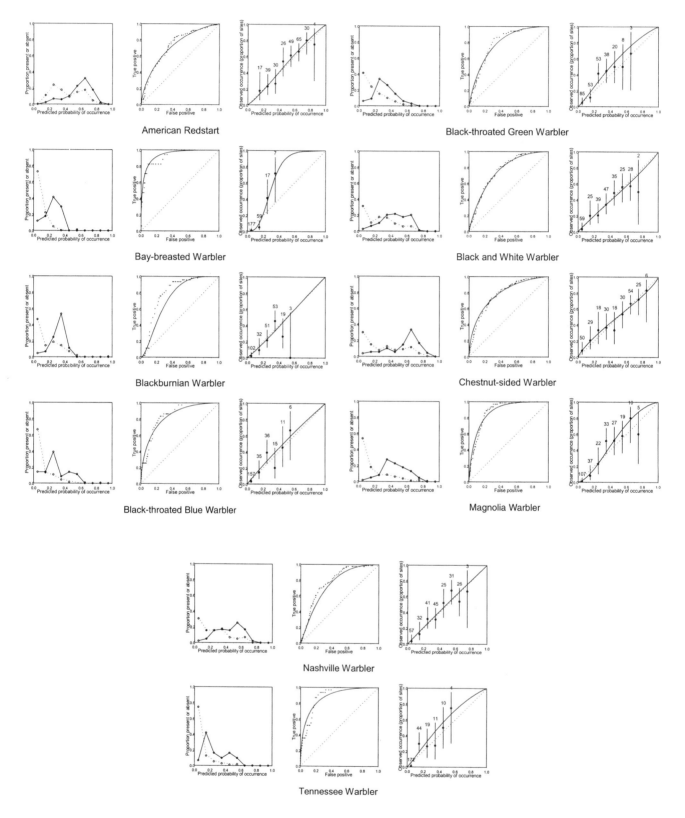

Figure 32.1. Graphical representation of predictive performance of models for ten species of forest songbirds. The first two graphs for each species depict discrimination ability using a discrimination histogram and an ROC curve. The third graph for each species is a calibration plot. Scientific names for all species listed can be found in Table 32.1.

as occupied. The false positive fraction (or commission error) is the proportion of unoccupied sites incorrectly predicted occupied by the model. The area under this curve, expressed as a proportion of the area yielded by a model with perfect accuracy, provides a measure of discrimination ability. An area of 0.5 suggests that the discrimination ability of a model is equivalent to that obtained by a random model (i.e., a random assignment of predicted values to sites). As discrimination ability improves, the area under the curve increases up to a maximum of one.

ROC curves were developed for each of the ten models evaluated, by plotting the sensitivity and false positive values obtained for one hundred evenly spaced probability thresholds. The thresholds spanned the range of predicted values within the evaluation data. The ROC plots for each species are presented in Figure 32.1. In each graph, discrimination performance equivalent to random is indicated by the 45° line. As discrimination accuracy increases, the curve tends toward the upper lefthand corner of the graph. The ROC curves in Figure 32.1 indicate fair to good discrimination performance for all ten species models.

An equivalent measure of discrimination ability to the area under the ROC curve is the Mann-Whitney-Wilcoxon test (Hanley and McNeil 1982). In this test, all possible pairs of sites are compared, and the ability of the model to accurately rank occupied sites as more likely to support the species than unoccupied sites is evaluated. This measure is easily calculated by any statistical package. However, it doesn't preclude the need to examine the ROC curve itself, especially if two equivalent models are being compared. The curve provides significant information on model performance at any given threshold level and describes the relationship between sensitivity and false positive values as the threshold level is changed (see Pearce and Ferrier 2000 for further discussion of the interpretation of the ROC curve). For each model, we calculated an index of discrimination accuracy using the Mann-Whitney-Wilcoxon statistic. A standard error for this index was calculated using bootstrapping (two hundred samples). As for the ROC area, this discrimination index ranges between 0.5 and 1, with 0.5 indicating discrimination performance equivalent to a random model and 1 indicating complete separation between the predictions for occupied and unoccupied sites. Pearce and Ferrier (2000) provide guidelines for interpreting the 0.5–1 value range. They suggest that values greater than 0.9 indicate an excellent level of discrimination because the sensitivity values are high relative to the false positive values. Values between 0.7 and 0.9 indicate a reasonable level of discrimination, while values between 0.5 and 0.7 indicate poor to marginal discrimination ability because the sensitivity rate is not much higher than the false positive rate. Based on these criteria, models for all ten species show acceptable (ROC area greater than 0.7) levels of discrimination, with discrimination indices in the 0.73-to-0.93 range (Table 32.2). Each of these ten models therefore provides a good index of species occurrence within the Great Lakes region.

Does a Model Provide an Accurate Estimate of the Probability of Detecting a Species at Each Site?

The second approach considered by Pearce and Ferrier places greater emphasis on the predictions than on the observations. That is, the researcher is interested in how accurately the model can provide estimates of the probability of a species being detected at a site. For presence/absence data, this can be examined graphically by deriving a calibration plot. This plot is derived by dividing the predicted probability range into classes (in this case, ten evenly spaced classes between zero and one) and plotting the proportion of occupied sites within each class (y-axis) against the median predicted value for the class (x-axis). For a well-calibrated model (as shown for the Nashville warbler in Fig. 32.1), the points should be distributed along a 45° line, where the observed proportion of occupied sites equals the median predicted value for each class. The relationship between model predictions and observations can be modeled using logistic regression to relate the observed values to the logit of the predictions. This line can be added to the calibration plot to further describe the distribution of the points around the 45° line.

A calibration plot was developed for each of the ten species models (Fig. 32.1). The fitted regression line in each of these plots follows the 45° line fairly closely

TABLE 32.2.

Results of statistical evaluation of predictive performance.

Species[a]	ROC (se)	Cox model: a+b (logitp)	Null deviance (0,1)	Deviance (a,b)	Cox statistic D(0,1)– D(a,b)	Deviance (a,1)	Bias statistic D(0,1) – D(a,1)	Spread statistic (a,1) – D(a,b)
American redstart	0.7335 (0.0023)	0.083 + 1.104	315.14	314.41	0.73			
Bay-breasted warbler	0.9228 (0.0019)	2.329 + 2.587	87.44	76.04	11.40 (0.01)	242.16	−154.72 (0.01)	166.11 (0.01)
Blackburnian warbler	0.7962 (0.0023)	−0.044 + 1.054	197.04	196.66	0.37			
Black-throated blue warbler	0.8454 (0.0022)	0.180 + 0.993	157.49	156.63	0.86			
Black-throated green warbler	0.7972 (0.0019)	0.473 + 1.127	241.03	236.30	4.73			
Black-and-white warbler	0.7604 (0.0021)	−0.082 + 0.938	268.38	268.13	0.25			
Chestnut-sided warbler	0.7839 (0.0021)	0.040 + 0.832	288.05	286.22	1.83			
Magnolia warbler	0.8813 (0.0015)	0.385 + 1.436	189.07	183.86	5.21			
Nashville warbler	0.7792 (0.0022)	0.054 + 0.990	268.57	268.41	0.16			
Tennessee warbler	0.8861 (0.0016)	0.510 + 1.117	129.59	127.01	2.58			

[a]See Table 32.1 for scientific names.

for all species except the bay-breasted warbler (*Dendroica castanea*). This indicates that nine out of the ten models are well calibrated.

To quantify the degree of calibration for each of the models, three statistics can be calculated. The first statistic, called the Cox statistic, describes whether there is systematic departure from the 45° line in the calibration plot. Calibration error can then be partitioned further into two measurable components of systematic error for which statistics can be calculated, bias and spread, and a third component, unexplained error. These components can be best explained in terms of the regression line fitted to observations and predictions in a calibration plot (Fig. 32.1). Bias and spread can be thought of as the intercept and slope, respectively, of this line. If a model is well calibrated, then the points in a calibration plot should lie along the 45° line. Bias describes

a consistent overestimate or underestimate of the probability of occurrence, resulting in an upward or downward shift in the regression line across the entire predicted probability range. Spread describes the systematic departure of the regression line from a gradient of 45 degrees. A gradient between 0 and 1 implies that the predicted values less than 0.5 are underestimating the occurrence of the species and that predicted values greater than 0.5 are overestimating the occurrence of the species. A gradient greater than 1 indicates that the converse is occurring.

Bias usually indicates that the prevalence of the species in the evaluation sample is greater or less than that in the model development sample due to, for example, the use of different survey techniques, or seasonal variation in the abundance or detectability of the species. Bias may also occur when a model developed in one region is applied to another region where the species is more or less prevalent. Spread error indicates misspecification of the model.

The third component of calibration, unexplained error, relates to the variability of individual records around the regression line fitted to predicted and observed values and describes variation not accounted for by the bias and spread of a model. Some of this variation may arise because particular covariate patterns or habitat types are not well represented in the development of the model. These sources of error may be identified through an analysis of residuals. These results are not presented here.

To partition the calibration into its error components, bias and spread, the observations were modeled as a function of the logit of the predictions, using logistic regression. The regression equation obtained for each species is listed in Table 32.2. In this equation, the *a* and *b* coefficients represent bias and spread respectively. The significance of bias and spread error in the predictions can be examined by calculating the deviance of the regression model fitted to observations and predictions, termed the Cox model. This deviance value, denoted deviance(a,b), describes the deviance of observations from predictions after accounting for any systematic bias or spread in the predictions. Therefore, if the deviance(a,b) is significantly different from the null deviance(0,1) of this Cox model, then predictions from the original model exhibit a significant level of bias or spread error. This change in deviance can be compared to a chi-squared distribution with two degrees of freedom. As shown in Table 32.2, the change in deviance (Cox statistic) is not significant for all models except for that of the bay-breasted warbler. Nine of the ten models evaluated therefore generate well-calibrated predictions that can be used confidently at face value as estimated probabilities of occurrence.

The predictions for the bay-breasted warbler were examined further to determine whether the poor calibration result obtained for this model was due to bias error or spread error or both. To this end, bias and spread statistics were calculated. The bias statistic is calculated as the difference between the deviance(0,1) of the original model and the deviance obtained from the Cox model when the *b* coefficient is held at 1, deviance(a,1). This tests the hypothesis that there is sig-

nificant bias error in model predictions. This change in deviance, when compared to a chi-squared distribution with one degree of freedom was highly significant. The high value (2.329) for the intercept in the Cox model suggests that the predictions are, on average, underestimating the occurrence of the species within the validation sample.

The significance of spread error was calculated as the difference between the deviance(a,b) of the Cox model, and the deviance(a,1) of the modified version of this model in which the *b* coefficient is held at 1. This change in deviance tests the hypothesis that there is no spread error, given the presence of bias error. This statistic was highly significant when compared to a chi-squared distribution with one degree of freedom. The positive *b* coefficient in the Cox model suggests that the underestimation problem revealed by the test for bias is more pronounced for high predicted probabilities than it is for low predicted probabilities. Predictions from this model are not well calibrated and therefore cannot be used confidently as estimated probabilities of occurrence. However, the evaluation of discrimination ability indicated that predictions from the model nevertheless perform well as a rank index of likelihood of occurrence.

Conclusion

Graphical and statistical evaluation of the predictive performance of the ten models for the Great Lakes Basin suggests that these models are suited to a wide range of applications, such as reserve design, exploring relationships between regional occurrence and environmental factors, or locating potential survey sites. They all provide predictions that have good to excellent discrimination ability, thereby providing good rank indices of occurrence across the region. All the models except that of the bay-breasted warbler are also well calibrated. The predictions from these models can therefore generally be relied on to provide an accurate estimate of the probability of detecting a species at a given site. The evaluation has given us confidence in using the models for regional conservation planning.

The techniques presented here are readily applied using any statistical package that can accommodate

TABLE 32.3.

Summary of evaluation objectives and graphical and statistical tests.

	Evaluation objective 1	**Evaluation objective 2**
Definition	The model is to be used to provide a rank index of species occurrence.	The predictions are to be used at face value.
Attribute of prediction quality	Discrimination	1. Calibration 2. Bias 3. Spread
Graphical evaluation	1. Discrimination histogram 2. ROC curve	Calibration graph
Statistical evaluation	Mann Whitney-Wilcoxon test	1. Cox statistic 2. Bias measure 3. Spread measure

logistic regression analysis. Therefore, the ease of application and interpretation of these evaluation techniques suggest that they should be more widely applied to test the performance of models of species distribution. Table 32.3 provides a summary of the evaluation approaches described by Pearce and Ferrier (2000) and applied in the current study. As a minimum, an evaluation of discrimination ability should be routinely performed before applying distributional models in any conservation planning or management exercise.

Toward Better Atlases:
Improving Presence-Absence Information

Douglas H. Johnson and Glen A. Sargeant

Atlases are effective tools for illustrating the spatial distribution of species (Bibby et al. 1992). A typical atlas represents a geographical area as a grid of cells. The entry in each cell indicates whether a particular species has been recorded in the area represented by that cell. Sometimes more than presence or absence is indicated, such as the type or strength of evidence for presence (e.g., recent versus historic records, or museum specimens versus field observations). An example of an atlas for breeding birds (western meadowlark, *Sturnella neglecta*) in North Dakota is given in Figure 33.1a in the color section. Squares denote confirmed (e.g., nests or dependent young observed) or suspected (e.g., singing male detected) evidence of breeding during 1950–1972.

Atlases serve many purposes. State breeding-bird atlases (e.g., Peterson 1995; Jacobs and Wilson 1997), for example, contribute information used in field guides. Atlases can be used to show habitat affinities of birds at a broad scale and can illustrate sympatry or allopatry of closely related or competitive species. Atlases also distinguish species that have broad ecological tolerances, and hence wide geographical distributions, from those with narrow tolerances and correspondingly limited distributions. Atlases help suggest areas where habitat changes, such as those caused by urbanization, can affect certain species. Gap analysis projects make effective use of atlases in developing layers in geographic information systems and identifying areas of high biotic diversity that are not protected by, for example, parks and nature reserves (Scott et al. 1993).

Ideally, an atlas would be based on exhaustive surveys of all habitats in each cell. Rarely is this perfection achieved, however, especially for large areas. More often, atlases are based on systematic surveys of a subsample of cells, on data gathered opportunistically, or on some combination of these. In the latter two cases, the amount of effort expended by observers in cells is likely to vary widely. This disparity in effort results in larger numbers of species recorded near human population centers, especially university towns, than in more remote locations, regardless of the true numbers of species that actually occur in each type of area.

Developing an atlas is complicated by variability in scale and time (extent and grain) (Fielding, Chapter 21). Atlases are subject to two kinds of errors: false positives (sometimes called commission errors)—cells with reported occurrences of a species that does not really occupy those cells; and false negatives (omission errors)—cells with no indication of a species that really is there. False positives are relatively uncommon: mistakes made due to misidentification, observations of escaped animals, or individuals blown off course during migration, for example. False negatives are much more likely to occur. Most such errors reflect insufficient effort by observers, especially for species that are rare, secretive, or occur only sporadically in some cells.

Recent advances in three fields facilitate the improvement of atlases. First, a considerable amount of effort has gone into developing models to predict species occurrences from habitat information (e.g., Verner et al. 1986a,b) and other explanatory variables, such as climatic features (e.g., Nicholls 1989; Walker 1990). By incorporating spatial information as well as other explanatory variables, such models can be useful for improving atlases (e.g., Le Duc et al. 1992; Buckland and Elston 1993; Smith 1994; Högmander and Møller 1995; Augustin et al. 1996). Second, in image processing, observed images are sometimes treated as degraded views of true images. Methods of image reconstruction have been developed (e.g., Besag 1986; Besag et al. 1991); these can be adapted to atlas mapping by considering an observed atlas to be an imperfect representation of the true distribution of a species (Heikkinen and Högmander 1994). Third, the widespread availability of powerful computers has permitted more-intensive statistical analyses, including methods such as Markov chain Monte Carlo (Geman and Geman 1984; Gelfand et al. 1990; Gelfand and Smith 1990; Besag et al. 1995). Some of these methods are complex, however, and currently not readily accessible to many biologists.

Our objective is to demonstrate some relatively simple and intuitive smoothing methods that can be applied to observed atlases to improve their accuracy. We use presence/absence information from nearby cells, as well as auxiliary data, such as habitat information, from each target cell. We propose a proxy for the effort expended by observers in a cell and incorporate that variable as well.

We illustrate the methods on both real and artificial data. The real data are breeding-bird atlases for western meadowlark (Fig. 33.1a), Baird's sparrow (*Ammodramus bairdii*, see Fig. 33.2a in color section), and blue jay (*Cyanocitta cristata*, see Fig. 33.3a in color section) in North Dakota (Stewart 1975). Cells in the atlas grids are legal townships, each about 6 miles square (total area is 9,324 hectares). Artificial data (see Figs. 33.4a, 33.5a in color section) were generated on a similar grid to exemplify a species—like the blue

jay, which is associated with woodland—that has a complex distribution that depends on a particular habitat, which is known and mapped. Two artificial distributions—one of a highly detectable species, another of a less-detectable species—are treated as known (Appendix). Our methods are applied to subsamples drawn from the distributions to demonstrate improvements in accuracy, that is, to validate the methods.

We were able to test the predictions generated by the methods developed here with an independent data set. Browder (1998) conducted point counts of birds at 885 points in 1995 and/or 1996. Points were clustered into forty-four roadside routes. Her methods were analogous to those used in the North American Breeding Bird Survey (Robbins et al. 1986).

Methods

The original data (Stewart 1975) are mapped on a grid of cells. For each cell, the indicator variable I assumes a value of 1 if the particular species was recorded there and 0 if it was not. We wish to estimate for each cell the probability (J) that the species actually occurs there. Thus, each entry will be a value between 0 and 1.

Simple Spatial Smoothing

The first model we describe simply replaces I in each cell with $I = 0$ with a weighted average of values of I from adjacent and diagonal townships. Weights (w) are inverses of mean distances (d) between the target cell and neighboring cells, standardized to sum to 1. This method of smoothing is illustrated for a single target cell and its eight neighbors or near-neighbors in Figure 33.6 (see color section). It produces a value in the target cell ranging from $J = 0$ (if the species was not recorded in any neighboring cell) to $J = 0.84$ (if the species was recorded in every neighboring cell but not in the target cell).

Habitat Commonality

The assumption underlying the simple spatial smoothing method, that a species probably occurs in a target cell if it occurs in most neighboring cells, may not be true if the target cell is lacking essential habitat that is

TABLE 33.1.

Weights used in the habitat commonality smoothing method, based on the presence or absence of the species in a neighbor cell and on the presence or absence of suitable habitat in the target cell and the neighbor cell.

Species present in neighbor cell		
	Suitable habitat in neighbor cell	
Suitable habitat in target cell	Yes	No
Yes	1	1[a]
No	0	1
Species absent in neighbor cell		
	Suitable habitat in neighbor cell	
Suitable habitat in target cell	Yes	No
Yes	1	0
No	1[a]	1

[a]Denotes nonintuitive weights (weights of 1 even though cells do not share the required habitat). The first indicates that a species is present in a neighbor cell that lacks appropriate habitat, so it is reasonable to deduce that the species might be present in the target cell, which does have the habitat. This assumes an error either in the habitat database or in the assumption that the habitat actually is required. The second says that the species is absent from a neighboring cell with the requisite habitat, so it is reasonable to assume it will be absent from the target cell, which does not have suitable habitat.

available in neighboring cells. We incorporated this concept by using habitat weights (Table 33.1) and multiplying the initial simple smoothing weights (w) by the habitat weights before standardizing them to sum to 1. Habitat weights are 1 if the target cell and the neighbor cell both have the habitat required by the species. They are also 1 if the required habitat is available in the target cell and the species is present in a neighbor cell without that habitat; or if the target cell lacks the requisite habitat and the species is absent in a neighboring cell with the habitat. This forms the basis for the second model.

The three species we illustrate have different habitat affinities. The western meadowlark is a grassland generalist, which uses a variety of short to tall, native or introduced, grassy to brushy habitats (Dechant et al. 1999a). Baird's sparrow is a grassland specialist; its favored habitat for breeding is grasslands of medium height, usually composed of native plant species, with little woody vegetation, and preferably in large

patches (Dechant et al. 1999b). The blue jay is a woodland edge species, rarely found in open habitats (Stewart 1975).

In North Dakota, suitable habitat for the western meadowlark exists in virtually every cell (township), so the habitat commonality method will offer no improvement over simple spatial smoothing. And, although Baird's sparrow is more restricted in its selection of grassland, there is no extant database that would outline suitable habitat for the species. For the blue jay, a database that provided useful information about one component of its favored habitat (forestland) was the land-use and land-cover maps (Anderson et al. 1976) developed by the USGS (1986) from color-infrared aerial photography.

Search Effort

With the third model, we account for variable search efforts among cells. For instance, we would not want to "borrow" information about a species being absent in a neighboring cell if very little observation had taken place there. Because the effort expended by observers in each cell was not recorded (cf. Osborne and Tigar 1992), we developed a proxy for it based on the number of species recorded. We used the double-log function,

$$E = \frac{\log(\log(Nspecies)+1)+1}{\log(\log(\max(Nspecies)+1)+1)} \quad (33.1)$$

where E is a proxy for effort and $Nspecies$ is the number of species recorded in a cell. This function incorporates the notion that cells with greater effort will have more species reported, but that, once the effort becomes substantial, new species will be added only slowly. Also note that this function estimates effort to be zero for cells with no species observed; we believe that, in our example, all townships in North Dakota have breeding birds, which would have been detected with even the slightest effort. In our North Dakota examples, max $(Nspecies) = 124$. E attains the value of 0.30 with but a single species recorded, reaches 0.60 with six species, and reaches 0.90 with forty-eight species.

This proxy for effort is based on the implicit assumption that each cell has the same number of species that could be detected. Although this assumption is no doubt false in general, it is likely to be approximately true for cells that are nearby, especially if they share similar habitats.

Testing the Models

For use here, we aggregated the points at which Browder (1998) surveyed birds into the legal townships that contained the points. The 885 were ultimately grouped into ninety-five legal townships, with the number of points per township ranging from one to twenty-three (mean = 9.3). We then determined for each township whether or not each species was detected on any point count in either year. We compared resulting presence/absence values to the values predicted from our model as well as to the presence/absence values in the original atlas. For the meadowlark and Baird's sparrow, we used the models that accounted for search effort; for the blue jay, we used the model that incorporated habitat information as well as search effort.

To evaluate the models, we computed the sum of squared errors as well as the sum of absolute values of the errors. We additionally conducted logistic regression analyses with observed presence/absence (Browder's results) as the binary response variable. Explanatory variables were predicted values from each model and from the presence/absence values in the original atlas.

Results

Spatial smoothing basically "spread" the information on occurrences in a cell into neighbor cells, or, from another point of view, each target cell "borrowed" information from neighbor cells. For the ubiquitous western meadowlark, spatial smoothing increased the number of cells with positive probabilities of occurrence (Fig. 33.1b), which is an entirely reasonable outcome. Similar changes occurred for the Baird's sparrow and blue jay (Figs. 33.2b, 33.3b), but we cannot be sure that those changes represent actual improvements in the atlases.

For the artificial data, for which we know the true distribution, simple spatial smoothing tended to fill in gaps in the distribution that resulted from insufficient effort, leading to improved atlases (Figs. 33.4b, 33.5b). Offsetting that gain in accuracy was incorrect expansion of the range caused by smoothing informa-

tion from cells at the border of the range into cells outside the range. Nonetheless, the overall error rate for the high-detectability situation decreased from 22 percent for the actual data to 7 percent after simple spatial smoothing. With low detectability, the error rate improved from 28 to 13 percent.

Habitat Commonality

We had useful habitat information only for the blue jay. By incorporating habitat availability in the weighting, information was not simply spread out; such spreading occurred only if suitable habitat was available (Fig. 33.3c). Compare, for example, the top center of the state in Figures 33.3b and 33.3c.

With the artificial data, incorporating habitat information (knowing there was no suitable habitat outside of the occupied range) enhanced the atlases (Figs. 33.4c, 33.5c). The improved accuracy was primarily due to the trimming of the smoothed atlases outside the boundary. Error rates decreased from 22 to 2 percent for the high-detectability scenario and from 28 to 11 percent for the low-detectability situation.

Search Effort

Adjusting for survey effort improved the atlas for the western meadowlark (Fig. 33.1c). The atlas has become nearly completely filled in, as we believe is appropriate for this species. We suspect that the Baird's sparrow results (Fig. 33.2c) are likewise improved, but because we lack knowledge of the true distribution, we cannot be sure. Adding information about search effort made little apparent difference for the blue jay atlas (Fig. 33.3d).

For the artificial data, the additional information about search effort did not improve error rates beyond what the habitat commonality method did (Figs. 33.4d, 33.5d); error rates remained at 2 and 11 percent for the high- and low-detectability scenarios, respectively. Nonetheless, the extra step did reduce the apparent patchiness of the resulting atlas, which was an appropriate change.

Testing the Model

The western meadowlark is ubiquitous in North Dakota. A perfect atlas would have every cell filled.

Browder recorded the species in 83.2 percent of the cells she surveyed (Table 33.2). In the original atlas, only 69.5 percent of those cells were filled. Our model produced an average value across all cells of 79.9 percent, closer to Browder's result and to the presumed true value of 100 percent. It could not attain that high a value because values smoothed from adjacent cells are less than 1. Clearly though, the model indicates that western meadowlarks likely occur in each cell (Fig. 33.1). Both the sum of squared errors and sum of absolute errors were smaller for our model values than for original atlas data (Table 33.2), suggesting the superiority of the model values. The logistic regression indicated that neither the original atlas nor our modeled value was useful in predicting where Browder would find the species; that result is consistent with the species' ubiquity.

Baird's sparrows are most common in the western two-thirds of North Dakota. They are erratic in their occurrence in most locations, however, and are easily overlooked. Browder detected Baird's sparrows on six (6.3 percent) of the ninety-five townships she visited. Only two of those townships had records in the original atlas data, suggesting that the atlas greatly underestimates the distribution of that species. Baird's sparrows were recorded in 12.6 percent of the cells in the original atlas (Table 33.2), lower than the average model value of 20.0 percent. The sum of squared errors and sum of absolute errors gave conflicting results (Table 33.2). Logistic regression indicated that our model values were useful in predicting where Browder would find Baird's sparrows ($P = 0.035$), even after accounting for the effects of the original atlas value ($P = 0.10$).

Blue jays are distributed widely throughout North Dakota where woodland occurs. Their numbers doubled between 1967, near the time when the original atlas was completed, and 1992–1993, near the time of Browder's survey (Igl and Johnson 1997). The increase is likely due at least in part to increases in woody vegetation such as shelterbelts (L. D. Igl and D. H. Johnson unpublished data) and possibly to increased feeding of birds, especially in winter. The sum of squared errors and sum of absolute errors were inconsistent (Table 33.2). Browder recorded the species in 20.0 percent of the cells she visited, whereas only 10.5 per-

TABLE 33.2.

Criteria comparing predictive ability of original atlas and atlas developed from models described in the text, when applied to an independent data set (Browder 1998).

	Western meadowlark	Baird's sparrow	Blue jay
Sum of squared errors			
Original atlas	33.0	14.0	17.0
Model values	15.2	10.9	13.2
Sum of absolute errors			
Original atlas	33.0	14.0	17.0
Model values	27.7	19.7	20.1
Average value			
Browder	0.832	0.063	0.200
Original atlas	0.695	0.126	0.105
Model values	0.799	0.200	0.114
Logistic regression deviance			
Original atlas	0.43	1.88	8.89[a]
Model values	0.56	4.43[b]	14.40[c]
Model values/original	0.18	2.70[d]	5.63[b]
Original/model values	0.05	0.16	0.10

[a] $P \leq 0.01$
[b] $P \leq 0.05$
[c] $P \leq 0.001$
[d] $P \leq 0.10$

cent of cells were filled in the original atlas. Our model improved that only to 11.4 percent, likely because it was based on the original data, which do not reflect recent increases in the number and distribution of blue jays. Both the original atlas value and our model value were useful predictors of where Browder would find blue jays ($P < 0.003$). Our model values were useful even after atlas values were included in the logistic regression ($P = 0.018$); the converse did not hold true ($P = 0.75$).

Discussion

We have demonstrated that relatively simple computational methods can markedly improve the accuracy of atlases. Without any additional information, simple spatial smoothing can improve an atlas, especially for a species with a widespread distribution. Knowledge of the habitat affinities of a species, in combination with a spatial representation of the availability of that habitat, can further improve the accuracy (as with the blue jay). Improvements may be greater for species with narrow ecological requirements, especially if the spatial distribution of those requirements is known. Finally, knowledge of the amount of effort expended in each cell can be exploited to improve atlases. Such knowledge may not be available, but the proxy we offered for effort appeared to work well in our examples.

Browder's (1998) data provided a useful test of the models developed here. The model for western meadowlark better represented the ubiquity of that species in North Dakota than did the original atlas. For the Baird's sparrow and blue jay, model values were better predictors than were the original data in logistic regression.

Data from the examples we presented (Stewart 1975) were not gathered according to any particular design. The resulting atlases, even for ubiquitous species such as the western meadowlark, have fairly large "holes" in them. The methods we presented cannot satisfactorily fill such holes. Moreover, they probably should not, for such an outcome would be based far more on assumptions that underlie the analytic method than on the data themselves. All statistical results are a product of, in various combinations, the data employed and the assumptions on which the analytic method is based. More sophisticated methods yield more definitive results, but only if the underlying assumptions are reasonable.

The design of an atlas project merits serious consideration. The optimal design will depend not only on the intended uses for the resulting atlas, but also on the analytic method to be used on the results. If smoothing procedures are not to be used, it is likely that the best design for gaining an overall impression of the distributions of the various species would involve samples spaced regularly throughout the grid. If, as in our examples, there is an extant database of opportunistically gathered presence-absence data, an optimal design for enhancing an atlas might involve drawing a new sample with the probability of including a cell based on the inverse of the effort already expended in that cell (or the inverse of the proxy we described).

Improved mappings of habitat availability could dramatically increase the value of the habitat commonality method we presented. Such maps are becoming more widely accessible with improved remote-sensing capabilities and efforts such as gap analysis. Also likely to be helpful is new information or better syntheses of information on the habitat affinities of species.

Other improvements might result from using other kinds of auxiliary information. One avenue we intend to explore is the possibility of using information about certain species to predict the presence or absence of other species. A particularly promising application of that approach would involve the use of information from a highly detectable species to predict the occurrence of a more secretive species that has very similar habitat requirements.

Atlases are becoming increasingly common, not only for birds but also for other vertebrates, invertebrates, and plants. Improved computer graphics capabilities facilitate the economical and widespread dissemination of full-color maps and other visual products. The World Wide Web offers great potential not only to display atlases, but also to encourage individuals to provide information to improve them. Considerable effort often goes into gathering the base data, verifying and processing them, and presenting the results. As important as the final products are, it behooves us to develop and use the best analytical methods for them.

Acknowledgments

We are grateful to Sharon F. Browder for the use of her data, to Laurence L. Strong of the North Dakota Gap Program, Northern Prairie Wildlife Research Center, for access to habitat information, and to Betty R. Euliss for numerous contributions. The manuscript benefited from reviews by Robert R. Cox Jr., Harri Högmander, and Laurence L. Strong.

Appendix: Generation of Artificial Data

We considered two species, one with high detectability and one with low detectability. We assumed that for the maximum effort expended in any cell, the probability (R) of recording the species was $R = 0.50$ for the highly detectable species and $R = 0.20$ for the less-detectable species.

We defined a grid of sixty-five by thirty-five cells, about the size of North Dakota used in the other examples. Within that grid, we assumed suitable habitat existed, and the species actually was present, in the cross-shaped region outlined in Figure 33.4. The area outside that region was deemed unsuitable for and unoccupied by the species.

To develop effort values for each cell, we used our proxy based on the number of species (*Nspecies*) recorded in the cell. We used a beta distribution (with exponents α = 1.2 and β = 5) to approximate the distribution of *Nspecies* recorded in the townships of North Dakota. For each cell in the grid, we randomly selected a value from that beta distribution, multiplied it by 125, and rounded down to next lowest integer to get a value for *Nspecies*. That value was used in Equation 33.1 to calculate the effort proxy, *E*. That value was multiplied by the detectability value to obtain a probability of recording the species (*P*) for each cell:

$$P = R \times E$$

We next drew a random variate from a binomial distribution with probability *P*. If the outcome was a "success," we recorded a detection for that cell (*I* = 1); otherwise, the species was not listed as recorded (*I* = 0). That set of values for the entire grid represented the observed or raw data, which we wished to smooth using the methods we present here.

To obtain effort data for using with the search effort method, we drew a random number (*Nspecies'*) from a Poisson distribution with parameter *Nspecies* obtained above. This step added random noise, so that we did not use the same value of effort to smooth the data as we used to create them.

We then applied the various smoothing methods, using the *I* and *Nspecies'* values generated above. We obtained two grids, one for a highly detectable species, another for a less-detectable species.

Predicting the Distributions of Songbirds in Forests of Central Wisconsin

Margaret J. Robertsen, Stanley A. Temple, and John Coleman

Many studies have demonstrated the importance of fine-grained habitat features, typically within plots the size of a bird's territory, on predicting the distribution of songbirds (Cody 1985a,b; Block and Brennan 1993). Others have considered the importance of coarse-scale variables, measured over landscapes much larger than a territory or home range (Shriner et al., Chapter 47; Boulinier et al. 1997; Lauga and Joachim 1992; Gustafson and Crow 1994; Pearson 1993). Habitat selection is believed to be a multiscale hierarchical phenomenon spanning selection of a geographic range (first-order selection) to patterns of habitat utilization (third- or fourth-order selection) (Johnson 1980). Combining fine- and coarse-grained levels of habitat selection requires a multilevel approach.

Geographic information systems (GIS) can be used to draw relationships between bird species and environmental variables and map distributions across large geographic areas. Palmeirim (1988) used a GIS to display the predicted distribution of songbirds by extracting cover-type information from Landsat satellite imagery (Thematic Mapper) and by incorporating rules on patch size requirements. Green et al. (1987) combined bird survey data with Landsat land-cover information to predict the abundance of hooded warblers across a two-million-hectare area of the Yucatan Peninsula. Information on slope, aspect, geology, and land cover extracted from a GIS database were used to map habitat areas in Texas for two endangered songbird species, the golden-cheeked warbler (*Dendroica chrysoparia*), and the black-capped vireo (*Vireo atricapillus*) (Shaw and Atkinson 1990).

GISs and categorical models can be combined to quantify habitat use across a landscape. Pereira and Itami (1991) applied a logistic regression model in a GIS environment to predict the distribution of Mt. Graham red squirrels based on environmental and locational variables. Mladenoff et al. (1995) used a similar approach to predict the presence or absence of wolf packs as a function of spatial indices and landscape variables.

Forest songbirds have been identified as high-priority indicators for ecosystem management in the Baraboo Hills of Wisconsin (Baraboo Hills Working Group 1994). The purpose of the study described in this chapter was to apply a multidisciplinary tool (GIS) in developing predictive habitat models that could be used for the conservation of forest songbirds across a 580-square-kilometer landscape. Our first objective was to determine whether we could reliably predict the distribution of six forest songbird species across the landscape by combining a GIS and a categorical modeling procedure (logistic regression). As a result we wanted to identify important forest stand and landscape-scale habitat components and to predict the distribution of songbirds based on these variables. Our second objective was to demonstrate the usefulness of the resulting models for long-term management of the forest landscape by projecting how bird habitats would change under realistic future scenarios.

Study Area

The Baraboo Hills ecosystem, located in south-central Wisconsin, is an area renowned for its biological diversity. This status was formalized in 1994 when it was declared a "bioreserve" by The Nature Conservancy. Twenty-seven natural communities, seventy-seven rare or sensitive plant and animal species, and 135 bird species have been documented within this area (Baraboo Hills Working Group 1994). This exceptional biodiversity can be partially explained by high topographic relief and the area's location at the junction of three ecoregions: the Moraines District, the Driftless District, and the Central Sand Plains District (Curtis 1959). Habitats of this ecosystem range from dry ridgetops covered with scrubby oaks and glade species to cool moist valleys harboring species such as white pine and northern hemlock.

The Baraboo Hills contains 22,260 hectares of forest surrounded by a largely agricultural landscape. The shallow, rocky soils of the Baraboo Hills made much of this area unsuitable for farming. Nearly 6,475 hectares of continuous forest remain in the Hills in contrast to other southern forests in the Midwest, which have undergone extensive fragmentation. However, human-related pressures increase daily, and only 11 percent of the area is currently protected from development. The forest community types we studied for this project are not considered rare on a national level but are highly significant in the context of statewide conservation efforts. In other words, the Baraboo Hills is the only remaining large tract of deciduous forest in south-central Wisconsin.

Methods

To develop our models, we used available GIS data layers and collected field information on the distribution of birds. The following sources of data were available: (1) vegetation data from a forest-stand in-ventory, (2) songbird data collected during point-count censuses, and (3) coarser landscape measures and spatial indices derived from GIS data (1:100,000 scale).

Forest-stand Inventory Data

Data collected by The Nature Conservancy from 1991 to 1993 provided information on stand-scale habitat variables that we believed would be important in determining the distribution of forest songbird species (Robertsen 1995). Forest stands were defined as relatively homogeneous units of forest with similar plant species composition and structure. Our research took place in "southern forest" communities (sensu Curtis 1959) in the Baraboo Hills because of the extent and importance of these types on the landscape (Clark et al. 1993a). We chose six forest-stand variables to use in model development: dominant tree type, dominant tree size class, dominant tree stocking density, understory type, understory size class, and disturbance class (Table 34.1; Robertsen 1995).

Forest Songbird Data: Selection of Plots

Point counts are an efficient censusing technique when the objective is to identify habitat use across a large geographic area (Bibby et al. 1992). A total of 550 point-count census plots were established within 233 southern forest stands to gather information on the distribution and habitat preference of songbirds (Robertsen 1995). In 1993, 468 plots were established to collect data necessary for the development of models. In 1994, eighty-two new plots were censused for the purpose of validating the predictive model for one species, the Acadian flycatcher (*Empidonax virescens*) (Robertsen 1995).

Census plots were visited once, and censuses were conducted within three hours of sunrise from June 1 to July 2 in 1993 and from June 12 to June 19 in 1994. This period covers the breeding season and the peak singing hours for male songbirds of most species in the Baraboo Hills (Mossman and Lange 1982). The number of singing males of each species was recorded within a 50-meter radius of the observer during a five-minute time period.

TABLE 34.1.

Forest songbird species of concern and eleven forest-stand and landscape variables used in the development of predictive songbird habitat models.

Forest songbirds	Forest-stand variables	Landscape (coarse-scale) variables
Veery	Dominant type: northern hardwoods, central hardwoods, oak	Distance to edge (meters): < 100, 100–200, > 200
Ovenbird	Dominant tree size class (dbh): 5–11 in., 11–15 in., 15–24 in.	Distance to water (meters): <100, 100–200, > 200
Chestnut-sided warbler	Understory type: central hardwoods, northern hardwoods	Percent forest cover within 200 meters (%): < 80, 80–99, 100
Eastern wood-pewee	Understory size class (dbh): 0–5 in., 5–11 in.	Slope (%): <10, 10–25, >25
Great crested flycatcher	Dominant tree stocking density (sq.ft/acre): 20–50, 51–85, 86+	Aspect: north, east/west, south
Acadian flycatcher	Disturbance class: grazed, unknown	

Landscape-scale Data

We expected that the following landscape-scale variables might be important in a songbird habitat analyses: distance to edge, percent forest cover, distance to water, slope, and aspect (Table 34.1). These variables were measured by using the spatial analytical capabilities of ArcInfo GIS software. The first step in obtaining these measures involved creating a point coverage that mapped our bird census plots. A vegetation-community coverage mapped the distribution of forest and nonforest and allowed us to calculate percent forest cover and distance-to-edge measures for each census plot. ArcInfo GRID was used to obtain percent forest measures, which were grouped into three equal-sized categories: less than 80 percent cover, 80–99 percent cover, and 100 percent cover (Robertsen 1995). A stream data layer (USGS digital line graph data, 1:100,000 scale, and USGS 7.5-minute quads, 1:24,000 scale) was used as the source of information for measuring distance to water. Based on a digitized contour map of the area, a triangulated irregular network (TIN) was created to determine slope (less than 10, 10–25, greater than 25 percent) and aspect (Robertsen 1995).

Model Development

We began building our models by selecting songbird species (dependent variables) and habitat features (independent variables). We chose songbird species if they were of special concern in the region (Thompson et al. 1992) and if we had detected at least thirty individuals during our censuses. By choosing this as our cutoff, we made the decision to exclude rare and accidental species to the study area. We were most interested in species such as the Acadian flycatcher that are still abundant in the Hills but considered less abundant or rare on a statewide scale. One or two years of surveys may be insufficient for sampling specialized or rare species, and such species may require specific survey methods (Karl et al., Chapter 51).

We selected eleven habitat variables based on evidence of biological significance and independence from other variables using a Pearson correlation analysis (Table 34.1, Robertsen 1995). Variables that were considered but excluded from analysis based on a correlation with another variable of 0.5 or greater included community type, understory density, and percent forest cover within 100 and 300 meters of a plot.

Songbird and habitat data were noncontinuous and required a categorical approach to modeling. We ran a univariate analysis on the 1993 census data to select habitat variables for inclusion in models (SAS Institute 1985). For each species, we evaluated the significance of each of the eleven habitat variables using a procedure outlined by Hosmer and Lemeshow (1989) and excluded those with p greater than 0.25 from further consideration.

In another premodeling step, cross-tabulations were run between all combinations of dependent and independent variables to check for marginal zeroes and to consider possible interactions between habitat variables. Because zero-cell frequencies may lead to erroneous results in categorical analyses due to the development of a singular (noninverted) covariance matrix (SAS Institute 1985), when possible, we combined categories where zero cells appeared to be a problem.

For each species and its remaining set of habitat variables, we ran a multivariate analysis to derive a final equation. Models were developed using the procedure CATMOD and followed a step-down method (SAS Institute 1985). We tested the importance of each variable removed using a likelihood ratio test (G statistic). Final models included no more than two habitat variables due to general sparseness of the data (few bird observations/total number of counts) and for ease of application in a GIS environment. For these same reasons, models considered only main effects, or, in other words, no interactions between variables.

The final model for each species included significant variables and showed a high probability of goodness of fit (Smith and Conners 1986; Hosmer and Lemeshow 1989). Model adequacy was determined by considering the residual (likelihood ratio) chi-square value and associated p value with a lack-of-fit test. The smaller the chi-square (and the higher the p) the better the model fits. Probabilities of a species being present or absent in each habitat class were generated (SAS Institute 1985).

Review and Validation of Models

To assess each model, we considered whether the final variables for each songbird made sense based on what we knew of the biology of a species (Mossman and Lange 1982; Robertsen 1995). Evaluations based on biological intuitiveness are a critical component of model validation (Burnham and Anderson 1991; Van Horne 1991). Using insights gleaned from a literature review, we deduced how songbirds would respond to stand-level variables based on their response to plot-level variables (Robertsen 1995).

We evaluated the Acadian flycatcher model using an independent data set collected in 1994. Testing a model with an independent data set is believed to be the best measure of accuracy (Fielding, Chapter 21; Capen et al. 1986). Karl et al. (Chapter 51) found that it is possible in some cases to develop and test models for more abundant species, with data from only one or two years. Censuses in 1994 were conducted at new plots within the six habitat types included in the 1993 Acadian flycatcher model. Because of the large number of censuses required to statistically test the model, the greatest sampling effort occurred at the extremes (i.e., sites with highest and lowest probability of occurrence). We compared observations for the Acadian flycatcher in each habitat class in 1994 to those predicted by the 1993 model.

Models for the other five species were not tested with independent data. However, confidence intervals on the predicted probability of occurrence in habitat of highest preference were calculated as an indication of model precision and are dependent on sample size (Table 34.2). Further application of these five models should be preceded by accuracy testing.

Model Application

One advantage of a GIS is its ability to create, display, and analyze the consequences of hypothetical changes to current landscape conditions. Using ArcInfo, we created scenarios for a 93-square-kilometer area of the Baraboo Hills and evaluated the implications of each scenario on songbird distributions. The scenarios were chosen to address likely land-use changes: (1) the combined effects of harvesting, succession, and fire management in the southern dry mesic forest communities, and (2) fragmentation of all forest types due to housing development. The assumptions required to create future conditions were based on expert opinion and existing literature (Clark et al. 1993a).

Our primary interest was in the future condition of relatively undisturbed mature stands of southern dry mesic forest and how changes in these stands would affect the songbird community. Southern dry mesic forests are dominated by a mixture of oaks (*Quercus alba* and *Quercus borealis*), basswood (*Tilia americana*) and sugar maple (*Acer saccharum*). These communities are considered unstable with dominance by one species likely to last one generation (Clark et al. 1993a). Due to the economic importance of oak, intense harvesting could rapidly

TABLE 34.2.

Results of categorical models for six species, including residual chi-square *P* for goodness-of-fit test, significant variables, habitat of highest preference, and probability of occurrence.

Species	Residual chi-square P	Variables and p-value	Habitat of highest preference	Predicted probability of occurrence in habitat of highest preference	95% confidence interval for the predicted probability
Acadian flycatcher	0.88	distance to edge (0.006) understory type (0.000)	> 200 m from edge, northern hardwoods	0.45	0.31–0.58
Veery	0.12	distance to edge (0.046) understory type (0.009)	> 200 m from edge, northern hardwoods	0.23	0.13–0.37
Ovenbird	0.38	dominant tree density (0.000) understory type (0.045)	stand basal area 86+, northern hardwoods	0.64	0.49–0.77
Chestnut-sided warbler	0.54	dominant tree size (0.002) dominant tree density (0.000)	5–11 in. (dbh), stand basal area 20–50	0.34	0.17–0.56
Eastern wood pewee	—	dominant tree density (0.019)	stand basal area 86+	0.50	0.40–0.60
Great crested flycatcher	—	dominant type (0.030)	central hardwoods	0.17	0.08–0.30

change the future composition and structure of southern dry mesic forests in the Baraboo Hills. Changes in the forest vegetation will affect habitat availability for forest songbirds.

To simulate a future condition, we changed composition and structure of 50 percent of the existing mature stands by making assumptions about timber harvesting, succession, and fire. A thirty-year timber projection for the state of Wisconsin estimates a 36.6-percent reduction in the volume of oak in the state (Spencer et al. 1988). Assuming a similar loss of oak in the Baraboo Hills—in other words, a 40 percent loss, we projected that half the reduction (20 percent) would occur due to harvesting and half (20 percent) through succession. Over thirty years, we further projected that 10 percent of the existing stands would retain or gain an oak component due to impacts of controlled burning. This projection was based on the current interest in using prescribed burns in the area. We assumed that dominant tree type, dominant tree size, dominant stocking density, and understory type changed as a result of harvesting, succession, and fire.

Fragmentation as a result of housing development has become a growing threat to the large blocks of forest critically important to forest interior songbirds

of the Baraboo Hills. Plans are underway to improve the main highway between the Baraboo Hills and Madison, Wisconsin, (a growing metropolitan area) less than 50 kilometers away. Easier commuting and the appeal of the scenic views from the Hills make them an attractive place to build new homes. Openings created by housing developments create hard edge types and lead to fragmentation of the forested landscape. We created two development scenarios, random and clustered, each with total area of new openings equal to 10 percent of the forest area.

We evaluated impacts on songbirds by calculating the area of "preferred occupied habitat" remaining under each scenario. "Preferred habitat" is defined here as the conditions with the highest probability of occurrence for a given species. We calculated number of preferred territories by multiplying the area of preferred habitat by the probability of occurrence and assuming territory size to be 1 hectare (Temple and Cary 1988).

Results

Based on regional concern and abundance on our census plots, six forest songbirds were selected for the habitat-model-building process. The Acadian fly-catcher (*Empidonax virescens*), veery (*Catharus fuscescens*), chestnut-sided warbler (*Dendroica pensylvanica*), eastern wood-pewee (*Contopus virens*), and great crested flycatcher (*Myiarchus crinitus*) are of high regional conservation concern (Thompson et al. 1992) and were relatively abundant (more than thirty individuals detected during our censuses). The oven-bird (*Seiurus aurocapillus*) is of lower regional concern, but it was chosen for our project because it was the most frequently recorded species.

Final models for each of the six species contained one to two variables (Table 34.2). Stand-scale variables were significant in all developed models. Landscape-scale (coarse-scale) variables were significant in two of the six models: the Acadian flycatcher and the veery. The equation for each of the final models is of the following form:

$$\ln(P/1 - P) = \text{baseline value} + \text{VAR1 effect} + \text{VAR2 effect} + \ldots$$

where

P = Probability that the site is occupied.

Highest probability of occurrence ranged from 0.17 to 0.64 depending on the species. Residual chi-square *P* ranged from 0.12 for the veery to 0.88 for the Acadian flycatcher (Table 34.2). Residual chi-square *P* could not be calculated for the two models with only one habitat variable.

In Table 34.3, an example of a final model for the Acadian flycatcher is summarized. A difference in coefficients between northern and central hardwoods of 2.02 indicates that northern hardwood understory

TABLE 34.3.

Logistic regression coefficients (B_i) for the effect of understory type and distance to edge on the occurrence of Acadian flycatchers (*Empidonax virescens*).

Variable and variable level	1993 (n = 486)	1994 (n = 82)
Understory type		
Central hardwoods	−1.01	−0.65
Northern hardwoods	+1.01	+0.65
Distance to edge		
Less than 100 meters	−0.75	−7.51
100–200 meters	+0.06	+3.81
More than 200 meters	+0.69	+3.70

types have increased odds of Acadian flycatcher occurrence of exp(2.02), or 7.5 times over central hardwoods. Similarly, the odds of finding an Acadian flycatcher in an area far from edge (more than 200 meters) is exp(1.44), or 4.2 times greater than the odds of finding this species close to edge (less than 100 meters) (Table 34.3; Robertsen 1995). The predicted probabilities of occurrence (P) for the final Acadian flycatcher model ranged from a low of 0.02 (central hardwood understory, close to edge) to a high of 0.45 (northern hardwood understory, far from edge).

We found that the probability of occurrence for Acadian flycatchers in each habitat class in 1994 closely followed what was predicted based on the 1993 model (Table 34.4). A chi-squared analysis indicated that differences between the two years were nonsignificant but that the test lacked the power re-

TABLE 34.4.

Observed number of 1994 plots with Acadian flycatchers (*Empidonax virescens*) present and absent for six habitat classes compared to predicted numbers based on the 1993 model.

Understory type and distance to edge	Predicted present	Observed present	Predicted absent	Observed absent
Central hardwoods, < 100 m from edge	0.4	0	17.6	18
Central hardwoods, 100–200 m from edge	0.2	0	3.8	4
Central hardwoods, > 200 m from edge	1.5	3	15.5	14
Northern hardwoods, < 100 m from edge	3.2	0	16.8	20
Northern hardwoods, 100–200 m from edge	1.6	3	4.4	3
Northern hardwoods, > 200 m from edge	6.8	5	8.2	10

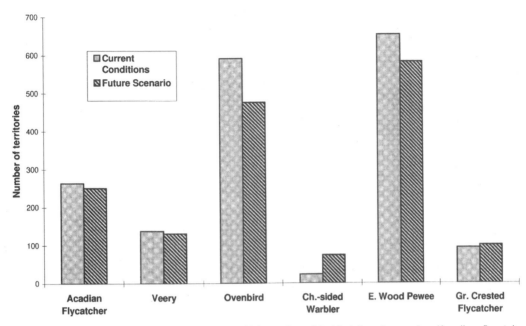

Figure 34.1. Predicted change in number of territories within preferred habitat for six species (Acadian flycatcher [*Empidonax virescens*], veery [*Catharus fuscescens*], ovenbird [*Seiurus aurocapillus*], chestnut-sided warbler [*Dendroica pensylvanica*], eastern wood-pewee [*Contopus virens*], great crested flycatcher [*Myiarchus crinitus*]) after thirty years of timber harvest, fire, and succession in the Baraboo Hills of Wisconsin.

quired to make a definite conclusion due to small sample size in 1994 (X^2 = 8.78, df = 7, p > 0.25).

Model Application

As a result of our harvesting/succession/fire scenario, four of the six species showed a decrease in abundance under projected future conditions: Acadian flycatcher (−5 percent), veery (−5 percent), ovenbird (−19 percent), and eastern wood-pewee (−11 percent) (Fig.

TABLE 34.5.

Predicted percent change in preferred territories for six forest songbird species after thirty years of timber harvest, fire, and succession in the Baraboo Hills of Wisconsin.

Common name/*Scientific name*	Predicted percent change in preferred territories (%)
Acadian flycatcher/*Empidonax virescens*	−5
Veery/*Catharus fuscescens*	−5
Ovenbird/*Seiurus aurocapillus*	−19
Chestnut-sided warbler/ *Dendroica pensylvanica*	+222
Eastern wood-pewee/*Contopus virens*	−11
Great crested flycatcher/*Myiarchus crinitus*	+8

34.1, Table 34.5). Two of the six increased in abundance: chestnut-sided warbler (+222 percent) and great crested flycatcher (+8 percent). The greatest benefit of this future scenario was realized by the chestnut-sided warbler, a species that currently has much less preferred habitat across the landscape than the other five. The greatest negative effect (a 19 percent loss) occurred for the ovenbird, the species currently having the largest amount of "preferred" habitat.

Five of the six songbird species experienced losses in total number of preferred territories (average 13-percent loss) under the random development scenario (Table 34.6). The losses resulting from a clustered development scenario ranged from 4 to 14 percent (average 8 percent loss) (Table 34.7). The losses of preferred territories greater than 200 meters from an edge was much higher under a random development scenario (average 25 percent loss) than under a clustered development scenario (average 11 percent loss) (Tables 34.6 and 34.7). The Acadian flycatcher and the veery showed a habitat preference for interior forest and experienced the greatest loss in territories (Tables 34.6 and 34.7). The ovenbird and the eastern wood-pewee did not show a large percentage loss in total

TABLE 34.6.

Percent reduction in preferred territories for six forest songbirds following a random housing development scenario.

Common name/*Scientific name*	Percent change in preferred territories (%)	Percent change in preferred territories more than 200 meters from an edge (%)
Acadian flycatcher/*Empidonax virescens*	–27	–27
Veery/*Catharus fuscescens*	–27	–27
Ovenbird/*Seiurus aurocapillus*	–5	–30
Chestnut-sided warbler/*Dendroica pensylvanica*	0	0
Eastern wood-pewee/*Contopus virens*	–7	–31
Great crested flycatcher/*Myiarchus crinitus*	–10	–33

TABLE 34.7.

Percent reduction in preferred territories for six forest songbirds following a clustered housing development scenario.

Common name/*Scientific name*	Percent change in preferred territories (%)	Percent change in preferred territories more than 200 meters from an edge (%)
Acadian flycatcher/*Empidonax virescens*	–14	–14
Veery/*Catharus fuscescens*	–14	–14
Ovenbird/*Seiurus aurocapillus*	–4	–13
Chestnut-sided warbler/*Dendroica pensylvanica*	–4	0
Eastern wood-pewee/*Contopus virens*	–5	–12
Great crested flycatcher/*Myiarchus crinitus*	–8	–13

number of preferred territories but did show a large decrease in the number of preferred territories more than 200 meters from an edge.

Discussion

Several studies have indicated the importance of landscape-scale features in determining the distribution of birds (Lauga and Joachim 1992; Robbins et al. 1989a; Lynch and Whigham 1984). Many of the landscape-scale (coarse grain) measures we expected would be important were not statistically significant in any of the models. These included slope, aspect, distance to water, and percent forest cover. However, significance tests on data may mask "preference" of resources due to the subjectiveness of the researcher in defining available habitat (Johnson 1980). Although they are similar measures, distance to edge was a better predictor than percent forest cover within 200 meters and

was a significant component in the Acadian flycatcher and veery models.

On the other hand, forest-stand variables were important predictors of songbird occurrence in all of our models (Table 34.2). Our forest-stand variables may have contained more useful information on habitat than can be measured by landscape-scale variables. Shriner et al. (Chapter 47) found topographic variables to be more important than habitat variables in predicting songbird distribution but for a largely undisturbed landscape. Ongoing timber harvesting and other developments in the Baraboo Hills may reduce the effectiveness of topographic variables in predicting vegetation types. Understory type and dominant tree-stocking density factored into more models than other forest-stand variables. Understory type was better at differentiating bird habitats than either dominant tree species or distance to water. Dominant tree size was a variable of significance for the chestnut-sided warbler—a species that utilizes early succes-

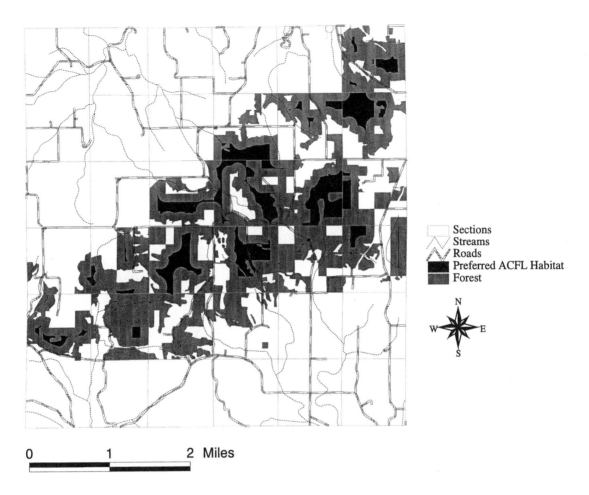

Figure 34.2. Preferred habitat for the Acadian flycatcher (*Empidonax virescens*) under a random development scenario.

sional stands. This variable was also highly significant for the ovenbird, which prefers closed-canopy, mature forests. Our results support the notion that it is important to consider macro- as well as micro-scale habitat features (Block and Brennan 1993). However, researchers should not neglect the importance of stand-level features in favor of coarser landscape-scale measures, especially in disturbed areas.

The ability to develop predictive models for species that have a strong association with stand or plot-scale variables may be limited by the cost of collecting such data and by the resolution of GIS layers. Collecting forest stand inventory data for the Baraboo Hills was an extensive effort and required skill and experience on the part of the field crew. Observer bias may have affected the accuracy of this data layer whereas

coarser-grain measures such as elevation could be obtained more objectively. In addition, forest stand inventory data is dynamic by nature and corresponding GIS layers must be updated on a regular basis. Because of the time and resources required to obtain and maintain landscape-level resource data, it is critical to evaluate the ability of a shared database to meet multiple objectives (Donovan et al. 1987). We were unable to develop a model for the wood thrush (*Hylocichla mustelina*), possibly because this species prefers gaps within a closed canopy, which was not a forest-stand measure available in our database. If different songbird habitat measures had been collected during the forest-stand inventory, they might have improved the predictability of models or permitted development of models for additional species.

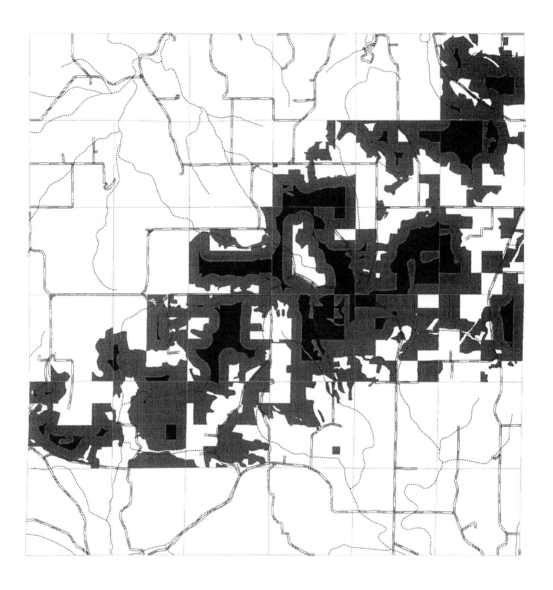

Figure 34.3. Preferred habitat for the Acadian flycatcher (*Empidonax virescens*) under a clustered development scenario.

Modeling Approach and Accuracy

We applied an empirical modeling approach based on the results of 550 songbird censuses. Habitat capability models (a theoretical approach) produce an index to habitat suitability ranging from zero (unsuitable) to one (highly suitable). It is difficult to know whether these indices reflect environmental conditions or population response (Morrison et al. 1992). Our approach directly measures and quantifies population response (probability of occurrence) within a given habitat type.

The final models met our expectations with regard to habitat selection by each species, but the goodness-of-fit test varied from poor for the veery to good for the Acadian flycatcher (Table 34.2). As might be expected, the two species with the highest goodness of fit are both habitat specialists, the Acadian flycatcher and the chestnut-sided warbler. The Acadian flycatcher's preference for closed-canopy, southern-mesic forest stands (Mossman and Lange 1982) is revealed by the importance of the northern understory variable. The chestnut-sided warbler's association with early successional habitats is revealed by the importance of tree size and density in that model. On the other hand, the veery is considered a habitat generalist with regard to forest structure, which may explain the relatively poor fit of that model. Dettmers and Bart (1999) found this same pattern in their development of GIS models for songbirds.

For all six species modeled, probability of occurrence in the most preferred habitat was relatively low (Table 34.2). Improved measures of habitat may increase these probabilities. Karl et al. (Chapter 51) found that models developed for heterogeneous landscapes (like the Baraboo Hills) required higher-resolution habitat data. More likely, this pattern is a result of factors other than habitat playing a role in the distribution of a species, such as competition, predation, stochastic events, weather, and regional population status. Others have found that predictive habitat-based models often only account for half the variation observed in species abundance or density (Morrison et al. 1992). The advantage to our approach is that with our 550 censuses we directly measured this variability rather than assuming that all "preferred habitat" was occupied.

One might expect the best models to be developed with species that are habitat specialists yet are abundant enough to provide adequate sample size. Our ability to predict the occurrence of a species using this method did not appear to be closely tied to abundance, with the exception of very rare species. The Acadian flycatcher and chestnut-sided warbler were detected on less than 12 percent of our census plots yet produced the best models. The ovenbird was detected on 38 percent of our plots, but the final model has a lower p value (Table 34.2). This lack of a close relationship between abundance and our ability to model preference may be tied to the scale of our study area. The accuracy of model predictions generally improves at very coarse scales, in other words, the state of Wisconsin versus the Baraboo Hills (Boone and Krohn 1999). In addition, all of the six species we developed models for could be classified as "abundant" as compared to those we chose to exclude (those species present on less than thirty stops). Five of the six models require additional tests to determine their true accuracy since accuracy is largely independent of goodness of fit (Fielding, Chapter 21).

As would be expected (Karl et al., Chapter 51), we were unable to develop models for several species of high concern that occurred in very low densities, such as the mourning warbler (*Oporornis philadelphia*) and hooded warbler (*Wilsonia citrina*), due to a high number of zero occurrences in some habitat types. In addition, our methodology would not work well for species with a low level of detectability since it is based on identifying bird presence in preferred habitats.

Predictive models should not be used to forecast into the future until they have been validated against an independent data set (Morrison et al. 1992). The Acadian flycatcher model is the only model that we tested with an independent data set. Results of this test indicate that the 1994 observations were not significantly different from the 1993 model predictions, but the test lacked the power to make a definite conclusion (Robertsen 1995). Better models could possibly be derived with increased sample size (Dettmers and Bart 1999). The performance of these models is unlikely to be as high as the performance of models at using coarser grain-habitat features over larger areas

(Karl et al., Chapter 51; Fielding, Chapter 21). However, these models are still useful for land managers in the Baraboo Hills by identifying potential habitats for protection. Models are unlikely to perform at 100 percent, and their usefulness should be judged based on the desired management application and scale (Van Horne, Chapter 4; Fielding, Chapter 21).

Management Applications

The fire/harvesting/succession and housing development scenarios demonstrate how these models can be applied by conservationists. If harvesting and successional trends continue, the chestnut-sided warbler will benefit the most, whereas the ovenbird will be the most negatively affected (Fig. 34.1). Most timber harvesting in the Baraboo Hills takes place in central hardwoods stands; however, those stands with an understory of northern hardwoods have the highest habitat value for the ovenbird, Acadian flycatcher, and veery. Impacts to habitat can be reduced in such cases by directing timber harvesting to stands with a central hardwoods understory or applying selection harvesting methods. Selection versus clearcut logging would also benefit the eastern wood-pewee by retaining more canopy cover. The great crested flycatcher showed a small increase in number of territories correlated with the succession of oak stands. The chestnut-sided warbler may benefit from created openings, but nesting success in these areas needs to be examined. Clustering housing development helped reduce loss in territories for most species, especially for the Acadian flycatcher and veery, which were significantly more likely to occur far from an edge (Figs. 34.2, 34.3).

Conclusion

We looked at the current and future distribution of six forest songbirds of conservation concern across a regionally important landscape. All models included forest-stand variables, thus emphasizing the importance of habitat measures at this scale. Only the Acadian fly-

catcher and the veery models contained a landscape-scale (coarse-scale) measure, distance to edge, as a significant predictor. We applied these models to look at the future impacts of timber harvesting, forest succession, prescribed burning, and housing developments. Under our timber harvest/succession/prescribed-burn scenario, preferred habitat increased for two of the six species, the chestnut-sided warbler and the great crested flycatcher but decreased for the Acadian flycatcher, veery, ovenbird, and eastern wood-pewee. Under a development scenario, random housing development resulted in a 13 percent average decrease in abundance as compared to an 8 percent average decrease for clustered housing development. These average losses were greater for forest interior habitats: 25 percent for random development and 11 percent for clustered development. Based on a test of the Acadian flycatcher model in 1994, the models we developed using 1993 data looked promising.

Acknowledgments

This project was funded by the North Central Research Station, USDA Forest Service, St. Paul, Minn. We thank the following individuals for their support of the project and editing of this manuscript, especially Tom Nicholls, Richard Buech, Kay Burcar, Mark Nelson, Leanne Egeland, and Steve Robertsen. We also wish to acknowledge the University of Wisconsin, Madison, Department of Wildlife Ecology and Land Information Computer Graphics Facility for technical advice. Other agencies that supported this project either through advice or funding include The Nature Conservancy of Wisconsin, the Wisconsin Department of Natural Resources, The Zoological Society of Milwaukee County, The USDI Fish and Wildlife Service, and the Wisconsin Society for Ornithology. The USDA Forest Service, Tongass National Forest, Wrangell Ranger District in Alaska provided support in the final phases of this project for which we are grateful.

Poisson Regression: A Better Approach to Modeling Abundance Data?

Malcolm T. Jones, Gerald J. Niemi, JoAnn M. Hanowski, and Ronald R. Regal

In many ecological studies researchers undertake what may be an elusive search for relationships between the occurrence and/or abundance of a particular species and its environment. The majority of these studies each use some form of regression technique in an attempt to develop a statistically valid model that subsequently can be used for predicting or elucidating an ecological process. In reviewing the last five years (1995–1999) of *Ecology, Ecological Applications, Ecological Monographs,* and *Conservation Biology,* we found that 43 percent of the articles discussing the development of models for predicting the occurrence or abundance of a species used multiple linear regression, 27 percent used logistic regression, 34 percent used other parametric techniques (e.g., principal component analysis), and 2 percent used classification and regression trees. Thus, we agree with Morrison et al. (1992) and Trexler and Travis (1993) that one of the most commonly used statistical techniques to develop predictive models relating habitat to abundance has been multiple linear regression. We suspect that this phenomenon is due to (1) the historical use and acceptance of this technique in the scientific literature, (2) the fact that more ecologists are exposed to this technique in statistics courses than to the more modern regression techniques, and (3) the ease of interpreting the results. It is interesting to note that in our literature review, we found no articles describing the use of Poisson regression in developing predictive models of species abundance.

With the continually increasing power and sophistication of personal computers and statistical software, ecologists have greater access to more modern regression techniques than ever before. The most notable modern regression techniques are those based on generalized linear models (GLMs), including logistic (Hosmer and Lemeshow 1989; Agresti 1990; Menard 1995; Long 1997), Poisson (Hastie and Pregibon 1997; Long 1997; Venables and Ripley 1997), and negative binomial (Long 1997; Venables and Ripley 1997). All three of these regression models are parametric because they are predicated on the assumption that the data conform to a particular frequency distribution. The most commonly used of these techniques is logistic regression, with numerous examples in the ecological literature (Osborne and Tigar 1992; Trexler and Travis 1993; Smith 1994). The use of Poisson regression in ecological studies is uncommon (Welsh et al. 1996; Vernier et al., Chapter 50). Another statistical approach is to use an adaptive or nonparametric technique (i.e., no *a priori* assumption about the underlying distribution) such as classification and regression trees (Breiman et al. 1984) or neural networks (Ripley 1996). Recently, several papers have discussed classification and regression tree models (Michaelsen et al. 1987; Walker 1990; Moore et al. 1991; Walker and Cocks 1991; Baker 1993; Michaelsen et al. 1994;

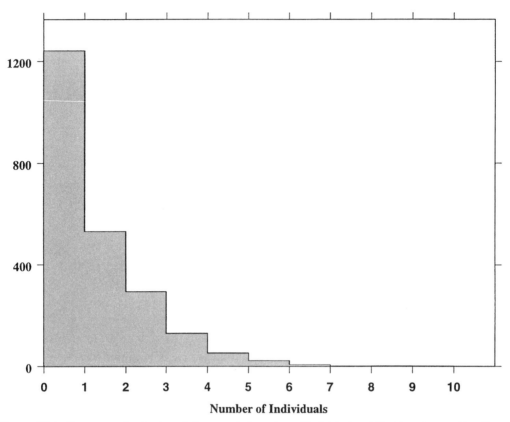

Figure 35.1. Frequency distribution of the sampled number of individuals of Blackburnian warbler (*Dendroica fusca*) from a long-term population monitoring study in northeastern Minnesota and northwestern Wisconsin. For details of this study, see Hawrot et al. (1998).

Hernandez et al. 1997; O'Connor et al. 1996; O'Connor and Jones 1997; Fertig and Reiners, Chapter 42; Thomas et al., Chapter 10).

The selection of any technique must be based on the type of data to be analyzed. First, the researcher must decide whether the data are measured on a ratio, interval, ordinal, or nominal scale, and whether these data are discrete or continuous (Zar 1984). Second, how well do these data conform to the underlying assumptions of the particular statistical procedure (e.g., normal distribution, linear relationships, homoscedasticity)? It is not uncommon for ecological data, such as abundance data, to show a highly skewed frequency distribution (Fig. 35.1). Although testing of the underlying assumptions of statistical tests is relatively common, it is likely that less emphasis has been placed on matching the technique to the type of data that has been collected.

A large proportion of floral and faunal surveys use count data or density estimates, standardized to a given areal unit (e.g., number of individuals per 10 hectares). Thus, it is not uncommon for ecologists to work with data that are discrete and were measured on a ratio scale (i.e., count data). By making this rather simple determination, we can use techniques that have been developed for analysis of count data, such as Poisson regression. A least-squares linear regression model could give us inefficient and biased parameter estimates for this type of data (Long 1997). Although Preston (1948, 1962a,b) introduced the concept of the veiled normal distribution to explain highly skewed abundance data, use of linear regression model with count data results in the possibility of predicting a negative abundance estimate. Biologically, such an outcome is unfeasible and only complicates the interpretation of the regression model. Logistic regression, on the other hand, requires one to discard meaningful abundance information.

Poisson Regression

The Poisson regression model (PRM) has several features that make its use appealing for ecological applications. PRM assumes an underlying Poisson distribution, which is defined in equation 35.1 where *P(X)* is the probability of *X* occurrences and *X* is the count of events.

$$P(X) = \frac{e^{-\mu}\mu^{x}}{X!} \qquad (35.1)$$

Although it is common to see μ referred to as the rate at which *X* events occur, it also can be expressed as the expected count (i.e., population mean count) (Long 1997). As μ increases, the Poisson distribution approximates a normal distribution (Fig. 35.2). This allows one to fit a wide range of apparent distributions with a unified modeling approach. Additionally, PRM is constrained by its underlying distribution to be bounded by zero at its lower end. Lastly, we can obtain predictions from PRM either on the scale of the response variable (i.e., an actual count) or as a probability of occurrence.

Most modern statistical software (e.g., SAS, S-Plus, and Systat) have routines for performing GLMs, and thus PRM. GLMs are fit using maximum likelihood estimation instead of ordinary least squares (Hastie and Pregibon 1997; Long 1997). The major assumptions of PRM are that (1) the data follow a Poisson distribution, and (2) the events are independent. Lack of fit between the data and the Poisson distribution is often attributed to overdisperson, an inequality of the variance and mean, or to the rate of the count variable varying between individuals (i.e., heterogeneity) (Long 1997). These potential problems have led to the development of many extensions of PRM, such as zero-inflated Poisson regression (Lambert 1992; Welsh et al.

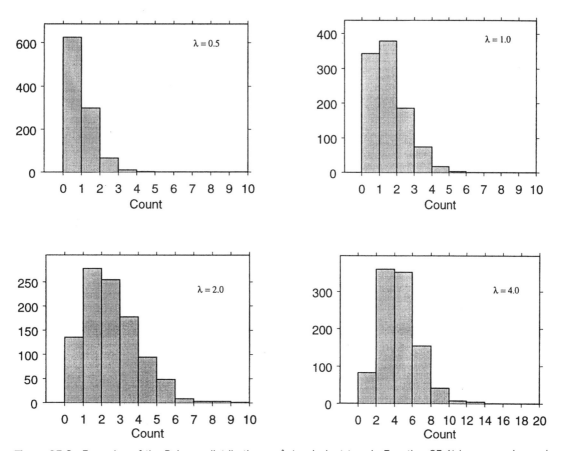

Figure 35.2. Examples of the Poisson distribution as λ (equivalent to μ in Equation 35.1) increases. Low values of λ, upper left histogram, result in right-skewed frequency distributions. As λ increases, the distributions begin to approximate the normal distribution.

1996; Long 1997), truncated Poisson regression (Long 1997), and even negative binomial regression (Long 1997; Venables and Ripley 1997). For a very readable discussion of PRM and the other methods mentioned above, see Long (1997).

The main purpose of this chapter is to increase awareness about regression techniques more appropriate than linear regression for ecological analyses, without necessarily reducing relative abundance information to presence/absence categories as required by logistic regression. Secondarily, we compare the performance of Poisson regression with that of logistic regression in a simulation exercise.

Methods

Wildlife surveys typically provide relative abundance data assumed to represent true densities of the organism being studied but that in fact do not necessarily do so. The accuracy of surveys depends on the detectability of individuals, which is influenced by the life-history traits of the organism as well as the habitat being sampled. Using real-world data to compare the performance of these two regression techniques would, under the best scenario, give us results confounded with the sampling design and, under the worst scenario, could give us false or misleading results. Therefore, we decided to use a Monte Carlo simulation to generate data for a single species so that we would be certain of the "truth" (i.e., the actual number of birds in a given sampling unit) and accurately compare the two regression procedures. We chose to generate our virtual birds in a real landscape that is a predominantly forested 1.1-million-hectare region located in St. Louis and Lake Counties in northeastern Minnesota (see Fig. 35.3 in color section). The land cover was classified using a Thematic Mapper (TM) satellite image with a ground resolution of 30 by 30 meters. The satellite imagery and resultant classification were obtained as part of an ongoing, long-term study designed to assess the distribution and abundance of breeding birds in forested regions of Minnesota (Hanowski and Niemi 1994, 1995). The original forty-two land-cover classes were reduced to six classes for this analysis to simplify the bird/habitat relationships we would be simulating.

We used ARC/INFO GIS software (version 7.2, ESRI) to identify individual patches of each cover type.

The birds in our simulation use conifer patches larger than 0.16 hectares (40×40 meters), which excluded all conifer patches too small to support one individual. For each iteration of the simulation, information on the 71,695 conifer patches was used as input for allocating a random number of birds to each patch. The number of birds randomly assigned to each patch was determined in the following manner.

Two variables, *TerrSize* (size of an individual territory) and *Occ_Rate* (number of birds per territory) were provided for each run of the simulation. *TerrSize* was allowed to vary from 0.5 hectare, representing high population density, to 10 hectares, representing low population density. To reduce the complexity of the comparisons, *Occ_Rate* was held constant at 0.6 for all simulations. A measure of density per hectare

$$Bird\_Den = \frac{Occ\_Rate}{TerrSize} \qquad (35.2)$$

was calculated by dividing the occupancy rate, *Occ_Rate*, by territory size, *TerrSize* (Eq. 35.2).

For each conifer patch we calculated the expected number of birds, *Expbirds*, based on the area of each patch, *Patch_Size* (Eq. 35.3). This was used to generate a random Poisson deviate using a random number generator from Press et al. (1992). The resulting value was the *true* number of birds, *Nbirds*, in a given patch.

$$Expbirds = Bird\_Den * Patch\_Size \qquad (35.3)$$

After populating all patches with a known number of birds, we sampled the landscape using 2,500 randomly located, nonoverlapping 25-hectare sample units. For each sample unit, the sampled number of birds was calculated by weighting the true number of birds in a patch by the proportion of the patch actually sampled (Eq. 35.4),

$$Samp\_Birds = \sum_{n=1}^{i} Nbirds_i * Samp\_Area \qquad (35.4)$$

where *Samp_Birds* is the number of birds in the sample, *n* is the number of patches in the sample grid, and *Samp_Area* is the proportion of each conifer patch in the sample grid. The number of birds sampled,

Samp_Birds, was constrained to be an integer, rounding fractions of birds to the nearest integer. If the number of birds sampled was greater than zero, then we recorded that the species was present in that patch; otherwise, the species was recorded as absent.

Both Poisson and logistic regressions were performed on the resultant data sets from our simulations. We randomly selected two hundred grids (10 percent) from each run of the simulation to be included in a training, or test, set. For the presence/absence data, we fit a GLM in which we specified a binomial distribution with a logit link (Eq. 35.5). Similarly, we fit a GLM in which we used a Poisson distribution with a log link to fit the actual count of sampled birds (Eq. 35.6).

$$PA = a + b * Samp\_Area_{tot} \qquad (35.5)$$

$$Samp\_Birds = a + b * Samp\_Area_{tot} \qquad (35.6)$$

In the logistic regression mode, *PA* indicated whether the species was present or absent and *Samp_area_{tot}* was the total conifer area in the sampled grids. *Samp_Birds* was the total number of birds sampled in each grid for the Poisson regression model. Statistical significance for the logistic and Poisson regression model was assessed using the difference between the null deviance and the residual deviance and calculating a P-value by assuming a chi-square distribution (Venables and Ripley 1997). Statistical significance was assessed for all simulations at an alpha level of 0.05. All statistical analyses were conducted using S-plus (S-Plus 2000 for Windows, MathSoft).

For each run of the simulation, we used the estimated regression coefficients from both regression models to predict the number of individuals or probability of occurrence expected in each conifer patch in our study area. Goodness of fit for each model to the simulated data was assessed by calculating the root mean squared error. For the logistic regression model (LRM), the root mean squared error is given by Equation 35.7, where $Pr(P)$ is the predicted probability of occurrence for each patch. This probability is subtracted from the true probability (0.0 or 1.0) of an occurrence obtained from each simulation. In order to compare the root mean squared error of the LRM with that of the PRM, we had to convert the predicted

number of individuals into the probability of obtaining a count of zero (i.e., absence). Thus, for the PRM, the root mean squared error is given by Equation 35.8 where $Pr(0) = e^{-m}$ is the predicted probability of an absence for each patch and m is the predicted count. This probability is subtracted from the true probability (0.0 or 1.0) of an absence obtained from each simulation.

$$\sum_{i=1}^{n} y_i$$
$$y_i = \begin{cases} (1 - Pr(P))^2 \forall Nbirds > 0 \\ (0 - Pr(P))^2 \forall Nbirds = 0 \end{cases} \qquad (35.7)$$

$$\sum_{i=1}^{n} y_i$$
$$y_i = \begin{cases} (0 - e^{-m})^2 \forall Nbirds > 0 \\ (1 - e^{-m})^2 \forall Nbirds = 0 \end{cases} \qquad (35.8)$$

Results

Ten simulations were conducted for each of six territory sizes, 0.5, 1, 2, 3, 6, and 10 hectares, while the occupancy rate, *Occ_Rate*, was held constant. Samples obtained from the simulated data had frequency distributions similar to those obtained in actual population monitoring studies (Fig. 35.1; Fig. 35.4). All sixty logistic regressions were statistically significant, with P-values less than or equal to 0.05. Similarly, all sixty Poisson regressions were statistically significant, with P-values less than or equal to 0.05.

The two regression models performed similarly across the range of simulated densities (Fig. 35.5). The difference between the root mean squared error ranged from a high of 25 percent at a territory size of 0.5 hectare to a low of 0 percent at a territory size of 10 hectares ($\bar{x} = 9.5$, s.e. = 3.7, $N = 6$). There was an unexpected curvature in the graph of root mean squared error, which suggests that both the Poisson and logistic regressions perform better when individuals are either always present or absent (Fig. 35.5). Examination of the residuals revealed that Poisson regression tended to overpredict abundance when the true number of birds was zero and underpredict abundance when the true abundance greater than zero (Fig. 35.6).

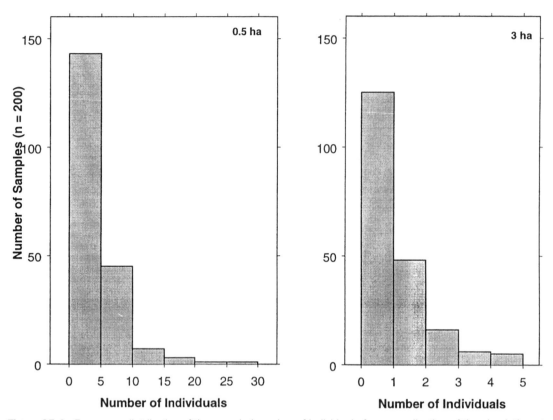

Figure 35.4. Frequency distribution of the sampled number of individuals from a realization of the simulation at two density levels. Each histogram is based on two hundred randomly placed 25-hectare sample grids.

Discussion

Poisson and logistic regression models were found to perform similarly on our simulated data sets. Both types of regression models had root mean squared errors that averaged less than 0.40 (logistic = 0.363 and Poisson = 0.397) across all simulated bird densities. However, Poisson regression tended to overpredict abundance at sites known to have low densities and underpredict abundance at sites known to have high densities. This error may be acceptable, however, given that logistic regression is only able to predict the probability of occurrence, which is difficult to relate to abundance.

There are several other advantages to using PRM. First, no transformations are required in order to meet normality assumptions. Transformations tend to complicate the interpretation of the results. Second, PRM is a nonlinear regression technique but the models are specified as if it were a linear model. Thus, errors in specifying the correct form of the model are avoided.

As with any simulation study, there are limits to

our study due to the assumptions of our design. Since we derived our simulated data from an underlying random Poisson process, it is not surprising that we obtained relatively good fits to the Poisson regression model. However, our results are applicable when the data being analyzed meet the assumptions of Poisson regression, namely that the events are independent and the data are generated by a Poisson process. Unfortunately, little is known about the true, underlying distribution of bird abundance in nature (see Smallwood, Chapter 6).

So why did our regression models not perform better than they did? We suspect that this behavior could be due to one or a combination of the following factors. First, we ignored the effects of autocorrelation that may occur due to the spatial configuration of the conifer patches in our landscape. Although we did not quantify the degree of spatial autocorrelation in the size of conifer patches, there is evidence that such nonrandomness exists in this landscape. This would be a

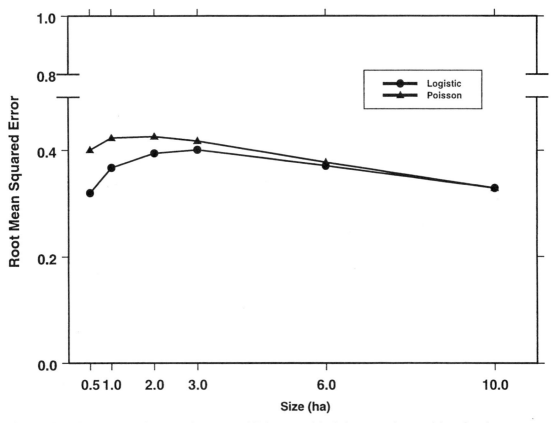

Figure 35.5. Comparison of the performance of Poisson and logistic regression models using the root mean squared error. For the Poisson regression, the error is the difference between the predicted probability of a zero count and the true probability of a zero count for each sample. Likewise, the error for the logistic regression is the difference between the predicted probability of occurrence and true probability of occurrence. The root mean squared error for each realization of the simulation is calculated using 1,709 sample grids not used during model building.

violation of the assumption of Poisson regression that all events are independent. Second, we chose to use a completely random sampling scheme that might not have produced a truly representative training set given the nonrandom distribution of patch sizes in the landscape. If a training set does not include the full range of possible values, it is unlikely that any statistical technique will be able to make accurate predictions outside the range of values encountered during model building. Third, our sampling scheme may have adequately sampled the full range of possible values but differentially sampled the response variable so that the resultant frequency distribution did not match that of the process that generated the data (i.e., the sampled data might be overdispersed).

How might one deal with these possibilities? One can explicitly model the spatial autocorrelation as sug-

gested by Klute et al. (Chapter 27) or Smith (1994). Overdispersion of these data can be handled in one of two ways. First, one can fit a Poisson regression model and account for the overdispersion by specifying an additional parameter that must be specified *a priori* (Long 1997; Venables and Ripley 1997). A second option is to fit a regression model based on a negative binomial distribution (Long 1997), which is a specialized case of Poisson regression. A third option, and perhaps the most intriguing but technically challenging, is to fit a zero-inflated Poisson (ZIP) regression (Lambert 1992; Welsh et al. 1996). ZIP regression is a mixture of logistic regression for the zero data and a Poisson regression for the positive integer data. The technical details of ZIP regression are beyond the scope of this chapter, but the reader is directed toward

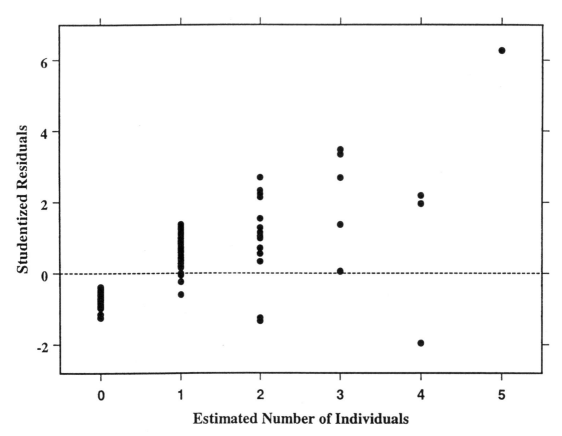

Figure 35.6. Studentized residuals for a Poisson regression model. The data used were generated for one realization of the simulation with a territory size equal to 3 hectares. Note that there is a tendency for Poisson regression to overpredict the number of individuals when the true number of birds is equal to zero and to underpredict when the true abundance is greater than zero.

Lambert's (1992) original paper or Long's (1997) discussion of this technique.

In summary, we have shown that Poisson regression performs similarly to logistic regression when using data that has a frequency distribution similar to real abundance data. Based on the results of our simulation, the mean difference in assessing the probability of occurrence was only 10 percent. When compared to the loss of information that occurs when converting abundance data to presence/absence categories, we believe this difference is not biologically significant. However, we should caution managers that predictive models developed using Poisson regression may be overly optimistic when used to model data from rare species and somewhat conservative when used for common species.

Acknowledgments

We would like to thank Cathy Johnson, Nick Danz, and three anonymous reviewers for their helpful comments on earlier versions of this manuscript. Financial assistance was provided by the Minnesota state legislature from the Environmental Trust Fund as recommended by the Legislative Commission on Minnesota Resources. This is contribution number 269 from the Center for Water and the Environment of the Natural Resources Research Institute.

Predicting Vertebrate Occurrences from Species Habitat Associations: Improving the Interpretation of Commission Error Rates

Sandra M. Schaefer and William B. Krohn

A crucial step in conservation is determining where animal and plant species occur. However, conducting complete field inventories of vertebrate occurrences is generally infeasible. So, wildlife-habitat relationship models are often used to predict species presence, absence, or relative abundance. Since our knowledge of species habitat use is limited, validation of these models is essential (Morrison et al. 1992; Csuti 1996; Krohn 1996). One common testing method is to compare the predicted occurrences to species lists obtained from test sites having long-term field inventories. Omission error (percentage of species present but not predicted), commission error (percentage of species predicted but not present), and the percentage of species matched (percentage of species present that are predicted) can then be used to evaluate model reliability (e.g., Scott et al. 1993; Edwards et al. 1996; Fielding and Bell 1997).

Problems with the above validation metrics are often encountered in the interpretation stage (Krohn 1996). There are many factors associated with species biology and the methods used that can influence errors and complicate their interpretation, including the presence of species on tests sites that go unsurveyed (Nichols et al. 1998b; Boone and Krohn 1999; Karl et al., Chapter 51). Further, size of test sites and definitions of species presence on a site can influence commission and omission errors (S. Schaefer in prepara-

tion). Generally, a close examination of the species being omitted and the data layers used in the prediction process will often identify the cause of the omission error. In contrast, commission error is more troublesome, with a key issue being the need to assess if the error reported is an *actual* error (the species is not present on the site) or if it is an *apparent* error (the species is present on the site but has not been recorded as a result of incomplete field inventories). For example, since the publication of the predicted distributions from the Idaho Gap Analysis Project, the sharp-tailed grouse (*Tympanuchus phasianellus*) has been confirmed to be present in areas where it had never before been recorded (Scott et al. 1993). In this case, commission error could be viewed as an apparent error of the prediction and not an actual error. (The species was present at the time of the GAP prediction.)

Rare and reclusive species can be difficult to detect during standard field surveys designed to inventory a wide variety of species. Thus, these species are likely to have higher estimates of commission error when predicted occurrences are compared to known field observations. Boone and Krohn (1999) recognized that biological characteristics of species can influence detectability and proposed that an *a priori* ranking system based upon the likelihood of detection could be related to commission error. Using avian occurrences from the Maine Breeding Bird Atlas (MBBA) (Adamus 1987), they established a ranking system called Likelihood of Occurrence Ranks (LOORs), which ranked all of the birds known to breed in Maine based upon how frequently they occurred in towns within their range limit (see below). In a gap-like analysis, they observed a strong correlation between LOORs and commission error on five of six test sites (ρ = 0.86–0.93, P = 0.002; Boone and Krohn 1999).

The purpose of the analysis discussed in this chapter was to determine if the *a priori* ranking system of Boone and Krohn (1999) improves the interpretation of the commission errors resulting from species-habitat models designed to predict presence or absence for gap analysis. The Gap Analysis Program (GAP) is a nationwide effort of the U.S. Geological Survey (USGS), Biological Resources Division (BRD), designed to assess some elements of biodiversity (Scott et al. 1993; Scott et al. 1996). GAP uses models primarily based on species-habitat associations, along with other data, such as range limits and vegetation, in a geographic information system (GIS), to predict the presence of terrestrial vertebrates that breed in a state (Scott et al. 1993). Data for this analysis came from the Maine Gap Analysis Project (Maine GAP) (Krohn et al. 1998), and our objective was to determine if LOORs and commission error were correlated. If rates of commission are constant across LOORs, then overprediction by the habitat models would be suggested (i.e., actual errors).

Methods

For this study, avian LOORs were calculated by Boone and Krohn (1999) and herptile LOORs were tabulated by Krohn et al. (1998). In both studies, atlas occurrence information was used to generate a spatial incidence for all species. Mammals were not included in this analysis because no unbiased surveys of incidence existed for Maine. To calculate the avian LOORs, Boone and Krohn (1999) used occurrence data from the MBBA (Adamus 1987). The spatial incidence was calculated by dividing the number of MBBA blocks having confirmed or potential breeding occurrences by the number of MBBA blocks within the species range. Because the MBBA was more than fifteen years old, they updated the spatial incidences

using logistic regression to model a suite of avian species-specific variables, including data taken from the USGS, BRD Breeding Bird Survey (BBS) during the period of the MBBA. The outdated MBBA data was then replaced with the new information, giving updated incidences. These incidences were sorted and assigned a rank, which became the species LOORs. Low ranks indicate the species has low detectability, and, conversely, high-ranking species are those that are easier to detect in the field. Species with inadequate data available to assign spatial incidences were given a rank of zero and excluded from the correlation analysis (Boone and Krohn 1999).

Krohn et al. (1998) used occurrence information from Maine Amphibian and Reptiles (Hunter et al. 1999) to calculate herptile LOORs. Since the information in the amphibian and reptile atlas was recent, there was no need to conduct additional modeling to update the data, as was done for avian species. Incidences for amphibians and reptiles were combined into one list of herptiles, sorted, and then ranked, giving the LOORs for each species. Combining the two taxonomic classes was done to increase the sample size used in the correlation analysis.

Predicted occurrences from the Maine Gap Analysis Project (Boone and Krohn 1998a,b) were compared to records from nine sites in Maine having field surveys. Amphibian, reptile, and bird occurrences came from checklists complied by National Park Service (NPS 1990, 1996) and the U.S. Fish and Wildlife Service (USFWS) (USFWS 1989, 1994a,b, 1995, 1996; Table 36.1, Fig. 36.1). Additional avian occurrences were also obtained from field inventory and research records from the White Mountains National Forest (D. Capen, Univ. of Vermont personal communication), Baxter State Park (Oliveri 1993), and two privately owned areas (Hagan et al. 1997; J. Witham, Univ. of Maine personal communication; Table 36.1, Fig. 36.1).

For each site, the number of species correctly predicted and the number in commission were tabulated and compared to five groups of species for birds and three groups of species for herptiles. Species were assigned to groups based upon LOORs (ranging from low to high) with equal number of species per group (as much as possible). This was done to remove any

TABLE 36.1.

Test-site names, data type, and available information used in testing the accuracy of the vertebrate predictions from Maine Gap Analysis.

Site no.	Name of test site	Size (ha)	Survey length	Amphibians and reptiles	Birds
1	North Maine Forestlands Study, Moosehead Lake Area	293	1[a]		x
2	Nesowadnehunk Field, Baxter State Park	177	3[a]		x
3	White Mountains National Forest	181	5[a]		x
4	Sunkhaze Meadows National Wildlife Refuge	3,833	10[b]		x
5	Holt Research Forest	172	15[a]	x	x
6	Petit Manan National Wildlife Refuge	993	22[b]		x
7	Rachel Carson National Wildlife Refuge	1,768	32[b]	x	x
8	Moosehorn National Wildlife Refuge	9,297	61[b]		x
9	Mount Desert Island/Acadia National Park	28,033	79[b]	x	x

[a]Number of years the area has been surveyed.

[b]Actual number of survey years unknown, so the number of years in existence is instead reported.

possibility of bias that might have occurred by including species that do not occur on a particular site due to range limits in the state. Spearman's Rho (alpha = 0.05) was used to quantify the relationship between species counts and the LOORs group for each taxonomic class on each site.

Results

Overall, the mean commission error for amphibians and reptiles was low ($\bar{x}$ = 12.3 percent, range 0 to 36.8 percent) and the mean percentage of species matched was high ($\bar{x}$ = 97.3 percent, range 92 to 100 percent; Table 36.2). No trend was apparent when combined amphibian and reptile errors were plotted for each LOOR group (Fig. 36.2). A valid Spearman's Rho analysis on commission error could only be conducted for the Holt Research Forest, because on Rachel Carson National Wildlife Refuge there was no commission error, and Mount Desert Island had too many ties in the number of species matched to conduct a rank correlation test (Table 36.3). The Spearman's rho for the Holt Research Forest indicated that there was not a significant relationship between commission error and the LOOR groups (ρ = –0.5, P ≤ 0.704; Table 36.3). The correlation between the number of species matched and LOORs was also not significant (ρ = 0.5, P ≤ 0.704; Table 36.3). Commission error was much higher for birds than it was for

herptiles ($\bar{x}$ = 76.8 percent ± 45.8, compared to 12.3 percent ± 21.2; Table 36.2). The number of bird species matched and the LOOR groups were positively correlated (ρ = 0.6 to 1.0) on all sites (n = 9; Fig. 36.3). Relationships were significant (P ≤ 0.05) for all sites except for the White Mountains National Forest (P ≤ 0.291) and Acadia National Park (P ≤ 0.059) (Table 36.4). An inverse relationship was observed between commission error and the LOORs (ρ = –0.87 to –1.0) (Fig. 36.3). The Spearman's rho tests confirmed the significance of this relationship on all sites except Acadia National Park (P ≤ 0.059; Table 36.4).

Discussion

Field surveys are often incomplete inventories of the species present in a given area (e.g., Nichols et al. 1998a). Factors such as species detectability, the number of years a survey has been conducted, and the amount of effort placed in searching for species all influence which species will be recorded and which will be missed during a survey (Boone and Krohn 1999; Karl et al., Chapter 51; Fielding, Chapter 21). We found that *a priori* ranking species based upon how likely a species is to be observed during a field inventory helps to detect the effects of incomplete field surveys on model validation.

An initial interpretation of the commission errors reported for Maine GAP, without correcting for

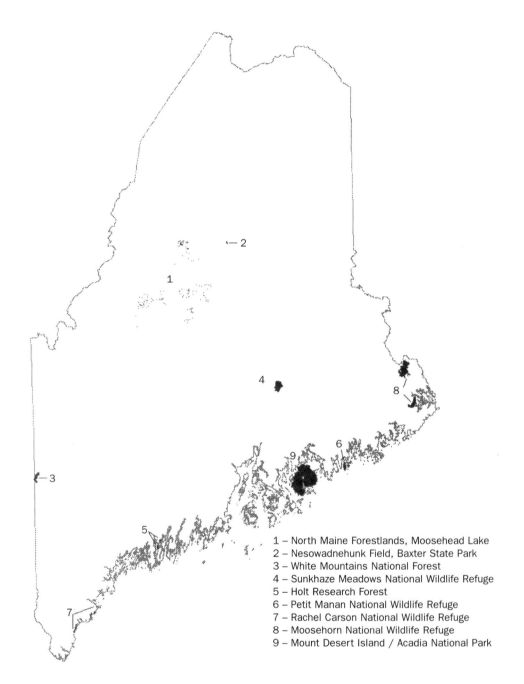

Figure 36.1. Locations of test sites used in the accuracy assessment of predicted distributions of terrestrial vertebrates from Maine Gap Analysis.

incompleteness of the inventories, would indicate that the models are overpredicting about 76 percent of the bird species in the state (Table 36.2). Block et al. (1994) faced a similar problem with their predictive models. They reported commission errors ranging from 29 to 44 percent and felt that this level of error was unacceptable. In both studies, these findings could

lead researchers to believe that the species-habitat association models have not been correctly constructed. However, the inverse correlation we observed between commission error and LOORs indicates that many of the errors reported in the Maine GAP predictions are related to the species detectability. Thus, by examining the models within an *a priori* ecological context of

(a) Temperature

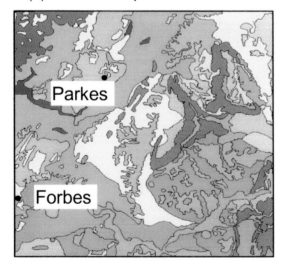

	13 - 14 °C		15 - 16 °C
	14 - 15 °C		16 - 17 °C

(b) Rainfall

	475 - 550 mm
	550 - 625 mm
	625 - 700 mm

(c) Soil landscapes

(d) Environmental stratification units

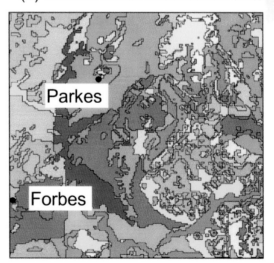

Figure 5.2. Environmental data used to derive the environmental stratification units (ESU) for the Parkes 1:100,000 mapsheet include (a) mean annual temperature classes, (b) mean annual rainfall classes, and (c) soil landscape types. The combinations of temperature (four classes), rainfall (three classes), and soil landscape types (thirty-one types) give rise to 101 environmental stratification units (d).

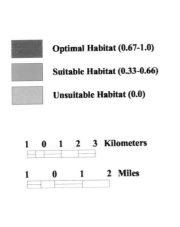

Optimal Habitat (0.67-1.0)

Suitable Habitat (0.33-0.66)

Unsuitable Habitat (0.0)

1   0   1   2   3   Kilometers

1   0   1   2   Miles

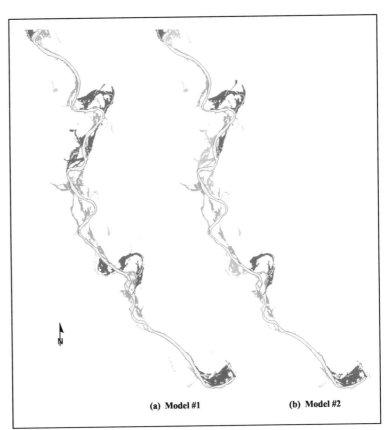

(a) Model #1   (b) Model #2

Figure 14.4. Geographic output of habitat suitability index (HSI) cartographic modeling for the yellow-billed cuckoo (*Coccyzus americanus*), (a) HSI reduced model (using three variables), (b) HSI full model (using five variables).

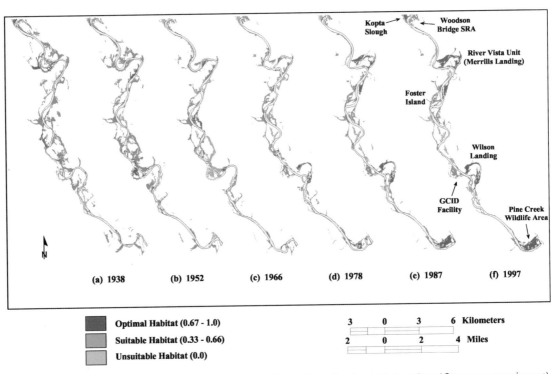

Kopta Slough
Woodson Bridge SRA
River Vista Unit (Merrills Landing)
Foster Island
Wilson Landing
GCID Facility
Pine Creek Wildlife Area

(a) 1938   (b) 1952   (c) 1966   (d) 1978   (e) 1987   (f) 1997

Optimal Habitat (0.67 - 1.0)

Suitable Habitat (0.33 - 0.66)

Unsuitable Habitat (0.0)

3   0   3   6   Kilometers

2   0   2   4   Miles

Figure 14.5. (a–f) Variation in the quality and spatial distribution of yellow-billed cuckoo (*Coccyzus americanus*) habitat from 1938 to 1997 on the Sacramento River, river-miles 196–219.

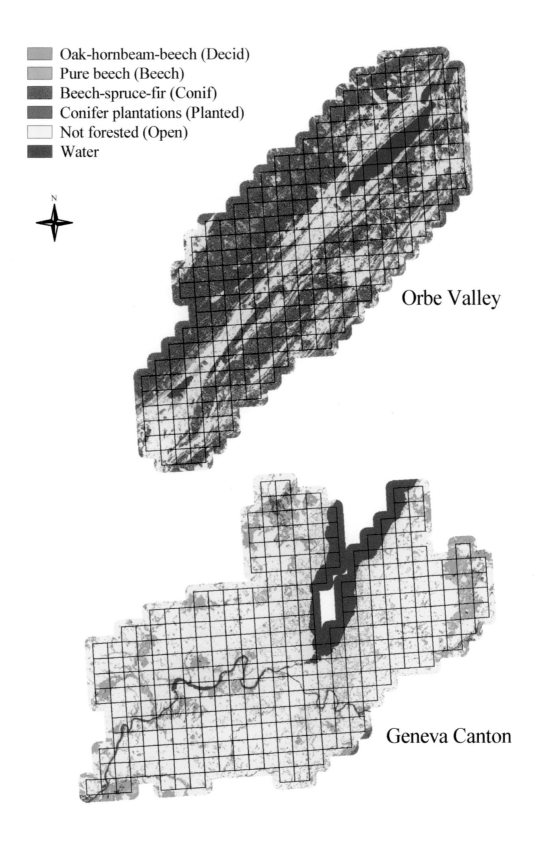

Figure 15.2. Distribution of land-cover types in the two study areas.

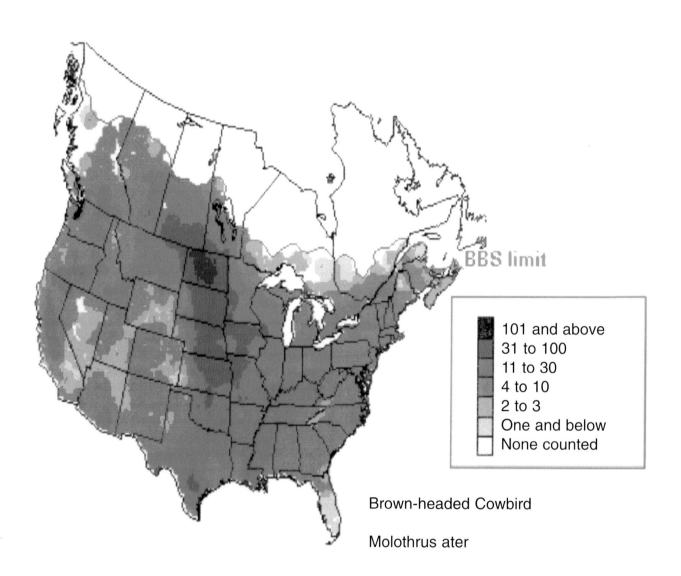

101 and above
31 to 100
11 to 30
4 to 10
2 to 3
One and below
None counted

BBS limit

Brown-headed Cowbird

Molothrus ater

Figure 17.1. Summer distribution map of brown-headed cowbird (*Molothrus ater*).

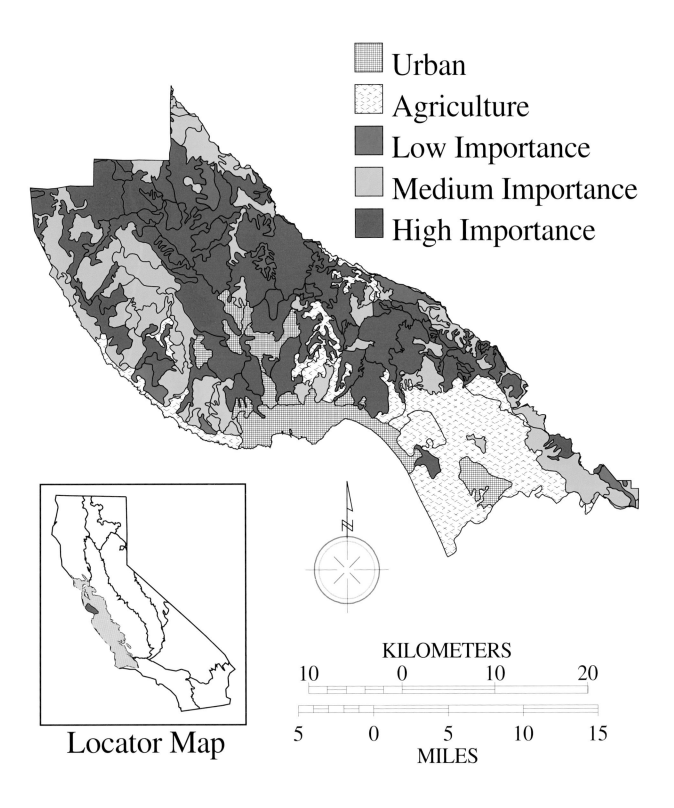

Figure 18.2. Three-level ecoregional importance valuation for Santa Cruz County, California. Central areas of high importance are influenced by the distribution of redwood forest across the ecoregion. Urban and agriculture areas are included for reference. Locator map shows Santa Cruz County within the Jepson Central West ecoregion.

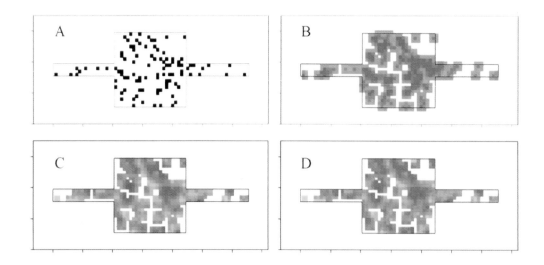

Figure 33.5. Simulated (known) distribution of hypothetical species with low detectability used to evaluate methods: (a) Original data; filled squares denote presence of species. (b) Results of simple spatial smoothing. (c) Results of smoothing based on habitat information. (d) Results of smoothing based on habitat information and search effort.

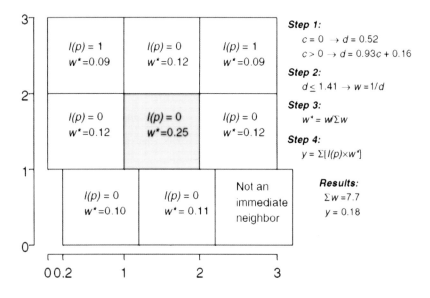

Figure 33.6. Weights ($w*$) used in simple spatial smoothing method, and steps taken to determine weights. Target cell (shaded) and immediate neighbors contribute data for determining the probability of a species' occurrence in the target cell. $I(p) = 1$ if the species is known to occur in a cell: otherwise, $I(p) = 0$. The variable c denotes the distance between the center of a data cell and the center of the target cell and is used to estimate d, the average distance between points in the two cells.

Location of Detail

N

Land Cover
Upland Conifer
Wetland
Hardwoods
Grass/Brush
Developed
Water

Figure 35.3. Location of landscape used in simulation, and detail of land cover classification.

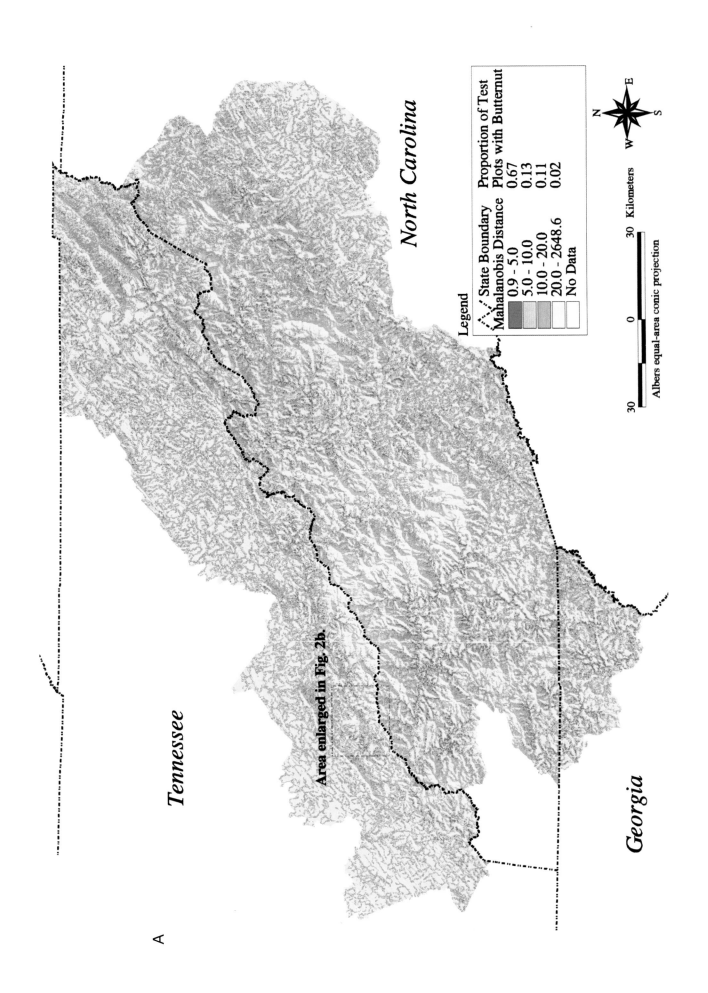

A

Tennessee

North Carolina

Georgia

Area enlarged in Fig. 2b.

Legend

State Boundary
Mahalanobis Distance

	Proportion of Test Plots with Butternut
0.9 - 5.0	0.67
5.0 - 10.0	0.13
10.0 - 20.0	0.11
20.0 - 2648.6	0.02
No Data	

N
W E
S

30   0   30   Kilometers

Albers equal-area conic projection

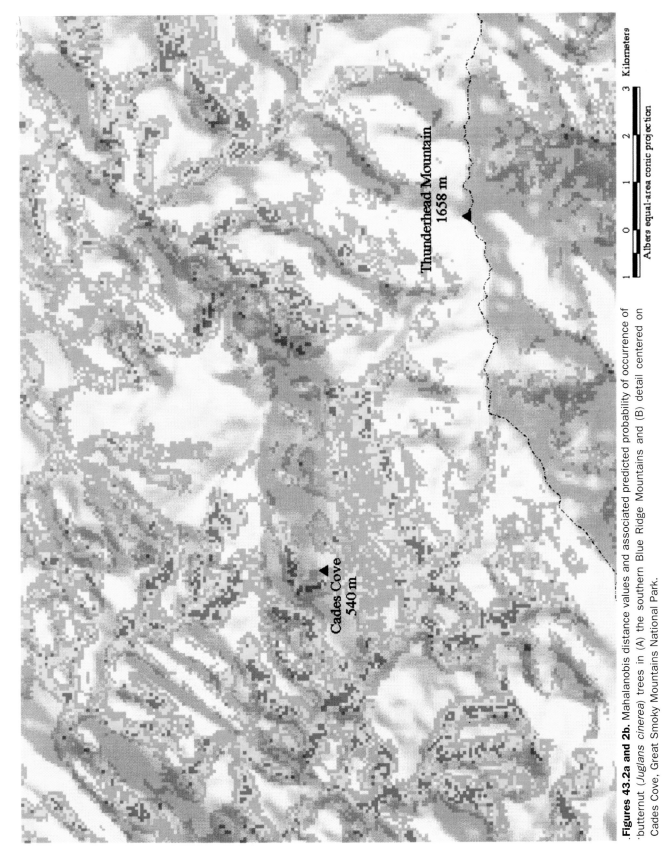

B

Figures 43.2a and 2b. Mahalanobis distance values and associated predicted probability of occurrence of butternut (*Juglans cinerea*) trees in (A) the southern Blue Ridge Mountains and (B) detail centered on Cades Cove, Great Smoky Mountains National Park.

Thunderhead Mountain
1658 m

Cades Cove
540 m

1 0 1 2 3 **Kilometers**

Albers equal-area conic projection

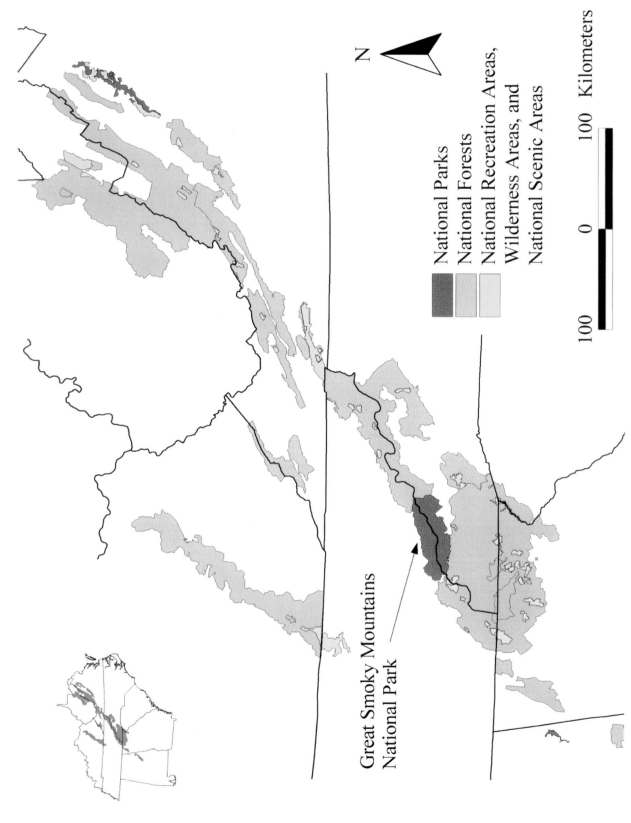

Great Smoky Mountains
National Park

National Parks
National Forests
National Recreation Areas,
Wilderness Areas, and
National Scenic Areas

N

100   0   100   Kilometers

Figure 47.1. Great Smoky Mountains National Park as part of the network of protected areas in the southern Appalachians.

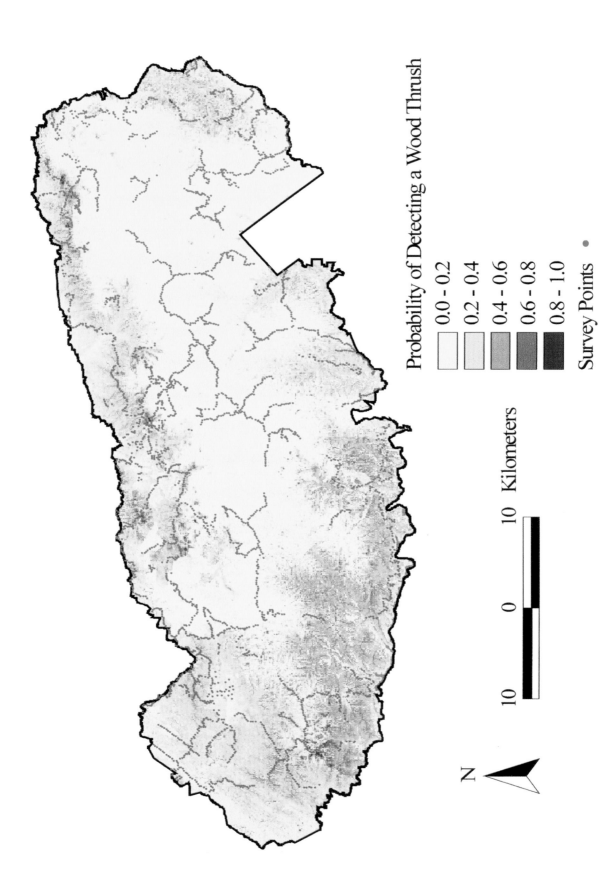

Probability of Detecting a Wood Thrush

- 0.0 - 0.2
- 0.2 - 0.4
- 0.4 - 0.6
- 0.6 - 0.8
- 0.8 - 1.0

Survey Points •

Figure 47.2. The probability of detecting a wood thrush in Great Smoky Mountains National Park.

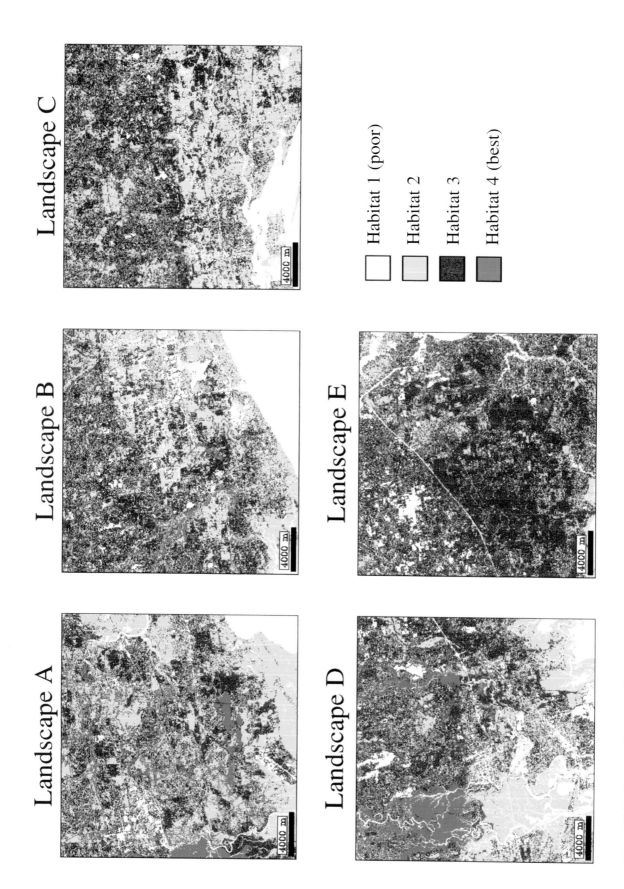

Landscape A

Landscape B

Landscape C

Landscape D

Landscape E

Habitat 1 (poor)

Habitat 2

Habitat 3

Habitat 4 (best)

Figure 52.1. The spatial pattern of habitats in five study landscapes of the Gulf Coast. Each study landscape is 25 kilometers by 25 kilometers.

A.

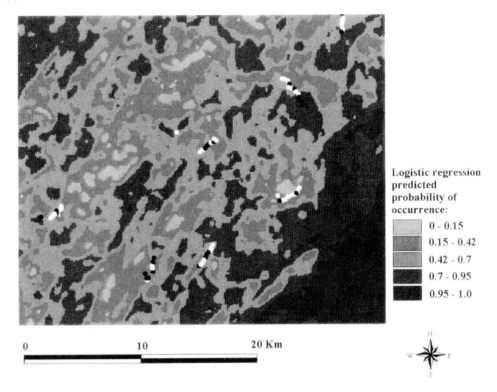

Logistic regression
predicted
probability of
occurrence:

	0 - 0.15
	0.15 - 0.42
	0.42 - 0.7
	0.7 - 0.95
	0.95 - 1.0

0   10   20 Km

B.

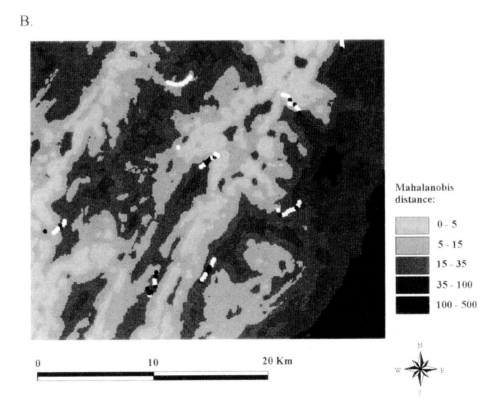

Mahalanobis
distance:

	0 - 5
	5 - 15
	15 - 35
	35 - 100
	100 - 500

0   10   20 Km

Figure 54.2. Model output and observed bird survey data for yellow-billed cuckoo (*Coccyzus americanus*) on the Deerfield Ranger District of the George Washington and Jefferson National Forest in Virginia. Portions of eight bird survey routes are pictured, with black dots indicating locations where yellow-billed cuckoos were present and white dots where they were absent. (*continues*)

Spatial random effect predictions

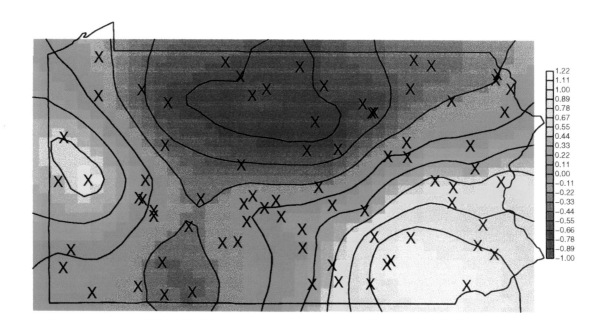

Prediction standard errors of spatial effect predictions

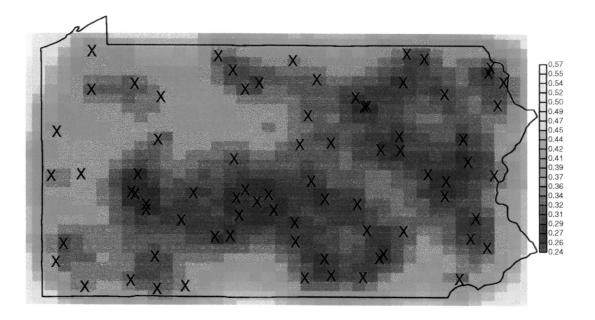

Figure 56.4. Top: Map of predicted random spatial effects at the grid locations. Bottom: Prediction standard errors (marginal posterior standard deviations). Route locations are indicated by an X. Contours in the top panel are from –0.75 to 1, incremented by 0.25.

Expected route count predictions

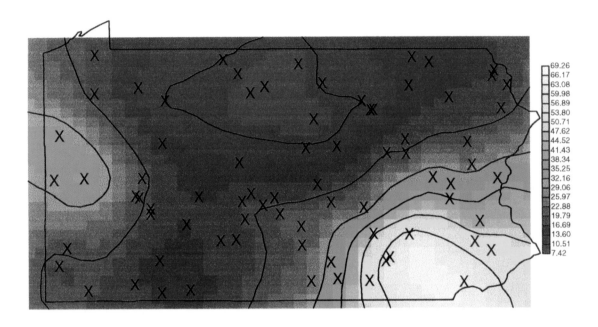

Prediction standard errors of expected route count predictions

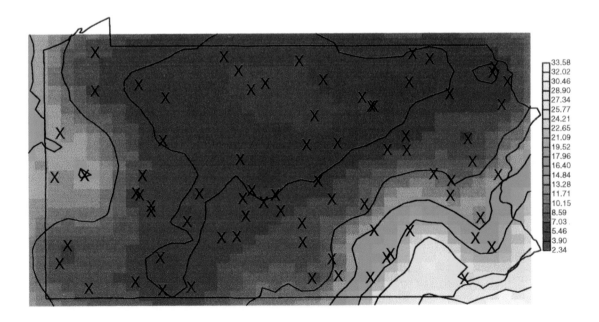

Figure 56.5. Top: Map of estimated expected route count at the grid locations. Bottom: Prediction standard errors. Route locations are indicated by an X. Contours in the top panel are from 10 to 60, incremented by 10. Contours in the bottom panel are from 5 to 30, incremented by 5.

Percent forest cover

Spatial random effect predictions

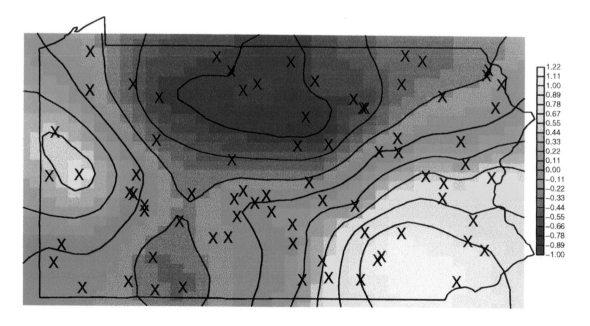

Figure 56.6. Top: Percent forest cover. Bottom: Map of predicted route effects (shown again to facilitate comparison).

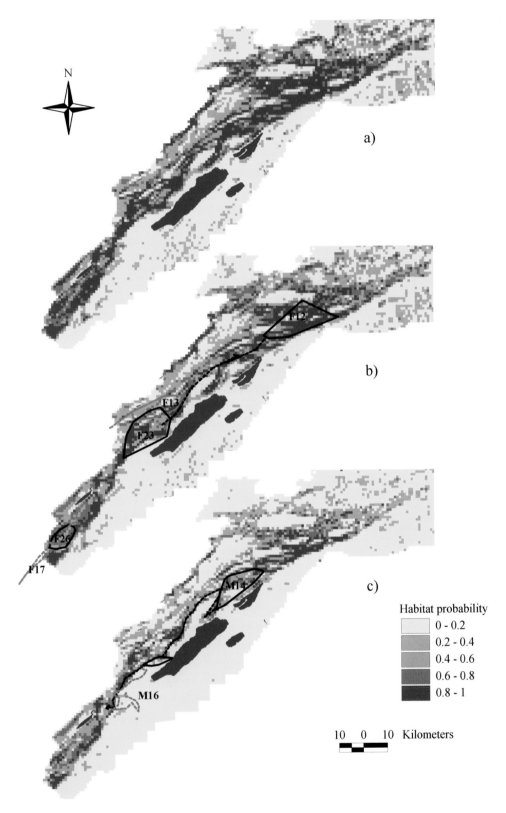

Figure 58.2. Potential distribution of the lynx in the Swiss Jura Mountains according to the model derived from the three sets of response variables (a = both sexes combined, b = females alone, and c = males alone). Lines represent the dispersal route of subadult lynx (F = females, M = males); polygons represent transient or definitive home ranges. Subadult lynx that survived the first year of independence are shown in black, and those that died are shown in red.

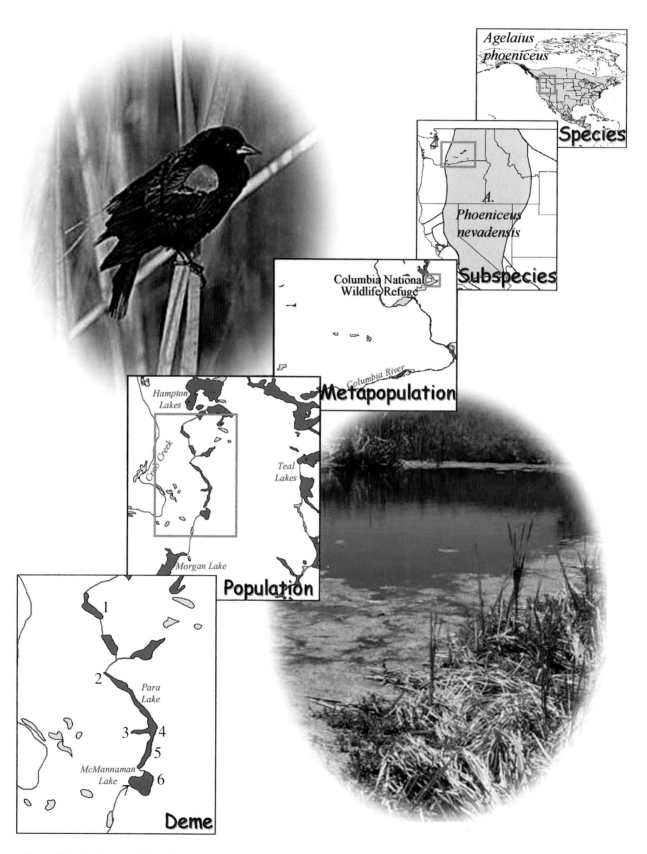

Figure I5.1. Red-winged Blackbird (*Agelaius phoeniceus*) hierarchy of spatial population units from deme to species at Columbia National Wildlife Refuge, WA.

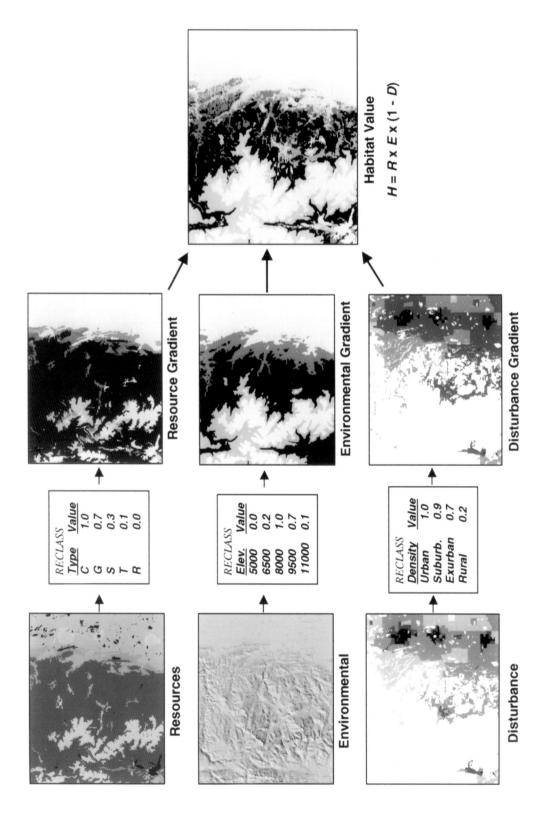

Figure 59.2. Dataflow diagram of inputs and steps to create the habitat-quality map.

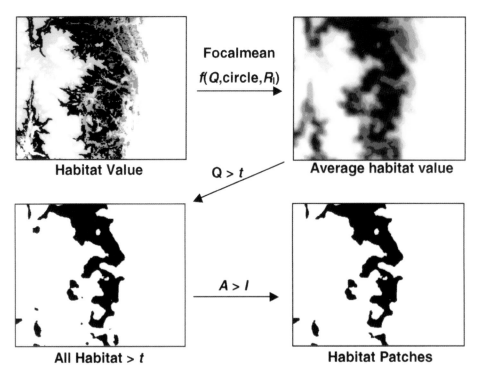

Figure 59.3. Habitat patches for both individual and population level are derived from the habitat-value map, using a neighborhood averaging function.

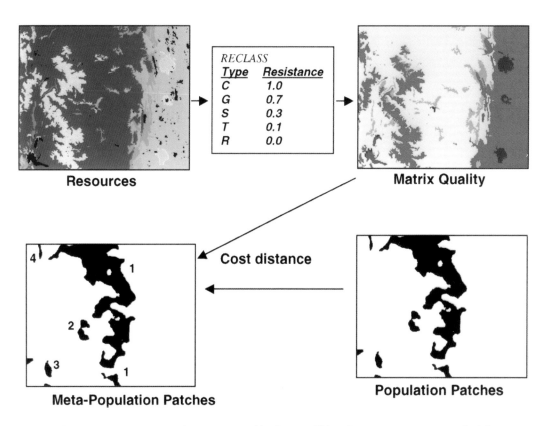

Figure 59.4. Meta-population patches are created by first modifying the resources map to reflect the matrix quality as a surrogate to resistance to movement between population patches.

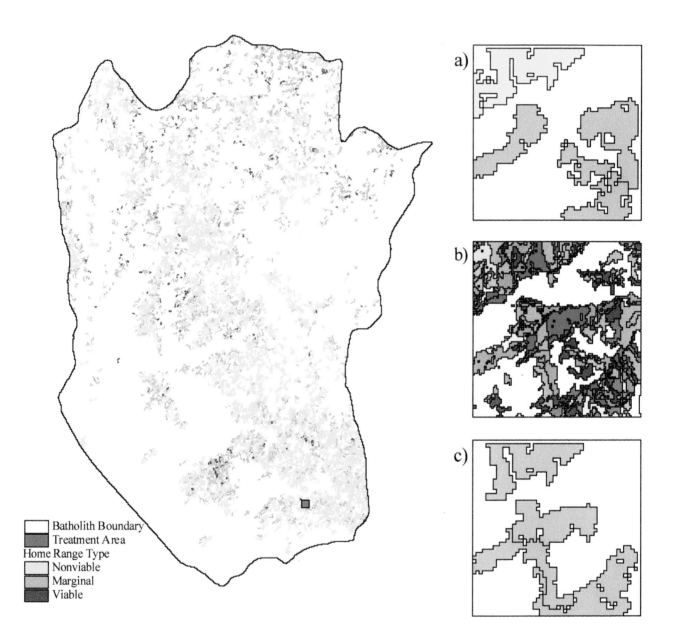

Figure 60.3. The Idaho Southern Batholith and location of 890-hectare area for silvicultural treatment, pre-treatment white-headed woodpecker (*Picoides albolarvatus*) habitat assessment (a), 52 hectares of forests subjected to treatment (b), and post-treatment white-headed woodpecker habitat assessment (c).

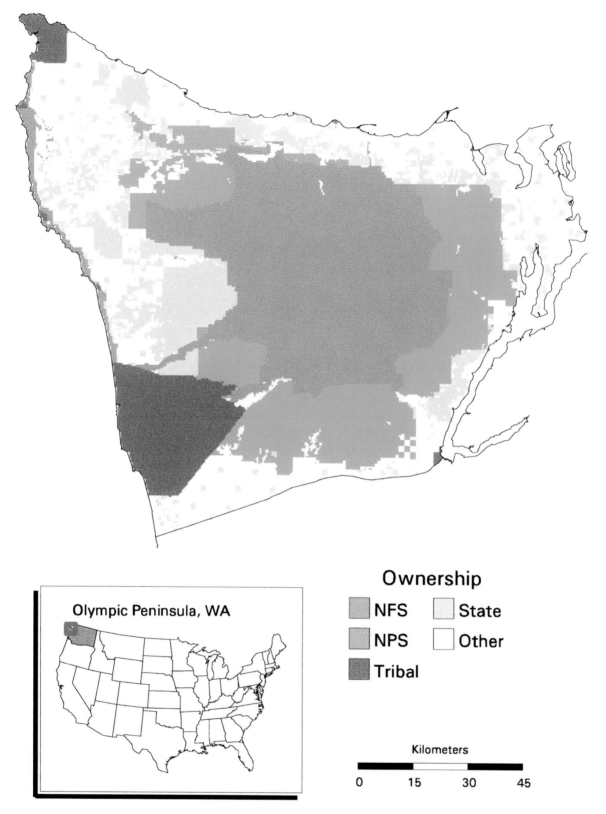

Figure 62.1. Map of the Olympic Peninsula, Washington, showing ownership patterns. Federal lands include National Forest system (NFS) lands and National Park Service (NPS) lands. Lands designated as "Other" include industrial and other privately owned areas.

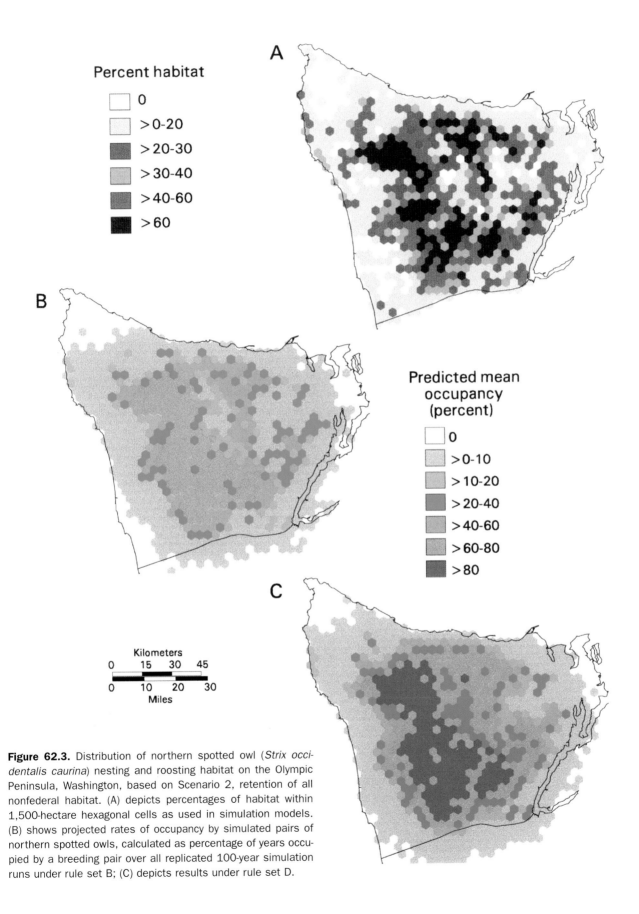

Figure 62.3. Distribution of northern spotted owl (*Strix occidentalis caurina*) nesting and roosting habitat on the Olympic Peninsula, Washington, based on Scenario 2, retention of all nonfederal habitat. (A) depicts percentages of habitat within 1,500-hectare hexagonal cells as used in simulation models. (B) shows projected rates of occupancy by simulated pairs of northern spotted owls, calculated as percentage of years occupied by a breeding pair over all replicated 100-year simulation runs under rule set B; (C) depicts results under rule set D.

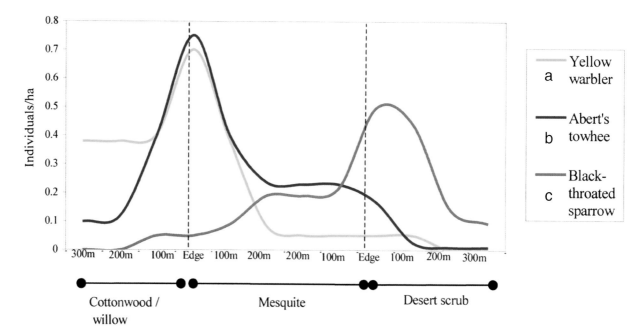

Figure 63.1. Changes in species densities across a riparian habitat gradient in southeastern Arizona. Cottonwood/willow gallery forest predominates within 200 meters of the river channel, giving way to mesquite woodland in the floodplain and desert scrub outside the riparian zone. The curves illustrate variation in species density for three bird species (a) yellow warbler (*Dendroica petechia*), a riparian specialist that favors cottonwood/willow habitat and exploits the edge with mesquite woodland; (b) Abert's towhee (*Pipilo aberti*), an edge exploiter that favors mesquite-dominated habitats; and (c) black-throated sparrow (*Amphispiza bilineata*), which prefers desert-scrub habitats and exploits the edge with mesquite woodland.

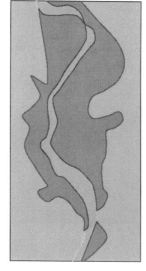

Figure 63.4. Thematic Mapper Simulator (TMS) image and simplified habitat grid for the modeled landscape along the upper San Pedro River corridor in southeastern Arizona. The landscape has been classified using three habitat classes that correspond to the most abundant habitat types: cottonwood/willow gallery forest, mesquite woodland, and desert scrub.

Habitat classes

Cottonwood/willow
Mesquite
Desert scrub

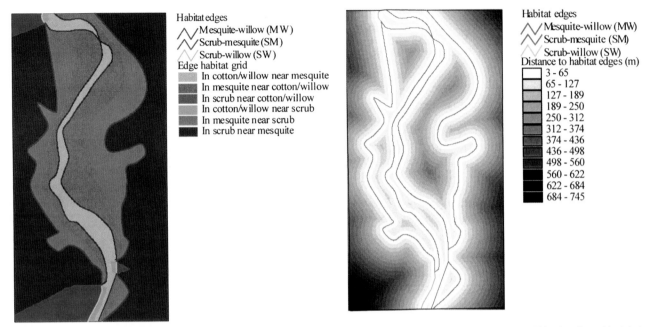

Figure 63.5. Edge habitat grid. The Effective Area Model creates a grid discriminating areas within each habitat class, based upon the habitat type that forms the nearest edge. This information is used to select the appropriate edge response curve (Figure 63.2) to apply to each pixel when estimating animal density across the landscape.

Figure 63.6. Edge proximity grid. The Effective Area Model determines the distance to the nearest edge for each pixel in the habitat grid. This grid serves as the template for projecting the appropriate edge response curve onto the habitat grid.

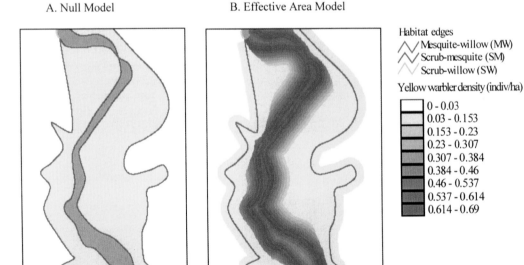

Figure 63.7. Animal density grids. Panel (A) shows the results of the null model. Panel (B) depicts the results of the Effective Area Model's projection of edge response functions onto the combined edge habitat grid, generating a map of expected animal density across the modeled landscape. Note that, for our example, yellow warbler (*Dendroica petechia*) densities are highest along the cottonwood/willow–mesquite edge.

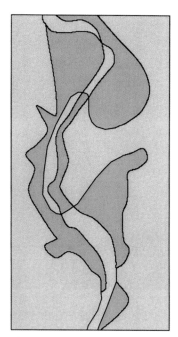

Habitat classes

☐ Cottonwood/willow
■ Mesquite
☐ Desert scrub

Figure 63.8. Habitat type conversion following an aquifer draw-down scenario, resulting in a reduction in river flow through a desert riparian landscape. In the case illustrated, 10 hectares of cottonwood/willow habitat have shifted to mesquite woodland, and 40 hectares of mesquite woodland have shifted to desert scrub.

A. Initial Habitat

B. After Habitat Conversion

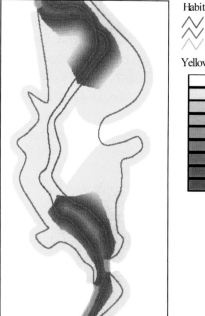

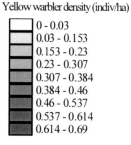

Habitat edges
/\/ Mesquite-willow (MW)
/\/ Scrub-mesquite (SM)
/\/ Scrub-willow (SW)

Yellow warbler density (indiv/ha)

☐ 0 - 0.03
☐ 0.03 - 0.153
☐ 0.153 - 0.23
☐ 0.23 - 0.307
☐ 0.307 - 0.384
☐ 0.384 - 0.46
☐ 0.46 - 0.537
☐ 0.537 - 0.614
☐ 0.614 - 0.69

Figure 63.9. The Effective Area Model applied to the riparian habitat depicted in Figures 63.4 and 63.8 to estimate predicted animal density before and after habitat-type conversion following a hypothetical reduction in river-flow volume. Panel (A) illustrates predicted density for the yellow warbler (*Dendroica petechia*) in the existing habitat, while panel (B) shows predicted bird densities for the same reach following conversion of riparian habitats to more xeric types.

TABLE 36.2.

Percentage and number of species matched and in commission for test site used in the predicted vertebrate accuracy assessment; sites are ordered by length of field inventory.

Test Site	Matches[a]			Commission error[b]	
	No. present	Count	Percent	Count	Percent
Amphibians and reptiles					
Holt Research Forest	19	19	100	7	36.8
Rachel Carson National Wildlife Refuge	32	32	100	0	0
Mount Desert Island/Acadia National Park	25	23	92	1	0.04
Mean (± st. dev)			97.3 (± 4.6)		12.3 (± 21.2)
Birds					
North Maine Forestland	72	72	100	67	93.1
Nesowadnehunk Field, Baxter State Park	55	55	100	76	138.2
White Mountains National Park	74	74	100	68	91.9
Sunkhaze Meadows National Wildlife Refuge	114	111	97.4	39	34.2
Holt Research Forest	60	57	95.0	81	135.0
Petit Manan National Wildlife Refuge	92	92	100	64	69.6
Rachel Carson National Wildlife Refuge	79	79	100	74	93.6
Moosehorn National Wildlife Refuge	137	133	97.1	25	18.2
Mount Desert Island /Acadia National Park	135	134	99.3	23	17.4
Mean (± St. Dev)			98.7 (± 1.83)		76.8 (± 45.8)

[a]Percent matched = [number of species predicted present that were present / number of species present on the site] * 100.
[b]Commission error = [number of species in predicted but not present on the site / number of species present] * 100.

TABLE 36.3.

Results of tests of Likelihood of Occurrence Ranks[a] for each test site having amphibian and reptile surveys in Maine; number of years the site has been potentially surveyed shown in parenthesis.

	LOOR				
	Low		High	ρ	*P*
	1	2	3		
Site 5: Holt Research Forest (15 years)					
Number of species predicted	8	9	9		
Number in commission	4	1	2	−0.5	0.704
Number of predicted species present	4	8	7	0.5	0.704
Site 8: Rachel Carson National Wildlife Refuge (32 years)					
Number of species predicted	10	11	11		
Number in commission	0	0	0	—	—
Number of predicted species present	10	11	11	—	—
Site 9: Mount Desert Island/Acadia National Park (79 years)					
Number of species predicted	8	8	8		
Number in commission	0	1	0	0.0	1.0
Number of predicted species present	8	7	8	0.0	1.0

[a]LOORs, defined by Boone and Krohn (1999)

TABLE 36.4.

Results of tests of Likelihood of Occurrence Ranks[a] for each test site having bird surveys in Maine; number of years the site has been potentially surveyed shown in parenthesis.

| | LOORs | | | | | | | |
| | Low | | | | High | | | |
	0	1	2	3	4	5	ρ	*P*
Site 1: North Maine Forestlands, Moosehead Lake area (1 year)								
Number of species predicted	3	27	27	28	27	27		
Number in commission	3	20	16	16	7	5	−0.97	0.006
Number of predicted species present	0	7	11	12	20	22	1.0	0.01
Site 2: Nesowadnehunk Field, Baxter State Park (3 years)								
Number of species predicted	5	25	25	26	25	25		
Number in commission	5	22	16	15	10	8	−1.0	0.01
Number of predicted species present	0	3	9	11	15	17	1.0	0.001
Site 3: White Mountains National Forest (5 years)								
Number of species predicted	2	28	28	28	28	28		
Number in commission	2	22	24	10	8	2	−0.9	0.042
Number of predicted species present	0	6	4	18	20	16	0.60	0.291
Site 4: Sunkhaze Meadows National Wildlife Refuge (10 years)								
Number of species predicted	3	29	29	31	29	29		
Number in commission	3	14	11	5	5	1	−0.97	0.006
Number of predicted species present	0	15	18	26	24	28	0.90	0.042
Site 5: Holt Research Forest (15 years)								
Number of species predicted	2	27	27	28	27	27		
Number in commission	2	25	24	17	9	4	−1.0	0.01
Number of predicted species present	0	2	3	11	18	23	1.0	0.001
Site 6: Petit Manan National Wildlife Refuge (22 years)								
Number of species predicted	4	30	30	32	30	30		
Number in commission	4	17	20	13	9	1	−1.0	0.01
Number of predicted species present	0	13	10	19	21	29	0.90	0.042
Site 7: Rachel Carson National Wildlife Refuge (32 years)								
Number of species predicted	8	29	29	29	29	29		
Number in commission	8	19	18	16	9	4	−1.0	0.001
Number of predicted species present	0	10	11	13	20	25	1.0	0.001
Site 8: Moosehorn National Wildlife Refuge (61 years)								
Number of species predicted	4	30	31	32	31	30		
Number in commission	3	6	6	6	3	1	−0.89	0.045
Number of predicted species present	1	24	25	26	28	29	1.0	0.001
Site 9: Mount Desert Island/Acadia National Park (79 years)								
Number of species predicted	5	30	30	32	30	30		
Number in commission	3	7	8	5	0	0	−0.87	0.059
Number of predicted species present	2	23	22	27	30	30	0.87	0.059

[a]LOORs, defined by Boone and Krohn (1999).

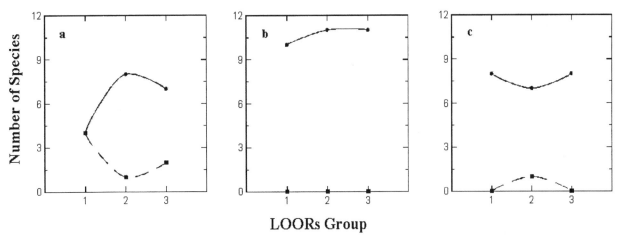

Figure 36.2. For each test site with amphibian and reptile data, the number of species correctly modeled (denoted by a circle and solid line) and the number of species in commission (denoted by a dashed line and a square). Sites are ordered from smallest to largest: (a) Holt Research Forest, (b) Rachel Carson National Wildlife Refuge, and (c) Mount Desert Island/Acadia National Park.

species detectability we were able to determine that much of our commission was due to apparent rather than actual errors in the models. We therefore suspect that the models are adequately predicting the presence of species in Maine. Although, additional effort needs to be put into surveying for those species with low LOORs to be fully confident.

The highest correlations between LOORs and the number of species in commission occurred on smaller sites with shorter surveys, which also indicates that the errors are apparent rather then actual (Fig. 36.3). Surveys such as those for the North Maine Forestlands and White Mountains National Park (conducted one year and five years, respectively) have not been established for a period of time long enough to capture the presence of the more uncommon and reclusive species (Schaefer in prep.). In addition, these surveys came from research projects having a specific objective of surveying forest songbirds (Hagan et al. 1997; D. Capen, Univ. of Vermont personal communication). These factors in an *a posterior* evaluation of error may lead researchers to incorrectly conclude that actual errors are present in the models (Edwards et al. 1996) when it is more likely that many of the errors are related to incomplete field surveys.

A more or less constant rate of commission error across LOORs on test sites with long histories of field inventories would indicate real overpredictions are being reported on the site (Boone and Krohn 1999).

This can be seen in the number of species in commission on Moosehorn National Wildlife Refuge and Acadia National Park (Fig. 36.3). The moderate correlation for these sites ($\bar{x}$ = −0.89 and −0.87) suggests that the species lists for these areas are relatively complete. Given that these sites have been surveyed for extended periods of time (sixty-one and seventy-nine years respectively; Table 36.1) it is reasonable to conclude that the species occurring on the sites have been well documented. However, even with a constant rate of error, care in the interpretation process must still be taken. The lack of correlation might be due to having too small of a sample size or by having too many LOORs groups (data spread too thinly). For example, too small of a sample size was problematic in this analysis with the herptile data. On all sites, predictions were relatively accurate, having a high percentage of species matched ($\bar{x}$ = 97.3 percent) and relatively little commission error ($\bar{x}$ = 12.3 percent; Table 36.2). On the site where a significant amount of commission was reported (36.8 percent), the number of species separated into the LOORs groups in the rank correlation test was extremely small, and thus our ability to detect a significant correlation between LOORs and commission was weak (Table 36.2 and 36.3).

Because of similarities between this analysis and that of Boone and Krohn (1999), further investigations still need to be made into the use of *a priori*

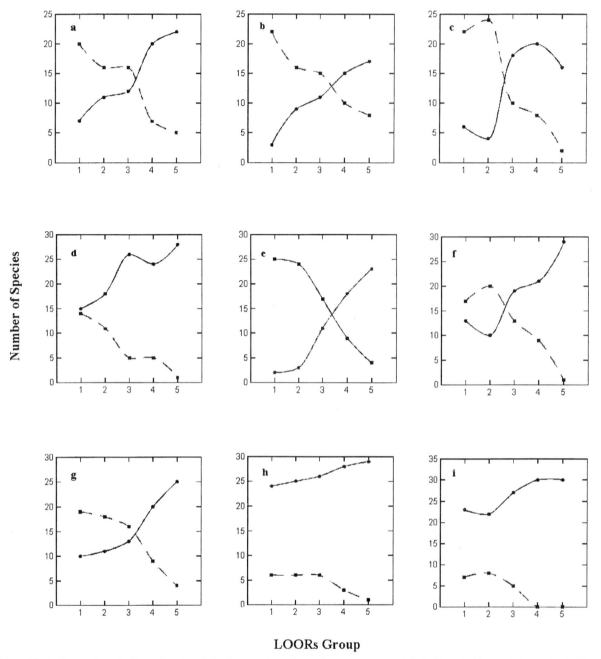

Figure 36.3. For each test site with avian data, the number of species correctly modeled (denoted by a circle and solid line) and the number of species in commission (denoted by a dashed line and a square). Sites are ordered from smallest to largest: (a) North Maine Forestlands, Moosehead Lake Area; (b) Nesowadnehunk Field, Baxter State Park; (c) White Mountains National Park; (d) Sunkhaze Meadows National Wildlife Refuge; (e) Holt Research Forest; (f) Petit Manan National Wildlife Refuge; (g) Rachel Carson National Wildlife Refuge; (h) Moosehorn National Wildlife Refuge; (i) Mount Desert Island/Acadia National Park.

ranking of species detectability to separate apparent from actual error in species predictions. There are major differences between these two studies, however, which are worthy of mention. First, the predictions of vertebrate occurrences we used were based upon an operational gap analysis, meaning that a statewide

vegetation and land-cover map created with remotely sensed data (Hepinstall et al. 1999) was the main data layer underlying species predictions (Boone and Krohn 1998a,b). In contrast, the vertebrate predictions reported in Boone and Krohn (1999) were not based on a statewide vegetation map but instead relied on the data available for each site, which in some cases included lists of vegetation cover (Boone and Krohn 1999). Unlike Boone and Krohn (1999), we wanted to study amphibians and reptiles as well as birds. Because there are few herptiles breeding in Maine, the number of LOOR groups was reduced to three. The number of LOOR groups identified by Boone and Krohn (1999) for birds was reduced from ten to five. This change tended to smooth some of the graphs of the number of species correctly modeled verses LOORs (e.g., note Moosehorn National Wildlife Refuge in Fig. 3b of Boone and Krohn [1999] versus our Fig. 36.3h) but did not change the overall patterns. Finally, this study reports data from nine test sites, three more then were used by Boone and Krohn (1999).

Conclusions

Having an ecological context in which to evaluate commission error is important and will help investigators to have greater confidence in their predictive models (Fielding and Bell 1997). The ideal situation in validating habitat-association models designed to predict the presence/absence of terrestrial vertebrates would be to have standardized field censuses capturing the presence of all species on test sites to compare to the predicted distributions. However, until such detailed surveys are available an *a priori* ranking system, such as LOORs, will permit fuller interpretation of rates of commission but helping to distinguish commission errors that are actual verses those resulting from incomplete field data. An ability to distinguish actual errors of overprediction from incomplete test data is critical to understanding the level of confidence model users should have in their findings.

Acknowledgments

We thank R. B. Boone, D. J. Harrison, and S. A. Sader for their insights and manuscript reviews. Our thanks to J. M. Hagan and S. L. Grove, Manomet Center for Conservation Science, who provided survey locations and observation information for the North Maine Forestlands study area used in this research. Thanks to J. W. Witham and M. L. Hunter, University of Maine (UM), who provided survey information for the Holt Research Forest, and to D. E. Capen, University of Vermont, and his staff who provided observation locations and data for the White Mountains National Forest. Many thanks to W. A. Halteman, UM, for advice on statistical methods. This research was supported by the Gap Analysis Program of the USGS, BRD through Cooperative Agreement No. 14-16,009-1557, Research Work Order Number 39 to the Maine Cooperative Fish and Wildlife Research Unit. The Maine Unit is supported by the BRD, University of Maine, Maine Department of Inland Fisheries and Wildlife, and the Wildlife Management Institute. This is scientific contribution number 2415 of the Maine Agricultural and Forest Experiment Station.

Assessment of Spatial Autocorrelation in Empirical Models in Ecology

Mary Cablk, Denis White, and A. Ross Kiester

A Brief Overview of Error and Error Sources

The existence of error in spatial analyses is a well-known occurrence (Veregin 1989). In statistical models, for example, error is introduced by simply calculating and using a mean value. We can estimate the error associated with a population mean by calculating the standard deviation or setting confidence limits. The resulting loss of accuracy is termed *random error* because there is no regularity in the direction or magnitude of the error. *Systematic errors*, on the other hand, are those that result from the introduction of a fixed and consistent difference from the true value. Errors are generated with successive iterations of analytical processing and as such become compounded. This compounding of error from multiple processing iterations is termed *error propagation*.

At the most fundamental level, error is introduced into an analysis by limiting precision. Rounding numbers from four decimal points to two, for example, introduces error. Additional calculations involving the parameter that has been rounded may or may not be precise, but they will be somewhat erroneous. At this level, we assume that such introduced error is within acceptable bounds of what we consider noise. However, as the analysis continues, we continue to round and thus continue to propagate error. As analyses increase in complexity, additional error types may be in-

curred. Although error and error propagation has been studied in detail (Chrisman 1982; Walsh et al. 1987; Lunetta et al. 1991; Lanter and Veregin 1992; Haining and Arbia 1993; Fielding, Chapter 21; Karl et al., Chapter 51), tracking error propagation in practice is difficult, particularly in spatial data.

Spatial analyses have undetectable errors that are generated at the basic production level. Thickness of pen, the wobble of a hand-drawn line, the number of nodes, or any other fundamental step in the mapmaking process introduces error to a thematic or other data layer. In an effort to introduce some level of standardization and quality, the U.S. Geological Survey (USGS) adopted the 1947 Bureau of Budget National Map Accuracy Standards (NMAS) to produce standard map products with a known degree of certainty, or uncertainty, in horizontal and vertical map dimensions. Map errors are usually not accounted for in spatial analyses because they are difficult to identify, quantify, and rectify. Veregin (1989) identified five dimensions, or categories, of opportunities for error in spatial databases: thematic, cartometric estimates, data compilation, geographic information system (GIS) operations, and other general issues. Of these dimensions, the easiest to quantify are those of classification of continuous data, such as remotely sensed imagery. This error, thematic error, can be accounted for and in many instances can be resolved with concentrated and guided effort.

Thematic error is simply whether or not a given area, pixel, polygon, or other spatial map feature, is an accurate representation of the landscape or mapped feature. This does not account for whether or not the classification scheme itself is an accurate representation of the landscape. Error matrices, sometimes referred to as confusion matrices, are generated for classifications with the purpose of evaluating a map product for thematic accuracy. That is to say, does this point on the map accurately represent what is on the ground? The associated error matrix provides the user with values indicating how true the map product is with respect to actual conditions on the ground. Error is calculated for each class and for the overall mapped area. Additionally, errors of omission and commission can be calculated to give the user information to update and refine the classification or for estimating confidence on a per-class basis in additional analyses. However, unless each minimum mapping unit, which may be a pixel, polygon, or any other feature with spatial properties, has been verified on the ground, there is still unknown and unquantified error. Furthermore, this error is unknown geographically.

An understanding of accuracy and error in classified or other thematic maps is critical when incorporating these data into a greater analysis. In studies of biodiversity, for example, thematic maps representing life zones, ecoregions, land cover, or vegetation composition are used to design field sampling strategies, conduct surveys, estimate habitat availability, and relate presence or absence to geographic locations. If these data are inaccurate, then subsequent analyses incorporating these data will be flawed as well. The difficulty in estimating and tracking these errors approximates the impossible.

An alternative to using thematic maps for correlation analyses of ecological data is to use continuous data, such as time series. Continuous data that retains its spatial component can be collected at any scale (grain) and analyzed using geostatistical methods to quantify pattern. Continuous data can be collected from many sources, such as field data, time series instrument recordings, or even satellite imagery.

Implication of Error in Ecological Modeling

There will always be error in modeling and computing because perfection does not exist where approximation or estimation is involved. To avoid errors associated with the process of creating thematic maps and to circumvent propagating unknown vectors and magnitudes of error, we might choose to use continuous data in analyses. Standard statistical methods can be used on continuous data and error terms can be estimated (i.e., standard deviation, [ε], etc.). The difficulty of analyzing spatial data with standard statistical techniques for hypothesis testing is that the assumption of independence is violated (Fortin et al. 1989; Legendre and Fortin 1989). Spatial data are inherently autocorrelated. In nature, ecological phenomena or discrete features influence neighbors, are influenced by neighbors, or both. The level of neighbor influence may vary with distance and other factors, making the autocorrelation function nonlinear. In general, the closer the neighbors, the greater the level of correlation, which distorts statistical tests of significance in analyses such as correlation, regression, or analysis of variance (Cliff and Ord 1981).

In ecology, we use this inherent autocorrelation to our advantage to make predictions where sampling does not or cannot occur. Therefore, we face a dilemma in ecological studies: we seek to explain ecological patterns but we have yet to develop tools that allow us to do so without violating our own set of terms. The inherent structure of the distribution of the natural world prohibits us from accurately explaining or evaluating the underlying pattern-process phenomena. Because of this paradigm, spatial analyses tend to be conducted either on continuous data analyzed using geostatistics or on a combination of thematic data operated on in a GIS. Associative wildlife habitat relationship models are an example of where true spatial methods would greatly advance our ability to quantify pattern-process relationships.

Quantifying Pattern-Process Relationships with Both Discrete and Thematic Data

Traditional habitat models are based on the concept or technique of categorization. For example, certain species are known to occur or are correlated with specific habitats. If we want to quantify a species-habitat relationship, we can mathematically relate the presence or absence of the particular species with the occurrence, amount, or distribution of a specific habitat type or types. When a statistically significant relationship is derived, the results can be explained based on an empirically derived statistical model. In reality, species are not related to class type or habitat type, rather, there exists an underlying relationship between the species and compositional landscape features or other characteristics of a certain class or habitat type. These components may include productivity, vertical vegetative complexity, temperature or photo gradients, among others. The reification of habitat by categorization in these circumstances thus serves to introduce a level of subjective interpretation that may mask the true correlations that exist between species and their environments.

If we further refine the analysis of species-habitat relationships and pose the question *"what mechanisms shape the patterns of biodiversity,"* we encounter more difficulties using categorically based methods. For example, the relationship between species diversity and a landscape comprising of sixty-six categories will be different from the relationship between diversity and the same landscape classified into thirteen categories (see Fig. 37.1 in color section). Therefore, the modeled relationships for species-habitat interactions or presence/absence predictions will vary depending on the land-cover base map used. The relationships found between an individual species and a sixty-six-class base map may be interpreted differently than if a base map with fewer or a greater number of categories were used.

A level of uncertainty exists in the use of categorical maps regardless of the number of interpreted classes. No two independent expert assessments will interpret a given landscape in exactly the same fashion because there is no universally accepted standard or set of rules that define land-cover types. Even given a set of criteria, there will exist variability in land-cover interpretation due to human-based differences regardless of interpreted media such as aerial photographs, hardcopy satellite image prints, or digital imagery. Another and perhaps more abstract issue with using

land-cover types as a basis for defining habitat is the assumption that fauna interpret or respond to the landscape in the same way as humans. Finally, land-cover classes do not represent temporal elements of a landscape. Temporal components might include length of photosynthetic productivity, timing of greenup or senescence, or duration of snowpack. These factors are not included in species-habitat assessments that are based on single-date derived land-cover maps. Wolff (1995) presents other criticisms of species-habitat association studies. He discusses ten points that address shortcomings of correlational analyses that include caveats of short-term localized studies, lack of independence in data, lack of replication, and potential differences in landscape interpretation by humans versus the individual animals.

The situation becomes more complicated when more than one species is evaluated. Groups of similar species may aggregate on the landscape in a similar pattern but with overlap in range while other species from different taxa may exhibit very different patterns of distribution with or without overlap in range. The pattern-process mechanisms become increasingly complex and nested and the results of these analyses are difficult to interpret. Because of the complexity of interactions between pattern and process of multiple species analyses, standard statistical techniques again fall short. Independent variable interactions may not be linear or known *a priori* and therefore may be excluded from a final empirical model. The assumption of independence remains violated. Finally, it is difficult to reconstruct spatial data once it has been despatialized, or reorganized into a structure for standard statistical analysis. Therefore, one place to begin an analysis of the spatial pattern of multiple species with multiple interacting independent variables is with exploratory data analysis.

Exploratory Data Analysis: An Alternative to Standard Statistical Techniques

Exploratory data analysis (EDA) is a means to quantify the inherent structure and variable interactions within a data set rather than forcing the data to fit a predefined or derived model. The fundamental philosophy is to use as much of the data as possible rather than creating summaries (such as means) that discard data. Exploratory techniques that are data-driven, nonparametric, and computer-intensive are alternatives to traditional Gaussian models for quantifying nonlinear structures in data (Tukey 1977; Miller 1994b), in contrast to Neyman-Pearson hypothesis testing. The computational intensity of large data sets at one time limited researchers' abilities for analysis, but with modern computers data volume is less of a limitation. With exploratory techniques, we can apply more-complicated statistical analyses to explore and describe data and to draw valid statistically based inferences on very large data sets (Efron and Tibshirani 1991). Exploratory methods allow us to uncover and quantify the structure inherent in data, free of traditional assumptions of normality.

In Oregon, the process-pattern relationship between vertebrate richness and environmental heterogeneity was expected to have a hierarchical structure of complexity and for our study, interactions between variables were unknown. We were interested in quantifying this hierarchical structure inherent in the data to provide insight for understanding process-pattern relationships rather than fitting the data to a predefined curve. With the use of exploratory data analysis and graphical display tools, these complex structures were described and interpreted.

Study Objectives

This study investigated the relationships between vertebrate species richness of birds, mammals, reptiles, and amphibians, respectively, with vegetation phenology, terrain, and climate for the state of Oregon. We attempted to maximize the number of species investigated and to directly correlate vertebrate richness with meaningful spatial and temporal environmental variables. Although the goal of this study was to determine whether satellite imagery, digital elevation model (DEM) data, and climate variables could be used directly to better explain observed pattern-process interactions of vertebrate species richness with the landscape, we present in this chapter our methods for evaluating the predictive and explanatory models derived in this process.

Data Compiled and Reviewed

A total of twenty-seven variables were used in this analysis (Table 37.1). These variables included indices derived from 1992 satellite imagery, DEM variables, and climate variables. Descriptions of each of these variables are given in Table 37.2. Data were collected at two different grains: the 1-km^2 pixel data, including Advanced Very High Resolution Radiometer (AVHRR)-derived parameters and DEM variables, and climate and richness data at 646-km^2 hexagons. The grain of the study was 646-km^2 hexagons and pixel data were aggregated up to the hexagon scale. Two summary statistics, median and variance, were calculated for variables within each of the 434 hexagons in Oregon. Median was not calculated for aspect and median values were not available for climate variables so mean values were substituted. Diversity, the number of different pixel values within a hexagon, was calculated for elevation, slope, and aspect as an additional measure of terrain variability.

Species richness data was compiled on a statewide tessellation of 646-km^2 hexagons by The Nature Conservancy (TNC) for native mammals, breeding birds, reptiles, and amphibians, and were reviewed by experts throughout the state (Master 1996). Each species

TABLE 37.1.

List of variables evaluated for correlational analyses with vertebrate diversity by taxa (birds, mammals, amphibians, and reptiles).

Summary statistic	PC	Greenness	DEM	Climate
Median and variance	PC2	maxv	elev	
	PC3	tot	slope	
		onv		
		sdn		
		range		
		sup		
Variance			aspect	seas
Mean				seas
				precip
Diversity			elev	
			slope	
			aspect	
Total	4	12	8	3

TABLE 37.2.

Interpretation of variables for each of the metrics used to model vertebrate species richness by taxa.

Variable	Abbrev.	Interpretation
Principal component 2	PC2	Seasonal pattern of phenology
Principal component 3	PC3	Nongrowing-season photosynthesis
Maximum NDVI value during growing season[a]	maxv	Level of maximum photosynthetic activity
Total integrated NDVI[a]	tot	Net primary production during the growing season
NDVI value at start of growing season[a]	onv	Level of potential photosynthetic activity at beginning of growing season
Rate of senescence[a]	sdn	Rate of senescence
Range of NDVI[a]	range	Range of annual photosynthetic activity
Rate of greenup[a]	sup	Rate of greenup
Elevation	elev	Elevation
Slope	slope	Slope
Aspect	aspect	Cardinal direction (aspect)
Seasonal mean temperature difference	seas	Difference between monthly mean July and January temperatures
Precipitation	precip	Rainfall

[a]*Adapted from Reed* et al. *(1994).*

was assigned an occurrence probability ranking for each hexagon. Recorded sighting and specimen collection locations were registered within the hexagon grid. These hexagons were termed "confirmed" and assigned a rank of 96–100 percent certainty that a given species occurred in that hexagon. The second ranking, "probable," was defined as 80–95 percent confidence that a given species occurred in that hexagon. Based on habitat and expert opinion, a species was assigned to a hexagon with the "probable" ranking if there was not a recorded specimen in that hexagon. This analysis included species given either "confirmed" or "probable" ranking.

The seasonal metrics derived from AVHRR time-series normalized difference vegetation index (NDVI) satellite imagery (Table 37.2) quantitatively characterized seasonal phenological phenomena within the course of one year. The six metrics in this analysis were chosen based on phenological relevance, units of measure, and whether or not each was a continuous versus discrete measure. Each of these metrics quantified a component of the annual NDVI curve over the course of one year thus capturing ecosystem dynamics (Reed et al. 1994).

Principal component analysis (PCA) was conducted on 1992 time series NDVI data. PCA was calculated

for twenty-one biweekly composites of National Oceanic and Atmospheric Administration (NOAA)–AVHRR 1-km^2 pixel data for the state of Oregon for the year 1992 to capture the greatest spatial and temporal variability in the twenty-one-scene data set. This type of analysis is well documented by Cicone and Olsenholler (1997), Eastman and Fulk (1993), Fung and LeDrew (1987), and Tucker et al. (1985). Principal component 2 (PC2) represented seasonal vegetation growth independent of the first principal component and accounted for 4.5 percent of the annual variation in NDVI. Principal component 3 (PC3) was interpreted to be baseline photosynthesis occurring in primarily coniferous-dominated or evergreen broadleaf forests during the winter months and as nongrowing season vegetation characteristics.

Terrain variables were calculated from 1-km^2 DEM obtained from the U.S. Geological Survey (USGS) EROS Data Center (EDC). Variables included were elevation (meters), slope (degrees), and aspect (degrees). Climate data were acquired for the state of Oregon as a subset from a larger database for the conterminous U.S. January and July temperatures and annual precipitation were compiled to a 1-km^2 rectangular grid. January and July mean temperature data were modeled and compiled using the method of

Marks (1990). Monthly averages of forty-year means, from approximately 1948 to 1988, were calculated at approximately 1,200 stations in the Historical Climate Network database. These values were first corrected to potential temperatures at a reference air pressure of 1,000 Mb using the station elevations and assuming a normal adiabatic lapse rate. The potential temperatures were then interpolated to the 1-km^2 grid using a linear model. These interpolated values were then converted to estimated actual temperatures from the adiabatic lapse rate correction using corresponding elevation values at each grid point. Annual precipitation data were compiled from the 10-kilometer resolution data to 1 kilometer (Daly et al. 1994). Mean precipitation and seasonal differences defined as (mean July–mean January temperatures) calculated on a per-pixel basis were summarized by hexagon as mean and variance. Seasonal difference and precipitation were also characterized by range, defined as (mean maximum value–mean minimum value) within a hexagon.

Data for the response, climate, and topographic predictor variables were assembled as part of a biodiversity research program that investigated mechanisms influencing the distributions of biodiversity, objective methods for prioritization places for conservation of biodiversity, and consequences for biodiversity of future landscape change (White et al. 1999).

Spatial Statistical Methods

Two spatial statistical methods were used: semivariance and Moran's I. Semivariance (γ) is a model of the average degree of similarity between observations as a function of distance (Rossi et al. 1992). Semivariance values range between 0, which indicates complete autocorrelation, and ∞, which indicates complete randomness in the data. A semivariogram shows autocorrelation as a function of distance and when plotted represents spatial variability (Cohen et al. 1990; Legendre and Fortin 1989). The two parameters used to describe the spatial pattern of a data set from a semivariogram are the *sill*, which is the value at which the curve levels off, and the *range*, which is the lag distance corresponding to the sill. Moran's I, I(d), is a spatial autocorrelation coefficient that indicates significant patch size pattern in spatial data. The two meth-

ods are complementary for evaluating the spatial structure of autocorrelated data. Values of I(d) range between −1 and 1 with positive values corresponding to positive autocorrelation, zero indicating randomness, and negative values representing negative autocorrelation. Positive significant values indicate similarity at the scale of the lag distance, or distance between pairs of points. Negative significant values show the distance between peaks and troughs. Spatial richness patterns are thus indicated by a series of significant positive and negative values (Legendre and Fortin 1989). Standardized semivariance and Moran's I with 95 percent confidence intervals were calculated for each of the taxa and for residuals from each of the regression tree models. The 95 percent confidence interval computed at each lag distance was a test for significance where the null hypothesis was that the coefficient at a given lag distance was not significantly different from zero.

Exploratory Statistics–Classification and Regression Tree Analysis

Classification and regression tree analysis (CART) is a tree-based exploratory data analysis method that has been shown to be effective in identifying and estimating complex hierarchical relationships in multivariate data such as satellite-derived indices, DEM, and other digital data (Rathert et al. 1999; Michaelsen et al. 1994). The data set is repeatedly partitioned into homogenous subsets using binary recursive partitioning until the entire data set has been evaluated. Node splits are determined by deviance, where a split occurs at a given node such that the change in deviance is maximized and the variance of all resulting subsets is minimized. CART is useful where there are expected but unknown nonlinear or nonadditive predictor-response interactions (Michaelsen et al. 1994). Regression tree analysis is a data-driven, nonparametric, computer-intensive method (Miller 1994b) that selects variables and values for splitting which best discriminate among responses (Efron and Tibshirani 1991). Because this method results in overfit models, meaning all data are categorized, regression trees must be pruned to select the most parsimonious subset that best explains the relationship between response and predictors. Cross-validation was used to select subtrees that have optimal predictive performance with lowest complexity (Breiman et al. 1984; Clark and Pregibon 1992). In a cross-validation, regression trees are computed on nine-tenths of the data and are checked with the one-tenth of the data withheld (Miller 1994b). Three cross-validation methods were used to determine the appropriate number of terminal nodes for the pruned trees: cost-complexity pruning, one standard error rule (one SE), and adjusted minimum risk (AMR). Each method was run on ten iterations where the suggested number of terminal nodes was calculated from each method and the complexity penalty was 0.01. Cross validation was run ten times for each of the five taxa. For each taxon, the median value of the ten cross validations was calculated. Tree length was based on median number of suggested nodes from the ten iterations.

Analyses resulted in numeric and graphical means for evaluating final CART models. Final models from each CART were indicated node splits, values at node splits, predicted average number of species per hexagon at terminal nodes, and number of hexagons at each terminal node. Semivariance and Moran's I were calculated on residuals from each CART model by taxa as a means to evaluate randomness in residuals similar to using a residual plot to check the fit of a regression model. In this manner, we integrated nonspatial statistical methods with spatial analyses as a check on how well the regression trees fit the data and to determine how well the models dealt with autocorrelated richness data.

CART Results for Predicting Richness

CART explained between 55 and 91 percent of the variation in taxonomic richness for vertebrates in Ore-

gon. Table 37.3 shows deviance explained—a measure similar to the familiar R^2 value in standard statistics—for each regression tree model by taxa and lists the optimal number of terminal nodes for each tree after pruning. Semivariance and Moran's I were calculated for the richness data by taxa and for the residuals of the regression tree models. This served to check how well the regression trees fit the data and to determine how well the individual models dealt with autocorrelated richness data. Like residual plots from classical regression methods, no pattern was expected from models that fit the data well. Spatial plots of residual values for regression tree models exhibited no pattern. Results from the spatial statistics on both the original data and residuals indicated that there was significant spatial autocorrelation to the vertebrate data but that CART was able to effectively model correlative relationships despite autocorrelation in the original data.

The semivariograms and corresponding correlograms for the spatial pattern of mammal, bird, reptile, and amphibian distributions, respectively, are shown in Figures 37.2–37.5. Semivariance and Moran's I calculated on the residuals (the difference between the actual species richness and predicted species richness) are also shown in Figures 37.2–37.5. As with residual plots from classical regression methods, no pattern was expected from models that fit the data well.

For mammals (Fig. 37.2), there was significant spatial autocorrelation in the data to a lag of 180 kilometers, or an approximate neighborhood of six hexagons. A secondary peak in autocorrelation appeared at longer lags. Moran's I was significant and positive to a lag of 60 kilometers, a distance about equal to two hexagon center-to-center distances. At the maximum lag distance, there was some indication of significant negative autocorrelation. The range of the semivariogram for mammal CART residuals was 120

TABLE 37.3.

Deviance explained for regression tree models by taxa and the number of terminal nodes for each data set.

	Mammals	Birds	Reptiles	Amphibians
Deviance explained	0.55	0.67	0.57	0.91
No. terminal nodes	11	6	9	5

MAMMALS

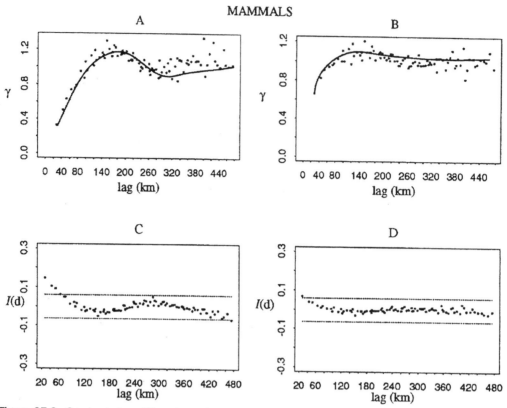

Figure 37.2. Semivariogram (A) and correlogram (C) for mammal richness and semivariogram (B) and correlogram (D) for residuals of mammal CART model. Dashed lines on correlograms indicate 95% confidence intervals. The fitted line on semivariograms is a spherical fitted model based on estimates using nonlinear least squares.

kilometers and the correlogram indicated a weak positive significance at a lag equivalent to one hexagon center-to-center distance.

Bird richness (Fig. 37.3) exhibited spatial autocorrelation to 340 kilometers, or nearly twelve adjacent hexagons. Moran's I coefficients were significant and positive to a lag of 100 kilometers. The residuals from the bird CART semivariogram and correlogram showed autocorrelation to lag distances of 200 kilometers and 27 kilometers, respectively. Some coefficients on the correlogram were weakly and negatively significant at a 95 percent confidence interval. This was not apparent in the correlogram of bird CART model residuals.

Semivariance of reptile richness (Fig. 37.4) had a range of 240 kilometers and Moran's I coefficients were positive and significant to a lag of 80 kilometers, or three adjacent hexagons. The CART residual semivariogram had a range of 140 kilometers. The corre-

sponding correlogram indicated CART residuals were positive and significant to one adjacent hexagon. The increasing trend in semivariance, which was apparent in the richness data, was not seen in the semivariogram of the reptile CART model residuals. Moran's I coefficients for reptile residuals were consistently closer to 0 than the coefficients for the corresponding reptile richness data.

The amphibian semivariogram in Figure 37.5 shows a significant positive correlation to a lag of 260 kilometers and significant negative correlations begin at a lag of 320 kilometers. The semivariogram and correlogram of CART residuals for amphibians showed no indication of spatial trend. The correlogram indicated significant positive correlation at smallest lag distances of 27 kilometers. A detailed explanation of the difference between the amphibian semivariogram and those of the other taxa follows.

BIRD

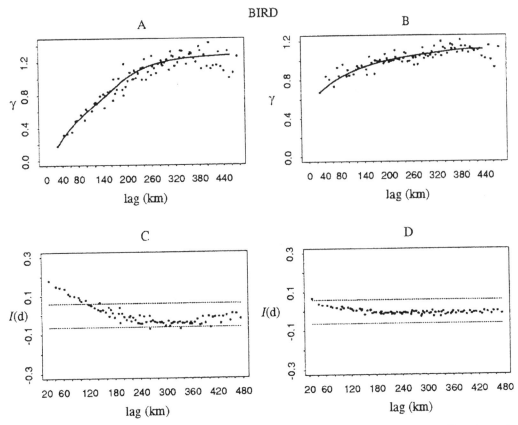

Figure 37.3. Semivariogram (A) and correlogram (C) for bird richness and semivariogram (B) and correlogram (D) for residuals of bird CART model. Dashed lines on correlograms indicate 95% confidence intervals. The fitted line on semivariograms is a spherical fitted model based on estimates using nonlinear least squares.

Discussion

We used semivariance and Moran's I in the same way that residual plots are evaluated in regression analysis: as a means of determining goodness of fit for our CART models. If the models fit well, we did not expect to see spatial autocorrelation in the semivariograms or correlograms of residuals. We did not expect to see a pattern because the over- or underestimation of the number of species in each hexagonal cell should not be systematic. We did expect to see a spatial pattern in the species richness data because ecological data are autocorrelated. Where CART excelled over standard regression modeling techniques was in the ability to make richness predictions based on spatially autocorrelated data that also exhibited nonlinear interactions. So, the use of CART as an exploratory method and as a predictive modeling tool combined with a goodness-of-fit test adapted

for spatial data was found to be successful. The success of CART to explain the spatial variance of vertebrate richness in Oregon showed promise but was somewhat dependent on the underlying spatial pattern of the dependent data (richness).

The spatial autocorrelation of the species data itself presented an interesting challenge for statistical fitting, and CART effectively modeled much of the spatial variability and autocorrelation of the original data. For any given explanatory variable by hexagon the value in a neighboring hexagon was expected to have a similar value. In most cases, CART models captured this spatial variability indicated by the fact that model residuals reflected low level noise and were spatially random. This was supported by the deviance explained, in combination with interpreted semivariograms and correlograms of the individual taxonomic distributions. All taxa had significant positive autocorrelation and

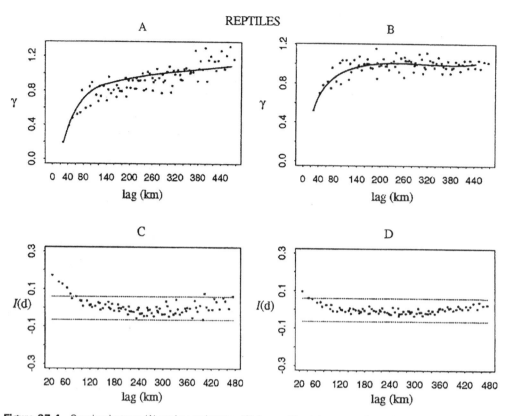

Figure 37.4. Semivariogram (A) and correlogram (C) for reptile richness and semivariogram (B) and correlogram (D) for residuals of reptile CART model. Dashed lines on correlograms indicate 95% confidence intervals. The fitted line on semivariograms is a spherical fitted model based on estimates using nonlinear least squares.

CART effectively modeled these groups evidenced by random semivariograms and nonsignificant values of Moran's I of model residuals. Statewide amphibian distribution, for example, was the simplest pattern, had the most variability accounted for in the final model (0.91), and had a random spatial residual plot. Mammal and reptile distributions exhibited the most complex spatial patterns and were likewise the most difficult to model. CART explained a little over half of the variability in mammal (0.55) and reptile (0.57) richness data. Unlike the amphibian distribution, mammals exhibited correlation at two different lag distances and, as a result, were the most difficult pattern to explain. The two different lag distances are apparent in the shape of the semivariogram.

Across taxa, results from semivariance and Moran's I indicated the CART methods reduced significant autocorrelation in the richness data, but there still existed some pattern to the residuals at shortest lag distances. Taxa with more complex spatial patterns were more difficult to model and this was reflected in corresponding lower values of deviance explained. The deviance explained for birds (0.67) was higher than for reptiles (0.57), and likewise reptiles had a more significant positive value of Moran's I for short lag distances. The case of amphibians is different because the pattern of amphibian distribution was not one of heterogeneous richness patterns like the other taxa.

The amphibian semivariogram is unlike those for the other taxa and was fit with an *exponential* model. This type of semivariogram does not have a sill and likewise lacks a range. Exponential semivariograms indicate a spatial gradient, which in the case of amphibians in Oregon runs east-west, with lowest richness in the east and highest richness in the west. The existence of a gradient is also supported by the significant positive and negative coefficients in the corresponding correlogram. Despite this strong gradient

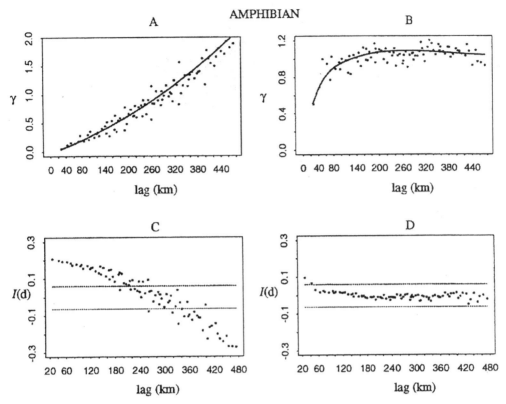

Figure 37.5. Semivariogram (A) and correlogram (C) for amphibian richness richness and semiovari-ogram (B) and correlogram (D) for residuals of amphibian CART model. Dashed lines on correlograms indicate 95% confidence intervals. The fitted line on semivariograms is a spherical fitted model based on estimates using nonlinear least squares.

pattern in amphibian richness, the CART residuals did not show a spatial pattern.

An argument could be made that, because the underlying richness data and independent variables have spatial dependence, we should expect to see a corresponding pattern in the residuals. The existence of such a pattern in residuals would indicate systematic error. Semivariance and I(d) indicate spatial dependence across distance, that is, how similar neighboring values are at regular distance intervals. A significant pattern in residuals indicates that a model consistently over- or underpredicts species richness in a spatially dependent manner. Therefore, pattern in residuals from the CART models is an indication of some sort of bias, such as inoptimal sample- or grid-cell size. There is evidence in our results to indicate our hexagon size was not optimal, seen for all taxa, in that the semivariance curves did not pass through zero. This *nugget effect* indicates either measurement error or variation at a finer grain size. The results of the spatial

statistics calculated on the amphibian residuals provide the strongest evidence to support both the results of the CART models and our method for evaluating the goodness of fit of those models. Amphibian richness had the most prominent spatial pattern and the highest deviance explained (0.91), and the analysis of residuals showed no spatial pattern.

Expert Judgment Required When Selecting a Statistical Model

The results from this study did not imply cause-and-effect relationships but did suggest relationships between diversity of taxa and specific phenologic, climatic, and topographically related forces. It is impossible to determine what factor(s) directly caused the patterns of diversity on the landscape today for several reasons, the most obvious being that we are restricted to retrospective studies given the relatively short life span of humans on an evolutionary time scale. We are also somewhat encumbered by our need to aggregate and characterize, evidenced in this study by the fact that none of the semivariograms of the taxa passed through the origin. This indicated an undetected pattern at a different spatial scale due to inappropriate sample size. Regardless of grain and extent, there will always be interpretation of the initial response variable (i.e., number of species per given area) and the explanatory variables (i.e., interpolated climate or averaged reflectance). These interpretations may create artifacts before any analysis is conducted. To effectively manage and interpret data and results from biodiversity analyses, however, we must aggregate, interpolate, and categorize.

This study not only showed some general trends that support findings of previous work, it also added new insight because of the methods used to assess suitability of the regression trees for predicting vertebrate richness. The development of spatial statistics has created tremendous opportunity for researchers to quantify and explain patterns in nature, but these new tools must be used with caution and thoughtfulness. With these new methods, researchers must question more than just the ecological or statistical significance of results. When expert judgment is required to select "the best" result, subjective interpretation plays an increasingly important role. Ecological interpretation of two closely "best" fitting models may be very different. We must have faith in our own knowledge to select the model that is closest to "truth." The use of parametric measures as a substitute for classified land-cover data for modeling vertebrate species richness may allow us to better relate patterns of diversity with correlated processes. In this manner, we have minimized subjective interpretation in terms of landscape function and habitat distribution.

Acknowledgments

The authors would like to thank all who contributed directly and indirectly to this work. Jeannie Sifneos, Manuela Huso, and Lisa Ganio provided statistical expertise. Brad Reed's assistance with phenology and greenness metrics was invaluable. Thanks to Bill Ripple, Chaur-Fong Chen, Ed Starkey, Jon Kimerling, and Kimberly Dunn for their excellent guidance as well.

Ranked Modeling of Small Mammals Based on Capture Data

Vickie J. Smith and Jonathan A. Jenks

Presence/absence models can be used to predict species occurrence and to map distributions (Scott et al. 1993; Krohn 1996; Smith and Catanzaro 1996). These models are based on habitats used by the animal and can be generated by season or year. When developing these models, all habitats are weighted equally and assumed to contain similar densities of species. For ubiquitous species occupying a number of habitat types, presence/absence models may be less useful for predicting habitats that are selected or avoided. Under these conditions, ranked models may provide a more accurate depiction of species occurrence and distribution than presence/absence models. Ranked models are formulated from capture data and habitat types are ranked, or weighted, based on higher capture densities in various habitat types.

The objective of the study described in this chapter was to compare presence/absence models to ranked models that account for variation in the number of small mammals captured within habitats. A more accurate depiction of habitat selection ranked by species density would enhance knowledge of distribution patterns and conservation status.

Although this study focuses on the use of capture data to assess the abundance of species in given habitat types, habitat quality is difficult to determine from trapping data without a study encompassing all seasons and reproductive success associated with habitat

types. To accurately assess habitat quality, extensive studies must be conducted that would measure reproductive state. Sources may occur where a high-density population has produced a large number of offspring. These offspring are forced into less-suitable habitat, where they occur at relatively high densities. In this case, a large number of individuals would not indicate high-quality habitat (Van Horne 1983). In spite of this knowledge, we have predicted occurrences based on the density of species in given habitat types. Reference to high-density habitats does not necessarily imply high-quality habitat.

Study Area and Methods

To generate ranked models, vegetation communities were sampled throughout eastern South Dakota, which consisted mostly of state-owned game production areas (GPA) and federally owned waterfowl production areas (WPA). These public areas contain remnant patches of mixed-grass prairie characterized by western wheatgrass (*Agropyron smithii*), needle-and-thread (*Stipa comata*), and sideoats grama (*Bouteloua curtipendula*) and are invaded by smooth brome (*Bromus inermis*). However, agricultural fields (i.e., corn, soybeans, wheat) dominate the landscape. Wetlands and shelterbelts also are common (Luttschwager et al. 1994).

Two satellite images from April 1992 were

obtained from the Multi-resolution Land Characteristics Consortium (MRLC) for path 30, row 30 (scene 7) of south-central South Dakota and georeferenced at EROS Data Center, Sioux Falls, South Dakota. Selected scenes were chosen based on availability, clarity, and low percent cloud cover. Agricultural lands and wetlands were masked from the scene based on an unsupervised classification (Vogelmann et al. 1998) and accuracy assessment. An unsupervised classification was performed on perennial vegetation resulting in one hundred clusters. Clusters were evaluated using known land cover to perform a supervised classification throughout the remaining unclassified pixels in the scene. Once general land-cover categories were determined, digital coverages of ancillary data, such as elevation and soils, were used to reduce confusion among clusters (Lauver and Whistler 1993; Egbert et al. 1998; Vogelmann et al. 1998).

Using IMAGINE software (ERDAS, Atlanta, Ga.), land-cover types from the general classification of Scene 7 were recoded into the appropriate ranked category (1–3). National Wetlands Inventory (NWI) basins were buffered by 90 meters. If a species occurred in wetlands, this coverage was added to the ranked land-cover coverage, resulting in a ranked prediction of density.

Small mammals were captured using snap traps (Woodstream Museum Special and Victor) and live traps (Woodstream Hav-a-Hart and Sherman) throughout eastern South Dakota from 16 May to 28 August 1998. Four trap lines were set per week and checked each morning for three days. Trap lines contained four to five traps (various combinations of live and snap traps) at each of twenty-five stations. Traps were baited with a mixture of oatmeal and peanut butter. Sampling areas were chosen based on the Environmental Protection Agency's (EPA) Environmental Monitoring and Assessment Program (EMAP) hexagons (Csuti and Crist 2000). GPAs and WPAs were chosen within the hexagons and traplines placed in wetlands, pasturelands, shelterbelts, deciduous trees, and grasslands.

Presence/Absence Models

National GAP standards (Csuti and Crist 2000) for creating presence/absence models were applied for the three species using literature on habitat use (deer mouse [*Peromyscus maniculatus*] [Rumble 1982; Forde et al. 1984; Hull-Sieg et al. 1984], meadow vole [*Microtus pennsylvanicus*] [Lokemoen and Duebbert 1976; Wilhelm et al. 1981], and arctic shrew [*Sorex arcticus*] [Gruebele and Steuter 1988]) and available capture information. Presence/absence models were created for the deer mouse, meadow vole, and arctic shrew, giving equal weight to each of the habitats within which the species was captured. Using IMAGINE software, each habitat type was assigned a 1 if the species was present and a 0 if the species was absent. Species that occurred within wetland habitat types were assigned a 1 inside the buffered wetland coverage. However, if a given pixel in the coverage contained suitable habitat in the land-cover image, the wetland coverage, or both, it was assigned a code of 1 (for presence).

Ranked Models

Capture data for three small mammal species, the deer mouse, meadow vole, and arctic shrew, were evaluated for each habitat type within trap lines to determine species density. Habitat types included wetlands, pasture, shelterbelts, deciduous trees, grasslands, and Conservation Reserve Program (CRP) lands. To allow comparison between habitats, the number of small mammals captured in each habitat was divided by trap nights and multiplied by one thousand (captures per thousand trap nights). A 1 was assigned to the habitat type with the lowest capture rate for each species. Ratios between the other capture rates were calculated for the remaining habitat types in which species were captured. These ratios were ranked from 1 to 3 (3 being the highest) to differentiate between capture abundance (Table 38.1).

Comparison of Models

Pixels contained in each category were converted to hectares for analysis. Models were compared by subtracting the value of each pixel of the presence/absence model from the ranked model. If no difference occurred between the two models, both methods were effective at determining density. Comparisons of hectares present in each rank may determine the areas with greater density for each species.

TABLE 38.1.

Captures per thousand trap-nights and rank (in parenthesis) by habitat type for deer mouse (*Peromyscus maniculatus*), meadow vole (*Microtus pennsylvanicus*), and arctic shrew (*Sorex arcticus*).

Species	Wetlands	Pasture	Shelterbelts	Deciduous trees	Grasslands	Conservation reserve program
Deer mouse	15.0	6.7	24.1	11.9	18.2	13.3
	(2)	(1)	(3)	(2)	(3)	(2)
Meadow vole	4.4	0.0	7.8	11.9	7.7	10.4
	(1)	(0)	(2)	(3)	(2)	(3)
Arctic shrew	0.0	0.0	0.3	0.0	0.3	0.0
	(0)	(0)	(1)	(0)	(1)	(0)

Results

Presence/Absence Models

Presence/absence models for the deer mouse and the meadow vole gave similar results (see Figs. 38.1 and 38.2 in color section). Both species used all habitats (grassland, including pasture and CRP [Rumble 1982; Forde et al. 1984; Agnew et al. 1986; Apa et al. 1991], shelterbelts [Hull-Sieg et al. 1984; Hodorff et al. 1988], and wetlands [Wilhelm et al. 1981]) except agriculture and were present in 900,708 hectares (60.3 percent of the 1,493,554-hectare area in Scene 7). The arctic shrew occurred in grassland, including pasture and CRP (Gruebele and Steuter 1988) and was present on 230,213 hectares (15.4 percent of total area in Scene 7) (Fig. 38.3).

Ranked Models

Ranked models for the deer mouse, meadow vole, and arctic shrew predicted the use of 900,708 hectares, 591,359 hectares, and 230,213 hectares of land, respectively (Table 38.2). Each model was unique based on the ranked habitat data for that species. Density ranks were calculated individually for each species. The deer mouse model contained five density ranks (with 5 having the highest predicted density) (Fig. 38.1). The habitat with the lowest density included 301,675 hectares (33.5 percent) and the habitat with the highest density included 72,421 hectares (8.0 percent) of habitat (Table 38.2). The meadow vole model contained four density ranks (with 4 having the highest predicted density) (Fig. 38.2). The habitat with the lowest density included 355,391 hectares (60.1

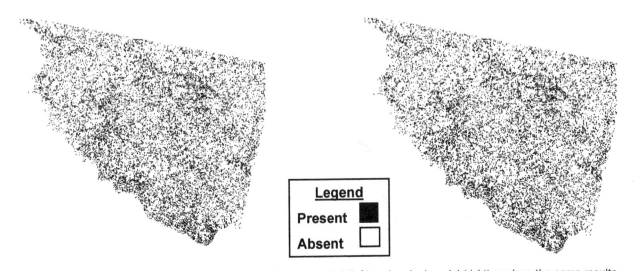

Figure 38.3. Arctic shrew (*Sorex arcticus*) presence/absence model (left) and ranked model (right) produce the same results.

TABLE 38.2.

Hectares present in ranked suitability models for deer mouse (*Peromyscus maniculatus*), meadow vole (*Microtus pennsylvanicus*), and arctic shrew (*Sorex arcticus*).

Rank	Deer mouse		Meadow vole		Arctic shrew	
	Hectares	Percent	Hectares	Percent	Hectares	Percent
1	301,674.7	33.5	355,391.0	60.1	230,212.8	100.0
2	249,397.2	27.7	22,062.1	3.7		
3	272,898.1	30.3	150,597.5	25.5		
4	4,316.5	0.5	63,308.5	10.7		
5	72,421.1	8.0				
Total	900,707.6	100.0	591,359.1	100.0	230,212.8	100.0

percent) and the habitat with the highest density included 63,309 hectares (10.7 percent) of habitat (Table 38.2). The arctic shrew model included one density rank (Fig. 38.3); this model included 230,213 hectares of habitat (Table 38.2).

In Scene 7, deer mice used 60.3 percent of the available habitat. Additionally, 38.8 percent of that habitat was classed as average to above average based on density (rank 3–5). Meadow voles used 39.6 percent of the available habitat in Scene 7. Of that, 36.2 percent was classed as above average based on density (rank 3–4). In Scene 7, arctic shrews utilized 15.4 percent of the available habitat. Only one habitat category was present in this model.

Comparison of Models

Presence/absence and ranked models for deer mice included the same area (900,708 hectares) of Scene 7, although differences occurred between the density rankings. The ranked model predicted a higher density in 66.5 percent of the area, or 599,033 hectares. Thus, our results indicated that ranked models might be more appropriate for predicting occurrences of deer mice.

The meadow vole presence/absence model included more area than the ranked model. This was caused by a discrepancy between information found in Lokemoen and Duebbert (1976) and our trapping information. Lokemoen and Duebbert (1976) reported use of grasslands by meadow voles, including pasture. However, meadow voles were not captured in pasture in our study, and, therefore, this habitat was not included in our model. The presence/absence model in-

cluded 900,708 hectares of suitable habitat. The ranked model predicted a higher density on 236,927 hectares, or 26.3 percent of the presence/absence model. Our results indicated that ranked models might be more appropriate for predicting occurrences of the meadow vole.

For the arctic shrew, presence/absence models and ranked models included 230,213 hectares. No differences were found between the two models. Arctic shrews were captured in low densities (0.3 individuals per thousand trap nights) in grasslands and shelterbelts, and, therefore, were given a rank of 1 for both habitat types. Results of this model did not differ from the presence/absence model. Therefore, both models were considered effective at predicting arctic shrew occurrence.

Discussion

Ranked density models were most informative when modeling habitat relationships for ubiquitous species. Species that occur in every habitat available at varying densities may be difficult to manage because the species' exact requirements may not be known. Ranked models can focus attention on habitats with higher or lower density, depending on the objective of the study or management. Management can be conducted at the local level to increase habitat quality in areas of low density or to maintain habitat quality in areas of high density. Conversely, presence/absence models were similar to ranked models for rare or low-density species. When a species occurs in only one or two habitat types, such as arctic shrews, one of the two available management scenarios may be restoration or maintenance of current habitat.

Competition for resources, such as food and living space, may limit the density of one species in the presence of another. Because our objective was to compare modeling results from presence/absence and ranked models, we attempted to minimize competition bias. Therefore, we reported results for three species, an herbivore (meadow vole), omnivore (deer mouse), and an insectivore (arctic shrew). Thus, for our study, competition among those species was assumed to be negligible. Nevertheless, presence of other small mammals not included in our study likely affected species distributions.

Due to a lack of information on small mammals in South Dakota, models could not be assessed for accuracy. Nonetheless, our objective was to illustrate differences in distributions of small mammals resulting from the presence/absence and density information. However, information from ranked models can give light to some issues of accuracy. For example, errors of omission (failure to capture a species where it was predicted to occur) is an issue when determining presence or absence of a species. Ranked models can help reduce errors of omission by determining the suitability of the habitat before trapping and by determining the number of trap nights necessary to capture a species. Areas with lower densities of a given species may require additional trapping to detect presence in that area. High-density species may require as few as fifty trap nights to detect presence. Consequently, ranked models can reduce time spent in high-density areas and focus effort on areas with relatively low detection probabilities.

Acknowledgments

S. K. Aker, M. A. Ohm, G. A. Wolbrink, and C. C. Zell conducted field work for this study throughout the summer of 1998. Thanks to J. M. Scott and P. J. Heglund for reviewing this manuscript. Funding was provided by the United States Geological Survey, Biological Resources Division, National Gap Analysis Program through the South Dakota Cooperative Fish and Wildlife Unit.

Calibration Methodology for an Individual-based, Spatially Explicit Simulation Model: Case Study of White-tailed Deer in the Florida Everglades

Christine S. Hartless, Ronald F. Labisky, and Kenneth M. Portier

Historically, simulation models have been important tools in wildlife ecology and conservation, and their use continues to increase. Simulations enable scientists to model the effects of environmental catastrophes or management strategies on target populations without conducting expensive or difficult experiments. Early population models ranged from simple models, such as logistic growth (Pearl and Reed 1920; Renshaw 1990) to stage- or age-based matrix models (Leslie 1945; Lefkovitch 1965). More recent population models have incorporated a spatial component, such as spatial dispersion models (Skellam 1951) and metapopulation models (Levins 1969; Hanski and Gilpin 1997). In the continuing evolution of simulation models, individual-based models—those using the individual as the basic unit (DeAngelis and Gross 1992; Grimm 1999)—are the current state of the art and represent a shift toward more mechanistic models (Maurer, Chapter 9). Within this large class of models, individual-based spatially explicit (IBSE) models are used to simulate individual movement processes and interactions over a heterogeneous landscape. This ability to model interactions among individuals and interactions between individuals and their environment has provided insight into many ecological processes (Huston et al. 1988). As computing power and speed increase, and computer costs decrease, more-complex IBSE models become feasible. The in-

creasing reliance of management decisions on simulation models drives the necessity for development of tools to adequately calibrate and validate these models.

Most individual-based models incorporate a large number of parameters (Grimm 1999). Some model parameters, such as the number of offspring or the survival rate, are estimated from published values. In contrast, parameters characterizing movement patterns of individuals in a simulation rarely can be estimated from published values or even derived from other measurable variables on study animals. These parameters (e.g., the distance an individual can "see" when making a movement decision, or the effect of previous movements on the current movement decision) are either difficult or impossible to estimate. However, the best-fitting movement algorithms and associated parameter values can be determined by evaluating discrepancies in measured outcomes (e.g., home range size) between study animals and simulated animals.

Bart (1995a) and Conroy et al. (1995) suggest guidelines for model development and testing that include the need for clearly stated model objectives, a description of the model structure, and a sensitivity analysis to assess effects of parameter uncertainty on model outputs. In addition, model development also requires verification, calibration, and validation. As defined by Rykiel (1996), verification is a demonstration that the model form is correct; calibration is the estimation and adjustment of model parameters to improve agreement between model output and observed data; and validation is a demonstration that the model output possesses the accuracy required for intended applications.

Although an individual-based model ought to be more testable than a state-variable model (i.e., a model with a population, community, or ecosystem as the basic unit; Murdoch et al. 1992), only 36 percent of the fifty individual-based modeling papers reviewed by Grimm (1999) explicitly discussed validation or corroboration of the presented models. Statistical tools for rigorous calibration and validation of simulation models do exist, and they generally fall into two groups. Some are described using ecological process and population models (e.g., Van der Molen and Pintér 1993; Rykiel 1996), and others are described in operations research and industrial engineering settings (e.g., Sargent 1984; Kleijnen 1987). In this chapter, statistical tools that aid in model verification and calibration in the context of IBSE models are presented. These tools are demonstrated with the calibration of a simulation model of the movement patterns of white-tailed deer (*Odocoileus virginianus seminolus*) in the Florida Everglades.

Methods

Determining correct model form (verification) involves evaluating the conceptual structure and the transformation of the structure into computer algorithms. Model structure is often visualized through flow charts. Detailed literature review and analyses of additional data aid in the verification of correct conceptual model structure. Assuming the model structure is correct, the calibration process consists of altering parameter values until the modeled system is represented adequately. Calibration also may reveal algorithms in the simulation model that need further modification if the optimum parameterization of the algorithm is not sufficient. Updating algorithms and parameter values is an iterative process requiring constant reevaluation of the simulation model.

The first issue to address in model calibration was the amount of time (i.e., number of iterations) the sim-

ulation must run before evaluation becomes meaningful. A second evaluation issue, quantitative comparison of the simulated data and observed data, was addressed with discrepancy measures. A third issue, optimization of simulation parameter settings, was accomplished by conducting experiments using the designs and iterative techniques of response surface methodology. Qualitative evaluation (e.g., visual comparison) was also an important component of model calibration.

Simulation Burn-in Time

The *burn-in* period, an artifact of computer simulation, was defined as the number of iterations required for the simulation to reach a steady state (Kleijnen 1987). Estimation of burn-in time was accomplished by allowing the simulation to run for extended periods of time and examining temporal trends and autocorrelations of simulation outcomes (e.g., annual home range size in an animal movement model). To avoid the confounding of burn-in and temporal algorithms and parameters, simulation burn-in time was evaluated before and after temporal effects were included in the simulation.

To estimate burn-in time, a simulation was run for an extended period of time and summary outcomes were calculated for all time intervals for each individual in the simulation (i.e., annual summary outcomes for a simulation run for fifteen years). A repeated measures analysis was performed to identify significant time trends in the summary outcome (Diggle et al. 1994; Littell et al. 1996; Vonesh and Chinchilli 1997). If no significant linear time trend was present, burn-in time did not affect that particular outcome. A significant linear time trend was evidence that simulation burn-in time did affect the outcome. If that occurred, the test for a linear trend was repeated using all intervals except the first. If the second test for time trend was not statistically significant, burn-in time was established at one interval; otherwise, the test for a linear trend was repeated excluding the first and second time intervals. These steps continued until the time trend in the remaining intervals was no longer significant, indicating simulation burn-in was completed. If the simulation did not reach a steady state until the end of the simulated time period or never reached a steady state (i.e., simulation burn-in time was equal to or longer than the time period of the simulation), further exploration of burn-in time and the simulation algorithms was necessary.

Discrepancy Measures

Discrepancy measures (DMs) quantify the difference between a simulated data set and an observed data set (Van der Molen and Pintér 1993). The general form of the discrepancy measure was

$$D(x) = f(P(x), O) \qquad (39.1)$$

where $P(x)$ was a summary statistic for a run of the simulation with parameter set x, x was an element of **X** (the set of all feasible model parameters), and O was the same summary statistic computed for the observed data. Examples of summary statistics for IBSE simulations of animal movement were mean home range size or the mean percentage of time individuals were located in a specific habitat.

One family of discrepancy measures had the form:

$$D(x) = |P(x) - O|^\beta \qquad (39.2)$$

where $1 \le \beta \le \infty$. For $\beta = 1$, $D(x)$ was the absolute deviation between simulated and observed data, and for $\beta = 2$, $D(x)$ was the squared deviation between simulated and observed data. Evaluation of the summary statistics such as mean annual home range size or mean distance between consecutive home range centers was accomplished with this family of DMs.

A discrepancy function useful for evaluating a set of n dependent outcomes, such as the percentage of observations (i.e., radio locations) in different habitats, had the form

$$D(x) = \sum_{i=1}^{n} \frac{(P_i(x) - O_i)^2}{O_i} \qquad (39.3)$$

where O_i was the summary statistic for the observed data for the ith outcome, $P_i(x)$ was the ith summary statistic for a run of the simulation with parameter set x, and x was an element of **X** (the set of all feasible model parameters). This DM approximated a chi-square goodness-of-fit statistic with n-1 degrees of freedom, where $P_i(x)$ and O_i were the percentage of locations in habitat i based on simulated and field data, respectively. Mayer and Butler (1993), Power

(1993), and Van der Molen and Pintér (1993) provided additional discrepancy measures.

Experimental Design and Analysis

Simulation experiments were conducted to determine the appropriate set of algorithms and parameter values that minimized discrepancies between simulated and observed data. Parameters from the simulation model were factors in the experiment, and one run of the simulation constituted an experimental unit (EU). Once the experimental design was selected and the simulation runs completed, burn-in time was evaluated for each summary outcome for each EU. Burn-in time for the experiment was the maximum burn-in time of all evaluated summary outcomes for all EUs. Statistical analyses were performed on the DMs, calculated using data from the time interval following the burn-in time for the experiment.

When evaluating large numbers of model parameters, effects of each parameter on simulation outcomes are often complicated and difficult to identify. Factorial experiments accomplish the task of simultaneous investigation of the effects of many factors (i.e., simulation model parameters). Moreover, the ensuing analysis of variance (ANOVA) of these experiments can include interaction terms that explain interrelationships among the parameters of the simulation. However, as the number of investigated factors increases, the number of EUs required to examine all possible factor combinations increases rapidly. For example, one replicate of a factorial experiment with p factors, each with k levels, requires k^p EUs. Other experimental designs (i.e., fractional factorials) make more efficient use of resources by requiring a minimal number of EUs. Response surface methodology provides a collection of specific experimental designs and statistical techniques to facilitate the estimation of factor settings that optimize a response variable (Khuri and Cornell 1987; Montgomery 1991).

Typically, a sequence of experiments is necessary to optimize the simulation model parameters, with the analysis results of each experiment dictating the particulars of the following experiment. First-order designs are used as initial screening experiments to estimate and test main effects and interactions among the factors. Common first-order designs are 2^p factorials and fractions of 2^p factorials (Cochran and Cox 1957; Montgomery 1991). Fractional factorials reduce the number of required EUs with the assumption that higher-order interactions (i.e., three- and four-way interactions) are negligible. This first-order experiment would determine if an optimum was attained inside the experimental region. If an optimum was not attained inside the initial experimental region, the method of steepest descent was used to establish the parameter settings for another experiment more likely to contain the optimum (Khuri and Cornell 1987; Montgomery 1991). This process was repeated, resulting in a sequence of experiments.

When statistical analyses indicated that the design settings were close to an optimum (i.e., the DM is close to zero inside the experimental region), additional experimentation was necessary to identify a set of parameter settings that minimizes the DM (a local minimum). More specifically, a second-order design, estimating main effects, first-order interactions, and quadratic effects, was often required to approximate the curvature of the true response surface. Common second-order designs are central composite designs, 3^p factorials, and 3^p fractional factorials (Cochran and Cox 1957; Khuri and Cornell 1987).

Visual Assessment of the Simulation Model

An additional component of model verification and calibration was the visual comparison of simulation output and observed data. Current geographic information systems (GIS) software is easy to use, accessible, and very powerful. Even if the DMs based on summary statistics demonstrate that simulation output is comparable to observed data, movement patterns also must be visually realistic.

Case Study: White-tailed Deer in the Florida Everglades

An IBSE simulation model of movement patterns of adult white-tailed deer in the Florida Everglades was developed. The model provided a means of exploring the patterns of habitat use of deer in response to frequent environmental catastrophes (e.g., tropical storms) and to different management regimes (e.g., water control). Maintenance of robust deer popula-

tions in this system is important because deer are the major prey of the endangered Florida panther (*Puma concolor coryi*) (Maehr et al. 1990) and the bobcat (*Lynx rufus*) (Maehr and Brady 1986), as well as a resource for human recreation. Furthermore, the statistical methods used to calibrate this IBSE model provide a foundation for the development of future simulation models. The simulation model was implemented in C++ (Borland C++ Builder 3.0, Inprise), using object-oriented programming techniques.

The 250-square-kilometer study area (Fig. 39.1) is located in the wet prairie/tree island ecosystem that extended from the Stairsteps Unit of the Big Cypress National Preserve (BCNP) south into Everglades National Park (ENP). It is bounded on the north by Loop Road, on the west by Lostmans and Dayhoff Sloughs, and on the east and south by Shark River Slough.

The Everglades ecosystem is characterized by a subtropical climate with alternating dry winters (November–April) and wet summers (May–October). Mean monthly temperature ranges from 14 degrees Celsius in January to 28 degrees Celsius in August. Mean annual precipitation is 136 centimeters, two-thirds of which falls between May and October (Duever et al. 1986). The onset and duration of the wet seasons are highly variable; thus, periods of either drought or flooding are common. Tropical cyclones (hurricanes and tropical storms) occur in this region of Florida at a frequency of one every three years (Gentry 1984) and often exacerbate the severity of floods.

Topographically the region is nearly flat and is characterized by a southwestward sheet flow of water. The major plant communities on the study area are wet prairie (87 percent), typified by a hydroperiod of 50–150 days, and small, widely dispersed, and slightly elevated hardwood tree islands (7 percent) (Duever et al. 1986; Miller 1993). The wet prairie is characterized by a complex of grasses and sedges; the tree islands contain both temperate and tropical hardwoods.

This white-tailed deer population was studied from 1989–1995 (Boulay 1992; Sargent 1992; Zultowsky 1992; Miller 1993; Sargent and Labisky 1995; MacDonald 1997; Labisky et al. 1999). Estimated densities for 1990–1992 averaged 3.65 (SE = 1.47) deer per square kilometer for the hunted BCNP population and

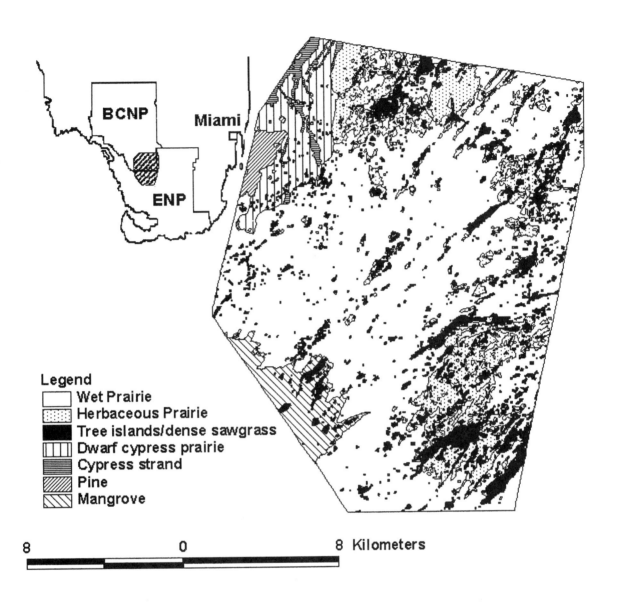

Figure 39.1. Habitat within the white-tailed deer study area in Big Cypress National Preserve (BCNP) and Everglades National Park (ENP). Map developed by Miller (1993).

4.68 (SE = 1.00) deer per square kilometer for the nonhunted ENP population (R. F. Labisky unpublished data). These deer are nonmigratory (Loveless 1959) and exhibit a high degree of site fidelity (MacDonald 1997). The cyclic rising and falling of water levels influences movements (Sargent and Labisky 1995; Zultowsky 1992), habitat use (Hunter 1990a; Miller 1993), and reproductive phenology (Loveless 1959; Richter and Labisky 1985; Boulay 1992).

Data Collection

The data set used for model calibration included yearling and adult deer that were captured, radio-collared, and monitored between 1989 and 1992. Due to the inaccessibility of the study area, all radio-monitoring was conducted during daylight hours from a fixed-wing aircraft. To obtain unbiased temporal monitoring, radio locations for each deer were evenly distributed among four daylight periods: sunrise to two hours post-sunrise, two hours post-sunrise to noon, noon to two hours pre-sunset, and two hours pre-sunset to sunset. Each deer was located, on average, once every five days. The location error associated with aerial-based telemetry, estimated from blind placement of dummy radio collars, was equal to 30 meters (Miller 1993).

Data on forty-six yearling or adult deer, characterized by a minimum of one year of radio-locations and no dispersal movements, were used for initial model calibration (Table 39.1). Twenty-four deer were radio-monitored for one year, eighteen deer for two years, and four deer for three years. For each deer, annual home range size was calculated using the 95 percent fixed kernel estimator with least squares cross validation (Silverman 1986; Worton 1989; Seaman and Powell 1996). Average distance between consecutive radio-locations served as an indicator for the minimum distance a deer traveled during a five-day interval. Distance between centers of consecutive annual home ranges was calculated to access the degree of site fidelity. Percentage of radio locations occurring in each habitat was used to examine habitat use.

Several raster maps of the study area were used, each with 20-meter×20-meter pixels (referred to as "20-meter pixels"). The habitat map, which had an estimated accuracy of 80.4 percent, contained seven habitat classes (Fig. 39.1). A spatiotemporal map of water depths was created using relative elevations of different habitats and mean monthly water depths recorded at the hydrological station (P-34) located in the wet prairie near the center of the study area.

Model Parameterization

The experiments presented in this chapter are two of a series of experiments used to calibrate the deer movement model. They illustrate the estimation of burn-in time and the use of experimental design to optimize model parameters. The focus of these two particular experiments, restricted to female deer, are scale of movement and home range parameters. Six model parameters were evaluated in the following calibration experiments. Temporal factors (water levels), survival and fecundity parameters, and deer interactions were not included in the presented simulation experiments.

Simulated deer made multiple daily movements, and

TABLE 39.1.

Average annual outcome measures for observed white-tailed deer in Big Cypress National Preserve and Everglades National Park, April 1989 to March 1992.

Observed outcome	Females (N = 29)	Males (N = 17)
Home range size	271 ha (20)[a]	316 ha (49)
Distance between consecutive locations	686 m (29)	779 m (79)
Distance between consecutive centers[b]	218 m (53)	243 m (56)
Percent observations in each habitat (%)		
Wet prairie	53% (4)	25% (4)
Herbaceous prairie	17% (3)	20% (3)
Tree islands/scrub	17% (2)	33% (4)
Willow/sawgrass	9% (1)	18% (2)
Dwarf cypress prairie	1% (1)	2% (2)
Cypress strand	< 1% (< 1)	1% (1)
Pine	< 1% (< 1)	< 1% (< 1)
Mangrove	2% (2)	1% (1)

[a]Standard error in parentheses.
[b]For females, N = 15, and for males, N = 7.

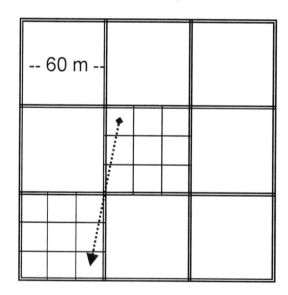

Figure 39.2. Two-stage movement process of a simulated deer using 60-meter pixels for the first stage. This hypothetical deer moved from the upper left 20-meter pixel in the center 60-meter pixel to the lower right 20-meter pixel in the lower left 60-meter pixel.

the number of steps over a five-day interval was one of the simulation parameters assessed in the following calibration experiments. Each step consisted of two stages: first, selection of a 40-meter by 40-meter pixel ("40-

TABLE 39.2.

Initial habitat relative affinity scores for simulated adult females.

Habitat	Symbol	Relative affinity
Wet prairie	A_{WPR}	10
Herbaceous prairie	A_{HPR}	20
Tree island	A_{TRE}	50
Willow/dense sawgrass	A_{WSA}	50
Cypress prairie/strand	A_{CYS}	10
Pine	A_{PIN}	10
Mangrove	A_{MAN}	1

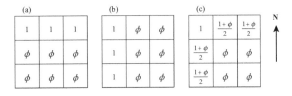

Figure 39.3. Illustration of homing beacon relative affinity calculations, with the homing beacon located southeast of current location (center pixel). Affinity scores to move (a) toward the south and (b) toward the east were averaged to give (c) the relative affinity scores of moving to each of nine possible pixels.

meter pixel") or a 60-meter by 60-meter pixel ("60-meter pixel"), then selection of a 20-meter pixel within the new 40-meter or 60-meter pixel (Fig. 39.2). The size of the pixel in the first stage of movement was the second parameter evaluated in these calibration experiments. In the first step of each movement iteration, deer moved a maximum of one pixel in any direction, using either 40- or 60-meter pixels. During this step, a deer evaluated its surroundings and determined the probability of moving to each pixel based on habitat and relative location inside its home range.

Each 20-meter pixel was assigned a relative affinity score based on habitat contained inside the pixel. Initial values for the relative affinities (Table 39.2) were updated throughout the calibration process. Relative affinity scores for the 40- and 60-meter pixels were determined using the mean relative affinity of their component 20-meter pixels.

The formation and maintenance of home ranges was controlled by two algorithms that incorporate the previous locations of the simulated deer. The length of this memory was the third parameter evaluated in the experiments.

The homing beacon algorithm gave simulated deer an affinity for pixels closer to their home-range center. The location of the homing beacon, based on k previous radio location coordinates (one every five days), was updated every five days using the moving average:

$$\bar{x}_{\text{home}} = \frac{1}{k}\left(x_t + x_{t-1} + x_{t-2} + \ldots + x_{t-k+1}\right)$$
$$\bar{y}_{\text{home}} = \frac{1}{k}\left(y_t + y_{t-1} + y_{t-2} + \ldots + y_{t-k+1}\right) \quad (39.4)$$

where (x_t, y_t) were the coordinates of the most recent radio-location and the moving window was k radio-locations wide (i.e., the length of the deer's memory). The relative affinity scores for the nine pixels to which the deer could move were based on the direction of travel from the current location of the deer to the homing beacon (Fig. 39.3). To avoid simulated deer gradually shrinking their home ranges with movements concentrated around the homing beacon, the strength of the beacon, δ (equal to $1, \frac{1+\phi}{2}$, or ϕ), was reduced exponentially as a deer moved closer to its beacon:

$$affinity_j = \begin{cases} \delta^{z/\mu} & \text{if } z < \mu \\ \delta & \text{otherwise} \end{cases} \quad (39.5)$$

for $j = 1, 2, 3, \ldots, 9$, and where δ was the relative affinity score, μ was the distance from homing beacon at which the relative affinity was constant, and z was the distance from current location to homing beacon. The fourth and fifth parameters to be accessed in the calibration experiments were ϕ and μ.

Simulated deer had a stronger affinity for previously visited pixels than for unfamiliar pixels. Relative affinity for a pixel j was defined as

$$affinity_j = \begin{cases} \lambda & \text{if pixel } j \text{ visited during known memory} \\ 1 & \text{otherwise} \end{cases} \quad (39.6)$$

for $j = 1, 2, \ldots, 9$ and where λ (> 1) was the relative affinity based on previous locations and the sixth simulation parameter evaluated in the calibration experiments.

After a simulated deer evaluated the eight surrounding pixels and its current location for these three factors, the probability of moving to each pixel was calculated using the relative affinity scores. The relative affinities for each factor were standardized by converting to probabilities using

$$p_{ij} = \frac{affinity}{\sum\limits_{j=1}^{9} affinity_{ij}} \qquad (39.7)$$

where p_{ij} was the probability of moving to pixel j for factor i ($i = 1,2,3$: habitat, homing beacon, and location memory, respectively) and $affinity_{ij}$ was the relative affinity score for pixel j for factor i. A weighted average of these probabilities was computed for each pixel using:

$$\pi_j = \frac{1}{2} p_{1j} + \frac{1}{4} p_{2j} + \frac{1}{4} p_{3j} \qquad (39.8)$$

where p_{1j}, p_{2j}, and p_{3j} were the probabilities of moving to pixel j based on habitat, homing beacon, and previous locations, respectively. The deer chose a pixel for its next location based on a random draw from the multinomial distribution (π_1, π_2, π_3, . . . , π_9), completing the first stage of the movement step.

The s 20-meter pixels ($s = 4$ if 40-meter pixels were used or $s = 9$ if 60-meter pixels were used for the first stage) contained inside the pixel representing the current location of the deer were evaluated based on habitat using the relative affinity scores. The scores were converted to probabilities:

$$\pi_j{}' = \frac{affinity_j}{\sum\limits_{j=1}^{s} affinity_j} \qquad (39.9)$$

where π'_1, π'_2, π'_3, . . . , π'_s were the probabilities of moving to pixel j based on habitat. The deer chose a 20-meter pixel based on a random draw from the multinomial distribution (π'_1, π'_2, π'_3 , . . . , π'_s).

Model Calibration

The first experiment was conducted using a half fraction of a 2^6 factorial design (thirty-two EUs) (Tables 39.3, 39.4). This design allowed estimation of all six main effects and all ten first-order interactions, as well as ten higher-order interactions; the higher-order interactions were assumed negligible and used as an esti-

TABLE 39.3.

Parameter settings for the first simulation experiment for adult females.

Experiment factor	Low	High
Number of steps[a]	150	500
First-stage pixel size	40 meters	60 meters
Memory length	2 months	6 months
λ[b]	2	10
ϕ[c]	2	4
μ[d]	1,000 meters	2,000 meters

[a]Number of movement steps per five-day interval.
[b]Relative affinity score for previously visited pixels.
[c]Maximum affinity used in equation 39.5.
[d]Distance from homing beacon at which affinity is constant.

mate of experimental error. Each EU consisted of thirty deer simulated over a fifteen-year time interval. Seventy-two locations per year (one per five days) were used to calculate summary statistics. The outcome summary statistics were annual average home range size, distance between consecutive locations, distance between consecutive annual centers, and percentage of observations in each habitat. Burn-in time for each EU for each outcome was estimated using repeated measures analyses with tests for linear time trends. Based on an α-level of 0.01, the maximum estimated burn-in time was four years, so summary data from the fifth year of the simulations were used to evaluate model parameters using ANOVA.

Discrepancy measures (DMs) were calculated for annual home range size, distance between consecutive radio locations, and distance between annual centers using Equation 39.2 with $\beta = 1$. The DM for habitat use was calculated using Equation 39.3. Significance of main effects and interactions was based on relative importance, using the magnitude of the F-statistic and the effect size, with the goal of focusing on factors with the largest impact on the DMs. The home range size DM was reduced with fewer steps, 40-meter pixels for the first stage of the step, larger λ and ϕ, and smaller μ. For distance between consecutive measurements, there was a significant interaction between first-stage pixel size and number of steps. If the first-stage pixels were 40 meters, more steps reduced the DM; however, if first-stage pixels were 60 meters, fewer steps reduced the DM (Fig. 39.4). The only

TABLE 39.4.

Factor levels and outcomes of home range size and distance between consecutive locations from the fifth simulation year of the first calibration experiment.

EU	Number of steps[a]	Pixel size (m)[b]	Memory length (mo)	λ	φ	μ(m)	Home range (ha)	Distance (m)
1	150	40	2	2	2	1,000	595	414
2	150	40	2	2	4	2,000	523	431
3	150	40	2	10	2	2,000	470	388
4	150	40	2	10	4	1,000	269	393
5	150	40	6	2	2	2,000	686	434
6	150	40	6	2	4	1,000	336	416
7	150	40	6	10	2	1,000	429	386
8	150	40	6	10	4	2,000	403	417
9	150	60	2	2	2	2,000	1,508	644
10	150	60	2	2	4	1,000	688	623
11	150	60	2	10	2	1,000	821	565
12	150	60	2	10	4	2,000	762	612
13	150	60	6	2	2	1,000	1,037	653
14	150	60	6	2	4	2,000	882	641
15	150	60	6	10	2	2,000	1,001	577
16	150	60	6	10	4	1,000	460	582
17	500	40	2	2	2	2,000	1,205	721
18	500	40	2	2	4	1,000	414	650
19	500	40	2	10	2	1,000	634	632
20	500	40	2	10	4	2,000	600	682
21	500	40	6	2	2	1,000	750	737
22	500	40	6	2	4	2,000	723	713
23	500	40	6	10	2	2,000	836	648
24	500	40	6	10	4	1,000	363	624
25	500	60	2	2	2	1,000	1,929	1039
26	500	60	2	2	4	2,000	1,232	1017
27	500	60	2	10	2	2,000	1,605	982
28	500	60	2	10	4	1,000	711	894
29	500	60	6	2	2	2,000	1,830	1106
30	500	60	6	2	4	1,000	810	929
31	500	60	6	10	2	1,000	1,185	928
32	500	60	6	10	4	2,000	981	962

[a]Number of movement iterations per five-day interval.

[b]Size of pixels for the first stage of the movement step.

model parameter affecting the DM for distance between consecutive centers was φ; a larger φ reduced the DM. The DM for habitat use was reduced with fewer steps.

Based on these results, the factor levels for the next experiment were determined. Because the interaction between number of movement iterations and first-stage pixel size indicated two conflicting directions in the parameter space for minimizing the DM for distance between consecutive locations, one direction was chosen for the next experiment. A pixel size of 60 meters and fewer steps was selected for several reasons. Although 60-meter pixels did increase the DM for home range size, reducing the number of steps

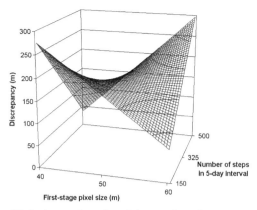

Figure 39.4. Interaction plot of the number of steps and first-stage pixel size for the distance between consecutive locations discrepancy measure for the first calibration experiment.

TABLE 39.5.

Parameter settings for the second simulation experiment for adult females[a].

Experiment factor	Low	High
Number of steps[b]	100	300
Memory length	2 months	6 months
λ[c]	4	12
ϕ[d]	3	5
μ[e]	750 meters	1,750 meters

[a]First-stage pixel size was fixed at 60 meters for this experiment.
[b]Number of steps per five-day interval.
[c]Relative affinity score for previously visited pixels.
[d]Maximum relative affinity used in equation 39.5.
[e]Distance from homing beacon at which affinity is constant.

decreased the DMs for both home range size and habitat use. Furthermore, run time for each simulation, an important consideration in computationally intensive simulations, was shorter with fewer steps. Levels of the four other factors were altered according to the observed trends in this experiment. Relative to the first experiment, memory length remained the same, λ and ϕ were increased, and μ was decreased (Table 39.5).

The second experiment was conducted using a half fraction of a 2^5 factorial design (sixteen EUs) (Table 39.6). This design allowed estimation of all main effects and first-order interactions. Three EUs at the center point of the design (number of steps = 200, memory length = 4 months, $\lambda = 8$, $\phi = 4$, and $\mu = 1,250$ meters) were run to estimate experimental error because no higher-order interactions were estimable with this design. Again, each EU consisted of thirty deer simulated over a fifteen-year time interval, and the summary outcomes and DMs were calculated. Based on an α-level of 0.01, the maximum estimated burn-in time for each outcome for each EU was four years; therefore, summary data from the fifth year of simulation was used to evaluate model parameters.

There were no significant interactions among model parameters ($p > 0.05$) for all four DMs. Discrepancies for home range size were reduced significantly with fewer steps, larger ϕ, and smaller μ. For distance between consecutive locations, changes in factor levels did not significantly affect the DM. Mean distance between consecutive locations ranged from 458 to 842 meters, approaching the observed mean of

671 meters. DMs for distance between consecutive home range centers decreased as ϕ increased. The DM for habitat use was not significantly affected by the model factors; however, simulated deer were located more often in tree islands and less often in the wet prairie than observed deer, indicating a need to reevaluate the habitat affinity scores.

In addition to quantitative analyses of simulation results, qualitative observations also aided in the verification and calibration processes. Movement paths of simulated deer and observed deer residing in the same geographic area were plotted and compared. For example, during a one-year interval, a simulated deer (parameters: number of steps over five days = 300, memory length = 6 months, $\lambda = 12$, $\phi = 3$, and $\mu = 750$ meters) had a realistic movement pattern when compared to an observed deer (three-year old female) in the same geographic area (Fig. 39.5). However, of the thirty deer in this EU, only 50 percent had movement paths and habitat use patterns that were similarly realistic, indicating further need for the refinement of the simulation model.

In the subsequent experiment, number of steps was decreased, memory length and λ were kept at the same settings, ϕ was increased, and μ was decreased. Habitat affinity scores were included as factors in the experiment. An initial estimate of burn-in time was now available, so seasonally fluctuating factors (i.e., water levels) were added to the model in later calibration experiments, and DMs also were calculated and evaluated for hydrologic seasons.

TABLE 39.6.

Factor levels and outcomes of home range size and distance between consecutive locations from the fifth simulation year in the second calibration experiment.[a]

EU	Number of steps[b]	Memory length (mo)	λ	ϕ	μ(m)	Home range (ha)	Distance (m)
1	100	2	4	3	1,750	792	499
2	100	2	4	5	750	428	496
3	100	2	12	3	750	436	458
4	100	2	12	5	1,750	487	487
5	100	6	4	3	750	502	503
6	100	6	4	5	1,750	528	493
7	100	6	12	3	1,750	638	470
8	100	6	12	5	750	356	481
9[c]	200	4	8	4	1,250	640	648
10[c]	200	4	8	4	1,250	598	653
11[c]	200	4	8	4	1,250	720	635
12	300	2	4	3	750	709	737
13	300	2	4	5	1,750	831	826
14	300	2	12	3	1,750	953	741
15	300	2	12	5	750	474	724
16	300	6	4	3	1,750	1,044	842
17	300	6	4	5	750	538	734
18	300	6	12	3	750	539	685
19	300	6	12	5	1,750	785	790

[a]First-stage pixel size was fixed at 60 meters for this experiment.

[b]Number of steps per five-day measurement interval.

[c]Center point of the experimental design.

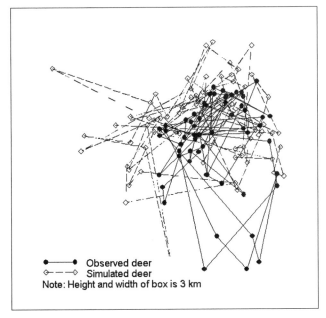

●————● Observed deer
◇– – –◇ Simulated deer
Note: Height and width of box is 3 km

Figure 39.5. Plot of radio locations of an observed adult female and locations of a simulated deer over one-year interval in the same geographic region.

Conclusions

The set of calibration tools and the iterative approach developed and demonstrated in this chapter start a framework for the more rigorous evaluation of simulation models. The general approach and techniques could be used for development of other classes of simulation models as well as IBSE models. Simulation model building is an iterative process; the developer must constantly reevaluate the model and update algorithms and parameter values to obtain the best fit possible. Burn-in time is explored as a nuisance parameter of simulations, and a technique for its estimation is presented. The degree of realism of the simulation is quantified by the use of discrepancy measures for each outcome summary statistic of the simulation.

Experimental designs and analysis techniques of response surface methods are presented as tools for the calibration of IBSE simulation models. The use of these techniques aids in the search for the optimum model parameter settings that will best represent the system of interest. This systematic, iterative approach logically attacks this multifactor optimization problem and provides opportunities for improvement in the model with each simulation experiment.

The experimentation revealed that the movement patterns of these deer can be modeled by the combination of several algorithms, including a two-stage movement step and a memory that is updated over time. Our example demonstrates that the described techniques facilitate the attainment of the model algorithms and parameter values that minimize the discrepancies between observed and simulated data. The final simulation model was tested for its predictive accuracy using deer radio-telemetry data collected from the same geographic area during the flood of 1994–1995 (Hartless 2000). The simulation model can be used by South Florida planners as they work toward restoration of the Everglades to better understand the impacts of altering water control regimes on the white-tailed deer population.

Acknowledgments

The National Park Service (BCNP and ENP) and the University of Florida provided funding. O. L. Bass (ENP) and D. K. Jansen (BCNP) provided liaison support. M. C. Boulay, K. E. Miller, R. A. Sargent Jr., and J. M. Zultowsky (University of Florida) assisted in data collection. K. E. Miller prepared the GIS habitat map of the study area. J. C. Allen (University of Florida) provided computer resources. The work on this project was done as part of a Ph.D. dissertation at the University of Florida and this chapter is a contribution (Journal Series R-07504) of the Florida Agricultural Experiment Station, Gainesville.

PART 4

Predicting Species Presence and Abundance

Predicting Species Presence and Abundance

Frances C. James and Charles E. McCulloch

In spite of the important environmental achievements it has permitted, the provisions of the Endangered Species Act cover only a limited range of biodiversity and resource problems. Managers would welcome practical ways to address additional problems and advice about how to choose among management options (Guikema and Milke 1999). Even progress toward predicting species distributions efficiently would be a step in the right direction.

This chapter reports on state-of-the-art examples of ways to accomplish that reduced objective, to produce maps for the probable presence, abundance, or absence of terrestrial species across large geographic areas, in cases when detailed local data are not available. Two chapters (Shapiro et al., Chapter 49; Pearson and Simons, Chapter 52) apply simulation modeling to related questions, and one (Angermeier et al., Chapter 46) discusses processes regulating distributions of stream fishes. With the recent advent of geographic information systems (GIS) for storing geospatial data, the availability of large data sets for climate variables, and satellite imagery and aerial photography for vegetation and landform variables, the subject of modeling the distributions of organisms has become a major enterprise within the field of landscape ecology. In this sphere, predicting species occurrences is less a matter of identifying limiting resources and more a matter of finding broad-scale associations between the distributions of taxa and combinations of values of readily available environmental variables. Some broad-scale environmental variables, such as those for landform, undoubtedly do limit distributions indirectly, but the motivation behind these studies seems tied more closely to the magnitude of biodiversity problems and the need for distribution maps.

An important example of the concept described above is gap analysis, a habitat-association method that uses distributions of vegetation classes and GIS technology to predict distributions of vertebrates and other taxa (Scott et al. 1993). In addition to vegetation and land-cover data, various landscape metrics taken from satellite imagery and climate data can be added as filters. Overlays, made with ArcView software, allow the identification of areas of high species richness or endemicity. Such areas that are not already protected are gaps, priority areas for ground surveys, and potential sites for new reserves. Gap analysis began in Hawaii and Idaho but has been extended by the U.S. Geological Survey to every state (Loomis and Echohawk 1999).

GIS predictions about the locations of species within their geographic ranges can be made not only from analysis of overlays but also with statistical models that work from data for the distribution of a focal taxon, combinations of associated environmental variables, and mapping routines. Probabilistic methods for estimating the extent of errors of omission (prediction of absence when a species is present) and commission (prediction of presence when it is absent) can be assessed and biases accounted for by various statistical approaches. Sometimes data are for individual marked animals recorded at more than one location. A sample or full data set for accompanying environmental variables can be measured at the site or derived from other sources. Because a set of correlated environmental variables is considered simultaneously, the challenge of the analysis is to reduce complex multivariate relationships and express them in clever ways that provide reliable predictions.

Traditional multivariate methods such as multiple linear regression, principal components analysis, discriminant function analysis, and Mahalanobis distance, which look for linear relationships, are being used. However, generalized linear models have gained in popularity as an alternative. They have the advantages of allowing some nonlinear responses, allowing a variety of distributions for the outcome variable, and accommodating prediction from a set of variables that may be continuous or discrete (Nicholls 1989). Generalized linear modeling includes multiple regression, logistic regression, and Poisson regression as special cases. Austin et al. (1990) used quadratic logistic regression for predicting presence and absence. Vernier et al. (Chapter 50) used Poisson regression for predicting abundance. Two compound nonlinear methods are classification and regression tree analysis (CART) (see Breiman et al. 1984) and genetic algorithm for rule-set prediction (GARP) (see, e.g., Lees and Ritman 1991; Stockwell and Noble 1992). CART derives flexible, nonlinear regressions and attempts to find the minimal ranges of environmental variables that fit the data (see, e.g., Nix 1986 for BIOCLIM to describe an environmental envelope). GARP is a decision tree and rule-induction approach, an artificial intelligence rule-set method that uses expert-system rules, logistic regression, CART, and envelope-type models and produces distribution maps.

Peterson et al. (Chapter 55) begin with known points of occurrence of a species and a set of environmental data for both these points and other places, where the species was not recorded. Then they apply "ecological niche modeling," which uses the GARP rule-set procedure (Stockwell and Noble 1992) to produce distribution maps.

Collectively the chapters in this section show the various ways researchers are predicting broad-scale species distributions. Dreisbach et al. (Chapter 41) used habitat suitability index scores (USFWS 1981b) and GIS rules to predict distributions of species of fungi in a gap-type system, pointing out the dietary links between fungi, flying squirrels, and northern spotted owls in the Pacific Northwest. Fertig and Reiners (Chapter 42) applied both nonlinear logistic regression and classification and regression tree analysis to a combination of herbarium data for plant localities and environmental data. They produced probabilistic range maps for the entire state of Wyoming. Van Manen et al. (Chapter 43), with data for surviving canker-resistant butternut trees in the Great Smoky Mountains National Park and associated habitat characteristics, used Mahalanobis distances and logistic regression to predict the locations of other potential sites and used jackknifing and bootstrapping for validation. Debinski et al. (Chapter 44) used field information and multispectral satellite data in especially small mapping units (0.25 hectare) for a set of montane meadows in the greater Yellowstone ecosystem. They used classification and regression tree analysis, discriminant analysis, and regression to map four types of plant communities. Various birds and butterflies were associated with each type. Gonzalez-Rebeles et al. (Chapter 57) removed filter variables one at a time from data for vertebrates in New Mexico to see how the predicted distributions would change. Zimmerman and Breitenmoser (Chapter 58) had radiotelemetry data for the locations of Eurasian lynx in the Jura Mountains of Switzerland. They used repeated runs of discriminant function analysis to predict the potential distribution of lynx and found that the distribution of prey (roe deer and chamois) was more important to the predictions than were habitat variables. Shriner et al. (Chapter 47) had data for the presence and absence of the wood thrush from point counts in Great Smoky Mountains National Park. They used logistic regression to predict occurrence throughout the park. In this case, topographic and landform indices were more useful than were local habitat variables. Fleischman et al. (Chapter 45) used data for butterflies in the Great Basin to demonstrate their recommended two-stage modeling process. They used multiple linear regression to predict species richness from environmental variables and then logistic regression to identify environmental variables that accounted for the presence of individual species. Vernier et al. (Chapter 50) used Poisson regression to predict the abundances of birds from forest inventory data in the boreal forest of Alberta, Canada. They used Akaike's information criterion (AIC, see Burnham and Anderson 1998) to select the final model. The data were taken at two scales, both of which were relevant to forestry practices, and links to simulation models for management options are planned.

The best way to check the accuracy of prediction methods is to try them first in a place where the distributions are known and predictions can be checked. That was tried by Karl et al. (Chapter 51), who used simulations to explore whether rarity has an effect on model accuracy over and above the effect of small sample size. They concluded that it does not. The availability of extensive field data for seven species of birds in Montana was used to validate their simulated results. Dettmers et al. (Chapter 54) began with point-count data for six species of birds and values of remotely sensed spatial data at the same sites in Tennessee. They compared predictions made by logistic regression, Mahalanobis distance, classification and regression trees, and discriminant function analysis for each species and decided that the method of classification and regression trees is best. Their judgment was based on jackknife tests and on how well the final, reduced models predicted occurrences in Georgia and Virginia. Hepinstall et al. (Chapter 53) used field data for birds in Maine to make a comparison between gap-type habitat-association modeling and a statistical approach that used Bayes' theorem to model associations with data from satellite imagery, a vegetation and land-cover map, and two derived layers of heterogeneity. They found that both methods overestimated the distributions of generalist-type species and that the overestimation was greater with the gap-type analysis.

Stockwell and Peterson (Chapter 48) are proponents of combining new data-mining techniques with flexible bias management within the GARP rule-set procedure (Stockwell and Noble 1992). The example here relates Breeding Bird Survey data (Sauer et al. 1999) for the wood thrush to data for mean annual temperature and precipitation. They explore ways to control three types of potential bias in predictions made from such data.

Of the two chapters reporting simulation work, Pearson and Simons (Chapter 52) used landscape-level metrics and an individual rule-based simulation model to compare the relative likelihood of successful stopovers during migration of birds that are habitat specialists and generalists. The objective here was to generate hypotheses for future work. Shapiro et al. (Chapter 49) were interested in the probability of parasitism by brown-headed cowbirds in nests of the endangered golden-cheeked warbler and black-capped vireo in Texas. They applied an individual rule-based simulation approach to female cowbird movement. Such models can be constructed to mimic the behavior of individual animals as they move among spatial units (pixels) on a grid having different environmental values.

Recently, researchers have begun to combine submodels using the spatially explicit rule-based approach with submodels using the statistical approaches and even submodels using state-variable approaches. An example is the paper by Gross and DeAngelis (Chapter 40), which describes the ambitious Across Trophic Level System Simulation for the Everglades. It begins with a broad-scale state model for the hydrological system and uses regional GIS and remote sensing to assess the impact of alternate management scenarios. Submodels have higher levels of resolution, the highest being detailed individual-based models for species of special interest. This project is squarely in the new field of computational ecology.

The only chapter in this section that addresses the analysis of causes of discontinuous distributions is the one by Angermeier et al. (Chapter 46) on stream fishes in Virginia. It discusses the complicating roles of population dynamics, density dependence, and temporal variation in multiscale analyses. For example, if population density is low, habitat associations may appear to be weak, even if they are in fact strong. This important paper does not present a quantitative analysis, but it highlights issues not mentioned elsewhere in the section. See also Hobbs and Hanley (1990).

Biometric approaches to modeling distributions of species and communities based on environmental variables began thirty years ago with studies of the distribution of birds in relation to the structure of the vegetation in their territories (Cody 1968; James 1971; Anderson and Shugart 1974). Wildlife biologists and land-use planners at that time became hopeful that such work would have practical applications for assessing the effects of habitat treatments, but in the 1970s communication between researchers and managers was poor (Verner 1981). Many of the habitat-suitability-index models and habitat-evaluation procedures that followed had low predictive power (Stauffer and Best 1986). There were complaints about the reliability of the results of applications of multivariate analysis reported in habitat modeling studies (Johnson 1981a; Rexstad et al. 1988) and concerns about the lack of attention to issues of scale (Wiens 1981b). Managers were not in fact finding such models as useful as they had hoped (Marcot 1986).

Data Could Be Improved in Future Work: Six Issues of Accuracy and Scale

The new generation of models in the 1990s uses GIS hardware and software to create spatial analyses and makes use of new broad-scale environmental databases (Morrison et al. 1998). Remote sensing, vegetation classification, and landform and climate data are often used to link the environment with the focal species or community so that predictions can be made. Even though access to such information is exciting and opens new possibilities (see, e.g., Sillett et al. 2000), whether landscape-scale predictors of biodiversity will provide useful data to managers is still uncertain (Short and Hestbeck 1995), and software for spatial statistical analysis of environmental data in a space-time framework that properly accommodates spatial correlation in the data is not yet available (Cressie and Ver Hoef 1993). In the meantime, here are six, by no means exhaustive, ways we think issues of accuracy and scale in predictions of presence and abundance of species from environmental data could be improved in future work. These suggestions do not originate with us. The same scientific principles and statistical requirements that applied to former generations of models (Johnson 1981a) still hold. Attention

to the first four ideas will improve predictions, but it will take all six to make the predictions useful to managers.

1. Use more *a priori* thinking.

   Develop a short list of candidate models for testing. Automated statistical techniques cannot be expected to sort out complicated multidimensional relationships without biological input (Box and Hill 1967; Burnham and Anderson 1998). As a result, the final model often includes too many terms, and both the composition of terms in the model and its predictions may be unstable. Think hard about the choice of variables and the likelihood that they constrain the distribution or functioning of organisms. For example, thermometers and rain gauges are cheap, but psychrometric variables that are a function of temperature and moisture jointly, like absolute humidity and wet-bulb temperature, are more likely to affect processes in living organisms than are the more conventional factors, dry-bulb temperature and precipitation (James 1991).

2. Pay more attention to model selection.

   The process of model selection often has as much to do with the final interpretations as it does with the fit of the decided-upon model. Nonparametric assessments of the model like the bootstrap (Efron and Gong 1983; Manly 1997) should be used to assess stability honestly. By that we mean that the model-selection process should be replicated, not just the fit of the final model, especially when complicated model-building processes are used and when the ratio of data to number of model terms is small. Burnham and Anderson (1998) advocate selection of the approximately best model, the one that simultaneously accounts for the most variation in the data with the fewest terms. The addition of significant terms in the model may not improve Akaike's information criterion.

3. Validate the final model.

   Habitat-based models have rarely been validated in the past (Chalk 1986; Raphael and Marcot 1986; Hansen et al. 1999), but validation is crucial if we are to understand the accuracy of predictions. The best validation is accomplished by repetition of the modeling exercise with newly obtained data from the population of interest (Taylor 1990; Chatfield 1995). This method often is not feasible, and then we must resort to resampling methods (Verbyla and Litvaitis 1989) or methods that split the data. If the data are split for validation purposes, they should still all be used to produce robust classification rules (Rencher 1995).

4. State the limitations of the proper interpretation of the results.

Remember that tests of regression coefficients do not tell you that the variables in the final model are in fact important to the species, just that they are jointly related to the presence of the species. If the objective is to produce cost-efficient range maps and locate sites of high conservation value for future on-the-ground surveys and management, then the job is simply to improve confidence about their accuracy. Reliable inference about causes is a false expectation from empirical modeling work such as the studies reported in this section.

5. Distinguish between hypothesis testing as statistical inference and hypothesis testing as analyzing the causes of processes.

With insightful analyses involving substantial *a priori* thinking, some progress can be made toward the discovery of environmental factors that managers might change to help a species. However, if the real challenge for managers is to assure the long-term health of representative ecosystems, then the task is much more difficult than evaluating statistically significant associations from empirical

models. To be useful to managers, predictions will have to be more focused on analyses of those environmental factors that are directly limiting and those that can be manipulated. The most obvious factors are those related to ecological succession, such as the history of fire. Inferences about processes that constrain distributions will be weak until they are set up as hypothesis tests that compare distributions among sites having different combinations of levels of environmental factors in an experimental design. The entire step of working experimental design into observational studies (Cochran 1983; James and McCulloch 1985, 1995; Rosenbaum 1995), which should come between the empirical predictions described in this chapter and making management plans, is not currently being addressed. Then adaptive management, as described by Kish (1987), can begin.

6. Think about the scales of processes that may be constraining distributions horizontally as well as vertically.

We manage places in patches, which are defined vertically, but processes that limit populations must be operating horizontally across their geographic ranges. Saving biodiversity by setting aside protected areas may succeed, but only by the accumulated influence of a large number of populations of individual species with individual requirements across broad horizontal scales. Even if single-species management is an inadequate management approach (Moss 2000), single-species research and on-the-ground fieldwork are still needed for progress in understanding how nature works.

Multimodeling: New Approaches for Linking Ecological Models

Louis J. Gross and Donald L. DeAngelis

The Everglades region of South Florida presents one of the major natural system management challenges facing the United States. With its assortment of alligators, crocodiles, manatees, panthers, large mixed flocks of wading birds, highly diverse subtropical flora, and sea of sawgrass, the ecosystem is unique in this country (Davis and Ogden 1994). The region is also perhaps the largest human-controlled system on the planet in that the major environmental factor influencing the region is water, and water flows are managed on a daily basis—subject to the vagaries of rainfall—by a massive system of locks, pumps, canals, and levees constructed over the past century. The changes brought about by such control have led to extensive modifications of historical patterns and magnitudes of flow, causing large declines in many native species, extensive changes in nutrient cycling and vegetation across south Florida, and great increases in pollutants such as mercury. Constrained by the conflicting demands of agriculture, urban human populations, and wildlife for control of water resources, and the varying agendas of hosts of government agencies and nongovernmental organizations, there is now an ongoing effort to plan for major changes to the system with expenditure estimates of eight billion dollars or more over the next several decades (USACOE 1999). Carrying out such planning, particularly as it impacts the natural systems of the region, provides one of the

major challenges to the new field of computational ecology.

Computational Ecology and Regional Assessment

Computational ecology is an emerging multidisciplinary field, similar in concept to the cell and molecular emphasis of bioinformatics, which applies modern computational methodology to key problems at higher levels of biological organization. The goal of computational ecology is to combine realistic models of ecological systems with the often-large data sets available to aid in analyzing these systems, utilizing techniques of modern computational science to manage the data, visualize model behavior, and statistically examine the complex dynamics that arise (Helly et al. 1995). Success in applying this new tool to regions such as the Everglades requires expertise in diverse areas such as field biology, complex systems theory, computational science, remote sensing, and mathematical modeling, as well as the development of mechanisms to link approaches that are at the forefront of research in many of these fields.

The vast majority of theory in ecology has started from very simple differential equations in which a single variable represents population densities; solutions of these are analyzed mathematically and may be compared to abundance estimates from field or lab observations. Although these models have had great influence on theory in ecology, their aggregated form is particularly difficult to relate to observational biology. Their application to complex natural systems with spatially and temporally varying environmental factors that force the system leads to models that are not analytically tractable and which must be investigated numerically. The study of equilibrium and stability behavior typical in mathematical ecology is of little relevance for natural systems with strong abiotic forcing.

One of the most important functions of the environmental sciences today is to analyze the impacts of human actions on ecosystems and to provide management recommendations in order to ameliorate these impacts. In all parts of the world, ecosystems are affected by the shrinkage and dissection of natural areas, disruptions of natural cycles, and the input of pollutants. Ecological assessment includes the determination of the impacts of various anthropogenic influences on a natural system. Common components of such an assessment would include the following:

- Changes in population densities of species considered "important" for either cultural or economic reasons, including endangered species such as the West Indian manatee (*Trichechus manatus*) and the Florida subspecies of the cougar (*Puma concolor coryi*) (hereafter the Florida panther).
- Changes in species composition of the system, particularly introductions of non-native species and associated issues of hybridization with species currently present, an example being the rapid spread of invading melaleuca trees in the Everglades.
- Changes in community structure (which may not necessarily be associated with biodiversity changes), such as the shift from sawgrass- to cattail-dominated wetlands.

Stressors requiring assessment include:

- Effects of pollutant inputs, an example being the high mercury levels in fish throughout the Everglades.
- Direct effects of human actions on the system, such as hunting, deforestation, and sewage/waste dis-

posal, an example being the effect of poaching on the Florida panther.
- Indirect effects of human actions, including habitat fragmentation, soil erosion, and salinity changes, an example being the increase in salinity in Florida Bay due to reduced fresh-water flows through Everglades National Park.

The spatial extent of the effects of anthropogenic impacts range from local (tens of meters) to regional (hundreds of kilometers) and therefore require assessments that can span these scales as well. Regional problems involve controls and dynamics changing over time periods far longer than the one- to five-year periods typical of academic research projects. In addition to Everglades restoration, projects involving extensive computational components are ongoing in the Columbia River Basin, the Sacramento River Basin, and the Southern Appalachians, to name just a few of the efforts in the United States. Dealing with such regional problems has led to the use of remote sensing data and geographic information systems (GIS). Still very much in their infancy are methods to couple GIS with information on tracking of animal movements (now feasible for large numbers of organisms due to the availability of inexpensive and highly reliable radio tags), site-specific studies of animal behavior, and dynamic models. Yet, there are immediate needs for rigorous scientific methods to couple available data with realistic ecological models to assess the potential impacts of alternative management scenarios.

Multimodeling and the Everglades

It is exactly this need for assessment of the long-term (e.g., over thirty years) impacts of alternative hydrologic scenarios on biotic components of the natural systems in South Florida that has led a large group of collaborators to develop an approach we call Across Trophic Level System Simulation, or ATLSS (DeAngelis et al. 1998). Just as the Atlas of mythology bore the world on his shoulders, we expect the methodologies we are developing to provide a firm basis for bearing the weight of ecological impact analysis for many natural systems across the planet. Key to this is our use of a multimodeling methodology (Fig. 40.1) in which we

Across Trophic Level Modeling

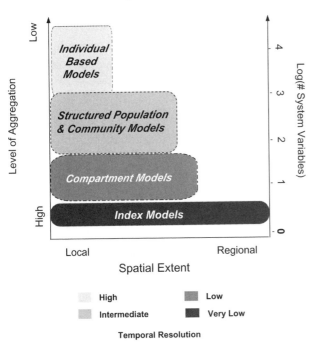

Figure 40.1. A general approach to across-trophic-level modeling is illustrated with four different modeling approaches, each with a range of spatial extents over which the variables utilized within the models act. Individual-based models are the most localized, involving the largest numbers of variables, with associated low levels of aggregation. Individual behavior depends inherently upon local conditions, not just on an average at larger spatial extents. Index models typically produce a single index value per location and may be constructed at a wide variety of spatial extents depending upon the level of spatial resolution available for the input data and the model assumptions. Compartment models and structured population/community models are intermediate in the spatial extents appropriate for their variables. Structured population and community models take account of the details of size, age, and physiological state structure within a population or community by breaking down populations and communities into discrete classes. Within a localized grid cell, such models can include 100–1,000 state variables. The hatching illustrates the temporal resolution appropriate for these models. In the ATLSS case, the index models operate on yearly time steps while the individual-based models have time steps of less than a day.

argue that the use of any single modeling approach is inappropriate for problems spanning a wide variety of temporal, spatial (grain and extent), and organismal scales.

The application of mathematical models in most fields involves determining whether the key features of the problem under consideration require a discrete or continuous formulation, for example populations with discrete life stages as compared to those with overlapping generations. A second dichotomy concerns whether stochastic factors should be included or ignored, that is if unpredictable abiotic environmental factors must be explicitly modeled or whether the averages of such variations are sufficient to describe and analyze the situation. Deciding these basic modeling issues helps to specify the most appropriate single modeling approach for a particular situation, such as a matrix approach for structured populations in discrete time, an ordinary differential equation system for discretely structured populations in continuous time, or a partial differential equation for a continuously structured population in continuous time. It is quite atypical for the modeling framework used for a particular problem to utilize more than one modeling approach. Rather, as the modeling process proceeds, and additional complexities of the system under study are being considered, a new approach might be taken. Thus, a simple population model for abundance dynamics could be phrased as a single differential equation, but when age structure is added, the model formulation would then involve a system of differential equations.

Unlike the situation described above, ATLSS relies upon a variety of different mathematical approaches, a process known as multimodeling. This mixture of approaches is based upon the inherent temporal and spatial resolution and extent of various trophic components linked together by spatially explicit information about underlying environmental (e.g., water, soil structure, etc.), biotic (e.g., vegetation), and anthropogenic (e.g., land-use) factors. The approaches currently in use include spatially explicit indices, compartment models, differential equations for structured populations and communities, and individual-based models. Linking models that operate at very different spatial and temporal extents has been a major challenge, requiring a variety of spatial interpolation methods (Luh et al. 1997) and careful design of model interfaces (Duke-Sylvester and Gross 2001). Due to the modular nature of the project and the desire for long-term flexibility in what submodels may be included, we chose from the start an object-oriented design, utilizing C++ software.

Figure 40.1 illustrates the variety of dynamic approaches utilized in ATLSS relative to the spatial and temporal resolutions of the trophic levels involved. The hierarchical pattern in Figure 40.1 for the components of ATLSS arises from the association of different model types with different species or groups of species in the ecosystem that have different typical spatial extents appropriate for the variables in the models. The spatial extent indicated ranges from the finest resolution currently available for certain system components, 30 meters from Landsat image information, to regional extents on the orders of hundreds of kilometers. The horizontal edge of one of the boxes indicates the range of spatial extents over which variables for that type of model may reasonably be considered important. Models that follow individuals necessarily require localized information about those individuals in order to appropriately assess their interactions between individuals that are inherently local. Although variables in these individual-based models may be averaged over larger spatial extents, these averaged variables cannot then be used to assess the basic interactions within the model. Other approaches have variables, such as those indicated as index models, for which the underlying environmental factors can be estimated at a wide range of spatial extents without affecting the basic assumptions of the model. Models such as those for structured populations have variables that may be defined at an intermediate range of spatial extents over which it may be assumed that within this extent the spatial variability of environmental conditions may be reasonably averaged.

The vertical axis of Figure 40.1 illustrates the complexity of the associated models, with the number of system variables used as a measure of such complexity. Alternatively, this complexity might indicate for the biotic system of interest the aggregation utilized in the description of its components. Some submodels, here indicated as compartment models typically applied to lower trophic levels, including algae and zooplankton, can for practical purposes be well described by simple coupled differential equations. These describe interactions between functional groups, defined here as groups of species that perform similar roles (on the same trophic level), with the forcing functions in these equations representing the effects of abiotic

influences. Thus, a reasonable description of a zooplankton community can be reduced to one or a few state variables when the focus is on how zooplankton contribute to biomass fluxes at higher trophic levels. Other components of the biotic community may have a more complex organization that requires a structured population or mixed-species community description. Additionally, this structure may directly impact higher trophic components. For example, fish survival and growth are typically size dependent, implying that a single state variable for a fish population may be inappropriate for many situations. The number of system variables needed to describe a structured population is higher than those needed when simply following population abundances or densities, thus leading to a lower level of aggregation.

Higher trophic-level organisms are generally much more behaviorally complex and are able to display a repertoire of responses to environmental conditions and other organisms that cannot be captured in a small number of state variables. Each individual may behave differently based upon its current condition and current environment. For such organisms, an individual-based approach may be indispensable for determining how the individual actions affect population-level phenomena. Adequate description of a single individual can easily require a number of variables detailing its current age, size, location, physiological status, and other information pertinent to its future actions. When considering populations with hundreds or thousands of individuals, the total number of state variables can easily be in the tens of thousands.

Although temporal resolution also has great importance in this modeling scheme, it is not as straightforward as spatial extent. In part, this is due to the fact that all populations are made up of individuals that carry out actions within short time frames. The relevant temporal resolution of importance here is that associated with the system variables of interest. Thus, if one is dealing with zooplankton density rather than actions of an individual zooplanktor, the appropriate temporal resolution is not the seconds upon which individuals move, but rather the many-day time frame over which significant changes in plankton density occur at the spatial extent chosen for the model. In Figure 40.1, the relevant temporal resolution should be viewed as the longest period for which a temporal description of the model makes sense. At the individual level, the relevant temporal resolution is set by the unit of time on which an individual can execute actions that affect its survival and reproduction. Thus, for individual birds, weekly time steps are inappropriate since within a week an individual bird's actions can have great impact on its survival or growth. For fish populations, however, in which the relevant state variables are size or age distributions, an upper bound on temporal effects may be weeks or longer.

Clearly, the temporal and spatial resolutions chosen for any particular trophic level depend greatly upon the questions being addressed. Any of the trophic components described above can certainly be investigated at much finer resolutions than those indicated in Figure 40.1. There are, for example, questions of great interest to ecophysiologists regarding algal responses to fine-resolution environmental fluctuations. However, if the focus is on the linkages between trophic levels, and in particular the effects of lower trophic levels upon the dynamics at higher levels, then we argue that it is appropriate, as illustrated in Figure 40.1, to consider different spatial and temporal resolutions for different trophic levels.

The ATLSS Models

The ATLSS hierarchy (see http://atlss.org) starts with models that translate coarse-resolution hydrologic information to a finer resolution appropriate for biotic components that operate at spatial extents much smaller than the resolution of the main hydrologic model. The development of ATLSS has been motivated to great extent by the efforts to analyze alternative restoration plans for south Florida through the Central and Southern Florida Project Restudy (the Restudy, see http://www.evergladesplan.org). Throughout the development of ATLSS, there has been extensive consultation between modelers and field biologists with experience in the Everglades. Indeed, the initial set of models included within ATLSS were chosen following discussions with field biologists as to which species would be important in evaluation of any restoration plan.

The main hydrologic model used for the Restudy, a product of the South Florida Management District, has been utilized to produce all of the alternative plans for water management and typically provides a thirty-year plan of daily water depths across approximately 1,700 spatial locations, each of which represents a 2×2-mile (3.2×3.2-kilometer) grid cell. A complete description of the methods used to produce a high-resolution hydrology model is available on the ATLSS Project Home Page (ATLSS). This model relies upon vegetation maps and associated limitations on hydroperiod for each vegetation type to characterize a 28.5-meter-resolution topography that preserves the volumes of water derived from the 2-mile (3.2-kilometer)-resolution hydrology model.

The ATLSS hierarchy next includes Spatially Explicit Species Index (SESI) models, which make use of the spatially explicit, within-year dynamics of hydrology to compare the relative potential for breeding and/or foraging across the landscape (Curnutt et al. 2000). These models produce yearly comparisons of the spatial extent of the index across the landscape. SESI models are viewed as approximations that are useful in coarse evaluations of scenarios and aid in interpreting the more-detailed models. SESI models have been constructed and applied during the Restudy to the Cape Sable seaside sparrow (*Ammodramus maritimus mirabilis*), the snail kite (*Rostrhamus sociabilis*), short- and long-legged wading birds, and white-tailed deer (*Odocoileus virginianus*).

The compartment models involve biotic components for which it is reasonable to model variation across a landscape by means of many local uncoupled spatial unit cell models. The cell size chosen is small enough to represent a tract relatively homogeneous in substrate and elevation. A cell's spatial extent might be several hundreds of meters in a relatively flat landscape such as the Everglades. This is particularly appropriate for the primary producers (e.g., periphyton, aquatic macrophytes, terrestrial macrophytes), and meso- and macroinvertebrate consumers and detritivores, which can be combined into a few main functional groups. These models had not yet been developed enough to adequately link them to hydrology, and so they were not applied to the Restudy. This low-trophic-level web is an important food resource for the higher trophic levels and for restoring higher-consumer populations. The validation procedure for fish models has shown that it is critical to account for the temporal variability in these components (Gaff 1999).

The age- and size-structured population and community models represent intermediate trophic levels, such as fish, macroinvertebrates, and small nonflying vertebrates. These populations may move short distances in response to changes in water levels. Thus the model spatial domains of interaction are larger than for SESI models. A typical extent is up to one square kilometer, thus encompassing many unit cells coupled to allow population movements. Detailed models have been developed and applied in the Restudy to estimate responses of functional groups of fish across the freshwater landscape (Gaff et al. 2000). These models consider the size distribution of large and small fish to be important to the basic food web that supports wading birds. It has been applied in order to assess the spatial and temporal distribution of availability of fish prey for wading birds.

Individual-based models are employed to represent populations of top predators or other large-bodied species. Individuals of these species may move over large areas, with movements over short periods of time spanning areas of thousands of spatial unit cells. Such models are also used for organisms of particular interest, such as endangered species, for which detailed behavioral information is available and there is particular interest about the detailed response of the population to alternative plans. These individual-based models, an ecological form of agent-based models (Duke-Sylvester and Gross 2001), are rule-based approaches that can track the growth, movement, and reproduction of many thousands of individuals across the landscape. Adequate description of a single individual can easily require a number of variables detailing its current age, size, location, physiological status, and other information pertinent to its future actions. ATLSS models of this type include the Cape Sable seaside sparrow, snail kite, white-tailed deer, Florida panther, and various species of wading birds. The models include great mechanistic detail, and their outputs may be compared to the wide variety of organism distribution data available, including that from radio-collared individuals. An advantage of these more-detailed models is that they link each individual animal to specific environmental conditions on the landscape. These conditions (e.g., water depth, food availability) can change dramatically through time and from one location to another, and they can determine when and where particular species will be able to survive and reproduce.

How ATLSS Was Applied during the Restudy

The U.S. Army Corps of Engineers was charged in 1996 by Congress to produce a restoration plan for the Everglades region of South Florida by 1 July 1999 (see USACOE 1999 for details). The production of such a plan involved a massive effort on the part of numerous federal and state agencies. We cannot begin to describe here the full procedures for the development of the final plan presented to Congress, but we do want to describe the procedures used to apply ATLSS. The essential goal was to provide a rational, scientific basis for developing relative rankings of the biotic impacts of the proposed hydrologic scenarios as input to the planning process.

A key feature of the ATLSS application is the reliance on what we call relative assessment. By this, we mean that we do not claim that the variety of models within ATLSS can produce quantitatively accurate forecasts of population responses across the Everglades over the thirty-year plan. There are simply too many uncertainties in the models and the data used to estimate the parameters to attach any great confidence to the exact spatial variation in population estimates provided by the models. Rather, what we provided during the Restudy was a ranking of how various alternative plans caused the models to respond, relative to a base plan chosen by the agencies involved.

The procedure used to apply ATLSS in the Restudy can be described as follows: First, a hydrologic plan was developed. Next, the ATLSS team (comprising eight individuals) had one week to run the model, evaluate them, and provide written comparisons of the results of each model to the base scenario and post the results and summaries on the ATLSS home page on the World Wide Web. There would follow several days of conversations with various individuals from agencies involved in assessing the scenarios (the

Alternative Evaluation Team) who would then make a set of recommendations to the Alternative Design Team (ADT). It was then the responsibility of the ADT to revise the hydrologic plan, run a new hydrology model based upon the design changes, and provide a new plan. The turnaround time between hydrologic plans was typically three weeks, and the ATLSS team carried out over fifteen complete evaluations of plans during a fourteen-month period.

The Future

The focus of ATLSS to date has been on the freshwater systems with an emphasis on the intermediate and upper trophic levels. The ATLSS structure was purposely formulated to provide for extension to estuarine and near-shore dynamic models once physical system models for these regions are completed. This would involve the construction of a variety of additional models for the biotic components. A further effort at lower trophic levels in the freshwater regions is needed to account for the impact of hydrologic plans on vegetation change and associated nutrient fluxes. Closely linked with these would be models for the effects of major disturbances to the system, including fire and hurricanes. Finally, ecotoxicological models coupled to transport models for toxicants such as mercury may be readily incorporated into the biotic components already constructed.

In the more general context of future developments, computational ecology will allow us to investigate ecological models far more realistic than before. Its practitioners are driven by the need to improve predictive capabilities—to more accurately assess the future impact of human actions on natural systems. The difficulty of manipulating and replicating experiments at regional extents raises issues about experimental design, the regularity and longevity of sampling, and the integration and storage of data. Associated with this is the need for methods to archive complex data structures arising from spatially explicit dynamic models, which we have only begun to address by utilizing object-oriented databases. Computational ecology also has the capability to suggest appropriate long-term monitoring plans for natural systems to aid adaptive management.

A central issue in computational ecology is the need to link dynamic processes that operate across differing spatial regions and at different rates. How do we link natural and anthropogenic forces that influence the demand for biological resources with the dynamics of those resources? How much averaging and smoothing of high-resolution biological data must be done to match the lower resolution of geophysical data while still preserving the predictive capabilities of the approach for the underlying natural systems? Much depends on what we wish to predict and what level of accuracy is needed for such a prediction to be useful. As always for such complex models, issues of model validation and error propagation are difficult to address. We argue, based upon our experience with the Restudy, that in many cases what we wish to produce are relative assessments of different scenarios rather than exact quantitative predictions for any particular scenario. Given the assumption that errors in parameter estimates and functional forms within such complex models do not interact differentially with changes in scenarios (a reasonable assumption for scenarios produced by methods external to the models being used to assess them), it is likewise reasonable to assume that errors propagate similarly in model runs for different scenarios. This may justify the use of highly complex models for relative assessments, though we have had little success mathematically proving such assertions.

In addition to providing the ability to develop multimodeling methods and the computationally efficient means of carrying them out, computational ecology offers the opportunity to combine spatially explicit ecological models with models that assess economic and social impacts. A related need is that of addressing problems of spatially explicit control (Hof and Bevers 1998). Landscape-level management (e.g., forest harvesting, water-flow management, conservation preserve design, etc.) is not an all-or-nothing affair that occurs uniformly in space. Rather, realistic management scenarios must take into account spatial heterogeneity in underlying resources as well as how such heterogeneity interacts with management through time (local ecological succession for example). Given that there are many potential criteria affecting the system management, and that the underlying nonspatial issue might be viewed as a multiple-criteria-optimization problem, how should the "control" of the system be applied spatially in order to carry out the optimization? This is a little-developed area of applied mathematics, particularly in systems in which stochastic factors interact with the management scheme. Yet, such control problems are at the heart of much of applied ecology today.

Acknowledgments

The efforts described here have been supported by the U.S. Geological Survey, Biological Resources Division, through Cooperative Agreement No. 1445-CA09-95-0094 with the University of Tennessee. The statements, findings, conclusions, recommendations, and other data in this report are solely those of the authors and do not necessarily reflect the views of the U.S. Geological Survey. A number of individuals have contributed greatly to the empirical work that underlies the ATLSS project, notably Sonny Bass, John Chick, John Curnutt, William Loftus, and Joel Trexler. Several individuals have contributed greatly to the ongoing computer modeling for this project, notably Jane Comiskey, Scott Duke-Sylvester, Holly Gaff, Wolf Mooij, Phil Nott, Mark Palmer, Michael Peek, and Rene Salinas. We acknowledge the assistance of all these individuals in bringing the ATLSS project to its current stage of development.

Challenges of Modeling Fungal Habitat: When and Where Do You Find Chanterelles?

Tina A. Dreisbach, Jane E. Smith, and Randy Molina

Fungal Habitat Modeling in the Pacific Northwest States

In the past decade, land-use patterns have changed dramatically in the Pacific Northwest. As a result, land management practices are being scrutinized for impacts on valued species other than those that provide timber. In the early 1990s, the Northwest Forest Plan was developed to conserve biodiversity and species viability by maintaining appropriate habitat. The record of decision (ROD) lists more than four hundred species of concern occurring in the range of the northern spotted owl (*Strix occidentalis caurina*). These organisms include mammals, birds, amphibians, plants, lichens, mosses, and fungi. Like the northern spotted owl, the species of concern are considered old-growth dependent and are collectively referred to as "survey and manage" species. Of the four hundred species, the largest proportion (36 percent) are fungi (USDA/USDI 1994a).

Within existing guidelines for the systematic evaluation of forest conditions, few methods exist for assessing species diversity and viability of fungi, yet fungi are important components of nutrient cycling, plant health, food webs, and ecosystem resiliency in forests (O'Dell et al. 1996). Furthermore, many are commercially valuable (Molina et al. 1993). Although fungal biodiversity has been studied in relation to particular forest conditions (Bills et al. 1986; Villeneuve

et al. 1989; Luoma et al. 1991; Nantel and Neumann 1992; Keizer and Arnolds 1994; Zhou and Sharik 1997; O'Dell et al. 1999; J. Smith unpublished data), little scientific information is available regarding the basic biology or habitat requirements for particular fungal species. This lack of information is a major impediment to the formulation of management recommendations and sampling procedures.

Information needed for developing species conservation plans includes availability of habitat, response of the organism to habitat change, and population dynamics. As a first step toward incorporating modeling into the study of forest fungi in the Pacific Northwest, we are developing habitat models. In the past several decades, habitat models have become decision-making tools for land managers, providing insight into ecosystem behavior over long periods and with hypothetical management scenarios.

Modeling as a tool has had limited use in the field of mycology. Plant pathologists have recently used a variety of modeling techniques to investigate host-pathogen interactions (Thrall et al. 1997; Taylor et al. 1998), pathogen fitness (Lannou and Mundt 1997; Newton et al. 1998), the effects of environmental factors on fungal pathogens (Gumpertz et al. 1997), and to simulate consequences of disease (Frankel 1998). The primary goal for modeling plant pathogenic fungi is to be able to predict disease and plan control strategies. In contrast, our interest in modeling is for conservation of forest fungi. Mathematical models have examined establishment and colonization of beneficial mycorrhizal fungi in roots, primarily for vesicular-arbuscular mycorrhizae in agricultural settings (Menge 1985; Tinker 1985; Walker and Smith 1985). We believe that modeling also can serve useful purposes in the study of forest fungi, particularly in ecosystems where conservation management is an issue.

Modeling can help mycologists predict the consequences of basic fungal processes such as dispersal, colonization, and reproduction. For forest fungi, we know virtually nothing about these processes. Modeling can provide a formal organizing framework for generating ideas, conducting data analysis, and assisting in determining research directions by helping to generate hypotheses and define problems. Models can simulate and project over time and under a variety of disturbance and land-management scenarios to predict species occurrence and interactions. Models can help prioritize data collection efforts by predicting areas where centers of biodiversity exist or where more-intensive sampling may be necessary (Kiester et al. 1996). In the Pacific Northwest, gap analysis is currently being explored as a method for helping design regional survey strategies for forest fungi (T. O'Dell and R. Kiester personal communication). To that end, models also can provide a method of accountability and support to land-management decisions. Using models as ecosystem-based management tools will allow land managers to calculate risks of decisions and actions. This will be particularly helpful in the development of sustainable harvest strategies for commercially valuable fungi and in predicting the impact of human intervention on fungal survival and productivity.

One of the primary needs of the USDA Forest Service and USDI Bureau of Land Management is development of habitat-based conservation plans. An important question in fungal biology is, for a given habitat, what are the odds of finding a particular species or group of species? Underlying this question are three assumptions: (1) species are tightly linked biologically to their habitats; (2) habitats can be detected, measured, and quantified; and (3) we are able to detect the species itself. Our work investigates habi-

tat requirements and modeling of several different fungal species displaying various life strategies. Several are considered habitat specialists, possibly restricted to one or a few host species or a unique identifiable vegetation type. Others are habitat generalists, able to associate with a variety of host plants and vegetation types. Our long-range plan is to develop habitat models encompassing multiple fungal species and life strategies, thereby fostering transition from single-species land management to multiple-species habitat management.

We have begun by modeling several fungal species commonly referred to as chanterelles. The species in the chanterelle group contain three genera: *Cantharellus*, *Gomphus*, and *Polyozellus*. We chose this group for the following reasons: (1) All species are included in the ROD survey-and-manage listing (USDA/USDI 1994a), but habitat information is minimal or lacking. (2) All may be considered "charismatic macrofungi"; they are easily recognized and identified, thereby increasing the possibilities of detection by field crews and others working in the forest. (3) These fungi cover the entire range from rare to weedy in occurrence, implying that habitat needs may be less stringent for some than for others. (4) The different life strategies displayed by our selected species are representative of other survey-and-manage fungi such that we can transfer results to developing models for other fungi. (5) Several species within this group are of interest as highly sought-after edibles and commercially valuable nontimber forest products. The objectives of our research are as follows:

- Identify ecological factors that determine fungal habitat.
- Develop spatial and statistical models to predict the occurrence of chanterelle species. In particular, the models are intended to predict the impact of environmental change at multiple geographic scales.
- Apply these models across a broad geographic area in the Pacific Northwest. This will determine the ability to evaluate and predict desired forest conditions for maintaining viability of cantharelloid species while meeting ecosystem management objectives.
- Assess how we might apply the models to other sur-

vey-and-manage fungi. Ultimately, this is necessary for multiple species-environment and species-habitat understanding.

Modeling Approach

Patton (1997) defines habitat as the environment and specific location where an organism lives, including the combination of factors where an organism can survive and reproduce. Guidelines for development of habitat models are given by the U.S. Fish and Wildlife Service (USFWS 1981b) and have been used to generate models for mammals, fish, and birds. In this approach, habitat variables are selected according to the following criteria: (1) Is the variable important to survival and reproduction of the species? (2) Is there a basic understanding of the habitat-species relation for the variable? (3) Are data available and is the variable practical to measure? Unfortunately, few studies of fungi have directly assessed any of these criteria.

Our most comprehensive database for survey-and-manage fungi consists of approximately five thousand records and includes information gleaned from historical material housed in herbaria as well as collections made by the USDA Forest Service Fungal Survey and Manage Team and cooperators (T. O'Dell personal communication). Although at first glance there appears to be an abundance of data; in fact, we lack the crucial information needed to develop models predicting fungal occurrences based on habitat. Because these data were collected by many different persons, statistical analysis is problematic if not impossible. The main difficulty arises from the manner in which data were collected. Collection labels may contain little to no information on habitat. For example, a subsample of 386 chanterelle records revealed that less than 50 percent include information on the type of substrate on which a mushroom was found growing. Even when information is included it is often vague, for example, habitat listed only as dense woods or mixed conifers. Consequently, our options of modeling methods are limited. We do not know enough about life processes of forest fungi to develop quantitative simulation models. We also have no idea of how population dynamics operate for most fungi, precluding population viability analysis.

We therefore must make assumptions about our study organisms by basing habitat parameter estimations on published studies of related organisms. In some cases, anecdotal information or expert opinion also is used. Scientific researchers, commercial mushroom harvesters, and recreational forest users have a great deal of experiential knowledge as to when and where particular mushroom species may be found. These experts often can describe optimal habitat in extensive detail. Accordingly, our approach is primarily qualitative: we use knowledge- and rules-based methods and expert systems for predicting expected distributions. These modeling methods are useful in that both quantitative and qualitative information can be synthesized (Starfield and Bleloch 1991). In addition, the availability of digital maps and geographic information systems (GIS) provides opportunities for linking our modeling rules with spatial databases and simulation models for other biological systems. In the following review, we discuss the development of rules for predicting the occurrence of suitable habitat for forest fungi in general and for chanterelles of the Pacific Northwest in particular.

Species of Interest

The chanterelle species all produce mushrooms with distinctive funnel shapes and spore-bearing folds. Table 41.1 lists the species chosen for the current study and their abundance in Pacific Northwest forests.

TABLE 41.1.

Species of interest: Cantharelloid fungi from the ROD (record of decision) list.

Species	Common name	Abundance[a]
Cantharellus formosus	yellow chanterelle	weedy
C. subalbidus	white chanterelle	common
C. tubaeformis	winter chanterelle	common
Gomphus bonari	scaly chanterelle	rare
G. floccosus	scaly chanterelle	weedy
G. kauffmanii	scaly chanterelle	common
G. clavatus	pig's ears	common
Polyozellus multiplex	blue chanterelle	rare

[a]Abundance: based on number of existing survey and manage and herbarium databases: Weedy = more than 100 records; common = 50–100 records; rare = less than 50 records.

Geographic Scope

Our efforts are concentrated in areas defined as potential northern spotted owl habitat, including U.S. Forest Service lands in northern California, western Oregon, and western Washington. Ideally, the extent of our models would include this entire area, for which forest inventory databases are collected and maintained by state and federal agencies. In addition, GIS coverages of vegetation, soils, and climate are available for much of this region. In the 1990s, specific areas, such as the Coast Range of Oregon, the H. J. Andrews Experimental Forest (on the western slope of the Cascade Range), and the Willamette National Forest, have been the sites of many ecological and mycological studies. Our first models will focus primarily on forest types within these geographic areas.

Difficulty arises from attempts to use existing data for construction of models with such a large extent. Many studies of fungi are conducted at a level of resolution of less than 1 meter and measure microhabitat factors such as the number of fruiting bodies occurring in a cluster or proximity to the nearest piece of down wood. In terms of grain, these studies may not be adequate to evaluate factors at the appropriate level of resolution for building large-area models. In a given forest type, habitat consists of a series of many discontinuous patches, and important biological processes to consider are host specialization, migration, and extinction. At the local scale, one habitat area may equal one of a few patches, and processes such as competition, aggregation, and the match between species niche and habitat becomes important.

Temporal Scope

Fungi live as aggregations (mycelia) of microscopic strands (hyphae). Mycelia inhabit soils, wood, litter, and living plants. Mycelial colonies can range in size from microscopic to many hectares and can persist for years (Smith et al. 1992; De la Bastide et al. 1994; Dahlberg and Stenlid 1995). Fungal reproductive structures are the parts readily seen in the forest in the form of cups, truffles, conks, and mushrooms. Timing of mushroom formation (and hence organism detection) is species specific and generally occurs only when nutritional requirements and environmental condi-

tions (temperature, light, pH, moisture) are appropriate during particular seasons of the year (Hunt and Trappe 1987; Luoma 1991). Therefore, it is necessary to define the season of the year for which the model is applicable for each species. Year-to-year variability in mushroom production also complicates modeling. Over a three-year period, Luoma (1991) documented a similar number of truffle species occurring each year; however, the proportions of those species differed greatly from year to year.

Our survey-and-manage records indicate that mushrooms of all chanterelle species occur most often in fall. At northern locations or higher elevations, or both, fruiting begins as early as late July or early August. In the southern part of our modeling area, and at lower elevations, fruiting extends from October through January. One species, *Cantharellus tubaeformis* (winter chanterelle), is documented from many locations in nearly all months. Anecdotally, mushrooms fruited prolifically in fall 1997 and poorly in fall 1998. The exact causes for this variability are unknown, but it is clear that this unpredictability requires incorporation of additional variables or stochasticity into the models. The purpose is not to define the suitability of habitat per se, but to give us an idea of the probability of detection.

Determination of Important Ecological Factors

Ecological factors often associated with fungi fall into three broad categories: vegetation, topography and soils, and climate.

Vegetation may be the primary factor contributing to fungal habitat, primarily because of the levels of host specificity required for many fungi. Our study includes several ectomycorrhizal species as well as species that serve a function as decomposers of dead material, thereby cycling nutrients. Ectomycorrhizal fungi (EMF) form mutually beneficial relationships with host plants by directly supplying nutrients in exchange for carbon. Molina and Trappe (1982) recognize three groups of EMF based on the relationships to host species, ranging from highly specific to nonspecific. Decomposer fungi also show various levels of host-substrate specificity (Swift 1982). We therefore can use the presence or absence of particular host tree species as an initial indicator for potential occurrence of many species of fungi.

Herbarium and collection records provide some information regarding the presence of dominant tree species in stands where chanterelles occur. *Cantharellus formosus* (Pacific golden chanterelle) is documented from forests dominated by many coniferous hosts, including *Abies*, *Picea*, *Pinus*, *Pseudotsuga*, *Thuja*, and *Tsuga*; *C. subalbidus* (white chanterelle) appears somewhat more limited in that no collection records thus far have indicated association with *Abies* or *Picea*. Recent research suggests that collections identified as yellow chanterelles may in fact represent at least two distinct species, with possible host and habitat differences (Feibelman et al. 1994; Redhead et al. 1997; S. Dunham personal communication).

Forest stand age and structure may play a critical role in species occurrence (Crites and Dale 1998). Disturbances such as harvesting or fire cause changes in fungal diversity and productivity (Amaranthus et al. 1994; Clarkson and Mills 1994; Stendell et al. 1999). Studies of EMF communities suggest that species succession occurs following wildfire (Visser 1995; Jonsson et al. 1999; Stendell et al. 1999). Tree harvest also affects the community of forest fungi (Pilz and Perry 1984; Colgan et al. 1999). The ROD assumes that listed fungi are old-growth dependent; however, chanterelles also may occur in younger stands. Although Danielson (1984) indicates that *Cantharellus* could not be found in a six-year-old jack pine stand, regenerating after wildfire, Pilz et al. (1998) found *Cantharellus* in coastal hemlock stands as young as twenty years in Washington. Research by S. Dunham (personal communication) indicates that one species of yellow chanterelle appears in both old-growth and rotation-age stands of forty-five to sixty years. Another species of yellow chanterelle and the white chanterelle are more likely to be found in old-growth forests. Data for effects of fire and harvest on other chanterelle species are not currently available. For modeling purposes, however, we will assume that the presence of fire or harvest within approximately twenty years precludes the occurrence of chanterelles.

Soil organic matter, including humus and coarse woody debris (CWD), especially advanced stages of decay in the form of brown cubical rot, is an important substrate for EMF (Harvey et al. 1976) and decomposer fungi. The CWD may be particularly important as a moisture-retaining substrate, allowing root tips to support active ectomycorrhizae in times of seasonal dryness. This is important on dry sites or following fire (Harvey et al. 1978; Amaranthus et al. 1989; Harmon and Sexton 1995). These fallen tree "reservoirs" may provide refugia for seedlings and mycorrhizal fungi, particularly in more arid forests. For dry forests in western Montana, Harvey et al. (1981) estimated that about 25–37 tons per hectare of CWD are needed to support ectomycorrhizal activity needed for a developing ecosystem. Currently, no data exist for quantities of CWD necessary to support viability of fungi in most forest ecosystems of the Pacific Northwest. After disturbance, colonization of CWD by ectomycorrhizal fungi may be limited in the early stages of stand development. As stands mature, the availability of CWD nevertheless seems to be crucial for the establishment of fungi as well as seedlings (Kropp 1982; Luoma et al. 1996). The CWD may be a good predictor of chanterelles, and we therefore assume that these fungi require a minimal volume of CWD for habitat maintenance. Our first parameter estimates will specify no less than 10 percent cover by CWD (Amaranthus et al. 1994). Detailed field studies evaluating relationship between CWD and chanterelle occurrence and production are now underway, and these data will be incorporated as they become available.

Topographic and soil factors that may be important to chanterelle occurrence include elevation, slope, aspect, soil properties, and local microtopography. The survey-and-manage database indicates approximate elevation ranges for each of the study fungi, and these are included as preliminary parameters. Factors such as slope, aspect, landform type, topographic moisture index, and soil texture are only now being investigated.

Climate seems to be a complex factor in fungal distribution. Seasonal and ecological distribution of fungi is partly determined by temperature and moisture. Wilkins and Harris (1946) contend that moisture may be the most important single environmental factor controlling fungal reproduction. Climatological databases are readily available from individual states. We know little, however, about the effects of climate on long-term survival of fungi or on timing of mushroom formation. From the survey-and-manage records we do know that all our study species form mushrooms in fall. Norvell (1995) found that amount of precipitation does not affect chanterelle productivity in moist forests of western Oregon. Chanterelle productivity was positively correlated with average mean summer temperature. All other factors being equal, years in which the average mean summer temperature was high showed greater productivity. Neither precipitation nor temperature have been directly linked to occurrence. In theory, however, increased productivity should increase our likelihood of detecting chanterelle occurrence.

The Models

Two types of models are being developed to predict occurrence of chanterelle species in the lowland mixed conifer forest type extending across the Oregon Coast Range and the west side of the Cascade Mountains in central Oregon. The first is based on ecological factors that contribute to fungal habitat over large areas (e.g., ecoregions), primarily plant association, dominant tree species present, forest structure, temperature, and precipitation. The habitat evaluation procedures (HEP) originally developed by the U.S. Fish and Wildlife Service (USFWS 1981a) are being modified to develop habitat suitability index (HSI) scores. These will correspond to the probability of occurrence of a species and will range in value from 0 (low probability of occurrence = poor habitat) to 1.00 (high probability of occurrence = optimal habitat). A rules-based approach, linked to a GIS in the form of queries, will provide predictions given current conditions of habitat variables as shown in composite GIS coverages for selected parameters. Figure 41.1 presents an example of rules we are developing for *Polyozellus multiplex*, the blue chanterelle. We will generate maps from these predictions to give a first approximation of where a particular species occurs and how likely we are to detect that species. Presumably, weedy species will generate more and larger areas of higher HSI value than those that are more specific in habitat requirements. The second type of model being developed to examine microhabitat variables, operable within a particular forest stand. Microhabitat factors such as amount and

IF	Longitude is < 123 degrees
AND	Elevation is > 1,000 meters, < 1,500 meters
AND	*Abies* is present
OR	*Pinus* is present
AND	Stand age is > 100 years
THEN	The probability of *Polyozellus multiplex* occurrence is 25 percent

Figure 41.1. Example of rules-based modeling technique for *Polyozellus multiplex*, the blue chanterelle.

quality of coarse woody debris present, are used to develop rules defining the probability of occurrence for a given species within a limited geographic area. For this type of modeling, we are using an expert-system approach in order to utilize both quantitative and qualitative data.

Validation and Accuracy Assessment

The Pacific Northwest has an abundance of experts on fungi and on chanterelles in particular. Among Oregon State University, USDA Forest Service Pacific Northwest Research Station, and other nearby agencies and institutions, about thirty mycologists will voice an opinion on model accuracy. Members of the North American Truffling Society, the Oregon Mycological Society, and commercial mushroom harvesters will contribute knowledge gained from many years of data collection and observations that will be invaluable in model development, although these individuals are often secretive about their picking territories.

Ground surveys will also be employed to determine accuracy of predictions. By visiting areas of both high and low probability of occurrence, we will quantify errors of omission (detection where not predicted) and errors of commission (failure to detect where predicted to occur). Concurrently, environmental factors will be measured in survey areas. These additional data will be incorporated into the habitat models to refine parameters and increase model accuracy. We need to keep in mind, however, that above-ground indicators (presence of mushrooms) may poorly reflect the composition of below-ground fungal communities (Gardes and Bruns 1996). Therefore, ground surveys will be conducted several times a year over several years.

Habitat Modeling—Opportunity and Challenge

By modeling fungal habitat, we have the opportunity to learn a great deal about the basic biology of these organisms. In our preliminary models, we are making major assumptions about selected fungi and their role in the forest ecosystem. We assume that as heterotrophic organisms, fungi are directly or indirectly dependent on plants and plant communities. We know mycorrhizal fungi form obligate associations with and obtain nutrients from one to several living host species and that decomposer fungi obtain nutrition from wood (living and dead), leaves and twigs in the litter layer, or freshly burned material. Some species also may require distinct ecological niches, such as forest gaps or undisturbed areas. Distribution of nutrient sources (plants, woody debris) partly determines fungal distribution. Fungal species have different tolerances to changing environment and plant communities. As plant communities change over ecological time under the influence of soil, climate, topography, and organisms, and in present time as a result of natural catastrophes or human activities, fungal species composition is altered. Fungal species composition also influences plant community structure, providing a complex feedback mechanism (van der Heijden et al. 1998).

Our models are conceptual and qualitative in nature, due to the current scarcity of statistically analyzable data. Figure 41.2 provides a flow chart of our generalized modeling scheme. Rules-based and mapping models are only the first step in understanding habitat and predicting fungal species occurrences. Exploratory data analyses are useful in determining trends. Bayesian and multivariate statistical techniques

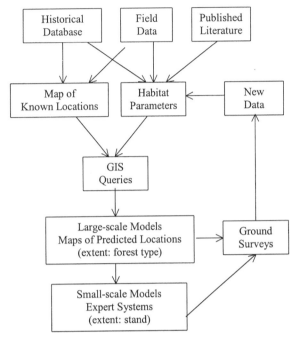

Figure 41.2. Generalized scheme for modeling fungal habitat.

(such as principal components analysis, correspondence analysis, cartographic and regression trees, multiple regression) provide approaches for analyzing large data sets and evaluating the contribution of many variables to habitat (Grubb 1988; Morrison et al. 1992).

As knowledge gaps are identified, the challenge to mycologists will be to address these gaps with studies designed to integrate with each other and with modeling efforts. Considerations in designing future studies include determining how temporal variability can be assessed and incorporated into models, and collecting data appropriate to the regional scale (grain and extent).

Predicting Presence/Absence of Plant Species for Range Mapping: A Case Study from Wyoming

Walter Fertig and William A. Reiners

The range of a plant species is shaped by its phylogeny, ecological adaptations, interactions with other species, and historical events (Daubenmire 1978). Detailed knowledge of a plant's range is increasingly important for land managers and conservation biologists. Distribution maps are one of the best tools for depicting information about a species' range and its environmental requirements.

Traditional range maps depict species' distributions as irregular polygons (outline maps), assemblages of points (dot maps), or a combination of both (Brown and Lomolino 1998). Outline maps are typically drawn by hand based on expert knowledge. The assumptions of the mapper are rarely stated (and thus are difficult to test), and such maps usually overestimate the actual range of a species, particularly for large areas. Dot maps show documented species locations but may represent only a fraction of the species' actual range or may reflect sampling bias (Brown and Lomolino 1998). Combination outline and dot maps may be an improvement but are still likely to overestimate distribution over large geographic areas.

The development of geographic information systems (GIS) has revolutionized the art and science of range mapping. GIS, in conjunction with large environmental data sets and computerized geostatistical methods, allows models of the interactions between a plant and its environment to be incorporated into the mapping process (Franklin 1995). Model-based range maps are superior to traditional ones because the assumptions of the map are explicit and readily testable with new data.

Plant distributions can be modeled mechanistically or empirically. Mechanistic models employ ecophysiological information about the modeled species and detailed fine-scale environmental measurements to predict a plant's potential range in a localized area. Such an approach is often difficult for most vascular plant species over geographic areas exceeding hundreds to thousands of kilometers because the requisite ecophysiological studies have not been conducted and fine-grained environmental data at these scales are often unavailable. Models based strictly on ecophysiology may also exclude the effects of competition and other biotic interactions, thereby generating a model of a species' potential niche rather than its realized niche.

Empirical (or correlational) models are based on correlations between selected environmental variables that directly influence a species or are surrogates for direct gradients (Austin et al. 1990; Franklin 1995). The resulting environmental envelope can then be used to generate a potential range map that approximates the species' realized niche. An important limitation of empirical models is that causal factors are not determined, and so these models will be less successful than mechanistic models at predicting range shifts due to changes in climatic or biotic variables. Empirical modeling may also be inadequate for rare species whose distributions do not reflect the full extent of their realized niche because of incomplete dispersal, recent origin, localized extinction, or historical accidents. Despite these limitations, empirical modeling can be an efficient technique for predicting ranges of relatively common species over scales of hundreds to thousands of square kilometers.

To date, most plant modeling studies have been applied to localized areas ranging in size from 1-square-meter plots (Wiser et al. 1998) to regions up to 70,000 square kilometers (Franklin 1998). These studies have focused on individual species (Davis and Goetz 1990; Guisan et al. 1998), guilds of species (Franklin 1998; Wiser et al. 1998), or plant communities (Lees and Ritman 1991; Moore et al. 1991; and reviewed by Franklin 1995). Those modelers studying larger spatial extents (regional or subcontinental scales exceeding several thousand square kilometers) have primarily addressed potential effects of climate change on distributions of ecologically important species or communities (Brzeziecki et al. 1995; Huntley et al. 1995; Iverson and Prasad 1998; Iverson et al. 1999).

We describe a prototype empirical modeling procedure for individual plant species centered in the state of Wyoming. Unlike many of the preceding studies, our approach covers a larger geographic area (approximately 252,000 square kilometers) and utilizes state and regionwide digital environmental and herbarium-based plant location data rather than localized, plot-derived environmental and presence/absence data sets. Our state-scale predictive models will ultimately be used for a gap analysis of selected elements of the vascular flora of Wyoming (Fertig et al. 1998) but could also be used as a baseline for studying future range shifts due to climate change or to identify new areas to survey for plants of management or conservation interest.

Methods

The vascular flora of Wyoming contains more than 2,700 native and introduced taxa, making the modeling of all species for range mapping impractical. We have randomly selected two hundred plant taxa that represent a cross-section of growth form, abundance,

biome affinity, and geographic distribution patterns in the flora and that were represented by at least twenty-five known locations in Wyoming (Fertig et al. 1998). We have chosen one of these target species for our pilot study to demonstrate the suitability of different statistical modeling approaches that will then be applied to the remaining study pool.

Mentzelia pumila (Nutt.) T. & G. var. *pumila* (Loasaceae), a short-lived perennial forb restricted to semibarren desert plains and slopes, occurs from south-central Montana and western North Dakota to central Wyoming, northeastern Utah, and northwestern Colorado. Based on collections at the Rocky Mountain Herbarium (RM) and reports from Hill (1975), *M. pumila* is known from approximately fifty locations in Wyoming (Fig. 42.1). Only thirty-five of these locations were chosen for modeling because they had sufficiently precise label data to place them within 0.5–2.5 kilometers of their presumed collection site.

Absence data have not been routinely collected for *M. pumila*, or virtually any other plant species, on a statewide basis. As a surrogate for confirmed absence points, we used the RM's database of over nine thousand sampling localities to identify areas where this species has not been collected (and is thus presumed absent). Since 1977, RM researchers have systematically established these sampling sites across the entire state of Wyoming for the purpose of collecting all plant taxa present in the immediate area (Hartman 1992). We stratified these putative absence sites by their environmental attributes and randomly selected 1,270 points that would reflect the full range of variation in each of the environmental variables selected for model development. By using such a large data pool, we were able to produce an approximately uniform distribution of absence location points across the state, although denser concentrations of points are present in mountainous areas with steep environmental gradients (Fig. 42.2). Any presumed absence sites that overlapped with known presence locations were removed from the final dataset.

We selected climatic, topographic, and edaphic variables for modeling based on their utility in describing the environmental space occupied by *M. pumila* and their availability in statewide digital format. In some cases, selected environmental variables

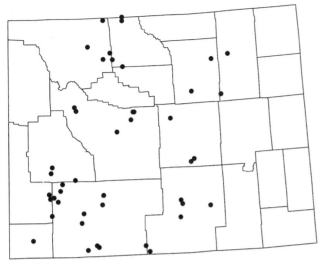

Figure 42.1. Dot distribution map of the known distribution of *Mentzelia pumila* in Wyoming based on records of the Rocky Mountain Herbarium.

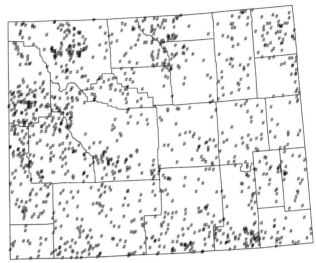

Figure 42.2. Dot distribution map of putative absence sites for *Mentzelia pumila* in Wyoming.

(such as terrain position or bedrock geology) have no direct physiological effect on this species but serve as useful surrogates for those factors that do have an influence (such as microclimate, soil texture, or soil pH).

For climate data, we used PRISM mean monthly precipitation available in 4-kilometer raster format (Daly et al. 1994, 1997) and unpublished PRISM mean monthly temperature data available in 2-kilometer raster format (Ken Driese personal communication). Topographic data, including elevation, slope, and aspect, were calculated from 30-meter digital elevation model (DEM) coverages for the state. An index of landscape position for each 30-meter pixel was calculated following the protocol of Fels and Matson (1996) and then reclassified into four terrain-position categories based on overall slope and shape (concave, slope, flat, and convex) (Ken Driese personal communication). We also used vector coverages of bedrock geology (Love and Christiansen 1985) and family-level soil type (Munn and Arneson 1998) derived from 1:500,000-scale maps. Bedrock geology units of comparable age and mineralogy were aggregated to form a coverage that approximated the statewide geologic highway map (Geological Survey of Wyoming 1991). Finally, land use and general vegetation type were derived from the 1:500,000-scale Wyoming gap land-cover map (Driese et al. 1997). Values for all environmental attributes were assigned to each presence and

absence point. To minimize potential errors based on imprecise locations, we derived terrain position from herbarium specimen labels for known presence locations.

Potential distribution models were constructed from our environmental data sets using logistic regression (Minitab, version 11) and classification tree analysis (S-plus, version 1.1) (Breiman et al. 1984; Hosmer and Lemeshow 1989). For both techniques, our data set was randomly divided into nineteen present and 626 absent locations for model building and sixteen present and 644 absent locations for model testing. In each model, presence and absence were chosen as response variables, and selected environmental attributes were used as predictors.

For the logistic regression analysis, we chose predictors based on their statistical significance in trial runs ($P < 0.05$). We selected the simplest model with the best goodness of fit for mapping in GIS (ArcView, version 3.1) at a resolution of 30 meters across the state domain. Using ArcView's map-calculator function, we determined the probability of presence or absence of *M. pumila* for each 30-meter pixel in the state. We then selected a cut-off probability for presence that represented the midpoint between the mean probabilities of the known present and absent points (Fielding and Haworth 1995). Points above this value were designated present in our final map, while those below were deemed absent.

In our classification tree model, we used box plots to visually identify the environmental predictors that best discriminated between presence and absence of *M. pumila*. We developed several initial classification tree models using six environmental predictors (terrain position; bedrock geology; July mean temperature; and January, April, and July mean precipitation) and a cut-off of five observations per terminal node beyond which no additional splitting occurred. For mapping, we selected the model with the fewest number of terminal nodes (14) and highest residual mean deviance (0.081). This model used only four of the six predictor variables (April mean monthly precipitation, terrain position, bedrock geology, and January mean precipitation) and was pruned to predict the presence of *M. pumila* at four terminal nodes. For each of these nodes, we developed a predicted range map in Arc-View based on the intersection of their composite environmental attributes. These individual layers were then merged to form the final statewide predicted range map at a resolution of 30 meters.

Results

Our logistic regression model was constructed with four environmental variables: April mean precipitation, terrain position, July mean temperature, and elevation. This model was strongly significant (G = 59.1, df = 4, P = 0.000) with good fit (Pearson deviance chi-square = 180.0, df = 607, P = 1.00). The multiple logistic regression equation was

$$\ln(Y / 1 - Y) = -23.04 - 1.0424 \text{ (terrain position)} + 0.002966 \text{ (elevation)} + 0.03252 \text{ (July mean temperature)} - 0.0015948 \text{ (April mean precipitation)}.$$

Of the four predictors, April precipitation had the strongest influence on the model (Z = −4.34, P = 0.000), followed closely by terrain position (Z = −3.29, P = 0.001), July mean monthly temperature (Z = 2.33, P = 0.020), and elevation (Z = 2.10, P = 0.036).

The model successfully classified thirteen (68.4 percent) of the known present points and 576 (92 percent) of the known absent points in the modeling data set, for an overall correct classification rate of 92 percent (Table 42.1). For the validation data set, the

model correctly predicted eight (50 percent) of the known present points and 573 (89 percent) of the known absent points for an overall success of 88 percent (Table 42.1). The distribution map produced from this model (Fig. 42.3) predicted presence for this species over an area of 49,482 square kilometers. Figure 42.4 illustrates the probability of occurrence at each pixel.

The final classification tree model selected April mean monthly precipitation as the most important predictor for the initial subdivision of the model-building data set followed by terrain position, geologic substrate, and January mean monthly precipitation. The model predicted likely presence of *M. pumila* under four different sets of environmental conditions (Table 42.2). The map derived from this model (Fig. 42.5) predicted a total statewide range of 15,962 square kilometers.

This model correctly classified seventeen of the known presence points (89.5 percent) and 562 of the known absence points (89.8 percent) from the modeling data set for a total classification success rate of 89.8 percent (Table 42.1). From the validation data set, the model correctly classified nine of the known present points (56.3 percent) and 548 (85 percent) of the known negative points (Table 42.1).

Model Performance

Both the logistic regression and classification tree models had overall classification success rates of 89.8–92 percent for the model-building data sets and 84.4–88 percent for the validation data set. Fielding (Chapter 21), however, has suggested that the overall rate of classification success is a poor indicator of prediction accuracy because this measure does not distinguish between errors of omission (missed present points) and commission (false positives), or weight these errors equally, when in most situations missed present points have more serious management consequences. The omission error rate was significantly higher (31.6 percent) for the model-building data set in our logistic regression model than in the comparable data set for the classification tree model (10.5 percent), (Table 42.1) although the commission error rates for both were similar (8 percent in the logistic re-

TABLE 42.1.

Comparison of classification success rates for logistic regression and classification tree models.

Logistic regression model Model data set			Logistic regression model Validation data set		
	Model present	Model absent		Model present	Model absent
Known present	13/19 (68.4%)	6/19 (31.6%)	Known present	8/16 (50%)	8/16 (50%)
Known absent	50/626 (8%)	576/626 (92%)	Known absent	71/644 (11%)	573/644 (89%)
Classification tree model Model data set			Classification tree model Validation data set		
	Model present	Model absent		Model present	Model absent
Known present	17/19 (89.5%)	2/19 (10.5%)	Known present	9/16 (56.2%)	7/16 (43.8%)
Known absent	64/626 (10.2%)	562/626 (89.8%)	Known absent	96/644 (15%)	548/644 (85%)

gression model and 10.2 percent for the classification tree model). Error rates were also comparable in both models for the validation data sets, although in this case, the logistic regression model performed slightly better in terms of commission errors.

Variation in prediction success rates may be related in part to differences in the choice of predictor variables between the two models. Both utilized April mean precipitation and terrain position, but only the classification tree model distinguished between levels

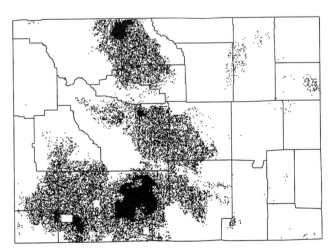

Figure 42.3. Predicted distribution of *Mentzelia pumila* in Wyoming based on a logistic regression model. Black area = predicted present, white area = predicted absent or no data.

of precipitation (selecting values between 27.25 and 33 millimeters) or terrain values (selecting only slopes or swales). Likewise, the classification tree model also selected categories or ranges of values for bedrock geology and January mean precipitation in model development, but the logistic regression model made no distinction between values of elevation or July mean temperature. The higher omission error rates in the logistic regression model may result from that model being too simplistic or from using parameters that are too conservative (Fielding, Chapter 21).

Differences in prediction success rates may also reflect methodological differences between the two modeling techniques. The logistic regression model uses a single regression equation to identify the optimal condition under which this species is present, whereas the classification tree model identifies multiple conditions under which *M. pumila* may occur. The environmental attributes used by the four pathways in the latter model are quantifiable and readily testable in the field (Franklin 1995), whereas the coefficients of the logistic regression equation are difficult to interpret. Logistic regression models are better suited for assessing probability of occurrence on a per-pixel basis than classification tree models (Dettmers et al., Chapter 54) but may be more prone to overestimation of potential

TABLE 42.2.

Combinations of environmental attributes used to model *Mentzelia pumila*.

Group	Mean precipitation	Terrain position	Bedrock geology
1	April: < 33 mm	Slopes or swales	Late Paleozoic/early Mesozoic sediments, Middle Cretaceous shales or Miocene sandstone/conglomerates
2	April: < 27.25 mm January: 7.25–8.85 mm	Slopes or swales	Upper Cretaceous shales, late Eocene interbedded sandstones, or Quaternary alluvium
3	April: 27.25–33 mm January: < 12.53 mm	Slopes or swales	Upper Cretaceous shales, late Eocene interbedded sandstones, or Quaternary alluvium
4	April: < 33 mm January: 12.53–16.1 mm	Slopes or swales	Jurassic/early Cretaceous sediments, Lower Cretaceous shales, Upper Cretaceous shales, Tertiary intrusive volcanics, Paleocene interbedded sandstones, late Eocene interbedded sandstones, Quaternary alluvium, Quaternary lacustrine deposits, or Quaternary till

range if cutoff probabilities are low (Fielding, Chapter 21).

Sources of Error

Errors in classification success may also be affected by imprecision in the plant location data set. Location points have an error range of 0.5 to 2.5 kilometers, depending on the label precision of the original specimens. Present locations may be ruled as absent if they fall within this error range on the projected range map. Including a buffer around each present and absent point could improve prediction accuracy. In the case of *M. pumila*, however, we found no difference between prediction success for present points with a buffer of 2.5 kilometers and unbuffered points.

Some error may derive from mistakes inherent in the environmental data sets. The conversion of small-scale cartographic maps to digital format may introduce spatial errors of up to 1 kilometer (Munn and Arneson 1998). DEMs frequently contain errors that can become amplified when they are used to derive slope, aspect, and terrain position, as we have done (Davis and Goetz 1990; Franklin 1995). Precipitation and temperature data derived by averaging values across each grid cell have a spatial precision no better than one-half the resolution of the cell (Daly et al. 1997). Due to the errors in all of these variables, digital versions of our modeled range maps should not be used at a scale below 2 kilometers.

Other sources of omission error in our models include an inadequate number of location points for *M. pumila*, the lack of digital data sets for important biotic factors (such as distribution of pollinators or seed dispersal vectors), and spatial autocorrelation. Commission errors may result from stochastic events, historical accidents, or incomplete dispersal into otherwise suitable environments (Fielding, Chapter 21).

Management Implications

Although the two models produced range maps encompassing similar regions of the state (Figs. 42.3 and 42.5), the logistic regression model predicted a total area nearly three times larger than the classification tree model. This disparity in size is probably due to the low cut-off level (6 percent) used to assign the probability of presence/absence to each pixel in the logistic regression model. When this map was recalculated based on higher probability cutoffs (greater than 25 percent), the resulting map depicts a smaller predicted range than the classification tree map does (Fig. 42.4). It is interesting to note that both modeling methods predicted the potential occurrence of *M. pumila* in northeastern Wyoming (Weston County), an area where this species has not previously been documented (Figs. 42.1, 42.3, 42.5). Under either mapping scenario, *M. pumila* is probably more widespread in Wyoming than current dot maps would suggest (Fig. 42.1).

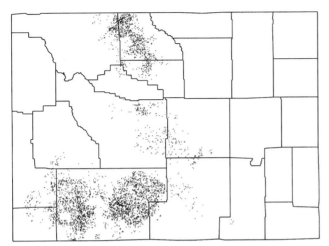

Figure 42.4. Probability of occurrence of *Mentzelia pumila* in Wyoming based on a logistic regression model. Black area = probability of occurrence greater than 50 percent, gray area = probability of occurrence 25–50 percent, white area = probability of occurrence less than 25 percent, or no data.

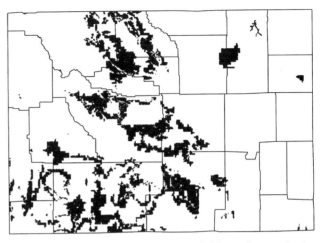

Figure 42.5. Predicted distribution of *Mentzelia pumila* in Wyoming based on a classification tree model. Black area = predicted presence, white area = predicted absence.

Range maps based on correlative models are useful for land managers and conservation biologists faced with the need to identify potential locations for rare or economically important plant species while constrained by limited budgets, time, or manpower. These models can be used to prioritize areas for survey or to identify potential preserves or reintroduction sites (Elith and Burgman, Chapter 24), to perform gap analyses (Fertig et al. 1998), or to determine baseline conditions for climate-change studies (Iverson et al. 1999). More important, the construction of these predictive maps is becoming increasingly easy and affordable as GIS and geostatistical programs improve and large-scale environmental and herbarium point-location data sets become more readily available.

Acknowledgments

We wish to thank Kenneth L. Driese, Philip Polzer, Robert Thurston, and Ellen Axtmann of the University of Wyoming for their help with compiling state and regional environmental data sets and computer assistance. Ronald L. Hartman and B. Ernie Nelson of the Rocky Mountain Herbarium provided point coverages of species locations and collection sites. Kenneth L. Gerow, Gary P. Beauvais, and Sharon Stewart of the University of Wyoming provided advice on statistical methods. Two anonymous reviewers provided many useful suggestions. Funding was provided by the U.S. Geological Survey, Biological Resources Division, National Gap Analysis Program, and NASA, through the University of Wyoming Planetary and Space Science Program.

A Model to Predict the Occurrence of Surviving Butternut Trees in the Southern Blue Ridge Mountains

Frank T. van Manen, Joseph D. Clark, Scott E. Schlarbaum, Kristine Johnson, and Glenn Taylor

Butternut, or white walnut (*Juglans cinerea*), is a highly valued hardwood species native to eastern North America. The tree is closely related to black walnut (*Juglans nigra*) and occurs on cove hardwood sites, but it can also grow on poorer and drier sites. The wood of butternut is used for veneer and lumber, and the mast is eaten by a variety of wildlife species.

Many butternut populations are currently being decimated by an exotic fungus, *Sirococcus clavigignenti-juglandacearum*. This fungus causes multiple branch and stem cankers with a characteristic black color. Main stem cankers eventually girdle the tree resulting in death. The disease was discovered in 1967 in southwestern Wisconsin (Renlund 1971), although it is believed to have first arrived on the eastern coast of the United States at least forty to sixty years ago (Anderson and LaMadeleine 1978). The canker has spread throughout most of the species' range, with greatest impacts on southeastern populations, where an estimated 77 percent of the butternuts are dead (USDA 1995). Because of this rapid decline of butternut populations, the U.S. Fish and Wildlife Service is currently reviewing the status of the species.

Surviving butternut trees in the southern Blue Ridge Mountains are primarily found in close proximity to streams. Because the disease can affect young trees and even seeds, genetic material of infected populations can be permanently lost. No effective strategies exist to protect butternut trees from the canker. However, putatively resistant trees have been located on the Daniel Boone National Forest in Kentucky and the Pisgah National Forest in North Carolina. If resistance is confirmed, a breeding program could develop disease-resistant trees. This strategy could eventually be used to return butternut to the southern Blue Ridge Mountains, but this would be predicated upon the ability to transfer resistance using germplasm with a sufficiently broad genetic base. Therefore, the ability to efficiently locate enough canker-resistant stock for evaluation and subsequent breeding is essential. At present, resistant stock is slowly being located by extensive searches, but that requires substantial time and resources.

Predicting the occurrence of organisms requires knowledge of resource conditions that lead to occupancy by a particular species. Some indicators of habitat where butternut is likely to occur can be discerned based on field experience, but other complex ecological relationships cannot. Moreover, habitat associations consistent with butternut occurrences can be easily overlooked in areas where access is prohibited, when field surveyors have limited experience or training, or when occurrence does not fit the surveyor's established "search image." GIS-based modeling, using multivariate statistics, can be an efficacious approach to predicting species occurrence across relatively large areas (Scott et al. 1993). A reliable model of butternut occurrence would enable managers to identify priority areas to survey for putatively resistant trees. Additionally, a spatial modeling approach may provide insights into habitat conditions and processes contributing to survival of butternut trees exposed to the fungus. Before such a model should be used, however, a thorough evaluation of its reliability is necessary. The objectives of this study were to develop a GIS-based occurrence model using habitat data with locations of surviving butternut trees and to test model predictions with independent field data.

Methods

We used 134 butternut locations from Great Smoky Mountains National Park (hereafter, GSMNP) to develop the model (Fig. 43.1). These locations were collected from 1990 to 1997 and compiled by Natural Heritage programs administered by North Carolina, Tennessee, and GSMNP. These 134 locations represented the entire database of known butternut trees within GSMNP. Butternut locations were identified by National Park Service botanists during surveys and inventories for other species throughout GSMNP, and, as such, should represent a relatively unbiased sample. We used a GIS database of eleven variables (coverages) to characterize habitat conditions of these butternut locations (Table 43.1). These variables were selected based on documented habitat associations of the species, field experience of botanists, and earlier experience developing models for other plant species. All GIS coverages were continuous variables based on 92.9×92.9-meter pixels.

Mahalanobis distance modeling. Mahalanobis distance (D^2) is a multivariate statistic that describes a measure of dissimilarity (Rao 1952). This statistic,

$$D^2 = (\underline{x} - \hat{\underline{u}})' \, \hat{\Sigma}^{-1} \, (\underline{x} - \hat{\underline{u}})$$

was calculated for each pixel in the GIS coverage of the study area by combining the information from the eleven habitat coverages, where $\underline{x}$ is a vector of habitat characteristics for each cell in the GIS grid, $\hat{\underline{u}}$ is the mean vector of habitat characteristics of the original sampling points, and $\hat{\Sigma}^{-1}$ is the inverse of the variance-covariance matrix calculated from the sampling points. This statistic represents the standard squared distance between a set of sample variates, $\underline{x}$, and "ideal" habitat represented by $\hat{\underline{u}}$. Thus, low values indicate conditions similar to those of the original sampling points, with greater D^2 values indicating increasingly dissimilar conditions. Mahalanobis distance is dimensionless because it is a function of standardized variables, despite the different measurement scales among the original variables. There is no one "best" combination of variables that results in the lowest D^2 values; a variety of habitat combinations can result in identical distance values (Clark et al. 1993b).

We calculated $\hat{\underline{u}}$ and $\hat{\Sigma}^{-1}$ with SAS (SAS Institute 1990a) based on the habitat characteristics of the 134 sample locations in GSMNP. Subsequently, we used this information to calculate D^2 in ArcInfo GRID (ESRI, Redlands, Calif.) for a 31,235-square-kilometer area in and around GSMNP; we refer to this area as the southern Blue Ridge Mountains (Fig. 43.1). This area was within the species' range (Rink 1990) and included counties where butternuts were surveyed for resistance during 1997. We used the Southern Appalachian Assessment database (Hermann 1996), a multiagency cooperative effort to assess and map the region's natural resources, to generate a GIS database for this area with the same eleven variables used to characterize the butternut sites upon which the model was based (Table 43.1). The pixel size of the regional GIS coverages also was set to 92.9×92.9 meters and D^2 was calculated for each pixel.

To test the hypothesis that the probability of encountering butternut increases with decreasing values of D^2, we generated 130 random sample points within the southern Blue Ridge Mountains (Fig. 43.1), but only in areas less than 500 meters from improved roads on publicly owned land. We restricted model testing to this smaller portion of the study area because of rugged terrain and the logistical difficulties of acquiring permits to access private property. We assumed the test locations were representative because the distribution of D^2 values of the test locations did not differ from the distribution for the entire southern Blue Ridge region (asymptotic Kolmogorov-Smirnov statistic = 1.12, P = 0.164).

Test locations were visited during spring 1999 and were coded so that field personnel had no knowledge of model predictions for that site. Fourteen of the 130

TABLE 43.1.

Geographic information system variables used to calculate Mahalanobis distance values in the southern Blue Ridge Mountains based on butternut (*Juglans cinerea*) locations in Great Smoky Mountains National Park (GSMNP), 1990–1997.

Variable	Description	Value Range[a]	Source
Aspect	Aspect transformed: 1 + cos(45 – aspect)	0.0–2.0	Calculated from aspect (Hermann 1996) based on Beers et al. (1966)
Elevation	Elevation (m)	308–1,470	U.S. Geological Survey digital elevation model from Hermann (1996)
Proximity to streams	Proximity to nearest stream (m)	0–956	Calculated from streams coverage (Hermann 1996) with the EUCDISTANCE Command (ArcInfo GRID)
Planform curvature[b]	Slope curvature in horizontal plane (divergence and convergence of water flow)	–0.21–0.35	Calculated from elevation with the CURVATURE command (ArcInfo GRID)
Profile curvature[c]	Slope curvature in vertical plane (acceleration and deceleration of water flow)	–0.19–0.31	Calculated from elevation with the CURVATURE command (ArcInfo GRID)
Relative slope position	Relative slope position (%)	0–100	Calculated from elevation based on Wilds (1996)
Slope	Slope steepness (degrees)	1–26	Calculated from elevation with the SLOPE command (ArcInfo GRID)
Solar insolation	Index of exposure to sunlight; approximated for the solar equinox	188–378	Calculated from elevation with the HILLSHADE command (ArcInfo GRID)
Topographic complexity	Shannon-Wiener index of topographic complexity considering elevation, aspect, and slope	18.7–33.8	Calculated based on Miller (1986)
Topographic convergence index	Simulates the flow accumulation of water; TCI = ln(A/tan B), where A is drained surface area and B is drained surface slope	1.0–13.2	Calculated based on Beven and Kirkby (1979), Wolock and McCabe (1995), and Halpin (1995)
Topographic relative moisture index	Index of moisture considering the effects of slope position, aspect, and elevation	13–59	Calculated based on Parker (1982)

[a]Value ranges are based on the 134 butternut locations sampled in GSMNP.
[b]Negative planform curvature indicates an upwardly concave curvature perpendicular to the direction of the slope; positive planform curvature indicates an upwardly convex curvature perpendicular to the direction of the slope.
[c]Negative profile curvature indicates an upwardly concave curvature of the surface in the direction of the slope; positive profile curvature indicates an upwardly convex curvature of the surface in the direction of the slope.

test plots were in GSMNP and were located with a global positioning system (GPS) (PLGR+96, Rockwell International, Cedar Rapids, Iowa) with unassisted military Y-code signal (3.5-meter accuracy). The remaining 116 test plots were located with a Garmin 12 XL GPS receiver (Navtech GPS, Alexandria, Va.). Because we could not achieve real-time differential corrections during location of these 116 test plots, this accuracy was subjected to the U.S. Department of Defense accuracy degradation up to 100 meters. Once each test point was identified, an area of 92.9×92.9 meters centered on this point was sampled for presence or absence of butternut trees. Location coordinates of butternut trees encountered in the vicinity (less than 500 meters) of the test plots also were recorded.

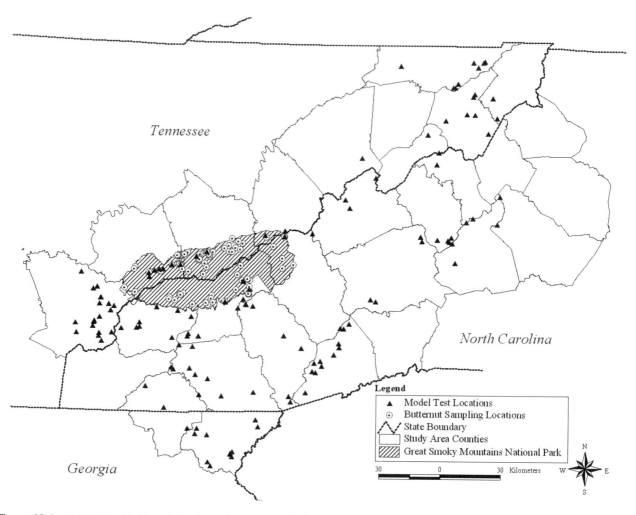

Figure 43.1. Locations of butternut (*Juglans cinerea*) trees in Great Smoky Mountains National Park used for model development (1990–1997) and locations of plots for model testing in the southern Blue Ridge Mountains (1999).

To assess model validity, we determined presence or absence of butternut trees for each test plot and used these observations as the dependent binomial variable in a logistic regression (Hosmer and Lemeshow 1989). We chose logistic regression to determine the relationship between butternut occurrence and the D^2 values of the pixels corresponding to the test plots. We tested all models for the assumption of linearity in the logit by plotting the D^2 values against associated logit values. We tested the fit of the model by use of the Hosmer-Lemeshow goodness-of-fit test (Hosmer and Lemeshow 1989). To determine whether statistical relationships were affected by scaling, we calculated mean D^2 values for areas surrounding the pixels of the test plots. We calculated this mean value based on

square "windows" of 279, 465, 650, 836, and 1,022 meters on a side.

Results

We calculated D^2 values for each pixel in the southern Blue Ridge Mountains based on the model developed from the original 134 butternut locations; these values ranged from 1.0 to 2,648.6, with a mean of 33.1 (sd = 31.7; see Fig. 43.2 in color section). D^2 values corresponding to the 134 model input positions (i.e., known butternut locations in GSMNP) ranged from 2.5 to 110.3 (mean = 11.2, median = 8.4, sd = 11.4) and the regional test points had D^2 values ranging from 3.1 to 366.2 (mean = 19.8, median = 13.2, sd =

TABLE 43.2.

Estimated parameters (coefficients) of a logistic regression model to determine the relationship between Mahalanobis distance values and presence or absence of butternut (*Juglans cinerea*) in 130 test plots, southern Blue Ridge Mountains, 1999.[a]

Variable	Parameter estimate	Standard error	Wald χ^2	$P > \chi^2$ [b]	Standardized estimate
Intercept	−0.037	0.625	0.004	0.952	
Mahalanobis distance	0.162	0.060	7.322	0.007	−3.001

[a]Model statistics: Hosmer-Lemeshow goodness-of-fit statistic = 11.11, 8 df, P = 0.196; maximum rescaled R^2 = 0.193.
[b]P-value indicating the probability of a greater value based on the Wald χ^2 statistic.

33.6). Sixteen of the 130 test plots contained butternut trees. Seventy-five percent of the butternut occurrences were in test plots with D^2 values less than 10.0, and 94 percent were in pixels with values less than 13.6 (Fig. 43.3). Conversely, the proportion of test plots containing butternut declined with increasing values of D^2: six out of nine (0.67) of the plots with values less than 5.0, five out of thirty-eight (0.133) with values = 5.0 and less than 10.0, four out of thirty-seven (0.11) with values = 10.0 and less than 20.0, and one out of forty-six (0.02) with values ≥ 20.0 (Fig. 43.2). The logistic regression analysis of the test data indicated a significant negative association between D^2 and presence of butternut (parameter estimate = −0.162, P = 0.007) and explained 19.3 percent of the variation (Table 43.2). The model fit the data (Hosmer and Lemeshow goodness-of-fit statistic = 11.1, 8 df, P = 0.196) and was linear in the logit. The

TABLE 43.3.

Estimated parameters (coefficients) of a logistic regression model to determine the relationships between Mahalanobis distance values and presence or absence of butternut (*Juglans cinerea*) in areas of different sizes surrounding 130 test plots, southern Blue Ridge Mountains, 1999.

Window size (m)	Parameter estimate	P[a]
93 × 93	−0.162	0.007
279 × 279	−0.082	0.059
465 × 465	−0.043	0.197
650 × 650	−0.016	0.525
836 × 836	−0.002	0.859
1,022 × 1,022	0.001	0.918

[a]P-value indicating the probability of a greater value based on the Wald χ^2 statistic.

relationship between D^2 and butternut occurrence was not significant for mean D^2 values in windows ≥ 279 meters (Table 43.3).

Discussion

Our analysis indicates that the Mahalanobis distance statistic can be an effective measure to delineate potential butternut habitat. For the southern Blue Ridge Mountains, the presence of butternut trees was more likely with decreasing values of D^2. However, this relationship explained only a small portion of the variation (19.3 percent); other factors that we did not incorporate in the model likely contributed to the presence or absence of butternut. The high false-positive rate shows that low D^2 values do not necessarily indicate species presence. However, one would not expect that species always occupy all suitable habitats. Because of dispersion limitations, competition, and other life-history phenomena, its absence at one point in time does not imply that the species was absent in the past or that it will not be present in the future (Andrewartha and Birch 1954; Hanski and Simberloff 1997). In addition, a possible confounding factor is the decline of butternut populations. This decline may result in species absence despite ideal environmental conditions.

When evaluating model performance, spatial errors, such as locational errors associated with the original butternut records, GIS mapping errors, and GPS errors should be considered (Fertig and Reiners, Chapter 42). Although these errors are unlikely to result in biases, they may affect the power to test for model performance. For example, test plots likely did

Cumulative Frequency of Butternut Presence

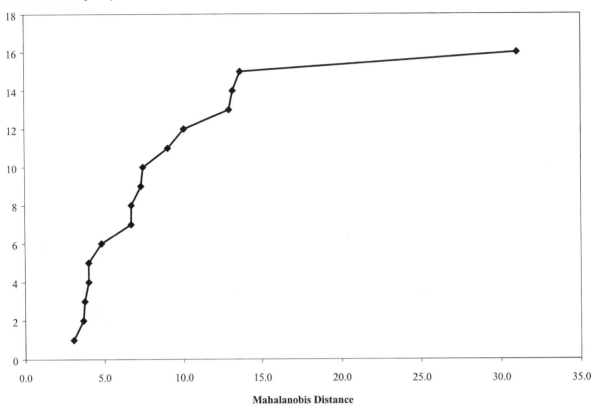

Figure 43.3. Cumulative frequency of butternut (*Juglans cinerea*) presence in test plots and associated Mahalanobis distance values, southern Blue Ridge Mountains, 1999.

not exactly coincide with the targeted pixel but instead were at least partially located within neighboring pixels. Despite the probable presence of such errors, however, we found that D^2 values successfully predicted butternut occurrence.

The data we used to develop the model came from a small spatial subset of the area to which predictions were made. Therefore, one primary concern is whether a model based on such data can be appropriately extrapolated to larger areas. We devised the independent test to specifically address this issue, and we conclude that our model was a good descriptor of butternut habitat in the southern Blue Ridge Mountains. Furthermore, the median value of Mahalanobis distance was similar for the original GSMNP locations (median = 8.4, n = 134) compared with the test locations where butternut was found elsewhere in the region (median = 6.7, n = 13).

Butternut occurrence did not show significant relationships with mean D^2 values calculated for square

areas ≥ 279 meters on a side centered on the targeted test pixel; with increasing area of the squares, the relationship tended toward a random model. We conclude that the model predictions were site specific and little affected by surrounding habitat conditions. A model developed using finer-grained habitat attributes than we used might be more accurate than ours is.

Many modeling efforts have not employed field tests to evaluate model predictions. Although various quasi-validation procedures have been used (e.g., jackknifing or bootstrapping), they do not eliminate possible biases in the collection of the original samples. We submit that the original butternut locations may not represent a completely unbiased sample. However, the field tests we conducted were independent of the observations used to develop the model and allowed us to more accurately quantify model performance, which can then be used to improve the model. These independent test results seem to indicate that the original butternut locations were relatively unbiased.

Six of nine (67 percent) test plots with D^2 values less than 5.0 contained butternut trees. Thus, when visiting areas represented by these pixels, resource managers can expect a relatively high probability of finding butternuts. Because these areas can easily be identified from the digital map (Fig. 43.2) and comprise only a small portion of the study area (1.2 percent in this example), searches can be conducted more efficiently and areas unfamiliar to surveyors can be searched. Although experts could find butternuts based on their knowledge of habitat indicators, their "search images" may differ or be prejudiced, and other habitat areas may be overlooked. Thus, applications of GIS technology coupled with multivariate statistics, as shown here, can be particularly useful for management of species that are in decline and may help guide restoration efforts or assist in the identification of high-quality habitat.

Biometric models of plant-habitat relationships are less common than wildlife habitat models, the latter being an active area of research in recent decades (see this volume and Verner et al. 1986b). In many ways, however, plants may be better subjects for habitat modeling. For example, plants are sessile and are more affected by microclimate, topography, and other typical GIS-based variables than most animals. Habitat conditions associated with an individual plant or tree are relatively static, whereas animals are mobile and habitat characteristics at their location at any one point in time may not be critical to survival. The testing of animal-habitat relationships is particularly challenging because false positive rates will likely be much greater than for plant models. Finally, field testing is more straightforward for plant models because presence is easier to detect.

Acknowledgments

Funding for this study was provided by the U.S. Geological Survey. We thank the many field surveyors from the National Park Service and the University of Tennessee who collected information on butternut occurrences. We extend a special thanks to G. Johnston and D. Griffin for surveying the test locations. We thank D. Buehler, M. Boersen, C. Dees, C. Parker, and J. Young for review of earlier drafts of the manuscript.

Predicting Meadow Communities and Species Occurrences in the Greater Yellowstone Ecosystem

Diane M. Debinski, Mark E. Jakubauskas, Kelly Kindscher, Erika H. Saveraid, and Maria G. Borgognone

For the past eight years, our research team has been developing predictive models of montane meadow communities in the Greater Yellowstone Ecosystem (e.g., Debinski 1996; Debinski and Humphrey 1997; Jakubauskas et al. 1998; Kindscher et al. 1998; Debinski et al. 1999). Our models are based upon landscape-level habitat analysis using geographic information systems (GIS) and remotely sensed data (Urban et al. 1987; Turner 1989; Davis et al. 1990; Hobbs 1990; Stoms and Estes 1993; Edwards et al. 1996) for small areas (0.25-hectare mapping unit), combined with field-sampled data of the biotic communities. We have focused on predicting species and communities along a gradient of montane meadow communities (spectrally distinct regions labeled M1-M7) ranging from extremely hydric wetlands (M1 meadows) to mesic, forb-rich meadows (M3 meadows) to xeric, sagebrush flats (M6 meadows), and to sites where bare ground is the predominant cover (M7).

The flora and fauna of these meadows were studied within two regions of the Greater Yellowstone Ecosystem: Grand Teton National Park and the Bridger-Teton National Forest of northwestern Wyoming (hereafter termed "Tetons") and the Gallatin National Forest and Yellowstone National Park in Montana (hereafter termed "Gallatins"). The taxonomic groups we investigated include birds, butterflies, and plants. Our assumptions were that the landscape was relatively pristine, and thus community structure should have been more predictable than in a recently altered habitat (e.g., Hanski et al. 1996b), and that species-habitat relationships could be predicted from abiotic factors such as a moisture gradient. Further, less than 1 percent of the species examined had range edges within our sampling area. Thus, if the habitat was present, the species could have potentially been present at that site. We limited our research to lower-elevation montane meadows (2,000–2,500 meters) to reduce the number of variables affecting the predictability of species distributions.

Our objectives were (1) to use remotely sensed imagery to map different montane meadow communities and develop spectrally based spatially explicit models for predicting plant and animal species distribution patterns in montane meadows, and (2) to test the predictive models in two areas of the ecosystem. We investigated the potential of remote sensing to map and predict montane meadows in general, and we examined specific subsets of these meadow types, such as montane wetland (Kindscher et al. 1998) and sagebrush (Jakubauskas et al. 1998) vegetation communities. We also examined the associations of specific animal species with our meadow types. In this chapter, we present a synthesis of these approaches.

Focal Taxa

Plants can be viewed both as a component of species diversity and as a component of habitat diversity. The presence of a particular plant species in a specific site can be highly indicative of the microhabitat of that site. Because vegetation is a major contributor to spectral reflectance patterns measured by satellite sensors, we believe that it is imperative that we test the relationship between remotely sensed meadow types and plant communities. If plant species distribution patterns can be predicted using remotely sensed data, relationships between remotely sensed data and animal taxa would be more probable. Thus, a plant survey is the critical link between remotely sensed data, habitat, and other species distribution patterns.

Butterflies are suitable species for testing the hypothesis that remotely sensed data can be used to predict species distributions because they are moderately to highly host-specific herbivores, and their diversity may be correlated with underlying plant diversity. Butterflies are also well known taxonomically and can be reliably identified in the field (Kremen 1992). Over one hundred species reside in the Greater Yellowstone Ecosystem (Bowser 1988; Brussard 1989). Birds are suitable because they are ecologically diverse and use a wide variety of food and other resources and reflect the condition of many aspects of the ecosystem. Birds often respond to spatial and temporal variation in a species-specific fashion (Wiens and Rotenberry 1981b; Steele et al. 1984; Taper et al. 1995). Birds are also conspicuous, ubiquitous, intensively studied, and often appear to be more sensitive to environmental changes than other vertebrates (Morrison 1986). Over two hundred different bird species reside in the Yellowstone National Park (McEneaney 1988), and most of these species also can be found in Grand Teton National Park. Our research focused primarily on songbirds and woodpeckers.

Field Sampling and Data Analysis

As a way of explaining our multiple projects, we've summarized in Table 44.1 meadow type, sample size, and region examined for each analysis. We began our study by defining meadow spectral classification, moved to specific subsets of montane meadows (sagebrush communities and wetland assessment), and then addressed biodiversity data. In our research, we tested for associations of specific animal species with our meadow types and finally analyzed the community similarity between the two regions of the ecosystem.

Meadow Classification

Computer analysis of multispectral satellite imagery was used to create maps of spectrally distinct montane meadows within the study regions. The maps were used to identify different meadow types and to guide plant and animal field sampling. Indian IRS LISS-II imagery from summer 1995 was used to produce maps for the 1996 sampling (Kindscher et al. 1998; Jakubauskas et al. 1998). Image data from the French SPOT (Système Pour l'Observation de la Terre) satellite for June and September 1996 were used to produce maps for the 1997 and 1998 fieldwork. Both the IRS and the SPOT systems are multispectral scanners that acquire data in green, red, and near-infrared bands of the electromagnetic spectrum, with spatial

TABLE 44.1.

Analysis type, regions included, meadow types, sample sizes, taxa, and years examined for each analysis described in this paper. M1–M7 meadows represent spectrally distinct montane meadows (M1= hydric, M6 and M7 = xeric) within Grand Teton National Park and the Bridger-Teton National Forest of northwestern Wyoming (Tetons), and the Gallatin National Forest and Yellowstone National Park (Gallatins).

Analysis	Region	Mtype	No. sites	Reps/ year
1996				
Sagebrush communities	Tetons	M6	51	1
Wetland assessment	Tetons	M1–M3; M5–M7	30	1
1997–1998				
Bird species	Gallatins	M1–M6	30	3
	Tetons	M1–M3; M5–M6	25	3
Butterfly species	Gallatins	M1–M6	30	4
	Tetons	M1–M3; M5–M6	25	4
Plant species	Gallatins	M1–M6	30	1
	Tetons	M1–M3; M5–M6	25	1

resolutions of 36.5 meters (IRS) and 20 meters (SPOT). Each image was georeferenced to a Universal Transverse Mercator (UTM) coordinate system. An unsupervised classification procedure was used to produce a map of spectrally distinct meadow types representing a hydric (M1) to xeric (M6) gradient, (and M7 for bare ground and talus). In order to reduce the number of variables affecting species distribution, we focused on low elevation sites (2,000–2,500 meters) with minimal slope (less than 30 degrees). To facilitate location of study sites during fieldwork, maps were plotted on translucent Mylar sheets, allowing overlay onto 1:24,000-scale USGS topographic maps of the study region.

Sagebrush Community Classification

In order to explore the limits of using multispectral, multi-temporal satellite imagery to identify and map specific community types, we carried out a second mapping procedure focused specifically on the sagebrush flats of Grand Teton National Park, Wyoming (meadow type M6). Fifty-one field plots were sampled in 1996, equally distributed among the four sagebrush communities (low sagebrush [*Artemesia arbuscula*], mountain big sagebrush [*Artemesia tridentata* ssp. *vaseyana*], mixed low sagebrush/big sagebrush, and bitterbrush [*Purshia tridentata*]/big sagebrush) as defined by Sabinske and Knight (1978). Shrub cover and height by species were surveyed as per Knight (1978); percent cover by forbs, grasses, shrubs, live stems, and mosses/lichens were recorded using the Daubenmire technique (Daubenmire 1959) within twenty 0.5×0.5-meter quadrats. An agglomerative hierarchical iterative clustering algorithm was used to assign each of the fifty-one sites to one of the four Sabinske and Knight (1978) community types (Jakubauskas et al. 1998). Variables used in the cluster analysis were height and percent shrub cover by big sagebrush, low sagebrush, and bitterbrush; and percent cover by five general understory classes (grasses, forbs, litter, persistent litter, and rock/soil cover). Multi-temporal satellite imagery was combined into a single image file, and groups of spectrally similar pixels were identified using an unsupervised classification clustering procedure. Data from the fifty-one field sites sampled in 1996 were used to identify the probable vegetation community represented by each cluster of spectrally similar pixels, producing a map of the four sagebrush communities described above. Accuracy assessment was carried out by comparing our 1996 classifications to points classified by Sabinske (1972) within the four sagebrush classes.

Wetland Assessment

Analysis of wetland classification was conducted across all meadow types in the Tetons (M1–M3, M5–M7) to determine whether our M1 and M2 meadows showed up distinctly as "wetlands" based upon an index of average wetland value. This index is a metric used by researchers studying wetland vegetation dynamics (Atkinson et al. 1993; Wilson and Mitsch 1996) and is based on the classification of all wetland plant species into one of five wetland classes (Reed 1988). Plant species are assigned a score from 1 to 5, depending on their wetland affinity. The wetland affinity value for each species is multiplied by the percent cover of each species at a sample site, yielding a numeric score (1.00–5.00), where values closer to 1.00 have more cover by obligate wetland vegetation and those closer to 5.00 have more cover of upland plant species. As an indicator of wetlands, any value under 3.00 indicates that the area has a wetland identity. We surveyed thirty sampling sites in Grand Teton National Park, five of each of six meadow types. Vegetation was sampled in a 20×20-meter plot for canopy cover (Daubenmire 1959) of all species during 1996 and an index of wetland value was computed for each sampling site (Kindscher et al. 1998). Data were analyzed using a nonparametric Kruskal-Wallis test in the SPSS/PC+ software package (SPSS 1988).

Biodiversity Assessment

A biodiversity assessment across all montane meadow types (M1–M6) was conducted for birds, butterflies, and plants in two regions of the ecosystem: the Tetons and the Gallatins. Our intention was to build models of species-habitat relations in one region of the ecosystem (e.g., Tetons) and then test them in the other region of the ecosystem (e.g., Gallatins). Field data were collected in 1997 and 1998 at five sites in each of the meadow classes (twenty-five sites in the Tetons, thirty sites in the Gallatins). All sites were a minimum of

100×100 meters in size, a distance of 500 meters or farther from other sites, and within 8 kilometers of a road or trail. Vegetation data were collected as described in the wetland section, but in a 100×100-meter plot. Abundance data were collected for butterflies building upon previously developed methods by Debinski and Brussard (1992). Taxonomy followed Scott (1986). Butterflies were surveyed between 1000 and 1630 hours on sunny days by netting and releasing for twenty minutes in a randomly selected 50×50-meter plot within the 100×100-meter sampling site. Surveys were repeated at each sampling site four times in each of the two years. Abundance data were collected for birds using 50-meter-radius point-count surveys. Surveys were performed three times at each site per season during 0530–1030 hours and between 1 June and 17 July. Each survey involved two observers for fifteen minutes. Landscape data, including percent willow cover, percent sagebrush cover, meadow size, distance to meadow of the same type, distance to treeline, tree density, vegetation biomass, and leaf area index, were also collected for examining relationships between birds and habitat (Saveraid 1999).

The species-habitat relationships were analyzed using a variety of multivariate statistical techniques. In order to assess the level of species detection, we first constructed a species accumulation curve for each taxon in each landscape by year using PC ORD (McCune and Mefford 1997) and estimated total species richness for each landscape using a first-order jackknife estimator (Heltshe and Forrester 1983; Palmer 1992). An important characteristic of this type of community data is that the number of observations (sampled sites) is small with respect to the number of variables (plant, bird, or butterfly species). A considerable number of species were represented by less than ten individuals and were dropped from the data sets. This approach restricted the analysis to those species with abundance measures large enough for rigorous statistical analysis. Multiple regression models were developed for eleven bird species using both meadow type and landscape variables (Saveraid 1999). Classification and regression tree analysis (CART; Steinberg and Colla 1997) was used to examine butterfly species relationships with meadow types. The latter analysis does not allow one to predict species presence based upon meadow type; in effect, our data (species distributions and abundances) constrain us to predict meadow type (a discrete category) from species abundance data. However, the results do give information regarding which species show affinities for specific meadow types. For example, the abundance of species X may differentiate the M5 meadows from the M6 meadows, and CART gives an importance score to species X and abundance value that allows for that split. In all of these analyses, each year, region, and taxa were tested separately.

Similarity between Communities

Differences between communities in the two regions of the ecosystem were examined using a Bray-Curtis measure of community distance (Bray and Curtis 1957). Bray-Curtis measures were calculated within each meadow type (excluding M4) for each taxon during 1997 and 1998 using each temporal replicate as a subsample and pooling all sites from the Gallatins to be compared with all sites from the Tetons. Tests for significant differences in community distance were conducted using an MDS on Bray-Curtis Distance (SAS Institute 1990b).

Results

Our results are divided into three categories: (1) classification of meadows into wetland or sagebrush community types, (2) tests for association of specific animal species with our meadow types, and (3) analysis of similarity between the two regions of the ecosystem. We should preface these comments by noting that we felt relatively confident that our analyses were based on a large proportion of the total species present in the area. The number of bird and butterfly species observed relative to the estimated total species richness (first-order jackknife estimator) in 1997 and 1998 showed that our sampling efficiency averaged just over 80 percent for butterflies and plants and 74 percent for birds (Table 44.2). The birds in the Gallatin landscape showed the lowest percentage of observed species relative to the total predicted (69 percent) but nonetheless represented the majority of the species in the landscape.

TABLE 44.2.

Species accumulation results for each taxon in each landscape by year. Total species richness was estimated for each landscape using a first-order jackknife estimator. Species accumulation curves were plotted across all meadow types using each sampling date as a separate replicate. For each year, there were three temporal replicates for birds, four temporal replicates for butterflies, and one temporal replicate for plants.

		No. spp. observed	First-order jackknife estimate	% Species observed relative to number predicted
Gallatins				
Bird	1997	33	47.8	69
Bird	1998	26	37.8	69
Butterfly	1997	45	55.9	81
Butterfly	1998	50	61.9	81
Plant	1997	253	307.1	82
Plant	1998	236	303.3	78
Tetons				
Bird	1997	40	48.9	82
Bird	1998	39	51.9	75
Butterfly	1997	59	70.9	83
Butterfly	1998	59	74.8	79
Plant	1997	257	320.8	80
Plant	1998	247	307.5	80

Sagebrush Community Classification

Overall accuracy of our sagebrush community classification was 65 percent (thirty-three of fifty-one sites) and highest for the mixed big sagebrush/low sagebrush community at 86 percent (Jakubauskas et al. 1998). Results indicate that the four sagebrush communities can be best differentiated using visible bands of reflected light recorded by a satellite sensor, independent of image date. Near-infrared reflectance was of value for distinguishing between sites only in late fall (October). Analysis of the accuracy assessment indicates that the pure big sagebrush and low sagebrush communities are consistently misclassified with the mixed big/low sagebrush community. Misclassification of the bitterbrush and big sagebrush communities as forb-dominated areas occurred primarily as a result of a fire east of Blacktail Butte in Grand Teton National

Park in 1994. Both the bitterbrush and the big sagebrush communities were burned during this fire, which regenerated to forb and grass cover by the date of the satellite overpass. Accuracy for low sagebrush may be higher than warranted, as areas of sparse cover within mixed and big sagebrush communities were mapped as low sagebrush. In the eastern part of the study area, where this community type does not occur, areas mapped as low sagebrush most likely represent areas of less-dense big sagebrush rather than low sagebrush.

Wetland Assessment

We were able to accurately classify 70 percent (seven out of ten) of the wetlands we studied into two wetland groups (M1 and M2) and to predict the location of 1,258 hectares of M1 wetland meadows and 1,711 hectares of M2 wetland meadows within Grand Teton National Park (Kindscher et al. 1998). One hundred and eighty-three plant species were found in the meadows surveyed, including ten obligate wetland species. Eight out of ten obligate wetland plant species had their greatest cover in M1 meadows and had significant cover differences among meadow types using the nonparametric Kruskal-Wallis test. Percent cover data showed that M1 and M2 meadows were dominated by obligate wetland vegetation, whereas other meadow types had less than 0.1 percent obligate wetland vegetation. Facultative wetland plant coverage showed similar trends.

Biodiversity Assessment

Eleven of the thirty-seven bird species observed were used to test predictability of species occurrence in both 1997 and 1998. We compared the accuracy of predicting bird occurrence using both spectrally defined meadow types and landscape data (Saveraid 1999). The selection criteria for the birds used in the analysis was a total abundance greater than or equal to fourteen in each year of the study. Meadow type, as determined from the satellite data, was significantly correlated with the abundance of six of the eleven bird species (R^2 range of 0.275 to 0.49). The abundance of generalist species (American robin [*Turdus migratorius*], dark-eyed junco [*Junco hyemalis*], white-crowned sparrow [*Zonotrichia leucophrys*], Brewer's blackbird [*Euphagus cyanocephalus*], and chipping sparrow [*Spizella passerina*]) was not strongly correlated with meadow type. We then tested the use of a combination of meadow type and landscape variables (e.g., meadow area, percent graminoid cover, percent willow cover, percent sagebrush cover, distance to treeline; see Saveraid 1999 for details) to increase the predictability of the models. Significant landscape variables were selected using a stepwise multiple regression for each bird species. These variables were then used in a multiple regression analysis with the classification variable meadow type. Ten out of the eleven bird species showed a significant correlation with one or more variables when both landscape variables and meadow type were used in the models (R^2 range of 0.189 to 0.811). Abundances of the species commonly associated with hydric meadows (common snipe [*Gallinago gallinago*], common yellowthroat [*Geothlypis trichas*], Lincoln's sparrow [*Melospiza lincolnii*], Savannah sparrow *[Passerculus sandwichensis]*, and yellow warbler [*Dendroica petechia*]) were significantly correlated with meadow type and landscape variables such as percent willow cover and percent woody vegetation. There were fewer species in the xeric meadows, but the most commonly observed species, the vesper sparrow (*Pooecetes gramineus*), was highly correlated with meadow type and percent sagebrush cover.

Butterfly distributions were much more strongly associated with specific meadow types in the Teton landscape relative to the Gallatin landscape. Using species abundance data for each meadow type, classification tree analysis was used to identify species that were most important in distinguishing among meadow types. Of the sixty-seven butterfly species found during our surveys in the Tetons, twenty species in 1997 and twenty-seven species in 1998 were used in the statistical analysis. Fourteen species were defined as important in distinguishing among meadow types, and these species could be used to classify sampling sites into one of five different meadow types with 96 percent accuracy in 1998 and a 92 percent accuracy in 1997. Six species showed high importance scores for both years in the Tetons. Each species is listed followed by the meadow type for which it showed a high affinity: *Coenonympha orchracea* (M6), *Lycaena het-*

eronea (M6), *Coenonympha haydenii* (M5), *Cercyonis oetus* (M3), *Speyeria mormonia* (M3), and *Boloria frigga* (M2). Predictability of species was much lower (63 percent) in the Gallatins, where patch size is much smaller, but many of the same species showed high importance scores in differentiating the meadows (Borgognone 1999).

Similarity between Communities

Community differences between the Gallatin and Teton meadows were significantly larger than expected via random variation in the majority of the comparisons examined (Table 44.3). Two of the five meadow types examined showed significant differences for birds and plants and four of the five meadow types showed significant differences for butterflies. We report results here from 1998; similar results were observed in 1997. Given these differences, we did not pursue testing models developed in one region of the ecosystem on the species observed in the alternate region.

Discussion

Each of these sub-projects has taken a similar approach to testing the predictability of communities and species patterns in the Greater Yellowstone Ecosystem. In each case, remotely sensed data were analyzed to create meadow classification maps that guided field surveys. Biotic communities were tested for their association with these remotely sensed meadow types. In general, predicting meadow type (and the major plant species associated with that meadow type) is somewhat easier than predicting a long list of specific species occurrences. This is not surprising given that meadow type is a more general classification and comprises both plant communities and abiotic factors that play a role in determining what reflectance patterns are measured by a satellite.

When we create predictive models of species occurrence, one of the major limitations is that rare species are often such a small component of the data set that they cannot be used to build predictive models (Hepinstall et al., Chapter 53; Karl et al., Chapter 51). An inherent statistical problem exists when the number of variables (in this case species) describing a site is more numerous than the number of sampling sites. In both

TABLE 44.3.

Bray Curtis distances for birds, butterflies, and plants in Gallatin versus Teton meadows, 1998. Five spatial replicates per meadow type were compared for differences in species composition. Distance value is noted above and the P value is noted below. Significant values (at P < .05) are noted in boldface type and indicate that between group distance is greater than within group distance.

	Taxon	M1	M2	M3	M5	M6
Birds	Distance	**0.636**	0.72	**0.782**	0.879	0.657
	P	0.01	0.09	0.04	0.46	0.51
Butterflies	Distance	**0.715**	**0.569**	**0.489**	0.488	**0.463**
	P	0.01	0.01	0.01	0.14	0.01
Plants	Distance	0.734	0.647	0.682	**0.602**	**0.527**
	P	0.31	0.07	0.18	0.01	0.01

the bird and the butterfly community analyses, we ended up analyzing only a small fraction of the total number of species for their associations with specific meadow types. The species we used in our models, however, gave us relatively strong results. The species that were used in these models may not include all of the rare species, but they are important indicators of each of the meadow types examined in this research. A future step in this research arena might be to test how often rare species are associated with more common "indicator" species. Interestingly, our ability to build models in one part of the ecosystem and test them in another part of the ecosystem was not successful. Even a distance of less than 200 kilometers showed significant differences between the Teton and Gallatin meadows with respect to the majority of the bird, butterfly, and plant communities.

In building models of community composition, it is important to examine the effects of species ranges, niche widths, seasonality, sampling intensity, landscape history, patch size, and landscape context. These variables may explain why we find a species in what could be considered unsuitable habitat, or alternatively why we might not find a species in what would be considered suitable habitat. Only one species (*Coenonympha haydenii*, the Hayden's ringlet butterfly) of the three taxa we surveyed would be considered endemic to the region. There were no range edges that should have significantly affected species distribution patterns within the ecosystem. Predictability may be affected by niche width (Hepinstall et al., Chapter 53),

and thus it may not be surprising that generalist birds showed different patterns than specialists in our study. We should note that many of our species predictions are seasonally limited. Because some of the bird species are migratory and many of the plant and butterfly species are not visible during the winter, these predictions are intended for the summer season only.

Our data concur with Karl et al. (Chapter 51) that intense sampling is necessary to develop predictive models of species-habitat relationships. Historical factors (e.g., fire in the sagebrush communities) may also affect patterns of species distribution. Two variables affecting predictability that we are just beginning to examine include patch size and landscape context. If habitat is present but present in small patches relative to the home range of an organism, the predictability of species-habitat relationships may be less reliable. Similarly, landscape context may complicate the predictability of species occurrence models, and community spillover (Holt 1997) from one patch to another may confound species-habitat relationships. We suspect that patch-size effects and spillover may have contributed to the lower predictability of butterfly species distributions in the Gallatin landscape. Several of the meadow types in the Gallatins are much smaller on average than their counterparts in the Tetons, even by a magnitude in size (Debinski unpublished data).

The methods described here have management applications from a short-term as well as a long-term planning perspective. In the short term, agencies such as

The Nature Conservancy are using these types of techniques in their ecoregional planning (The Nature Conservancy 2000). Satellite data may increase our efficiency in locating areas of high biodiversity or with high probabilities of supporting rare species. From a longer-term planning perspective, one of the more interesting applications of these models will be that of monitoring communities for effects of global climate change. Detection and characterization of changes in vegetation extent and condition at multiple spatial and temporal scales provides important information about the variability within a habitat patch. Recent advances in remote sensing technology and theory have expanded opportunities to characterize the seasonal and interannual dynamics of vegetation communities. Analysis of changing spectral patterns can provide precursor measurements of terrestrial ecosystem dynamics (Waring et al. 1986; Ustin et al. 1993; Lancaster et al. 1996). The temporal domain of multispectral data frequently provides more information about land cover and condition than do the spatial, spectral, or radiometric domains (Kremer and Running 1993; Eastman and Fulk 1993; Samson 1993; Reed et al. 1994; Jensen et al. 1993). Quantification of this inherent landscape variability at a regional scale using remotely sensed data, in concert with predictive models of community type and species distributions, will provide the foundation for modeling the effects of environmental change and consequent effects upon plant and animal biodiversity.

Our long-term goal is to develop a suite of remotely sensed and field-based indices of community identity and species composition that will serve as indicators of environmental change. Montane communities are likely to be some of the most susceptible to climate change (Peters and Lovejoy 1992). Changes in hydrology and plant species composition, themselves driven by seasonal and interannual changes in climate, determine cover and vigor of individual plant species and their availability and use by animals. Naturally occurring plant communities occupy specific geographic sites based on narrowly defined adaptations to gradients of temperature and moisture. Short-term changes in environmental conditions are manifested as changes in vegetation condition, while long-term, directional shifts in temperature and moisture regimes drive changes in species composition and diversity (e.g., Harte and Shaw 1995). Although the experts are still debating whether montane ecosystems will become more wet or more dry, there is virtually no disagreement that the earth's surface will become warmer (Schneider 1993). Because the meadows we have surveyed are arranged along a moisture gradient, we expect to be able to detect changes in the landscape and associated communities at either end of the spectrum.

Conclusions

The results from our study indicate that satellite imagery is applicable for mapping wetland and sagebrush communities and for predicting dominant vegetation in each of these communities with a high level of accuracy. We were also able to show significant correlation between specific meadow types and the distribution patterns of a subset of the bird and butterfly species in montane meadow types. Butterflies showed stronger relationships with meadow type than birds did, especially in the Teton landscape. Because birds respond to habitat structure, adding landscape variables such as percent cover of the dominant woody vegetation increased the predictability of our models.

Acknowledgments

Funding for this research was provided by the U.S. Environmental Protection Agency, National Center for Environmental Research and Quality Assurance (NCERQA), STAR Grant R825155. Although the research described in this chapter has been funded in part by the EPA, it has not been subjected to the Agency's peer review and therefore does not necessarily reflect the views of the Agency, and no official endorsement should be inferred. This is Journal Paper No. J-18518 of the Iowa Agriculture and Home Economics Experiment Station, Ames, Iowa, Project 3377, and supported by Hatch Act and State of Iowa funds. Field assistance was provided by E. Saveraid, T. Saveraid, C. Blodgett, A. Fraser, J. Auckland, D. Friederick, T. Aschenbach, S. Ashworth, H. Loring, J. Hanlon, J. Pritchard, and others. We thank J. Su and J. M. Scott for comments on previous versions of the manuscript.

Modeling Species Richness and Habitat Suitability for Taxa of Conservation Interest

Erica Fleishman, Dennis D. Murphy, and Per Sjögren-Gulve

Models that reliably quantify relationships between environmental variables, species occurrences, and population dynamics can enhance our capacity for land-use planning substantially. Attempting to develop such models is critical, for managers simultaneously face daunting obligations and stringent resource constraints (Stohlgren et al. 1995; Oliver and Beattie 1996; Longino and Colwell 1997; Niemi et al. 1997; Simberloff 1998; Maurer, Chapter 9). Analytic and predictive models of species richness, occurrence, and viability may assist practitioners in meeting numerous objectives, including conservation of relatively rich or unusual species assemblages and protection or eradication of individual species (or other taxonomic levels). If a subset of the significant predictive variables can be modified by human activities, then the models also may be used to forecast the biological effects of alternative management strategies.

Species of concern often include native plants and animals that are threatened by anthropogenic or natural factors. Introduced or invasive species represent a second focal group. Yet other species become the target of planning efforts because their measurement is believed to provide a scientifically reliable and cost-effective surrogate measure of a distinct ecological parameter that is difficult to assess directly. Indicator species, for example, exhibit distributions, abundances, or population dynamics that can serve as sub-stitute measures of the status of other species or environmental attributes (Noss 1990b). Potential umbrella species, taxa whose conservation confers a protective umbrella to numerous cooccurring species, and keystone species, taxa with a disproportionate impact on the dynamics of their ecosystems, also have attracted considerable attention from biologists and practitioners (Simberloff 1998). Because few organisms have proved to be dependable and affordable gauges of variables that characterize communities or ecosystems (Scott 1998), we urge ecologists to test the efficacy of potential surrogate species rigorously before employing them for on-the-ground management.

Because the status of and threats to a given species may depend upon the spatial extent at which those phenomena are considered, models ideally should be able to address factors that influence species distributions and viability at multiple spatial extents. The balance between benign neglect and substantive human intervention that best will protect a certain assemblage or species likewise may fluctuate in space and time (New et al. 1995; Thomas and Hanski 1997).

We recently introduced a modeling approach that can be utilized at different grains and extents to examine species richness in relation to environmental variables and to recognize independent variables that influence the occurrence and population dynamics of selected species (Fleishman et al. 2000). The framework initially was developed to identify species that might be used as indicators of high species richness. In this chapter, we apply the approach in the context of regional land-use planning and management. As presented here, our model has three steps. First, we use multiple linear regression to predict species richness as a function of easily quantified environmental variables (e.g., derived from geographic information systems [GIS]). The validity of the species richness model is evaluated by calculating the percentage of variance in species richness that it explains and the probability that it correctly classifies locations with respect to chosen levels of species richness. These tests are performed not only upon the data used to build the model but also upon independent sets of data. Second, we employ logistic regression to identify easily quantified environmental variables that help explain occupancy patterns of individual species over relatively large areas. Third, we use data on species occurrences and local resources to develop a spatially realistic population model that generates testable probabilities of persistence (see Hanski and Simberloff 1997). Our modeling approach is a general procedure that can be used for diverse taxonomic groups and ecoregions. Here we present an overview of the approach and illustrate its application with case studies of butterflies in the Great Basin.

Modeling Framework

Land-use planning benefits from a general understanding of which locations have relatively high or low concentrations of native species and where individual species are likely to occur. Clearly, comprehensive field inventories address these issues directly, and empirical data are essential for developing effective models and management strategies. Nonetheless, tried-and-proven predictive models have tremendous value. For example, models sometimes can be used to assign tentative species richness or occurrence values to planning units based on efficiently derived biophysical variables (Angermeier and Winston 1999; Mac Nally et al. 2002; Fleishman et al. 2001). These projections may help prioritize groundtruthing and more-detailed field studies. Knowledge of circumstances in which predictive ability tends to be low can be helpful too—empirical data collection may be most critical in those situations. Moreover, even when ample inventory data exist, predictive models may help practitioners weigh the potential benefits and risks of alternative management approaches.

Environmental variables that may affect species distributions across a landscape include topography, climate, frequency and magnitude of disturbance events, and patterns of human land use. Over large areas, multiple populations of some species, especially those regulated largely by density-independent factors, tend to fluctuate in synchrony (Pollard 1991). However, long-term trends in occupancy or abundance at a particular location sometimes deviate from regional trends. Models may help elucidate the extent to which divergent population dynamics are functions of local environmental characteristics or human influences. In a similar vein, habitat-based models may clarify whether correlates of occupancy or viability differ across a species' range (Thomas et al. 1998; Maurer, Chapter 9).

Predicting Species Richness

The first step of our modeling framework uses multiple linear regression to predict species richness across a landscape as a function of easily derived environmental variables. We assume that relatively complete lists of resident species have been compiled, using standard methods (e.g., Pollard and Yates 1993; Heyer et al. 1994; Wilson et al. 1996), for a representative sample of locations to be managed in the future (i.e., spanning the range of major environmental gradients and/or vegetational communities). Is this assumption realistic? We acknowledge that inventory data for many landscapes are sparse, but we argue that these data are prerequisite for developing productive models.

Predictive variables entered into the species richness model should be tractable to measure, resistant to observer bias, and should be expected to reasonably influence species distributions in the focal assemblage. Many relevant variables, such as elevation or aspect, can be derived from existing electronic sources of data such as digital elevation models (DEMs) and digital line graphs (DLGs). Various DEM and DLG coverages of the United States are currently available for no charge from the U.S. Geological Survey's EROS Data Center, http://edc.usgs.gov/doc/edchome/ndcdb/ndcdb.html. Other variables, including climate parameters like temperature or precipitation, are relatively simple to quantify at numerous points in space and time via field measurements, remote data loggers, or realistic models (e.g., the PRISM family of orographic climate models, Daly et al. 1994, 1997). If possible, a subset of the independent variables should be responsive to management.

To qualify as valid and practical, a predictive model of species richness should explain a significant percentage of variation in the number of species. The model also should correctly classify locations with respect to species richness. In other words, the model should accurately identify locations with a species richness greater than a certain value and should not erroneously predict that other locations meet that criterion. Because using a model that is not demonstrably successful offers no advantages for land-use planning, its validity should be assessed not only with the data used to build the model, but also with independent sets of data (Fielding, Chapter 21). If managers are most interested in predicting how species richness in a single system will respond to human activities or ecological change, then existing data may be divided into two separate sets for model building and model validation. Alternatively, data from later time steps may be used for validation. If predicting species richness patterns in ecologically similar but poorly inventoried systems is a higher priority, then data from new locations can be used to test the model's accuracy.

Predicting Species Occurrence

Species richness is one consideration in delineating land uses. In many cases, the distributional patterns of certain species also help managers evaluate locations for activities ranging from resource extraction to treatment with herbicides or prescribed fire (Van Horne, Chapter 4). Predictive models of species occurrence, like predictive models of species richness, not only are valuable tools in the absence of complete inventory data, but also may clarify whether environmental changes will affect the probability that a target species will inhabit a specified location.

Therefore, the second step of our modeling approach tests whether easily quantified environmental variables explain significant variation in the occurrence patterns of individual species. In this phase, we use logistic regression to analyze simultaneously correlations between each of a suite of predictive variables and the distribution (presence/absence) of a species of interest. Models that are based upon correlations do not necessarily identify causes of ecological patterns. Nonetheless, they are useful for planning purposes because they link landscape variables with species distributions and do not require overwhelming quantities of field data. Furthermore, it often is possible to draw strong biological inferences about why certain variables have significant analytic or predictive ability.

Predicting Persistence

Especially in a dynamic system, managers may want to know not only whether a species is likely to be present in a given location, but also what ecological conditions might promote its persistence. Identifying variables that are associated with turnover (colonization and extinction) is especially helpful if management options include restoration of habitat and/or reintroduction. It may be possible to test whether turnover events are correlated with changes in environmental parameters that are affected by human activities and whether management can minimize variation in some of those factors. Detailed population modeling is usually data-intensive and therefore restricted to species of considerable management concern. Fortunately, the logistic investments necessary to obtain spatially explicit data on dependent and independent variables (often key resources such as nesting sites or food base) can yield high returns. If validation tests demonstrate that models parameterized with empirical data accurately predict turnover events, then it also may be possible to generate realistic forecasts using projections of environmental change.

Habitat-based occupancy and viability models are applicable to species with diverse population structures. Knowledge of the population structure of the target species will help guide selection of predictive variables. In species with open populations, for example, individuals usually are relatively mobile. Therefore, assessment of relatively coarse-grained environmental attributes is likely most relevant (New et al. 1995). In species that have closed populations, by contrast, local habitat features or resources (e.g., vegetation composition and structure, water chemistry) may have a more substantial bearing on demographic parameters such as birth and death rates.

Some other species persist as metapopulations, sets of local populations that to some degree are linked by dispersal (Hanski and Gilpin 1997; Hanski 1999). Although each patch of habitat in a metapopulation can support a local breeding population, no single population is adequate to ensure the long-term viability of the entire metapopulation (Hanski et al. 1995). Measurement of a suite of environmental variables that collectively characterize a range of spatial grains may be necessary to accurately model habitat requirements for metapopulations. Local habitat features tend to affect demographic parameters within each patch; larger-area attributes such as isolation or climate often affect regional demographic trends and population dynamics (Hanski and Gilpin 1997). Conservation of a metapopulation demands that multiple patches of habitat be maintained, but assessing the importance of each patch is complicated by the fact that not all suitable patches are occupied at any given time. Thus, metapopulation models serve at least two purposes. First, they can distinguish between locations that are suitable but unoccupied and locations that probably are not suitable for the focal species. Second, they can identify environmental correlates of extinction and colonization within networks of suitable habitat patches.

There appears to be a trend to classify any spatially structured population as a metapopulation. Although logistic regression models of population dynamics can be applied to species with a broad range of population structures, we caution against automatically assuming that any "patchy" population functions as a metapopulation (Hanski and Simberloff 1997; Brommer and Fred 1999). If distinct populations are not interdependent, then it probably is more appropriate to manage each one separately rather than as members of a regional system in which local extinction, colonization, and dispersal are fundamental considerations.

Which spatially realistic model is likely to be most effective depends in part upon the population dynamics of the target species. For example, incidence function models are useful for management purposes because they can be parameterized with data on species occupancy from a single time step, and they yield location-specific extinction and colonization probabilities (Hanski et al. 1996b). However, incidence function models assume that species occurrence is equilibrial at the regional level (Hanski 1994a,b; Hanski et al. 1996b; Sjögren-Gulve and Hanski 2000). Repeated field inventories are essential for evaluating whether these assumptions are met. If the proportion of locations occupied by the target species changes significantly over time, it is more appropriate to model extinction and colonization patterns with logistic regression (Sjögren-Gulve and Hanski 2000).

Case Study

The Great Basin of western North America includes nearly 430,000 square kilometers of internal drainage extending from the east slope of the Sierra Nevada and southern Cascades to the west, the west slope of the Wasatch Range to the east, the Columbia River to the north, and the Colorado River to the south (Grayson 1993). More than 75 percent of the ecoregion is federally owned and managed under multiple-use mandates. Managers face considerable resource constraints; even the most fundamental data on species distributions frequently are not available.

Topographically, the Great Basin is dominated by more than two hundred mountain ranges. After the Pleistocene, these ranges were isolated from the surrounding valleys as the regional climate became warmer and drier (Brown 1978; Grayson 1993). Individual mountain ranges, and the canyons that deeply incise many of them, essentially function as islands of habitat for numerous taxa that either are restricted to montane vegetation types or have relatively low mobility, including butterflies (McDonald and Brown 1992; Murphy and Weiss 1992). Federal oversight agencies generally develop management plans on a range-by-range basis. Within mountain ranges, land uses commonly are delineated at the level of individual or several adjacent canyons.

We chose to focus on butterflies for several reasons. They are well understood biologically, are fairly easy to study and monitor, and have relatively short generation times, thus possibly exhibit rapid responses to management (e.g., Ehrlich and Davidson 1960; Pollard 1977; Scott 1986; New 1991; Kremen 1992; Pollard and Yates 1993; Harding et al. 1995; Shapiro 1996). In addition, the presence of some butterflies has been proposed to convey information on other taxonomic groups or ecosystem attributes of concern (e.g., Pyle et al. 1981; Erhardt 1985; Ehrlich and Murphy 1987; Eyre and Rushton 1989; Morris et al. 1989; Viejo et al. 1989; Hafernik 1992; Prendergast et al. 1993; Kremen 1994; Nelson and Andersen 1994; Holl 1995; Thomas 1995; Blair and Launer 1997).

From 1994 through 1996, we conducted comprehensive inventories of butterflies in nineteen canyons in the Toiyabe Range, a large mountain range in the central Great Basin. We used walking transects, an established technique that reliably detects species presence (Kremen 1992; Pollard and Yates 1993; Harding et al. 1995). It is unlikely that we failed to detect species that actually were present in a given location. Field personnel were familiar with the regional butterfly fauna, and we restricted our inventories to times when the weather was most favorable for flight (Shapiro 1975; Pollard 1977; Swengel 1990; Thomas and Mallorie 1985; Pollard and Yates 1993). It is reasonable to interpret that a given butterfly species is absent if the area has been searched with these methods during the appropriate season and weather conditions (Pullin 1995; Reed 1996). During the study period, we recorded sixty-eight resident species (those that complete their entire life cycle in the Toiyabe Range), 98 percent of the number expected under a Michaelis-Menten model (Clench 1979; Raguso and Llorente-Bousquets 1990; Soberón and Llorente 1993).

Species Richness of Butterflies

Identifying and predicting where native species richness is relatively high or low holds considerable value for development of management strategies. Using our data on butterflies in the Toiyabe Range as a case study, we used multiple linear regression to build a predictive model of species richness as a function of GIS-derived environmental variables. Managers could potentially employ this model in planning efforts not only for the Toiyabe Range, but also for ecologically similar mountain ranges in the Great Basin.

Our regression analysis was based on inventory data and environmental variables for sixty-eight canyon segments, each extending for approximately one hundred vertical meters, in fifteen Toiyabe Range canyons (see Fleishman et al. 1998). Data from an additional thirty-four canyon segments were used to validate the regression model. Species richness of these 102 canyon segments ranged from seven to fifty-three. We included forty-eight environmental variables in our analysis (Table 45.1). All are either static or amenable to repeated measurement and were relatively easy to quantify. To obtain values, we first recorded the endpoints of each canyon segment with differentially corrected global positioning systems (GPS). We overlaid the GPS points on a thirty-meter DEM and buffered each canyon segment to a width of 100 meters. Next, we used ArcView and ArcInfo (GIS software packages, Environmental Systems Research Institute, Redlands, Calif., http://www.esri.com) and a small suite of existing algorithms (e.g., Dubayah and Rich 1996) to derive the environmental variables. Values for this set of independent variables easily can be obtained for locations that have not yet been inventoried.

Our analysis of butterfly species richness in the Toiyabe Range was highly statistically significant (Fleishman et al. 2000). The full model, $S = -129.78 + 5.476$ Llength $- 0.1246$ NORMEAN $+ 0.1382$ NORMAX $- 0.0032$ EQINMAX $+ 7.084$ Leqinstd $+ 0.00892$ SSINMAX $- 4.608$ SSHRMAX $- 0.285$ EX15MEAN, explained a considerable percentage (77 percent) of the variation in species richness (see Table 45.1 for a complete explanation of the independent variables). Species richness of a canyon segment tended to increase with increasing area and with increasing heterogeneity in aspect and solar insolation. These correlations are not surprising given the life history of many butterfly species. For example, area may be correlated with diversity of larval host plants and adult nectar sources and with topographic heterogeneity. Variation in aspect and solar insolation may function as surrogate measures of overall topographic diversity. Not only is topographic heterogeneity often associated with vegetational diversity, but also the adults of many butterfly species locate mates on

TABLE 45.1.

Environmental variables for each canyon segment included in step one of the case study, how each was derived, and the data and software needed for their derivation.[a]

Variable	Definition	Data and software needs[b]
EASTMEAN	Mean "eastness" on a scale from –100 (west-facing) to 100 (east-facing). Derivation for each cell: 100*sine(aspect in degrees)	1, 2, 3
EASTMIN	Minimum eastness	1, 2, 3
EASTMAX	Maximum eastness	1, 2, 3
EASTSTD	Standard deviation of eastness	1, 2, 3
NORMEAN	Mean "northness" on a scale from –100 (south-facing) to 100 (north-facing). Derivation for each cell: 100*cosine (aspect in degrees)	1, 2, 3
NORMIN	Minimum northness	1, 2, 3
NORMAX	Maximum northness	1, 2, 3
NORSTD	Standard deviation of northness	1, 2, 3
Lelevmea	Mean elevation in meters (ln-transformed). Derived directly from DEM cell values	1, 2, 3
Lelevmin	Minimum elevation (ln-transformed)	1, 2, 3
Lelevmax	Maximum elevation (ln-transformed)	1, 2, 3
Lelevstd	Standard deviation of elevation(ln-transformed)	1, 2, 3
Lelevmid	Midpoint elevation (ln-transformed)	1, 2, 3
Lgrad	Gradient (ln-transformed). Derivation: 57.3*arctan[(maximum elevation – minimum elevation)/length]	1, 2, 3
LNA	Surface area (ln-transformed). Derivation for each cell: cell size/cosine(slope)	1, 2, 3
SLOPMEAN	Mean slope in degrees. Derived directly from the DEM with the ArcView slope command	1, 2, 3
Lslopmin	Minimum slope (ln-transformed)	1, 2, 3
SLOPMAX	Maximum slope	1, 2, 3
SLOPSTD	Standard deviation of slope	1, 2, 3
Llength	Length in meters (ln-transformed). Derived by measuring the length of the DLG road or trail vector bisected by the GPS points	GPS points (see text), DLG transportation coverage
PRECIP	Mean annual precipitation in mm for the 4x4-kilometer cell in which the canyon segment falls or weighted mean of the cells in which the canyon segment falls. Derived directly from PRISM cell values (Daly et al. 1994)	2, 3
EQINMEAN	Mean solar insolation in kilojoules at the vernal equinox	1, 4, 5
EQINMIN	Minimum solar insolation at the vernal equinox	1, 4, 5
EQINMAX	Maximum solar insolation at the vernal equinox	1, 4, 5
Leqinstd	Standard deviation of solar insolation at the vernal equinox (ln-transformed)	1, 4, 5
SSINMEAN	Mean solar insolation in kilojoules at the summer solstice	1, 4, 5
SSINMIN	Minimum solar insolation at the summer solstice	1, 4, 5
SSINMAX	Maximum solar insolation at the summer solstice	1, 4, 5
Lssinstd	Standard deviation of solar insolation at the summer solstice (ln-transformed)	1, 4, 5
EQHRMEAN	Mean duration of direct sunlight in hr at the vernal equinox	1, 4, 5
EQHRMIN	Minimum duration of direct sunlight at the vernal equinox	1, 4, 5
EQHRMAX	Maximum duration of direct sunlight at the vernal equinox	1, 4, 5
EQHRSTD	Standard deviation of duration of direct sunlight at the vernal equinox	1, 4, 5
SSHRMEAN	Mean duration of direct sunlight in hr at the summer solstice	1, 4, 5
SSHRMIN	Minimum duration of direct sunlight at the summer solstice	1, 4, 5
SSHRMAX	Maximum duration of direct sunlight at the summer solstice	1, 4, 5
SSHRSTD	Standard deviation of duration of direct sunlight at the summer solstice	1, 4, 5

TABLE 45.1. (Continued)

Environmental variables for each canyon segment included in step one of the case study, how each was derived, and the data and software needed for their derivation.[a]

Variable	Definition	Data and software needs[b]
EX3MEAN	Mean topographic exposure within a 300-meter radius. Compares the elevation of the canyon segment with the mean elevation of a specified neighborhood around that segment. If the segment is in a valley, value < mean; if on a ridge, value > mean; if on an open slope, value = mean, slope ≠ 0; if flat, value = mean, slope = 0. Derivation: elevation of a given cell − (mean elevation of all cells within 300 m)	1, 2, 3
EX3MIN	Minimum topographic exposure within a 300-meter radius	1, 2, 3
EX3MAX	Maximum topographic exposure within a 300-meter radius	1, 2, 3
Lex3std	Standard deviation of topographic exposure within a 300-meter radius (ln-transformed)	1, 2, 3
EX15MEAN	Mean topographic exposure within a 150-meter radius	1, 2, 3
EX15MIN	Minimum topographic exposure within a 150-meter radius	1, 2, 3
EX15MAX	Maximum topographic exposure within a 150-meter radius	1, 2, 3
Lex15std	Standard deviation of topographic exposure within a 150-meter radius (ln-transformed)	1, 2, 3
H20MEAN	Mean distance in meters from the centroid of the canyon segment to running or standing water minimum = 0, maximum set at 500. Derivation: distance from the centroid to the nearest water on the DLG hydrology coverage	2, 3, 6
H20MIN	Minimum distance to running or standing water	2, 3, 6
H20MAX	Maximum distance to running or standing water	2, 3, 6

[a]In the digital elevation model (DEM), cells are 30 horizontal meters on each side. All insolation and sunlight values were derived with the SolarFlux AML script.
[b]Data and software needs: 1 = DEM, 2 = ArcView, 3 = Spatial Analyst (a plug-in extension for ArcView), 4 = ArcInfo, 5 = SolarFlux Arc Macro Language (AML) script (developed and distributed by the GIS and Environmental Modeling Laboratory at the University of Kansas, http://www.gemlab.ukans.edu/gemlab/software.htm), 6 = digital line graph (DLG) hydrology coverage.

hilltops or other prominent topographic features (Scott 1975).

Our multiple regression model performed fairly well at classifying locations used to build the model with respect to species richness. For example, the model correctly classified 85 percent of the thirty-three canyon segments with at least twenty-seven butterfly species (i.e., more than 50 percent of the maximum number of species recorded from a canyon segment), and 50 percent of the ten canyon segments with forty or more species (i.e., more than 75 percent of the maximum number of species recorded). The model's error rate was relatively low. It overestimated species richness in seven (20 percent) of the thirty-five canyons segments with twenty-six or fewer species; in no instance did the model erroneously predict that a canyon segment was occupied by forty or more species.

Species Richness: Independent Tests

We conducted two independent assessments of the accuracy of our predictive model. First, we examined its ability to predict species richness of additional locations within the same geographic planning unit (i.e., other canyon segments in the Toiyabe Range). The model was highly statistically significant ($F_{1,33} = 37.5$, $P < 0.001$) and explained 53 percent of the variance in species richness. It correctly classified fourteen (93 percent) of the fifteen segments with twenty-seven or more species and misclassified four locations with fewer than twenty-seven species (21 percent). Although only one of five locations with forty or more species was classified correctly, the misidentification rate was a mere 3 percent.

Managers in our study region would like to predict species richness patterns in other planning units (mountain ranges) that are remote and poorly inventoried but that are nonetheless expected to support multiple uses and objectives. Therefore, we conducted a second assessment of the accuracy of our model using data from forty-three canyon segments in the neighboring Toquima Range, a 1,750-square-kilometer mountain range that lies roughly 10 kilometers to the east of the Toiyabe Range. Methods for butterfly inventories and segment delineation were identical to those used in the Toiyabe Range. For each canyon segment, we used GIS to derive the model's predictive variables. We then predicted species richness of butterflies using the Toiyabe Range model and compared those values to our observations.

The relationship between predicted and observed species richness was not statistically significant ($F_{1,42}$ = 0.38, P = 0.54). Nor could the model be used for qualitative predictions, in other words, to identify which locations have relatively many or relatively few species of butterflies (Spearman rank-correlation: ρ = 0.05, P = 0.73). We can draw at least some ecological inferences for these results. For example, the Toquima Range is smaller and, on average, lower in elevation and more arid than the Toiyabe Range. Thus, it spans a different range of values with respect to several predictive variables. It probably would be more appropriate to apply the Toiyabe-based model to mountain ranges with relatively similar topography and moisture gradients. The results of our independent tests emphasize the importance of conducting validation assessments before employing a species richness model in management planning for locations that were not used to build the model.

Occurrence of the Apache Silverspot Butterfly

Many species that inhabit the Great Basin face substantial threats to their viability. Managers of public lands are eager to develop conservation plans for these taxa that are compatible with multiple-use mandates and will preclude the need to invoke the federal Endangered Species Act. To facilitate these goals, it would be expedient to predict whether (1) species of concern occupy areas that have not yet been inventoried, (2) locations that are not currently occupied nevertheless appear to be suitable habitat, and (3) man-

agement actions could render a location suitable for the target species.

In the second step of our case study, we used logistic regression to determine whether environmental variables explained significant variance in the occurrence pattern of *Speyeria nokomis apacheana* (*S. nokomis*), the Apache silverspot butterfly. Breeding populations of this rare butterfly are confined to seeps, springs, and riparian areas in the western and central Great Basin. We chose to model the occurrence of this taxon because not only the butterfly itself but also its habitat is the focus of conservation attention. In xeric ecoregions, riparian areas both provide water and have plant communities with relatively diverse composition and structure. Therefore, they tend to receive disproportionately heavy use from numerous faunal groups, including humans (Kauffman and Krueger 1984; USGAO 1988; Armour et al. 1991; Thomas 1991; Dawson 1992; Chaney et al. 1993). Protection of riparian-obligate plants and animals is frequently a high priority of resource agencies in the Great Basin.

Step two included all of the 102 locations and independent variables described in step one and incorporated one additional variable, distance to the nearest canyon segment occupied by *S. nokomis*. The logistic regression model had high statistical significance (deviance goodness-of-fit χ^2 = 22.09, df = 97, P = 1.000) and identified five variables that were significantly associated with the presence of the species. The variable that was most strongly correlated with occurrence of *S. nokomis*, distance to nearest occupied location (improvement χ^2 = 34.74, df = 1, P < 0.0001), suggested that the spatial distribution of the butterfly may affect its viability. This result is consistent with our determination that *S. nokomis* is distributed as a metapopulation in the Toiyabe Range (see "Persistence of the Apache Silverspot Butterfly"). Probability of occurrence, not surprisingly, also was greater in locations that are relatively close to water (χ^2 = 23.38, df = 1, P < 0.0001), and in locations that have high vernal equinox insolation (χ^2 = 11.57, df = 1, P < 0.001), face east (χ^2 = 4.95, df = 1, P < 0.05), and are relatively flat (χ^2 = 3.43, df = 1, P < 0.10).

Managers obviously cannot affect the insolation or aspect of a given location, but our results indicate that they can take several steps to help protect the butterfly. For example, locations with isolated populations might receive a lower priority for species conservation, in other words, a higher priority for other land uses, than locations that are close to occupied areas. Another potential management action, particularly in areas with extremely wet soils, would be to switch the period of heaviest livestock use from early summer to late summer or early autumn. This would help prevent soil compaction, depression of the water table, and, ultimately, conversion of mesic meadows to dry meadows. If resources (and ecological circumstances) allow for restoration of degraded riparian areas, then our analyses also may give managers some guidance as to which locations, once rehabilitated, are most likely to support *S. nokomis*.

Because we have searched virtually all accessible areas in the Toiyabe Range that might have suitable habitat for *S. nokomis*, and because historical records and recent searches indicate that the butterfly is not present in any nearby (within 100 kilometers) mountain ranges, we have not yet conducted an independent test of our occurrence model. However, we expect to initiate surveys for the butterfly in one or more distant Great Basin mountain ranges in upcoming field seasons. At that point, we plan to use the model in conjunction with GIS-based maps and computational tools to identify promising locations for field surveys. The outcome of those surveys will provide data with which to test the accuracy of the model.

Persistence of the Apache Silverspot Butterfly

Step two of our model demonstrated that strong correlations exist between easily measured environmental variables and the distribution of *Speyeria nokomis apacheana*. Step two thus yielded a means tentatively to classify locations as suitable or unsuitable for the species. As a third step, we developed spatially realistic models to assess probabilities of turnover (colonization or extinction) at locations that appear to be suitable for the taxon. We defined a patch as suitable if adult butterflies were recaptured there and/or their larval host plants and adult nectar sources were present. Calculating turnover probabilities for individual

patches of habitat may help practitioners prioritize which patches are most critical for viability.

Field data collected between 1994 and 1999 indicate that in the Toiyabe Range, *S. nokomis* exists as a traditionally defined metapopulation (many small and ephemeral populations) (Hanski 1994a,b; Hanski et al. 1995; Harrison and Taylor 1997). Therefore, we included both local habitat features (e.g., area, percent cover of larval host plants) and larger-scale attributes (e.g., isolation) as predictive variables (Table 45.2). Values were obtained from recent field measurements or GIS. Surface area was the only variable common to both step two and step three. In step two, however, area was measured as the area of the canyon segment, which often included both suitable and unsuitable habitat. Patch area, included in step three, encompassed only suitable habitat.

Because we had data from multiple time steps (annual censuses), and because those data demonstrated that occupancy fluctuates in time, we used logistic regression to model extinction and colonization patterns pooled over four years. We found that, relative to patches that remained vacant, patches that were colonized had significantly ($P < 0.05$) greater host plant abundance. In addition, the bull thistle *Cirsium vulgare* tended to be present in patches that were colonized. State transitions for unoccupied patches were predicted moderately well with the resulting colonization logit (deviance goodness of fit: $\chi^2 = 19.35$, df = 22, $P = 0.624$). Patches that went extinct were nearer to other extinction patches, lacked the lavender thistle *Cirsium neomexicanum*, and had a greater percent cover of live vegetation and litter. The logistic regression based upon the extinction logit predicted extinctions quite well (deviance goodness of fit: $\chi^2 = 9.89$, df = 34, $P = 1.00$).

Because steps two and three included different environmental variables, the significant predictors of extinction and colonization in the logistic models (step three) were distinct from the significant predictors of occupancy in the multiple regression model (step two). Step two essentially produced a quantitative definition of suitable habitat; step three provided additional information about potential gradients in the quality of suitable habitat (Harrison and Taylor 1997; Thomas and Hanski 1997). Thus, step two might provide

TABLE 45.2.

Environmental variables included in logistic regression models of extinction and colonization probabilities for *Speyeria nokomis apacheana* (case study, step three).

Variable	Description
Vegetation variables[a]	
VINEct	Number of host plants (the violet *Viola nephrophylla*)
THISTct	Number of adult nectar sources (various species of thistles, see below)
VINEcov	Percent cover of host plants
THISTcov	Percent cover of thistles
TVC	Percent cover of live vegetation
LITTER	Percent cover of dead vegetation
BARE	Percentage of ground not covered by live vegetation or litter
VEGht	Mean height of live and standing dead vegetation
CANU	Presence of the thistle *Carduus nutans*
CIVU	Presence of the thistle *Cirsium vulgare*
CISC	Presence of the thistle *Cirsium scariosum*
CINE	Presence of the thistle *Cirsium neomexicanum*
Other habitat attributes	
AREA	Area in square meters (ln-transformed)
Dloc	Distance in meters to the closest occupied patch for each year 1995–1998, taken from censuses in the preceding year
Dext	Distance in meters to the closest neighboring extinction site for each year 1995–1998, taken from same-year censuses
DIST	Human disturbance (including livestock grazing and recreational use) for each year 1994–1998, categorized as minimal (0), moderate (1), or heavy (2)
PERIM	Perimeter in meters
UTMX	Latitudinal coordinate
UTMY	Longitudinal coordinate

[a]Data collected from 1-square-meter quadrats spaced evenly throughout each patch. In most patches, the number of quadrats sampled was proportional to patch area. Abundance was recorded as the midpoint of one of four abundance classes (1–3, 4–10, 11–30, 31–100; Fleishman et al. 1996). Percent cover was recorded as the midpoint of one of twelve percent cover classes (0–0.99, 1–4.99, 5–14.99, 15–24.99, 25–34.99, 35–44.99, 45–54.99, 55–64.99, 65–74.99, 75–84.99, 85–94.99, 95–100). Absolute cover values were recorded. In other words, 20 percent cover means 20 percent of the quadrat, not 20 percent of the total vegetation cover in that quadrat.

managers with a relatively coarse filter through which to evaluate whether certain areas could support a focal species, while step three could guide strategies for maintaining or improving habitat quality over time.

To test the accuracy of the turnover models for *S. nokomis*, we will use the colonization and extinction logits to calculate probabilities of colonization and extinction for each of the thirty-nine patches of suitable habitat. We will then compare those "predicted" probabilities to turnover events that are observed in 2000 and beyond and test whether significantly more predictions are accurate than would be expected by chance. We are optimistic about the outcome of these tests in light of the models' ability to account for variation in the data used for their construction. Some potential deviations between predicted and observed states may be explained by our recent finding that in the Toiyabe Range metapopulation of *S. nokomis*, different aspects of habitat quality affect turnover patterns in different years. We have also discovered that factors associated with turnover in single years are not always reflected in multiple-year analyses. We suggest that there is likely a spectrum of relative variation in habitat quality along which metapopulations fall. Toiyabe Range *S. nokomis* appear to lie at the higher end of this gradient, opposite from several systems that have laid the groundwork for current metapopulation paradigms (Thomas and Hanski 1997). Nonetheless, our models may be quite useful for identifying habitat attributes that may affect turnover in some years, and we believe that our overall modeling approach holds considerable promise for management of diverse "patchy" populations.

Discussion

Management at the regional level has multiple objectives, including maintenance of native species diversity, conservation of rare species, and tracking the effects of ecological changes on biological communities. No single tool or method is appropriate for all management challenges (Maurer, Chapter 9; Van Horne, Chapter 4). Instead, it is important to develop and validate a range of complementary modeling approaches that can be employed in various scenarios. From a technical standpoint, it can be extremely difficult to develop a model that successfully predicts species richness patterns in a geographic region other than that used to construct the model, yet the potential value of such a model is manifest. Validation is essential if one hopes to apply a single model across many planning units.

Our modeling framework takes somewhat different approaches to predicting species richness and to modeling occupancy and, particularly, to modeling persistence probabilities of individual taxa. Our species richness and occupancy models rely especially heavily upon variables that can be derived efficiently with GIS. The reasons for this are largely pragmatic. Although modeling species distributions as a function of location-specific variables related to the resource requirements of the focal assemblage frequently yields successful results (Hanski and Gilpin 1997), obtaining the necessary data can present substantial logistic obstacles. Furthermore, planning and decision-making for federal lands is often based in a central office—a "headquarters." Personnel may have less-detailed knowledge of local environmental conditions or species distributions than do staff in field offices, but they are more likely to have access to computerized mapping and planning tools, including digital and remotely sensed sources of data. A central office also may be the best location for tracking and archiving data on landscape features such as percent cover of major vegetational communities and road density.

Managers can overlay predictions of species richness with predicted occurrences of species of interest and data on land-use patterns. The resulting maps may be utilized for gap analysis (e.g., Caicco et al. 1995; Edwards et al. 1995), to prioritize field investigations, and to guide land-use planning (Cogan, Chapter 18). If road closures are politically contentious, for example, then it may be possible to minimize vehicle restrictions in locations with a baseline topography that is unlikely to harbor a relatively high number of native species.

The multiple regression analysis in step one of our model obviously does not examine the requirements of individual species of management interest, nor does it address the probabilities of persistence of those species at certain locations. Environmental variables that influence species richness may have little effect on the distribution of a particular rare taxon (Cody 1986; Thomas 1995; Fagan and Kareiva 1997; Freitag et al. 1997), and even locations with relatively low species diversity may serve as key supports for populations in species-rich areas (Fleishman et al. 2000). In addition, management for individual species often must rely more heavily upon local and/or species-specific data than upon remotely sensed variables. Step two of our model begins to address the needs of target species using a readily available set of environmental variables. It highlights some ecological conditions that are associated with the presence of the species and by extension suggests management actions that may promote occupancy. Step three incorporates relatively detailed habitat and species data in order to generate more-refined predictions and associated management recommendations.

As we noted, some species are of interest to practitioners because of their potential to serve as surrogate measures of high species richness or ecosystem function. Some species also may be useful for quantifying the biological effects of known ecological changes, particularly if those changes can be modified by management. One practical obstacle to using so-called indicator or umbrella species is that their occupancy can fluctuate independently of variation in human-influenced environmental parameters. By identifying some of the factors that may affect occupancy and persistence of potential indicator species, steps two and three of our modeling framework can help test whether the species is likely to function as a scientifically reliable surrogate.

Land-use planners rely upon predictive models and other tools to inform management strategies in the absence of detailed data. Models that can address a range of ecological phenomena, from species richness to occurrence patterns of rare or invasive taxa, may assist practitioners in delineating land uses, prioritizing field research, and anticipating the outcome of various management alternatives.

Discontinuity in Stream-fish Distributions: Implications for Assessing and Predicting Species Occurrence

Paul L. Angermeier, Kirk L. Krueger, and C. Andrew Dolloff

Overview of Species Occurrence

Understanding the patterns and causes of species distributions through time and space is a central goal of ecology. Development of models to predict species occurrence (i.e., presence or absence) enhances ecological knowledge and conservation effectiveness. Predicting occurrence for unsampled times or places requires the ability to explain alternations between presence and absence along environmental continua within a species' geographic range. Moreover, such explanations must distinguish between real discontinuity and the perceived discontinuity stemming from modeling or sampling errors.

Based on a review of the literature and our experiences, we propose that discontinuity in species distributions is regulated by environmental suitability and colonist availability. Environmental suitability comprises both abiotic and biotic factors. A broad array of physical and chemical features determines habitat (abiotic) suitability; greater suitability facilitates continuity in distributions. In contrast, intense biotic interactions promote discontinuous distributions. Colonist availability comprises three factors: population size, fluctuation in population size, and dispersal ability. We define dispersal as movement beyond the home range, including migratory and nonmigratory movement. Greater population size and dispersal ability facilitate continuity in species distributions, whereas

wide fluctuations in population size promote discontinuity. Explicitly incorporating these five factors into explanations of species distributions should enhance the reliability of predictive models.

The factors regulating discontinuity interact in complex ways. For example, habitat suitability may not predict occurrence if species-habitat associations are density-dependent. Both chance and biotic interactions can cause density-dependence, but their relative importance is rarely examined. Correlations between species abundance and occurrence are expected by chance, even with no habitat selection or biotic interactions (Wright 1991). At low population densities, the association between a preferred habitat configuration and species presence may be weak, even if preference is strong. Conversely, at high densities, preferred configurations may not accommodate all individuals, and a species might occupy a broad range of habitats, including less-preferred configurations. Similar distribution patterns commonly are predicted by models that assume animals make informed, optimal choices based on habitat suitability (e.g., Fretwell and Lucas 1969). That is, intraspecific competition could expand the range of suitable patches, and habitat selectivity (as measured by presence/absence) would appear reduced. However, for suitable habitat to be used, it must be accessible (i.e., within a species' dispersal ability). Thus, occurrence is also related to proximity to centers of abundance (Legendre 1993; Hanski 1994b), where proximity is defined in terms of dispersal ability.

Models to predict species occurrence are built from sample observations of presence versus absence. Because observed absences are more difficult to interpret than observed presences, the interpretation of observed absences is more likely to control model reliability. False absences are caused by three types of error: (1) failure to sample appropriate spatial or temporal strata, (2) inadequate sampling effort, and (3) ineffective or inappropriate survey technique. Herein, we focus on the first two error types because we believe that they are most limiting to the understanding of discontinuity in species distributions.

Although observations of occurrence are instantaneous, the predictions of interest typically involve extrapolation to much longer time frames (e.g., years). To be predictive in "new" places and times, models must incorporate knowledge of spatial and temporal variation in occurrence, which is accomplished via assessment of occurrence over ranges of the values taken by regulating factors. Incorporation of larger ranges should produce more generally applicable models. Relations between occurrence and some regulating factors (e.g., habitat suitability) are commonly incorporated into predictive models, whereas influences of other factors (e.g., population fluctuation) on occurrence are rarely modeled. In part, this disparity reflects the greater feasibility of measuring physicochemical conditions relative to estimating population parameters and biotic processes. In this chapter, we summarize knowledge of spatial and temporal variation in habitat suitability for stream fishes as might be used in building models to predict species occurrence. Where possible, we also discuss documented and hypothesized influences of other regulating factors on occurrence.

Stream-fish Distributions

Models to predict occurrence of stream fishes are interesting because stream features are highly variable through time and space and because modern sampling technology often allows confident assessment of occurrence. Local assemblages of stream fishes are subsets of regional species pools; local species have been "filtered" by a hierarchical array of limiting factors related to dispersal ability, physiological tolerance, life-history, and biotic interactions (Angermeier and Winston 1998). Spatiotemporal variation in the filtering process generates discontinuous distributions over a broad range of spatial scales (extent and grain). Confidence in assessments of species absence accrues because system boundaries (i.e., streambanks) are frequently obvious and because sampling effectiveness is well studied (Hankin and Reeves 1988; Bayley et al. 1989; Angermeier et al. 1991; Dolloff et al. 1993; Ensign et al. 1995; Simonson and Lyons 1995; Thurow and Schill 1996). Many sampling protocols provide the accuracy and known variance needed to develop predictive models.

The grain at which discontinuity in a species' distribution is relevant depends on the question being asked. Distribution within a pool might be especially relevant to individual survival, whereas distribution across a landscape might be especially relevant to species conservation. Detectability and pattern of occurrence are functions of study grain (Wiens 1989a). Moreover, the range of relevant scales is species-specific, bounded by the smallest scale at which patches are differentiated (species grain) and the largest scale of heterogeneity perceived (species extent; Kotliar and Wiens 1990). Wide-ranging, mobile species (great dispersal ability) tend to perceive mosaics at a given scale as more homogeneous than do more sedentary species (Kotliar and Wiens 1990). Multiscale analyses generally offer the most powerful approach for understanding stream-fish occurrence patterns (Lohr and Fausch 1997; Watson and Hillman 1997; Wiley et al. 1997; Angermeier and Winston 1998; Dunham and Rieman 1999; Torgersen et al. 1999).

Frissell et al. (1986) provide a hierarchical framework for stream habitat; levels increase in grain from microhabitat to stream system and reflect basic geomorphic processes. We adopted their two largest-grain levels, which we call reach and watershed. A reach is a continuous array of pool-riffle configurations (10^1 to 10^3 meters) and is bounded by major geomorphic (e.g., waterfall) or hydrologic (e.g., confluence with another reach) discontinuities. A watershed is a network of reaches and intervening land (10^6 to 10^8

square meters) with a single outflowing reach of third- or fourth-order (Strahler 1957). We also consider a larger-grain level: landscape, which comprises a cluster of watersheds (10^7 to 10^9 square meters) with a single outflow. Landscapes are delineated by boundaries of major river drainages or physiographic provinces; they are relatively homogeneous with respect to phylogenetic history (Hocutt and Wiley 1986; Mayden 1992) and habitat availability. Geographic ranges of many fish species coincide with drainage and/or physiographic boundaries. We believe that the three study grains considered in this chapter represent the scales most relevant to the conservation of stream fishes.

We address three major objectives. First, we review major spatiotemporal strata related to discontinuity in stream-fish distributions at three study grains (reach, watershed, and landscape). Samples from key strata (the next-smaller grain-level) are critical to assessing whether a species is present or absent in a study unit at the grain of interest. Because patterns of occurrence among sampled units provide a basis for predicting occurrence in unsampled places or times, we also review, for each grain, how models are used to predict stream-fish occurrence. Third, we develop a conceptual model, including several hypotheses, for understanding the relative importance, over a range of scales, of the five factors regulating discontinuity in species distributions.

Patterns of Occurrence in a Reach

Reliable assessments of species occurrence must be based on samples from appropriate spatial and temporal strata. Stratified sampling of habitat increases the probability of detecting species relative to the probability of detection with random sampling (McArdle 1990). The most commonly studied strata in reaches are riffles and pools (Table 46.1), which are spatially distinctive configurations of depth, velocity, substrate type, and cover availability. Most fish species and life stages regularly use only a subset of the available configurations. Occurrence may also be influenced by habitat-unit size or proximity to dispersers. Biotic interactions (e.g., competition, predation) commonly induce shifts in habitat use. For example, benthic species

may exclude each other from riffles (Baltz et al. 1982; Taylor 1996) or pool-dwelling piscivores may restrict their prey to riffles or pool margins (Power et al. 1985; Schlosser 1987).

Although less studied than spatial strata, there are several important temporal strata in stream reaches (Table 46.1). Environmental suitability of a reach may fluctuate seasonally or vary sporadically with major disturbances (e.g., flood, drought). Annual variation in flow regime and longer-term variation associated with succession can strongly influence stream-fish reproductive success and occurrence. In addition, some species (e.g., anadromous fishes) may use a reach only seasonally for a particular life-history function (e.g., spawning), whereas others may be present only during periods of high abundance. Multiyear studies indicate that breadth of habitat use by some fishes is correlated with population density, presumably because of intraspecific competition.

The relationship between abundance and distribution strongly influences the sampling effort needed to reliably assess species occurrence. The probability of detecting a species is generally correlated with population size (Nichols et al. 1998a). Increasing the sampling extent decreases the likelihood of a false absence, especially for uncommon species. Because many fish species are distributed sporadically along stream reaches, estimates of species richness are low and imprecise when sampling effort is small (Lyons 1992; Angermeier and Smogor 1995). The effort needed to detect most species varies among streams and is probably related to species-specific habitat selectivities, population densities, and habitat complexity. Although largely undocumented, analogous relations between number of reaches or watersheds sampled and species richness estimates for watersheds and landscapes, respectively, also probably exist. For accurate assessments of occurrence, we generally expect study units with many sporadically occurring species to require more sampling effort than do units with fewer such species.

Predicting Occurrence in a Reach

Predicting species occurrence depends on estimates of environmental suitability and colonist availability, as

TABLE 46.1.

Important strata associated with stream-fish occurrence at three study grains. Fish distribution is also influenced by a broad array of water chemistry variables, such as pH, temperature, dissolved oxygen, and toxins, but these are not summarized here. Selected references are provided for most strata.

Grain	Spatial strata	Temporal strata
Reach	• Riffles, pools, and subtypes (Hawkins et al. 1993) Near to versus far from dispersers (Schlosser 1995, 1998) Large versus small habitat unit (Taylor 1997)	Seasons (Matthews and Hill 1979) Before versus after disturbance (Sedell et al. 1990; Schlosser 1995) Before versus after atypical flow regime (Freeman et al. 1988; Strange et al. 1993) Successional stages (Snodgrass and Meffe 1998; Schlosser and Kallemeyn 2000) Life-history stages High versus low population size (Fraser and Sise 1980; Moyle and Vondracek 1985)
Watershed	• Large versus small reach (Burton and Odum 1945; Sheldon 1968) High versus low position (Gorman 1986; Osborne and Wiley 1992) Flow regime types (Poff and Allan 1995)	Seasons (anadromous species) Before versus after catastrophic disturbance (Reeves et al. 1995) Life-history stages High versus low population size
Landscape	• Large versus small watershed (Rieman et al. 1997; Dunham and Rieman 1999) High- versus low-elevation watershed (Thurow et al. 1997; Dunham and Rieman 1999) Flow regime types (Thurow et al. 1997) Many versus few human modifications (Thurow et al. 1997; Dunham and Rieman 1999) Connected versus isolated watershed (Dunham et al. 1997; Dunham and Rieman 1999))	Seasons (long-distance anadromous species) Before versus after anthropogenic transformation (Moyle and Williams 1990; Angermeier 1995) Before versus after climate change (Matthews and Zimmerman 1990; Poff et al. 2001)

well as knowledge of the temporal variation in these estimates. Environmental suitability of a stream reach can be described by variables related to physiological tolerance (e.g., pH, temperature), habitat preference (e.g., depth, patch configuration), biotic interactions (e.g., abundance of predators or competitors), or any combination of these. Colonist availability in a reach is a function of past and current population abundance at the watershed scale and of accessibility to dispersers.

Although abundance of species or life stages is often estimated (Baltz 1990), we know of no predictive models that incorporate most of the factors regulating discontinuity in the distribution of fishes. The models most commonly used to infer fish occurrence in reaches are those developed via the instream flow incremental methodology (IFIM), a complex protocol that uses models of habitat suitability and hydraulics to predict how habitat quality changes with variation in discharge (Stalnaker et al. 1995). Users of the IFIM often assume that fish abundance is proportional to habitat availability, thereby implicitly assuming that other factors do not influence abundance (or occurrence). Sampling protocols such as the basinwide visual estimation technique (BVET; Hankin and Reeves 1988; Dolloff et al. 1993; Fig. 46.1) can generate empirical estimates of fish abundance with known variances for reaches in a watershed. This two-stage stratified-random protocol

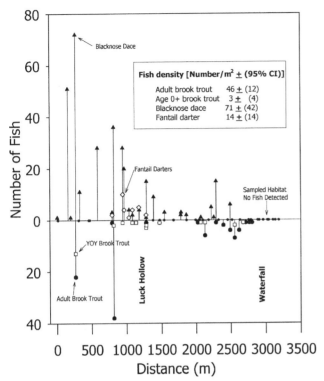

Figure 46.1. Number of fish counted by a diver in systematically selected pools and riffles in three reaches of White Oak Run, Shenandoah National Park. Counts were obtained using the basinwide visual estimation technique. Reaches are bounded by the confluence with Luck Hollow and a waterfall. Inset shows the density of fish over the entire survey distance. Densities for riffles or pools or reaches could also be estimated.

has mostly been used to estimate salmonid abundance but is also effective in estimating abundance of other stream fishes (e.g., Leftwich et al. 1997). The BVET focuses on spatial strata, but analogous protocols could be developed to sample temporal strata.

Many studies have addressed whether species-habitat associations observed in reaches of one system can predict fish occurrence in reaches of another system. Broadly transferable models would obviate the need to develop many system-specific models. A few models of fish-habitat associations are broadly transferable (e.g., Belaud et al. 1989) but most are not, especially across major drainages (Layher et al. 1987; Leftwich et al. 1997) or regions (Bowlby and Roff 1986; Hubert and Rahel 1989). Similarly, fish-habitat associations in a reach can vary among years (Angermeier 1987; Bozek and Rahel 1992). To enhance model transferability, researchers will need to explicitly in-

corporate factors that limit fish distribution and abundance. Rather than relying on population density to indicate habitat quality (Van Horne 1983), mechanistic links between habitat configurations and fitness measures (e.g., survival, reproduction) should be identified (e.g., Dolloff 1987; Schlosser 1998).

Patterns of Occurrence in a Watershed

The most commonly studied strata in watersheds are related to stream size (Table 46.1), which integrates variation in a suite of physicochemical features, such as temperature and flow regimes, and depth and velocity frequencies. Other spatial strata that influence stream fish distribution include position in the watershed and flow regime type (Table 46.1). Biotic interactions that exclude species from reaches are scarcely documented, but complementary distributions of some species-pairs are attributable to competition (Winston 1995; Taylor 1996; Clark and Rose 1997) or predation (Fraser et al. 1995).

Environmental suitability of a watershed may fluctuate seasonally or vary sporadically with catastrophic disturbances (Table 46.1). Watershed-wide extinctions induced by disturbances have long recurrence intervals (e.g., centuries) but may reverse themselves if suitable habitat remains or develops and if connections to disperser sources exist (Reeves et al. 1995). In addition, some species (e.g., anadromous fishes) may use a watershed only seasonally for a particular life-history function (e.g., spawning). Density-dependence in the occurrence of fishes among watersheds is undocumented, but relations between occurrence and watershed size and isolation (see Occurrence in a Landscape) suggest influences of population stability and dispersal ability, respectively (i.e., colonist availability). Density-dependent occurrence may be especially likely for species that have declined dramatically because of human impacts. Long-term dynamics of occurrence among watersheds might be expected for species with metapopulation structure, but empirical evidence of such structure is scarce (Schlosser and Angermeier 1995; but see Rieman and Dunham 2000). In any case, the ecological significance of a species' observed absence from a watershed (in a landscape where the species in present) is equivocal because long-term dynamics may obscure the watershed's suitability.

Predicting Occurrence in a Watershed

Predictions of species occurrence in watersheds are rare. Environmental suitability of a watershed might be described by the same variables used in reach-level models, but variables would be "scaled up" to represent cumulative suitability of component reaches by using frequencies of various reach-types in the watershed. In addition, suitability may be affected by key ecosystem properties such as flow, nutrient, sediment, and thermal regimes. The importance of certain reach-type combinations or juxtapositions to stream-fish occurrence is largely unexplored (but see Labbe and Fausch 2000) despite recognition that many species require multiple habitat configurations to complete their life cycles (Schlosser and Angermeier 1995). In part, this information gap reflects the common belief that most stream fish move little (less than 500 meters) during their lifetimes (Gerking 1959; but see Gowan et al. 1994). However, recent findings of extensive movement suggest that some fish select and use habitat at large spatial scales (Fausch and Young 1995; Schlosser 1995; Gowan and Fausch 1996b). A few models based on watershed features that reflect both environmental suitability and colonist availability predict occurrence of salmonids in Pacific Northwest watersheds (Rieman et al. 1997; Thurow et al. 1997; Dunham et al., Chapter 26). We anticipate that analogous models will be useful for other taxa and regions.

Some predictions of fish occurrence in watersheds are based on iterative assessments of reaches rather than on watershed descriptors (i.e., models with reach-level grain are extended to watersheds). However, the influence of watershed-level land use is increasingly recognized in characterizations of reach-level suitability (Richards et al. 1996; Roth et al. 1996). In reach-grain models (e.g., IFIM applications), predicted presence in any reach means predicted presence in the watershed. The reliability of such predictions depends on the transferability of fish-habitat associations from the reaches where a model was developed to the reaches of the watershed being assessed. Thus, this approach has the same transferability constraints as discussed for reach-level predictions.

Predicting occurrence in an unsampled watershed should be based on models developed from samples of similar watersheds in the landscape. Estimating prediction accuracy requires a probabilistic sampling protocol (e.g., stratified-random). Presumably, a protocol analogous to the BVET could be developed for sampling watersheds in a landscape. However, because rare fishes are often distributed capriciously across landscapes, highly accurate predictions of their occurrence in watersheds are unlikely (Boone and Krohn 1999).

Occurrence in a Landscape

Spatial and temporal strata associated with fish occurrence in landscapes are scarcely studied. Occurrence of some species is related to watershed size, elevation, or flow regime and may reflect occurrence in nearby watersheds (Table 46.1). Occurrence is also affected by human modifications such as dams and roads. We know of no species-pairs with complementary distributions among watersheds attributable to biotic interactions. Some long-distance migrants may use a landscape only seasonally, but the most obvious temporal strata are related to human impacts on environmental suitability (Table 46.1). Specific mechanisms are poorly understood, but anthropogenic changes in flow, nutrient, sediment, and thermal regimes across landscapes are primarily responsible for the pervasive endangerment and extinction of North American stream fishes.

Analyses of fish occurrence in landscapes are descriptive rather than predictive. Explanations for discontinuous distributions of fishes among landscapes are actively debated by zoogeographers and systematists (Hocutt and Wiley 1986; Mayden 1992), but we know of no models that predict fish occurrence in unsampled landscapes. However, some authors have tested predictions of occurrence in landscapes based on reach-gain models (Bowlby and Roff 1986; Layher et al. 1987; Hubert and Rahel 1989; Leftwich et al. 1997). We speculate that certain combinations or juxtapositions of watersheds may regulate metapopulation persistence in some landscapes.

Synthesis and Hypotheses

Stream-fish distributions can be discontinuous at any spatial scale, but the influence of each regulating factor on species absence probably varies greatly with study grain (Fig. 46.2). We expect a factor's influence on absence to be proportional to the range of variation at a given grain, and we hypothesize that these influences exhibit general patterns from microhabitat to regional scales. Habitat suitability, the focus of most occurrence models, strongly influences species absence at all scales (Fig. 46.2A). For stream fishes, habitat suitability typically is based on structural variables such as water depth and velocity, substrate and cover type, and patch size and juxtaposition as well as on a suite of water chemistry variables such as temperature, pH, and dissolved oxygen concentration. In contrast, biotic interactions such as competition, predation, and disease are much more likely to cause species absence at small scales than at large scales (Fig. 46.2B).

Population size and variability have opposing effects on species discontinuity, but we expect the influence of both to decrease with increasing study grain (Fig. 46.2C). Population size at a given grain will generally be correlated with the number of spatial strata (next-smaller grain) occupied. However, as grain increases, optimal patch choice becomes increasingly infeasible because of the escalating cost/risk of gathering relevant information (Lima and Zollner 1996). Furthermore, in habitat units large enough to support multiple populations, asynchrony in population fluctuations could preclude unit-wide density-dependence in occurrence. At smaller grains, we expect patch choice driven by availability of resources (e.g., food, spawning sites) to be more density-dependent than that driven by physiological tolerance. Population instability promotes species absence via the recurrence of small population size; fluctuations at large study grains are likely to be less severe or frequent than fluctuations at small grains. We expect occurrence models for species with stable populations to be more transferable than models for species with highly variable populations.

Barriers to movement can cause species absence at any study grain. Barrier permeability, which is a

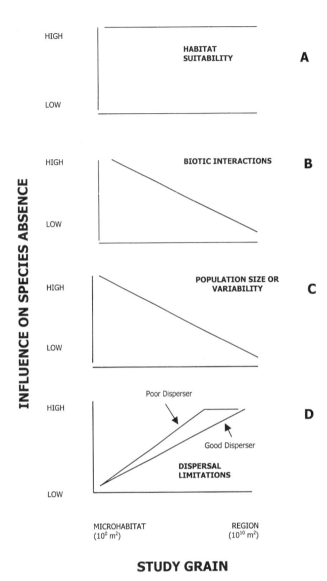

Figure 46.2. Hypothesized relations between study grain (size of habitat units) and the influence of habitat suitability (A), biotic interactions (B), population size or variability (C), and dispersal limitations (D) on species absence from a habitat unit. Relations are based on the assumption that the species is present in the next-larger study grain.

function of behavior, size, shape, dispersal frequency and distance, and other factors, is poorly understood for most fishes. Because barriers at small study grains tend to be more temporary than barriers at large grains (e.g., Schlosser and Kallemeyn 2000), we expect the influence of dispersal limitations to increase with grain (Fig. 46.2D). However, the grain at which this influence is maximal is smaller for poor dispersers than for good dispersers. Effects of artificial barriers (e.g., dams) on distributions of

TABLE 46.2.

Values (and probabilities) for Kendall's tau concordance between proportions of habitat units from which fish species were absent at four study grains.

Study grains[a]	Channel units[b]	Reaches[c]	HPUs[d]	Drainages[e]
Beaver Creek (31 species)				
N	18	32	9	6
Channel units		0.499 (0.0002)	0.027 (0.847)	0.122 (0.380)
Reaches			0.431 (0.002)	0.238 (0.091)
HPUs				0.387 (0.008)
Bernards Creek (21 species)				
N	17	15	9	6
Channel units		0.284 (0.098)	0.114 (0.510)	0.281 (0.122)
Reaches			0.390 (0.026)	−0.170 (0.354)
HPUs				−0.032 (0.863)

[a]The species lists compared came from samples of two streams in the James River drainage, Beaver Creek and Bernards Creek.

[b]Data for channel units came from samples of contiguous riffles and pools (Angermeier and Smogor 1995).

[c]Data for reaches came from a large statewide database of fish collections. Only those reaches associated with the same HPU and a fish collection were included in analyses for Beaver Creek and Bernards Creek.

[d]Hydrologic-physiographic units (HPUs) are the portions of U.S. Geological Survey 8-digit hydrologic units that intersect a single physiographic province (see Angermeier and Winston 1998). Data for HPUs came from a large statewide database of fish collections. Only those HPUs with twenty or more fish collections were considered.

[e]Data for drainages came from Jenkins and Burkhead (1994) and are limited to native species. Only those Atlantic-slope drainages well represented in Virginia were considered.

migratory fishes are well documented, but effects are likely underestimated for nonmigrants (but see Winston et al. 1991), which need to track spatiotemporal variation in resource availability and habitat suitability to persist.

Our conceptual framework builds on the notion (e.g., Watson and Hillman 1997; Dunham and Rieman 1999) that the most useful models of species occurrence are hierarchical, beginning with patterns occurring over large areas. Occurrence at a given study grain is uninteresting unless presence at the next-larger grain is known or expected. The hypothesized relations in Figure 46.2 suggest which regulating factors may be most influential at a given grain. We expect occurrence among landscapes to be regulated primarily by habitat suitability and dispersal barriers, but occurrence among reaches to be regulated by habitat suitability, biotic interactions, and population size and stability.

Factors may interact at any grain, but we expect probability of occurrence generally to be independent across grains. That is, infrequent presence at one grain does not imply infrequent presence at another grain. As a preliminary test of this hypothesis, we assessed concordance of fish occurrences across four study grains in Virginia (Table 46.2). Grain size ranged from channel unit (e.g., riffle, pool) to major river drainage. Overall, there was some concordance between species' tendency to be rare (or widespread) at one grain and their tendency to be rare (or widespread) at another grain (four of twelve Kendall's tau values > 0.35; Table 46.2). However, all significant concordance occurred between the most-similar-sized grains, a pattern that may be an artifact of spatial autocorrelation (Legendre 1993).

Predictive models of species occurrence are important conservation tools. They help managers to (1) identify likely occupied areas for poorly surveyed species, (2) assess status of species relative to historical distribution, (3) identify the best areas for re-introducing extirpated species, (4) justify protecting areas that are suitable for, but currently unoccupied by, valued species, and (5) predict the extent of invasion by colonizing species. In addition, hierarchical models of occurrence help managers focus on the most relevant scales and regulating factors thereby enhancing cost-effectiveness. For most species, occurrence in reaches and watersheds should be of greatest interest to managers because the time frames typically associated with population dynamics at those scales (years to decades) are the most tractable. However, highly migratory species (e.g., diadromous fishes) do choose among landscapes; understanding occurrence of such species among landscapes will also facilitate effective management. Moreover, if metapopulation dynamics are important to the persistence of most stream fishes, understanding occurrence in landscapes will be critical to long-term conservation. In this context, predictive models can shed light on the conservation significance of particular presences and absences. An area that is primarily a population sink (i.e., species is present but not self-sustaining) is less conservation-worthy than a population-source area (Schlosser and Angermeier 1995), and perhaps less conservation-worthy than some unoccupied, but environmentally suitable, areas (Rieman and Dunham 2000). Similarly, because stream-fish populations typically transcend individual reaches, occurrence at reach grains, where most models focus, has less conservation significance than occurrence at larger study grains.

Research Needs

The reliability of assessing or predicting stream-fish occurrence could be enhanced substantially by additional research in three interrelated areas. First, better understanding of temporal variation in species distribution and abundance is needed, including analyses of variation in habitat use in the context of population fluctuations. Such analyses would clarify the importance of density-dependence but would likely require much longer-term data than are typically collected (e.g., Wiley et al. 1997). Failure to incorporate temporal variation into fish-habitat models can cause them

to be overfitted because temporal variation erroneously appears as spatial variation (Wiley et al. 1997). Second, more insight into the spatial and temporal scales most relevant to stream fish population dynamics, including those of metapopulations, is needed to help identify key strata to include in occurrence models. Especially lacking is knowledge of spatiotemporal dynamics of fish dispersal, including estimates of movement frequency and distance and barrier permeability. Finally, additional conceptions of hierarchical, multiscale models are needed to understand how the relative importance of various factors that regulate species occurrence changes with study grain and to understand whether occurrence patterns observed at one grain influence those, or can be expected, at other grain sizes.

Our ability to assess or predict occurrence has important implications for how we prioritize areas for conservation and for our use of fish communities as environmental monitors. Acquiring knowledge of occurrence is a small but essential step toward achieving conservation goals. Intense societal demands on ecosystems dictate that only the most irrefutable knowledge of occurrence and persistence will be tolerated as justification for pursuing aggressive conservation of fish species or biotic integrity. In this context, the research areas noted above become crucial. Without basic understanding of the factors regulating distribution, scientifically sound recommendations on how to conserve fish populations and communities are unlikely. As our understanding of species discontinuity advances, so will the effectiveness of our conservation efforts.

Acknowledgments

We thank M. B. Bain, J. B. Dunham, W. A. Hubert, D. J. Orth, B. E. Rieman, and four anonymous reviewers for many helpful comments on an earlier draft. K. L. Krueger was supported by a Federal Aid grant from Virginia Department of Game and Inland Fisheries.

A GIS-based Habitat Model for Wood Thrush, *Hylocichla mustelina,* in Great Smoky Mountains National Park

Susan A. Shriner, Theodore R. Simons, and George L. Farnsworth

Management of wildlife populations requires knowledge of the ecology and habitat preferences of the species of interest. Typically, the majority of habitat information available for Neotropical migratory birds has been collected at a microhabitat scale. However, managers interested in identifying key areas for a given species rarely have the resources to collect detailed microhabitat data over broad regions. Therefore, identifying habitat variables that are both readily available over large areas and correlated to species occurrence is essential to effective management of large areas (Simberloff 1988). Topographic indexes based on readily available data (digital elevation models) have been successfully applied to models developed to predict plant occurrence (Wilds 1996; Wiser et al. 1998) but are not often used to predict animal occurrence. Nonetheless, these indexes may be useful predictors of animal occurrence over broad regions. In this chapter, we present a logistic regression model for wood thrush (*Hylocichla mustelina*) in Great Smoky Mountains National Park based on topographic and other readily available habitat variables.

As the result of several studies showing possible population declines (Robbins et al. 1989b; Askins et al. 1990), Neotropical migratory songbirds have become major research species in conservation biology. In particular, the wood thrush has been the focus of many studies examining the impact of fragmentation

and land-use change on population abundance, distribution, and productivity (Robinson 1988; Hoover and Brittingham 1993; Robinson et al. 1993; Hoover et al. 1995; Robinson et al. 1995a). Most efforts to develop predictive models to assess the location of suitable habitat over large areas for Neotropical migratory songbirds have given rise to models based on patch size and configuration and/or habitat fragmentation indexes. Although these types of models may be a useful first attempt at identifying potential habitat, they do not provide information about which habitat features within large forested landscapes are important predictors of Neotropical migratory songbird occurrence. Therefore, these models may not be usefully applied to large areas of contiguous forests. On the other hand, models that do incorporate variables associated with habitat use are usually based on fine-grained microhabitat data that require extensive data collection and cannot feasibly be applied to large areas.

The southeastern United States, in particular, Great Smoky Mountains National Park, is a species-rich region with a complex biota. The park is the largest tract of contiguous forest within the southeastern United States and the second-largest national park in the eastern United States. As part of a network of protected areas in the southern Appalachians, the park is an important reserve for many Neotropical migratory land birds. Surveying the physiological, genetic, and life-history characteristics of each species in such a region is a near-impossible task (Martin et al. 1993). Models based on variables readily available in a GIS may offer a solution to such an intractable problem. Therefore, we investigated the utility of using commonly available GIS data to assess several topographic indexes and other habitat variables as predictors of species presence or absence for wood thrush in Great Smoky Mountains National Park.

Great Smoky Mountains National Park represents a unique opportunity to examine eastern forest birds in a setting free from recent human disturbance. A relatively large portion of the park has been undisturbed, and most of the areas that have been previously altered have had sixty-five to one hundred years to recover. This lack of human disturbance enhances the likelihood that topographic indexes correlated with vegetation features will be useful predictors of bird presence or absence. In disturbed areas, variables such as stand age and landscape configuration may be more-important predictors of vegetation features, but in undisturbed areas, topographic indexes are likely to be highly correlated with vegetation. Several researchers (Austin et al. 1984; Yee and Mitchell 1991) have been able to successfully predict vegetation characteristics based on site conditions (Wiser et al. 1998). We hypothesized that for areas primarily undisturbed by recent human impacts, topographic variables may be useful predictors of wood thrush presence or absence, since topographic indexes are likely to act as predictors of vegetation features.

Methods

Great Smoky Mountains National Park comprises 205,665 hectares of primarily contiguous forest straddling the Appalachian Trail along the Tennessee–North Carolina border. The park serves as the nucleus of a group of protected areas in the southern Appalachians that includes national forests, federally designated wilderness areas, state lands, Tennessee Valley Authority reservoirs, and National Park Service lands (see Fig. 47.1 in color section). This region consists of 15.1 million hectares, including more than 2 million hectares of public land. More than 70 percent of the region is currently forested (SAMAB 1996), providing the largest extent of forested landscape in the eastern United States.

Great Smoky Mountains National Park is characterized by a wide elevation range (575–1,830 meters) and complex topography, creating a rich diversity of habitat and vegetation types. MacKenzie (1993) used Landsat imagery to develop a vegetation map of the park based on thirteen major forest types. MacKenzie patterned his vegetation scheme after the original classification system of Whittaker (1956). The MacKenzie vegetation types range from pine and pine oak forests at the lowest elevations of the park to northern hardwoods and spruce fir at the highest elevations. Seven of these vegetation types are regularly used by breeding wood thrush. In general, these habitats occur along a wet-to-dry moisture gradient and include cove hardwood, mixed mesic hardwood, tulip poplar, mesic oak, xeric oak, pine-oak, and pine.

The Great Smoky Mountains was established as a national park in 1934. At that time, more than half of the land had been disturbed, primarily by different methods of logging. Other disturbances were caused by large-scale fires and clearing of the land for homesteads. Current disturbances include extensive loss of Fraser fir (*Abies fraseri*) due to infestation by the exotic balsam woolly adelgid (*Adelges piceae*) and decline of many pine species (*Pinus* spp.) due to southern pine beetle (*Dendroctonus frontalis*) infestations. The hemlock woolly adelgid (*Adelges tsugae*) is also expected to spread to the park within the next decade (SAMAB 1996). This exotic pest will likely affect eastern hemlock (*Tsuga canadensis*) populations and therefore impact many of the cove hardwood and mixed mesic hardwood forests. Bird species such as the wood thrush may be seriously impacted if the park experiences major losses of eastern hemlock. More than 90 percent of wood thrush nests found in the park have been located in eastern hemlocks (Farnsworth and Simons 1999).

Field Data

We conducted variable circular plot point counts (Reynolds et al. 1980) at more than four thousand locations throughout our study area during May and June of 1996–1999. Due to the large number of observers (n = 40), we employed several strategies to minimize observer variability. We trained and tested all observers prior to the initiation of fieldwork. We also provided a review training session at the midpoint of the field season. In addition, we rotated observers throughout the different areas of the park so that the efforts of a single observer were not restricted to a particular area of the park. We conducted counts for a ten-minute interval between dawn and 10:15 A.M. and only in good weather (no rain or excessive wind). Our count protocol is consistent with the majority of the recommendations for point count methodology detailed by Ralph et al. (1995b).

During each ten-minute count, we recorded the number of breeding pairs of each bird species present. A single observer collected data at any given point. Before each count, observers estimated a 50-meter radius circle by spotting landmarks using a laser range finder and began the count immediately thereafter. At each point, we recorded bird detections in all directions for an unlimited radius plot. We mapped the location and movement of all individual birds detected in order to avoid double counting.

We established the majority of points systematically along trails with some points located on minor roads and some points located off-trail along transects. The majority of trails in Great Smoky Mountains National Park are low impact and do not result in a gap in the tree canopy. Comparison of counts from points located on- and off-trail showed no significant differences (S. Shriner unpublished data). We spaced points 250 meters apart to avoid double-counting birds. We sited points primarily by pacing and occasionally by using a laser range-finder. The meandering nature of the trails resulted in an average horizontal distance between points of approximately 175 meters. In areas where we could discern that the trail was very windy, we paced an additional 50 to 500 meters between points. We sited a small number of points in Cades Cove, a large, open area. We spaced these points 500 meters apart to account for an increase in detectability due to the sparse vegetation. In the event that an individual bird was heard at more than one point, we only recorded the bird at the point with the smaller detection distance.

We stratified points throughout the park with respect to the availability of each MacKenzie vegetation type. We used Trimble GeoExplorer II global positioning system (GPS) units to collect the geographic coordinates of each point.

Digital Data

We obtained habitat and topographic variable data from a GIS database provided by the Inventory and Management Division of Great Smoky Mountains National Park. All queries were performed using ArcInfo (ESRI 1997) software. The Great Smoky Mountains National Park GIS database contains 90×90-meter grid data for vegetation type, bedrock geology, and disturbance history. Vegetation type information is based on the groundtruthed analysis of satellite imagery produced by MacKenzie (1993); it includes thirteen vegetation types. The bedrock geology data include twenty-four classes of bedrock. The disturbance history data are based on an analysis of park records and include five categories of human disturbance: undisturbed, selective cut, light cut, heavy cut, and settlement (Pyle 1985).

Data for elevation and several topographic indexes (including topographic convergence index, terrain shape index, landform index, topographic complexity, relative moisture, and relative slope position) are available in the database as 30×30-meter grids. Elevation for each point was calculated from a digital elevation model (DEM) using an interpolation function based on the nearest-neighbor cells. Slope and aspect were computed in ArcInfo using the DEM. Aspect was transformed into north/south and east/west components using sin(aspect) and cos(aspect), respectively. The different topographic variables available in the park GIS were developed by several different researchers to characterize the shape of the landscape and local moisture regimes at various spatial scales. These indexes are coded in Arc Macro Language (AML) and are available as coverages in the park GIS database. The topographic convergence index (TCI) is an index of potential soil moisture developed by Beven and Kirkby (1979) and was developed to simulate runoff saturation and infiltration. This index has been successfully used in spatial models of vegetation distribution. The underlying formula is TCI = ln[(A/tanB)] where A is the surface area of each grid cell providing drainage and B is the surface slope of the grid cell.

The terrain shape index (TSI) was developed by McNab (1989) to characterize the topographical curvature of the landscape. The TSI distinguishes between ridges/exposed areas and coves/protected areas by calculating the average difference in elevation between the center of a plot and its boundary. The landform index (LFI) was also developed by McNab (1993) and is a large-area parameter that describes general classes of protection at a site, or in other words, cove, slope, or ridge. The LFI is the mean of eight different slope gradients (N, NE, E, SE, S, SW, W, NW) calculated from the center of a plot to the skyline.

Topographic complexity was computed as the Shannon-Weaver Index of Topographic Complexity (SWI), an index developed by Miller (1986) to explain the distribution of rare and endemic plant species. The SWI index is a fine-scale index of the topographic diversity of a 150×150-meter plot. Topographic complexity is calculated for elevation, slope, and aspect and then combined into a single measure (Wilds 1996). The SWI is calculated as SWI = $-\Sigma(p_i*\log2(p_i))$, where p = the proportion of area for each elevation/slope/aspect category.

Relative moisture is described by the topographic relative moisture index (TRMI) developed by Parker (1982) to model vegetation distribution in the western United States. The TRMI was developed to describe local moisture regimes for areas with diverse topography. It is based on aspect, steepness, topographic position, and curvature. Relative slope is also a measure of relative moisture and is the distance from the point to the bottom of the slope divided by the total distance from the bottom of the slope to the nearest ridge. The distance to the bottom of the slope is measured to the nearest stream or the nearest topographic concavity from the point, perpendicular to contour lines. The distance to the top of the slope is measured to the nearest ridge or topographic convexity, perpendicular to contour lines.

The Model

The model is a logistic regression model that returns the probability of detecting a wood thrush as a function of habitat and topographic variables. Logistic regression is a statistical technique used to predict the probability of an event occurrence (P_0) and has been used to predict the probability that an organism will occur based on the conditions present at a particular site (e.g., Margules and Stein 1989; Wiser et al. 1998). Probability values are constrained to range between 0 and 1 and therefore cannot be modeled using linear functions. In logistic regression, explanatory variables that can range from positive to negative infinity are transformed to a probability using a logit link function. The logistic regression equation models the logit transformation as a linear function of the explanatory variables (Christensen 1997). The result is a probability value in the 0 to 1 range.

We used presence/absence data for wood thrush derived from the point counts as the dependent variable in the model and the eleven habitat and topographic variables derived from the GIS as the explanatory variables. We ran the model using the PROC LOGISTIC procedure in SAS (SAS Institute 1990b). We used the backward elimination and forward selection procedures (threshold p-values = 0.1) in SAS to compare different logistic regression models based on the eleven variables, the squared values of the topographic variables, and interactions between the topographic variables. The backward elimination procedure analyzes the full model and then removes variables one at a time as they fail to meet the specified significance level for staying in the model. The forward selection procedure begins with an intercept-only model and adds variables one at a time based on adjusted chi-square statistics. A variable is added to the model if it meets the specified significance level for staying in the model.

The topographic variables (but not the squared terms) were standardized (mean = 0, variance = 1) prior to analysis to aid in the interpretation of the parameter estimates. If a squared variable or an interaction variable was significant (p-value less than or equal to 0.1), then the main variable was included in the model without regard to its significance level. A significant interaction term indicates that one of the variables has a modifying effect on the impact of another variable.

We treated the thirteen vegetation types, twenty-three bedrock geology types, and five disturbance classes as class variables. We only included type variables that were represented by a minimum of thirty observations such that the final analysis was performed on ten vegetation types and eighteen geology types. All counts with missing data or type variables not included in the analysis were deleted from the data set, which resulted in 3,743 points available for analysis. Each of these data points was randomly assigned to either a model development data set (n = 1,833) or a validation data set (n = 1,910); only the model development data set was used for model selection.

Final model selection was based on concordance scores (SAS Institute 1990a,b). Concordance is a measure of rank correlation between the predicted probability that a wood thrush is present at a site and the actual presence/absence of wood thrush at that site (Bolger et al. 1997). Concordance is calculated as a percentage of all possible pairs of observations where one member of the pair represents the presence of a wood thrush and the other member of a pair represents an absence of a wood thrush. A pair is concordant if an observation representing an absence of wood thrush has a lower predicted event probability than does the observation representing the presence of wood thrush. A pair is discordant if the observation representing an absence of wood thrush has a higher predicted event probability than the observation representing the presence of wood thrush. If a pair is neither concordant nor discordant, then it is a tie. A model with a higher concordance score is more likely to correctly classify the presence or absence of wood thrush at a particular location. We compared the concordance scores for the models determined by the backward elimination and forward selection procedures to choose the final model.

Validation and Model Assessment

We applied the final model to the validation data set to compare the performance of the validation data with the model development data. Although model selection was based on concordance scores, we also evaluated the final model according to its ability to correctly classify presence and absence for the ob-

served data. We classified points with a probability greater than or equal to 0.3 as present. We chose this probability because it represents a balance between model sensitivity and specificity (see Dettmers et al., Chapter 54).

The Probability Map

We used the logistic regression model to develop a probability map for Great Smoky Mountains National Park. The probability map represents the probability of detecting a wood thrush at any location in the Park. We created a 90×90-meter grid cell coverage of the park by programming the logistic regression model in ArcInfo map algebra language. Because the vegetation type, bedrock geology, and disturbance history coverages are only available as 90×90-meter grids, we coded the probability map at that same grain. For each grid cell, the values of each of the explanatory variables were determined by querying the appropriate GIS coverage. The probability for each grid cell was then calculated based on the logistic regression equation to create a probability coverage of detecting a wood thrush.

Results and Discussion

Two of the topographic indexes (SWI and LFI) tested were significantly correlated with wood thrush occurrence (Table 47.1). The model with the highest concordance score includes these two indexes as well as elevation, disturbance history, and geology type. Several squared and interaction terms were also significantly associated with wood thrush presence/absence. All of the variables included in the model were significant at the p less than 0.05 level. The overall model had a relatively high concordance of 77.9 percent and correctly classified 83 percent of the observed data points. Application of the best-fit model to the validation data set resulted in a concordance of 78.9 percent and correct classification of 86 percent of the observed data.

The parameter estimates indicate that elevation had the strongest predictive power of the variables included in the model. This result is consistent with the elevation range of the wood thrush, which is primarily re-

TABLE 47.1.

Result of logistic regression model selection for wood thrush presence/absence using model selection data.

Variable[a]	DF	Parameter estimate	Standard error	Chi-square Wald	Pr
INTERCPT	1	−12.3752	19.2466	4.9268	0.0264
ELEV	1	9.0385	1.9063	22.4803	< 0.0001
ELEV2	1	< −0.001	< 0.001	43.3424	< 0.0001
SWI	1	−2.8090	0.9866	8.1065	0.0044
SWI2	1	0.0231	0.0075	9.5938	0.0020
LFI	1	0.8238	0.3433	5.7578	0.0164
ELLFI	1	−1.2637	0.4531	7.7668	0.0053
GEOL	21	—	—	40.4717	0.0065
DISTURB	4	—	—	10.3836	0.0344

Concordance: Concordant = 77.9 percent; Discordant = 21.7; Tied = 0.4 percent; (356,040 pairs)
[a]Variable abbreviations are as follows: ELEV = elevation, ELEV2 = elevation squared, SWI = Shannon-Wiener Index of Topographic Complexity, SWI2 = SWI squared, LFI = Landform Index, ELLFI = elevation * LFI, GEOL = bedrock type, DISTURB = disturbance history type.

stricted to elevations below 1,200 meters while elevations in the park range up to 1,800 meters. This elevation boundary is evident in the probability map (see Fig. 47.2 in color section), which shows low probability of detecting a wood thrush in the center (highest elevations) of the park. The Shannon-Weaver Index of Topographic Complexity (SWI) was negatively associated with the likelihood of detecting a wood thrush and had the next-greatest explanatory power. The SWI is a relatively fine-scale (150-meter) measure of land form that describes the diversity of elevations, slopes, and aspects around a center point. The model also includes a positive association with the coarse-scale Landform Index, which is a broad measure of site protection on the scale of kilometers. This variable indicates that the probability of detecting a wood thrush increases with increasing protection. The disturbance history and geology type class variables were also significantly associated with the probability of detecting a wood thrush. Parameter estimates (Table 47.2) for individual disturbance types were calculated relative to the undisturbed type. Negative parameter estimates for areas that experienced logging in the past indicate these areas may be less likely to be associated with wood thrush occurrence than undisturbed areas. It is interesting that the vegetation type variable was

not identified as being significantly correlated with wood thrush presence/absence. It is possible that the vegetation type variable did not have any explanatory power beyond the variation explained by the topographic variables.

The specific results of this model are unlikely to be applicable outside of Great Smoky Mountains National Park, which is largely undisturbed. However, this research highlights the potential for topographic indexes to be useful predictors of songbird occurrence and they should be more commonly tested in habitat models. In addition to their possible predictive value, topographic indexes are attractive because many of them can be easily calculated from DEM data, which is often readily available for large areas. These vari-

TABLE 47.2.

Parameter estimates for the disturbance history type class variable. Estimates are relative to the undisturbed type.

Type	Parameter estimate
Selective cut	−0.0222
Light commercial cut	−0.3675
Heavy cut	−0.3603
Settlement area	0.4152

ables are also appealing because they may be useful surrogates for vegetation information that might otherwise have to be collected in the field.

Conclusion

In order to develop a management strategy for the conservation of wood thrush, it is important to identify variables associated with species distribution. Traditionally, ecologists have sought to identify fine-scale microhabitat features associated with occurrence. Microhabitat data is typically only available for small areas and can be quite costly and time intensive to collect, making large-area assessments difficult. On the other hand, the data needed to develop models based on topographic variables are often readily available in GIS databases and can be applied to large areas.

Acknowledgments

This work was funded by grants to Ted Simons from the Friends of Great Smoky Mountains National Park and the U.S. Geological Survey, Biological Resources Division. We would like to thank Jeremy Lichstein for helpful comments and statistical advice. We thank the staff at Great Smoky Mountains National Park for technical and logistical support. We also thank the many members of our field crew who made this work possible.

Controlling Bias in Biodiversity Data

David R. B. Stockwell and A. Townsend Peterson

Advances in information technology are revolutionizing the way science is being performed. One noticeable change is the growth of data mining—finding useful information among ad hoc collections of data (Weiss and Indurkhya 1997). In the biodiversity realm, museum specimen data are seeing increased attention (Peterson et al. 1998). Automated methods of analysis are being applied to such data via the World Wide Web (Stockwell and Peters 1999) for modeling species distributions. These models relate the occurrence of the species with environmental variables in what has been termed a phenomenological model (Maurer, Chapter 9). There is concern with the quality of the biological data used in these models, particularly the fact that museum data is not based on a comprehensive and random survey design. A large component of these concerns can be categorized as concerns about bias.

Randomized and comprehensive biological surveys such as the Maryland Biological Stream Survey are suitable for analysis using a range of statistical methods (Southerland and Weisberg 1995). However, randomized surveys of large areas are rare. More commonly, data are collected in an ad hoc or opportunistic fashion and without random sampling. Because these data violate assumptions of common multivariate statistical methods, their use is deprecated as a source of reliable results (James and McCulloch 1990). But does this mean that the data are flawed and unusable?

Another potential problem of museum collections data is that the spatial resolution of the records is rarely more precise than 0.5 kilometer. Although this may limit their use for smaller areas, it is adequate for use at resolutions greater than 1 kilometer. The scale of modeling is largely determined by the scale of the environmental variables correlated with the species data. As climate variables (temperature and precipitation) are usually the main factors in these models, and the resolution available for these variables is usually much greater than 0.5 kilometer (e.g., 2.5 kilometer minimum), the resolution of the museum collections data is not a practical limitation.

Another concern with museum collections data is that they are composed of records-of-presence of the species but no records-of-absence of the species. This characteristic can limit the application of chi-squared tests and other parametric statistical methods simply because they require more than one value. However, it has not prevented analysis, as shown by the BIOCLIM method, which develops an ecological niche model by fitting an environmental envelope at the climatic extremes of the data set (Nix 1986). A solution is to generate "pseudo-absences," a set of data points selected at random and used as absences that then allows parametric statistics to be applied to the data (Stockwell and Peters 1999). In the view of this chapter, presence-only data is simply the result of an extreme form of bias.

Therefore, the common concerns with the deficiencies of museum data do not necessarily mean they are unusable in analysis. Additionally, when museum collections databases are combined, museum specimen data represent a vast storehouse of species' occurrences, placing particular taxa in a complete geographic and hence ecological context (Peterson et. al. 1998). Thus, questions can be answered using these data, but inherent bias can undermine confidence in the results. These biases include focus of sampling activities in areas or habitats of easy access and concentration of sampling on taxa that are easily detected, captured, or preserved.

We should remember that problems of bias are also not restricted to museum data. Surveys crossing many habitats can have different observation rates due to the density of the habitat. If estimates of abundance are based on these survey results, patterns of abundance can be distorted by the relative difficulty of surveying the habitat. Methods of controlling bias are thus of interest, not only for analyzing museum data, but also because they are potentially applicable to any type of data.

There is sometimes the possibility of modifying the survey design to reduce bias, or augmenting records with field notes where species were seen but not collected. In this chapter, however, we are concerned with strategies for dealing with bias after data collection. Developing analytical methods less affected by bias, or methods that can reduce bias prior to analysis, presents many advantages. Control of bias allows use of data that could not otherwise be used. Hence, flexible bias management should be an important feature of learning systems in applications for automated modeling, based on data mining. More generally, a better understanding of bias in ad hoc data sets, such as most biodiversity information, is needed.

The analysis of biodiversity data we are concerned with is the ecological niche modeling of species, which quantifies probability of occurrence as a response to environmental variables. To determine the effects of bias on model performance, we develop models of species' ecological niches using two major environmental dimensions, temperature and rainfall. We define an experimental protocol for incorporating bias and then controlling its effects on predictions of species distribution. We show the effect of methods of bias control on accuracy and predicted distribution maps of a well-surveyed species.

The analysis in this study uses methods incorporated into the GARP (genetic algorithm for rule-set prediction) modeling system (Stockwell and Noble 1992; Stockwell 1999; Stockwell and Peters 1999). The performance of the methods on biased and unbiased data is compared with and without methods for controlling bias. The primary concern is to demonstrate that the methods do indeed control bias in data sets.

Forms of Bias

A number of meanings exist for the word *bias*. Clarification of terminology is essential if workers are to communicate effectively in ecology (Morrison and Hall, Chapter 2). Unfortunately, the semantics of bias has not been treated as extensively in the ecological literature. Bias can be interpreted in at least three ways: statistical bias, inductive bias, and sampling bias.

Statistical Bias

The tendency of a statistical parameter, such as a mean or variance, to converge on a value not equal to its true value (Mendenhall et al. 1981) is referred to as *statistical bias*. For example, if the mean of a population parameter, such as abundance or probability, is P_1, a measure of that population parameter is biased if given a large number of samples, it gives a value P_2, where $P_2 \neq P_1$.

Inductive Bias

Algorithms and statistical models are often used to explore patterns in data. For example, stepwise elimination of variables in linear regression is an iterative algorithm for exploring possible models with a minimal number of variables. This process of developing a model through fitting a range of models to data is called *induction*. However, the effectiveness of such approaches depends critically on algorithmic design choices, such as representation language and performance criteria (Forsyth 1981). Little is known about the effects of these choices and their relationships to types of data (DeJong and Mooney 1986). One of the most difficult problems in practical induction is assessing the way the design choices interact with data sets analyzed (Turney 1991). In practice, factors are manipulated experimentally until the algorithm meets set criteria of adequacy in the application domain. When done informally, this practice is known as "tweaking." Because the biases in an algorithm are rarely made explicit, researchers struggle with the application of vast numbers of subtle biases (Michalski 1983).

Sampling Bias

A characteristic of data relative to the population from which they are drawn is *sampling bias*, which can cause statistical bias if the set of data is a nonrandom subset of the entire population. The most typical example for museum data is data collected from sites in a limited area rather than from over the whole area. However, sampling bias can also occur where data come from a set of representative sites in which some of the sites consist of multiple samples, such as collections focused in one area but only a few in another.

Statistical, inductive, and sampling bias can be present in any empirical study. Owing to the difficulty of obtaining truly random samples and robust methods that would be free of these problems, it is possible that bias is the greatest impediment to the reliable analysis of natural data sets.

Controlling Bias

Where random sampling is not possible because data have already been collected, an effective *a posteriori* approach to controlling bias is needed. This includes preparation of data for the analysis, both through selection of controlled sets of data to be modeled and through the selection of the independent variables used to develop the model. We now examine sources of bias more peculiar to biodiversity data and explore approaches to controlling them through data manipulation.

Presence-only Bias

Much of biodiversity data—especially data associated with scientific specimens—present an odd situation in which the location of specimens collected is recorded

(positive data) but absence of species is not. These data represent only a subset of possible information: points of presence only. The approach for controlling this source of bias is to generate *pseudo-absences*: points that provide a contrast in prior probabilities of presence to the given data. One method of generating the set of data for pseudo-absences is to draw points at random anywhere the species has not been recorded, called *background* points. This method is used in the GARP system (Stockwell and Peters 1999). Alternatively, these points could be drawn from specific areas known not to hold the species, based on field notes of collectors or the recollection of experienced field workers.

Abundance Bias

Two examples of abundance bias are rarity and hotspots. Species can be recorded from very few locations or many owing to a number of factors: inherent ease of encounter or capture, intrinsic rarity, or difficulty of capture or preparation. These imbalances can be regarded as a form of bias, since frequencies are affected by observational factors. In hotspots, large numbers of specimens may be recorded at some locations because a particular collector focused activities at particular sites, or because many collectors worked particular sites more than others did. The approach here is to map the data into a regular grid (rasterizing data), allowing only one data point to be selected from each grid cell. We then control this source of bias by generating data sets for analysis with even distributions, using resampling.

Correlation Bias

Bias can be introduced through the tendency of sampling to be more frequent in certain geographic features. Collections along roadsides are a particularly frequent example. Another is the tendency of inaccessible habitat types such as aquatic vegetation or montane areas to be underrepresented. The approach developed here is to identify those variables correlated with the form of bias and eliminate them from the set of possible predictor variables in the model. Certain methods of analysis may be more affected by correlation bias than others might.

Methods

The environmental data employed for predicting the distributions consisted of two climate layers: average annual temperature (Fig. 48.1A) and average annual precipitation (Fig. 48.1B). These data layers were taken from the NOAA-EPA (1992) data set, comprising the continental United States at a resolution of 0.5 raster grid. In practical modeling applications, many other variables may be required to obtain highly accurate predictions of species' distributions. For simplicity, though, as the purpose of this study is illustrative rather than predictive, only these two variables were used.

The species modeled was *Hylocichla mustelina*, the wood thrush. The point occurrence data used in this study were drawn from the U.S. Breeding Bird Survey (BBS) data set (Sauer et al. 1997). This data set presently consists of approximately 6×10^5 records of species of bird species sighted in yearly census at sites across North America. The surveys are conducted along prescribed transects approximately 45 kilometers in length. The grid cells at 0.5-degree resolution containing survey paths are shown in Fig. 48.1D.

Experimental Protocol

Our experimental protocol was designed to allow comparison of the accuracy of models and distributions developed on biased data with and without the

method of controlling bias. The exact approach depends on the particular bias. In general, the protocol consisted of:

1. Take a large data set with almost-complete data about the distribution of a species.
2. Sub-sample those data in a controlled way to develop data with the bias being considered.
3. Develop models on the biased data with and without the method of controlling bias.
4. Evaluate the accuracy of the model by comparison with the known distribution.

Bias in the data used to develop the model should lead to inaccuracy in modeling niches and in resulting geographic distributions. Using biased data, the control method should lead to increased accuracy, compared with analysis without bias control.

In this protocol, the comparison data consist of all grid cells that have been surveyed by the BBS—determined to be approximately 50 percent of the grid cells in the continental United States. Data sets for analysis are random samples of (1) sites where birds have been observed (presences), and (2) sites where they have never been observed (absences). Data sets for analysis contain 2,500 sampled points.

We used three modeling methods. The first, *logit*, was a quadratic logistic regression model for predict-

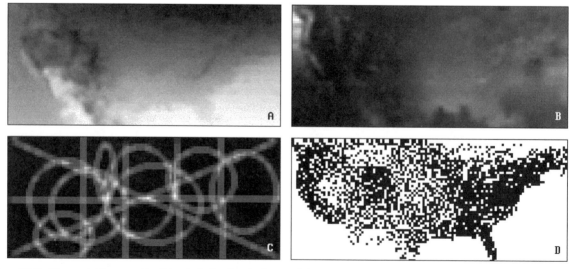

Figure 48.1. The variables used in this study for developing models: (A) the average annual temperature and (B) average annual precipitation for the continental United States at a resolution of 0.5 degrees derived from interpolated weather station data, (C) a simulated road variable, and (D) a mask consisting of cells containing survey locations. Legend: white = masked.

ing binary variables that has been applied to predicting the probability of occurrence of species (Austin et al. 1990). The model was developed to fit probabilities to a linear combination of variables where the response could be either linear or unimodal. Examples of species' responses to environment that would be well modeled by this method are probability increases with increasing precipitation (linear), or particular temperature range (unimodal). The probability of occurrence P is

$$P = \frac{1}{1 + e^{-y}}$$

$$y = a_1 x_1 + a_2 x_1 x_1 + a_3 x_2 + a_4 x_2 x_2 + \ldots + a_{2n} x_n x_n$$

where x_i represents the value of a particular environmental parameter I and a represents a fitted coefficient.

The second method, *envelope*, is an envelope-fitting method similar to BIOCLIM (Nix 1986). An envelope uses only presence points by fitting an envelope around the range of the set of points. In contrast, the logit model cannot be used for presence-only data as it requires both presence and absence points for fitting. This model is described as

$$P = 1 \text{ if } x_1 \text{ in } (a_1, a_2) \wedge x_2 \text{ in } (a_3, a_4) \wedge \ldots \wedge x_n \text{ in } (a_{2n-1}, a_n)$$
$$0 \text{ otherwise}$$

The parameters $a_i \ldots a_n$ in a presence-only model are the minimal range of the parameters that fits all the data between them. The values for each pair of parameters are calculated variable by variable.

The GARP *rule-set* method is used in most other analyses (Stockwell and Noble 1992; Stockwell and Peters 1999). This artificial intelligence method develops a set of models, including regression and envelope models as described above. The models are optimized in an iterative fashion by incremental modification and testing in an evolutionary algorithm. The resulting set of models is output and available for prediction. In using the set of models for prediction, the best model at each prediction point (grid cell) is selected and the result used as the predicted value. This method has shown greater robustness in a variety of applications due to its capacity to draw from a set of possible models. These and other algorithms for species distribution modeling are implemented in the GARP analysis package (http://biodi.sdsc.edu).

Results

The accuracy of the results is quoted as the 95th percentile range in the mean accuracy over six trials. Using this method of comparison, the actual accuracy is expected to fall between this range 95 percent of the time. Accuracies are not significantly different at the 95 percent confidence level if the quoted ranges of the accuracy overlap. If the range of accuracies does not overlap, we may be confident in asserting that the expected accuracy of the results is significantly different.

The predicted distribution of the experiments shown in Figures 48.2 to 48.4 should also be considered along with accuracy figures, as they are a more sensitive indicator of the results. The panels in the figures are referred to in a row-column order, with A in the top righthand corner and D in the bottom lefthand corner.

Presence-only Bias

In this protocol, we examined the bias toward presence-only data by comparing the results of different modeling methods. The envelope modeling method used only data from sites where the species occurred by fitting a climatic envelope around the presence data points. The envelope model minimizes error of presence, using the GARP option "envelope" to produce a model capable of predicting presence and absences. The presence rule encloses all the data points and the absence rule is the negation of that rule, predicting absence for all points outside the range of the variables. We then added pseudo-absences by adding the pseudo-absence "background" information into the

TABLE 48.1.

The accuracy of trials using presence-only bias control (see Fig. 48.2).

B. Bias	C. Bias control	D. Bias control
Presence-only data, Envelope method	Presence and background data, Logit method	Presence and background data, GARP method
n = 6	n = 6	n = 6
mean = 0.685	mean = 0.748	mean = 0.778
s.d. = 0.014	s.d. = 0.010	s.d. = 0.015

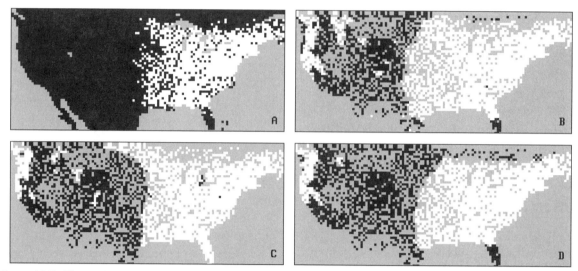

Figure 48.2. The predicted distributions of *Hylocichla mustelina*, or wood thrush, using modeling methods that make use of presence data only, and presence and absence data: (A) the actual distribution of the wood thrush, (B) the (biased) predicted distribution using the bioclimatic envelope method, (C) the (bias controlled) predicted distribution using the GARP rule-set method, and (D) the (bias controlled) predicted distribution using the logistic regression method. Legend: black = absent, white = present, gray = unpredicted.

data. Models that minimize error of presences and absences are the rule-set method and logistic regression (logit).

The accuracy in each of these experiments is shown on Table 48.1. The accuracy of predictions of the species' distribution in the envelope method is between 0.67 and 0.69. The accuracy with background data and the logit method is between 0.74 and 0.75.

This is a significant improvement in accuracy. The GARP rule-set method gave the highest accuracy (0.77 to 0.78) of the three methods.

The comparison of the actual predicted distributions is shown in Figure 48.2. The envelope model developed using presences only (B) predicts a broader distribution than the rule-set model (D) that uses presence and absence data. Providing pseudo-absences

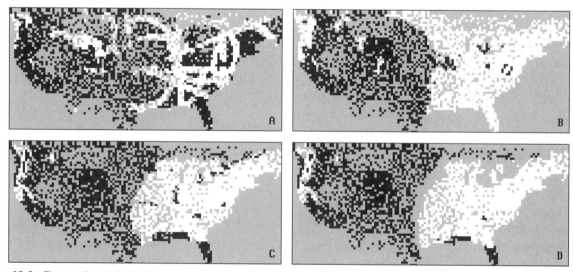

Figure 48.3. The predicted distributions of *Hylocichla mustelina*, or wood thrush, using different proportions of presence and absence data. The actual distribution (A) was sampled according to proportions in A, to produce (biased) predicted distribution B. Sampling according to strongly biased proportions in original data produced predicted distribution C. Sampling from A using even proportions (bias controlled) produced predicted distribution D. Legend: black = absent, white = present, gray = unpredicted.

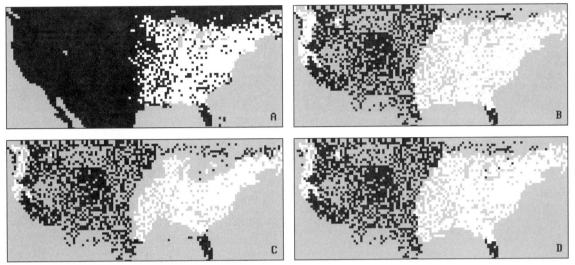

Figure 48.4. The predicted distribution of *Hylocichla mustelina*, or wood thrush, from presence and absence data selected for proximity to simulated roads. The predicted distribution of the bird in A and B uses the logistic regression method. In A, the (biased) model was developed using climate and road proximity variables and shows the pattern of the roads. In B, the (bias controlled) model was developed without the road variable and shows a more accurate predicted distribution. Maps C and D are the predicted distributions of the bird using the GARP (genetic algorithm for rule-set prediction) model. In C, the (biased) models used the climate and road variables. In D (bias controlled), only climate variables were used. The predicted distribution using the GARP model did not show a pattern of roads in either case. Legend: black = absent, white = present, gray = unpredicted.

based on a random background data population helps to constrain the model and increase accuracy of prediction of the absence of species. The distribution predicted by the logistic model is also shown (C). The distribution predicted using this method considerably overpredicts the distribution of the species.

Abundance Bias

In this protocol, we develop a biased data set by using data points with varying proportions of presence and absences. For example, a form of hotspot bias in the BBS data could be due to sampling of the same area over many years. The method of controlling this bias is "rasterizing"—mapping the temporal data onto a grid and allowing just a single point to represent each grid cell.

Although rasterizing reduced some of the redundancy in the data, models are still exposed to rarity bias. For example, the proportion of cells occupied by the species influences the proportions in the data used to develop the model. If all data points from the grid are used, the biased set has a small proportion of presences as compared with absences. We controlled this

source of data by sampling to an even proportion of data points among both presence and absences.

In the example species, the cells where the species were found constitute 44 percent of the total cells. The biased data was modeled with 44 percent presences and 56 percent background (Fig. 48.3B). The data were again biased to a proportion of 5 percent presences and 95 percent absences (Fig. 48.3C). The bias controlled data sets had a proportion of 50 percent presences and 50 percent background (Fig 48.3D). The accuracy expected using the rule-set method on data sets of 44/56 proportions was between 0.78 and 0.79 (Table 48.2). The accuracy achieved on datasets of 5/95 proportions was between 0.80 and 0.81. The expected accuracy using even proportion of data was between 0.78 and 0.79.

The accuracy of the models is similar in each case, indicating that the rule-set method is not sensitive to the relative proportions of presence and absence data used to develop the model. Examination of the predicted distributions (Fig. 48.3) shows the most extreme abundance bias (Fig. 48.3C) predicted less area

TABLE 48.2.

Accuracy of trials of abundance bias control (see Fig. 48.3).

B. Bias	C. Bias control	D. Bias control
Uneven (44/56) proportions of presences and absences, GARP method	Uneven (5/95) proportions of presences and absences, GARP method	Even proportions of presences and absences, GARP method
n = 6 mean = 0.787 s.d. = 0.015	n = 6 mean = 0.812 s.d. = 0.012	n = 6 mean = 0.783 s.d. = 0.008

for the distribution of the species relative to other models (Fig. 48.3A).

Correlation Bias

In this protocol, we develop a biased data set by selecting the data preferentially with proximity to a given set of lines crossing the study area. The bias is created by a mask that simulates roads that might cross the region (Fig. 48.1C). This protocol simulates a collection strategy where the presence or absence of specimens is recorded along roadsides.

The control of this form of bias includes both methods and the strategy for inclusion of predictor variables. In the evaluation of logit and GARP methods, the road variable is first included as a potential predictor variable. The method of bias control is omission of the biased variable in development of the model. The biased data set is then analyzed without the simulated road variable.

The accuracy of the biased logit model was between 0.67 and 0.68 (Table 48.3). The accuracy of the bias controlled model was significantly higher, between 0.75 and 0.76. Figure 48.4A clearly shows that logit analysis of biased data with the simulated road variable present predicts a species distribution containing patterns of the road distribution. When a logistic regression analysis is performed without the road variable using the biased data, the pattern of roads is not evident (Fig. 48.4B). Thus, removing the variable that is correlated with the bias in the data helps to control the effects of the bias in the logistic regression method.

When the analysis is performed using the rule-set method and the road proximity variable, the accuracy is between 0.76 and 0.77, and the resulting predicted distribution closely resembles the actual distribution of the bird (Fig. 48.4C). A slightly more accurate result was achieved without the simulated road variable (between 0.79 and 0.79). This demonstrates clearly that the rule-set method is capable of predicting the correct distribution, even in the presence of the variable correlated with the bias in the data.

Discussion

The above results include the distribution predictions and accuracy of models of the wood thrush developed with and without bias control methods. The presence-only and correlation bias control methods showed improvement in the accuracy of predicted distributions at dealing with the respective forms of biased data.

TABLE 48.3.

The accuracy of trials using correlation bias control (see Fig. 48.4).

A. Bias	B. Bias control	C. Bias	D. Bias control
Climate and simulated road predictors, Logit method	Climate and no simulated road predictors, Logit method	Climate and simulated road predictors, GARP method	Climate and no simulated road predictors, GARP method
n = 6 mean = 0.675 s.d. = 0.005	n = 6 mean = 0.758 s.d. = 0.008	n = 6 mean = 0.773 s.d. = 0.018	n = 6 mean = 0.795 s.d. = 0.008

The use of background data to provide pseudo-absences for the GARP rule-set model had greater accuracy than an envelope model developed using only presence data (Table 48.1). This demonstrates that only having positive information about presence of a species and no information about absences of that species does not necessarily present an obstacle to accurate prediction.

Abundance bias did not show a strong effect on the accuracy of models developed using differing proportions of presence and absence data, even when the proportion varies between 50/50 and 5/95. However, the effect could be seen on the predicted distributions, where the uneven proportions resulted in contraction of the predicted distribution (Fig. 48.3C).

The effect of correlation bias on predicted distributions was clearly shown using the simulated road variable in a logistic regression analysis (Fig. 48.4A). The removal of this variable as a potential predictor produced accurate distribution patterns on data strongly biased with a simulated road mask. The rule-set method also produced accurate distributions with or without the biasing road proximity variable.

In general, the bias control methods improve the predicted distributions of species, particularly in the case of the logistic regression method. The GARP method was less susceptible to bias in the data but did show quantitative improvements when the bias control methods were applied. In the case of logistic regression modeling, elimination of the biasing variable when analyzing correlation bias allowed a dramatic improvement in the accuracy of the predicted distribution. These results demonstrate the potential of bias control methods to overcome problems posed in analysis of museum data. These improvements were quantified by accuracy measures and, even more clearly, through visual examination of the predicted distributions.

In this study, we have categorized biases into three types based on where the bias is present. Abundance bias occurs in the proportions of data presented to the analytical method. Presence-only bias can be regarded as an extreme form of abundance bias in which the presence data are highly abundant. The problem of these biases is that bias results in less information being effectively present. In the case of

abundance biases (hotspots and rarity), the samples of biased data contain few points providing information about presences. In the case of presence-only bias, no information about absences is present. Resampling to even proportions of presences and absences corrects this by putting more information into the training data set.

In presence-only bias and correlation bias, the bias in the data sample pushes methods into suboptimal solutions. In presence-only bias, the model contains all presence points; an optimal solution if only presence points were important is to the detriment of accuracy by predicting too many absences as presences. In the case of correlation bias and the logistic regression algorithm, the correlation of presence data with the biasing (road) variable ensures that the dummy variable figures highly in the model. The variable is actually irrelevant to the species and its ecology.

These are cases of inductive bias because the assumptions of the analytical system force the suboptimal solutions. The removal of the biasing variable from the set of predictor variables in susceptible analysis methods such as logistic regression helps to control correlation bias. The redundancy in the GARP rule-set model makes it less susceptible to correlation bias, as models in the rule set that incorporate the biasing variable are not necessarily used in predicting the presence or absence of the species.

The biases described above represent the main biases we have encountered in analyzing museum data, although others certainly exist. With the exception of correlation bias, methods of controlling bias can be incorporated into the standard operation of the modeling system: rasterizing, use of pseudo-absences, and resampling in even proportions. That is, the control method operates without human intervention and can therefore be part of the standard methodology.

In the case of correlation bias, we need to determine which variables are potentially likely to cause increased error due to correlation with the sampling effort of the survey. Roadsides have been mentioned, and consequently variables based on proximity to roads should never be used. Biasing variables may be detectable using jackknife and other resampling approaches applied to environmental data sets (Peterson and Cohoon 1999).

The logistic regression method requires the identification and removal of potential biasing variables to control bias in the data. The GARP rule-set method shows the capacity, in this example, to control correlation bias without the need to identify biasing predictor variables. These results show the great potential of the GARP rule-set method for utilizing data containing unknown biases, such as museum collections data.

Conclusions

Managers could use these methods to improve the ac-curacy of predicting species distributions when using museum data or other ad hoc sampled data. They could reduce potential errors due to bias by

1. Evaluating the potential for bias and looking for patterns of bias in their biodiversity data.
2. Identifying and removing variables that correlate with bias from the analysis.
3. Applying the strategies implemented in the GARP modeling system for reducing the effect of bias: rasterizing and, where there are no absence observations, augmenting the data set with randomly selected background data.

Modeling Cowbird Occurrences and Parasitism Rates: Statistical and Individual-based Approaches

Ann-Marie Shapiro, Steven J. Harper, and James Westervelt

Fort Hood, Texas, a large army installation, actively manages significant breeding populations of the black-capped vireo (*Vireo atricapillus*) and the golden-cheeked warbler (*Dendroica chrysoparia*) within its boundaries (Weinberg et al. 1998; Jette et al. 1998). Both species are susceptible to brood parasitism by the brown-headed cowbird (*Molothrus ater*). Long-term increases in cowbird populations and resultant parasitism pressure have been implicated as significant factors in the decline of a large number of passerines (Robinson and Wilcove 1994; Robinson et al. 1995b). On the Fort Hood landscape, the black-capped vireo appears to be particularly vulnerable to reduced reproductive success from cowbird parasitism (Hayden et al. 2000). Low productivity due to cowbird parasitism has been identified as a major reason for the endangered status of this species (USFWS 1991). As a result, an important component of endangered species management on Fort Hood is the systematic live trapping and removal of female cowbirds from the installation landscape (Eckrich et al. 1999). In 1998, over 3,800 female cowbirds were trapped and 159 were shot within the breeding season on Fort Hood (Eckrich and Koloszar 1998). In comparison, there were fewer than 250 black-capped vireos and only eighty-nine territorial male golden-cheeked warblers documented on the installation (although these data come from intensively studied sites and do not reflect total

installation populations; Koloszar and Bailey 1998; Craft et al. 1998).

Cowbird control has proved successful and has generated a large increase in black-capped vireo nesting success. For example, over 90 percent of black-capped vireo nests were parasitized prior to initiation of the control program, while in recent years fewer than 15 percent of nests have been parasitized (Eckrich et al. 1999). This comprises fairly strong evidence that until recently cowbird parasitism has been a limiting factor in the reproductive success of this species. Without landscape-level changes in factors supporting cowbird populations on Fort Hood, control of cowbirds through trapping and shooting must be conducted indefinitely for continued benefits to be realized. An experiment at Fort Hood showed that cessation of the control program resulted in increases in cowbird densities, increases in parasitism levels, and decreases in the reproductive success of black-capped vireos (Cook et al. 1998). Therefore, one goal of land managers at Fort Hood is to optimize the placement of traps in space and time in order to capture the most cowbirds with the least effort.

Our research team has developed and applied simulation models to assess landscape- and population-level implications of various endangered species management efforts, including the cowbird control program (Trame et al. 1997, 1998). After efforts to model cowbird parasitism using statistical approaches within a dynamic landscape simulation model, the Fort Hood Avian Simulation Model (FHASM), failed to produce spatially explicit results (Trame et al. 1997), we developed the individual cowbird behavior model (ICBM; Trame et al. 1998; Harper et al. 2001) using an individual-based approach. Despite a wealth of empirical data related to cowbird parasitism, cowbird control efforts, and reproductive success in the two endangered species, and somewhat less knowledge about the vegetation on Fort Hood, a "classical," statistical approach to modeling cowbird parasitism proved unsatisfactory. The simple, rule-based approach of individual-based modeling appears to have served better for this particular application.

Individual-based models differ from aggregated, state-variable approaches in both philosophy and content (Grimm 1999). Most classical approaches in ecology, whether static or dynamic, rely on averaging the attributes of a population and projecting the emerging characteristics, which may not represent the variability exhibited by the natural system. Model projections can be improved, to a certain extent, through the addition of greater complexity in model structure and the incorporation of greater detail in input data. A relatively recent, alternative approach is to capture the different attributes and behavior of each individual organism in question, keep track of individual variation throughout a simulation, and eventually aggregate the resulting patterns at a higher level of organization for analysis and interpretation. The individual-based modeling approach supports the use of fairly simple rule sets while creating aggregated results that reflect higher levels of complexity (Huston et al. 1988). The challenge of keeping track of individual attributes of many organisms in a single simulation is no longer a limitation, since greater computational power has become readily available. The development of hundreds of individual-based models (Grimm 1999) supports the assertion that the individual organism is the logical unit for modeling certain ecological questions (Judson 1994). As this approach gains acceptance, new simulation environments provide the tools needed to develop individual-based ecological models. Swarm (Minar et al. 1996), Echo (Jones and Forrest 1993, http://www.santafe.edu/projects/echo/echo.html), Gecko, a simulator developed within Swarm (Booth

1999), the Model of Animal Behavior (MOAB; http://www.usgs.gov/tech-transfer/factsheets/FS-056-97.html), XRaptor (Bruns et al. 1999), and even the educational tool, Ecobeaker (http://www.ecobeaker.com/index.html), all provide development environments for individual-based ecological models. Topping et al. (1998) have presented a new "biological programming language" to facilitate programming of individual-based ecological models.

This chapter will compare the development and content of two models of cowbird parasitism. First, we list the data sources available throughout both modeling efforts. Then, we briefly sketch the approach and content of FHASM, our statistical landscape model, focusing on the portion of the model that addressed cowbird parasitism. Then, we detail the approach and content of the ICBM, our individual-based cowbird behavior model. Last, we compare the ability of the two models to predict brown-headed cowbird occurrences and parasitism, and briefly discuss the importance of scale in studying and modeling brown-headed cowbird behavior.

Data Sources

During model development, we had access to the results of years of field research and management efforts on Fort Hood. Nest-specific parasitism and reproductive success measures were available from 1994 to 1996, trapping results were available for 1993–1996, trap locations were available for 1995 and 1996, and an avian community study documented the distribution and abundance of potential cowbird hosts within warbler breeding habitat in 1994. In addition, summarized data on nesting success and trapping efforts were available for all years since 1987. A 1987 vegetation map was available (Trame et al. 1997), but confidence in the map was very low and we did not feel comfortable using it to develop predictive relationships between vegetation and cowbird or host parameters. An extensive land-cover trends analysis (LCTA; Diersing et al. 1992) database was provided for 1989 through 1995 and contained data from randomly placed transects, including vegetation, vertebrates, and land uses across the entire installation. A GIS database was also available from various efforts on the installation since the mid-1980s. When we initiated the individual-based approach, it became obvious that we would need to simulate the behavior of cattle herds, and we sought additional data to support that requirement. We were provided with the locations of supplemental feeding stations and salting areas used in cattle herd management on Fort Hood, and we attained 1996 Landsat imagery of the installation. Two vegetation categories, grassland and non-grassland, were determined from the imagery using neural network classification. The results of two neural network calculations were combined to create a 50-meter-resolution vegetation map. Only cells classified as grassland in both calculations were accepted as grassland cells. The neural network was trained with 152 LCTA points from 1995. Settings for the two calculations were percent trained: 0.75, 0.75; learn: 0.20, 0.60; momentum: 0.90, 0.50; epochs: 100, 100; and iterations: 100, 1,000. Error for the two calculations were 0.0979 and 0.0094.

The Statistical Approach

Our initial modeling effort produced the Fort Hood Avian Simulation Model (FHASM), a dynamic, spatially explicit model of ecosystem processes and the population dynamics of the black-capped vireo and the golden-cheeked warbler (Trame et al. 1997). Our first effort to simulate cowbird parasitism was embedded within the development of this complex model. The model simulated changes in vegetation and in golden-cheeked warbler and black-capped vireo populations across the installation landscape over 100-year periods. We designated management policies, or altered a variety of input variables; FHASM generated maps of vegetation, fires, management activities, and bird populations for each year of the simulation. The relative results of different management scenarios were compared, something that could not be assessed easily nor be accomplished quickly through field studies.

FHASM was created using the Spatial Modeling Environment (SME) software, developed at the University of Maryland (http://kabir.cbl.umces.edu/SME3/index.html), GRASS GIS software (http://www.baylor.edu/~grass/), and STELLA modeling software

(High Performance Systems 1994). The general model of FHASM was constructed using the graphical, user-friendly STELLA program on a desktop computer. The propagation of the general FHASM model across a grid landscape was accomplished by the SME software, running on a single Unix workstation.

The landscape of Fort Hood was divided into 21,540 square raster cells, each representing a 4-hectare area (200×200 meters). This cell size provided a reasonable representation of the average breeding territory for either endangered species on Fort Hood (Weinberg et al. 1996; Jette et al. 1998). FHASM proceeded in a three-month timestep, allowing the use of summary statistics that described recruitment by breeding birds over an entire breeding season (quarter 2). FHASM was composed of five submodels—habitat, avian, management, and two submodels that managed spatial data. The habitat submodel simulated successional changes among fifteen mapped plant communities. Plant community type then was used by the avian submodel, which determined the quality of the vegetation as breeding habitat for vireos and warblers. Territory choices were made based on habitat characteristics and the history of occupation of a site. The management submodel simulated the activities of endangered species managers to promote and/ or protect endangered species habitat (affecting the habitat submodel), and to reduce the risk of cowbird parasitism (affecting the reproductive parameters in the avian submodel). The avian submodel calculated reproductive success from habitat quality and the probability of parasitism by cowbirds. The map input submodel informed SME of the proper GRASS input maps from which to attain values for mapped variables. Finally, the simulation submodel stored spatial variables (in the form of GRASS maps) from each time step and made them available to other sectors of the model, as needed.

We modeled cowbird parasitism in FHASM based on a strong empirical relationship between cowbird control efforts and installation-wide parasitism rates. As cowbird control efforts (days trapping or days shooting), and numbers of females killed, increased from 1988 through 1995, installation-wide parasitism rates dropped dramatically from 90.91 percent in 1987 to a low of 12.59 percent in 1994 and 15.17

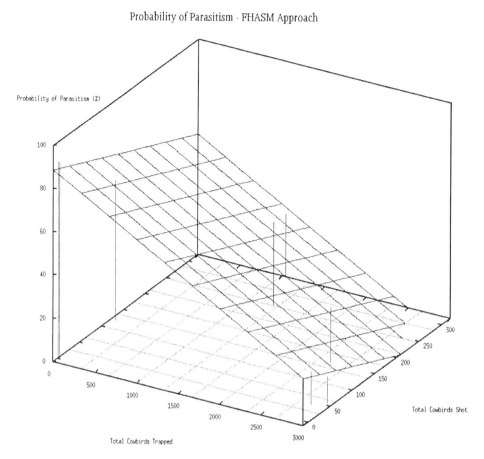

Figure 49.1. The empirical relationship between total numbers of female brown-headed cowbirds (*Molothrus ater*) trapped and shot on Fort Hood and installation-wide nest-specific probabilities of parasitism.

percent in 1995 (Hayden and Tazik 1991; Bolsinger and Hayden 1992, 1994; Tazik and Cornelius 1993; Weinberg et al. 1995, 1996; Fig. 49.1). The same data sources revealed a coarse spatial pattern based on accessibility to different areas on the installation. Due to the nature of the military mission, there were areas on post that were inaccessible to regular civilian access, and thus no consistent cowbird control was possible. For these inaccessible areas, the number of parasitized vireo nests was strongly related (adjusted R^2 = 0.94839) to the number of females captured by trapping (F = 129.6201, df = 1, P < 0.0000). In accessible areas, percentage of parasitized vireo nests was strongly related (adjusted R^2 = 0.980507) to the number of females killed by shooting (F = 59.5913, df = 1, P < 0.0015), the number of females captured by trapping (F = 166.2290, df = 1, P < 0.0002), and their interaction (F = 16.8250, df = 1, P < 0.0148). We

searched for similar relationships at a finer spatial scale, across the five historically recognized subregions of the 87,000-hectare installation, without success. Although parasitism was influenced by total number of females captured, the relationship did not show a spatial effect. This suggested that differences in parasitism rates were due to total trapping efforts across the entire installation.

Based on these analyses, the risk of cowbird parasitism was modeled as a function of control effort and the efficiency (success rate) of the control effort (Table 49.1; Fig. 49.2). Effort was defined as the number of trap days or shooting excursions for the second quarter of each simulation year (the breeding season). Efficiency was defined as the number of females trapped per trap day or the number of females shot per shooting excursion. Since there were differences among regions in the efficiency of traps, trapping effort and ef-

TABLE 49.1.

Input data sources used for the brown-headed cowbird (*Molothrus ater*) control section of Fort Hood Avian Simulation Model (FHASM).

Variable name	Content	Data source	Value used[a]
M TRAPDAYS LF	Sum of number of traps open by number of days open within the region of Fort Hood known as the "Live fire area"	1995 data, provided by Mr. Gil Eckrich in 1996	0
M TRAPDAYS EARA	Sum of number of traps open by number of days open within the region of Fort Hood known as the "eastern ranges"	1995 data, provided by Mr. Gil Eckrich in 1996	1890
M TRAPDAYS WERA	Sum of number of traps open by number of days open within the region of Fort Hood known as the "western ranges"	1995 data, provided by Mr. Gil Eckrich in 1996	2074
M TRAPDAYS WEFH	Sum of number of traps open by number of days open within the region of Fort Hood known as "West Fort Hood"	1995 data, provided by Mr. Gil Eckrich in 1996	0
M TRAPDAYS CANT	Sum of number of traps open by number of days open within the region of Fort Hood known as the "cantonment"	1995 data, provided by Mr. Gil Eckrich in 1996	153
M TRAP EFFIC LF	Mean number of females trapped per trapday within the region of Fort Hood known as the "life fire area"	1995 data, provided by Mr. Gil Eckrich in 1996	0.0664
M TRAP EFFIC EARA	Mean number of females trapped per trapday within the region of Fort Hood known as the "eastern ranges"	1995 data, provided by Mr. Gil Eckrich in 1996	0.4894
M TRAP EFFIC WERA	Mean number of females trapped per trapday within the region of Fort Hood known as the "western ranges"	1995 data, provided by Mr. Gil Eckrich in 1996	0.8120
M TRAP EFFIC WEFH	Mean number of females trapped per trapday within the region of Fort Hood known as "West Fort Hood"	1995 data, provided by Mr. Gil Eckrich in 1996	0.0614
M TRAP EFFIC CANT	Mean number of females trapped per trapday within the region of Fort Hood known as the "cantonment"	1995 data, provided by Mr. Gil Eckrich in 1996	1.3660
M SHOOTDAYS	Sum of number of days in which a cowbird technician patrols endangered species breeding habitat and shoots individual female cowbirds	Estimated by Mr. Gil Eckrich to represent shooting effort in 1995	27
M SHOOT EFFIC	Mean number of female cowbirds killed per shooting day	1989 data, provided by Mr. Gil Eckrich in 1996	1.5682

[a]Can be altered by the user.

ficiency were input at the regional scale, while effort and efficiency for shooting was modeled at the scale of the entire installation. The products of effort and efficiency across the installation determined a single rate of parasitism for the entire landscape.

As a result, the trapping of cowbirds affected simulated parasitism on an installation-wide scale (i.e., the effects of trapping were experienced by nests uniformly across the installation, rather than in the local area surrounding a trap). Although this uniformity was not believed to be accurate, it was the only result

successfully developed using statistical approaches (see Discussion for explanation of other analyses we attempted).

The Rule-Based, Individual Approach

In response to the limited success of modeling cowbird parasitism in FHASM, we developed a second model to simulate cowbird behavior and trapping efforts on Fort Hood. The individual cowbird behavior model (ICBM) was a two-stage individual-based model that

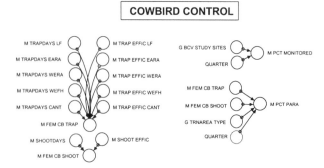

Figure 49.2. STELLA diagram of the Cowbird Control subsection of FHASM. Explanations of input variables are found in Table 49.1.

simulated female cowbird movement in response to breeding habitat quality, the locations of feeding cattle, and movement decision rules (Table 49.2). The first stage of the model produced maps of cumulative species occurrences through time for both individual female cowbirds and for cattle herds. If desired, output could be analyzed within the second stage of the ICBM to ascertain optimal trapping strategies for endangered species management (see Harper et al. 2001 for a discussion of the trapping component of the model and application of the two-stage model). Otherwise, output from the first stage could be used directly for applications such as risk-analysis of endangered species breeding habitat. This chapter focuses on the technical development of the first stage of the ICBM.

The ICBM was developed using GRASS GIS (http://www.baylor.edu/~grass/) and Swarm (Minar et al. 1996), an individual (agent-based) simulation envi-

ronment available from the Santa Fe Institute (http://www.santafe.edu/projects/swarm/swarmdoc/swarmdoc.html). Westervelt (2001) has described the integration of the two modeling tools for development of the ICBM. Swarm employed an object-oriented approach to simulate independent entities (agents) interacting via discrete events controlled by an activity schedule. The ICBM was composed of feeding area agents, (the equivalent of cells on a grid landscape), agents representing locations known to influence cattle behavior (such as persistent water sources and supplemental cattle feeding stations), and agents representing the organisms of interest (i.e., cattle herds and female cowbirds). GIS was required to assess spatial features of vegetation, such as distance to edge, and to prepare spatially explicit data for input to the simulation. The dynamic model simulated movement behavior of both female cowbirds and herds of free-ranging cattle on Fort Hood. Telemetry studies of female cowbirds on Fort Hood during the breeding seasons of 1995 and 1996 revealed that over 90 percent of the afternoon sightings of feeding cowbirds were of birds in the presence of at least thirty cattle (Cook et al. 1997). Early experiences with cowbird control demonstrated that trapping is most successful when efforts targeted feeding cowbirds in the afternoon hours (Hayden et al. 2000). As a result, simulated cattle herd movement was an important component of the ICBM. The landscape of Fort Hood was divided into a grid with a resolution of 750×750 meters, corresponding to a cell area of 56.25 hectares. This scale was a reasonable match to the median (50 hectares)

TABLE 49.2.

Input data sources used for the first stage of the individual cowbird behavior model (ICBM).

Data source	Resolution	Year	Other
Landsat TM imagery	30 meter	1996	EOSAT
			Seven bands
Corral Locations	Point data	1998	Trame et al. 1999
Warbler and vireo breeding habitat raster map	50 meter	1991[a]	Probably hand digitized from 1:50,000-scale map
Number of cattle herds	N/A	1996	Equal to the legal permit for number of Animal Units
Water map	50 meter	1990[a]	Originally from a 1:75,000-scale map

[a]Raster map completed.

and mean (65 hectares) size of female cowbird breeding territories from 1995 telemetry studies on Fort Hood (Cook et al. 1995), although mean breeding territory size in 1996 was 99 hectares, (Cook et al. 1997). The UTM (universal transverse mercator) coordinate for each agent within the Swarm simulation was recorded at each timestep and was "known" to the other agents.

The dynamic model ran with a daily timestep for ninety-day simulations, corresponding to a single breeding season of the endangered species on the installation. Output reflected cumulative movement decisions over this length of time. We assumed that vegetation characteristics of habitat did not change significantly during a single breeding season and that female cowbirds maintained a single breeding territory, equivalent to one grid cell in the model, throughout a single breeding season. Thus, our primary goal was to simulate the daily movements of cowbirds as they traveled from their breeding territories to new feeding sites each day.

Modeling Cowbird Locations and Parasitism

Breeding-habitat quality for female cowbirds, and feeding habitat for both cowbirds and cattle, were evaluated using GRASS. Using a neural network, we classified 1996 Landsat TM imagery into grassland and non-grassland categories at a resolution of 50 meters. We then calculated the quantity and continuity of grassland within grid cells at the resolution of the ICBM, 750 meters (details can be found in Trame et al. 1998).

Breeding Habitat

Using GRASS, we determined the amount of non-grassland habitat within delineated endangered species areas on Fort Hood, and we found that endangered species habitat was characterized by at least 30 percent non-grassland cover. Since we intended the model to focus on parasitism of the black-capped vireo and golden-cheeked warbler in particular, we initialized the ICBM with cowbirds in grid cells of at least 30-percent non-grassland vegetation.

Decreasing mean distances to grassland edge within each grid cell contributed to higher breeding habitat quality. Three levels of breeding habitat quality were recognized. Grid cells containing a mean distance between grassland and non-grassland areas of less than or equal to 100 meters were considered highest quality based on the results of Brittingham and Temple (1983). Grid cells containing a mean distance between grassland and non-grassland areas greater than 300 meters were considered lowest quality (again, based on Brittingham and Temple 1983), while intermediate mean distance values were considered of intermediate quality. Using GRASS, female cowbirds were allocated through a weighted random algorithm. The three quality levels were weighted in the following way: low-quality cells = 1.0; intermediate-quality cells = 2.3, and the high-quality cells = 3.6. Weights were based on Brittingham and Temple's (1983) observation that the proportion of nests parasitized in edge habitats is 3.6 times higher than the proportion of nests parasitized in nonedge habitats (65 percent versus 18 percent), while the intermediate weight of 2.3 was the mean of the extremes.

The actual number of breeding cowbirds on Fort Hood has not been determined. The ICBM was designed to be initialized by placing a single female cowbird into any grid cell with potential breeding habitat characteristics. The proportion of breeding habitat utilized by cowbirds could vary from 1 to 100 percent breeding habitat "saturation." For the simulations illustrated here, breeding capacity was equal to 50 percent, for a total of 624 females in each simulation.

Cattle-grazing Habitat

Female cowbirds on Fort Hood were documented on their breeding territories during the morning hours and feeding or resting in the presence of domestic cattle herds during the afternoon (Cook et al. 1998). To model cowbird movement between breeding and feeding habitats, it was necessary to determine the location of cattle herds (thirty or more individuals) for the majority of time in the afternoon period (grazing, resting, and drinking combined).

Grazing-habitat quality was determined using GRASS. The amount and continuity of grasslands, as well as the distance to water and the distance to supplemental feed or salt equally contributed to the grazing-habitat quality of each grid cell. Habitat quality was represented with a value from 0 to 1 for each of these factors, and their product yielded the measure of grazing habitat quality.

Grazing quality increased as proportion of grassland increased to 50 percent, at which point quality reached a maximal plateau. Greater habitat-quality values were assigned to areas of contiguous grassland compared with areas containing numerous disjunct patches of grassland (details on the mathematical calculations for habitat quality are found in Trame et al. 1998).

Research on range management has documented a solid relationship between abiotic factors, including the location of water, and cattle movement over large areas (Senft et al. 1987; Coughenour 1991). Valentine (1947) recommended that stocking rates be adjusted for distance to water, since vegetation far from water was underutilized compared to vegetation near water. The influence of distance to water on cattle grazing also was found to be highly significant by numerous other researchers (Cook 1966; Senft et al. 1983; Stafford Smith 1988). GRASS layers showing sources of water expected to be available during the avian breeding season on Fort Hood were available through installation GIS databases. Simulated cattle remained close to water sources, since grazing quality dropped as much as 50 percent at distances of only 1,500 meters (Trame et al. 1998).

No quantitative information was available describing cattle foraging response to locations of supplemental feeding, although it has been shown to be significant (McDougald et al. 1989). Personnel at Fort Hood believed a majority of cattle remained within 1.5 to 2 kilometers of supplemental feeding stations most of the time (T. L. Cook, Director of the Fort Hood Project of the Texas Nature Conservancy, Fort Hood, Texas; A-M Shapiro and S. J. Harper personal communication 1996; J. D. Cornelius, Branch Chief, Endangered Species Branch, Fort Hood, Texas). However, field work in 1998 frequently documented cattle herds in locations greater than 2 kilometers from feeding stations, and the model was altered to better match the new data (Trame et al. 1999). More-accurate locations of supplemental feeding stations and salt blocks also were recorded during 1998 and

entered into a GRASS layer to improve model accuracy.

Cattle Movement

Simulated cattle movement was largely based on day-to-day movement patterns recorded by Bailey et al. (1990) in a 248-hectare Texas pasture of fairly homogenous forage. Their experimental pasture was initially divided into sixty-three unfenced units. However, this was too fine-scale to capture actual cattle movement patterns, and the final analysis divided the pasture into five large sections to calculate transition probabilities among different areas on a daily time scale. The average area of the five sections was 49.6 hectares, approximated in the ICBM with grid-cell sizes of 56.25 hectares. Bailey et al. (1990) concluded that cattle display a "win-switch" strategy when foraging. Rather than stay in a productive area until the value drops, as predicted by optimal foraging theory (Charnov 1976), they switch grazing areas before forage quality drops. In addition, cattle may utilize spatial memory (Bailey 1995, Bailey et al. 1996) to avoid recently grazed areas. Behavioral research has shown that cattle recall locations of food depletion for at least eight days (Bailey et al. 1996). Such mechanisms may be important in homogeneous environments; studies in heterogeneous landscapes may show different patterns arising from alternative mechanisms (Bailey et al. 1990). In the ICBM, we assumed that forage quality was relatively homogeneous and cattle movement of Fort Hood was comparable to that recorded by Bailey et al. (1990).

Cattle were modeled as small herds, with each agent representing thirty individuals. Supplemental feeding and salting locations served as "home base" to the herds of cattle and will be referred to by that name. Grid cells with GIS-generated habitat characteristics (described earlier) were referred to as "feeding areas" in the Swarm model, since that is the primary function for which they were evaluated by both the cowbirds and the cattle herds (but note that many of these "feeding areas" actually served as breeding territories for cowbirds). To approximate the current grazing capacity on Fort Hood, four cow herd agents were placed into each home base upon initialization of Swarm. Each cow herd agent then identified the feeding areas within a fixed distance (3.75 kilometers) from its home base and created cow-memory objects for each. During simulations, cow herds only considered moving to these nearby feeding areas to minimize computational overhead.

During simulation, each cow herd agent chose a feeding area in which to feed at the beginning of each timestep. The choice was based on an attractiveness value composed of three factors: (1) its grazing quality, (2) time since first occupied by the herd, and (3) distance from the herd's current location.

When a cow herd first visited a feeding area, both the cow herd and the feeding area agents recorded the move. The time since first occupied was then factored in during subsequent movement decisions. The value of the visited feeding area decreased daily through the fifth day following initial visitation. By that day, the probability of re-visiting dropped to zero. The value of the feeding area then increased over the following eight days. No reduction in value was created due to recent occupation by a different herd, since avoidance was thought to be based on memory of the cattle's own occupation, and not environmental cues (Bailey et al. 1996). The distance from a herd's current location was factored into subsequent moves as well. Attractiveness was not reduced for the feeding area currently occupied, nor for immediately adjacent feeding areas, but the value declined in a linear fashion for feeding areas at distances of two, three, and four cells. Beyond distances of four cells, the attractiveness of feeding areas dropped to zero. Each of these three attractiveness factors (grazing quality, distance from the previous herd location, and time since first occupied) was represented by values ranging from 0 to 1, which were multiplied together with equal weighting to calculate the overall attractiveness value. Once the grazing quality of each feeding area was updated in each new time step, the movement rule instructed each cow herd to move to the feeding area with the highest grazing quality. Percentage of cattle herds moving zero, one, two, and three or more cells distance is output for each time step during simulation.

Cowbird Movement

Female cowbird agents were placed into breeding territories upon Swarm initialization, as described above.

To follow movements of those cowbirds most likely to parasitize endangered species, we labeled each cowbird agent as breeding in black-capped vireo protection areas, golden-cheeked warbler protection areas, or locations not within protection areas. These labels were available in model output of cowbird behavior patterns.

Feeding cowbirds at Fort Hood nearly always associate with small herds of cattle, so movement rules in ICBM rely on cowbird agents "perceiving" the locations of cattle herd agents, within limits. We know of no field research describing how cowbirds locate feeding sites with cattle, areas that can be more than 18 kilometers from their breeding territories (Cook et al. 1998). We developed four behavioral algorithms, or movement rules, to simulate the daily search for a suitable feeding area after leaving the breeding territory, and compared the daily movement distances resulting from each rule with daily movement data from telemetry on Fort Hood. The best-performing rule (for comparisons see Harper et al. 2001) directed each cowbird agent to move, in reverse chronological sequence, directly to each of the three most recently utilized feeding areas until a cattle herd was located. If no cattle herd was found at any of these previously successful locations, the cowbird would move randomly to the nearest feeding area, then to the next nearest feeding area, and so forth, until a cattle herd was located. Along the search path, the cowbird had the ability to assess feeding areas en route to the three previously successful locations. We defined the perception distance (i.e., distance away from the straight-line travel path) within which cowbirds could locate cattle herds to be 1,000 meters while in flight, a conservative distance considering the open landscape of Fort Hood grasslands. The decision to feed in a feeding area was registered by the feeding area agent and the summed counts were used to produce output of cumulative occurrences in each feeding area on the landscape.

Additionally, the status of each female cowbird, whether trapped or not yet trapped, was recorded for each time step by the feeding area agent serving as the breeding territory, and the summed counts provided the total number of days that a simulated cowbird occupied her breeding territory. From this output,

Probability of Parasitism - ICBM Approach

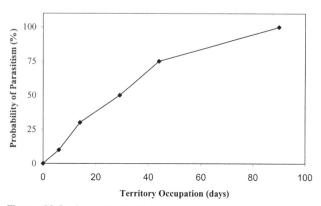

Figure 49.3. An example of how the probability of parasitism within a single ICBM (individual cowbird behavior model) grid cell could be based on the length of occupation of the grid cell by a breeding female brown-headed cowbird (*Molothrus ater*). This example is not based on empirical data, but on professional judgment. Figure was generated using spreadsheet software Microsoft Excel.

spatially explicit estimates of cowbird parasitism probabilities can be calculated by assigning probability of parasitism according to how many days a female occupied a territory (Fig. 49.3).

Discussion

Altogether, three separate analytical efforts were made to predict cowbird occurrences and/or parasitism rates. Two attempts at modeling spatially explicit parasitism rates failed. Due to the rich data sources we had available, and to our inability to develop spatially explicit predictions of cowbird occurrences, we feel it is worthwhile to explain the two analyses attempted and rejected during model development. First, we examined available data on northern cardinal (*Cardinalis cardinalis*) densities, preferred habitat of cardinals, and nest parasitism of black-capped vireos, based on Barber's (1993) findings that numbers of nearby cardinals are strongly correlated to high parasitism in the vireo. This approach failed because the value needed for our model was a probability for parasitism, whereas the categorical nature of individual nest information (parasitized versus not parasitized) was inappropriate. We attempted to aggregate individual nests into groups for calculations of probability for parasitism and then to subsequently compare

those groups to the independent variables of interest. (It is important to realize that determining probability of parasitism was essential; we modeled the impact of parasitism on black-capped vireo fecundity with Pease and Grzybowski's [1995] model, which required probability of parasitism as an input.)

We also attempted to generate parasitism probabilities based on distance to traps, amount of nearby disturbed habitat, and categorical (parasitized versus not parasitized) nest data. For each grid cell in the model, the numbers of parasitized vireo nests and total nests in a 5-kilometer radius were used to calculate the percent parasitism in the neighborhood. The weighted counts were generated by a circular matrix filter, then summed, and the neighborhood value for percent parasitism was assigned to the centroid (focal) cell. The same approach was used to calculate weighted counts of nearby disturbed cells and weighted counts of nearby traps. Our regression analyses revealed no significant relationship between numbers of nearby traps, amount of nearby disturbed habitat and the probability of parasitism assigned to focal cells.

Comparison of Inputs and Outputs

The FHASM approach required input values for management effort and efficiency (success rate) of simulated cowbird trapping and shooting programs. These values were specified for each of five subregions of the installation, but numbers of birds killed were added together to produce a single cowbird result variable for the entire installation. A single value for probability of parasitism across the whole installation was assigned to each cell of the model. Probability of parasitism was used, along with other factors (for details see Trame et al. 1997), in Pease and Grzybowski's (1995) mathematical model to generate breeding season total number of offspring for each vireo breeding pair. FHASM did not generate maps of cowbird locations or cumulative occurrences through time. It was not possible to assess the risk of parasitism in different regions of the installation. However, the approach used in FHASM incorporated the influence of cowbird parasitism in simulations of vireo and warbler breeding populations. In FHASM simulations, overall parasitism and its effects on fecundity responded to different trapping strategies, which allowed modeling of

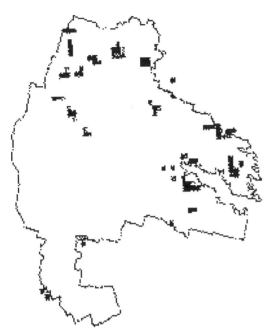

Figure 49.4. Example GIS layer output from the first stage of the individual cowbird behavior model (ICBM), illustrating simulated cumulative occurrences (in days) of feeding brown-headed cowbirds (*Molothrus ater*) on Fort Hood. Darker cells experienced higher cumulative occurrences. Cells outside boundary represent cowbird feeding on nearby private lands. Figure created using GRASS GIS software.

Figure 49.5. Example output following the application of the second stage of the individual cowbird behavior model (ICBM), in which some female brown-headed cowbirds (*Molothrus ater*) have been trapped. This GIS layer illustrates the total length of territory occupation (in days) for each female cowbird on Fort Hood. Darker cells experienced longer lengths of occupation. Figure created using GRASS.

comparative scenarios by endangered species managers on Fort Hood.

The ICBM was developed to provide spatially explicit parasitism risks (and to simulate alternative trap management scenarios, an extension that is beyond the scope of this chapter). We specified the proportion of all potential cowbird breeding territories that received a cowbird upon model initialization. We provided spatially explicit habitat-quality levels for both cowbirds and cattle herds and for the sites of landscape features, such as water sources and supplemental feeding or salting stations, that significantly affect cattle behavior. The first stage of the ICBM produced spatially explicit cumulative occurrence counts of feeding female cowbirds (Fig. 49.4). When the second stage of the ICBM was applied, some of the female cowbirds were trapped and killed, and the ICBM generated spatially explicit territory occupation data, showing the cumulative number of days each breeding territory was inhabited by a breeding female cowbird (Fig. 49.5). Using GRASS for further analysis, this output was used to generate probabilities of parasitism for each grid cell and supported

risk-analysis of parasitism of endangered species. As with FHASM output, endangered species managers can specify different trapping scenarios (in much greater detail using the ICBM, see Harper et al. 2001) and compare between scenarios for effectiveness in removing female cowbirds and in reducing parasitism rates across different regions of the installation. Field validation was conducted to test ICBM predictions of cattle and cowbird occurrences, and in a separate validation study, the ICBM was parameterized and applied to a landscape that has not yet experienced any cowbird control. Results from these studies are currently being documented (A.-M. Shapiro in preparation). Recent work created a link between the FHASM and ICBM models. In the future, FHASM will be able to access spatially explicit parasitism probabilities from the ICBM instead of using a single value for every cell across the landscape (Trame et al. 1999).

Issues of Scale

The ecology and management of cowbirds is well suited to landscape simulation since the cowbird responds to landscape-level patterns in feeding and breeding habitats. Because they do not protect their young or a nest, female cowbirds can range large distances in search of suitable feeding areas. Researchers have reported maximum daily movements between breeding territories and feeding sites as much as 7 to 18 kilometers (Rothstein et al. 1984; Cook et al. 1998, respectively). Management and control of cowbirds must target the entire landscape.

Recent telemetry studies on Fort Hood have demonstrated that livestock pastures and bird feeders located on lands outside the installation boundary influenced the behavior of female breeding cowbirds on Fort Hood (Cook et al. 1998); in response, the ICBM included off-post landscape and livestock (Trame et al. 1999). It seems likely that cattle herds and endangered species nests located relatively close to places with abundant livestock experience more occurrences of cowbirds. However, the need for a female cowbird to return to her breeding territory may scatter individuals more than expected by livestock patterns alone. The spatial scale at which patterns develop depends on the landscape context of all available breeding habitat and feeding habitat but are not currently understood. Theoretically, output from the ICBM could be analyzed to search for and quantify the scale at which cowbirds congregate in feeding areas on a landscape. Such an analysis would be desirable in the future and should further elucidate the ability of individual-based models to generate landscape-level insights that are not possible using aggregated approaches such as that applied in FHASM.

Modeling Bird Abundance from Forest Inventory Data in the Boreal Mixed-wood Forests of Canada

Pierre R. Vernier, Fiona K. A. Schmiegelow, and Steve G. Cumming

Effective wildlife conservation in forested landscapes managed for multiple objectives increasingly relies on models to predict the outcome of alternative management scenarios on the distribution and abundance of focal species. Habitat models based on remotely sensed data such as forest inventories or satellite imagery are inexpensive to develop compared to models based on detailed vegetation data collected in the field (e.g., Venier and Mackey 1997) and may be as effective in predicting abundance at the spatial scales considered here (F. K. A. Schmiegelow et al. unpublished analysis). As our goal is prediction at spatial extents commensurate with forest management planning, candidate independent variables should be derivable from available spatial data. We intend to use the models such as those presented here within a spatially explicit landscape simulation model of wildfire and stand dynamics (Cumming et al. 1998; see also He and Mladenoff 1999), which is initialized from digital forest inventories. Thus, our objective was to develop habitat models using only such data. Once developed and validated, these models will be used to evaluate the consequences of alternative management activities and policies over spatial extents of several thousand square kilometers and time horizons of at least one hundred years (e.g., Hansen et al. 1993, 1995).

In this study, we used bird survey and forest in-

ventory data from the mixedwood region of the boreal forest in Alberta, Canada, to model abundances of six bird species as functions of habitat characteristics measured at two spatial scales (see Maurer 1985; Addicott et al. 1987; Karl et al., Chapter 51). We refer to these scales as the local and neighborhood scales, which are intended to represent a habitat patch within a spatial context. At the local scale, commensurate with the territory sizes of species considered here (F. K. A. Schmiegelow unpublished data), and with the resolution at which bird observations were recorded (see Methods), we measured forest characteristics such as stand height and crown closure in a 100-meter-radius buffer. We defined a neighborhood as a 400-meter-wide buffer surrounding the local habitat patch, where we measured the abundance and configuration of different forest cover types and anthropogenic features. The neighborhood size was selected, in part, to be consistent with the extent at which other ecological phenomena, such as fire ignition and spread, are represented in our landscape simulator. In some recent assessments of the effects of habitat loss and fragmentation on forest birds, our neighborhood scale is considered a landscape (Edenius and Sjöberg 1997; Drolet et al. 1999). However, we considered both our spatial extents to be consistent with a broad interpretation of Johnson's (1980) second-order habitat selection in which habitat composition and configuration are characterized at multiple spatial extents, at the level of territories. Our specific objectives here were to develop statistical habitat models relating observed bird abundances to local and neighborhood habitat characteristics measured from forest inventory data and then to estimate the predictive performance of the habitat models using a statistical cross-validation approach. We did not attempt to determine the relative (and unique) contributions of local and neighborhood habitat characteristics on bird species abundance. We are exploring this issue further.

Methods

Our study area encompassed about 140 square kilometers of boreal mixedwood forest near Calling Lake in north-central Alberta, Canada (55°N, 113°W; Figs. 50.1 and 50.2). Mean summer (early June through mid-August) precipitation in the region is about 320 millimeters, accounting for more than 70 percent of the total yearly precipitation; July is generally the wettest month. The mean summer temperature is 12.0°C, and the mean freeze-free period is eighty-five days (Strong and Leggat 1981). Trembling aspen (*Populus tremuloides*), balsam poplar (*P. balsamifera*), and white spruce (*Picea glauca*) are the most abundant upland tree species, often occurring together in old, mixed stands, whereas black spruce (*P. mariana*) characterizes wetter sites (Strong and Leggat 1981). The dominant understory shrubs are alder species (*Alnus tenuifolia, A. crispa*) with lesser amounts of willow (*Salix* spp.). Various fruiting shrubs (*Rubus, Rosa* spp.), sarsaparilla (*Aralia nudicaulis*), and other herbaceous plants dominate the lower strata.

Bird Surveys

We used bird abundance data collected by point-count surveys conducted between 1993 and 1998 as part of the Calling Lake Fragmentation Experiment and related studies (e.g., Schmiegelow et al. 1997). A total of 406 permanent sampling stations were located within sixty-five sites, which we define as contiguous areas of similar forest type and age. Site types included areas clearcut in 1993 as part of the experimental design, young and old deciduous forests, mature coniferous

Figure 50.1. Location of the Calling Lake Study Area in north-central Alberta, Canada.

forests, and mixedwood forests. There was at least 200 meters between each sampling station. In every year that a station was sampled, point counts were conducted five times during the breeding season at ten-day intervals from the third week in May through early July. Upon arrival at a station, observers waited for one minute and then recorded all birds seen and heard during a five-minute sampling interval, within 50-meter and 100-meter distance classes. All individuals recorded at a station during a given visit were mapped, and movements during the five-minute sampling period noted, ensuring that individuals were recorded only once (see Ralph et al. 1993). Sampling effort, in years, ranged from one to six years across stations, as resources allowed additional survey points to be added to the main experimental design described by Schmiegelow et al. (1997).

We developed models for six bird species (Table 50.1): black-capped chickadee (BCCH; *Poecile atricapilla*), black-throated green warbler (BGNW; *Dendroica virens*), red-breasted nuthatch (RBNU; *Sitta canadensis*), white-throated sparrow (WTSP; *Zonotrichia albicollis*), yellow-rumped warbler (YRWA; *D. coronata*), and yellow warbler (YWAR; *D. petechia*). This suite of species follows Schmiegelow and Hannon (1999), representing a range of observed abundances and expected responses to forest fragmentation. For each bird species, we calculated

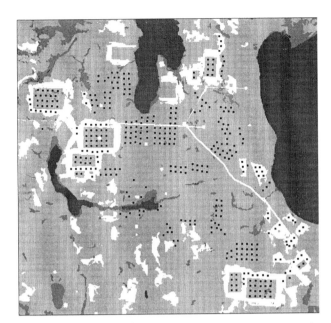

Figure 50.2. Distribution of forest habitat (light gray), clearcuts (white), lakes (dark gray), nonforest habitats (medium gray), and bird sampling stations in the Calling Lake Study Area.

the mean abundance per station per year (after Schmiegelow et al. 1997) and multiplied (weighted) this by the number of years a station was sampled (between one and six years). We used this aggregated count value as our response variable in subsequent statistical modeling. We did not include 1994 data from experimental areas where transient effects had been documented. This applied to ninety stations surrounded by the 1993 clearcuts, where temporary crowding of birds occurred (see Schmiegelow et al. 1997 for details).

Habitat Variables

Habitat patterns around each bird sampling stations were quantified using 1:20,000-scale digital Alberta Vegetation Inventory (AVI) maps. These maps were produced by interpretation of 1:15,000-scale aerial photography, updated to 1994. The digital maps contain several data layers that are potentially useful for modeling wildlife habitat relationships. The forest cover layer links stand polygons to forest attributes such as species composition, crown closure, height, and estimated stand age. This layer also includes information on nonforest cover types such as permanent

clearings, lakes, and wetlands. Two other map layers described the location of streams, logging roads, and seismic cutlines. We developed a habitat classification system based on the overstory and understory tree species (or genus in the case of *Populus*), stand age, and management history (Table 50.2). The classification system was used to create a generalized map of forest and nonforest habitat classes within the study area. The point count stations were georeferenced and linked to the AVI spatial database.

We used the original and derived map layers to measure habitat characteristics around each bird-sampling station at two spatial scales (Table 50.3, Fig. 50.3): the local scale, which matched the size and shape of the circular bird-sampling stations (inner buffer of 100-meter radius, or 3.14 hectares), and the neighborhood scale, which extended from 100 to 500 meters beyond the sampling stations (outer buffer, 75.4 hectares). The habitat characteristics we chose have either previously been used in the literature or were hypothesized correlates of species abundance based on the ecology of the species.

Seven variables characterized the structure and composition of the inner buffers (Table 50.3). A categorical variable (having discrete, unordered values) specified the habitat class at the origin of the station. Four continuous variables quantified the size of the habitat patch containing the origin and the area weighted means of canopy height, crown closure, and proportion of deciduous species in the canopy, for forested habitats intersecting the buffer. Two index variables coded the presence/absence of streams and anthropogenic edges within the buffer.

Twelve variables characterized the structure and composition of neighborhoods, or outer buffers. Five of these variables measured the proportional areas of deciduous forest, mixedwood forest, clearcuts, mid-seral forest (fifteen to ninety years), and late-seral forest (more than ninety years). Four variables indexed the presence of white spruce, black spruce, water, and anthropogenic habitat. We also derived three variables descriptive of neighborhood spatial structure. Simpson's diversity index (N_SIMP) measures the number of patch types and their relative abundance. The index represents the probability that two randomly selected patches belong to different patch types. The higher the

TABLE 50.1.

Distribution and abundance of bird species at Calling Lake, Alberta, Canada, from 1993 to 1998.

Code	Common name/Scientific name[a]	No. of stations	Mean detections	Mantel signif.[b]
WTSP	White-throated sparrow			
	Zonotrichia albicollis	356	3.96	0.000
YRWA	Yellow-rumped warbler			
	Dendroica coronata	327	2.11	0.000
RBNU	Red-breasted nuthatch			
	Sitta canadensis	266	0.61	0.000
BGNW	Black-throated green warbler			
	Dendroica virens	224	0.92	0.000
YWAR	Yellow warbler			
	Dendroica petechia	172	0.13	0.000
BCCH	Black-capped chickadee			
	Poecile atricapilla	151	0.14	0.285

[a]Species are listed from most common to least common based on the number of stations in which they were detected.

[b]Mantel signif. indicates the significance level of a randomization procedure used to test for spatial autocorrelation in bird counts.

value, the greater the structural diversity. After gridding the classified map of habitat classes to 0.01 hectare, we used FRAGSTATS (McGarigal and Marks 1995) to calculate two contrast-weighted measures of edge density to characterize the heterogeneity and fragmentation of the neighborhoods. N_EDGEN and N_EDGEA measured the density in meters per hectare of natural and anthropogenic edges, respectively. Both measures relied on separate edge contrast matrices

that assigned weights, ranging from 0 to 1, to adjacent habitat classes (Table 50.4). The weights were estimated subjectively, based on our knowledge and field experience. To N_EDGEA, we added the density of logging roads and cutlines; these were linear features in the underlying AVI data.

Prior to model development, we checked the distributional assumptions of our candidate predictor variables. N_PATCH was log transformed to improve normality. Numeric variables with an excess of zeroes were converted to binary variables (e.g., N_SB). There were no highly correlated pairs of predictor variables (Pearson's $r > 0.75$). We checked for nonlinear relationships between bird abundances and our continuous variables using scatterplots with lowess smoothers. We found no evidence of nonlinearity.

TABLE 50.2.

Habitat classification system used to calculate several local and neighborhood-level habitat variables.

Class	Description
WATER	Water (lake, ice, river)
NONFOR	Nonforest and wetland
Y_DECID	> 70% deciduous and ≤ 90 years
O_DECID	> 70% deciduous and > 90 years
W_SPRUCE	> 70% white spruce
B_SPRUCE	Leading black spruce
PINE	Leading pine
MIXED	Mixed deciduous/white spruce
CCUT	Clearcuts < 15 years
ANTHRO	Anthropogenic (wellsites, large cutlines, etc.)

Statistical Analysis

We used generalized linear models (GLM; McCullagh and Nelder 1989) to model the response of bird species to local and neighborhood habitat characteristics. GLMs can represent a greater variety of relationships between response and explanatory variables than can linear regression models, and do not

TABLE 50.3.

AVI-based habitat variables. Local habitat variables were measured within a 100-meter radius while neighborhood variables were measured in a 400-meter radius beyond each local (inner) buffer.

Variable	Variable type	Range of values	Description
Local			
L_CCUT, L_MIXED, L_ODEC, L_PINE, L_SB, L_SW, L_YDEC	Dummy coded	7 classes	Habitat types in which stations were located (see Table 50.2 for descriptions)
L_SIZE	Numeric	0.5–703.4 ha	Patch size; relies on a habitat classification system (Table 50.2)
L_DIST	Numeric	0–1238.9 m	Distance of station center to nearest anthropogenic edge (habitat classes 9 and 10)
L_CROWN	Numeric	0–85.5 %	Mean crown closure among forested polygons
L_DEC	Numeric	0–1.0	Mean deciduous proportion of forested polygons
L_HT	Numeric	0–31.0 m	Mean stand height of forested polygons L_STREAM Binary 0 or 1 Presence of streams or lakes
Neighborhood			
N_CUT	Numeric	0–0.66	Proportion of neighborhood in a clearcut
N_MID	Numeric	0–0.99	Proportion of neighborhood in mid seral forest (15–90 years)
N_LATE	Numeric	0–1.00	Proportion of neighborhood in late seral forest (more than 90 years)
N_DEC	Numeric	0–1.00	Proportion of neighborhood in deciduous forest
N_MIXED	Numeric	0–0.77	Proportion of neighborhood in mixedwood forest
N_SB	Binary	0 or 1	Presence of black spruce forest
N_SW	Binary	0 or 1	Presence of white spruce forest
N_ANTHRO	Binary	0 or 1	Presence of anthropogenic features (well sites, clearings, gravel pits, highways, etc.)
N_WATER	Binary	0 or 1	Presence of lakes, ponds, etc.
N_SIMP	Numeric	0–0.83	Habitat patch diversity measured using Simpson's index
N_EDGEA	Numeric	0–319.2 m/ha	Anthropogenic edge density calculated using habitat classification system (Table 50.2) and edge contrast matrix (Table 50.4)
N_EDGEN	Numeric	0–85.3 m/ha	Natural edge density calculated using habitat classification system (Table 50.2) and edge contrast matrix (Table 50.4)

assume constant variance. Because our response variables are counts, we assumed a Poisson error distribution (McCullagh and Nelder 1989; Jones et al., Chapter 35). Thus for a set of *n* explanatory variables, the Poisson regression model specifies that the distribution of responses is Poisson and that the log of the mean is linear in the regression coefficients:

$$\log(expected\ count) = \log(effort) + b_0 + b_1x_1 + b_2x_2 + \ldots + b_nx_n$$

where *effort*, the number of years a station was sampled, is an offset (StataCorp 1999) to correct for variable sampling effort between stations. This model of sampling effort is formally equivalent to assuming

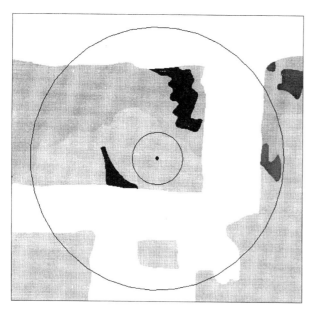

Figure 50.3. Example of inner and outer buffers placed around one of the bird survey stations. White areas represent clearcuts and shades of gray represent different forest cover types.

that, within stations and species, annual counts are independent, identically distributed Poisson random variates. As an informal test of model specification, we used the deviance-based dispersion statistic (sum-of-squared deviance residuals divided by degrees of freedom), or mean deviance per sample. Values greater than 1.5 indicate overdispersion, implying, for count data, that a negative binomial link may be more appropriate.

For each bird species, we selected explanatory variables from the candidate set by a backward stepwise procedure (P-to-enter < 0.001, P-to-remove < 0.0015). A conservative level of significance was chosen as a correction for multiple tests. Significance levels were based on standard likelihood ratio tests. Model strength was measured using the percent of deviance explained. This measure is analogous to the multiple coefficient of determination (R^2) and measures the proportion of the deviance in the independent

TABLE 50.4.

Anthropogenic (top number) and natural (bottom number) edge contrast values used to calculate N_EDGEA and N_EDGEN, respectively.

thro	water	nonfor	ydec	odec	sw	sb	pine	mixed	ccut	an
water	0.00									
	0.00									
nonfor	0.00	0.00								
	1.00	0.00								
ydec	0.00	0.00	0.00							
	1.00	0.75	0.00							
odec	0.00	0.00	0.00	0.00						
	1.00	0.75	0.50	0.00						
sw	0.00	0.00	0.00	0.00	0.00					
	1.00	0.75	0.75	0.50	0.00					
bs	0.00	0.00	0.00	0.00	0.00	0.00				
	1.00	0.25	0.50	0.75	0.50	0.00				
pine	0.00	0.00	0.00	0.00	0.00	0.00	0.00			
	1.00	0.75	0.50	0.50	0.50	0.50	0.00			
mixed	0.00	0.00	0.00	0.00	0.00	0.00	0.00	0.00		
	1.00	0.75	0.50	0.25	0.25	0.75	0.50	0.00		
ccut	1.00	0.50	0.50	0.75	0.75	0.50	0.50	0.75	0.00	
	0.00	0.00	0.00	0.00	0.00	0.00	0.00	0.00	0.00	
anthro	1.00	1.00	1.00	1.00	1.00	1.00	1.00	1.00	0.00	0.00
	0.00	0.00	0.00	0.00	0.00	0.00	0.00	0.00	0.00	0.00

variables associated with the deviance in the dependent variable (Cameron and Windmeijer 1997).

To evaluate the relative influence of local and neighborhood habitat variables on each species, we compared five alternative habitat models. At the ends of the spectrum were the null and full models. The null model was simply the mean count over all stations while the full model included all local and neighborhood variables. The three intermediate models used subsets of the variables selected by the backward stepwise procedure: local variables only, neighborhood variables only, or both sets of variables. We used Aikaike's Information Criteria (AIC, Akaike 1974) to select the best of the five models. AIC measures the tradeoff between model goodness of fit (measured as the log-likelihood) and model parsimony (measured by the number of parameters included in the model).

We assessed model assumptions by examining diagnostic plots and maps of the response variables and the model residuals. Plots of deviance residuals against the linear predictor, and a normal scores plot of standardized deviance residuals, were used to identify skewness and outliers and to assess the overall behavior of the model. Plots of approximate Cook statistics against leverage/(1 − leverage) and case plots of Cook statistics allowed us to identify potentially influential observations and their location in the dataset (i.e., case number), respectively. These diagnostic methods follow Scott (1997). The maps allowed us to visually examine the assumption that counts were independent between stations. This assumption may fail, as counts at stations within the same site are likely to be correlated. This is because the mean distance between such stations is low and because sites are, by assignment, relatively homogeneous. Preliminary inspection of the mapped counts indicated the need to test for spatial autocorrelation. We used Mantel's test (Mantel 1967), which is based on a scalar measure of the association between two distance matrices, one describing the absolute difference in counts between stations and the other the Euclidean distance between stations. Significance is tested with respect to a simulated distribution, obtained by re-computing the statistic under five thousand random permutations of one of the matrices (Manly 1997). Where there was evidence of spatial autocorrelation in counts, we repeated Mantel's test on model residuals (after building regression models).

The diagnostic plots and maps revealed that the most obvious problems were the presence of a few large residuals and influential observations, as well as some unexplained spatial autocorrelation (see Figs. 50.4 and 50.5 for examples). Visual inspection of the maps indicated that spatial dependence was primarily the result of within-site correlation, as expected. Our solution to both of these problems (influential observations and spatial autocorrelation) was to use STATA's cluster option to calculate variance estimates that are robust to both influential observations and within site correlation (StataCorp. 1999). Using the cluster option has the additional benefit of adjusting for any undetected overdispersion.

We estimated model uncertainty using a *leave-one-out* cross-validation approach (Burt and Barber 1996). This was done by eliminating observations one at a time and predicting this dependent value with a regression model estimated from the remaining observations. We then used the prediction residuals (n = 406) to calculate two criteria to assess the performance of our habitat models: the percent deviance explained (described earlier) and s^2, the sum of squared deviance residuals divided by the residual degrees of freedom, which measured the mean prediction error. The bias in each criterion was then calculated as the difference between the mean of the 406 leave-one-out statistics (e.g., σ^2) and the actual value obtained when all observations are included.

Results

The overall distribution and abundance of each species is summarized in Table 50.1. Of 102 bird species recorded during point count surveys, the six species considered here rank among the thirty most abundant, with white-throated sparrow and yellow-rumped warbler being the most and second-most abundant species, respectively. The ranges of values of our predictor variables are summarized in Table 50.3. The contrast between sites for our continuous variables is nearly as high as possible, given the overall composition of the study area.

Statistically significant regression models were

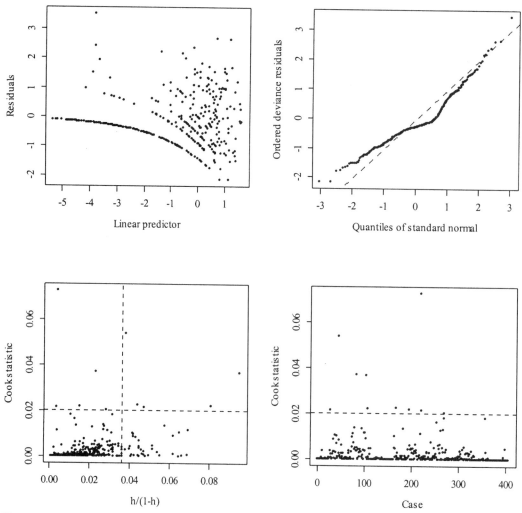

Figure 50.4. Diagnostic plots for the black-throated green warbler showing the existence of some large residuals (top panels) and influential observations (bottom panels). Top left: plot of deviance residuals against linear predictor; top right: normal scores plots of standardized deviance residuals; bottom left: plot of approximate Cook statistics against leverage/(1 − leverage); bottom right: case plot of Cook statistic.

developed for five out of six species (Tables 50.5 and 50.6), the exception being the black-capped chickadee. None of these models had deviance statistics suggesting overdispersion. We conclude that our choices of link and variance functions are appropriate and that the models are correctly specified.

For three species, the best of the five alternate models we considered were those including both local and neighborhood habitat variables. The models explained between 54 and 73 percent of the variation in abundances of the black-throated green warbler, yellow-rumped warbler, and white-throated sparrow. For the red-breasted nuthatch and yellow warbler, the best models included only local habitat variables and ex-

plained only 43 and 37 percent of the variation in bird abundance, respectively. The model comparison process is summarized for the black-throated green warbler in Table 50.7. Although the three stepwise models (local, neighborhood, and local plus neighborhood variables) are all highly significant, the third is clearly the best. BGNW abundance was positively related to local canopy height (L_HT) and to features of the neighborhood (N_DEC, N_SIMP, N_LATE, N_SW, N_CUT). Of these, the first three were most significant. High abundances of this species are associated with areas of older, structurally diverse, deciduous dominated forest, containing at least some white spruce. The relatively weak positive association with

Spatial Distribution of BGNW Counts

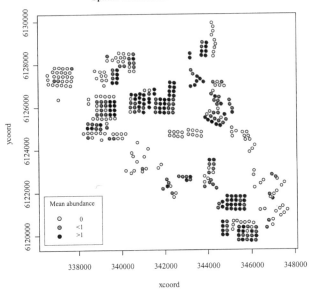

Figure 50.5. Simplified spatial distribution of counts for the black-throated green warbler. Empty circles indicate absence of BGNW (n = 182), gray circles indicate mean abundance is less than or equal to 1 (n = 112), and black circles indicate mean abundance is greater than 1 (n = 112).

clearcuts may be spurious due to the distribution of available habitat relative to harvested areas (see Discussion).

The white-throated sparrow was positively related to two local (L_CCUT, L_DEC) and two neighborhood (N_DEC, N_SIMP) variables, and negatively related to one local (L_PINE) and two neighborhood (N_EDGEN, N_MID) variables. The most important

predictors were the presence clearcuts, the local abundance of deciduous forest, and the proportion of mid-seral forest in the neighborhood. This species was most abundant in clearcuts adjacent to deciduous forest, with a low proportion of midseral forest in the surrounding area.

The yellow-rumped warbler was negatively associated with four local variables (L_CCUT, L_ODEC, L_SIZE, L_YDEC) and positively associated with two neighborhood variables (N_LATE, N_MID). The presence of both clearcuts and old deciduous forest were the most important predictors at the local level, while the proportion of mid- and late-seral forest were influential at the neighborhood level. Stations located in either clearcuts or old deciduous forests had the lowest abundance of YRWA whereas stations with a high proportion of young and mature forest in the surrounding area had the highest abundance.

Both the red-breasted nuthatch and yellow warbler were best predicted by models consisting of only two local habitat variables. RBNU was associated with tall coniferous or coniferous dominated stands; YWAR was associated with old, relatively open patches of deciduous forest. Neither species was sensitive to any of the neighborhood variables we measured, after accounting for variability in local habitat characteristics.

Our diagnostic plots and maps are illustrated for one species (BGNW, Figs. 50.4 and 50.5). Influential observations and potential outliers (indicated by high residual values) were present. Similarly, the map of the

TABLE 50.5.

Summary statistics for each bird species' best model (i.e., lowest AIC) indicating the number of local and neighborhood variables selected, degrees of freedom, the null model deviance (assuming only a mean effect), the residual deviance unaccounted for by the model, the percent deviance explained, and AIC values.

Summary statistic	BGNW[a]	RBNU[a]	WTSP[a]	YRWA[a]	YWAR[a]
Local variables	1	2	3	4	2
Neighborhood variables	5	0	4	2	0
Degrees of freedom	399	403	398	399	403
Null model deviance	582.73	252.08	984.60	520.65	122.15
Residual deviance	266.43	143.29	267.69	220.91	77.45
% deviance explained	54.28	43.16	72.81	57.57	36.59
AIC	751.49	619.35	1364.67	1105.30	248.90

[a]See Table 50.1 for bird code definitions

TABLE 50.6.

Coefficients, robust standard errors, and a measure of variable importance for habitat variables selected in the best model developed for each species.

Species	Variable	Coefficient	Robust standard error[a]	Importance[b] (%)
BGNW	L_HT	0.105	0.021	24.7
	N_CUT	1.348	0.533	3.9
	N_LATE	2.333	0.461	17.2
	N_DEC	2.195	0.300	21.7
	N_SW	0.459	0.136	4.4
	N_SIMP	3.344	0.500	14.4
	Constant	−8.853	0.650	
RBNU	L_DEC	−0.875	0.148	43.2
	L_HT	0.099	0.009	11.5
	Constant	−3.440	0.246	
WTSP	L_CCUT	2.463	0.229	38.0
	L_PINE	−1.356	0.366	7.3
	L_DEC	2.366	0.226	32.8
	N_MID	−1.555	0.160	31.7
	N_DEC	0.915	0.265	9.4
	N_SIMP	1.475	0.289	13.9
	N_EDGEN	−0.008	0.002	4.2
	Constant	−2.862	0.282	
YRWA	L_CCUT	−2.638	0.438	36.7
	L_ODEC	−0.336	0.063	7.8
	L_YDEC	−0.511	0.179	6.4
	L_SIZE	−0.001	0.000	8.6
	N_MID	1.055	0.303	8.6
	N_LATE	1.000	0.284	9.1
	Constant	−0.932	0.238	
YWAR	L_ODEC	1.768	0.276	18.6
	L_CROWN	−0.029	0.005	30.1
	Constant	−3.081	0.255	

[a] All variables are significant at the 5 percent level using robust standard errors.

[b] The importance of a variable within the final model is measured by 1 minus the ratio of the deviance explained by the final model and the deviance explained by a reduced model where the variable was excluded.

spatial distribution of counts provided some visual evidence of spatial autocorrelation (Fig. 50.5). This was confirmed using Mantel's test (Table 50.1). Spatial autocorrelation of the model residuals was less striking, but Mantel's test statistic was still significant ($0.05 > p > 0.01$). Similar patterns were observed for the other species, except the black-capped chickadee, whose counts were not spatially correlated. These results justify our use of robust variance estimators (Table 50.6).

Cross-validation analysis revealed that the predictive ability of the Poisson regression models were neither under- nor overestimated. The strengths of the original models versus the cross-validated models, as measured by percent deviance explained, were within 0.2 percent of each other. Similarly, estimates of prediction error using the sum-of-squared deviance residuals divided by the residual degrees of freedom (s^2) agreed to three significant digits. There was no evidence of bias in either of the two measures that we used for assessing the predictive ability of the models.

Discussion

In general, our quantitative models are consistent with qualitative accounts of habitat requirements for the selected species (e.g., Ehrlich et al. 1988; Semenchuck 1992; Kaufman 1996; Fisher and Acorn 1998), and we do not dwell on specific interpretations of variables here. A notable anomaly, however, was inclusion of the proportion of clearcut in the neighborhood as a positive predictor of black-throated green warbler abundance. As the species has been identified as one at risk due to forest harvesting (Schmiegelow and Hannon 1999; Norton 1999), the result was surprising. We believe it to be an artifact of the nonrandom harvest of suitable BGNW habitat in areas adjacent to sampled sites in order to satisfy an experimental design. Many sampling sites at which BGNWs were recorded were previously contiguous stands of older, deciduous-dominated forest fragmented by harvesting. Of the five species we modeled, the BGNW showed the greatest affinity for these older stands, and thus occupied sites were closer in proximity to harvested areas than expected by chance. Nevertheless, such post hoc explanations emphasize the importance of model validation and will be discussed later in the text in some detail.

An interesting outcome of our analyses was the variation in the inclusion of local and neighborhood habitat descriptors in species models. We quantified habitat composition and configuration at two scales in order to test whether the spatial context of habitat

TABLE 50.7.

Summary of alternative habitat models for the black-throated green warbler.

Model	Habitat variables	Df	Deviance[a]	% Deviance explained[a]	AIC[a,b]
Null		405	582.7		
Local	L_MIXED, L_ODEC, L_HT	402	333.5	42.8	812.6
Neighborhood	N_LATE, N_DEC, N_SIMP	402	364.7	37.4	843.8
Local + neighborhood	*L_HT, N_CUT, N_LATE, N_DEC, N_SW, N_SIMP*	*399*	*266.4*	*54.3*	*751.5*
Full	All local + neighborhood variables	380	252.5	56.7	775.6

[a]Deviance, percent deviance explained, and AIC are explained in the text.

[b]The model with the lowest AIC is italicized.

patches affected habitat selection at the level of territories. Our results suggest that, for some species, habitat quality at the level of territories (as approximated by long-term mean species abundances) is mediated by characteristics of the surrounding area. In other cases, only local characteristics explained variation in abundances. In the latter cases, explained deviance of our models was lower than for species models that included habitat descriptors at multiple scales. We present two contrasting interpretations of these results.

First, species responding only to local habitat characteristics may have selected specific features, regardless of their spatial context, and these features were not well described by relatively coarse-scale habitat data. Second, such species represent generalists relatively insensitive to both habitat composition and configuration, regardless of scale. We can test hypotheses arising from the first interpretation with extensive, detailed vegetation data collected in our study area over the same time period during which birds were sampled. The second interpretation is testable in the analytical framework presented here, with a larger suite of species. Our selection of species for preliminary model development was not made on this basis. Regardless, we emphasize that the species models containing only local habitat descriptors, although poorer than those containing both local and neighborhood variables, were still able to account for about 40 percent of the variation in abundance using, in each case, only two habitat variables derived from forest inventory data.

A fundamental criticism of habitat-based models is that they are rarely validated (Hansen et al. 1999). As a first step, we employed a statistical cross-validation approach. However, because we are ultimately interested in testing hypotheses about landscape dynamics and wildlife response, our predictive models must be validated across a range of spatial scales and geographic locations. We plan to use spatially referenced bird data from other localized studies in the boreal mixedwood for geographic validation. At a coarser scale, survey data in the form of Breeding Bird Survey routes and the Alberta Breeding Bird Atlas permit us to test whether our finer-scale models can be generalized, or in other words, whether the spatial scale of response to habitat varies.

Another criticism of habitat-based abundance models is that abundance in a given habitat is not necessarily indicative of quality, as measured by reproductive success (Van Horne 1983). If occupation of sink habitats (those where expected reproduction is below replacement) is limited by immigration from nearby source habitats (e.g., Pulliam 1988), then projections of population abundance may fail to predict actual population persistence. In the absence of productivity measures, we use long-term mean abundance as a proxy for habitat quality (see also arguments in Boyce and McDonald 1999). We assume that the system is dynamic and unsaturated, and that optimal habitat will be most frequently occupied, subject to variation imposed by environmental stochasticity, individual mortality, and dispersal limitation, or in other words, an ideal free distribution (Fretwell and Lucas 1969).

Occupancy of suboptimal habitats will also be a function of their density. In regions where suboptimal habitats make up a large portion of the landscape, surplus production from the rare nearby source habitats is unlikely to provide colonists to occupy sink habitats with a frequency sufficient to permit estimation of their relative suitability. In such areas, observed abundances in areas of relatively low predicted quality should be less than expected, given our models. Thus, our models can be used to design extensive sampling efforts to test metapopulation models and identify sink habitats. In such an application, source habitats would be identified as areas of both high predicted quality and high observed abundances.

This again points to the necessity for model validation across multiple sites and scales (extent and grain), as source habitat detection is dependent on predicted habitat quality. In the landscape from which the present models were derived, the contrast in continuous variables (at local and neighborhood scales) was nearly as high as possible (see Table 50.3) given the overall composition of the study area, but this area is dominated by older forest. An efficient way of proceeding may be to use our existing models to locate new sampling sites in areas with high contrast in independent variables that are highly significant in existing models, or where uncertainty (prediction variance) is high. Criteria based on the expected influence of new observations on parameter estimates or other measures based on Fisher's information matrix can also be devised. In addition, future sampling sites should target areas with contrast in independent variables that are anticipated to change most as industrial development in the region proceeds (e.g., the density of anthropogenic edges).

In Alberta, forest management planning is largely based on forest inventory information, but the ability of such information to predict species abundances has not previously been evaluated. We attempted such an evaluation, using Poisson regression analysis to model the relationship between bird species abundances observed in the field and habitat characteristics derived from forest inventory data. Poisson regression deals explicitly with characteristics of count data and is generally more efficient and consistent, and less biased than linear regression models using the same data (Scott 1997). For these reasons, it is receiving increasing attention in the ecological literature (e.g., Nicholls 1989; Bustamante 1997). However, it is still necessary to check model assumptions, including statistical independence of observations, correct specification of the link and variance functions, correct scale for measurement of the explanatory variables, and lack of undue influence of individual observations on the fitted model (Scott 1997). In our case, residual diagnostics revealed several influential observations, as well as spatial dependence in the model residuals. Hence, we used robust variance estimates and corrected for spatial dependence within bird survey clusters (sites). Our final models demonstrated good predictive ability with no evidence of bias. We conclude that their use in landscape simulations is justified pending their validation against independent data sets.

We believe the approach to modeling abundance presented here is robust and appropriate to the questions at hand, namely, assessing the potential ecological outcomes of various forest management scenarios in the boreal mixedwood forests of Alberta. We demonstrated that predictive models of bird abundance could be generated from forest inventory data, permitting evaluation of activities at a resolution and extent commensurate with management planning. The models presented here are a subset of those we have developed, representing species with a range of observed abundances and expected responses to forest fragmentation (Schmiegelow and Hannon 1999). Our final selection of species to model for scenario evaluations requires identification of species most at risk from land-use practices (primarily forestry and energy sector development) that are resulting in widespread habitat modification in Alberta's boreal forests (see Hansen et al. 1999 for a summary of approaches).

Adaptive resource management requires evaluation, prediction, and monitoring. Uncertainties associated with policy options are exposed as alternative hypotheses, implemented management strategies are treated as experiments, and observed conse quences are used to test hypotheses and refine management (Walters and Holling 1990). Land managers in Alberta have professed the advantages of an adaptive approach to resource management. We hope our work will provide some of the tools necessary for this process, through identification of measurable parameters, and development of analytical techniques, for assessment and monitoring of management activities.

Acknowledgments

We thank R. Pelletier, D. Demarchi, and T. Morcos for computer and GIS assistance, S. Hannon for her collaboration in the Calling Lake Fragmentation Experiment, and the numerous field assistants who collected the point count data. This research was supported by the Sustainable Forest Management Network and by Alberta Pacific Forest Industries.

Species Commonness and the Accuracy of Habitat-relationship Models

Jason W. Karl, Leona K. Svancara, Patricia J. Heglund, Nancy M. Wright, and J. Michael Scott

Two types of error are possible when assessing the accuracy of models predicting species presence or absence: omission error (failure to predict species occurrence in an occupied area) and commission error (prediction of species occurrence in unoccupied areas)(see Fielding, Chapter 21). Of these two, omission errors are relatively easy to measure (Krohn 1996; Karl et al. 2000) because observation of a species in an unpredicted area necessitates an omission error. Conversely, failure to observe a species in a predicted area, while necessary to the definition, is not sufficient to classify it as a commission error (Krohn 1996; Boone and Krohn 1999; Karl et al. 2000). This can be due to inefficient or inappropriate sampling, species life history characteristics (e.g., avoids humans, cryptic nature, episodic), or temporal and spatial variation in species distributions (Karl et al. 2000; Fielding, Chapter 21; Schaefer and Krohn, Chapter 36). Thus, field measures of commission error contain both true error and apparent error (Karl et al. 2000; Schaefer and Krohn, Chapter 36).

Attributes of species biology can affect our estimates of model accuracy, but the effect of rarity on model accuracy is not well defined. It has been proposed that the presence of "species with high spatial and temporal evenness" (Krohn 1996) (e.g., common species) would be easier to predict with habitat-relationship models (e.g., gap analysis models) than

species with low evenness (Boone and Krohn 1999) for most modeling applications. Karl et al. (2000) reported a significant decline in commission error accompanied with a slight increase in omission error with number of species detections on two study areas in north Idaho. As such, apparent error decreased with increased sample size. However, it was unclear whether high error rates at low numbers of detections were a result of differences in model accuracy between rare and common species or an artifact of sample size used to estimate model performance.

A rarity effect would exist if the models for species less-frequently encountered were less accurate than those for common species. Lower model accuracy for rare species in one situation could be caused by incomplete knowledge of the species' range or habitat associations, or the species responding to habitat features that cannot be measured (or mapped). Alternatively, because large numbers of rare species detections often take a large investment of time and money, model accuracy is assessed with few data points (if done at all). Depending on the statistics used, accuracy assessment with small sample sizes could lead to erroneous measures.

We investigated whether the pattern described by Karl et al. (2000) was due to a rarity effect or to an artifact of sample size. We simulated small sample sizes by randomly subsampling our data set for the most common species and using the subset of observations to test model accuracy. By doing this, we held the biological attributes of species constant, varying only the sample size. If models developed for rare species (i.e., those with few detections) have poorer prediction accuracy than common ones (rarity effect), then the slope of regression lines from a plot of error rates against number of detections for field data set should be steeper than that obtained by simulation. Although this approach did not consider reasons for rarity and may not appropriately approximate distribution of rare species, it was adequate for examining the effects of sample size on model accuracy.

Study Area

Our study area encompassed most of the Idaho portion of U.S. Forest Service (USFS) Northern Region (the Idaho Panhandle, Clearwater, and Nez Perce National Forests) as well as land owned by the Potlatch Corporation (Fig. 51.1). This area (2.75 million hectares) begins just north of the Clearwater River, extending northward to the tip of the Idaho panhandle, but excluding the dry grasslands of the Snake River Valley and the Palouse agriculture lands. Most of this area is dominated by mixed coniferous forests in various stages of timber management.

Methods

Breeding birds were surveyed on the U.S. Forest Service Northern Region in 1994 to 1996 (R. L. Hutto and U.S. Forest Service unpublished data; P. J. Heglund, Potlatch Corporation unpublished data) using a variable-radius circular plot technique (Ralph et al. 1995a). Each of 1,628 survey points was surveyed one time per year for up to three years following the methods described by Hutto and Hoffland (1996).

We eliminated from the data set all birds that were flying when detected, except for those birds whose detections are mostly restricted to aerial foraging (i.e., swallows, swifts, hawks). We further truncated the data set to only those observations occurring within 50 meters of the survey point for two reasons. First, the ability to accurately judge the distance of an observation and the cover type in which it occurred decreases with distance from the survey point (Hutto

and Hoffland 1996; see also Scott et al. 1981). Second, limiting the area of analysis around the survey point reduces the potential for variation in the values of the geographic information system (GIS) data layers around the survey point.

We received GIS coordinates for the survey points from the U.S. Forest Service Northern Region's Landbird Monitoring Program. These coordinates were digitized from geo-registered aerial photographs of the study area. We then converted the vector point coverages from each study area to raster grids with a 0.09-hectare cell size.

We used models developed by Scott et al. (unpublished data) for the Idaho Gap Analysis Project to predict the presence/absence of the species detected in the breeding bird surveys. These models were built using methods proposed by Scott et al. (1993) (see also Butterfield et al. 1994; Csuti 1996; Smith and Catanzaro 1996) consisting of four major steps: (1) establishing a species list, (2) defining species range limits, (3) collecting species habitat information and determining habitat relationships, and (4) modeling the species habitat in a GIS using the information gathered.

To assess model accuracy, we compared the model predictions with survey data for each species detected. We tallied the number of omission errors (observed, not predicted) and commission errors (predicted, not observed) and calculated percent omission (number of omissions divided by the total number of observations) and commission error (number of commissions divided by the total number of survey points), respectively. All species measures were combined into one data set. We plotted omission and commission error by the number of species detections for all species. An inverse relationship existed between omission and commission errors (Karl et al. 2000); but, this relationship was not easily quantifiable. For this reason, we treated omission and commission error separately. We separately regressed omission and commission error rates against number of detections to achieve a regression coefficient and standard error describing the relationship between model error and number of detections.

We selected the seven species with more than five hundred detections and subjected their accuracy assessment to a simulation designed to approximate

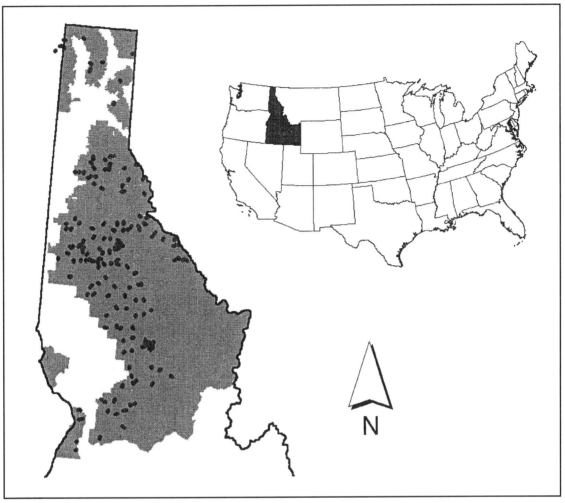

Figure 51.1. The Idaho portion of the U.S. Forest Service Northern Region consists of 2.75 million hectares dominated by coniferous forest land cover types, interspersed with dry grasslands and shrublands.

rarity. Exploratory data analysis indicated variability of omission error estimates was small for species with more than five hundred detections. Additionally, the seven species selected shared similar life history attributes (i.e., broadly distributed, similar habitat associations). For each species, we randomly selected a subset of its observations and estimated accuracy with this subset. Subset size was varied from five to the full number of observations for that species by increments of five (e.g., 5, 10, 15, . . .). We repeated this procedure for each of the seven species. Simulation data for all seven species were combined into one data set. Once the simulations were run, we plotted the simulated accuracy data against the number of observations included in each subset. We separately regressed omission and commission error rates against number

of detections to achieve a regression coefficient and standard error describing the relationship between model error types and number of detections.

If the observed pattern of change in error rates with number of species detections is an artifact of sample size, the slope of a linear regression line for the field data should be the same as that obtained by simulation. However, if there is a rarity effect, causing the models of less-common species to have lower accuracy than more common ones, then the slope of the field data regression line should be greater. To test for this, we used a student's t-test with the following null hypotheses:

$$H0: \beta_{cf} = \beta_{cs} \qquad (51.1)$$

$$H0: \beta_{of} = \beta_{os} \qquad (51.2)$$

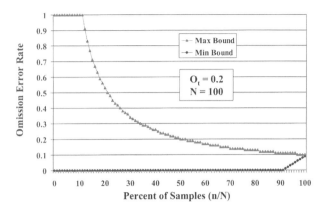

Figure 51.2. Maximum and minimum bounds on the possible values that an estimate of omission error assumes are dependent on the sample size. Sample size has been scaled to the percentage of points necessary to sample every individual in the study area.

where b_{cf} is the slope of the regression line from the field observations for commission error, b_{cs} is the slope of the regression line from simulation for commission error, b_{of} is the slope of the regression line from field observations for omission error, and b_{os} is the slope of the regression line from simulation for omission error. Because the simulations yielded large amounts of data behaving in mostly predictable patterns, the standard errors for the simulation regression coefficients were very small with respect to the parameter estimates. Thus, for the purpose of comparison, we constructed our statistical tests treating the simulation results as constants (Ramsey and Schafer 1997).

Plotting the possibilities that an estimate of omission or commission error could attain for a given sample size gave insight into the bounds within which error rates must be. To see how upper and lower bounds for omission error rates changed (Fig. 51.2), we assumed that a given model had a true omission error (O_t), that there were a definite number of individuals within the modeling area at a given time (N), and at some maximum amount of effort all individuals (N) were sampled and O_t obtained. For all detections of n individuals (where n is less than or equal to N), omission error rates were bounded by 0.0 and 1.0 as long at n/N is less than or equal to O_t. When the proportion of sampled individuals (n) to the total number of individuals on the study area (N) exceeded

the true omission error of the model, the upper bound decreased as

$$O_{max} = O_t N/n \qquad (51.3)$$

The minimum bound for omission errors remained 0.0 as long as $n/N \leq 1 - O_t$. When the proportion of samples individuals (n) to total individuals (N) exceeded one minus the true omission error rate (O_t), the lower bound increased as

$$O_{min} = O_t - (N - n)/N \qquad (51.4)$$

When n reached N, the only value that could be obtained for estimated omission error is O_t.

To see how upper and lower limits of commission error rates changed with sample size (Fig. 51.3), the same types assumptions for omission error rate bounds were made (i.e., actual number of sampling units and true commission error rate [C_t] that could be attained with some maximum effort). Additionally, the total number of predictions made (P) and the true omission error rate (O_t) must be known. The minimum bound for commission error rates originated at 1.0 for n = 0 and decreased linearly until estimated commission error reached C_t. The maximum commission error rate bound was 1.0 until n/N exceeds the omission error rate when it decreased linearly at the same rate as the minimum bound until C_t was reached. The greatest difference between the maximum and minimum bounds for commission error rates was O_t.

Results

The graph of commission error by number of detections (Fig. 51.4a) showed a strong negative trend as sample sizes increased across all species ($R^2 = 0.9861$: $P \ll 0.0001$) and behaved as predicted (Fig. 51.3). The regression line intercept was approximately equal to 1 (i.e., no observations necessitates total commission error). Commission error rates decreased 0.1 (or 10 percent) for every 167 observations. Five species had commission error rates less than predicted by the regression line (western meadowlark [see Appendix for scientific names and number of detections], spotted towhee, yellow warbler, song sparrow, warbling vireo). Omitting the seven species included in the sim-

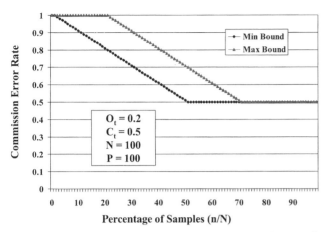

Figure 51.3. Maximum and minimum values that estimates of commission error assume are dependent on sample size and the actual omission error rate of the model. Sample size has been scaled to the percentage of points necessary to sample every individual in the study area.

ulation did not significantly change the regression coefficient ($\beta_{cf} = -0.0007$; $R^2 = 0.9577$; $P \ll 0.0001$)

Omission error rates showed a statistically significant decrease with changes in number of detections (Fig. 51.4b; $R^2 = 0.0716$; $P = 0.0051$). Given the low correlation, however, we did not consider this biologically significant because the change was less than 0.025 across the range of sample sizes 5 to 899. Variation in the values of omission error rates decreased as sample size increased. This was in line with our prediction (Fig. 51.2). Four species had significantly higher omission error rates than other species with similar numbers of detections (yellow warbler, song sparrow, black-capped chickadee, warbling vireo). Omitting the seven species included in the simulation significantly changed the regression coefficient ($\beta_{of} = -0.0009$; $R^2 = 0.0617$; $P = 0.0123$). We also did not consider this biologically significant.

In our simulation studies, commission error rates decreased predictably as sample size increased (Fig. 51.5a; $R^2 = 0.9973$; $P \ll 0.0001$). The regression line intercept was equal to one. Omission error was generally low and showed no correlation with respect to sample size but was statistically significant due to the large sample size (Fig. 51.5b; $R^2 = 0.0283$; $P < 0.0001$). Given the low correlation, we did not consider it biologically significant. Variation in the simulated omission error rates tended to decrease as sample size increased.

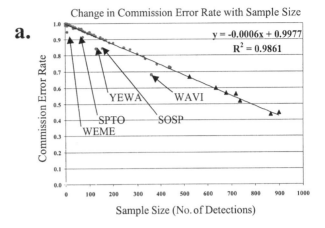

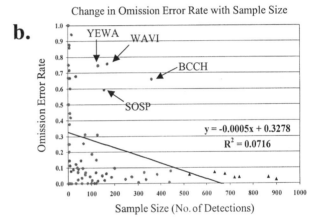

Figure 51.4. Change in error rates with number of detections for 108 bird species detected on the Idaho portion of U.S. Forest Service Northern Region. The seven species with more than five hundred detections (marked with dark triangles) were used in the simulation exercise. The dispersed nature of estimated commission (a) and omission (b) error rates obscured trends in the data due to sample sizes. Given that models with high commission error rates had low omission error and vice versa (indicating either over- or underprediction, correspondingly), we averaged commission and omission error rates for each model. Black triangles indicate the seven species included in the simulation. BCCH = black-capped chickadee (See Appendix for scientific names), WAVI = warbling vireo, YEWA = yellow warbler, SOSP = song sparrow, SPTO = spotted towhee, and WEME = western meadowlark.

Field and Simulation Comparison

Field estimates of commission error change with number of detections were not significantly different from simulation estimates ($P = 0.1747$). The slope of the regression line for change in field estimates of omission error with number of detections was significantly less than that of simulation estimates ($P = 0.0065$). Given the variability in the omission error

Change in Simulated Commission Error Rate with Sample Size

a.

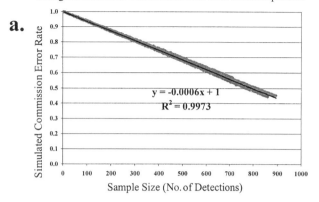

$y = -0.0006x + 1$
$R^2 = 0.9973$

Change in Simulated Omission Error Rate with Sample Size

b.

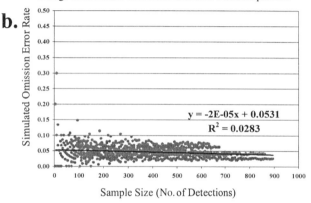

$y = -2E{-}05x + 0.0531$
$R^2 = 0.0283$

Figure 51.5. Change in error rates with simulated number of detections for the seven most common species detected on the U.S. Forest Service Northern Region. Because random subsets of observations were selected from the total observation set for each species, commission error decreased in a predictable manner (a). Omission error rates were low and exhibited more variability (b). Mean error rates for the simulations indicated similar patterns in error rate change with number of detections as the field observations.

data, we do not believe that this difference is biologically significant.

Error Rate Possibilities

We found it was possible to account for the pattern in model error by changing the number of species detections. Error rates at small sample sizes were characterized by high estimates of commission error and high variability in omission error estimates. Commission error rates declined predictably with increasing number of observations. Variability in omission error estimates also decreased with increased observations. For predicting presence and absence of the seven simulated species, we can estimate the true versus apparent

error at sample sizes less than the full number of detections (assuming that commission error at the full number of detections is the actual commission error of the model). At the smallest sample size (five detections), apparent error accounted for as much as 55 percent of measured commission error when averaged over the seven simulated species.

Simulation results suggest for forest songbirds on our study area, approximately 167 observations are needed to decrease commission error estimates by 10 percent. Potentially, more data would be needed for highly confident accuracy measures than was necessary for constructing the model. However, this is undoubtedly related to number of survey points versus area modeled. Still, this is a significant finding, as most accuracy assessments for wildlife-habitat models are either carried out with a very small number of field observations or not conducted at all (Salwasser and Krohn 1982; Morrison et al. 1998; Verbyla and Litvaitis 1989; T. C. Edwards personal communication). Project goals and the precision of results may need to be modified to fit within budgetary constraints. Thus, the additional expense in getting a test set of sufficient size may not always be possible to managers operating with small budgets.

Rabinowitz et al. (1986; see also Rabinowitz 1981) described rarity in terms of the interaction of geographic range, habitat specificity, and local density. Under this hypothesis, a species that occurred over a large region and in a variety of ecological conditions but had naturally low densities can be distinguished from a narrow endemic species that was strongly associated with localized habitat features but occurred in dense populations. This has important implications for assessing the accuracy of wildlife-habitat models. For habitat-general species that occurred in low densities over large regions, commission error rates at low sample sizes would contain a large apparent error component. However, for habitat-specific species occurring in high densities over small areas, true commission error may be much greater than apparent model error. Boone and Krohn (1999) attempted to quantify the attributes associated with rarity in Maine birds to predict whether wildlife-habitat models could be expected to have high apparent error components.

The intermountain northwest of the United States

has relatively few endemic bird species (AOU 1998). Therefore, the species that we detected infrequently would most likely fit into the category of broad-range, low-density species (after Rabinowitz et al. 1986). Additionally, simulation of rarity by random subsampling of a data set would tend to produce distributions equivalent to that of a broad-range, low-density species. We then would not expect the models for most species we detected infrequently to perform any worse than more-abundant species. However, more research should be directed toward the effects of other factors contributing to rarity (i.e., geographic range, habitat specificity).

Given that the presence or absence of a species is related to habitat features that are easily mapped, it is plausible that the ability to correctly model species occurrence could be as much a function of how much is known about the species as it is a function of factors contributing to rarity. In the case of a species with a limited geographic range, incomplete knowledge as to the extent of its range could result in higher commission error. For widely distributed species occurring at low densities, apparent model error is likely very high given the difficulty in collecting sufficient observations. However, often more is known about the habitat associations and ranges of the rarest species than many common ones. Therefore, small sample sizes preclude reliable estimates of accuracy of habitat-relationship models for many rare species.

To the manager using habitat-relationship models to aid decision-making, this means that reported accuracies could be misleading. We do not advocate that effort should not be spent toward assessing model accuracy. Assessment with even the smallest sample size can give some information about model performance. However, the results of such calculations should be viewed with extreme caution since actual error rates could by above or below what is estimated.

Appendix

Common and scientific names for species detected on U.S. Forest Service's Northern Region (Region 1), and the number of sites at which each species was detected.

Common name	Scientific name	No. of detection sites
Mallard	*Anas platyrhynchos*	2
Common merganser	*Mergus merganser*	2
Osprey	*Pandion haliaetus*	2
Sharp-shinned hawk	*Accipiter striatus*	4
Cooper's hawk	*Accipiter cooperii*	1
Northern goshawk	*Accipiter gentilis*	5
Red-tailed hawk	*Buteo jamaicensis*	8
American kestrel	*Falco sparverius*	12
Blue grouse	*Dendragapus obscurus*	2
Ruffed grouse	*Bonasa umbellus*	84
Wild turkey	*Meleagris gallopavo*	3
California quail	*Callipepla californica*	1
Spotted sandpiper	*Actitis macularia*	1
Common snipe	*Gallinago gallinago*	3
Mourning dove	*Zenaida macroura*	9
Barn owl	*Tyto alba*	1
Common poorwill	*Phalaenoptilus nuttallii*	1
Vaux's swift	*Chaetura vauxi*	2
White-throated swift	*Aeronautes saxatalis*	1
Calliope hummingbird	*Stellula calliope*	9
Broad-tailed hummingbird	*Selasphorus platycercus*	2
Rufous hummingbird	*Selasphorus rufus*	59
Belted kingfisher	*Ceryle alcyon*	7
Lewis's woodpecker	*Melanerpes lewis*	4
Williamson's sapsucker	*Sphyrapicus thyroideus*	8
Red-naped sapsucker	*Sphyrapicus nuchalis*	122
Downy woodpecker	*Picoides pubescens*	7
Hairy woodpecker	*Picoides villosus*	63
Three-toed woodpecker	*Picoides tridactylus*	8
Black-backed woodpecker	*Picoides arcticus*	1
Northern flicker	*Colaptes auratus*	112
Pileated woodpecker	*Dryocopus pileatus*	42
Olive-sided flycatcher	*Contopus cooperi*	56
Western wood-pewee	*Contopus sordidulus*	11
Willow flycatcher	*Empidonax traillii*	34
Hammond's flycatcher	*Empidonax hammondii*	302
Dusky flycatcher	*Empidonax oberholseri*	247
Cordilleran flycatcher	*Empidonax occidentalis*	23
Violet-green swallow	*Tachycineta thalassina*	1
Barn swallow	*Hirundo rustica*	1
Gray jay	*Perisoreus canadensis*	94
Steller's jay	*Cyanocitta stelleri*	67
Clark's nutcracker	*Nucifraga columbiana*	5
American crow	*Corvus brachyrhynchos*	1
Common raven	*Corvus corax*	15
Black-capped chickadee	*Poecile atricapilla*	171

(continues)

Appendix. (Continued)

Common name	Scientific name	No. of detection sites
Mountain chickadee	*Poecile gambeli*	165
Boreal chickadee	*Poecile hudsonica*	1
Chestnut-backed chickadee	*Poecile rufescens*	385
Red-breasted nuthatch	*Sitta canadensis*	635
White-breasted nuthatch	*Sitta carolinensis*	40
Pygmy nuthatch	*Sitta pygmaea*	4
Brown creeper	*Certhia americana*	74
Rock wren	*Salpinctes obsoletus*	1
Canyon wren	*Catherpes mexicanus*	1
House wren	*Troglodytes aedon*	102
Winter wren	*Troglodytes troglodytes*	335
American dipper	*Cinclus mexicanus*	14
Golden-crowned kinglet	*Regulus satrapa*	740
Ruby-crowned kinglet	*Regulus calendula*	120
Western bluebird	*Sialia mexicana*	1
Mountain bluebird	*Sialia currucoides*	13
Townsend's solitaire	*Myadestes townsendi*	75
Veery	*Catharus fuscescens*	1
Swainson's thrush	*Catharus ustulatus*	524
Hermit thrush	*Catharus guttatus*	33
American robin	*Turdus migratorius*	439
Varied thrush	*Ixoreus naevius*	187
Gray catbird	*Dumetella carolinensis*	4
Cedar waxwing	*Bombycilla cedrorum*	34
European starling	*Sturnus vulgaris*	1
Plumbeous vireo	*Vireo cassinii*	327
Warbling vireo	*Vireo gilvus*	361
Red-eyed vireo	*Vireo olivaceus*	17
Orange-crowned warbler	*Vermivora celata*	127
Nashville warbler	*Vermivora ruficapilla*	100
Yellow warbler	*Dendroica petechia*	129
Yellow-rumped warbler	*Dendroica coronata*	678
Townsend's warbler	*Dendroica townsendi*	865

Common name	Scientific name	No. of detection sites
American redstart	*Setophaga ruticilla*	10
Northern waterthrush	*Seiurus noveboracensis*	7
MacGillivray's warbler	*Oporornis tolmiei*	719
Common yellowthroat	*Geothlypis trichas*	8
Wilson's warbler	*Wilsonia pusilla*	158
Western tanager	*Piranga ludoviciana*	444
Black-headed grosbeak	*Pheucticus melanocephalus*	128
Lazuli bunting	*Passerina amoena*	104
Spotted towhee	*Pipilo maculatus*	74
Chipping sparrow	*Spizella passerina*	273
Savannah sparrow	*Passerculus sandwichensis*	3
Fox sparrow	*Passerella iliaca*	147
Song sparrow	*Melospiza melodia*	156
Lincoln's sparrow	*Melospiza lincolnii*	7
White-crowned sparrow	*Zonotrichia leucophrys*	8
Dark-eyed junco	*Junco hyemalis*	899
Red-winged blackbird	*Agelaius phoeniceus*	1
Western meadowlark	*Sturnella neglecta*	9
Brewer's blackbird	*Euphagus cyanocephalus*	2
Brown-headed cowbird	*Molothrus ater*	120
Bullock's oriole	*Icterus bullockii*	1
Pine grosbeak	*Pinicola enucleator*	5
Cassin's finch	*Carpodacus cassinii*	43
Red crossbill	*Loxia curvirostra*	45
White-winged crossbill	*Loxia leucoptera*	4
Pine siskin	*Carduelis pinus*	202
American goldfinch	*Carduelis tristis*	4
Evening grosbeak	*Coccothraustes vespertinus*	59
House sparrow	*Passer domesticus*	2

Spatial Analysis of Stopover Habitats of Neotropical Migrant Birds

Scott M. Pearson and Theodore R. Simons

Recent declines in populations of Neotropical landbird migrants (Robbins et al. 1989b; Askins et al. 1990; Finch 1991) have prompted a wave of new research into the factors affecting populations of these birds on their breeding and wintering grounds (Hagan and Johnston 1992; Finch and Stangel 1993). Although breeding and wintering activities are essential for population persistence, migration represents a significant event in the yearly life cycle of these birds because it requires a large energetic investment and can represent a period of high mortality. Moreover, areas used by migrating birds for rest and foraging (i.e., stopover areas) are experiencing rapid changes due to increasing urban, residential, and industrial/agricultural development. Increases in the abundance of these habitats may be detrimental to migrants (Yong et al. 1998). Some studies have focused on the factors affecting birds during migration (Moore and Simons 1992; Winker et al. 1992; Watts and Mabey 1993; Morris et al. 1994; Simons et al. 2000; Moore et al. 1995) and are shedding light on how stopover areas can be critical for many migratory species. Designing conservation-oriented studies of the stopover ecology of migrants is complicated by the fact that migration occurs over a broad geographic area, but over a relatively short time period. It is often difficult for field researchers to be in the right place at the right time.

Remote-sensing technology and spatial-modeling techniques are providing new research tools for investigating how the distribution and abundance of habitats may affect wildlife populations. Metrics of landscape pattern provide a means to quantify differences between landscapes. Landscape indices and spatial models (e.g., habitat suitability, individual-based models) are tools that allow biologists to assess how differences in the abundance and spatial arrangement of habitats may affect the suitability of landscapes for selected species. Moreover, these models can be modified to assess the quality of landscapes for species with different habitat needs, physiological requirements, or foraging strategies (Dunning et al. 1995; Turner et al. 1995). Conclusions gleaned from these models, however, need to be interpreted within the context of the model design and assumptions.

We have recently discussed how spatial models can be applied to questions about the stopover ecology of trans-Gulf migrants (Simons et al. 2000). We have shown how models incorporating available data on the arrival condition of migrants, energetic and morphological constraints on movement, and species-specific habitat preferences can provide insights into how the abundance, quality, and spatial pattern of habitats interact with the arrival energetic state of migrants to determine the suitability of migratory stopover habitats along the northern Gulf Coast. Our goal in this chapter is to explore further how an analysis of landscape composition and spatial models that distinguish between habitat specialists and generalists can improve our understanding of the factors constraining migrants at stopover sites. We hope that the results of this analysis will provide insights that are useful in setting priorities for future research and conservation.

This research will compare five study landscapes located on the northern coast of the Gulf of Mexico using (1) landscape-level metrics of spatial pattern, and (2) output from an individual-based model of stopover habitat use. Landscape-level metrics provide a means to quantify the abundance and spatial pattern of habitat types in study landscapes (Turner and Gardner 1991). The most straightforward measure is the area of suitable habitat types. The spatial arrangement of habitats can also be measured. For example, habitat fragmentation is often quantified using a combination of the number of habitat patches, mean patch size, and area of the largest patch. The juxtaposition of habitat types can be measured with indices of edge density or contagion (Hargis et al. 1997). Landscape-level metrics provide objective measures of habitat patterns; however, interpreting these metrics from the perspective of a species' biology remains challenging.

Field studies have shown that the arrival energetic state of migrants at trans-Gulf stopover sites is highly variable (Moore et al. 1990; Moore and Simons 1992; Morris et al. 1994; Moore et al. 1995; Fransson and Jakobsson 1998). Energy reserves can constrain long- and short-range movements of migrants (Jenni and Jenni-Eiermann 1998). Measurements of fat reserves of birds arriving along the northern Gulf Coast have been used with flight performance models to estimate the potential flight ranges of migrants at coastal stopover sites (Simons et al. 2000). Flight ranges may be as little as a few kilometers in extreme cases and average from tens to several hundred kilometers in most species. Thus, the distribution, abundance, and quality of suitable habitat within the range of migrants at coastal stopover sites can be viewed as an important constraint on the likelihood of a successful migration. Habitat preferences have been demonstrated for migrants during stopover (e.g., Weisbrod et al. 1993; Russell et al. 1994). Along the Gulf Coast, field studies have shown that migrants appear to pre-

fer forests with a well-developed understory and riparian bottomlands over other habitats available at stopover sites (Moore et al. 1990; Simons et al. 2000). These observations suggest habitat specialization may also constrain the suitability of stopover sites for some species.

We have developed an individual-based model that simulates habitat usage and energy gain by birds during migration stopover events. The model is simplistic and makes minimal assumptions about the details of habitat use. It is designed to provide information about the consequences of foraging in landscapes that vary with respect to the abundance, spatial arrangement, and quality of habitats that differ with respect to their foraging returns. Changing the model parameters also allows us to simulate birds that vary in their energy states and habitat specificity (e.g., generalists versus specialists).

The specific objectives of this work were (1) to investigate the relative importance of landscape pattern of habitats, energetic state of arriving birds, and habitat specialization for successful stopover, and (2) to identify the types of landscapes that will provide suitable stopover habitat. This second objective was implemented by ranking a set of real landscapes according to their suitability for stopover habitat use. To achieve these objectives, we analyzed data from remote sensing using modeling and analytical approaches based on an understanding of stopover ecology gained from field studies.

Study Area and Methods

The study landscapes were five 25×25-kilometer regions located along the northern Gulf Coast in the states of Texas, Louisiana, and Mississippi (see Fig. 52.1 in color section). Habitat data for these landscapes were derived from a supervised classification of two 1990 Landsat Thematic Mapper images. The classification was performed by the U.S. Geological Survey's Southern Science Center in Lafayette, Louisiana. The original map consisted of eighteen cover types in raster format having square 28.5×28.5-meter cells. These eighteen habitat types were reclassified into four habitat categories that represent four classes of habitat quality (Table 52.1). Category 1 rep-

TABLE 52.1.

Habitat categories used for landscape suitability analyses.

	Category 1[a]	Category 2	Category 3	Category 4[b]
	Unclassified	Emergent marsh	Mixed shrub-scrub	Deciduous forest
	Water	Residential	Evergreen shrub-scrub	Bottomland forest
	Excavated soil	Pine forest	Mixed forest	
	Beach/sand	Cropland		
	Sand bar	Orchards		
	Commercial			
	Transportation			
	Industrial			
% Total[c]	7.5	53.6	31.4	7.5

[a]Category 1 habitats offer few opportunities for foraging; therefore, this category represents the poorest habitats for migrants.
[b]Category 4 habitats were assumed to be the best habitats for migrants.
[c]The percent representation of each of the categories, summed over all landscapes, is reported.

resented the poorest-quality habitat class. These habitats provide few if any opportunities for foraging by migrating landbirds. Category 4 represents the highest-quality habitats that provide abundant food resources. Assignment of habitats to these classes was based on field studies of habitat usage and habitat-specific weight gain conducted during 1987–1994 (Moore et al. 1990; Kuenzi et al. 1991; Simons et al. 2000).

Landscape Metrics

For each study landscape, the spatial pattern of the four habitat categories was quantified using FRAGSTATS (McGarigal and Marks 1995). For each habitat category, the following metrics were recorded: percentage of total area of map occupied by each habitat category, total number of patches of all habitat types, patch density (number of patches per hectare), mean patch size (hectare), edge density (meters of edge per hectare), Simpson's index, and contagion. Proportion of total area serves a measure of the abundance of the habitat category. Collectively, the number of patches, patch density, and mean patch size provide a means to compare the relative fragmentation or connectivity of a given habitat type among the five study landscapes. Landscapes in which habitats are more fragmented have a greater number of patches, higher

patch density, and smaller mean patch size. Edge density and contagion measure the degree of interspersion and contact between habitat types. Edge density will increase in landscapes in which patch sizes are small and patches have complex or elongated shapes (e.g., skinny rectangles or dendritic shapes) rather than compact shapes (e.g., circles or squares). Contagion also measures the degree of habitat fragmentation and contact between habitat types. The contagion value represents the probability that two adjacent cells, chosen at random, will be of the same habitat type. Thus, contagion will be greater for landscapes in which habitats are highly clumped and lower for maps in which habitats are fragmented and highly interspersed. See McGarigal and Marks (1995) for a more complete description of the calculation and interpretation of these landscape metrics.

In addition to the metrics provided by FRAGSTATS, we calculated an index of general landscape quality using the following formula:

$$Quality = P_1 + 2*P_2 + 3*P_3 + 4*P_4$$

where P_x represents the proportion of the total landscape area occupied by habitat category x. Landscapes with a greater proportion of high-quality habitat types will receive a greater-quality score. Although this index will permit the ranking of landscapes with respect to the abundance of habitat categories, it does not indicate anything about the spatial arrangement of habitats within the landscape. This index uses an econometric approach similar to that of a habitat suitability index (HSI). In HSI techniques, each habitat unit (usually an areal unit) is multiplied by an ordinal index of habitat quality.

Individual-based Model

This model uses an energy state index (ESI) to indicate the relative energetic state of birds during migratory stopover. We assume that birds arrive at the stopover site with low energy reserves. During stopover, an individual bird forages to improve its energetic state, rebuilding energy reserves that will fuel its next migratory flight (e.g., Alerstam and Lindstrom 1990; Fransson 1998). The species that inspired this study typically migrate by flying long distances (usually hundreds of kilometers per flight) at night. At the end of one of these long-distance flights, the birds "stop over" and spend one or more days foraging to refuel before the next long-distance flight. The availability of habitats that provide foraging opportunities at the stopover site will determine the rate at which energy reserves can be rebuilt. Birds that land in relatively rich sites will be able to refuel quickly; those that land in poorer sites will take longer to store enough energy to make another long-distance, nocturnal flight (Yong et al. 1998). In a worst case, a migrant in a very poor site may starve because it cannot ingest enough energy to satisfy its immediate energetic needs.

The model incorporates these assumptions about energetic state in the following way. Arriving birds with a given ESI land in a randomly determined cell in the habitat map. By changing the initial ESI value, we simulated birds that have varying levels of energy upon arriving at the stopover site. The bird then foraged by moving from cell to cell. At each cell, ESI is updated using this equation:

$$\text{ESI}_j = (\text{energy gained in cell}_j) -$$
$$(\text{energy cost of movement and foraging}).$$

Foraging cost was held constant, and foraging gain accrued by the birds as they moved across the landscape depended on the habitat category of each cell

TABLE 52.2.

Energetic gain values by habitat categories for habitat generalists and specialists. The habitat categories are defined in Table 52.1.

	Habitat category			
	1	**2**	**3**	**4**
Generalist	0.10	0.55	0.75	0.75
Specialist	0.00	0.10	0.45	1.50

encountered. In productive habitats, migrants experienced a net energy gain (i.e., gain is greater than cost). In the poorest habitats, there was a net loss (gain is less than cost).

The degree of ESI gain or loss depended on whether the bird was classified as a habitat *generalist* or a habitat *specialist* (Table 52.2). For both generalist and specialist, the cost of foraging and moving was fixed at 0.50 ESI units for all habitat types. The habitat-specific gain values were adjusted so that both generalists and specialists would receive the same gain if they encountered equal proportions of the four habitat categories. However, the specialist did better than the generalist on category 4 cells and worse on category 2 and 3 cells.

In the model, the bird continued to forage until its ESI crossed one of two thresholds. If the individual gained enough energy, it left the study landscape on another long-range migratory movement. In contrast, individuals that failed to find productive habitat continually lost energy and died (if the ESI dropped too low). When an individual migrated or died, the number of cells visited was recorded in the model output. For these simulations, the migration threshold was fixed at an ESI of 30.0. The death threshold was set at an ESI of 2.0. Thresholds were the same for both habitat generalists and specialists.

A subroutine that governs movement from cell to cell was used that incorporated knowledge of adjacent cells, ability to choose among cells based on habitat quality, and a northerly bias to movement. When moving from cell to cell, the individual was assumed aware of the habitat types of the adjacent eight cells. The bird would choose cells with higher ESI gain values (i.e., higher habitat-quality category) over cells with lower ESI gain values. Laboratory studies have

TABLE 52.3.

Coefficients used to incorporate a northerly bias to bird movement during stopover.

		North		
	0.90	1.00	0.90	
West	0.75	**Focal Cell**	0.75	**East**
	0.60	0.50	0.60	
		South		

demonstrated a tendency for birds migrating through the northern Gulf in the spring to orient and move northward (Gauthreaux 1971; Emlen 1975). Therefore, we assumed that given two cells of equal habitat quality, a bird is more likely to move to the more northerly cell. This bias was accomplished by discounting the gain value for each cell by its position relative to north. Thus, the bird would calculate an attractiveness value for each cell:

$$Attractiveness = (ESI\ gain)*(nbias)$$

where *nbias* is this discounting coefficient. Table 52.3 shows the *nbias* coefficients used in these simulations. Birds would move to the cell with the greatest attractiveness value. If two cells were equal in greatest attractiveness, choice between them would be made randomly. Birds were not allowed to return to cells that were previously visited. Arriving birds were randomly located in the southern portion of the map to a cell within 1.0 kilometer of the Gulf Coast. Data for birds that wandered to the edge of the map without migrating or dying were discarded, and the simulation was reinitiated for that individual.

Simulation Experiment

A set of simulations was designed to assess the performance of birds using the five study landscapes as potential stopover sites. The goal of this experiment was to determine the relative importance of landscape pattern of habitats, arrival energetic state of arriving birds, and species' habitat specialization to stopover performance. Performance was measured by (1) the proportion of birds that survived and migrated, and (2) the number of cells visited by migrating birds. In highly suitable landscapes, it is expected that a greater proportion of birds will migrate and that fewer cells

will be visited during the stopover time because energy gain per cell will be greater on average. However, the actual performance of the birds could be affected by the patchiness and interspersion of habitat types.

A factorial design was used with the following levels: (1) five study landscapes, (2) two levels of habitat specialization, and (3) four levels of arriving ESI. The levels of arrival ESI were set at 5, 10, 15, and 20. For each of forty treatment combinations, five hundred replicate birds were simulated. The relative influence of landscape, arrival ESI, and habitat specialization were compared by examining the magnitude of F scores from the analyses of variance. ANOVAs were conducted using SAS (1985). To improve normality, the proportions of migrants were arcsine–square-root transformed and counts of cells were square-root transformed before conducting the ANOVAs (Sokal and Rohlf 1995) .

Results

Differences among the five study landscapes evident in Figure 52.1 were also revealed in the landscape metrics, although variation in the selected metrics was not striking. Comparing the relative abundance of each habitat type is the simplest way to compare the five landscapes. Landscape A had the greatest amount of the highest-quality habitat, category 4 (Fig. 52.2).

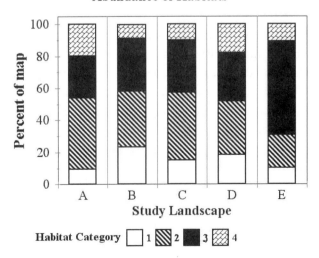

Figure 52.2. The relative abundance of four habitat types in each of the five study landscapes.

TABLE 52.4.

Landscape metrics for five study landscapes.

Metric[a]	Landscape				
	A	B	C	D	E
Number of patches	32,426	31,333	31,059	29,491	31,463
Patch density (no./ha)	51.9	50.2	49.7	47.2	50.4
Mean patch size (ha)	1.93	1.99	2.01	2.12	1.97
Edge density (m/ha)	196.7	194.7	199.3	188.7	192.0
Simpson's index	0.69	0.71	0.68	0.73	0.60
Contagion (%)	34.8	34.0	35.5	31.8	40.9
Quality index[b]	2.57	2.27	2.37	2.47	2.70

[a]See McGarigal and Marks (1995) for a complete description of the first six metrics.
[b]The quality index is described in the methods section of this chapter.

Landscape E had the greatest amount of habitat category 3 and the greatest amount of habitats 3 and 4 combined. There was little difference in the abundance of high-quality habitat among landscapes B, C, and D. Among those three study areas, landscape C had the least amount of poor-quality, category 1 habitat. Landscape D had the greatest value of Simpson's index (Table 52.4) indicating that the habitat categories were most evenly distributed in this map compared to the other five maps. Landscape E had the lowest value of this index due the greater relative abundance of category 3 habitats.

The remaining metrics provide measures of the spatial pattern of the four habitats and their interspersion. The number of patches, patch density, and mean patch size provide a means to compare the relative levels of fragmentation among the five maps. Landscape A had the greatest number of patches, largest patch density, and smallest mean patch size (Table 52.4), which indicates that this map had the greatest degree of habitat fragmentation. In contrast, landscape D had the lowest number of patches, least patch density, and largest mean patch size, indicating that habitats in this landscape tended to be more clumped in larger patches (landscapes A and D, Fig. 52.1).

The interspersion of habitats may be important for birds as they move across the landscapes. Although interspersion is related to measures of habitat fragmentation, the metrics of edge density and contagion provide information on the likelihood that a moving bird will encounter different habitat types during stopover.

Landscape C had the greatest edge density (Table 52.4) due to the high interspersion of habitats and the complex shapes of habitat patches evident in northern portion of this map (Fig. 52.1). This landscape had an intermediate level of contagion (Table 52.4) due to the differences between the northern and southern sections of this map. Landscapes B, C, and D show a sharp gradient in the pattern of category 4 habitats. These habitats, which include deciduous and bottomland forests, become more abundant in wetland areas that are farther away from the brackish water influence of the Gulf of Mexico. Landscape E had the highest level of contagion and an intermediate value of edge density (Table 52.4) caused by the increased coverage of category 3 habitats in this map (Fig. 52.2). This habitat dominates much of the middle portions of this landscape (Fig. 52.1).

Probability of Successful Stopover

Highly suitable landscapes would be expected to have a high proportion of birds that acquire enough energy to migrate after stopover. The relative influence of landscape, arrival ESI, and habitat specialization were compared by examining the magnitude of F scores from the analysis of variance. Habitat specialization had the greatest effect on probability of successful migration (Table 52.5). Habitat generalists had consistently higher probabilities of successful stopover than did specialists (Fig. 52.3). The success of specialists varied among landscapes (Fig. 52.3), resulting in a weaker, although significant landscape

TABLE 52.5.

Analysis of variance of the proportion of birds surviving to migrate.

Source	df	Type III SS	F	P
Landscape	4	0.458	34.5	< 0.001
Arrival ESI	1	0.200	60.0	< 0.001
Habitat specialization	1	0.659	198.2	< 0.001
Landscp.×ESI	4	0.012	0.9	0.457
Landscp.×Specialization	4	0.679	51.1	< 0.001
ESI×Specialization	1	0.001	0.4	0.550
Total	39	6.686		

Note: Proportions were arcsine–square root transformed before analysis.

main effect and landscape × specialization interaction. Among the main effects, arrival ESI had an influence of intermediate magnitude. Higher values of arrival ESI resulted in a greater chance of successful stopover, especially for habitat specialists (Fig. 52.3). There were no significant interactions involving arrival ESI.

Number of Cells Visited by Migrants

Among the main effects, the strongest influences were arrival ESI and landscape pattern (Table 52.6). In-

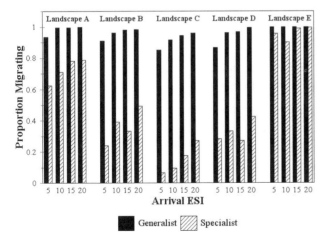

Figure 52.3. The proportion of simulated birds obtaining enough energy to survive and migrate out of the study landscape. Results for the five landscapes are plotted in separate groups of bars. Two types of birds were simulated: habitat specialists and habitat generalists. Each bird began its stopover habitat use with one of four levels of energy reserves. These reserves were tracked in the individual-based model using a energy state index (ESI).

creasing arrival ESI tended to reduce the number of cells visited by successful migrants (Fig. 52.4). Landscapes A and E (mean ± SD respectively: 96.0 ± 74.9, 103 ± 778.2) required fewer cells for successful migration than landscapes B, C, and D (136.7 ± 144.4, 136.6 ± 110.1, 159.3 ± 146.0, respectively). The effect of habitat specialization was of intermediate magnitude. However, there were strong interactions between this factor, landscape, and arrival ESI. Arrival ESI had a stronger influence on generalists than on specialists (Fig. 52.4). The number of cells visited was consistently reduced for generalists with higher levels of energy on arrival; the influence of this factor on specialists was less consistent among the landscapes. Specialists visited more cells than generalists did in four of the five landscapes. Specialists visited fewer cells than generalists did in landscape A. The higher efficiency of specialists in this landscape was likely due to the greater abundance and dispersion of category 4 habitats in this map.

Ranking Study Landscapes

The landscape metrics and output from the individual-based model allowed the ranking of landscapes. The most straightforward means was to use the quality index (Table 52.4) that is similar to HSI approaches. Although this measure is not spatially explicit, it provided an initial intuitive assessment of the relative abundance of high-quality habitats among the alternative landscapes. Based on the quality index, landscapes E and A ranked highest; landscapes C and B

TABLE 52.6.

Analysis of variance of the number of cells visited by successful migrants.[a]

Source	df	Type III SS	F	P
Landscape	4	1,751.2	37.4	< 0.001
Arrival ESI	1	11,544.5	985.6	< 0.001
Habitat specialization	1	1,260.4	107.6	< 0.001
Landscp.×ESI	4	885.0	18.9	< 0.001
Landscp.×Specialization	4	20,044.1	427.8	< 0.001
ESI×Specialization	1	4,766.2	406.9	< 0.001
Total	14,679	234,342.3		

[a]Cell counts were square root transformed before analysis.

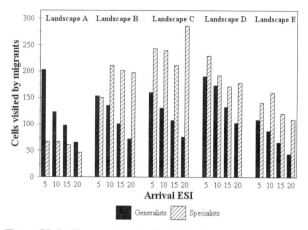

Figure 52.4. The number of cells visited by birds that survived and migrated from study landscape. See Figure 52.3 caption for explanation of format.

ranked the worst (Table 52.7). The high rank of landscape E was produced by its greater relative abundance of category 3 habitats.

Output from the individual-based model provided information on the relative success rates of migrant birds using these landscapes. The proportion of individuals surviving to migrate and the number of cells visited by successful migrants served as measures of success and were used to rank the landscapes. High proportions of successful migrants and low numbers of cells visited were taken to represent the best landscapes. Based on proportion of migrants, landscapes E and A ranked best; landscape C ranked worst (Table 52.7). Landscapes B and D had similar proportions of

TABLE 52.7.

Ranking of landscapes based on quality index, proportion of successful migrants, and number of cells visited by successful migrants.

	Landscape designation[a]		
Rank	**Quality index[b]**	**Proportion migrants**	**Cells visited**
1 (best)	highest E	most E (0.980)	least A (95.0)
2	A	A (0.854)	E (103.7)
3	D	B (0.662)	C (136.6)
4	C	D (0.638)	B (136.7)
5 (worst)	lowest B	least C (0.535)	most D (159.3)

[a]Mean proportion of successful migrants and mean number of cells visited are reported in parentheses after landscape number in the respective columns.

[b]The quality index is reported in Table 52.4.

migrants and were intermediate in rank. Based on the number of cells visited, landscapes A and E again ranked best; landscape D was worst. Landscapes C and B had almost identical numbers of cells visited and had an intermediate rank.

In summary, the precise ranking of landscapes depended on the method employed. Although landscapes A and E consistently ranked as the best, the ranking of remaining landscapes varied with method. For example, landscape D ranks as third best with respect to the quality index but fourth or worst according to the individual-based model.

Discussion

Transcontinental migration is a brief period in the annual cycle of a Neotropical migrant bird (i.e., a couple weeks during the spring and autumn). However, successful completion of migration between wintering and breeding grounds is essential to an individual's evolutionary fitness and for the persistence of the population as a whole. Successful migration depends on the existence of suitable stopover habitat along migration routes. The availability and quality of this habitat are as crucial as the quality of wintering and breeding sites. Although stopover sites are used only during a brief period of the year, the existence of high-quality sites is necessary for the continued persistence of these avian species.

Conservation of Stopover Sites:
A Challenging Problem

The conservation strategy that led to the present system of waterfowl reserves along the Gulf Coast may not work for migrant songbirds. There are fifteen national wildlife refuges and two national seashores that protect coastal habitat along the northern Gulf Coast. These areas were established primarily for the purposes of waterfowl conservation and recreation. In large part, the remaining coastal cheniers and riparian woodlands that are important to trans-Gulf migrants are unprotected and not in the public trust. The waterfowl refuges were selected because specific locations used by these birds could be identified. Providing habitat for migrating songbirds is not as straightforward.

The geographic extent and stochastic nature of

stopover events presents a challenge for the purposes of maintaining populations of migrating songbirds. For example, on the Gulf Coast the exact location of "fall out" events is determined by onshore weather encountered when the birds reach the coast and the weather patterns experienced during their flight over the Gulf of Mexico. Groups of migrant birds may land just about anywhere in a given landscape used for stopover during the course of several seasons. The conservation manager is then faced with the difficulty of devising ways to protect a set of species that are dependent on a particular landscape but whose use of the landscape is geographically random and seasonally ephemeral. Moreover, the manager has little or no control over a vast portion of the landscape that is owned and used by a large number of private landowners. These challenges to protection efforts may seem insurmountable. Nevertheless, we are optimistic that feasible strategies for preserving an essential amount and geographic distribution of stopover habitat can be developed if scientists develop a better understanding of stopover ecology.

The same characteristics of stopover events that discourage managers (i.e., geographically extensive, randomly located, temporally ephemeral) also confound research scientists seeking to conduct field studies. Some of these difficulties may be overcome by the increasing availability and accuracy of remotely sensed data covering large areas. Maps derived from these data allow scientists to examine the relative abundance and spatial pattern of habitats over broad geographic areas. By combining these maps with field studies, researchers should develop a better understanding of the characteristics of landscapes that provide quality stopover sites.

Field Studies at Stopover Sites and Landscape-level Analyses

The analytical and modeling approaches used in this study were developed from the findings of field studies. Although limited in geographic and temporal extent, field studies provide information about the ecology of individual and small groups of birds using stopover habitat. These studies provide information on the important features of habitat use such as coarse-grained and microhabitat preferences, duration of stay at a given site, patterns of movement within a site, energetic condition of arriving birds, and qualitative comparison of relative rates of energy gain among alternative sites. The vegetation types and microhabitats available at a given study site may be quantified, but researchers must rely on accurate maps of surrounding areas to learn about the abundance of habitats in the landscape surrounding the site. Habitat maps permit the study of research questions related to the influence of the surrounding landscape on the use of a given stopover site (e.g., Pearson 1993). For example, does the use of a given 5-hectare study plot of high-quality habitat depend on the abundance of that habitat in the surrounding landscape?

Other questions about landscape-level habitat uses are more difficult (but not impossible) to address with field studies. For example, what is the scale of habitat selection conducted by arriving birds? Once the bird lands, what aspects of the landscape affect its pattern of movement among habitat patches? The geographic extent of migration makes it difficult to design comprehensive field studies. However, we should be able to explore these questions using landscape analyses and spatial models that are based on our understanding of stopover ecology gained from past and present field studies. These explorations should lead to predictions and hypotheses that could be tested in future, carefully designed field studies. The approach taken in our study was to use a model to rank landscapes according to the "success" of migrants during stopover. The measures of success included the duration of stay (i.e., number of cells visited by migrants).

By using the model, we gained insights into how alternative landscape configurations may affect this aspect of stopover ecology, and we generated predictions about the duration of stay that can be tested in future field studies. Analyzing the predictions of a general model can help identify aspects of stopover ecology that need addressing in future field studies. For the conservation manager, a general model could be used to evaluate management alternatives. For example, the performance of migrants could be simulated on alternative landscape patterns produced by different policy options that affect land use on private lands (e.g., conservation easements) or by the creation of new protected areas under public ownership. Given the limited resources for conservation, an approach like this can help managers select strategies to achieve the greatest positive change in the landscape with limited fiscal resources. Thus, landscape analyses and modeling studies provide a powerful tool to complement field studies.

Results from This Study

Habitat specialization emerged as an important influence on the probability of successful stopover and the amount of time needed to rebuild energy reserves (Tables 52.5 and 52.6). Our results indicate that the effects of habitat pattern and abundance on migrants will be amplified for habitat specialists and when birds arrive at stopover sites with their energy reserves depleted. The difference in the relative abundance of habitats in these landscapes is not striking (Fig. 52.2). Although some differences in spatial arrangement of habitats are apparent in Figure 52.1, it is not clear from a cursory inspection of these maps which landscape should be better or worse. A more thorough analysis was needed. While landscape pattern ranked the lowest among main effects in both ANOVAs, these analyses revealed that there were strong interactions among landscape pattern, habitat specialization, and arrival ESI for number of cells visited (Table 52.6). These results suggest that the differences in the abundance and spatial arrangement of habitats matter more for species with specialized habitat needs and less for generalists. Admittedly, this is an intuitive result, but the degree of difference that habitat specialization makes, and the relative ranking of these landscapes, would not have been possible without the use of the model.

Ranking landscapes depends on the method for quantifying habitat suitability. Table 52.7 shows that the exact ranking depends on the schema being used. Landscape C ranks third, fourth, or fifth, depending on which ranking method is employed. The differences in the ranking of this landscape based on proportion of successful migrants versus number of cells visited is due to differences in performance between habitat generalists and specialists. Although the performance of generalists on this landscape was comparable to landscapes B and C, the performance of spe-

cialists was much worse. Specialists had the lowest success rates on landscape C (Fig. 52.3) and required a higher number of cells visited (Fig. 52.4) than on any of the other landscapes. Choices about the most desirable criteria for ranking landscapes must be made by knowledgeable managers within the context of specific conservation goals. Spatial models and landscape analyses can provide information useful for comparing alternative criteria.

The results caused us to reevaluate the effects of some landscape features. *A priori*, we expected landscape D to rank high because of the presence of two large river corridors dominated by high-quality deciduous bottomland forest. However, this landscape ranked in the middle or lower half of the rankings (Table 52.7). The presence of large patches of bottomland forest was counteracted by the abundance of large patches of low-quality habitat in the southern portion of the map. The southern region provides none of the category 4 habitats that provide high returns for the habitat specialists. Migrants landing in the southern half had to contend with these relatively poor habitats before encountering the richer sites to the north.

The model output also assisted in interpreting the landscape metrics, such as contagion. While the relative abundance of habitats affects the mean habitat quality, patch size, and interspersion of habitats affect the variance in foraging returns experienced by a bird as it moves across the landscape. Higher levels of interspersion (i.e., low contagion) reduce the variance in foraging returns for a bird visiting a fixed number of cells. Thus, in landscapes with the same levels of interspersion, mean habitat quality (calculated at a scale relevant to a single bird during stopover) is most important. In a landscape with the same mean quality but lower levels of interspersion (i.e., high contagion), the variance among birds will be higher because some will encounter large patches of high-quality habitat while others encounter patches of low-quality habitat.

Changes in interspersion can be good or bad depending on mean habitat quality. If the average habitat quality is great enough for most birds to be successful, then increasing interspersion will require birds to visit a larger area (i.e., more cells), because they will inevitably encounter low-quality sites, although almost all are assured to gather enough energy to migrate. Decreasing interspersion (increasing contagion) would mean that some birds would encounter large patches of rich and poor habitats. More birds would die without migrating because they landed in large patches of low-quality habitat. In contrast, decreasing interspersion could be a good thing for landscapes in which the mean habitat quality is so low that the average foraging return is less than that needed for successful migration. If interspersion is high, practically no birds will be able to obtain enough energy during stopover because they will be receiving the mean return for foraging on this landscape. Alternatively, if rich habitats are more clumped, then at least some birds will land in these areas and become successful, although most will perish because they landed in poor areas. This is a spatially explicit example of issues addressed by the topic of risk-sensitive foraging (e.g., Caraco et al 1980; Stephens and Charnov 1982) investigated in the field of optimal foraging theory. These issues are have been addressed in other systems where prey are cryptic and have heterogeneous spatial distributions by using a combination of field data and simulation modeling (e.g., ungulates; Turner et al. 1994; Pearson et al. 1995).

This understanding helps interpret differences in the performance of migrants on different landscapes. Landscape E consistently ranked high because it has the highest abundance of category 3 and 4 habitats (Table 52.4). Moreover, this map has the highest level of contagion (Table 52.4), driven by the extensive well-connected regions of category 3 habitat. The number of successful migrants is highest on this map (Table 52.7, Fig. 52.3). Habitat generalists do better than specialists do in this landscape because of their higher foraging returns in category 3 habitats (Table 52.2, Fig. 52.4). In contrast, compare landscapes A, B, and C. These three landscapes have a similar level of contagion (Table 52.4). However, the greater abundance of high-quality habitats in landscape A increases the average quality of cells encountered during stopover. This difference results in a greater chance of successful stopover (Fig. 52.3) and fewer cells being visited by successful birds (Fig. 52.4) than in landscapes B and C. Whereas successful habitat generalists visit fewer cells than specialists in landscapes B and C (Fig. 52.4), the greater abundance of category 4 habitats in landscape A (Fig. 52.2) allowed specialists to visited fewer cells than specialists in this landscape. Specialists also had a higher probability of successful migration in landscape A relative to B and C (Fig. 52.3). Thus, this simulation model provided a means to evaluate the relative quality of landscapes from the perspective of birds that have different habitat needs. It also provides a mechanistic understanding of stopover habitat use that enhances our ability to interpret metrics of habitat pattern.

The strengths and weaknesses of a given model should be known by the users. The main value of the simulation model used in this study is to illustrate the complex interactions that shape the process of songbird migration. The factors shaping that process include the pattern, abundance, and quality of stopover habitats, and the mobility and foraging ecology of individual migrants. The individual-based model provides a means to compare landscapes from a less-anthropocentric perspective. The specific weaknesses of this model include the fact that (1) we have little knowledge about how variations in habitat quality at stopover sites translate into different rates of energy gain for migrants even though data on relative abundance of migrants, residency times, and fat condition in different habitats are available; (2) we have little data on movement patterns of migrants during at stopover sites (but see Aborn and Moore 1997); and (3) we know little about the settling patterns of migrants at migratory stopover sites.

Important Considerations in the Use of Metrics and Models

At present, approaches to measuring landscape characteristics include (1) metrics of landscape patterns, and (2) implementation of spatial models. Both of these methods allow the researcher or manager to rank a series of real or hypothetical landscapes based on the abundance and spatial arrangement of habitats. Ranking landscapes with respect to their suitability for a given population or suite of species can be challenging. Obviously, the ranking will depend on the method for quantifying landscape suitability. Researchers and managers should ideally use methods that are both realistic and appropriate for the species of interest. Landscape metrics are useful if they can be readily interpreted—that is, if variation in a metric can be directly related to an important aspect of the species' biology. Some progress has been made in this area, but better links between metrics and specific ecological processes need to be forged. Approaches such as the habitat suitability index move a step beyond the use of landscape metrics because these indices can incorporate the positive and negative influences of diverse habitat measurements from a variety of scales. Spatial models can be more realistic because they can incorporate more-complicated aspects of the species' habitat use, such as the consequences of movement and habitat use on survival and reproduction.

Although spatial models can be realistic, tradeoffs will exist between the general use of a model and the number of assumptions it makes about the species' ecology. The design of any complex model will involve making assumptions about habitat selection or other aspects of habitat use. These assumptions should be testable, or models should be based on our best understanding of species biology. Models involving many precise parameters and detailed mechanisms (e.g., fine-grained habitat selection, movement between cells) are fraught with many more assumptions than models that have fewer details and are more general in design.

Conservation programs often raise questions about the ecology of species, reserve design, or other management issues for which definitive data are not available and/or are difficult to collect. Spatial analyses and modeling are tools that can be useful in making the best of these difficult situations, as long as the model results are viewed within a context of model design, assumptions, parameter values, and spatial (or other) data. Models and analyses based on the current best understanding of a species' ecology can make it possible to pose meaningful "what if" questions about management alternatives or ecological processes. These explorations often generate a better understanding of the system in question, and they can produce hypotheses that can be tested with empirical data from field studies.

Summary

Stopover habitat use presents challenges to research and management due to its broad spatial extent and seasonal, ephemeral time span. A landscape-level approach is essential. Understanding gained from field studies can guide landscape-level analyses that in turn can be used to develop testable hypotheses for future research or to inform difficult management decisions. This study produced findings relevant to management. Landscapes with the greatest amount of high-quality habitat were most suitable, but the spatial arrangement of habitats modified suitability. For example, the fragmentation of habitats as measured by interspersion can be good or bad depending on mean habitat quality experienced by migrants. Decrease interspersion if mean quality is too low. Spatial arrangement becomes more important as landscape-wide average habitat quality declines. Moreover, landscape suitability depends on habitat specialization; the relative performance of generalists and specialists depends on the details of the relative abundance and spatial arrangements of habitats. Given the broad spatial extent of bird migration, policies that favor the protection of high-quality habitats throughout the landscape, including private lands, would be more beneficial than the purchase and management of a few preserves of limited spatial extent.

Acknowledgments

This work received support from a grant from the National Science Foundation DEB 9416803, the U.S. Fish and Wildlife Service, the National Park Service, and the U.S. Geological Survey, Biological Resources Division. The habitat maps were derived from a Landsat image classified by P. O'Neil, U.S. Geological Survey. J. M. Scott and two anonymous reviewers provided valuable comments on the original manuscript.

Effects of Niche Width on the Performance and Agreement of Avian Habitat Models

Jeffrey A. Hepinstall, William B. Krohn, and Steven A. Sader

Conservation assessment and planning require knowledge about the regional occurrence of species as well as information about trends in species abundance (Morrison et al. 1992). Many methods exist to correlate species presence with their environment, to derive habitat associations, and to use these associations to build predictive models of vertebrate species occurrence. Two general groups of habitat models are statistical models (e.g., Hepinstall and Sader 1997; Tucker et al. 1997; Schulte and Niemi 1998; Dettmers et al., Chapter 54; Vernier et al., Chapter 50) and those models derived from matrices documenting species-habitat associations (Salwasser et al. 1980; Scott et al. 1993; Boone and Krohn 2000a,b).

Because habitat association models and statistical models differ in their basic premises, one would expect predictions of species presence to differ for models using the different methods. Species-habitat association matrices are derived from the literature and from expert review and, therefore, presumably represent the range of habitats used, at least in well-studied species. Species models derived from these habitat associations, therefore, predict potential rather than actual habitat for a species. Statistical models are derived from mathematical relationships between species data and field data and, therefore, represent at best the current habitat of a species. Given the differences in the input data used to build habitat association matrix

models and statistical models, statistical methods would be expected, in general, to underpredict species presence with respect to the more general habitat association models.

One factor that may affect the results of comparisons of model predictions derived from statistical models or habitat association models is breadth of niche (here used in the Grinnellian [Grinnell 1917] sense of niche as the range of environmental attributes enabling individuals to survive and reproduce) that a species can occupy. Species that are generalists (eurytopes) and use many different habitats could be predicted to occur everywhere by habitat association methods. Species with narrower niches are more likely to be accurately predicted by both methods. However, habitat specialists (stenotopes) will be modeled well by either method only if the required habitat is mapped and correctly delineated.

Testing model output is important in determining model performance and accuracy, but test approaches are limited by what they are compared against (Krohn 1992; Fielding and Bell 1997). For example, model predictions for some gap analysis projects have been tested against species lists from national parks and national wildlife refuges (Scott et al. 1993; Edwards et al. 1996; Krohn et al. 1998). Species lists may contain records of species that no longer occur at the site or may not record species that have recently moved into an area. Confusion matrices can be used to calculate a variety of measures of agreement between observed species occurrences and predicted species occurrences (Fielding and Bell 1997). Errors of omission (not predicting a species that was present) and commission (predicting a species that was not present) typically are calculated, as is correct classification rate (correctly predicted species presence and species absence). Understanding the ecological context of model errors is essential to understanding model performance (Fielding and Bell 1997). For example, when identifying areas of conservation concern for reserve creation, commission errors may be more detrimental than omission errors, because overpredicting many species may lead to incorrect assessments of which areas are of higher species richness. In contrast, reducing omission errors may be important when predicting the distribution of an endangered species.

Interpreting the results of accuracy assessments requires an understanding of the biases inherent in the modeling process. Predictions based on habitat association models use general vegetation types as surrogates for a species' habitat and thus assume that required microhabitat elements will be present at least at some locations; as a result, commission errors of species presence will be higher than omission errors (Krohn 1996). Higher commission errors have been observed in several studies (Scott et al. 1993; Edwards et al. 1996; Block et al. 1994; Krohn et al. 1998; Karl et al., Chapter 51). However, if the objective of a study is to predict potential use of areas versus currently used areas, these commission errors are not necessarily serious errors (Edwards et al. 1996; Fielding and Bell 1997).

The objectives of this study were to (1) compare spatially explicit predictions of species occurrence for land bird species in Maine derived from two modeling methods (one statistical and the other based on species-habitat association matrices), (2) test for the effect of species niche width on agreement between species predictions for each method, and (3) test for the effect of species niche width on model accuracy.

Modeling Paradigms and Species Selection

The selection of species modeled in this study was based on availability of survey data at a sufficient

grain to build statistical models (n = 60) and validate model results (n = 20). Sufficient data were available to build and test models for twenty-eight species (Table 53.1). To rate species niche width, species were ranked by the number of vegetation and land-cover types used by each species according to the habitat associations models used in the Maine Gap Analysis Project (ME-GAP; Boone and Krohn 1998a,b).

Maine Gap Analysis

ME-GAP (Krohn et al. 1998) used a thirty-seven-class vegetation and land-cover map (Hepinstall et al. 1999) along with ancillary geographic information system (GIS) data to delineate habitat for each species. Bird species range limits were modified from DeGraaf and Rudis (1986), with modifications from several published sources (Adamus 1987; Erskine 1992; Foss 1994; Gauthier and Aubry 1996), and expert review (Krohn et al. 1998). Species-habitat association matrices (heuristic measures) began with DeGraaf and Rudis (1986) and were modified with other published data and expert review (Krohn et al. 1998). The association matrices formed look-up tables to recode the vegetation and land-cover types into used and unused types (Boone and Krohn 1998b).

Normally, appropriate habitat polygons extending beyond a species' range limit are included as predicted species presence (Scott et al. 1993). Because binary species predictions at the edges of a species' range do not accurately reflect how a species occurs at range limits, a random feathering that eliminated an increasing proportion of the appropriate habitat toward the edge of a species' range (Krohn 1996) was used to approximate a theoretical range edge (Krohn et al. 1998; Boone and Krohn 1998b). Feathering was done 3 to 50 kilometers from the edge of a species range, depending on each species' mobility (Boone and Krohn 1998c). Feathering of species predictions at the edge of their ranges will increase the disagreement between species predictions from each modeling method because Bayesian predictions were made statewide. However, only four of the modeled species discussed in this chapter (*Regulus satrapa*, *Parula americana*, *Sitta canadensis*, and *Sphyrapicus varius*) had range limits in the state (all reach their southern limits in

TABLE 53.1.

Species common and scientific names, species code, and the percentage of ME-GAP vegetation and land-cover types used by each species (based on Boone and Krohn 1998b).

Common name	Scientific name	Species code	% of Types used[a]
American robin	*Turdus migratorius*	AMRO	73
American crow	*Corvus brachyrhynchos*	AMCR	70
Common yellowthroat	*Geothlypis trichas*	COYE	65
Song sparrow	*Melospiza melodia*	SOSP	65
Eastern wood-pewee	*Contopus virens*	EWPE	59
White-throated sparrow	*Zonotrichia albicollis*	WTSP	54
Nashville warbler	*Vermivora ruficapilla*	NAWA	51
Black-capped chickadee	*Poecile atricapilla*	BCCH	49
American redstart	*Setophaga ruticilla*	AMRE	49
Chestnut-sided warbler	*Dendroica pensylvanica*	CSWA	46
Rose-breasted grosbeak	*Pheucticus ludovicianus*	RBGR	46
Blue jay	*Cyanocitta cristata*	BLJA	46
Hermit thrush	*Catharus guttatus*	HETH	46
Winter wren	*Troglodytes troglodytes*	WIWR	46
Yellow-bellied sapsucker	*Sphyrapicus varius*	YBSA	46
Red-eyed vireo	*Vireo olivaceus*	REVI	43
Purple finch	*Carpodacus purpureus*	PUFI	41
Magnolia warbler	*Dendroica magnolia*	MAGW	41
Least flycatcher	*Empidonax minimus*	LEFL	38
Veery	*Catharus fuscescens*	VEER	38
Northern parula	*Parula americana*	NOPA	35
Yellow-rumped warbler	*Dendroica coronata*	YRWA	32
Black-throated green warbler	*Dendroica virens*	BTNW	32
Black-and-white warbler	*Mniotilta varia*	BAWW	30
Ovenbird	*Seiurus aurocapillus*	OVEN	30
Red-breasted nuthatch	*Sitta canadensis*	RBNU	27
Blackburnian warbler	*Dendroica fusca*	BLBW	27
Golden-crowned kinglet	*Regulus satrapa*	GCKI	16

[a]Defined as the percentage of vegetation and land-cover types (n = 37) used by a species in the habitat association models used in Maine Gap Analysis Project species models.

extreme southern Maine), making concern over this potential source of error minimal.

Bayesian Methods

Species associations with environmental variables were based on species records in 1990 Breeding Bird Survey (BBS) data (J. A. Hepinstall et al. unpublished data). In 1990, thirty-nine BBS routes were run in Maine. Data were gathered during fifty three-minute point counts at 0.8-kilometer intervals along 39.4-kilometer road routes. All bird species seen or heard within 0.4 kilometer of stop locations were recorded.

To predict species occurrences, J. A. Hepinstall et al. (unpublished data) used six explanatory data layers. Three were derived directly from unclassified 1991 Landsat Thematic Mapper[(TM)] imagery: band 4 (near-infrared), band 5 (mid-infrared), and a texture measure derived from the variance of normalized difference vegetation index (NDVI) values within a 210×210-meter window (forty-nine pixels). TM imagery and variance texture data were stored as 8-bit data (0–255). We grouped (binned) the data values by sets of five values to reduce the total number of classes in a data layer. Data values for each pixel within 400 meters of BBS stops were extracted from the binned data sets. The remaining three layers were derived from the 1993 ME-GAP vegetation and land-cover map (Hepinstall et al. 1999): classes within 400 meters of BBS stop locations; classes within 200 meters of BBS stop locations; and vegetation class richness within 400 meters of BBS stop locations. The vegetation class richness was calculated as the number of classes in a 210×210-meter window.

Bayes' Theorem provides a method for combining frequencies of association (conditional probabilities) between species presence and values in each explanatory data layer with *a priori* (subjective) probabilities of occurrence to estimate posterior probabilities of species presence. Conditional probabilities given species presence or absence were derived for each value in each data layer (J. A. Hepinstall et al. unpublished data). Data layers were recoded using mean estimates (based on one hundred bootstrapped samples) of conditional probabilities that were determined to be significantly different for species presence and absence through a contingency table analysis. Probabilities from each data-layer value were combined using Bayes' Theorem to produce a probability of species presence using Equation 53.3 from Hepinstall and Sader (1997). Model output ranged from 0 to 1.0, with values greater than 0.5 indicating predicted species presence.

There were forty-seven possible model permutations for each species (six possible data layers, but with no models combining the two data layers of the ME-GAP vegetation and land-cover map). Model performance was evaluated through a number of tests using 1990 BBS data, change from *a priori*, and difference from a random model. The best Bayesian model for each species was defined as the satisfactory model with the highest agreement with BBS verification data.

Between Model Comparisons

The best Bayesian model for each species was compared with the prediction from ME-GAP. Because both methods produce spatially explicit measures of species presence for the state of Maine, a complete cross-tabulation of each model's prediction was possible. The statewide correct classification rate (CCR; Eq. 53.1) was calculated for each species (Fielding and Bell 1997).

$$\text{Correct Classification Rate} = \frac{(a+d)}{(a+b+c+d)} \quad (53.1)$$

where a, b, c, and d are taken from the agreement matrix in Figure 53.1.

The McNemar test (Conover 1980) was used to test if ME-GAP "overpredicted" species occurrence for significantly more area than Bayesian predictions (b and c in Fig. 53.1) (Eq. 53.2).

$$R = \frac{(b-c)^2}{(b+c)} \quad (53.2)$$

where b and c are taken from the agreement matrix in Figure 53.1. The test statistic would be positive if more area was predicted as species presence in ME-GAP models and species absence in Bayesian models and would be negative if the opposite were true. A

1)   Bayesian Prediction

		Present	Absent
ME-GAP Prediction	Present	a	b
	Absent	c	d

2)   Field Data (observed)

		Present	Absent
Predicted	Present	a	b
	Absent	c	d

Figure 53.1. (1) Example agreement matrix comparing Maine Gap Analysis Project (ME-GAP) and Bayesian predictions and (2) confusion matrix comparing observed to predicted.

simplified measure, the off-diagonal ratio (Eq. 53.3), which varies from 0 to 1 and equals 0.5 when the errors are balanced between b and c, was used to evaluate the directionality of model prediction errors.

$$\text{Off - diagonal Ratio} = \frac{b}{(b+c)} \qquad (53.3)$$

where b and c are taken from the agreement matrix in Figure 53.1.

Linear regression analysis was used to test for a significant relationship between (1) measures of agreement of ME-GAP and Bayesian species predictions (CCR and off-diagonal ratio) and (2) species niche width (percentage of vegetation and land-cover types used by a species from ME-GAP).

Spatial autocorrelation, which is the tendency of nearer objects to be more similar (or more dissimilar) than expected by chance, may have affected the direct comparisons of model predictions from both methods. Measurements of agreement, such as the correct classification rate, should be larger within smaller distances from random points if positive spatial autocorrelation exists. Although we did not measure spatial autocorrelation directly, the comparison of CCR and the off-diagonal ratio at various distances from random points can be used to index the spatial juxtaposition of model predictions. To determine if there were biases associated with comparing the species predictions for each method on a statewide, pixel-by-pixel basis, we calculated CCR and off-diagonal ratio for five hundred points randomly generated throughout the state. We buffered these five hundred points at distances of 50, 200, and 400 meters. We calculated the model agreement between Bayesian predictions and ME-GAP predictions for the buffer strips of 0–50 meters, 51–200 meters, and 201–400 meters. If the measured variables reached an asymptote at 51–200 meters, it would potentially indicate a spatial limit to the autocorrelation of agreement measures.

Model Verification: Agreement with BBS Data

We calculated the agreement between model predictions for each method and BBS 1990 stop data. Because the BBS data were used to build the Bayesian

models, measures comparing Bayesian predictions with species observed in the BBS record are only verification of model formulation (see Conroy and Moore, Chapter 16) and not of model validation (accuracy of model at predicting species presence or absence at new sites).

We assessed agreement between predicted and observed species presence using two methods. For the first test, we scored a species as predicted to be present at a BBS stop location if any model output value within 50 meters of a BBS stop was greater than 0.5. Our rationale for querying only pixels within 50 meters of a survey point location was to limit random agreement between the model prediction and the BBS record. However, this measure would be positively biased toward ubiquitous species. We calculated the number of stops where a species was observed in the 1990 BBS data and was predicted to occur by each method. Because BBS survey data do not explicitly measure species absence on a site, measurements of agreement that incorporate b or d (Fig. 53.1) were inappropriate (see discussion of commission error rates in Schaefer and Krohn, Chapter 36). Instead, we calculated the positive agreement of model predictions with BBS data according to Equation 53.4.

$$\text{Positive Agreement} = \frac{a}{(a+c)} \qquad (53.4)$$

where a, b, and c are taken from the confusion matrix in Figure 53.1. Positive agreement is defined as the conditional probability that a site correctly classified species occurrence given observed species presence (Fielding and Bell's [1997] "sensitivity"). We also calculated the CCR for each species using the data within 50 meters of BBS stops where the species was observed (Eq. 53.1). Because field data do not actually record all species present on a site (Boone and Krohn 1999), positive agreement is a less-biased measure than CCR, which is likely to be negatively biased.

As a comparison of the above measures of positive agreement and CCR, we tested for a significant difference in the mean proportion of predicted species occurrence for the area within 200 meters of point locations where a species was observed and where a species was not observed. We used a one-tailed t-test with unequal variances (Welch's approximation; Zar 1996:129) to test for differences. This test was less biased toward ubiquitous species than the measures of positive agreement and CCR within 50 meters were.

Model Validation

We also tested our model predictions against field data from Manomet Center for Conservation Sciences (J. Hagan personal communication) to validate model predictions (Conroy and Moore, Chapter 16). Ten-minute point counts (n = 387) were run twice during the breeding season (1992 and 1993) in west-central Maine (Hagan et al. 1997). We calculated the same measures of agreement between model output and the Manomet field data as we calculated for the BBS data (i.e., positive agreement and CCR within 50 meters and significant difference of means within 200 meters of survey point locations).

Linear regression analysis was used to test for a significant relationship between (1) measures of agreement between model predictions and field data (positive agreement and CCR) and (2) species niche width (percentage of vegetation and land-cover types used by a species from ME-GAP).

Results

Agreement between Methods

Three of the top four eurytopic species (as ranked by their use of vegetation types in the ME-GAP species-habitat association matrices: *Turdus migratorius*, *Geothlypis trichas*, and *Melospiza melodia*) had low (less than 30 percent) CCR between the two prediction methods (Table 53.2). Only one other species, *Regulus satrapa*, had a CCR value of less than 50 percent. Only three species (*Troglodytes troglodytes*, *Sphyrapicus varius*, and *Dendroica magnolia*) had greater than 70 percent CCR. These three species were modeled by ME-GAP to use 41–46 percent (ranked fourteenth, fifteenth, and eighteenth of the twenty-eight species) of the available vegetation and land-cover types (Table 53.1). Maps of the Bayesian and ME-GAP predictions for thee species (*Corvus brachyrhynchos*, *Troglodytes troglodytes*, and *Dendroica fusca*) clearly show the

TABLE 53.2.

Correct classification rate (CCR), off-diagonal ratio, and McNemar's test statistic calculated for statewide Bayesian predictions and Maine Gap Analysis predictions (Krohn et al. 1998) for each of the twenty-eight species modeled.

Species name	CCR[a]	Off-diagonal ratio[b]	McNemar's test statistic[c]
Turdus migratorius	27.8	0.935	2440
Corvus brachyrhynchos	51.2	0.946	2024
Geothlypis trichas	28.4	0.955	2495
Melospiza melodia	32.2	0.957	2442
Contopus virens	52.3	0.862	1625
Zonotrichia albicollis	61.6	0.895	1580
Vermivora ruficapilla	54.8	0.863	1578
Poecile atricapilla	58.7	0.838	1401
Setophaga ruticilla	52.4	0.766	1184
Dendroica pensylvanica	52.4	0.337	-728
Pheucticus ludovicianus	55.4	0.882	1824
Cyanocitta cristata	55.3	0.811	1342
Catharus guttatus	61.7	0.538	153
Troglodytes troglodytes	72.0	0.710	719
Sphyrapicus varius	71.6	0.810	1068
Vireo olivaceus	63.4	0.697	775
Carpodacus purpureus	53.6	0.727	997
Dendroica magnolia	72.2	0.647	499
Empidonax minimus	70.8	0.557	200
Catharus fuscescens	65.4	0.643	543
Parula americana	57.5	0.649	628
Dendroica coronata	59.1	0.401	-409
Dendroica virens	64.8	0.706	794
Mniotilta varia	56.7	0.431	-293
Seiurus aurocapillus	64.3	0.610	428
Sitta canadensis	59.7	0.341	-651
Dendroica fusca	58.9	0.653	632
Regulus satrapa	48.2	0.864	1691

[a]Correct classification rate is the percentage of overall agreement between species predictions from Bayesian and Maine Gap Analysis models (joint predicted species presence and joint predicted absence); Equation 53.1.

[b]Off-diagonal ratio is a measure of the directionality of disagreement between predictions from Bayesian and Maine Gap Analysis methods (Equation 53.3). Values above 0.5 indicate ME-GAP models overpredict species presence with respect to Bayesian models.

[c]Conover (1980); Equation 53.2, positive measures indicate Maine Gap Analysis models overpredict species presence with respect to Bayesian models.

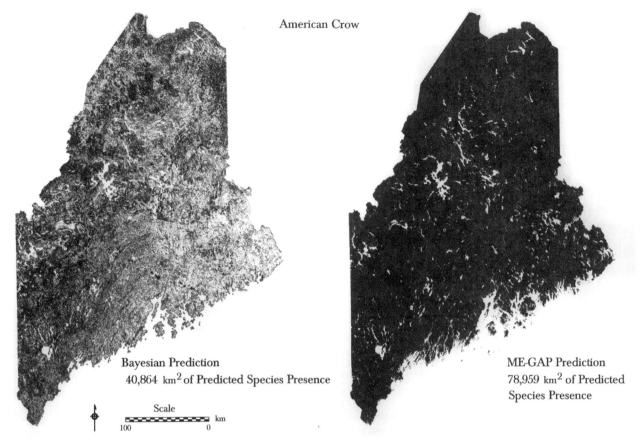

Figure 53.2. Predicted occurrences (black) of the American crow (*Corvus brachyrhynchos*) in Maine from the Bayesian method and the Maine Gap Analysis Project (ME-GAP) method. This species uses 70 percent (second-widest niche of the twenty-eight species modeled) of the vegetation and land-cover types present in the ME-GAP map (Boone and Krohn 1998b).

trend between model agreement and species niche width (Figs. 53.2, 53.3, and 53.4).

Species niche width did have an effect on the agreement between model predictions from the two methods. The values for CCR, off-diagonal ratio, and Mc-Nemar's test statistic were generally poorer for the more-generalist species than for the more-specialist species (Table 53.2). Correct classification rate decreased significantly (P = 0.0139, r^2 = 0.21, slope = –0.416) as the percentage of vegetation and land-cover types used increased. The off-diagonal ratio increased significantly (P = 0.001, r^2 = 0.34, slope = 0.0021) with increased use of vegetation and land-cover types.

The potential effects of spatial autocorrelation, as indexed by the CCR and off-diagonal ratio measured for five hundred random points at three distances (50, 200, and 400 meters) may be limited to less than 400 meters (Fig. 53.5). For fifteen of the twenty-eight

species modeled, CCR rates were maximized at 200 meters, potentially indicating a limit to the direct spatial coincidence of predictions for the two methods. Such patterns were not as clear with the measures of off-diagonal ratio for each buffer-strip distance. However, the off-diagonal values tended to decrease for the more-specialized species (right side of graphs in Fig. 53.5). Two species, *Empidonax minimus* and *Regulus satrapa*, had off-diagonal ratios below 0.5 for all three distances, but off-diagonal values above 0.5 when species predictions were compared statewide (Table 53.2).

Model Verification and Validation

Species predictions from the Bayesian models, by definition of a satisfactory model, were all equal to or greater than 70 percent (Table 53.3). ME-GAP predictions for two species (*Regulus satrapa* and *Dendroica pensylvanica*) were the only species with positive

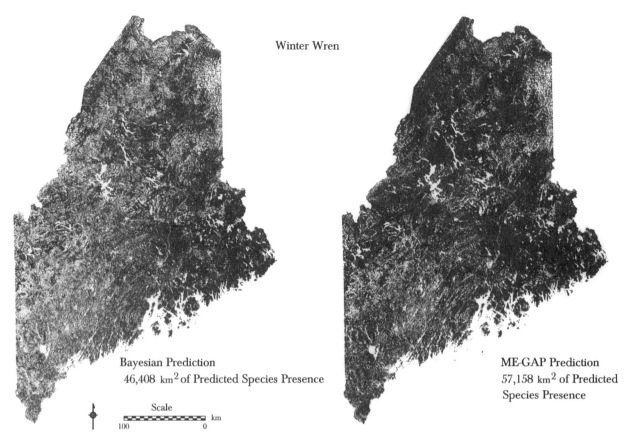

Figure 53.3. Predicted occurrences (black) of the winter wren (*Troglodytes troglodytes*) in Maine from the Bayesian method and the Maine Gap Analysis Project (ME-GAP) method. This species uses 46 percent (fourteenth-widest niche of the twenty-eight species modeled) of the vegetation and land-cover types present in the ME-GAP map (Boone and Krohn 1998b).

agreement less than 70 percent with the BBS data. Positive agreement of ME-GAP predictions with Manomet data was less than 70 percent for two species (*Mniotilta varia* and *Regulus satrapa*). For seven species, generally the more stenotopic species of the species modeled, Bayesian model predictions had higher positive agreement with Manomet data than ME-GAP model predictions did (Table 53.3).

The CCR was lower than the positive agreement for both modeling methods for all species except for ME-GAP predictions for *Dendroica pensylvanica* (BBS) and *Regulus satrapa* (BBS and Manomet) and Bayesian predictions for *Turdus migratorius* and *Geothlypis trichas* (Manomet) (Table 53.3). CCR was much lower than positive agreement for ME-GAP predictions for six species (Table 53.3), dropping from more than 90 percent positive agreement with BBS to less than 35 percent CCR. The same trend was seen for seven species when compared with Manomet data.

All of these species were the more eurytopic species modeled. The differences between positive agreement and CCR were generally much less for Bayesian predictions.

Agreement between predicted and observed species presence within 200 meters of survey locations differed from agreement within 50 meters of survey locations (Table 53.3). No relationship was observed between the measure of significant agreement at 200 meters and species niche width for either Bayesian or ME-GAP predictions, although species with significant ME-GAP agreement with BBS data were skewed toward the more specialist species.

Significant trends existed between species niche width (percentage of vegetation and land-cover types used) and measures of positive agreement and CCR for BBS data and Manomet data for ME-GAP predictions (Table 53.4). Positive agreement with field data increased with increased number of vegetation types

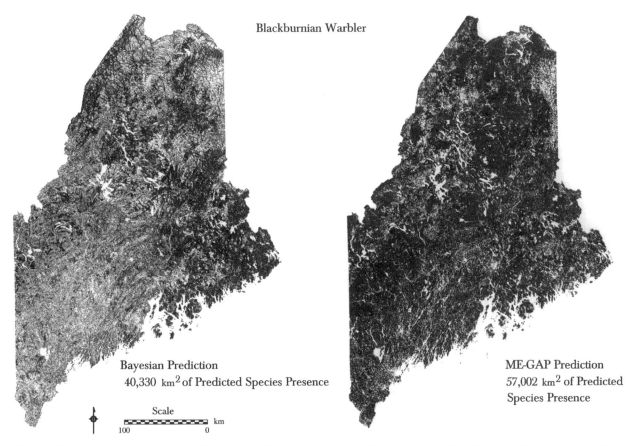

Blackburnian Warbler

Bayesian Prediction
40,330 km² of Predicted Species Presence

Scale

100   0 km

ME-GAP Prediction
57,002 km² of Predicted
Species Presence

Figure 53.4. Predicted occurrences (black) of the Blackburnian warbler (*Dendroica fusca*) in Maine from the Bayesian method and the Maine Gap Analysis Project (ME-GAP) method. This species uses 27 percent (twenty-seventh widest niche of the twenty-eight species modeled) of the vegetation and land cover types present in the ME-GAP map (Boone and Krohn 1998b).

used by a species; the opposite trend was observed with measures of CCR, indicating the increase in positive agreement is likely an artifact of increased area of predicted species presence rather than an increase in the accuracy of model predictions. No significant trends were observed with measures of agreement (CCR and positive agreement) between Bayesian predictions and field data.

Discussion

Habitat association models and statistical models differ in their basic premises. Habitat association methods, being based on literature review and expert knowledge, will more likely model potential rather than actual habitat for a species. Statistical models, being derived from field data, at best will model the current (at the time the field data were gathered) distribution of a species.

Habitat association methods based on associations observed over larger temporal and spatial ranges, yield binary responses predicting the areas that could possibly be suitable for a species based on the presence or absence of habitat elements. Rule-based models derived from simple heuristics, such as species-habitat association matrix models, will tend to integrate over the variability inherent in statistical models. Therefore, rule-based models are expected to be more general in their predictions of species occurrence and will be less influenced by yearly variation in habitat use patterns.

The advantages of using species-habitat associations to predict species presence are that the predictions can be made for large spatial extents (but see Austin 1999b for an example of statistical models for large spatial extents) and for as many species as there exists adequate knowledge for general rules of association. The disadvantages are that (1) species

A) Correct Classification Rate (CCR)

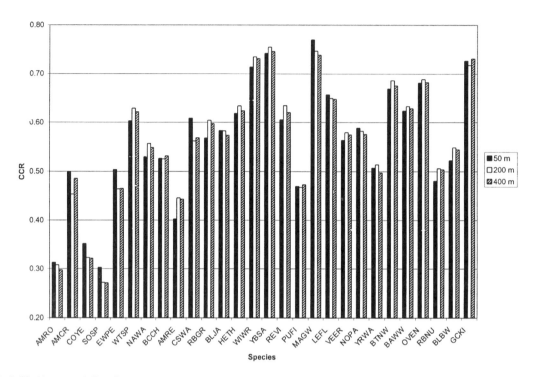

B) Off-diagonal Ratio

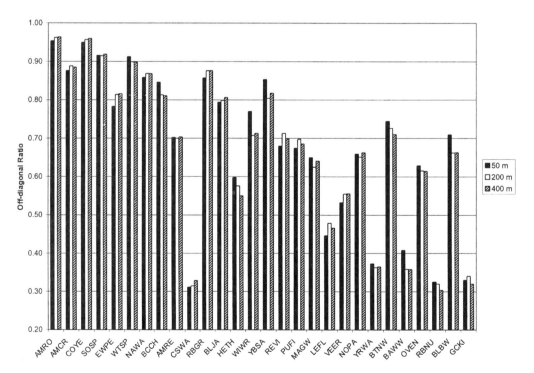

Figure 53.5. (A) Correct classification rate (CCR) and (B) off-diagonal ratio values for agreement between Bayesian and Maine Gap Analysis Project (ME-GAP) species predictions calculated for five hundred random points located throughout Maine for three distances from point locations (0–50 meters, 51–200 meters, and 201–400 meters). Species are ordered by decreasing number of vegetation and land-cover types used from left to right. Species AOU codes given in Table 53.1.

TABLE 53.3.

Percent agreement (positive agreement and correct classification rate [CCR]) and significant agreement of Maine Gap Analysis Project (ME-GAP) predictions and Bayesian predictions with verification data (Breeding Bird Survey [BBS]) and validation data (Manomet Center for Conservation Sciences field data [Manomet]).

| | BBS | | | | | | Manomet | | | | | |
| | Positive agreement[a] | | CCR[b] | | Difference: t-test value[c] | | Positive agreement[a] | | CCR[b] | | Difference: t-test value[c] | |
	ME-GAP	BAYESIAN	ME-GAP	BAYESIAN	ME-GAP	BAYESIAN	ME-GAP	BAYESIAN	ME-GAP	BAYESIAN	ME-GAP	BAYESIAN
Turdus migratorius	0.98	0.93	0.27	0.41	−2.6	**4.0[d]**	1.00	0.43	0.17	0.67	−0.27	0.10
Corvus brachyrhynchos	0.99	0.92	0.17	0.50	−3.5	**9.4**	1.00	0.83	0.07	0.41	0.46	0.15
Geothlypis trichas	0.97	0.91	0.23	0.40	0.3	**6.6**	1.00	0.63	0.31	0.66	**1.90**	**4.73**
Melospiza melodia	0.97	0.94	0.17	0.61	−3.9	**16.6**	1.00	0.74	0.11	0.69	0.96	**4.46**
Contopus virens	0.78	0.92	0.39	0.55	−0.1	**3.8**	1.00	0.70	0.23	0.60	1.12	-0.97
Zonotrichia albicollis	0.91	0.88	0.44	0.60	**8.9**	**9.3**	1.00	0.87	0.44	0.62	0.81	**3.48**
Vermivora ruficapilla	0.95	0.92	0.13	0.61	−1.5	**5.4**	1.00	0.86	0.34	0.58	0.67	**5.37**
Poecile atricapilla	0.80	0.90	0.39	0.51	−1.5	0.2	0.97	0.73	0.28	0.51	0.82	−0.60
Setophaga ruticilla	0.80	0.80	0.41	0.57	1.5	**3.3**	0.95	0.58	0.31	0.48	1.00	−1.77
Dendroica pensylvanica	0.53	0.96	0.70	0.32	0.1	**4.6**	0.76	0.62	0.61	0.62	**4.00**	1.57
Pheucticus ludovicianus	0.77	0.81	0.59	0.59	1.3	**2.2**	0.67	0.59	0.61	0.39	−0.32	-3.87
Cyanocitta cristata	0.82	0.95	0.39	0.37	−0.8	**4.3**	0.97	0.69	0.27	0.62	0.45	−0.81
Catharus guttatus	0.94	0.87	0.25	0.58	0.0	**7.6**	0.89	0.95	0.51	0.50	−0.60	−1.75
Troglodytes troglodytes	0.87	0.89	0.52	0.64	**9.2**	**8.2**	0.94	0.74	0.65	0.65	**2.00**	0.79
Sphyrapicus varius	0.96	0.95	0.38	0.51	**6.0**	**7.9**	1.00	0.86	0.27	0.46	0.28	−1.12
Vireo olivaceus	0.82	0.89	0.33	0.49	**5.6**	1.8	1.00	0.75	0.11	0.73	**3.00**	**3.29**
Carpodacus purpureus	0.75	0.85	0.44	0.63	−0.6	**4.2**	0.86	0.75	0.28	0.53	−1.21	0.09
Dendroica magnolia	0.89	0.95	0.47	0.62	**7.8**	**11.1**	0.89	0.83	0.62	0.67	**2.10**	**2.36**
Empidonax minimus	0.72	0.85	0.48	0.65	**1.8**	**5.3**	0.72	0.78	0.62	0.53	**2.65**	0.89
Catharus fuscescens	0.75	0.80	0.54	0.68	0.3	**6.4**	0.86	0.89	0.52	0.39	**1.80**	0.35
Parula americana	0.82	0.92	0.55	0.59	**5.7**	**8.4**	0.78	0.75	0.44	0.52	0.90	0.87
Dendroica coronata	0.70	0.87	0.55	0.61	−1.4	**5.8**	0.70	0.93	0.62	0.59	−0.18	−2.24
Dendroica virens	0.81	0.92	0.52	0.58	**5.4**	**7.6**	0.88	0.86	0.63	0.64	**4.50**	−0.14
Mniotilta varia	0.71	0.91	0.57	0.49	1.1	**4.1**	0.69	0.86	0.54	0.40	−0.08	−0.12
Seiurus aurocapillus	0.75	0.87	0.55	0.48	**4.5**	**5.1**	0.90	0.82	0.67	0.73	**5.90**	**7.06**
Sitta canadensis	0.84	0.94	0.51	0.57	**1.8**	**7.5**	0.90	0.94	0.62	0.56	**2.00**	−2.69
Dendroica fusca	0.77	0.95	0.59	0.54	**4.4**	**6.2**	0.84	0.71	0.58	0.53	0.90	−1.37
Regulus satrapa	0.49	0.93	0.78	0.60	**1.6**	**5.7**	0.47	0.72	0.63	0.64	0.90	**2.16**

[a]Positive agreement is the percentage the joint predicted and observed is of the total observed within 50 meters of field survey points (Equation 53.4).

[b]Correct classification rate is the percentage of overall agreement between species predictions and field observations (joint predicted presence and joint predicted absence); Equation 53.1.

[c]The t-test value for difference in the proportion of the area with predicted species presence within 200 meters of survey points where the species was observed where the species was not observed.

[d]Boldface type indicates one-tailed (positive) significant (P < 0.05) difference.

TABLE 53.4.

Results for agreement measures between model predictions and verification data (Breeding Bird Survey [BBS]) and validation data (Manomet Center for Conservation Sciences field data) regressed against the percentage of vegetation and land-cover types used by each of the twenty-eight species modeled.

	BBS		Manomet	
	Positive agreement	CCR	Positive agreement	CCR
Slope	0.0062	–0.009	0.0068	–0.011
P-value	0.0001	< 0.0001	< 0.0001	< 0.0001
r²	0.461	0.601	0.480	0.542

habitat relationships are based on descriptive rather than statistical relationships, (2) predictions are binary (presence or absence) or categorical (e.g., rarely used, sometimes used, often used) with no associated probability of finding a species, and (3) predictions are of large spatial resolution (grain), usually greater than 90 meters and often as large as 1 kilometers. Rule-based approaches also may be more limited than statistical methods in their inclusion of data layers other than vegetation and land-cover types because of the lack of known relationships between species presence and derived environmental data layers.

Limitations of the statistical method described in this chapter include (1) incomplete associations between explanatory variables and species presence due to incomplete species counts (Boone and Krohn 1999), (2) the ability to model only species with sufficient field data to build valid statistical relationships, and (3) yearly variation in species distribution affected by the particular spatiotemporal dynamics functioning at both local and regional scales (Wiens 1981b, 1989c; Maurer, Chapter 9). The number of species successfully modeled using a statistical approach potentially could be increased by pooling multiple years of BBS data for species that are less common, but issues of pseudo-replication would have to be addressed.

The ability to make precise predictions about what is appropriate habitat for a species will be directly related to the breadth of habitats used by that species

(Csuti 1996). It is unlikely that species with narrow habitat requirements would be clearly correlated with unclassified satellite imagery or vegetation cover types derived from satellite imagery. The probability of a landscape containing microhabitat resources required by a species increases as spatial extent increases (Csuti 1996). However, if relationships between species presence and vegetation types are too general, associations will be so broad as to yield predictions of species presence everywhere, which may not accurately depict even species potential distribution (see also Karl et al., Chapter 51 and Dettmers et al., Chapter 54). Johnson and Krohn (Chapter 13) also found differences in the error rates of predictions for three species of seabirds and related this variability to differences in species niche width.

Agreement between the predictions for the habitat association method and statistical method tested in this study was variable, but, as expected, the statistical models tended to underpredict species presence relative to habitat association model output. Gap predictions are known to overestimate (commission errors) species presence (Scott et al. 1993; Edwards et al. 1996; Krohn et al. 1998; Garrison and Lupo, Chapter 30) more often than underestimate presence (omission errors), although high commission error can result from incomplete field surveys (Boone and Krohn 1999). As predicted, ME-GAP methods predicted more occurrences for the more eurytopic species than were predicted by the Bayesian methods. For four species (*Dendroica pensylvanica*, *Dendroica coronata*, *Mniotilta varia*, and *Sitta canadensis*), Bayesian models tended to overpredict species occurrence with respect to ME-GAP predictions. These species, with the exception of the *Dendroica pensylvanica*, tended to be the more stenotopic species.

ME-GAP methods clearly overestimated species occurrence when compared against field data. The positive agreement of ME-GAP predictions with BBS and Manomet field data generally increased from species habitat specialists to habitat generalists. The trend in correct classification rate was exactly the opposite: ME-GAP predictions for the generalist species potentially greatly overpredict the current distribution of generalist species. However, the CCR metric is dependent on the assumption that field data is complete (i.e., few to no nondetected birds).

The correct classification rate for ME-GAP models between observed and predicted species occurrences was much lower for most species than the positive agreement (correctly predicted presence) was. However, because field test data were survey data, species absence was not explicitly recorded. Therefore, comparing the positive agreement of model predictions with field data is more appropriate than testing the correct classification rate, which will tend to inflate errors. Also, because the purpose of GAP prediction is to attempt to predict potential habitat versus actual habitat and to aid in conservation planning, Edwards et al. (1996) argues that errors of commission are preferable to errors of omission. However, because species predictions from GAP models are ultimately pooled to create maps of species richness, commission errors may mask true areas of species richness if the patterns of commission errors exhibit positive spatial autocorrelation. Testing the spatial pattern of commission errors through the inclusion of neighborhood values (e.g., Augustin et al. 1996; Fielding and Bell 1997; Klute et al., Chapter 27) potentially could be used to correct for spatial patterns of errors. Ultimately, field verification will be needed to test any model prediction, whether attempting to protect species richness or a single endangered species.

Because birds more often respond to structural elements than to the floristic composition of their environment (Cody 1985b), the vegetation and land-cover map created for ME-GAP concentrated on delineating general vegetation types. The general vegetation and land-cover types, while sufficient for ME-GAP predictions, may have been too general in classification, yielding nonsignificant relationships in the Bayesian model parameterization. Habitat generalists (e.g., *Geothlypis trichas*), which will more likely be overpredicted than underpredicted by habitat association methods, would likely not be modeled well by Bayesian methods unless the species was abundant and clear differences in presence/absence existed between vegetation and land-cover types.

ME-GAP predictions for one species (*Regulus*

satrapa) had poor agreement with BBS and Manomet data. The ME-GAP model predicted much less area of habitat for *Regulus satrapa*s than Bayesian models predicted (28,355 versus 36,098 square kilometers, respectively). *Regulus satrapa*s used the fewest vegetation and land-cover classes of the species discussed in this study. It is likely that the mapped classes did not correspond well to the habitat used by *Regulus satrapa*, and the decision to label each vegetation and land-cover class as used or unused—an example of misclassified cases (see Fielding, Chapter 21)—underestimated the available habitat for *Regulus satrapa*.

Changes in vegetation due to conversion or forest harvest operations will change available habitat for some species. Species increasing in range or decreasing in number should have more potential habitat versus occupied habitat. Species with stable populations that have occupied all potential habitat should have similar values for potential and occupied habitat. Krohn (1996) postulated that species-habitat relationships measured at time t_1 and predicted for time t_2 will lead to models with poor fit or erroneous predictions unless the populations under consideration are at or near their carrying capacity. Low populations will lead to underestimating the full range of habitats the species will potentially use, whereas high population levels will have species present in sub-optimal habitat (Brown 1969a; Fretwell and Lucas 1969; O'Connor 1986). It is likely that population status influenced both the comparisons of model predictions for each method and agreement of model predictions with field data.

Problems can arise in interpreting the results of accuracy assessments for at least five reasons: (1) incomplete or unrepresentative reference data sets (e.g., Edwards et al. 1996; Boone and Krohn 1999; Fielding, Chapter 21); (2) unsaturated populations; (3) site fidelity; (4) temporal and spatial changes; and (5) incorrect accuracy measurements (see Fielding, Chapter 21). Incomplete reference data sets will inflate measures of commission errors (e.g., Edwards et al. 1996) and may deflate measures of overall prediction accuracy. If the local population has not saturated the available optimum habitat, individuals may not be observed in all available areas. In this scenario, models may either fail to predict suitable habitat or the commission error rate may be increased above the "true" rate (Fielding and Bell 1997). Species that are philopatric will remain in suboptimal habitat even when optimal habitat becomes available. If a species' local population is declining from a once-high level where many sub-optimal areas were occupied, models predicting species occurrence may actually measure the "wrong" (i.e., sub-optimal) associations between a species and its environment. Temporal and spatial changes in species distribution will most likely follow changes on the landscape, although studies have observed lags in population changes following environmental changes (e.g., Wiens 1989c). Krohn (1996) showed that while the number of forest bird and small mammal species did not change with time at one site in Maine, the cumulative number of species increased over the same time period. Because data sets for testing species model predictions are rarely gathered for the same time period for which predictions are based, temporal changes in species composition will lead to calculated errors in model accuracy that may not be on-the-ground errors.

If populations were fluctuating during the time period being modeled, agreement between the two modeling approaches may reflect this variation as discussed above. If, for example, a species was below carrying capacity, then habitat association methods, when based on robust literature covering many environmental conditions, would presumably predict a wider range of potentially used habitats than the statistical models would predict. In this instance, statistical models would likely predict currently used habitats and would underestimate potential species distributions. When populations are close to K, model predictions may be similar because the patterns of recorded and observed use of vegetation and land-cover type would be similar (Krohn 1996). This agreement will hold only if populations remain relatively stable.

Conclusions and Management Implications

The results of the study presented in this chapter suggest the importance in comparing the results from different methods of predicting species occurrences. Specifically, statistical methods generally will be more conservative than habitat association models in the area of predicted habitat for a species. Additionally, species niche width will affect the agreement between predicted species occurrence and measured presence; generalists will generally be more overpredicted by habitat association methods with respect to statistical methods than will specialists. Finally, a single view of a species' distribution will always be constrained by the data used to build the model and the assumptions inherent in each attempted modeling approach. Comparison of the predictions from highly different methodologies can be used to better understand natural processes and to aid managers in decision making. Ultimately current field data will be required to accurately gauge predictions from any type of model.

Acknowledgments

We thank J. M. Hagan and S. L. Grove, Manomet Center for Conservation Sciences, who provided field observations and survey locations from west central Maine used in testing model output in this study. We thank R. J. O'Connor for comments on earlier drafts and for the use of digital Breeding Bird Survey 1990 stop data. This manuscript benefited from critical comments by B. Compton, A. Fuller, and A. Guerry. We thank R. B. Boone and the other participants in the Maine Gap Analysis Project. This is scientific contribution number 2413 of the Maine Agriculture and Forest Experiment Station.

A Test and Comparison of Wildlife-habitat Modeling Techniques for Predicting Bird Occurrence at a Regional Scale

Randy Dettmers, David A. Buehler, and John B. Bartlett

Empirically based wildlife-habitat models that can predict species occurrence over large spatial extents (e.g., regional areas of millions of hectares) can be very useful in developing regionally based or ecosystem-level management plans for wildlife resources. Numerous statistical modeling techniques exist for developing habitat-based models for predicting wildlife species occurrence. These methods include logistic regression (Straw et al. 1986; Nadeau et al. 1995), discriminant function analysis (Livingston et al. 1990; Fielding and Haworth 1995), Mahalanobis distance statistic (Clark et al. 1993b; Knick and Dyer 1997), classification and regression tree analysis (O'Connor et al. 1996), cumulative distribution analysis (Dettmers and Bart 1999), and principal component analysis (Debinski and Brussard 1994). However, models produced with these techniques often are not validated over large extents or they perform poorly when tested (Dedon et al. 1986; Raphael and Marcot 1986; Johnson et al. 1989). Furthermore, these techniques are rarely compared with one another to provide information on their relative ability to produce accurate predictive models. Little information exists regarding whether some of these techniques are more appropriate for certain types of modeling applications beyond what the statistical constraints of the data might be.

In this study, we developed habitat models for predicting the occurrence of breeding birds in the southern Appalachian Mountains region using several different modeling techniques. We present validation results on the ability of the different models to accurately predict bird occurrence at two different locations within the region. We used logistic regression, Mahalanobis distance method, classification and regression tree analysis (Breiman et al. 1984; Clark and Pregibon 1992), and discriminant function analysis to develop predictive models based on bird survey data in Tennessee, and we tested the models on similar survey data from Georgia and Virginia.

Study Area

Our study area was the same as the area chosen for the Southern Appalachian Assessment (SAMAB 1996), which extends from western Virginia south to northern Georgia (Fig. 54.1) and encompasses over 22 million hectares. An extensive spatial database, including land-cover types and landscape metrics, was assembled for the Southern Appalachian Assessment (SAA). We used this collection of spatial data as the source of predictor variables for our habitat models.

Bird Data

We used data collected from three point-count survey projects on three national forests within our study

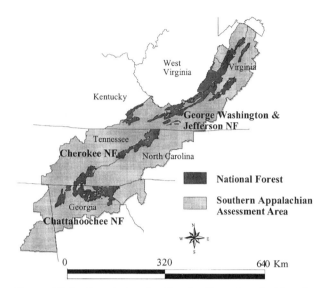

Figure 54.1. Boundaries for the Southern Appalachian Assessment and the three national forests from which bird survey data were collected.

area: the George Washington and Jefferson National Forests in Virginia, Cherokee National Forest in Tennessee, and Chattahootchee National Forest in Georgia. All point counts were conducted following the same protocol, which conforms to the point-count standards recommended by Ralph et al. (1995a). The surveys were ten-minute, unlimited-radius point counts, with detections noted as either 50 meters or less, or more than 50 meters from the point. We used only the 50-meters-or-less data for the development and testing of models reported in this chapter. All counts were conducted between the hours of 0600 and 1000. Within each national forest, the point count locations were selected in a stratified random manner such that approximately equal numbers of points were placed in stands representing combinations of six major forest types (yellow pine, mixed hardwood–yellow pine, oak–hickory, eastern hemlock–white pine, cove hardwood, and northern hardwood) and three size-class categories (seedling–sapling, pole timber, saw timber). Points were surveyed once annually between mid-May and mid-June. We used data collected from 1992 to 1996 in Georgia, where a total of 650 points were surveyed, from 1992 to 1996 in Tennessee, where a total of 215 points were surveyed, and from 1993 to 1997 in Virginia, where a total of 764 points were surveyed.

Model Development and Testing

We developed models for twenty-five common bird species, but we will present results from six representative species (black-throated blue warbler [*Dendroica caerulescens*], black-throated green warbler [*Dendroica virens*], Carolina chickadee [*Poecile carolinensis*], ovenbird [*Seiurus aurocapillus*], veery [*Catharus fuscescens*], yellow-billed cuckoo [*Coccyzus americanus*]), which we will use to generalize the results in a more manageable form. These six species were chosen as a representative sample of the entire twenty-five species set. They included species that nest on the ground (ovenbird), at the shrub and mid-story levels (black-throated blue warbler, veery, yellow-billed cuckoo), in the canopy (black-throated green warbler), and in cavities (Carolina chickadee). This group of species also had several habitat "generalists" (i.e., used numerous forest types: Carolina chickadee, ovenbird, yellow-billed cuckoo) and several habitat "specialists" (i.e., restricted to a small number of forest types: black-throated blue warbler, black-throated green warbler, veery).

For each species, we used the Tennessee data to develop models for predicting species occurrence based on a suite of twenty-six habitat-related variables (Table 54.1), all of which were derived from remotely sensed data and taken from the spatial databases of the SAA (SAMAB 1996). We selected these twenty-six variables based partly on availability in SAA database and partly on past modeling experience, which suggested these variables can be useful in predicting bird occurrence. Using this suite of explanatory variables, we developed four models for each species—one from each of the modeling techniques we were interested in comparing: logistic regression, Mahalanobis distance method, classification and regression tree (CART) analysis, and discriminant function. We then used the Georgia and Virginia data as separate, independent tests of the models' ability to accurately predict the occurrence of the six bird species across the Southern Blue Ridge physiographic province.

For logistic regression, we used Proc Logistic (SAS Institute 1990a) to complete a step-wise variable selection followed by a best-subsets procedure. This procedure found the model that resulted in the lowest Akaike Information Criterion (AIC) value, had a

TABLE 54.1.

Predictor variables used to develop the bird-habitat models. All variables were derived from remotely sensed data collected for the Southern Appalachian Assessment.

Elevation	Water flow accumulation downhill	Forest type
Slope	Distance from nearest stream	Dominant forest type—1 km
Aspect	Relative slope position	Dominant forest type—5 km
Planiform curvature	Shannon-Weaver topographic index	Dominant forest type—10 km
Profile curvature	Topographic similarity index	Land-cover diversity—1 km
Solar exposure	Topographic relative moisture index	Land-cover diversity—5 km
	Topographic convergence index	Land-cover diversity—10 km
	% Forest cover—1 km	Forest diversity—1 km
	% Forest cover—5 km	Forest diversity—5 km
	% Forest cover—10 km	Forest diversity—10 km

significant fit to the data according the Hosmer and Lemeshow (1989) goodness-of-fit test, and provided the highest-overall correct classification on jackknife tests when sensitivity (percentage of occurrences correctly classified) and specificity (percentage of absences correctly classified) were equally balanced. For the Mahalanobis distance method, we used the same procedure described in Clark et al. (1993b) and then performed a best subsets procedure to select the model that resulted in the largest correct classification rate from jackknife tests when sensitivity and specificity were balanced equally. For the CART models, we followed the same procedure described in O'Connor and Jones (1996), which was to overfit an initial tree to the data and then prune the tree to an optimum fit based on cross-validation analysis, which is a form of jackknife test. We used Proc Discrim (SAS Institute 1990a) to develop discriminant functions for the presence/absence of each species, and we used the cross-validation option to perform a jackknife test. We used a best subsets procedure, and the final discriminant function model was chosen as the one for which all variables remaining in the model were significant and that resulted in the greatest correct classification rate from the jackknife test.

All of the forms of jackknife tests used in these different methods are calculated in similar fashion by recursively removing one observation (i.e., results from one point count station) at a time from the total data set, recalculating the model parameters, testing whether the model correctly predicts the removed observation, replacing that observation, removing the

next, and so on through the entire data set. The cumulative percent correct classification over the entire data set provides an indication of a model's predictive ability for a new data set.

We tested the best models from the procedures just described by using the Georgia and Virginia data sets as independent tests. For each of these data sets, we had global positioning system (GPS) locations for the point count stations. We used these locations to sample the SAA spatial data sets containing the explanatory habitat variables and to obtain the habitat values associated with each point count location. We then ran the various models on these habitat data sets to get predictions of presence/absence (CART and discriminant function) or probability of occurrence (logistic regression and Mahalanobis distance). For all four of the modeling methods, we calculated the percent correct classification of the observations in each of the new data sets (Georgia and Virginia). We termed this procedure the classification test. For logistic regression and the Mahalanobis distance method, we had to select a cut-off probability value in order to convert the probabilities of occurrence predicted from these methods into predictions of presence/absence. We used the same cut-off values that resulted in the best models from the model development phase using the Tennessee data set. These cut-off values were selected to achieve the best overall correct classification rate for the jackknife tests on the Tennessee data while maintaining a balance between sensitivity and specificity.

In addition to calculating the percent correct

TABLE 54.2.

Percentage of correct classifications for jackknife tests on the Tennessee avian point count data for species-specific models developed using four different modeling methods.

Species	% points occupied[a]	Logistic regression	Mahalanobis distance	CART[b]	Discriminant function
Black-throated blue warbler					
Dendroica caerulescens	26	85.7	84.3	90.7	94.4
Black-throated green warbler					
Dendroica virens	48	70.7	63.7	80.9	84.2
Carolina chickadee					
Poecile carolinensis	44	58.6	56.3	80.5	94.9
Ovenbird					
Seiurus aurocapillus	53	68.4	61.4	79.5	91.2
Veery					
Catharus fuscescens	17	94.4	94.4	95.8	96.7
Yellow-billed cuckoo					
Coccyzus americanus	19	71.6	66.5	87.9	90.2

[a]A total of 215 points were surveyed. This column represents the percentage of the total sample of points in Tennessee on which a species was detected.

[b]Classification and regression tree analysis.

classification using these methods, we also used the actual probabilities of occurrence predicted from logistic regression and Mahalanobis distance to test the usefulness of these models in another manner. For this second test, we used logistic regression analysis to test for a significant relationship across an entire data set between the model-predicted probabilities of occurrence for each location and the observed presence/absence at that location. We termed this procedure the association test. A significant relationship ($P < 0.05$) in the correct direction indicated that the models were performing well because higher predicted probabilities of occurrence were associated with locations where the given species was actually observed to occur. For logistic regression, the correct direction of this relationship was positive (higher predicted probabilities of occurrence associated with observed occurrence). The correct direction was negative for the Mahalanobis distance models because we used the actual Mahalanobis distance values, for which smaller values indicated a higher probability of occurrence (Clark et al. 1993b).

Model Development Results

Results from the jackknife tests on the data set used for developing the models showed variability, both among species and among modeling techniques (Table 54.2). These results provide an indication of how well the models should perform when tested on new data. Discriminant function models performed the best on the jackknife tests for all species, correctly classifying over 80 percent of the observations in the Tennessee data set for all species and over 90 percent for many of the species. However, this high rate of correct classification is somewhat misleading, because the jackknife test of the discriminant function procedure does not balance sensitivity and specificity but rather seeks the best total correct classification. CART models also performed well on the jackknife tests, with results that were better than logistic regression and Mahalanobis distance but slightly worse than discriminant function models for all species. CART models typically correctly classified 80 percent or more of the observations, with some of these models correctly classifying more than 90 percent of the observations. The logistic regression and Mahalanobis distance models did not perform as well on the jackknife test for all species, although they both did well (over 80 percent correct) on black-throated blue warbler and veery (species that are restricted to higher elevations in our study area). The logistic regression models also correctly classified over 70 percent of the observations for black-throated green warbler and yellow-billed cuckoo. Both modeling techniques produced models that correctly classified less than 70 percent of the Tennessee observations for ovenbird and less than 60 percent for Carolina chickadee, suggesting that the models for these two species, which tend to be widely distributed across forest types and stand conditions, were not likely to be as successful at accurately predicting presence/absence in new locations.

Model Testing Results

As an example, we provide a graphical representation of output from the four models in comparison to observed field data for yellow-billed cuckoo in the Deerfield Ranger District of the George Washington and

Jefferson National Forests in Virginia (see Fig. 54.2 in color section). Although only one species and a small portion of the Virginia test data used are shown, this figure provides a visual reference for how the test-data points were spatially configured and what typical model output was like for each of the modeling techniques.

A noticeable decrease occurred in the percentage of observations in the Georgia and Virginia test data sets that were correctly classified by the discriminant function and CART models compared to their level of performance on the jackknife tests with the Tennessee data. The discriminant function models, in particular, performed much worse on the test data sets and did not achieve a correct classification rate over 70 percent for any of the species (Table 54.3). The CART

TABLE 54.3.

Results from testing bird-habitat models for six species on two independent data sets. The models were developed using four different modeling methods and covered the southern Appalachian Mountains.

Species[c]	Logistic regression				Mahalanobis distance				CART[a]	DF[b]
	% correct	Sign	Wald χ^2	*P* value	% correct	Sign	Wald χ^2	*P* value	% correct	% correct
Georgia										
BTBW	65.5	+	36.49	0.001	65.2	—	17.64	0.001	92.0	65.6
BTNW	54.8	+	11.48	0.001	57.1	—	16.38	0.001	55.4	27.1
CACH	44.8	—	4.85	0.028	45.7	+	5.69	0.015	49.2	26.5
OVEN	52.9	+	0.86	0.354	54.9	—	3.57	0.059	53.1	40.9
VEER[d]										
YBCU	61.8	+	15.58	0.001	34.0	+	1.09	0.298	54.6	37.7
Virginia										
BTBW	65.4	+	25.18	0.001	58.6	—	6.77	0.009	94.9	56.7
BTNW	54.6	+	5.96	0.015	61.4	—	7.99	0.005	67.1	35.5
CACH	44.5	—	0.01	0.907	52.2	—	0.06	0.802	50.7	23.6
OVEN	53.0	+	1.05	0.305	56.2	—	4.18	0.041	56.0	34.6
VEER	62.9	+	9.19	0.002	62.4	—	15.84	0.001	85.6	62.4
YBCU	61.8	+	24.25	0.001	55.2	—	1.52	0.217	67.3	47.1

Note: The final models developed from each of the modeling methods were tested on similar point count data sets from Georgia and Virginia. "% correct" indicates the percentage of observations in the test data sets that were correctly classified by the model for a species. The rest of the statistics indicate the strength of the relationship between the probability of occurrence predicted by a model and the observed occurrence at each point count location. Logistic regression was used to test for the significance of these relationships. We considered a significant positive relationship to be an indication that logistic regression models performed acceptably well and a significant negative relationship to be an indication of acceptable performance for the Mahalanobis distance models.

[a]Classification and regression tree analysis.

[b]Discriminant function analysis.

[c]Species codes (see Table 54.2 for scientific names): BTBW = black-throated blue warbler, BTNW = black-throated green warbler, CACH = Carolina chickadee, OVEN = ovenbird, VEER = veery, YBCU = yellow-billed cuckoo.

[d]An insufficient number of veeries were detected on the Georgia sites (N = 4) for testing the models of this species.

models performed well (over 85 percent correct) on the classification tests for black-throated blue warbler and veery, which were the species restricted to high elevations and thus should be the ones most easily predicted.

The logistic regression and Mahalanobis distance models generally performed rather poorly (less than 70 percent correct) on the classification tests for all species but performed quite well on the association tests for a majority of the species (Table 54.3). Black-throated blue warbler, veery, and yellow-billed cuckoo were the only species for which the logistic regression models correctly classified 60 percent or more of the observations in the Georgia and Virginia data sets. Black-throated blue warbler, black-throated green warbler, and veery were the only species for which the Mahalanobis distance models achieved more than 60 percent correct classification on the test data sets. However, since both of these modeling techniques are more appropriate for predicting probabilities of occurrence than presence/absence, we expected these two methods to perform better on the association tests. The results from the association tests indicated that the logistic regression models for black-throated blue warbler, black-throated green warbler, veery, and yellow-billed cuckoo performed well (Table 54.3). The Mahalanobis distance models for black-throated blue warbler, black-throated green warbler, ovenbird (marginally on the Georgia data set), and veery performed well on the association tests. Neither the logistic regression nor the Mahalanobis distance models for Carolina chickadees performed well on the association tests or the classification tests.

Comparison of Model Performance

Discriminant function models for all the species appeared to perform very well when assessed by jackknife tests, but they performed very poorly for all species when their predictive ability was tested on independent data. All of the discriminant function models correctly classified over 90 percent of the observations in the development data set using the jackknife test, with the exception of the black-throated green warbler model, which still performed well (over 80 percent correct). However, when these models were

tested on data from Georgia and Virginia, their performance dropped to less than 70 percent correct for all of the species tested, with most species less than 50 percent. We believe this poor performance on the classification test resulted from discriminant function models being chosen on the basis of the best overall classification rate, rather than on the basis of balancing sensitivity and specificity, resulting in models biased toward either presences or absences and thus unable to accurately predict new observations.

Results of the classification tests were mixed for the three other modeling techniques. CART models consistently performed similarly or better than the other models on the classification tests for both the Georgia and Virginia data sets across all the species. Performance of CART models for black-throated blue warbler (over 90 percent correct classification for both test data sets) and veery (85.6 percent correct classification on the Virginia data) was exceptionally good and considerably better than logistic regression and Mahalanobis distance models (65 percent or less correct classification). Since the occurrence of these two species was restricted to higher elevations and could be predicted fairly well by elevation alone, our results suggest CART might be the most suitable method for modeling species whose occurrences are distributed in an approximately binomial fashion with regard to a key environmental factor (e.g., restricted to occurring above a given elevation).

Performance on the classification tests was considerably lower and very similar among these three methods (logistic regression, Mahalanobis distance, and CART) for the rest of the species tested in this chapter, with none of the models for the remaining species correctly classifying over 70 percent of either test data set (and most less than 60 percent). Our results suggested that for these three modeling methods, their ability to correctly predict presence/absence might be fairly similar for many species. However, the CART method was the only one that proved capable of highly accurate models (over 80 percent correct) for at least some species.

Correct classification of 75–80 percent has been suggested as a level of accuracy that both researchers and managers consider to be acceptable levels of model performance (Chalk 1986; Hurley 1986). Our models for most of the species fell somewhat below this level of performance for the classification tests. However, Morrison et al. (1992:258–262) cautioned that even good habitat models typically account for less than 50 percent of the variation in species occurrence. Most of our models performed at least this well.

One of the factors likely contributing to the low levels of model performance was our testing of Tennessee-based models on data from Georgia and Virginia. Although these locations all fell within the southern Appalachian region, some of the natural variation among these locations undoubtedly was not incorporated in a model based solely on the data from Tennessee. Thus, the somewhat low levels of performance for many of the models were not entirely unexpected, as species responses to habitat conditions certainly could vary across this large geographic region, especially along the north-south gradient. As suggested by Heglund (Chapter 1), where steep environmental gradients exist, stratified surveys along the gradients are useful for capturing the variability in species responses and reducing bias to responses at more local scales. The data from Georgia, Tennessee, and Virginia represent a stratified sample of the north-south gradient, so our models for the southern Appalachians might be improved by developing the models from data across the entire region. Combining all the data from the three locations and then dividing the entire set into training and testing groups may have improved the predictive ability of the models across the region.

Due to the subjectivity involved in selecting cut-off values to predict presence/absence (see further discussion to follow) from the probabilities of occurrence generated by logistic regression and Mahalanobis distance models, we considered the association tests to be a more appropriate method of assessing the performance of models developed from these two methods. For the association tests, a significant association (P < 0.05) in the correct direction between the predicted probabilities of occurrence (or predicted distance from the ideal value, in the case of Mahalanobis distance models) and the observed presence or absence across all samples in the test data sets suggested the models performed well at what they were designed to do—

predict which locations were more likely to have the species of interest present.

Based on the association tests, both the logistic regression and Mahalanobis distance models performed well (P < 0.05) for most of the species tested. Ovenbird and Carolina chickadee were the only species for which the logistic regression method did not perform well, while yellow-billed cuckoo and Carolina chickadee were the only species for which the Mahalanobis distance method did not perform well. As suggested by the large Wald χ^2 values, both methods produced models that performed very well for black-throated blue warbler and veery, which are species restricted to higher elevations in the Southern Appalachians, as well as black-throated green warbler. The logistic regression model for yellow-billed cuckoo also performed well, while the Mahalanobis distance model for ovenbird was acceptable, but neither method produced good models for both of these species. Developing models that perform well for these two species could be difficult because they tend to be fairly general in their habitat associations and thus occur in a wide variety of forest types and condition classes. We have no clear explanation why the different methods would produce a model that worked well for one of these species but not the other. Carolina chickadee was another species that was widespread and occurred in a variety of forest types and condition classes, but neither of the methods produced a model that performed well for that species.

Analytical Method

The classification tests were perhaps a conservative means of testing the predictive abilities of the logistic regression and Mahalanobis distance models. We consider the need to choose a cut-off probability level, above which the model output is considered to be equivalent to a prediction of presence, to be a limiting factor in the correct prediction of presence/absence, particularly for models that are designed to predict the probability of occurrence rather than absolute presence or absence. No standardized rules exist for choosing a cut-off value, and the cut-off value ultimately influences the total correct classification rate, thus adding an element of subjectivity to the results of classification tests.

A common approach to choosing a cut-off value is to select the probability value that results in a balance between the sensitivity of the model (proportion of the "presence" observations correctly predicted) and the specificity of the model (proportion of the "absence" observations correctly predicted). Depending on the balance between the number of occurrences and absences in the data, this approach to selecting the cut-off value may not result in the highest-possible total correct classification rate. For instance, if the data set includes few occurrences and many absences, then selecting a cut-off value that maximizes specificity would result in a much higher total correct classification than selecting the cut-off value that balances sensitivity and specificity. Thus, our method of always selecting the cut-off value that balances sensitivity and specificity often resulted in a total correct classification that was below what the model could achieve.

We believe that balancing sensitivity and specificity is the most appropriate method because we are interested in maintaining reasonable levels (i.e., over 50 percent) of sensitivity and specificity for all models, regardless of the distribution of occurrences and absences in the data sets for a given species. If the distribution of occurrence is skewed and the cut-off value is chosen to maximize either sensitivity or specificity, then the other measure will usually be extremely low. Furthermore, maximizing either sensitivity or specificity to achieve a high correct classification of the developmental data is likely to bias the models and limit their ability to make accurate predictions for new locations, as suggested by the results from testing the discriminant function models.

Management Implications and Conclusions

Our development and comparison of modeling methods was part of a large effort to provide a series of regional wildlife-habitat models for the southern Appalachians. The U.S. Forest Service plans to use these models in an effort to coordinate national forest planning across this region. Having bird models developed from the same set of habitat data across the entire region will allow the Forest Service to consistently assess how alternative management options for the different national forests will affect the availability of bird habitat across the region. The results from this study indicate that reasonable models can be developed for some species over large spatial extents. Thus, in areas where resource management agencies have an interest in coordinating planning efforts over large areas, useful predictive models can be developed to assist in evaluating management alternatives.

Our results suggest several generalizations regarding the ability of the four modeling methods compared in this study to produce good predictive wildlife-habitat models for large spatial extents when using data similar to those used in this study. First, while discriminant function models are likely to distinguish between presence and absence locations within the original data set, such models are unlikely to perform well when tested on new data sets. Second, both logistic regression and Mahalanobis distance models appear to be good general methods for predicting probabilities of occurrence for new locations, although they may not perform as well predicting absolute presence/absence. These two methods performed well on our association tests for most of the species that we evaluated. CART analysis may be the best method when the correct prediction of species presence and absence at given locations is the desired product from the models.

The difficulties in developing successful predictive models for some species were also clear from our results. Generalist species that occur over a wide range of habitat types seem to be particularly problematic (Hepinstall et al., Chapter 53). The models for some species in our study might have been improved by developing the models with data from across the entire region rather than just from Tennessee, but some generalist species may simply be difficult to model. Species that are more limited in their habitat utilization (e.g., elevation, forest type, condition class) are more likely to be modeled well. This trend was especially true for the CART models, which performed extremely well on the classification tests for the species that were elevation limited (e.g., black-throated blue warbler, veery) but did not perform very well on the species that were more general in their habitat use. Both logistic regression and Mahalanobis distance models performed well on our association tests for many of the species and thus both should be considered good options for general wildlife-habitat modeling. Logistic regression is a more completely developed technique that has been extensively described by statisticians (e.g., Hosmer and Lemeshow 1989), including appropriate methods for model selection and determining the goodness-of-fit of models. The Mahalanobis distance method is not yet a commonly used method and lacks well-defined procedures for model assessment and determination of significant variables in a given model.

Finally, while the test results from our models may not have been as good as we would have liked for many of the species, our results indicate that acceptable predictive models covering large spatial extents can be developed, at least for some species. For the species in our study that were sufficiently restricted to specific habitat components (e.g., northern hardwood forests associated with high elevations in the mountains), we were able to develop models from data collected in Tennessee and then correctly predict the occurrence of those species in Georgia and Virginia with a high degree of accuracy. Thus, some species will lend themselves to this kind of modeling process much better than others will, but our results show that successful modeling of this type is possible over large extents.

Acknowledgments

Funding was provided by the USDA Forest Service (Southern Experiment Station and Cherokee National Forest) and the University of Tennessee Institute of Agriculture. Thanks to Eddie Morris for kindly sharing his Georgia data with us, as well as all the personnel at the George Washington and Jefferson National Forest for helping to make their data available to us. We also thank the many field assistants who made the collection of all the data possible. Frank van Manen was an indispensable source of information and feedback for this project, particularly regarding Mahalanobis distance.

Distributional Prediction Based on Ecological Niche Modeling of Primary Occurrence Data

A. Townsend Peterson, David R. B. Stockwell, and Daniel A. Kluza

Knowledge of biodiversity is incomplete (Wilson 1988). The magnitude of numbers of species (10^6–10^7) combined with relatively small numbers of systematic biologists, suggest that scientific documentation of world biodiversity (currently about 3×10^9 specimens) is fragmentary (Anonymous 1994). Explorations of the completeness of biodiversity data on regional scales (e.g., Peterson et al. 1998) have found *J*-shaped frequency distributions—only a few species or sites are well documented, and most are poorly known and underdocumented.

Many broadly based biodiversity programs, such as the U.S. National Gap Analysis Program and others (e.g., ICBP 1992; Scott et al. 1996), focus on geographic distributions of species. Predictions of geographic distributions, however, depend on the incomplete sampling available and therefore would be improved by some inference or modeling. Inferential procedures offer the possibility of using existing knowledge to predict into the gaps in knowledge. The nature of these inferential procedures, which carry their own assumptions and biases, affects the characteristics of any products of such programs.

The purpose of this chapter is to explore some of the base constructs underlying modeling procedures (algorithms) used to infer species' distributions from incomplete data and to make recommendations regarding these algorithms. Throughout this discussion, we distinguish between primary information (direct observation or documentation, often in the form of a specimen, of a particular species taken at a particular place and time) and secondary information (a synthesized product based on primary information, often in the form of a species biography, range map, or description of habitat use). This distinction offers insights into the types of algorithms that are desirable for such analyses.

Distributional Modeling

The procedure used to convert primary distributional data (i.e., known points of occurrence) into spatially continuous information (i.e., predictions of presence and absence across a landscape for a particular species) is critical to the value of biodiversity products. These procedures, however, are not all alike, and the differences among them have important implications. Understanding these differences may allow development of criteria for improved methodologies.

One or Two Steps

A first choice is whether the algorithm to be employed is of one or two steps: one-step approaches (e.g., Hollander et al. 1994) focus on modeling the geographic distribution directly from the spatial arrangement of known occurrence points. Two-step approaches (e.g.,

Austin et al. 1990) attempt to model the ecological requirements of species (their ecological niches) and project those requirements onto maps to produce a potential geographic distribution.

One-step models are convenient because they focus directly on prediction of known geographic distributions and are often computationally much less expensive. However, they usually require restrictive assumptions regarding existing information, such as that sampling is sufficient to represent geographic limits, or that sampling is random or uniform such that ranges may be estimated from central tendencies (Hollander et al. 1994). These assumptions are rarely evaluated prior to application of such approaches (Csuti 1996) and are almost never realistic given the odd characteristics of existing biodiversity information—many species and sites are poorly sampled, and few species and sites are well sampled (see Peterson et al. 1998c for a quantitative analysis). Their weakness, to be more precise, lies in the fact that no special quality of a species is linked to its *known* geographic distribution, and that species' true geographic distributions result from complex interactions of ecological and historical factors (Peterson et al. 1999).

Two-step models, on the other hand, offer a direct tie to species' biology. An ecological determinant of a species' distribution, its ecological niche (MacArthur 1972b), can be modeled to hypothesize environmental conditions under which it is likely to be able to maintain populations. Then, geographic areas with those conditions can be identified as a potential distribution for the species. This procedure could potentially overcome some of the problems of sampling bias given that the ecological niche may be identifiable without a complete or balanced sampling of the species' geographic distribution. In essence, the question of bias in sampling passes from geographic space, where spatial biases are well known (e.g., Peterson et al. 1998c), to ecological space, where bias may be reduced. A niche model also allows additional inferences, such as changes in the species' potential distribution under scenarios of environmental change (Peterson et al. 2001).

Error Components and Decision Criteria

In developing a model predicting a species' geographic distribution, two types of error are possible. *Omission*

is leaving out areas actually inhabited, whereas *commission* is including areas not actually inhabited (Scott et al. 1993; Krohn 1996). Although simple solutions can minimize *either* (predicting the whole map as present reduces omission error to zero; predicting only the known occurrence points reduces commission error to zero), an ideal algorithm would minimize *both* simultaneously.

A typical, testable implementation of ecological or distributional modeling procedures involves subsetting available information to provide tests of model accuracy. For instance, a model can be developed based on 60 percent of the distributional points available ("training data") and tested based on the remaining 40 percent ("test data"), or a standard number of points can be chosen at random and set aside for model tests (Fielding and Bell 1997); similar test data sets can be developed for absence based on known absences or on points chosen from the background (Stockwell and Peters 1999). In this case, omission error is directly estimable as the proportion of the test presence data set not predicted as "present." Commission errors are more problematic with presence-only data, as no direct estimator is available (Krohn 1996). The overall area of the distribution predicted includes both actual distributional area (unknown but real, and generally larger than the known distribution) and the commission error component. Under this view, geographic predictions of smaller areas will generally be better at reducing commission error. Hence, a reasonable decision rule is to reduce predicted areas as much as possible (reducing commission error) without increasing omission error.

Given these arguments, the following ranked criteria are recommended for evaluating the potential of inferential approaches to modeling species' geographic distributions:

1. *Minimize omission.* Minimal omission of known occurrence points.
2. *Minimize commission.* Find the smallest distributional prediction that does not fail on Criterion 1.
3. *Emphasize data economy.* With insensitivity to small sample sizes, elimination of potential biases in geographic or ecological coverage, and ability to use inexpensive data (i.e., no detailed physiological

or bibliographical data necessary), modeling can be applied to distributions of large numbers of species, not just the small, well-studied minority.

4. *Model the niche.* Although not a requirement for many simple applications, development of ecological niche models makes possible extension of biodiversity analyses to more complex situations, such as scenarios of climate change, species invasions, human activity shifts, and the like.

5. *Check repeatability and emphasize renewability.* The ability to replicate a particular result via known, quantitative procedures lends scientific credibility to products. Especially useful would be approaches that are perpetually renewed and thus do not get old or out of date (e.g., published faunas or floras).

Although relative importances of these criteria may vary according to their associated costs, the above criteria offer a general framework within which potential approaches may be viewed.

The Primacy of Point Occurrence Data

Throughout this discussion, the importance of basing biodiversity products (synthetic analyses such as gap analyses, conservation prioritizations, and richness maps) on primary point occurrence data is emphasized. Secondary data are often in temptingly convenient form—published range maps, species accounts, or ecological summaries. However, secondary data usually carry with them several problems. First, they insert elements of subjectivity—development of a range map usually involves an expert drawing a polygon or polygons around known areas of occurrence and guessing at which of the inevitable unsampled areas are likely to hold or not to hold populations. Second, availability of secondary data depends on *publication* of knowledge and for that reason often lags behind the *existence* of that knowledge. Data for particular taxonomic groups or regions may exist, but secondary data may be unavailable (compare numbers of publications on birds with those on nematodes, or publications about Massachusetts with those about Vietnam).

Finally, and perhaps most critically, secondary data are usually one-time products (e.g., publication of a book), or at best products that are renewed or revised periodically, which makes the biodiversity products based on them degrade over time. For example, if range maps published in a regional field guide are used to develop a biodiversity product, those maps were probably somewhat out of date even when the guide finished the yearlong publication process—habitat modification may have extinguished populations or permitted invasion of new areas, or new distributional points may have been discovered. Publication of such a regional summary usually stimulates a rapid spate of additions and modifications to the state of knowledge as well, making the range maps degrade in quality and completeness even more rapidly. The biodiversity product degrades along with the underlying information, and soon is out of date.

Primary data, on the other hand, offer solutions to most of these problems. These data, including occurrences vouchered by scientific specimens as well as observational data, are abundant (Peterson et al. 1998c). Although some primary data are decades or even centuries old, and precision of locality information is clearly reduced with older records, when modeling procedures are properly designed, model results may have excellent predictive ability (Godown and Peterson 2000, Peterson et al. 2002, Peterson 2001). Moreover, primary data, when aggregated from world holdings, are renewable and *improve* over time, as more and more information is added to the storehouse.

Recently, a multidisciplinary, inter-institutional effort has laid a technological and political foundation for making primary biodiversity data from world scientific institutions available for such applications. The North American Biodiversity Information Network of the Commission for Environmental Cooperation (Montreal, Canada) and the National Science Foundation have funded the development of an Internet-based distributed database system called *The Species Analyst* (speciesanalyst.net). This facility uses the Z39.50 information transfer protocol to integrate institutional holdings of primary biodiversity data into a virtual "world database" that provides the information on which various biodiversity products can be based. The system presently serves about fifteen million data points housed at twenty-two institutions and provides a direct connection to facilities for analysis of data via the tools described below. A combination of this distributed store of primary biodiversity data and rapid computational approaches is expected to provide continually renewed and improved analytical abilities.

Ecological Niche Modeling and Distributional Prediction

The niche, as we use the term, is a set of tolerances and limits that define where a species is potentially able to maintain populations (Grinnell 1917), later visualized as an *N*-dimensional polyhedron in ecological space (Hutchinson 1957). Given the spatial scale of our analyses, we focus on niche dimensions relevant to geographic distributions rather than to local distributional issues such as microhabitat or substrate selection. Hence, the niche dimensions employed are those usually considered in geographic limitation of species—temperature, precipitation, elevation, vegetation, and so forth, and not the finer-resolution information that may be necessary to resolve more detailed issues, such as local substrate selection.

Ecological niches are generally divided into fundamental and realized ecological niches: the former represents the base ecological capacity of the species, and the latter incorporates the effects of interactions with other species (MacArthur 1972b). Because community composition varies greatly over space, its impacts should vary as well (Dunson and Travis 1991). Models at the level of entire species' distributions may allow identification of broader ranges of environments potentially suitable for a species, analogous to a fundamental niche, although on coarse spatial scales. In some senses, most of the approaches used in distributional modeling approximate a niche as they are defined in multidimensional ecological space; however, the one-step approaches discussed above intermingle modeling of the fundamental and realized niches and therefore stray from a simple list of environmental factors important to a species' presence or absence.

Several two-step approaches have been used to approximate species' ecological niches; three are treated herein. The simplest is BIOCLIM (Nix 1986), which involves tallying species' occurrences in categories along each environmental dimension, trimming marginal portions of distributions, and taking the niche as the conjunction of the trimmed ranges (e.g., annual mean temperature between 4°C and 6°C, annual mean precipitation between 1,000 and 1,200 millimeters, etc.). Although easy to implement and conceptually attractive, BIOCLIM suffers reduction in efficacy when many environmental dimensions are included (D. R. B. Stockwell unpublished data; B. Loiselle personal communication).

A second class of approaches is based on logistic multiple regression, a class of statistical techniques aimed at predicting the probability of "yes" versus "no" in the independent variable (e.g., Mladenoff et al. 1995). Although complex in its application to biodiversity data for which absence data are not common, thus requiring sampling of the background landscape as a substitute, this idea combines well with the concept of physiological tolerances determining species' presences along continuous climate axes. It, however, is less well suited to incorporation of categorical information (e.g., vegetation type, soil type). In effect, logistic regression divides environmental space into two portions ("habitable" and "uninhabitable") at a particular probability threshold, an approach that may be useful under some circumstances. More-recent implementations of this approach have included improvements such as relaxation of assumptions regarding distributions of errors in the regression (e.g., Austin et al. 1990).

Finally, the *Genetic Algorithm for Rule-set Prediction* (GARP) includes simplified versions of both of the methods described above, as well as other set-based approaches in an iterative, artificial-intelligence approach (Stockwell and Noble 1992; Stockwell and Peters 1999; http://biodi.sdsc.edu/). Individual algorithms (e.g., BIOCLIM, logistic regression) are used to produce component "rules" in a broader rule-set, and hence portions of the species' distribution may be determined as inside or outside of the niche based on different rules from several algorithms. As such, GARP is a superset of the other approaches and should always perform as well as or better than any one of them. Extensive testing of GARP has indicated excellent predictive ability and relative insensitivity to small sample sizes—even down to 10–20 sample points (Peterson et al. 2002; Peterson 2001; Stockwell and Peterson in press)—and insensitivity to BIOCLIM's problems with environmental data density (Peterson and Cohoon 1999), although precision of predictions clearly depends on the minimum resolution available in both environmental and species' occurrence data (Peterson and Kluza unpublished data).

GARP models generally predict geographic distributions of species accurately (Peterson et al. 1999; Peterson and Cohoon 1999; Peterson et al. 2002; Peterson 2001). Within the region in which a particular species occurs, it does an excellent job of including and excluding areas that are inhabited and not. Two phenomena cause deviations from accurate predictions. The first involves insufficient data dimensions to delimit ecological niches sufficiently (Peterson and Cohoon 1999). The effect of including too few environmental variables in a modeling effort is that factors critical to limiting species' distributions in space may be omitted, and for that reason predicted geographic distributions are too large. Tests of these methodological challenges indicate that most models stabilize with four to five environmental dimensions (Peterson and Cohoon 1999); however, clearly, ecological models will be more predictive with more and more environmental dimensions that are included in the analysis.

The second type of error in GARP involves prediction of occurrence in biogeographic regions not inhabited by the species in question. Although initially worrisome, this phenomenon makes considerable sense: species are limited to particular geographic areas not solely by ecological factors (modeled by GARP), but also by historical phenomena. These historical factors include speciation events (producing sister species in the "other" area), extinction events, and limited colonization ability (why are hummingbirds not in Africa?), as well as possible historical species' interactions (R. Anderson et al. submitted). In fact, a recent study demonstrated that ecological niches of many species accurately predict geographic distributions of their allopatric sister species because ecological niches apparently evolve more slowly than speciation events occur (Peterson et al. 1999). The

high degree of predictability for species invasions achieved by applying GARP to native distributions and projecting to invaded ranges (Peterson and Vieglais 2001) is further indication that species commonly do not inhabit the entire geographic extent of their niches. This category of error is eliminated by restriction to biogeographic regions known to be inhabited by the particular species under study (requiring assumptions about the thoroughness of sampling at the level of biogeographic regions), but may also suggest possible undiscovered isolated populations that should be investigated.

A Test

GARP is a quantitative method for modeling ecological niches and predicting geographic distributions from primary point occurrence data. This approach makes point-occurrence data relevant to biodiversity applications that require information on geographic distributions of species. The GARP approach is an alternative to the approach used (for example) in most Gap applications (Scott et al. 1996), so we developed a direct comparison of the two methods; we report initial results herein and will provide a complete summary elsewhere (Peterson and Kluza submitted).

In consultation with Gap programs around the country, we selected the Maine Gap Analysis Program as an ideal testbed, being recently completed and replete with both fine-scale bird distributional data and relevant environmental coverages. We used stop-level U.S. Breeding Bird Survey (BBS) data (http://www.mp2-pwrc.usgs.gov/bbs/index.htm) for thirty forest bird species from 1990 (kindly provided by R. O'Connor and W. Krohn), combined with environmental coverages including elevation; slope; aspect; annual mean precipitation; vegetation type; and average maximum, absolute maximum, average minimum, and absolute minimum temperatures for winter, spring, summer, fall, and the entire year. Twenty known occurrence points for each species were set aside for testing models' predictions of presence: the remaining known occurrence points were used to build models. All BBS points from which particular species were *not* known were used to test predictions of absence, although these points clearly included some undetected presences as well.

As an example of test results, we discuss Gap and GARP predictions for the wood thrush (*Hylocichla mustelina*) (see Fig. 55.1 in color section). Of the twenty test presence points, both Gap and GARP models correctly predicted eleven; of the 1,152 test absence points, however, the GARP model successfully predicted 806, whereas the Gap model successfully predicted only 471. These results indicate that the two approaches are generally comparable on false-negative (omission) error (Fielding and Bell 1997), but GARP performed better on avoiding false-positive (commission) error (Fielding and Bell 1997), at least given the test absence data that were available to us. Direct statistical comparison of the performance of the two approaches (Fielding and Bell 1997) indicated that the GARP model was highly statistically significantly more accurate (McNemar's test, $X^2 = 215$, $P < 10^{-48}$). Indeed, of the thirty species tested, twenty-eight were significantly better predicted by GARP than Gap, in all cases with greatly reduced false-positive (commission) error rates (Peterson and Kluza submitted).

The Maine study is a direct test of predictive efficiency using GARP and Gap methodologies. Our interpretation of its results is that Gap models identified species' habitat associations reasonably well but failed to predict regional restriction within the state (hence the high false-positive error rates). GARP models, on the other hand, performed similarly or somewhat worse at identifying habitat associations but were better able to restrict regional distributions. Referring to the criteria outlined above, while GARP and Gap models both avoided omission (criterion 1) about as well (or Gap models performed slightly better), GARP models avoided commission considerably better than the Gap models (criterion 2). (It is worthy of note that our measures of commission error are subject to the criticism that our test absence data set is certainly a mixture of *real* absences and apparent absences.) GARP models function based on the minimum-quality data available (point-occurrence data) and do not require further information (criterion 3), such as literature on habitat preferences. GARP models produce a niche model (criterion 4) and are readily repeatable (at

least in a statistical sense, as they are able to be rerun and reproduced at will) and renewable (criterion 5). Hence, we suggest further exploration of the possibilities of GARP for improving the predictivity of distributional models used in gap analyses.

Possibilities and Conclusions

Development of accurate, quantitative, and repeatable methods for inferring ecological niches and geographic distributions from primary point-occurrence data opens doors to new advances in studying biodiversity. A series of products becomes possible, including the following:

1. *Single species distributional predictions*. Distributional predictions for single species can be useful, for example, in localizing rare and poorly known species, designing reintroduction programs, and protecting endangered species.
2. *Community predictions*. Overlaying results for numerous species, estimation of assemblages of species at particular points in space becomes feasible, providing a "rapid inventory" approach to understanding local species distribution patterns.
3. *Conservation prioritization and impact assessment*. Once local assemblages are understood, combinations of species present and absent at sites can be analyzed; if interpreted as conjunctions of species of concern, then concentrations represent areas for conservation, whereas areas in which such species are absent can be interpreted as areas of reduced concern (Godown and Peterson 2000).
4. *Climate change/environmental change*. Development of niche models opens the possibility of projecting an ecological niche onto other landscape scenarios besides the present set of conditions; these approaches can be illuminating for applications such as understanding the effects of global climate change (Peterson et al. 2001), predicting species invasions (Peterson and Vieglais 2001), or projecting effects of human population trends.

Exploration of these possibilities should bring about a phase of rapid development of improved products and qualitatively new products. Gap analysis, for example, could be moved from a one-time product that must be laboriously updated from time to time to a continuously updated, never-out-of-date, growing database that builds products in real time, thus taking advantage of a maximum of information for every result.

Acknowledgements

We thank David A. Vieglais for his development of the data access tools discussed herein, and the Commission on Environmental Cooperation and the National Science Foundation for financial support of these efforts.

Statistical Mapping of Count Survey Data

J. Andrew Royle, William A. Link, and John R. Sauer

The North American Breeding Bird Survey (BBS; Robbins et al. 1986) is conducted during the breeding season of each year by volunteer observers. The sampling unit in the BBS is a roadside route 24.5 miles (39.2 kilometers) in length, containing fifty stops. At each stop along a route, birds are counted by sight and sound for a period of three minutes. Over four thousand routes have been surveyed in North America. The BBS is incapable of providing measures of absolute abundance because the proportion of the population counted (the count proportion) is unknown and the effective sampling area and relevant local populations are poorly defined. Thus, the BBS produces counts along transects that may be loosely interpreted as indices to local abundance, or relative abundance (Link and Sauer 1998). Although data from the BBS are most often used for monitoring and assessment of temporal trends in bird abundance, the data are also useful for providing information on the spatial distribution of relative abundance. Such information can be used, for example, to relate abundance to habitat attributes, to examine changes in the spatial distribution of abundance, and to assess the relative abundance in sparsely sampled regions.

In this chapter, we consider the problem of spatial prediction of relative abundance from BBS count data. Nominally, our goal is mapping; that is, predicting at many points within a region for the purpose of pro-

ducing a relative abundance map for a particular species. Mapping may be approached in a number of ways using various more or less ad hoc procedures such as inverse-distance squared interpolation, spline or kernel-density smoothing, and kriging (e.g., Sauer et al. 1995). However, we believe that mapping should be based on a biologically reasonable statistical model. Thus, we approach the mapping problem within a model-based framework, assuming that BBS route counts are observations of a spatially indexed Poisson random variable.

Because count data are discrete, positive valued, and often exhibit strong mean-variance relationships, Poisson models are a natural starting point for modeling such data. Our objective here is to demonstrate the application of a mapping procedure based on a Poisson model with spatially correlated mean. We adopt the Poisson modeling approach proposed by Diggle et al. (1998), which they used for mapping radiation radionuclide concentrations. The Poisson mean may depend on fixed covariates, as in standard Poisson regression (e.g., Jones, Chapter 35), which is common throughout statistics. However, because the model allows for the Poisson mean to be spatially correlated, it departs from the standard Poisson regression framework. It is perhaps best viewed as a Poisson-analog of the autologistic model for modeling a spatially correlated binary variable (e.g., Klute, Chapter 27). More generally, this model is itself a special case of the generalized linear mixed models (GLMM) that have been widely adopted in all fields of statistical practice.

There are several important reasons for pursuing a model-based strategy for mapping BBS data, and count data in general. First, in order to conduct formal statistical inference, a statistical model is required. Map-based inference problems that might be of interest include comparison of maps over time (such as to assess the impact of climate events or landscape changes); comparison of areal averages over arbitrary regions, thus avoiding the problem of subjective and ad hoc stratification; model-based design under which routes are added and deleted, or moved in a manner that is optimal with respect to a variance-based criterion; and inference about particular predicted values. Of course, such maps may still serve as a purely descriptive assessment of the spatial distribution of a bird species. Second, because count data are nonnegative and often exhibit strong mean-variance relationships, a proper predictor should accommodate these features. Putative model-free procedures do not address either of these concerns and can produce unreasonable negative predictions. Third, model-based procedures produce a natural measure of prediction uncertainty (i.e., the prediction variance), which can be used to qualify prediction maps. Finally, proper modeling of count data allows a much more efficient use of data than in analyses based on reductions of counts to presence/absence (e.g., "range maps"), as pointed out by Jones et al. (2001).

In this chapter, we propose a framework for mapping count data from the BBS, based on a Poisson model with spatial correlation. In the next section, we define the general mapping problem with some discussion of the common technique known as kriging. In the Breeding Bird Survey Data section, we discuss mapping issues relevant to the BBS data. The Poisson mixed model is introduced in the section that follows. Because we adopt a Bayesian approach to analysis of the model, some issues pertaining to this are then discussed. In this section, we also introduce a general method of fitting complex Bayesian models known as Markov chain Monte Carlo (MCMC), although details of the algorithm that we used is deferred to the Appendix. Finally, results from applying the model to mapping the relative abundance of mourning doves

(*Zenaida macroura*) are given, and the chapter concludes with discussion of issues pertaining to extension of the model.

Mapping

Mapping is inherently a problem of spatial prediction. That is, given a set of observations $y = (y_1, y_2, \ldots, y_n)$, we wish to predict the value of a new observation, say y_0. Predictions on a grid of new locations produce a map. There are many techniques commonly used for mapping data. These include variations on "inverse-distance" interpolation, kernel smoothing, thin-plate splines, and kriging. Most of these are essentially "black-box" procedures that do not rely on specification of a statistical model for the underlying process. As such, quantification of the prediction uncertainty is generally difficult for most of these procedures. We briefly discuss kriging here because it is one of the more common techniques employed for mapping. Kriging can be viewed in a model-based context, but it is most often viewed as a "model-free" procedure.

Let $(s_1, s_2, \ldots, s_n)$ denote a set of sampling locations at which observations $y(s_1), y(s_2), \ldots, y(s_n)$ of some spatial variable are made. Kriging assumes that the data are a realization of a random spatial process, $\{Y(s) : s \in D\}$, where D is a fixed subset of R^2 (i.e., D is some spatial region). Predictions are produced based on first and second moments (i.e., mean and covariance) of the underlying random process. For clarity, we will present what is known as "simple kriging" here (Cressie 1991, 110), which assumes that the mean of the process is known. Many generalizations of kriging are possible, and the interested reader is referred to Cressie (1991) for details. The traditional development of kriging specifies the mean and covariance structure as

$$E[Y(s)] = \mu \quad Var[Y(s)] = \sigma^2$$

$$Corr(Y(s), Y(s')) = k_\theta(\| s - s' \|)$$

Here, $k_\theta(\| s - s' \|)$ is the correlation function, which depends on parameter θ (possibly a vector) and the distance between locations s and s'. Thus, it is assumed that the correlation function is stationary and isotropic; in other words, it depends only on the distance between points and not on their absolute locations. A common correlation model is the single parameter exponential function given by

$$k_\theta(\| s - s' \|) = e^{-\| s - s' \|/\theta}.$$

Under this model, the correlation between two points decays exponentially as the distance between those points increases. The rate of decay is controlled by θ, which is loosely interpreted as a correlation range parameter (i.e., larger values of θ lead to higher correlation between points located further apart). Although there are many other parametric models for the correlation function, including multiparameter models, there is often little theoretical justification for use of one over the other. Thus, given the concise description of correlation provided by the exponential model, we feel this simple model to be a reasonable choice for many applications involving mapping. Occasionally, it makes sense to parameterize a nugget effect in the covariance function, which amounts to a discontinuity at the origin of the covariance function and arises due to measurement error and uncorrelated "small-scale" variation in the process.

Kriging seeks to predict the process Y at an unmonitored location s_0, say $y(s_0)$, using a linear function of the observed data. Thus, the kriging predictor is of the form

$$\hat{y}(s_0) = \sum_{i=1}^{n} \lambda_i y(s_i)$$

where the λ_i are the kriging weights. The kriging weights are selected so as to minimize the mean-squared prediction error

$$E\left[\left(y(s_0) - \sum_{i=1}^{n} \lambda_i y(s_i) \right)^2\right] \qquad (56.1)$$

subject to the predictor being unbiased. Under the assumption of constant mean, the requirement of unbiasedness is equivalent to $\sum \lambda_i = 1$. The kriging weights are computed as the solution to minimizing Equation 56.1 subject to the unbiased constraint. This is a simple optimization problem solved by taking derivatives, setting them to zero, and solving the resulting linear

system of equations. See Cressie (1991, 120) for details.

In practice, parameters of the kriging model must be estimated. This is often done in an ad hoc fashion. Typically, mean parameters are estimated by generalized least-squares, the correlation parameter is estimated by fitting the parametric model to "empirical" estimates computed using residuals from this fit, and the procedure is iterated. Likelihood estimation is apparently neglected due to the absence of a likelihood.

Kriging leads to predictions that are optimal in the sense that they have minimum variance among all linear, unbiased predictors. That is, the kriging predictor is the Best Linear Unbiased Predictor, or BLUP, regardless of distributional assumptions beyond existence of the first two moments (hence the "model-free" interpretation). Although this is often seen as a benefit of kriging, we believe it to be a major deficiency, particularly when the random variable is clearly non-normal, such as with count data. In general, for non-normal data, there is little theory to suggest that the kriging predictor is reasonable. Indeed, for non-normal data, distributions are not completely specified by the first and second moments and therefore use of a kriging predictor is somewhat ambiguous because it is based on specification of first and second moments alone. Moreover, estimates of prediction variance computed under kriging procedures tend to underestimate true uncertainty in the prediction since error in estimation of covariance parameters is not accounted for. Recent developments (e.g., Handcock and Stein 1993; Diggle et al. 1998) toward Bayesian formulations of kriging alleviate this problem and are more naturally applied to non-normal data problems.

Under distributional assumptions on the data, a predictor that is optimal in a stronger sense can be computed. The Best Unbiased Predictor (BUP) is defined to be the conditional expectation of the quantity to be predicted given the data, or in other words $E[y(s_0)| \mathbf{y}]$ (in general, this conditional expectation follows directly from specification of the joint distribution of $y(s_0)$ and $\mathbf{y}$). For the special case when the spatial process has a normal distribution, it can be shown that the BUP is linear and therefore equivalent to the BLUP. Thus, the kriging predictor is equivalent to the BUP derived under a normality assumption; use of kriging can therefore be viewed as an implicit assumption of normality.

Count data are clearly non-normal. Because count data cannot be negative, one should use a predictor that guarantees positivity of the predictions. Additionally, count data typically exhibit strong mean-variance relationships that are ignored by kriging procedures. Although one can mitigate some of these problems by use of a transformation (e.g., log-normal kriging), there is little theoretical basis for this in the context of modeling counts. Therefore, we believe there is a need for a model-based mapping procedure for count data that produces accurate predictions of relative abundance and associated measures of prediction uncertainty while at the same time imposing realistic distributional assumptions on the data.

Breeding Bird Survey Data

The data used for illustration here consists of mourning dove counts made on eighty-seven routes in Pennsylvania during 1990 (Fig. 56.3, top panel). The grid of points on this map are those locations at which predictions are desired in order to construct the map. We arbitrarily chose a 30×40 grid of points for prediction. For convenience, we associate the total count for each BBS route with the spatial location of the route midpoint.

A major consideration in the analysis of BBS data is observer biases (Sauer et al. 1994). In the analysis of trends, observer effects are typically parameterized as nuisance effects and not modeled directly (e.g., Link and Sauer, 1997a). For mapping BBS data from a single year, a simple treatment of observer effects is possible. Generally, each route count is made by a different observer, and it is reasonable to assume that the observer effects are independent across space. Our solution then, is to parameterize the observer variation as measurement error, which is discussed in the following section.

We expect BBS counts made at different routes to be spatially correlated for (at least) two reasons. First, correlation arises due to similarities (or differences) in underlying habitat structure. That is, all other things being equal, counts made on routes sharing similar habitat characteristics should be more similar than counts made on routes with dissimilar habitat characteristics. Second, even conditional on habitat structure, there are broad-scale differences in abundance of a species within that species' range that are likely to vary smoothly over space. The model that we propose in the following section accommodates spatial correlation among counts in a very general fashion and can be modified slightly to model both sources individually.

A Poisson Mixed Model for Mapping Bird Counts

Let $y(s)$ be the count made at some route centered at location s. We assume that $y(s)$ has a Poisson distribution with mean $\lambda(s)$, which is denoted as

$$y(s) \mid \lambda(s) \sim Poisson\ (\lambda(s))$$

The "$\mid \lambda\ (s)$" notation merely makes explicit that this model for $y(s)$ is *conditional on* λ(s). Though generally ignored, this notation is more rigorous for our subsequent treatment of $\lambda(s)$ as itself being a random variable, as we discuss shortly. Thus, two observations, say $y(s)$ and $y(s')$, are independent *conditional on* their means λ(s) and $\lambda(s')$; that is, any association in the counts at two sites is in their expected values. The Poisson variation in counts—that due to chance events in the observation process—is assumed to be independent from site to site. As is natural for Poisson data, we then specify a model for the logarithm of the mean of this Poisson variable as

$$log\ (\lambda(s)) = \mu + Z(s) + \eta(s)$$

Here, μ is a constant and $Z(s)$ and $\eta(s)$ are random effects, the meanings of which will be described shortly. Generally, m could be replaced with an arbitrary regression function such as $\mu(s) = \sum_{j=1}^{p} \beta_j x_j(s)$ where the $x_j(s)$ are spatially indexed covariates, such as habitat, and β_j determine the change in abundance (on the log scale) per unit change in covariate j. This model is more generally useful, but the added generality is not necessary in our development. To reduce clutter in formulae given below, we will define $u(s) \equiv log\ (\lambda(s))$.

The random effects in this model consist of a smooth, spatially correlated "map," $Z(s)$, and an uncorrelated "measurement error" term, due to observer differences, $\eta(s)$. More specifically, we assume that $Z(s) \sim N(0, \sigma_z^2)$ with $Corr(Z(s), Z(s')) = k_\theta(\| s - s' \|)$, and $\eta(s) \sim N(0, \sigma_\eta^2)$ with $Corr(\eta(s), \eta(s')) = 0$ for $s \neq s'$. Thus, the correlation between the route effects at any two sites, say s and s', is given by the correlation function $k_\theta(\| s - s' \|)$. We used the exponential function discussed in the section on mapping, so $k_\theta(\| s - s' \|) = e^{-\| s - s' \|/\theta}$. Other correlation functions could be considered (see Cressie 1991 for other choices). The fact that the route effects are correlated allows for prediction of $Z(s)$ (and $\lambda(s)$) at arbitrary, unsampled, locations.

Variation that is not explicitly accounted for in the mean term is absorbed into the two random effects, one being a surrogate for spatial sources of variation that have been neglected (such as omitted habitat covariates) and that may be of interest (for example, in prediction), the latter representing unpredictable (i.e., observer, protocol, etc.) variation. In other words, the set of $\eta(s)$ parameters are merely nuisance parameters not of direct interest for any inferential problem. The interpretation of $Z(s)$ depends on whether μ contains habitat covariates. The variance components σ_z^2 and σ_η^2 are the variances attributable to the smooth spatial process and the observer differences, respectively. Thus, when μ contains habitat covariates, we would expect s to be smaller than when habitat information was omitted from μ.

Since the $\eta(s)$ terms are assumed to be independent nuisance parameters, it is clear that, conditional on μ and $Z(s)$, $log(\lambda(s))$ has a normal distribution:

$$log(\lambda(s)) | \mu, Z(s) \sim (\mu + Z(s), \sigma_\eta^2)$$

This is convenient for employing the estimation algorithm discussed later in this chapter.

To simplify presentation of estimation and prediction procedures given below, we will make use of the notation [•] to represent the "distribution of •." The symbol "|" means conditional upon and is read as "given." For clarity, we will neglect the convention of indexing quantities by s and instead use the more compact notation of subscripting the observations and corresponding random effects by i; in other words, $y(s_i) \equiv y_i$. Thus, we will denote the $N \times 1$ vector of observed BBS counts (on the N BBS routes) as $\mathbf{y} = (y_1, y_2,$

$. . . , y_N)$. Similarly, the corresponding Poisson mean vector, the vector of log-means, and the random effects are the $N \times 1$ vectors λ, $\mathbf{u} = \log(\lambda)$, and $\mathbf{Z}$, respectively.

The joint likelihood of all observations for our problem is then

$$[\mathbf{y}|\lambda] = \prod_{i=1}^N \text{Poisson } (\lambda_i) \qquad (56.2)$$

in other words, the product of N Poisson likelihoods. The joint distributions of $\mathbf{u} \equiv \log(\lambda)$ and $\mathbf{Z}$ are multivariate normal distributions:

$$[\mathbf{u}|\mu, \mathbf{Z}, \sigma_\eta^2] = MVN(\mu\mathbf{1} + \mathbf{Z}, \sigma_\eta^2) \qquad (56.3)$$

where $\mathbf{1}$ is an $N \times 1$ vector of ones, and $\mathbf{I}$ is the $N \times N$ identity matrix, and

$$[\mathbf{Z}|\theta, \sigma_z^2] = MVN(0, \sigma_z^2 \mathbf{K}_\theta) \qquad (56.4)$$

where $\mathbf{K}_\theta = Corr (\mathbf{Z}, \mathbf{Z})$.

Note that the Z-process does not depend on sand so σ_z^2 does not appear after the "|" in Equation 56.4. Similarly, conditional on $\mathbf{Z}$, $\mathbf{u}$ in Equation 56.3 is independent of σ_z^2 and θ.

Our goals under this model are *estimation* of the parameters, μ, θ, σ_z^2, σ_η^2 and perhaps the vector of route effects at the data locations, $\mathbf{Z}$. Of more interest is *prediction* of values of $Z(s)$ on a grid, say $\mathbf{Z}_g$ (a vector), and estimating appropriate variances for these predictions. The vector of predictions of $\mathbf{Z}_g$, plotted with respect to their spatial attribution, forms a map depicting spatial variation in relative abundance, albeit on the log scale. We may also be interested in the expected count at location s; in other words, $E[\lambda(s)] = E[e^{\mu + Z(s) + \eta(s)}]$, which may be more meaningful for interpretation. One implication of the Poisson model (and, indeed, for most non-normal models), is that closed-form expressions for parameter estimates and predictions are not obtainable. Instead, solutions must be obtained numerically, or by simulation, which is the approach that we take in the following section.

Finally, although we have presented the model in the absence of habitat covariate information, we would generally recommend that important habitat attributes be conditioned on (that is, incorporated into the μ term as discussed above) where they are available.

Model Fitting and Prediction by Markov Chain Monte Carlo

The Poisson mixed model described above is a special case of a generalized linear mixed model—that is, a model that contains both *fixed* and *random* terms. Under such models, traditional estimation is difficult. Since the random effects in the model are essentially high-dimensional nuisance parameters, the classical approach is to integrate them from the conditional likelihood, producing a simplified unconditional likelihood that is only a function of parameters μ, θ, σ_z^2 and σ_η^2. The resulting *n*-fold integration problem is analytically intractable and must be tackled using numerical or simulation-based integration. We do this below, though in a more formal Bayesian framework. Strictly speaking, there are no fixed effects in a Bayesian analysis. Instead, we use this term to distinguish between those parameters in the model that are given flat priors and those that are related through a prior distribution. Traditional prediction is equally difficult, requiring the numerical calculation of conditional expectations such as $E[Z(s)|y]$. In a Bayesian setting, this quantity is just the posterior mean of $Z(s)$ and thus is simple to compute using techniques more common in Bayesian analysis, described below. We adopt several Markov chain Monte Carlo (MCMC) techniques for sampling from the relevant posterior distribution. Before discussing the MCMC algorithm, we provide a Bayesian formulation of our mapping problem.

Bayesian Formulation of the Mapping Problem

The heart of the Poisson mixed model is specified by the three probability distributions (Eqn. 56.2–56.4). To complete the model specification, prior distributions are required for the parameters. We assume that the parameters are independent of one another and express the joint prior distribution as the product of the prior distributions for each parameter: $[\mu, \theta, \sigma_\eta^2, \sigma_z^2] = [\mu][\theta][\sigma_\eta^2][\sigma_z^2]$. In our analysis, uniform priors were assigned to μ and θ so as to reflect little prior information about them. Specifically, [μ] was taken to be constant (a so-called improper uniform prior) and

[θ] was taken to be uniform on [0,1000], a proper uniform prior. The implication of choosing a uniform prior for a parameter is that the resulting estimates are essentially maximum likelihood estimates. In the case of the uniform prior on θ, there is some ambiguity in restricting the range to be on [0,1000] since, strictly speaking, any positive value is possible. This was a choice made more for pragmatic reasons having to do with the simulation-based estimation of parameters. We felt this range of values to be a reasonable range of probable values for this parameter. Indeed, the estimates were well below the 1,000 upper limit. See Appendix for further discussion. We parameterized the variance components in terms of precision: $\tau_\eta = 1/\sigma_\eta^2$ and $\tau_z = 1/\sigma_z^2$. Diffuse gamma priors were assigned to both of these precision parameters: $\tau_\eta = Gamma(a_\eta, b_\eta)$ and $\tau_z = Gamma(a_z, b_z)$. Parameters of prior distributions are known as hyperparameters. In our case the hyperparameters are a_η, b_η, a_z, and b_z. Often, their values are fixed in the analysis, for example at values that indicate virtually no prior knowledge of the value of τ_η and τ_z. Discussion of this can be found in the Appendix.

Bayesian analysis focuses on analysis of appropriate posterior distributions—that is, the conditional distribution of the unknown quantities (e.g., parameters, predictands) given the data. Typically, one is interested in marginal posterior distributions of a particular unknown. For example, the marginal posterior of the random effect at a location, say $Z(s)$, is simply the conditional distribution $[Z(s)|y]$. Generally, it is this distribution upon which one would conduct inference regarding $Z(s)$. For example, the Bayes estimate is the mean of this posterior distribution, the posterior variance quantifies uncertainty, and so forth.

Unfortunately, in our problem, these posterior distributions are analytically intractable in the sense that they are not of a "known," convenient form. However, to within a normalizing constant, the joint posterior distribution is merely the product of the likelihood (Poisson) and prior distributions (multivariate normals, gammas, and uniforms). For our spatial model of BBS data, the joint posterior (of all unknowns in the model) is

$$[u, Z, \mu, \theta, \tau_\eta, \tau_z | y] \propto [y|u][u|\mu, Z, \tau_\eta][Z|\theta, \tau_z][\mu, \theta, \tau_\eta \tau_z] \quad (56.5)$$

Although this distribution cannot be analyzed directly, simple algorithms exist that allow us to draw samples from distributions that are specified up to a normalizing constant. Our general strategy for analysis of this model then, is to draw a large number of samples from the joint posterior distribution given by Equation 56.5 and then to estimate the quantities of interest from the simulated values. For example, to estimate μ, we sample a large number of values from Equation 56.5, isolate those simulated values of μ, say $\mu^{(1)}$, $\mu^{(2)}$, . . ., and then use these values to estimate interesting features of the marginal posterior of μ (e.g., the mean and variance). The general method for accomplishing this simulation task is known as Markov chain Monte Carlo; we describe this in the following section. Since our Poisson model is similar to that of Diggle et al. (1998), the interested reader is referred to their paper for further details on MCMC-based estimation.

Markov Chain Monte Carlo

Markov chain Monte Carlo is a collection of methods for simulating from complicated multivariate distributions. The idea behind MCMC is to simulate a random walk through the parameter space that converges to the target distribution (that is, the distribution of the samples converges to the target distribution). Formally, we simulate the sequence θ^t, $t = 1, 2 \ldots$ by starting at some point θ^0 and then, for each t, drawing θ^t from a transition distribution, say $P_t(\theta^t | \theta^{t-1})$ that depends on the previous value and that may vary from one iteration to the next, hence the subscript t on P. The key is construction of the transition probability distributions so that the Markov chain converges to a stationary distribution that is the desired target distribution. Upon convergence of the chain, say at iteration T, subsequent samples θ^{T+1}, θ^{T+2}, ... are correlated observations from the appropriate distribution and relevant quantities of interest (e.g., means, variances) may be computed from them. In the present context, we use MCMC to generate samples from the posterior distribution in Equation 56.5. Although MCMC is relatively "new" in statistical science, the literature on MCMC techniques is vast, and even a simple presentation of the details is beyond the scope of this chapter. The interested reader is referred to Gelman et al. (1995) and Gilks et al. (1996) for recent expositions on the subject.

In our application, we use Gibbs sampling (Gelfand et al. 1990; Zeger and Karim 1991; Casella and George 1992) and the Metropolis-Hastings algorithm (Chib and Greenberg 1995). Detailed discussion of both methods, in addition to background and other related material, can also be found in the texts by Gelman et al. (1995) and Gilks et al. (1996). In Gibbs sampling, the transition distributions are chosen to be the full conditional distributions of each parameter or vector of parameters. The full conditional distribution of a parameter is defined as the conditional distribution of that parameter given all other quantities in the model. These are typically easy to construct and are nominally proportional to the product of the likelihood, or the component model in which the parameter appears, and the prior distribution for that parameter. The key advantage of Gibbs sampling is that the full conditional distributions are typically of simple form and can often be sampled directly. However, sometimes they too are only known up to a normalizing constant. In this case, one must rely on other methods to sample from them. A common method, and that which we used where necessary, is the Metropolis-Hastings algorithm, which requires sampling from an approximating distribution.

In theory, samples from the full conditional distributions produce a Markov chain whose stationary distribution is the target posterior. Posterior quantities of interest are then computed from the resulting simulated data after convergence is judged to have occurred. There are many issues that one must address in applying MCMC algorithms, including selecting approximating distributions for the Metropolis-Hastings algorithm, assessing convergence of the chain, and so forth. For details, interested reader may consult Gilks et al. (1996).

We present our MCMC algorithm in the Appendix. This is of sufficient detail so as to permit the interested reader to fit our model, but technical details as to how the algorithm is derived are omitted. The interested reader is referred to Gilks (1996) for discussion of this.

Results

In our analysis, the Markov chain was run for fifty thousand iterations at which point convergence was judged to have occurred. We then ran the chain for another ten thousand iterations, sampling Z_g every fifth iteration in order to reduce serial correlation in the simulated values. Estimates of parameters other than Z_g were based on all ten thousand "converged" samples.

To illustrate what "output" from an MCMC-based analysis looks like, the simulated values for the structural parameters of the model, μ, θ, τ_z, and τ_η, are shown in Figure 56.1. Each point depicted on these plots (which are connected by lines) represents a single simulated value of the relevant parameter (and thus there are ten thousand points). The corresponding posterior distribution estimates, formed by computing a histogram of the simulated values, are shown in Figure 56.2. The simulation history plots are indicative of fairly rapid convergence, the values fluctuating within a narrow range of their mean, with no slowly varying departures or rapid shifts in the mean. Parameter estimates were computed as the means of the marginal posterior distributions, and error estimates from the standard deviations of the marginal posteriors. Posterior means and standard deviations of μ, θ, τ_z, and τ_η were $\hat{\mu} = 2.96$ (0.0605), $\hat{\theta} = 310.2$ (169.6), $\hat{\tau}_z = 1.03$ (0.414), and $\hat{\tau}_\eta = 4.47$ (1.413). Notably, the estimate of θ is very imprecise. This is not surprising given the relatively small sample size ($N = 87$) for a spatial statistical problem.

Inference, Prediction Uncertainty, and Sampling Design

Inference within the context of a Bayesian model is always relative to the posterior distribution of the quantity of interest. For example, if one were interested in inference concerning μ in our model, then one can construct a 95 percent credible interval (a Bayesian "confidence interval") from the simulated values by locating the 2.5 and 97.5 percentiles. The resulting interval is (2.85, 3.094). Of course, a point estimate is the posterior mean (2.96), and an assessment of variation is the posterior variance. Other interesting quantities are easily estimated from the simulated values, for example, the median, mode, and arbitrary percentiles. For a model containing habitat covariates, one would likely be interested in carrying out inference regarding the effect of those habitat covariates.

The general strategy is the same when interest centers on assessment of prediction uncertainty. In this case, the marginal posterior distribution of each point at which a prediction is desired can be used to obtain a measure of uncertainty of the prediction at that point. We illustrate this while at the same time showing the implication of spatial correlation in the model. Figure 56.3 (top) shows the locations of two grid points; one, labeled "B" is located relatively far from sample locations, and thus we expect less-accurate prediction of its route effect, $Z(s_B)$. The other, labeled "G" is located very near several sample locations, and so we expect more accurate prediction of its route effect, $Z(s_G)$. Figure 56.3 (bottom) shows the estimated marginal posterior distributions for these two route effects. The posterior for the first point is noticeably more diffuse than that of the latter. The posterior standard deviation of point B was 0.375, whereas the posterior standard deviation of point G was 0.258. One could also obtain credible intervals for these predictions, and other interesting quantities, from the values simulated from the posterior distribution. Ecologists may be interested in other types of inference problems. For example, comparison of averages over geographic strata formed by watershed boundaries, land-use considerations, and the like. Such comparisons can be made by examining the posterior distribution of the quantity of interest (an areal average).

Of course, our ultimate aim with this work was to produce a map depicting the relative abundance of birds over space while at the same time qualifying that map with an assessment of uncertainty. A map of the predicted route effects (the marginal posterior means) on the 1,200 grid points is shown in Figure 56.4 (top) and the corresponding prediction standard error map (the marginal posterior standard deviations) is shown in Figure 56.4 (bottom) (see color section). Maps of the expected route count predictions, $\lambda(s) = e^{u(s)}$ and associated standard errors are shown in Figure 56.5 (see color section). Of note here is that the standard errors of the expected count exhibit a strong relationship with the mean, as is expected under a Poisson model. Under many traditional methods of mapping data such as these, assessment of uncertainty is difficult, and often ad hoc at best. We feel that the ability to do this within a rigorous statistical framework is the primary benefit of the Poisson model that we have proposed.

Finally, the model also provides a framework for

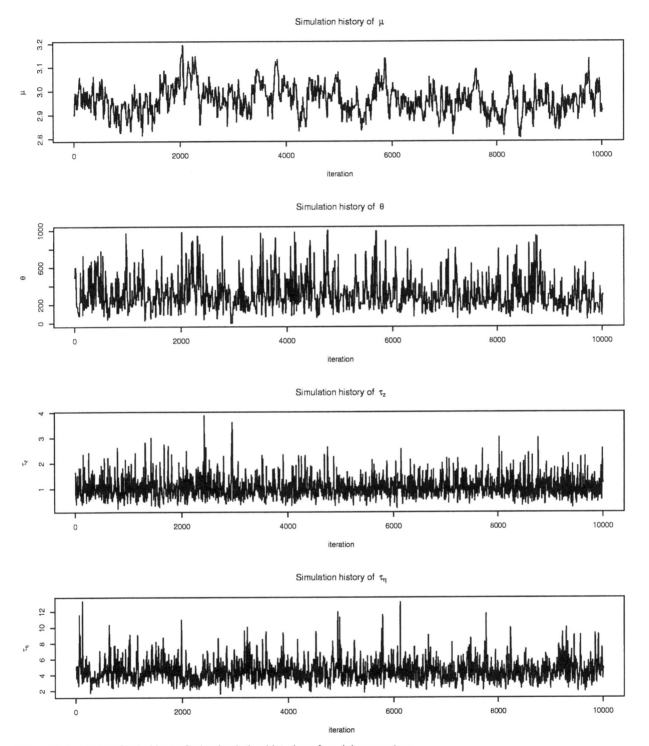

Figure 56.1. Markov Chain Monte Carlo simulation histories of model parameters.

evaluating issues of spatial sample allocation, allowing for sampling design in order to better estimate abundance at specific locations or over regions of particular interest. For example, in Figure 56.4, we observe regions of relative high prediction error, and other regions of relatively low prediction error. Intuitively, such information can be used in redesign of the sampling plan by moving clustered routes into regions of high prediction error. More formally, one can use the estimated covariance structure of the spatial process

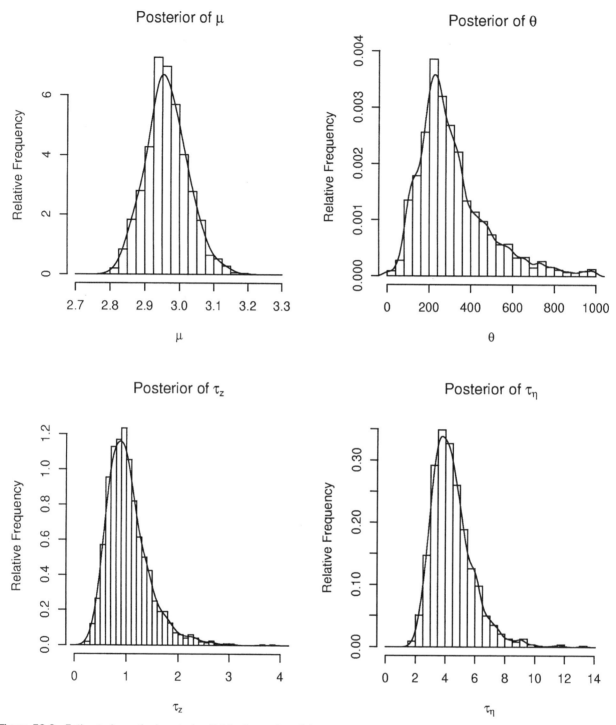

Figure 56.2. Estimated marginal posterior distributions of model parameters.

to address design issues. For example, identification of that subset of routes which, when deleted, or moved, produces the best change in average prediction variance of the route effects. This is a relatively straightforward optimization exercise. Such problems are well studied in the context of air pollution monitoring networks (see Cox et al. [1995] and Nychka and Saltzman [1998] for reviews), yet this work has not been adopted in the development of sampling plans for ecological processes, to the best of our knowledge.

Variation in Prediction Precision

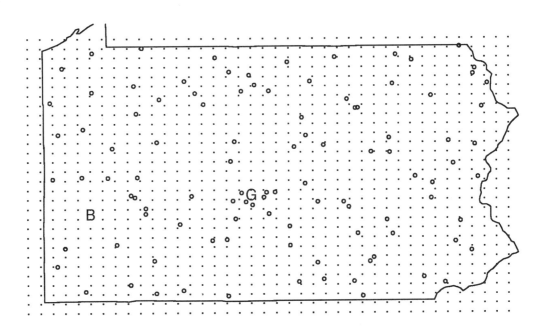

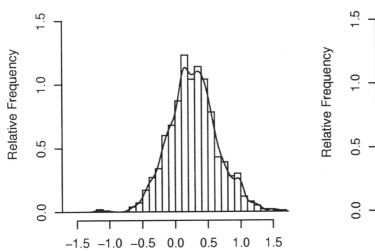

Posterior of prediction at point B

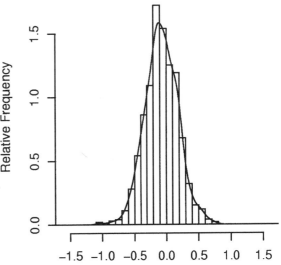

Posterior of prediction at point G

Figure 56.3. Top: BBS route locations and prediction grid, indicating a well (point G) and poorly (point B) predicted grid point. Bottom: Posterior distributions of predication at points labeled G and B.

Discussion and Further Investigations

The Poisson model presented here represents an advance over inverse-distance and kriging-based approaches to spatial summary of count data in that an appropriate statistical model is fit to the data using realistic distributional assumptions. Using this model, estimates with appropriate standard errors can be used to evaluate questions about spatial differences in relative abundance, inference regarding model parameters and predictions, and sampling design issues. Fitting and prediction under the Poisson model was based on Markov chain Monte Carlo methods. These methods, while computationally intensive, provide a powerful tool for fitting complicated models, and we expect that this approach will have great applicability for a variety of ecological modeling exercises.

The final model (Figs. 56.4 and 56.5) provides a reasonable spatial smooth of mourning dove relative abundance in Pennsylvania. Spatial patterns exist in Pennsylvania mourning dove populations, with lowest abundances in the north-central part of the state, and higher abundances in the southeast and west-central parts of the state. These patterns appear to be closely associated with the distribution of forest cover in the state (see Fig. 56.6 in color section), with high dove relative abundance being associated with low forest cover, although we have not yet addressed these issues formally within the context of our model. These patterns raise an important question. Namely, what is the genesis of spatial correlation in BBS data? Clearly, habitat plays a major role and we feel that models that address mapping and other prediction problems should accommodate habitat information when possible. The Poisson model is easily extended to include information on habitat covariates. The large variety of remotely sensed databases on habitat (e.g., Jones et al. 1997 for Pennsylvania) greatly facilitate these sorts of analyses, and we are presently working to include habitat variables from remotely sensed data in our analyses of BBS abundances. However, since habitat structure is generally sampled infrequently in time, is measured with error, and is often ambiguously defined, it leads to many interesting and complex statistical problems that we hope to address in extensions of this work. Nevertheless, one could potentially condition on the "right" set of habitat variables, thereby diminishing the spatial correlation remaining to be modeled. But, since spatial structure from other sources may exist (such as general patterns in abundance within a species' range), conditioning on habitat might not entirely account for all of the spatial variation that is present. We also computed predictions under a model lacking observer variation. The impact of neglecting observer variation was obvious; "rougher" maps with more local maxima and minima were produced, since the model did not have the ability to smooth unusually small and large observations. Although these results were interesting insofar that they lend support for the need to accommodate observer variation, space does not permit their presentation here.

Of course, the BBS is an ongoing survey, and data exist for over thirty years of surveys on BBS routes in Pennsylvania. The temporal structure of the BBS can also be included in a Poisson model, although incorporating the limitations of the BBS design poses interesting problems in most analyses of population change (e.g., Link and Sauer 1998). For example, it has been documented that temporal patterns exist in observer quality, and hence this covariate must be accommodated in analyses of temporal change in counts. Any analysis of count survey data must accommodate factors that influence detectability of birds among BBS routes, and simple analyses that do not include these factors can lead to biased estimates of spatial and temporal change in populations (e.g., Link and Sauer 1997b).

Finally, this general modeling approach extends easily to other non-normal data. For example, in mapping the range of a bird species using presence/absence (or atlas) data, one might assume that data follow a Bernoulli distribution where the probability of occurrence is spatially correlated. The same idea can be applied to spatially indexed binomial counts, as Diggle et al. (1998) illustrated. Presence/absence mapping may also be handled using the autologistic model (as in Klute, Chapter 27), but the latter requires that the model be defined with respect to a lattice, so care must be taken when dealing with data collected in continuous space.

Acknowledgments

The authors thank Jane Fallon and Ian Thomas for GIS support in digitizing BBS routes and summarizing habitat data along routes; an anonymous reviewer for many helpful comments; Patricia J. Heglund and J. Michael Scott for finding space for us to present this work at the presymposium workshop, and the many BBS participants for collecting the data used in our analyses.

Appendix: The MCMC Algorithm

In the following description of the MCMC algorithm that we employed, the notation $[x| \bullet]$ denotes the full conditional of the parameter x, and superscripts index the iteration; for example, $\theta^{(t)}$ is the value of at iteration t. Before beginning the MCMC simulation, initial values must be chosen for all parameters. There is not a general rule of thumb for choosing initial values; however estimates based on traditional methods perform well when they can be computed. For example, initial values for $\mathbf{Z}$ might be computed by smoothing the log-transformed data whereas the initial value for μ can be based on a GLM fit. The reader should keep in mind that this algorithm generates a sequence of samples from the joint posterior, say, $\{(\mathbf{u}^{(t)}, \mathbf{Z}^{(t)}, \mu^{(t)}, \ldots) : t = 1, 2, \ldots, M\}$. These samples can then be used to estimate important features of the posterior distribution, as illustrated in Discussion and Further Readings.

Each iteration of the MCMC algorithm consists of the following six steps (comments and explanation follow):

Step (1) For $i = 1, 2, \ldots N$ sample from $[u_i| \bullet] \propto [y_i| u_i][u_i| \mu, Z_i, \tau_\eta]$ as follows:

(1a) Generate $u_i^* \sim N (u_i^{(t-1)}, \delta_1^2)$ and compute the ratio

$$r = \frac{[y_i|u_i^*][u_i^*|\mu^{(t-1)}, Z_i^{(t-1)}, \tau_\eta^{(t-1)}]}{[y_i|u_i^{(t-1)}][u_i^{(t-1)}|\mu^{(t-1)}, Z_i^{(t-1)}, \tau_\eta^{(t-1)}]}$$

(1b) Set $u_i^{(t)} = u_i^*$ with probability $\min(r, 1)$

(1c) Otherwise, set $u_i^{(t)} = u_i^{(t-1)}$

Step (2) Sample $\mathbf{Z}^{(t)}$ directly from the full conditional:

$$Z^{(t)} \sim MVN\big((\tau_\eta^{(t-1)}\mathbf{I} + \tau_z^{(t-1)}\mathbf{K}^{-1})^{-1} \tau_\eta^{(t-1)}(\mathbf{u}^{(t)} - \mu^{(t-1)}),$$
$$(\tau_\eta^{(t-1)}\mathbf{I} + \tau_\eta^{(t-1)}\mathbf{K}^{-1})^{-1}\big)$$

Step (3) Sample $\theta^{(t)}$ from $[\theta | \bullet] \propto [\mathbf{Z} | \theta, \tau_z]$ as follows:

(3a) Generate $\theta^* \sim DU(0,1000)$ and compute the ratio

$$r = \frac{[Z^{(T)}|\theta^*, \tau_z^{(t-1)}]}{[Z^{(t)}|\theta^{(t-1)}, \tau_z^{(t-1)}]}$$

(3b) Set $\theta^{(t)} = \theta^*$ with probability $\min(r,1)$

(3c) Otherwise, set $\theta^{(t)} = \theta^{(t-1)}$

Step (4) Sample $\mu^{(t)}$ directly from its full conditional distribution:

$$\mu^{(t)} \sim Normal\left(\overline{u}^{(t)}, \frac{1}{N\tau_\eta^{(t-1)}}\right)$$

where $\overline{u}^{(t)} = (1/N) \sum u_i^{(t)}$ is the mean of the u_i values generated in step (1) above.

Step (5) Sample τ_z directly from its full conditional

$$[\tau_z| \bullet] = Gamma\left(a_z + (N/2), \frac{1}{(1/b_z) + (1/2)\mathbf{Z}'\mathbf{K}^{-1}\mathbf{Z}}\right)$$

Step (6) Sample τ_η directly from its full conditional

$$[\tau_\eta| \bullet]=$$
$$Gamma\left(a_\eta + (N/2), \frac{1}{(1/b_\eta) + \Sigma(1/2)(u_i^{(t)} - (\mu^{(t)} + Z_i^{(t)})^2}\right)$$

Comments

The matrix $\mathbf{K}$ depends explicitly on θ, though this dependence has been suppressed. As a consequence, as θ is updated, so must $\mathbf{K}$ be. The notation DU(0,1000) indicates that candidate values of θ were drawn from a *discrete uniform* distribution on [0,1000]. Because simulating from the full conditional of θ requires computing both the determinant of $\mathbf{K}$ and its inverse, and since $\mathbf{K}$ depends on θ, performing this computation is very computationally demanding. Therefore, the determinants and inverses were computed for 100 values on the interval [0,1000] and stored for subsequent use, thereby avoiding on-the-fly computation at every iteration of the Gibbs sampler. Although this induces some loss of precision in estimating θ, we do not feel this to be of significant enough concern to warrant a more precise candidate distribution, such as continuous uniform on the same interval.

The mean and variance of a Gamma(a,b) distribution are ab and ab^2, respectively. In our analysis, we set $a_z = a_\eta = 1/10$ and $b_z = b_\eta = 10$ so that the priors on t_z and τ_η both had mean 1 and variance 10, which are highly diffuse priors for the variance components, expressing little prior information about them.

The parameters δ_1^2 and δ_2^2 of Steps (1a) and (4) are merely *tuning* parameters of the MCMC algorithm; that is, they have no effect on the estimates obtained, but only on the manner in which they are obtained. Large values ensure that the parameter space is explored more rapidly, at the expense of rejecting a higher proportion of the draws. Conversely, small values lead to less rapid movement about the parameter space but a higher acceptance rate of the draws. Of course, "large" and "small" are relative to the distribution of mass in the posterior, and for our analysis $\delta_1^2 = 1$ and $\delta_2^2 = 1/8$ appeared to provide a reasonable compromise between exploration of the parameter space and acceptance rate.

To compute the prediction of Z_g, the above algorithm is run until convergence at which point (following Diggle et al. [1998]) we add the following step to the algorithm:

Step (7) Sample $Z_g^{(t)}$ directly from the full conditional:

$$[Z_g | Z, \theta, \sigma_z^2] = MVN(S_1 K^{-1} Z, S_2 - S_1 K^{-1} S_{1'})$$

where and $S_1 = \text{Cov}^*(Z_g, Z)$ and $S_2 = \text{Var}(Z_g)$. Also, it may be desirable to estimate the vector λ_g, the Poisson mean at the grid of prediction points. To do this, we include the following step into the algorithm:

Step (8) Sample $u_g^{(t)}$ directly from:

$$[\mathbf{u}_g | \mu, Z, \theta, \sigma_z^2] = MVN\left(\mu^{(t)} + Z_g^{(t)}, \frac{1}{\tau_\eta^{(t)}} I\right).$$

These simulated values are then exponentiated to obtain the sample of λ_g. This simulation is then run until enough samples have been drawn to achieve precise estimates of the model parameters and predictions.

Sampling from the full conditional in Step (7) is, in general, the computationally limiting stage of the MCMC simulation since it requires manipulating the multivariate normal variance-covariance matrix, which is of dimension $N_g \times N_g$ where N_g is the number of grid points at which predictions are desired (in our Mourning Dove illustration, $N_g = 1,200$). Since successive iterates of the Markov chain tend to be highly correlated, and since Step (7) is computationally expensive, it is recommended that this step be performed only every few iterations. We sampled Z_g every 5th iteration, as in Diggle et al. (1998), but for problems such as ours in which N is relatively small compared with N_g, less frequent sampling may be more efficient. As a consequence, draws of Z_g will be much less correlated, and fewer samples will be required to produce an adequate estimate.

Influence of Selected Environmental Variables on GIS-habitat Models Used for Gap Analysis

Carlos Gonzalez-Rebeles, Bruce C. Thompson, and Fred C. Bryant

Gap analysis is a geographic approach to assessing biodiversity distribution and its conservation status based on a series of digital information layers that are combined and analyzed in a geographic information system (GIS). The analysis requires the following data sets: (1) present distribution of land cover (produced from the classification of satellite imagery together with ancillary information), (2) distributions for vertebrate species (predicted from knowledge of their habitat associations and the spatial representation of those habitats), and (3) maps of land stewardship (to differentiate management status relative to conservation potential). Within the GIS, distribution of plant and animal species are analyzed relative to location of areas devoted to conservation. Sites with selected species or significant vegetation communities not adequately represented by current conservation systems will constitute "gaps" requiring priority attention. A detailed description of gap analysis concepts and specific procedures are presented in Scott et al. (1993) and Gap Analysis Program (1998).

Modeling wildlife species distributions is a key element in the gap analysis process and a variety of other conservation endeavors (Morrison et al. 1998). The definition of priority areas and their actual mapping is based on the combination of all individual species distribution estimates for a project area (Butterfield et al. 1994). Wildlife distribution predictions, within gap

analysis context, are mainly based on two types of information: location data (geographic unit or point data) and species associations with land cover (habitat indicator) (Butterfield et al. 1994; Csuti and Crist 1998). Within the GIS, species are identified as being present in unique polygons where the geographic location characteristics and the preferred land cover overlap. The process selectively filters out unsuitable habitat from the original coarse distribution and adds other potential habitat sites by extrapolating known distributions to the boundaries set by suitable land-cover associations (Scott et al. 1993; Scott and Jennings 1998).

However, these types of "basic" models tend to overestimate distributions (Stoms and Estes 1993; Stoms 1994). Thus, different environmental variables based on specific habitat characteristics (e.g., soil, elevation, temperature, slope) also are included with the basic model as "filters" to provide further detail for the distribution estimates. It is expected that the addition of each variable will result in a cumulative restriction in the spatial distribution of wildlife species such that an "adjustment" is made to the model (Butterfield et al. 1994; Csuti and Crist 1998). It is assumed that a different, more simplified distribution model would produce a different but less-accurate estimate if fewer or different variables were used. Notwithstanding, it is impossible to know *a priori* how each variable will be spatially represented in the habitat map or about its relationships with other variables. Some of these variables may be spatially correlated and may not actually provide additional information.

Our research addressed questions related to the efficacy of using these filter variables in the model (i.e., if their use produced a difference, and the level of this influence was determined). Our work was based on the premise that GIS-habitat models used for gap analysis are structured from a basic model of habitat associations represented by land cover and location data (base variables) and that additional sets of different environmental variables (filter variables) are used to adjust the distribution estimate. For the analysis, we used New Mexico Gap Analysis Project (NMGAP) predictive models for animal distribution. Objectives attempted for this research were (1) to quantify the level of response (spatial changes) reflected by altered distribution estimates to different levels of model perturbation (systematic reduction of filter variables) as an indirect measure of the value of the information contributed by including different combinations of filter variables in the distribution models, and (2) to examine the response patterns relative to differences in model types (number of filter variables used).

Methods

Data to be used in the analysis were obtained from preliminary NMGAP vertebrate distribution models. They integrated a species-habitat relationship database for a total of 584 species (26 amphibians, 96 reptiles, 324 birds, and 138 mammals); see NMGAP Final Report (Thompson et al. 1996). By the time the study was developed, distribution models were defined from nine major habitat association categories. Base variables always included in all models were (1) counties where the species is known to occur, (2) presence by county within watershed limits, and (3) land-cover types with which species are associated. Depending on the species, different combinations were applied for the remaining six filter variables: (1) soil associations, (2) elevation limits (minimum and maximum), (3) association with aquatic features, (4) slope affinities, (5) temperature limits, and (6) distribution by mountain ranges. Number and type of these filter variables could

coincide or differ among species, but specific descriptions of habitat elements represented by variable categories were unique for each of the species.

Our sample was representative of the different types of NMGAP distribution models based on how they were structured by the number of filter variables included and not specifically directed to represent any taxa in particular. We did not consider special modeling cases defined by an aquatic variable (i.e., amphibians, marsh birds, shore birds, and waterfowl models) or preliminary models formed only by base variables. From the remaining 423 models, we selected a stratified random sample to represent three basic model type groups. The groups were composed of base variables and either one, two, or three filter variables. An additional restriction was applied to proportionally represent models for species with widespread and restricted distributions (distinction defined in spatial extent relative to the size of New Mexico: less than one-fourth or more than one-fourth up to approximately one-half of state's surface).

The final sample contained a total of thirty-one habitat models representing different species of mammals, birds, and reptiles subdivided into three groups: Group 1 had twelve distribution models with only one filter variable, Group 2 had twelve distribution models with sets of two filter variables, and Group 3 had seven distribution models with sets of three filter variables.

Data were generated following NMGAP modeling procedures (see Gonzalez-Rebeles 1996 and Thompson et al. 1996). Models were perturbed by systematically removing filter variables one at a time and also by combined sets (two or three at a time) to complete all possible permutations depending on the number of filter variables contained by model type (Groups 1, 2, and 3). By nature of the modeling process, changes produced from removing variables were expected to be in one direction, progressively expanding the distribution estimates (i.e., loss of detail). Thus, size of the response measured was considered indirectly to be an indicator of the amount of information, the adjustment value, provided by the variables tested (removed). Every time a filter variable or set of filter variables was removed from the model, other filter variables in the model (if present) were maintained.

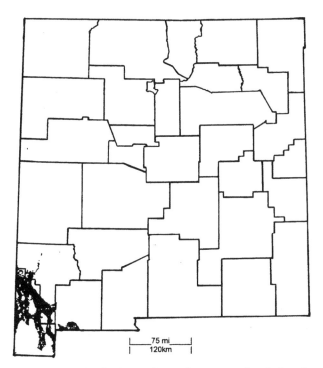

Figure 57.1. Distribution estimate for western banded gecko (*Coleonyx variegatus*) using the full set of model variables (Source: New Mexico Gap Analysis Project, NM-CFWRU).

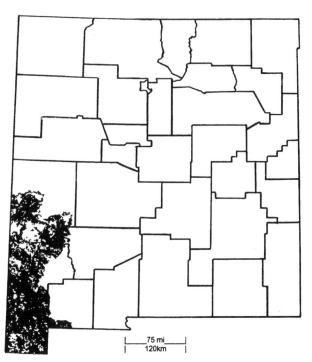

Figure 57.2. Distribution estimate for western banded gecko (*Coleonyx variegatus*) after removing elevation (filter variable) from the model (Source: New Mexico Gap Analysis Project, NM-CFWRU).

Base variables were always maintained. Response measured from altered distribution estimates was the percent difference in total area (square kilometers) of species distribution covered by the estimate as compared to original distributions with the full complement of filter variables plus base variables (Figs. 57.1 and 57.2).

Response to model perturbation was examined across all thirty-one models by types of filter variables removed (individually or combined sets). For these initial tests, the effect of removing filter variables was evaluated directly without considering a potential influence by the presence or absence of other filter variables in the model.

To quantify the level of information contributed by filter variables, we examined the degree of change produced from reduction of the individual filter variables with the least effect. Statistical significance was defined by comparing to a constant value defined as a 5 percent threshold of change. The magnitude for this threshold value was an arbitrary value chosen to set a minimum limit beyond which we considered that a filter variable contributed substan-

tive information (i.e., a meaningful degree of model adjustment).

Finally, the effect by type of filter variable removed (individually), relative to the presence or absence of other filter variables within the same model, was also evaluated. For this test, the responses from initial model perturbations were used to selectively subtract the effect of every other filter variable from a particular combination. This was done to isolate their individual effects when removed from the model and to estimate variations in their effect relative to a potential presence or absence of the other filter variables when combined in the model (depending on model types: Groups 1, 2, or 3). Three kinds of potential combination cases were evaluated: (1) removal of one filter variable given no other is present (R[X]), (2) removal of one filter variable given another is present (R[X|Y] and R[Y|X], and (3) removal of one filter variable given two others are present (R[X|YZ], R[Y|XZ], R[Z|YX]). In all cases, base variables were present in the models. For a complete description of methods followed for the whole analysis, see Gonzalez-Rebeles (1996).

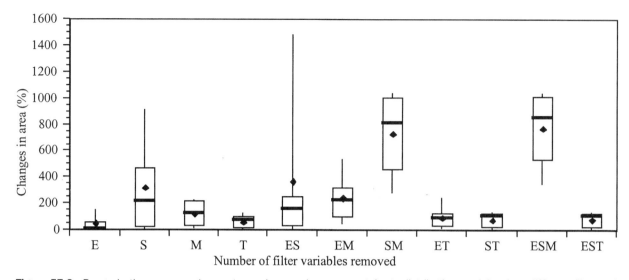

Figure 57.3. Range in the response (percent area increase) across vertebrate-distribution models when different filter variables were removed (E = elevation, S = soil, T = temperature, and M = mountain range). Boxes represent the interquartile range (50 percent of sample data found within top and lower quartile), top whiskers extend down from the ninetieth percentile (top decile) and bottom whiskers extend up from the tenth percentile (bottom decile), horizontal lines represent median and diamond symbols the mean. Letter n indicates number of models evaluated for each variable or combination set across the thirty-one models, when n = 27 (E), n = 15 (S), n = 8 (M), n = 7 (T), n = 11 (ES), n = 8 (EM), n = 4 (SM), n = 7 (ET), n = 4 (ST), n = 4 (ESM), and n = 3 (EST).

Results

Response to the removal of different types of filter variables (individually or combined) across all thirty-one species distribution models was highly variable, between extremes of zero to very high values (see interquartile ranges, Fig. 57.3). Median of the response across the distribution models showed that some filter variables or combinations were more influential than others were. When single filter variables were removed, soil was the most influential (216 percent increase, n = 15), while elevation was the least influential (11 percent area increase, n = 27). Mountain range and temperature showed an intermediate influence (129 and 79 percent, n = 8 and n = 7, respectively). When sets of two were removed (i.e., across models in Groups 2 and 3), soil and mountain range produced the most difference (810 percent, n = 4), whereas elevation and temperature were the least influential (90 percent, n = 7). When models permitted the removal of sets of three filter variables combined (i.e., across Group 3), the most influential combination was the set of variables elevation-soil-mountain range (856 percent increase, n = 4) (Fig. 57.3).

Generally, we observed that response values of larger magnitude occurred after perturbing models that represented species with an original small distribution area (predicted from the full set of variables). In contrast, response values of smaller magnitude were observed after perturbing models representing species with larger original distribution areas (Fig. 57.4). For example, when a single filter variable was removed, the largest response observed among all models was from the desert pocket mouse (*Chaetodipus penicillatus*) distribution model, a 1,355.4 percent area increase (Table 57.1). The original distribution predicted for this species (with the full set of variables) covered an area of approximately 4,811 square kilometers. By contrast, considering species for which removal of one filter variable produced no response, such as American kestrel (*Falco sparverius*), Lincoln's sparrow (*Melospiza lincolnii*), green-tailed towhee (*Pipilo chlorurus*), and dark-eyed junco (*Junco hyemalis*), their original distribution estimates covered 139,268 square kilometers, 104,119 square kilometers, 102,726 square kilometers, and 227,241 square kilometers, respectively. These areas were all much larger relative to the area of the original distribution from the first species cited (Table 57.1 and Fig. 57.4).

TABLE 57.1.

List of all individual response values observed (percent area increase) organized by level of model perturbation (removal of different single filter variables, or by different sets of two or three).

Distribution models

No.	Range	Spp code	Change in area (%) by removal of different filter variables						
Group 1			**1 removed**						
1.	(R)	Piab	3.7						
2.	(W)	Caga	57.6						
3.	(W)	Vuma	73.0						
4.	(R)	Peta	160.2						
5.	(R)	Meur	14.0						
6.	(W)	Coco	93.2						
7.	(W)	Thel	3.8						
8.	(W)	Lage	66.1						
9.	(W)	Chin	266.5						
10.	(R)	Typa	916.4						
11.	(R)	Thum	360.7						
12.	(R)	Pepe	520.6						
Group 2			**1 removed**			**2 removed**			
1.	(R)	Cova	430.2	509.7		1442.9			
2.	(W)	Came	3.3	215.9		226.6			
3.	(R)	Chpe	24.7	1355.4		1546.6			
4.	(W)	Locu	1.0	130.0		132.9			
5.	(R)	Casi	11.4	79.1		270.7			
6.	(W)	Sipy	14.3	27.4		45.7			
7.	(W)	Stma	4.7	11.6		36.3			
8.	(W)	Fasp	0.0	89.6		89.6			
9.	(R)	Synu	1.1	217.4		248.7			
10.	(W)	Vece	46.5	123.5		600.9			
11.	(R)	Clga	5.4	32.4		39.6			
12.	(R)	Lale	2.1	211.8		390.7			
Group 3			**1 removed**			**2 removed**			**3 Removed**
1.	(R)	Scpo	1.7	159.3	226.0	163.5	227.4	1,036.3	1,042.8
2.	(R)	Ocpr	3.4	29.3	179.8	46.2	186.3	278.8	345.7
3.	(R)	Crle	13.5	96.0	135.3	121.5	159.8	640.2	721.5
4.	(R)	Sona	2.0	223.1	246.2	224.2	260.0	980.2	990.5
5.	(W)	Meli	0.0	2.4	103.1	2.4	103.1	106.5	106.5
6.	(W)	Pich	0.0	3.1	129.4	3.1	129.9	133.7	134.2
7.	(W)	Juhy	0.0	1.2	1.7	1.2	1.7	2.9	2.9

Notes: Model groups 1, 2, and 3 represent the number of filter variables in each model.

Different response values of some models (groups 2 and 3) at same perturbation level were organized by increasing magnitude (across row) within each model.

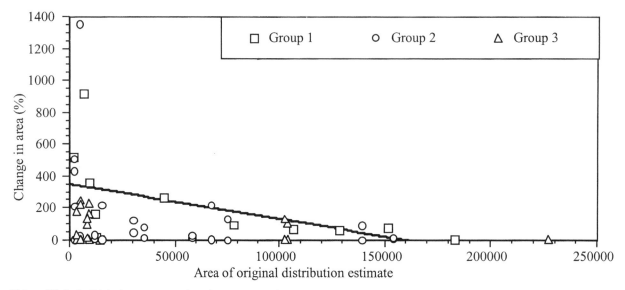

Figure 57.4. Individual response values (percent area increase) observed when one filter variable was removed by model in relation to the original distribution predicted for each species (with full set of variables). Where for Group 1, n = 12 (single filter variable removed from twelve models); for Group 2, n = 24 (one of two filter variables removed at a time from twelve models); and for Group 3, n = 21 (one of three filter variables removed at a time from seven models).

Similar tendency was observed when two or three filter variables were removed. See Gonzalez-Rebeles (1996) for full details.

Considering the lowest response values observed from the removal of single-filter variables across the thirty-one different animal distribution models, values ranged between extremes of null (American kestrel, Lincoln's sparrow, green-tailed towhee, and dark-eyed junco) to a 916 percent change in area (lesser prairie-chicken, *Tympanuchus pallidicinctus*) (see first column of data in Table 57.1). However, most of the response observed across models at this minimum perturbation level presented values nearer to the lower end of this range. Nineteen of the thirty-one models were less than a 15 percent change in area (Table 57.1, first column).

Median of the response across all models (n = 31) when the least-influential filter variable was removed was an 11 percent increase in area, which did not differ (p > 0.05) (Median test; Iman and Conover 1989) from the 5 percent threshold value set as the minimum acceptable level of change. Median response across each individual model type group (Groups 1, 2, or 3) when the least-influential variable was removed showed that only Group 1 (83 percent area increase, n = 12) was different (p < 0.05) from the 5 percent

threshold value, whereas the median responses estimated across Group 2 and Group 3 (5 and 2 percent, n = 12 and n = 7, respectively) were not different (p > 0.05) from the 5 percent threshold (Median test; Iman and Conover 1989). At other levels of model perturbation (i.e., removal of other single-filter variables or by sets of two or three combined), we observed much-higher response values that consequently were statistically different from the 5 percent threshold. Group 3 was an exception, where due to the combination of small sample size (n = 7) and extreme range in responses (3 to 1,048 percent area increase), the test was not sufficiently powerful to detect differences (see Gonzalez-Rebeles 1996).

Effects of individual variables differed relative to the presence or absence of other filter variables in the particular models. Results by type of filter variable removed for three potential combinations (R[X], or R[X|Y] and R[Y|X], or R[X|YZ], R[Y|XZ], and R[Z|YX]), was highly variable (between extremes of null to very high response) (Fig. 57.5). Median of the response values, by type of filter variable, was larger for mountain range (401 percent increase in area, n = 8), when it was considered the only filter variable in the model and it was removed. The next-largest response was observed when soil was removed (361 per-

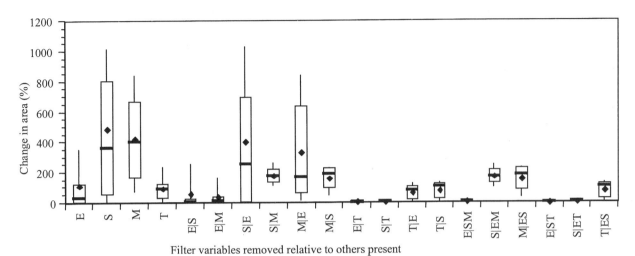

Figure 57.5. Range in the response (percent area increase) across vertebrate-distribution models when single filter variables were removed relative to others present in model (E = elevation, S = soil, T = temperature, and M = mountain range). Boxes represent the interquartile range (50 percent of sample data found within top and lower quartile), top whiskers extend down from the ninetieth percentile (top decile) and bottom whiskers extend up from the tenth percentile (bottom decile), horizontal lines represent median, and diamond symbols represent the mean. Letter n refers to number of models evaluated for each potential case depending on combinations that their filter variables permitted, when n = 27 (E), n = 15 (S), n = 8 (M), n = 7 (T), n = 11 (E|S), n = 8 (E|M), n = 11 (S|E), n = 4 (S|M), n = 8 (M|E), n = 4 (M|S), n = 7 (E|T), n = 3 (S|T), n = 3 (T|E), n = 3 (T|S), n = 4 (E|SM), n = 4 (S|EM), n = 4 (M|ES), n = 3 (E|ST), n = 3 (S|ET), n = 3 (T|ES).

cent increase, n = 15), while elevation produced the least response (31 percent increase, n = 27). Temperature produced an intermediate effect (90 percent increase in area, n = 7). Median of the response, when other filter variables were considered, showed a similar relationship in size of the effects among the different filter variables removed. We observed higher values when soil was removed given the presence of elevation (241 percent increase in area, n = 11) and when mountain range was removed given the presence of soil (184 percent increase, n = 4). The exceptions were two cases with very low response: (1) when soil was removed given the presence of temperature (2 percent increase, n = 3), and (2) when elevation and temperature were combined (2 percent increase, n = 3). We observed intermediate values with the removal of temperature for all potential combinations of other filter variables (1 or 2) considered present in respective models. Lowest response values were observed when elevation was removed (Fig. 57.5).

We observed a variation in the effect with filter variables present individually versus when the variable was removed given the number of other filter variables (1 or 2) were present in the model. For example, when soil was considered the only filter variable in models

in which it was used and removed, median response across models produced a 361 percent increase in area (n = 15). When the soil variable was removed given the presence of elevation or mountain-range variables, median response across corresponding models was a 250 percent (n = 11) and 172 percent increase (n = 4), respectively. When the soil variable was removed, given the presence of both elevation and mountain-range variables, median response was 170 percent (n = 4) (Fig. 57.5).

After polling all response values across each potential combination case by the number of filter variables present or not in the model (R[1], R[1|1], and R [1|2]), median of the response by combination case showed a general decline for cases when a larger number of filter variables were present (i.e., 93 percent [n = 57], 26 percent [n = 66], and 13.5 percent [n = 21] increases in area, respectively). Considering only the data estimated from Group 2, the removal of single filter variables when no other was present, R(1), and removal of single filter variables when another is present, R(1|1), were different (p < 0.05) (Sign test; Conover 1980). Median response estimated by removal case was a 155 percent increase (n = 24) and a 30 percent increase (n = 24), respectively. Considering the data from Group 3 as well, contrast between removal cases R(1), R(1|1), and R(1|2) showed differences (p < 0.05) among all three cases (Friedman test; Conover 1980). Median response by removal case was 81 percent (n = 21), 25 percent (n = 42), and 13 percent (n = 21), respectively. Consistency of these results with the previous analysis (contrasted with a 5 percent threshold) provides further indication of potential interaction when adding more than one filter variable.

This analysis of the effect by filter variable relative to the presence or absence of other variables provided insights useful for identifying interacting variables and combinations. For example, in comparing Figure 57.3 with Figure 57.5 it can be seen that the largest median response was produced when the variable set elevation/soil/mountain range were removed in combination, followed in magnitude by the removal of the variable set soil/mountain range (Fig. 57.3). This pattern suggested the fewest problems of correlation when these variables were used in combination. Similar results (high response) were observed when the effects of the variables soil and mountain range were estimated relative to the potential presence of respective variables with each other (i.e., S|M, M|S) and with elevation (i.e., S|E, M|E, S|EM, and M|ES) (Fig. 57.5). This pattern further indicated that these variables produced less correlation whenever they were combined.

Conversely, the combination sets of filter variables assumed to present more problems with correlation among themselves were elevation/temperature, soil/temperature, and elevation/soil/temperature (Fig. 57.3). Considering the effect of the variable soil relative to the presence of the variable temperature, or of elevation and temperature combined, we found extremely low response values, indicating a high degree of spatial correlation between them (Fig. 57.5). The same was observed with elevation in the presence of temperature, or soil and temperature (Fig. 57.5). However, the effect of temperature relative to the presence of elevation, soil, or elevation and soil combined, median response suggested no correlation in general of temperature with these two variables (Fig. 57.5). This in turn, suggested that the response observed from perturbation cases when the sets soil/temperature and elevation/soil/temperature were removed (Fig. 57.3) was mainly produced by the sole effect of

temperature, given that the other variables were correlated with this. Full data sets and analyses have been described elsewhere (Gonzalez-Rebeles 1996).

Discussion

Modeling vertebrate distributions is subject to uncertainty, with potential errors in sources of information, the relationships defined, or model structure (Marcot et al. 1983; Morrison et al. 1998). Factors and interactions that determine habitat suitability for wildlife species are generally not well known. Parameters used are biased toward those easily measured and for which the species are assumed to be most responsive (Marcot et al. 1983; Schamberger and O'Neil 1986; Verner et al. 1986a). Additionally, sources of error, how they are transferred through the modeling process, and how they are expressed in the final products are difficult to determine. This is especially true for GIS-habitat models (such as those used for gap analysis) that combine spatial or nonspatial information from different sources through several operations of overlaying and map transformations (Lodwick et al. 1990; Stoms et al. 1992; Edwards et al. 1995).

Our research reported here should not be considered an attempt to validate NMGAP vertebrate distribution models. Model validation involves intensive field sampling procedures that are expensive and time consuming. For a review of issues about habitat-model uncertainties and validation theory, see Marcot et al. (1983), Verner et al. (1986b), and Morrison et al. (1998). See also Karl et al. (Chapter 51), Schaefer and Krohn (Chapter 36), and Fielding (Chapter 21) for specific discussions on validation problems related to detection and interpretation between apparent and actual errors. Model reliability can also be assessed through sensitivity analysis, when other validation procedures are not practical or possible (Heinen and Lyon 1989; Lodwick et al. 1990; Lyon et al. 1987; Stoms 1992; Stoms et al. 1992). Our work involved a "sensitivity analysis;" however, it was not addressed in the traditional sense of evaluating model robustness (or specific sensitivities) to controlled alteration of model parameters or modification of relationships among model elements (Grant 1986) but instead evaluated the effect of excluding different information layers (see Lodwick et al. 1990 for a description of the different approaches for geographic analyses). For this specific case, we strove to evaluate the effect of filter variables used for distribution models as an indirect measure of their contribution.

Animal distribution models in NMGAP are based on extensive literature review and expert consultation. From the information available on species-habitat associations, only those variables that can be represented spatially (digitized) or considered appropriate to characterize species habitats are considered for modeling (Thompson et al. 1996). Based on systematic selection of variables and expert review at different stages, we expected the modeling approach would produce reasonably accurate distribution estimates (at the desired landscape scale and assuming a correct model). The effect measured from the different combination of filter variables was considered as the level of "additional adjustment" provided to the "basic model" (i.e., within known occurrences and land-cover associations). In our example using the western banded gecko (*Coleonyx variegatus*) (Figs. 57.1 and 57.2), the final model entailed a 430 percent change in area versus removal of elevation, yet it accurately predicted all thirteen known field locations for the species (Degenhardt et al. 1996). In this case, the contribution of elevation combined with base variables and the other filter variable in the model (land cover, presence by county within watershed limits, and soil), provided the expected refinement and adjustment (assuming that known associations for base and filter variables are correct). Another example is that of the distribution model for desert pocket mouse. The removal of the variables soil and elevation produced 1,355.4 percent (maximum response when a single filter variable is removed, see Results above) and 24.7 percent of change in area respectively, indirectly meaning it resulted in a final model that included thirty of thirty-one locations reported by Findley et al. (1975).

Results should be interpreted relative to the efficacy of using those different combinations of variables. All filter variables included in a model should produce an effect to be useful, and this is expected whether one, two, or three filter variables are used. The contrary would indicate some sort of interaction among variables that consequently could be interpreted as the variable or combination not being useful. Our comparisons among model groups were intended to identify combinations with potential problems (interactions) and the pattern of response was examined for insights about whether the model performed as expected. Nevertheless, real direction of this effect (correct adjustment) and the optimum combination of filter variables can only be evaluated from field verification.

For example, white-tailed ptarmigan (*Lagopus leucurus*) and the dwarf shrew (*Sorex nanus*) have been associated with montane habitats (Bailey 1928; Ligon 1961; Findley et al. 1975). However, the first species modeled (with two filter variables) showed a large response when elevation was removed (212 percent). Response was almost null to the removal of mountain range (2 percent). By contrast, the dwarf shrew model (with three filter variables) showed large responses when soil (246 percent) or mountain range (223 percent) was removed. In this case, the least response was observed when elevation was removed (2 percent). For the first species, the smaller effect of mountain range was potentially due to some sort of correlation of this variable with elevation and/or other base variables. It is probable that limits defining the mountain-range polygons, used to predict its distribution, were located at lower elevations than those represented by the actual elevation variable used (possibly nested within mountain-range polygons). In the case of the dwarf shrew model, the combination of two filter variables—suitable soil associations within mountain-range polygons—were the ones that appeared to control this estimate relative to elevation.

Traditionally, various parameters have been considered biologically significant for spatial analysis of the landscape, such as area measurements, fractal dimension, and indices to assess shape, contiguity, and dispersion patterns (O'Neill et al. 1988b; Turner et al. 1989b; LaGro 1991). We did not prepare maps for all perturbation levels applied to all species, thus restricting some spatial analyses. Area is considered a significant measure because it directly reflects the degree of spatial alteration produced from the original distribution patch predicted (full set of variables) when a perturbation is applied to the model (removal of filter variables).

However, not much is known about the size of the effects that different filter variables contribute to distribution models. Initial gap analysis research evaluated the effect of modifying different variables using several trial models for different species (Scott et al. 1993; Butterfield et al. 1994; Csuti and Crist 1998). In particular, Butterfield et al. (1994) reported how the use of filter variables improves distribution estimates for different taxa. However, none of these research efforts measured the size of changes produced. Other studies examined the sensitivity of richness mapping to habitat generalization and errors in vegetation classification (Stoms 1991), to variations in the resolution of mapped habitats with fixed sampling unit sizes (Stoms 1992), and to variations in sampling unit size with fixed resolutions for the habitats mapped (Stoms 1994). In these cases Stoms (1991, 1992, 1994), simplified the models in the traditional sense of "sensitivity analysis" (as mentioned above) by altering information about base variables (not removing complete subsets of filter variables). In other words, the author assessed taxonomic and resolution sensitivities (see Lodwick et al. 1990) but did not quantify levels of change in area of the estimates. Thus, there is little published information for comparative evaluation of our findings.

Magnitudes of the measured responses indicated some relationship with size of the original distribution estimates (not perturbed model) of the species sampled. This was a function of the arithmetic relationship established by calculating proportions of change, which is unavoidable in this kind of data. However, in addition to the arithmetic relationship, the effect of interest (by filter variable, relative to combination used and species modeled) was still considered measurable by the fact that both large and small values were observed for species models with similar original distribution size (with exception at the extremes of larger distributions) (Fig. 57.4). For example, with one variable removed, maximum response (1,355 percent) observed was for a species (desert pocket mouse) whose original distribution was small (4,811 square kilometers). At the other extreme, two species that also had relatively small distributions, the white-tailed ptarmigan (2,326 square kilometers) and the dwarf shrew (5,325 square kilo-

meters), presented very small responses (2 percent) when the mountain range and elevation variables, respectively, were removed.

The large variability in the response of the models to perturbation was expected because of the nature of model structure and function. A variety of potential interrelationships may occur by combining different filter variables, each corresponding to particular environmental elements (represented spatially as specific polygons) that defined the habitat associations. However, our results indicated that some filter variables were more influential than others across the models.

Our selection of a 5 percent threshold in area change to evaluate the level of contribution from filter variables can be questioned, but the 5 percent threshold has some biological significance in conservation planning considering the size of the original vertebrate distribution estimates (complete set of variables). Sizes ranged from a minimum area of 1,984 square kilometers for the white-ankled mouse (*Peromyscus pectoralis*) to a maximum of 227,241 square kilometers for the dark-eyed junco. Five percent from each of these areas will then represent values ranging from 99 to 11,365 square kilometers. The minimum reserve size required to maintain a viable population for small mammals is estimated to be from 10 to 100 square kilometers and for larger mammals from 10,000 to 100,000 square kilometers (Schonewald-Cox et al. 1983). This means that 5 percent of the smallest distribution (1,984 square kilometers) represents an area that is approximately the minimum reserve size required for small mammals (upper limit) and 5 percent from the largest distribution (227,241 square kilometers) represents the minimum size required for large animals (lower limit).

However, the 5 percent threshold was an arbitrary value determined to set a minimum limit to evaluate if filter variables were useful. It could have been set at any level, depending on the efficacy attempted and taking into consideration typical trade-offs between efficiency and the costs inherent to the project. (Each variable added would represent a new thematic layer in the GIS that would require time and cost for development and processing.) Filter variables producing a response value equal to or below this threshold were assumed to contribute insufficient information to improve the model. By contrast, all levels of response above the threshold were assumed to indicate an adequate contribution.

Specific weights of relevance between differences in magnitudes of response above this level (5 percent) were not considered further because of the arithmetic relationship between the magnitude of response and area of distribution, as discussed before. Thus, contribution was evaluated by differentiating between levels of response versus no response (above or below the threshold). This 5 percent threshold was useful for determining that some of the filter variables did not contribute valuable information for some models (assumed adjustment) and that this problem mainly occurred for model types formed by two- or three-filter variables, due to potential spatial correlation.

A relevant issue for modeling is that of balancing the cost and effort of model building with the quality of the output. Stoms (1991, 1992) reported that these types of GIS-habitat models are not robust enough in regard to variations in the information about habitat (as represented by land cover, a base variable), reflected as variations of species richness mapping. Our results further indicated a limit to the usefulness of filter variables when applied in increasing numbers, suggesting that special care be taken regarding the quality of information about habitat associations and the selection of filter variables.

Conclusions

Results indicated that most filter variables presented adequate levels of influence when used in the models (required adjustment), although some were more influential than others. Our findings also suggested that special care must be taken when applying more than two filter variables due to potential spatial correlation as the number of filter variables increase in the model. However, model developers will not know at the outset which of the variables they plan to use will be best and therefore cannot remove less-influential variables beforehand. There is no way to know *a priori* how these variables will interact spatially. Due to the nature of this modeling approach, it

is clear that all filter variables combined produce a cumulative effect by model (e.g., see the effect of combining two- and three-filter variables in Table 57.1). However, individual effects are difficult to predict and will depend on the particular spatial extent each one depicts.

A practical recommendation is to apply a test similar to this type of sensitivity analysis to evaluation of distribution models at the stage when the preliminary maps are being reviewed by experts. This simple method will provide enlightening information by detecting all potential types of models and variables for which major problems may arise. For example, in the case of NMGAP, the variable elevation, followed by temperature, showed the most correlation problems, as did models with three filter variables. Temperature was among the filter variables eliminated from final vertebrate distribution modeling and maps production (Thompson et al. 1996). For a different study area, problem detection will be concurrent with the spatial representation of habitat features in that particular region and model specific characteristics.

We undertook our research in the context of gap analysis, but we evaluated a standard modeling approach that is used in many species modeling endeavors. Our results provide quantitative evidence for the influence of filter variable combinations. This insight can be useful to other researchers, serving as a reference about the potential effects of the variables used for their own distribution estimates. Our intent was not to estimate absolute accuracy of such modeling, but to illustrate the relative efficacy and efficiency inherent in using filter variables in such predictive modeling. We hope our experiences with trying to understand predictive modeling of animal distribution for biodiversity can aid in developing the philosophy of biodiversity in conservation as encouraged by Callicott et al. (1999).

Acknowledgments

The study was developed by Carlos Gonzalez-Rebeles while under a scholarship from CONACYT, Mexico, and the Fulbright Program, IIE, USIA. Logistic support was provided by Texas Tech University (TTU).

Special appreciation goes to the New Mexico Cooperative Fish and Wildlife Research Unit, at New Mexico State University (NMSU) (especially Mary Ann Hughes and Robert A. Dietner), and the Geographic Applications Research Lab, Dept. of Geography, NMSU (especially David L. Garber), for permitting use of their data on wildlife species distributions and computing equipment respectively. New Mexico State University's Agricultural Experiment Station provided additional financial support. We thank Dr. William J. J. Conover from TTU for his advice during statistical analysis and Dr. J. Michael Scott for editorial advice.

Appendix.

Species distribution models sampled from New Mexico Gap Analysis Project. Model groups 1, 2, and 3 indicate the corresponding numbers of filter variables each include.

Range		Name	Base variables			Filter variables	
Group 1							
1.	(R)	Pipilo aberti	Cnty	LaCo	Elvt		
2.	(W)	Callipepla gambelii	Cnty/Wshd	LaCo	Elvt		
3.	(W)	Vulpes macrotis	Cnty/Wshd	LaCo	Elvt		
4.	(R)	Peucedramus taeniatus	Cnty/Wshd	LaCo	Elvt		
5.	(R)	Melanerpes uropygialis	Cnty/Wshd	LaCo	Elvt		
6.	(W)	Coluber constrictor	Cnty/Wshd	LaCo	Elvt		
7.	(W)	Thamnophis elegans	Cnty/Wshd	LaCo	Elvt		
8.	(W)	Lampropeltis getulus	Cnty/Wshd	LaCo	Elvt		
9.	(W)	Chaetodipus intermedius	Cnty/Wshd	LaCo	Soil		
10.	(R)	Tympanuchus pallidicinctus	Cnty/Wshd	LaCo	Soil		
11.	(R)	Thomomys umbrinus	Cnty/Wshd	LaCo	Soil		
12.	(R)	Peromyscus pectoralis	Cnty/Wshd	LaCo	Soil		
Group 2							
1.	(R)	Coleonyx variegatus	Cnty/Wshd	LaCo	Elvt	Soil	
2.	(W)	Catherpes mexicanus	Cnty/Wshd	LaCo	Elvt	Soil	
3.	(R)	Chaetodipus penicillatus	Cnty/Wshd	LaCo	Elvt	Soil	
4.	(W)	Loxia curvirostra	Cnty/Wshd	LaCo	Elvt	Soil	
5.	(R)	Cardinalis sinuatus	Cnty/Wshd	LaCo	Elvt	Temp	
6.	(W)	Sitta pygmaea	Cnty/Wshd	LaCo	Elvt	Temp	
7.	(W)	Sturnella magna	Cnty/Wshd	LaCo	Elvt	Temp	
8.	(W)	Falco sparverius	Cnty/Wshd	LaCo	Elvt	Temp	
9.	(R)	Sylvilagus nuttallii	Cnty	LaCo	Elvt	MntR	
10.	(W)	Vermivora celata	Cnty	LaCo	Elvt	MntR	
11.	(R)	Clethrionomys gapperi	Cnty	LaCo	Elvt	MntR	
12.	(R)	Lagopus leucurus	Cnty	LaCo	Elvt	MntR	
Group 3							
1.	(R)	Sceloporus poinsettii	Cnty	LaCo	Elvt	Soil	MntR
2.	(R)	Ochotona princeps	Cnty	LaCo	Elvt	Soil	MntR
3.	(R)	Crotalus lepidus	Cnty	LaCo	Elvt	Soil	MntR
4.	(R)	Sorex nanus	Cnty	LaCo	Elvt	Soil	MntR
5.	(W)	Melospiza lincolnii	Cnty/Wshd	LaCo	Elvt	Soil	Temp

Appendix. (Continued)

Species distribution models sampled from New Mexico Gap Analysis Project. Model groups 1, 2, and 3 indicate the corresponding numbers of filter variables each include.

Species			Habitat variable categories				
			Base variables			Filter variables	
Range		Name					
6.	(W)	*Pipilo chlorurus*	Cnty/Wshd	LaCo	Elvt	Soil	Temp
7.	(W)	*Junco hyemalis*	Cnty/Wshd	LaCo	Elvt	Soil	Temp

Variable key

Cnty	=	Species presence by county
Cnty/WShd	=	Presence by county extrapolated to limits within all intersected watersheds
LaCo	=	Land cover associations
Elvt	=	Association with elevation
Soil	=	Association with soil type
Temp	=	Association with temperature gradient
MntR	=	Association with mountain range

Species range

W	=	Widespread
R	=	Restricted

A Distribution Model for the Eurasian Lynx (*Lynx lynx*) in the Jura Mountains, Switzerland

Fridolin Zimmermann and Urs Breitenmoser

The lynx (*Lynx lynx*) populations of western and southern Europe disappeared during the eighteenth and nineteenth centuries as a consequence of direct persecution, alteration of the ecosystem (forest destruction and expansion of cultivated land), and excessive reduction of wild ungulates (Breitenmoser 1998). Since the end of the nineteenth century, forests have regenerated in many mountainous region of Europe, and the wild ungulate populations have recovered quickly. This improvement in the ecological conditions also inspired the idea to bring back large predators. Lynx were re-introduced to the Swiss Alps and the Swiss Jura Mountains in the 1970s (Breitenmoser et al. 1998). Although the Swiss reintroductions are considered to be rare examples of successful translocations of large predators (Yalden 1993), these small populations cannot yet be regarded as viable. In the Swiss Jura Mountains, only the southern half of the range is permanently occupied by lynx. The reasons for the lack of vitality are not known; they may include ecological, anthropogenic, and intrinsic (genetic) factors. However, habitat suitability analyses were never carried out for the Jura Mountains, although such a tool is recognized to be important for reintroduction programs.

The purpose of this study is to assess small-scale habitat variables and their importance to lynx recolonization of the whole Jura Mountains and to estimate available lynx habitat throughout the mountain range. We used a geographic information system (GIS) to determine if easily available spatial data can successfully describe lynx habitat and contribute to a predictive spatial model (see Guisan and Zimmermann [2000] for a review). The model was built using data from adult, resident lynx that were followed by means of radiotelemetry in the southern part of the Swiss Jura Mountains. We then extrapolated the model over the entire Swiss Jura Mountains and evaluated the reliability of the resulting maps using radio fixes from dispersing subadult lynx. Such a spatial model permits prediction of the future distribution and the potential size of the lynx population in the Jura Mountains and could be of use in drafting a lynx conservation plan for this mountain range.

Study Area

The study was performed in the Jura Mountains, a secondary limestone mountain chain forming the northwestern border of Switzerland with France (Fig. 58.1). The altitude varies from 372 meters (Lake of Geneva) to 1,679 meters (Mont Tendre). The main study area (680 square kilometers) was confined to the northern part of the canton of Vaud (VD). Lynx were also followed into the adjoining areas of the canton of Neuchâtel (NE) and into France; this total area is approximately 3,000 square kilometers. Deciduous forests

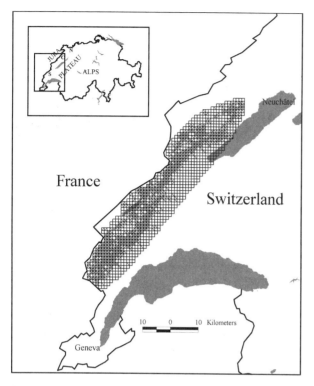

Figure 58.1. Situation of the study area in the Jura Mountains of Switzerland (small map) and France. The grid (large map) shows the 1,085 quadrats where the predictor and response variables have been sampled. Cells are hatched as follows: \ = cells visited by female lynx; / = visited by male lynx; X = visited by both sexes. In addition, S = southern, C = central, and N = northern part of the Swiss Jura Mountains.

along slopes and coniferous forests on the ridges cover 53 percent of the highlands. Cultivated areas are typically pastures. The human population reaches a density of 120 per square kilometer in most parts of the Jura Mountains, and people living on the Swiss Plateau use the highlands intensively for recreation. The center of the study area is crossed by two railways, a highway, and some additional roads with dense traffic. As in the Swiss Alps (Breitenmoser and Haller 1987; Haller 1992), Western roe deer (*Capreolus capreolus*), and chamois (*Rupicapra rupicapra*) are the main prey of lynx in the Jura Mountains (Jobin et al. 2000).

Method

From April 1988 to June 1998 a total of twenty-nine lynx were surveyed by means of radiotelemetry in the Jura Mountains (Breitenmoser et al. 1993; Swiss Lynx

Project unpublished data). Some of the radio-tagged lynx roamed into the French part of the mountain range. All analyses for this study were done using the radio fixes from the Swiss part of the study area, because environmental data for France were unavailable.

We used a total of 6,282 radio fixes of eleven resident lynx followed from 1988 to 1998 to generate the models on the assumption that these adult, territorial individuals (Breitenmoser et al. 1993) would occupy the best habitat. The sample unit was a 1×1-kilometer quadrat. The sampling area was restricted to all the quadrats intercepted by the minimum convex polygon (MCP), including all the fixes of the resident lynx. Quadrats falling within France were disregarded. A total of 1,085 quadrats remained for the analyses (Fig. 58.1). We split the data in two subsets. One was used for calibration of the model, and the other was used to evaluate the model predictions (split sample approach; see Guisan and Zimmermann 2000).

We compared the results from different sample sizes ranging from two hundred to one thousand quadrats for the calibration of the model in order to test the consistency (stability) of our model. Since all samples greater than three hundred quadrats produced the same parameters, we decided on a sample of four hundred quadrats to calibrate our model. The quadrats were chosen randomly with a distance constraint between them in order to reduce spatial autocorrelation. The remaining 685 quadrats were then used to evaluate the model.

The response variable is the presence/absence of lynx in each quadrat. Lynx were considered to be present in each quadrat containing one or more telemetry fixes. From this set of quadrats, three sets were prepared (Fig. 58.1), using radio fixes of (1) all lynx (females and males), (2) females only, and (3) males only.

The eighteen predictor variables (Table 58.1) were selected from among all statistical parameters available according to our empirical knowledge of the lynx's ecological requirements, but also with respect to their availability in digital form. A previous study of lynx recolonization of the Jura Mountains (Breitenmoser and Baettig 1992), based on random observations, had shown that the lynx distribution was *a priori* determined by the extension of the forest and human activity. Each of these factors can be described in terms of many concurrent environmental predictors and can be correlated to the lynx distribution. The environmental predictors came from the database of the Federal Office of Statistics and from the database of the Federal Office of Topography. Both databases had an accuracy of 1 hectare and were in digital form and ready to be analyzed using GIS ArcView software (ESRI 1996a,b,c). From the hectare information, we then computed a summary statistics to each square-kilometer quadrat: (1) the proportion of the different land-use predictors, and (2) the mean value in the case of the quantitative predictors(fringe length, elevation, declivity, human population density, exposure of the slope [predictors 1–18 in Table 58.1]).

General linear models (GLM; McCullagh and Nelder 1989; see Nicholls 1989) were used to select those predictors that best explained the presence/absence of lynx. All the analyses were computed in S-PLUS (MathSoft) according to the method described in Guisan et al. (1999). To facilitate the final ecological discussion of the model, we did not orthogonalize the predictors (e.g., through principal component analysis) prior to the model calibration. Predictors were only selected when they significantly contributed to the deviance reduction, as attested by a χ^2-test (p-value ≤ 0.05). In addition, we did not retain the predictors that explained less than one percent of the total deviance to avoid having predictors with few or no biological meaning appearing in the final model.

We used the receiver operating characteristic (ROC; see Fielding, Chapter 21), a threshold-independent measure of accuracy, to evaluate our models. An ROC plot is obtained by plotting the true positive proportion on the y-axis against the false positive proportion on the x-axis. The area under the ROC function (AUC) is usually taken as the index of performance because it provides a single measure of overall accuracy independent of any particular threshold in the training data (Fielding, Chapter 21). Final GLMs were fitted and evaluated using custom S-Plus functions (written by A. Guisan).

We compared the three lynx distribution maps by subtracting the computed probabilities of lynx presence for each quadrat in the GIS: (1) total (both sexes combined) minus female, (2) total minus male, and (3) female minus male. Values close to −1 or +1 indicate a high discrepancy between corresponding grid cells, whereas values close to 0 indicate a high conformity.

We extrapolated the resulting model over the entire Swiss Jura Mountains in the GIS. GLM models are readily implemented in a GIS by building a single formula in which each coefficient multiplies its related predictor variable (Guisan et al. 1999). The results of the calculations are obtained to the scale of the linear predictor so that the inverse logistic transformation is then necessary to obtain probability values between 0 and 1 at every quadrat of the grid. Finally, we evaluated the resulting models with the spatial behavior of dispersing subadult lynx.

Results

The proportion of deviance significantly explained (adj-D^2) in the models ranged from 0.39 to 0.44, corresponding to a medium fit of the models (both sexes

TABLE 58.1.

Sources of the eighteen predictors used in the logistic regression analysis.

Predictor[a]	Units	Sources[b]
Forest areas	ha/km^2	GEOSTAT
Other wooded areas	ha/km^2	GEOSTAT
Fringe length	meter	FOT
Horticulture, viticulture	ha/km^2	GEOSTAT
Arable land, meadows	ha/km^2	GEOSTAT
Pastures	ha/km^2	GEOSTAT
Pastures in mountain areas	ha/km^2	GEOSTAT
Lakes and rivers	ha/km^2	GEOSTAT
Nonproductive vegetation	ha/km^2	GEOSTAT
Areas without vegetation	ha/km^2	GEOSTAT
Built-up areas	ha/km^2	GEOSTAT
Rest areas, parks	ha/km^2	GEOSTAT
Roads and railways	ha/km^2	GEOSTAT
Elevation	meter	GEOSTAT
Slope	degree	GEOSTAT
Eastness	(cosinus)	GEOSTAT
Northness	(sinus)	GEOSTAT
Human population density	ind./ha	GEOSTAT

[a]Predictors had an accuracy of 1 hectare.
[b]GEOSTAT database of the Federal Office of Statistics and FOT = Vector 200 database of the Federal Office of Topography.

TABLE 58.2.

Results of the GLM analyses with the three different sets of response variables.

Response variables	Calibration			Evaluation
	GLM formulas	Proportion of explained variance	AUC	AUC
Presence/absence of lynx	elev^2 + slo + forest	0.435	0.89	0.88
Presence/absence of females	elev^2 + slo + forest	0.386	0.87	0.84
Presence/absence of males	elev + slo + forest + roads	0.423	0.90	0.89

pooled, females, males). The AUC at calibration and evaluation ranged between 0.84 and 0.90 (Table 58.2). Three out of eighteen predictors were selected in the final model when presence/absence data of both sexes were used. They were elevation (second-order polynomial; 20 percent of the deviance explained), slope (20.5 percent), and forest (3.4 percent). The same predictors were retained when presence/absence data from female lynx were used to build the model (elevation second-order polynomial 16.3 percent,

TABLE 58.3.

Survival of subadult lynx (F = females, M = males) according to their habitat use.

Model[a]	Lynx	Probability class[b]					Destiny
		1	2	3	4	5	
Both sexes pooled	F12	0.0	4.4	2.2	3.3	90.1	Survived
	M14	13.3	3.5	12.1	12.1	59.0	Survived
	F23	12.1	4.1	19.7	7.6	56.5	Survived
	F26	0.0	1.8	21.4	25.0	51.8	Survived
	F13	0.0	17.1	17.2	28.6	37.1	Died
	F17	0.0	0.0	11.1	66.7	22.2	Died
	M16	66.8	4.8	11.9	2.3	14.2	Died
Females	F12	0.0	2.3	4.3	6.6	86.8	Survived
	M14	8.1	6.9	6.9	14.5	63.6	Survived
	F23	3.6	5.8	10.3	8.5	71.8	Survived
	F26	0.0	2.4	7.8	38.0	51.8	Survived
	F13	8.6	5.7	20.0	37.1	28.6	Died
	F17	0.0	22.2	44.4	33.3	0.0	Died
	M16	61.9	7.1	16.7	2.4	11.9	Died
Males	F12	4.4	3.3	6.6	6.6	79.1	Survived
	M14	14.5	9.8	13.3	16.2	46.2	Survived
	F23	21.1	8.5	1.8	20.6	48.0	Survived
	F26	2.4	6.0	19.6	44.0	28.0	Survived
	F13	8.6	17.1	34.3	11.4	28.7	Died
	F17	11.2	22.2	44.4	22.2	0.0	Died
	M16	69.1	19.0	0.0	4.8	7.1	Died

[a]The percentage of radio fixes of the subadults during their first year of independence falling into the different lynx habitat probability categories is shown for each response variable set.
[b]Probability class: 1: 0–0.2; 2: 0.2–0.4; 3: 0.4–0.6; 4: 0.6–0.8; 5: 0.8–1.

slope 19.9 percent, and forest 3.4 percent, respectively). Four predictors were selected when we used presence/absence data from males. Here, elevation explained 17.4 percent, slope 20.2 percent, forest 4.2 percent, and roads 1.4 percent of the deviance.

The comparison of the resulting probabilities showed a high conformity between the three distribution maps. The subtractions of the probability values of most grid cells gave results close to zero. The differences between the probabilities of the 1,085 grid cells from the female versus the male distribution ranged from –0.34 to 0.33. More than 90 percent (978 cells), however, had values between –0.2 and 0.2. When subtracting the female lynx distribution probability map from the total map, all grid cells had a positive value less than 0.12. The differences for the male versus total map comparison ranged from –0.31 to 0.45, with 68 percent (735 cells) falling into the class from –0.2 to 0.2.

We then extrapolated the outcome of the three distribution probabilities over the entire Swiss Jura Mountains (see Fig. 58.2 in color section). The map of the potential distribution for males shows the most restrictive potential distribution, whereas the map for the females and for both sexes combined showed larger areas in the higher probability classes.

Maps of potential lynx distribution were based on the telemetry locations of resident lynx. As a supplementary evaluation of the models, we investigated the survival of young, dispersing lynx according to their habitat use. The lynx is a solitary, territorial species, and subadult lynx have to leave the parental home range at the age of about ten months (Breitenmoser et al. 1993). One can predict that subadult lynx can only establish a permanent home range if they find free space; otherwise, they would be driven into suboptimal habitat. Each subadult lynx revealed an individual fate, although the tendency observed was consistent with our habitat model: two subadult lynx (M14, F12 in Fig. 58.2) dispersed north from our study area into the still-unoccupied part of the Jura Mountains (Breitenmoser and Baettig 1992; Capt et al. 1998). They traveled along the corridors predicted from the habitat model (Fig. 58.2), and both settled in good-quality habitat (Table 58.3). F12 was poached one year after independence. Two other young lynx (F23, F26) were able to establish home ranges in

good-quality habitat inside the study area (Table 58.3). Both had taken over the home ranges of their mothers after the deaths of the latter. The subadult female (F17) was killed by a car during her dispersal. Finally, the locations of F13 and M16 showed a high share of suboptimal habitat (Table 58.3). Both lynx died from a natural death during the dispersal—F13 after she had left a temporary home range in marginal habitat (Fig. 58.2).

Discussion

Our models do not identify single variables but rather the combination of variables limiting lynx distribution. Different combinations of variables can result in the same probability of presence. Slope and elevation were the most powerful variables predicting lynx presence/absence in the three models. This is not so much typical for the lynx, which lives in a large part of its distribution area in lowland forests, but was for our study area, where forested areas are correlated with elevation and slope as a result of human activities. This observation underlines the local nature of our models and shows that the selected variables do not necessarily have a biological value for the species in question, as discussed by Guisan and Zimmermann (2000). Consequently, such models should only be applied to regions similar to those where the basic data were originally gathered. The human impact on carnivores is extremely difficult to evaluate, although today this is the main factor limiting their distribution (Boitani and Cuicci 1993; Mladenoff et al. 1995; Corsi et al. 1998). It is not a simple variable, nor can its distribution be easily mapped. In our model, we suppose that the human impact is included in other variables such as road density, human population density, or land use. Even in areas of generally good habitat, roads, which have a limited spatial extension and seem not to reduce the habitat quality considerably, can be a risk factor, as demonstrated by the fate of F17 (Fig. 58.2). Failure to incorporate such spot-like or linear, but critical, habitat features or ecological factors such as prey availability, competition, predation (Pearce et al., Chapter 32) and the like can lead to prediction errors. Data on number and distribution of roe deer and chamois, the main prey of lynx in the study area (Jobin et al. 2000), are presently not available in adequate form or precision to be incorporated into a habitat model. However, as ungulate distribution is habitat dependent, too, we can assume that the presence/absence approach of lynx at least partly reflects prey availability.

Most classifiers assume that class membership is known without errors (Fielding, Chapter 21). Lynx were not located in all favorable lynx zones within the study area, because peripheral spots of good habitat (1) might not be connected to the lynx zone, (2) might be occupied by neighboring lynx, or (3) surveillance density might have been insufficient. It is a shortcoming of our method that the defined categories (presence/absence) are not exclusive. Assuming absence of a species where it was actually present is a type II error that could be corrected with adequate sample size and monitoring duration to increase the power of the statistical evaluation (Morrison et al. 1998). To minimize this error, we restricted our sample area to the southern portion of Swiss Jura Mountains, where lynx were followed most intensively by telemetry.

The model provides a tool for the conservation and the management of the lynx in the Jura Mountains. An early study by Breitenmoser and Baettig (1992), based on random observations of lynx gathered from 1972 to 1987, revealed the discontinuous distribution of the lynx in the Jura Mountains and a lack of observations in the central part of the range (Figs. 58.1, 58.2). Our model confirms that the central part, especially for males, seems to be a suboptimal habitat (Fig. 58.2). So far, the AUC values from the evaluation as well as the anecdotal observations of dispersing subadult lynx seem to confirm the validity of the model for the Jura Mountains. None of the subadults settled in the central part of the Jura Mountains but continued on to adjacent areas (Figs. 58.1, 58.2). All subadult lynx dispersed through corridors predicted by the model. The final test for our model, however, will be the future spread of the lynx population through the northern part of the Swiss Jura Mountains. The model can predict the potential distribution of the lynx in the Jura Mountains and, when based on knowledge of the land tenure system of the resident lynx (Breitenmoser et al. 1993), allows estimation of the possible population size. Such knowledge will be

crucial for the conservation and management of this large carnivore species living in such close proximity to intensive human activities. Since large-carnivore populations are difficult to census over vast areas, a modeling approach based on high-quality, local data from telemetry may be more efficient. Decisions will have to be made in regard to the choice of the model and the threshold value. We prefer the model built from presence/absence data of both sexes, because this had the best fit (Table 58.2) and represents best the need of the population as a whole.

In conservation-oriented models, the overestimation of false-positive locations (the model predicts presence of a species when in fact it is absent) versus the overestimation of false negative locations includes different conservation risks (see also Fielding, Chapter 21). The balance between false positives and false negatives is defined through the threshold value and must be set according to the question to be answered. The lower the threshold value, the higher the percentage of all quadrats containing lynx fixes included, but also the higher the share of quadrats without any locations.

Another practical use of the model will be the evaluation of potential connections of the Jura population with neighboring lynx populations in the Alps or in the Vosges Mountains. For this purpose, however, we will have to expand our model into France and test its capability to predict corridors or barriers.

Acknowledgments

We thank A. Guisan from the Swiss Center for Faunal Cartography (CSCF) for his help with the GLM analyses in S-PLUS and for his valuable comments to our manuscript. We thank Ch. Breitenmoser-Würsten, S. Capt, A. Jobin, and P. Molinari for valuable discussions and—including many more unnamed colleagues—for gathering field data over the past ten years. We appreciated the GIS knowledge of U. Müller and the assistance in the analyses by J. Hausser and P. Vogel. We thank R. Mace and M. L. Morrison for critical comments and P. Jackson for help with the language. The study in the Jura Mountains was supported by the Federal Office of Environment, Forestry and Landscape (FOEFL), the canton of Vaud, the Swiss League for Protection of Nature (Pro Natura), WWF Switzerland, the Agassiz Foundation and, in the early phase, by the Swiss Foundation for Scientific Research (project no. 3.119-0.85). Sources of the digital geographical database were: highways and forest, VECTOR 200, Federal Office of Topography; rivers, BFS GEOSTAT, Federal Office of Topography; land use 1992–1997, BFS GEOSTAT; digital terrain model RIMINI, BFS GEOSTAT.

PART 5

Predicting Species: Populations and Productivity

Mapping a Chimera?

Edward O. Garton

Can we map abundance and productivity of a population from ecological-habitat maps of characteristics such as soil type, canopy cover, land use, and land-type morphology? Unfortunately, populations of animals are dynamic, changing, and even chimeric. Presence-absence maps oversimplify complex patterns of continually expanding and shrinking distributions. Perhaps a better approach would be to use our ecological maps as the basis for sampling to estimate population characteristics. The most extensive effort to estimate abundance and production for any population in the world is the continental waterfowl survey conducted annually by the U.S. Fish and Wildlife Service and Canadian Wildlife Service (1987). Biologists from federal, state, and provincial resource management agencies sample wetlands using aerial and ground counts of waterfowl on more than 3.6 million square kilometers of breeding habitat in Canada and the United States. These sample counts produce annual estimates of population size and production of various species of ducks (Smith 1995). This approach uses maps of wetlands to delineate geographic units of potential habitat and then sample the geographic units within species distributions. Finite sampling methods (Hayek and Buzas 1997) are applied to these samples to estimate the characteristics for the entire population. Another approach is a hybrid strategy that combines current sampling data with previous studies and ecological maps to model population characteristics. This

modeling approach is the most powerful approach to predicting population characteristics. A majority of chapters in this section take this modeling approach to predicting species-population characteristics.

Whether we choose to map, sample, or model population characteristics of species, we must immediately confront the issue of variability. Can we treat population characteristics as constant and make deterministic predictions, or must we take a probabilistic approach? Assuming constancy of populations is conceptually simple but unrealistic, and it makes validation with new data difficult. A better approach would be to embrace the variability by predicting probabilistically. There are two principal types of variability that we must address: (1) variation through time within spatial units, and (2) spatial variability among units.

Temporal Variation within Spatial Units

Populations within individual spatial units can vary substantially, both seasonally and yearly, due to genetic, demographic, and environmental variability (Shaffer 1981). Census data for such small populations are best described with means and variances of the characteristics such as abundance, survival rate, annual rate of change, and so forth, or by the trend or periodicity for populations showing long-term trends or cycles. At larger spatial extents, populations or metapopulations typically show less variation through time because genetic, demographic, and environmental variation of the subpopulations are uncorrelated or weakly correlated with each other over larger areas. These independent random deviations tend to cancel each other out over larger areas.

Variation through time threatens the long-term persistence of small populations, leading to a major focus on population viability analysis (PVA; Soulé 1987a,b). Roloff and Haufler (Chapter 60) propose a habitat-based PVA that hierarchically structures population modeling from individual to subpopulation to population level. They use this habitat-based PVA to predict population numbers, distributions, and probability of persistence for white-headed woodpeckers (*Picoides albolarvatus*) in an area subject to silvicultural treatments. In a similar way, Raphael and Holthausen (Chapter 62) model the effects of habitat management using a spatially explicit population model for the northern spotted owl (*Strix occidentalis caurina*). They used locally collected demographic information to model the results of proposed management choices, and, in the process, found several approaches superior to those planned originally by managers.

Variability between Spatial Units

Some of the differences between habitat patches (or larger-scale spatial units) in natural systems are obvious, such as size, shape, resource productivity, amount of edge, and distance to adjacent habitat patches. Other differences are more subtle, such as influence of adjacent matrices, abundance of predators and competitors, and connectivity to nearby source habitats. The only realistic approach possible is to estimate, through sampling or modeling, the mean and variance of characteristics of populations occupying particular spatial units. Sisk, Noon, and Hampton (Chapter 63) demonstrated the power of an effective area model that predicted spatially explicit animal density as a function of maps of habitat type, adjacent habitat, and distance from edge. This was done in a probabilistic manner, generating maps of animal abundance with an associated measure of certainty. Hunsaker, et al. (Chapter 61) utilized the variability between populations of California spotted owls to identify habitat

characteristics that were statistically different for successful, productive pairs.

Hierarchy of Spatial Units

The classic definition of a population as "a group of organisms of the same species occupying a particular space at a particular time" (Krebs 1994:151) is so general that it could describe everything from a single deme occupying a single patch of habitat to a metapopulation distributed across great regions containing enormous areas of unoccupied nonhabit in addition to areas of high-quality habitat occupied by dense concentrations of highly productive individuals. A better approach would be to delineate a series of hierarchical spatial units containing groupings of individuals significant for our understanding, estimation, and prediction of future population conditions. I suggest delineating five levels of spatial aggregation on the basis of demography, movement, geography, and genetics (see Fig. I5.1 in color section) as follows:

1. *Deme*: The smallest grouping of individuals showing random breeding (within the constraints of the species' social system) where it is reasonable to estimate birth, death, immigration, and emigration rates. Animals in this grouping are ideally distributed continuously in one patch of habitat, and their movements within this patch of habitat are restricted to home ranges for breeders during the breeding season. The size of this patch ideally would be related to the dispersal distance of juveniles or perhaps equal an area twenty to fifty times the size of a home range. For example, red-winged blackbirds (*Agelaius phoeniceus*) occupy territories variable in size but averaging 0.05 hectare for males in marsh habitat (n = 868 territories) and 0.3 hectare for males in upland habitat (n = 97 territories, Beletsky 1996:182), with each male territory holder guarding a harem averaging 3.3 females (n = 2389, Beletsky 1996:136). Males disperse an average distance of 1.4 kilometers from their natal nest to their first breeding territory and females disperse 1 kilometer on core marshes (Beletsky 1996:28). This suggests that demes of this species may cover only 3–5 square kilometers in areas such

as Columbia National Wildlife Refuge. This size area constituted Beletsky and Orians' (1996) core study area, and supported seventy to eighty male redwing territories each year (Beletsky 1996:13) and three times that many breeding females.

2. *Population*: A collection of demes with strong connections demographically (very high correlations), genetically, and through frequent dispersal. The population occupies a collection of patches of habitat, without large areas (relative to dispersal distance) of nonhabitat intervening. The area is typically less than one hundred times the size of an average home range and not larger than the dispersal distance of 95 percent of initial dispersers, but it may be much larger if the habitat patches are linear in shape or widely dispersed. For example, the maximum known dispersal distance for red-winged blackbirds from Beletsky and Orians' (1996) study at Columbia National Wildlife Refuge was 7 kilometers for males and 2.8 kilometers for females (Beletsky 1996:13). Delineating a redwing population to include all the patches of marsh habitat within 5 kilometers of their 3-square-kilometer study area would identify a population of approximately seven hundred male territory holders (Beletsky and Orians 1996:152) and 2,700 accompanying females occupying an area of patchy marsh habitat bordering lakes and streams distributed throughout a rolling desert grass/shrubland of 150 square kilometers (Fig. I5.1).

3. *Metapopulation*: A collection of populations sufficiently close together that dispersing individuals from source populations readily colonize empty habitat resulting from local population extinction. Populations within a single metapopulation may or may not show correlations in demographic rates but the low rates of dispersal are sufficient to maintain substantial genetic similarity. For example, red-winged blackbird populations distributed among the seven national wildlife refuges along 200 kilometers of the Columbia River in the south-central part of the state of Washington may constitute a metapopulation (Fig. I5.1).

4. *Subspecies*: A collection of metapopulations in a geographic region where very rare dispersals maintain genetic similarity but populations and meta-

populations occupy habitat patches that may be separated by enormous distances or by large areas of nonhabitat resulting in substantial demographic independence among metapopulations. The metapopulations of red-winged blackbirds occupying the intermountain region east of the Sierra Nevada and Cascade Mountains and west of the Rocky Mountains in southern British Columbia, eastern Washington, eastern Oregon, eastern California, and in Idaho and Nevada are together categorized as the subspecies *A. phoeniceus nevadensis* (Beletsky 1996:21).

5. *Species*: The collection of subspecies encompassing the entire distribution and geographic range of the species. The species may encompass substantial differences in phenotypes (habitat, physiology, behavior) and genotypes. For example, red-winged blackbirds breed extensively across North and Central America from east-central Alaska and the Yukon to Costa Rica and Cuba (Fig. I5.1). The twenty-two recognized subspecies vary in size, shape, and plumage, yet numerous genetic studies have repeatedly observed a remarkably high degree of genetic similarity and a lack of genetic differentiation among subspecies (Beletsky 1996:22). Dolbeer (1982) documented numerous movements between winter roosts by banded redwings of hundreds or thousands of kilometers, suggesting that even a small amount of straying from breeding areas due to males or females following flock members from winter roosts would lead to frequent genetic exchange among populations spread widely throughout the species distribution.

Theobald and Hobbs (Chapter 59) outlined an approach to habitat delineation, based on habitat quality, that incorporates functional relationships of species to resources, environmental factors (e.g., elevation, aspect), and disturbance. They propose delineation at three of the scales identified above based on allometric relationships: individual scale based on foraging behavior, population scale based on minimum viable populations, and metapopulation scale dependent on dispersal. They argue that if such an approach to delineation of spatial aggregates proves successful for specific animal species, it would provide a very useful tool for identifying appropriate patches for purposes of prediction and estimation.

Delineation of spatial units appropriate for each of the five levels in this hierarchy would lead to dramatic improvements in our understanding of the processes driving population changes and in our ability to predict the consequences of management activities. Efforts to increase our understanding of population processes beyond those operating primarily on single isolated populations in independent patches of habitat depend upon successfully viewing populations from this hierarchical perspective. Without this perspective, we may fail to understand the causes and consequences of interactions between populations and incorporate them into our thinking. Further advances in understanding the processes determining population change depend critically upon this next step. Otherwise, we may simply be trying to map a chimera.

Functional Definition of Landscape Structure Using a Gradient-based Approach

David M. Theobald and N. Thompson Hobbs

A standard approach to conceptualizing landscape patterns requires distinguishing between patches and the matrix surrounding them (Dramstad et al. 1996; Forman and Godron 1986; Fig. 59.1). This distinction is rooted in island biogeography theory (MacArthur and Wilson 1967), where patches are metaphors for islands and the matrix is an inhospitable "sea" (Wiens 1995; Wiens 1996b). The approach is straightforward and accessible because it usually corresponds to the scale of human perception of the landscape and resembles traditional cartographic representations of landscapes using categorical maps (Gustafson 1998). Furthermore, analysis of these landscape maps can be accomplished easily through standard geographic information systems (GIS) methods (e.g., Haines-Young et al. 1993). As a result, the patch/matrix (hereafter PM) approach has been a dominant approach to quantifying landscape structure to date.

However, we argue here that the PM approach is limited because it cannot easily incorporate functional characteristics of species or ecological processes in representations of landscape structure. In particular, biological mechanisms that operate on a landscape are often poorly represented by PM models. An important challenge for the field of landscape ecology is to move beyond simple representations of pattern that fail to

reveal the consequences of pattern for ecological processes (Wiens 1999).

For example, the basis for most studies in landscape ecology is a land-cover or vegetation map. Patches of vegetation are commonly represented by polygons in a categorical map. Such maps emerge from photointerpretation of aerial photography or aggregation of adjacent (either four or eight) cells from a classified remotely sensed image, and often patches are not defined consistently or are not biologically based (Paton 1994). Representing a patch as a discrete entity (e.g., a polygon) ignores both the fuzziness of the patch boundary and heterogeneity within the patch (Gustafson 1998). Noncontiguous patches of vegetation can be functionally integrated if a species or process of interest operates at a scale that can span patches (With and Crist 1995; Hobbs 1999). Uncertainties associated with vegetation data are seldom reflected in habitat maps modeled from vegetation and other elements (Flather et al. 1997), even though all maps have some level of inaccuracy (Goodchild and Gopal 1989) and there are well-established methods of error assessment in land-cover mapping (e.g., Congalton 1991).

An important recent finding demonstrates that models that use the core area of a patch, rather than the entire patch, are better predictors of species presence/absence for species that avoid patch edges and utilize patch interiors (Temple 1986a). The reduction in habitat quality at patch edges is caused by processes such as

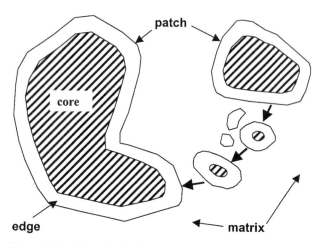

Figure 59.1. The standard approach to conceptualizing landscape structure distinguishes among patches, core areas, edge areas, and the matrix.

changes in microclimate regimes (Chen et al. 1995; Stevens and Husband 1998) and increased predation rates (Paton 1994). Although the depth of this edge effect is typically modeled using a uniform distance (e.g., Reed et al. 1996), processes that create edge are known to vary considerably, for example by aspect and edge contrast (Chen et al. 1995). Methods have been developed to vary the edge distance (e.g., Baskent and Jordan 1995) but these techniques are rarely used in landscape analysis. Moreover, edges are typically modeled as a step function, where all habitat value within the edge-effect distance is lost, despite the knowledge that for most processes the magnitude of the edge effect decreases with increasing distance from the edge (e.g., Chen et al. 1992; Chen et al. 1995; Paton 1994).

Patches are typically considered to be imbedded in a matrix of an inhospitable intervening landscape, and this matrix is generally considered biologically inert or even hostile. As a result, isolation of patches are thought of simply in terms of Euclidean distance between patches (e.g., Schumaker 1996; Keitt et al. 1997). However, the matrix is not ecologically homogeneous and movement through the landscape is affected, not only by the probability of encountering patch edges but also by how species perceive and respond to heterogeneity within the matrix (Wiens et al. 1993). Matrix properties such as edge contrast, vegetation structure, and land use clearly influence species movement (Fagan et al. 1999).

Although the limitations of the PM approach are

generally recognized, they are often overlooked because there are no widely available alternatives for modeling landscape structure. We assert that it will remain difficult to demonstrate linkages between landscape structure and ecological processes until alternative approaches are developed that explicitly include understanding of process in the representation of habitat pattern.

In this chapter, we describe an approach that develops two innovations beyond the typical PM approach. First, although the typical approach maps a species' habitat, or the physical space within which a species lives, we recognize that landscapes include a full range of habitat quality that better describes the ability of an area to provide the appropriate conditions. Second, we define patches on a functional basis by identifying components of landscape structure by scaling to ecological distances. We define patches at three levels of organization: individual, population, and metapopulation. Each of these levels is defined by a corresponding process: daily foraging, movement within home ranges, and dispersal between habitat. We represent these processes with simple allometric models that scale model parameters to body mass. We then use these models to develop functional representations of landscape structure. After developing this algorithm, we illustrate the differences between this gradient-based approach and the typical PM approach to landscape characterization. A broader framework that explicitly incorporates biological knowledge is needed to better understand the consequences of changes in landscape structure on species distributions and to understand the consequences of land-use change on population viability and distribution.

Methodology

Our primary goal is to develop an approach to defining landscape structure in functional terms. To do so, we first develop a model that quantifies habitat quality. We then identify patches of habitat based on scaling of the behavior of a species at three levels of organization: individual, population, and metapopulation. A major challenge in modeling landscape structure is to better incorporate biological mechanisms, yet not exceed our ability to parameterize a model nor overwhelm the interpreter with overly complex modeling results. We chose to use allometric relationships to illustrate our functional approach (Peters 1987), because these relationships provide a useful way to parameterize models for a broad range of species with a minimum of detail. Because differences in life history characteristics of species are important, the allometric relationships should only be used in a general way and with an understanding of the error terms of the allometric relationship. Modeling habitat for a specific species requires detailed understanding of the species and careful parameterization of the model.

Habitat Quality

We identify areas that provide resources for a given species. Ideally, resources are defined in terms of the individual elements (e.g., flora, fauna, soil, water, etc.) used by a species at a given location (Morrison and Hall, Chapter 2). However, resources are most commonly defined in terms of overstory vegetation or land-cover types. Typically, this is done through a species-vegetation affinity table that contains the vegetation types utilized by a species (e.g., Caicco et al. 1995; Edwards et al. 1996; White et al. 1997a). We extend this approach by allowing the affinity value to range from 0.0 to 1.0, describing the strength of the association. If an error matrix has been built for the vegetation image, uncertainty can be incorporated into the habitat-quality model by multiplying the affinity score by the probability that a location mapped as a particular type is correct. Fuzzy logic can be used to further incorporate fine-scale information into distribution maps by computing the degree of membership in each vegetation type for each location (pixel) on a map (e.g., Hill and Binford, Chapter 7). Other factors such as stand structure, canopy cover, and soil type can be incorporated to adjust estimates of resource availability. Although there are practical limitations to resource data, landscape analysts should strive to narrow the gap between the ideal definition of resources and the limitations of available data.

Second, we identify environmental gradients that modify a species' ability to utilize resources, such as elevation, slope/aspect, precipitation, or solar exposure. These environmental gradients are used to adjust the predicted habitat for a species. Typically, environmental gradients are represented by a binary map that describes the limits of a range, but we extend the values to range from 0.0 to 1.0. For example, species distributions are usually limited by upper and lower elevation ranges, yet these limits create artificial boundaries.

Third, the quality of habitat resources for certain species can be modified by either in situ or nearby natural or human-caused disturbance. For example, some species have lower population densities in areas adjacent to land-cover conversions such as forest clearcuts or urban areas (Harrison 1997). Typically, habitat loss caused by in situ impacts on habitat is represented by a change in land-cover type (e.g., forested to urban land conversion) at a particular location, whereas habitat loss from nearby land-cover types is modeled by removing the patch edge, leaving the patch core (Temple 1986a). However, we represent both in situ and nearby modifications to habitat quality by developing a relationship between the disturbance (e.g., road or housing density) and the impact on habitat use (e.g., Theobald et al. 1997). This allows us to model the reduction of habitat quality caused by in situ impacts that may not be captured by typical land-cover maps. For example, habitat quality can be reduced by human activities in suburban and rural areas (Harrison 1997), yet these locations are not represented by urban land-cover types. Typically, edge effects reduce patch quality near the edge, but we extend our approach to allow both lowered and increased habitat quality as a function of distance to edge. For interior species, habitat quality near the patch edge can be lowered, but for edge species, patch core areas may be poorer habitat than at the edge.

Thus, we define (Q) as an index of habitat quality, which is based on surrogate measures of habitat quality. Q is measured in terms of area (e.g., hectares) and is a function of three components (see Fig. 59.2 in color section):

$$Q = f(R, S, D) \qquad (59.1)$$

where resources, environmental factors, and disturbances are denoted by R, S, and D, respectively.

Individual

Within areas smaller than a home range, movement is characterized by foraging, when species are maximizing their energetic return through foraging behavior (Hobbs 1999). Species integrate resources over a local area through foraging and so definition of patches should reflect this scale of behavior (Addicott et al. 1987).

We define patches at the individual level based on home-range requirements. The area of home range (I) is based on an allometric relationship to body size (B) (Eq. 59.2) (Harestad and Bunnell 1979):

$$I = 1.166M^{1.06} \quad \text{where } S_{xy} = 1.12 \quad (59.2)$$

where M is body mass in kilograms and I is in square kilometers. Next, we derive the radius required (R_I) to fulfill the area requirement using Equation 59.3. For every cell in the habitat map, we calculate the average index of habitat quality within the radius R_I using Equation 59.4 (see Fig. 59.3 in color section).

$$R_I = (I / 3.1415)^{1/2} \quad (59.3)$$

$$P_I = focalmean(Q, circle, R_I) > t \quad (59.4)$$

The resulting map depicts a gradient of values, or the proportion of a cell that contributes to habitat. To identify contiguous areas of habitat adequate for an individual, we threshold the gradient values, where the average quality P_I exceeds a user-specified efficiency threshold (t). Ultimately, the appropriate threshold value depends on the behavior of the species or process in consideration. We then aggregate adjacent (eight neighborhood) cells to form patches that contribute to individual's habitat.

Population

Population patches are identified in a similar fashion as the individual level, but we identify the area required to support minimum viable population patches based on allometric scaling from minimal mammal densities (Silva and Downing 1994) (Eqs. 59.5 and 59.6):

$$D = e^{(-0.68*\ln(M) + 2.1414)} \quad (59.5)$$

$$P = A / D \quad (59.6)$$

where D is the minimum viable density in animals per square kilometer and P is the area required to support A animals. The population patch size based on minimal mammal densities is sensitive to uncertainty and

variability in population densities (Bowers and Matter 1997; Van Horne et al. 1997).

Metapopulation

A third level of organization is the metapopulation level. At this spatial extent, species response to landscape structure is characterized by dispersal. Here, the goal is to identify clusters of population patches that are within the dispersal distance of one another. Conversely, population patches are considered isolated when they are beyond the dispersal distance. The typical approach is to consider patch isolation in terms of the Euclidean distance from one patch to another (e.g., Keitt et al. 1997). However, dispersal is affected not only by the location of habitat patches, but also by the characteristics of the intervening patches that make up the matrix.

The relative ease or difficulty a species has in moving through the matrix is largely influenced by the land use/cover type at any given location. We conceptualize the matrix in terms of impedance to movement and represent it as a cost surface. This allows Euclidean distance to be modified by the relative ease or difficulty to travel from one location to another. Also, human land-use and natural-disturbance regimes in the matrix interact with land-use/cover and can further restrict dispersal. The relative resistance to movement needs to be specified for each land use/cover type.

Again, we use allometric scaling to parameterize dispersal ability based on body size:

$$\beta = 0.001M^{-0.91} \quad (59.7)$$

$$L = \ln Z / -\beta \quad (59.8)$$

where β is the probability of successful dispersal (D. Malkinson and N. T. Hobbs unpublished data), and L is the dispersal distance (m) for a probability of Z. We then calculate the distance based on matrix quality using a cost-distance function, where the population patches are the source patches for the cost-distance function. Typically, we reclassify land use/cover maps to reflect how the cover types would impede the movement through that area (see Fig. 59.4 in color section). The impedance values could also reflect other data layers, such as road or housing density, and very high

TABLE 59.1.

Parameters for a range of body sizes based on allometric relationships for mammals.

Body size (kg)	Individual patch			Population patch			Meta-population dispersal	
	Area (km²)	Radius (m)	No. patches	Area (km²)	Radius (m)	No. patches	Distance (m) at p = 0.5	No. patches
1	1.17	609	481	11.75	1,934	119	693	113
10	13.39	2,064	89	56.23	4,231	33	5,634	16
50	73.72	4,844	31	168.00	7,313	27	24,372	3

values could cause a "barrier" to movement. Population patches are then grouped if they are within the distance *L*.

We illustrate our methodology by examining individual, population, and metapopulation patches defined for a representative range of mammals in Colorado (Table 59.1). We use the vegetation map from the Colorado Gap Analysis Project, which was produced by interpretation (using a 100-hectare minimum mapping unit) of a Landsat Thematic Mapper image (30-meter resolution). We identified forested-type habitats (including coniferous, deciduous, and mixed). We then found the individual and population patches for mammals of 1, 10, and 50 kilograms in size and reclassified the land-use/cover map to reflect matrix quality and its influence on the interpopulation-patch movement.

Results

Over 568,600 square kilometers of habitat are defined using the typical binary approach, and there are over 5,900 patches. As is typical of habitat maps produced using the binary approach, the distribution of patch sizes is highly skewed. For instance, over 45 percent of the patches are two cells or less in size (8 hectares), and 67 percent of patches are less than ten cells in size. The average patch size is 45 square kilometers.

In contrast, the habitat maps produced using the functional algorithm have fewer patches (Table 59.1), and patches are more evenly distributed by size. The smallest average patch size is 166 square kilometers for individual population patches for 1-kilogram body size. The number of patches decreases rapidly (nonlinearly) with body size.

Conclusions

Assessments that examine the consequences of development and land-use change for habitat quality and fragmentation should be based on analyses that explicitly incorporate the functional response of a species or process. Without incorporating these responses, evaluations of landscape change quantify structural changes that are simply an artifact of the data and interpretation scale. This requires biological parameters to be incorporated, and we find that parameterization of these models using allometric scaling of body size for three levels of organization is a useful way to incorporate biological realism into modeling habitat fragmentation.

We do not contend that our approach is necessarily more accurate than other modeling approaches. Indeed, testing predicted versus observed patch occupancy is fraught with its own challenges (Fielding, Chapter 21, Karl et al., Chapter 51), including the difficulty of understanding errors of commission (i.e., interpreting predictions of occupied habitat versus the lack of field data that demonstrates unoccupied habitat). However, we do believe that the approach we offer is useful because it provides results that are more easily interpreted than results from traditional methods. That is, the linkage between the assumptions made about the mechanisms affecting habitat use and the resulting landscape pattern is explicit, ecologically based, and repeatable. At minimum, it provides a starting point for managers to understand how individual mechanisms might contribute to habitat loss and/or fragmentation.

We see three immediate applications of this approach for wildlife managers. First, this methodology has been used to develop refined maps of potentially suitable habitat for given species of interest. These refined maps are more defensible because they better reflect known life-history characteristics and can better incorporate data uncertainty. Second, the potential consequences of management actions or of changes in landscape context on local-scale conservation sites could be assessed using this approach. For example, would a significant loss of habitat occur if a road were built through a conservation site? What if five roads were developed through a site? Third, this approach offers a way to screen, at ecoregional scales, potentially imperiled species by identifying critical thresholds and characteristics landscape patterns that result from known or predicted landscape changes. For example, the consequences of urban growth on sensitive species could be assessed by comparing habitat maps that reflect the extent and intensity of urban and suburban development at two time periods (e.g., 1990 versus 2020).

Modeling Habitat-based Viability from Organism to Population

Gary J. Roloff and Jonathan B. Haufler

Population viability is a primary management issue in the United States as directed by the Endangered Species Act (ESA) and National Forest Management Act (NFMA). An underlying premise of the ESA is to ensure that distinct population segments receive appropriate levels of protection and that their persistence is ensured into the future (Endangered Species Act of 1973). Similarly, the NFMA (and pursuant regulations) requires the United States Forest Service to "maintain viable populations of all native vertebrate species." The NFMA does not require the Forest Service to identify minimum viable numbers per se, but rather to ensure that species persist over time (Marcot and Holthausen 1987). Population viability analysis (PVA) has most often been used to portray the effects of genetic, demographic, environmental, and catastrophic stochasticities on long-term population trends (Table 60.1). However, one of the greatest challenges facing resource managers on lands throughout the United States is how to evaluate the effects of individual management projects on population viability and biodiversity (see Cogan, Chapter 18), especially over larger areas (e.g., national forests, species' range). To this end, we developed a modeling framework that permits evaluation of operational-level projects in the context of overall population viability.

In this chapter, we discuss population viability as a planning goal, review the role that habitat resources (e.g., vegetation, patch configuration; see Morrison and Hall, Chapter 2) have played in viability analyses, review a process for incorporating habitat resources into spatially explicit viability assessments, and demonstrate use of this mechanism for three species. In this chapter, a population is defined as consisting of all individual locations (Morrison and Hall, Chapter 2) and subpopulations of conspecifics that are demographically, genetically, or spatially disjunct (Wells and Richmond 1995). We refer to a subpopulation as a set of individuals that are not spatially isolated from other individuals (Wells and Richmond 1995), synonymous to Smallwood's (Chapter 6) "constrained aggregations." Our example explicitly addresses three components of population viability analyses: demographic stochasticity, environmental stochasticity, and population spatial structure. Demographic stochasticity is the variation in birth and death rates observed from an independent sample of individuals that make up a population (Miller and Lacy 1999). Environmental stochasticity is variation in the population mean itself (Miller and Lacy 1999). Population structure refers to the spatial organization of organisms and subpopulations and how those entities interact. Our discussion emphasizes PVA on terrestrial fauna (excluding insects).

TABLE 60.1.

A review of population viability analyses.

Study and species[a]	Conservation issue	Temporal scale (yrs)	Treatment of habitat quality	Modeling approach
Akçakaya et al. (1995) Helmeted honeyeater	Effectiveness of translocations on species viability	50	Explicitly modeled as patches	RAMAS[b]
Beier (1993) Cougar	Prediction of minimum areas and levels of immigration needed	100	Implied in stochastic simulation	Leslie matrix
Beissinger (1995) Snail kite	Effects of periodic environments on species viability	50–100	Explicitly modeled as environmental states	Random walk model
Bustamante (1996) Bearded vulture	Use of reintroduction to ensure species viability	200	Implied in stochastic simulation	VORTEX[c]
Doak et al. (1994) Desert tortoise	Effects of different management scenarios on species viability	150	Implied in stochastic simulation	Stage-based models
Elliott (1996) Mohua	Effects of breeding strategy and predation on species viability	100	Implied in stochastic simulation	Stochastic simulation modeling
Foin and Brenchley-Jackson (1991) Light-footed clapper rail	Effects of estuary type and management strategy on species viability	1	Explicitly modeled as habitat patches	Simulation modeling
Forys and Humphrey (1999) Lower Keys marsh rabbit	Effects of current conditions and management scenarios on species viability	50	Explicitly modeled as habitat patches	VORTEX
Gaona et al. (1998) Iberian lynx	Effects of demographics and spatial structure on species viability	100	Implied in stochastic simulation	Age and space
Goldingay and Possingham (1995) Yellow-bellied glider	Effect of habitat area on species viability	100	Implied in stochastic simulation	Simulation model ALEX[d]
Green et al. (1996) White-tailed eagle	Feasibility of reintroduction program on species viability	100	Implied in stochastic simulation	Stochastic simulation modeling
Hamilton and Moller (1995) Sooty shearwaters	Effects of management scenarios on species viability	100	Implied in stochastic simulation	VORTEX
Howells and Edwards-Jones (1997) Wild boar	Feasibility of reintroduction program on species viability	50	Implied in stochastic simulation	VORTEX
Lamberson et al. (1994) Northern spotted owl	Reserve design for species viability	100	Explicitly modeled to identify the number of suitable sites per cluster	Stage projection and territory cluster models
Li and Li (1998) Crested ibis	Effects of external and intrinsic factors on species viability	100–500	Implied in stochastic simulation	VORTEX
Lindenmayer and Lacy (1995a) Mountain brushtail possum Greater glider	Effects of forestry practices on species viability	100	Implied in stochastic simulation	VORTEX

Reference / Species	Purpose	Years	Catastrophes	Method
Lindenmayer and Lacy (1995b) Leadbeater's possum	Effects of forestry practices on species viability	100	Explicitly modeled as patches	VORTEX
Lindenmayer and Possingham (1995) Leadbeater's possum	Effect of wildfire on species viability	35–120	Implied in stochastic simulation	ALEX
Lindenmayer and Possingham (1996) Leadbeater's possum	Effects of timber management alternatives on species viability	750	Explicitly modeled as patches	ALEX
Maguire et al. (1987) Sumatran rhino	Effects of management alternatives and associated costs on species viability	30	Explicitly modeled in decision tree	Decision analysis
Maguire et al. (1995) Red-cockaded woodpecker	Effect of demographic uncertainty on species viability	100	Implied in stochastic simulation	RAMAS-STAGE
Marcot and Holthausen (1987) Northern spotted owl	Evaluation of species viability in context of forest plan	150	Explicitly modeled as patches	Stochastic Leslie matrix
Marmontel et al. (1997) Florida manatee	Effect of demographic uncertainty on species viability	1,000	Implied in stochastic simulation	VORTEX
McCarthy et al. (1995) Helmeted honeyeater	Effects of fecundity and initial non-breeder density on species viability	50	Implied in stochastic simulation	Individual-based stochastic simulation
Miller and Botkin (1974) Whooping crane Sandhill crane	Effects of management alternatives on species viability	100	Implied in stochastic simulation	Stochastic modeling
Mills et al. (1996) Grizzly bear	Evaluation of species viability predictions from four different models	48	Implied in stochastic simulation	GAPPS,[e] INMAT,[f] RAMAS-AGE, VORTEX
Nolet and Baveco (1996) Beaver	Species viability for a translocated population	100	Explicitly modeled in the metapopulation structure	Individual-based simulation model
Pascual et al. (1997) Wildebeest	Effect of harvest on species viability	100	Implied in stochastic simulation	Generalized difference equations
Possingham et al. (1994) Greater glider	Effect of forestry and wildfire on species viability	200	Explicitly modeled as patches	ALEX
Root (1998) Florida scrub jay	Effects of habitat loss and modification on species ability	60	Explicitly modeled as patches	RAMAS
Saether et al. (1998) Brown bear	Effect of population size on species viability	150–300	Explicitly modeled according to subpopulation location	Stochastic diffusion process
Saltz (1996) Persian fallow deer	Feasibility of reintroduction program on species viability	100	Not modeled	Monte Carlo Leslie matrix
Song (1996) Haianan Eld's deer	Effects of stochasticities and translocation on species viability	200	Implied in stochastic simulation	VORTEX
Southgate and Possingham (1995) Greater bilby	Effect of reintroduction on species viability	20–100	Explicitly modeled as patches	ALEX

(Continued)

TABLE 60.1. (Continued)

A review of population viability analyses.

Study and species[a]	Conservation issue	Temporal scale (yrs)	Treatment of habitat quality	Modeling approach
Suchy et al. (1985) Grizzly bear	Effect of population size on species viability	100	Implied in stochastic simulations	Stochastic simulation modeling
Swart et al. (1993) Samango monkey	Effect of low population size on species viability	60	Not modeled	Density-dependent demographic model
Vucetich et al. (1997) Gray wolf	Effects of social structure and prey dynamics on species viability	200	Implied in modeling prey dynamics	Stochastic simulation modeling
Wootton and Bell (1992) Peregrine falcon	Effects of alternative management strategies on species viability	100	Implied in stochastic simulations	Modified Lefkovitch stage matrix
Zhou and Pan (1997) Giant panda	Effects of changing conditions on species viability	200	Not modeled	Leslie matrix models

[a]Scientific names:

Bearded vulture, *Gypaetus barbatus*

Beaver, *Castor fiber*

Brown bear, *Ursus arctos*

Cougar, *Felis concolor*

Crested ibis, *Nipponia nippon*

Desert tortoise, *Gopherus agassizii*

Florida manatee, *Trichechus manatus latirostris*

Florida scrub jay, *Aphelocoma coerulescens*

Giant panda, *Ailuropoda melanoleuca*

Gray wolf, *Canis lupus*

Greater bilby, *Macrotis lagotis*

Greater glider, *Petauroides volans*

Grizzly bear, *Ursus arctos horribilis*

Haianan Eld's deer, *Cervus eldi*

Helmeted honeyeater, *Lichenostomus melanops*

Iberian lynx, *Lynx pardinus*

Leadbeater's possum, *Gymnobelideus leadbeateri*

Light-footed clapper rail, *Rallus longirostris levipes*

Lower Keys marsh rabbit, *Sylvilagus palustris hefneri*

Mohua, *Mohoua ochrocephala*

Mountain bushtail possum, *Trichosurus caninus*

Northern spotted owl, *Strix occidentalis caurina*

Peregrine falcon, *Falco peregrinus*

Persian fallow deer, *Dama dama mesopotamica*

Red-cockaded woodpecker, *Picoides borealis*

Samango monkey, *Cercopithecus mitis*

Sandhill crane, *Grus canadensis*

Snail kite, *Rostrhamus sociabilis*

Sooty shearwaters, *Puffinus griseus*

Sumatran rhino, *Dicerorhinus sumatrensis*

White-tailed eagle, *Haliaeetus albicilla*

Whooping crane, *Grus americana*

Wild boar, *Sus scrofa*

Wildebeest, *Connochaetes taurinus*

Yellow-bellied glider, *Petaurus australis*

[b]RAMAS (Ferson 1993, Akçakaya 1994).

[c]VORTEX (Miller and Lacy 1999).

[d]ALEX (Possingham et al. 1992).

[e]GAPPS (Downer 1993).

[f]INMAT (Mills and Smouse 1994).

Population Viability As a Planning Goal

In the NFMA, a viable population is defined as "one which has the estimated numbers and distribution of reproductive individuals to ensure its continued existence is well distributed in the planning area" (Code of Federal Regulations, Title 36, Volume 2, Section 219.19). The definition, consistent with other definitions (Soulé 1987a; Boyce 1992), implies that "continued existence" is the management goal. PVA was developed as a tool for assisting land managers in understanding and portraying the effects of change on this management goal; however, the number of species that must be considered under NFMA often make the use of PVAs restrictive in resource planning (Raphael and Marcot 1994). In addition, lack of information about species vital rates, movement capabilities, and habitat relationships often prohibit the rigorous application of PVA. Coarse-filter approaches have been advocated to help alleviate this problem (Hunter 1990; Raphael and Marcot 1994; Haufler et al. 1996, 1999b). These approaches assume that the proper distribution and amounts of ecological communities inherent to landscapes will directly relate to the viability requirements for most species. Recently, Haufler et al. (1996, 1999b) proposed a strategy for ecosystem management that used "adequate ecological representation" as a coarse-filter planning goal to satisfy population viability. Adequate ecological representation refers to threshold amounts of ecological communities that are described and distributed according to potential vegetation, existing vegetation successional stage, and historic disturbance regimes (Haufler et al. 1996, 1999b). Adequate ecological representation is not advocated as a management target; rather, it should be viewed as a threshold level that is never compromised. If adequate ecological representation is not compromised, then sufficient amounts of ecological communities inherent to the landscape are provided, and it is assumed that organisms associated with these communities are viable (Haufler et al. 1996, 1999b). According to the coarse-filter strategy proposed by Haufler et al. (1996, 1999b), the viability assumption must be tested using an appropriate PVA for a select group of species that represent the diversity of ecological communities found in the planning landscape (Haufler et

al. 1996, 1999b). They called this process a "coarse filter–fine filter approach to ecosystem management." As a means of fulfilling the PVA portion of Haufler et al.'s (1996, 1999b) approach, we developed a method for assessing viability (Roloff and Haufler 1997) that can be used to address viability goals in a variety of planning scenarios.

The tangible components of the viability goal in NFMA include organism density and the spatial arrangement of those organisms. Thus, consistent with other efforts that linked demographic and environmental stochasticities, habitat, and metapopulation models in a PVA (e.g., Akçakaya et al. 1995; Root 1998), our viability approach is based on these components. We developed a habitat-based framework for indexing organism density and for spatially portraying population structure (Roloff and Haufler 1997). In this approach, planning units are home ranges of varying quality and sizes, information we use as input to PVAs. We define home range quality as the ability of an area traversed by an organism during a breeding cycle to provide conditions appropriate for individual persistence (Morrison and Hall, Chapter 2) and successful reproduction. In comparison to habitat quality (see Morrison and Hall, Chapter 2), home range quality also explicitly defines the area of use and indexes the reproductive contribution of that area to population persistence. Our approach portrays the effects of resource conditions on population viability and offers resource planners a unit of measure (the contribution of individual home ranges) that can be mapped and linked to population viability planning through time (Roloff and Haufler 1997).

In our approach, the basis of PVA is the quality and size of home ranges and their locations (Roloff and Haufler 1997). Numerous factors influence home range density and spatial arrangement in viability analyses, and these factors have been categorized as genetic, demographic, environmental, and catastrophic (Shaffer 1981; Gilpin 1987). Ideally, PVA should include all of these factors (Gilpin and Soulé 1986); however, complete data sets are rare, even for the most-studied wildlife species (Boyce 1992; Sæther et al. 1998). Thus, when faced with the daunting task of accounting for the viability of all species in a planning area, the challenge in conducting PVAs is to create a model that captures important aspects of a species' ecology while only using realistically measurable parameters (Eberhardt 1987; Boyce 1992). Often, the most realistically measurable data available for population viability planning across large areas are habitat based. This conforms to the notion that loss or degradation of habitat is the most significant factor threatening the future extinction of species (Wilcox and Murphy 1985; Pimm and Gilpin 1989).

A Perspective on Previous Viability Analyses and the Role of Habitat

Except for situations in which population numbers are excessively small, the effects of demographic and environmental stochasticities on population vital rates have been identified as arguably the most important factors influencing population viability (Shaffer 1987; Lande 1988a,b; Boyce 1992; Wissel and Zaschke 1994). Demographic and environmental stochasticities are incorporated into viability analyses by first deriving the best estimates of individual and subpopulation vital rates and then simulating randomness around those rates based on the breadth of uncertainty surrounding the estimates. Most often, the portrayed vital rates involve birth and death processes (Table 60.1). In population viability analysis, these stochastic events represent a variety of factors, including variation in individual reproduction and survival rates, weather, prey and predator distributions, competition, and parasites and diseases (Hunter 1990; Caughley and Sinclair 1994). We concur that the randomization process is essential for portraying the complete range of responses that organisms may exhibit to chance events; however, the importance of accurate vital rate estimates should not be overlooked. To that end, we encourage the use of habitat as a means to refine vital rate estimation and to constrain the stochastic models.

We reviewed population viability analyses that were published in the readily accessible literature to understand the conservation issues driving the analyses, the temporal scale of analyses, how habitat was incorporated, and the modeling approaches used (Table 60.1). We reviewed thirty-nine papers that described forty-one population viability analyses on thirty-five species (Table 60.1). Most (79 percent)

PVAs were performed on species that were at least locally rare or exhibited range reductions. In papers we reviewed, PVAs were most often conducted to evaluate the effects of management alternatives (Table 60.1). Other common conservation issues addressed were the effects of demographic uncertainty and reintroduction or translocations on species viability (Table 60.1). PVAs were also conducted to evaluate reserve design (Lamberson et al. 1994; Goldingay and Possingham 1995), the effects of life-history strategies on species viability (Elliott 1996; Vucetich et al. 1997), and to compare different modeling approaches (Mills et al. 1996). Some PVAs were used as exploratory data analysis to provide general insight on the factors influencing species' persistence (Li and Li 1998; Sæther et al. 1998).

The utility of PVAs for resource planning and management has been questioned. One problem with PVA is that it often depicts species viability over time frames beyond operational-level planning horizons (Boyce et al. 1994). In addition, the output of PVA (a probability of persistence for some time period) is difficult to frame as a tangible management objective. In a typical natural resource planning scenario, strategic goals are established for fifty to hundred years and operational activities for achieving that goal are projected in five- to ten-year increments. In an ideal situation, progress toward strategic goals is reevaluated after the five- to ten-year time period and revised accordingly. More realistically, strategic goals change on five- to ten-year intervals in response to changes in ecological, economic, or social influences, and knowledge and operational activities are adjusted accordingly. Most viability assessments reviewed for this study projected population dynamics for one hundred years or more (Table 60.1) conforming to the notion that longer time frames are required to portray population trajectories. Thus, although the temporal scales of strategic planning and PVA are theoretically comparable, in reality the modeling requirements of PVA dictate that individual operational-level activities get "lost" in the stochastic simulations.

The mismatch between PVA output and operational-level decision making should not preclude the use of PVA in planning. Rather, PVA is a tool that can be used to assess the long-term cumulative effects of operational-level management activities on species viability. This view is consistent with Thomas et al. (1990), who recommended the use of PVA for extrapolating long-term population sizes that are used to frame management objectives. Thus, the challenge of using PVA for operational-level decision making lies in our ability to align the temporal and spatial scales used for both processes. This can be accomplished by using home ranges as the planning unit, if the home ranges are understood in the context of population dispersion and regional pattern of distribution (Smallwood, Chapter 6).

Frameworks for integrating habitat quality, quantity, and spatial arrangement into PVA have been demonstrated (e.g., Akçakaya et al. 1995; Lindenmayer and Lacy 1995a; Lindenmayer and Possingham 1996; Root 1998; Forys and Humphrey 1999). Of the thirty-nine papers reviewed for this study, twenty-six (67 percent) either implied habitat effects via stochastic modeling or did not model habitat. Of the thirteen (33 percent) papers that explicitly incorporated habitat quality (i.e., some differentiation between resource areas as to their fitness contribution, see Morrison and Hall, Chapter 2) and location into the PVA, nine (23 percent) explicitly modeled habitat patches (Table 60.1). These PVAs used habitat to estimate carrying capacity (Forys and Humphrey 1999), model different demographics (Foin and Brenchley-Jackson 1991; Akçakaya et al. 1995; Root 1998), and assess metapopulation structure (Lamberson et al. 1994; Akçakaya et al. 1995; Lindenmayer and Lacy 1995a; Lindenmayer and Possingham 1996). The findings from our review of PVA models are consistent with Beissinger and Westphal (1998), who noted that relatively few modeling approaches incorporated habitat- (or patch) specific information.

Using Habitat and Home Ranges for Viability Analyses

Reasons for species peril are varied (Table 60.1), although habitat loss and fragmentation are frequently cited as primary contributors to population decline. The ESA and NFMA viability guidelines explicitly identify habitat as a risk or limiting factor important for most species managed under these regulations. For

example, Temple (1986b) demonstrated that habitat loss was responsible for 82 percent of the avian taxa endangered by extinction. Although Boyce (1992) recommended that viability analyses include genetic, demographic, environmental, and catastrophic stochasticities, Foin and Brenchley-Jackson (1991) demonstrated that useful estimates of population dynamics could be derived from simple, habitat-based models. However, describing habitat is by no means a simple task. Habitat is defined as the physical space within which an organism lives (Morrison and Hall, Chapter 2) and as such can include a wide range of environmental resources that often form complex interactions (Belovsky 1987; Hunter 1990; Caughley and Sinclair 1994). Thus, even though the effects of habitat on demographic and environmental stochasticities have been identified as an important aspect of population viability, the question remains as to what expression of habitat should be quantified in viability planning.

By definition, habitat quality is the suite of resources and environmental conditions that determine the presence, survival, and reproduction of a population (Morrison and Hall, Chapter 2), and thus it makes sense to use habitat quality in viability assessments. In using habitat quality as a framework for PVA, one assumes that home ranges or habitat patches of similar quality exhibit more-predictable demographics and are exposed to similar levels of environmental variation (e.g., rates of predation, forage success; Akçakaya et al. 1995; Beissinger 1995). For example, Root (1998) showed that Florida scrub-jays (*Aphelocoma coerulescens*) in different vegetation patches exhibited different probabilities for survival and breeding. She attributed these differences in vital rates in part to habitat quality. The importance of stochasticities, particularly as applied to birth and death rates in viability analyses can be of overwhelming importance (Wissel and Zaschke 1994). Thus, techniques that refine stochastic modeling are encouraged. PVAs that used habitat patches of differing quality to stratify stochastic simulations have been conducted (Table 60.1), and the process is analogous to stratifying a sample to reduce statistical variance. Similar to these studies, our approach uses an organism-based scalar hierarchy linked to habitat (see Roloff and

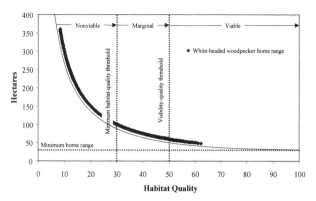

Figure 60.1. Viability relationship and associated home ranges for white-headed woodpeckers (*Picoides albolarvatus*) in the Idaho Southern Batholith. See Roloff and Haufler (1997) for details on the development of the viability relationship.

Haufler 1997) for refining the stochastic simulations of PVA. Scales in the hierarchy include home ranges, subpopulations, and populations (Roloff and Haufler 1997).

As part of the coarse filter–fine filter ecosystem management strategy, Haufler et al. (1996, 1999a) contended that the basic unit of the fine filter check for species viability was the individual home range (Roloff and Haufler 1997). Briefly, our habitat-based framework for PVA consists of four steps. The first three steps involve home-range-level analyses and include the identification and mapping of habitat based on the organism's life requisites, the development of criteria used for identifying home range viability, and application of the viability criteria to the habitat assessment to map home ranges (Roloff and Haufler 1997). The contribution of these home ranges to viability is based on an estimate of habitat quality with the assumption that higher-quality home ranges as scored and mapped in this process will offer more-favorable conditions for survival and reproduction (Roloff and Haufler 1997). The final step of the framework involves spatial analyses at the subpopulation and population levels. The framework as a whole offers the opportunity to integrate organism demographics and habitat quality in the PVA (Boyce 1992).

Scoring (as to viability potential) and mapping individual home ranges provides a means to link operational-level decision making to population viability for planning. Given that resource management affects

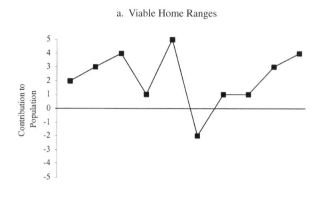

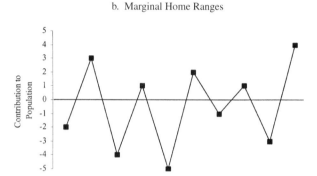

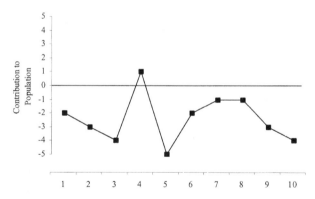

Figure 60.2. Relationships between home range types and contribution to population viability.

home range quality, size, and location, a method to index the magnitude and direction of these management effects was desired. To that end, Roloff and Haufler (1997) proposed that home ranges could be plotted as a viability relationship (Fig. 60.1). Viability relationships are species specific, and details of their development are provided elsewhere (Roloff and Haufler 1997). The axes of the viability relationship are habitat quality and home range size (Fig. 60.1). In addition, the quality axis is divided into viable, marginal, and nonviable areas (Roloff and Haufler 1997; Fig. 60.1). Viable home ranges are assumed to have the lowest adult mortality and to consistently produce young (Fig. 60.2; Roloff and Haufler 1997). Marginal home ranges are assumed to exhibit variable adult mortality and reproductive rates depending on the availability of resources. During good resource years, these home ranges will contribute to population viability similar to viable ranges; however, during lean resource years these ranges will not contribute to population viability (Fig. 60.2; Roloff and Haufler 1997). Nonviable home ranges are assumed to rarely contribute to population viability regardless of resource availability (see Fig. 60.3 in color section; Roloff and Haufler 1997). Operational-level manipulations to habitat will alter the quality and area of home ranges, and, thus, their relative locations in the viability relationship will change. The magnitude, direction, and number of home range shifts can be used to estimate the effects of management activities on population viability.

An Example:
White-headed Woodpecker Viability

To demonstrate the process of identifying, scoring, and mapping home ranges as an index to population viability, we conducted a habitat-based viability analysis for white-headed woodpeckers (*Picoides albolarvatus*) according to the framework of Roloff and Haufler (1997) for the Idaho Southern Batholith (Fig. 60.3). The Idaho Southern Batholith is 2.3 million hectares, a size that corresponds to some ecosystem management efforts (Haufler et al. 1996, 1999a,b). These large areas are useful for assessing population viability of most species and for establishing strategic planning goals; however, it is difficult to demonstrate and quantify the effects of individual management projects on viability over areas that large. We used habitat potential modeling (Roloff and Haufler 1997) to score white-headed woodpecker nesting and foraging habitat quality on a scale of 0 to 100, with 100 representing optimum habitat conditions. Scores were based on a mathematical representation of the vegetation resources required by white-headed woodpeckers to per-

sist and reproduce. Nesting and foraging scores were assigned to map pixels in the assessment area. Subsequently, nesting and foraging map pixels were aggregated into home ranges according to Roloff and Haufler (1997). Each home range was ranked on the same scale of 0 to 100 according to the quality and amounts of nesting and foraging habitats that were used to delineate the home range.

Individual home ranges were plotted on the viability relationship depicted in Figure 60.1 according to their size (ordinate) and mean habitat quality score (abscissa) for the entire 2.3-million-hectare assessment area. For purposes of this demonstration, we assumed that scores greater than or equal to 50 represented viable home ranges, scores from 30 to 49 represented marginal home ranges, and scores from 1 to 29 represented nonviable home ranges (Fig. 60.1). Thus, all home ranges with an average habitat-quality score greater than 50 in Figure 60.1 were considered viable. In addition, to understand the precision of our model projections, we calculated 90 percent confidence intervals for our habitat model output according to Bender et al. (1996). The process used by Bender et al. (1996) quantifies the variability associated with model input data (i.e., the measurement of habitat attributes). Also, the map pixel aggregation process includes randomness (Roloff and Haufler 1997), and, thus, different spatial patterns can result from different model runs. We spatially aggregated each bootstrap iteration three times and calculated the mean number of home ranges. Thus, output from our home range analysis includes estimates of vegetation sampling error and randomness in the home range aggregation process. It is important to note that the bootstrap process only accounts for the error associated with model input data (Bender et al. 1996); it does not represent the validity of the habitat model.

Using the viability thresholds in Figure 60.1, the model identified a total of 4,640 (4,091–4,915) (mean and 90 percent confidence interval) white-headed woodpecker home ranges in the Batholith landscape: 348 (172–616) viable, 1,965 (1,632–2,165) marginal, and 2,312 (2,161–2,465) nonviable. Figure 60.1 shows the results from one of the model iterations.

To simulate the effects of an operational-level management activity on both localized and population-level viability, we selected 890 hectares centered on an area of marginal to low-quality white-headed woodpecker home ranges in the Batholith (Fig. 60.3). The 890-hectare area is characterized as a mosaic of open shrubland and mid- to high-elevation dry forest communities. Dominant habitat type (sensu Daubenmire 1952) classes included cool, moist Douglas-fir (*Pseudotsuga menziesii*); cool, dry Douglas-fir; warm, dry, subalpine fir (*Abies lasiocarpa*); and high-elevation subalpine fir (Haufler et al. 1996, 1999b). Forest structure was dominated by medium (30.5–50.8 centimeters diameter at breast height [dbh]) and large (greater than 50.8 centimeters dbh) single- and multistoried stands. We identified 52 hectares of large-tree, densely stocked, single-storied forests on the warm, dry subalpine fir habitat type class for treatment (Fig. 60.3b). These forests consisted primarily of mature Douglas-fir and subalpine fir and thus were not considered high-quality white-headed woodpecker habitat (Garrett et al. 1996). Therefore, we predicted that the proposed treatment would have a minimal effect on white-headed woodpecker viability, and we used our framework to quantify and document this effect.

We simulated a historical range of variability harvest prescription on the selected 52 hectares and generated pre- and post-treatment viability relationships for the 890-hectare assessment area (Figs. 60.4a,b). Historically, these forest types in central Idaho were subjected to a fifty- to ninety-year fire mosaic that maintained sparsely stocked stands of Douglas-fir and lodgepole pine (*Pinus contorta*). We simulated a prescription that thinned the subalpine fir from below and retained patches of large Douglas-fir in the overstory. Pretreatment white-headed woodpecker habitat analysis indicated that zero (0–1; 90 percent confidence interval) viable, three (0–4) marginal, and one (0–6) nonviable home ranges were located within the 890-hectare assessment area (Fig. 60.4a). Post-treatment habitat analysis indicated that zero (0–1) viable, four (1–6) marginal, and zero nonviable (0–5) home ranges were located in the assessment area (Fig. 60.4b). Consistent with our prediction, treatment appeared to have a negligible effect on localized white-headed woodpecker viability. Similarly, in the context of the entire Idaho

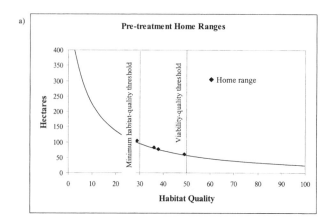

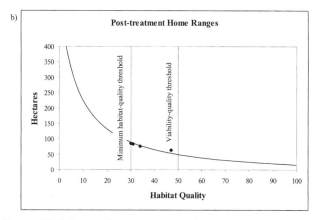

Figure 60.4. Pre- and post-treatment viability relationships for white-headed woodpeckers (*Picoides albolarvatus*) in the 890-hectare treatment assessment area ([a] and [b], respectively).

Southern Batholith (Fig. 60.1), the treatment had a negligible effect on population-level viability.

Assessing the Viability of Multiple Species

For individual species, the habitat-based PVA process outlined above will provide planners and managers a means to evaluate the effects of alternative management scenarios on viability. For example, instead of simulating only an historical range of variability timber harvest as described above, one could also simulate a clearcut prescription and compare the effects of the different prescriptions on white-headed woodpecker viability. But what about the hundreds of other species potentially impacted by proposed projects? As noted earlier, the magnitude of species numbers and computational complexity of the problem necessitates a coarse-filter approach. Only by properly addressing a coarse-filter approach can maintenance and enhancement of both biological diversity and ecosystem integrity (see Haufler et al. 1996) be attained. However, to assure proper functioning of the coarse filter, viability of selected individual species needs to be checked (Haufler et al. 1996, 1999b). To keep the number of PVAs required for a legitimate check of the coarse filter manageable, species should be selected using ecological rationale (Haufler et al. 1996, 1999b). Thus, this process uses an ecological indicator species approach (Morrison et al. 1998), where species are used to represent ecological communities found in the assessment area (Haufler et al. 1996, 1999b). The limitations of using indicator species have been discussed elsewhere (Mannan et al. 1984; Patton 1987; Landres et al. 1988); however, indicator species are likely useful for monitoring general ecosystem conditions (Kremen 1992).

In our example, we selected the white-headed woodpecker as a species that represented the sparsely stocked, large-tree, understory-fire-maintained ponderosa pine (*Pinus ponderosa*) ecological community that was historically a dominant feature of low-elevation sites in the Idaho Southern Batholith landscape. To demonstrate our approach for species representing a diversity of coarse-filter conditions, we also conducted habitat-based viability assessments for pileated woodpeckers (*Dryocopus pileatus*) and dusky flycatchers (*Empidonax oberholseri*) in the 890-hectare assessment area. In areas representative of the Idaho Southern Batholith, pileated woodpeckers have been shown to represent a more closed-canopy, large-tree community with an abundance of snags, downed

wood, and tree defects (Bull 1987). In the Idaho Southern Batholith, this vegetation community historically occurred in riparian areas and on mesic sites that were less susceptible to a recurring understory fire regime. Dusky flycatchers are considered habitat generalists and represent more open-forest conditions with moderately dense understories and ground cover (Dobkin 1994). This type of community is more typical of potential vegetation types (sensu Daubenmire 1952) that support dense shrubs and that are not subjected to periodic understory burns.

Habitat-based viability results for the 890-hectare treatment area indicated that pileated woodpecker habitat was scarce for both pre- and post-treatment conditions (Table 60.2). Only one nonviable pileated woodpecker home range was identified (Table 60.2). Similarly, no viable dusky flycatcher home ranges were identified both pre- and post-treatment in the assessment area; however, the model identified twenty-eight and twenty-five marginal home ranges pre- and post-treatment, respectively (Table 60.2). Although the treatment resulted in one additional dusky flycatcher home range, it caused a reduction in the number of marginal home ranges (Table 60.2). Overlapping 90-percent confidence intervals for the dusky flycatcher data suggest that these differences are not significant (Bender et al. 1996). Thus, using the habitat-based approach for assessing viability, the silvicultural treatment was estimated to have negligible effect on the three species (Table 60.2).

Our multispecies demonstration used species that represent a portion of the ecological complexity found in the Idaho Southern Batholith. For the Batholith, we

TABLE 60.2.

Pre- and post-treatment viability relationships for white-headed woodpecker (*Picoides albolarvatus*), pileated woodpecker (*Dryocopus pileatus*), and dusky flycatcher (*Empidonax oberholseri*) for the 890-hectare treatment area in the Idaho Southern Batholith.[a]

| Home range type | Species | | | | | |
| | White-headed woodpecker | | Pileated woodpecker | | Dusky flycatcher | |
	Pre	Post	Pre	Post	Pre	Post
Viable	0(0–1)	0(0–1)	0	0	0(0–8)	0(0–7)
Marginal	3(0–4)	4(1–6)	0	0	28(21–38)	25(14–31)
Nonviable	1(0–6)	0(0–5)	1(0–1)	1(0–1)	6(0–18)	10(5–16)

[a]Values represent mean and 90% confidence intervals.

suspect that the number of species-specific PVAs required in a legitimate check of the coarse filter in a resource plan would range from twenty to forty. The actual number would depend on landscape complexity, the number of protected species that must be considered, and the variety of management alternatives considered.

Integrating Home Ranges into PVA Stochastic Sampling

For most species managed under the NFMA, the habitat-based level of analysis discussed above will suffice for planning. However, some species that are rare or protected under ESA will require a more comprehensive PVA. A comprehensive PVA considers all aspects of viability, including habitat; genetic; demographic, environmental, and catastrophic stochasticities; community interactions; and spatial structure. In this chapter, we focus on demographic and environmental stochasticities and spatial structure. Genetics, catastrophes, and community structure are important in some PVAs (e.g., Hedrick and Miller 1992; Haig et al. 1993; Lacy 1997); however, extinction is usually more affected by demographic and environmental stochasticities (Lande 1988a,b; Boyce 1992). Genetic and catastrophic stochasticities and community interaction models (e.g., competition models, predator-prey models) may be integrated at the appropriate level in the process described herein (Fig. 60.5).

We recommend the use of PVAs consistent with previous efforts that have used habitat to stratify stochastic modeling (e.g., Akçakaya et al. 1995; Lindenmayer and Lacy 1995b; Root 1998), but we advocate the use of home ranges instead of habitat patches as the planning unit. The distinction between demographic and environmental stochasticities in PVA is subtle and conforms to the organism-subpopulation-population hierarchy discussed earlier (Fig. 60.5). We propose stratifying demographic simulations in PVAs by home range types (viable, marginal, and nonviable) that form a subpopulation. This approach is analogous to the "environmental states" modeling described by Beissinger (1995). Classifying environments into "states" has been used as a method by demographers to measure environmental predictability (re-

viewed by Beissinger 1995). By definition (Roloff and Haufler 1997), the demographics of viable and nonviable home ranges will be less variable than marginal ranges (Fig. 60.2). The purpose of stratifying demographic variation by home range type is to more consistently represent these differences in the stochastic simulation (Beissinger 1995).

Environmental stochasticity applies to subpopulations as delineated by groups of home ranges. The key to expressing subpopulation averages for vital rates that retain the home-range-specific demographics is to portray the subpopulation mean and associated data distribution in conformity with home range composition. For example, if a subpopulation is dominated by viable home ranges, environmental simulations should occur from a data distribution that conforms to this skewness. Several PVA computer programs automate this simulation process (Table 60.1), but we are not aware of any that accommodate data distributions that deviate from normal. Monte Carlo simulation works well for normally distributed data and small sample sizes (Miller and Lacy 1999). For non-normal distributions, bootstrapping has also been used (Sæther et al. 1998), but bootstrapped simulations based on low sample sizes are suspect.

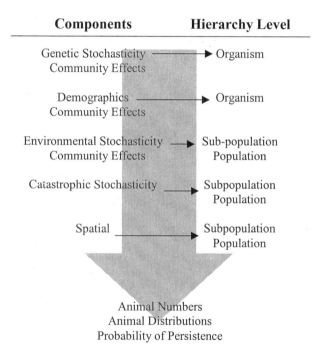

Figure 60.5. The components and hierarchical levels of population viability analysis (PVA).

After simulating the effects of demographic and environmental stochasticities on the population, the final step in the PVA is to analyze the spatial structure of subpopulations or conduct a metapopulation analysis. Metapopulation analysis is based on the notion that subpopulations are spatially structured into assemblages of local breeding populations and that immigration and emigration among subpopulations have an effect on local population dynamics (Hanski and Simberloff 1997). Some metapopulation parameters (e.g., number of emigrants over time; dispersal potential) for the subpopulations can be estimated from the habitat-based and stochastic analyses discussed earlier. Subpopulations that represent "local breeding populations" can be identified from mapped viable home ranges. It is important to note that we are not *a priori* recognizing the population as a metapopulation. Metapopulations have distinct characteristics (under the classical definition) and as such are rare in real systems (Harrison and Taylor 1997). Rather, we recommend using the tools associated with metapopulation analyses (Harrison 1993; Hanski 1996; McCullough 1996; Hanski and Gilpin 1997) to identify and describe the spatial structure of the subpopulations and population.

Spatial structure in PVA influences species viability at two scales: within subpopulations and between subpopulations (if more than one exists). Within a subpopulation, managers must consider the quality of habitat between viable home ranges, distance between viable home ranges, movement capabilities of the organism, organism susceptibility to mortality while in transit, and life history spacing strategies (Roloff and Haufler 1997). Between subpopulations, the size, shape, and relative contribution of each subpopulation have important consequences for population persistence and stability (Maurer 1994). The questions facing resource managers include whether these subpopulations interchange, the contribution of individual subpopulations to overall population structure and stability, and the resiliency of the system to landscape-level dynamics (e.g., fire, drought). Metapopulation analysis was developed to assist managers in addressing these questions and software has been developed to aid in these analyses (e.g., RAMAS/space; Akçakaya and Ferson 1990; Possingham et al. 1992; RAMAS-GIS; Akçakaya 1994).

Conclusions

As outlined above, we envision our habitat-based PVA approach as a hierarchically structured process for indexing the persistence of organisms over time. Components of the hierarchy include the organism, subpopulations, and population (Fig. 60.5). Although the components of a PVA can be associated to specific levels in the hierarchy, the effects of each component cascade through all levels and are ultimately expressed as animal numbers and distributions and probability of persistence (Fig. 60.5). Habitat pervades all of these levels and is an ultimate determinant of viability (Gilpin 1987; Lande 1987; Lawton et al. 1994; Hanski et al. 1996a; Fahrig 1997; Drechsler and Wissel 1998). Here, we presented a spatially explicit approach to population viability modeling that accounts for individuals, subpopulations, and populations within a landscape and that offers a means for defining spatial relations between habitat patches.

In a complex system such as the environment that influences organism viability, validation of the assumptions, models, linkages, and processes described here is critical (Morrison et al. 1998). It is naïve to think we can consistently model organism viability with complete accuracy. Nonetheless, land management decisions that influence viability are made on a daily basis, and, thus, we as wildlife professionals must offer the tools for making informed, defensible decisions. Validation is the key to credibility, and it should be used to understand and quantify sources of error. Validation should occur at all levels in the process described above, starting with the assumptions, variables, relationships, and outputs of the species habitat model, and transcending through all levels of the stochastic hierarchy (Fig. 60.5). Model errors propagate through a process like the one described here (Bender et al. 1996; Morrison et al. 1998), and, thus, it is critical to minimize and incorporate that error into outputs. We encourage model developers and users to rigorously validate their modeling processes. In addition, decision makers should scrutinize model error and understand the limitations of using model output in an inherently variable environment.

Relations between Canopy Cover and the Occurrence and Productivity of California Spotted Owls

Carolyn T. Hunsaker, Brian B. Boroski, and George N. Steger

Effective land management for multiple uses such as wildlife habitat, timber production, and fuels management requires a multiscaled landscape planning approach (Raphael and Holthausen, Chapter 62). This chapter addresses forest structural characteristics associated with the California spotted owl (*Strix occidentalis occidentalis*) in the southern Sierra Nevada of California. The California spotted owl was designated as a Sensitive Species on national forests throughout California in the late 1970s, and the northern spotted owl (*Strix occidentalis caurina*) was federally listed in 1990 as a threatened species. Previous studies have shown a strong association between both subspecies of spotted owls and mature and old-growth forests at the scales of (1) home ranges, (Blakesley et al. 1992; Verner et al. 1992; Hunter et al. 1995; Bart 1995b; Franklin et al. 2000) and (2) habitat components within home ranges (LaHaye 1988; Solis and Gutiérrez 1990; Verner et al. 1992). Franklin et al. (2000) found that climate and habitat models could account for 88.5 percent of the total observed process variation in owl reproduction.

This chapter evaluates three hypotheses: (1) no difference between the composition of canopy-cover classes in the study area as a whole and sites used by spotted owls, (2) no difference in the composition of canopy-cover classes among analysis areas exhibiting different levels of occupancy and productivity, and (3)

no difference in conclusions about relations between owl occupancy and productivity and canopy-cover classes based on aerial photography or Landsat Thematic Mapper imagery. First, we tested for differences in the proportion of canopy-cover classes in the study area as a whole and sites used by spotted owls. If owls are selecting areas based on canopy cover, the next questions to answer are what classes relate to occurrence over time and what classes correspond to different levels of productivity? We therefore tested for differences in the composition of canopy-cover classes among analysis areas exhibiting different levels of occupancy and owl productivity. Last, we recognized that different sources of canopy-cover data might result in a different composition of the canopy-cover classes within the landscape. Thus, we tested whether the rejection of the first two hypotheses was influenced by the data source (aerial photography and satellite imagery) used to derive canopy-cover values.

The study was done in the Sierra National Forest in the Sierra Nevada of California (Fig. 61.1). The study area encompassed about 60,600 hectares and ranged in elevation from 853 to 2,743 meters. Five vegetation types, as described in Mayer and Laudenslayer (1988), occur in the study area: montane hardwood-conifer, ponderosa pine (*Pinus ponderosa*), Sierran mixed-conifer, white fir (*Abies concolor*), and red fir (*Abies magnifica*).

Figure 61.1. Location of spotted owl study area within the Sierra National Forest and the state of California.

Methods

Spotted owls were monitored from 1990 through 1998. Owl activity centers are areas within which owls find suitable nesting sites and several suitable roosts and in which they do a substantial amount of their foraging (Zabel et al. 1992). Activity centers are smaller than both territories and home ranges. The mean home-range size was estimated by the minimum convex polygon (MCP) method (Mohr 1947; Odum and Kuenzler 1955; and Jennrich and Turner 1969) to be 728 ± 319 hectares for individual owls during the breeding season in conifer forests of the Sierra National Forest (Verner et al. 1992). For our study, the central location of an activity center is a single geographic location where owls were observed roosting or nesting repeatedly over time; we refer to this location as a site.

Owl Demographics

Roosts and nests were located using methods described by Forsman (1983) and based on actual owl observations. Sex, age class, pair status, nesting status, and reproductive success of each owl were determined (Forsman 1981, 1983). Reproductive success at each site was determined through repeat observations (minimum of six) of adult owls and their fledglings (Steger et al. 1993).

A productivity score was assigned to each site every year it was surveyed to protocol standards (Steger et al. 1993). Values ranged from 0 to 9 based on the presence of owls, their attempt to nest, and the number of fledglings produced. A nonsequential series was used to represent the extra energy required to nest and produce fledglings. Sites that were sampled and determined to be empty were assigned a value of zero while a score of 1 was given to those containing a single bird. Sites with non-nesting pairs and pairs that nested but failed to fledge young were given scores of 2 and 4, respectively. Lastly, scores of 7, 8, and 9 were assigned to sites producing one, two, and three fledglings. A summary productivity index was calculated for each site across the years it was surveyed by summing the yearly scores for the site and dividing by the number of years it was surveyed (Table 61.1).

Estimation of Owl Activity Centers

Telemetry data were not available for owls at every site, but a subset of spotted owls was captured, fitted with radio transmitters, and tracked from spring 1987 through fall 1990. The number of locations per owl ranged from thirty-two to sixty-seven (means and standard deviations by year: 1987: 45.6, 6.9; 1988: 56.6, 12.4; 1989: 38.5, 11.0) and typically represented single nightly locations that were mapped on 7.5-minute-series maps (1:24,000-scale U.S. Geological Survey, Menlo Park, CA 94025). We used data from these owls to calculate generic owl analysis areas. We considered it reasonable to represent the generic owl analysis areas as circles because owls fit a central-place foraging pattern. To calculate these circles, we first estimated 50 percent, 70 percent, and 90 percent MCP home-range estimates for each radio-tagged owl and represented home range areas as circles having the radii r_{50}, r_{70}, and r_{90}. Mean values for these radii were calculated from all radio-tagged owls having at least twenty-five locations between sunset and sunrise throughout the breeding season. Using a geographic information system (GIS) (ARC/INFO Version 7.0, Environmental Systems Research Institute, Redlands, CA 92373), these mean radii were used to generate three concentric circles of 72, 168, and 430 hectares around the central location of each owl activity center. Here, we call these circles owl analysis areas and use them to represent areas likely to be used by owls.

Vegetation Classification

Tree canopy-cover values came from two different vegetation data sets, one based on aerial photography and the other based on Landsat Thematic Mapper imagery. For the aerial photography (hereafter referred to as photography), vegetation was classified by G. N. Steger into habitat communities representing a homogeneous unit of vegetation for overstory, understory, and base material (soil or rock). The key features used to determine homogeneity in the stands were size and spacing of trees and the species and density of the understory vegetation. A minimum mapping unit of 1 hectare was used for delineating vegetation class polygons, although some features such as meadows, lakes, and houses were

TABLE 61.1.

Productivity indices for California spotted owls (*Strix occidentalis occidentalis*) on the Sierra National Forest in California.

	Score by year[a]									Summary score[b]	Years surveyed
Site ID	1990	1991	1992	1993	1994	1995	1996	1997	1998		
3	2			1	1	0	0	0	0	0.57	7
4	7	2	8	2	8	2	2	2	4	4.11	9
5			0	1		1	0	1		0.60	5
6	7			1	1	2	2	2	4	2.71	7
9					2	1	0	0		0.75	4
15	2				0	0	0	0		0.40	5
25	8	7	7	2	4	4	2	1	7	4.67	9
31						0	1	0	1	0.50	4
33	8		7	7	2	2	2	0		4.00	7
35	2	2	9	2	7	2	7	2	2	3.89	9
36	2	1		0	0	0	0	0	0	0.38	8
38	2	2	8	8	2	2	2	2	2	3.33	9
41	2	2	4	7	4	4	4	2	4	3.67	9
43	2		9	2	2	2	2	0	2	2.63	8
48	1	2	7	2	2	2	2	2	0	2.22	9
49	8	2	9	4	4	4	2	4	0	4.11	9
53	8	2	8	8	2	1	2	2	1	3.78	9
57		7	8	2	0	0	0	0		2.43	7
58	2	7	8	2	2	2	2	2	2	3.22	9
61	7	7	8	8	4	2	2	2	7	5.22	9
62	8	2	2	8	2		0	0		3.14	7
64		1	9	2	4		2	8	4	4.29	7
65	7	2	7	8	2	4	4	2	1	4.11	9
67	2	2	8	4	7	2	2	2	2	3.44	9
70	2	1	1	1	2	1	0	1		1.13	8
77	8	2	9	7	2	2	2	2	1	3.89	9
80	8	2	0	1	1	0	0	0		1.50	8
83	1	2	9	2	2	0	0	0	0	1.78	9
84	2	2	8	2	2	2	2	2	2	2.67	9
87	2		7	2	7	2	8	8	2	4.75	8
91a	8	4	8	2	4	4	2	2	4	4.22	9
91b									4	4.00	1
100	1	2	8	2	8	2	2	2	2	3.22	9
219					2	2	1	2	2	1.80	5
221					7	2	2	8	2	4.20	5
225						2	0	0		0.67	3
227					2	2	2	2		2.00	4
228					2	2	2	2	2	2.00	5
229a					8	2	2	1	2	3.00	5
229b					1		2	2	2	1.75	4
230					2	2	2	2	2	2.00	5
234					2	2	2	2	2	2.00	5
239					8	2	8	2	0	4.00	5

TABLE 61.1. (Continued)

Productivity indices for California spotted owls (*Strix occidentalis occidentalis*) on the Sierra National Forest in California.

Site ID	Score by year[a]									Summary score[b]	Years surveyed
	1990	1991	1992	1993	1994	1995	1996	1997	1998		
241					8	2	2	2	8	4.40	5
244					2	7	2	2	1	2.80	5
245					8	2	2	2	2	3.20	5
247					8	2	2	8	7	5.40	5
257							2	0		1.00	2
266					2		0	0		0.67	3

[a]0 = no owls; 1 = single owl; 2 = non-nesting pair; 4 = failed nest; 7 = one fledgling; 8 = two fledglings; 9 = three fledglings.

[b]Summary score = (sum of annual scores / years surveyed).

portrayed using a resolution of 0.5 hectare. Vegetation typing from aerial photographs followed the guidelines from Avery (1978). Using a stereoscope, lines delineating polygons were drawn on 22.9 by 22.9 centimeter vertical photographs flown in 1996. These images were geocorrected when transferred onto orthophotoquads (U.S. Geological Survey, Menlo Park, CA 94025) and then scanned into a GIS (CartaLinx, Clark Lab, Clark University, Worcester, MA 01610-1477).

The structural characteristics derived from the photography included crown cover of trees with a crown diameter greater than 8 meters, crown cover of all trees, and crown cover of all vegetation with each being subdivided into five canopy-cover categories (State of California Resource Agency 1969). The type of vegetative cover and dominance of land-cover types—conifer, hardwood, shrub, grass, ground, rock, or cultivated land—were recorded in rank order from highest to lowest within the polygon. Crown-cover increments of 5 percent were used for conifer, hardwood, and shrub types, and our classes were based on these values. For this analysis, the vegetation and structural data from the photography data have not been extensively groundtruthed; however, this is being done in 2000–2001.

The second data set for canopy cover was a 1995 classified image from the Landsat Thematic Mapper satellite (hereafter referred to as Landsat); the classified image was produced by San Diego State University (J. Franklin personal communication). The classified image had a pixel size of 30 meters and represented canopy cover in 10 percent increments using a canopy reflectance model. We ran a 7×7 modal filter followed by a 3×3 modal filter on this image so the mean patch size per class was similar to the photography data ($F = 0.87$, df = 4,6426, $P = 0.48$). Averaging attributes over areas larger than the single grid cell tends to reduce errors in the estimates of those geographic phenomena that are spatially autocorrelated. Although formal accuracy assessment is not available for the Landsat data, the producers of these data believe that the canopy cover falls within two cover classes (20 percent) of the actual cover most of the time and within one cover class much of the time (J. Franklin personal communication).

For this analysis, forest structure was represented as five categories of canopy closure: 0–19 percent, 20–39 percent, 40–49 percent, 50–69 percent, and 70–100 percent. The planimetric proportions of each canopy-cover class were derived for the three concentric owl analysis areas around each owl activity center.

Canopy Cover of Owl Analysis Areas

We examined second-order habitat selection (Johnson 1980) by testing the null hypothesis that spotted owls arbitrarily selected areas to use based on the available composition of canopy-cover classes between the elevations of 853 and 2,743 meters within the study area. The maximum likelihood statistics resulting from the compositional analysis method described by Aebischer et al. (1993) were used to determine significant departures from random for the concentric analysis areas within the 60,600-hectare study area.

The above approach was also used to examine third-order selection (Johnson 1980) by testing the null hypothesis that the composition of canopy-cover classes within the 72-hectare analysis area was no different from the composition within the 430-hectare analysis area. An additional compositional analysis tested the null hypothesis that the central locations of the analysis areas for our middle size (168-hectare) analysis areas reflected the classes' availabilities within these areas, ranking the classes based on their relative occurrence as central locations (Aebischer et al. 1993).

The occupancy of activity centers by owls varied throughout the study period (Table 61.1). Using a general linear model (SPSS 1998), we examined the relationship between the occupancy of a site (no birds, single bird, or pair) and yearly variation, the canopy-cover class at the center of the activity center, and the proportion of moderate to dense (50–100 percent canopy cover) and sparse to open habitat (0–39 percent canopy cover) within the 72-hectare analysis area. At this stage of analysis, multivariate models are being used to explore relations between habitats and owl occurrence and productivity.

Productivity scores were compared to the proportions of canopy-cover classes within the analysis areas to determine whether productivity correlated well with the classes. We calculated Pearson correlation coefficients for groups of classes: classes 1 and 2, and 3 through 5. Using a generalized additive model for this binary data (S-PLUS 4 1997; Hastie and Tibshirani 1990), we evaluated the relations between observed site characteristics and the success of activity centers (a successful activity center was defined to mean that it includes a nesting pair of owls). Explanatory variables included in the model were year, the canopy-cover class at the center of the activity center, and a smooth function of the percentage of the analysis area within the 40–49 percent, 50–69 percent, and 70–100 percent canopy-cover classes. Using the same variables, we used a generalized additive model to explore the relations between the canopy-cover classes in an analysis area and the number of fledglings at a successful site.

Results

Nine owls in 1987 and eight each in 1988 and 1989 met the criteria of having at least twenty-five nighttime locations. The mean size of estimated home ranges for owls was similar across years for MCP estimates based on 50 percent, 70 percent, and 90 percent of the radio locations (Table 61.2). The three concentric owl analysis areas calculated from the mean radii for individual home ranges were 72, 168, and 430

TABLE 61.2.

Minimum convex polygon home ranges (hectares) of California spotted owls in the Sierra National Forest, Fresno County, California.

		Means[a]								
		50% of locations			70% of locations			90% of locations		
Year	n	area	SD	radius	area	SD	radius	area	SD	radius
1987	9	73.8	44.3	467	194.2	99.6	763	442.9	193.4	1159
1988	8	73.3	36.9	468	169.4	97.2	713	457.0	230.5	1175
1989	8	83.4	37.4	503	165.3	62.9	714	458.5	228.9	1176
All years		76.7	38.5	479	177.0	86.0	731	452.4	208.1	1170
Test statistic		0.80			0.57			0.001		
Probability		0.67			0.75			0.99		

[a]Mean radius (m) represents the average radius of circles having areas equal to the individual home-range estimates. Test statistics comparing home-range area across the three years are from Kruskal-Wallis one-way analysis of variance tests, assuming a chi-square distribution with two degrees of freedom.

hectares. Figure 61.2 depicts these analysis areas for site 61 with respect to the canopy-cover classes derived from photography and from Landsat.

Forty-nine owl activity centers were surveyed within the study area from 1990 to 1998. The number of years surveyed during the nine-year period varied but averaged seven years per site (range one to nine years). The distributions for occurrence and productivity values were non-normal and reflected the tendency for sites to be occupied by non-nesting pairs of birds and pairs that nested but failed to fledge young (median summary score for productivity = 3, range 0.38 to 5.40). Pairs of owls were observed at the forty-nine sites nearly 80 percent of the time (median occurrence = 2, sd = 0.72).

Vegetation Composition

Although the proportion of a given canopy-cover class within the study area varied depending on the method used to characterize the landscape (Table 61.3), both methods resulted in similar distributions of available habitat (Kolmogorov-Smirnov two-sided probability = 0.82) (Fig. 61.3). Landsat data depicted more of the landscape in canopy-cover classes of 50–69 percent and 70–100 percent as compared to photography data. Mean polygon size differed significantly among classes for both the photography (F = 4.80, df = 42,710, P < 0.001) and Landsat data (F = 4.23, df = 43,716, P = 0.002). The mean size of polygons depicting areas with at least 70 percent canopy cover was significantly greater than for canopy cover classes from 0 to 49 percent when photography data were used and significantly greater than the 20–39 percent and 40–49 percent classes when Landsat data were used (Table 61.3). The 70–100 percent and 0–19 percent canopy-cover classes were dominant in the photography data set as compared to the Landsat data set, in which the 50–69 percent and 70–100 percent classes were dominant.

Compositional analyses based on both photography and Landsat data sets for canopy cover lead us to reject the hypothesis that the selection of sites used by owls was random within the study area. The likelihood ratio statistics were 41, 30, and 34 for the photography data and 69, 45, and 43 for the Landsat data with reference to the 72-hectare, 168-hectare, and

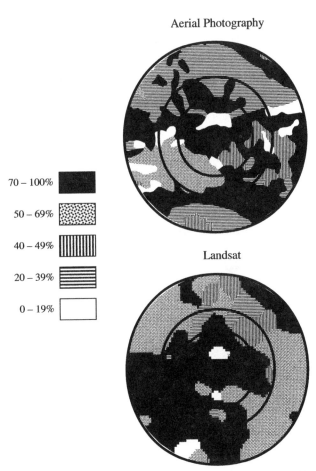

Figure 61.2. The canopy-cover classes estimated from aerial photography and Landsat Thematic Mapper data are shown for the three different sizes of owl analysis areas around site 61.

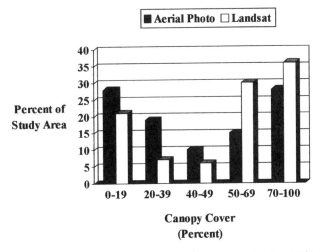

Figure 61.3. Distribution of canopy-cover classes for the study area by data source: aerial photography and Landsat Thematic Mapper.

TABLE 61.3.

Data on available habitat classes derived from aerial photography and Landsat Thematic Mapper data for a 60,600-hectare study area on the Sierra National Forest.

Class	% Canopy closure	Area (ha)	% of landscape	Number of polygons	$\bar{x}$ polygon size (ha)	Standard deviation	Class differences[a]
Aerial photo							
1	70–100	16,894	28	453	37.3	123.1	3,4,5
2	50–69	9,459	15	410	23.1	50.7	
3	40–49	5,940	10	341	17.4	34.8	1
4	20–39	11,556	19	561	20.6	73.6	1
5	0–19	16,732	28	950	17.6	88.3	1
Landsat Thematic Mapper							
1	70–100	21,784	36	514	42.4	471.7	3,4
2	50–69	17,987	30	871	20.7	78.1	
3	40–49	3,734	6	815	4.6	7.3	1
4	20–39	4,481	7	751	5.9	11.0	1
5	0–19	12,540	21	770	16.3	62.8	

[a]Class differences based on mean polygon size were determined by an analysis of variance test followed by a post hoc Tukey multiple comparison test. Significant differences between classes were based on an alpha level of 5 percent.

430-hectare analysis areas (in all cases df = 4, $P < 0.00001$). Results from both data sets also rejected the null hypothesis that the composition of habitat classes within the 72-hectare area was not different from that in the 430-hectare area (likelihood ratio statistics equaled 47 (photography) and 56 (Landsat), df = 4, $P < 0.00001$). Last, we examined the canopy-cover classes for the polygon at the center of the analysis area and determined that the selection of these central locations within the 168-hectare areas was significantly nonrandom (likelihood ratio statistics equaled 28 (photography) and 19 (Landsat), df = 4, $P < 0.001$). Rankings of the classes based on their relative use within the analysis areas were different for the photography and Landsat data. The photography data resulted in the ranked order of $\underline{1 > 3 > 2 > 5} > 4$ while the Landsat data returned a ranking of $\underline{1 > 4 > 3 > 5} > \underline{2}$, where underlined classes reflect no significant difference at an alpha level of 5 percent.

Owl Occupancy

The results from general linear models examining the relations between occurrence, yearly variation, and canopy cover were similar for both data sets, with the exception of the factor that described the canopy-cover class at the activity center. In both models, year was significant (photography $F = 2.8$, df = 8, 315, $P = 0.005$, Landsat $F = 3.5$, df = 8, 312, $P = 0.001$) as was the proportion of the 72-hectare area comprising habitat with canopy cover between 0 and 39 percent (photography $F = 49.4$, df = 1, 315, $P < 0.001$; Landsat $F = 5.3$, df = 1, 312, $P = 0.022$). In both models, the occupancy of sites increased as the proportion of the area having canopy-cover values between 0 and 39 percent declined, but the proportion affecting occupancy differed for photography and Landsat data sets (Fig. 61.4). In the model with Landsat data, the canopy-cover class at the activity center was also a significant variable ($F = 6.8$, df = 3, 312, $P < 0.001$). When the canopy cover at the activity center was between 20 and 39 percent, occupancy averaged 0.75 birds (sd = 0.87), which was significantly less (Tukey multiple comparison test $P < 0.02$) than when the crown cover at the activity center was greater than 50 percent (cover 50–69 percent, mean = 1.51 birds, sd = 0.83; cover 70–100 percent, mean = 1.71 birds, sd = 0.63).

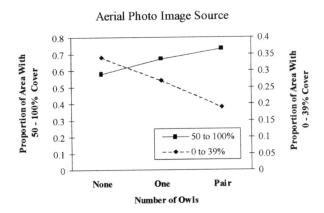

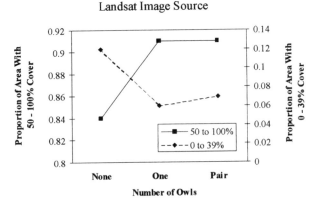

Figure 61.4. The relations between canopy cover and the occurrence of spotted owls within 72-hectare areas surrounding known owl activity centers for two image sources. Data on the annual occupancy of the activity centers was based on nine years of data from the Sierra National Forest, Fresno County, California.

Owl Productivity

For both image types and across the three analysis areas, scores of productivity were positively correlated with the proportion of the analysis area having *greater than or equal to* 50 percent canopy-cover and negatively correlated with those having less than 50 percent cover (Table 61.4). The median proportions for these classes for unproductive sites (productivity scores *less than or equal to* 2) and productive sites (productivity scores greater than 2) were different for estimates based on the aerial photography and Landsat imagery (Fig. 61.5). Mann-Whitney U test statistics for each analysis area by breeding status combination returned probability values of less than 0.0001 for comparisons between proportions derived from photography and Landsat imagery. The proportions of the analysis areas in habitat classes with greater than

or equal to 50 percent canopy cover were substantially greater for sites with productivity scores greater than 2 (Table 61.5). As analysis areas increased in size, the trend was for a greater proportion of the areas to be in classes 3, 4, and 5 irrespective of the productivity score and type of imagery. The median proportions for classes 1 and 2 were consistently larger for analysis areas having a productivity score greater than 2 for photography data and larger or almost the same for Landsat data. For all analysis areas, the median proportions of the 0–19 percent canopy-cover class (unsuitable owl habitat) were larger in the areas having productivity scores less than 2.

Models of productivity using canopy-cover estimates from photography data had more significant class determinants than models using data from Landsat. The generalized additive model for binary data (owl[s] present or owl[s] productive) using photography identified year ($P < 0.001$) and the proportion of the area within 70–100 percent ($P = 0.047$) and 40–49 percent ($P = 0.036$) canopy-cover classes as significant factors affecting the probability of the site consistently supporting breeding activity. The logistic model using Landsat identified only year ($P < 0.001$) and the proportion of the area in the 40–49-percent canopy-cover class ($P = 0.011$) as significant factors.

The generalized additive model for the number of fledglings at a site, which used canopy-cover estimates from photography, indicated that the canopy-

TABLE 61.4.

Pearson correlation coefficients between productivity scores and the proportion of canopy-cover classes derived from aerial photography and Landsat data.

Image source	Analysis area[a] (ha)	Classes 1, 2	P	Classes 3, 4, 5	P
Photography	72	0.30	0.04	−0.30	0.04
Landsat	72	0.29	0.04	−0.29	0.04
Photography	168	0.35	0.01	−0.35	0.01
Landsat	168	0.36	0.01	−0.36	0.01
Photography	430	0.37	0.01	−0.37	0.01
Landsat	430	0.33	0.02	−0.33	0.02

[a]Probability values are given for three owl-analysis areas (72 hectares, 168 hectares, and 430 hectares) centered on owl nest sites or primary roost sites and are based on Bartlett's chi-square statistic with one degree of freedom.

TABLE 61.5.

Proportion of canopy-cover classes within owl analysis areas[a] on the Sierra National Forest derived from aerial photography and Landsat imagery.

Canopy cover class	Imagery[b]	Productivity scores[c] ≤ 2				Productivity scores[d] > 2			
		%[e]	Median	Range	SD	%[e]	Median	Range	SD
72-hectare area									
70–100	A	94	0.52	0.00–0.94	0.24	100	0.53	0.06–0.92	0.23
	L	100	0.61	0.12–0.98	0.25	100	0.67	0.09–1.00	0.26
50–69	A	72	0.04	0.00–0.62	0.18	84	0.13	0.00–0.78	0.20
	L	100	0.22	0.02–0.77	0.22	94	0.20	0.00–0.68	0.22
40–49	A	72	0.03	0.00–0.29	0.08	68	0.04	0.00–0.56	0.14
	L	61	0.02	0.00–0.14	0.04	58	0.00	0.00–0.16	0.04
20–39	A	94	0.16	0.00–0.53	0.17	87	0.07	0.00–0.41	0.10
	L	50	0.00	0.00–0.25	0.07	55	0.00	0.00–0.13	0.03
0–19	A	89	0.13	0.00–0.42	0.11	90	0.07	0.00–0.38	0.08
	L	83	0.03	0.00–0.42	0.10	65	0.01	0.00–0.30	0.08
168–hectare area									
70–100	A	94	0.46	0.00–0.89	0.21	100	0.48	0.16–0.84	0.18
	L	100	0.61	0.09–0.93	0.24	100	0.59	0.10–0.98	0.23
50–69	A	89	0.09	0.00–0.44	0.13	87	0.15	0.00–0.62	0.16
	L	100	0.17	0.05–0.63	0.18	100	0.24	0.01–0.58	0.18
40–49	A	89	0.04	0.00–0.33	0.09	77	0.06	0.00–0.48	0.12
	L	83	0.02	0.00–0.22	0.06	87	0.02	0.00–0.13	0.03
20–39	A	100	0.20	0.00–0.47	0.13	100	0.13	0.01–0.33	0.09
	L	83	0.03	0.00–0.31	0.09	77	0.01	0.00–0.15	0.04
0–19	A	100	0.19	0.00–0.35	0.10	97	0.10	0.00–0.31	0.07
	L	89	0.09	0.00–0.34	0.09	90	0.06	0.00–0.29	0.07
430–hectare area									
70–100	A	100	0.39	0.00–0.73	0.19	100	0.43	0.14–0.74	0.14
	L	100	0.57	0.14–0.80	0.20	100	0.53	0.15–0.85	0.20
50–69	A	100	0.11	0.00–0.43	0.11	100	0.17	0.00–0.47	0.11
	L	100	0.26	0.07–0.53	0.13	100	0.29	0.05–0.55	0.14
40–49	A	94	0.08	0.00–0.38	0.10	97	0.08	0.00–0.34	0.09
	L	100	0.03	0.00–0.18	0.05	100	0.03	0.00–0.15	0.03
20–39	A	100	0.21	0.03–0.45	0.11	100	0.16	0.03–0.32	0.07
	L	94	0.04	0.00–0.18	0.06	97	0.02	0.00–0.16	0.04
0–19	A	100	0.19	0.05–0.36	0.10	100	0.11	0.03–0.26	0.07
	L	100	0.11	0.02–0.34	0.08	100	0.09	0.00–0.23	0.06

[a]n = 49.

[b]Aerial photography = A; Landsat = L.

[c]Eighteen analysis areas had productivity scores less than or equal to two.

[d]Thirty-one analysis areas had scores greater than two.

[e]The number of times a class type was present in an analysis area is expressed as the percentage (%) of the total number of areas.

Aerial Photo Image Source

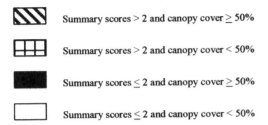

Summary scores > 2 and canopy cover ≥ 50%

Summary scores > 2 and canopy cover < 50%

Summary scores ≤ 2 and canopy cover ≥ 50%

Summary scores ≤ 2 and canopy cover < 50%

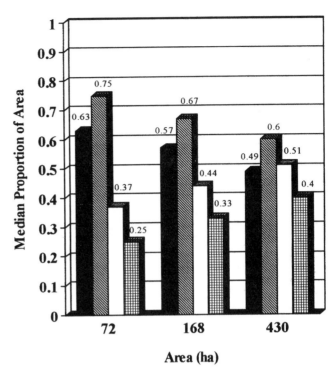

Landsat Image Source

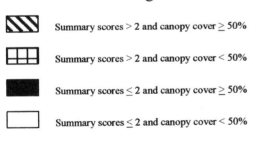

Summary scores > 2 and canopy cover ≥ 50%

Summary scores > 2 and canopy cover < 50%

Summary scores ≤ 2 and canopy cover ≥ 50%

Summary scores ≤ 2 and canopy cover < 50%

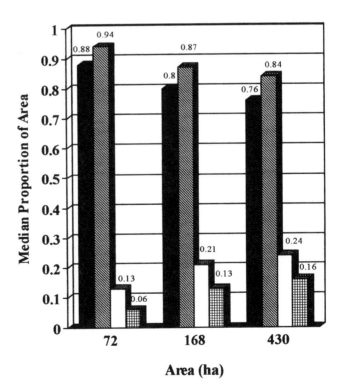

Figure 61.5. Median proportion in each owl analysis area for nesting and good foraging habitat (50–100 percent canopy cover, classes 1 and 2 combined) and for foraging and unsuitable habitat (50 percent canopy cover or less, classes 3, 4, and 5 combined) by productivity score groups. Productivity scores greater than 2 indicate breeding success while scores less than or equal to 2 indicate nonbreeding situations.

cover of the habitat containing the center of the analysis area ($P = 0.026$), year ($P < 0.001$), and the proportion of the area in the 50–69-percent ($P = 0.026$) and 40–49 percent ($P = 0.010$) canopy-cover classes were significant. In contrast, results of the generalized additive model using Landsat indicated that year ($P < 0.001$) was the only significant factor explaining the number of fledglings produced at a site. The proportion of the area in the 50–69 percent cover class was not significant ($P = 0.692$), but the proportion of the area in the 40–49 percent class ($P = 0.081$) and the habitat containing the center of the analysis area ($P = 0.053$) would be significant at an alpha level of 10 percent.

Discussion and Conclusions

The main focus of this research was to evaluate the influence of different vegetation data sets on the relations between canopy cover and the occurrence and productivity of the owls. Our research builds on the work of Verner et al. (1992), who evaluated forest characteristics for roost, nest, and home-range locations with respect to all available habitats in the Sierra Nevada. We analyzed relations between forest canopies and the occupancy and productivity of owl activity centers over the past nine years. Results tested the following three hypotheses and rejected them all:

- No difference in the proportion of canopy-cover classes in the study area as a whole and sites used by spotted owls.
- No difference in the composition of canopy-cover classes among analysis areas exhibiting different levels of owl productivity.
- No difference in conclusions about relations between owl occurrence and productivity and canopy-cover classes based on aerial photography or Landsat Thematic Mapper imagery.

Owls selected use areas at three scales: (1) activity centers as represented by our analysis areas within the study area, (2) core areas within activity centers, and (3) roosting or nesting habitat. Results show that the California spotted owl is selecting for specific forest structure, especially near the central location of their activity center (72-hectare and 168-hectare areas) and,

furthermore, that the composition of the area relates to the occupancy of the site by owls through time. Verner et al. (1992) found the mean size of nest stands to be 40 hectares in the Sierra Nevada. Our results corroborate the findings of Verner et al. (1992): that California spotted owls are habitat specialists and, for nesting, they select stands with relatively closed canopies (greater than 70 percent). They also found that owls foraged significantly more than expected in stands with greater than or equal to 70 percent canopy cover and significantly less than expected in stands with 0–39 percent canopy closure. Laymon's (1988) and Call's (1990) studies suggest that spotted owls in the Sierra Nevada tended to forage in stands of intermediate to older ages.

Hunter et al. (1995) and Meyer et al. (1998) found that landscape characteristics had the highest levels of significance between random sites and sites used by northern spotted owls in the Klamath province when circles with radii of 0.8 kilometer were used for analysis areas as compared to larger circles centered around the same central locations. Franklin et al. (2000) used a 0.71-kilometer-radius circle (one-half the median nearest-neighbor distance among thirty-seven territory centers) to represent spotted owl territories—similar to our 168-hectare analysis area with a radius of 0.73 kilometer. Meyer et al. (1998) suggested that characteristics of these core areas may be most influential in determining territory locations for northern spotted owls.

We conclude that canopy-cover relates to owl occurrence and productivity within our study area. Productivity scores were significantly correlated with canopy closure, and canopy-cover classes were significant in the multivariate models. For sites that consistently produced young, the median proportion of good habitat (canopy cover greater than 50 percent) was usually about 10 percent higher than for unproductive sites (based on photography data in Fig. 61.5). The values ranged from 75 percent of the 72-hectare analysis area to 60 percent of the largest analysis area. When canopy cover was based on Landsat data, the difference between productive and unproductive sites remained, but the magnitude decreased to a 5–7 percent difference. Landsat data suggest that a higher proportion (80 to 90 percent) of the owl analysis area needs to be in good habitat. Recall, however, that Landsat depicted 23 percent more of the landscape as having 60–100 percent canopy cover. Our results from the photography data agree with Bart and Forsman (1992), who reported that areas with less than 40 percent suitable owl habitat supported lower densities of spotted owls, and pairs had lower reproduction than in areas with greater than 60 percent suitable owl habitat.

Although correlations between reproduction and canopy cover in our data were not large, they were consistent. Classes 1 and 2 were positively correlated and classes 3, 4, and 5 were negatively correlated with our productivity index for all three sizes of analysis area. From these correlations, one would conclude that the threshold between canopy cover values that contribute to or detract from occurrence and productivity is a value near 50 percent. Indeed, the occurrence of owl pairs at sites declined as the proportion of habitat with 0–39 percent canopy cover increased in the area surrounding the activity center. A difference of 10 percent in cover (say between 40 and 50 percent, or between 50 and 60 percent) is small and can easily be similar to the magnitude of uncertainty in canopy cover from remotely sensed data and in field measurements made by different people or measurement devices.

Results from our multivariate models are logical given the spotted owl's strong association with late seral-stage forests for nesting and roosting and with early seral stages for prey sources. Zabel et al. (1995) found spotted owls foraging near edges of late and early seral-stage forests, and Ward et al. (1998) reported that woodrat abundance at sites where owls foraged was greatest at the ecotone between late and early seral stages. Franklin et al. (2000) found reproductive output to depend on a substantial amount of edge habitat, a low amount of core area, and a certain amount of habitat fragmentation. His models indicated that changes in reproductive output were most sensitive to changes in edge, while our results indicate that changes in canopy-cover composition of less than 10 percent can significantly affect occupancy.

Franklin et al. (2000) stress the importance of climate for temporal variation in life-history traits of the northern spotted owl. Unseasonably cold and wet events in late spring are thought to be a major factor in annual productivity. The year variable was always significant in our models, and we assume that it is a surrogate for such events. A more-refined analysis could use the actual number and magnitude of such climate events.

The results of compositional analyses and correlations between owl habitat and productivity were similar for the two data types. Aerial photography and Landsat estimates for percentage of the landscape in each of the five canopy classes (Fig. 61.5) differed; moreover, conclusions about occurrence and what constitutes productive owl habitat would be different (Figs. 61.3 and 61.5). Landsat data suggest that a greater proportion of the landscape around an activity center needs to have canopy cover values above 50 percent.

Conclusions about the importance of the canopy-cover class in the area immediately surrounding a nest site or primary roost site differed between the Landsat and photography data. The habitat at the activity center was a significant factor when examining occurrence with the Landsat data set but not with the photography data. The opposite was found, however, when the generalized additive model was used to test relations between canopy cover and the number of fledglings produced. The models developed from the Landsat data concerning productivity typically contained fewer variables relating to canopy cover. At this stage of analysis, multivariate models are being used to explore relations between habitats and owl occurrence and productivity, and not for predictive purposes. The photography data seem to provide more insight into the relations that we examined, but formal accuracy assessments are needed on the vegetation data sets before we can conclude which data source is more accurate and whether differences between results are real.

Our results provide further insights into the types of data required for long-term maintenance of spotted owls and the potential bias of using different sources of data for management decisions. Results such as ours will be used to set standards and guidelines for vegetation management within the distribution of the California spotted owl. Empirical relations between spatial habitat data and spotted owl occurrence/ productivity could be used to model suitable habitat or predict responses of the owls to forest management. Further research needs to address additional forest structural attributes, landscape patterns (especially edge), and the uncertainty in the spatial habitat data prior to such model development.

Acknowledgments

The authors benefited from the guidance of Jared Verner during the development and execution of this research, and we express our gratitude. We thank Haiganoush Preisler, Pacific Southwest Research Station, Albany, for assistance with the statistical model aspects of this research. We also thank Gary Eberlein, computer specialist, Tom Munton and Ken Johnson, crew leaders, and all the people who performed fieldwork over the nine years of owl data collection. Region 5 and the Pacific Southwest Research Station, USDA Forest Service, provided support for this work.

Using a Spatially Explicit Model to Analyze Effects of Habitat Management on Northern Spotted Owls

Martin G. Raphael and Richard S. Holthausen

Conservation of the northern spotted owl (*Strix occidentalis caurina*) has been a central issue in struggles over forest management in the Pacific Northwest (Thomas et al. 1990; USDI 1992; FEMAT 1993; Hunsaker et al., Chapter 61). A recent demographic analysis (Forsman et al. 1996a) indicated that the population of the northern spotted owl has declined over large portions of its range, but this report does not relate rates of population change to variation in habitat quality over the species' range. Because of the finding that the owl population has declined over the past ten years, likely in response to habitat loss (Murphy and Noon 1992), the effect of further harvest of habitat is of great interest. A major step in managing federal lands to conserve the owl was the development of a series of options for ecosystem management on federal forest lands by the Forest Ecosystem Management Assessment Team (FEMAT 1993). Adoption of one of those options as a strategy for managing late-successional and old-growth forests on federal forestlands in the range of the owl has become known as the Northwest Forest Plan. Over the long term, the habitat reserve design that is part of this plan will likely support stable and well-distributed populations of owls when unsuitable habitat within the reserves has matured (Thomas et al. 1990; USDI 1992; Murphy and Noon 1992; Thomas et al. 1993; Raphael et al. 1994, 1998).

However, the Northwest Forest Plan is focused on management of spotted owl habitat on federal lands. It is of interest to consider the degree to which additional contributions of habitat on nonfederal lands might further support spotted owl populations. The analysis presented in this chapter was originally requested by the U.S. Fish and Wildlife Service to provide background for development of a section 4(d) rule that might authorize incidental take of some northern spotted owls on nonfederal forests on the Olympic Peninsula, Washington (see Fig. 62.1 in color section). This analysis describes likely patterns of distribution and persistence of owls on the Olympic Peninsula under the provisions of the Northwest Forest Plan, and benefits to the owl population of varying levels of habitat contribution from nonfederal lands. This chapter also extends earlier results reported elsewhere (Holthausen et al. 1995). Holthausen et al. (1995) reported on simulated owl population responses to a series of hypothetical scenarios that might be considered in the development of a possible 4(d) rule for spotted owl management on nonfederal lands. One of the actions that had been considered by the U. S. Fish and Wildlife Service (FWS) was adoption of a special rule under section 4(d) of the Endangered Species Act that addressed the conservation of northern spotted owls while reducing the prohibition against incidental take of owls in the course of timber harvest and related activities on nonfederal lands. A Notice of Intent was published on

29 December 1993 that identified a series of Special Emphasis Areas where the FWS believed that maintenance of nonfederal habitat was necessary. The FWS was further evaluating those areas and additional alternatives to develop a proposed 4(d) rule. The question posed for this analysis was "Can the contribution of nonfederal lands be made more efficiently than through the current take guidelines?" The analysis focused on alternative scenarios for retention of nonfederal habitat throughout the Olympic Peninsula and within the western Special Emphasis Area (SEA). Here, a more efficient scenario was defined as one that required less contribution of nonfederal habitat for similar benefits to owl conservation compared with current take guidelines.

Methods

We based our analysis on the same digital map of nesting, roosting, and foraging (NRF) habitat used by Holthausen et al. (1995). This map was assembled by the Washington Department of Natural Resources (WDNR 1997) and is an aggregation of the best available information from a variety of different sources, including the FEMAT (1993) habitat classification for federal lands and data from the Washington Department of Fish and Wildlife for lands administered by the WDNR (Fig. 62.2). On remaining state lands, the WDNR used inventoried stands (land use/land cover database) that met NRF habitat definitions. For all other lands, the WDNR used a satellite-derived forest-

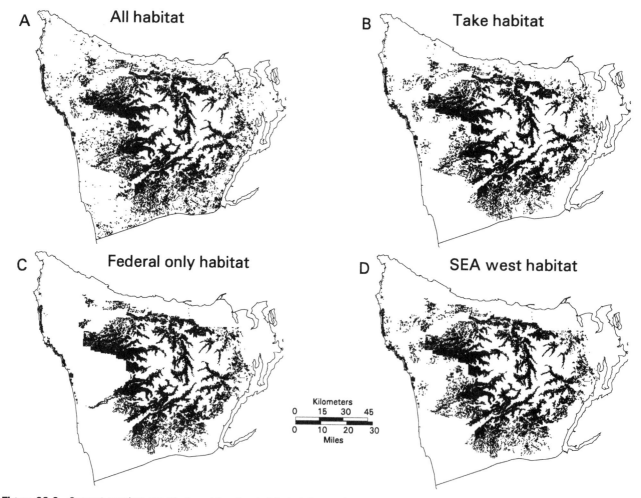

Figure 62.2. Current nesting, roosting, and foraging habitat of the northern spotted owl (*Strix occidentalis caurina*) on the Olympic Peninsula, Washington. Figure (A) depicts habitat under Scenario 2, retention of habitat on all lands; (B) depicts habitat remaining under Scenario 3, application of take guidelines; (C) depicts habitat remaining under Scenario 1, removal of all habitat on nonfederal lands; and (D) depicts Scenario 4, retention of habitat within the western Special Emphasis Area (SEA) (see text).

cover classification (Green et al. 1993) and selected as habitat either the late-successional category (generally stands with greater than 10 percent cover of trees greater than 53 centimeters [21 inches] in diameter at breast height [dbh]) or a combination of the late-successional and mid-successional (less than 10 percent cover of 53-centimeter-dbh trees) categories, or whichever had the best fit to classified owl habitat within a particular planning unit. We assumed static habitat conditions, that is, we did not model growth in habitat over time, nor did we model the effects of catastrophic fire. In addition, for each of the new land-management scenarios, we assumed, based on advice provided by the FWS, that no habitat would be retained on tribal lands.

Spotted Owl Population Simulation

To evaluate the relative likelihood of persistence of the northern spotted owl under alternative land allocation scenarios, we used a spatially explicit life-history simulator (OWL [version 2.01] McKelvey et al. 1992). This model is a single-organism simulator that is based largely on models developed by Lande (1987, 1988a) and Lamberson (Thomas et al. 1990; Lamberson et al. 1992, 1994) and is similar to Pulliam's BACHMAP model (Pulliam et al. 1992). The model is sensitive to the shape and location of high-quality habitat, which we mapped as described above. The model should be viewed as a tool for landscape design that allows a logical framework in which to assess qualitative differences in various land management plans in regard to population dynamics of the northern spotted owl.

We parameterized the model as described in Holthausen et al. (1995). The primary relationships built into the simulation model were variation in fecundity and adult survivorship in relation to amount of NRF habitat. These relationships were derived from regressions reported by Bart (1995b):

$$f = 0.32 + 0.54p$$

$$s_a = 0.63 + 0.39p$$

where f is fecundity, s_a is adult survival, and p is the proportion of the area surrounding an owl activity center that is composed of NRF habitat. We used these regressions to establish rule sets relating percent-

TABLE 62.1

Adult survival and fecundity in relation to percent suitable habitat within hexagonal cells (spotted owl sites).

| Parameter | Percent suitable habitat | | | | |
	0–20	>20–30	>30–40	>40–60	> 60
Adult survival	0.76	0.82	0.86	0.92	0.92
Fecundity	0.24	0.33	0.38	0.46	0.46

age of suitable habitat in owl territories to demographic parameters (see Fig. 62.3 in color section). These relationships were adjusted to match demographic data reported by Forsman et al. (1996b) from their study on the Olympic Peninsula. For this adjustment, we retained the slopes of the regression reported above but adjusted the intercepts to match the Olympic results. In addition, to avoid implausibly high rates, we truncated the survival and fecundity parameters in the highest-quality cells (Table 62.1).

These parameters (and others listed by Holthausen et al. 1995), were used as input to the simulation model and were labeled "rule set B." In this rule set, we used an estimate of juvenile survival of 0.29, the median from the values in Burnham et al. (1996) for eleven study areas throughout the owl's range. Rule set A was identical to rule set B except that the parameters were shifted to the right (into the next-highest habitat category); the parameters of rule set C were shifted to the left (into the lower habitat category). A fourth rule set, rule set D, was identical to rule set B except that juvenile survival was increased to 0.38, the estimate from Burnham et al. (1996) that was adjusted for emigration of juveniles out of study areas.

As in these previous analyses, we defined hexagonal cells of 1,500 hectares and initialized the model with pairs of owls wherever habitat exceeded 30 percent of a hexagonal cell. The cell size of 1,500 hectares was selected based upon field data on the observed density of owls and to achieve a carrying capacity in line with estimated population size on the Peninsula (see Holthausen et al. 1995 for details). To describe relative variability of simulation runs (each run consisting of fifty separate simulations), we replicated an entire sequence of runs ten times and tabulated variation in simulation results for rule sets B and D. Variation is due to stochastic elements of the simulation model itself. To gain some insight into validity of the model projections, we compared projected occupancy within each cell to observed locations of spotted owls as identified in agency databases. We performed this comparison with occupancy from rule sets B and D.

Land Management Scenarios

Most scenarios were reproduced from Holthausen et al. (1995) and represented the likely retention of nonfederal NRF habitat within various alternatives (Fig. 62.2). Scenario 1 represented the retention of only federal habitat. Scenario 2 represented the retention of all currently existing nonfederal and federal NRF habitat. Scenario 3 represented current take guidelines where nonfederal habitat was retained only within owl circles of 40 percent habitat or less. This scenario serves as an appropriate benchmark against which to compare alternative Peninsula-wide approaches. Scenario 4W represented retention of nonfederal habitat following take guidelines within the western SEA.

Scenario 5 was also taken from Holthausen et al. (1995), where it was presented as one approach toward a more efficient mechanism for retention of nonfederal habitat. Efficiency was evaluated as the ratio of the gain in owl occupancy or population size to the gain in amount of nonfederal habitat retained under a scenario relative to the gain expected between Scenarios 1 and 2. As seen in Holthausen et al. (1995), Scenarios 1, 2, and 3 fall along an essentially straight line, where more habitat translated to greater numbers of owls. Any scenario falling above that line can be considered more efficient; such a scenario would indicate a relatively greater contribution to the population of owls per unit of nonfederal habitat. As noted by Holthausen et al. (1995), Scenario 5 was developed by identifying habitat that occurred in hexagonal cells with successively greater predicted occupancy by pairs of owls as evaluated at the conclusion of 100-year simulation runs. Any nonfederal habitat in a cell supporting less than or equal to 10 percent mean occupancy by pairs of owls was removed, then nonfederal habitat in a cell with less than or equal to 20 percent occupancy, less than or equal to 40 percent occupancy, and less than or equal to 60 percent occupancy was successively removed. After each removal, a map of

remaining habitat was created, and a new simulation run was completed.

Scenario 6 is a new set based on removal of habitat without regard to occupancy by owls. For this set, we first eliminated any habitat that occurred on tribal land. We then eliminated nonfederal habitat in any hexagonal cell where habitat was less than or equal to 20 percent of that cell and ran the owl population simulation on the resulting habitat map. Next, we identified cells with less than or equal to 40 percent habitat, removed all nonfederal habitat from those cells, and ran another simulation. Similarly, we successively removed nonfederal habitat at the 60 and 80 percent levels, running simulations after each step.

Results

The amount of nonfederal habitat that might be retained under each of the scenarios varied from 0 hectares under Scenario 1 to 64,600 hectares under Scenario 2 over the entire Peninsula; 32,000 hectares were retained under Scenario 3 (Table 62.2). Removal of habitat from tribal lands reduced total nonfederal habitat by 6,800 hectares.

Our model validation test revealed general agreement between occupancy from the simulation model and occupancy by observed birds. The owl location database contains 169 sites occupied by spotted owls. Under rule set B, 168 of the 169 hexagons containing known locations had predicted occupancy greater than 0 (errors of omission less than 0.1 percent). However, a larger number of hexagons were predicted to be occupied at some level but were not actually occupied (errors of commission = 56 percent). Under rule set D, all 169 sites were predicted to be occupied at some level (errors of omission = 0 percent); 73 percent of sites with predicted occupancy were not actually occupied (errors of commission). For both rule sets, the errors of commission were most prevalent at low levels of occupancy (i.e., over half of the errors involved cells where expected occupancy was less than 20 percent). Observed rates of occupancy increased with increasing predicted rates of occupancy under both rule sets (Fig. 62.4). We believe that errors of omission are of greater concern than errors of

TABLE 62.2.

Total area of suitable northern spotted owl (*Strix occidentalis caurina*) habitat (hectares) under each management scenario, Olympic Peninsula.[a]

Scenario	Total habitat	Nonfederal habitat
1. Federal habitat only[b]	265,900	0
2. Federal and nonfederal	330,600	64,600
3. Current take guidelines	297,900	32,000
4. Realized Special Emphasis Area: West	287,100	21,100
5. Selective removal:[c]		
Rule A		
≤ 10%	278,900	13,000
≤ 20%	272,000	6,100
≤ 40%	266,600	700
≤ 60%	266,000	0
Rule B		
≤ 10%	295,700	29,700
≤ 20%	288,600	22,600
≤ 40%	280,600	14,600
≤ 60%	271,500	5,500
Rule C		
≤ 10%	310,400	44,400
≤ 20%	300,300	34,300
≤ 40%	291,500	25,600
≤ 60%	285,000	19,100
Rule D		
≤ 10%	306,200	40,300
≤ 20%	298,900	32,900
≤ 40%	290,100	24,100
≤ 60%	283,800	17,800
6. Federal and nonfederal, not including tribal [d]	323,800	57,900
≤ 20%	300,800	34,800
≤ 40%	277,300	11,300
≤ 60%	269,600	3,700
≤ 80%	267,800	1,800

[a]All values rounded to the nearest 100 hectares.
[b]Includes national forest and national park land only (does not include minor contributions of other federal land).
[c]Selective removal of nonfederal habitat in cells based on successively higher mean occupancy (Scenario 5, see text for details).
[d]Selective removal of nonfederal habitat based on successively higher percentage of habitat in cells (Scenario 6, see text for details).

commission: if the model predicted absence at a site and we actually observed a pair of owls at that site, we would be concerned that our model was missing important habitat attributes. That the model predicted low rates of occupancy in sites where no owls were present in the current database may simply mean we did not observe that site over enough years to record an owl there. The simulation model ran for one

TABLE 62.3.

Range[a] of results associated with simulating northern spotted owl population dynamics using rule sets B and D under Scenario 2, federal and nonfederal habitat, Olympic Peninsula.

Criterion	Rule B	Rule D
Number of cells > 60–80% mean occupancy	120–125 (5)	109–121 (12)
Number of cells > 80% mean occupancy	0	138–144 (6)
Total of cells > 60% mean occupancy	120–125 (5)	248–260 (12)
Mean number of pairs	173–177 (4)	282–287 (5)
Pairs at year 60	149–153 (4)	277–283 (6)
Pairs at year 100	127–137 (10)	274–282 (8)
λ[b] Evaluated for years 60–100	0.9960–0.9973 (0.0013)	0.9992–1.0003 (0.0011)

[a]Ranges are the extreme values generated from ten separate simulation runs for each rule set (each run consisted of fifty separate simulations). Simulations assumed that current vegetation patterns were static (no regrowth), and all habitat currently on the Olympic Peninsula was retained. Values in parentheses are ranges.

[b]Finite rate of population change.

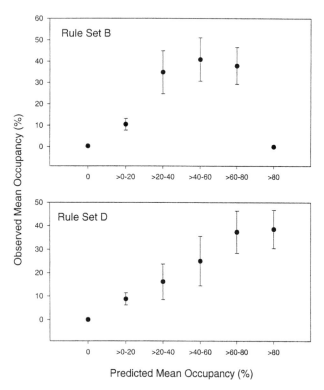

Figure 62.4. Observed occupancy (known northern spotted owls [*Strix occidentalis caurina*]) in relation to predicted occupancy under two rule sets on the Olympic Peninsula, Washington. Predicted occupancy is the percentage of years in which a hexagonal cell was occupied by a pair of owls during a set of simulation runs (see text for details).

hundred years whereas our owl database represents only five years of observation.

The results of population simulations based on each of the scenarios show that some scenarios do result in more-efficient habitat retention. All comparisons should be made bearing in mind the expected levels of variability among runs (Table 62.3). As summarized in Figure 62.5, Scenario 5 and its subparts seem to result in greater mean numbers of owls per unit of nonfederal habitat. For example, under rule set B, Scenario 5A (removal based on cells with less than or equal to 10 percent occupancy) resulted in about the same mean number of pairs as Scenario 2 (all nonfederal habitat, 64,600 hectares), while retaining only half the nonfederal habitat (29,700 hectares, Table 62.2, Fig. 62.5). This scenario also supported more owls per unit nonfederal habitat than Scenario 3 (current take guidelines) under all rule sets except rule set A (Fig. 62.5).

Removal of habitat based on percent habitat within each cell regardless of occupancy was not as efficient as Scenario 5. As shown in Fig. 62.5, for any given level of nonfederal habitat, mean number of pairs was greater under Scenario 5 than under this scenario. However, removal of habitat based on its percentage within cells was more efficient than cur-

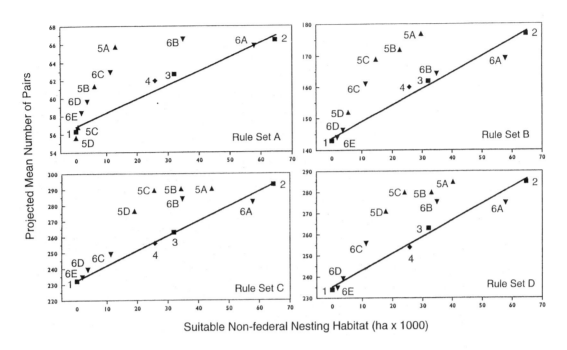

Figure 62.5. The relation between nonfederal habitat for each scenario and the expected average number of pairs, evaluated over the entire Olympic Peninsula. Scenario 1 models federal habitat only, without regrowth. Scenario 2 includes all existing federal and nonfederal habitat without regrowth or harvest. Scenario 3 retains habitat following the current take guidelines. Scenario 4 retains nonfederal habitat that occurs only within a western Special Emphasis Area. Scenario 5 saves nonfederal habitat based on its simulated contribution to population size; 5A results from removal of nonfederal habitat from cells with mean occupancy of less than or equal to 10 percent, 5B from cells of less than or equal to 20 percent, 5C from cells of less than or equal to 40 percent, and 5D from cells of less than or equal to 60 percent. Scenario 6 is the same as Scenario 2 except that we assumed no habitat on tribal lands; 6A is the retention of all nonfederal habitat, 6B is removal of nonfederal habitat from any cell with less than or equal to 20 percent habitat, 6C is removal of nonfederal habitat from any cell with less than or equal to 40 percent habitat, 6D and 6E result from removal of habitat in cells with less than or equal to 60 percent or 80 percent habitat, respectively. Occupancy patterns and mean numbers of pairs were based on simulations using rule sets A, B, C, and D. The diagonal line is a regression of mean number of owl pairs against amount of nonfederal habitat for Scenarios 1, 2, and 3.

rent take guidelines (Table 62.4). For example, under rule sets A and B, removal of habitat from cells with less than or equal to 40 percent habitat (6B) resulted in mean pairs of about the same level as Scenario 3 but with about one-third the total nonfederal habitat (11,300 hectares, 27,900 acres) versus 32,000 hectares, Table 62.2, Fig. 62.5). Rates of occupancy were also greater under this approach (n = 112 cells with more than 60 percent occupancy under the 40 percent removal scenario and using rule set B versus 108 cells under Scenario 3; Table 62.5).

The location of habitat that would be retained under these approaches (Fig. 62.3) reflects the current distribution of higher-quality patches where habitat occurs at higher concentrations. For example, Scenario 5C (selective removal when occupancy was less than or equal to 40 percent) under rule set B resulted in reten-

tion of habitat within about 5 kilometers of federal lands along the western boundary of the Olympic National Park, the southwestern boundary of the Olympic National Forest, and small patches along the northern boundary of the national forest. Many of the scenarios resulted in a similar distribution of retained habitat, although the amounts and exact distributions varied among the scenarios and rule sets.

Discussion

Discussion of this analysis is organized around the following questions:

1. Can the contribution of nonfederal lands to owl conservation be made more efficiently than through the current take guidelines?

TABLE 62.4.

Summary of simulation results under Scenario 5, selective removal of low-occupancy habitat, using rule sets A, B, C, and D, Olympic Peninsula.

Criterion	Percent mean occupancy[a]			
	≤ 10%	≤ 20%	≤ 40%	≤ 60%
Rule set A:				
Amount of habitat (ha × 1,000)	279	272	267	266
Number of cells > 60–80% mean occupancy	0	0	0	0
Number of cells > 80% mean occupancy	0	0	0	0
Total of cells > 60% mean occupancy	0	0	0	0
Mean number of pairs	66	61	˙57	56
Pairs at year 60	36	32	27	25
Pairs at year 100	15	12	9	9
λ[b] evaluated for years 60–100	0.9784	0.9755	0.9731	0.9745
Rule set B:				
Amount of habitat (ha × 1,000)	296	289	281	271
Number of cells > 60–80% mean occupancy	121	122	124	113
Number of cells > 80% mean occupancy	0	0	0	0
Total of cells > 60% mean occupancy	121	122	124	113
Mean number of pairs	177	172	169	152
Pairs at year 60	154	152	149	134
Pairs at year 100	139	131	128	115
λ evaluated for years 60–100	0.9975	0.9962	0.9964	0.9961
Rule set C:				
Amount of habitat (ha × 1,000)	310	300	292	285
Number of cells > 60–80% mean occupancy	136	134	133	136
Number of cells > 80% mean occupancy	127	130	135	125
Total of cells > 60% mean occupancy	263	264	268	261
Mean number of pairs	290	290	290	276
Pairs at year 60	289	287	287	273
Pairs at year 100	286	280	285	267
λ evaluated for years 60–100	0.9998	0.9994	0.9998	0.9995
Rule set D:				
Amount of habitat (ha × 1,000)	306	299	290	284
Number of cells 60–80% mean occupancy	116	119	110	117
Number of cells > 80% mean occupancy	141	134	146	137
Total of cells > 60% mean occupancy	257	253	256	254
Mean number of pairs	285	280	280	271
Pairs at year 60	283	276	274	266
Pairs at year 100	277	278	277	262
λ evaluated for years 60–100	0.9995	1.0002	1.0003	0.9996

[a]All nonfederal habitat was removed from cells with an indicated mean occupancy percentage based on earlier simulations of Scenario 2 (all habitat).
[b]Finite rate of population change.

TABLE 62.5.

Summary of simulation results under Scenario 6, selective removal of habitat from cells based on percent habitat within each cell, using rule sets A, B, C, and D, Olympic Peninsula.

Criterion	Percent habitat removed[b]				
	0%	≤ 20%	≤ 40%	≤ 60%	≤ 80%
Rule set A:					
Amount of habitat (ha × 1,000)	324	301	277	270	268
Number of cells > 60–80% mean occupancy	0	0	0	0	0
Number of cells > 80% mean occupancy	0	0	0	0	0
Total of cells > 60% mean occupancy	0	0	0	0	0
Mean number of pairs	66	67	63	60	58
Pairs at year 60	33	34	32	30	28
Pairs at year 100	12	14	14	12	11
λ[b] evaluated for years 60–100	0.9763	0.9787	0.9791	0.9781	0.9776
Rule set B:					
Amount of habitat (ha × 1,000)	324	301	277	270	268
Number of cells > 60–80% mean occupancy	117	109	112	94	86
Number of cells > 80% mean occupancy	0	0	0	0	0
Total of cells > 60% mean occupancy	117	109	112	94	86
Mean number of pairs	169	164	161	146	144
Pairs at year 60	148	139	141	126	123
Pairs at year 100	119	119	123	105	100
λ evaluated for years 60–100	0.9947	0.9963	0.9966	0.9955	0.9948
Rule set C:					
Amount of habitat (ha × 1,000)	324	301	277	270	268
Number of cells > 60–80% mean occupancy	140	144	151	151	169
Number of cells > 80% mean occupancy	113	115	81	69	49
Total of cells > 60% mean occupancy	253	259	232	220	218
Mean number of pairs	282	284	249	239	235
Pairs at year 60	274	280	242	236	227
Pairs at year 100	275	271	241	226	215
λ evaluated for years 60–100	1.0001	0.9991	0.9999	0.9990	0.9986
Rule set D:					
Amount of habitat (ha × 1,000)	324	301	277	270	268
Number of cells 60–80% mean occupancy	110	115	107	108	104
Number of cells > 80% mean occupancy	138	131	125	109	110
Total of cells > 60% mean occupancy	248	246	232	217	214
Mean number of pairs	275	275	256	239	235
Pairs at year 60	270	269	251	232	228
Pairs at year 100	268	267	250	232	226
λ evaluated for years 60–100	0.9998	0.9998	0.9998	1.0000	0.9998

[a]All nonfederal habitat was removed in cells with the indicated habitat percentage.

[b]Finite rate of population change.

2. If there is a more efficient way to provide a nonfederal contribution, what are the characteristics of habitat that contribute to it?
3. To what degree do different levels of nonfederal contribution contribute to the persistence of the owl population on the Olympic Peninsula?
4. How could the results of this analysis be used appropriately in policy decisions?

Throughout this discussion, any conclusions are tempered by general cautions concerning the interpretation of model results (Holthausen et al. 1995). Because the relationship between model runs and reality is not known, the relative differences observed among land management scenarios are of most interest. The different rule sets are used to show how those relative differences change based on a change in biological understandings or assumptions.

Can Nonfederal Lands Contribute to Owl Conservation more Efficiently Than through the Current Take Guidelines?

Scenario 5 was developed to investigate this question as part of the initial analysis done in Holthausen et al. (1995). The results for Scenario 5 suggested that more-efficient nonfederal contributions could be crafted. That scenario selectively removed habitat from the simulated landscape based on the occupancy of the habitat observed in model results. The results of the scenario were viewed with some caution because they used the results of one model run to improve the performance of the population in a second model run. The new analyses reported here were designed in part to further validate the conclusions drawn from Scenario 5.

The current analyses confirm that configurations of habitat different from the take guidelines are more efficient in contributing to owl persistence on the Olympic Peninsula. This effect can be observed in Figure 62.5, which shows the mean number of owl pairs maintained through one hundred years in each of the scenarios and the hectares of nonfederal owl habitat that were retained in the scenarios. Scenario 3 represents application of take guidelines throughout the Peninsula, Scenario 2 includes all existing federal and nonfederal habitat, and Scenario 1 includes only federal habitat. The regression line shows the expected relationship between nonfederal habitat and owl pairs based on these three scenarios. Scenarios are judged more efficient than the regression line if they fall above it, indicating that they support a greater increase in number of owl pairs per hectare of habitat. To be conclusive, the difference in number of pairs supported must be greater than the differences expected from the stochastic nature of the model (Table 62.3). The specific scenarios shown to be more efficient differ depending on the rule set used.

Among the scenarios applied to nonfederal land across the entire peninsula (Fig. 62.5), the following are more efficient than the take guideline under at least one rule set: 5A through 5D, 6A, and 6B. In all these scenarios, decisions to retain habitat in the landscape are based on either the amount of habitat present in hexagonal cells or the occupancy of cells observed in model runs that included all existing habitat. Thus, these scenarios all selectively retain areas where habitat is concentrated and eliminate smaller, scattered patches.

What Are the Characteristics of Habitat That Provide for an Efficient Contribution to Owl Conservation?

The model results indicate that cells containing larger amounts of habitat make a more-efficient contribution to owl persistence than do cells containing lesser amounts. In addition, the location of those cells relative to other large concentrations of habitat apparently influences efficiency of the habitat in providing for owls. The effect of location is observed most strongly in Scenario 5 and its subparts. Because both the amount of habitat in cells and the location of that habitat contributed to occupancy of habitat cells in initial runs, the selection of cells for retention was influenced by both amount of habitat and the location of cells relative to the large concentrations of federal habitat.

The conclusions that amount of habitat in cells and location of cells affect relative efficiency of contribution seems relatively robust. However, it is much more difficult to identify critical values of those characteristics. Under all rule sets except B, a gain in efficiency resulted from selecting cells with more than 20 percent suitable habitat rather than selecting cells based on the current take guidelines (i.e., based on proximity to owl activity centers). However, use of 20 percent suitable habitat as a critical value should be viewed with skepticism, since it is dependent on the structure of the demographic rule sets developed for the model (Holthausen et al. 1995). Likewise, the scenarios identified as being efficient (5A, 5B, and 5C under rule set B, 5A, 5B, 5C, and 5D under rule set D, 6A, and 6B) consistently include nonfederal habitat that falls within approximately 10 kilometers of federal land. However, the specific value of 10 kilometers must be viewed with caution, since that value is dependent on the structure of the model.

Examination of the distribution of retained habitat under these efficient scenarios (e.g., Fig. 62.2) suggests that lands on the western side of the Peninsula provide the greatest opportunity for efficient contribution to owl conservation. This is likely due to the location of those lands relative to federal lands and is consistent with findings of Holthausen et al. (1995). However, the specific boundary of the most efficient contribution is again difficult to identify. Some lands in all portions of the Peninsula are identified in some of these scenarios, but the nonfederal contribution becomes more concentrated in the western and southwestern portion of the Peninsula as the selection criteria are tightened.

To What Degree Do Different Levels of Nonfederal Contribution Affect Owl Persistence on the Olympic Peninsula?

In the initial analysis of owls on the Olympic Peninsula, Holthausen et al. (1995) concluded that retention of nonfederal habitat could provide a biologically significant contribution to the maintenance of a stable population of owls on the Peninsula. However, no specific thresholds could be identified in this benefit. Population performance simply increased steadily with nonfederal habitat contribution (Fig. 62.5). Analyses completed for this report suggest that the effects of several of the efficiently designed scenarios (e.g., 5A) may be similar to the effects of retaining all nonfederal habitat (Fig. 62.5). This suggests that retention of only one-half to two-thirds of the existing nonfederal habitat may be highly effective in providing for owl persistence. However, this finding should

again be viewed with caution because it varies across demographic rule sets and is dependent on the structure and assumptions in the model. The finding also does not reflect potential value of habitat in providing a connection between the population in the center of the Peninsula and the population on the western coast of the Peninsula, especially the coastal strip of the Olympic National Park.

What Is the Appropriate Use of the Analysis in Policy Decisions?

The link between the analysis reported here and policy formulation is not straightforward. This analysis does not use the existing take guideline as an operating assumption, and thus it cannot be applied directly to decisions being made as part of the 4(d) rule process. Rather, the analysis assumes that some nonfederal habitat may be retained in areas that are not in proximity to owl activity centers, that the take circles around other activity centers might be managed to retain more than 40 percent suitable habitat, and that still other take circles might be managed to retain less than 40 percent suitable habitat.

The analysis could be used to identify areas where the contribution of nonfederal habitat appears to be most significant. We found that areas on the western and southwestern sides of the Peninsula within approximately 10 kilometers of federal land are common to all of the scenarios that were identified as being efficient. As the amount of nonfederal land to be contributed increases, other geographic areas are also identified in some of the scenarios (Scenario 5A, rule set B; Scenario 5A, rule set D; Scenario 6A, Fig. 62.5). Thus, nonfederal contributions might emphasize the western and southwestern portions of the Peninsula, but other lands cannot be ruled out based on this analysis. All increments of habitat within these scenarios appear to provide positive benefits.

Within the areas identified for nonfederal habitat contribution, the objective should be to retain habitat that is most concentrated and occurs in the largest blocks. The model suggests that home range areas containing less than 20 percent suitable habitat make a much smaller contribution to owl persistence than do areas with more habitat. Although this result is dependent on the structure and assumptions in the model, it could be used as a management hypothesis and further tested through time. Application of this guideline to nonfederal habitat management would require some mechanism other than the current take guideline. Habitat Conservation Planning might provide such a mechanism, allowing for retention of habitat in configurations other than the simple maintenance of 40 percent suitable habitat within home ranges. Unless a specific mechanism is identified to provide the types of habitat configurations described here, the results of this analysis will have little value to the establishment of a 4(d) rule. If this analysis were used to locate areas for nonfederal contributions, but those areas were subsequently managed with the existing take guidelines, the benefit to owls predicted here would not be realized.

Specific results of these analyses should not be extrapolated to other geographic regions within the range of the northern spotted owl. Conditions on the Olympic Peninsula are unique, with a large block of federal habitat surrounded by nonfederal habitat. In other parts of the range, the relative contributions of federal and nonfederal habitat to owl persistence may differ from the patterns observed in this analysis.

Acknowledgments

This chapter was derived from earlier work to which K. S. McKelvey, E. D. Forsman, E. E. Starkey, and D. E. Seaman contributed. We thank C. Ogden and J. Michaels for useful discussion of ideas in this chapter, and we thank B. M. Galleher for analytical support and graphics preparation. This work was supported by the U. S. Fish and Wildlife Service and by the Pacific Northwest Research Station.

Estimating the Effective Area of Habitat Patches in Heterogeneous Landscapes

Thomas D. Sisk, Barry R. Noon, and Haydee M. Hampton

Animal abundance and population viability have often been inferred via direct extrapolation from estimates of the area of suitable habitat. However, such estimates fail to account for changes in the spatial distribution of habitats such as those that may arise from fragmentation (e.g., Saunders et al. 1991; Wilcove et al. 1986; Faaborg et al. 1995). Consequently, it is now widely recognized that indirect estimates of animal abundance based on simple area metrics, which do not include information about the spatial distribution of habitat, can be misleading. Added to the effects of spatial pattern of habitat is the recognition that source:sink dynamics (Pulliam 1988) and other subtle differences in habitat quality may result in strong gradients in population density.

Collectively, these insights indicate that inferences to animal abundance based on habitat assessments should be done at a landscape scale and that they should explicitly include information about spatial pattern and gradients in habitat quality (Beutel et al. 1999). Recently added to the list of habitat factors known to affect animal abundance is variation in the character of the matrix that surrounds patches of habitat (Wiens 1996a,b, 1999; Gascon et al. 1999). Animal abundance, as well as survival and reproductive rates, may be governed most strongly by landscape-scale factors, including edge and matrix effects. This observation suggests that habitat quality is as much a product of landscape-scale factors as it is of proximate factors within a given habitat patch (Forman and Godron 1981; Temple 1986a; Murcia 1995; Gascon et al. 1999; Mesquita et al. 1999). As terrestrial landscapes become increasingly fragmented by human activities, a predictive approach that accounts for spatially extensive ecological factors in a spatially explicit context is needed to guide habitat management and conservation.

The detrimental effects of changes in habitat area, spatial patterning, and landscape context are now widely accepted. However, the development of practical tools to predict the effects of factors and design appropriate mitigation efforts has progressed relatively slowly. Temple (1986a) and Temple and Cary (1988) developed a "core area" model to explore possible effects of forest fragmentation on interior-habitat bird species, and Laurance and Yensen (1991) extended the capabilities of the model and applied it to other fragmentation scenarios (e.g., Laurance 1991). In this chapter, we present a spatially explicit landscape model that builds upon these and other efforts to account for edge effects and patch context (i.e., matrix effects), given changes in landscape pattern and context. Our approach extends the Effective Area Model (EAM; Sisk et al. 1997), a spatial model that adjusts animal density estimates based on a species' response to habitat edges. The model "works" by linking field data and remotely sensed imagery through a geographic information system (GIS) interface with maps of actual landscapes. The eventual goal of our research is to develop a tool that will help managers select among alternative management actions, each generating a characteristic landscape configuration based on their effects on animal populations.

Overview

Many authors have emphasized the need for landscape-scale assessments of the relationship between variation in animal abundance and variation in habitat pattern and quality (Forman and Godron 1986; Turner and Gardner 1991; Wiens 1995; Morrison et al. 1998). This need has been recognized for decades. For example, Sisk and Battin (in press) point out that Leopold's discussion of edge effects (Leopold 1933) constitutes an early conceptualization of matrix effects and a call for greater attention to landscape-scale factors influencing habitat selection. Since that early beginning, attempts to deal with edge, matrix, and other landscape-scale factors have received much attention, but primarily in a descriptive context. Sisk and Battin (in press) note that "edge effects" has become a poorly defined concept that is routinely employed to "explain away" unidentified or poorly quantified processes that influence measured attributes within the focal habitat patch. As habitats become increasingly fragmented, large, homogeneous patches that were formerly dominated by internal processes may become highly influenced by adjacent habitats and thus be strongly affected by processes external to the boundaries of the patch (Harris 1988; Wiens 1989a; Laurance 1991; Sisk and Battin in press)

We believe that understanding patterns in animal abundance and population persistence in rapidly changing landscapes will require an explicit, process-oriented approach to address edge effects and the juxtaposition of patches within increasingly heterogeneous landscapes. Such an approach requires methods for estimating the differing responses of animals to spatial patterning of their environment—that is, the type of habitat patch in which the animal is located, as well as the context of that patch within the larger landscape. Collectively, information about location and context, coupled with empirical measures of dis-tribution, abundance, and demography, will allow us to predict various population attributes, such as expected abundance or reproductive success, in changing landscapes.

Effective Area Models: Scaling Issues and Spatial Variability in Animal-Habitat Relationships

Wildlife biologists often struggle with scaling issues when attempting to model the relationships between animals and their habitats. The scale at which we collect information on species distribution, abundance, and fecundity, for example, are often much finer (typically on the order of tens of hectares) than the scale at which we wish to draw inference (typically on the order of hundreds or thousands of hectares) (Wiens et al. 1985; Withers and Meentemeyer 1999). Associated with this difference in scale is a difference in the degree to which spatial variability is captured and quantified (Wilkie and Finn 1996). Field studies typically employ study designs that use replication to capture between-site variability in response variables. However, logistic constraints limit the number of habitat types that can be studied with adequate replication. At the landscape scale, the number of habitat types and the variation among patches of similar habitat is often much greater than what can be measured in the field. Similarly, empirical studies usually must focus on only one or a few focal species, yet variation in responses among species is often large, making it difficult to generalize animal responses to habitat fragmentation and other changes in landscape structure (Noss 1991; Sisk et al. 1997).

The issues that arise with mismatched scales have led to two contrasting attitudes that are prevalent in animal ecology studies. The first, widely held among field biologists, emphasizes the variability in nature. Extensive observation often suggests that every study site and each species have unique characteristics and, therefore, the animal:habitat relationship cannot be generalized. For our purposes, this translates to the assumption that every species scales its environment uniquely, and developing a useful modeling approach for predicting the effects of landscape change would require a unique, data-based model for each species: habitat combination. For many ecologists, this is tantamount to stating that modeling approaches are of limited utility. A second perspective emphasizes general patterns in animal:habitat responses and focuses more on comparing likely outcomes rather than attempting to predict with great precision the responses of individual species. For example, a suite of "habitat interior species" may be identified from field data or life-history characteristics, and a general model of their sensitivity to predators and parasites can be used to estimate the effects of habitat fragmentation (Temple and Cary 1988; Laurance 1991; Thompson 1993). This approach embraces the goal of most ecological models: to provide a useful simplification of complex phenomena in order to increase understanding and predictive power. Of course, theorists and empirical ecologists often differ on what is a "useful simplification."

Models that attempt to fully incorporate the fine-scale heterogeneity that characterizes most terrestrial landscapes, and the great variability and complexity of population responses occuring at local scales, often become so complex as to be of limited use to wildlife and resource managers (e.g., Malcolm 1994). In attempting to address directly the complexities of heterogeneous, dynamic landscapes, and the suite of animal responses to landscape-scale influences, we suggest two guiding principles of model development. First, the approach must be empirically based. Purely theoretical models often sidestep the critical problems of data collection and parameter estimation—particularly challenging issues when managing for multiple species. Elegant theoretical models often provide useful insights but have limited application when real-world management issues arise. Second, useful management-oriented models should allow the user to compare the expected outcome of various management alternatives (e.g., Noon and McKelvey 1996b). Although optimization approaches may prove useful in some applications, managers seldom have the luxury of selecting the optimal management alternative, even if it were possible to identify one *a priori*. Instead, managers often are faced with a constrained set of options, each of which may have different impacts on animal populations, and the task is to select the best of the alternative actions. Therefore, management-oriented landscape models should allow the user to evaluate the alternative land-

scape patterns projected to arise from the different management options and to select the alternative that "best" addresses the management objectives. From a decision-support perspective, a simple ecological model that will allow managers to correctly rank the relative values of competing alternatives is often more useful than a more elegant, complex model. There is often a pronounced trade-off between model utility and model complexity. If a simple model is adequate to discriminate among and rank the outcomes from possible alternatives, it obviously will be easier and cheaper to implement and thus be more likely to be used to help solve actual problems.

Below, we describe a flexible modeling approach that is empirically based and designed to capture many of the influences that emerge from heterogeneous landscapes and affect the distribution and abundance of animal populations. The model, as presented here, is parameterized by measuring the density response of target organisms as a function of their location in the landscape. The value of a population attribute, such as abundance, is affected by the quality of the patch in which the individuals reside as well as by the context of the patch within the landscape. Our approach is map-based and assumes that abundance can be characterized as a graded response dependent upon location relative to various habitat patches and patch boundaries. The term *edges* is used generically in our discussion to refer to abrupt patch boundaries that arise from many management activities. However, the model structure described below is flexible enough to include abrupt or gradual habitat transitions.

Conceptual Approach

The Effective Area Model (EAM) is a straightforward approach that combines field-based measures of species' responses to habitat edges with landscape-scale habitat maps derived from remotely sensed data. The integration of these two sources of information allows us to predict variations in animal abundance across heterogeneous landscapes and to explicitly account for the spatial context of habitat patches. The method incorporates among-species variability in response to landscape boundaries by relating multiple species' responses to a common, classified habitat map. Depending on the spatial resolution of the map, using a single map may result in some unavoidable "smoothing" of habitat relationships. That is, there is a danger of excluding some habitat attributes that are associated with a given species' spatial distribution. However, we believe the EAM approach achieves the proper balance between model complexity and management utility.

By selecting a response variable (e.g., density, reproductive success) expected to vary with landscape heterogeneity, one can conduct the empirical research necessary for parameterizing the model. The association of the edge response function with a detailed habitat map combines a simplified biological response, adequately described in the response function, with the detailed spatial information available from remotely sensed data and GIS technology.

Taxonomic Foci

Our objective is to develop a map-based modeling approach that is practical for use as a management tool and useful in addressing the habitat needs of diverse species. Our current model parameterization efforts focus on birds and butterflies because these taxa represent diverse life histories and ecologies, providing an appropriate set of species for model development and testing. Both taxa are highly mobile and, thus, are able to respond rapidly to changes in habitat quality. In addition, we assume that their vagility allows them to select habitats based on behavioral cues rather than on limitations imposed by dispersal ability. Both groups are rich in species and easy to identify and survey in the field, and there is abundant natural history information available to permit analysis of the links between habitat selection and life history traits. To illustrate our methods, we restrict our discussion in this chapter to birds of desert riparian ecosystems. Other systems (Sisk et al. 1997; Haddad and Baum 1999) and taxa (Sisk and Haddad in press) are considered elsewhere.

Landscape Characterization

A chief limitation of most spatial models in ecology is their failure to capture key components of variability in landscape pattern—patch attributes such as size, shape, number, distribution, and spatial context. A typical modeling approach starts with a raster-based map of vegetation community types—that is, a classification based on some combination of structural and compositional attributes (see, e.g., Avery and Berlin 1992; Scott et al. 1996). Based on prior knowledge of the focal taxon, the map is categorized into habitat types by imposing a grain size and discriminating vegetation communities on assumed differences in quality to the focal taxon. By use of some neighborhood rule, individual habitat cells are then aggregated to define patches of habitat. From that point forward, each habitat patch is assumed to be homogeneous, and in most cases the ecological dynamics that occur within the patch are assumed to occur in virtual isolation from the rest of the landscape, perhaps with a dispersal rule connecting some patches genetically and/or demographically. In other words, the landscape-scale analysis is reduced to a series of exercises whose goal is to classify the landscape into patches in a parsimonious manner. Unless the spatial location of individual patches is explicitly considered, such as in some metapopulation models (e.g., Hanski 1994a), the models no longer contain explicit spatial information. Even when movement is simulated across these mosaic landscapes, the dynamics are restricted to simple functions that describe the likelihood of successful movement based on between-patch distances.

The expected quality of a patch, however, is not just a function of its size and type but also of its landscape context. The influence of the surrounding habitats on the quality of a focal habitat patch, often called the matrix effect, is an important but often-overlooked influence on animal demography (Forman and Godron 1986). Brittingham and Temple (1983) noted that proximity to agricultural habitats could lead to novel interspecific interactions in forest patches and the decline of songbirds, and others have pursued this issue in greater detail (e.g., Temple and Cary 1988; Thompson 1993; Faaborg et al. 1995; Robinson et al. 1995; Williams-Linera et al. 1998). Our approach to modeling the effects of landscape heterogeneity on animal abundance incorporates the effects of patch context into assessment of habitat quality. That is, we compute the effective area of a habitat patch by adjusting estimates of animal abundance upward or downward to account for patch context and edge effects. Our efforts attempt to integrate consideration of within-patch habitat quality with the influence of surrounding habitats.

Methods

Collection of appropriate field data for model parameterization depends on locating sites for bird surveys that are suitable for drawing inference over the modeled landscape. Where possible, we advocate selection of study sites that span habitat edges between all well-represented habitat types. In practice, this may involve an impractically large number of habitats and edge combinations, and the selection of study sites will have to be prioritized. In such cases, we prioritize edge types based on their commonness of occurrence in the modeled landscape and their sensitivity to proposed management activities. Common edges receive higher priority that rare edge types, and those that are most likely to be altered by management actions receive greater attention that those less sensitive to human activities. We have found that the number of edge types that are abundant in real landscapes is typically much smaller than the number of possible combinations of all habitat types. Based on field experiences in California, Arizona, and Costa Rica, the number of edges that must be examined empirically rarely exceeds five, and in many cases it is lower. Although field efforts to parameterize edge responses of multiple species at up to five different edge types requires a large field effort, this typically is much less demanding than the requirements for field data imposed by demographic models. In fact, we believe that the effort required to parameterize the EAM is modest compared to most other empirically driven landscape models and that the necessary resources are likely to be available for many pressing habitat management issues. For example, in the application presented below from the San Pedro River in southeastern Arizona, we are focusing on three predominate edge types in a complex mosaic of desert riparian habitats. Funding limitations prevent the collection of empirical data across all possible edge combinations; however, these data gaps affect model predictions over only a small area of the landscape

(see Figs. 63.7, 63.9 in color section). This field effort, while considerable, has been carried out by a four-person crew over two field seasons, a modest commitment of resources in comparison to several other ongoing bird population studies in this watershed.

Density Estimates

We first established multiple survey points along transects running orthogonal to transitions between different habitat patches. For each transect, we randomly selected a survey point along the habitat edge, then extended the transect 200 meters into both of the adjoining habitats. Transects spanned the full range of habitat types, with multiple sample points located within each habitat type at different distances from the patch boundary. At each survey point, we record the location (distance and direction from survey point) and identity of each bird detected. Because we expect the likelihood of detection to decrease with distance from the point, we used variable distance point sampling, recording the distance from each sample point to the detection. To estimate the probability of detection (P_a) out to some effective distance (w) from the survey point, these data are analyzed separately by species with program DISTANCE (Buckland et al. 1993). Detection probabilities are estimated for each species in each major habitat type, and these estimates are used to adjust the number of individuals counted (n) at each survey point. Density at each survey point is estimated by

$$\hat{D} = \frac{n}{\pi w^2 P_a}$$

Finally, each survey point, with its corresponding density estimate for species i (D_i), is given precise spatial coordinates using global positioning system (GPS) technology.

Density Response Functions

Each survey point that provides a density estimate for species i is assigned a habitat type label, and its spatial coordinates allow an estimate of the distance from the point to the patch boundary (edge). These data allow us to fit a regression of density (dependent variable) on distance from the edge (independent variable) for each edge type (see Fig. 63.1 in color section). The

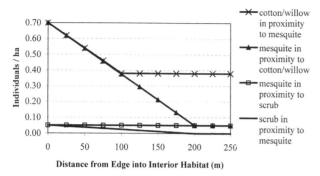

Figure 63.2. Species-specific responses to landscape heterogeneity can be expressed as a family of edge response curves (see text). Here, we fit linear response functions to empirical data quantifying the yellow warbler (*Dendroica petechia*) response to four unique edge types in the riparian landscape. This family of curves describes the influence of the adjacent habitats on the quality of the three basic habitat types.

methods used to estimate response functions and the terminology used to describe the functions are similar to Malcolm (1994). Various regression models can be assumed for species showing either declines or increases in abundance as edges are approached, and both linear and nonlinear regression models can be used. In practice, however, we have found that most response curves can be described adequately by a limited set of functions, including simple linear, exponential, half-normal, power, and second-degree polynomials.

We consider the edge to be the origin in our regression models—that is, distance 0 represents the edge. As a consequence, two separate response functions are estimated for each species for each edge type—the density response as location is moved from the edge into each of the adjoining habitat types (habitat type A and habitat type B). The regression models provide estimates of the density at the edge (y-intercept), the distance over which density is dependent upon distance from the edge (d_{max}), the constant density that occurs beyond $d_{max} = k$, the basal density in a given habitat type). It is important to note that the two functions need not be mirror images of each other and that d_{max} and basal density may differ between adjacent habitat types. An example of density response functions for a warbler breeding along the San Pedro River in southeastern Arizona is shown in Figure 63.2 (A. Brand unpublished data).

Landscape Maps and Habitat Classification

Defining and mapping the habitat types relevant to the focal taxon is a difficult and time-consuming task. Since this process has been described extensively in the published literature (e.g., Avery and Berlin 1992; Wilkie and Finn 1996; Turner and Gardner 1991) and elsewhere in this volume, we do not go into detail here. With the increase in availability of spatial data, such as gap analysis project (GAP; Scott et al. 1996) coverages and numerous sources for classified remotely sensed data, users may opt to use previously prepared vegetation or habitat maps in the model.

To create our landscape-scale maps, we relied extensively on remotely sensed data, including Thematic Mapper Simulator (TMS) data. Researchers familiar with the site classified the habitat maps we use in our example, drawing on georeferenced TMS imagery and referring to hardcopy aerial photographs. We also use parametric and nonparametric classification algorithms, such as parallelepiped and minimum distance to means, which are available in off-the-shelf image processing software, for generating habitat data for the model. The particular technique chosen by the user will depend on several factors, including the availability of spatial data, digital-image processing capabilities, and expertise. Accuracy assessment of the resulting maps is itself an involved process and may vary from map product to map product. The effect of mapping error in habitat-relationship models is an area of active research (see Garrison and Lupo, Chapter 30; Gonzalez-Rebeles et al., Chapter 57; Karl et al., Chapter 51). In cases where model error has been partitioned among mapping error and error associated with the empirical estimation of population parameters, the latter has proven to be of greater importance (A. King personal communication). The number of case studies is, however, quite limited, and further investigation is needed before general conclusions can be drawn about the relative importance of mapping error and error in population parameters.

Model Structure

In this raster-based spatial model, a species-specific density grid is created by evaluating the response func-

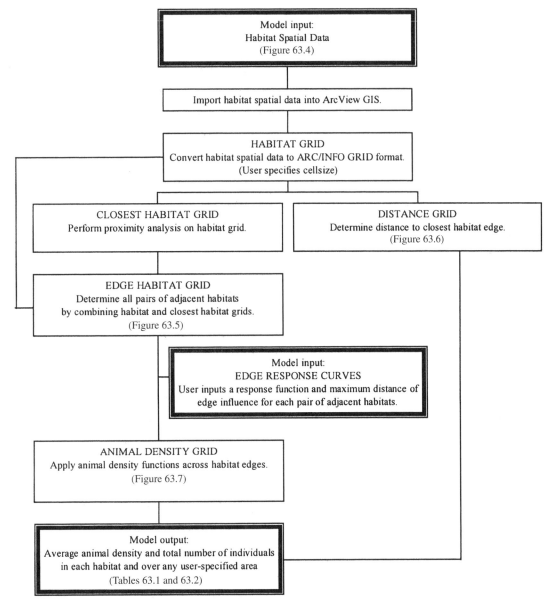

Figure 63.3. Effective Area Model flowchart. Key steps in model development are illustrated. They are described in greater detail in the text.

tions relative to the nearest edge for each pixel in the habitat map. The number of individuals in any region or habitat type is then calculated from the density grid. The model was developed for ArcView GIS (ESRI, Redlands, Calif.) using the Spatial Analyst extension and Avenue scripting language. The current developmental version has been tested under ArcView 3.2 (Windows 95/98/NT) but should also work under ArcView 3.1. The step-down process we used to develop the EAM is summarized in Figure 63.3.

The EAM requires two classes of model input:

characterization of each species density response to habitat types and edges, and a detailed habitat map (see Fig. 63.4 in color section). The habitat map, portrayed at a landscape scale, is developed from remotely sensed and/or field data. Two decisions are made at this point: (1) the number and characteristics of habitat types to use in classifying the map, and (2) the size of the minimum mapping unit. In practice, these decisions are often constrained by data availability and the management objectives. It is important, however, that the habitat classes and minimum-

TABLE 63.1.

Projected population size for three bird species, based on animal density grids (e.g., Figure 63.7).

Species	Null model				Effective area model				% Difference			
	C/W[a]	Mes[b]	DS[c]	Total	C/W[a]	Mes[b]	DS[c]	Total	C/W[a]	Mes[b]	DS[c]	Total
Yellow warbler (*Dendroica petechia*)	14.8	8.4	0.0	23.2	22.9	41.7	3.8	68.4	35	80	100	66
Abert's towhee (*Pipilo aberti*)	3.9	16.8	2.6	23.3	23.9	52.4	8.9	85.2	84	68	70	73
Black-throated sparrow (*Amphispiza bilineata*)	0.0	31.9	26.4	58.3	1.5	35.4	56.7	93.6	100	10	53	38

Note: Numbers of individuals are broken down by habitat type as estimated by a null model ignoring edge effects and the Effective Area Model. Percent differences illustrate the important influences that edge responses may have in determining species abundances at the landscape scale.
[a]C/W = cottonwood/willow gallery forest.
[b]Mes = mesquite woodland.
[c]DS = desert scrub.

mapping unit be appropriate for the focal species with the finest-grained response to edge and matrix effects.

In order to apply the density response curves, the model first determines which types of contiguous habitat exist and where they are located. This is accomplished by combining the habitat map with a map of the closest habitat type. The resulting edge habitat map (see Fig. 63.5 in color section) divides the landscape into classes of edge habitat. The model then prompts the user to input species-specific response functions and maximum distance of edge influence (d_{max}) for each type of edge habitat. For example, we found the yellow warbler (*Dendroica petechia*) to have a basal density of 0.38 individuals per hectare in the interior of the cottonwood/willow habitat, rising to 0.70 individuals per hectare along the boundary with mesquite (Fig. 63.2). This increasing trend begins 100 meters from the habitat boundary. Based on preliminary data, we assumed this trend was linear, but nonlinear response curves can also be used.

Once the model locates and characterizes the edge habitat classes, a distance-to-edge grid (see Fig. 63.6 in color section) is created to supply distance values for the density response curves. The model then projects values from the appropriate density response curves to their respective edge habitats, forming a grid of den-

sity values for a given species (see Fig. 63.7 in color section). The EAM's adjustment for edge effects can significantly change population estimates. For example, when edge effects were included, the predicted population size of yellow warblers in the study area increased by 66 percent over the null model (Table 63.1). Because the yellow warbler is an edge exploiter at each edge type, its estimated population size is higher in each habitat when edge effects are considered. In this example, the Abert's towhee (*Pipilo aberti*) and black-throated sparrow (*Amphispiza bilineata*) exhibit different edge-exploiting behavior, influencing predicted population sizes (Table 63.1).

As noted above, time, cost, and other constraints are likely to limit the amount of field data on edge responses that will be available or can be collected across the model landscape in a timely manner. Various assumptions regarding animal density surrounding these edges can be specified in the EAM. Here we assume that densities near edges for which we have no empirical edge response are equivalent to the values of the interior habitats. That is, the EAM reverts to a null model that ignores edge and matrix effects in locations where edge responses are not quantified. In our example, it is assumed that edge response curves are not available for the habitat boundary between cottonwood/willow and desert scrub habitats. Thus, the null model and EAM return the same animal density values adjacent to these edge types.

Although many sources of error influence EAM output, we believe that the most important potential source of prediction error is associated with the empirical estimation of edge response functions. We are currently examining two approaches for incorporating error estimates into the EAM. One approach assumes that the dependent variable (e.g., animal density) is normally distributed given any value of the independent variable (i.e. distance to the habitat edge) and calculates expected error based on the mean squared error. An alternative approach employs Monte Carlo methods to sample from the empirically derived distribution of the response variable and constructs confidence intervals for estimates at various distances from the edge. Ongoing efforts to test and refine these methods will lead to incorporation of error estimation algorithms into the first version of the model to be widely distributed to managers.

Application: Assessing the Impacts of Habitat Change on Birds in a Desert Riparian Ecosystem

The San Pedro River in southeastern Arizona is the last free-flowing river in the Southwest. It supports very high species diversity for many taxonomic groups, and it is an important seasonal habitat for Neotropical migrant birds (Skagen et al. 1998). The San Pedro is a threatened river due to increased demand for water to supply agriculture and rapid urban growth. Groundwater depletion may have dramatic effects on riparian vegetation leading to the decline in recruitment of riparian flora and increased mortality of mature vegetation (Stromberg et al. 1992, 1996; Auble et al. 1994). Excessive pumping of groundwater resources has led to the degradation and de-watering of arid streams and springs worldwide (Gremmen et al. 1990), threatening rare riparian habitats and compromising the viability of species dependent upon them. The conflict over groundwater pumping and instream flow of the San Pedro River, and associated impacts on riparian habitats and the biological diversity

they support, is one of the most controversial in the nation, involving county, state, and international law.

Change in habitat type and landscape composition has been prevalent over the past several decades on most desert rivers, and the San Pedro is no exception. The first European explorers described the San Pedro watershed as an expansive wetland system, with meandering small streams and only sparse woody vegetation (Hendrickson and Minkeley 1984). In the early twentieth century, channel incision and the establishment of cottonwood-willow gallery forest along the river channel dramatically changed the vegetation of the riparian corridor (Hastings and Turner 1980). Mining activities placed increased demands on limited water resources in southeastern Arizona during the first half of the twentieth century, and rapid urban expansion over the past twenty years has dramatically increased groundwater pumping rates. Water extraction has lowered the water table in some locations, threatening cottonwood-willow, mesquite woodland, and grassland habitats that compose the corridor of riparian habitat that occupies the river floodplain. As demands for groundwater continue to increase, shifts in habitat type and in the spatial configuration of patches of riparian vegetation become more likely (Auble et al. 1994; Stromberg et al. 1996), and the fate of the riparian habitats that support the majority of the region's biodiversity hangs in the balance. The complex relationship between hydrology, vegetation, and biodiversity—poorly understood and laced with scientific uncertainty—has emerged as a controversial management issue.

We used a prototype of the EAM to examine avian responses to habitat heterogeneity and landscape change along a 1.5×3-kilometer section of the upper San Pedro River. We tested the ability of the model to capture edge and matrix effects by contrasting predictions with those of a null model that ignores edge effects and uses a single mean density to predict animal abundance in each habitat patch. We also demonstrate use of the model to predict the consequences of habitat change (in this case habitat type conversion from more mesic to more xeric types) resulting from a hypothetical drawdown of the aquifer maintaining flows in this particular segment of the San Pedro River.

Drawdown Scenario

The riparian vegetation along the river corridor stands in stark visual contrast to the neighboring desert scrub habitat that predominates across much of the surrounding region, where the water table is too deep for the establishment of woody riparian species.

Cottonwood/willow gallery forest dominates along the perennial stretches of river, while mesquite bosque and *Sacaton* grasslands occupy more xeric sites in the river floodplain. The edges between these habitat types tend to be sharp and obvious in this hydrologically driven landscape structure (Fig. 63.4). In the hypothetical scenario that we simulated for the purpose of demonstrating the EAM, we assumed a simple type conversion from more mesic to more xeric habitat types following a significant drawdown of the water table. Approximately 10 hectares of cottonwood/willow habitat are converted to mesquite bosque, and 40 hectares of mesquite are converted to desert scrub (see Fig. 63.8 in color section). Although the response of vegetation to hydrological change is certain to be more complex, similar landscape-scale changes have been observed in some desert riparian habitats following water diversion or prolonged drought.

Using the methodology described above, we created a habitat map for the drawdown scenario and generated new density grids for each of three bird species—yellow warbler, Abert's towhee, and black-throated sparrow. This allowed us to estimate the effects of simulated habitat change by comparing existing conditions with those following the drawdown.

Assessing the Importance of Edge and Matrix Effects

Comparisons of the EAM and the null model suggest that the influences of habitat edges on avian distribution and density can be large in heterogeneous landscapes such as desert riparian systems. Where species show strong affinity for particular habitat edges, such as was seen for the yellow warbler at the cottonwood/willow/mesquite edge (Fig. 63.1) the resulting differences in predicted avian densities in the adjoining habitats may be quite large. For example, the EAM predicts a density of 22.9 yellow warblers per

hectare in cottonwood/willow habitat within the study areas compared to 14.8 birds per hectare in the null model (Table 63.1). Mean density of this species in mesquite habitats is expected to be 41.7 birds per hectare, compared to only 8.4 birds per hectare in the null model. For other species, the contrast between predictions of the EAM and null model are less pronounced but still large enough to suggest that consideration of edge responses in heterogeneous riparian landscapes is important when assessing habitat quality and avian distributions (Table 63.1).

Model results for this hypothetical scenario demonstrate possible applications of the Effective Area Model for predicting the outcomes of alternative management scenarios. Although it is inappropriate to attempt to extract detailed management lessons from this hypothetical scenario that is based on preliminary field data, several salient points emerge from this exercise that reflect on the utility of the EAM approach, in general.

First, model results demonstrate that the EAM is sensitive to habitat fragmentation, as the amount, type, and location of habitat edges change. Given realistic species-specific edge responses, the EAM provides a rapid, automated means of assessing the simultaneous shifts in species abundance, or other edge-sensitive response variables, given a relatively small set of clear assumptions regarding habitat use and landscape structure. Figure 63.9 (see in color section) illustrates how changes in subsurface hydrological properties may impact animal populations by driving complex changes in the amount of available riparian habitat and the spatial distribution and juxtaposition of habitat patches. Clearly, the relationship between hydrological processes and vegetation change lie at the heart of any predictions of avian responses. Such modeling efforts are currently underway along the San Pedro River (e.g., Stromberg et al. 1996), and the possibility of linking the EAM to hydrological models is an intriguing concept that may soon be tractable.

Finally, it is apparent from our analysis of the drawdown scenario that differences between EAM and null model predictions were less in the original landscape than for the drawdown scenario (Tables 63.1 and 63.2). Since more complex landscape struc-

TABLE 63.2.

Projected population sizes for three bird species before and after habitat type conversion resulting from a hypothetical aquifer drawdown scenario along a desert riparian river.

	Null model	Effective area model	% Difference
Yellow warbler (*Dendroica petechia*)			
Initial landscape	23.2	68.4	195
Following habitat type conversion	17.9	48	168
Predicted change	–5.3	–20.4	
Abert's towhee (*Pipilo aberti*)			
Initial landscape	23.3	85.2	266
Following habitat type conversion	19.9	65.7	230
Predicted change	–3.4	–19.5	
Black-throated sparrow (*Amphispiza bilineata*)			
Initial landscape	58.3	93.6	61
Following habitat type conversion	56.8	109.4	93
Predicted change	–1.5	15.8	

Note: Type conversion increases the relative abundance of edge-influenced habitat, increasing the differences between estimates from the EAM and null models.

tures generally contain more edge (Sisk and Margules 1993), this result is not unexpected. However, it suggests a more general rule: that the consideration of edge effects becomes more important as landscape complexity increases.

Discussion

A number of mechanistic explanations for edge effects on birds have been demonstrated, and additional mechanisms have been proposed. These include increases in parasitism and predation rates (e.g., Robinson et al. 1995; Arango-Velez and Kattan 1997), changes in microclimate (Kuitunen and Makinen 1993), and changes in the distribution of food resources resulting from changes in vegetation structure or cross-boundary subsidies (e.g., Restrepo and Gomez 1998). Because a diversity of mechanisms may be operative at patch edges, and because of great difference in species' life histories, responses to these factors range from positive, to neutral, to negative. Furthermore, the response of a particular species may differ at different types of habitat edges (Sisk and

Haddad in press). Despite this complexity in edge responses, the EAM approach has proven practical in several studies of animal:habitat relationships (see Sisk and Haddad in press), and the EAM performed significantly better than a null model ignoring edge effects when predicting the abundances of breeding birds in patches of oak woodland habitat in coastal California (Sisk et al. 1997).

The EAM does not explicitly incorporate the mechanisms that give rise to edge effects. Rather, the consequences of edge effects are assumed to be expressed in some demographic variable, such as density or reproductive success. In addition, the value of this variable is assumed to vary continuously as a function of distance from a habitat edge. Though some ecologists may view the EAM approach as deficient because it is more phenomenological than mechanistic, we believe the model structure can be defended on purely pragmatic grounds. The realities of conservation planning and multispecies management preclude the luxury of developing detailed mechanistic models for most species.

Management Implications

Which habitats are most important for sustaining biological diversity? What management activities can be implemented to improve the quality of existing habitats for species of concern? These are common, fundamental questions frequently asked of researchers by land and wildlife managers (see, e.g., Dasmann 1981; Forman and Godron 1986). Although inherently difficult to answer, these questions are unfortunately becoming more urgent and intractable as a consequence of rapid rates of habitat loss and changes in land cover. The result is a displacement of ecological systems outside their historic ranges of variability, a consequence of introducing novel conditions of land use, water diversion and disruption of hydrological cycles, shifting climate, and other forms of global and regional change. The significance of a landscape ecological perspective is widely acknowledged by land managers. However, at the same time that managers are discovering the relevance of concepts such as landscape pattern, heterogeneity, connectivity, and edge effects, they are discovering that few practical tools have been developed for linking landscape-scale ecological theory with practical management issues.

The EAM is an attempt to estimate the effects of habitat transitions and boundaries on the distribution and abundance of mobile organisms in heterogeneous landscapes. Predictions from our prototype model suggest that the responses of species to habitat edges may exert profound influence on such landscape-scale patterns. These effects may be particularly pronounced in the desert riparian ecosystems explored here because they are naturally linear, heterogeneous, disturbance-prone, and characterized by abrupt habitat edges. However, the differences between predictions of the EAM and a null model that excludes edge effects suggest that edges should not be ignored in other landscapes when assessing the effects of landscape pattern and heterogeneity on distribution and abundance.

Although animal abundance may not always be a reliable predictor of population status (Van Horne 1983; Pulliam 1988), we believe that it is often the only tractable response variable when many species are of interest and more-detailed demographic approaches are impractical. Where demographic data permit detailed analysis of population viability, the EAM can provide important refinements in the assessment of habitat quality in mapped habitat patches. Furthermore, the model structure of the EAM is sufficiently general to allow the modeling of any ecological process or variable that can be represented as an edge response, that is, as a continuous function across an ecological gradient running orthogonal to a habitat edge. We are currently using the same model structure to address microclimatic variation and butterfly abundance across forest structural types in ponderosa pine forests on the Colorado Plateau (Meyer and Sisk in press) and to model avian nest productivity across heterogeneous forest and riparian ecosystems (J. Battin, A. Brand unpublished data). Given the flexibility of the EAM in modeling edge-related processes at the landscape scale, we believe that the model will provide a versatile new tool for assessing animal:habitat relationships at the landscape scale.

As discussed above, a significant limitation to wide application of the EAM approach is its requirement for robust edge-response data for all species of management interest. Although the cost and time investment for obtaining the relevant empirical data are much less than those of intensive demographic approaches (e.g., the northern spotted owl (*Strix occidentalis caurina*) research program, see Forsman et al. 1996a; Noon and McKelvey 1996a), field efforts required for estimation of edge responses for rare taxa may, nevertheless, be significant. Several lines of evidence, however, suggest that efficiencies in data acquisition may arise. For example, multi-taxon sampling by the co-location of survey points is recommended. (See Fig. 63.1 for edge responses of three of several dozen species sampled along the San Pedro River, using the methodology described above.) In addition, preliminary data suggest that density response functions are restricted to a small number of possible shapes. An active area of research is to determine if the shape of edge response curves can be inferred as a general function of taxonomy, life history attributes, habitat types, and landscape characteristics.

In its prototype stage, described here, the EAM links field and remotely sensed data in a landscape model that permits comparison of the effects of landscape structure on wildlife populations. In this context, we believe the EAM provides decision makers with a useful tool for comparing the impacts of alternative land management plans.

Acknowledgments

We thank Arriana Brand for providing the preliminary field data used to generate the edge response functions presented here. Leslie Ries offered insight on landscape composition and structure on the upper San Pedro River, and John Fay, Center for Conservation Biology, Stanford University, provided helpful advice on Avenue scripting. Finally, comments and suggestions from James Battin, Nick Haddad, J. Michael Scott, and one anonymous reviewer greatly improved this chapter. This work was supported by the Strategic Environmental Research and Development Program (Project CS-1100), and aided by collaborators in the Semi-Arid Land-Surface-Atmosphere Program, coordinated by the USDA Agricultural Research Service, Tucson, Ariz.

Demographic Monitoring and the Identification of Transients in Mark-recapture Models

M. Philip Nott and David F. DeSante

M ethods for monitoring bird populations fall into two categories: population monitoring and demographic monitoring. Population monitoring includes techniques such as point count surveys and area searches that can provide inferences regarding species richness and abundance. Over time these techniques can detect change in population size but may not be able to identify whether the cause(s) of that change are associated with birth or death processes if they do not discriminate between breeding and nonbreeding individuals. The demographic causes of population change may be identified by demographic monitoring techniques, including banding and nest monitoring, that provide inferences regarding productivity and survivorship parameters. Accurate estimates of these parameters are essential to the construction of predictive population models. Constant-effort mist netting provides spatially explicit estimates of survival rates from mark-recapture modeling of banding data (Buckland and Baillie 1987; Rosenberg et al. 1999) and indices of productivity obtained from the ratio of young to adult birds captured (DeSante 1992; DeSante et al. 1995; Peach et al. 1996). This technique provides both within-year and between-year information, allowing modifications to be made to mark-recapture models that consider the existence of transient individuals and provide estimates of demo-

graphic parameters for the resident proportion of the population.

Pradel et al. (1997) proposed a method for identifying transients and adjusting survival rate estimates accordingly. This method is incorporated into a modification of the Cormack-Jolly-Seber mark-recapture model (Cormack 1964; Jolly 1965; Seber 1965) called TMSURVIV (Pradel et al. 1997). This model provides estimates of the proportion of residents (τ) in addition to estimates of survival rate (ϕ) and recapture probability (p). Although this method allows that a bird caught in only one year may be a resident, based on the probability of between-year recaptures being less than unity, it ignores within-year information inherent in banding data that is derived from constant effort mist-netting. Analyzing these data requires a more sophisticated mark-recapture model that considers a length-of-stay criterion for individual birds (Pradel et al. 1997). This is provided by LOSSURVIV, a recent modification by TMSURVIV and J. D. Nichols and J. Hines.

In this chapter, we describe the flexibility of constant-effort mist-netting data and outline methods by which the accuracy of estimating demographic parameters can be improved. We utilize banding data from ten bird species (three temperate-wintering, three temperate-tropical, and four tropical-wintering species) captured in the northwestern United States at thirty-six constant-effort banding stations operated by the Institute for Bird Populations as part of the Monitoring Avian Productivity and Survivorship (MAPS) program (DeSante 1992; DeSante et al. 1995). We analyze banding data for the period 1992 to 1998 to provide temporal patterns of the resident adult, transient adult, and juvenile portions of the population. For each species, we compare these trends with those obtained from population monitoring data for the western states provided by the North American Breeding Bird Survey (BBS; Peterjohn et al. 1995). We compare estimates of adult survival rates, recapture probabilities, and proportions of residents obtained from three different mark-recapture models (Pollock et al. 1990; Lebreton et al. 1992). These include LOSSURV, a recent modification of SURVIV (White 1983, 1986) that considers individuals categorized as transients or residents according to a length-of-stay criterion. Finally, we explore the relationship between survival rate and productivity as a function of migratory strategy with respect to accepted life-history theory for temperate, temperate/tropical, and tropical migrants.

Methods

The Monitoring Avian Productivity and Survivorship (MAPS) program collects data from over five hundred field stations across the North American continent (DeSante et al. 1998, 1999). This program adopts a "constant-effort mist-netting" protocol to provide survivorship estimates and productivity indices for passerines. Typically, at each station, bird-banding teams operate ten mist nets located within the central 8 hectares of a 20-hectare study plot for six hours following sunrise. Each station is visited on one day within sequential ten-day periods throughout the breeding season (May to August) up to a maximum of ten periods. Normally, six stations constitute a MAPS location, which represents a monitoring effort in a national forest, national park, or other managed land area. The protocol assumes that captures include adults and young from both the vicinity of the monitoring station early in the breeding season and from the surrounding landscape later in the season as breeding activity ceases and the birds begin to disperse.

TABLE 64.1.

Common names, scientific names, and Breeding Bird Laboratory (BBL) abbreviations for ten bird species represented by over five hundred individuals in the MAPS database for USDA Forest Service Region 6.

Common name	Scientific name	BBL abbr.
Western flycatcher	*Empidonax difficilis* and *E. occidentalis*	WEFL
Winter wren	*Troglodytes troglodytes*	WIWR
Swainson's thrush	*Catharus ustulatus*	SWTH
Yellow-rumped (Audubon's) warbler	*Dendroica coronata audubonii*	AUWA
Townsend's warbler	*Dendroica townsendi*	TOWA
MacGillivray's warbler	*Oporornis tolmiei*	MGWA
Wilson's warbler	*Wilsonia pusilla*	WIWA
Song sparrow	*Melospiza melodia*	SOSP
Lincoln's sparrow	*Melospiza lincolnii*	LISP
Dark-eyed (Oregon) junco	*Junco hyemalis oregonus*	ORJU

Here, we consider a group of thirty-six stations that allow us to monitor demographic parameters in forested lands under the stewardship of the USDA Forest Service Region 6, which covers the Pacific Northwest region of the United States. Specifically, the locations included are Mount Baker National Forest and Wenatchee National Forest in the state of Washington and Willamette National Forest, Siuslaw National Forest, Umatilla National Forest, and Fremont National Forest in the state of Oregon. These forests are typically heavily managed and share many plant, animal, and bird species. Data were pooled for ten target species (top ten species ranked by total number of captures in each case, represented by over five hundred individuals captured) captured at these stations to provide regional survival rate estimates for adult birds.

We selected banding data for the ten most-captured species (Table 64.1) during the seven-year period from 1992 through 1998 inclusive. Dates of operation vary by station dependent upon latitude and elevation. We only considered captures made between MAPS periods 4 and 10 representing the seven ten-day periods between May 31 and August 8 that were common to all stations.

Demographic Groups and Indices of Productivity

Birds that are caught and banded belong to one of three demographic groups—residents, transients, and juveniles. Early in the breeding season the catch includes philopatric individuals returning to breeding territories within the boundaries of the banding station, many of which are caught year after year. The remaining portion of the early-season catch is made up of transient individuals passing through the station on their way to distant territories, or seeking habitat in which to establish new territories. In the middle of the breeding season the catch consists of resident breeders (whose activity space includes a mist net location) and unpaired adults (floaters) that may be queuing for available territories or passing through in search of otherwise unoccupied breeding habitat. Although some individuals are only caught in one year, they may be caught more than once in that year and could be considered as resident birds. Later in the season, as young birds fledge and adult territoriality relaxes, the catch includes dispersing juveniles and adults from the surrounding area.

For each species, we constructed temporal activity patterns by categorizing individuals as resident, transient, or juvenile individuals according to their capture histories:

1. **Adults Seen Once (ASO).** Transient adults include those individuals captured once, and once only, and those individuals caught in only one year but more than once within a period of less than seven days.
2. **Between-Year Residents (BYRES).** Those individuals caught in at least two different years. It is assumed that the probability of any these individuals are merely passing through the site and are caught in more than one year is very low.
3. **Within-Year Residents (WYRES).** Those individuals caught in only one year but caught more than once in that year. They are only classified as WYRES if the maximum "length-of-stay" between captures exceeded six days; otherwise, they were assigned to the ASO category.
4. **Known Residents (KNRES).** Those individuals classified as BYRES or WYRES; we assume these

birds represent the group that are resident at the monitoring site.

5. **Young (YNG).** Those individuals identified in the database as juvenile birds that may have come from the site or from the surrounding landscape.

For each species, we estimated linear temporal trends for the annual numbers of all adults, known residents, adults seen once, and juveniles. We indexed productivity as the proportion of young in the catch both annually and as a mean annual index for the whole period. We recorded the frequency of captures of birds (pooled across years) in each category by ten-day period, calculated the abundance of each category as a proportion of the total number of captures, and plotted the results as a histogram for each species. Furthermore, to look at the diurnal patterns of activity, we plotted the frequency of captures by hour after sunrise. Henceforth, we refer to these histograms as "seasonal capture profiles" and "hourly capture profiles," respectively.

Comparing Population Trends in MAPS with Breeding Bird Survey Data

The North American Breeding Bird Survey (BBS) provides trends of the numbers of birds of all species seen and heard at a number of stops along routes distributed across North America (Peterjohn et al. 1995). We obtained the regional BBS abundance trends (James et al. 1996; Link and Sauer 1998) for ten target species (Table 64.1) of British Columbia, Washington, Oregon, and California for the period 1992–1998. For each species, we compared these trends with corresponding trends in the number of adults calculated from MAPS data.

Estimates of Survivorship and Transience

Birds caught only once may belong to any of the three demographic groups. Adults may be transient individuals, residents that died or left the area by the next year, or residents caught in the latest year of banding that may be caught in future years. The transient individuals cause survivorship estimates to be biased low in closed-population models. Pradel et al. (1997) approached the problem of identifying transients using an ad hoc approach to produce unbiased estimates of resident proportions that effectively ignores the first year of capture for all individuals. This approach is incorporated into TMSURVIV, which produces estimates of survivorship and capture probabilities for residents, as well as proportions of residents. This approach, although unbiased with regard to the recapture probabilities, does not take into account the important "within-year" information that can indicate an individual captured in only one year is a resident and not a transient. Pradel et al. (1997) suggest that the critical parameter in identifying transients is the length-of-stay period and that transients can be detected with a suitable study design in which the interval between capture sessions exceeds that of the maximum length-of-stay of a transient individual. Constant-effort banding studies represent a sampling design appropriate for detecting transients using a length-of-stay method because banding sessions can be separated by an interval that exceeds the period of time a transient might be expected to stay in the area.

For each species, we constructed capture histories for all adult birds. We obtained time-constant estimates of adult survival probability (ϕ), recapture probability (p), and proportion of residents (τ) by entering the capture histories into three different mark-recapture models. The first and oldest model, SURVIV (White 1983), assumes a closed population and does not distinguish between transient and resident individuals. The second model, TMSURVIV, is a modification of SURVIV that provides an unbiased estimate of the resident proportion based on between-year information (Pradel et al. 1997). The final model, LOSSURVIV, is a modification of TMSURVIV that considers additional within-year information. This model requires categorizing capture histories as (1) unmarked individuals caught only once in the first year of capture, regardless of how many times they were caught in subsequent years, and (2) marked individuals caught more than once seven or more days apart in their first year of capture. We processed these capture histories for each species through a number of permutations of time-constant and time-independent (with respect to ϕ, p, and τ) submodels to ascertain whether the time-constant $\phi p \tau$ submodel represented the best (or equivalent) submodel for estimating demographic parameters. We compared and contrasted the values of demographic parameters resulting from these models.

The Relationship between Survival Rate and Productivity

DeSante et al. (1998) presented evidence for the existence of the trade-off between survival rate and productivity suggested by Martin (1995) and found the relationship to be a function of longitude and migration strategy. The underlying theory is that the longer a species lives the fewer offspring it needs to produce to maintain stable population levels. Avian migration to climatically more stable tropical wintering areas may lead to higher overwintering survival rates relative to those of temperate-wintering species. On the other hand, temperate-wintering species can exploit available breeding habitat sooner than tropical-wintering species and potentially produce more clutches. All else being equal, survival rates and productivity indices should correlate negatively and be a function of migration strategy. To explore this relationship, we plotted the relationship between survivorship and productivity for the ten species of this study.

Results

Seasonal capture profiles (Fig. 64.1) show the number known resident individuals (KNRES = BYRES + WYRES), individual adults seen once only (ASO), and individual young birds as a function of sequential visits (periods) to the stations throughout the breeding season. Typically, the peak of KNRES captures occurs in period 6 or 7 corresponding to late June or early July, with the exception of dark-eyed juncos (*Junco hyemalis*) for whom the peak capture period occurs in period 4 at the beginning of June. Note that the seasonal capture profiles of KNRES include, for many individuals, multiple captures of the same birds. The peak of ASO captures relative to the peak of resident captures varies across species and occurs in the same or a later period, never earlier. The ASO profile for the western flycatcher (Pacific-slope flycatcher [*Empidonax difficilis*], Cordilleran [*E. occidentalis*]) takes a sudden jump from fifty to eighty individuals in period 7 whereas that for the dark-eyed junco (*Junco hyemalis*) decreases gently over the entire season.

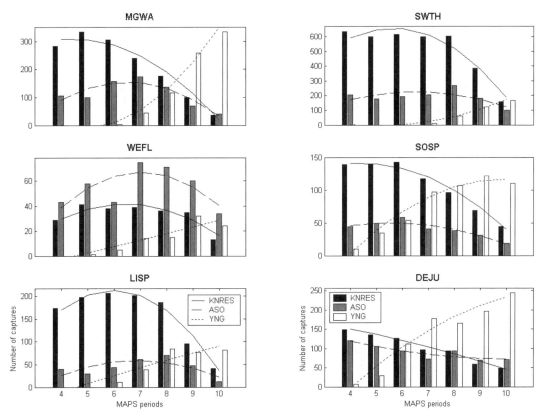

Figure 64.1. Seasonal capture profiles of the number of individuals caught in mist nets by ten-day periods between May 31 and August 8. These data are pooled from thirty-six banding stations across a seven-year period (1992–1998). Six species are represented: MacGillivray's warbler (MGWA), Swainson's thrush (SWTH), western flycatcher (WEFL), song sparrow (SOSP), Lincoln's sparrow (LISP), and dark-eyed (Oregon) junco (ORJU) Individuals are categorized into three groups, and second-order polynomials are fitted to each set of histogram bars. These categories are known residents (black bars, solid line), adults seen once only (gray bars, dashed line), and young birds (white bars, dotted line).

Although the date varies by year, the appearance of young first occurs in period 6 or 7 and generally increases to a peak in periods 9 and 10 for all species except song sparrow (*Melospiza melodia*) and Lincoln's sparrow (*Melospiza lincolnii*) for which it occurs in periods 9 and 8, respectively.

The hourly capture profiles depicted in Figure 64.2 also show differences between species but not across demographic groups within a species. Generally, the proportion of total captures decreases as the day progresses. Approximately 45 percent of MacGillivray's warbler (*Oporornis tolmiei*), Swainson's thrush (*Catharus ustulatus*), and song sparrow captures occur in the first two hours (peaking in the second hour), followed by a gradual decline in the hourly capture proportion thereafter. Lincoln's sparrow captures

occur mainly in the first three hours (60 percent of total) and decline more rapidly thereafter. Hourly capture profiles for western flycatcher and dark-eyed junco show a more even distribution except that the proportion of western flycatcher captures made in the first hour is very low.

Population Trends in Maps and Breeding Bird Survey Data

Six of the ten species show a negative trend in the total number of adults caught during the period 1992–1998 (Table 64.2) but only Townsend's warbler (*Dendroica townsendi*) shows a significant trend ($P < 0.05$). Generally, these trends are reflected in the trends reported for the KNRES and ASO portions of the captures as well for the annual numbers of young

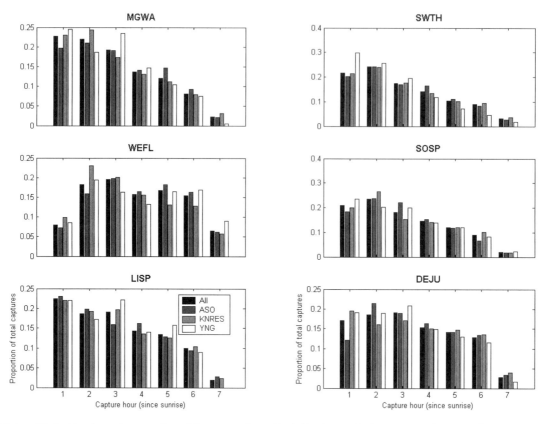

Figure 64.2. Hourly capture profiles of the proportion of individuals caught in mist nets in each of seven hours after sunrise. These data are pooled from thirty-six banding stations visited annually between May 31 and August 8, across a seven-year period (1992–1998). Six species are represented: MacGillivray's warbler (MGWA), Swainson's thrush (SWTH), western flycatcher (WEFL), song sparrow (SOSP), Lincoln's sparrow (LISP), and dark-eyed (Oregon) junco (ORJU). Individuals are categorized into four groups: all adults (black bars), adults seen once only (dark gray bars), known residents (light gray bars), and young birds (white bars).

captured. The increases in the total number of adult song sparrows and winter wrens (*Troglodytes troglodytes*) appears to be associated with significant increases in the number of adults only seen once. Similarly, the decline in western flycatcher adults is associated with a significant decline in the numbers of adults seen once. Conversely, the declines in Townsend's warbler and dark-eyed junco adults are associated with significant declines in the numbers of known residents. For the dark-eyed junco population at least, the overall decline may be driven by a significant decline in the numbers of young.

Breeding Bird Survey trends for these species in the western coastal states region agree with the MAPS adult trends in nine out of ten cases. The MAPS trend for adult Swainson's thrush shows an increase and the BBS trend a decline.

Competing Mark-recapture Models

For each species, comparisons of the survivorship probability, capture probability, and resident proportion from three competing mark-recapture models—SURVIV, TMSURVIV, and LOSSURVIV—are shown in Table 64.3. In all cases, the time-independent transient model represents the best (or equivalent) model reported by TMSURVIV or LOSSURVIV based on the values of Akaike Information Criteria (Akaike 1981; Burnham and Anderson 1992) associated with each combination of time dependent and time-independent ϕ, P, and τ models. The values of both ϕ and P parameter estimates are significantly greater for both TMSURVIV and LOSSURVIV than for SURVIV ($P < 0.01$, two-tailed t-test), whereas exactly half are greater and half are smaller when comparing

TABLE 64.2.

Total number and short-term trends (1992–1998) in all adults (Adult), known residents (KNRES), adults seen once (ASO), and young (Young) for ten bird species well represented in the MAPS data covering USDA Forest Service Region 6.

Code[a]	Adult	Trend	KNRES	Trend	ASO	Trend	Young	Trend	BBS[b]	PI[c]
WEFL	470	–4.0	92	–2.2	378	–1.8[d]	88	0.3	–2.79[d]	0.14
WIWR	534	1.2	138	–2.8	396	4.1[d]	266	2.5	5.57[d]	0.31
SWTH	2213	5.6	947	0.5	1266	5.2	342	–3.2	–2.25[d]	0.09
AUWA	379	–1.2	55	–2.0	324	0.8	186	–4.0	–1.08	0.29
TOWA	447	–8.5[d]	68	–7.1[d]	379	–1.4	469	–6.6	–2.26	0.45
MGWA	1158	–4.8	407	–5.6	751	–0.8	720	–5.9	–0.9	0.32
WIWA	776	–3.5	199	–2.9	557	–0.6	312	–5.7	–1.46	0.28
SOSP	484	5.0	219	0.5	265	4.5[d]	492	2.5	1.31	0.45
LISP	550	0.9	279	–0.7	271	1.6	267	0.0	2.37[d]	0.25
ORJU	889	–2.1	277	–3.7[d]	612	1.7	939	–18.3[d]	–1.22[d]	0.45[e]

[a]Species abbreviations: WEFL—*Empidonax difficilis* and *E. occidentalis*; WIWR—*Troglodytes troglodytes*; SWTH—*Catherus ustulatus*; AUWA—*Dendroica coronata auduboni*; TOWA—*Dendroica townsendi*; MGWA—*Oporornis tolmiei*; WIWA—*Wilsonia pusilla*; SOSP—*Melospiza melodia*; LISP—*Melospiza lincolnii*; and ORJU—*Junco hyemalis oregonus*.
[b]North American Breeding Bird Survey (BBS) regional population trends for British Columbia, Canada, and Washington, Oregon, and California, USA.
[c]The mean annual proportion of young (PI) in the total catch (pooled across all stations) expressed as (Young/[Adults + Young]).
[d]Statistically significant ($P < 0.05$).
[e]Significantly decreasing productivity ($P = 0.01$).

TMSURVIV with LOSSURVIV for ϕ and P (non-significant). The ϕ and P increases that result from using LOSSURVIV over TMSURVIV are associated with decreases in the estimates of the resident proportion (τ). Conversely, the ϕ and p decreases that result from using LOSSURVIV rather than TMSURVIV are associated with increases in τ for only three of five species. The percent change in the precision of the estimate of survivorship (ϕ) results from comparing the coefficient of variation (standard error of the estimate of ϕ / estimate of ϕ) from LOSSURVIV with that from TMSURVIV. Precision increased for all species, with values ranging from 1 to 29.3 percent and with a mean improvement of about 16 percent ($P < 0.01$, two-tailed t-test), indicating that estimates of survival rate produced by LOSSURVIV are generally more precise than those produced by TMSURVIV.

A plot of the mean annual productivity indices (proportion of young in the catch) derived from Table 64.2, against the survival estimates from LOSSURV in Table 64.3, is shown in Figure 64.3. A regression line reveals a negative relationship in which a high survival rate is associated with a low productivity index and a low survival rate is associated with a high productivity index. Importantly, survival rates of temperate-wintering species tend to be lower than those associated with species with intermediate or tropical-wintering strategies. In turn, survival rates for these species is lower than those associated with tropical-wintering species.

Discussion

Analysis of constant-effort mist-netting data can detect annual changes in population size and structure as well as seasonal or diurnal patterns in the capture rates of resident adult, transient adult, and juvenile portions of avian populations. This information can be used to increase the accuracy and precision of demographic parameter estimates. Capture histories derived from these data show interspecific differences in temporal patterns of activity. For some species, the probability of capture peaks during the first few hours of the morning, whereas for others it remains relatively constant over a larger part of the morning.

Similarly, the probabilities of capturing young and adults vary by species throughout the season. This

TABLE 64.3.

Comparison of estimates of survivorship probability (φ), capture probability (P), and resident proportion (τ) from SURVIV, TMSURVIV, and LOSSURVIV for ten bird species[a] utilizing MAPS data for the period 1992–1998 in USFS Region 6.

	SURVIV				TMSURVIV						LOSSURVIV						
Species[a]	φ	CV (φ)	P	CV(P)	φ	CV (φ)	P	CV(P)	τ	CV(τ)	φ	CV (φ)	P	CV(P)	τ	CV(τ)	ΔCV (φ)%
WEFL	0.505	12.7	0.150	24.0	0.553	13.0	0.215	31.2	0.592	34.8	0.561	11.2	0.232	24.1	0.461	25.8	13.7
WIWR	0.233	18.5	0.460	22.8	0.364	19.8	0.648	16.8	0.399	29.3	0.314	16.2	0.593	17.9	0.430	24.4	17.9
SWTH	0.494	2.6	0.533	3.9	0.585	3.2	0.601	3.7	0.660	5.6	0.577	2.6	0.597	3.5	0.546	5.9	20.0
AUWA	0.481	18.5	0.111	34.2	0.582	17.4	0.251	38.2	0.320	44.1	0.549	17.1	0.194	35.1	0.403	37.7	1.3
TOWA	0.349	16.9	0.236	25.8	0.427	17.1	0.370	27.6	0.475	34.9	0.390	16.7	0.315	28.3	0.609	31.9	2.5
MGWA	0.398	10.1	0.523	7.3	0.525	5.9	0.633	6.0	0.533	9.9	0.495	4.8	0.612	6.0	0.487	9.9	17.9
WIWA	0.359	10.0	0.343	15.2	0.425	11.3	0.442	15.8	0.607	21.4	0.445	9.0	0.467	12.8	0.423	16.8	20.4
SOSP	0.371	8.4	0.604	10.6	0.394	11.7	0.626	11.0	0.884	17.9	0.436	8.3	0.658	9.3	0.584	15.6	29.3
LISP	0.418	6.0	0.654	7.5	0.427	8.9	0.662	8.0	0.957	13.4	0.439	6.8	0.670	7.3	0.857	11.9	23.2
ORJU	0.413	7.5	0.345	11.9	0.415	9.6	0.347	15.9	0.989	19.3	0.433	8.1	0.373	12.6	0.832	14.5	16.1
Mean	0.402		0.396		0.470		0.480		0.642		0.464		0.471		0.563		16.2

Note: In all cases the time-independent transient model represents the best (or equivalent) model reported by TMSURVIV or LOSSURVIV. Coefficients of variation (standard error/estimate) are calculated for each parameter and shown as percentages (in boldface italic type). The percent change in the precision of the estimate of survivorship (f) results from comparing the coefficients of variation from LOSSURVIV with that from TMSURVIV. A positive value indicates that the estimate of survival rate from LOSSURVIV is more precise than that from TMSURVIV.
[a]Species abbreviations: WEFL—*Empidonax difficilis* and *E. occidentalis*; WIWR—*Troglodytes troglodytes*; SWTH—*Catherus ustulatus*; AUWA—*Dendroica coronata audubonii*; TOWA—*Dendroica townsendi*; MGWA—*Oporornis tolmiei*; WIWA—*Wilsonia pusilla*; SOSP—*Melospiza melodia*; LISP—*Melospiza lincolnii*; and ORJU—*Junco hyemalis oregonus*.

strongly suggests that foraging strategies vary by species. Obviously, when indexing productivity by the proportion of young in the total catch, the seasonal timing of missing effort can bias the results. If effort is missed early in the season, the number of adult captures may be underestimated and bias productivity high. Conversely, if effort is missed late in the season, the number of young may be underestimated and bias productivity low. Let us assume, for a given species, one hundred adults and forty young should have been captured, 20 percent of the adults are normally captured in the first banding period and 40 percent of the young are normally captured in period 10. If half of the effort in the first period were missed, then the productivity index (proportion of young in the catch) would be biased 17 percent higher than expected. If half of the effort in period 10 were missed, the productivity index would be 15 percent lower than expected. Because adult and young captures are a function of both the time of day and the banding period, species-specific temporal patterns of capture probabil-

ity must be used to adjust productivity indices when banding effort is missing.

In fact, most mist-netting effort is missed early or late in the season due to unfavorable weather, is missed early in the day due to logistic or weather problems, or is missed late in the day due to high ambient temperatures forcing nets to be closed (P. Nott unpublished data). Because interspecific diurnal activity patterns vary, effort missed in a particular hour may differentially underestimate numbers of captures among species. It is also likely that these temporal activity patterns vary geographically and with patterns of environmental conditions (e.g., temperature and humidity). We propose that to provide comparisons of annual indices of productivity over many years, corrections for missing effort should be based by region and on the expected proportion of the catch by both hour and by banding period.

The first step in this process involves constructing matrices (hours by periods) expressing the expected proportion of resident, transient, and young captures in each time slot for each species. Young birds, except-

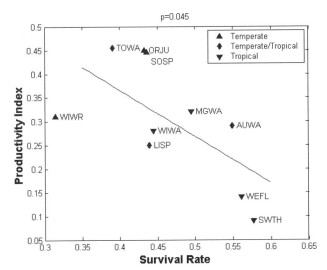

Figure 64.3. Relationship between mean annual productivity and survival derived from banding data for ten species of passerines breeding in the Pacific Northwest region of the United States. Productivity is expressed as the annual mean of the proportion of young in the total catch from banding data pooled across thirty-six stations. Survival rates are estimated from a modified Cormack-Jolly-Seber mark-recapture model that considers transient individuals (see text). Three temperate-wintering species (upward pointing triangles), three temperate/tropical-wintering species (downward pointing triangles) and four tropical-wintering species (diamonds) are shown. Linear regression shows a significant negative relationship ($P < 0.05$).

ing those subsequently recruited into the population, are generally caught only once and therefore generally require only a simple proportional correction as do the number of adults seen once. The relationship between missing effort and changes in the resident proportion of the population is more complex. Many between-year captures are also caught multiple times within a year and so would remain classified as residents until considerably more effort is missed and they become classified as adults seen once or even transients. Conversely, as effort increases, a portion of the adults seen once may be recaptured in another period and reclassified as within-year residents. The proposed method to obtain species-specific correction rates for each of the time slots and demographic groups is to apply a simple Monte Carlo method. Small increments of effort may be removed randomly (and repeatedly) from existing data, pooled across a number of stations, by period and hour. This process facilitates the formulation of time-specific (diurnal and seasonal) rate equations expressed as the proportion of individuals lost per unit of

missing effort. These relationships can then be used to correct for missed effort by adjusting the numbers of individuals caught at individual banding stations.

Although MAPS adult population trends generally agree with BBS trends, it is important to note that the two protocols are very different. BBS point count data is collected from 50-kilometer-long roadside routes representative of a number of different habitat types and environmental conditions. As such, BBS provides a good indicator of overall numbers of birds of many species. On the other hand, MAPS data is collected from spatially restricted forest interior or offroad forest edge plots, and, because it distinguishes between adults and young, can provide indices of productivity, estimates of survivorship, or estimates of the proportion of transient individuals in the numbers of adults detected. In this way, constant-effort bird banding can identify the proximal causes of population changes. In the case of the Oregon junco (*J. hyemalis oregonus*), it seems likely from banding data that the number of young produced in a year are declining rapidly. For Swainson's thrush, the numbers of detections in BBS data are increasing as the numbers of adults seen once are increasing, but the numbers of resident adults captured show no temporal trend.

Perhaps, in cases where there is a great disparity between trends of adults seen once and resident adults, the source-sink dynamics of those populations are changing. Fortunately, morphological data collected from captured individuals allow banders to identify the age (Pyle 1997) and breeding status (identified by brood patch and cloacal protuberance) of captured birds. It is possible that further analysis of the banding data may reveal stations at which, for one or more species, the age of recruited breeding birds has changed over time. In turn, this might suggest a shift from source to sink population or *vice versa*.

Obviously, open-population mark-recapture models (TMSURVIV and LOSSURVIV) that estimate demographic parameters based on within-year and/or between-year information provide higher values for survival rates and recapture probabilities than does the closed-population model SURVIV. Although the net overall difference in the estimates of survival rate produced by TMSURVIV and LOSSURVIV is minimal (about 1 percent per species) and statistically insignificant, LOSSURVIV provides a significant improvement in the precision of those estimates (about 16 percent). LOSSURVIV also conserves more of the information in the capture histories, which may be responsible for the improvement in precision surrounding survival rate estimates.

Across species, we detected a negative correlation between survival rates and productivity indices, which adheres to generally accepted life history theory (e.g., Martin 1995), whereby species with higher survival rates (e.g., long-distance migrants) produce fewer young than do species with lower survival rates (e.g., short-distance migrants). Within species, however, an analysis of banding data collected across a range of latitudes (and elevations) will facilitate testing of alternate life-history theories. One such theory, the "*time-allocation hypothesis*" (Greenberg 1980) predicts that within species long-distance migrants survive better at higher latitudes (or elevations) than at lower latitudes (or elevations) because they have compensated for having less time to devote to reproduction by evolving to live longer.

For several target species, genetic analyses (Milá 2000) and isotope analyses (J. F. Kelly pers. comm.) of feather and blood samples taken from both North American populations in the breeding season (MAPS program) and from populations wintering in North America, Mexico, and Central America make it possible to identify discrete breeding and wintering populations of individual species (e.g., MacGillivray's warbler and Wilson's warbler). Then for those species that exhibit spatial demographic structure, such as that caused by a leap-frog migration strategy (e.g., Bell 1997), it may be possible to quantify the effect of environmental stressors (e.g., weather or land use changes) on various stage(s) of species' life cycles and their relative contributions to trends in population sizes. Further research will therefore explore the relationships among regional variations in age structure, morphometrics, survivorship, and productivity toward a better understanding of source-sink population dynamics, phylogeography, and life-history theory.

Acknowledgments

We thank the DoD Legacy Resources Management Program for financial support for this work. We thank Peter Boice and Alison Dalsimer of the Legacy Program, Joe Hautzenroder and Chris Eberly of the DoD Partners in Flight Program, and Robert Johnson of the U.S. Army Corps of Engineers, Huntsville Division for logistical support. We also acknowledge the staff of the Patuxent Wildlife Research Center (USGS/BRD Resources Division), especially Jim Nichols and William Kendall for their help in designing TMSURVIV and LOSSURVIV and Jim Hines for programming these models and subsequent technical support. Furthermore, we would like to thank Rodney Siegel, Amy McAndrews, and Nichole Michel at IBP for their help in crafting this chapter, along with the many biologists, interns, and especially volunteers who diligently operate the monitoring stations every year to provide these valuable data. This is contribution number 109 of The Institute for Bird Populations.

PART 6

Future Directions

Predicting Species Occurrences: Progress, Problems, and Prospects

John A. Wiens

In one way or another, predicting the occurrence of species in space and time comes down to dealing with habitats. The issues that we face in predicting occurrences therefore are issues of habitat—how we define it; how we measure, map, and model it; how we analyze it; what it means to organisms; and, ultimately, how we can use knowledge about habitat to manage natural resources in a sustainable and balanced way.

These are not novel insights. Awareness of "habitat" and its importance has been with us for a long time. It was the foundation of the observations of early naturalists, such as Gilbert White, Henry David Thoreau, or Alexander von Humboldt, and it became formalized in the writings of Charles Darwin, Alfred Russell Wallace, John Muir, and Aldo Leopold. It continues as a thread that runs through the thinking and work of all modern ecologists. For some, however, it is buried deep in their subconscious. This is especially true of some theoretical ecologists, although certainly not all. After all, Robert MacArthur, perhaps the most influential theoretical ecologist of the twentieth century, really knew his birds. And "habitat" has become the foundation on which management of species and communities is often based.

All of these approaches, from the observations of early naturalists to the applications of modern land managers, are founded on the premise that there is a close relationship between the occurrence and abundance of species and characteristics of their habitats, that we can predict what will happen to species by knowing their habitat relationships. There have been times when we thought we had it right, that we could actually predict species occurrences with a high degree of certainty and accuracy. During the 1950s and 1960s, for example, niche theory was thought to provide a reasonable abstraction of the critical features of habitat. This approach gave way in the 1970s and 1980s to the construction of models based on multivariate statistics—principal components analysis (PCA), discriminant function analysis (DFA), and a variety of others—what Dean Stauffer (Chapter 3) has referred to as the "multivariate muddle" period, although it might more appropriately be termed "multivariate madness." Yet, every time, as we learned more ecology, we realized that critical components were missing from these approaches. I know. I began my career basing my analyses of habitat relationships on niche theory (Wiens 1969) and then went through a multivariate phase—thanks to John Rotenberry (Wiens and Rotenberry 1981b). Each time I thought I had uncovered the "true" habitat relationships only to realize that my "other things being equal" assumption contained too much interesting ecology to ignore. Now I'm in a spatially explicit landscape phase; we'll see where that leads.

So thinking about and assessing "habitat" has evolved. Perhaps the surest way to see where progress has been made and problems remain—for the detection of new, unforeseen problems is an inevitable consequence of progress—is to compare the present volume with the *Wildlife 2000* collection published sixteen years ago (Verner et al. 1986b). My intent here is to offer a collection of thoughts, perspectives, and observations that have been prompted by these books, by hallway discussions at the Snowbird conference, and by my own tortuous pathway toward understanding "habitat." My presentation will be more in the style of an extended essay than a comprehensive synthesis or review. Thence, I will refer to the published literature only sparingly.

Progress

Dean Stauffer (Chapter 3) has provided an enlightening and entertaining historical perspective on how we got where we are. Clearly, approaches to wildlife-habitat evaluation have changed a lot, and we'd like to think that most of the change represents progress. Looking over *Wildlife 2000*, one finds considerable attention devoted to describing and (occasionally) testing habitat models, many of them based on a habitat suitability index (HSI) protocol. Statistical analyses of habitat relationships highlighted regression approaches and multivariate procedures, and although the emphasis was on linear relationships, the HSI formulations often included strong response thresholds. Few of the papers in that volume considered temporal or spatial variation explicitly (although habitat fragmentation was a major topic of concern), and even fewer mentioned issues of scale.

Have things changed all that much? In their introductory remarks to *Wildlife 2000*, Jerry Verner, Mike Morrison, and C. J. Ralph observed that "What we need now are some appropriate ways to organize our information and get it into computer files, and suitable methods for analyzing the information in those files." In that respect, the past sixteen years have certainly seen remarkable progress. Progress is evident in other areas as well. I'll comment on this progress in three areas: tools, concepts, and sociology.

Tools

The most obvious evidence of progress in dealing with wildlife-habitat relationships is in the size and sophis-

tication of the toolbox. One of the distinguishing features of many of the chapters in this volume, for example, is the size of the data sets analyzed. Dealing with thousands of sampling points, or hundreds of thousands of pixels, would have been unimaginable a decade ago. Moore's Law, named after Intel founder Gordon Moore, suggests that transistor density, and thence microprocessor speed, doubles every eighteen months. There are theoretical limits to this function, but we don't seem near them yet. The crunching capacity (CC) of computers continues to increase exponentially, and progress is being made in dealing with the problem of mismatches in the scale of data related to different components of an ecosystem or of a model—variously known as the transmutation problem or the Modifiable Areal Unit Problem (Henebry and Merchant, Chapter 23). In many cases, the wealth of data derives from our capacity to gather information by remote sensing and analyze the data using geographic information systems (GIS). The development of GIS has had a tremendous influence on assessing species' occurrences, enabling us to map (with apparent precision) the spatial pattern and distribution of multiple habitat and environmental features. Virtually all of the empirical chapters in this book make use of GIS in one way or another. Gap analysis, and its application to mapping biodiversity, land use, and stewardship, would be impossible without GIS. Beyond these obvious uses, GIS has led to closer examination of classification and categorization procedures and to the development of numerous metrics to describe landscape patterns. It is difficult to underestimate the importance of this tool to what we do.

Progress in the area of data analysis, while not so obvious, has been just as remarkable. Advances in regression modeling, such as generalized linear modeling (GLM), generalized additive modeling (GAM), and Poisson regression have increased analytical power, and regression has been coupled with hierarchical classification approaches in classification and regression tree (CART) analyses. (I suppose that a useful index of progress might be the proliferation of acronyms in a science.) New approaches developed in other disciplines, such as genetic algorithm for rule-set prediction (GARP) modeling or neural network modeling, are just beginning to be applied to habitat evaluation, but their potential appears to be great. Progress has also been evident in how we deal with space. Ecologists, modelers, and managers have long realized that natural systems vary spatially and that this variation affects the dynamics of natural systems. It can also confound conventional statistical analysis. A rapidly growing suite of spatial statistics, such as autologistic regression, semivariance analysis, kriging, and correlograms (see Fortin [1999] or Dale [1999] for useful and understandable reviews), is providing the tools necessary to describe spatial patterns quantitatively as well as to assess the magnitude of spatial autocorrelation and thereby evaluate its potential effects on more conventional, nonspatial approaches.

In the era of *Wildlife 2000*, statistics and modeling (primarily simulation modeling) were largely separate activities: one analyzed the data and then used the results to specify parameter values (usually only means) as inputs to simulation models. Increasingly, statistical analysis blends into modeling, and both are being linked to spatial and cartographic analyses via GIS. In addition, rather than seeking a single wildlife-habitat model that works and labeling it "best," modelers are now developing multiple models to address a particular situation and are using statistical tools such as Akaike's information criterion (AIC) to select among alternatives. Considerable attention is being given to evaluating model performance, both in terms of the logics of the model structure and functions, and the accuracy of the predictions about species occurrence. Perhaps the greatest change in modeling, however, is the level of detail and complexity that can be included in models as a consequence of increased computer capacity and speed. This has led to the development of models in which the dynamics and interrelationships among numerous locations (spatially explicit models, or SEMs) or individuals (individual-based models, or IBMs) can be tracked in mindboggling detail. Inevitably, these approaches have given rise to integrated spatially explicit individual-based models (SEIBMs). Or, in a delightful play on acronyms, the individual cowbird behavior model (ICBM; Shapiro et al., Chapter 49). We are reaching the point where we can incorporate additional levels of complexity into models as fast as we can recognize them. The potential development of technologies such as optical or molecular

computing or self-organizing computational networks suggests that modeling complexity may be limited more by our ingenuity than by our technology—as, perhaps, it has always been.

Concepts

Ultimately, our ability to use tools intelligently depends largely on how we think about Nature or, more proximately, wildlife-habitat relationships and species occurrences. Theory and concepts provide the intellectual context for our efforts. As in any discipline, theory and concepts in ecology develop in fits and starts, largely because the advances that foster breakthroughs tend then to establish a tenacious hold that may slow further development—the continued dominance of the assumption-laden logistic model in population biology is a good example. Kingsland (1985) provides a splendid review of this process, while Peters' (1991) treatment is more caustic. Nonetheless, ecological thinking has advanced considerably over the past sixteen years, largely due to a shedding (at least in part) of concepts and theories founded on assumptions of temporal equilibrium and spatial homogeneity in favor of a recognition of the importance of variation in time and space. These changes have affected both how we think about organisms and populations and how we view the environments they occupy. On the biological side of the ledger, concepts such as keystone species, umbrella species, functional types, and ecological redundancy have influenced how we regard the role or importance of particular species in an ecological context. Metapopulation theory and population viability analysis (PVA) have opened new avenues for assessing the spatial, or demographic, structure of populations and its consequences. And, of course, "biodiversity" (which traces its roots directly to the focus on species diversity that dominated community ecology during the 1960s, or thinking about species-abundance distributions that developed decades earlier) has spawned a plethora of papers, books, web sites, and videos, even though there is not a clear consensus about what the word really means.

Then there is the matter of how we deal with environmental variation. More than a generation ago, Robert MacArthur (1972a,b) advised us to study homogeneous areas of habitat, where ecological patterns should emerge most clearly. For a decade (or two), ecologists followed this advice, ignoring the heterogeneity that confronted them both within and among habitats. Heterogeneity is now receiving much closer attention, both in terms of description and measurement and of its consequences for population dynamics and community structure. We are beginning to consider the linkages that exist between temporal variation and spatial variation and how they can affect both how we model and how we manage habitat for wildlife. But perhaps the greatest shift has occurred in our thinking about scale (extent and grain), from ignoring it almost completely (as in *Wildlife 2000*) to realizing that everything we observe and model is sensitive to scale. (The collection assembled by Peterson and Parker [1998b] provides an excellent introduction to the manifold ways of viewing scale and its consequences.) The emergence of landscape ecology has done much to focus attention directly on issues of heterogeneity and scale.

Sociology

There is more to science and its applications than theories, concepts, models, and data. People are involved. Without going deeply into the demographic and cultural changes that have affected science and management in the last sixteen years, let me call attention to two significant changes, one of which represents real progress, the other, perhaps not.

If one takes the contributions and contributors to *Wildlife 2000* as reasonable indicators of the field of wildlife-habitat evaluation and management in the mid-1980s, it is apparent that women were not well represented. By my count, only 9 percent of the authors of chapters in that volume were women. In the present volume, 28 percent of the authors are women, a three-fold increase. Although this proportion does not yet come close to matching the representation of women in graduate programs in wildlife management and ecology, it nonetheless indicates substantial progress. Less progress has been made in involving ethnic minorities in the task of evaluating and applying wildlife-habitat relationships.

The other sociological aspect has to do with the transfer of new knowledge, approaches, and technologies into management practice and policy. Applied ecology is often used to refer to ecological studies that have some real-world relevance, but the term has real meaning only if the results of the science actually affect the implementation of management at some level. There is abundant evidence that many of the buzzwords of modern ecology, such as biodiversity, ecosystem management, indicator species, and landscape, have made their way into the lexicon of policy and legislation, but the science behind these terms has not permeated practice to an equivalent extent. This requires that the findings and thinking of science be communicated to managers and policy makers in understandable terms and that the needs and priorities of management and policy be communicated to scientists in equally understandable terms. *Wildlife 2000* was explicitly focused on fostering this communication and was at least moderately successful. Since then, however, the gulf between science and management has widened instead of closing. The very innovations in technology and concepts that have contributed so much to our progress in modeling wildlife-habitat relationships have made it increasingly difficult for managers, whose training has been in other areas, to keep up. Moreover, ecologists and researchers still seem more interested in talking to one another than in communicating with practitioners and policymakers. I'll return later to this aspect of the sociology of applied science.

Problems

Given all of the progress we've made, one would think we would be closer to solutions, to achieving a real understanding of wildlife-habitat relationships, so we could generate robust and reliable predictions. We are. Through no fault of their own (except history), many of the contributions to *Wildlife 2000* now seem overly simplistic or downright naive. Yet, the new tools and concepts that have encouraged all this progress have also exposed new problems, many of which we were only dimly aware of, or actively swept aside or suppressed, as we defended our beloved paradigms and procedures. Equilibrium thinking and linear regression come quickly to mind. Perhaps one additional sign of progress is that many of the contributions to this volume consider these problems in one way or another.

But the concerns are so basic to what we want to do that, at the risk of generating a litany of criticisms, I'll highlight those that to me seem most critical, touching on all of the key words in the title of this book (and then some).

The Use and Misuse of Tools

Models of one sort or another have become our primary means of assessing wildlife-habitat relationships and generating predictions of the consequences of habitat change. Ultimately, the strength or weakness of models is related to the tools used to construct or provide information to the models, the data used to drive the models, and the internal structure of the models themselves. Technological advances have given us tools that are immensely powerful, but behind this power lies the peril of indiscriminate use, of mindless application. We can calculate the fractal dimension of all sorts of things, for example, and there is a rich body of theory and application of fractal geometry in other sciences. But what do fractals really mean, ecologically? We have precious little theory to suggest what we might expect from a movement pathway or a landscape of a given fractal dimension in terms of ecologically important consequences. It would be premature, however, to disregard fractal measures simply because we haven't yet developed fractal ecology theory. Milne (1997) provides a nice perspective on the uses of fractals in wildlife biology.

GIS analysis has become a major bandwagon in ecology, and justly so. Yet, the ability to generate multilevel, brilliantly colored maps and to link these with spatially explicit computer models does not ensure that we have a better understanding of ecologically important process-pattern linkages. Because our choice of variables and scales is often fixed by the technology (e.g., particular spectral bandwidths or 30-meter pixel sizes in remote sensing), the "truth" that emerges in a computer-generated map is as much a product of the constraints as of the data. One can lie—inadvertently, I'm sure—with maps just as easily as one can lie with statistics. Van Horne (Chapter 4) and Henebry and Merchant (Chapter 23) provide some additional cautionary words about the overly enthusiastic use of GIS in wildlife-habitat analysis and modeling.

In some cases, we aren't making sufficient use of the tools that are available. All environments are heterogeneous and all organisms are aggregated at one scale or another. We know that this spatial heterogeneity can have profound ecological consequences (for a sampling see Kolasa and Pickett 1991; Tilman and Kareiva 1997; Maudsley and Marshall 1999; Hutchings et al. 2000), yet the expanding array of spatial statistics that can be used to analyze spatial patterns is still largely ignored in wildlife-habitat investigations. For example, of the chapters in this volume that include specific, empirical analyses, hardly any use spatial statistics. As Austin (Chapter 5) has warned, however, it is risky to ignore spatial autocorrelation. Beyond this, there is a wealth of information buried in the spatial patterns of organisms and habitats, and spatial statistics provide a way to distill that information.

Variable Selection

It is trite to say that models (at least empirical ones) are only as good as the data that drive them. Our capacity to gather data has never been greater. But what sorts of data should we gather? Ecologists, wildlife biologists, and modelers have grappled for decades with the issue of which habitat variables to measure (Morrison et al. 1998 review this history). To some degree, the selection of variables has been based on a knowledge of the natural history of the target organisms, but it has also been strongly influenced by tradition (i.e., we measure what those before us have measured) and by the availability of tools (e.g., if we are interested in measures of landscape structure, we go to FRAGSTATS and follow the directions). As the collection of candidate variables for habitat analysis continues to grow, it becomes more difficult to choose among them and more tempting to employ a "shotgun" approach, hoping that computer models or statistical analyses will, in the end, tell us which ones are important. Moreover, every variable we measure has an associated variance—that is, after all, why they're called variables. Although consideration of this variance is part of any statistical analysis, we still tend to regard it as "sampling error" or "noise," something to be controlled for so we can get at the real meaning of the means. But, as Dan Simberloff once observed, what is noise to a physicist (or a statistician) is music to an ecologist. (Dan failed to specify the kind of music; given contemporary tastes, this could span a wide array of discordant harmonies [cf. Botkin 1990].) In other words, the variance contains important biological information in addition to an inevitable stochastic component, and we would do well not to ignore it. What does it tell us about habitat quality, for example, if a species occurs at low density in one area and high density in another, even though densities vary among years much more in the latter area? Not much, perhaps, but the contrast is worth considering, and by focusing on mean values we may miss important insights.

Model Evaluation

We must also be concerned about model structure and evaluation. It is well and good to insist that models be "logically consistent," but what, exactly, do we mean by this? That the model operations are founded on proper mathematics? That circularity is avoided? That the model functions portray biologically feasible processes? Ideally, a "good" model includes all of these. We usually evaluate models, however, by judging their goodness of fit (i.e., their accuracy), either in a comparison with the original data from which the model was generated—a logical no-no for any purposes other than determining whether the model works (i.e., validation)—or with freshly minted data. But what is the standard for such evaluations? Perhaps the model could be compared with a null or neutral model, except that we know some such models are so unrealistic that a departure from them only indicates that the model under consideration may have some degree of biological realism (Gotelli and Graves 1996; With and King 1997). Or, perhaps we could develop multiple, alternative models and pick the "best" model using an objective measure (e.g., AIC), but such comparisons are only as good as the set of models considered. Being the "best" among a series of unrealistic or poor models does not make a model "good." Model evaluation is a central focus of many of the contributions to this volume, so the problem is clearly receiving a lot of attention.

Prediction and Accuracy: The Holy Grail of Wildlife Biology

Certainly one of the best measures with which to evaluate models, and the one in which we are ultimately most interested, is predictiveness. In the realm of wildlife-habitat management, the most elegant and logically consistent model remains an esoteric toy unless it can actually predict something useful. Predictability, however, is itself problematic. What, exactly, is it that we want to predict? The simple presence of a species? Its abundance? The annual variance in abundance? Its sensitivity to habitat change? Its population dynamics? Clearly, different questions demand different approaches, and much of the variety of models represented in this volume undoubtedly reflects differences among authors in what they wanted to predict. We also aim for accuracy in our predictions. Indeed, accuracy is the motherhood-and-apple-pie of modeling and quantitative ecology. Because of this, there may be a temptation to measure those things that we can measure with great accuracy, hoping that accuracy in measurement will translate into accuracy of model predictions. Again, we need to ask whether these are the right variables to measure.

Instead of endlessly pursuing ever-greater accuracy, we should also ask how much accuracy we really need. Will a marginally significant model—$P > 0.05$, of course—suffice if our objective is to assess coarse associations between wildlife and habitats (even though such a model would have a low probability of acceptance ($P < 0.05$) in a mainstream scientific journal)? Conversely, is a model that generates a significant relationship between variables really useful if the amount of variance it explains (the oft-overlooked R^2 value) is low? We have all seen (and some of us have published) scatter plots containing a large cloud of points intersected by a "significant" regression line. As our capacity to gather data (and therefore generate large sample sizes) increases, the likelihood of finding such significant but meaningless relationships also increases. And what do we do when a seemingly good wildlife-habitat model (as judged by a high R^2, for example) fails to predict what it should? This was the problem that John Rotenberry posed in his contribution to the *Wildlife 2000* volume ("Habitat Relationships of Shrubsteppe Birds: Even 'Good' Models

Cannot Predict the Future." In fact, an entire section of that volume was devoted to the problem of the failure of habitat models as predictors.). John and I have recently confirmed the accuracy of his title in a resoundingly discouraging way. Beginning in 1977, we surveyed bird populations and measured habitat features over six years at fourteen sites in the shrubsteppe of Oregon and Nevada. Despite annual variation, we were able to develop significant regression models ($P < 0.01$) of bird abundance and habitat variables that had a reasonably good fit. The multiple regression model for sage sparrows (*Amphispiza belli*), for example, had an R^2 value of 0.60. John and I returned to the shrubsteppe in the summer of 1997 and resurveyed the same transects using the same methods (but with older observers). We used the 1997 habitat data as input to the 1977–1982 model to generate predicted sage sparrow abundances in 1997. The match with the observed abundances was, to say the least, poor (Fig. 65.1). Being good ecologists, we can come up with several *post facto* explanations for why the model predictions failed. For example, events elsewhere—such as El Niño, which we tend to blame for everything we can't otherwise explain—or events in the past may have affected current abundances and may have done so differently in different areas. Departures from habitat occupancy based on an ideal-free distribution, or due to individual variation in habitat selection, or associated with stochastic factors—the other favorite explanation for variations we can't otherwise explain—could also erode the fit between observations and predictions. Perhaps more sophisticated modeling approaches might yield better predictions. The basic problem, however, is that we are dealing with a dynamic, nonequilibrial system in which different factors, acting at different scales, with varying time lags, determine abundances at the scale of local survey plots in a given year (or decade). Studies of Townsend's ground squirrels (*Spermophilus townsendii* [now Piute ground squirrel, *Spermophilus molis*]) by Van Horne et al. (1997) illustrate how "good" habitat (as judged by demographic measures) in a wet year can become "poor" habitat during a drought, and vice versa. We are trying to model a moving target, and an erratically moving one at that.

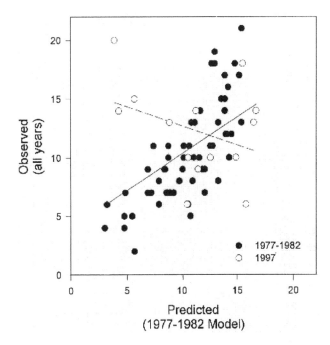

Figure 65.1. The problem in a nutshell: abundances of sage sparrows (*Amphispiza belli*) on survey plots in western shrubsteppe predicted on the basis of a model incorporating several habitat measures are displayed versus abundances actually observed on those plots. The solid circles are the data points used to develop the model (1977–1982), and the close fit between predicted and observed values ($P < 0.01$, $R^2 = 0.60$) suggests that the model might have reasonable predictive power. The open circles are from surveys conducted in 1997, in which the 1997 habitat measures were used in the 1977–1982 model to generate predictions of 1997 sparrow abundances. Clearly, it's not working.

Species Occurrences

The intent of wildlife-habitat modeling and analysis, in general, is to understand and predict the occurrence of species in space and time. The simplest approach (at least in terms of data collection) seemingly is to record species presence and absence. Quite apart from the problem of model efficacy in predicting presence and absence based on habitat measures (errors of omission and commission, discussed often in this volume) there is the problem of interpreting "presence" or "absence" in ecologically meaningful terms. If a species is present in an area, does that mean that the habitat is suitable? Suitable for what? If areas are surveyed only once (as is often the case), records of presence may include transients that don't really "belong there." (This is my chance to refer you to Wiens

(1981c), a [perhaps justly] long-neglected paper.) Are species that are present in few sites actually rare, or simply undetectable? Or, if a species is absent from a location or habitat, does that mean that it isn't really "habitat" for that species, or that it was really there but wasn't detected, or that it is normally there but was absent when the survey was conducted? I think we often assume that these uncertainties will be minimized if we accumulate a lot of presence-absence observations. Even if that is done, however, there is an implicit assumption that presence is matched with "suitable" habitat and absence with "unsuitable" habitat. In other words, habitat occupancy is (with a certain amount of "noise") equilibrial. But we know from numerous studies that events elsewhere (e.g., on the wintering grounds of migratory birds) can affect patterns of habitat occupancy, or that there may be time lags in the response of a species to a change in habitat suitability (e.g., species loss following forest fragmentation). We also know (or at least think) that habitat occupancy is related to density. The Fretwell-Lucas ideal-free distribution (IFD) model, for example, predicts that the range of habitat occupancy will increase as regional population density increases, and this formulation has influenced much of our thinking about habitat occupancy.

Ah, density! Surely there is no more persistent notion in the annals of wildlife-habitat analysis than the expectation that local population density should be positively related to habitat quality. After all, we are part of a culture that generally adheres to the premise that more is better, so why shouldn't this apply to habitat occupancy? The problem is that (at least with respect to wildlife and habitat) the premise is both logically and empirically flawed—as Van Horne (1983) pointed out nearly two decades ago, Maurer (1986) reinforced, and Garshelis (2000) has forcefully reiterated. More resources (i.e., better-quality habitat) do not necessarily translate into more individuals. It is hard to escape from the intuitive notion that abundance must indicate something important—we just aren't sure what it is.

The issue of habitat occupancy and what it means leads to another, more basic concern. Our approach to "habitat" is strongly categorical. Although we may take all sorts of quantitative measures of habitat vari-

ables, we tend to do so only to characterize a habitat type that has been determined *a priori* (often on the basis of dominant vegetation). Habitat types are regarded as discrete entities, and while we may array them in nested hierarchies or use Boolean or fuzzy set approaches to define them (see Hill and Binford, Chapter 7), they end up being categories nonetheless. Tools such as GIS simply reinforce this perspective, for what is a habitat map without discrete boundaries on the habitat types? Managers like habitat maps because their decisions are usually made with reference to land-use blocks, which are administratively discrete. The assignment of "habitats" to categories also dovetails nicely with categorical forms of statistical analysis, such as ANOVA (or, less conspicuously, CART). Yet, we know that there is substantial variation within any defined habitat type, and their boundaries in the real world are often gradual and indistinct, especially in relatively "natural" environments. Rather than thinking discretely about habitat, it might be worthwhile to think of gradients in habitat conditions, to let topographic maps rather than land-cover maps be our conceptual guides. Mike Austin (Chapter 5) has provided some guidance about how we might do this.

Scale

Scale has rapidly become a central issue in ecology. It was scarcely mentioned in *Wildlife 2000*, yet it figures prominently in many of the contributions to this volume. But an awareness of scale does not necessarily translate into effective ways of dealing with it. There are three aspects to the "scaling problem." First, it has become apparent that virtually everything in ecology depends on the scale at which it is viewed. Patterns that are obvious at fine spatial scales or over short time periods dissolve at broader scales of space and time, perhaps to be replaced by startlingly different patterns. The associations of sage sparrows with big sagebrush (*Artemisia tridentata*) cover, for example, are either positive or negative depending on the spatial scale at which they are assessed. Statements about habitat relationships or projections of wildlife-habitat models are useless unless they are accompanied by qualifiers linking them with a particular scale of measurement or application.

The second aspect of scaling that is relevant here is

how scale should be specified. It has become customary to bracket the range of scale encompassed by a study (or a model) by *grain* (the finest scale of resolution of the data, generally equivalent to survey-plot or pixel size) and *extent* (the area or region from which observations are made or to which a model is applied). Grain and extent define the window through which we view patterns—by definition, we cannot detect patterns at scales finer than the grain or broader than the extent. When we assess variance or autocorrelation in habitat features or species occurrence or abundance, the variance measures, like the means (or whatever other metrics we derive), apply within but not beyond the grain-extent window. The criteria used to determine grain and extent vary tremendously among studies, depending on their objectives and constraints. This makes comparisons among studies difficult, for investigations conducted with different windows on the world are likely to see different things simply because their windows differ in size. More problematic, however, are differences in grain and extent that occur *within* a study. It is not unusual, for example, to see species surveyed using 50-meter radius circular plots, vegetation sampled using 10-square-meter quadrats, and GIS layers developed using 30-square-meter pixels, and all of them then combined in an analysis as if the scales used were the most appropriate, biologically, and as if the differences in scales didn't matter. Perhaps they don't. Given the pervasiveness of scale dependence in ecological systems, however, it would be good to be sure.

The third problematic aspect of scale has to do with extrapolation—what is in the technical literature known more cryptically as the transmutation problem, which Bob O'Neill (1979) presciently called attention to decades ago, in another (unjustifiably) neglected paper. How does one use the results of studies conducted at one (grain-extent) scale to draw inferences or make predictions at other scales? Simply stated, one can't. Yet, we are generally constrained to conduct ecological studies at scales that are often considerably finer than their intended or desired scales of application. This is surely one of the most vexing, yet most critical, problems in applied ecology. Initial approaches have been made in both empirical and modeling studies. Empirically, the favored method seems

to be to conduct a multiscale investigation in which habitat features and species responses are assayed at several scales of resolution (i.e., several discrete window sizes). Roughly half of the empirical studies reported in this book used this approach. Although this represents an improvement on single-scale studies, most such studies consider only a few (generally, two) scales, and the scales are selected based on arbitrary (and rarely justified) criteria. A few modeling studies (e.g., Cogan, Chapter 18; Gross and DeAngelis, Chapter 40) are exploring the possibilities of multimodeling, in which models operating at different scales are operationally linked.

Prospects

Where does this leave us? Certainly with no dearth of problems to address. I'd like to conclude, however, by drawing attention to several considerations that I think lie at the foundation of our efforts to (as the book title suggests) predict species occurrences.

First, scale (again). The problem is not only that ecologists deal with a wide array of scales that are often incompatible with one another. When the time comes to apply the results of these studies to management or policy problems, we are confronted with the reality that management and policy are frequently exercised at quite different scales from the scales of ecology. Management scales are determined by administrative boundaries, land ownership, and resource-extraction practices, not ecology, and policy is developed at even broader scales. To think that Nature, which follows its own scaling rules, can somehow be made to fit within the arbitrary scales of management and policy is fantasy.

This relates to my second point: organisms follow their own algorithms in responding to "habitat," and this determines the scales at which they operate and over which variation in environmental conditions may be relevant to them. John Addicott and his students recognized the importance of organism-determined scaling in their concept of ecological neighborhoods, in a paper (Addicott et al. 1987) that has (thankfully) not been neglected. To understand scaling relationships, and indeed to understand anything about wildlife-habitat relationships, we need to adopt a view of "habitat" that is centered on how organisms might perceive and respond to it, not how humans think of it. Obviously, this is not easy to do, and it becomes more difficult as one considers organisms that are not mammals or birds, but insects or plants. The key, as several contributors to this volume have emphasized, is to focus on *process*, on the aspects of physiology, behavior, life history, and demography that actually produce the patterns we so meticulously document. To understand why a model does or does not predict well, for example, requires that we ask "why?" This comes down to an understanding of process, of mechanisms, rather than a further refinement of patterns. Going back to the now-ancient notion of ecological niches (e.g., Austin, Chapter 5), developing "envirograms" (Andrewartha and Birch 1984), or employing knowledge-based approaches (e.g., Van Horne, Chapter 4) may help us to incorporate process more directly into our thinking.

Thinking about niches may also free us from the constraints of viewing habitats as categories. My third point, then, is to emphasize the value of viewing ecological systems as arrayed over gradients and considering the response functions of species to these gradients, be they unimodal, bimodal, or amodal. Environmental gradients may steepen in places, which we may then define as boundaries (or thresholds, or nonlinearities). It is the form of the gradient, and of the species response, that will ultimately tell us what we need to know to model wildlife-habitat relationships.

Fourth, it is imperative that we be clear about our goals. Why do we conduct a particular study or develop a certain model? In my own experience, the objectives of a research project may be clear at the beginning and at the end, but along the way they have probably undergone a transformation—I prefer to think of it as "maturation." It's important to be clear about goals and objectives both within a study and, particularly, among studies. A model developed for one purpose is not immediately transferable to some other purpose, even if it is a "good" model. Investigations may be conducted for a variety of purposes and at a variety of levels. Many of the contributions to this volume, for example, focus on individual species, often in only a portion of their geographic ranges. As a consequence of the conservation focus on biodiver-

sity and policy dictates to implement ecosystem management, however, attention is increasingly shifting to multispecies investigations. This shift carries with it a host of new challenges, not the least of which is reconciling the multifarious scaling responses of different species with one another. (As an aside, I should note that there is some disagreement among reputable ecologists about the value of investigations of ecological communities. John Lawton [1999], for example, has concluded that "community ecology is a mess, with so much contingency that useful generalisations are hard to find," that "community ecology may have the worst of all worlds," and that "the time has come to move on." On the other hand, Ed Wilson [2000] observes that "community ecology . . . is about to emerge as one of the most significant intellectual frontiers of the twenty-first century," and that "it stands intellectually in the front rank with astrophysics, genomics, and neuroscience." You decide.)

Finally, if any of this is to make any difference whatsoever, it is essential that the widening gulf between scientists and managers be bridged, and bridged quickly. Ecologists must learn that the variability, complexity, and contingencies of Nature that so enthrall them (Simberloff's "music") are not so appealing to managers, who of necessity must seek simple solutions. Statements that "it all depends" and calls for "more research" may satisfy ecologists, but they ring hollow to the manager. At the same time, managers must come to realize that the systems they wish to manage, and the species occurrences they wish to predict, *are* variable, and any conclusions or actions are bound to be accompanied by uncertainty (see Bradshaw and Borchers 2000). They also should realize that the arbitrary scales of management they have applied in the past may not be appropriate for attaining goals that relate to ecological systems that scale things differently. Progress is being made in addressing the issue of how much variability or complexity we really need to consider in addressing particular problems (e.g., Gross and DeAngelis, Chapter 40), and adaptive management is widely touted as an antidote for uncertainty. Such approaches need to be tested and expanded. Most importantly, however, both ecologists and managers need to talk with one another, free of their disciplinary defenses and jargon. *Wildlife 2000* made a good start on this; we need desperately to resurrect it.

Coda

It's interesting, and of more than passing importance, I think, to contemplate from whom we might draw inspiration in dealing with these issues in wildlife-habitat relationships. G. Evelyn Hutchinson (1959), for example, drew his inspiration for contemplating the sources of what we now term biodiversity from Santa Rosalia. More recently, O'Neill and King (1998) adopted St. Michael as the patron for their exploration of scaling issues. Several other possibilities come to mind. Charles Elton developed early thinking about ecological niches, laid some of the empirical foundation for modern studies of population dynamics, and in his later years focused increasingly on the importance of "habitat." David Lack did much to influence our thinking about life histories and evolutionary adaptations in ecology. Robert MacArthur catalyzed the growth of theoretical ecology during the last half of the twentieth century. Through his sensitivity and understated elegance, Aldo Leopold gave both rigor and passion to wildlife ecology, and thence to conservation biology.

I could go on. I, however, take my inspiration from the Australian ecologists Harry Andrewartha and Charles Birch. Before ecology became a modern science (i.e., 1954), they recognized that, when all is said and done, our central task is explaining the distribution and abundance of organisms. The key to doing this, they held, lies in an understanding of how organisms respond to their environments—what we now call mechanistic or process-based ecology. These are the themes that echo through this book and that must underlie our efforts to understand and model wildlife-habitat relationships.

Acknowledgments

No one should be forced to take credit for these musings except me. However, my long association with John Rotenberry, and his influence on my thinking, cannot be ignored. In no lesser way, Mike Scott has always forced me to deal with new problems—long ago by introducing me to the world of seabirds (and ships) and most recently by offering me the challenge to summarize the Snowbird conference. And Bea Van Horne has brought me back to earth, to the realities of mechanisms, when my thinking has become too ethereal. At least most of the time.

Literature Cited _____

Aaris-Sørensen, J. 1987. Past and present distribution of badgers *Meles meles* in the Copenhagen area. *Biological Conservation* 41:159–165.

Abbott, I. 2000. Improving the conservation of threatened and rare mammal species through translocation to islands: Case study Western Australia. *Biological Conservation* 93:195–201.

Aborn, D. A., and F. R. Moore. 1997. Pattern of movement by summer tanagers during stopover: A telemetry study. *Behaviour* 134:1077–1100.

Abramsky, Z., and M. L. Rosenzweig. 1983. Predicted productivity-diversity relationship shown by desert rodents. *Nature* 309:150–151.

Adamus, P. R., ed. 1987. *Atlas of breeding birds in Maine, 1978–1983*. Augusta: Maine Department of Inland Fisheries and Wildlife.

Adamus, P. R. 1995. Validating a habitat evaluation method for predicting avian richness. *Wildlife Society Bulletin* 23:743–749.

Addicott, J. F., J. M. Aho, M. F. Antolin, D. K. Padilla, J. S. Richardson, and D. A. Soluk. 1987. Ecological neighborhoods: Scaling environmental patterns. *Oikos* 49:340–346.

Aebischer, N. J., P. A. Robertson, and R. E. Kenward. 1993. Compositional analysis of habitat use from animal radio-tracking data. *Ecology* 74:1313–1325.

Agnew, W. D., D. W. Uresk, and R. M. Hansen. 1986. Flora and fauna associated with prairie dog colonies and adjacent ungrazed mixed-grass prairie in western South Dakota. *Journal of Range Management* 39:135–139.

Agresti, A. 1990. *Categorical data analysis*. New York: John Wiley & Sons.

———. 1996. *An introduction to categorical data analysis*. New York: John Wiley & Sons.

Airola, D. A. 1988. *Guide to the California wildlife habitat relationships system*. Sacramento: Jones and Stokes Associates.

Airola, D. A., and R. H. Barrett. 1985. Foraging and habitat relationships of insect-gleaning birds in a Sierra Nevada mixed-conifer forest. *Condor* 87:205–216.

Akaike, H. 1974. A new look at statistical model identification. *IEEE Transactions on Automatic Control* AC-19:716–723.

———. 1978a. A new look at the Bayes procedure. *Biometrika* 65:53–59.

———. 1978b. A Bayesian analysis of the minimum AIC procedure. *Annals of the Institute of Statistical Mathematics* 30:9–14.

———. 1981a. Likelihood of a model and information criteria. *Journal of Econometrics* 16:3–14.

———. 1981b. Modern development of statistical methods. In *Trends and progress in system identification*, ed. P. Eykhoff. 169–184. Paris: Pergamon Press.

Akçakaya, H. R. 1994. *RAMAS GIS: Linking landscape data with population viability analyses*. Version 1.0. Setauket, N.Y.: Applied Biomathematics.

Akçakaya, H. R., and J. L. Atwood. 1997. A habitat-based metapopulation model of the California gnatcatcher. *Conservation Biology* 11:422–434.

Akçakaya, H. R., and S. Ferson. 1990. *RAMAS/space user manual: Spatially structured population models for conservation biology*. New York: Applied Biomathematics.

Akçakaya, H. R., M. A. McCarthy, and J. L. Pearce. 1995. Linking landscape data with population viability analysis: Management options for the helmeted honeyeater *Lichenostomus melanops cassidix*. *Biological Conservation* 73:169–176.

Alatalo, R. V., A. Lundberg, and S. Ulfstrand. 1985. Habitat selection in the pied flycatcher *Ficedula hypoleuca*. In *Habitat selection in birds*, ed. M. L. Cody, 59–83. Orlando: Academic Press.

Aldrich, R. C., R. A. Harding, and D. P. Paine. 1984. Remote sensing. In *Forestry handbook*, ed. K. E. Wegner, 1117–1163. 2nd ed. New York: John Wiley & Sons.

Alerstam, T., and A. Lindstrom. 1990. Optimal bird migration: The relative importance of time, energy and safety. In *Bird migration: Physiology and ecophysiology*, ed. E. Gwinner, 331–351. Berlin: Springer-Verlag.

Alford, R. A. 1999. Ecology—Resource use, competition, and predation. In *Tadpoles: The biology of anuran larvae*, ed. R. W. McDiarmid, and R. Altig, 240–278. Chicago: University of Chicago Press.

Alldredge, J. R., and J. T. Ratti. 1986. Comparison of some statistical techniques for analysis of resource selection. *Journal of Wildlife Management* 50:157–165.

Alldredge, J. R., D. L. Thomas, and L. L. McDonald. 1998. Survey and comparison of methods for study of resource selection. *Journal of Agricultural, Biological, and Environmental Statistics* 3:237–253.

Allen, T. F. H., and T. W. Hoekstra. 1991. Role of heterogeneity in scaling in ecological systems under analysis. In *Ecological heterogeneity*, ed. J. Kolasa and S. T. A. Pickett, 47–68. New York: Springer-Verlag.

Allen, T. F. H., and T. W. Hoekstra. 1992. *Toward a unified ecology*. New York: Columbia University Press.

Allen, T. F. H., and T. B. Starr. 1982. *Hierarchy: Perspectives for ecological complexity*. Chicago: University of Chicago Press.

Al-Mufti, M. M., C. L. Sydes, S. B. Furness, J. P. Grime, and S. R. Band. 1977. A quantitative analysis of shoot phenology and dominance in herbaceous vegetation. *Journal of Ecology* 65:759–791.

Alpin, P., P. M. Atkinson, and P. J. Curran. 1997. Fine spatial resolution satellite sensors for the next decade. *International Journal of Remote Sensing* 18:3873–3881.

Amaranthus, M., J. M. Trappe, L. Bednar, and D. Arthur. 1994. Hypogeous fungal production in mature Douglas-fir forest fragments and surrounding plantations and its relation to coarse woody debris and animal mycophagy. *Canadian Journal of Botany* 24:2157–2165.

Amaranthus, M. P., D. S. Parrish, and D. A. Perry. 1989. Decaying logs as moisture reservoirs after drought and wildfire. In *Stewardship of soil, air and water resources*, ed. E. B. Alexander, 191–194. Proceedings of Watershed 89. R10-MB-77. Juneau: USDA Forest Service, Alaska Region.

Analytical Software. 1996. *Statistix*. Analytical Software, Inc.

Anderson, D. R., K. P. Burnham, and W. L. Thompson. 2000. Null hypothesis testing: Problems, prevalence, and an alternative. *Journal of Wildlife Management* 64: 912–923.

Anderson, J. A. 1984. Regression and ordered categorical variables. *Journal of the Royal Statistical Society Series B* 46:1–30.

Anderson, J. R., E. E. Hardy, J. T. Roach, and R. E. Witmer. 1976. A land use and land cover classification system for use with remote sensor data. Professional Paper 964. Reston, Va.: U.S. Geological Survey.

Anderson, R. L., and L. A. LaMadeleine. 1978. The distribution of butternut decline in the eastern United States. Report S-3-78. Radnor, Pa.: USDA Forest Service, Northeastern Area States and Private Forestry.

Anderson, S. H. 1981. Correlating habitat variables and birds. *Studies in Avian Biology* 6:538–542.

Anderson, S. H., and H. H. Shugart Jr. 1974. Habitat selection of breeding birds in an east Tennessee deciduous forest. *Ecology* 55:828–837.

Anderson, T. W. 1958. *An introduction to multivariate statistical analysis*. New York: John Wiley & Sons.

Andrewartha, H. G., and L. C. Birch. 1954. *The distribution and abundance of animals*. Chicago: University of Chicago Press.

———. 1984. *The ecological web: More on the distribution and abundance of animals*. Chicago: University of Chicago Press.

Angelstam, P. 1990. Factors determining the composition and persistence of local woodpecker assemblages in taiga forest in Sweden—a case for landscape ecological studies. In *Conservation and management of woodpecker populations*, ed. A. Carlson and G. Aulén, 147–164. Report 17. Uppsala: Swedish University of Agricultural Science.

Angelstam, P., and G. Mikusinski. 1994. Woodpecker assemblages in natural and managed boreal and hemiboreal forest—a review. *Annales Zoologici Fennici* 31:157–172.

Angermeier, P. L. 1987. Spatiotemporal variation in habitat selection by fishes in small Illinois streams. In *Community and evolutionary ecology of North American stream fishes*, ed. W. J. Matthews, and D. C. Heins, 52–60. Norman: University of Oklahoma Press.

———. 1995. Ecological attributes of extinction-prone species: Loss of freshwater fishes of Virginia. *Conservation Biology* 9:143–158.

Angermeier, P. L., and R. A. Smogor. 1995. Estimating number of species and relative abundances in stream-fish communities: Effects of sampling effort and discontinuous spatial distributions. *Canadian Journal of Fisheries and Aquatic Sciences* 52:936–949.

Angermeier, P. L., R. A. Smogor, and S. D. Steele. 1991. An electric seine for collecting fish in streams. *North American Journal of Fisheries Management* 11:352–357.

Angermeier, P. L., and M. R. Winston. 1998. Local vs. regional influences on local diversity in stream fish communities of Virginia. *Ecology* 79:911–927.

———. 1999. Characterizing fish community diversity across Virginia landscapes: Prerequisite for conservation. *Ecological Applications* 9:335–349.

Anonymous 1994. *Systematics agenda 2000: Charting the biosphere*. Systematics Agenda 2000, New York.

Anselin, A., P. M. Meire, and L. Anselin. 1989. Multicriteria techniques in ecological evaluation: An example using the analytical hierarchy process. *Biological Conservation* 49:215–229.

Anselin, L. 1988. *Spatial econometrics: Methods and models*. Boston: Kluwer Academic Publishers.

AOU. American Ornithologists' Union. 1895. *Check-list of North American birds*. 2nd and rev. ed. New York.

———. 1910. *Check-list of North American birds*. 3rd ed. New York.

———. 1931. *Check-list of North American birds*. 4th ed. Lancaster: Lancaster Press

———. 1957. *Check-list of North American birds*. 5th ed. Baltimore: Lord Baltimore Press.

———. 1998. *Check-list of North American birds*. 7th ed. Lawrence: Allen Press.

Apa, A. D., D. W. Uresk, and R. L. Linder. 1991. Impacts of black-tailed prairie dog rodenticides on non-target passerines. *Great Basin Naturalist* 51:301–309.

Apostle, H., trans. 1980. *Aristotle's categories and propositions*. Grinnell, Iowa: Peripatetic Press.

Arango-Velez, N., and G. H. Kattan. 1997. Effects of forest fragmentation on experimental nest predation in an Andean cloud forest. *Biological Conservation* 81:137–143.

Arbia, G. 1989. *Spatial data configuration in statistical analysis of regional economic and related problems*. Boston: Kluwer.

Arbia, G., R. Haining, and D. Griffith. 1998. Modelling error propagation in GIS overlay operations. *International Journal of Geographical Information Systems* 12:145–167.

Armour, C.L, D. A. Duff, and W. Elmore. 1991. The effects of livestock grazing on riparian and stream ecosystems. *Fisheries* 16:7–11.

Arnett, E. B., and R. Sallabanks. 1998. Land manager perceptions of avian research and information needs: A case study. In *Avian conservation*, ed. J. M. Marzluff and R. Sallabanks, 399–413. Washington, D.C.: Island Press.

Arthur, W. 1987. *The niche in competition and evolution*. New York: John Wiley & Sons.

Ashford, J. R. 1959. An approach to the analysis of data for semi-quantal responses in biological assay. *Biometrics* 15:573–581.

Ashmole, N. P. 1963. The regulation of numbers of tropical oceanic birds. *Ibis* 103:458–473.

Askins, R. A., J. F. Lynch, and R. Greenberg. 1990. Population declines in migratory birds in eastern North America. In *Current ornithology*, ed. D. M. Power, Vol. 7, 1–57. New York: Plenum Press.

Aspinall, R. 1992. An inductive modelling procedure based on Bayes' theorem for analysis of pattern in spatial data. *International Journal of Geographical Information Systems* 6:105–121.

Aspinall, R., and N. Veitch. 1993. Habitat mapping from satellite imagery and wildlife survey data using a Bayesian modeling procedure in a GIS. *Photogrammetric Engineering and Remote Sensing* 59:537–543.

Atkinson, P. M., P. J. Curran, and R. Webster. 1990. Sampling remotely sensed imagery for storage, retrieval, and reconstruction. *Professional Geographer* 42:345–353.

Atkinson, R. B., J. E. Perry, E. Smith, and J. Cairns Jr. 1993. Use of created wetland delineation and weighted averages as a component of assessment. *Wetlands* 13:185–193.

ATLSS. Across Trophic Level System Simulation. 1999. http://www.atlss.org (assessed 23 June 1999).

Auble, G., J. M. Friedman, and M. L. Scott. 1994. Relating riparian vegetation to present and future streamflows. *Ecological Applications* 4:544–554.

Augspurger, C.K. 1983. Offspring recruitment around tropical trees: Changes in cohort distance with time. *Oikos* 40:189–196.

Augustin, N. H., M. A. Mugglestone, and S. T. Buckland. 1996. An autologistic model for the spatial distribution of wildlife. *Journal of Applied Ecology* 33:339–347.

Austin, M. P. 1976. On non-linear species response models in ordination. *Vegetatio* 33:33–41.

———. 1980. Searching for a model for use in vegetation analysis. *Vegetatio* 423:11–21.

———. 1991. Vegetation theory in relation to cost-effective surveys. In *Nature conservation: Cost effective biological surveys and data analysis*, ed. C. R. Margules and M. P. Austin, 17–21. Melbourne: CSIRO Australia.

———. 1992. Modelling the environmental niche of plants: Implications for plant community response to elevated CO_2 levels. *Australian Journal of Botany* 40:615–630.

———. 1994. *Data capability, subproject 3, modelling of landscape patterns and processes using biological data*. Melbourne: CSIRO Division of Wildlife and Ecology.

———. 1998. An ecological perspective on biodiversity investigations: Examples from Australian eucalypt forests. *Annals of the Missouri Botanical Garden* 85:2–17.

———. 1999a. A silent clash of paradigms: Some inconsistencies in community ecology. *Oikos* 86:170–178.

———. 1999b. The potential contribution of vegetation ecology to biodiversity research. *Ecography* 22:465–484.

Austin, M. P., E. M. Cawsey, B. L. Baker, M. M. Yialelogou, D. J. Grice, and S. V. Briggs. 2000. *Predicted Vegetation Cover in the Central Lachlan Region*. Final report of the Natural Heritage Trust Project AA1368.97. CSRIO

Wildlife and Ecology, Canberra, Australia. http://www.cse.csiro.au/research/SL/Resources/Lachlan_veg.htm.

Austin, M. P., R. B. Cunningham, and P. M. Fleming. 1984. New approaches to direct gradient analysis using environmental scalars and statistical curve-fitting procedures. *Vegetatio* 55:11–27.

Austin, M. P., R. B. Cunningham, and R. B. Good. 1983. Altitudinal distribution in relation to other environmental factors of several eucalypt species in southern New South Wales. *Australian Journal of Ecology* 8:169–180.

Austin, M. P., and P. C. Heyligers. 1989. Vegetation survey design for conservation: Gradsect sampling of forests in north-eastern New South Wales. *Biological Conservation* 50:13–32.

———. 1991. New approach to vegetation survey design: Gradsect sampling. In *Nature conservation: Cost effective biological surveys and data analysis*, ed. C. R. Margules and M. P. Austin, 31–36. Melbourne, Australia: CSIRO Australia.

Austin, M. P., and J. A. Meyers. 1996. Current approaches to modelling the environmental niche of eucalypts: Implication for management of forest biodiversity. *Forest Ecology and Management* 85:95–106.

Austin, M. P., J. A. Meyers, D. L. Belbin, and M. D. Doherty. 1995. Modeling of landscape patterns and processes using biological data. Subproject 5: Simulated data case study. Consultancy report for ERIN. Melbourne, Australia: CSIRO Division of Wildlife and Ecology.

Austin, M. P., A. O. Nicholls, M. D. Doherty, and J. A. Meyers. 1994. Determining species response functions to an environmental gradient by means of a beta-function. *Journal of Vegetation Science* 5:215–228.

Austin, M. P., A. O. Nicholls, and C. R. Margules. 1990. Measurement of the realized qualitative niche: Environmental niches of five Eucalyptus species. *Ecological Monographs* 60:161–177.

Austin, M. P., J. G. Pausas, and I. R. Noble. 1997. Modelling environmental and temporal niches of *Eucalyptus*. In *Eucalypt ecology: Individuals to ecosystems*, ed. J. E. Williams and J. C. Z. Woinarski. Cambridge: Cambridge University Press.

Austin, M. P., and T. M. Smith. 1989. A new model for the continuum concept. *Vegetatio* 83:35–47.

Avery, B. W. 1980. Soil classification for England and Wales (higher categories). Technical Monograph No. 14. Harpenden, UK: Soil Survey.

Avery, T. E. 1978. Forester's guide to aerial photo interpretation. Agriculture Handbook 308. Washington, D.C.: USDA Forest Service.

Avery, T. E., and G. L. Berlin. 1992. *Fundamentals of remote sensing and airphoto interpretation.* 5th ed. Upper Saddle River, N.J.: Prentice Hall.

Ayyub, B., and R. McCuen. 1987. Quality and uncertainty assessment of wildlife habitat with fuzzy sets. *Journal of Water Resources Planning and Management* 113:95–109.

Azuma, D. L., J. A. Baldwin, and B. R. Noon. 1990. Estimating the occupancy of spotted owl habitat areas by sampling and adjusting for bias. General Technical Report PSW-124. Berkeley, Calif.: USDA Forest Service, Pacific Southwest Research Station.

Bailey, D. W. 1995. Daily selection of feeding areas by cattle in homogenous and heterogenous environments. *Applied Animal Behavioral Science* 45:183–200.

Bailey, D. W., J. E. Gross, E. A. Laca, L. R. Rittenhouse, M. B. Coughenour, D. M. Swift, and P. L. Sims. 1996. Mechanisms that result in large herbivore grazing distribution patterns. *Journal of Range Management* 49:386–400.

Bailey, D. W., J. W. Walker, and R. L. Rittenhouse. 1990. Sequential analysis of cattle location: Day-to-day movement patterns. *Applied Animal Behavioral Science* 25:137–148.

Bailey, F. M. 1928. *Birds of New Mexico.* Santa Fe: New Mexico Department of Game and Fish, State Game Protective Association, and the Bureau of Biological Survey.

Bailey, R. G. 1983. Delineation of ecosystem regions. *Environmental Management* 7:365–373.

Baker, F. A. 1993. Classification and regression tree analysis for assessing hazard of pine mortality caused by *Heterobasidion annosum. Plant Disease* 77:136–139.

Baker, W., and Y. Cai. 1992. The r.le programs for multiscale analysis of landscape structure using the GRASS geographical information system. *Landscape Ecology* 7:291–302.

Bakker, J. P. 1989. *Nature management by grazing and cutting.* Geobotany 14. Dordrecht, Netherlands: Kluwer.

Ballinger, R. E., J. D. Lynch, and P. C. Cole. 1979. Distribution and natural history of amphibians and reptiles in western Nebraska with ecological notes on the herpetiles of Arapaho Prairie. *Prairie Naturalist* 11:65–74.

Baltz, D. M. 1990. Autecology. In *Methods for fish biology*, ed. C. B. Schreck and P. B. Moyle, 585–607. Bethesda, Md.: American Fisheries Society.

Baltz, D. M., P. B. Moyle, and N. J. Knight. 1982. Competitive interactions between benthic stream fishes, riffle sculpin *Cottus gulosus* and speckled dace *Rhinichthys osculus. Canadian Journal of Fisheries and Aquatic Sciences* 39:1502–1511.

Banks, R. C. 1988. Geographic variation in the yellow-billed cuckoo. *Condor* 90:473–477.

Baraboo Hills Working Group. 1994. Baraboo Hills Plan. Madison: Wisconsin Chapter of The Nature Conservancy.

Baran, P., S. Lek, M. Delacoste, and A. Belaud. 1996. Stochastic models that predict trout population density or biomass on a mesohabitat scale. *Hydrobiologia* 337:1–9.

Barber, D. R. 1993. Effects of alternate host densities on brown-headed cowbird parasitism rates in black-capped vireos. M.S. thesis, University of Arkansas, Fayetteville.

Barkman, J. J., H. Doing, and S. Segal. 1964. Kritische Bemerkungen und Vorlschläge zur quantitativen Vegetationsanalyse. *Acta Botanica Neerlandica* 13:394–419.

Bart, J. 1995a. Acceptance criteria for using individual-based models to make management decisions. *Ecological Applications* 5:411–420.

———. 1995b. Amount of suitable habitat and viability of northern spotted owls. *Conservation Biology* 9:943–946.

Bart, J, and E. D. Forsman. 1992. Dependence of northern spotted owls *Strix occidentalis caurina* on old-growth forests in the western USA. *Conservation Biology* 6:95–100.

Baskent, E. Z., and G. A. Jordan. 1995. Characterizing spatial structure of forest landscapes. *Canadian Journal of Forest Research* 25:1830–1849.

Batty, M. 1998. Urban evolution on the desktop: Simulation with the use of extended cellular automata. *Environment and Planning A* 30:1943–1967.

Baxter, C. V., C. A. Frissell, and F. R. Hauer. 1999. Geomorphology, logging roads, and the distribution of bull trout in a forested river basin: Implications for management and conservation. *Transactions of the American Fisheries Society* 128:854–867.

Bayer, M, and W. Porter. 1988. Evaluation of a guild approach to habitat assessment for forest-dwelling birds. *Environmental Management* 12: 797–801.

Bay Institute. 1998. *From Sierra to the sea: The Ecological history of the San Francisco Bay-Delta watershed.* San Rafael, Calif.: Bay Institute of San Francisco.

Bayley, P. B., R. W. Larimore, and D. C. Dowling. 1989. Electric seine as a fish-sampling gear in streams. *Transactions of the American Fisheries Society* 118:447–453.

Beard, M. K., and B. P. Buttenfield. 1999. Detecting and evaluating errors by graphical methods. In *Geographical*

Information Systems: Principles, Techniques, Applications and Management, ed. P. A. Longley, M. F. Goodchild, D. J. Maguire, and D. W. Rhind, Vol. 1, Principles and Technical Issues 2/e, 219–233. New York: Wiley.

Beers, T. W., P. E. Dress, and L. C. Wensel. 1966. Aspect transformation in site productivity research. *Journal of Forestry* 64:691–692.

Begon, M., J. L. Harper, and C. R. Townsend. 1990. *Ecology: Individuals, populations, and communities.* 2nd ed. Sunderland, Mass.: Sinauer.

Beier, P. 1993. Determining minimum habitat areas and habitat corridors for cougars. *Conservation Biology* 7:94–108.

Beissinger, S. R. 1995. Modeling extinction in periodic environments: Everglades water levels and snail kite population viability. *Ecological Applications* 5:618–631.

Beissinger, S. R., and M. I. Westphal. 1998. On the use of demographic models of population viability in endangered species management. *Journal of Wildlife Management* 62:821–841.

Belaud, H., P. Chaveroche, P. Lim, and C. Sabaton. 1989. Probability-of-use curves applied to brown trout (*Salmo trutta fario* L.) in rivers of southern France. *Regulated rivers: Research and management* 3:321–336.

Belding, L. 1890. *Land birds of the Pacific district.* California Academy of Sciences. Occasional Papers Number 2. San Francisco: California Academy of Sciences.

Beletsky, L. 1996. *The red-winged blackbird: The biology of a strongly polygynous songbird.* New York: Academic Press.

Beletsky, L. and L. D. Orians. 1996. *Red-winged blackbirds: Decision-making and reproductive success.* Chicago: University of Chicago Press.

Bell, C. P. 1997. Leap-frog migration in the fox sparrow: Minimizing the cost of spring migration. *Condor* 99:470–477.

Bell, J. F. 1999. Tree based methods. In *Machine learning methods for ecological applications*, ed. A. Fielding, 89–106. Boston: Kluwer Academic Publishers.

Belovsky, G. E. 1987. Extinction models and mammalian persistence. In *Viable populations for conservation*, ed. M. E. Soulé, 35–57. Cambridge: Cambridge University Press.

Bendel, R. B., and A. A. Afifi. 1977. Comparison of stopping rules in forward regression. *Journal of the American Statistical Association* 72:46–53.

Bendell, J. F., and P. W. Elliott. 1966. Habitat selection in blue grouse. *Condor* 68:431–446.

Bender, D. J., T. A. Contreras, and L. Fahrig. 1998. Habitat loss and population decline: A meta-analysis of the patch size effect. *Ecology* 79(2):517–533.

Bender, L. C., G. J. Roloff, and J. B. Haufler. 1996. Evaluating confidence intervals for habitat suitability models. *Wildlife Society Bulletin* 24:347–352.

Bent, A. C. 1940. *Life histories of North American cuckoos, goatsuckers, hummingbirds and their allies.* Smithsonian Institution United States National Museum Bulletin 176. Washington, D.C.: United States Printing Office.

Berendse, F. 1994. Competition between plant populations at low and high nutrient supplies. *Oikos* 71:253–260.

Bergin, T. M. 1992. Habitat selection by the western kingbird in western Nebraska: A hierarchical analysis. *Condor* 94:903–911.

Berry, J. K. 1991. Assessing variation, shape, and pattern of map patches. *GIS World* August–October issues.

Besag, J., J. York, and A. Mollié. 1991. Bayesian image restoration, with two applications in spatial statistics (with discussion). *Annals of the Institute of Statistical Mathematics* 43:1–59.

Besag, J. E. 1986. On the statistical analysis of dirty pictures (with discussion). *Journal of the Royal Statistical Society Series B* 48:259–302.

Besag, J. E., P. Green, D. Higdon, and K. Mengersen. 1995. Bayesian computation and stochastic systems. *Statistical Science* 10:3–66.

Best, L. G., and D. F. Stauffer. 1986. Factors confounding evaluation of bird-habitat relationships. In *Wildlife 2000: Modeling habitat relationships of terrestrial vertebrates*, ed. J. Verner, M. L. Morrison, and C. J. Ralph, 209–216. Madison: University of Wisconsin Press.

Beutel, T. S., R. J. S. Beeton, and G. S. Baxter. 1999. Building better wildlife-habitat models. *Ecography* 22: 219–223.

Beven, K. J., and M. J. Kirkby. 1979. A physically based, variable contributing area model of basin hydrology. *Hydrological Science Bulletin* 24:43–69.

Bian, L. 1997. Multiscale nature of spatial data in scaling up environmental models. In *Scale in remote sensing and GIS*, ed. D. A. Quattrochi and M. F. Goodchild, 13–26. Boca Raton, Fla.: CRC Press.

Bibby, C. J., N. D. Burgess, and D. A. Hill. 1992. *Bird census techniques*. London: Academic Press.

Bierregaard, R. O., Jr., T. E. Lovejoy, V. Kapos, A. A. dos Santos, and R. W. Hutchings. 1992. The biological dynamics of tropical rainforest fragments. *BioScience* 42: 859–866.

Bills, G. F., G. I. Holtzman, and O. K. Miller Jr. 1986. Comparison of ectomycorrhizal-basidiomycete communities in red spruce versus northern hardwood forests of West Virginia. *Canadian Journal of Botany* 64:760–768.

Bio, A. M. F., R. Alkemande, and A. Barendregt. 1998. Determining alternative models for vegetation response analysis—a non-parametric approach. *Journal of Vegetation Science* 9:5–16.

Birkhead, T. R., and R. W. Furness. 1985. Regulation of seabird populations. In *Behavioural ecology: Ecological consequences of adaptive behaviour* ed. R. M. Sibly, and R. H. Smith, 145–167. British Ecological Society Symposium 25. Oxford: Blackwell Scientific Publications.

Bissonette, J. A. 1997a. Scale-sensitive ecological properties: Historical context, current meaning. In *Wildlife and landscape ecology: Effects of pattern and scale*, ed. J. A. Bissonette, 3–31. New York: Springer-Verlag.

———. ed. 1997b. *Wildlife and landscape ecology: Effects of pattern and scale.* New York: Springer-Verlag.

Bissonette, J. A., R. J. Fredrickson, and B. J. Tucker. 1989. American marten: A case for landscape-level management. *Transactions of the North American wildlife and natural resources conference* 54:89–101.

Bissonette, J. A., D. J. Harrison, C. D. Hargis, and T. G. Chapin. 1997. The influence of spatial scale and scale-sensitive properties on habitat selection by American marten. In *Wildlife and landscape ecology: Effects of pattern and scale*, ed. J. A. Bissonette, 368–385. New York: Springer-Verlag.

Bivand, R. 1998. A review of spatial statistical techniques for location studies. http://www.nhh.no/geo/gib/gib1998/gib98-3/lund.html. Current 2001/06/01.

Blackburn, T. M., and K. J. Gaston. 1996. Abundance-body size relationships: The area you census tells you more. *Oikos* 75:303–309.

Blackwell, B. F., W. B. Krohn, and R. B. Allen. 1995. Foods of double-crested cormorants in Penobscot Bay, Maine, USA: Temporal and spatial comparisons. *Colonial Waterbirds* 18:199–208.

Blair, R. B., and A. E. Launer. 1997. Butterfly diversity and human land use: Species assemblages along an urban gradient. *Biological Conservation* 80:113–125.

Blakesley, J. A., A. B. Franklin, and R. J. Gutiérrez. 1992. Spotted owl roost and nest site selection in northwestern California. *Journal of Wildlife Management* 56:388–392.

Blaustein, A. R., and D. B. Wake. 1990. Declining amphibian populations: A new global phenomenon? *Trends in Ecology and Evolution* 5:203–204.

Blaustein, A. R., D. B. Wake, and W. P. Sousa. 1994. Amphibian declines: Judging stability, persistence, and susceptibility of populations to local and global extinctions. *Conservation Biology* 8:60–71.

Block, W. M., and L. A. Brennan. 1993. The habitat concept in ornithology: Theory and applications. In *Current ornithology*, ed. D. M. Power, 35–91. New York: Plenum Press.

Block, W. M., M. L. Morrison, J. Verner, and P. N. Manley. 1994. Assessing wildlife-habitat-relationships models: A case study with California oak woodlands. *Wildlife Society Bulletin* 22:549–561.

Blumton, A. K., R. B. Owen Jr., and W. B. Krohn. 1988. Habitat suitability models: American Eider (breeding). Biological Report 82. Washington, D.C.: U.S. Department of the Interior, U.S. Fish and Wildlife Service.

Boal, C. W., and R. W. Mannan. 1998. Nest-site selection by Cooper's hawks in an urban environment. *Journal of Wildlife Management* 62:864–871.

Bock, W., and A. Salski. 1998. A fuzzy knowledge-based model of population dynamics of the yellow-necked mouse (*Apodemus flavicollis*) in a beech forest. *Ecological Modelling* 108:155–161.

Boddy, L., and C. W. Morris. 1999. Artificial neural networks for pattern recognition. In *Machine learning methods for ecological applications*, ed. A. Fielding, 37–88. Boston: Kluwer Academic Publishers.

Boitani, L., and P. Cuicci. 1993. Wolves in Italy: Critical issues for their conservation. In *Wolves in Europe: Status and perspectives*, ed. C. Promberger and W. Schröder, 74–79. Ettal, Germany: Munich Wildlife Society.

Bolger, D. T., T. A. Scott, and J. T. Rotenberry. 1997. Breeding bird abundance in an urbanizing landscape in coastal southern California. *Conservation Biology* 11:406–421.

Bolsinger, J. S., and T. J. Hayden. 1992. Project status report: 1992 field studies of two endangered species (the black-capped vireo and the golden-cheeked warbler) and the cowbird control program on Fort Hood, Texas. Report submitted to HQIII Corps and Fort Hood, Texas. Champaign, Ill.: U.S. Army Corps of Engineers Construction Engineering Research Lab.

———. 1994. Project status report: 1993 field studies of two endangered species (the black-capped vireo and the golden-cheeked warbler) and the cowbird control program on Fort Hood, Texas. Report submitted to HQIII Corps and Fort Hood, Texas. Champaign, Ill.: U.S. Army Corps of Engineers Construction Engineering Research Lab.

Boone, R. B. 1997. An assessment of terrestrial vertebrate diversity in Maine. Ph.D. dissertation, University of Maine, Orono.

Boone, R. B., and W. B. Krohn. 1998a. Maine gap analysis vertebrate data—Part 1: Distribution, habitat relations, and status of amphibians, reptiles, and mammals in Maine. Orono: Maine Cooperative Fish and Wildlife Research Unit, University of Maine.

———. 1998b. Maine gap analysis vertebrate data—Part 2: Distribution, habitat relations, and status of breeding birds in Maine. Orono: Maine Cooperative Fish and Wildlife Research Unit, University of Maine.

———. 1998c. A technique for representing diminishing habitat occupation: Feathering predicted species distributions near range limits in Maine. In *GAP Analysis Bulletin*, ed. E. S. Brackney and M. D. Jennings, 41–43. Moscow, Idaho: Bulletin No. 7, USGS BRD Gap Analysis Program.

———. 1999. Modeling the occurrence of bird species: Are the errors predictable? *Ecological Applications* 9:835–848.

———. 2000a. Partitioning sources of variation in vertebrate species richness. *Journal of Biogeography* 27:457–470.

———. 2000b. Predicting broad-scale occurrences of vertebrates in patchy landscapes. *Landscape Ecology* 15:63–74.

Booth, G. 1999. Gecko: A continuous 2-D world for ecological modeling. http://peaplant.biology.yale.edu:8001/papers/swarmgecko/rewrite.html.

Booth, G. D., M. J. Niccolucci, and E. G. Schuster. 1994. Identifying proxy sets in multiple linear regression: An aid to better coefficient interpretation. Research Paper INT-470. Ogden, Utah: Intermountain Research Station.

Borcard, D., P. Legendre, and P. Drapeau. 1992. Partialling out the spatial component of ecological variation. *Ecology* 73:1045–1055.

Borgognone, M. G. 1999. Statistical methods for habitat classification using butterfly abundance data. M.S. thesis, Iowa State University, Ames.

Bormann, B. T., J. R. Martin, F. H. Wagner, G. Wood, J. Alegria, P. G. Cunningham, M. H. Brookes, P. Friesema, J. Berg, and J. Henshaw. 1999. Adaptive management: Common ground where managers, scientists, and citizens can try to learn to meet society's needs and wants while maintaining ecological capacity. In *Ecological stewardship: A common reference for ecosystem management*, ed. N. C. Johnson, A. J. Malk, W. Sexton, and R. Szaro, 505–534. Kidlington: Elsevier Science.

Bormann, F. H., and G. E. Likens. 1979. *Pattern and*

process in a forested ecosystem. New York: Springer-Verlag.

Botkin, D. B. 1990. *Discordant harmonies: A new ecology for the twenty-first century.* Oxford: Oxford University Press.

Boulay, M. C. 1992. Mortality and recruitment of white-tailed deer fawns in the wet prairie/tree island habitat of the Everglades. M.S. thesis, University of Florida, Gainesville.

Boulinier, T., C. H. Flather, J. D. Nichols, K. H. Pollock, J. R. Sauer, and J. E. Hines. 1997. Using landscape ecology to test hypotheses about large-scale changes in bird communities. *Bulletin of the Ecological Society of America* (suppl.) 78:57.

Bowers, M. A., and S. F. Matter. 1997. Landscape ecology of mammals: Relationships between density and patch size. *Journal of Mammalogy* 78:999–1013.

Bowlby, J. N., and J. C. Roff. 1986. Trout biomass and habitat relationships in southern Ontario streams. *Transactions of the American Fisheries Society* 115:503–514.

Bowser, G. 1988. Phenology of butterflies' (Lepidoptera–Rhopalocera) adult food plants in Yellowstone National Park. M.S. thesis, University of Vermont, Burlington.

Box, G. E. P. 1976. Science and statistics. *Journal of the American Statistical Association* 71:791–799.

Box, G. E. P., and W. J. Hill. 1967. Discrimination among mechanistic models. *Technometrics* 9:57–71.

Boyce, M. S. 1992. Population viability analysis. In *Annual review of ecology and systematics* ed. D. G. Fautin, Vol. 23, 481–506. Palo Alto, Calif.: Annual Reviews.

Boyce, M. S., and L. L. McDonald. 1999. Relating populations to habitats using resource selection functions. *Trends in Ecology and Evolution* 14:268–272.

Boyce, M. S., J. S. Meyer, and L. L. Irwin. 1994. Habitat-based PVA for the northern spotted owl. In *Statistics in ecology and environmental monitoring,* ed. D. J. Fletcher and B. F. J. Manly, 63–85. Series 2. Dunedin, New Zealand: University of Otago.

Bozek, M. A., and F. J. Rahel. 1992. Generality of micro-habitat suitability models for young Colorado River cut-throat trout (*Oncorhynchus clarki pleuriticus*) across sites and among years in Wyoming streams. *Canadian Journal of Fisheries and Aquatic Sciences* 49:552–564.

Bradley, A. P. 1997. The use of the area under the ROC curve in the estimation of machine learning algorithms. *Pattern Recognition* 30:1145–1159.

Bradshaw, G. A., and J. G. Borchers. 2000. Uncertainty as information: Narrowing the science-policy gap. *Conser-vation Ecology,* [online] 4(1):7 http://www.consecol.org/vol4/iss1/art7.

Bragg, A. N. 1940. Observations on the ecology and natural history of anura I. Habits, habitat and breeding of *Bufo cognatus* Say. *American Naturalist* 74:322–349, 424–438.

Braithwaite, L. W. 1983. Studies on the arboreal marsupial fauna of eucalypt forests being harvested for wood pulp at Eden, New South Wales. I. The species distribution of animals. *Australian Wildlife Research* 10:219–229.

———. 1984. The identification of conservation areas for possums and gliders within the Eden woodpulp concession district. In *Possums and gliders,* ed. A. P. Smith and I. D. Hume, 501–508. Chipping Norton, NSW: Surrey Beatty & Sons in association with the Australian Mammal Society.

Braithwaite, L. W., M. P. Austin, M. Clayton, J. Turner, and A. O. Nicholls. 1989. On predicting the presence of birds in *Eucalyptus* forest types. *Biological Conservation* 50:33–50.

Braithwaite, L. W., M. L. Dudzinski, and J. Turner 1983. Studies on the arboreal marsupial fauna of eucalypt forests being harvested for wood pulp at Eden, New South Wales. II. Relationships between the faunal density richness and diversity, and measured variables of habitat. *Australian Wildlife Research* 10:231–247.

Braithwaite, L. W., J. Turner, and J. Kelly. 1984. Studies on the arboreal marsupial fauna of eucalypt forests being harvested for woodpulp at Eden, New South Wales. III. Relationships between faunal densities, eucalypt occurrence and foliage nutrients, and soil parent materials. *Australian Wildlife Research* 11:41–48.

Braithwaite, R. W., and W. J. Muller. 1997. Rainfall ground-water and refuges: Predicting extinctions of Australian tropical mammal species. *Australian Journal of Ecology* 22:57–67.

Braun-Blanquet, J. 1964. *Pflanzensoziologie.* 3rd ed. Vienna: Springer-Verlag.

Bray, J. R., and C. T. Curtis. 1957. An ordination of the upland forest communities of southern Wisconsin. *Ecological Monographs* 27:325–349.

Breiman, L., J. H. Friedman, R. A. Olshen, and C. J. Stone. 1984. *Classification and regression trees.* Belmont, Calif.: Wadsworth International Group.

Breitenmoser, U. 1998. Large predators in the Alps: The fall and rise of man's competitors. *Biological Conservation* 83:279–289.

Breitenmoser, U., and M. Baettig. 1992. Wiederansiedlung und Ausbreitung des Luchses (*Lynx lynx*) im Schweizer Jura. *Revue Suisse de Zoologie* 99:163–176.

Breitenmoser, U., C. Breitenmoser-Würsten, and S. Capt. 1998. Re-introduction and present status of the lynx in Switzerland. *Hystrix* 10:17–30.

Breitenmoser, U., and H. Haller. 1987. Zur Nahrungsökologie des Luchses *Lynx lynx* in den schweizerischen Nordalpen. *Zeitschrift für Säugetierkunde* 52:168–191.

Breitenmoser, U., P. Kaczensky, M. Dötterer, F. Bernhart, C. Breitenmoser-Würsten, S. Capt, and M. Liberek. 1993. Spatial organization and recruitment of Lynx (*Lynx lynx*) in a reintroduced population in the Swiss Jura Mts. *Journal of Zoology* 231:449–464.

Brennan, L. A. 1991. How can we reverse the northern bobwhite population decline? *Wildlife Society Bulletin* 19:544–555.

Brittingham, M. C., and S. A. Temple. 1983. Have cowbirds caused forest songbirds to decline? *BioScience* 33:31–35.

Brodman, R. 1998. Biogeography of Midwestern amphibians. In *Status and conservation of Midwestern amphibians*, ed. M. J. Lannoo, 24–30. Iowa City: University of Iowa Press.

Brommer, J. E., and M. S. Fred. 1999. Movement of the Apollo Butterfly *Parnassius apollo* related to host plant and nectar plant patches. *Ecological Entomology* 24:125–131.

Browder, S. F. 1998. Assemblages of grassland birds as indicators of environmental condition. M.S. thesis, University of Montana, Missoula.

Brown, J. 1978. The theory of insular biogeography and the distribution of boreal mammals and birds. *Great Basin Naturalist Memoirs* 2:209–228.

Brown, J. H. 1973. Species diversity of seed-eating desert rodents in sand dune habitats. *Ecology* 54:775–787.

———. 1995. *Macroecology*. Chicago: University of Chicago Press.

Brown, J. H., and M. V. Lomolino. 1998. *Biogeography*. Sunderland, Mass.: Sinauer Associates.

Brown, J. H., and B. A. Maurer. 1987. Evolution of species assemblages: Effects of energetic constraints and species dynamics on the diversification of the North American avifauna. *American Naturalist* 130:1–17.

———. 1989. Macroecology: The division of food and space among species on continents. *Science* 243:1145–1150.

Brown, J. H., D. W. Mehlman, and G. C. Stevens. 1995. Spatial variation in abundance. *Ecology* 76:2028–2043.

Brown, J. L. 1969a. Territorial behavior and population regulation in birds: A review and re-evaluation. *Wilson Bulletin* 81:293–329.

———. 1969b. The buffer effect and productivity in tit populations. *American Naturalist* 103:347–354.

Bruggink, J. G. 1997. American Woodcock Harvest and Breeding Population Status, 1996. Washington, D.C.: U.S. Department of the Interior, U.S. Fish and Wildlife Service.

Brunckhorst, D. J. 2000. *Bioregional planning: Resource management beyond the new millennium*. Amsterdam, The Netherlands: Harwood Academic Publishers.

Bruns, G., P. Mössinger, D. Polani, R. Schmitt, R. Spalt, T. Uthmann, and S. Weber. 1999. Xraptor: A simulation environment for continuous virtual multi-agent systems. Users manual. http://www.informatik.uni-mainz.de/~polani/XRaptor/XRaptor.html.

Brussard, P. F. 1989. Butterflies of the Greater Yellowstone Ecosystem. In *Rare, sensitive, and threatened species of the Greater Yellowstone Ecosystem*, ed. T. Clark and A. H. Harvey, 28–32. Jackson Hole: Northern Rockies Conservation Cooperative, Montana Natural Heritage Program, The Nature Conservancy, and Mountain West Environmental Services.

Brzeziecki, B., F. Kienast, and O. Wildi. 1995. Modelling potential impacts of climate change on the spatial distribution of zonal forest communities in Switzerland. *Journal of Vegetation Science* 6:257–268.

Buckland, S. T., D. R. Anderson, K. P. Burnham, and J. L. Laake. 1993. *Distance sampling: Estimating abundances of biological populations*. London: Chapman and Hill.

Buckland, S. T., and S. R. Baillie. 1987. Estimating bird survival rates from organized mist-netting programs. *Acta Ornithologica* 23:89–100.

Buckland, S. T., and D. A. Elston. 1993. Empirical models for the spatial distribution of wildlife. *Journal of Applied Ecology* 30:478–495.

Buckley, N. J. 1990. Diet and feeding ecology of great black-backed gulls (*Larus marinus*) at a southern Irish breeding colony. *Journal of Zoology* 222:363–373.

Buckley, P. A., and F. G. Buckley. 1984. Seabirds of the north and middle Atlantic coast of the United States: Their status and conservation. In *Status and conservation of the world's seabirds*, ed. J. P. Croxall, P. G. H. Evans, and R. W. Schreiber, 101–133. Technical Publication 2. Cambridge, England: International Council for Bird Preservation.

Buckley, P. A., and R. Downer. 1992. Modeling metapopulation dynamics for single species of seabirds. In *Wildlife 2001: Populations*, ed. D. R. McCullough and R. H. Barrett, 563–585. London: Elsevier Applied Science.

Bull, E. L. 1987. Ecology of the pileated woodpecker in northeastern Oregon. *Journal of Wildlife Management* 51:472–481.

Bunnell, F. L. 1989. Alchemy and uncertainty: What good are models? General Technical Report PNW-GTR-232. Portland, Ore.: USDA Forest Service, Pacific Northwest Research Station.

Burgess, R. L., and D. M. Sharpe, eds. 1981. *Forest Island dynamics in man-dominated landscapes.* New York: Springer-Verlag.

Burgman, M. A., and D. B. Lindenmayer. 1998. *Conservation biology for the Australian environment.* Chipping Norton, NSW: Surrey Beatty & Sons.

Burley, F. W. 1988. Monitoring biological diversity for setting priorities in conservation. In *Biodiversity*, ed. E. O. Wilson, 227–230. Washington: National Academy Press.

Burnham, K. P., and D. R. Anderson. 1992. Data-based selection of an appropriate biological model: The key to modern data analysis. In *Wildlife 2001: Populations*, ed. D. R. McCullough and R. H. Barrett, 16–30. London: Elsevier Applied Science.

———. 1998. *Model selection and inference: A practical information-theoretic approach.* New York: Springer-Verlag.

Burnham, K. P., D. R. Anderson, and J. L. Laake. 1980. Estimation of density from line transect sampling of biological populations. *Wildlife Monographs* 72:1–202.

Burnham, K. P., D. R. Anderson, and G. C. White. 1996. Meta-analysis of vital rates of the northern spotted owl. *Studies in Avian Biology* 17:92–101.

Burnham, K. P., D. R. Anderson, G. C. White, C. Brownie, and K. H. Pollock. 1987. Design and analysis methods for fish survival experiments based on release-recapture. *American Fisheries Society Monograph* 5:1–437.

Burrough, P. 1989. Fuzzy mathematical methods for soil survey and land evaluation. *Journal of Soil Science* 40:477–492.

———. 1992. Fuzzy classification methods for determining land suitability from soil profile observations and topography. *Journal of Soil Science* 43:193–210.

Burrough, P., and R. McDonnell. 1998. *Principles of geographic information systems.* New York: Oxford University Press.

Burrough, P. A., and A. U. Frank. 1995. Concepts and paradigms in spatial information: Are current geographical

information systems truly generic? *International Journal of Geographical Information Systems* 9:101–116.

Burroughs, R. H., and T. W. Clark. 1995. Ecosystem management: A comparison of Greater Yellowstone and Georges Bank. *Environmental Management* 19:649–663.

Burt, J. E., and G. M. Barber. 1996. *Elementary statistics for geographers.* 2nd ed. New York: Guilford Press.

Burt, T. P. 1992. The hydrology of headwater catchments. 3–28 In *The rivers handbook: Hydrological and ecological principles,* ed. P. Calow and G. Petts. Oxford: Blackwell Scientific.

Burton, G. W., and E. P. Odum. 1945. The distribution of stream fish in the vicinity of Mountain Lake, Virginia. *Ecology* 26:182–194.

Burvill, G. H., ed. 1979. *Agriculture in Western Australia, 150 years of development and achievement, 1829–1979.* Sesquicentenary Celebration Series. Nedlands: University of Western Australia Press.

Busby, J. R. 1986. A biogeographic analysis of *Nothofagus cunninghamii* (Hook.) Oerst. in south-eastern Australia. *Australian Journal of Ecology* 11:1–7.

———. 1991. BIOCLIM: A bioclimate analysis and prediction system. In *Nature conservation: Cost effective biological surveys and data analysis*, ed. C. R. Margules and M. P. Austin, 64–68. Melbourne: CSIRO Australia.

Busby, W. H., and J. R. Parmelee. 1996. Historical changes in a herpetofaunal assemblage in the Flint Hills of Kansas. *American Midland Naturalist* 135:81–91.

Bustamante, J. 1996. Population viability analysis of captive and released bearded vulture populations. *Conservation Biology* 10:822–831.

———. 1997. Predictive models for lesser kestrel *Falco naumanni* distribution, abundance and extinction in southern Spain. *Biological Conservation* 80:153–160.

Butterfield, B. R., B. Csuti, and J. M. Scott. 1994. Modeling vertebrate distributions for gap analysis. In *Mapping the diversity of nature*, ed. R. I. Miller, 53–68. London: Chapman and Hall.

Cade, B. S., J. W. Terrell, and R. L. Schroeder. 1999. Estimating effects of limiting factors with regression quantiles. *Ecology* 80:311–323.

Caicco, S. L., J. M. Scott, B. Butterfield, and B. Csuti. 1995. A gap analysis of the management status of the vegetation of Idaho (USA). *Conservation Biology* 9:498–511.

Cairns, D. K. 1992. Population regulation of seabird colonies. In *Current ornithology,* ed. D. M. Powers, Vol. 9, 37–61. New York: Plenum Press.

Cale, W. G., and P. L. Odell. 1980. Behavior of aggregate state variables in ecosystem models. *Mathematical Biosciences* 46:121–137.

Cale, W. G., R. V. O'Neill, and H. H. Shugart. 1983. Development and application of desirable ecological models. *Ecological Modelling* 18:171–186.

Caley, M. J., and D. Schluter. 1997. The relationship between local and regional diversity. *Ecology* 78:70–80.

California Department of Fish and Game. 1998. *State and Federally Listed Endangered and Threatened Animals of California.* Sacramento: California Department of Fish and Game, Natural Heritage Division.

California Resources Agency. 1998. *Sacramento River conservation area handbook.* Sacramento: California Resources Agency.

Call, D. R. 1990. Home range and habitat use by California spotted owls in the central Sierra Nevada. M.S. thesis, Humboldt State University, Arcata, Calif.

Callicott, J. B., L. B. Crowder, and K. Mumford. 1999. Current normative concepts in conservation. *Conservation Biology* 13:22–35.

Callicott, J. B., and K. Mumford. 1997. Ecological sustainability as a conservation concept. *Conservation Biology* 11:32–40.

Cameron, A. C., and F. A. G. Windmeijer. 1997. An R-squared measure of goodness of fit for some common nonlinear regression models. *Journal of Econometrics* 77:329–342.

Campbell, H. 1968. Seasonal precipitation and scaled quail in eastern New Mexico. *Journal of Wildlife Management* 32: 641–644.

Campbell, H., D. K. Martin, P. E. Ferkovich, and B. K. Harris. 1973. Effects of hunting and some other environmental factors on scaled quail in New Mexico. *Wildlife Monographs* 34:1–49.

Candy, S. G. 1991. Modeling insect phenology using ordinal regression and continuation ratio models. *Environmental Entomology* 20:190–195.

Cantin, M., J. Bedard, and H. Milne. 1974. The food and feeding of common eiders in the St. Lawrence Estuary in summer. *Canadian Journal of Zoology* 52:319–334.

Cao, G. 1995. The definition of the niche by fuzzy set theory. *Ecological Modelling* 77:65–71.

Capen, D. E., ed. 1981. The use of multivariate statistics in studies of wildlife habitat. General Technical Report RM-87. Fort Collins, Colo.: USDA Forest Service, Rocky Mountain Forest and Range Experiment Station.

Capen, D. E., J. W. Fenwick, D. B. Inkley, and A. C. Boyn-ton. 1986. Multivariate models of songbird habitat in New England forests. In *Wildlife 2000: Modeling habitat relationships of terrestrial vertebrates,* ed. J. A. Verner, M. L. Morrison, and C. J. Ralph, 171–175. Madison: University of Wisconsin Press.

Capt, S., U. Breitenmoser, and C. Breitenmoser-Würsten. 1998. Monitoring the lynx population in Switzerland. In *The Re-introduction of lynx into the Alps,* ed. C. Breitenmoser-Würsten, C. Rohner, and U. Breitenmoser, Vol. 38, 105–108. Strasbourg, Germany: Council of Europe Publishing.

Caraco, T., S. Martindale, and T. S. Whittam. 1980. An empirical demonstration of risk-sensitive foraging preferences. *Animal Behaviour* 28:820–830.

Carey, A. B., J. M. Calhoun, B. Dick, K. O'Halloran, L. S. Young, R. E. Bigley, S. Chan, C. A. Harrington, J. P. Hayes, and J. Marzluff. 1999. Reverse technology transfer: Obtaining feedback from managers. *Western Journal of Applied Forestry* 14:153–163.

Carpenter, S. R. 1998. Keystone species and academic-agency collaboration. *Conservation Ecology* [online] 2: R2. http://www.consecol.org/Journal/vol2/iss1/resp2.

Casella, G., and E. I. George. 1992. Explaining the Gibbs sampler. *The American Statistician* 46:167–174.

Casper, G. S. 1996. Geographic distributions of the amphibians and reptiles of Wisconsin. An interim report of the Wisconsin Herpetological Atlas. Milwaukee, Wis.: Milwaukee Public Museum.

Catling, P. C., and R. J. Burt. 1994. Studies of the ground-dwelling mammals of eucalypt forests in south-eastern New South Wales: The species, their abundance and distribution. *Wildlife Research* 21:219–239.

———. 1995a. Studies of the ground-dwelling mammals of eucalypt forests in south-eastern New South Wales: The effect of habitat variables on distribution and abundance. *Wildlife Research* 22:271–288.

———. 1995b. Studies of the ground-dwelling mammals of eucalypt forests in south-eastern New South Wales: The effect of environmental variables on distribution and abundance. *Wildlife Research* 22:669–685.

———. 1997. Studies of the ground-dwelling mammals of eucalypt forests in northeastern New South Wales: The species, their abundance and distribution. *Wildlife Research* 24:1–19.

Catling, P. C., R. J. Burt, and R. I. Forrester. 1998. Models of the distribution and abundance of ground-dwelling mammals in the eucalypt forests of south-eastern New South Wales. *Wildlife Research* 25:449–466.

Caudill, M. 1990. *AI expert: Neural networks primer.* San Francisco: Miller Freeman Publications.

Caughley, G. 1994. Directions in conservation biology. *Journal of Animal Ecology* 63:215–244.

Caughley, G., and A. R. E. Sinclair. 1994. *Wildlife ecology and management.* Boston: Blackwell Scientific Publications.

Chalk, D. E. 1986. Summary: Development, testing, and application of wildlife-habitat models—the researcher's viewpoint. In *Wildlife 2000: Modeling habitat relationships of terrestrial vertebrates* ed. J. Verner, M. L. Morrison, and C. J. Ralph, 155–156. Madison: University of Wisconsin Press.

Chaney, E., W. Elmore, and W. S. Platts. 1993. *Livestock grazing on western riparian areas.* Washington, D.C.: U.S. Government Printing Office.

Chapin, F. S., III, B. H. Walker, R. J. Hobbs, D. U. Hooper, J. H. Lawton, O. E. Sala, and D. Tilman. 1997. Biotic control over the functioning of ecosystems. *Science* 277:500–504.

Charnov, E. L. 1976. Optimal foraging: The marginal value theorem. *Theoretical Population Biology* 9:129–136.

Chatfield, C. 1995. Model uncertainty, data mining and statistical inference. *Journal of the Royal Statistical Society Series A* 158:419–466.

Chauhan, N. S. 1997. Soil moisture estimation under a vegetation cover: Combined active and passive microwave remote sensing approach. *International Journal of Remote Sensing* 18:1079–1097.

Chen, Jiquan, J. F. Franklin, and T. A. Spies. 1992. Vegetation responses to edge environments and old-growth Douglas-fir forests. *Ecological Applications* 2:387–396.

———. 1995. Growing-season microclimatic gradients from clearcut edges into old-growth Douglas-fir forests. *Ecological Applications* 5: 74–86.

Chernoff, H., and L. E. Moses. 1959. *Elementary decision theory.* New York: John Wiley & Sons.

Chesson, P. 1986. Environmental variation and the coexistence of species. In *Community ecology,* eds. J. Diamond, and T. J. Case, 240–256. New York: Harper and Row.

Chib, S., and E. Greenberg. 1995. Understanding the Metropolis-Hastings algorithm. *The American Statistician* 49:327–335.

Choate, J. S. 1967. Factors influencing nesting success of eiders in Penobscot Bay, Maine. *Journal of Wildlife Management* 31:769–777.

Chomicki, J, and P. Revesz. 1999a. A geometric framework for specifying spatiotemporal objects. In *Proceedings of the 6th international workshop on time representation and reasoning, TIME-99,* ed. C. Dixon and M. Fisher, 41–46. Los Alamitos, Calif.: IEEE Computer Society.

———. 1999b. Constraint-based interoperability of spatiotemporal databases. *GeoInformatica* 3:211–243.

Chomicki, J., and D. Toman. 1998. Temporal logic in information systems. In *Logics for databases and information systems,* ed. J. Chomicki and G. Saake, 31–70. Boston: Kluwer Academic Publishers.

Chou, Y. H. 1991. Map resolution and spatial autocorrelation. *Geographical Analysis* 23:228–246.

Chrisman, N. R. 1982. Beyond accuracy assessment: Correction of missclassification. In *Fifth international symposium on computer-assisted cartography and international society for photogrammetry and remote sensing commission IV,* ed. J. Foreman, 123–132. Falls Church, Va.: Society of Photogrammetry and American Congress on Surveying and Mapping.

Christensen, R. 1997. *Log-linear models and logistic regression.* New York: Springer-Verlag.

Christiansen, J. L. 1981. Population trends among Iowa's amphibians and reptiles. *Proceedings of the Iowa Academy of Science* 88:24–27.

Christiansen, J. L., and C. M. Mabry. 1985. The amphibians and reptiles of Iowa's Loess Hills. *Proceedings of the Iowa Academy of Science* 82:159–163.

Church, K. E., J. R. Sauer, and S. Droege. 1993. Population trends of quails in North America. In *Quail 3: National quail symposium,* ed. K. E. Church and T. V. Dailey, 44–54. Pratt, Kans.: Kansas Department of Wildlife and Parks.

Cicone, R. C., and J. A. Olsenholler. 1997. A summary of Asian vegetation using annual vegetation dynamic indicators. *Geocarto International* 12:13–25.

Clark, F., B. Isenring, and M. Mossman. 1993a. The Baraboo Hills Inventory Final Report. Madison: Nature Conservancy, Wisconsin Chapter.

Clark, D. A., and D. B. Clark. 1984. Spacing dynamics of a tropical rainforest tree: Evaluation of the Janzen-Connell model. *American Naturalist* 124:769–788.

Clark, J. D., J. E. Dunn, and K. G. Smith. 1993b. A multivariate model of female black bear habitat use for a geographic information system. *Journal of Wildlife Management* 57:519–526.

Clark, L. A., and D. Pregibon. 1992. Tree-based models. In *Statistical models in S,* ed. J. M. Chambers and T. J. Hastie, 377–419. Pacific Grove, Calif.: Wadsworth & Brooks, Cole Advanced Books and Software.

Clark, M. E., and K. A. Rose. 1997. An individual-based modeling analysis of management strategies for enhancing brook trout populations in southern Appalachian streams. *North American Journal of Fisheries Management* 17:54–76.

Clark, P. J., and F. C. Evans. 1954. Distance to nearest neighbor as a measure of spatial relationships in populations. *Ecology* 35:445–453.

Clark, R. G., and D. Shutler. 1999. Avian habitat selection: Pattern from process in nest-site use by ducks? *Ecology* 80:272–287.

Clark, W. R., R. A. Schmitz, and T. R. Bogenschutz. 1999. Site selection and nest success of ring-necked pheasants as a function of location in Iowa landscapes. *Journal of Wildlife Management* 63:976–989.

Clarke, K. C., and L. J. Gaydos. 1998. Loose-coupling a cellular automaton model and GIS: Long-term urban growth prediction for San Francisco and Washington/Baltimore. *International Journal of Geographical Information Science* 12:699–714.

Clarkson, D. A., and L. S. Mills. 1994. Hypogeous sporocarps in forest remnants and clearcuts in southwest Oregon. *Northwest Science* 68:259–265.

Clench, H. K. 1979. How to make regional lists of butterflies: Some thoughts. *Journal of the Lepidopterists' Society* 33:216–231.

Cleveland, W. S. 1979. Robust locally weighted regression and smoothing scatterplots. *Journal of the American Statistical Association* 74:829–836.

Cliff, A. D., and J. K. Ord. 1981. *Spatial processes: Models and applications.* London: Pion Limited.

Cochran, W. 1983. *Planning and analysis of observational studies.* New York: Wiley.

Cochran, W. G. 1977. *Sampling techniques.* 3rd ed. New York: Wiley.

Cochran, W. G., and G. M. Cox. 1957. *Experimental designs.* 2nd ed. New York: John Wiley & Sons.

Cody, M. L., 1968. On the methods of resource division in grassland bird communities. *American Naturalist* 102:107–147.

———. 1985a. An introduction to habitat selection in birds. In *Habitat selection in birds*, ed. M. L. Cody, 3–56. Orlando: Academic Press.

———. ed. 1985b. *Habitat selection in birds.* Orlando: Academic Press.

———. 1986. Diversity, rarity, and conservation in Mediterranean-climate regions. In *Conservation biology: The science of scarcity and diversity*, ed. M. E. Soulé, 122–152. Sunderland, Mass.: Sinauer Associates.

Cohen, J. 1960. A coefficient of agreement for nominal scales. *Educational and Psychological Measurement* 20:37–46.

Cohen, W. B., T. A. Spies, and G. A. Bradshaw. 1990. Semivariograms of digital imagery for analysis of conifer canopy structure. *Remote Sensing of Environment* 34:167–178.

Colasanti, R. L. 1991. Discussions of the possible use of neural network algorithms in ecological modelling. *Binary* 3:13–15.

Colgan, W., III., A. B. Carey, J. M. Trappe, R. Molina, and D. Thysell. 1999. Diversity and productivity of hypogeous fungal sporocarps in a variably thinned Douglas-fir forest. *Canadian Journal of Forest Research* 29:1259–1268.

Collett, D. 1991. *Modelling binary data.* London: Chapman and Hall.

Collins, S. L. 1983. Geographic variation in habitat structure of the black-throated green warbler (*Dendroica virens*). *Auk* 100:382–389.

Common, M. S., and D. W. McKenney. 1994. Investigating the reliability of a hedonic travel cost model: A Monte Carlo approach. *Canadian Journal of Forest Research* 24:358–363.

Conant R., and J. T. Collins. 1991. *A field guide to reptiles and amphibians: Eastern and central North America.* 3rd ed. Boston: Houghton Mifflin.

Concannon, J. A., C. L. Shafer, R. L. DeVelice, R. M. Sauvajot, S. L. Boudreau, T. E. DeMeo, and J. Dryden. 1999. Describing landscape diversity: A fundamental tool for ecosystem management. In *Ecological stewardship: A common reference for ecosystem management*, ed. N. C. Johnson, A. J. Malk, W. Sexton, and R. Szaro, 195–218. Kidlington: Elsevier Science.

Congalton, R. G. 1991. A review of assessing the accuracy of classifications of remotely sensed data. *Remote Sensing of Environment* 37:35–46.

Congalton, R. G., M. Balogh, C. Bell, K. Green, J. Milliken, and R. Ottman. 1998. Mapping and monitoring agricultural crops and other land cover in the Lower Colorado Basin. *Photogrammetric Engineering and Remote Sensing* 64:1107–1113.

Congalton, R. G., and K. Green. 1999. *Assessing the accuracy of remotely sensed data: Principles and practices.* Boca Raton, Fla.: Lewis.

Conkling, P. W. 1995. *From Cape Cod to the Bay of Fundy:*

An environmental atlas of the Gulf of Maine. Cambridge: MIT Press.

Connell, J. H. 1978. Diversity in tropical rainforests and coral reefs. *Science* 199:1302–1310.

———. 1980. Diversity and coevolution of competitors, or the ghost of competition past. *Oikos* 35:131–138.

Conover, W. J. 1980. *Practical nonparametric statistics.* 2nd ed. New York: John Wiley & Sons.

Conroy, M. J. 1993. The use of models in natural resource management: Prediction, not prescription. *Transactions of the North American Wildlife and Natural Resource Conference* 58:509–519.

Conroy, M. J., and B. R. Noon. 1996. Mapping species richness for conservation of biological diversity: Conceptual and methodological issues. *Ecological Applications* 6:763–773.

Conroy, M. J., Y. Cohen, F. C. James, Y. G. Matsinos, and B. A. Maurer. 1995. Parameter estimation, reliability, and model improvement for spatially explicit models of animal populations. *Ecological Applications* 5:17–19.

Cook, C. W. 1966. Factors affecting utilization of mountain slopes by cattle. *Journal of Range Management* 19:200–204.

Cook, T. L., J. A. Koloszar, and M. D. Goering. 1995. 1995 annual report: Behavior and movement of the brown-headed cowbird (*Molothrus ater*) on Fort Hood, Texas. Fort Hood: Nature Conservancy of Texas.

Cook, T. L., J. A. Koloszar, M. D. Goering, and L. L. Sanchez. 1997. Brown-headed cowbird (*Molothrus ater*) radio-tracking study on Fort Hood, Texas. In *The Nature Conservancy summary of 1996 research activities,* 11–50. Fort Hood: Texas Conservation Data Center, The Nature Conservancy.

———. 1998. The spatial and temporal response of brown-headed cowbirds (*Molothrus ater*) to cattle removal. In *The Nature Conservancy summary of 1997 research activities,* 76–96. Fort Hood: Texas Conservation Data Center, The Nature Conservancy.

Cooke, J. L. 1997. A spatial view of population dynamics. In *Wildlife and landscape ecology: Effects of pattern and scale,* ed. J. A. Bissonette, 288–309. New York: Springer-Verlag.

Cooperrider, A. Y. 1986. Habitat evaluation systems. In *Inventory and monitoring of wildlife habitat,* ed. A. Y. Cooperrider, R. J. Boyd, and H. R. Stuart, 757–776. Denver: U.S. Department of the Interior, Bureau of Land Management.

Cork, S. J., and P. C. Catling. 1996. Modelling distributions of arboreal and ground-dwelling mammals in relation to climate, nutrients, plant chemical defences and vegetation structure in eucalypt forests of southeastern Australia. *Forest Ecology and Management* 85:163–176.

Cormack, R. M. 1964. Estimates of survival from the sighting of marked animals. *Biometrika* 51:429–438.

———. 1968. The statistics of capture-recapture methods. *Oceanographic Marine Biology Annual Review* 6:455–506.

———. 1979. Models for capture-recapture. *Statistical Ecology Series* 5:217–255.

Corn, P. S., and C. R. Peterson. 1996. Prairie legacies—amphibians and reptiles. In *Prairie conservation: Preserving North America's most endangered ecosystem,* ed. F. B. Samson and F. L. Knopf, 125–134. Washington, D.C.: Island Press.

Cornell, H. V., and R. H. Karlson. 1996. Species richness of reef-building corals determined by local and regional processes. *Journal of Animal Ecology* 65:233–241.

Cornell, H. V., and J. H. Lawton. 1992. Species interactions, local and regional processes, and limits to the richness of ecological communities: A theoretical perspective. *Journal of Animal Ecology* 61:1–12.

Corsi, F., E. Duprè, and L. Boitani. 1999. A large-scale model of wolf distribution in Italy for conservation planning. *Conservation Biology* 13:150–159.

Corsi, F., I. Sinibaldi, and L. Boitani. 1998. *Large carnivores conservation areas in Europe.* Rome: Istituto Ecologia Applicata.

Costanza, R. 1992. Toward an operational definition of ecosystem health. In *Ecosystem health: New goals for environmental management,* ed. R. Costanza, B. G. Norton, and B. D. Haskell, 239–256. Washington, D.C.: Island Press.

Coss, R. G., and R. O. Goldthwaite. 1995. The persistence of old designs for perception. In *Perspectives in ethology,* ed. N. S. Thompson, Vol. 2, 83–148. New York: Plenum Press.

Couclelis, H. 1992. People manipulate objects (but cultivate fields): Beyond the raster-vector debate in GIS. In *Theories and methods of spatio-temporal reasoning in geographic space,* ed. A. U. Frank, I. Campari and U. Formentini, 65–77. LNCS 639. New York: Springer-Verlag.

———. 1997. From cellular automata to urban models: New principles for model development and implementation. *Environment and Planning B: Planning and Design* 24:165–174.

———. 1999. Space, time, geography. In *Geographical infor-*

mation systems, ed. P. A. Longley, M. F. Goodchild, D. J. Maguire, and D. W. Rhind, Vol. 1, Principles and technical issues. 29–38. 2nd ed. New York: Wiley.

Coughenour, M. B. 1991. Spatial components of plant-herbivore interactions in pastoral, ranching, and native ungulate ecosystems. *Journal of Range Management* 44: 530–542.

Coughlan, B. A. K., and C. L. Armour. 1992. Group decision-making techniques for natural resources management applications. Resource Publication 185. Washington, D.C.: U.S. Department of the Interior, U.S. Fish and Wildlife Service.

Cousins, S. 1977. Sample size and edge effect on community measures of farm bird populations. *Polish Ecological Studies* 3:27–35.

Cowger, J. 1976. Alcid nesting habitat on the Maine coast and its relevance to the Critical Areas Program. Augusta: Maine State Planning Office.

Cox, D. D., L. H. Cox, and K. B Ensor. 1995. Spatial sampling and the environment. Technical Report 18. Research Triangle Park, N.C.: National Institute of Statistical Sciences.

Cracknell, A. P. 1998. Review article. Synergy in remote sensing—what's in a pixel? *International Journal of Remote Sensing* 19:2025–2047.

Craft, R. A., J. R. Zelenak, and M. M. Stake. 1998. Monitoring the golden-cheeked warbler (*Dendroica chrysoparia*) during 1998 on Fort Hood, Texas. In *1998 annual report*, 55–97. Fort Hood: Nature Conservancy of Texas.

Cramp, S. 1985. *The birds of western palaearctic*. Vol. 4. Oxford: Oxford University Press.

CRES. Centre for Resource and Environmental Studies. 1998. ANUCLIM documentation. http://cres.anu.edu.au/outputs/anuclim.html.

Cressie, N., and J. M. Ver Hoef. 1993. Spatial statistical analysis of environmental and ecological data. In *Environmental modeling with GIS*, ed. M. F. Goodchild, B. O. Parks, and L. T. Steyaert, 404–413. New York: Oxford University Press.

Cressie, N. A. C. 1991. *Statistics for spatial data*. New York: John Wiley & Sons

———. 1993. *Statistics for spatial data*. Rev. ed. New York: John Wiley & Sons.

Cresswell, P., S. Harris, and D. J. Jefferies. 1990. *The history, distribution, status and habitat requirements of the badger in Britain*. Peterborough, UK: Nature Conservancy Council.

Crist, P. J., T. W. Kohley, and J. Oakleaf. 2000. Assessing land-use impacts on biodiversity using an expert systems tool. *Landscape Ecology* 15:47–62.

Crites, S., and M. R. T. Dale. 1998. Diversity and abundance of bryophytes, lichens, and fungi in relation to woody substrate and successional stage in aspen mixed wood boreal forests. *Canadian Journal of Botany* 76:641–651.

Crump, M. L., and N. J. Scott Jr. 1994. Visual encounter surveys. In *Measuring and monitoring biological diversity: Standard methods for amphibians*, ed. W. R. Heyer, M. A. Donnelly, R. W. McDiarmid, L. C. Hayek, and M. S. Foster, 84–92. Washington, D.C.: Smithsonian Institution Press.

Csillag, F., M. J. Fortin, and J. L. Dungan. 2000. On the limits and extensions of the definition of scale. *Bulletin of the Ecology Society of America* 81:230–232.

Csuti, B. 1996. Mapping animal distribution areas for gap analysis. In *Gap analysis: A landscape approach to biodiversity planning*, ed. J. M. Scott, T. H. Tear, and F. W. Davis, 135–146. Bethesda, Md.: American Society for Photogrammetry and Remote Sensing.

Csuti, B., and P. Crist. 1998. Methods for developing terrestrial vertebrate distribution maps for gap analysis. Gap Analysis Program: A Handbook for Gap Analysis. Revised ed. Moscow, Idaho: USGS Gap Analysis Program. http://www.gap.uidaho.edu/handbook/default.htm.

———. 2000. Methods for developing terrestrial vertebrate distribution models for GAP Analysis. Version 2.0.0. A Handbook for Conducting Gap Analysis. Moscow, Idaho: USGS Gap Analysis Program. http://www.gap.uidaho.edu/handbook/VertebrateDistributionModeling/history.htm.

Cuff, D. J., W. J. Young, E. K. Muller, W. Zelinsky, and R. F. Abler. 1989. *The atlas of Pennsylvania*. Philadelphia: Temple University Press.

Cullinan, V. I., and J. M. Thomas. 1992. A comparison of quantitative methods for examining landscape pattern and scale. *Landscape Ecology* 7: 211–227.

Cumming, S. G., D. A. Demarchi, and C. Walters. 1998. A grid-based spatial model of forest dynamics applied to the boreal mixedwood region. Sustainable Forest Management Network Working Paper 1998-8. http://sfm-1.biology.ualberta.ca/english/pubs/epubframe.htm.

Curnutt, J. L., J. Comiskey, M. P. Nott, and L. J. Gross. 2000. Landscape-based spatially explicit species index models for natural Everglades restoration. *Ecological Applications* 10:1849–1860.

Curran, P. J. 1988. The semivariogram in remote sensing:

An introduction. *Remote Sensing Environment* 24: 493–507.

Currie, D. J. 1991. Energy and large-scale patterns of animal- and plant-species richness. *American Naturalist* 137:27–49.

Curtis, J. R. 1959. *The vegetation of Wisconsin: An ordination of plant communities.* Madison: University of Wisconsin Press.

Cyr, H. 1997. Does inter-annual variability in population density increase with time? *Oikos* 79:549–558.

Dahl, T. E. 1990. *Wetland losses in the United States 1780s to 1980s.* Washington, D.C.: U.S. Department of the Interior, Fish and Wildlife Service.

Dahlberg, A., and J. Stenlid. 1995. Spatiotemporal patterns in ectomycorrhizal populations. *Canadian Journal of Botany* 73 (suppl. 1):1222-1230.

Dale, M. R. T. 1999. *Spatial pattern analysis in plant ecology.* Cambridge: Cambridge University Press.

Dale, V. H., R. Gardner, and M. Turner. 1989. Predicting across scales: Comments of the guest editors of landscape ecology. *Landscape Ecology* 3:147–151.

Daly, C., R. P. Neilson, and D. L. Phillips. 1994. A statistical-topographic model for mapping climatological precipitation over mountainous terrain. *Journal of Applied Meteorology* 33:140–158.

Daly, C., G. H. Taylor, and W. P. Gibson. 1997. The PRISM approach to mapping precipitation and temperature. Tenth Conference on Applied Climatology. Reno, Nev.: American Meteorological Society.

Danielson, B. J. 1992. Habitat selection, interspecific interactions and landscape composition. *Evolutionary Ecology* 6:399–411.

Danielson, R. M. 1984. Ectomycorrhizal associations in jack pine stands in northeastern Alberta. *Canadian Journal of Botany* 62:932–939.

Danko, D. M. 1992. The digital chart of the world. *GeoInfo Systems* 2:29–36.

Dasgupta, N., and J. R. Alldredge. 1998. A multivariate χ^2 analysis of resource selection data. *Journal of Agricultural, Biological, and Environmental Statistics* 3: 323–334.

Dasmann, R. F. 1981. *Wildlife biology.* 2nd ed. New York: John Wiley & Sons.

Daubenmire, R. 1952. Forest vegetation of northern Idaho and adjacent Washington, and its bearing on concepts of vegetation classification. *Ecological Monographs* 22: 301–330.

———. 1959. A canopy-coverage method of vegetational analysis. *Northwest Science* 33:43–66.

———. 1978. *Plant geography, with special reference to North America.* New York: Academic Press.

David, F. N., and P. G. Moore. 1954. Notes on contagious distributions in plant populations. *Annals of Botany* 18:47–53.

Davis, F. W., and S. Goetz. 1990. Modeling vegetation pattern using digital terrain data. *Landscape Ecology* 4:69–80.

Davis, F. W., and D. M. Stoms. 1996. A spatial analytical hierarchy for GAP analysis. In *Gap analysis: A landscape approach to biodiversity planning,* ed. J. M. Scott, T. H. Tear, and F. W. Davis, 15–24. Bethesda, Md.: American Society for Photogrammetry and Remote Sensing.

Davis, F. W., D. M. Stoms, J. E. Estes, and J. M. Scott. 1990. An information systems approach to the preservation of biological diversity. *International Journal of Geographical Information Science* 4:55–78.

Davis, F. W., D. M. Stoms, A. D. Hollander, K. A. Thomas, P. A. Stine, D. Odion, M. I. Borchert, J. H. Thorne, M. V. Gray, R. E. Walker, K. Warner, and J. Graae. 1998. *The California gap analysis project: Final report.* Santa Barbara: University of California.

Davis, S. M., and J. C. Ogden, eds. 1994. *Everglades: The ecosystem and its restoration.* Delray Beach, Fla.: St. Lucie Press.

Dawson, W. R. 1992. Physiological responses of animals to higher temperatures. In *Global warming and biological diversity,* ed. R. L. Peters and T. E. Lovejoy, 158–170. New Haven, Conn.: Yale University Press.

DeAngelis, D. L., and L. J. Gross, eds. 1992. *Individual-based models and approaches in ecology: Populations, communities, and ecosystems.* New York: Chapman and Hall.

DeAngelis, D. L., L. J. Gross, M. A. Huston, W. F. Wolff, D. M. Fleming, E. J. Comiskey, and S. M. Sylvester. 1998. Landscape modeling for Everglades ecosystem restoration. *Ecosystems* 1:64–75.

Debinski, D. M. 1996. Using satellite data to support field-work: Can species distributions be predicted? *Yellowstone Science* 4:2–5.

Debinski, D. M., and P. F. Brussard. 1992. Biological diversity assessment in Glacier National Park, Montana: I. Sampling design. In *Proceedings from the international symposium on ecological indicators,* ed. D. H. McKenzie, D. E. Hyatt, and V. J. McDonald, 393–407. Essex, England: Elsevier Publishing.

_____. 1994. Using biodiversity data to access species-habitat relationships in Glacier National Park, Montana. *Ecological Applications* 4:833–843.

Debinski, D. M., and P. S. Humphrey. 1997. An integrated approach to biodiversity assessment. *Natural Areas Journal* 17:355–365.

Debinski, D. M., M. E. Jakubauskas, and K. Kindscher. 1999. A remote sensing and GIS-based model of habitats and biodiversity in the Greater Yellowstone Ecosystem. *International Journal of Remote Sensing* 20:3281–3291.

Dechant, J. A., M. L. Sondreal, D. H. Johnson, L. D. Igl, C. M. Goldade, M. P. Nenneman, and B. R. Euliss. 1999a. *Effects of management practices on grassland birds: Baird's sparrow.* Jamestown, N.D.: Northern Prairie Wildlife Research Center. http://www.npwrc.usgs.gov/resource/literatr/grasbird/bairds/bairds.htm (Version 19FEB99).

Dechant, J. A., M. L. Sondreal, D. H. Johnson, L. D. Igl, C. M. Goldade, A. L. Zimmerman, and B. R. Euliss. 1999b. *Effects of management practices on grassland birds: Western meadowlark.* Jamestown, N.D.: Northern Prairie Wildlife Research Center. http://www.npwrc.usgs.gov/resource/literatr/grasbird/weme/weme.htm (Version 06MAY99).

Dedon, M. F., S. A. Laymon, and R. H. Barrett. 1986. Evaluating models of wildlife-habitat relationships of birds in black oak and mixed-conifer habitats. In *Wildlife 2000: Modeling habitat relationships of terrestrial vertebrates*, ed. J. Verner, M. L. Morrison, and C. J. Ralph, 115–119. Madison: University of Wisconsin Press.

Degenhardt, W. G., C. W. Painter, and A. H. Price. 1996. *Amphibians and reptiles of New Mexico.* Albuquerque: University of New Mexico Press.

DeGraaf, R. M., and D. D. Rudis. 1986. New England wildlife: Habitat, natural history, and distribution. General Technical Report NE-108. Broomall, Pa.: USDA Forest Service, Northeastern Forest Experiment Station.

DeJong, G., and R. Mooney. 1986. Explanation-based learning: An alternative view. *Machine Learning* 1: 145–176.

De la Bastide, P. Y., B. R. Kropp, and Y. Piche 1994. Spatial distribution and temporal persistence of discrete genotypes of the ectomycorrhizal fungus *Laccaria bicolor* (Maire) Orton. *New Phytologist* 127:547–556.

Deleo, J. M. 1993. Receiver operating characteristic laboratory (ROCLAB): Software for developing decision strategies that account for uncertainty. In *Proceedings of the Second International Symposium on Uncertainty Modelling and Analysis*, ed. B. M. Ayyub, 318–325. Los Alamitos, Calif.: IEEE Computer Society Press.

Deleo, J. M., and G. Campbell. 1990. The fuzzy receiver operating characteristic function and medical decisions with uncertainty. In *Proceedings of the First International Symposium on Uncertainty Modelling and Analysis*, ed. B. M. Ayyub, 694–699. Los Alamitos, Calif.: IEEE Computer Society Press.

DeLong, E. R., D. M. DeLong, and D. L. Clarke-Pearson. 1988. Comparing the area under two or more receiver operating characteristic curves: A non-parametric approach. *Biometrics* 44:837–845.

deMaynadier, P. G., and M. L. Hunter Jr. 1998. Effects of silvicultural edges on the distribution and abundance of amphibians in Maine. *Conservation Biology* 12: 340–352.

den Boer, P. J. 1981. On the survival of populations in a heterogeneous and variable environment. *Oecologia* 50: 39–53.

DeSante, D. F. 1992. Monitoring avian productivity and survivorship (MAPS): A sharp, rather than blunt, tool for monitoring and assessing landbird populations. In *Wildlife 2001: Populations*, ed. D. R. McCullough and R. H. Barrett, 511–521. London: Elsevier Applied Science.

DeSante, D. F., K. M. Burton, J. F. Saracco, and B. L. Walker. 1995. Productivity indices and survival rate estimates from MAPS, a continent-wide programme of constant effort mist-netting in North America. *Journal of Applied Statistics* 22:935–947.

DeSante, D. F., M. P. Nott, and D. R. O'Grady. 2001. Identifying the proximate demographic cause(s) of population change by modelling spatial variation in productivity, survivorship and population trends. *Ardea* 89:185–208.

DeSante, D. F., D. R. O'Grady, K. M. Burton, P. Velez, D. Froehlich, E. E. Feuss, H. Smith, and E. D. Ruhlen. 1998. The Monitoring Avian Productivity and Survivorship (MAPS) program sixth and seventh annual report (1995 and 1996). *Bird Populations* 4:69–122.

DeSante, D. F., D. R. O'Grady, and P. Pyle. 1999. Measures of productivity and survival derived from standardized mist netting are consistent with observed population changes. *Bird Study* (suppl.) 46: s178–s188.

Desrochers, A. 1989. Sex, dominance, and microhabitat use in wintering black-capped chickadees: A field experiment. *Ecology* 70:636–645.

De Staso, J. III., and F. J. Rahel. 1994. Influence of water temperature on interactions between juvenile Colorado River cutthroat trout and brook trout in a laboratory stream. *Transactions of the American Fisheries Society* 123:289–297.

Dettmers, R., and J. Bart. 1999. A GIS modeling method applied to predicting forest songbird habitat. *Ecological Applications* 9:152–163.

Diersing, V. E., R. B. Shaw, and D. J. Tazik. 1992. U.S. Army land condition-trend analysis (LCTA) program. *Environmental Management* 16:405–414.

Diggle, P. J., K.-Y. Liang, and S. L. Zeger. 1994. *Analysis of Longitudinal Data*. Oxford: Oxford Science Publications, Clarendon Press.

Diggle, P. J., J. A. Tawn, and R. A. Moyeed. 1998. Model-based geostatistics (with discussion). *Applied Statistics* 47:299–350.

Doak, D., P. Kareiva, and B. Klepetka. 1994. Modeling population viability for the desert tortoise in the western Mojave Desert. *Ecological Applications* 4:446–460.

Dobkin, D. S. 1994. *Conservation and management of Neotropical migrant landbirds in the northern Rockies and Great Plains*. Moscow: University of Idaho Press.

Dobson, A. P., A. D. Bradshaw, and A. J. M. Baker. 1997. Hopes for the future: Restoration ecology and conservation biology. *Science* 277:515–522.

Dodson, S. I., S. E. Arnott, and K. L. Cottingham. 2000. The relationship in lake communities between primary production and species richness. *Ecology* 81:2662–2679.

Dolbeer, R. A. 1982. Migration patterns for age and sex classes of blackbirds and starlings. *Journal of Field Ornithology* 53:28–46.

Dolloff, C. A. 1987. Seasonal population characteristics and habitat use by juvenile coho salmon in a small southeast Alaska stream. *Transactions of the American Fisheries Society* 116:829–838.

Dolloff, C. A., D. G. Hankin, and G. H. Reeves. 1993. Basinwide estimation of habitat and fish populations in streams. General Technical Report SE-83. Asheville, N.C.: USDA Forest Service, Southern Research Station.

Donovan, M. L., D. L. Rabe, and C. E. Olson Jr. 1987. Use of geographic information systems to develop habitat suitability models. *Wildlife Society Bulletin* 15:574–579.

Downer, R. 1993. *GAPPS II user manual. Version 1.3*. Setauket, N.Y.: Applied Biomathematics.

Doyle, T. W. 1981. The role of disturbance in the gap dynamics of a montane rain forest: An application of a tropical forest succession model. In *Forest succession concepts and applications*, ed. D. C. West, H. H. Shugart, and D. B. Botkin, 56–73. New York: Springer-Velag.

Dramstad, W. E., J. D. Olson, and R. T. T. Forman. 1996. *Landscape ecology principles in landscape architecture and land-use planning*. Washington, D.C.: Island Press.

Drechsler, M., and C. Wissel. 1998. Trade-offs between local and regional scale management of metapopulations. *Biological Conservation* 83:31–41.

Driese, K. L., W. A. Reiners, E. H. Merrill, and K. G. Gerow. 1997. A digital land cover map of Wyoming, USA: A tool for vegetation analysis. *Journal of Vegetation Science* 8:133–146.

Drolet, B. and A. Desrochers. 1999. Effects of landscape structure on nesting songbird distribution in a harvested boreal forest. *Condor* 101:699–704.

Drury, W. H. 1973. Population changes in New England seabirds. *Bird-banding* 44:267–313.

Dubayah, R., and P. M. Rich. 1996. GIS-based solar radiation modeling. 129–134. In *GIS and environmental modeling: Progress and research issues*, ed. M. F. Goodchild, L. T. Steyaert, B. O. Parks, C. Johnston, D. Maidment, M. Crane, and S. Glendinning. Fort Collins, Colo.: GIS World Books.

Dubois, D., and H. Prade. 1980. *Fuzzy sets and systems: Theory and applications*. New York: Academic Press.

Dueser, R. D., and H. H. Shugart Jr. 1978. Microhabitats in a forest-floor small mammal fauna. *Ecology* 59:89–98.

Duever, M. J., J. E. Carlson, J. F. Meeder, L. C. Duever, L. H. Gunderson, L. A. Riopelle, T. R. Alexander, R. L. Myers, and D. P. Spangler. 1986. *The Big Cypress National Preserve*. Research Report Number 8. New York: National Audubon Society.

Duke, G. E. 1966. Reliability of censuses of singing male woodcocks. *Journal of Wildlife Management* 30: 697–707.

Duke-Sylvester, S., and L. J. Gross. 2001. Integrating spatial data into an agent-based modeling system: Ideas and lessons from the development of the across trophic level system simulation (ATLSS). In *Integrating Geographic Information Systems and Agent-based Modeling Techniques for Stimulating Social and Ecological Processes*, ed. H. R. Gimblett. Oxford: Oxford University Press.

Dunham, J. B., M. M. Peacock, B. E. Rieman, R. E. Schroeter, and G. L. Vinyard. 1999. Local and geographic variability in the distribution of stream-living Lahontan cutthroat trout. *Transactions of the American Fisheries Society* 128:875–889.

Dunham, J. B., and B. E. Rieman. 1999. Metapopulation structure of bull trout: Influences of physical, biotic, and geometrical landscape characteristics. *Ecological Applications* 9:642–655.

Dunham, J. B., G. L. Vinyard, and B. E. Rieman. 1997.

Habitat fragmentation and extinction risk of Lahontan cutthroat trout (*Oncorhynchus clarki henshawi*). *North American Journal of Fisheries Management* 17: 1126–1133.

Dunn, C. P., D. M. Sharpe, G. R. Guntenspergen, F. Stearns, and Z. Yang. 1991. Methods for analyzing temporal changes in landscape pattern. In *Quantitative methods in landscape ecology*, ed. M. G. Turner and R. H. Gardner, 173–198. New York: Springer-Verlag.

Dunning, J. B., Jr., D. J. Stewart, B. J. Danielson, B. R. Noon, T. L. Root, R. H. Lamberson, and E. E. Stevens. 1995. Spatially explicit population models: Current forms and future uses. *Ecological Applications* 5:3–11.

Dunson, W. A., and J. Travis. 1991. The role of abiotic factors in community organization. *American Naturalist* 138:1067–1091.

du Toit, J. T. 1995. Determinants of the composition and distribution of wildlife communities in southern Africa. *Ambio* 24:2–6.

Dwyer, T. J., G. F. Sepik, E. L. Derleth, and D. G. McAuley. 1988. Demographic characteristics of a Maine woodcock population and effects of habitat management. Research Report 4. Washington, D.C.: U.S. Department of the Interior, U.S. Fish and Wildlife Service.

Eastman, J. R. 1997. *Idrisi.* Worcester: Clark Labs for Cartographic Technology and Geographic Analysis.

Eastman, J. R., and M. Fulk. 1993. Long sequence time series evaluation using standardized principal components. *Photogrammetric Engineering and Remote Sensing* 59:1307–1312.

Eastman, J. R., W. G. Jin, P. A. K. Kyem, and J. Toledano. 1995. Raster procedures for multi-criteria multi-objective decisions. *Photogrammetric Engineering and Remote Sensing* 61:539–547.

Eaton, J. G., J. H. McCormick, B. E. Goodno, D. G. O'Brien, H. G. Stefany, M. Hondzo, and R. M. Scheller. 1995. A field information-based system for estimating fish temperature tolerances. *Fisheries* 20:10–18.

Eberhardt, L. L. 1987. Population projections from simple models. *Journal of Applied Ecology* 24:103–188.

Eckrich, G. H., and T. E. Koloszar. 1998. Brown-headed cowbird (*Molothrus ater*) control program on Fort Hood, Texas. In *1998 Annual report*, 151–167. Fort Hood: Nature Conservancy of Texas.

Eckrich, G. H., T. E. Koloszar, and M. D. Goering. 1999. Effective landscape management of brown-headed cowbirds at Fort Hood, Texas. *Studies in Avian Biology* 18:267–274.

Edenius, L., and K. Sjöberg. 1997. Distribution of birds in natural landscape mosaics of old-growth forests in northern Sweden: Relations to habitat area and landscape context. *Ecography* 20:425–431.

Edwards, M., and D. R. Morse. 1995. The potential for computer-aided identification in biodiversity research. *Trends in Ecology and Evolution* 10:153–158.

Edwards, T. C., Jr., E. T. Deshler, D. Foster, and G. G. Moisen. 1996. Adequacy of wildlife habitat relation models for estimating spatial distributions of terrestrial vertebrates. *Conservation Biology* 10:263–270.

Edwards, T. C., Jr., C. G. Homer, S. D. Bassett, A. Falconer, R. D. Ramsey, and D. W. Wight. 1995. Utah Gap Analysis: An environmental information system. Final Project Report 95-1. Logan: Utah Cooperative Fish and Wildlife Research Unit, Utah State University.

Edwards, W. R. 1972. Quail, land use, and weather in Illinois 1956–70. In *Proceedings of the first national bobwhite quail symposium*, ed. J. A. Morrison and J. C. Lewis, 174–183. Stillwater, Okla.: Oklahoma State University.

Efron, B. 1982. *The jackknife, the bootstrap and other resampling plans.* Philadelphia: Society for Industrial and Applied Mathematics.

———. 1983. Estimating the error rate of a prediction rule: Improvement on cross-validation. *Journal of the American Statistical Association* 78:316–331.

Efron, B., and G. Gong. 1983. A leisurely look at the bootstrap, the jackknife, and cross validation. *American Statistician* 37:36–48.

Efron, B., and R. Tibshirani. 1997. Improvements on cross-validation: The .632+ bootstrap method. *Journal of the American Statistical Association* 92:548–506.

———. 1991. Statistical data analysis in the computer age. *Science* 253:390–395.

Egbert, S. L., R. Y. Lee, K. P. Price, R. Boyce, and M. D. Nellis. 1998. Mapping Conservation Reserve Program (CRP) grasslands using multi-seasonal Thematic Mapper imagery. *GeoCarto International* 13:17–24.

Ehrlich, D., J. E. Estes, and A. Singh. 1994. Applications of NOAA-AVHRR 1 km Data for Environmental Monitoring. *International Journal of Remote Sensing* 15: 145–161.

Ehrlich, P. R., and S. E. Davidson. 1960. Techniques for capture-recapture studies of Lepidoptera populations. *Journal of the Lepidopterists' Society* 14:227–229.

Ehrlich, P. R., D. S. Dobkin, and D. Wheye. 1988. *The*

birder's handbook: A field guide to the natural history of North American birds. New York: Simon and Schuster.

——. 1992. *Birds in jeopardy: The imperiled and extinct birds of the United States and Puerto Rico.* Stanford: Stanford University Press.

Ehrlich, P. R., and D. D. Murphy. 1987. Conservation lessons from long-term studies of checkerspot butterflies. *Conservation Biology* 1:122–131.

EIP-Associates. 1997. Yolo County habitat conservation plan: A plan to mitigate biological impacts from urban development in Yolo County. Sacramento, Calif.: EIP-Associates.

Elith, J., M. A. Burgman, and P. Minchin. 1998. Improved protection strategies for rare plants. Consultancy report for Environment Australia. Project No. FB-NP22. Canberra: Environment Australia.

Ellenberg, H. 1956. Ausgaben und Methoden der Vegetationskunde. In *Einfuhrung in die Phtologie, IV Grundlagen der Vegetations gliederung*, ed. H. Walter and I. Teil. Stuttgart, Germany: Ulmer.

Ellingwood, M. R., B. P. Shissler, and D. E. Samuel. 1993. Evidence of leks in the mating strategy of the American woodcock. In *Proceedings of the Eighth Woodcock Symposium*, ed. J. R. Longcore and G. F. Sepik, 109–115. Biological Report 16. Washington, D.C.: U.S. Department of the Interior, U.S. Fish and Wildlife Service.

Elliott, G. P. 1996. Mohua and stoats: A population viability analysis. *New Zealand Journal of Zoology* 23:239–247.

Elton, C. 1927. *Animal ecology.* London: Sidgwick and Jackson.

——. 1930. *Animal ecology and evolution.* Oxford: Clarendon Press.

Emlen, S. T. 1975. The stellar-orientation system of a migratory bird. *Scientific American* 233:102–111.

Ensign, W. E., P. L. Angermeier, and C. A. Dolloff. 1995. Use of line transect methods to estimate abundance of benthic stream fishes. *Canadian Journal of Fisheries and Aquatic Sciences* 52:213–222.

Erhardt, A. 1985. Diurnal Lepidoptera: Sensitive indicators of cultivated and abandoned grassland. *Journal of Applied Ecology* 22:849–861.

Erskine, A. J. 1992. *Atlas of breeding birds of the maritime provinces.* Nova Scotia: Nimbus Publications, Nova Scotia Museum.

Erwig, M., R. H. Güting, M. Schneider, and M. Vazirgiannis. 1999. Spatio-temporal data types: An approach to modeling and querying moving objects in databases. *GeoInformatica* 3:269–296.

Erwig, M., M. Schneider, and R. H. Güting. 1998. Temporal and spatio-temporal data models and their expressive power. In *Advances in database technologies (ER '98 Workshops on Data Warehousing and Data Mining, Mobile Data Access, and Collaborative Work Support and Spatio-Temporal Data Management), LNCS 1552,* ed. Y. Kambayashi, D. L. Lee, E.-P. Lim, M. K. Mohania, Y. Masunaga, 454–465. Lecture Notes in Computer Science 1552. New York: Springer-Verlag.

ESRI. Environmental Systems Research Institute. 1995. *Arc/Info 7.0.3.* Redlands, Calif.: Environmental Systems Research Institute.

——. 1996a. *Using ArcView GIS: User manual.* Redlands, Calif.: Environmental Systems Research Institute.

——. 1996b. *Using ArcView GIS Spatial Analyst.* Redlands, Calif.: Environmental Systems Research Institute.

——. 1996c. *Using Avenue.* Redlands, Calif.: Environmental Systems Research Institute.

——. 1997. *Understanding GIS: The Arc/Info method. Version 7.1 for Unix and Windows NT.* 4th Edition. Redlands, Calif.: Environmental Systems Research Institute.

——. 1998. *ARC/INFO Version 7.2.* 1982–1998. Redlands, Calif.: Environmental Systems Research Institute.

Everitt, B. S. 1977. *The analysis of contingency tables.* London: Chapman and Hall.

Eyre, M. D., and S. P. Rushton. 1989. Quantification of conservation criteria using invertebrates. *Journal of Applied Ecology* 26:159–171.

Faaborg, J., M. Brittingham, T. Donovan, and J. Blake. 1995. Habitat fragmentation in the temperate zone. In *Ecology and management of Neotropical migratory birds,* ed. T. E. Martin and D. M. Finch, 357–380. Oxford: Oxford University Press.

Fagan, W. F., R. S. Cantrell, and C. Cosner. 1999. How habitat edges change species interactions. *American Naturalist* 153:165–182.

Fagan, W. F., and P. M. Kareiva. 1997. Using compiled species lists to make biodiversity comparisons among regions: A test case using Oregon butterflies. *Biological Conservation* 80:249–259.

Fahrig, L. 1997. Relative effects of habitat loss and fragmentation on population extinction. *Journal of Wildlife Management* 61:603–610.

Farina, A. 1997. Landscape structure and breeding bird distribution in a sub-Mediterranean ecosystem. *Landscape Ecology* 12:365–378.

Farmer, A. H. and J. A. Wiens. 1999. Models and reality:

Time-energy trade-offs in pectoral sandpiper (*Calidris melanotos*) migration. *Ecology* 80:2566–2580.

Farnsworth, G. L., and T. R. Simons. 1999. Factors affecting nesting success of wood thrushes in Great Smoky Mountains National Park. *Auk* 116:1075–1082.

Fausch, K. D., C. L. Hawkes, and M. G. Parsons. 1988. Models that predict standing crop of stream fish from habitat variables: 1950–85. General Technical Report PNW-GTR-213. Portland, Ore.: USDA Forest Service, Pacific Northwest Research Station.

Fausch, K. D., and M. K. Young. 1995. Evolutionarily significant units and movement of resident stream fishes: A cautionary tale. *American Fisheries Society Symposium* 17:360–370.

Fauth, J. E. 1997. Working toward operational definitions in ecology: Putting the system back into ecosystem. *Bulletin of the Ecological Society of America* 78:295–297.

Fauth, J. E., J. Bernardo, M. Camara, and S. A. McCollum. 1996. Simplifying the jargon of community ecology: A conceptual approach. *American Naturalist* 147:282–286.

Feibelman, T., P. Bayman, and W. G. Cibula. 1994. Length variation in the internal transcribed spacer of ribosomal DNA in chanterelles. *Mycological Research* 98:614–618.

Feller, W. 1968. *An introduction to probability theory and its applications*. 3rd ed. New York: Wiley.

Fels, J. E., and K. C. Matson. 1996. A cognitively-based approach for hydrogeomorphic land classification using digital terrain models. In *Third International Conference/Workshop on Integrating GIS and Environmental Modeling CDROM*. Santa Barbara, Calif.: National Center for Geographic Information and Analysis.

FEMAT. Forest Ecosystem Management Assessment Team. 1993. Forest ecosystem management: An ecological, economic, and social assessment. Report of the Forest Ecosystem Management Assessment Team. Washington, D.C.: U.S. Department of Agriculture, U.S. Department of Commerce, U.S. Department of the Interior, Environmental Protection Agency.

Feminella, J. W., and C. P. Hawkins. 1995. Interactions between stream herbivores and periphyton: A quantitative analysis of past experiments. *Journal of the North American Benthological Society* 14:465–509.

Ferrier, S., and G. Watson. 1996. An evaluation of the effectiveness of environmental surrogates and modelling techniques in predicting the distribution of biological diversity. Consultancy report from New South Wales National Parks and Wildlife Service to the Biodiversity Convention and Strategy Section of the Biodiversity Group. Armidale, New South Wales: Environment Australia.

Ferson, S. 1993. *RAMAS STAGE: Generalized stage-based modeling for population dynamics*. Setauket, N.Y.: Applied Biomathematics.

Fertig, W., W. A. Reiners, and R. L. Hartman. 1998. Gap analysis for plant species. *Gap Analysis Bulletin* 7:24–25.

FGDC. Federal Geographic Data Committee. 1997. Vegetation classification standard, FGDC-STD-005. http://www.fgdc.gov/standards/status/sub2_1.html.

Fielding, A. H. 1999. An introduction to machine learning methods. In *Machine learning methods for ecological applications*, ed. A. Fielding, 1–35. Boston: Kluwer Academic Publishers.

Fielding, A. H., and J. F. Bell. 1997. A review of methods for the assessment of prediction errors in conservation presence/absence models. *Environmental Conservation* 24:38–49.

Fielding, A. H., and P. F. Haworth. 1995. Testing the generality of bird-habitat models. *Conservation Biology* 9:1466–1481.

Finch, D. M. 1991. Population ecology, habitat requirements, and conservation of Neotropical migratory birds. General Technical Report RM-205. Fort Collins: USDA Forest Service, Rocky Mountain Forest and Range Experiment Station.

Finch, D. M., and P. W. Stangel. 1993. Status and management of Neotropical migratory birds. General Technical Report RM-229. Fort Collins: USDA Forest Service, Rocky Mountain Forest and Range Experiment Station.

Findley, J. S., A. H. Harris, D. E. Wilson, and C. Jones. 1975. *Mammals of New Mexico*. Albuquerque: University of New Mexico Press.

Fingleton, B. 1983. Log-linear models with dependent spatial data. *Environment and Planning A* 15:801–813.

Fisher, C., and J. Acorn. 1998. *Birds of Alberta*. Edmonton: Lone Pine Publishing.

Fisher, P. 1997. The pixel: A snare and a delusion. *International Journal of Remote Sensing* 18:679–685.

Fisher, R. A. 1924. The conditions under which χ^2 measures the discrepancy between observation and hypothesis. *Journal of the Royal Statistical Society Series A* 87:442–50.

———. 1950. The significance of deviations from expectation in a Poisson series. *Biometrics* 6:17–24.

Fitzgerald, R. W., and B. G. Lees. 1992. The application of neural networks to the floristic classification of remote sensing and GIS data in complex terrain. In *Proceedings of Seventeenth International Archives of the International Society for Photogrammetry and Remote Sensing*, ed. L. W. Fritz and J. R. Lucas, 570–582. Commission 7. Bethesda, Md.: Committee of the Seventeenth International Congress for Photogrammetry and Remote Sensing.

Flather, C. H., and R. M. King. 1992. Evaluating performance of regional wildlife habitat models: Implications to resource planning. *Journal of Environmental Management* 34:31–46.

Flather, C. H., K. R. Wilson, D. J. Dean, and W. C. McComb. 1997. Identifying gaps in conservation networks: Of indicators and uncertainty in geographic-based analyses. *Ecological Applications* 7:531–542.

Flebbe, P. A. 1994. A regional view of the margin: Salmonid abundance and distribution in the southern Appalachian Mountains of North Carolina and Virginia. *Transactions of the American Fisheries Society* 123:657–667.

Fleishman, E. 1997. Mesoscale patterns in butterfly communities of the central Great Basin and their implications for conservation. Ph.D. dissertation, University of Nevada, Reno.

Fleishman, E., G. T. Austin, and A. D. Weiss. 1998. An empirical test of Rapoport's rule: Elevational gradients in montane butterfly communities. *Ecology* 79:2482–2493.

Fleishman, E., B. G. Jonsson, and P. Sjögren-Gulve. 2000. Focal species modeling for biodiversity conservation. In *The use of population viability analyses in conservation planning*, ed. P. Sjögren-Gulve and T. Ebenhard. *Ecological Bulletins* 85–99.

Fleishman, E., A. E. Launer, K. R. Switky, and U. Yandell. 1996. Development of a long-term monitoring plan for the endangered plant *Cordylanthus palmatus*. In *Selected proceedings of the 1994 Conference of the Society of Wetland Scientists, western chapter*, ed. D. M. Kent, J. J. Zentner, and K. D. Whitney, 45–57. Berkeley, Calif.: Society of Wetland Scientists, Western Chapter.

Fleishman, E., R. Mac Nally, J. P. Fay, and D. D. Murphy. 2001. Modeling and predicting species occurrences using broad-scale environmental variables: An example with butterflies of the Great Basin. *Conservation Biology* 15: in press.

Fogel, D. B. 2000. *Evolutionary computation: Toward a new philosophy of machine intelligence.* New York: IEEE Press.

Foin, T. C., and J. L. Brenchley-Jackson. 1991. Simulation model evaluation of potential recovery of endangered light-footed clapper rail populations. *Biological Conservation* 58:123–148.

Foody, G. M., and D. P. Cox. 1994. Sub-pixel land cover composition estimation using a linear mixture model and fuzzy membership functions. *International Journal of Remote Sensing* 15:619–631.

Forbes, A. D. 1995. Classification algorithm evaluation: Five performance measures based on confusion matrices. *Journal of Clinical Monitoring* 11:189–206.

Forde, J. D., N. F. Sloan, and D. A. Shown. 1984. Grassland habitat management using prescribed burning in Wind Cave National Park, South Dakota. *Prairie Naturalist* 16:97–110.

Foreman, D. B., and N. G. Walsh, eds. 1993. *Flora of Victoria. Vol. 1.* Melbourne: Inkata Press.

Forman, R. T. T., and M. Godron. 1981. Patches and structural components for a landscape ecology. *BioScience* 31:733–740.

———. 1986. *Landscape ecology.* New York: John Wiley & Sons.

Forrester, N. D., F. S. Guthery, S. D. Kopp, and W. E. Cohen. 1998. Operative temperature reduced habitat space for northern bobwhites. *Journal of Wildlife Management* 62:1506–1511.

Forsman, E. D. 1981. Molt of the spotted owl. *Auk* 98:735–742.

———. 1983. Methods and materials for locating and studying spotted owls. General Technical Report PNW-GTR-162. Portland: USDA Forest Service, Pacific Northwest Forest and Range Experiment Station.

Forsman, E. D., S. DeStafano, M. G. Raphael, and R. J. Gutiérrez, eds. 1996a. Demography of the northern spotted owl. *Studies in Avian Biology* 17. Los Angeles: Cooper Ornithological Society.

Forsman, E. D., S. G. Sovern, D. E. Seaman, K. J. Maurice, M. Taylor, and J. Zisa. 1996b. Demography of the northern spotted owl on the Olympic Peninsula and east slope of the Cascade Range, Washington. *Studies in Avian Biology* 17:21–30.

Forsyth, R. 1981. BEAGLE—A Darwinian approach to pattern recognition. *Kybernetes* 10:159–166.

Fortin, M., P. Drapeau, and P. Legendre. 1989. Spatial autocorrelation and sampling design in plant ecology. *Vegetatio* 83:209–222.

Fortin, M. J. 1999. Spatial statistics in landscape ecology. In *Landscape ecological analysis: Issues and applications*, ed. J. M. Klopatek and R. H. Gardner, 253–279. New York: Springer-Verlag.

Forys, E. A., and S. R. Humphrey. 1999. Use of population viability analysis to evaluate management options for the endangered lower Keys marsh rabbit. *Journal of Wildlife Management* 63:251–260.

Foss, C. R., ed. 1994. *Atlas of breeding birds in New Hampshire*. Dover, N.H.: Audubon Society of New Hampshire and Arcadia Publishing.

Fotheringham, A. S., M. Charlton, and C. Brunsdon. 1996. The geography of parameter space: An investigation of spatial non-stationarity. *International Journal of Geographic Information Systems* 10:605–627.

Fox, J. F. 1979. Intermediate disturbance hypothesis. *Science* 204:1344–1345.

Frankel, S. J. 1998. User's guide to the western root disease model, version 3.0. USDA Forest Service General Technical Report PSW-GTR-165. Albany: USDA Forest Service, Pacific Southwest Research Station.

Franklin, A. B. 1997. Factors affecting temporal and spatial variation in northern spotted owl populations in northwest California. Ph.D. dissertation, Colorado State University, Fort Collins.

Franklin, A. B., D. R. Anderson, R. J. Gutiérrez, and K. P. Burnham. 2000. Climate, habitat quality, and fitness in northern spotted owl populations in northwestern California. *Ecological Monographs* 70:539–590.

Franklin, J. 1995. Predictive vegetation mapping: Geographic modelling of biospatial patterns in relation to environmental gradients. *Progress in Physical Geography* 19:474–499.

———. 1998. Predicting the distribution of shrub species in southern California from climate and terrain-derived variables. *Journal of Vegetation Science* 9:733–748.

Franklin, J., T. Keeler-Wolf, K. Thomas, D. Shaari, P. Stine, J. Michaelsen, and J. Miller. In press. Stratified sampling for field survey of environmental gradients in the Mojave Desert ecoregion. In *GIS and remote sensing applications in biogeography and ecology*, ed. A. Millington, S. Walsh, and P. Osborne. Dordrecht, Netherlands: Kluwer Academic Publishers.

Franklin, J. F. 1988. Structural and functional diversity in temperature forests. In *Biodiversity*, ed. E. O. Wilson, 166–175. Washington, D.C.: National Academy Press.

Fransson, T. 1998. Physiology and stopover: Patterns of migratory fueling in whitethroats *Sylvia communis* in relation to departure. *Journal of Avian Biology* 29:569–573.

Fransson, T., and S. Jakobsson. 1998. Fat storage in male willow warblers in spring: Do residents arrive lean or fat? *Auk* 115:759–763.

Franzreb, K. E., and S. A. Laymon. 1993. A reassessment of the taxonomic status of the yellow-billed cuckoo. *Western Birds* 24:17–28.

Fraser, D. F., J. F. Gilliam, and T. Yip-Hoi. 1995. Predation as an agent of population fragmentation in a tropical watershed. *Ecology* 76:1461–1472.

Fraser, D. F., and T. E. Sise. 1980. Observations of stream minnows in a patchy environment: A test of a theory of habitat distribution. *Ecology* 61:790–797.

Freeman, M. C., M. K. Crawford, J. C. Barrett, D. E. Facey, M. G. Flood, J. Hill, D. J. Strouder, and G. D. Grossman. 1988. Fish assemblage stability in a southern Appalachian stream. *Canadian Journal of Fisheries and Aquatic Sciences* 45:1949–1958.

Freitag, S., A. S. van Jaarsveld, and H. C. Biggs. 1997. Ranking priority biodiversity areas: An iterative conservation value-based approach. *Biological Conservation* 82:263–272.

Fretwell, S. D. 1972. *Populations in a seasonal environment*. Princeton: Princeton University Press.

Fretwell, S. D., and H. L. Lucas Jr. 1969. On territorial behaviour and other factors influencing habitat distribution in birds. I. Theoretical development. *Acta Biotheoretica* 19:16–36.

Friedmann, H. 1963. Host relations of the Parasitic Cowbirds. Bulletin of the U.S. National Museum 233.

Frissell, C. A., W. J. Liss, C. E. Warren, and M. D. Hurley. 1986. A hierarchical framework for stream habitat classification: Viewing streams in a watershed context. *Environmental Management* 10:199–214.

Fritzell, E. K. 1987. Gray fox and island gray fox. In *Wild furbearer management and conservation in North America*, ed. M. Novak, J. Baker, M. Obbard, B. Malloch, 408–421. Ontario: Ministry of Natural Resources.

Fukunaga, K., and D. Kessell. 1971. Estimation of classification error. *IEEE Transactions on Computers* C-20: 1521–1527.

Fuller, T. K. 1989. Population dynamics of wolves in north-central Minnesota. *Wildlife Monographs* 105:1–41.

Fung, T., and E. LeDrew. 1987. Application of principal components analysis to change detection. *Photogrammetric Engineering and Remote Sensing* 53:1649–1658.

Gaff, H. D. 1999. Spatial heterogeneity in ecological models: Two case studies. Ph.D. dissertation, University of Tennessee, Knoxville.

Gaff, H. D., D. L. DeAngelis, L. J. Gross, R. Salinas, and M. Shorrosh 2000. A dynamic landscape model for fish in the Everglades and its application to restoration. *Ecological Modelling* 127:33–52.

Gaines, D. A. 1970. The nesting riparian avifauna of the Sacramento Valley, California, and the status of the yellow-billed cuckoo. M.S. thesis, University of California, Davis.

———. 1974. Review of the status of the yellow-billed cuckoo in California: Sacramento Valley populations. *Condor* 76:204–209.

———. 1977. The valley riparian forests of California: Their importance to bird populations. In *Riparian forests in California: Their ecology and conservation*, ed. A. Sands, 57–73. Publication 4101. Berkeley: Division of Agricultural Sciences, University of California.

Gaines, D. A., and S. A. Laymon. 1984. Decline, status and preservation of the yellow-billed cuckoo in California: Sacramento Valley populations. *Western Birds* 15:49–80.

Ganey, J. L., and W. M. Block. 1994. A comparison of two techniques for measuring canopy closure. *Western Journal of Applied Forestry* 9:21–23.

Ganey, J. L., and J. L. Dick Jr. 1995. Habitat relationships of the Mexican spotted owl: Current knowledge. Chapter 4 in *Recovery plan for the Mexican spotted owl*. Vol. 2. *USDI Fish and Wildlife Service*, 1–42. Albuquerque, New Mexico.

Gaona, P., P. Ferreras, and M. Delibes. 1998. Dynamics and viability of a metapopulation of the endangered Iberian lynx (*Lynx pardinus*). *Ecological Monographs* 68:349–370.

Gap Analysis Program. 1998. *A handbook for gap analysis.* Moscow: USGS Gap Analysis Program. http://www.gap.uidaho.edu/handbook/default.htm.

Gardes, M., and T. D. Bruns. 1996. Community structure of ectomycorrhizal fungi in a *Pinus muricata* forest: Above- and below-ground views. *Canadian Journal of Botany* 74:1572–1583.

Gardner, R. H., W. G. Cale, and R. V. O'Neill. 1982. Robust analysis of aggregation error. *Ecology* 63:1771–1779.

Garrett, K. L., M. G. Raphael, and R. D. Dixon. 1996. White-headed woodpecker (*Picoides albolarvatus*). In *The birds of North America*, ed. A. Poole and F. Gill, 1–24. No. 252. Philadelphia: Academy of Sciences and Washington: American Ornithologists' Union.

Garrison, B. A. 1993. Validation studies of the California wildlife habitat relationships system. *Fish and Wildlife Information Exchange Newsletter* 2:2–5.

Garrison, B. A., R. A. Erickson, M. A. Patten, and I. C. Timossi. 1999. California wildlife habitat relationships system: Effects of county attributes on prediction accuracy for bird species. *California Fish and Game* 85:87–101.

Garrison, B. A., and K. J. Sernka. 1997. User's manual for version 6.0 of the California wildlife habitat relationships

system database. CWHR Technical Report 29. Sacramento: California Department of Fish and Game.

Garshelis, D. L., and L. G. Visser. 1997. Enumerating megapopulations of wild bears with an ingested biomarker. *Journal of Wildlife Management* 61:466–480.

Garshelis, D. L. 2000. Delusions in habitat evaluation: Measuring use, selection, and importance. In *Research techniques in animal ecology: Controversies and consequences*, ed. L. Boitani and T. K. Fuller, 111–164. New York: Columbia University Press.

Gascoigne, J., and R. Wadsworth. 1999. Mapping misgivings: Monte Carlo modeling of uncertainty and the provision of spatial information for international policy. In *Spatial accuracy assessment: Land information uncertainty in natural resources*, ed. K. Lowell and A. Jaton, 53–60. Chelsea, Mich.: Ann Arbor Press.

Gascon, C., T. E. Lovejoy, R. O. Bierregaard, J. R. Malcolm, P. C. Stouffer, H. L. Vasconcelos, W. F. Laurence, B. Zimmerman, M. Tocher, and S. Borges. 1999. Matrix habitat and species richness in tropical forest remnants. *Biological Conservation* 91:223–229.

Gates, D. M. 1993. *Climate change and its biological consequences.* Sunderland, Mass.: Sinauer Associates.

Gates, S., D. W. Gibbons, P. C. Lack, and R. J. Fuller. 1994. Declining farmland bird species: Modelling geographical patterns of abundance in Britain. In *Large-scale ecology and conservation biology*, ed. P. J. Edwards, R. M. May, and N. R. Webb, 153–177. London: Blackwell Scientific Publications.

Gauch, H. G. 1982. *Multivariate analysis in community ecology.* Cambridge: Cambridge University Press.

———. 1993. Prediction, parsimony, and noise. *American Scientist* 81:468–478.

Gauch, H. G., Jr., and R. H. Whittaker. 1972. Coencline simulation. *Ecology* 53:446–451.

Gauch, H. G., Jr., and G. B. Chase. 1974. Fitting the Gaussian curve to ecological data. *Ecology* 55:1377–1381.

Gauthier, J., and Y. Aubry. 1996. *The Breeding birds of Quebec: Atlas of the breeding birds of southern Quebec.* Montreal: Province of Quebec Society for the Protection of Birds and the Canadian Wildlife Service.

Gauthreaux, S. A., Jr. 1971. A radar and direct visual study of passerine spring migration in Southern Louisiana. *Auk* 88:343–365.

Geist, D. R., and D. D. Dauble. 1998. Redd site selection and spawning habitat use by fall chinook salmon: The importance of geomorphic features in large rivers. *Environmental Management* 22:655–669.

Gelfand, A. E., S. E. Hills, A. Racine-Poon, and A. F. M. Smith. 1990. Illustration of Bayesian inference in normal data models using Gibbs sampling. *Journal of the American Statistical Association* 85:972–985.

Gelfand, A. E., and A. F. M. Smith. 1990. Sampling-based approaches to calculating marginal densities. *Journal of the American Statistical Association* 85:398–409.

Gelman, A., J. B. Carlin, H. S. Stern, and D. B. Rubin. 1995. *Bayesian data analysis.* London: Chapman and Hall.

Geman, S., and D. Geman. 1984. Stochastic relaxation, Gibbs distributions, and the Bayesian restoration of images. *IEEE Transactions on Pattern Analysis and Machine Intelligence* 6:721–741.

Gentry, R. C. 1984. Hurricanes in south Florida. In *Environments of south Florida: Past and present 2*, ed. P. J. Gleason, 510–519. Coral Gables, Fla.: Miami Geological Society.

Geographic Resource Solutions. 1997. *The GRS densitometer.* Arcata, Calif.: Geographic Resource Solutions.

Geological Survey of Wyoming. 1991. Wyoming geologic highway map. Canon City, Colo.: Western Geographics.

Gerking, S. D. 1959. The restricted movement of fish populations. *Biological Review* 34:221–242.

Géroudet, P., C. Guex, M. Maire, et callaborateurs. 1983. *Les Oiseaux Nicheurs du Canton de Genève.* Suisse: Muséum de Genève.

Gibbs, J. P. 1998. Distribution of woodland amphibians along a forest fragmentation gradient. *Landscape Ecology* 13:263–268.

Gilks, W. R. 1996. Full conditional distributions. In *Markov Chain Monte Carlo in practice*, ed. W. R. Gilks, S. Richardson, and D. J. Spiegelhalter, 75–87. London: Chapman and Hall.

Gilpin, M. E. 1987. Spatial structure and population vulnerability. In *Viable populations for conservation*, ed. M. E. Soulé, 125–139. Cambridge: Cambridge University Press.

Gilpin, M. E., and M. E. Soulé. 1986. Minimum viable populations: The processes of species extinctions. In *Conservation biology: The science of scarcity and diversity*, ed. M. E. Soulé, 13–34. Sunderland, Mass.: Sinauer Associates.

Girman, D. 2000. Phylogeography of the Wilson's warbler on breeding and wintering grounds: Evolutionary and conservation implications. Master's thesis in progress, San Francisco State University, California.

Giuliano, W. M., and R. S. Lutz. 1993. Quail and rain: What's the relationship? In *Quail 3: National Quail Symposium*, ed. K. E. Church and T. V. Dailey, 64–68. Pratt, Kans.: Kansas Department of Wildlife and Parks.

Glayre, D., and D. Magnenat. 1984. *Oiseaux Nicheurs de la Haute Vallée de l'Orbe.* Nos Oiseaux 398. Fascicule spécial du volume 37.

Godown, M. E., and A. I. Peterson. 2000. Preliminary distributional analysis of U.S. endangered bird species. *Biodiversity and Conservation* 9:1313–1322.

Goldingay, R., and H. Possingham. 1995. Area requirements for viable populations of the Australian gliding marsupial (*Petaurus australis*). *Biological Conservation* 73:161–167.

Goldstein, P. Z. 1999. Clarifying the role of species in ecosystem management: A reply. *Conservation Biology* 13:1515–1517.

Gonzalez, R., and T. O. Nelson. 1996. Measuring ordinal association in situations that contain tied scores. *Psychological Bulletin* 119:159–165.

Gonzalez-Rebeles, C. 1996. A sensitivity test for species distribution models used for gap analysis in New Mexico. Ph.D. dissertation. Texas Tech University, Lubbock.

Goodchild, M. F. 1997. Scale in a digital geographic world. *Geographical and Environmental Modelling* 1:5–23.

———. 2000. Communicating geographic information in a digital age. *Annals of the Association of American Geographers* 90:344–355.

Goodchild, M., and S. Gopal, eds. 1989. *The accuracy of spatial databases.* London: Taylor and Francis.

Goodchild, M. F., B. O. Parks, and L. T. Steyaert, eds. 1993. *Environmental modeling with GIS.* New York: Oxford University Press.

Goodchild, M. F., L. T. Steyaert, B. O. Parks, C. Johnston, D. Maidment, M. Crane, and S. Glendinning. 1996. *GIS and environmental modeling: Progress and research issues.* Fort Collins, Colo.: GIS World Books.

Goodman, D. 1987. The demography of chance extinction. In *Viable populations for conservation*, ed. M. E. Soulé, 2–34. Cambridge: Cambridge University Press.

Goodman, L. A., and W. H. Kruskal. 1954. Measures of association for cross-classifications. *Journal of the American Statistical Association* 86:1085–1111.

Goodwin, B. J., and L. Fahrig. 1998. Spatial scaling and animal population dynamics. In *Ecological scale: Theory and applications*, ed. D. L. Peterson and V. T. Parker, 193–206. New York: Columbia University Press.

Gordon, I. J. 1989. Vegetation community selection by ungulates on the Isle of Rhum. I. Food supply. *Journal of Applied Ecology* 26: 35–51.

Gorman, O. T. 1986. Assemblage organization of stream fishes: The effect of rivers on adventitious streams. *American Naturalist* 128:611–616.

Goss-Custard, J. D., R. W. G. Caldow, R. T. Clarke, S. E. A. le V. dit Durell, J. Urfi, and A. D. West. 1994. Consequences of habitat loss and change to populations of wintering migratory birds: Predicting the local and global effects from studies of individuals. *Ibis* (suppl.) 137: S56–S66.

Gotelli, N. J., and G. R. Graves. 1996. *Null models in ecology*. Washington: Smithsonian Institution Press.

Gowan, C., and K. D. Fausch. 1996a. Long-term demographic responses of trout populations to habitat manipulations in six Colorado streams. *Ecological Applications* 6:931–946.

_____. 1996b. Mobile brook trout in two high-elevation Colorado streams: Re-evaluating the concept of restricted movement. *Canadian Journal of Fisheries and Aquatic Sciences* 53:1370–1381.

Gowan, C., M. K. Young, K. D. Fausch, and S. C. Riley. 1994. Restricted movement in resident stream salmonids: A paradigm lost? *Canadian Journal of Fisheries and Aquatic Sciences* 51:2626–2637.

Goward, S. N., and D. L. Williams. 1997. Landsat and earth systems science: Development of terrestrial monitoring. *Photogrammetric Engineering and Remote Sensing* 63: 887–900.

Grace, J. B. 1999. The factors controlling species density in herbaceous plant communities: An assessment. *Perspectives in Plant Ecology, Evolution, and Systematics* 2: 1–28.

Graetz, D. 1990. Remote sensing of terrestrial ecosystem structure: An ecologist's pragmatic view. In *Remote sensing of biosphere functioning*, ed. R. J. Hobbs and H. A. Mooney, 5–30. New York: Springer-Verlag.

Grant, W. E. 1986. *Systems analysis and simulation in wildlife and fisheries sciences*. New York: John Wiley & Sons.

Grayson, D. K. 1993. *The desert's past: A natural prehistory of the Great Basin*. Washington, D.C.: Smithsonian Institution Press.

Greco, S. E. 1999. Monitoring riparian landscape change and modeling habitat dynamics of the yellow-billed cuckoo on the Sacramento River, California. Ph.D. dissertation, University of California, Davis.

Green, D. M. 1997. Perspectives on amphibian population declines: Defining the problem and searching for answers. In *Amphibians in decline: Canadian studies of a global problem*, ed. D. M. Green, 291–308. Herpetological Conservation 1. St. Louis, Mo.: Society for the Sudy of Amphibians and Reptiles.

Green, K., S. Bernath, L. Lackey, M. Brunengo, and S. Smith. 1993. Analyzing the cumulative effects of forest practices: Where do we start? *GeoInfo Systems* 3:31–41.

Green, K. M., J. Lynch, J. Sircar, and L. Greenberg. 1987. Landsat remote sensing to assess habitat for migratory birds in the Yucatan Peninsula, Mexico. *Vida Silvestre Neotropical* 1:27–38.

Green, R. E., M. W. Pienkowski, and J. A. Love. 1996. Long-term viability of the re-introduced population of the white-tailed eagle *Haliaeetus albicilla* in Scotland. *Journal of Applied Ecology* 33:357–368.

Green, R. H. 1979. *Sampling design and statistical methods for environmental biologists*. New York: John Wiley & Sons.

Greenberg, R. 1980. Demographic aspects of long-distance migration. In *Migrant birds in the Neotropics*, ed. A Keast and E. Morton, 493–504. Washington, D.C.: Smithsonian Institute Press.

Greenhouse, L. 1997. Court ruling expands list of species act challengers: Economic interest allowed as basis for suits. *New York Times* March 20, 146: A11, B10.

Gremmen, N. J. M., M. J. S. M. Reijnen, J. Wiertz, and G. van Wirdum. 1990. A model to predict and assess the effects of ground-water withdrawal on vegetation in the Pleistocene areas of the Netherlands. *Journal of Environmental Management* 31:143–155.

Griffith, D. A., R. P. Haining, and G. Arbia. 1999. Uncertainty and error propagation in map analyses involving arithmetic and overlay operations: Inventory and prospects. In *Spatial accuracy assessment: Land information uncertainty in natural resources*, ed. K. Lowell and A. Jaton, 11–25. Chelsea, Mich.: Ann Arbor Press.

Grime, J. P. 1973a. Competitive exclusion in herbaceous vegetation. *Nature* 242:344–347.

_____. 1973b. Control of species density in herbaceous vegetation. *Journal of Environmental Management* 1: 151–167.

_____. 1979. *Plant strategies and vegetation processes*. New York: Wiley.

Grimm, V. 1999. Ten years of individual-based modelling in ecology: What have we learned and what could we learn in the future? *Ecological Modelling* 115:129–148.

Grinnell, J. 1917. The niche-relationships of the California thrasher. *Auk* 34:427–433.

Grinnell, J., and A. H. Miller. 1944. *The distribution of the birds of California.* Pacific Coast Avifauna, Number 27. Berkeley, Calif.: Cooper Ornithological Club.

Gross, A. O. 1944a. The present status of the double-crested cormorant on the coast of Maine. *Auk* 61:513–537.

_____. 1944b. The present status of the American eider on the Maine coast. *Wilson Bulletin* 56:15–26.

_____. 1945. The present status of the great black-backed gull on the coast of Maine. *Auk* 62:241–257.

Grossman, D. H., D. Faber-Langendoen, A. S. Weakley, M. Anderson, P. Bourgeron, R. Crawford, K. Goodin, S. Landaal, K. Metzler, K. D. Patterson, M. Pyne, M. Reid, and L. Sneddon. 1998. *International classification of ecological communities: Terrestrial vegetation of the United States.* Vol. 1, The National Vegetation Classification System: Development, Status, and Applications. Arlington, Va.: The Nature Conservancy.

Grubb, T. G. 1988. Pattern recognition—a simple model for evaluating wildlife habitat. Research Note RM-487. Fort Collins, Colo.: USDA Forest Service, Rocky Mountain Forest and Range Experiment Station.

Gruebele, M. J., and A. A. Steuter. 1988. South Dakota records of pygmy and arctic shrews: Response to fire. *Prairie Naturalist* 20:95–98.

Guégan, J. F., S. Lek, and T. Oberdorff. 1998. Energy availability and habitat heterogeneity predict global riverine fish diversity. *Nature* 391:382–384.

Guerry, A. D. 2000. Amphibian distributions in an agricultural and forested mosaic: Forest extent, isolation, and landscape context. M.S. thesis, University of Maine, Orono.

Guikema, S., and M. Milke. 1999. Quantitative decision tools for conservation programme planning: Practice, theory, potential. *Environmental Conservation* 26:179–189.

Guisan, A. 1997. Distribution de taxons végétaux dans un environnement alpin: Application de modélisations statistiques dans un système d'information géographique. Thèse de doctorat. No. 2892. Université de Genève, Suisse.

Guisan, A., J.-P. Theurillat, and F. Kienast. 1998. Predicting the potential distribution of plant species in an alpine environment. *Journal of Vegetation Science* 9:65–74.

Guisan, A., S. B. Weiss, and A. D. Weiss. 1999. GLM versus CCA spatial modeling of plant species distribution. *Plant Ecology* 143:107–122.

Guisan, A., and F. E. Harrell. 2000. Ordinal response regression models in ecology. *Journal of Vegetation Science* 11:617–626.

Guisan, A., and J.-P. Theurillat. 2000. Equilibrium modeling of alpine plant distribution and climate change: How far can we go? *Phytocoenologia* 30:353–384.

Guisan, A., and N. E. Zimmermann. 2000. Predictive habitat distribution models in ecology. *Ecological Modelling* 135:147–186.

Gumpertz, M. L., J. M. Graham, and J. B. Ristaino. 1997. Autologistic model of spatial pattern of *Phytophthora* epidemic in bell pepper: Effects of soil variables on disease presence. *Journal of Agricultural, Biological, and Environmental Statistics* 2:131–156.

Gunderson, L. 1999. Resilience, flexibility, and adaptive management: Antidotes for spurious certitude? *Conservation Ecology* [online] 3:7. http://www.consecol.org/Journal/vol3/iss1/art7.

Gunderson, L. H., C. S. Holling, and S. S. Light. 1995. *Barriers and bridges to the renewal of ecosystems and institutions.* New York: Columbia University Press.

Guo, Q., and W. L Berry. 1998. Species richness and biomass: Dissection of the hump-shaped relationships. *Ecology* 79:2555–2559.

Gustafson, E. J. 1998. Quantifying landscape spatial pattern: What is the state of the art? *Ecosystems* 1:143–156.

Gustafson, E. J., and T. R. Crow. 1994. Modeling the effects of forest harvesting on landscape structure and the spatial distribution of cowbird brood parasitism. *Landscape Ecology* 9:237–248.

Gustafson, E. J., G. R. Parker, and S. E. Backs. 1994. Evaluating spatial pattern of wildlife habitat: A case study of the wild turkey (*Meleagris gallopavo*). *American Midland Naturalist* 131:24–33.

Guthery, F. S. 1997. A philosophy of habitat management for northern bobwhites. *Journal of Wildlife Management* 61:291–301.

Guthery, F. S., and R. L. Bingham. 1992. On Leopold's principle of edge. *Wildlife Society Bulletin* 20:340–344.

Guthery, F. S., N. D. Forrester, K. R. Nolte, W. E. Cohen, and W. P. Kuvlesky. 2000. Potential effects of global warming on quail populations. *Proceedings of the National Bobwhite Quail Symposium* 4:198–204.

Gutiérrez, R. J., A. B. Franklin, and W. S. LaHaye. 1995. Spotted owl (*Strix occidentalis*). In *The birds of North America,* No. 179, ed. A. Poole and F. Gill. Philadelphia: Academy of Natural Sciences, and Washington: American Ornithologists' Union.

Gutiérrez, R. J., J. E. Hunter, G. Chavez-Leon, and J. Price. 1998. Characteristics of spotted owl habitat in landscapes disturbed by timber harvest in northwestern California. *Journal of Raptor Research* 32:104–110.

Gutzwiller, K. J., K. R. Kinsley, G. L. Storm, W. M. Tzilkowski, and J. S. Wakeley. 1983. Relative value of vegetation structure and species composition for identifying American woodcock breeding habitat. *Journal of Wildlife Management* 47:535–540.

Hacking, I. 1975. *The emergence of probability*. Cambridge: Cambridge University Press.

Haddad, N. M., and K. A. Baum. 1999. An experimental test of corridor effects on butterfly densities. *Ecological Applications* 9:623–633.

Haeuber, R. 1996. Setting the environmental policy agenda: The case of ecosystem management. *Natural Resources Journal* 36:1–28.

Hafernik, J. E., Jr. 1992. Threats to invertebrate biodiversity: Implications for conservation strategies. In *Conservation biology: The theory and practice of nature conservation, preservation, and management*, ed. P. L. Fiedler and S. K. Jain, 172–195. New York: Chapman and Hall.

Hagan, J. M., and D. W. Johnston. 1992. *Ecology and conservation of Neotropical migrant landbirds*. Washington, D.C.: Smithsonian Institution Press.

Hagan, J. M., P. S. McKinley, A. L. Meehan, and S. L. Grove. 1997. Diversity and abundance of landbirds in northeastern industrial forest. *Journal of Wildlife Management* 61:718–735.

Hagan, M. T., H. B. Demuth, and M. Beale. 1996. *Neural network design*. Boston: PWS.

Hågvar, S., G. Hågvar, and E. Monness. 1990. Nest site selection in Norwegian woodpeckers. *Holarctic Ecology* 13:156–165.

Haig, S. M., J. R. Belthoff, and D. H. Allen. 1993. Population viability analysis for a small population of red-cockaded woodpeckers and an evaluation of enhancement strategies. *Conservation Biology* 7:289–301.

Haila, Y. I., I. K. Hanski, and S. Raivio. 1993. Turnover of breeding birds in small forest fragments: The "sampling" colonization hypothesis corroborated. *Ecology* 74: 714–725.

Haines-Young, R., D. R. Green, and S. H. Cousins. 1993. Landscape ecology and spatial information systems. In *Landscape ecology and GIS*, ed. R. Haines-Young, D. R. Green, and S. H. Cousins, 3–8. London: Taylor & Francis.

Haining, R., and G. Arbia. 1993. Error propagation through map operations. *Technometronics* 35:293–305.

Hall, F. G., D. B. Botkin, D. E. Strebel, K. D. Woods, and S. J. Goetz. 1991. Large-scale patterns of forest succession as determined by remote sensing. *Ecology* 72:628–640.

Hall, F. G., D. E. Knapp, and K. F. Huemmrich. 1997a. Physically based classification and satellite mapping of biophysical characteristics in the southern boreal forest. *Journal of Geophysical Research* 102 (D24):29, 567–29, 580.

Hall, L. S., P. R. Krausman, and M. L. Morrison. 1997b. The habitat concept and a plea for standard terminology. *Wildlife Society Bulletin* 25:173–182.

Haller, H. 1992. Zur ökologie des Luchses (*Lynx lynx*) im Verlauf seiner Wiederansiedlung in den Walliser Alpen. *Mammalia Depicta* 15:1–62.

Halpin, P. N. 1995. A cross-scale analysis of environmental gradients and forest patterns in the giant sequoia–mixed conifer forest of the Sierra Nevada. Ph.D. dissertation, University of Virginia, Charlottesville.

Halterman, M. D. 1991. Distribution and habitat use of the yellow-billed cuckoo (*Coccyzus americanus occidentalis*) on the Sacramento River, California, 1987–90. M.S. thesis, California State University, Chico.

Hamel, P. B. 1992. *Land manager's guide to the birds of the South*. Chapel Hill: Nature Conservancy.

Hamilton, S., and H. Moller. 1995. Can PVA models using computer packages offer useful conservation advice? Sooty shearwaters *Puffinus griseus* in New Zealand as a case study. *Biological Conservation* 73:107–117.

Hand, D. J. 1982. *Kernel discriminant analysis*. New York: Research Studies Press.

———. 1997. *Construction and assessment of classification rules*. London: John Wiley & Sons.

Handcock, M. S., and M. L. Stein. 1993. A Bayesian analysis of kriging. *Technometrics* 35:403–410.

Hankin, D. G., and G. H. Reeves. 1988. Estimating total fish abundance and total habitat area in small streams based on visual estimation. *Canadian Journal of Fisheries and Aquatic Sciences* 45:834–844.

Hanley, J. A., and B. J. McNeil. 1982. The meaning and use of the area under a receiver operating characteristic (ROC) curve. *Radiology* 143:29–36.

———. 1983. A method of comparing the areas under receiver operating characteristic curves derived from the same cases. *Radiology* 148:839–843.

Hanowski, J. M., and G. J. Niemi. 1994. Breeding bird abundance patterns in the Chippewa and Superior National Forests from 1991 to 1993. *The Loon* 66:64–70.

———. 1995. Experimental design considerations for establishing an off-road, habitat-specific bird monitoring program using point counts. In *Monitoring bird populations by point counts*, technical ed. C. J. Ralph, J. R. Sauer, and S. Droege, 145–150. General Technical Report PSW-GTR-149. Albany, Calif.: USDA Forest Service, Pacific Southwest Research Station.

Hansen, A. J., S. L. Garman, B. Marks, and D. L. Urban. 1993. An approach for managing vertebrate diversity across multiple-use landscapes. *Ecological Applications* 3:481–496.

Hansen, A. J., S. L. Garmen, J. F. Weigand, D. L. Urban, W. C. McComb, and M. G. Raphael. 1995. Alternative silvicultural regimes in the Pacific Northwest: Simulations of ecological and economic effects. *Ecological Applications* 5:535–554.

Hansen, A. J., J. J. Rotella, M. P. V. Kraska, and D. Brown. 1999. Dynamic habitat and population analysis: An approach to resolve the biodiversity manager's dilemma. *Ecological Applications* 9:1459–1476.

Hansen, A. J., and D. L. Urban. 1992. Avian response to landscape patterns: The role of species' life histories. *Landscape Ecology* 7:163–180.

Hansen, R. M., and E. E. Remmenga. 1961. Nearest neighbor concept applied to pocket gopher populations. *Ecology* 42:812–814.

Hanski, I. 1982. Dynamics of regional distribution: The core and satellite species hypothesis. *Oikos* 38:210–221.

———. 1994a. A practical model of metapopulation dynamics. *Journal of Animal Ecology* 63:151–162.

———. 1994b. Patch-occupancy dynamics in fragmented landscapes. *Trends in Ecology and Evolution* 9:131–135.

———. 1996. Metapopulation ecology. In *Population dynamics in ecological space and time*, ed. O. E. Rhodes Jr., R. K. Chesser, and M. H. Smith, 13–43. Chicago: University of Chicago.

Hanski, I., and M. Gilpin. 1991. Metapopulation dynamics: Brief history and conceptual domain. *Biological Journal of the Linnean Society* 42:3–16.

Hanksi, I., A. Moilanen, and M. Gyllenberg.1996a. Minimum viable metapopulation size. *American Naturalist* 147:527–541.

Hanski, I., A. Moilanen, T. Pakkala, and M. Kuussaari. 1996b. The quantitative incidence function model and persistence of an endangered butterfly metapopulation. *Conservation Biology* 10:578–590.

Hanski, I., T. Pakkala, M. Kuussaari, and G. Lei. 1995. Metapopulation persistence of an endangered butterfly in a fragmented landscape. *Oikos* 72:21–28.

Hanski, I., and D. Simberloff. 1997. The metapopulation approach, its history, conceptual domain, and application to conservation. In *Metapopulation biology: Ecology, genetics, and evolution*, ed. I. A. Hanski and M. E. Gilpin, 5–26. San Diego: Academic Press.

Hanski, I. A. 1999. *Metapopulation ecology*. Oxford: Oxford University Press.

Hanski, I. A., and M. E. Gilpin, eds. 1997. *Metapopulation biology: Ecology, genetics, and evolution*. San Diego: Academic Press.

Hanson, W. R., and R. J. Miller. 1961. Edge types and abundance of bobwhites in southern Illinois. *Journal of Wildlife Management* 25:71–76.

Hansteen, T. L., H. P. Andreassen, and R. A. Ims. 1997. Effects of spatiotemporal scale on autocorrelation and home range estimators. *Journal of Wildlife Management* 61:280–290.

Harding, J. H. 1997. *Amphibians and reptiles of the Great Lakes region*. Ann Arbor: University of Michigan Press.

Harding, P. T., J. Asher, and T. J. Yates. 1995. Butterfly monitoring 1—recording the changes. In *Ecology and conservation of butterflies*, ed. A. S. Pullin, 3–22. London: Chapman and Hall.

Harestad, A. S., and F. L. Bunnell. 1979. Home range and body weight: A reevaluation. *Ecology* 60:389–402.

Hargis, C. D., J. A. Bissonette, and J. L. David. 1997. Understanding measures of landscape pattern. In *Wildlife and landscape ecology: Effects of pattern and scale*, ed. J. A. Bissonette, 231–261. New York: Springer-Verlag.

Harmon, M. E., and J. Sexton. 1995. Water balance of conifer logs in early stages of decomposition. *Plant and Soil* 172:141–152.

Harper, S. J., J. Westervelt, and A. Trame. 2001. Management application of an agent-based model: Control of cowbirds at the landscape scale. In *Integrating geographic information system and agent-based modeling techniques for simulating social and ecological processes*, ed. H. R. Gimblett. Santa Fe: Santa Fe Institute.

Harrell, F. E. 1999. Design: S-Plus functions for biostatistical/epidemiologic modelling, testing, estimation, validation, graphics, prediction, and typesetting by storing enhanced model design attributes in the fit. Unix and Windows versions available from statlib@lib.stat. cmu.edu. http://www.stat.cmu.edu.

Harris, L. D. 1988. Edge effects and conservation of biotic diversity. *Conservation Biology* 2:330–332.

Harris, S., P. Morris, S. Wray, and D. Yalden. 1995. *A review of British mammals*. Peterborough, UK: Joint Nature Conservation Committee.

Harrison, R. L. 1997. A comparison of gray fox ecology between residential and undeveloped rural landscapes. *Journal of Wildlife Management* 61:112–122.

Harrison, S. 1991. Local extinction in a metapopulation context: An empirical evaluation. *Biological Journal of the Linnean Society* 42:73–88.

Harrison, S. 1993. Metapopulations and conservation. In *Large-scale ecology and conservation biology*, ed. P. J. Edwards, R. M. May, and N. R. Webb, 111–128. London: Blackwell Scientific Publications.

Harrison, S., and L. Fahrig. 1995. Landscape pattern and population conservation. In *Mosaic landscapes and ecological processes*, ed. L. Hansson, L. Fahrig, and G. Merriam, 293–308. New York: Chapman Hall.

Harrison, S., S. J. Ross, and J. H. Lawton. 1992. Beta diversity on geographic gradients in Britain. *Journal of Animal Ecology* 61:151–158.

Harrison, S., and A. D. Taylor. 1997. Empirical evidence for metapopulation dynamics. In *Metapopulation biology: Ecology, genetics, and evolution*, ed. I. A. Hanski and M. E. Gilpin, 27–42. San Diego: Academic Press.

Harte, J., and R. Shaw. 1995. Shifting dominance within a montane vegetation community: Results of a climate-warming experiment. *Science* 267:876–880.

Hartless, C. S. 2000. Modeling spatial use patterns of white-tailed deer in the Florida Everglades. Ph.D. dissertation, University of Florida, Gainesville.

Hartman, R. L. 1992. The Rocky Mountain herbarium, associated floristic inventory, and the flora of the Rocky Mountains project. *Journal Idaho Academy of Science* 28:22–43.

Harvey, A. E., M. F. Jurgensen, and M. J. Larsen. 1978. Seasonal distribution of ectomycorrhizae in a mature Douglas-fir/larch forest soil in western Montana. *Forest Science* 24:203–208.

———. 1981. Organic reserves: Importance to ectomycorrhizae in forest soils of western Montana. *Forest Science* 27:442–445.

Harvey, A. E., M. J. Larsen, and M. F. Jurgensen. 1976. Distribution of ectomycorrhizae in a mature Douglas-

fir/larch forest soil in western Montana. *Forest Science* 22:393–398.

Hassler, C.C., S.A. Sinclair, and E. Kallio. 1986. Logistic regression: A potentially useful tool for researchers. *Forest Products Journal* 36:16–18.

Hastie, T., and R. Tibshirani. 1990. *Generalized additive models*. London: Chapman and Hall.

Hastie, T. J., and D. Pregibon. 1997. Generalized linear models. In *Statistical models in S*, ed. M. J. Chambers and T. J. Hastie, 195–248. London: Chapman and Hall.

Hastings, J. R., and R. M. Turner. 1980. *The changing mile*. Tucson: University of Arizona Press.

Haufler, J. B., T. Crow, and D. Wilcove. 1999a. Scale considerations for ecosystem management. In *Ecological stewardship: A common reference for ecosystem management*, ed. R. C. Szaro, N. C. Johnson, W. T. Sexton, and A. J. Malk, Vol. 2, 331–342. New York: Elsevier Science.

Haufler, J. B., C. A. Mehl, and G. J. Roloff. 1996. Using a coarse-filter approach with species assessment for ecosystem management. *Wildlife Society Bulletin* 24:200–208.

———. 1999b. Conserving biological diversity using a coarse-filter approach with a species assessment. In *Practical approaches to the conservation of biological diversity*, ed. R. K. Baydack, H. Campa, III., and J. B. Haufler, 107–125. Washington, D.C.: Island Press.

Haug, E. A., B. A. Millsap, and M. S. Martell. 1993. Burrowing Owl (*Speotyto cunicularia*). In *The birds of North America*, No. 61, ed. A. Poole and F. Gill. Philadelphia: Academy of Natural Sciences, and Washington: American Ornithologists' Union.

Hawkins, C. P., J. L. Kershner, P. A. Bisson, M. D. Bryant, L. M. Decker, S. V. Gregory, D. A. McCullough, C. K. Overton, G. H. Reeves, R. J. Steedman, and M. K. Young. 1993. A hierarchical approach to classifying stream habitat features. *Fisheries* 18(6):3–12.

Hawkins, C. P., and J. A. MacMahon. 1989. Guilds: The multiple meanings of a concept. *Annual Review of Ecology and Systematics* 56:283–300.

Hawrot, R. Y., J. M. Hanowski, A. R. Lima, G. J. Niemi, and L. Pfannmuller. 1998. Bird population trends in Minnesota and northwestern Wisconsin forests, 1991–1997. *Loon* 70:130–137.

Hayden, T. J., and D. J. Tazik. 1991. Project status report: 1991 field studies of two endangered species (the black-capped vireo and the golden-cheeked warbler) and the cowbird control program on Fort Hood, Texas. Champaign, Ill.: Report submitted to HQIII Corps and Fort

Hood, Texas, U.S. Army Corps of Engineers Construction Engineering Research Lab.

Hayden, T. J., D. J. Tazik, R. H. Melton, and J. D. Cornelius. 2000. Cowbird control program on Fort Hood, Texas: Lessons for mitigation of cowbird parasitism on a landscape scale. In *The ecology and management of cowbirds and their hosts*, ed. J. N. M. Smith, T. L. Cook, S. I. Rothstein, S. K. Robinson, and S. G. Sealy, 357–370. Austin: University of Texas Press.

Hayek, L. C., and M. A. Buzas. 1997. *Surveying natural populations*. New York: Columbia University Press.

Haykin, S. 1999. *Neural networks: A comprehensive foundation*. 2nd ed. Upper Saddle River: Prentice Hall.

HCN. Historical Climatology Network. 1996. Monthly precipitation and temperature data. U.S. Department of Energy, Oak Ridge National Laboratory and National Oceanic and Atmospheric Administration, National Climatic Data Center. http://cdiac.esd.ornl.gov/cdiac/r3d/ushcn/ushcn.html#TEXT.

He, H. S., and D. J. Mladenoff. 1999. Spatially explicit and stochastic simulation of forest-landscape fire disturbance and succession. *Ecology* 80:81–99.

Heady, E. O., J. T. Pesek, and W. G. Brown. 1955. Crop response surfaces and economic optima in fertilizer use. Research Bulletin 424. Ames: Agricultural Experiment Station, Iowa State College.

Hecnar, S. J., and R. T. M'Closkey. 1996. Regional dynamics and the status of amphibians. *Ecology* 77:2091–2097.

———. 1997. Spatial scale and determination of species status of the green frog. *Conservation Biology* 11:670–682.

Hedrick, P. W., and P. S. Miller. 1992. Conservation genetics: Techniques and fundamentals. *Ecological Applications* 2:30–46.

Heffelfinger, J. R., F. S. Guthery, R. J. Olding, C. L. Cochran Jr., and C. M. McMullen. 1999. Influence of precipitation timing and summer temperatures on reproduction of Gambel's quail. *Journal of Wildlife Management* 63: 154–161.

Heglund, P. J., J. R. Jones, L. H. Frederickson, and M. S. Kaiser. 1994. Use of boreal forested wetlands by Pacific loons (*Gavia pacifica* Lawrence) and horned grebes (*Podiceps auritus* L.): Relations with limnological characteristics. *Hydrobiologia* 279/280: 171–183.

Heikkinen, J., and H. Högmander. 1994. Fully Bayesian approach to image restoration with an application in biogeography. *Applied Statistics* 43: 569–582.

Heikkinen, R. K. 1998. Can richness patterns of rarities be predicted from mesoscale atlas data? A case study of vas-

cular plants in the Kevo Reserve. *Biological Conservation* 83: 133–143.

Heinen, J. T., and J. G. Lyon. 1989. The effects of changing weighting factors on wildlife habitat index values: A sensitivity analysis. *Photogrammetric Engineering and Remote Sensing* 55:1445–1447.

Heinrich, M. L., and D. W. Kaufman. 1985. Herpetofauna of the Konza Prairie Research Natural Area, Kansas. *Prairie Naturalist* 17:101–112.

Hejl, S. J., and K. M. Granillo. 1998. What managers really need from avian researchers. In *Avian conservation: Research and management*, ed. J. M. Marzluff and R. Sallabanks, 431–437. Washington, D.C.: Island Press.

Helle, P., and O. Jarvinen. 1986. Population trends of north Finnish land birds in relation to their habitat selection and changes in forest structure. *Oikos* 46:107–115.

Helly, J., T. Case, F. Davis, S. Levin, and W. Michener, eds. 1995. *The state of computational ecology*. http://www.sdsc.edu/compeco_workshop/report/report.html (assessed 23 June 1999).

Heltshe, J. F., and N. E. Forrester. 1983. Estimating species richness using the jackknife procedure. *Biometrics* 39: 1–12.

Hemesath, L. M. 1998. Iowa's frog and toad survey: 1991–1994. In *Status and conservation of Midwestern amphibians*, ed. M. J. Lannoo, 206–216. Iowa City: University of Iowa Press.

Hendrickson, D. A., and W. L. Minckley. 1984. Cienegas—vanishing climax communities of the American Southwest. *Desert Plants* 6:131–175.

Henebry, G. M. 1993. Detecting change in grasslands using measures of spatial dependence with Landsat TM data. *Remote Sensing of Environment* 46:223–234.

———. 1995. Spatial model error analysis using autocorrelation indices. *Ecological Modelling* 82:75–91.

———. 1997. Advantages of principal components analysis for land cover segmentation from SAR image series. In *Proceedings of Third ERS Symposium on Space at the Service of our Environment*, 175–178. Noordwijk, The Netherlands: European Space Agency. http://earth.esrin.esa.int/symposia/papers/data/henebry3/index.html.

Henebry, G. M., and H. Su. 1993. Using landscape trajectories to assess the effects of radiometric rectification. *International Journal of Remote Sensing* 14:2417–2423.

Henery, R. J. 1994. Classification. In *Machine learning, neural and statistical classification*, ed. D. Michie, D. J. Spiegelhalter, and C. C. Taylor, 6–16. New York: Ellis Horwood.

Hengeveld, R. 1990. *Dynamic biogeography*. Cambridge: Cambridge University Press.

Hepinstall, J. A., and S. A. Sader. 1997. Using Bayesian statistics, thematic mapper satellite imagery, and breeding bird survey data to model bird species probability of occurrence in Maine. *Photogrammetric Engineering and Remote Sensing* 63:1231–1237.

Hepinstall, J. A., S. A. Sader, W. B. Krohn, R. B. Boone, and R. I. Bartlett. 1999. Development and testing of a vegetation and land cover map of Maine. Technical Bulletin 173. Orono: Maine Agriculture and Forest Experiment Station, University of Maine.

Heppell, S. S., J. R. Walters, and L. B. Crowder. 1994. Evaluating management alternatives for red-cockaded woodpeckers: A modeling approach. *Journal of Wildlife Management* 58:479–487.

Hermann, K. A. 1996. *The southern Appalachian assessment GIS data base CD ROM set*. Gatlinburg: Southern Appalachian Man and the Biosphere Program.

Hernandez, J. E., L. D. Epstein, M. H. Rodriguez, A. D. Rodriguez, E. Rejmankova, and D. R. Roberts. 1997. Use of generalized regression tree models to characterize vegetation favoring *Anopheles albimanus* breeding. *Journal of the American Mosquito Control Association* 13:28–34.

Herrera, C. M. 1978. On the breeding distribution pattern of European migrant birds: MacArthur's theme reexamined. *Auk* 95:496–509.

Hetrick, W. A., P. M. Rich, F. J. Barnes, and S. B. Weiss. 1993. GIS-based solar radiation flux models. In *American Society for Photogrammetry and Remote Sensing technical papers*, Vol. 3, 132–143. GIS Photogrammetry, and Modeling.

Heuvelink, G., and M. F. Goodchild. 1998. *Error propagation in environmental modelling with GIS*. London: Taylor & Francis.

Heyer, R. W., M. Donnelly, and L. C. Hayek, eds. 1994. *Measuring and monitoring biological diversity: Standard methods for amphibians*. Washington, D.C.: Smithsonian Institution Press.

Heywood, V. H. 1994. The measurement of biodiversity and the politics of implementation. In *Systematics and conservation evaluation*, ed. P. L. Forey, C. J. Humphries, R. I. Vane-Wright, 15–22. Oxford: Clarendon Press.

Hickman, J. C. ed. 1993. *The Jepson manual: Higher plants of California*. Berkeley: University of California Press.

Hiebeler, D. 2000. Populations on fragmented landscapes with spatially structured heterogeneities: Landscape generation and local dispersal. *Ecology* 81:1629–1641.

High Performance Systems. 1994. *STELLA 2, technical documentation*. Hanover: High Performance Systems.

Hilborn, R., and M. Mangel. 1997. *The ecological detective: Confronting models with data*. Princeton: Princeton University Press.

Hildén, O. 1965. Habitat selection in birds. *Annales Zoologici Fennici* 2:53–75.

Hill, K. 1997. The representation of categorical ambiguity: A comparison of fuzzy, probabilistic, boolean, and index approaches in suitability analysis. Ph.D. dissertation, Harvard University, Cambridge.

Hill, M. O. 1979. *DECORANA—A FORTRAN program for detrended correspondence analysis and reciprocal averaging*. Ithaca, N.Y.: Cornell University.

Hill, R. J. 1975. A biosystematic study of the genus *Mentzelia* (Loasaceae) in Wyoming and adjacent states. M.S. thesis, University of Wyoming, Laramie.

Hingston, F. J., G. M. Dimmock, and A. G. Turton. 1980. Nutrient distribution in a jarrah (*Eucalyptus marginata* Donn ex Sm.) ecosystem in south-west Western Australia. *Forest Ecology and Management* 3:183–207.

Hirons, G., R. B. Owen Jr. 1982. Comparative breeding behavior of European and American woodcock. In *Proceedings of the Seventh Woodcock Symposium*, technical coordinators T. J. Dwyer and G. L. Storm, 179–186. Research Report 14. U.S. Department of the Interior, U.S. Fish and Wildlife Service.

Hobbs, N. T. 1999. Responses of large herbivores to spatial heterogeneity in ecosystems. In *Nutritional ecology of herbivores: Proceedings of the Fifth International Symposium on the Nutrition of Herbivores*, ed. H. G. Jung, G. C. Fahey, 97–129. Savory, Ill.: American Society of Animal Science.

Hobbs, N. T., and T. A. Hanley. 1990. Habitat evaluation: Do use/availability data reflect carrying capacity? *Journal of Wildlife Management* 54:515–522.

Hobbs, R. J. 1990. Remote sensing of spatial and temporal dynamics of vegetation. In *Remote sensing of biosphere functioning*, ed. R. J. Hobbs and H. A. Mooney, 203–219. New York: Springer-Verlag.

———. 1994. Dynamics of vegetation mosaics: Can we predict responses to global change? *Écoscience* 1:346–356.

Hocutt, C. H, and E. O. Wiley, eds. 1986. *The zoogeography of North American freshwater fishes*. New York: John Wiley & Sons.

Hodorff, R. A., C. Hull-Sieg, and R. L. Linder. 1988. Wildlife response to stand structure of deciduous woodlands. *Journal of Wildlife Management* 52:667–673.

Hof, J., and M. Bevers. 1998. *Spatial optimization for managed ecosystems.* New York: Columbia University Press.

Hof, J. G. 1993. *Coactive forest management.* San Diego: Academic Press.

Högmander, H., and J. Møller. 1995. Estimating distribution maps from atlas data using methods of statistical image analysis. *Biometrics* 51:393–404.

Hole, F. D., and J. B. Campbell. 1985. *Soil landscape analysis.* Totowa, N.J.: Rowman and Allanheld.

Holl, K. D. 1995. Nectar resources and their influence on butterfly communities on reclaimed coal surface mines. *Restoration Ecology* 3:76–85.

Hollander, A. D., F. W. Davis, and D. M. Stoms. 1994. Hierarchical representations of species distributions using maps, images, and sighting data. In *Mapping the diversity of nature*, ed. R. I. Miller, 71–88. London: Chapman and Hall.

Holling, C. S. ed. 1978. *Adaptive environmental assessment and management.* New York: John Wiley & Sons.

———. 1992. Cross-scale morphology, geometry, and dynamics of ecosystems. *Ecological Monographs* 62: 447–502.

Holt, R. D. 1984. Spatial heterogeneity, indirect interactions, and the coexistence of prey species. *American Naturalist* 124:377–406.

———. 1997. From metapopulation dynamics to community structure—some consequences of spatial heterogeneity. In *Metapopulation biology: Ecology, genetics, and evolution*, ed. I. A. Hanski and M. E. Gilpin, 149–164. San Diego: Academic Press.

Holthausen, R. S., M. G. Raphael, K. S. McKelvey, E. D. Forsman, E. E. Starkey, and D. E. Seaman. 1995. The contribution of federal and non-federal habitat to persistence of the northern spotted owl on the Olympic Peninsula, Washington: Report of reanalysis team. General Technical Report PNW-GTR-352. Portland, Ore.: USDA Forest Service, Pacific Northwest Research Station.

Homer, C. G., T. C. Edwards Jr., R. D. Ramsey, and K. P. Price. 1993. Use of remote sensing methods in modelling sage grouse winter habitat. *Journal of Wildlife Management* 57:78–84.

Hoover, J. P., and M. C. Brittingham. 1993. Regional variation in cowbird parasitism of wood thrushes. *Wilson Bulletin* 105:28–238.

Hoover, J. P., M. C. Brittingham, and L. J. Goodrich. 1995. Effects of forest patch size on nesting success of wood thrushes. *Auk* 112:146–155.

Hoover, R. L., and D. L. Wills, eds. 1984. Managing forested lands for wildlife. Denver: Colorado Division of Wildlife in cooperation with USDA Forest Service, Rocky Mountain Region.

Hopkins, L. 1977. Methods for generating land suitability maps: A comparative evaluation. *Journal of the American Institute of Planners* 43:386–400.

Hoshovsky, M. 1988. The lands and natural areas project: Identifying and protecting California's significant natural areas. In *Plant biology of eastern California*, ed. C. A. Hall Jr. and V. Doyle-Jones, 199–206. Los Angeles: University of California, White Mountain Research Station.

Hosking, J. R. M., P. D. Edwin, and M. S. Pednault. 1997. A statistical perspective on data mining. *Future Generation Computer Systems* 13:117–134.

Hosmer, D. W., Jr., and S. Lemeshow. 1989. *Applied logistic regression.* New York: John Wiley & Sons.

Howells, O., and G. Edwards-Jones. 1997. A feasibility study of reintroducing wild boar *Sus scrofa* to Scotland: Are existing woodlands large enough to support minimum viable populations. *Biological Conservation* 81: 77–89.

Hubbell, S. P. 1980. Seed predation and the coexistence of tree species in tropical forests. *Oikos* 35:214–229.

Hubbell, S. P., R. B. Foster, S. T. O'Brien, K. E. Harms, R. Condit, B. Wechsler, S. J. Wright, and S. L. de Lao. 1999. Light-gap disturbances, recruitment limitation, and tree diversity in a Neotropical forest. *Science* 283:554–557.

Hubert, W. A., and F. J. Rahel. 1989. Relations of physical habitat to abundance of four nongame fishes in high plains streams: A test of habitat suitability index models. *North American Journal of Fisheries Management* 9: 332–340.

Huberty, C. J. 1994. *Applied discriminant analysis.* New York: John Wiley & Sons.

Hudgins, J. E., G. L. Storm, and J. S. Wakeley. 1985. Local movements and diurnal-habitat selection by male American woodcock in Pennsylvania. *Journal of Wildlife Management* 49:614–619.

Huffer, F. W., and H. Wu. 1998. Markov chain Monte Carlo for autologistic regression models with application to the distribution of plant species. *Biometrics* 54:509–524.

Hull-Sieg, C., R. A. Hodorff, and R. L. Linder. 1984. Stand condition as a variable influencing wildlife use of green ash woodlands. *Great Plains Agriculture Council* No. 11, 36–39.

Hunsaker, C. T. 1996. Estimating uncertainty in spatial data: Implications for landscape ecology. *Bulletin of the Ecological Society of America* 77 (3 suppl. part 2): 208.

Hunsaker C. T., M. F. Goodchild, M. A. Friedl, T. J. Case. 2001. *Spatial uncertainty for ecology: Implications for remote sensing and GIS applications.* New York: Springer.

Hunsaker, C. T., R. A. Nisbet, D. C. Lam, J. A. Browder, W. L. Baker, M. G. Turner, and D. B. Botkin. 1993. Spatial models of ecological systems and processes: The role of GIS. In *Environmental modeling with GIS*, ed. M. Goodchild, B. Parks, and L. Steyaert, 248–264. New York: Oxford University Press.

Hunsaker, C. T., R. V. O'Neill, S. P. Timmins, B. L. Jackson, D. A. Levine, and D. J. Norton. 1994. Sampling to characterize landscape pattern. *Landscape Ecology* 9: 207–226.

Hunt, C. B. 1966. Plant ecology of Death Valley, California. U.S. Geological Survey Professional Paper 509. Washington, D.C.: U.S. Government Printing Office.

Hunt, G. A., and J. M. Trappe. 1987. Seasonal hypogeous sporocarp production in a western Oregon Douglas-fir stand. *Canadian Journal of Botany* 65:438–445.

Hunter, C. 1990a. *Odocoileus virginianus seminolus*: The ecology of fawning in wet and dry prairies. M.S. thesis, University of Florida, Gainesville.

Hunter, J. E., R. J. Gutiérrez, and A. B. Franklin. 1995. Habitat configuration around spotted owl sites in northwestern California. *Condor* 97:684–693.

Hunter, J. E., K. N. Schmidt, H. B. Stauffer, S. L. Miller, C. J. Ralph, and L. Roberts. 1998. Status of the marbled murrelet in the inner north coast ranges of California. *Northwestern Naturalist* 79:92–103.

Hunter, M. L. Jr. 1990. *Wildlife, forests, and forestry: Principles of managing forests for biological diversity.* Englewood Cliffs, N.J.: Prentice-Hall.

Hunter, M. L., Jr., J. Albright, and J. Arbuckle, eds. 1999. *Maine amphibians and reptiles.* Orono: University of Maine Press.

Hunter, M. L., Jr., G. L. Jacobson Jr., and T. Webb, III. 1988. Paleocology and the coarse-filter approach to maintaining biological diversity. *Conservation Biology* 2:375–385.

Huntley, B., P. Berry, W. Cramer, and A. P. McDonald. 1995. Modelling present and potential future ranges of some European higher plants using climate response surfaces. *Journal of Biogeography* 22:967–1001.

Hurlbert, S. H. 1981. A gentle depilation of the niche: Dicean resource sets in resource hyperspace. *Evolutionary Theory* 5:177–184.

———. 1984. Pseudoreplication and the design of ecological field experiments. *Ecological Monographs* 54: 187–211.

Hurley, J. F. 1986. Summary: Development, testing, and application of wildlife-habitat models—the manager's viewpoint. In *Wildlife 2000: Modeling habitat relationships of terrestrial vertebrates*, ed. J. A. Verner, M. L. Morrison, C. J. Ralph, 151–153. Madison: University of Wisconsin Press.

Huston, M. A. 1979. A general hypothesis of species diversity *American Naturalist* 113:81–101.

———. 1980. Soil nutrients and tree species richness in Costa Rican forests. *Journal of Biogeography* 7:147–157.

———. 1985. Patterns of species diversity on coral reefs. *Annual Review of Ecology and Systematics* 16:149–177.

———. 1994. *Biological Diversity: The coexistence of species on changing landscapes.* Cambridge: Cambridge University Press.

———. 1997. Hidden treatments in ecological experiments: Re-evaluating the ecosystem function of biodiversity. *Oecologia* 110:449–460.

———. 1999. Local processes and regional patterns: Appropriate scales for understanding variation in the diversity of plants and animals. *Oikos* 86:393–401.

Huston, M. A., D. L. DeAngelis, and W. M. Post. 1988. New computer models unify ecological theory. *BioScience* 38:682–692.

Huston, M. A., and L. Gilbert. 1996. Consumer diversity and secondary production. In *Biodiversity and ecosystem processes in tropical rainforests*, ed. G. H. Orians, R. Dirzo, and J. H. Cushman, 33–47. New York: Springer-Verlag.

Huston, M. A., and T. M. Smith. 1987. Plant succession: Life history and competition. *American Naturalist* 130:168–198.

Hutchings, M. J., L. A. John, and A. J. A. Stewart, eds. 2000. *The ecological consequences of environmental heterogeneity.* Malden, Mass.: Blackwell Science.

Hutchinson, G. E. 1953. The concept of pattern in ecology. *Proceedings of the Academy of Natural Sciences of Philadelphia* 105:1–12.

———. 1957. Concluding remarks. *Cold Harbor symposium on quantitative biology.* 22: 415–427.

———. 1959. Homage to Santa Rosalia, or Why are there so many kinds of animals? *American Naturalist* 93: 145–159.

Hutchinson, M. F. 1987. Methods for generation of weather sequences. In *Agricultural environments: Characterisa-*

tion, classification and mapping, ed. A. H. Bunting, 149–157. Wallingford, UK: CAB International.

———. 1995. Interpolating mean rainfall using thin plate smoothing splines. *International Journal of GIS* 9: 385–403.

———. 1997. *Anusplin version 3.2 user guide*. Canberra: Centre for Resource and Environmental studies, Australian National University.

———. 1998. Interpolation of rainfall data with thin plate smoothing splines 2: Analysis of topographic dependence. *Journal of Geographic Information and Decision Analysis* 2:152–167.

Hutchinson, M. F., H. A. Nix, D. J. Houlder, and J. P. McMahon. undated. *ANUCLIM version 1.8 user guide*. Canberra: Centre for Resource and Environmental Studies, Australian National University.

Hutto, R. L. 1985. Habitat selection by nonbreeding, migratory land birds. In *Habitat selection in birds*, ed. M. L. Cody, 455–476. Orlando: Academic Press.

———. 1990. Measuring the availability of food resources. *Studies in Avian Biology* 13:20–28.

Hutto, R. L., and J. Hoffland. 1996. USDA Forest Service Northern Region Landbird Monitoring Project: Field Methods. Missoula, Mont.: U.S. Forest Service Report, Region 1.

Hutto, R. L., S. M. Pletschet, and P. Hendricks. 1986. A fixed-radius point count method for nonbreeding and breeding season use. *Auk* 103: 593–602.

Hutto, R. L., and J. S. Young. 1999. Habitat relationships of landbirds in the northern region, USDA Forest Service. General Technical Report RMRS-GTR-32. Ogden, Utah: USDA Forest Service, Rocky Mountain Research Station.

ICBP. International Council for Bird Preservation. 1992. *Putting biodiversity on the map*. Cambridge, England: International Council for Bird Preservation.

Igl, L. D., and D. H. Johnson. 1997. Changes in breeding bird populations in North Dakota: 1967 to 1992–93. *Auk* 114:74–92.

Illinois Department of Natural Resources. 1996. *Illinois land cover: An atlas on compact disc*. Springfield, Ill.: Illinois Department of Natural Resources, Office of Realty and Environmental Planning.

Iman, R. L., and W. J. Conover. 1989. *Modern business statistics*. 2nd ed. New York: John Wiley & Sons.

Imhof, J. G., J. Fitzgibbon, and W. K. Annable. 1996. A hierarchical evaluation system for characterizing watershed ecosystems for fish habitat. *Canadian Journal of Fisheries and Aquatic Sciences* 53 (suppl. 1):312–326.

Imhoff, M. L., T. D. Sisk, A. Milne, G. Morgan, and T. Orr. 1997. Remotely sensed indicators of habitat heterogeneity: Use of synthetic aperture radar in mapping vegetation structure and bird habitat. *Remote Sensing of Environment* 60:217–227.

Ims, R. A., and N. G. Yoccoz. 1997. Studying transfer processes in metapopulations: Emigration, migration, and colonization. In *Metapopulation biology: Ecology, genetics, and evolution*, ed. I. A. Hanski and M. E. Gilpin, 247–264. New York: Academic Press.

Iverson, L. R., and A. M. Prasad. 1998. Predicting abundance of eighty tree species following climate change in the eastern United States. *Ecological Monographs* 68:465–485.

Iverson, L. R., A. M. Prasad, B. J. Hale, and E. K. Sutherland. 1999. *Atlas of current and potential future distributions of common trees of the eastern United States*. Radnor, Pa.: USDA Forest Service Northeast Research Station.

Ivlev, V. S. 1961. *Experimental ecology of the feeding of fishes*. New Haven: Yale University Press.

Jackson, D. A. 1993. Stopping rules in principal components analysis: A comparison of heuristical and statistical approaches. *Ecology* 74:2204–2214.

Jacobs, B., and J. D. Wilson. 1997. *Missouri breeding bird atlas: 1986–1992*. Jefferson City, Mo.: Missouri Department of Conservation.

Jacobs, J. 1974. Quantitative measurement of food selection: A modification of the forage ratio and Ivlev's electivity index. *Oecologia* 14:413–417.

Jaksic, F. M. 1981. Abuse and misuse of the term "guild" in ecological studies. *Oikos* 37:397–400.

Jakubauskas, M. E., K. Kindscher, and D. M. Debinski. 1998. Multitemporal characterization and mapping of montane sagebrush communities using Indian IRS LISS-II imagery. *Geocarto International* 13:65–74.

James, F. C. 1971. Ordinations of habitat relationships among breeding birds. *Wilson Bulletin* 83:215–236.

———. 1991. Complementary descriptive and experimental studies of clinal variation in birds. *American Zoologist* 31:694–706.

James, F. C., and C. E. McCulloch. 1985. Data analysis and the design of experiments in ornithology. In *Current ornithology*, ed. R. J. Johnston, Vol. 2, 1–63. New York: Plenum.

———. 1990. Multivariate analysis in ecology and systematics: Panacea or Pandora's box? *Annual Review of Ecology and Systematics* 21:129–166.

———. 1995. The strength of inferences about causes of trends in populations. In *Ecology and management of Neotropical migratory birds: A synthesis and review of critical issues*, ed. T. E. Martin and D. M. Finch, 40–51. New York: Oxford University Press.

James, F. C., C. E. McCulloch, and D. A. Wiedenfeld. 1996. New approaches to the analysis of population trends in land birds. *Ecology* 77:13–27.

James, F. C., and H. H. Shugart. 1970. A quantitative method of habitat description. *Audubon Field Notes* 24:727–736.

Jankowski, P., T. L. Nyerges, A. Smith, T. J. Moore, and E. Horvath. 1997. Spatial group choice: A SDSS tool for collaborative spatial decision-making. *International Journal of Geographical Information Science* 11:577–602.

Jarvis, A. M., and A. Robertson. 1999. Predicting population sizes and priority conservation areas for ten endemic Namibian bird species. *Biological Conservation* 88:121–131.

Jelinski, D. E., and J. Wu. 1996. The modifiable areal unit problem and implications for landscape ecology. *Landscape Ecology* 11:129–140.

Jenerette, G. D., and J. Wu. 2000. On the definitions of scale. *Bulletin of the Ecological Society of America* 81:104–105.

Jenkins, R. E., and N. M. Burkhead. 1994. *Freshwater fishes of Virginia*. Bethesda: American Fisheries Society.

Jenni, L., and S. Jenni-Eiermann. 1998. Physiology and stopover: Fuel supply and metabolic constraints in migrating birds. *Journal of Avian Biology* 29:521–528.

Jennings, C. W. 1977. *Geologic map of California: 1969–1973*. Map No. 2. Sacramento, Calif.: Division of Mines and Geology Geologic Data.

Jennrich, R. I., and F. B. Turner. 1969. Measurement of non-circular home range. *Journal of Theoretical Biology* 22:227–237.

Jensen, J. R., S. Narumalani, O. Weatherbee, and H. E. Mackey. 1993. Measurement of seasonal and yearly cattail and waterlily changes using multidate SPOT panchromatic data. *Photogrammetric Engineering and Remote Sensing* 59:519–525.

Jette, L. A., T. J. Hayden, and J. D. Cornelius. 1998. Demographics of the golden-cheeked warbler (*Dendroica chrysoparia*) on Fort Hood, Texas. USACERL Technical Report 98/52. Champaign Ill.: U.S. Army Corps of Engineers, Construction Engineering Research Laboratory.

Jobin, A., P. Molinari, and U. Breitenmoser. 2000. Prey spectrum, prey preference and consumption rates of Eurasian lynx in the Swiss Jura Mountains. *Acta Theriologica* 45:243–252.

Johnson, A. R. 1996. Spatio-temporal hierarchies in ecological theory and modeling. In *GIS and environmental modeling: Progress and research issues*, ed. M. F. Goodchild, L. T. Steyaert, B. O. Parks, C. Johnston, D. Maidment, M. Crane, and S. Glendinning, 451–456. Fort Collins, Colo.: GIS World Books.

Johnson, B. L. 1999b. The role of adaptive management as an operational approach for resource management agencies. *Conservation Ecology* [online] 3:8. http://www.consecol.org/vol3/iss2/art8.

Johnson, C. M. 1998. Spatial and temporal considerations for identifying important seabird nesting habitats in Maine. Ph.D. dissertation, University of Maine, Orono.

Johnson, C. M., and W. B. Krohn. 1998. Coastal Maine: Island habitats and fauna. In *Status and trends of the nation's biological resources*, ed. M. J. Mac, P. A. Opler, C. E. Puckett Haecker, and P. D. Doran, 207–208. Fort Collins, Colo.: U.S. Department of the Interior, U.S. Geological Survey.

Johnson, D. H. 1980. The comparison of usage and availability measurements for evaluating resource preference. *Ecology* 61:65–71.

———. 1981a. How to measure habitat: A statistical perspective. In *The use of multivariate statistics in studies of wildlife habitat*, ed. D. E. Capen, 53–57. General Technical Report RM-87. Fort Collins, Colo.: USDA Forest Service, Rocky Mountain Forest and Range Experiment Station.

———. 1981b. The use and misuse of statistics in wildlife habitat studies. In *The use of multivariate statistics in studies of wildlife habitat* ed. D. E. Capen, 11–19. General Technical Report RM-87. Fort Collins, Colo.: USDA Forest Service, Rocky Mountain Forest and Range Experiment Station.

———. 1995. Statistical sirens: The allure of nonparametrics. *Ecology* 76:1998–2000.

———. 1999. The insignificance of statistical significance testing. *Journal of Wildlife Management* 63:763–772.

Johnson, D. H., M. C. Hammond, T. L. McDonald, C. L. Nustad, and M. D. Schwartz. 1989. Breeding canvasbacks: A test of a habitat model. *Prairie Naturalist* 21:193–202.

Johnson, F., and K. Williams. 1999. Protocol and practice in the adaptive management of waterfowl harvests. *Conservation Ecology* [online]3:8. http://www.consecol.org/vol3/iss1/art8.

Johnson, K. N., D. P. Jones, and B. Kent. 1980. *Forest planning model (FORPLAN): User's guide and operations manual.* Fort Collins, Colo.: USDA Forest Service, Systems Applications for Land Management Planning.

Johnson, N. K., F. Swanson, M. Herring, and S. Greene, eds. 1999. *Bioregional assessments.* Washington, D.C.: Island Press.

Johnson, N. L., and S. Kotz. 1969. *Discrete distributions.* New York: Houghton Mifflin.

Johnston, C. A. 1993. Introduction to quantitative methods and modeling in community, population, and landscape ecology. In *Environmental modeling with GIS*, ed. M. F. Goodchild, B. O. Parks, and L. T. Steyaert, 276–283. New York: Oxford University Press.

Johnston, C. A., Y. Cohen, and J. Pastor. 1996. Modeling of spatially static and dynamic ecological processes. In *GIS and environmental modeling: Progress and research issues*, ed. M. F. Goodchild, L. T. Steyaert, B. O. Parks, C. Johnston, D. Maidment, M. Crane, and S. Glendinning, 149–154. New York: Oxford University Press.

Jolly, G. M. 1965. Explicit estimates from capture-recapture data with both death and immigration-stochastic model. *Biometrika* 52:225–247.

———. 1982. Mark-recapture models with parameters constant in time. *Biometrics* 38:301–321.

Jones, K. B., K. H. Riitters, J. D. Wickham, R. D. Tankersley Jr., R. V. O'Neill, D. J. Chaloud, E. R. Smith, and A. C. Neale. 1997. *An ecological assessment of the United States mid-Atlantic region: A landscape atlas.* EPA/600/R-97/130. Washington, D.C.: U.S. Environment Protection Agency, Office of Research and Development.

Jones, L. T., and H. G. Elliott. 1944. Copper deficiency in the Busselton-Augusta district. *Journal of Agriculture, Western Australia* (2nd series) 21:342–357.

Jones, S. M., R. E. Ballinger, and J. W. Nietfeldt. 1981. Herpetofauna of Mormon Island Preserve, Hall County, Nebraska. *Prairie Naturalist* 13:33–41.

Jones, T., and S. Forrest. 1993. An introduction to SFI echo. Technical Report 93-12-074. Santa Fe: Santa Fe Institute. Available via anonymous ftp from ftp.santafe.edu: pub/echo/how-to.ps.Z.

Jongman, R. H. G., C. J. F. ter Braak, and O. F. R. Van Tongeren. 1987. *Data analysis in community and landscape ecology.* Wageningen, The Netherlands: Pudoc.

Jonsson, L., A. Dahlberg, M.-C. Wilsson, O. Zackrisson, and O. Kårén. 1999. Ectomycorrhizal fungal communities in late-successional Swedish boreal forests, and their composition following wildfire. *Molecular Ecology* 8:205–215.

Jorgensen, E. E., and S. Demarais. 1999. A comparison of modelling techniques for small mammal diversity. *Ecological Modelling* 120:1–8.

Jovéniaux, A. 1993. *Atlas des Oiseaux Nicheurs du Jura.* Lons-le-Saunier, France: Groupe Ornithologique du Jura.

Judson, O. P. 1994. The rise of the individual-based model in ecology. *Trends in Ecology and Evolution* 9:9–14.

Justice, C. O., E. Vermote, J. R. G. Townshend, R. DeFries, D. P. Roy, D. K. Hall, V. V. Salomonson, J. L. Privette, G. Riggs, A. Strahler, W. Lucht, R. B. Myneni, Y. Knyazikhin, S. W. Running, R. Nemani, Z. Wan, A. Huete, W. Leeuwen, R. Wolfe, L. Giglio, J.-P. Muller, P. Lewis, and M. Barnsley. 1998. The moderate resolution imaging spectroradiometer (MODIS): Land remote sensing for global change research. *IEEE Transactions on Geoscience and Remote Sensing* 36:1228–1249.

Kaiser, M. S., P. L. Speckman, and J. R. Jones. 1994. Statistical models for limiting nutrient relations in inland waters. *Journal of the American Statistical Association* 89:410–423.

Kalos, M. H., and P. A. Whitlock. 1986. *Monte Carlo methods.* New York: Wiley.

Kaluzny, S. P., S. C. Vega, T. P. Cardoso, and A. A. Shelly. 1996. *S-PLUSSPATIALSTATS user's manual.* Seattle: MathSoft.

Kanda, N. 1998. Genetics and conservation of bull trout: Comparison of population genetic structure among different markers and hybridization with brook trout. Ph.D. dissertation, University of Montana, Missoula.

Karl, J. W., N. M. Wright, P. J. Heglund, E. O. Garton, J. M. Scott, and R. L. Hutto. 2000. Sensitivity of species habitat relationship model performance to factors of scale. *Ecological Applications* 10:1690–1705.

Karl, J. W., N. M. Wright, P. J. Heglund, and J. M. Scott. 1999. Obtaining environmental measures to facilitate vertebrate habitat modeling. *Wildlife Society Bulletin* 27:357–365.

Karlson, R. H., and H. V. Cornell. In press. Species richness of coral assemblages: Detecting regional influences at local scales. *Ecology.*

Karr, J. R. 1980. History of the habitat concept in birds and the measurement of avian habitats. In *Acta Internationalis Congressus Ornithologici 27*, ed. R. Nöhring, 991–997. Berlin: Verlag der Deutschen Ornithologen-Gesellschaft.

Karr, J. R., and T. E. Martin. 1981. Random numbers and

principal components: Further searches for the unicorn? In *The use of multivariate statistics in studies of wildlife habitat*, ed. D. E. Capen, 20–24. General Technical Report RM-87. Fort Collins, Colo.: USDA Forest Service, Rocky Mountain Forest and Range Experiment Station.

Kauffman, J. B., and W. C. Krueger. 1984. Livestock impacts on riparian ecosystems and stream management implications: A review. *Journal of Range Management* 37:430–438.

Kaufman, K. 1996. *Lives of North American birds*. Boston: Houghton Mifflin.

Keddy, P., L. Twolan-Strutt, and B. Shipley. 1997. Experimental evidence that interspecific competitive asymmetry increases with soil productivity. *Oikos* 80:253–256.

Keddy, P. A. 1989. *Competition*. London: Chapman and Hall.

Keitt, T. H., D. L. Urban, and B. T. Milne. 1997. Detecting critical scales in fragmented landscapes. *Conservation Ecology* 1:4.

Keizer, P. J., and E. Arnolds. 1994. Succession of ectomycorrhizal fungi in roadside verges planted with common oak (*Quercus robur* L.) in Drenthe, The Netherlands. *Mycorrhiza* 4:147–159.

Keleher, C. J., and F. J. Rahel. 1996. Thermal limits to salmonid distributions in the Rocky Mountain region and potential habitat loss due to global warming: A geographic information system (GIS) approach. *Transactions of the American Fisheries Society* 125:1–13.

Keller, F. 1992. Automated mapping of mountain permafrost using the program PERMAKART within the geographical information system ARC/INFO. *Permafrost and Periglacial Processes* 3:133–138.

Kelly, J. F., and B. Van Horne. 1997. Effects of scale-dependent variation in ice cover on the distribution of wintering belted kingfishers. *Ecography* 20:506–512.

Kelly, J. P. 1993. The effect of nest predation on habitat selection by dusky flycatchers in limber pine-juniper woodland. *Condor* 95:83–93.

Kendall, K. C., L. H. Metzgar, D. A. Patterson, and B. M. Steele. 1992. Power of sign surveys to monitor population trends. *Ecological Applications* 2:422–430.

Kennedy, P. 1992. *A guide to econometrics*. Cambridge: MIT Press.

Khuri, A. I., and J. A. Cornell. 1987. *Response surfaces: Designs and analyses*. New York: Marcel Dekker.

Kiester, A. R., J. M. Scott, B. Csuti, and R. F. Noss, B. Butterfield, K. Sahr, and D. White. 1996. Conservation prioritization using GAP data. *Conservation Biology* 10: 1332–1342.

Kimes, D. S., R. F. Nelson, M. T. Manry, and A. K. Fung. 1998. Review article. Attributes of neural networks for extracting continuous vegetation variables from optical and radar measurements. *International Journal of Remote Sensing* 19:2639–2664.

Kindscher, K., A. Fraser, M. E. Jakubauskas, and D. M. Debinski. 1998. Identifying wetland meadows in Grand Teton National Park using remote sensing and average wetland values. *Wetlands Ecology and Management* 5:265–273.

Kineman, J. J. 1992. Global ecosystems database, version 1.0 (on CD-ROM): EPA Global Climate Research Program, NOAA/NGDC Global Change Database Program: user's guide. Boulder, Colo.: U.S. Department of Commerce, National Oceanic and Atmospheric Administration, National Geophysical Data Center.

King, A. W. 1991. Translating models across scales in the landscape. In *Quantitative methods in landscape ecology*, ed. M. G. Turner and R. H. Gardner, 479–517. New York: Springer-Verlag.

———. 1997. Hierarchy theory: A guide to system structure for wildlife biologists. In *Wildlife and landscape ecology: Effects of pattern and scale*, ed. J. A. Bissonette, 185–212. New York: Springer-Verlag.

King, A. W., A. R. Johnson, and R. V. O'Neill. 1991. Transmutation and functional representation of heterogeneous landscapes. *Landscape Ecology* 5:239–253.

Kingsland, S.E. 1985. *Modeling nature: Episodes in the history of population ecology*. Chicago: University of Chicago Press.

Kinsley, K. R., S. A. Liscinsky, and G. L. Storm. 1980. Changes in habitat structure on woodcock singing grounds in central Pennsylvania. In *Proceedings of the Seventh Woodcock Symposium*, technical coordinators. T. J. Dwyer, and G. L. Storm, 40–50. Research Report 14. Washington, D.C.: U.S. Department of the Interior, U.S. Fish and Wildlife Service.

Kish, L. 1987. *Statistical design for research*. New York: Wiley.

Klebenow, D. A. 1969. Sage grouse nesting and brood habitat in Idaho. *Journal of Wildlife Management* 33: 649–662.

Kleijnen, J. P. 1987. *Statistical tools for simulation practitioners*. New York: Marcel Dekker.

Klimstra, W. D., and J. L. Roseberry. 1975. Nesting ecology

of the bobwhite in southern Illinois. *Wildlife Monographs* 41:1–37.

Klute, D. S. 1999. Modeling habitat relationships for American woodcock in Pennsylvania: Effects of spatial autocorrelation and spatial scale. Ph.D. dissertation, Pennsylvania State University, University Park.

Klute, D. S., M. J. Lovallo, W. M. Tzilkowski, and G. L. Storm. 2000. Determining multi-scale habitat and landscape associations for American woodcock in Pennsylvania. In *Proceedings of the Ninth American Woodcock Symposium* ed. McAnley, D. G., J.G. Bruggink, and G. F. Sepik. Information Technology Report USGS/BRD/ITR-2000-0009. Laurel, Md.: U.S. Department of the Interior, U.S. Geological Survey.

Knick, S. T., and D. L. Dyer. 1997. Spatial distribution of black-tailed jackrabbit habitat determined by GIS in southwestern Idaho. *Journal of Wildlife Management* 61:75–85.

Knick, S. T., and J. T. Rotenberry. 1995. Landscape characteristics of fragmented shrubsteppe habitats and breeding passerine birds. *Conservation Biology* 9:1059–1071.

———. 1997. Landscape characteristics of disturbed shrubsteppe habitats in southwestern Idaho (USA). *Landscape Ecology* 12:287–297.

———. 1998. Limitations to mapping habitat use areas in changing landscapes using the Mahalanobis distance statistic. *Journal of Agricultural, Biological, and Environmental Statistics* 3:311–322.

———. 1999. Spatial distribution of breeding passerine bird habitats in southwestern Idaho. *Studies in Avian Biology* 19:104–111.

———. 2000. Ghosts of habits past: Contribution of landscape change to current habitats used by shrubland birds. *Ecology* 81:220–227.

Knick, S. T., J. T. Rotenberry, B. A. Hoover, and G. S. Olson. 1996. Simulation of raptors, prey, and vegetation dynamics: A spatial approach. Chapter 5 in *Effects of military training and fire in the Snake River Birds of Prey National Conservation Area*, Vol. 2. BLM/IDARNG Research Project Final Report. Boise, Idaho: Snake River Field Station.

Knick, S. T., J. T. Rotenberry, and T. J. Zarriello. 1997. Supervised classification of Landsat thematic mapper imagery in a semiarid rangeland by nonparametric discriminant analysis. *Photogrammetric Engineering and Remote Sensing* 63:79–86.

Knight, D. H. 1978. Methods for sampling vegetation. Unpublished manual. Laramie: Department of Botany, University of Wyoming.

Knight, T. W., and D. W. Morris. 1996. How many habitats do landscapes contain? *Ecology* 77:1756–1764.

Knopf, F. L. 1985. Significance of riparian vegetation to breeding birds across an altitudinal cline. In *Riparian ecosystems and their management: Reconciling conflicting uses*, technical coordinators R. R. Johnson, C. D. Ziebell, D. R. Patton, P. F. Ffolliot, and R. H. Hamre, 105–111. General Technical Report RM-120. Fort Collins, Colo.: USDA Forest Service, Rocky Mountain Forest and Range Experiment Station.

———. 1986. Changing landscapes and the cosmopolitism of the eastern Colorado avifauna. *Wildlife Society Bulletin* 14:132–142.

Knopf, F. L., J. A. Sedgwick, and D. B. Inkley. 1990. Regional correspondence among shrubsteppe bird habitats. *Condor* 92:45–53.

Kochert, M. N., and M. W. Collopy. 1998. Relevance of research to resource managers and policy makers. In *Avian conservation: Research and management*, ed. J. M. Marzluff and R. Sallabanks, 423–430. Washington, D.C.: Island Press.

Koerth, N. E., and F. S. Guthery. 1988. Relatioships between prior rainfall and current body fat for northern bobwhites in south Texas. *Texas Journal of Agriculture and Natural Resources* 2:10–12.

Koford, C. B. 1958. Prairie dogs, whitefaces, and blue grama. *Wildlife Monographs* 3:1–78.

Kolasa, J., and S. T. A. Pickett. 1989. Ecological systems and the concept of biological organization. *Proceedings of the National Academy of Sciences* 86:8837–8841.

———, eds. 1991. *Ecological heterogeneity*. New York: Springer-Verlag.

Koloszar, J. A., and J. W. Bailey. 1998. Monitoring of the black-capped vireo (*Vireo atricapillus*) during 1998 on Fort Hood, Texas. In *1998 annual report*, 13–49. Fort Hood: Nature Conservancy of Texas.

Korschgen, C. E. 1979. *Coastal waterbird colonies: Maine.* FWS/OBS-79/09. Washington, D.C.: U.S. Department of the Interior, U.S. Fish and Wildlife Service, Office of Biological Services, National Coastal Ecosystem Team.

Kosko, B. 1992. *Neural networks and fuzzy systems: A dynamical systems approach to machine intelligence.* Englewood Cliffs, N.J.: Prentice Hall.

Kotliar, N. B., and J. A. Wiens. 1990. Multiple scales of patchiness and patch structure: A hierarchical framework for the study of heterogeneity. *Oikos* 59:253–260.

Kovac, M., B. W. Murphy, and J. W. Lawrie. 1990. *Soil*

of edge effects in fragmented habitats. *Biological Conservation* 55:77–97.

Lauver, C. L., and J. L. Whistler. 1993. A hierarchical classification of Landsat TM imagery to identify natural grassland areas and rare species habitat. *Photogrammetric Engineering and Remote Sensing* 59:627–634.

Lawton, J. H. 1999. Are there general laws in ecology? *Oikos* 84:177–192.

Lawton, J. H., and C. G. Jones. 1995. Linking species and ecosystems: Organisms as ecosystem engineers. In *Linking species and ecosystems*, ed. C. G. Jones and J. H. Lawton, 141–150. London: Chapman and Hall.

Lawton, J. H., S. Nee, A. J. Letcher, and P. H. Harvey. 1994. Animal distributions: Patterns and processes. In *Large-scale ecology and conservation biology*, ed. P. J. Edwards, R. M. May, and N. R. Webb, 41–58. Boston: Blackwell Scientific Publications.

Layher, W. G., O. E. Maughan, and W. D. Warde. 1987. Spotted bass habitat suitability related to fish occurrence and biomass and measurements of physicochemical variables. *North American Journal of Fisheries Management* 7:238–251.

Laymon, S. A. 1980. Feeding and nesting behavior of the yellow-billed cuckoo in the Sacramento Valley. Wildlife Management Branch Administrative Report 80-2. Sacramento: State of California, Resources Agency, Department of Fish and Game.

———. 1988. Ecology of the spotted owl in the central Sierra Nevada, California. Ph.D. dissertation, University of California, Berkeley.

Laymon, S. A., and R. H. Barrett. 1986. Developing and testing habitat-capability models: Pitfalls and recommendations. In *Wildlife 2000: Modeling habitat relationships of terrestrial vertebrates*, ed. J. Verner, M. L. Morrison, and C. J. Ralph, 87–92. Madison: University of Wisconsin Press.

Laymon, S. A., and M. D. Halterman. 1987a. Can the western subspecies of the yellow-billed cuckoo be saved from extinction? *Western Birds* 18:19–25.

———. 1987b. Distribution and status of the yellow-billed cuckoo in California 1986–1987. Final Report to the Nongame Bird and Mammal Section, contract number C-1845. Sacramento: Wildlife Management Division, California Department of Fish and Game.

———. 1989. A proposed habitat management plan for yellow-billed cuckoos in California. In *Proceedings of the California Riparian Systems Conference: Protection, management and restoration for the 1990's*, technical coordinator D. L. Abell, 272–277. General Technical Re-

port PSW-110. Berkeley, Calif.: USDA Forest Servic, Pacific Southwest Forest and Range Experiment Station.

Laymon, S. A., and J. A. Reid. 1986. Effects of grid-cell size on tests of a spotted owl HSI model. In *Wildlife 2000: Modeling habitat relationships of terrestrial vertebrates*, ed. J. Verner, M. L. Morrison, and C. J. Ralph, 93–96. Madison: University of Wisconsin Press.

Laymon, S. A., P. L. Williams, and M. D. Halterman. 1997. Breeding status of the yellow-billed cuckoo in the South Fork Kern River Valley, Kern County, California: Summary report 1985–1996. Weldon, Calif.: USDA Forest Service, Sequoia National Forest, Cannel Meadow Ranger District. Kern River Research Center.

Leathwick, J. R. 1995. Climatic relationships of some New Zealand forest tree species. *Journal of Vegetation Science* 6:237–248.

———. 1998. Are New Zealand's *Nothofagus* species in equilibrium with their environment? *Journal of Vegetation Science* 9:719–732.

Leathwick, J. R., and N. D. Mitchell. 1992. Forest pattern, climate and vulcanism in central North Island, New Zealand. *Journal of Vegetation Science* 3:603–616.

Leathwick, J. R., D. Whitehead, and M. McLeod. 1996. Predicting changes in the composition of New Zealand's indigenous forests in response to global warming: A modelling approach. *Environmental Software* 11:81–90.

Lebreton, J. D., K. P. Burnham, J. Clobert, and D. R. Anderson. 1992. Modeling survival and testing biological hypotheses using marked animals: A unified approach with case studies. *Ecological Monographs* 62:67–118.

Le Duc, M. G., M. O. Hill, and T. H. Sparks. 1992. A method for predicting the probability of species occurrence using data from systematic surveys. *Watsonia* 19:97–105.

Lee, K. N. 1993. *Compass and gyroscope: Integrating science and politics for the environment*. Washington, D.C.: Island Press.

———. 1999. Appraising adaptive management. *Conservation Ecology* [online] 2:3. http://www.consecol.org/Journal/vol3/iss2/art3.

Lees, B. G., and K. Ritman. 1991. Decision-tree and rule-induction approach to integration of remotely sensed and GIS data in mapping vegetation in disturbed or hilly environments. *Environmental Management* 15:823–831.

Lefkovitch, L. P. 1965. The study of population growth in organisms grouped by stages. *Biometrics* 21:1–18.

Leftwich, K. N., P. L. Angermeier, and C. A. Dolloff. 1997. Factors influencing behavior and transferability of habitat models for a benthic stream fish. *Transactions of the American Fisheries Society* 126:725–734.

Legendre, P. 1993. Spatial autocorrelation: Trouble or new paradigm? *Ecology* 74:1659–1673.

Legendre, P., and M. Fortin. 1989. Spatial pattern and ecological analysis. *Vegetatio* 80:107–138.

Legendre, P., and L. Legendre. 1998. *Numerical ecology.* 2nd ed. New York : Elsevier Scientific Publishing.

Leger, D. W., D. H. Owings, and R. G. Coss. 1983. Behavioral ecology of time allocation in California ground squirrels (*Spermophilus beecheyi*): Microhabitat effects. *Journal of Comparative Psychology* 97:283–291.

Lehman, J. T. 1986. The goal of understanding in limnology. *Limnology and Oceanography* 31:1160–1166.

Lehmann, A. 1998. GIS modeling of submerged macrophyte distribution using generalized additive models. *Plant Ecology* 139:113–124.

Lehtinen, R. M., S. M. Galatowitsch, and J. R. Tester. 1999. Consequences of habitat loss and fragmentation for wetland amphibian assemblages. *Wetlands* 19:1–12.

Leigh, E. G. 1981. The average lifetime of a population in a varying environment. *Journal of Theoretical Biology* 90:213–239.

Lek, S., A. Belaud, P. Baran, I. Dimopoulos, and M. Delacoste. 1996a. Role of some environmental variables in trout abundance models using neural networks. *Aquatic Living Resources* 9:23–29.

Lek, S., M. Delacoste, P. Baran, I. Dimopoulos, J. Lauga, and S. Aulagnier. 1996b. Application of neural networks to modelling nonlinear relationships in ecology. *Ecological Modelling* 90:39–52.

Lek, S., I. Dimopoulos, and A. Fabre. 1996c. Predicting phosphorus concentrations and phosphorus loads from watershed characteristics using backpropagation neural networks. *Acta Ecologica* 17:43–53.

Lek, S., and J. F. Guégan. 1999. Artificial neural networks as a tool in ecological modelling, an introduction. *Ecological Modelling* 120:65–73.

Lele, S., M. L. Taper, and S. Gage. 1998. Statistical analysis of population dynamics in space and time using estimating functions. *Ecology* 79:1489–1502.

Lemons, J. 1996. The conservation of biodiversity: Scientific uncertainty and the burden of proof. In *Scientific uncertainty and environmental problem-solving*, ed. J. Lemons. Cambridge: Blackwell Science.

Lenat, D. R. 1988. Water quality assessment of streams using a qualitative collection method for benthic macroinvertebrates. *Journal of the North American Benthological Society* 7:222–233.

Leopold, A. 1933. *Game management.* New York: Charles Scribner's Sons.

Lescourret, F., and M. Genard. 1994. Habitat, landscape and bird composition in mountain forest fragments. *Journal of Environmental Management* 40:317–328.

Lesica, P., and F. W. Allendorf. 1992. Are small populations of plants worth preserving? *Conservation Biology* 6:135–139.

Leslie, P. H. 1945. On the use of matrices in certain population mathematics. *Biometrika* 33:183–212.

Levin, S. A. 1992. The problem of pattern and scale in ecology. *Ecology* 73:1943–1967.

Levins, R. 1966. The strategy of model building in population biology. *American Scientist* 54:421–431.

———. 1968. *Evolution in changing environments: Some theoretical explorations.* Princeton: Princeton University Press.

———. 1969. Some demographic and genetic consequences of environmental heterogeneity for biological control. *Bulletin of the Entomological Society of America* 15:237–240.

———. 1970. Extinction. In *Some mathematical problems in biology*, ed. M. Gerstenhaber, 75–107. Providence: American Mathematical Society.

Li, X., and D. Li. 1998. Current state and the future of the crested ibis (*Nipponia nippon*): A case study by population viability analysis. *Ecological Research* 13:323–333.

Lidicker, W. Z., Jr. 1975. The role of dispersal in the demography of small mammals. In *Small mammals: Their productivity and population dynamics*, ed. F. B. Golley, K. Petusewicz, and L. Ryszkowski, 103–128. New York: Cambridge University Press.

———. 1995. The landscape concept: Something old, something new. In *Landscape approaches in mammalian ecology and conservation*, ed. W. Z. Lidicker, 3–19. Minneapolis: University of Minnesota Press.

Ligon, J. S. 1961. *New Mexico birds, and where to find them.* Albuquerque: University of New Mexico Press.

Lillesand, T., J. Chipman, D. Nagel, H. Reese, M. Bobo, and R. Goldmann. 1998. Upper Midwest gap analysis program image processing protocol. EMTC 98-G001. Onalaska, Wis.: Report prepared for the U.S. Geological Survey, Environmental Management Technical Center.

Lima, S. L., and P. A. Zollner. 1996. Towards a behavioral ecology of ecological landscapes. *Trends in Ecology and Evolution* 11:131–135.

Lindenmayer, D. B., R. B. Cunningham, and C. F. Donnelly. 1993. The conservation of arboreal marsupials in the montane ash forests of the central highlands of Victoria, south-east Australia, IV. The presence and abundance of arboreal marsupials in retained linear habitats (wildlife corridors) within logged forest. *Biological Conservation* 66:207–221.

———. 1994. The conservation of arboreal marsupials in the montane ash forests of the central highlands of Victoria, south-eastern Australia, VI. The performance of statistical models of the nest tree and habitat requirements of arboreal marsupials applied to new survey data. *Biological Conservation* 70:143–147.

Lindenmayer, D. B., R. B. Cunningham and M. A. McCarthy. 1999. The conservation of arboreal marsupials in the montane ash forests of the central highlands of Victoria, southeastern Australia, VIII. Landscape analysis of the occurrence of arboreal marsupials. *Biological Conservation* 89:83–92.

Lindenmayer, D. B., R. B. Cunningham, H. A. Nix, M. T. Tanton, and A. P. Smith. 1991a. Predicting the abundance of hollow-bearing trees in montane forests of southeastern Australia. *Australian Journal of Ecology* 16:91–98.

Lindenmayer, D. B., R. B. Cunningham, M. T. Tanton, H. A. Nix, and A. P. Smith. 1991b. The conservation of arboreal marsupials in the montane ash forests of the central highlands of Victoria, south-east Australia, III. The habitat requirements of leadbeater's possum (*Gymnobelideus leadbeateri*) and models of the diversity and abundance of arboreal marsupials. *Biological Conservation* 56:295–315.

Lindenmayer, D. B., R. B. Cunningham, M. T. Tanton, A. P. Smith, and H. A. Nix. 1990a. The conservation of arboreal marsupials in the montane ash forests of the Central Highlands of Victoria, south-east Australia, I. Factors influencing the occupancy of trees with hollows. *Biological Conservation* 54:111–131.

———. 1990b. Habitat requirements of the mountain brushtail possum and the greater glider in the montane ash-type eucalypt forests of the central highlands of Victoria. *Australian Wildlife Research* 17:467–478.

———. 1991c. Characteristics of hollow-bearing trees occupied by arboreal marsupials in montane ash forests of the central highlands of Victoria, south-east Australia. *Forest Ecology and Management* 40:289–308.

Lindenmayer, D. B., and R. C. Lacy. 1995a. Metapopulation viability of arboreal marsupials in fragmented old-growth forests: Comparison among species. *Ecological Applications* 5:183–199.

———. 1995b. Metapopulation viability of Leadbeater's possum, *Gymnobelideus leadbeateri*, in fragmented old-growth forests. *Ecological Applications* 5:164–182.

Lindenmayer, D. B., H. A. Nix, J. P. McMahon, M. F. Hutchinson, and M. T. Tanton. 1991d. The conservation of Leadbeater's possum, *Gymnobelideus leadbeateri* (McCoy): A case study of the use of bioclimatic modelling. *Journal of Biogeography* 18:371–383.

Lindenmayer, D. B., and H. P. Possingham. 1995. Modelling the impacts of wildfire on the viability of metapopulations of the endangered Australian species of arboreal marsupial, Leadbeater's possum. *Forest Ecology and Management* 74:197–222.

———. 1996. Ranking conservation and timber management options for Leadbeater's possum in southeastern Australia using population viability analysis. *Conservation Biology* 10:235–251.

Lindenmayer, D. B., K. Ritman, R. B. Cunningham, J. D. B. Smith, and D. Horvath. 1995. A method for predicting the spatial distribution of arboreal marsupials. *Wildlife Research* 22:445–456.

Lindley, D. V. 1985. *Making decisions.* 2nd ed. London: John Wiley & Sons.

Link, W. A., R. J. Barker, J. R. Sauer, and S. Droege. 1994. Within-site variability in surveys of wildlife populations. *Ecology* 75:1097–1108.

Link, W. A., and J. R. Sauer. 1997a. Estimation of population trajectories from count data. *Biometrics* 53:63–72.

———. 1997b. New approaches to the analysis of population trends in land birds: A comment. *Ecology* 78:2632–2634.

———. 1998. Estimating population change from count data: Application to the North American breeding bird survey. *Ecological Applications* 8:258–268.

Lischke, H., A. Guisan, A. Fischlin, and H. Bugmann. 1998. Vegetation responses to climate change in the Alps—modeling studies. In *A view from the Alps: Regional perspectives on climate change*, ed. P. Cebon, U. Dahinden, H. Davies, D. Imboden, and C. Jaeger, 309–350. Boston: MIT Press.

Littell, R. C., G. A. Milliken, W. W. Stroup, and R. D. Wolfinger. 1996. *SAS system for mixed models.* Cary, N.C.: SAS Institute.

Litvaitis, J. A. 1993. Response of early successional vertebrates to historic changes in land use. *Conservation Biology* 7:866–873.

Livingston, S. A., C. S. Todd, W. B. Krohn, and R. B. Owen. 1990. Habitat models for nesting bald eagles in Maine. *Journal of Wildlife Management* 54:644–653.

Lloyd, M. 1967. Mean crowding. *Journal of Animal Ecology* 36:1–30.

Lodwick, W. A., W. Monson, and L. Svoboda. 1990. Attribute error and sensitivity analysis of map operations in geographical information systems: Suitability analysis. *International Journal of Geographical Information Science* 4:413–428.

Lohr, S. C., and K. D. Fausch. 1997. Multiscale analysis of natural variability in stream fish assemblages of a western Great Plains watershed. *Copeia* 1997:706–724.

Lokemoen, J. T., and H. F. Duebbert. 1976. Ferruginous hawk nesting ecology and raptor population in northern South Dakota. *Condor* 78:464–470.

Long, J. S. 1997. *Regression models for categorical and limited dependent variables*. Thousand Oaks, Calif.: Sage Publications.

Longino, J. T., and R. K. Colwell. 1997. Biodiversity assessment using structured inventory: Capturing the ant fauna of a tropical rain forest. *Ecological Applications* 7:1263–1277.

Loomis, J., and J. C. Echohawk. 1999. Using GIS to identify underrepresented ecosystems in the National Wilderness Preservation System in the USA. *Environmental Conservation* 26:53–58.

Lopez, W. S. 1998. Application of the HEP methodology and use of GIS to identify priority sites for the management of white-tailed deer. In *GIS methodologies for developing conservation strategies*, ed. B. G. Savitsky and T. E. Larcher Jr., 127–137. New York: Columbia University Press.

Lord, L. A., and T. D. Lee. 2001. Interactions of local and regional processes: Species richness in tussock sedge communities. *Ecology* 82:313–318.

Loucks, O. L. 1970. Evolution of diversity, efficiency, and community stability. *American Zoologist* 10:17–25.

Love, J. D., and A. C. Christiansen. 1985. *Geologic map of Wyoming*. Reston, Va.: U.S. Department of the Interior, U.S. Geological Survey.

Loveland, T. R., J. W. Merchant, J. F. Brown, D. O. Ohlen, B. C. Reed, P. Olsen, and J. Hutchinson. 1995. Seasonal land cover regions of the United States. *Annals of the Association of American Geographers* 85:339–355.

Loveland, T. R., J. W. Merchant, D. J. Ohlen, and J. F. Brown. 1991. Development of a land-cover characteristics database for the conterminous U.S. *Photogrammetry Engineering and Remote Sensing* 57:1453–1463.

Loveless, C. M. 1959. The Everglades deer herd: Life history and management. Technical Bulletin 6. Tallahassee: Florida Game and Fresh Water Fish Commission.

Lubchenco, J. 1978. Plant species diversity in a marine intertidal community: Importance of herbivore food preference and algal competitive ability. *American Naturalist* 112:23–39.

Luh, H.-K., C. Abbott, M. Berry, E. J. Comiskey, J. Dempsey, and L. J. Gross. 1997. Parallelization in a spatially explicit individual-based model (I) Spatial data interpolation. *Computers and Geosciences* 23:293–304.

Lunetta, R. S., R. G. Congalton, L. K. Fenstermaker, J. R. Jensen, K. C. McGwire, and L. R. Tinney. 1991. Remote sensing and geographic information system data integration: Error sources and research issues. *Photogrammetric Engineering and Remote Sensing* 57:677–687.

Luoma, D. L. 1991. Annual changes in seasonal production of hypogeous sporocarps in Oregon Douglas-fir forests. In *Wildlife and vegetation of unmanaged Douglas-fir forests*, technical coordinators L. F. Ruggiero, K. B. Aubry, A. B. Carey, and M. M. Huff, 83–89. General Technical Report PNW-GTR-285. Portland, Ore.: USDA Forest Service, Pacific Northwest Research Station.

Luoma, D. L., J. L. Eberhart, and M. P. Amaranthus. 1996. Response of ectomycorrhizal fungi to forest management treatments—sporocarp production. In *Mycorrhizas in integrated systems: From genes to plant development*, ed. C. Azcon-Aguilat and J. M. Barea, 553–556. Luxembourg: Office for Official Publications of the European Communities, European Commission.

Luoma, D. L., R. E. Frenkel, and J. M. Trappe. 1991. Fruiting of hypogeous fungi in Oregon Douglas-fir forests: Seasonal and habitat variation. *Mycologia* 83:335–353.

Lusk, J. J., F. S. Guthery, and S. J. DeMaso. 2001. Northern bobwhite (*Colinus virginianus*) abundance in relation to yearly weather and long-term climate patterns. *Ecological modelling*. In press.

Luttschwager, K. A., K. F. Higgins, and J. A. Jenks. 1994. Effects of emergency haying on duck nesting in conservation reserve program fields, South Dakota. *Wildlife Society Bulletin* 22:403–408.

Lynch, J. D. 1985. Annotated checklist of the amphibians and reptiles of Nebraska. *Transactions of the Nebraska Academy of Sciences* 13:33–57.

Lynch, J. F., and D. F. Whigham. 1984. Effect of forest fragmentation on breeding bird communities in Maryland, USA. *Biological Conservation* 28:287–324.

Lynn, H., C. L. Mohler, S. D. DeGloria, and C. E. McCulloch. 1995. Error assessment in decision-tree models applied to vegetation analysis. *Landscape Ecology* 10: 323–335.

Lyon, J. G., J. T. Heinen, R. A. Mead, and N. E. G. Roller. 1987. Spatial data for modelling wildlife habitat. *Journal of Surveying Engineering* 113:88–100.

Lyons, J. 1992. The length of stream to sample with a towed electrofishing unit when fish species richness is estimated. *North American Journal of Fisheries Management* 12:198–203.

Ma, Z. 1995. Using a rule-based merging algorithm to eliminate "salt/pepper" and small regions of classified image. In *Ninth annual symposium on geographic information systems in natural resources management*, 834–837. Vancouver, BC: GIS World.

Ma, Z., and R. L. Redmond. 1995. Tau coefficients for accuracy assessment of classifications of remote sensing data. *Photogrammetric Engineering and Remote Sensing* 61:435–439.

MacArthur, R. H. 1958. Population ecology of some warblers of northeastern coniferous forests. *Ecology* 39: 599–619.

———. 1964. Environmental factors affecting bird species diversity. *American Naturalist* 68:387–397.

———. 1968. The theory of the niche. In *Population biology and evolution*, ed. R. C. Lewontin, 159–176. Syracuse: Syracuse University Press.

———. 1972a. Coexistence of species. In *Challenging biological problems*, ed. J. Behnke, 253–259. Oxford: Oxford University Press.

———. 1972b. *Geographical ecology: Patterns in the distribution of species*. New York: Harper & Row.

MacArthur, R. H., and J. W. MacArthur. 1961. On bird species diversity. *Ecology* 42: 594–598.

MacArthur, R. H., and E. O. Wilson. 1967. *The theory of island biogeography*. Princeton: Princeton University Press.

Macdonald, D. W., F. Mitchelmore, and P. J. Bacon. 1996. Predicting badger sett numbers: Evaluating methods in East Sussex. *Journal of Biogeography* 23:649–655.

MacDonald, K. 1997. Site fidelity and its effects on survival of *Odocoileus virginianus seminolus* during a catastrophic flood in the Everglades. M.S. thesis, University of Florida, Gainesville.

MacFadyen, A. 1957. *Animal ecology*. London: Pitman.

Machlis, G. E. 1992. The contribution of sociology to biodiversity research and management. *Biological Conservation* 62:161–170.

MacKenzie, M. D. 1993. The vegetation of Great Smoky Mountains National Park: Past, present, and future. Ph.D. dissertation, University of Tennessee, Knoxville.

Mackey, B. G. 1994. Predicting the potential distribution of rainforest structural characteristics. *Journal of Vegetation Science* 5:43–54.

Mackey, B. G., I. Mullen, D. W. McKenney, and R. Cunningham. Unpublished. Issues associated with modelling and spatially predicting the climatic domain of tree species using logistic regression: A case study based on jack pine (*Pinus banksiana* Lamb.) in Ontario, Canada.

MacMahon, J. A., D. J. Schimpf, D. C. Anderson, K. G. Smith, and R. L. Bayn Jr. 1981. An organism-centered approach to some community and ecosystem concepts. *Journal of Theoretical Biology* 88:287–307.

Mac Nally, R. C. 1990a. An analysis of density responses of forest and woodland birds to composite physiognomic variables. *Australian Journal of Ecology* 15:267–275.

———. 1990b. The roles of floristics and physiognomy in avian community composition. *Australian Journal of Ecology* 15:321–327.

Mac Nally, R., and G. P. Quinn. 1998. Symposium introduction: The importance of scale in ecology. *Australian Journal of Ecology* 23:1–7.

Mac Nally, R., E. Fleishman, J. P. Fay, and D. D. Murphy. 2002. Modeling butterfly species richness using mesoscale environmental variables: Model construction and validation. *Biological Conservation*: in press.

Maehr, D. S., R. C. Belden, E. D. Land, and L. Wilkins. 1990. Food habits of panthers in southwest Florida. *Journal of Wildlife Management* 54:420–423.

Maehr, D. S., and J. R. Brady. 1986. Food habits of bobcats in Florida. *Journal of Mammalogy* 67:133–138.

Magder, L. S., and J. P. Hughes. 1997. Logistic regression when the outcome is measured with uncertainty. *Amercian Journal of Epidemiology* 146:195–203.

Magder, L. S., M. A. Sloan, S. H. Duh, J. F. Abate, and S. J. Kittner. 2000. Utilization of multiple imperfect assessments of the dependent variable in a logistic regression analysis. *Statistics in Medicine* 19:99–111.

Magnuson, J. J., W. M. Tonn, A. Banerjee, J. Toivonen, O. Sanchez, and M. Rask. 1998. Isolation vs. extinction in the assembly of fishes in small northern lakes. *Ecology* 79:2941–2956.

Maguire, L. A., U. S. Seal, and P. F. Brussard. 1987. Managing critically endangered species: The Sumatran rhino as a case study. In *Viable populations for conservation*, ed.

M. E. Soulé, 141–158. Cambridge: Cambridge University.

Maguire, L. A., G. F. Wilhere, and Q. Dong. 1995. Population viability analysis for red-cockaded woodpeckers in the Georgia piedmont. *Journal of Wildlife Management* 59:533–542.

Maier, H. R., G. C. Dandy, and M. D. Burch. 1998. Use of artificial neural networks for modelling cyanobacteria *Anabaena spp.* in the River Murray, South Australia. *Ecological Modelling* 105:257–272.

Maisongrande, P., A. Ruimy, G. Dedieu, and B. Saugier. 1995. Monitoring seasonal and interannual variations of gross primary productivity, net primary productivity and net ecosystem productivity using a diagnostic model and remotely sensed data. *Tellus* 47B:178–190.

Majumdar, S. K. F. J. Brenner, J. E. Lovich, J. F. Schalles, and E. W. Miller, eds. 1994. *Biological diversity: Problems and challenges.* Easton, Pa.: Pennsylvania Academy of Science.

Makse, H. A., J. S. Andrade Jr., M. Batty, S. Havlin, and H. E. Stanley. 1998. Modeling urban growth patterns with correlated percolation. *Physical Review E* 58: 7054–7062.

Makse, H. A., S. Havlin, and H. E. Stanley. 1995. Modelling urban growth patterns. *Nature* 377:608–612.

Malanson, G. P. 1993. *Riparian landscapes.* New York: Cambridge University Press.

Malcolm, J. R. 1994. Edge effects in central Amazonian forest fragments. *Ecology* 75: 2438–2445.

Malkinson, D., and N. T. Hobbs. 1995. Unpublished manuscript. Fort Collins, Colo.: Natural Resource Ecology Lab, Colorado State University.

Maller, R. A. 1990. Some aspects of a mixture model for estimating the boundary of a set of data. *Journal du Conseil International pour l'Exploration de la Mer* 46: 140–147.

Manel, S., and D. Debouzie. 1997. Modeling insect development time of two or more larval stages in the field under variable temperatures. *Environmental Entomology* 26:163–169.

Manel, S., J.-M. Dias, and S. J. Ormerod. 1999. Comparing discriminant analysis, neural networks and logistic regression for predicting species distributions: A case study with a Himalayan bird. *Ecological Modelling* 120: 337–347.

Manel, S., Ceri Williams, H., Ormerod, S. J. Evaluating presence-absence models in ecology: The need to account for prevalence. *Journal of Applied Ecology* (in press).

Mangel, M. and C. W. Clark. 1988. *Dynamic modeling in behavioral ecology.* Princeton, N.J.: Princeton University Press.

Manly, B. F. J. 1997. *Randomization, bootstrap, and Monte Carlo methods in biology.* 2nd ed. London: Chapman and Hall.

Manly, B. F. J., L. L. McDonald, and D. L. Thomas. 1993. *Resource selection by animals: Statistical design and analysis for field studies.* London: Chapman and Hall.

Mannan, R. W., M. L. Morrison, and E. C. Meslow. 1984. Comment: The use of guilds in forest bird management. *Wildlife Society Bulletin* 12:426–430.

Mantel, N. 1967. The detection of disease clustering and a generalized regression approach. *Cancer Research* 27: 209–220.

Marangio, M. S., and R. Morgan. 1987. The endangered sandhills plant communities of Santa Cruz County. In *Conservation and management of rare and endangered plants: Proceedings of a California conference on the conservation and management of rare and endangered plants,* ed. T. S. Elias, 267–273. Sacramento: California Native Plant Society.

Marcot, B. G. 1986. Biometric approaches to modeling—the manager's viewpoint. In *Wildlife 2000: Modeling habitat relationships of terrestrial vertebrates,* ed. J. Verner, M. L. Morrison, and C. J. Ralph, 203–204. Madison: University of Wisconsin Press.

Marcot, B. G., and R. Holthausen. 1987. Analyzing population viability of the spotted owl in the Pacific Northwest. *Transactions of the North American Wildlife and Natural Resources Conference* 52:333–347.

Marcot, B. G., M. G. Raphael, and K. H. Berry. 1983. Monitoring wildlife habitat and validation of wildlife-habitat relationships models. *Transactions of the North American Wildlife and Natural Resources Conference* 48: 315–329.

Margules, C. R., and M. P. Austin, eds. 1991. *Nature conservation: Cost effective biological surveys and data analysis.* Melbourne: CSIRO Australia.

Margules, C. R., and T. R. Redhead. 1995. *BioRap: Guidelines for using the BioRap methodology and tools.* Melbourne: CSIRO Australia.

Margules, C. R., and J. L. Stein. 1989. Patterns in the distributions of species and the selection of nature reserves: An example from *Eucalyptus* forests of south-eastern New South Wales. *Biological Conservation* 50:219–238.

Marks, D. 1990. The sensitivity of potential evapotranspiration to climate change over the continental United States. In *Biospheric feedbacks to climate change: The sensitivity of regional trace gas emissions, evapotranspiration, and energy balance to vegetation redistribution*, ed. H. Gucinski, D. Marks, and D. P. Turner, IV-1 to IV-31. EPA/600/3-90/078. Washington, D.C.: U.S. Environmental Protection Agency, Office of Research and Development.

Marmontel, M., S. R. Humphrey, and T. J. O'Shea. 1997. Population viability analysis of the Florida manatee (*Trichechus manatus latirostris*), 1976–1991. *Conservation Biology* 11:467–481.

Marsden, S., and A. H. Fielding. 1999. Habitat associations of parrots on the islands of Buru, Seram and Sumba. *Journal of Biogeography* 26:439–446.

Martin, A. C., H. S. Zim, and A. L. Nelson. 1951. *American wildlife and plants: A guide to wildlife food habits.* New York: Dover Publications.

Martin, J. W., and B. A. Carlson. 1998. Sage sparrow (*Amphispiza belli*). In *The birds of North America, No. 326*, ed. A. Poole and F. Gill. Philadelphia: The Birds of North America.

Martin, T. E. 1992. Breeding productivity considerations: What are the appropriate habitat features for management? In *Ecology and conservation of Neotropical migrant landbirds*, ed. J. M. Hagan and D. W. Johnston, 455–473. Washington, D.C.: Smithsonian Institution Press.

———. 1995. Avian life history evolution in relation nest sites, nest predation, and food. *Ecological Monographs* 65:101–127.

Martin, T. E., and G. R. Geupel. 1993. Nest-monitoring plots: Methods for locating nests and monitoring success. *Journal of Field Ornithology* 64:507–519.

Martin, W. H., S. G. Boyce, and A. C. Echternacht, eds. 1993. *Biodiversity of the southeastern United States*. Vol. 2, Upland Terrestrial Communities. New York: John Wiley & Sons.

Martinka, R. R. 1972. Structural characteristics of blue grouse territories in southwestern Montana. *Journal of Wildlife Management* 36:498–510.

Marzluff, J. M., and R. Sallabanks, eds. 1998. *Avian conservation*. Washington, D.C.: Island Press.

Master, L. 1996. Predicting distributions for vertebrate species: Some observations. In *Gap analysis: A landscape approach to biodiversity planning*, ed. J. M. Scott, T. H. Tear, and F. W. Davis, 171–176. Bethesda, Md.: American Society for Photogrammetry and Remote Sensing.

Mastrorillo, S., S. Lek, F. Dauba, and A. Belaud. 1997. The use of artificial neural networks to predict the presence of small-bodied fish in a river. *Freshwater Biology* 38: 237–246.

MathSoft. 1995. *S-PLUS guide to statistical and mathematical analysis, version 3.3*. Seattle: StatSci Division, MathSoft.

Matsumoto, T. A. 1999. Sampling threatened and endangered species with non-constant occurrence and detectability: A sensitivity analysis of power when sampling low-occurrence populations with varying probability parameters. M.S. thesis, Humboldt State University, Arcata, Calif.

Matthews, W. J., and L. G. Hill. 1979. Influence of physicochemical factors on habitat selection by red shiners, *Notropis lutrensis* (Pisces: Cyprinidae). *Copeia* 1979: 70–81.

Matthews, W. J., and E. G. Zimmerman. 1990. Potential effects of global warming on native fishes of the southern Great Plains and the Southwest. *Fisheries* 15(6):26–32.

Maudsley, M., and J. Marshall, eds. 1999. *Heterogeneity in landscape ecology: Pattern and scale*. Aberdeen, Scotland, UK: International Association of Landscape Ecology.

Maurer, B. A. 1985. Avian community dynamics in desert grasslands: Observational scale and hierarchial structure. *Ecological Monographs* 55:295–312.

———. 1986. Predicting habitat quality for grassland birds using density-habitat correlations. *Journal of Wildlife Management* 50:556–566.

———. 1990. The relationship between distribution and abundance in a patchy environment. *Oikos* 58:181–189.

———. 1993. Are cowbirds increasing in abundance and expanding their geographic range? Evidence from the breeding bird survey. Abstract presented at the North American Research Workshop on the Ecology and Management of Cowbirds, sponsored by The Nature Conservancy of Texas, Austin, Tex., November 4–5, 1993.

———. 1994. *Geographical population analysis: Tools for the analysis of biodiversity*. Boston: Blackwell Scientific Publications.

———. 1998. Ecological science and statistical paradigms: At the threshold. *Science* 279:502–503.

———. 1999. *Untangling ecological complexity: The macroscopic perspective*. Chicago: University of Chicago Press.

Maxwell, J. R., C. J. Edwards, M. E. Jensen, S. J. Paustian, H. Parrott, and D. M. Hill. 1995. A hierarchical framework of aquatic ecological units in North America (Nearctic Zone). General Technical Report NC-176. St.

Paul, Minn.: USDA Forest Service, North Central Forest Experiment Station.

May, R. M. 1986. The search for patterns in the balance of nature: Advances and retreats. *Ecology* 67:1115–1126.

Mayden, R. ed. 1992. *Systematics, historical ecology, and North American freshwater fishes.* Stanford, Calif.: Stanford University Press.

Mayer, D. G., and D. G. Butler. 1993. Statistical validation. *Ecological Modelling* 68:21–32.

Mayer, K. E., and W. F. Laudenslayer Jr., eds. 1988. *A guide to wildlife habitats of California.* Sacramento: California Department of Forestry and Fire Protection.

Mayr, E. 1982. *The growth of biological thought: Diversity, evolution, and inheritance.* Cambridge, Mass.: Harvard University Press.

McArdle, B. H. 1990. When are rare species not there? *Oikos* 57:276–277.

McArthur, W. M. 1991. *Reference soils of south-western Australia.* Perth, Western Australia: Department of Agriculture, Western Australia on behalf of the Australian Society of Soil Science.

McAuliffe, J. R. 1994. Landscape evolution, soil formation, and ecological patterns and processes in Sonoran Desert bajadas. *Ecological Monographs* 64:111–148.

McCarthy, M. A., M. A. Burgman, and S. Ferson. 1995. Sensitivity analysis for models of population viability. *Biological Conservation* 73:93–100.

McClelland, B. R. 1977. Relationships between hole-nesting birds, forest snags, and decay in western larch–Douglas-fir forests of the northern Rocky Mountains. Ph.D. dissertation, University of Montana, Missoula.

McCloy, K. R. 1995. *Resource management information systems.* London: Taylor and Francis.

McCoy, E. D., and S. S. Bell. 1991. Habitat structure: The evolution and diversification of a complex topic. In *Habitat structure: The physical arrangement of objects in space,* ed. S. S. Bell, E. D. McCoy, and H. R. Mushinsky, 3–21. New York: Chapman and Hall.

McCullagh, P., and J. A. Nelder. 1989. *Generalized linear models.* 2nd ed. London: Chapman and Hall.

McCullagh, P. 1980. Regression models for ordinal data. *Journal of the Royal Statistical Society Series B.* 42: 109–142.

McCullagh, P., and J. A. Nelder. 1983. *Generalized linear models: Monographs on statistics and applied probability.* London: Chapman and Hall.

_____. 1989. *Generalized linear models.* 2nd ed. London: Chapman and Hall.

McCullough, D. R. 1996. *Metapopulations and wildlife conservation.* Washington, D.C.: Island Press.

McCune, B., and M. J. Mefford. 1997. *PC-ORD for Windows. Version 3.0 for Windows.* Gleneden Beach, Ore.: MjM Software Design.

McDonald, K. A., and J. H. Brown. 1992. Using montane mammals to model extinctions due to global change. *Conservation Biology* 6:409–415.

McDougald, N. K., W. E. Frost, and D. E. Jones. 1989. Use of supplemental feeding locations to manage cattle use on riparian areas of hardwood rangelands. In *Proceedings of the California Riparian Systems Conference: Protection, management, restoration for the 1990s,* technical coordinator D. L. Abell, 124–126. General Technical Report PSW-110. Berkeley, Calif.: USDA Forest Service, Pacific Southwest Forest and Range Experiment Station.

McEneaney, T. 1988. *Birds of Yellowstone: A practical habitat guide to the birds of Yellowstone National Park—and where to find them.* Boulder, Colo.: Roberts Rinehart Publishers.

McGarigal, K., and B. J. Marks. 1995. FRAGSTATS: Spatial pattern analysis program for quantifying landscape structure. General Technical Report PNW GTR-351. Portland, Ore.: USDA Forest Service, Pacific Northwest Research Station.

McKelvey, K., B. R. Noon, and R. H. Lamberson. 1992. Conservation planning for species occupying fragmented landscapes: The case of the northern spotted owl. In *Biotic interactions and global change,* ed. P. M. Kareiva, J. G. Kingsolver, and R. B. Huey, 424–450. Sunderland, Mass.: Sinauer Associates.

McKenney, D. W., B. G. Mackey, and D. Joyce. 1999. SEEDWHERE: A computer tool to support seed transfer and ecological restoration decisions. *Environmental Modeling and Software* 14:589–595.

McKenney, D. W., R. S. Rempel, L. A. Venier, Y. Wang, and A. R. Bisset. 1998. Development and application of a spatially explicit moose population model. *Canadian Journal of Zoology* 76:1922–1931.

McNab, W. H. 1989. Terrain shape index: Quantifying effect of minor landforms on tree height. *Forestry Science* 35:91–104.

_____. 1993. A topographic index to quantify the effect of meso-scale landform on site productivity. *Canadian Journal of Forest Research* 23:1100–1179.

McQuarrie, A. D. R., and C. L. Tsai. 1998. *Regression and time series model selection*. Singapore: World Scientific.

Meentemeyer, V., and E. O. Box. 1987. Scale effects in landscape studies. In *Landscape heterogeneity and disturbance*, ed. M. G. Turner, 15–34. New York: Springer-Verlag.

Meents, J. K., J. Rice, B. W. Anderson, and R. D. Ohmart. 1983. Nonlinear relationships between birds and vegetation. *Ecology* 64:1022–1027.

Meffe, G. K., and C. R. Carroll. 1994. *Principles of conservation biology*. Sunderland, Mass.: Sinauer Associates.

Meisner, J. D. 1990. Effect of climate warming on the southern margins of the native range of brook trout, *Salvelinus fontinalis*. *Canadian Journal of Fisheries and Aquatic Sciences* 47:1065–1070.

Menard, S. 1995. *Applied logistic regression analysis*. Thousands Oaks, Calif.: Sage Publications.

Mendall, H. L. 1936. The home-life and economic status of the double-crested cormorant, *Phalacrocorax auritus auritus* (Lesson). The Maine Bulletin 39.3. University of Maine Studies, Second Series, No. 38.

———. 1976. Eider ducks, islands, and people. *Maine Fish and Wildlife* 18:4–7.

Mendenhall, W., R. L. Scheaffer, and D. D. Wackerly. 1981. *Mathematical statistics with applications*. Belmont, Calif.: Wadsworth.

Menge, J. A. 1985. The use and application of modeling systems to the study of VA mycorrhizal fungi. In *Proceedings of the Sixth North American Conference of Mycorrhizae*, ed. R. Molina, 142. Corvallis, Ore.: Forest Research Laboratory, Oregon State University.

Merchant, J. W., L. Yang, and W. Yang. 1994. Validation of continental-scale land cover databases derived from AVHRR data. In *Proceedings of the Pecora Twelve Symposium*, ed. L. Pettinger and R. H. Haas, 63–72. Bethesda, Md.: American Society for Photogrammetry and Remote Sensing.

Mesquita, R. C. G., P. Delamonica, and W. F. Laurance. 1999. Effects of surrounding vegetation on edge-related tree mortality in Amazonian forest fragments. *Biological Conservation* 91:129–134.

Metz, C. E. 1978. Basic principles of ROC analysis. *Seminars in Nuclear Medicine* 8:283–298.

Meyer, C. and T. D. Sisk. In press. Butterfly response to microclimatic changes following ponderosa pine restoration. *Restoration Ecology*.

Meyer, J. S., L. L. Irwin, and M. S. Boyce. 1998. Influence of habitat abundance and fragmentation on northern spotted owls in western Oregon. *Wildlife Monographs* 139:1–51.

Michaelsen, J., F. W. Davis, and M. Borchert. 1987. A nonparametric method for analyzing hierarchical relationships in ecological data. *Coenoses* 2:39–48.

Michaelsen, J., D. S. Schimel, M. A. Friedl, F. W. Davis, and R. C. Dubayah. 1994. Regression tree analysis of satellite and terrain data to guide vegetation sampling and surveys. *Journal of Vegetation Science* 5:673–686.

Michalski, R. S. 1983. A theory and methodology of inductive learning. In *Machine learning—an artificial intelligence approach*, ed. R. S. Michalski, J. G. Carbonell, and T. M. Mitchell, 83–133. Palo Alto, Calif.: Tioga.

Mickey, J., and S. Greenland. 1989. A study of the impact of confounder-selection criteria on effect estimation. *American Journal of Epidemiology* 129:125–137.

Mielke, P. W. 1986. Non-metric statistical analyses: Some metric alternatives. *Journal of Statistical Planning and Inference* 13:377–387.

Milà, B. D. Girman, M. Kimura, and T. B. Smith. 2000. Genetic evidence for the effect of a postglacial population expansion on the phylogeography of a North American songbird. *Proceedings of the Royal Academy of Science* 267:1033–1040.

Miller, K. E. 1993. Habitat use by white-tailed deer in the Everglades: Tree islands in a seasonally flooded landscape. M.S. thesis, University of Florida, Gainesville.

Miller, P. S., and R. C. Lacy. 1999. *VORTEX: A stochastic simulation of the extinction process*. Version 8 user's manual. Apple Valley, Minn.: Conservation Breeding Specialists Group.

Miller, R. 1986. Predicting rare plant distribution patterns in the southern Appalachians of the southeastern U.S.A. *Journal of Biogeography* 13:293–311.

Miller, R. I. ed. 1994a. *Mapping the diversity of nature*. London: Chapman and Hall.

Miller, R. S., and D. B. Botkin. 1974. Endangered species: Models and predictions. *American Scientist* 62:172–181.

Miller, T. W. 1994b. Model selection in tree-structured regression. In *Proceedings of the Statistical Computing Section*, 158–163. Alexandria, Va.: American Statistical Association.

Mills, L. S., and P. E. Smouse. 1994. Demographic consequences of inbreeding in remnant populations. *American Naturalist* 144:412–431.

Mills, L. S., S. G. Hayes, C. Baldwin, M. J. Wisdom, J. Citta, D. J. Mattson, and K. Murphy. 1996. Factors leading to different viability predictions for a grizzly bear data set. *Conservation Biology* 10:863–873.

Milne, B. T. 1991. Lessons from applying fractal models to landscape patterns. In *Quantitative methods in landscape ecology*, ed. M. G. Turner and R. H. Gardner, 199–235. New York: Springer-Verlag.

———. 1997. Applications of fractal geometry in wildlife biology. In *Wildlife and landscape ecology: Effects of pattern and scale*, ed. J. A. Bissonette, 32–69. New York: Springer-Verlag.

Milne, B. T., K. M. Johnston, and R. T. T. Forman. 1989. Scale-dependent proximity of wildlife habitat in a spatially neutral Bayesian model. *Landscape Ecology* 2: 101–110.

Milne, B. T., M. G. Turner, J. A. Wiens, and A. R. Johnson. 1992. Interactions between the fractal geometry of landscapes and allometric herbivory. *Theoretical Population Biology* 41:337–353.

Minar, N., R. Burkhart, C. Langton, and M. Askenazi. 1996. The swarm simulation system: A toolkit for building multi-agent simulations. http://www.santafe.edu/projects/swarm/swarmdoc/swarmdoc.html.

Minchin, P. R. 1987. Simulation of multidimensional community patterns: Towards a comprehensive model. *Vegetatio* 71:145–156.

Minnesota Department of Natural Resources. 1996. Minnesota Land Use and Cover: 1990s Census of the Land (http://mapserver.lmic.state.mn.us/landuse/). Land Management Information Center, St. Paul, Minn.

Mitchell, M. 1996. *An introduction to genetic algorithms.* Cambridge, Mass.: MIT Press.

Mitchell, N. D. 1992. The derivation of climate surfaces for New Zealand and their application to the bioclimatic analysis of the distribution of Kauri (*Agathis australis*). *Journal of the Royal Society of New Zealand* 21:13–24.

Mladenoff, D. J., and W. L. Baker. 1999. *Spatial modeling of forest landscape change: Approaches and applications.* Cambridge: Cambridge University Press.

Mladenoff, D. J., T. A. Sickley, R. G. Haight, and A. P. Wydeven. 1995. A regional landscape analysis and prediction of favorable gray wolf habitat in the northern Great Lakes region. *Conservation Biology* 9:279–294.

Mohr, C. O. 1947. Table of equivalent populations of North American small mammals. *American Midland Naturalist* 37:223–249.

Moilanen, A., and I. Hanski. 1998. Metapopulation dynamics: Effects of habitat quality and landscape structure. *Ecology* 79:2503–2515.

Molina, R., T. O'Dell, D. Luoma, M. Amaranthus, M. Castellano, and K. Russell. 1993. Biology, ecology, and social aspects of wild edible mushrooms in the forests of the Pacific Northwest: A preface to managing commercial harvest. General Technical Report PNW-GTR-309. Portland, Ore.: USDA Forest Service, Pacific Northwest Research Station.

Molina, R., and J. M. Trappe. 1982. Patterns of ectomycorrhizal host specificity and potential among Pacific Northwest conifers and fungi. *Forest Science* 28:423–458.

Monmonier, M. S. 1982. *Computer-assisted cartography: Principles and prospects.* Englewood Cliffs, N.J.: Prentice Hall.

Montana Bird Distribution Committee. 1996. *P. D. Skaar's Montana bird distribution.* 5th ed. Special Publication No. 3. Helena: Montana National Heritage Program.

Montgomery, D. C. 1991. *Design and analysis of experiments.* 3rd ed. New York: John Wiley & Sons.

Montgomery, D. R., and J. M. Buffington. 1998. Channel process, classification, and response. In *River ecology and management: Lessons from the Pacific coastal ecoregion*, ed. R. J. Naiman and R. E. Bilby, 13–42. New York: Springer-Verlag.

Montreal Process. Undated. http://www.mpci.org/whatis/criteria_e.html.

Moorcroft, P. R., G. C. Hurtt, and S. W. Pacala. In press. A method for scaling vegetation dynamics: The ecosystem demography model. *Ecological Monographs*.

Moore, D. M., B. G. Lees, and S. M. Davey. 1991. A new method for predicting vegetation distributions using decision tree analysis in a geographic information system. *Environmental Management* 15:59–71.

Moore, F. R., S. A. Gauthreaux Jr., P. Kerlinger, and T. R. Simons. 1995. Habitat requirements during migration: Important link in conservation. In *Ecology and management of Neotropical migratory birds: A synthesis and review of critical issues*, ed. T. E. Martin and D. M. Finch, 121–144. New York: Oxford University Press.

Moore, F. R., P. Kerlinger, and T. R. Simons. 1990. Stopover on a Gulf Coast barrier island by spring trans-Gulf migrants. *Wilson Bulletin* 102:487–500.

Moore, F. R., and T. R. Simons. 1992. Habitat suitability and the stopover ecology of Neotropical landbird migrants. In *Ecology and conservation of Neotropical migrant landbirds*, ed. J. M. Hagan and D. W. Johnston, 345–355. Washington: Smithsonian Institution Press.

Moore, I. D., P. E. Gessler, G. A. Nielsen, and G. A. Peterson. 1993. Soil attribute prediction using terrain analysis. *Soil Science Society of America Journal* 57:443–452.

Moors, P. J., and I. A. E. Atkinson. 1984. Predation on seabirds by introduced animals and factors affecting its severity. In *Status and conservation of the world's seabirds*, J. P. Croxall, P. G. H. Evans, and R. W. Schreiber, 667–690. Technical Publication 2. Cambridge, England: International Council for Bird Preservation.

Moran, P. A. P. 1950. Notes on continuous stochastic phenomena. *Biometrika* 37:17–23.

Morgan, J. N., and B. G. Savitsky. 1998. Error and the gap analysis model. In *GIS methodologies for developing conservation strategies*, ed. B. G. Savitsky and T. E. Larcher Jr., 170–178. New York: Columbia University Press.

Morris, D. W. 1989. Density-dependent habitat selection: Testing the theory with fitness data. *Evolutionary Ecology* 3:80–94.

———. 1990. Temporal variation, habitat selection and community structure. *Oikos* 59:303–312.

Morris, M. G., N. M. Collins, R. I. Vane-Wright, and J. Waage. 1989. The utilization and value of non-domesticated insects. In *The conservation of insects and their habitats*, ed. N. M. Collins and J. A. Thomas, 319–347. London: Academic Press.

Morris, S. R., M. E. Richmond, and D. W. Holmes. 1994. Patterns of stopover by warblers during spring and fall migration on Appledore Island, Maine. *Wilson Bulletin* 106:703–718.

Morrison, D. F. 1990. *Multivariate statistical methods*. 3rd ed. New York: McGraw-Hill.

Morrison, D. G. 1969. On the interpretation of discriminant analysis. *Journal of Marketing Research* 4:156–163.

Morrison, M. L. 1986. Bird populations as indicators of environmental change. *Current Ornithology* 3:429–451.

———. 1988. On sample sizes and reliable information. *Condor* 90:275–278.

Morrison, M. L., and D. C. Hahn. In press. Geographic variation in cowbird parasitism. In *Effects of habitat fragmentation on western bird populations*, ed. D. S. Dobkin and L. George. Studies in Avian Biology.

Morrison, M. L., L. S. Hall, S. K. Robinson, S. I. Rothstein, D. C. Hahn, and T. D. Rich, eds. 1999. Research and management of the brown-headed cowbird in western landscapes. *Studies in Avian Biology* 18. Camarillo, Calif.: Cooper Ornithological Society.

Morrison, M. L., and B. G. Marcot. 1995. An evaluation of resource inventory and monitoring program used in na-

tional forest planning. *Environmental Management* 19:147–156.

Morrison, M. L., B. G. Marcot, and R. W. Mannan. 1992. *Wildlife-habitat relationships: Concepts and applications*. Madison: University of Wisconsin Press.

———. 1998. *Wildlife-habitat relationships: Concepts and applications*. 2nd ed. Madison: University of Wisconsin Press.

Morrison, M. L., I. C. Timossi, and K. A. With. 1987. Development and testing of linear regression models predicting bird-habitat relationships. *Journal of Wildlife Management* 51:247–253.

Morton, S. R. 1990. The impact of European settlement on the vertebrate animals of arid Australia: A conceptual model. *Proceedings of the Ecological Society of Australia* 16:201–213.

Mosher, J. A., K. Titus, and M. R. Fuller. 1986. Developing a practical model to predict nesting habitat of woodland hawks. In *Wildlife 2000: Modeling habitat relationships of terrestrial vertebrates*, ed. J. Verner, M. L. Morrison, and C. J. Ralph, 31–35. Madison: University of Wisconsin Press.

Moss, B. 2000. Biodiversity in fresh waters—an issue of species preservation or ecosystem functioning. *Environmental Conservation* 27:1–4.

Mossman, M. J., and K. I. Lange. 1982. *Breeding birds of the Baraboo Hills, Wisconsin: Their history, distribution and ecology*. Hartland, Wis.: Wisconsin Department of Natural Resources and Wisconsin Society for Ornithology.

Mourell, C., and E. Ezcurra. 1996. Species richness of Argentine cacti: A test of biogeographic hypotheses. *Journal of Vegetation Science* 7:667–680.

Mowat, F. 1984. *Sea of slaughter*. Boston: Atlantic Monthly Press.

Mowrer, H. T. 1999. Accuracy (re)assurance: Selling uncertainty assessment to the uncertain. In *Spatial accuracy assessment: Land information uncertainty in natural resources*, ed. K. Lowell and A. Jaton, 3–10. Chelsea, Mich.: Ann Arbor Press.

Moyle, P. B., and B. Vondracek. 1985. Persistence and structure of the fish assemblage in a small California stream. *Ecology* 66:1–13.

Moyle, P. B., and J. E. Williams. 1990. Biodiversity loss in the temperate zone: Decline of the native fish fauna of California. *Conservation Biology* 4:275–284.

Mueller-Dombois, D., and H. Ellenberg. 1974. *Aims and methods of vegetation ecology*. New York: Wiley.

Munn, L. C., and C. S. Arneson. 1998. 1:500,000-scale Digital Soils Map of Wyoming. Laramie, Wyo.: University of Wyoming Agricultural Experiment Station.

Murcia, C. 1995. Edge effects in fragmented forests: Implications for conservation. *Trends in Ecology and Evolution* 10:58–62.

Murdoch, W. W., E. McCauley, R. M. Nisbet, S. C. Gurney, and A. M. de Roos. 1992. Individual-based models: Combining testability and generality. In *Individual-based models and approaches in ecology: Populations, communities, and ecosystems*, ed. D. L. DeAngelis and L. J. Gross, 18–35. New York: Chapman and Hall.

Murphy, D. D., and B. R. Noon. 1992. Integrating scientific methods with habitat conservation planning: Reserve design for the northern spotted owl. *Ecological Applications* 2:3–17.

Murphy, D. D., and S. B. Weiss. 1992. Effects of climate change on biological diversity in western North America: Species losses and mechanisms. In *Global warming and biological diversity*, ed. R. L. Peters and T. E. Lovejoy, 355–368. New Haven: Yale University Press.

Murphy, D. L. 1985. Estimating neighborhood variability with a binary comparison matrix. *Photogrammetric Engineering and Remote Sensing* 51:667–674.

Murray, B. G. 2000. Universal laws and predictive theory in ecology and evolution. *Oikos* 89:403–408.

Murray, R. W. 1958. The effects of food plantings, climatic conditions, and land use practices upon the quail population on an experimental area in northwest Florida. *Proceedings of the annual conference of the Southeastern Association of Game and Fish Commissioners* 12:269–274.

Musick, H. B., and H. D. Grover. 1991. Image textural measures as indices of landscape pattern. In *Quantitative methods in landscape ecology*, ed. M. G. Turner and R. H. Gardner, 77–103. New York: Springer-Verlag.

Nachlinger, J., and G. A. Reese. 1996. Plant community classification of the Spring Mountains National Recreation Area, Clark and Nye Counties, Nevada. Las Vegas: Unpublished report on file with Toiyabe National Forest, Spring Mountains National Recreation Area.

Nadeau, S., R. Décarie, D. Lambert, and M. St-Georges. 1995. Nonlinear modeling of muskrat use of habitat. *Journal of Wildlife Management* 59:110–117.

Naiman, R. J. 1998. Biotic stream classification. In *River ecology and management: Lessons from the Pacific coastal ecoregion*, ed. R. J. Naiman and R. E. Bilby, 97–119. New York: Springer-Verlag.

Nakano, S., F. Kitano, and K. Maekawa. 1996. Potential fragmentation and loss of thermal habitats for charrs in the Japanese archipelago due to climatic warming. *Freshwater Biology* 36:711–722.

Nantel, P., and P. Neumann. 1992. Ecology of ectomycorrhizal-basidiomycete communities on a local vegetation gradient. *Ecology* 73:99–117.

NAS. National Academy of Sciences, National Academy of Engineering, Institute of Medicine, and National Research Council. 1997. *Preparing for the Twenty-first Century: The Environment and the Human Future.*

Nathan, R., U. N. Safriel, I. Noy-Meir, and G. Schiller. 1999. Seed release without fire in *Pinus halapensis*, a Mediterranean serotinous wind-dispersed tree. *Journal of Ecology* 87:659–669.

Nathan, R., U. N. Safriel, and I. Noy-Meir. 2001. Field validation and sensitivity analysis of a mechanistic model for tree seed dispersal by wind. *Ecology* 82:374–388.

National Park Service. 1990. *Birds of Acadia National Park.* Bar Harbor, Maine: U.S. Department of the Interior, National Park Service.

———. 1996. *Acadia amphibians, reptiles, and mammals.* Bar Harbor, Maine: U.S. Department of the Interior, National Park Service.

National Research Council. 1995. *Science and the Endangered Species Act.* Washington: National Academy Press.

Neal, E. G. 1972. The national badger survey. *Mammal Review* 2:55–64.

Nekola, J. C., and P. S. White. 1999. The distance decay of similarity in biogeography and ecology. *Journal of Biogeography* 26:867–878.

Neldner, V. J., D. C. Crossley, and M. Cofinas. 1995. Using geographic information systems (GIS) to determine the adequacy of sampling in vegetation surveys. *Biological Conservation* 73:1–17.

Nelson, R. D., and H. Salwasser. 1982. The Forest Service wildlife and fish habitat relationships program. *Transactions of the North American Wildlife and Natural Resources Conference* 47:174–182.

Nelson, S. M., and D. C. Andersen. 1994. An assessment of riparian environmental quality by using butterflies and disturbance susceptibility scores. *The Southwestern Naturalist* 39:137–142.

Neter, J., M. H. Kutner, C. J. Nachtsheim, and W. Wasserman. 1996. *Applied linear statistical models.* 4th ed. Chicago: Irwin.

Nettleship, D. N., and T. R. Birkhead, eds. 1985. *The Atlantic alcidae.* Orlando: Academic Press.

Neu, C. W., C. R. Byers, and J. M. Peek. 1974. A technique

for analysis of utilization-availability data. *Journal of Wildlife Management* 38:541–545.

New, T. R. 1991. *Butterfly conservation.* Melbourne: Oxford University Press.

New, T. R., R. M. Pyle, J. A. Thomas, C. D. Thomas, and P. C. Hammond. 1995. Butterfly conservation management. *Annual Review of Entomology* 40:57–83.

Newsome, A. E., and P. C. Catling. 1979. Habitat preferences of mammals inhabiting heathlands of warm temperate coastal, montane and alpine regions of southeastern Australia. In *Ecosystems of the world: Heathlands and related shrublands of the world*, ed. R. L. Specht, Vol. 9A, 301–316. Amsterdam: Elsevier Scientific Publishing.

Newton, I. 1998. *Population limitation in birds.* San Diego: Academic Press.

Newton, M. R., L. L. Kinkel, and K. J. Leonard. 1998. Determinants of density- and frequency-dependent fitness in competing plant pathogens. *Phytopathology* 88:45–51.

Nicholls, A. O. 1989. How to make biological surveys go further with generalised linear models. *Biological Conservation* 50:51–75.

———. 1991. Examples of the use of generalized linear models in analysis of survey data for conservation evaluation. In *Nature conservation: Cost effective biological surveys and data analysis*, ed. C. R. Margules and M. P. Austin, 191–201. Melbourne: CSIRO Australia.

Nichols, J. D., T. Boulinier, J. E. Hines, K. H. Pollock, and J. R. Sauer. 1998a. Estimating rates of local species extinction, colonization, and turnover in animal communities. *Ecological Applications* 8:1213–1225.

———. 1998b. Inference methods for spatial variation in species richness and community composition when not all species are detected. *Conservation Biology* 12: 1390–1398.

Nichols, J. D., F. A. Johnson, and B. K. William. 1995. Managing North American waterfowl in the face of uncertainty. *Annual Review of Ecology and Systematics* 26:177–199.

Nichols, J. D., B. R. Noon, S. L. Stokes, and J. E. Hines. 1981. Remarks on the use of mark-recapture methodology in estimating avian population size. In *Estimating numbers of terrestrial birds*, ed. C. J. Ralph, and J. M. Scott, 426–435. Studies in Avian Biology 6. Los Angeles: Cooper Ornithological Society.

Nielsen, J. L. ed. 1995. *Evolution and the aquatic ecosystem: Defining unique units in population conservation.*

Symposium 17. Bethesda, Md.: American Fisheries Society.

Niemi, G., J. Hanowski, P. Helle, R. Howe, M. Monkkonen, L. Venier, and D. Welsh. 1998. Ecological sustainability of birds in boreal forests. *Conservation Ecology* 2(2):17.

Niemi, G. J., J. M. Hanowski, A. R. Lima, T. Nicholls, and N. Weiland. 1997. A critical analysis on the use of indicator species in management. *Journal of Wildlife Management* 61:1240–1252.

Nix, H. A. 1986. A biogeographic analysis of Australian elapid snakes. In *Atlas of elapid snakes of Australia*, ed. R. Longmore. *Australian flora and fauna series*, no. 7, 4–15. Canberra: Australian Government Publishing Service, Bureau of Flora and Fauna.

NOAA. National Oceanic and Atmospheric Administration. 1991. Our living oceans: The first annual report on the status of U.S. living marine resources. Technical Memorandum NMFS-F/SPO1. Washington, D.C.: U.S. Deparetment of Commerce, National Oceanic and Atmospheric Administration, National Marine Fisheries Service.

NOAA-EPA Global ecosystems database project. 1992. Global ecosystems database version 1.0. User's Guide, Documentation, Reprints, and Digital Data on CD-ROM. USDOC/NOAA National Geophysical Data Center, Boulder, CO.

Nolet, B. A., and J. M. Baveco. 1996. Development and viability of a translocated beaver *Castor fiber* population in the Netherlands. *Biological Conservation* 75:125–137.

Noon, B. R., and K. S. McKelvey. 1996a. A common framework for conservation planning: Linking individual and metapopulation models. In *Metapopulations and wildlife conservation*, ed. D. R. McCullough, 139–165. Washington, D.C.: Island Press.

———. 1996b. Management of the spotted owl: A case history in conservation biology. *Annual Review of Ecology and Systematics* 27:135–162.

Northcote, T. G. 1997. Potamodromy in salmonidae—living and moving in the fast lane. *North American Journal of Fisheries Management* 17:1029–1045.

Norton, B. G., and R. E. Ulanowicz. 1992. Scale and biodiversity policy: A hierarchical approach. *Ambio* 21: 244–249.

Norton, M. 1999. Status of the black-throated green warbler (*Dendroica virens*) in Alberta. Wildlife Status Report No. 23. Edmonton: Alberta Environment, Fisheries and Wildlife Management Division, and Alberta Conservation Association.

Norvell, L. 1995. Loving the chanterelle to death? The ten-year Oregon chanterelle project. *McIlvainea* 12:6–23.

Noss, R. F. 1987. From plant communities to landscapes in conservation inventories: A look at the Nature Conservancy (USA). *Biological Conservation* 41:11–37.

_____. 1990a. Can we maintain biological and ecological integrity? *Conservation Biology* 4:241–243.

_____. 1990b. Indicators for monitoring biodiversity: A hierarchical approach. *Conservation Biology* 4:355–364.

_____. 1991. Effects of edge and internal patchiness on avian habitat use in an old-growth Florida hammock. *Natural Areas Journal* 11:34–47.

Noss, R. F., E. T. LaRoe, III, and J. M. Scott. 1995. *Endangered ecosystems of the United States: A preliminary assessment of loss and degradation.* Washington, D.C.: U.S. Department of the Interior, National Biological Service. Biological Report 28.

Noss, R. F., M. A. O'Connell, and D. D. Murphy. 1997. *The science of conservation planning: Habitat conservation under the Endangered Species Act.* Washington, D.C.: Island Press.

NRCS. Natural Resources Conservation Service. 1999. The Plants Database. http://plants.usda.gov.

NSW NPWS. 1998. Eden Fauna Modelling—A report undertaken for the NSW CRA/RFA Steering Committee Number NE 24/EH. New South Wales National Parks and Wildlife Service, Canberra.

Nychka, D., and N. Saltzman. 1998. Design of air-quality monitoring networks. In *Case studies in environmental statistics*, ed. D. Nychka, W. W. Piegorsch, and L. H. Cox, 51–75. New York: Springer-Verlag.

O'Connor, R. J. 1986. Dynamical aspects of avian habitat use. In *Wildlife 2000: Modeling habitat relationships of terrestrial vertebrates*, ed. J. Verner, M. L. Morrison, and C. J. Ralph, 235–240. Madison: University of Wisconsin Press.

_____. 1987a. Environmental interest of field margins for birds. In *Field margins*, ed. M. J. Way and P. W. Greig-Smith, 35–48. Croydon, UK: British Crop Protection Council.

_____. 1987b. Organization of avian assemblages—the influence of intraspecific habitat dynamics. In *Organization of communities: Past and present*, ed. J. H. R. Gee and P. S. Geller, 163–183. London: Blackwell Scientific Publications.

O'Connor, R. J., and M. T. Jones. 1997. Using hierarchical models to index the ecological health of the nation. *Transactions of the North American Wildlife and Natural Resources Conference* 62:558–565.

O'Connor, R. J., M. T. Jones, R. B. Boone, and T. B. Lauber. 1999. Linking continental climate, land use, and land patterns with grassland bird distribution across the conterminous United States. *Studies in Avian Biology* 19:45–59.

O'Connor, R. J., M. T. Jones, D. White, C. Hunsaker, T. Loveland, B. Jones, and E. Preston. 1996. Spatial partitioning of environmental correlates of avian biodiversity in the conterminous United States. *Biodiversity Letters* 3:97–110.

O'Connor, R. J., and M. Shrubb. 1986. *Farming and birds.* Cambridge: Cambridge University Press.

O'Dell, T. E., J. F. Ammirati, and E. G. Schreiner. 1999. Species richness and abundance of ectomycorrhizal basidiomycete sporocarps on a moisture gradient in the Tsuga heterophylla zone. *Canadian Journal of Botany* 77:1699–1711.

O'Dell, T., J. E. Smith, M. A. Castellano, and D. L. Luoma. 1996. Diversity and conservation of forest fungi. In *Managing forest ecosystems to conserve fungus diversity and sustain wild mushroom harvests*, ed. D. Pilz and R. Molina, 5–18. General Technical Report PNW-GTR-371. Portland, Ore.: USDA Forest Service Pacific Northwest Research Station.

O'Donnell, B. P., N. L. Naslund, and C. J. Ralph. 1995. Patterns of seasonal variation of activity of marbled murrelets in forested stands. In *Ecology and conservation of the marbled murrelet*, technical ed. C. J. Ralph, G. L. Hunt Jr., M. G. Raphael, and J. F. Piatt, 117–128. General Technical Report PSW-GTR-152. Albany, Calif.: USDA Forest Service, Pacific Southwest Research Station.

Odum, E. P. 1959. *Fundamentals of ecology.* Philadelphia: W. B. Saunders.

Odum, E. P., and E. J. Kuenzler. 1955. Measurement of territory and home range size in birds. *Auk* 72:128–137.

Office of Population Census. 1991. *1991 Census county report: Greater Manchester.* London: HMSO.

Ojeda, F. P., and J. H. Dearborn. 1990. Diversity, abundance and spatial distribution of fishes and crustaceans in the rocky subtidal zone of the Gulf of Maine. *Fishery Bulletin* 88:403–410.

Oliver, I., and A. J. Beattie. 1996. Designing a cost-effective invertebrate survey: A test of methods for rapid assessment of biodiversity. *Ecological Applications* 6:594–607.

Oliver, M. A., and R. Webster. 1986. Semivariograms for

modeling the spatial pattern of landform and soil properties. *Earth surface processes and landforms* 11:491–504.

Oliveri, S. F. 1993. Bird responses to habitat changes in Baxter State Park, Maine. *Maine Naturalist* 1:145–154.

Olson, K. C., and R. C. Cochran. 1998. Radiometry for predicting tallgrass prairie biomass using regression and neural models. *Journal of Range Management* 51: 186–192.

Omernik, J. M. 1987. Ecoregions of the conterminous United States. *Annals of the Association of American Geographers* 77:118–125.

O'Neil, L.J., and A. B. Carey. 1986. Introduction: When habitats fail as predictors. In *Wildlife 2000: Modeling habitat relationships of terrestrial vertebrates*, ed. J. Verner, M. L. Morrison, and C. J. Ralph, 207–208. Madison: University of Wisconsin Press.

O'Neil, L. J., T. H. Roberts, J. S. Wakeley, and J. W. Teaford. 1988a. A procedure to modify habitat suitability index models. *Wildlife Society Bulletin* 16:33–36.

O'Neill, R. V. 1979. Transmutations across hierarchical levels. In *Systems analysis of ecosystems*, ed. G. S. Innis and R. V. O'Neill, 59–78. Fairlands, Md.: International Cooperative Publishing House.

———. 1989. Perspectives in hierarchy and scale. In *Perspectives in ecological theory*, ed. J. Roughgarden, R. M. May, and S. A. Levin, 140–156. Princeton, N.J.: Princeton University Press.

O'Neill, R. V., D. L. DeAngelis, J. B. Waide, and T. F. H. Allen. 1986. *A hierarchical concept of ecosystems*. Princeton, N.J.: Princeton University Press.

O'Neill, R. V., R. H. Gardner, B. T. Milne, M. G. Turner, and B. Jackson. 1991. Heterogeneity and spatial hierarchies. In *Ecological heterogeneity*, ed. J. Kolasa and S. T. A. Pickett, 85–96. New York: Springer-Verlag.

O'Neill, R. V., C. T. Hunsaker, K. B. Jones, K. H. Riitters, J. D. Wickham, P. M. Schwartz, I. A. Goodman, B. L. Jackson, and W. S. Baillargeon. 1997. Monitoring environmental quality at the landscape scale. *Bioscience* 47: 513–519.

O'Neill, R.V., and A.W. King. 1998. Homage to St. Michael: Or, why are there so many books on scale? In *Ecological scale: Theory and applications*, ed. D. L. Peterson and V. T. Parker, 3–15. New York: Columbia University Press.

O'Neill, R. V., J. R. Krummel, R. H. Gardner, G. Sugighara, B. Jackson, D. L. DeAngelis, B. T. Milne, M. G. Turner, B. Zygiment, S. W. Christensen, V. H. Dale, and R. L. Graham. 1988b. Indices of landscape pattern. *Landscape Ecology* 1:153–162.

Openshaw, S. 1984. *The modifiable areal unit problem: Concepts and techniques in modern geography*. Norwich, UK: Geo-Books.

Openshaw, S., and S. Alvanides. 1999. Applying geocomputation to the analysis of spatial distributions. In *Geographical information systems*, ed. P. A. Longley, M. F. Goodchild, D. J. Maguire, and D. W. Rhind, 267–282. Vol. 1, Principles and Technical Issues. 2nd ed. New York: Wiley.

Orians, G. H., and J. F. Wittenberger. 1991. Spatial and temporal scales in habitat selection. *American Naturalist* (suppl.) 137:S29–S49.

Oring, L. W. 1982. Avian mating systems. In *Avian biology*, ed. D. S. Farmer and J. R. King, Vol. 6, 1–92. New York: Academic Press.

Ortega, C. P. 1998. *Cowbirds and other brood parasites*. Tucson: University of Arizona Press.

Osborne, L. L., and M. J. Wiley. 1992. Influence of tributary spatial position on the structure of warm water fish communities. *Canadian Journal of Fisheries and Aquatic Sciences* 49:671–681.

Osborne, P. E., and B. J. Tigar. 1992. Interpreting bird atlas data using logistic models: An example from Lesotho, Southern Africa. *Journal of Applied Ecology* 29:55–62.

Osenberg, C. W., O. Sarnelle, S. D. Cooper, and R. D. Holt. 1999. Resolving ecological questions through meta-analysis: Goals, metrics, and models. *Ecology* 80: 1105–1117.

Otis, D. L. 1998. Analysis of the influence of spatial pattern in habitat selection studies. *Journal of Agricultural, Biological, and Environmental Statistics* 3:254–267.

Otis, D. L., K. P. Burnham, G. C. White, and D. R. Anderson. 1978. Statistical inference from capture data on closed animal populations. *Wildlife Monographs* 62: 1–135.

Owen, J. G. 1988. On productivity as a predictor of rodent and carnivore diversity. *Ecology* 69:1161–1165.

———. 1990. Patterns of mammalian species richness in relation to temperature, productivity, and variance in elevation. *Journal of Mammalogy* 71:1–13.

Özesmi, S. L., and U. Özesmi. 1999. An artificial neural network approach to spatial habitat modelling with interspecific interaction. *Ecological Modelling* 116:15–31.

Özesmi, U., and W. J. Mitsch. 1997. A spatial habitat model for the marsh-breeding red-wing blackbird (*Agelaius phoeniceus* L.) in coastal Lake Erie wetlands. *Ecological Modelling* 101:139–152.

Pace, R. K., and R. Barry. 1997. Performing large-scale spatial autoregressions. *Economics Letters* 54:283–291.

Paetkau, D. 1999. Using genetics to identify intraspecific conservation units: A critique of current methods. *Conservation Biology* 13:1507–1509.

Paine, R. T. 1966. Food web complexity and species diversity. *American Naturalist* 100:65–75.

Palmer, M. W. 1988. Fractal geometry: A tool for describing spatial patterns of plant communities. *Vegetatio* 75: 91–102.

———. 1992. The coexistence of species in fractal landscapes. *American Naturalist* 139:375–397.

Palmer, M. W., and E. Van der Maarel. 1995. Variance in species richness, species association, and niche limitation. *Oikos* 73:203–213.

Palmeirim, J. M. 1988. Automatic mapping of avian species habitat using satellite imagery. *Oikos* 52:59–68.

Paloheimo, J. E. 1979. Indices of food type preference by a predator. *Journal of Fisheries Research Board of Canada* 36:470–473.

Parker, A. J. 1982. The topographic relative moisture index: An approach to soil-moisture assessment in mountain terrain. *Physical Geography* 3:160–168.

Parker, J. L., and J. W. Thomas, eds. 1979. Wildlife habitats in managed forests: The Blue Mountains of Oregon and Washington. U.S. Department of Agriculture, Agricultural Handbook No. 553. Washington, D.C.: Wildlife Management Institute.

Paruelo, J. M., and F. Tomasel. 1997. Prediction of functional characteristics of ecosystems: A comparison of artificial neural networks and regression models. *Ecological Modelling* 98:173–186.

Pascual, M. A., P. Kareiva, and R. Hilborn. 1997. The influence of model structure on conclusions about the viability and harvesting of Serengeti wildebeest. *Conservation Biology* 11:966–976.

Pastor, J., J. Bonde, C. A. Johnston, and R. J. Naiman. 1993. Markovian analysis of the spatially dependent dynamics of beaver ponds. In *Predicting spatial effects in ecological systems, LMLS 23*, ed. R. Gardner, 5–27. Providence, R.I.: American Mathematical Society.

Paton, P. W. C. 1994. The effect of edge on avian nest success: How strong is the evidence? *Conservation Biology* 8:17–26.

Patton, D. R. 1978. RUN WILD: A storage and retrieval system for wildlife habitat information. General Technical Report RM-51. Fort Collins, Colo.: USDA Forest Service, Rocky Mountain Forest and Range Experiment Station.

———. 1987. Is the use of "management indicator species" feasible? *Western Journal of Applied Forestry* 2:33–34.

———. 1997. *Wildlife habitat relationships in forested ecosystems*. Portland, Ore.: Timber Press.

Pausas, J. G., and M. P. Austin. 1998. Potential impact of harvesting for the long-term conservation of arboreal marsupials. *Landscape Ecology* 13:103–109.

Pausas, J. G., M. P. Austin, and I. R. Noble. 1997. A forest simulation model for predicting eucalypt dynamics and habitat quality for arboreal marsupials. *Ecological Applications* 7:921–933.

Pausas, J. G., L. W. Braithwaite, and M. P. Austin. 1995. Modelling habitat quality for arboreal marsupials in the south coastal forests of New South Wales, Australia. *Forest Ecology and Management* 78:39–49.

Peach, W. J., S. T. Buckland, and S. R. Baillie. 1996. The use of constant mist-netting to measure between-year changes in the abundance and productivity of common passerines. *Bird Study* 43:142–156.

Peak, D. 1997. Taming chaos in the wild: A model-free technique for wildlife population control. In *Wildlife and landscape ecology: Effects of pattern and scale*, ed. J. A. Bissonette, 70–100. New York: Springer-Verlag.

Pearce, J. L., and S. Ferrier. 2000. Evaluating the predictive performance of habitat models developed using logistic regression. *Ecological Modelling* 133:225–245.

Pearl, R., and L. J. Reed. 1920. On the rate of growth of the population of the United States since 1790 and its mathematical representation. *Proceedings of the National Academy of Sciences of the United States of America* 6:275–288.

Pearson, K. 1900. On the criterion that a given system of deviations from the probable in the cause of a correlated system of variables is such that it can be reasonably supposed to have arisen from random sampling. *Philosophical Magazine* (ser. 5) 50:157–175.

———. 1901. On lines and planes of closest fit to systems of points in space. *Philosophical Magazine* 2:559–572.

Pearson, S. M. 1993. The spatial extent and relative influence of landscape-level factors on wintering bird populations. *Landscape Ecology* 8:3–18.

Pearson, S. M., M. G. Turner, L. L. Wallace, and W. H. Romme. 1995. Winter habitat use by large ungulates following fire in northern Yellowstone National Park. *Ecological Applications* 5:744–755.

Pease, C. M., and J. A. Grzybowski. 1995. Assessing the

consequences of brood parasitism and nest predation on seasonal fecundity in passerine birds. *Auk* 112:343–363.

Pechmann, J. H. K., D. E. Scott, R. D. Semlitsch, J. P. Caldwell, L. J. Vitt, and J. W. Gibbons. 1991. Declining amphibian populations: The problem of separating human impacts from natural fluctuations. *Science* 253:892–895.

Pechmann, J. H. K., and H. M. Wilbur. 1994. Putting declining amphibian populations in perspective: Natural fluctuations and human impacts. *Herpetologica* 50: 65–84.

Pedhazur, E. J. 1982. *Multiple regression in behavioral research: Explanation and prediction.* 2nd ed. Fort Worth, Tex.: Harcourt Brace College Publishers.

Pedlar, J. H., L. Fahrig, and H. G. Merriam. 1997. Raccoon habitat use at two spatial scales. *Journal of Wildlife Management* 61:102–112.

Peet, R. 1974. The measurement of species diversity. *Annual Review of Ecology and Systematics* 5:285–302.

Pendleton, G. W. 1995. Effects of sampling strategy, detection probability, and independence of counts on the use of point counts. In *Monitoring bird populations by point counts*, ed. C. J. Ralph, J. R. Sauer, and S. Droege, 131–133. USDA Forest Service General Technical Report PSW-149. Albany, Calif.: USDA Forest Service Pacific Southwest Research Station.

Penhollow, M. E., and D. F. Stauffer. 2000. Large-scale habitat relationships of Neotropical migratory birds in Virginia. *Journal of Wildlife Management* 64:362–373.

Peoples, A. 1991. Upland game investigations: Quail monitoring, performance report. Federal Aid in Wildlife Restoration Project W-82-R-30. Oklahoma City: Oklahoma Department of Wildlife Conservation.

Pereira, J. M. C., and R. M. Itami. 1991. GIS-based habitat modelling using logistic multiple regression: A study of the Mt. Graham red squirrel. *Photogrammetric Engineering and Remote Sensing* 57:1475–1486.

Perry, A. D. 1995. *Forest ecosystems.* Baltimore: John Hopkins University Press.

Peterjohn, B. G., J. R. Sauer, and C. S. Robbins. 1995. Population trends from the North American breeding bird survey. In *Ecology and management of Neotropical migratory birds*, ed. T. E. Martin and D. M. Finch, 357–380. New York: Oxford University Press.

Peterjohn, B. G., J. R. Sauer, and S. Schwartz. 2000. Temporal and geographic patterns in population trends of brown-headed cowbirds. In *The biology and management of cowbirds and their hosts*, ed. J. N. M. Smith, T.

L. Cook, S. I. Rothstein, S. K. Robinson, and S. G. Sealy, 21–34. Austin: University of Texas Press.

Peters, R. H. 1987. *The ecological implications of body size.* Cambridge: Cambridge University Press.

———. 1991. *A critique for ecology.* Cambridge: Cambridge University Press.

Peters, R. L. 1992. Conservation of biological diversity in the face of climate change. In *Global warming and biological diversity*, ed. R. L. Peters and T. E. Lovejoy, 15–30. New Haven, Conn.: Yale University Press.

Peters, R. L., and T. E. Lovejoy, eds. 1992. *Global warming and biological diversity.* New Haven, Conn.: Yale University Press.

Peterson, A. T. 2001. Predicting species' geographic distributions based on ecological niche modeling. *Condor* 103:599–605.

Peterson, A. T., L.G. Ball, and K. P. Cohoon. 2002. Predicting distributions of Mexican birds using ecological niche modeling methods. *Ibis.* In press.

Peterson, A. T., and K. P. Cohoon. 1999. Sensitivity of distributional prediction algorithms to geographic data completeness. *Ecological Modelling* 117:159–164.

Peterson, A. T., A. G. Navarro-Siguenza, and H. Benitez-Diez. 1998c. The need for continued scientific collecting: A geographic analysis of Mexican bird specimens. *Ibis* 140:288–294.

Peterson, A. T., V. Sánchez-Cordero, J. Soberón, J. Partley, R. W. Buddemeier, and A. G. Navarro-Sigüenza. 2001. Effects of global climate change on geographic distributions of Mexican Cracidae. *Ecological Modelling.* In press.

Peterson, A. T., J. Soberón, and V. Sánchez-Cordero. 1999. Stability of ecological niches in evolutionary time. *Science* 285:1265–1267.

Peterson, A. T., and D. A. Vieglais. 2001. Predicting species' invasions based on ecological niche modeling. *BioScience* 51:363–371.

Peterson, D. L., and V. T. Parker. 1998a. Dimensions of scale in ecology, resource management, and society. In *Ecological scale: Theory and applications*, ed. D. L. Peterson and V. T. Parker, 499–522. New York: Columbia University Press.

———. 1998b. *Ecological scale: Theory and applications.* New York: Columbia University Press.

Peterson, R. A. 1995. *The South Dakota breeding bird atlas.* Aberdeen, S. Dak.: South Dakota Ornithologists' Union.

Peuquet, D. J. 1994. It's about time: A conceptual framework for the representation of temporal dynamics in geographic information systems. *Annals of the American Association of Geographers* 84:375–394.

———. 1999. Time in GIS and geographical databases. In *Geographical information systems*, ed. P. A. Longley, M. F. Goodchild, D. J. Magurie, and D. W. Rhind, 91–103. Vol. 1, Principles and Technical Issues. 2nd ed. New York: Wiley.

Peuquet, D. J., and N. Duan. 1995. An event-based spatio-temporal data model (ESTDM) for temporal analysis of geographic data. *International Journal of Geographical Information Systems* 9:2–24.

Phipps, M. 1981. Entropy and community pattern analysis. *Journal of Theoretical Biology* 93:253–273.

Pianka, E.R. 1966. Latitudinal gradients in species diversity: A review of concepts. *American Naturalist* 100:33–46.

———. 1970. On r- and K-selection. *American Naturalist* 104:592–597.

Pickett, S. T. A., J. Kolasa, and C. G. Jones. 1994. *Ecological understanding*. San Diego: Academic Press.

Pickett, S. T. A., and K. H. Rogers. 1997. Patch dynamics: The transformation of landscape structure and function. In *Wildlife and landscape ecology: Effects of pattern and scale*, ed. J. A. Bissonette, 101–127. New York: Springer-Verlag.

Pickett, S. T. A., and J. N. Thompson. 1978. Patch dynamics and the design of nature reserves. *Biological Conservation* 13:27–37.

Pickett, S. T. A., and P. S. White, eds. 1985. *The ecology of natural disturbance*. Orlando: Academic Press.

Pielou, E. C. 1977. *Mathematical ecology*. New York: John Wiley & Sons.

———. 1984. *The interpretation of ecological data*. New York: John Wiley & Sons.

Pienkowski, M. W., E. M. Bignal, C. A. Galbraith, D. I. Mc-Cracken, R. A. Stillman, M. G. Boobyer, and D. J. Curtis. 1996. A simplified classification of land-type zones to assist the integration of biodiversity objects in land-use policies. *Biological Conservation* 75:11–25.

Pilz, D., R. Molina, and L. Liegel. 1998. Biological productivity of chanterelle mushrooms in and near the Olympic Peninsula biosphere reserve. *Ambio Special Report Number* 9:8–13.

Pilz, D. P., and D. A. Perry. 1984. Impact of clearcutting and slash burning on ectomycorrhizal associations of Douglas-fir seedlings. *Canadian Journal of Forest Research* 14:94–100.

Pimm, S. L., and M. E. Gilpin. 1989. Theoretical issues in conservation biology. In *Perspectives in ecological theory*,

ed. J. Roughgarden, R. M. May, and S. A. Levin, 287–205. Princeton, N.J.: Princeton University.

Pitt, D. G., and D. P. Kreutzweiser. 1998. Applications of computer-intensive statistical methods to environmental research. *Ecotoxicology and Environmental Safety* 39:78–97.

PNRHJ. Parc Naturel Régional du Haut-Jura. 1988. *Le Haut-Jura . . . De Crêts en Combes*. Lajoux, France: Parc Naturel Régional du Haut-Jura.

Poff, N. L., and J. D. Allan. 1995. Functional organization of stream fish assemblages in relation to hydrological variability. *Ecology* 76:606–627.

Poff, N. L., P. L. Angermeier, S. D. Cooper, P. S. Lake, K. D. Fausch, K. O. Winemiller, L. A. K. Mertes, M. W. Oswood, J. Reynolds, and F. J. Rahel. 2001. Fish diversity in streams and rivers. In *Global biodiversity in a changing environment: Scenarios for the 21st century*, ed. F. S. Chapin, O. E. Sala, and R. Huber-Sannwald, 315–349. New York: Springer-Verlag.

Pollard, E. 1977. A method for assessing changes in the abundance of butterflies. *Biological Conservation* 12:115–134.

———. 1991. Synchrony of population fluctuations: The dominant influence of widespread factors on local butterfly populations. *Oikos* 60:7–10.

Pollard, E., and T. J. Yates. 1993. *Monitoring butterflies for ecology and conservation*. London: Chapman and Hall.

Pollock, K. H. 1981. Capture-recapture models: A review of current methods, assumptions, and experimental design. In *Estimating numbers of terrestrial birds*, ed. C. J. Ralph and J. M. Scott, 426–435. Studies in Avian Biology 6. Los Angeles: Cooper Ornithological Society.

Pollock, K. H., J. D. Nichols, C. Brownie, and J. E. Hines. 1990. Statistical inference for capture-recapture experiments *Wildlife Monographs* 107:1–97.

Possingham, H. P., I. Davies, I. R. Noble, and T. W. Norton. 1992. A metapopulation simulation model for assessing the likelihood of plant and animal extinctions. *Mathematics and Computers in Simulation* 33:367–372.

Possingham, H. P., D. B. Lindenmayer, T. W. Norton, and I. Davies. 1994. Metapopulation viability analysis of the greater glider *Petauroides volans* in a wood production area. *Biological Conservation* 70:227–236.

Powell, L. A., J. D. Lang, M. J. Conroy, and D. G. Krementz. 2000. Effects of forest management on density, survival, and population growth of wood thrushes. *Journal of Wildlife Management* 64:11–23.

Power, M. 1993. The predictive validation of ecological and environmental models. *Ecological Modelling* 68:33–50.

Power, M. E. 1992. Top-down and bottom-up forces in food webs: Do plants have primacy? *Ecology* 73:733–746.

Power, M. E., W. J. Matthews, and A. J. Stewart. 1985. Grazing minnows, piscivorous bass, and stream algae: Dynamics of a strong interaction. *Ecology* 66: 1448–1456.

Pradel, R., J. Hines, J. D. Lebreton, and J. D. Nichols. 1997. Estimating survival probabilities and proportions of "transients" using capture-recapture data. *Biometrics* 53:60–72.

Preisler, H. K. 1993. Modelling spatial patterns of trees attacked by bark-beetles. *Applied Statistics* 42:501–514.

Prendergast, J. R., R. M. Quinn, J. H. Lawton, B. C. Eversham, and D. W. Gibbons. 1993. Rare species, the coincidence of diversity hotspots and conservation strategies. *Nature* 365:335–337.

Press, D., D. F. Doak, and P. Steinberg. 1996. The role of local government in the conservation of rare species. *Conservation Biology* 10:1538–1548.

Press, W. H., S. A. Teukolsky, W. T. Vetterling, and B. P. Flannery. 1992. *Numerical recipes in C: The art of scientific computing*. 2nd ed. New York: Cambridge University Press.

Preston, F. W. 1948. The commonness, and rarity, of species. *Ecology* 29:254–283.

———. 1962a. The canonical distribution of commonness and rarity: Part 1. *Ecology* 43:185–215.

———. 1962b. The canonical distribution of commonness and rarity: Part 2. *Ecology* 43:410–432.

Price, J., S. Droege, and A. Price. 1995. *The summer atlas of North American birds*. San Diego: Academic Press.

Pringle, C. M., R. J. Naiman, G. Bretschko, J. R. Karr, M. W. Oswood, J. R. Webster, R. L. Welcomme, and M. J. Winterbourne. 1988. Patch dynamics in lotic systems: The stream as a mosaic. *Journal of the North American Benthological Society* 7:503–524.

Proulx, M., and A. Mazumder. 1998. Reversal of grazing impact on plant species richness in nutrient poor vs. nutrient-rich ecosystems. *Ecology* 79:2581–2592.

Proulx, M., F. R. Pick, A. Mazumder, P. B. Hamilton, and D. R. S. Lean. 1996. Experimental evidence for interactive impacts of human activities on lake algal species richness. *Oikos* 76:191–195.

Provost, F., and T. Fawcett. 1997. Analysis and visualization of classifier performance: Comparison under imprecise class and cost distributions. In *Proceedings of the Third International Conference on Knowledge Discovery and Data Mining*, ed. D. E. Heckerman, H. Mannila, and D. Pregibon, 43–48. Menlo Park, Calif.: AAAI Press.

Pulliam, H. R. 1988. Sources, sinks and population regulation. *American Naturalist* 132:652–661.

Pulliam, H. R., and B. J. Danielson. 1991. Sources, sinks, and habitat selections: A landscape perspective on population dynamics. *American Naturalist* (suppl.) 137: S50–S66.

Pulliam, H. R., J. B. Dunning Jr., and J. Liu. 1992. Population dynamics in complex landscapes: A case study. *Ecological Applications* 2:165–177.

Pullin, A. S. ed. 1995. *Ecology and conservation of butterflies*. London: Chapman and Hall.

Pyle, C. 1985. Vegetation disturbance history of Great Smoky Mountains National Park: An analysis of archival maps and records. Research/Resources Management Report SER-77. Atlanta, Ga.: U.S. Department of the Interior, National Park Service.

Pyle, P. 1997. *Identification guide to North American birds*. Part 1. Bolinas, Calif.: Slate Creek Press.

Pyle, R., M. Bentzien, and P. Opler. 1981. Insect conservation. *Annual Review of Entomology* 26:233–258.

Quinlan, F. T., T. R. Karl, and C. N. Williams Jr. 1987. United States Historical Climatology Network (HCN) serial temperature and precipitation data. NDP019. Oak Ridge, Tenn.: Carbon Dioxide Information Analysis Center, Oak Ridge National Laboratory.

Quinn, T. P. 1993. A review of homing and straying of wild and hatchery-produced salmon. *Fisheries Research* 18:29–44.

Rabinowitz, D., S. Cairns, and T. Dillon. 1986. Seven forms of rarity and their frequency in the flora of the British Isles. In *Conservation biology: The science of scarcity and diversity*, ed. M. E. Soulé, 182–204. Sunderland, Mass.: Sinauer Associates.

Rabinowitz, D. S. 1981. Seven forms of rarity. In *The biological aspects of rare plant conservation*, ed. H. Synge, 205–217. Chichester, UK: Wiley.

Radeloff, V. C., A. M. Pidgeon, and P. Hostert. 1999. Habitat and population modelling of roe deer using an interactive geographic information system. *Ecological Modelling* 114:287–304.

Raguso, R. A., and J. Llorente-Bousquets. 1990. The butterflies (Lepidoptera) of the Tuxlas Mountains,, Veracruz, Mexico, revisited: Species-richness and habitat disturbance. *Journal of Research on the Lepidoptera* 29: 105–133.

Rahel, F. J., C. J. Keleher, and J. L. Anderson. 1996. Potential habitat loss and population fragmentation for coldwater fish in the North Platte River drainage of the Rocky Mountains: Response to climate warming. *Limnology and Oceanography* 41:1116–1123.

Ralph, C. J., S. Droege, and J. R. Sauer. 1995a. Managing and monitoring birds using point counts: Standards and applications. In *Monitoring bird populations by point counts*, ed. C. J. Ralph, J. R. Sauer, and S. Droege, 161–175. General Technical Report PSW-GTR-149. Albany, Calif.: USDA Forest Service, Pacific Southwest Research Station.

Ralph, C. J., G. R. Geupel, P. Pyle, T. E. Martin, and D. F. DeSante. 1993. Handbook of field methods for monitoring landbirds. General Technical Report PSW-GTR-144. Albany, Calif.: USDA Forest Service, Pacific Southwest Research Station.

Ralph, C. J., S. L. Miller, and B. P. O'Donnell. 1992. Capture and monitoring of foraging and breeding of the marbled murrelet in California during 1990: An interim report. Technical Report 1992-7. Sacramento: California Department of Fish and Game, Nongame Bird and Mammal Section.

Ralph, C. J., and S. K. Nelson, eds. 1992. Methods of surveying marbled murrelets at inland forested sites. Report on file. Corvallis, Ore.: Pacific Seabird Group, Oregon Cooperative Wildlife Research Unit, Oregon State University.

Ralph, C. J., S. K. Nelson, M. M. Shaughnessy, S. L. Miller, and T. E. Hamer. Pacific Seabird Group, Marbled Murrelet Technical Committee. 1994. Methods for surveying marbled murrelets in forests. Technical Paper No. 1, Revision. Corvallis, Ore.: Pacific Seabird Group and Oregon Cooperative Wildlife Research Unit, Oregon State University.

Ralph, C. J., J. R. Sauer, and S. Droege, eds. 1995b. Monitoring bird populations by point counts. General Technical Report PSW-GTR-149. Albany, Calif.: U.S. Department of Agriculture, Pacific Southwest Research Station, Forest Service.

Ralph, C. J., and J. M. Scott, eds. 1981. Estimating numbers of terrestrial birds. *Studies in Avian Biology* 6. Los Angeles: Cooper Ornithological Society.

Ramsey, F. L., and D. W. Schafer. 1997. *The statistical sleuth: A course in methods of data analysis*. Belmont, Calif.: Duxbury Press.

Ramsey, F. L., and J. M. Scott. 1981. Analysis of bird survey data using a modification of Emlen's method. In *Estimating numbers of terrestrial birds*, ed. C. J. Ralph and J. M.

Scott, 483–487. Studies in Avian Biology 6. Los Angeles: Cooper Ornithological Society.

Ranney, J. W., M. C. Bruner, and J. B. Levenson. 1981. The importance of edge in the structure and dynamics of forest islands. In *Forest island dynamics in man-dominated landscapes*, R. L. Burgess, and D. M. Sharpe, 67–96. New York: Springer-Verlag.

Rao, C. R. 1952. *Statistical methods in biometric research*. New York: John Wiley & Sons.

Raper, J., and D. Livingstone. 1994. Species and ecosystem viability. *Journal of Forestry* 92:45–47.

———. 1996. High-level coupling of GIS and environmental process modeling. In *GIS and environmental modeling: Progress and research issues*, ed. M. F. Goodchild, L. T. Steyaert, B. O. Parks, C. Johnston, D. Maidment, M. Crane, and S. Glendinning, 387–390. Fort Collins, Colo.: GIS World Books.

Raphael, M. G., and B. G. Marcot. 1986. Validation of a wildlife-habitat relationship model: Vertebrates in a Douglas-fir sere. In *Wildlife 2000: Modeling habitat relationships of terrestrial vertebrates*, ed. J. Verner, M. L. Morrison, and C. J. Ralph, 129–138. Madison: University of Wisconsin Press.

Raphael, M. G., K. S. McKelvey, and B. M. Galleher. 1998. Using geographic information systems and spatially explicit population models for avian conservation: A case study. In *Avian conservation: Research and management*, ed. J. M. Marzluff and R. Sallabanks, 65–74. Washington, D.C.: Island Press.

Raphael, M. G., J. A. Young, K. McKelvey, B. M. Galleher, and K. C. Peeler. 1994. A simulation analysis of population dynamics of the northern spotted owl in relation to forest management alternatives. In *Final environmental impact statement on management of habitat for late-successional and old-growth forest related species within the range of the northern spotted owl*, Vol. 2, App. J-3. Portland, Ore.: Interagency SEIS Team.

Rastetter, E. B., A. W. King, B. J. Cosby, G. M. Hornberger, R. V. O'Neill, and J. E. Hobbie. 1992. Aggregating fine-scale ecological knowledge to model coarser-scale attributes of ecosystems. *Ecological Applications* 2:55–70.

Ratcliffe, J. H., and M. J. McCullagh. 1998. Aoristic crime analysis. *International Journal of Geographical Information Science* 12:751–764.

Rathert, D., D. White, J. C. Sifneos, and R. M. Hughes. 1999. Environmental correlates of species richness for native freshwater fish in Oregon, USA. *Journal of Biogeography* 26:1–17.

Ratti, J. T., and J. M. Scott. 1991. Agricultural impacts on wildlife: Problem review and restoration needs. *The Environmental Professional* 13:263–274.

Raven, P. H., and E. O. Wilson. 1992. A fifty-year plan for biodiversity survey. *Science* 258:1099–1100.

Reader, R. J., and B. J. Best. 1989. Variation in competition along an environmental gradient: *Hieracium floribundum* in an abandoned pasture. *Journal of Ecology* 77:673–684.

Recher, H. F., J. Majer, and H. A. Ford. 1991. Temporal and spatial variation in the abundance of eucalypt canopy arthropods: The response of forest birds. *Proceedings of the International Ornithological Congress* 20: 1568–1575.

Reckhow, K. H. 1990. Bayesian inference in non-replicated ecological studies. *Ecology* 71:2053–2059.

Recknagel, F., M. French, P. Harkonen, and K.-I. Yabunaka. 1997. Artificial neural network approach for modelling and prediction of algal blooms. *Ecological Modelling* 96:11–28.

Redhead, S. A., L. L. Norvell, and E. Danell. 1997. *Cantharellus formosus* and the Pacific golden chanterelle harvest in western North America. *Mycotaxon* 65: 295–322.

Reed, B. C., J. F. Brown, D. VanderZee, T. R. Loveland, J. W. Merchant, and D. O. Ohlen. 1994. Measuring phenological variability from satellite imagery. *Journal of Vegetation Science* 5:703–714.

Reed, B. C., and L. Yang. 1997. Seasonal vegetation characteristics of the United States. *GeoCarto International* 12:65–71.

Reed, J. M. 1996. Using statistical probability to increase confidence of inferring species extinction. *Conservation Biology* 10:1283–1285.

Reed, P. B., Jr. 1988. National list of plant species that occur in wetlands: National summary. Biological Report 88-24. Washington, D.C.: U.S. Department of the Interior, U.S. Fish and Wildlife Service, Research and Development.

Reed, R. A., J. Johnson-Barnard, and W. L. Baker. 1996. Contribution of roads to forest fragmentation in the Rocky Mountains. *Conservation Biology* 10:1098–1106.

Reeves, G. H., L. E. Benda, K. M. Burnett, P. A. Bisson, and J. R. Sedell. 1995. A disturbance-based ecosystem approach to maintaining and restoring freshwater habitats of evolutionarily significant units of anadromous salmonids in the Pacific Northwest. *American Fisheries Society Symposium* 17:334–349.

Rencher, A. C. 1995. *Methods of multivariate analysis*. New York: Wiley.

Renlund, D. W. 1971. Forest pest conditions in Wisconsin: Annual report. Poynette, Wis.: Wisconsin Department of Natural Resources.

Renshaw, E. 1990. *Modelling biological populations in space and time*. Cambridge: Cambridge University Press.

Restrepo, C., and N. Gomez. 1998. Responses of understory birds to anthropogenic edges in a Neotropical montane forest. *Ecological Applications* 8:170–183.

Rexstad, E. A., D. D. Miller, C. H. Flather, E. M. Anderson, J. W. Hupp, and D. R. Anderson. 1988. Questionable multivariate statistical inference in wildlife habitat and community studies. *Journal of Wildlife Management* 52:794–798.

Reynolds, R. T., J. M. Scott, and R. A. Nussbaum. 1980. A variable circular-plot method for estimating bird numbers. *Condor* 82:309–313.

RFA Steering Committee. 1996. Central Highlands comprehensive regional assessment summary report. Melbourne, Australia: Joint Commonwealth and Victorian Regional Forest Agreement (RFA) Steering Committee.

Richards, C., L. B. Johnson, and G. E. Host. 1996. Landscape-scale influences on stream habitats and biota. *Canadian Journal of Fisheries and Aquatic Sciences* (suppl. 1) 53:295–311.

Richardson, D., R. Costa, and R. Boykin. 1998. Strategy and guidelines for the recovery and management of the red-cockaded woodpecker and its habitats on national wildlife refuges. Atlanta, Ga.: U.S. Fish and Wildlife Service Region 4.

Richardson, D. M., and J. P. McMahon. 1992. A bioclimatic analysis of *Eucalyptus nitens* to identify potential planting regions in southern Africa. *South African Journal of Science* 88:380–387.

Richter, A. R., and R. F. Labisky. 1985. Reproductive dynamics among disjunct white-tailed deer herds in Florida. *Journal of Wildlife Management* 49:964–971.

Rickers, J. R., L. P. Queen, and G. J. Arthaud. 1995. A proximity-based approach to assessing habitat. *Landscape Ecology* 10:309–321.

Ricklefs, R. E. 1980. Geographical variation in clutch size among passerine birds: Ashmole's hypothesis. *Auk* 97:38–49.

———. 1987. Community diversity: Relative roles of local and regional processes. *Science* 235:167–171.

———. 2000. Density dependence, evolutionary optimization, and the diversification of avian life histories. *Condor* 102:9–22.

Ricklefs, R. E., and D. Schluter. 1993. Species diversity: Regional and historical influences. In *Species diversity in ecological communities*, ed. R. E. Ricklefs and D. Schluter, 350–363. Chicago: University of Chicago Press.

Ridgway, R. 1887. *A manual of North American birds*. Philadelphia: J. B. Lippincott.

Rieman, B. E., and J. B. Dunham. 2000. Metapopulations and salmonids: A synthesis of life history patterns and empirical observations. *Ecology of Freshwater Fish* 9: 51–64.

Rieman, B. E., D. C. Lee, and R. F. Thurow. 1997. Distribution, status, and likely future trends of bull trout within the Columbia River and Klamath River Basins. *North American Journal of Fisheries Management* 17: 1111–1125.

Rieman, B. E., and J. D. McIntyre. 1995. Occurrence of bull trout in naturally fragmented habitat patches of varied size. *Transactions of the American Fisheries Society* 124:285–296.

Ringold, P. L., B. S. Mulder, J. Alegria, R. Czaplewski, and T. V. Tolle. 1999. Establishing a regional monitoring strategy: The Pacific Northwest Forest Plan. *Environmental Management* 23:179–192.

Rink, G. 1990. *Juglans cinerea* L. Butternut. In *Silvics of North America*, ed. R. M. Burns and B. H. Honkala, 386–390. Vol. 2, Hardwoods. Agriculture Handbook 654. Washington, D.C.: USDA Forest Service.

Riordan, P. 1998. Unsupervised recognition of individual tigers and snow leopards from their footprints. *Animal Conservation* 1:253–262.

Ripley, B. D. 1996. *Pattern recognition and neural networks*. New York: Cambridge University Press.

Ripple, W. J., G. A. Bradshaw, and T. A. Spies. 1991. Measuring landscape pattern in the Cascades Range of Oregon, USA. *Biological Conservation* 57:73–88.

Risbey, D. A., M. C. Calver, and J. Short. 1999. The impact of cats and foxes on the small vertebrate fauna of Heirisson Prong, Western Australia. 1. Exploring potential impact using diet analysis. *Wildlife Research* 26:621–630.

Robbins, C. S., D. Bystrak, and P. H. Geissler. 1986. The breeding bird survey: Its first fifteen years, 1965–1979. Resource Publication 157. Washington, D.C.: U.S. Department of the Interior, U.S. Fish and Wildlife.

Robbins, C. S., D. K. Dawson, and B. A. Dowell. 1989a. Habitat area requirements of breeding forest birds of the middle Atlantic states. *Wildlife Monographs* 103:1–34.

Robbins, C. S., J. R. Sauer, R. S. Greenberg, and S. Droege. 1989b. Population declines in North American birds that

migrate to the Neotropics. *Proceedings of the National Academy of Science* 86:7658–7662.

Robertsen, M. J. 1995. A landscape-scale approach to forest songbird management. M.S. thesis, University of Wisconsin, Madison.

Robinson, S. K. 1988. Reappraisal of the costs and benefits of habitat heterogeneity for nongame wildlife. *Transactions of the North American Wildlife Natural Resources Conference* 53:145–155.

Robinson, S. K. 1992. Population dynamics of breeding Neotropical migrants in a fragmented Illinois landscape. In *Ecology and conservation of Neotropical migrant landbirds*, ed. J. M. Hagan, III., and D. W. Johnston, 408–418. Washington, D.C.: Smithsonian Institution Press.

———. 1999. Cowbird ecology: Factors affecting the abundance and distribution of cowbirds. In *Research and management of the brown-headed cowbird in western landscapes*, ed. M. L. Morrison, L. S. Hall, S. K. Robinson, S. I. Rothstein, D. C. Hahn, and T. D. Rich, 4–9. Studies in Avian Biology 18. Camarillo, Calif.: Cooper Ornithological Society.

Robinson, S. K., S. I. Rothstein, M. C. Brittingham, L. J. Petit, and J. A. Grzybowski. 1995. Ecology and behavior of cowbirds and their impact on host populations. In *Ecology and management of Neotropical migratory birds: A synthesis and review of critical issues*, ed. T. E. Martin and D. M. Finch, 428–460. New York: Oxford University Press.

Robinson, S. K., F. R. Thompson, III., T. M. Donovon, D. R. Whitehead, and J. Faaborg. 1995. Regional forest fragmentation and the nesting success of migratory birds. *Science* 267:1987–1990.

Robinson, S. K., and D. S. Wilcove. 1994. Forest fragmentation in the temperate zone and its effects on migratory songbirds. *Bird Conservation International* 4:233–249.

Robinson, S. K., J. A. Grzybowski, S. I. Rothstein, M. C. Brittingham, L. J. Petit, and F. R. Thompson. 1993. Management implications of cowbird parasitism for Neotropical migrant songbirds. In *Status and management of Neotropical migratory birds*, ed. D. M. Finch and P. W. Stangel, 93–102. General Technical Report RM-229. Fort Collins, Colo.: USDA Forest Service, Rocky Mountain Forest and Range Experiment Station.

Robinson, V., and A. Frank. 1985. About different kinds of uncertainty in collections of spatial data. *Proceedings of AUTO-CARTO* 7:440–449.

Rodenhouse, N. L., L. B. Best, R. J. O'Connor, and E. K. Bollinger. 1993. Effects of temperate agriculture on

Neotropical migrant landbirds. In *Status and management of Neotropical migratory birds*, ed. D. M. Finch and P. W. Stengel, 280–295. General Technical Report RM-229. Fort Collins, Colo.: USDA Forest Service, Rocky Mountain Forest and Range Experiment Station.

———. 1995. Effects of agricultural practices and farmland structure. In *Status and management of Neotropical migratory birds*, ed. T. E. Martin and D. M. Finch, 269–293. New York: Oxford University Press.

Rogers, K. 1998. Managing science/management partnerships: A challenge of adaptive management. *Conservation Ecology* [online] 2: R1. http://www.consecol.org/Journal/vol2/iss2/resp1.

Roloff, G. J., and J. B. Haufler. 1997. Establishing population viability planning objectives based on habitat potentials. *Wildlife Society Bulletin* 25:895–904.

Roloff, G. J., and B. J. Kernohan. 1999. Evaluating reliability of habitat suitability index models. *Wildlife Society Bulletin* 27:973–985.

Romesburg, H. C. 1981. Wildlife science: Gaining reliable knowledge. *Journal of Wildlife Management* 45: 293–313.

Romme, W. H. 1982. Fire and landscape diversity in subalpine forests of Yellowstone National Park. *Ecological Monographs* 52:199–221.

Romme, W. H., and D. G. Despain. 1989. Historical perspectives on the Yellowstone fires of 1988. *BioScience* 39:695–699.

Root, K. V. 1998. Evaluating the effects of habitat quality, connectivity, and catastrophes on a threatened species. *Ecological Applications* 8:854–865.

Root, R. B. 1967. The niche exploitation pattern of the blue-gray gnatcatcher. *Ecological Monographs* 37: 317–350.

Root, T. 1988a. *Atlas of wintering North American birds: An analysis of Christmas bird count data.* Chicago: University of Chicago Press.

———. 1988c. Environmental factors associated with avian distributional boundaries. *Journal of Biogeography* 15:489–505.

Root, T. L. 1988b. Energy constraints on avian distributions and abundances. *Ecology* 69:330–339.

———. 1993. Effects of global climate change on North American birds and their communities. In *Biotic interactions and global change*, ed. P. M. Kareiva, J. G. Kingsolver, and R. B. Huey, 280–292. Sunderland, Mass.: Sinauer Associates.

Roper, T. J. 1993. Badger setts as a limiting resource. In *The badger*, ed. T. J. Hayden, 26–34. Dublin: Royal Irish Academy.

Rosch, E. 1975. Basic objects in natural categories. Working Paper No. 43. Berkeley: Language Behavior Research Laboratory, University of California.

———. 1978. *Cognition and categorization.* Hillsdale, N.J.: Erlbaum.

Roseberry, J. L., and W. D. Klimstra. 1984. *Population ecology of the bobwhite.* Carbondale, Ill.: Southern Illinois University Press.

Roseberry, J. L., and S. D. Sudkamp. 1998. Assessing the suitability of landscapes for northern bobwhites. *Journal of Wildlife Management* 62:895–902.

Rosenbaum, P. R. 1995. *Observational studies. Springer series in statistics.* New York: Springer-Verlag.

Rosenberg, D. K., D. F. DeSante, K. S. McKelvey, and J. E. Hines. In review. Monitoring survival rates of landbirds from a continental network of mist-net stations. *Ecological Applications.*

———. 1999. Monitoring survival rates of Swainson's thrush Catharus ustulatus at multiple spatial scales. *Bird Study* (suppl.) 46:s198–s208.

Rosenberg, D. K., and K. S. McKelvey. 1999. Estimation of habitat selection for central-place foraging animals. *Journal of Wildlife Management* 63:1028–1038.

Rosenberg, K. V., and M. G. Raphael. 1986. Effects of forest fragmentation in Douglas-fir forests. In *Wildlife 2000: Modeling habitat relationships of terrestrial vertebrates*, ed. J. Verner, M. L. Morrison, and C. J. Ralph, 263–272. Madison: University of Wisconsin Press.

Rosenzweig, M. 1985. Some theoretical aspects of habitat selection. In *Habitat selection in birds*, ed. M. L. Cody, 517–540. Orlando: Academic Press.

Rosenzweig, M. L. 1981. A theory of habitat selection. *Ecology* 62:327–335.

———. 1989. Habitat selection, community organization and small mammal studies. In *Patterns in the structure of mammalian communities*, ed. D. W. Morris, Z. Abrambsky, B. J. Fox, and M. R. Willig, 5–21. Special publications no. 28. Lubbock, Tex.: Texas Tech University Press.

———. 1991. Habitat selection and population interactions: The search for mechanism. *American Naturalist* (suppl.) 137:S5-S28.

Rosenzweig, M. L., and Z. Abramsky. 1993. How are diversity and productivity related? In *Species diversity in ecological communities*, ed. R. E. Ricklefs and D. Schluter, 52–65. Chicago: University of Chicago Press.

Ross, R. K. 1974. A comparison of the feeding and nesting

requirements of the great cormorant (*Phalocrocorax carbo*) and the double-crested cormorant (*P. auritus*) in Nova Scotia. *Proceedings of the Nova Scotian Institute of Science* 27:114–132.

Ross, T. J. 1995. *Fuzzy logic with engineering applications.* New York: McGraw-Hill.

Rossi, R. E., D. J. Mulla, A. G. Journel, and E. H. Franz. 1992. Geostatistical tools for modeling and interpreting ecological spatial dependence. *Ecological Monographs* 62:277–314.

Rotenberry, J. T. 1981. Why measure bird habitat? In *The use of multivariate statistics in studies of wildlife habitat,* ed. D. E. Capen, 29–32. General Technical Report RM-87. Fort Collins, Colo.: USDA Forest Service, Rocky Mountain Forest and Range Experiment Station.

––––––. 1985. The role of habitat in avian community composition: Physiognomy or floristics? *Oecologia* 67:213–217.

––––––. 1986. Habitat relationships of shrubsteppe birds: Even "good" models cannot predict the future. In *Wildlife 2000: Modeling habitat relationships of terrestrial vertebrates,* ed. J. Verner, M. L. Morrison, and C. J. Ralph, 217–221. Madison: University of Wisconsin Press.

––––––. 1998. Avian conservation research needs in western shrublands: Exotic invaders and the alteration of ecosystem processes. In *Avian conservation: Research and management,* ed. J. M. Marzluff and R. Sallabanks, 262–272. Washington, D.C.: Island Press.

Rotenberry, J. T., and S. T. Knick. 1999. From the individual to the landscape: Multiscale habitat associations of a shrubsteppe passerine and their implications for conservation biology. *Studies in Avian Biology* 19:95–103.

Rotenberry, J. T., and J. A. Wiens. 1980. Temporal variation in habitat structure and shrubsteppe bird dynamics. *Oecologia* 47:1–9.

Roth, N. E., J. D. Allan, and D. L. Erikson. 1996. Landscape influences on stream biotic integrity assessed at multiple spatial scales. *Landscape Ecology* 11:141–156.

Roth, R. R. 1976. Spatial heterogeneity and bird species diversity. *Ecology* 57:773–782.

Rothstein, J. 1951. Information, measurement, and quantum mechanics. *Science* 114:171–175.

Rothstein, S. I. 1994. The cowbird's invasion of the Far West: History, causes and consequences experienced by host species. *Studies in Avian Biology* 15:301–315.

Rothstein, S. I., and S. K. Robinson. 1998. The evolution and ecology of avian brood parasitism. In *Parasitic birds and their hosts,* ed. S. I. Rothstein and S. K. Robinson, 3–56. Oxford: Oxford University Press.

Rothstein, S. I., J. Verner, and E. Stevens. 1984. Radio-tracking confirms a unique diurnal pattern of spatial occurrence in the parasitic brown-headed cowbird. *Ecology* 65:77–88.

Rudis, V. A., and J. B. Tansey. 1995. Regional assessment of remote forests and black bear habitat from forest resource surveys. *Journal of Wildlife Management* 59:170–180.

Rumble, M. A. 1982. Biota of uranium mill tailings near the Black Hills. In *Proceedings of the annual conference of the Western Association of Fish and Wildlife Agencies,* 278–292. Las Vegas, Nev., Western Association of Fish and Wildlife Agencies.

Russell, R. W., F. L. Carpenter, M. A. Hixon, and D. C. Paton. 1994. The impact of variation in stopover habitat quality on migrant rufous hummingbirds. *Conservation Biology* 8:483–490.

Ruttiman, U. E. 1994. Statistical approaches to development and validation of predictive instruments. *Critical Care Clinics* 10:19–35.

Ryan, T. P. 1997. *Modern regression methods.* New York: John Wiley & Sons.

Rykiel, E. J., Jr. 1996. Testing ecological models: The meaning of validation. *Ecological Modelling* 90:229–244.

Saaty, T. L. 1980. *The analytic hierarchy process: Planning, priority setting, resource allocation.* New York: McGraw-Hill.

––––––. 1994. Highlights and critical points in the theory and application of the analytic hierarchy process. *European Journal of Operational Research* 74:426–447.

Saaty, T. L., and L. G. Vargas. 1994. *Decision making in economic, political, social, and technological environments with the analytic hierarchy process.* Pittsburgh: RWS Publications.

Sabinske, D. W. 1972. The sagebrush community of Grand Teton National Park: A vegetation analysis. M.S. thesis, University of Wyoming. Laramie.

Sabinske, D. W., and D. H. Knight. 1978. Variation within the sagebrush vegetation of Grand Teton National Park, Wyoming. *Northwest Science* 52:195–204.

Sæther, B.-E., S. Engen, J. E. Swenson, Ø. Bakke, and F. Sandegren. 1998. Assessing theviability of Scandinavian brown bear, *Ursus arctos*, populations: The effects of uncertain parameter estimates. *Oikos* 83:403–416.

Sall, J., and A. Lehman. 1996. *JMP Start Statistics: A guide to statistical and data analysis using JMP and JMPIN, software.* Pacific Grove, Calif.: Duxbury Press.

Saltz, D. 1996. Minimizing extinction probability due to demographic stochasticity in a reintroduced herd of Persian fallow deer *Dama dama mesopotamica. Biological Conservation* 75:27–33.

Salwasser, H. 1986. Conserving a regional spotted owl population. In *Ecological knowledge and environmental problem solving: Concepts and case studies*, ed. N. Grossblatt, 227–247. Washington, D.C.: National Academy Press.

Salwasser, H., H. Black Jr., and T. Hanley. 1980. The Forest Service fish and wildlife habitat relationships system. San Francisco: USDA Forest Service, Pacific Southwest Region.

Salwasser, H., and W. B. Krohn. 1982. Habitat classification—assessments for wildlife and fish: Closing remarks. *Transactions of the North American Wildlife and Natural Resources Conference* 47:184–185.

SAMAB. 1996. The southern Appalachian assessment summary report. Asheville, N.C.: Southern Appalachian Man and the Biosphere Cooperative.

Samson, F. B., and F. L. Knopf. 1982. In search of a diversity ethic for wildlife management. *Transactions North American Wildlife and Natural Resources Conference* 47:421–431.

Samson, S. A. 1993. Two indices to characterize temporal patterns in the spectral response of vegetation. *Photogrammetric Engineering and Remote Sensing* 59: 511–517.

Sargent, G. A., and D. H. Johnson. 1997. Carnivore scent-station surveys: Statistical considerations. *Proceedings of the North Dakota Academy of Science* 51:102–104.

Sargent, R. A., Jr. 1992. Movement ecology of adult male white-tailed deer in hunted and non-hunted populations in the wet prairie of the Everglades. M.S. thesis, University of Florida, Gainesville.

Sargent, R. A., and R. F. Labisky. 1995. Home range of male white-tailed deer in hunted and non-hunted populations. *Proceedings of the annual conference of the Southeastern Association of Fish and Wildlife Agencies* 49:389–398.

Sargent, R. E. 1984. A tutorial on verification and validation of simulation models. In *Proceedings of the 1984 Winter Simulation Conference*, ed. S. Sheppard, U. W. Pooch, and C. D. Pegden, 115–122. IEEE 84CH2098-2. La Jolla, Calif.: Society for Computer Simulation.

Sarkar, S. 1996. Ecological theory and anuran declines. *BioScience* 46:199–207.

Särndal, C. E., B. Swensson, and J. Wretman. 1992. *Model assisted survey sampling.* New York: Springer-Verlag.

SAS Institute. 1985. *SAS/STAT guide for personal computers.* Version 6. ed. Cary, N.C.: SAS Institute.

_____. 1990a. *SAS/STAT user's guide.* Version 6. 4th ed. Vols. 1 and 2. Cary, N.C.: SAS Institute.

_____. 1990b. *SAS procedures guide.* Version 6. 3rd ed. Cary, N.C.: SAS Institute.

_____. 1994. *JMP statistics and graphics guide.* Version 3. Cary, N.C.: SAS Institute.

_____. 1995. *Logistic regression examples using the SAS system.* Version 6. Cary, N.C.: SAS Statistical Institute.

Sauer, J. R., and S. Droege, eds. 1990. Survey designs and statistical methods for the estimation of avian population trends. Biological Report 90(1). Washington, D.C.: U.S. Department of the Interior, Fish and Wildlife Service.

Sauer, J. R., J. E. Hines, G. Gough, I. Thomas, and B. G. Peterjohn. 1997. The North American Breeding Bird Survey results and analysis. Version 96.3. Laurel, Md.: Patuxent Wildlife Research Center.

Sauer, J. R., J. E. Hines, G. Gough, I. Thomas, and B. G. Peterjohn. 1997. The North Amercian breeding bird survey results and analysis. Version 96.4. Laurel, Md.: Patuxent Wildlife Research Center.

Sauer, J. R., J. E. Hines, I. Thomas, J. Fallon, and G. Gough. 1999. The North American Breeding Bird Survey, Results and Analysis 1966–1998. Version 98.1. Laurel, Md.: USFS Patuxent Wildlife Research Center. http://www.mbr-pwrc.usgs.gov/bbs.

Sauer, J. R., G. W. Pendleton, and S. Orsillo. 1995. Mapping of bird distributions form point count surveys. In *Monitoring bird populations by point counts*, ed. C. J. Ralph, J. R. Sauer, and S. Droege, 151–160. General Technical Report PSW-GTR-149. Albany, Calif.: USDA Forest Service, Pacific Southwest Research Station.

Sauer, J. R., B. G. Peterjohn, and W. A. Link. 1994. Observer differences in the North American Breeding Bird Survey. *Auk* 111:50–62.

Saunders, D. A., R. J. Hobbs, and C. R. Margules. 1991. Biological consequences of ecosystem fragmentation: A review. *Conservation Biology* 5:18–32.

Saveraid, E. H. 1999. Using satellite imagery and landscape variables to predict bird communities in montane meadows of the Greater Yellowstone Ecosystem. M.S. thesis, Iowa State University, Ames.

Savitsky, B. G., J. Fallas, C. Vaughan, and T. E. Larcher Jr. 1998. Wildlife and habitat data collection and analysis. In *GIS methodologies for developing conservation strate-*

gies, ed. B. G. Savitsky and T. E. Larcher Jr., 158–169. New York: Columbia University Press.

Schaafsma, W., and G. N. van Vark. 1979. Classification and discrimination problems with applications. Part 2a. *Statistica Neerlandica* 33:91–126.

Schabenberger, O. 1995. The use of ordinal response methodology in forestry. *Forest Science* 41:321–336.

Schaefer, S. M. In preparation. Investigations into accuracy tests of predicted vertebrate occurrences from Maine Gap Analysis. M.S. thesis, University of Maine, Orono.

Schamberger, M., A. H. Farmer, and J. W. Terrell. 1982. Habitat suitability index models: Introduction. FWS/OBS-82/10. Washington, D.C.: U.S. Department of the Interior, U.S. Fish and Wildlife Service, Office of Biological Services and Division of Ecological Services.

Schamberger, M. L., and L. J. O'Neil. 1986. Concepts and constraints of habitat-model testing. In *Wildlife 2000: Modeling habitat relationships of terrestrial vertebrates*, ed. J. Verner, M. L. Morrison, and C. J. Ralph, 5–10. Madison: University of Wisconsin Press.

Schellenberg, J. A., J. N. Newell, R. W. Snow, V. Mung'ala, K. Marsh, P. G. Smith, and R. J. Hayes. 1998. An analysis of the geographical distribution of severe malaria in children in Kilifi District, Kenya. *International Journal of Epidemiology* 27:323–329.

Schemnitz, S. D. 1961. Ecology of the scaled quail in the Oklahoma panhandle. *Wildlife Monographs* 8:1–47.

Schiffers, J. 1997. A classification approach incorporating misclassification costs. Intelligent Data Analysis 1(1): an electronic journal. http://www-east.elsevier.com/ida/browse/96-8/ida96-8.htm.

Schlosser, I. J. 1987. The role of predation in age- and size-related habitat use by stream fishes. *Ecology* 68:651–659.

———. 1995. Dispersal, boundary processes, and trophic-level interactions in streams adjacent to beaver ponds. *Ecology* 76:908–925.

———. 1998. Fish recruitment, dispersal, and trophic interactions in a heterogeneous lotic environment. *Oecologia* 113:260–268.

Schlosser, I. J., and P. L. Angermeier. 1995. Spatial variation in demographic processes of lotic fishes: Conceptual models, empirical evidence, and implications for conservation. *American Fisheries Society Symposium* 17:392–401.

Schlosser, I. J., and L. W. Kallemeyn. 2000. Spatial variation in fish assemblages across a beaver-influenced successional landscape. *Ecology* 81:1371–1382.

Schluter, D., and R. E. Ricklefs. 1993. Species diversity: An introduction to the problem. In *Species diversity in ecological communities*, ed. R. E. Rickleffs and D. Schluter, 1–10. Chicago: University of Chicago Press.

Schmiegelow, F. K. A., and S. J. Hannon. 1999. Forest-level effects of fragmentation on boreal songbirds: The Calling Lake fragmentation studies. In *Forest fragmentation: Wildlife and management implications*, ed. J. A. Rochelle, L. A. Lehmann, and J. Wisniewski, 201–221. Leiden, The Netherlands: Brill.

Schmiegelow, F. K. A., C. S. Machtans, and S. J. Hannon. 1997. Are boreal birds resilient to forest fragmentation? An experimental study of short-term community responses. *Ecology* 78:1914–1932.

Schneider, D. C. 1994. *Quantitative ecology: Spatial and temporal scaling*. San Diego: Academic Press.

Schneider, S. H. 1993. Scenarios of global warming. In *Biotic interactions and global change*, ed. P. M. Kareiva, J. G. Kingsolver, and R. B. Huey, 9–23. Sunderland, Mass.: Sinauer Associates.

Schneider, S. H., L. Mearns, and P. H. Gleick. 1992. Climate-change scenarios for impact assessment. In *Global warming and biological diversity*, ed. R. L. Peters and T. E. Lovejoy, 38–55. New Haven, Conn.: Yale University Press.

Schoener, T. W. 1974. Resource partitioning in ecological communities. *Science* 185:27–39.

Schonewald-Cox, C. 1994. Protection of biological diversity: Missing connection between science and management. In *Biological diversity: Problems and challenges*, ed. S. K. Majumdar, F. J. Brenner, J. E. Lovich J. F. Schalles, and E. W. Miller, 170–183. Easton, Pa.: Pennsylvania Academy of Science.

Schonewald-Cox, C. M., S. M. Chambers, B. MacBryde, and W. L. Thomas, eds. 1983. *Genetics and conservation: A reference for managing wild animal and plant populations*. Menlo Park, Calif.: Benjamin/Cummings.

Schroeder, R. L. 1983a. Habitat suitability index models: Black-capped chickadee. U.S. Fish and Wildlife Service. FWS/OBS 82/10.37. Washington, D.C.: U.S. Department of the Interior, U.S. Fish and Wildlife Service, Western Energy and Land Use Team, Division of Biological Services, Research and Development.

———. 1983b. Habitat suitability index models: Pileated woodpecker. FWS/OBS 82/10.39. Washington, D.C.: U.S. Department of the Interior, U.S. Fish and Wildlife Service, Western Energy and Land Use Team, Division of Biological Services, Research and Development.

———. 1990. Tests of a habitat suitability model for black-capped chickadees. Biological Report 90 (10). Washington, D.C.: U.S. Department of the Interior, U.S. Fish and Wildlife Service.

Schulte, L. A, and G. J. Niemi. 1998. Bird communities of early successional burned and logged forest. *Journal of Wildlife Management* 62:1418–1429.

Schultz, T. T., and L. A. Joyce. 1992. A spatial application of a marten habitat model. *Wildlife Society Bulletin* 20:74–83.

Schumaker, N. H. 1996. Using landscape indices to predict habitat connectivity. *Ecology* 77:1210–1225.

Scott, C. T. 1998. Sampling methods for estimating change in forest resources. *Ecological Applications* 8:228–233.

Scott, J. A. 1975. Mate-locating behavior of western North American butterflies. *Journal of Research on the Lepidoptera* 14:1–40.

———. 1986. *Butterflies of North America.* Stanford: Stanford University Press.

Scott, J. M., G. T. Auble, and J. M. Friedman. 1997. Flood dependency of cottonwood establishment along the Missouri River, Montana, USA. *Ecological Applications* 7:677–690.

Scott J. M., B. Csuti, and S. Caicco. 1991b. Gap analysis: Assessing protection needs. In *Landscape linkages and biodiversity*, ed. W.E. Hudson, 15–26. Washington, D.C.: Defenders of Wildlife and Island Press.

Scott, J. M., B. Csuti, J. D. Jacobi, and J. E. Estes. 1987. Species richness: A geographic approach to protecting future biological diversity. *BioScience* 37:782–788.

Scott, J. M., B. Csuti, K. Smith, J. E. Estes, and S. Caicco. 1991a. Gap analysis of species richness and vegetation cover: An integrated biodiversity conservation strategy. In *Balancing on the brink of extinction: The Endangered Species Act and lessons for the future*, ed. K. A. Kohm, 282–297. Washington, D.C.: Island Press.

Scott, J. M., F. Davis, B. Csuti, R. Noss, B. Butterfield, C. Groves, H. Anderson, S. Caicco, F. D'Erchia, T. C. Edwards Jr., J. Ulliman, and R. G. Wright. 1993. Gap analysis: A geographic approach to protection of biological diversity. *Wildlife Monographs* 123:1–41.

Scott, J. M., and M. D. Jennings. 1997. A description of the national gap analysis program. Moscow, Idaho: USGS Biological Resources Division.

———. 1998. Large-area mapping for biodiversity. *Annals of the Missouri Botanical Garden* 85:34–47.

Scott, J. M., S. Mountainspring, F. L. Ramsey, and C. B. Kepler. 1986. *Forest bird communities of the Hawaiian Islands: Their dynamics, ecology, and conservation.* Studies in Avian Biology 9. Santa Barbara, Calif.: Cooper Ornithological Society.

Scott, J. M., F. L. Ramsey, and C. B. Kepler. 1981. Distance estimation as a variable in estimating bird numbers from vocalizations. In *Estimating numbers of terrestrial birds* ed. C. J. Ralph, and J. M. Scott, 334–340. Studies in Avian Biology 6. Los Angeles: Cooper Ornithological Society.

Scott, J. M., T. H. Tear, and F. W. Davis, eds. 1996. *GAP analysis: A landscape approach to biodiversity planning.* Bethesda, Md.: American Society of Photogrammetry and Remote Sensing.

Scott, J. S. 1997. Regression models for categorical and limited dependent variables. Advanced quantitative techniques in the social sciences 7. London: Sage Publications.

Seaman, D. E., and R. A. Powell. 1996. An evaluation of the accuracy of kernel density estimators for home range analysis. *Ecology* 77:2075–2085.

Seastedt, T. R., and A. K. Knapp. 1993. Consequences of non-equilibrium resource availability across multiple time scales: The transient maxima hypothesis. *American Naturalist* 141:621–633.

Seber, G. A. F. 1965. A note on the multiple-recapture census. *Biometrika* 52:249–259.

———. 1982. *The estimation of animal abundance and related parameters.* 2nd ed. New York: MacMillan.

———. 1984. *Multivariate observations.* New York: John Wiley & Sons.

———. 1986. A review of estimating animal abundance. *Biometrics* 42:267–292.

Seburn, C. N. L., D. C. Seburn, and C. A. Paszkowski. 1997. Northern leopard frog (*Rana pipiens*) dispersal in relation to habitat. In *Status and conservation of midwestern amphibians*, ed. M. J. Lannoo, 64–72. Iowa City: University of Iowa Press.

Sedell, J. R., G. H. Reeves, F. R. Hauer, J. A. Stanford, and C. P. Hawkins. 1990. Role of refugia in recovery from disturbances: Modern fragmented and disconnected river systems. *Environmental Management* 14:711–724.

Sellers, P. J., M. D. Heiser, and F. G. Hall. 1992. Relations between surface conductance and spectral vegetation indices at intermediate (100 m squared to 15 km squared) length scales. *Journal of Geophysical Research* 97(D17): 19033–19059.

Semenchuck, G. ed. 1992. *The atlas of breeding birds of Alberta.* Edmonton: Federation of Alberta Naturalists.

Senft, R. L., M. B. Coughenour, D. W. Bailey, L. R. Rittenhouse, O. E. Sala, and D. M. Swift. 1987. Large herbivore foraging and ecological hierarchies. *Bioscience* 37:789–799.

Senft, R. L., L. R. Rittenhouse, and R. G. Woodmansee. 1983. The use of regression models to predict spatial patterns of cattle behavior. *Journal of Range Management* 36:553–557.

Shaffer, M. L. 1981. Minimum population sizes for species conservation. *BioScience* 31:131–134.

———. 1987. Minimum viable populations: Coping with uncertainty. In *Viable populations for conservation*, ed. M. E. Soulé, 69–86. Cambridge: Cambridge University.

Shannon, C. E., and W. Weaver. 1949. *The mathematical theory of communication*. Urbana, Ill.: University of Illinois Press.

Shapiro, A. M. 1975. The temporal component of butterfly species diversity. In *Ecology and evolution of communities*, ed. M. L. Cody and J. M. Diamond, 181–195. Cambridge, Mass.: Belknap Press of Harvard University Press.

———. 1996. Status of butterflies. In *Sierra Nevada Ecosystem Project: Final report to congress*. Vol. 2, *Assessments and scientific basis for management options*, 743–757. Davis, Calif.: Centers for Water and Wildland Resources, University of California.

Shaw, D. M., and S. F. Atkinson. 1990. An introduction to the use of geographic information systems for ornithological research. *Condor* 92:564–570.

Sheldon, A. L. 1968. Species diversity and longitudinal succession in stream fishes. *Ecology* 49:193–198.

Short, H. L., and J. B. Hestbeck. 1995. National biotic resource inventories and gap analysis. *BioScience* 45: 535–539.

Shrader-Frechette, K. S., and E. D. McCoy. 1993. *Method in ecology: Strategies for conservation*. London: Cambridge University Press.

Shugart, H. H., T. M. Smith, and W. M. Post. 1992. The potential for application of individual-based simulation models for assessing the effects of global change. *Annual Review of Ecology and Systematics* 23:15–38.

Shy, E. 1984. Habitat shift and geographical variation in North American tanagers (Thraupinae: *Piranga*). *Oecologia* 63:281–285.

Sillett, T. S., R. T. Holmes, and T. W. Sherry. 2000. Impacts of a global climate cycle on population dynamics of a migratory songbird. *Science* 188:2040–2042.

Silva, M., and J. A. Downing. 1994. Allometric scaling of minimal mammal densities. *Conservation Biology* 8: 732–743.

Silverman, B. W. 1986. *Density estimation for statistics and data analysis*. London: Chapman and Hall.

Simberloff, D. 1988. The contribution of population and community biology to conservation science. *Annual Review of Ecology and Systematics* 19:473–511.

———. 1998. Flagships, umbrellas, and keystones: Is single-species management passé in the landcape era? *Biological Conservation* 83:247–257.

Simons, T. R., S. M. Pearson, and F. R. Moore. 2000. Application of spatial models to the stopover ecology of trans-Gulf migrants. *Studies in Avian Biology* 20:5–15.

Simonson, T. D., and J. Lyons. 1995. Comparison of catch per effort and removal procedures for sampling stream fish assemblages. *North American Journal of Fisheries Management* 15:419–427.

Simpson Timber Company. 1992. Habitat conservation plan for the northern spotted owl on the California timberlands of Simpson Timber Company. Arcata, Calif.: Simpson Timber Company.

Sisk, T. D., and W. J. Battin. In press. Understanding the influence of habitat edges on avian ecology: Geographic patterns and insights for western landscapes. *Studies in Avian Biology*.

Sisk, T. D., and N. M. Haddad. In press. Incorporating the effects of habitat edges into landscape models: Effective area models for management. In *Integrating landscape ecology into natural resource management*, ed. J. Liu and W. W. Taylor. Oxford: Oxford University Press.

Sisk, T. D., N. M. Haddad, and P. R. Ehrlich. 1997. Bird assemblages in patchy woodlands: Modeling the effects of edge and matrix habitats. *Ecological Applications* 7:1170–1180.

Sisk, T. D., and C. R. Margules. 1993. Habitat edges and restoration: Methods for quantifying edge effects and predicting the results of restoration efforts. In *Nature conservation 3: Reconstruction of fragmented ecosystems*, ed. D. A. Saunders, R. J. Hobbs, and P. R. Ehrlich, 57–69. Chipping Norton, N.S.W.: Surrey Beatty & Sons in association with Western Australian Laboratory of the Commonwealth Scientific and Industrial Research Organization, Division of Wildlife and Ecology.

Sjögren-Gulve, P. 1994. Distribution and extinction patterns within a northern metapopulation of the pool frog, *Rana lessonae*. *Ecology* 75:1357–1367.

Sjögren-Gulve, P., and I. Hanski. 2000. Metapopulation viability analysis using occupancy models. In *The use of population viability analyses in conservation planning*, ed. P. Sjögren-Gulve and T. Ebenhard. *Ecological Bulletins* 53–71.

Skagen, S. K., C. P. Melcher, W. H. Howe, and F. L. Knopf. 1998. Comparative use of riparian corridors and oases by migrating birds in southeast Arizona. *Conservation Biology* 12:896–909.

Skellam, J. G. 1951. Random dispersal in theoretical populations. *Biometrika* 38:196–218.

Skelly, D. K., E. E. Werner, and S. A. Cortwright. 1999. Long-term distributional dynamics of a Michigan amphibian assemblage. *Ecology* 80:2326–2337.

Skidmore, A. K., A. Gauld, and P. Walker. 1996. Classification of kangaroo habitat distribution using three GIS models. *International Journal of Geographic Information Science* 10:441–454.

Skinner, C. A., P. J. Skinner, and S. Harris. 1991. The past history and recent decline of badgers *Meles meles* in Essex: An analysis of some of the contributory factors. *Mammal Review* 21:67–80.

Small, A. 1994. *California birds: Their status and distribution.* Vista, Calif.: Ibis.

Smallwood, K. S. 1993. Understanding ecological pattern and process by association and order. *Acta Oecologica* 14:443–462.

———. 1995. Scaling Swainson's hawk population density for assessing habitat-use across an agricultural landscape. *Journal of Raptor Research* 29:172–178.

———. 1997. Interpreting puma (*Puma concolor*) density estimates for theory and management. *Environmental Conservation* 24:283–298.

———. 1998. On the evidence needed for listing northern goshawks (*Accipter gentilis*) under the Endangered Species Act: A reply to Kennedy. *Journal of Raptor Research* 32:323–329.

———. 1999. Scale domains of abundance among species of mammalian carnivora. *Environmental Conservation* 26:102–111.

———. 2001. Linking habitat restoration to meaningful units of animal demography. *Restoration Ecology.* 9:253–266.

Smallwood, K. S., and W. A. Erickson. 1995. Estimating gopher populations and their abatement in forest plantations. *Forest Science* 41:284–296.

Smallwood, K. S., and E. L. Fitzhugh. 1995. A track count for estimating mountain lion *Felis concolor californica* population trend. *Biological Conservation* 71:251–259.

Smallwood, K. S., and S. Geng. 1997. Multi-scale influences of gophers of alfalfa yield and quality. *Field Crops Research* 49:159–168.

Smallwood, K. S., and C. Schonewald. 1996. Scaling population density and spatial pattern for terrestrial, mammalian carnivores. *Oecologia* 105:329–335.

Smallwood, K. S., B. Wilcox, R. Leidy, and K. Yarris. 1998. Indicators assessment for habitat conservation plan of Yolo County, California, USA. *Environmental Management* 22:947–958.

Smith, A. P., and D. B. Lindenmayer. 1988. Tree hollow requirements of Leadbeater's possum and other possums and gliders in timber production ash forests of the Victorian central highlands. *Australian Wildlife Research* 15:347–362.

Smith, C. C., and A. N. Bragg. 1949. Observation on the ecology and natural history of anura, 7. Food and feeding habits of the common species of toad in Oklahoma. *Ecology* 30:333–349.

Smith, G. W. 1995. A critical review of the aerial and ground surveys of breeding waterfowl in North America. Biological Science Report 5. Washington: U.S. Department of the Interior, National Biological Service.

Smith, J. N. M., T. L. Cook, S. I. Rothstein, S. K. Robinson, and S. G. Sealy, eds. 2000. *The biology and management of cowbirds and their hosts.* Austin: University of Texas Press.

Smith, K., and P. G. Conners. 1986. Building predictive models of species occurrence from total-count transect data and habitat measurements. In *Wildlife 2000: Modeling habitat relationships of terrestrial vertebrates*, ed. J. Verner, M. L. Morrison, and C. J. Ralph, 45–50. Madison: University of Wisconsin Press.

Smith, K. G. 1977. Distribution of summer birds along a forest moisture gradient in an Ozark watershed. *Ecology* 58:810–819.

Smith, K. G., and D. G. Catanzaro. 1996. Predicting vertebrate distributions for gap analysis: Potential problems in constructing the models. In *Gap analysis: A landscape approach to biodiversity planning*, ed. J. M. Scott, T. H. Tear, and F. W. Davis, 163–169. Bethesda, Md.: American Society for Photogrammetry and Remote Sensing.

Smith, M. 1996. *Neural networks for statistical modeling.* London: International Thomson Computer Press.

Smith, M. L., J. N. Bruhn, and J. B. Anderson. 1992. The fungus *Armillaria bulbosa* is among the largest and oldest organisms. *Nature* 356:428–431.

Smith, M. S. 1988. Modeling three approaches to predicting how herbivore impact is distributed in rangelands. Research Report 628. Las Cruces: New Mexico State University Agricultural Experiment Station.

Smith, P. A. 1994. Autocorrelation in logistic regression

modelling of species' distributions. *Global Ecology and Biogeography Letters* 4:47–61.

Smith, R. D., A. Ammann, C. Bartoldus, and M. M. Brinson. 1995. An approach for assessing wetland functions using hydrogeomorphic classification, reference wetlands, and functional indices. Technical Report WRP-DE-9. Vicksburg, Miss.: U.S. Army Engineer Waterways Experiment Station.

Smith, T. M., and H. H. Shugart. 1987. Territory size variation in the ovenbird: The role of habitat structure. *Ecology* 68:695–704.

Smith, T. M., and M. A. Huston. 1989. A theory of the spatial and temporal dynamics of plant communities. *Vegetatio* 83:49–69.

Smith, T. M., and D. L. Urban. 1988. Scale and resolution of forest structural pattern. *Vegetatio* 74:143–150.

Snell, E. J. 1964. A scaling procedure for ordered categorical data. *Biometrics* 20:592–607.

Snodgrass, J. W., and G. K. Meffe. 1998. Influence of beavers on stream fish assemblages: Effects of pond age and watershed position. *Ecology* 79:928–942.

Snodgrass, T. R. 1992. Temporal databases. In *Theories and methods of spatio-temporal reasoning in geographic space*, ed. A. U. Frank, I. Campari, and U. Formentini, 2–64. New York: Springer-Verlag.

Soberón, J. M., and J. B. Llorente. 1993. The use of species accumulation functions for the prediction of species richness. *Conservation Biology* 7:480–488.

Soderquist, T. R., and R. Mac Nally. 2000. The conservation value of mesic gullies in dry forest landscapes: Mammal populations in the box-ironbark ecosystem of southern Australia. *Biological Conservation* 93:281–291.

Sodhi, N. S., C. A. Paszkowski, and S. Keehn. 1999. Scale-dependent habitat selection by American redstarts in aspen-dominated forest fragments. *Wilson Bulletin* 111:70–75.

Sokal, R. R., and N. L. Oden. 1978. Spatial autocorrelation in biology. 2. Some biological implications and four applications of the evolutionary and ecological interest. *Biological Journal of the Linnean Society* 10:229–249.

Sokal, R. R., and F. J. Rohlf. 1981. *Biometry: The principles and practice of statistics in biological research.* 2nd ed. San Francisco: W.H. Freeman.

———. 1995. *Biometry: The principles and practice of statistics in biological research.* 3rd ed. New York: W. H. Freeman.

Solis, D. M., Jr., and R. J. Gutiérrez. 1990. Summer habitat ecology of northern spotted owls in northwestern California. *Condor* 92:739–748.

Somers, R. H. 1962. A new asymmetric measure of association for ordinal variables. *American Sociological Review* 27:799–811.

Song, Y.-L. 1996. Population viability analysis for two isolated populations of Haianan Eld's deer. *Conservation Biology* 10:1467–1472.

Sonquist, J. A., E. L. Baker, and J. N. Morgan. 1973. *Searching for structure.* Revised ed. Ann Arbor: Institute for Social Research, University of Michigan.

Soulé, M. E.1987a. Introduction. In *Viable populations for conservation*, ed. M. E. Soulé, 1–10. Cambridge: Cambridge University Press.

———. 1987b. *Viable populations for conservation.* Cambridge: Cambridge University Press.

———. 1991. Conservation tactics for a constant crisis. *Science* 253:744–750.

Soulé, M. E., and L. S. Mills. 1998. No need to isolate genetics. *Science* 282:1658–1659.

Sousa, W. P. 1979. Disturbances in marine intertidal boulder fields: The nonequilibrium maintenance of species diversity. *Ecology* 60:1225–1239.

Southerland, M. T, and S. B. Weisberg. 1995. Maryland Biological Stream Survey: The 1995 workshop summary. Report to Chesapeake Bay Research and Monitoring Division. Anapolis: Maryland Department of Natural Resources.

Southgate, R., and H. Possingham. 1995. Modelling the reintroduction of the greater bilby *Macrotis lagotis* using the metapopulation model analysis of the likelihood of extinction (ALEX). *Biological Conservation* 73:151–160.

Southwood, T. R. E. 1977. Habitat, the templet for ecological strategies? *Journal of Animal Ecology* 46:337–365.

Sowls, L. K. 1960. Results of a banding study of Gambel's quail in southern Arizona. *Journal of Wildlife Management* 24:185–190.

Spears, G. S., F. S. Guthery, S. M. Rice, S. J. DeMaso, and B. Zaiglin. 1993. Optimum seral stage for northern bobwhites as influenced by site productivity. *Journal of Wildlife Management* 57:805–811.

Spencer, J. S., Jr., W. B. Smith, J. T. Hahn, and G. K. Raile. 1988. Wisconsin's fourth forestry inventory, 1983. Resource Bulletin NC-107. St. Paul, Minn.: USDA Forest Service, North Central Forest Experiment Station.

Spitz, F. and S. Lek. 1990. Environmental impact prediction using neural network modelling. An example in wildlife damage. *Journal of Applied Ecology* 36:317–326.

Spitznagel, A. 1990. The influence of forest management on woodpecker density and habitat use in floodplain forests of the Upper Rhine Valley. In *Conservation and management of woodpecker populations*, ed. A. Carlson and G. Aulén, 177–146. Report 17. Uppsala: Swedish University of Agricultural Science.

S-PLUS 4. 1997. *Guide to statistics*. Seattle: Data Analysis Products Division, MathSoft.

Springer, J. T. 1979. Some sources of bias and sampling error in radio triangulation. *Journal of Wildlife Management* 43:926–935.

Spruell, P., B. E. Rieman, K. L. Knudsen, F. M. Utter, and F. W. Allendorf. 1999. Genetic population structure within streams: Microsatellite analysis of bull trout populations. *Ecology of Freshwater Fish* 8:114–121.

SPSS. 1988. *Statistical package of the social sciences*. Chicago: SPSS.

———. 1990. *SPSS. Version 4.0.4. for the Macintosh*. Chicago: SPSS.

———. 1997. *SYSTAT 7.0.: for Windows*. Chicago: SPSS.

———. 1998. *SYSTAT 8.0: Statistics*. Chicago: SPSS.

Stalnaker, C., B. L. Lamb, J. Henriksen, K. Bovee, and J. Bartholow. 1995. The instream flow incremental methodology: A primer for IFIM. Biological Report 29. Washington, D.C.: U.S. Department of the Interior, National Biological Service.

Stamps, J. A. 1991. The effect of conspecifics on habitat selection in territorial species. *Behavioral Ecology and Sociobiology* 28:29–36.

Starfield, A. M. 1997. A pragmatic approach to modeling for wildlife management. *Journal of Wildlife Management* 61:261–270.

Starfield, A. M., and A. L. Bleloch. 1991. *Building models for conservation and wildlife management*. Edina, Minn.: Burgess International Group.

StataCorp. 1999. *Stata statistical software: Release 6.0*. College Station, Tex.: Stata Corporation.

State of California Resources Agency. 1969. Soil vegetation surveys of California. Rev. ed. Sacramento: Department of Conservation and Division of Forestry.

Stauffer, D. F., and L. B. Best. 1980. Habitat selection by birds of riparian communities: Evaluating effects of habitat alterations. *Journal of Wildlife Management* 44:1–15.

———. 1986. Effects of habitat type and sample size on habitat suitability index models. In *Wildlife 2000: Modeling habitat relationships of terrestrial vertebrates*, ed. J. Verner, M. L. Morrison, and C. J. Ralph, 71–77. Madison: University of Wisconsin Press.

Steele, B. B., R. L. Bayn Jr., and C. V. Grant. 1984. Environmental monitoring using populations of birds and small mammals: Analysis of sampling effort. *Biological Conservation* 30:157–172.

Steele, R. G. D., and J. H. Torrie. 1960. *Principles and procedures of statistics: A biometrical approach*. New York: McGraw Hill.

Steger, G. N., T. E. Munton, and J. Verner. 1993. Preliminary results from a demographic study of spotted owls in Sequoia and Kings Canyon National Parks, 1990–1991. In *Proceedings of the fourth conference on research in California's national parks*, ed. S. D. Veirs Jr., T. J. Stohlgren, and C. Schonewald-Cox, 83–92. NPS/NRUC/NRTP-93/9. Davis, Calif.: Cooperative Park Studies Unit, University of California.

Stehman, S. V. 2000. Practical implications of design-based sampling inferences for thematic map accuracy assessment. *Remote Sensing of Environment* 72:35–45.

Steidl, R. J., J. P. Hayes, and E. Schrauber. 1997. Statistical power analysis in wildlife research. *Journal of Wildlife Management* 61:270–279.

Stein, B. A., L. S. Kutner, and J. A. Adams. 2000. *Precious heritage: The status of biodiversity in the United States*. New York: Oxford University Press.

Steinberg, D., and P. Colla. 1997. *CART—Classification and regression trees*. San Diego: Salford Systems.

Steinhart, P. 1990. *California's wild heritage: Threatened and endangered animals in the golden state*. Sacramento: California Department of Fish and Game, California Academy of Sciences, Sierra Club Books.

Steinitz, C. 1969. *A comparative study of resource analysis methods*. Cambridge, Mass.: Dept. of Landscape Architecture Research Office, Graduate School of Design, Harvard University.

———. 1979. *Defensible processes for regional landscape design*. Washington, D.C.: Landscape Architecture Technical Information Series, American Society of Landscape Architects.

Steinman, A. D. 1996. Effects of grazers on freshwater benthic algae. In *Ecology of freshwater benthic algae*, ed. R. J. Stevenson, M. L. Bothwell, and R. L. Lowe, 341–374. New York: Academic Press.

Stendell, E. R., T. R. Horton, and T. D. Bruns. 1999. Early effects of prescribed fire on the structure of the ectomycorrhizal fungus community in a Sierra Nevada ponderosa pine forest. *Mycological Research* 103:1353–1359.

Stephens, D. W., and E. L. Charnov. 1982. Optimal foraging: Some simple stochastic models. *Behavioral Ecology and Sociobiology* 10:251–263.

Stevens, S. 1946. On the theory of scales of measurement. *Science* 103:677–680.

Stevens, S. M., and T. P. Husband. 1998. The influence of edge on small mammals: Evidence from Brazilian Atlantic forest fragments. *Biological Conservation* 85:1–8.

Stewart, R. E. 1975. *Breeding birds of North Dakota*. Fargo, N.D.: Tri-College Center for Environmental Studies.

Stith, B. M., J. W. Fitzpatrick, G. E. Woolfenden, and B. Pranty. 1996. Classification and conservation of metapopulations: A case study of the Florida Scrub Jay. In *Metapopulations and wildlife conservation*, ed. D. R. McCullough, 187–215. Washington, D.C.: Island Press.

Stockwell, D. R. B. 1999. Genetic algorithms 2. In *Machine learning methods for ecological applications*, ed. A. H. Fielding, 123–144. Boston: Kluwer Academic Publishers.

Stockwell, D. R. B., S. M. Davey, J. R. Davis, and I. R. Noble. 1990. Using induction of decision trees to predict greater glider density. *AI Applications* 4:33–43.

Stockwell, D. R. B., and I. R. Noble. 1992. Induction of sets of rules from animal distribution data: A robust and informative method of analysis. *Mathematics and Computers in Simulation* 33:385–390.

Stockwell, D. R. B., and D. P. Peters. 1999. The GARP modelling system: Problems and solutions to automated spatial prediction. *International Journal of Geographical Information Science* 13:143–158.

Stockwell, D. R. B., and A. T. Peterson. In press. Effects of sample size on accuracy of species distribution models. *Ecological Modelling.*

Stohlgren T. J., D. Binkley, G. W. Chong, M. A. Kalkhan, L. D. Schell, K. A. Bull, Y. Otuski, G. Newman, M. Bashkin, and Y. Son. 1999. Exotic plant species invade hot spots of native plant diversity. *Ecological Monographs* 69:25–46.

Stohlgren, T. J., J. F. Quinn, M. Ruggiero, and G. S. Waggoner. 1995. Status of biotic inventories in U.S. national parks. *Biological Conservation* 71:97–106.

Stoms, D. M. 1991. Mapping and monitoring regional patterns of species richness from geographic information. Ph.D. dissertation, University of California, Santa Barbara.

———. 1992. Effects of habitat map generalization in biodiversity assessment. *Photogrammetric Engineering and Remote Sensing* 58:1587–1592.

———. 1994. Scale dependence of species richness maps. *Professional Geographer* 46:346–358.

Stoms, D. M., F. W. Davis, and C. B. Cogan. 1992. Sensitivity of wildlife habitat models to uncertainties in GIS data. *Photogrammetric Engineering and Remote Sensing* 58:843–850.

Stoms, D. M., and J. E. Estes. 1993. A remote sensing research agenda for mapping and monitoring biodiversity. *International Journal of Remote Sensing* 14:1839–1860.

Storey, K. B., and J. M. Storey. 1986. Freeze tolerance and intolerance as strategies of winter survival in terrestrially hibernating amphibians. *Comparative Biochemistry and Physiology. A, Comparative Physiology* 83:613–617.

Strahler, A. N. 1957. Quantitative analysis of watershed geomorphology. *Transactions of the American Geophysical Union* 38:913–920.

Strange, E. M., P. B. Moyle, and T. C. Foin. 1992. Interactions between stochastic and deterministic processes in stream fish community assembly. *Environmental Biology of Fishes* 36:1–15.

Straw, J. A., Jr., J. S. Wakely, and J. E. Hudgins. 1986. A model for management of diurnal habitat for American woodcock in Pennsylvania. *Journal of Wildlife Management* 50:378–383.

Stromberg, J. C., R. Tiller, and B. Richter. 1996. Effects of groundwater decline on riparian vegetation of semiarid regions: The San Pedro, Arizona. *Ecological Applications* 6:113–131.

Stromberg, J. C., J. A. Tress, S. D. Wilkins, and S. Clark. 1992. Response of velvet mesquite to ground water decline. *Journal of Arid Environments* 23:45–58.

Strong, W. L., and K. R. Leggat. 1981. Ecoregions of Alberta. Technical Report T/4. Edmonton: Alberta Energy and Natural Resources.

Suchy, W. J., L. L. McDonald, M. D. Strickland, and S. H. Anderson. 1985. New estimates of minimum viable population size for grizzly bears of the Yellowstone ecosystem. *Wildlife Society Bulletin* 13:223–228.

Sumner, E. L., Jr. 1935. A life history study of the California quail with recommendations for conservation and management. *California Fish and Game* 21:167–256, 277–342.

Suter, G. W. 1981. Ecosystem theory and NEPA assessment. *Bulletin of the Ecological Society of America* 62: 186–192.

Svardson, G. 1949. Competition and habitat selection in birds. *Oikos* 1:157–174.

Swank, W. G., and S. Gallizioli. 1954. The influence of

hunting and of rainfall upon Gambel's quail populations. *Transactions of the North American Wildlife and Natural Resources Conference* 19:283–297.

Swart, J., M. J. Lawes, and M. R. Perrin. 1993. A mathematical model to investigate the demographic viability of low-density samango monkey (*Cercopithecus mitis*) populations in Natal, South Africa. *Ecological Modelling* 70:289–303.

Sweeney, J. M., and W. D. Dijak. 1985. Ovenbird habitat capability model for an oak-hickory forest. *Proceedings of the annual conference of the Southeastern Association of Fish and Wildlife Agencies* 39:430–438.

Swengel, A. G. 1990. Monitoring butterfly populations using the Fourth of July butterfly count. *American Midland Naturalist* 124:395–406.

Swets, J. A. 1988. Measuring the accuracy of diagnostic systems. *Science* 240:1285–1293.

Swift, M. J. 1982. Basidiomycetes as components of forest ecosystems. In *Decomposer basidiomycetes: Their biology and ecology*, ed. J. C. Frankland, J. N. Hedger, and M. J. Swift, 307–338. Cambridge: Cambridge University Press.

Takeyama, M., and H. Coucelis. 1997. Map dynamics: Integrating cellular automata and GIS through geo-algebra. *International Journal of Geographical Information Science* 11:73–91.

Taniguchi, Y., F. J. Rahel, D. C. Novinger, and K. G. Gerow. 1998. Temperature mediation of competitive interactions among three fish species that replace each other along longitudinal stream gradients. *Canadian Journal of Fisheries and Aquatic Sciences* 55:1894–1901.

Taper, M. L., K. B. Gaese, and J. H. Brown. 1995. Individualistic responses of bird species to environmental change. *Oecologia* 101:478–486.

Tautin, J. 1982. Assessment of some important factors affecting the singing-ground survey. In *Proceedings of the Seventh Woodcock Symposium*, technical coordinators. T. J. Dwyer and G. L. Storm, 6–11. Research Report 14. Washington, D.C.: U.S. Department of the Interior, U.S. Fish and Wildlife Service.

Tautin, J., P. H. Geissler, R. E. Munro, and R. S. Pospahala. 1983. Monitoring the population status of American woodcock. *Transactions of the North American Wildlife and Natural Resources Conference* 48:376–388.

Taylor, C. M. 1996. Abundance and distribution within a guild of benthic stream fishes: Local processes and regional patterns. *Freshwater Biology* 36:385–396.

———. 1997. Fish species richness and incidence patterns in

isolated and connected stream pools: Effects of pool volume and spatial position. *Oecologia* 110:560–566.

Taylor, D. R., A. M. Jarosz, R. E. Lenski, and D. W. Fulbright. 1998. The acquisition of hypovirulence in host-pathogen systems with three trophic levels. *American Naturalist* 151:343–353.

Taylor, J. 1990. Questionable multivariate statistical inference in wildlife habitat and community studies: A comment. *Journal of Wildlife Management* 54:186–189.

Taylor, L. R. 1961. Aggregation, variance and the mean. *Nature* 189:732–735.

Taylor, L. R., and R. A. J. Taylor. 1977. Aggregation, migration and population mechanics. *Nature* 265:415–421.

Taylor, L. R., I. P. Woiwood, and J. N. Perry. 1978. The density-dependence of spatial behaviour and the rarity of randomness. *Journal of Animal Ecology* 47:383–406.

Taylor, R. A. J., and L. R. Taylor. 1979. A behavioral model for the evolution of spatial dynamics. In *Population dynamics*, ed. R. M. Anderson, B. D. Turner, and L. R. Taylor, 1–28. Oxford, UK: Blackwell Scientific Publications.

Tazik, D. J., and J. D. Cornelius. 1993. Status of the black-capped vireo at Fort Hood, Texas. Vol. 3, Population and nesting ecology. Technical Report EN-94/01/ADA 277544. Champaign, Ill.: U.S. Army Construction Engineering Research Laboratory.

Temple, S. A. 1986a. Predicting impacts of habitat fragmentation on forest birds: A comparison of two models. In *Wildlife 2000: Modeling habitat relationships of terrestrial vertebrates*, J. Verner, M. L. Morrison, and C. J. Ralph, 301–304. Madison: University of Wisconsin Press.

———. 1986b. The problem of avian extinctions. *Current Ornithology* 3:453–485.

Temple, S. A., and J. R. Cary. 1988. Modeling dynamics of habitat-interior bird populations in fragmented landscapes. *Conservation Biology* 2:340–347.

ter Braak, C. J. F., and C. W. N. Looman. 1995. Regression. In *Data analysis in community and landscape ecology*, ed. R. H. G. Jongman, C. J. F. ter Braak, and O. F. R. van Tongeren, 29–77. 2nd ed. Cambridge: Cambridge University Press.

ter Braak, C. J. F., and I. C. Prentice. 1988. A theory of gradient analysis. *Advances in Ecological Research* 18: 271–317.

The Nature Conservancy. 2000. *Designing a geography of hope: A practitioner's handbook to ecoregional conservation planning*. Vol. 1. 2nd ed. Arlington, Va.: The Nature Conservancy.

The Pacific Lumber Company. 1999. Habitat conservation plan for the properties of The Pacific Lumber Company, Scotia Pacific Company LLC, and Salmon Creek Corporation. Scotia, Calif.: The Pacific Lumber Company.

Theobald, D. M., J. M. Miller, and N. T. Hobbs. 1997. Estimating the cumulative effects of development on wildlife habitat. *Landscape and Urban Planning* 39:25–36.

Theurillat, J.-P., and A. Schlüssel. 1998. Phenology of *Rhododendron ferrugineum* L. in two altitudinal gradients of the Valaisian Alps (Switzerland). *Ecologie* 29: 429–433.

Thomas, C. D., and J. C. G. Abery. 1995. Estimating rates of butterfly decline from distribution maps: The effect of scale. *Biological Conservation* 73:59–65.

Thomas, C. D., and I. Hanski. 1997. Butterfly metapopulations. In *Metapopulation biology*, ed. I. A. Hanski and M. E. Gilpin, 359–386. San Diego: Academic Press.

Thomas, C. D., and H. C. Mallorie. 1985. Rarity, species richness and conservation: Butterflies of the Atlas Mountains in Morocco. *Biological Conservation* 35:95–117.

Thomas, D. L, and E. J. Taylor. 1990. Study designs and tests for comparing resource use and availability. *Journal of Wildlife Management* 54:322–330.

Thomas, J. A. 1991. Rare species conservation: Case studies of European butterflies. In *The scientific management of temperate communities for conservation. Twenty-ninth Symposium of the British Ecological Society*, ed. I. F. Spellerberg, M. G. Morris, and F. B. Goldsmith, 141–197. Oxford, UK: Blackwell Scientific Publications.

———. 1995. The conservation of declining butterfly populations in Britain and Europe: Priorities, problems and successes. *Biological Journal of the Linnaean Society* 56 (suppl.):55–72.

Thomas, J. A., D. J. Simcox, J. C. Wardlaw, G. W. Elmes, M. E. Hochberg, and R. T. Clarke. 1998. Effects of latitude, altitude and climate on the habitat and conservation of the endangered butterfly *Maculinea arion* and its myrmica ant hosts. *Journal of Insect Conservation* 2:39–46.

Thomas, J. W. 1979. Wildlife habitats in managed forests: The Blue Mountains of Oregon and Washington. Washington, D.C.: USDA Forest Service Agricultural Handbook No. 553.

Thomas, J. W. 1982. Needs for and approaches to wildlife habitat assessment. *Transactions of the North American Wildlife and Natural Resources Conference* 47:35–46.

Thomas, J. W., E. D. Forsman, J. B. Lint, E. C. Meslow, B. R. Noon, and J. Verner. 1990. A conservation strategy for the northern spotted owl: Report of the Interagency Scientific Committee to address the conservation of the northern spotted owl. Portland, Ore.: U.S. Department of Agriculture Forest Service; U.S. Department of the Interior, Bureau of Land Management, U.S. Fish and Wildlife Service, National Park Service.

Thomas, J. W., M. G. Raphael, R. G. Anthony, E. D. Forsman, A. G. Gunderson, R. S. Holthausen, B. G. Marcot, G. H. Reeves, J. R. Sedell, and D. M. Solis. 1993. Viability assessments and management considerations for species associated with late-successional and old-growth forests of the Pacific Northwest. Portland, Ore.: USDA Forest Service.

Thome, D. M., C. J. Zabel, and L. V. Diller. 1999. Forest stand characteristics and reproduction of northern spotted owls in managed north-coastal California forest. *Journal of Wildlife Management* 63:44–59.

Thompson, B., P. J. Crist, J. S. Prior-Magee, R. A. Dietner, D. L. Garber, and M. A. Hughes. 1996. Gap analysis of biological diversity conservation in New Mexico using geographic information systems. Research Completion Report. Las Cruces: New Mexico Cooperative Fish and Wildlife Research Unit, New Mexico State University.

Thompson, F. R., III. 1993. Simulated responses of a forest-interior bird population to forest management options in central hardwood forests of the United States. *Conservation Biology* 7:325–333.

Thompson, F. R., S. J. Lewis, J. Green, and D. Ewert. 1992. Status of Neotropical migrant landbirds in the Midwest: Identifying species of management concern. In *Status and management of Neotropical migratory birds*, ed. D. M. Finch and P. W. Stangel, 145–158. General Technical Report RM-229. Fort Collins, Colo.: USDA Forest Service, Rocky Mountain Forest and Range Experiment Station.

Thompson, S. K. 1992. *Sampling*. New York: Wiley.

Thompson, W. L., G. C. White, and C. Gowan. 1998. *Monitoring vertebrate populations*. San Diego: Academic Press.

Thomson, J. D., G. Weiblen, B. A. Thomson, S. Alfaro, and P. Legendre. 1996. Untangling multiple factors in spatial distributions: Lilies, gophers, and rocks. *Ecology* 77: 1698–1715.

Thornton, P. S. 1988. Density and distribution of badgers in south-west England: A predictive model. *Mammal Review* 18:11–23.

Thrall, P. H., J. D. Bever, J. D. Mihail, and H. M. Alexander. 1997. The population dynamics of annual plants and soil-borne fungal pathogens. *Journal of Ecology* 85: 313–328.

Thurow, R. F., D. C. Lee, and B. E. Rieman. 1997. Distribution and status of seven native salmonids in the Interior Columbia River Basin and portions of the Klamath River and Great Basins. *North American Journal of Fisheries Management* 17:1094–1110.

Thurow, R. F., and D. J. Schill. 1996. Comparison of day snorkeling, night snorkeling, and electrofishing to estimate bull trout abundance and size structure in a second-order Idaho stream. *North American Journal of Fisheries Management* 16:314–323.

Tileston, J. V., and R. R. Lechleitner. 1966. Some comparisons of the black-tailed and white-tailed prairie dogs in north-central Colorado. *American Midland Naturalist* 75:292–316.

Tilman, D. 1982. *Resource competition and community structure*. Princeton, N.J.: Princeton University Press.

——. 1987. Secondary succession and the pattern of plant dominance along experimental nitrogen gradients. *Ecological Monographs* 57:189–214.

——. 1993. Species richness of experimental productivity gradients: How important is colonization limitation? *Ecology* 74:2179–2191.

——. 1996. Biodiversity: Population versus ecosystem stability. *Ecology* 77:350–363.

——. 1997. Community invasibility, recruitment limitation, and grassland biodiversity. *Ecology* 78:81–92.

Tilman, D., and J. A. Downing. 1994. Biodiversity and stability in grasslands. *Nature* 367:363–365.

Tilman, D., and P. Kareiva, eds. 1997. *Spatial ecology: The role of space in population dynamics and interspecific interactions*. Princeton, N.J.: Princeton University Press.

Timothy, K. G., and D. F. Stauffer. 1991. Reliability of selected avifauna information in a computerized information system. *Wildlife Society Bulletin* 19:80–88.

Tinker, P. B. 1985. Modelling mycorrhizal development. In *Proceedings of the Sixth North American Conference of Mycorrhizae*, ed. R. Molina, 140–141. Corvallis, Ore.: Forest Research Laboratory, Oregon State University.

Titus, K., J. A. Mosher, and B. K. Williams. 1984. Chance-corrected classification for use in discriminant analysis: Ecological applications. *American Midland Naturalist* 111:1–7.

Tobalske, C. 1998. Modeling the distribution of woodpecker species in the Jura, France, and in Switzerland, using atlas data. Ph.D. dissertation, University of Montana, Missoula.

Tobalske, C., and B. W. Tobalske. 1999. Using atlas data to model the distribution of woodpecker species in the Jura, France. *Condor* 101:472–483.

Toft, C. A. 1985. Resource partitioning in amphibians and reptiles. *Copeia* 1985:1–21.

Toft, C. A., and P. J. Shea. 1983. Detecting community-wide patterns: Estimating power strengthens statistical inference. *American Naturalist* 122:618–625.

Tomlin, C. 1990. *Geographic information systems and cartographic modeling*. Englewood Cliffs, N.J.: Prentice Hall.

Topping, C. J., M. J. Rehder, B. H. Mayoh, P. Odderskaer, and J. Reddersen. 1998. Biola: A new biological programming language for developing individual based models. http://acarus.entu.cas.cz/Abstbook/abstbook/node28.html.

Torgersen, C. E., D. M. Price, H. W. Li, and B. A. McIntosh. 1999. Multiscale thermal refugia and stream habitat associations of chinook salmon in northeastern Oregon. *Ecological Applications* 9:301–319.

Trame, A., S. J. Harper, J. Aycrigg, and J. Westervelt. 1997. The Fort Hood Avian Simulation Model: A dynamic model of ecological influences on two endangered species. USACERL Technical Report 97/88. Champaign, Ill.: U.S. Army Construction Engineering Research Laboratory.

Trame, A., S. J. Harper, and J. Westervelt. 1998. Management of cowbird traps on the landscape: An individual-based modeling approach for Fort Hood, Texas. USACERL Technical Report 98/121. Champaign, Ill.: U.S. Army Construction Engineering Research Laboratory.

Trame, A., A. Krzysik, S. Briggs, B. MacAllister, S. Harper, and W. Seybold. 1999. Report to Fort Hood: FY98 validation and improvement of the ICBM and FHASM models. USACERL Report to Fort Hood, Texas. Champaign, Ill.: U.S. Army Construction Engineering Research Lab.

Trani, M. K. 1996. Landscape pattern analysis related to forest wildlife resources. Ph.D. dissertation, Virginia Polytechnic Institute and State University, Blacksburg.

Trani, M. K., and R. H. Giles Jr. 1999. An analysis of deforestation. *Forest Ecology and Management* 114:459–470.

Trexler, J. C., and J. Travis. 1993. Nontraditional regression analyses. *Ecology* 74:1629–1637.

Trietz, P. M., P. J. Howarth, and P. Gong. 1992. Application of satellite and GIS technologies for land cover and land use mapping at the rural-urban fringe: A case study. *Photogrammetric Engineering and Remote Sensing* 58:439–448.

Tucker, C. J., J. R. G. Townshend, and T. E. Goff. 1985.

African land-cover classification using satellite data. *Science* 227:369–375.

Tucker, K., S. P. Rushton, R. A. Sanderson, E. B. Martin, and J. Blaiklock. 1997. Modelling bird distributions—a combined GIS and Bayesian rule-based approach. *Landscape Ecology* 12:77–93.

Tukey, J. W. 1977. *Exploratory data analysis*. Reading, Mass.: Addison-Wesley.

Turner, M. G. 1989. Landscape ecology: The effect of pattern on process. *Annual Review of Ecology and Systematics* 20:171–197.

Turner, M. G., G. J. Arthaud, R. T. Engstrom, S. J. Hejl, J. Liu, S. Loeb, and K. McKelvey. 1995. Usefulness of spatially explicit population models in land management. *Ecological Applications* 5:12–16.

Turner, M. G., V. H. Dale, and R. H. Gardner. 1989a. Predicting across scales: Theory development and testing. *Landscape Ecology* 3:245–252.

Turner, M. G., and R. H. Gardner, eds. 1991. *Quantitative methods in landscape ecology: The analysis and interpretation of landscape heterogeneity*. New York: Springer-Verlag.

Turner, M. G., R. V. O'Neill, R. H. Gardner, and B. T. Milne. 1989b. Effects of changing spatial scale on the analysis of landscape patterns. *Landscape Ecology* 3:153–162.

Turner, M. G., W. H. Romme, R. H. Gardner, R. V. O'Neill, and T. H. Kratz. 1993. A revised concept of landscape equilibrium: Disturbance and stability on scaled landscapes. *Landscape Ecology* 8:213–227.

Turner, M. G., Y. Wu, L. L. Wallace, W. H. Romme, and A. Brenkert. 1994. Simulating winter interactions among ungulates, vegetation, and fire in northern Yellowstone Park. *Ecological Applications* 4:472–496.

Turney, P. 1991. The gap between abstract and concrete results in machine learning. *Journal of Experimental and Theoretical Artificial Intelligence* 3:179–190.

Turney, P. D. 1995. Cost-sensitive classification: Empirical evaluation of a hybrid genetic decision tree induction algorithm. *Journal of Artificial Intelligence Research* 2:369–409.

University Consortium for Geographic Information Science. 1996. Research priorities for geographic information science. *Cartography and Geographic Information Systems* 23:115–127.

Urban, D. L., R. V. O'Neill, and H. H. Shugart Jr. 1987. Landscape ecology: A hierarchical perspective can help scientists understand spatial patterns. *BioScience* 37:119–127.

Urban, D. L., and H. H. Shugart Jr. 1986. Avian demography in mosaic landscapes: Modeling paradigm and preliminary results. In *Wildlife 2000: Modeling habitat relationships of terrestrial vertebrates*, ed. J. Verner, M. L. Morrison, and C. J. Ralph, 273–279. Madison: University of Wisconsin Press.

Urban, D. L., and T. M. Smith. 1989. Microhabitat pattern and the structure of forest bird communities. *American Naturalist* 133:811–829.

USACOE. United States Army Corps of Engineers. 1991. Sacramento River, sloughs, and tributaries. California 1991 Aerial Atlas, Collinsville to Shasta Dam. Sacramento: U.S. Army Corps of Engineers, Sacramento District.

———. 1993. *GRASS 4.1 reference manual*. Champaign, Ill.: United States Army Corps of Engineers Construction Engineering Research Laboratory.

———. 1999. Central and South Florida comprehensive review study. Final integrated feasibility report and programmatic environmental impact statement. http://www.restudy.org/overview.pdf (accessed 23 June 1999).

U.S. Census Bureau. TIGER. 1995. http://prome.snu.ac.kr/~ohrora/lis/tiger.htm.

USDA. United States Department of Agriculture, U.S. Department of the Interior, U.S. Department of Commerce, U.S. Environmental Protection Agency. 1993. Forest ecosystem management: An ecological, economic, and social assessment. USDA Forest Service, Washington, D.C.

USDA. United States Department of Agriculture, Forest Service. 1995. Forest Insect and Disease Conditions in the United States 1994. Washington, D.C.: USDA Forest Service, Forest Pest Management.

USDA/USDI. United States Department of Agriculture, Forest Service and United States Department of Interior, Bureau of Land Management. 1994a. Record of decision for amendments to Forest Service and BLM planning documents within the range of northern spotted owl; standards and guidelines for management of habitat for late-successional and old-growth forest related species within the range of the northern spotted owl. Washington, D.C.: USDA Forest Service and USDI Bureau of Land Management.

———. 1994b. Final supplemental environmental impact statement on management of habitat for late-successional and old-growth forest related species within the range of the northern spotted owl. Washington, D.C.: USDA Forest Service and USDI Bureau of Land Management.

USDI. United States Department of the Interior. 1992. Recovery plan for the northern spotted owl: Final draft. Washington, D.C.: U.S. Department of the Interior.

———. 1996. Effects of military training and fire in the Snake River Birds of Prey National Conservation Area. BLM/IDARNG Research Project Final Report. Boise, Idaho: U.S. Geological Survey, Biological Resources Division, Snake River Field Station.

USFWS. United States Fish and Wildlife Service. 1980. Habitat as a basis for environmental assessment. ESM 101. Washington, D.C.: U.S. Department of the Interior, U.S. Fish and Wildlife Service, Division of Ecological Services.

———. 1981a. Habitat evaluation procedures (HEP). ESM 102. Washington, D.C.: U.S. Department of the Interior, U.S. Fish and Wildlife Service, Division of Ecological Services.

———. 1981b. Standards for the development of habitat suitability index models. ESM 103. Washington, D.C.: U.S. Department of the Interior, U.S. Fish and Wildlife Service, Division of Ecological Services.

———. 1989. Birds: Moosehorn National Wildlife Refuge, Maine. Washington, D.C.: U.S. Department of the Interior, U.S. Fish and Wildlife Service.

———. 1990. Endangered and threatened wildlife and plants: Determination of threatened status for the northern spotted owl; final rule. *Federal Register* 55: 26114–26194.

———. 1991. Black-capped vireo (*Vireo atricapillus*) recovery plan. Austin, Tex.: U.S. Department of the Interior, U.S. Fish and Wildlife Service.

———. 1994a. Birds: Petit Manan National Wildlife Refuge, Milbridge, Maine. Washington, D.C.: U.S. Department of the Interior, U.S. Fish and Wildlife Service.

———. 1994b. Birds: Rachel Carson National Wildlife Refuge, Wells, Maine. Washington, D.C.: U.S. Department of the Interior, U.S. Fish and Wildlife Service.

———. 1995. Birds: Sunkhaze Meadows National Wildlife Refuge, Old Town, Maine. Washington, D.C.: U.S. Department of the Interior, U.S. Fish and Wildlife Service.

———. 1996. Reptiles and Amphibians: Rachel Carson National Wildlife Refuge, Wells, Maine. Washington, D.C.: U.S. Department of the Interior, U.S. Fish and Wildlife Service.

———. 1997. Recovery plan for the threatened marbled murrelet (*Brachyramphus marmoratus*) in Washington, Oregon, and California. Portland, Ore.: U.S. Department of the Interior, U.S. Fish and Wildlife Service.

U.S. Fish and Wildlife Service and Canadian Wildlife Service. 1987. Standard operation procedures [SOP] for aerial waterfowl breeding ground population and habitat surveys in North America; revised; unpublished report.

USGAO. United States General Accounting Office. 1988. Rangeland management: More emphasis needed on declining and overstocked grazing allotments. Report number GAO/RCED-88-80. Washington, D.C.: U.S. General Accounting Office.

USGS. United States Geological Survey. 1986. Land use and land cover digital data from 1:250,000- and 1:100,000-scale maps. Data Users Guide 4. Reston, Va.: U.S. Geological Survey.

U.S. Laws, Statutes, etc. Public Law 91–190. [S. 1075], Jan. 1, 1970. National Environmental Policy Act of 1069. An act to establish a national policy for the environment, to provide for the establishment of a Council on Environmental Quality, and for other purposes. In its United States statutes at large. 1969. Vol. 83, 852–856. U.S. Gov. Print. Off., Washington, D. C. 1970. [U.S. C. sec. 4321, et seq.(1970).]

Ustin, S. L., M. O. Smith, and J. B. Adams. 1993. Remote sensing of ecological processes: A strategy for developing and testing ecological models using spectral mixture analysis. In *Scaling physiological processes: Leaf to globe*, ed. J. R. Ehlringer and C. B. Field, 339–357. San Diego, Calif.: Academic Press.

Utter, F. M., J. E. Seeb, and L. W. Seeb. 1993. Complementary uses of ecological and biochemical genetic data in identifying and conserving salmon populations. *Fisheries Research* 18:59–76.

Utter, F. M., R. S. Waples, and D. J. Teel. 1992. Genetic isolation of previously indistinguishable chinook salmon populations of the Snake and Klamath Rivers: Limitations of negative data. *Fishery Bulletin* 90:770–777.

Valentine, K. A. 1947. Distance from water as a factor in grazing capacity of rangeland. *Journal of Forestry* 45:749–754.

Van der Heijden, M. G. A., J. N. Klironomos, M. Ursic, P. Moutoglis, R. Streitwold-Engel, T. Boller, A. Wiemken, and I. R. Sanders. 1998. Mycorrhizal fungal diversity determines plant biodiversity, ecosystem variability and productivity. *Nature* 396:69–72.

Van der Maarel, E. 1979. Transformation of cover-abundance values in phytosociology and its effects on community similarity. *Vegetatio* 39:97–114.

Van der Maarel, E., V. Noest, and M. W. Palmer. 1995. Variation in species richness on small grassland quadrants: Niche structure or small-scale plant mobility? *Journal of Vegetation Science* 6:741–752.

Van der Molen, D. T., and J. Pintér. 1993. Environmental model calibration under different specifications: An application to the model SED. *Ecological Modelling* 68:1–19.

Vanderwerf, E. A. 1993. Scales of habitat selection by foraging 'Elepaio in undisturbed and human-altered forests of Hawaii. *Condor* 95:980–989.

Van Deursen, W. P. A. 1995. *Geographical information systems and dynamic models: Development and application of a prototype spatial modelling language.* Nederlandse Geografische Studies 190. Utrecht, The Netherlands: Koninklijk Nederlands Aardrijkskundig Genntschap/Faculteit Ruimtelijke Wetenschappen Universiteit Utrecht.

Van Dorp, D., and P. F. M. Opdam. 1987. Effects of patch size, isolation, and regional abundance on forest bird communities. *Landscape Ecology* 1:59–73.

Van Horne, B. 1983. Density as a misleading indicator of habitat quality. *Journal of Wildlife Management* 47:893–901.

———. 1991. Spatial configuration of avian habitats. *Acta XX Congressus Internationalis Ornithologici* 4:2313–2319.

Van Horne, B., G. S. Olson, R. L. Schooley, J. G. Corn, and K. P. Burnham. 1997. Effects of drought and prolonged winter on Townsend's ground squirrel demography in shrubsteppe habitats. *Ecological Monographs* 67:295–315.

Van Horne, B., and J. A. Wiens. 1991. Forest bird habitat suitability models and the development of general habitat models. Research 8. Washington, D.C.: U.S. Department of the Interior, U.S. Fish and Wildlife Service.

Vaughan, V. 2000. Reinventing previous work. *The Scientist* 14 (11): 6.

Venables, W. N., and B. D. Ripley. 1994. *Modern applied statistics with S-PLUS.* New York: Springer-Verlag.

———. 1997. *Modern applied statistics with S-PLUS.* 2nd ed. New York: Springer-Verlag.

Venier, L. A., A. A. Hopkin, D. W. McKenney, and Y. Wang. 1998. Spatial, climate-determined risk rating for scleroderris disease of pines in Ontario. *Canadian Journal of Forest Research* 28:1398–1404.

Venier, L. A., and B. G. Mackey. 1997. A method for rapid, spatially explicit habitat assessment for forest songbirds. *Journal of Sustainable Forestry* 4:99–118.

Venier, L. A., D. W. McKenney, Y. Wang, and J. McKee.

1999. Models of large-scale breeding-bird distribution as a function of macro-climate in Ontario, Canada. *Journal of Biogeography* 26:315–328.

Verbyla, D. L., and J. A. Litvaitis. 1989. Resampling methods for evaluating classification accuracy of wildlife habitat models. *Environmental Management* 13:783–787.

Veregin, H. 1989. Taxonomy of error in spatial databases. Technical Report 89-12. Santa Barbara, Calif.: National Center for Geographic Information and Analysis.

Verner, J. 1981. Measuring responses of avian communities to habitat manipulation. In *Estimating numbers of terrestrial birds*, ed. C. J. Ralph and J. M. Scott, 543–547. Studies in Avian Biology 6. Los Angeles: Cooper Ornithological Society.

Verner, J., and A. S. Boss. 1980. California wildlife and their habitats: Western Sierra Nevada. General Technical Report PSW-37. Berkeley, Calif.: USDA Forest Service, Pacific Southwest Forest and Range Experiment Station.

Verner, J., K. S. McKelvey, B. R. Noon, R. J. Gutiérrez, G. I. Gould Jr., and T. W. Beck. technical coordinators. 1992. The California spotted owl: A technical assessment of its current status. General Technical Report PSW-GTR-133. Albany, Calif.: USDA Forest Service, Pacific Southwest Research Station.

Verner, J., M. L. Morrison, and C. J. Ralph. 1986a. Introduction. In *Wildlife 2000: Modeling habitat relationships of terrestrial vertebrates*, ed. J. Verner, M. L. Morrison, and C. J. Ralph, xi–xv. Madison: University of Wisconsin Press.

———, eds. 1986b. *Wildlife 2000: Modeling habitat relationships of terrestrial vertebrates.* Madison: University of Wisconsin Press.

Vickery, P. D., M. L. Hunter Jr., and J. V. Wells. 1992. Is density an indicator of breeding success? *Auk* 109:706–710.

Viejo, J. L., M. G. de Videma, and E. M. Falero. 1989. The importance of woodlands in the conservation of butterflies (Lep.: Papilionoidea and Hesperioidea) in the centre of the Iberian Peninsula. *Biological Conservation* 48:101–114.

Villard, M., and B. A. Maurer. 1996. Geostatistics as a tool for examining hypothesized declines in migratory songbirds. *Ecology* 77:59–68.

Villard, M. A., E. V. Schmidt, and B. A. Maurer. 1998. Contribution of spatial modeling to avian conservation. In *Avian conservation: Research and management*, ed. J. M. Marzluff and R. Sallabanks, 49–64. Washington, D.C.: Island Press.

Villeneuve, N., M. M. Grandtner, and J. A. Fortin. 1989. Frequency and diversity of ectomycorrhizal and saprophytic macrofungi in the Laurentide Mountains of Quebec. *Canadian Journal of Botany* 67:2616–2629.

Virkkala, R. 1991. Spatial and temporal variation in bird communities and populations in north-boreal coniferous forests: A multiscale approach. *Oikos* 62:59–66.

Virtanen, A., V. Kairisto, and E. Uusipaikka. 1998. Regression-based reference limits: Determination of sufficient sample size. *Clinical Chemistry* 44:2353–2358.

Visser, S. 1995. Ectomycorrhizal fungal succession in jack pine stands following wildfire. *New Phytologist* 129: 389–401.

Vitousek, P. M., H. A. Mooney, J. Lubchenco, and J. M. Melillo. 1997. Human domination of earth's ecosystems. *Science* 277:494–499.

Vogelmann, J. E., T. Sohl, and S. M. Howard. 1998. Regional characterization of land cover using multiple sources of data. *Photogrammetric Engineering and Remote Sensing* 64:45–57.

Vonesh, E. F., and V. M. Chinchilli. 1997. *Linear and nonlinear models for the analysis of repeated measurements.* New York: Marcel Dekker.

Vucetich, J. A., R. O. Peterson, and T. A. Waite. 1997. Effects of social structure and prey dynamics on extinction risk in gray wolves. *Conservation Biology* 11:957–965.

Vuillod, P. 1994. Paysage visible et aménagement: Modélisations cartographiques et test sur le Haut-Jura. Ph.D. dissertation, Université de Franche-Comté, BesanVon France.

Waide, R. B., M. R. Willig, C. F. Steiner, G. Mittelbach, L. Gough, S. I. Dodson, G. P. Juday, and R. Parmenter. 1999. The relationship between productivity and species richness. *Annual Review of Ecology and Systematics* 30:257–300.

Wake, D. B. 1991. Declining amphibian populations. *Science* 253:860.

Walker, M. A., and S. E. Smith. 1985. The usefulness of current models of VA mycorrhizal infection. In *Proceedings of the Sixth North American Conference of Mycorrhizae,* ed. R. Molina, 143–144. Corvallis, Ore.: Forestry Research Laboratory, Oregon State University.

Walker, P. A. 1990. Modelling wildlife distributions using a geographic information system: Kangaroos in relation to climate. *Journal of Biogeography* 17:279–289.

Walker, P. A., and R. J. Aspinall. 1997. *Evaluating spatial modelling techniques.* Canberra, Australia: CSIRO Wildlife and Ecology.

Walker, P. A., and K. D. Cocks. 1991. HABITAT: A procedure for modelling a disjoint environmental envelope for a plant or animal species. *Global Ecology and Biogeography Letters* 1:108–118.

Walker, R. E. 1992. Community models of species richness: Regional variation of plant community species composition on the west slope of the Sierra Nevada, California. M.A. thesis, University of California, Santa Barbara.

Walker, S. H., and D. B. Duncan. 1967. Estimation of the probability of an event as a function of several independent variables. *Biometrika* 54:167–178.

Wallace, A., E. M. Romney, and M. T. Mueller. 1982. Sodium relations in desert plants: Effects of sodium chloride on *Atriplex polycarpa* and *Atriplex canescens. Soil Science* 134:65–68.

Walsh, S. J., D. R. Lightfoot, and D. R. Butler. 1987. Recognition and assessment of error in geographic information systems. *American Society for Photogrammetry and Remote Sensing* 53:1423–1430.

Walter, H. 1964/1968. *Die Vegetation der Erde in ökophysiologischer Betrachtung.* Vol. 1 (1964) and Vol. 2 (1968). Stuttgart, Germany:Fischer.

———. 1970. *Vegetationszonen und Klima.* Stuttgart, Germany: Verlag Eulen Ulmer.

Walters, C. J. 1986. *Adaptive management of renewable resources.* New York: MacMillan.

———. 1997. Challenges in adaptive management of riparian and coastal ecosystems. *Conservation Ecology* [online]1:1. http://www.consecol.org/Journal/vol1/iss2/art1.

Walters, C. J., and C. S. Holling. 1990. Large-scale management experiments and learning by doing. *Ecology* 71:2060–2068.

Ward, J. P. Jr., R. J. Gutiérrez, and B. R. Noon. 1998. Habitat selection by northern spotted owls: The consequences of prey selection and distribution. *Condor* 100:79–92.

Waring, R. H., J. D. Aber, J. M. Melillo, and B. Moore, III. 1986. Precursors of change in terrestrial ecosystems. *BioScience* 36:433–438.

Warwick, J. J., and W. G. Cale. 1987. Determining the likelihood of obtaining a reliable model. *Journal of Environmenal Engineering* 113:1102–1119.

———. 1988. Estimating model reliability using data with uncertainty. *Ecological Modelling* 41:169–181.

Watson, G., and T. W. Hillman. 1997. Factors affecting the distribution and abundance of bull trout: An investigion at hierarchical scales. *North American Journal of Fisheries Management* 17:237–252.

Watts, B. D., and S. E. Mabey. 1993. Spatio-temporal patterns of landbird migration on the lower Delmarva Peninsula. Annual report to the Virginia Department of Environmental Quality. Williamsburg, Va.: Center for Conservation Biology, College of William and Mary.

WDNR. Washington Department of Natural Resources. 1997. Final habitat conservation plan. Olympia, Wash.: Washington State Department of Natural Resources.

Wehde, M. 1982. Grid cell size in relation to errors in maps and inventories produced by computerized map processing. *Photogrammetric Engineering and Remote Sensing* 48:1289–1298.

Weinberg, H. J., J. S. Bolsinger, and T. J. Hayden. 1995. Project status report: 1994 field studies of two endangered species (the black-capped vireo and the golden-cheeked warbler) and the cowbird control program on Fort Hood, Texas. Report submitted to HQIII Corps and Fort Hood, Texas. Champaign, Ill.: U.S. Army Construction Engineering Research Laboratory.

Weinberg, H. J., J. A. Jette, and J. D. Cornelius. 1996. Project status report: 1995 field studies of two endangered species (the black-capped vireo and the golden-cheeked warbler) and the cowbird control program on Fort Hood, Texas. Report submitted to HQIII Corps and Fort Hood, Texas. Champaign, Ill.: U.S. Army Construction Engineering Research Laboratory.

Weinberg, H. J., T. J. Hayden, and J. D. Cornelius. 1998. Local and installation-wide black capped vireo dynamics on the Fort Hood, Texas, military reservation. USACERL Technical Report 98/54. Champaign, Ill.: U.S. Army Construction Engineering Research Laboratory.

Weisbrod, A. R., C. J. Burnett, and J. G. Turner. 1993. Migrating birds at a stopover site in the Saint Croix River Valley. *Wilson Bulletin* 105:265–284.

Weishampel, J. F., D. L. Urban, H. H. Shugart, and J. B. Smith Jr. 1992. Semivariograms from a forest transect gap model compared with remotely sensed data. *Journal of Vegetation Science* 3:521–526.

Weiss, S. M. and N. Indurkhya. 1997. Predictive data mining: A practical guide. San Francisco, Calif.: Morgan Kaufmann Pub.

Weiss, S. B., and A. D. Weiss. 1998. Landscape-level phenology of a threatened butterfly: A GIS-based modeling approach. *Ecosystems* 1:299–309.

Wells, J. V., and M. E. Richmond. 1995. Populations, metapopulations, and species populations: What are they and who should care? *Wildlife Society Bulletin* 23: 458–462.

Wells, K. B. 1977. The social behaviour of anuran amphibians. *Animal Behavior* 25:666–693.

Welsh, A. H., R. B. Cunningham, C. F. Donnelly, and D. B. Lindenmayer. 1996. Modelling the abundance of rare species: Statistical models for counts with extra zeros. *Ecological Modelling* 88:297–308.

West, G. B., J. H. Brown, and B. J. Enquist. 1997. A general model for the origin of allometric scaling laws in biology. *Science* 276:122–126.

West, N. E. 1983. Intermountain salt-desert shrubland. In *Temperate deserts and semi-deserts*, ed. N. E. West. Ecosystems of the World series. New York: Elsevier Scientific. 5: 375–397.

Westervelt, J. D. 2001. Computational approach to integrating GIS and agent-based modeling. In *Integrating geographic information system and agent-based modeling techniques for simulating social and ecological processes*, ed. H. R. Gimblett. Santa Fe, N. Mex.: Santa Fe Institute.

Whitcomb, R. F., C. S. Robbins, J. F. Lynch, B. L. Whitcomb, M. K. Klimkiewicz, and D. Bystrak. 1981. Effects of forest fragmentation on avifauna of eastern deciduous forest. In *Forest island dynamics in man-dominated landscapes*, ed. R. L. Burgess and D. M. Sharpe, 125–205. New York: Springer-Verlag.

White, D., J. Kimmerling, and W. S. Overton. 1992. Cartographic and geometric components of a global design for environmental monitoring. *Cartography and geographic information systems* 19:5–22.

White, D., P. G. Minotti, M. J. Barczak, J. C. Sifneos, K. E. Freemark, M. V. Santelmann, C. F. Steinitz, A. R. Kiester, and E. M. Preston. 1997a. Assessing risks to biodiversity from future landscape change. *Conservation Biology* 11:349–360.

White, D, E. M. Preston, K. E. Freemark, and A. R. Kiester. 1999. A hierarchical framework for conserving biodiversity. In *Landscape ecological analysis: Issues and applications*, ed. J. M. Klopatek and R. H. Gardner, 127–153. New York: Springer-Verlag.

White, D. L., T. A. Waldrop, and S. M. Jones. 1991. Forty years of prescribed burning on the Santee fire plots: Effects on understory vegetation. In *Fire and the environment: Ecological and cultural perspectives*, ed. S. C. Nodvin and D. A. Waldrop, 45–59. Asheville, N.C.: Southeastern Forest Experiment Station.

White, G. C. 1983. Numerical estimation of survival rates from band-recovery and biotelemetry data. *Journal of Wildlife Management* 47:716–728.

———. 1986. *Program SURVIV User's Manual Version 1.2.* Los Alamos, N. Mex.: Environmental Science Group.

White, G. C., D. R. Anderson, K. P. Burnham, and D. L. Otis. 1982. Capture-recapture and removal methods for sampling closed populations. Technical Report LA 8787-NERP. Los Alamos, N. Mex.: Los Alamos National Laboratory.

White, G. C., and R. E. Bennetts. 1996. Analysis of frequency count data using the negative binomial distribution. *Ecology* 77:2549–2557.

White, G. C., and K. P. Burnham. 1999b. Program MARK-survival estimation from populations of marked animals. *Bird Study* (suppl.) 46: S14-S21.

White, P. S. 1979. Pattern, process, and natural disturbance in vegetation. *Botanical Review* 45:229–299.

White, R., G. Engelen, and I. Uljee. 1997b. The use of constrained cellular automata for high-resolution modelling of urban land-use dynamics. *Environment and Planning B-Planning and Design* 24:323–343.

Whittaker, R. H. 1956. Vegetation of the Great Smoky Mountains. *Ecological Monographs* 26:1–80.

———. 1960. Vegetation of the Siskiyou Mountains, Oregon and California. *Ecological Monographs* 30:279–338.

———. 1967. Gradient analysis of vegetation. *Biological Review* 42:207–264.

———. 1975. *Communities and ecosystems*. 2nd ed. New York: Macmillan.

———. 1977. Evolution of species diversity in land communities. *Evolutionary Biology* 10:1–67.

———. 1978. Direct gradient analysis. In *Ordination of plant communities*, ed. R. H. Whittaker, 7–51. The Hague: W. Junk.

Wickham, J. D., K. B. Jones, K. H. Riitters, T. G. Wade, and R. V. O'Neill. 1999. Transitions in forest fragmentation: Implications for restoration opportunities at regional scales. *Landscape Ecology* 14:137–145.

Wickham, J. D., J. Wu, and D. F. Bradford. 1997. A conceptual framework for selecting and analyzing stressor data to study species richness at large spatial scales. *Environmental Management* 21:247–257.

Wiedenfeld, D. A. 2000. Cowbird population changes and their relationship to changes in some host species. In *The biology and management of cowbirds and their hosts*, ed. J. N. M. Smith, T. L. Cook, S. I. Rothstein, S. K. Robinson, and S. G. Sealy, 35–46. Austin: University of Texas Press.

Wiens, J. A. 1969. An approach to the study of ecological relationships among grassland birds. *Ornithological Monographs* 8:1–93.

———. 1977. On competition and variable environments. *American Scientist* 65:590–597.

———. 1981a. Scale problems in avian censusing. In *Estimating numbers of terrestrial birds*, ed. C. J. Ralph and J. M. Scott, 513–521. Studies in Avian Biology 6. Los Angeles: Cooper Ornithological Society.

———. 1981b. Single-sample surveys of communities: Are the revealed patterns real? *American Naturalist* 117:90–98.

———. 1985. Habitat selection in variable environments: Shrub-steppe birds. In *Habitat selection in birds*, ed. M. L. Cody, 227–251. London: Academic Press.

———. 1989a. Spatial scaling in ecology. *Functional Ecology* 3:385–397.

———. 1989b. *The ecology of bird communities*. Vol. 1, Foundations and Patterns. Cambridge: Cambridge University Press.

———. 1989c. *The ecology of bird communities*. Vol. 2, Processes and Variations. Cambridge: Cambridge University Press.

———. 1995. Habitat fragmentation: Island vs. landscape perspectives on bird conservation. *Ibis* 137:97–104.

———. 1996a. Metapopulation dynamics and landscape ecology. In *Metapopulation biology: Ecology, genetics, and evolution*, ed. I. A. Hanski and M. E. Gilpin, 43–62. New York: Academic Press.

———. 1996b. Wildlife in patchy environments: Metapopulations, mosaics, and management. In *Metapopulations and wildlife conservation*, ed. D. R. McCullough, 53–84. Washington, D.C.: Island Press.

———. 1999. The science and practice of landscape ecology. In *Landscape ecological analysis: Issues and applications*, ed. J. M. Klopatek and R. H. Gardner, 371–383. New York: Springer-Verlag.

Wiens, J. A., J. F. Addicott, T. J. Case, and J. Diamond. 1986. Overview: The importance of spatial and temporal scale in ecological investigations. In *Community ecology*, ed. J. Diamond and T. J. Case, 145–153. New York: Harper & Row.

Wiens, J. A., C. S. Crawford, and J. R. Gosz. 1985. Boundary dynamics: A conceptual framework for studying landscape ecosystems. *Oikos* 45:421–427.

Wiens, J. A., and J. T. Rotenberry. 1981a. Censusing and the evaluation of avian habitat occupancy. *Studies in Avian Biology* 6:522–532.

———. 1981b. Habitat associations and community structure of birds in shrubsteppe environments. *Ecological Monographs* 51:21–41.

———. 1985. Response of breeding passerine birds to rangeland alteration in a North American shrub-steppe locality. *Journal of Applied Ecology* 22:655–668.

Wiens, J. A., J. T. Rotenberry, and B. Van Horne. 1987. Habitat occupancy patterns of North American shrub-steppe birds: The effect of spatial scale. *Oikos* 48: 132–147.

Wiens, J. A., N. C. Stenseth, B. Van Horne, and R. A. Ims. 1993. Ecological mechanisms and landscape ecology. *Oikos* 66:369–380.

Wiertz, J., and J. Vink. 1986. The present status of the badger *Meles meles* (L. 1758) in the Netherlands. *Lutra* 29:21–53.

Wilbur, H. M. 1980. Complex life cycles. *Annual Review of Ecology and Systematics* 11:67–93.

Wilcove, D. S., and T. Eisner. 2000. The impending extinction of natural history. Chronicle of Higher Education. September 15, 2000. B24.

Wilcove, D. S., C. J. McLellan, and A. P. Dobson. 1986. Habitat fragmentation in the temperate zone. In *Conservation biology: The science of scarcity and diversity*, ed. M. E. Soulé, 237–256. Sunderland, Mass.: Sinauer Associates.

Wilcove, D. S., D. Rothstein, J. Dubow, A. Phillips, and E. Losos. 1998. Quantifying threats to imperiled species in the United States. *BioScience* 48:607–615.

Wilcox, B. A., and D. D. Murphy. 1985. Conservation strategy: The effect of fragmentation on extinction. *American Naturalist* 125:879–887.

Wilds, S. P. 1996. Gradient analysis of the distribution of flowering dogwood (*Cornus florida* L.) and dogwood anthracnose (*Discula destructiva* Redlin.) in western Great Smoky Mountains National Park. M.A. thesis, University of North Carolina, Chapel Hill.

Wiley, M. J., S. L. Kohler, and P. W. Seelbach. 1997. Reconciling landscape and local views of aquatic communities: Lessons from Michigan trout streams. *Freshwater Biology* 37:133–148.

Wilhelm, R. B., J. R. Choate, and J. K. Jones Jr. 1981. Mammals of LaCreek National Wildlife Refuge, South Dakota. Special Publications of the Museum, Number 17. Lubbock: Texas Tech University.

Wilkie, D. S., and J. T. Finn. 1996. *Remote sensing imagery for natural resources monitoring*. New York: Columbia University Press.

Wilkins, W. H., and G. C. M. Harris. 1946. The ecology of the larger fungi. 5. An investigation into the influence of rainfall and temperature on the seasonal production of fungi in a beechwood and a pinewood. *Annals of Applied Biology* 33:179–188.

Wilkinson, L. 1998. *SYSTAT: Statistics*. Evanston, Ill.: SYSTAT.

Williams, G. L., K. R. Russell, and W. K. Seitz. 1978. Pattern recognition as a tool in the ecological analysis of habitat. In *Classification, inventory, and analysis of fish and wildlife habitat: The proceedings of a national symposium*, ed. A Marmelstein, 521–531. FWS/OBS-78/76. Washington, D.C.: U.S. Department of the Interior, U.S. Fish and Wildlife Service, Office of Biological Services.

Williams-Linera, G., V. DomRiguez-Gasted, and M. E. Garca-Zurita. 1998. Microenvironment and floristics of different edges in fragmented tropical rainforest. *Conservation Biology* 12:1091–1102.

Wilson, D. E., F. R. Cole, J. D. Nicholas, and R. Rudran, eds. 1996. *Measuring and monitoring biological diversity: Standard methods for mammals*. Washington, D.C.: Smithsonian Institution Press.

Wilson, E. O. 1988. *Biodiversity*. Washington, D.C.: National Academy Press.

———. 2000. On the future of conservation biology. *Conservation Biology* 14:1–3.

Wilson, M. F. 1974. Avian community organization and habitat structure. *Ecology* 55:1017–1029.

Wilson, R. F., and W. J. Mitsch. 1996. Functional assessment of five wetlands constructed to mitigate wetland loss in Ohio, USA. *Wetlands* 16:436–451.

Winker, K., D. W. Warner, and A. R. Weisbrod. 1992. Migration of woodland birds at a fragmented inland stopover site. *Wilson Bulletin* 104:580–598.

Winston, M. R. 1995. Co-occurrence of morphologically similar species of stream fishes. *American Naturalist* 145:527–545.

Winston, M. R., C. M. Taylor, and J. Pigg. 1991. Upstream extirpation of four minnow species due to damming of a prairie stream. *Transactions of the American Fisheries Society* 120:98–105.

Wise, S. M. 1998. The effect of GIS interpolation errors on the use of digital elevation models in geomorphology. In *Landform monitoring, modelling and analysis*, ed. S. N. Lane, K. S. Richards, and J. H. Chandler, 139–164. Chichester: John Wiley & Sons.

Wiser, S. K., R. K. Peet, and P. S. White. 1998. Prediction of rare-plant occurrence: A southern Appalachian example. *Ecological Applications* 8:909–920.

Wissel, C., and S.-H. Zaschke. 1994. Stochastic birth and

death processes describing minimum viable populations. *Ecological Modelling* 75/76:193–201.

With, K. A., and T. O. Crist. 1995. Critical thresholds in species' responses to landscape structure. *Ecology* 76:2446–2459.

With, K. A., and A. W. King. 1997. The use and misuse of neutral landscape models in ecology. *Oikos* 79:219–229.

Withers, M. A., and V. Meentemeyer. 1999. Concepts of scale in landscape ecology. In *Landscape ecological analysis: Issues and applications*, ed. J. M. Klopatek and R. H. Gardner, 205–252. New York: Springer-Verlag.

Woinarski, J. C. Z., H. F. Recher, and J. D. Majer. 1997. Vertebrates of eucalypt formations. In *Eucalypt ecology: Individuals to ecosystems*, ed. J. E. Williams and J. C. Z. Woinarski, 303–341. Cambridge: Cambridge University Press.

Wolf, A. T., R. W. Howe, and G. J. Davis. 1995. Detectability of forest birds from stationary points in northern Wisconsin. In *Monitoring bird populations by point counts*, ed. C. J. Ralph, J. R. Sauer, and S. Droege, 19–23. General Technical Report PSW-149. Albany, Calif.: USDA Forest Service, Pacific Southwest Research Station.

Wolff, J. O. 1995. On the limitations of species-habitat association studies. *Northwest Science* 69:72–76.

Wolock, D. M., and G. J. McCabe Jr. 1995. Comparison of single and multiple flow direction algorithms for computing topographic parameters in TOPMODEL. *Water Resources Research* 31:1315–1324.

Wolter, P. T., D. J. Mladenoff, G. E. Host, and T. R. Crow. 1995. Improved forest classification in the northern lake states using multi-temporal Landsat imagery. *Photogrammetric Engineering and Remote Sensing* 61:1129–1143.

Woodgate, P. W., W. D. Peel, K. T. Ritman, J. E. Coram, A. Brady, A. J. Rule, and J. C. G. Banks. 1994. *A study of the old-growth forests of East Gippsland*. East Melbourne, Australia: Department of Conservation and Natural Resources.

Wootton, J. T., and D. A. Bell. 1992. A metapopulation model of the peregrine falcon in California: Viability and management strategies. *Ecological Applications* 2:307–321.

Worboys, M. F. 1994. A unified model for spatial and temporal information. *Computer Journal* 37:26–34.

Worton, B. J. 1989. Kernel methods for estimating the utilization distribution in home-range studies. *Ecology* 70:164–168.

Wright, A. 1997. Predicting the distribution of Eurasian badger (*Meles meles*) setts. Ph.D. thesis, Manchester Metropolitan University, Manchester, UK.

Wright, A., A. H. Fielding, and C. P. Wheater. 2000. Predicting the distribution of Eurasian badger (*Meles meles*) setts over an urbanized landscape: A GIS approach. *Photogrammetric Engineering and Remote Sensing* 66:423–428.

Wright, D. H. 1983. Species-energy theory: An extension of species-area theory. *Oikos* 41:496–506.

———. 1991. Correlations between incidence and abundance are expected by chance. *Journal of Biogeography* 18:463–466.

Wright, H. E., and A. W. Bailey. 1982. *Fire ecology: United States and southern Canada*. New York: Wiley.

Yahner, R. H. 1988. Changes in wildlife communities near edges. *Conservation Biology* 2:333–339.

Yalden, D. W. 1993. The problems of reintroducing predators. *Symposia of the Zoological Society of London* 65:289–306.

Yang, Dongsheng, B. C. Pijanowski, and S. H. Gage. 1998a. Analysis of gypsy moth (Lepidoptera: Lymantriidae) population dynamics in Michigan using geographic information systems. *Environmental Entomology* 27:842–852.

Yang, L., B. K. Wylie, L. L. Tieszen, and B. C. Reed. 1998b. An analysis of relationships among climate forcing and time-integrated NDVI of grasslands over the U.S. northern and central Great Plains. *Remote Sensing of Environment* 65:25–37.

Yee, T. W., and N. D. Mitchell. 1991. Generalized additive models in plant ecology. *Journal of Vegetation Science* 2:587–602.

Yong, W., D. M. Finch, F. R. Moore, and J. F. Kelly. 1998. Stopover ecology and habitat use of migratory Wilson's warblers. *Auk* 115:829–842.

Yost, R. S., G. Uehara, and R. L. Fox. 1982. Geostatistical analysis of soil chemical properties of large land areas. 1. Semivariograms. *Soil Science Society American Journal* 46:1028–1032.

Young, J. S. 1996. Nonlinear bird-habitat relationships in managed forest of the Swan Valley, Montana. M.S. thesis, University of Montana, Missoula.

Young, L. S., and D. E. Varland. 1998. Making research meaningful to the manager. In *Avian conservation: Research and management*, ed. J. M. Marzluff and R. Sallabanks, 415–422. Washington, D.C.: Island Press.

Zabel, C. J., K. S. McKelvey, and J. P. Ward Jr. 1995. Influence of primary prey on home range size and habitat-use patterns of spotted owls (*Strix occidentalis caurina*). *Canadian Journal of Zoology* 73:433–439.

Zabel, C. J., G. N. Steger, K. S. McKelvey, G. P. Eberlein, B. R. Noon, and J. Verner. 1992. Home-range size and habitat-use patterns of California spotted owls in the Sierra Nevada. In *The California spotted owl: A technical assessment of its current status*, technical coordinators. J. Verner, K. S. McKelvey, B. R. Noon, R. J. Gutiérrez, G. I. Gould Jr., and T. W. Beck, 149–163. General Technical Report PSW-GTR-133. Albany, Calif.: USDA Forest Service, Pacific Southwest Research Station.

Zabel, C. J., J. R. Dunk, H. B. Stauffer, L. M. Roberts, B. S. Mulder, and A. Wright. In review. Northern spotted owl habitat selection models for research and management application in California. *Ecological Applications*.

Zadeh, L. 1965. Fuzzy sets. *Information and Control* 8:338–353.

Zar, J. H. 1984. *Biostatistical analysis*. 2nd ed. Englewood Cliffs, N.J.: Prentice Hall.

———. 1996. *Biostatistical analysis*. 3rd ed. Upper Saddle River, N.J.: Prentice Hall.

Zeger, S. L., and M. R. Karim. 1991. Generalized linear models with random effects: A Gibbs sampling approach. *Journal of the American Statistical Association* 86:79–86.

Zeiner, D. C., W. F. Laudenslayer Jr., K. E. Mayer, and M. White, eds. 1990. *California's wildlife: Birds*. Vol. 2. Sacramento: California Department of Fish and Game.

Zerger, A. 1995. The role of ground sample resolution and spatial dependence in the prediction of vegetation communities using digital elevation models. Master's thesis, Department of Geomatics, University of Melbourne.

Zhou, Z., and W. Pan. 1997. Analysis of the viability of a giant panda population. *Journal of Applied Ecology* 34:363–374.

Zhou, M., and T. L. Sharik. 1997. Ectomycorrhizal associations of northern red oak (*Quercus rubra*) seedlings along an environmental gradient. *Canadian Journal of Forest Research* 27:1705–1713.

Zielinski, W. J., and H. B. Stauffer. 1996. Monitoring martes populations in California: Survey design and power analysis. *Ecological Applications* 6:1254–1267.

Zimmerman, G. S., and W. E. Glanz. 2000. Habitat use of bats in eastern Maine. *Journal of Wildlife Management* 64:1032–1040.

Zimmermann, N. E., and F. Kienast. 1999. Predictive mapping of alpine grasslands in Switzerland: Comparing the species and community approach. *Journal of Vegetation Science* 10:469–482.

Zultowsky, J. M. 1992. Behavioral and spatial ecology of female white-tailed deer in the Everglades ecosystem. M.S. thesis, University of Florida, Gainesville.

Zweig, M. H., and G. Campbell. 1993. Receiver-Operating Characteristic (ROC) Plots: A fundamental evaluation tool in clinical medicine. *Clinical Chemistry* 39:561–577.

Contributing Authors

PAUL L. ANGERMEIER IS a research scientist and assistant unit leader with the U.S. Geological Survey and the Virginia Cooperative Fish and Wildlife Research Unit at Virginia Polytechnic Institute and State University. His current research interests include aquatic community ecology, conservation of aquatic biodiversity, and the use of biotic communities to assess environmental quality.

MICHAEL P. AUSTIN is a plant community ecologist at the CSIRO (Commonwealth Scientific and Industrial Research Organisation) Sustainable Ecosystems, Canberra, Australia. His research interests include statistical modeling of species distributions, theoretical community ecology, and analysis of plant competition experiments.

REGINALD H. BARRETT teaches courses in wildlife management, case histories in wildlife management, and wildlife management planning at the University of California, Berkeley. His research has included studies of alien species, responses of terrestrial vertebrates to forest management practices, and development and testing of wildlife habitat relationships models for California wildlife.

JOHN B. BARTLETT is a research associate at the U.S. Forest Service's Southern Global Change Program, where he works on the development of modeling strategies for interfacing environmental, demographic, and socioeconomic data.

VAL BEASLEY is a veterinary, wildlife, and ecological toxicologist, and director of the Envirovet Program in Wildlife and Ecosystem Health at the College of Veterinary Medicine, University of Illinois, Urbana. His current research interests include declines in amphibian populations, mass die-offs in flamingo populations, and the sources and effects (direct and indirect) of heavy metals, pesticides, and blue-green algal toxins in wildlife.

MICHAEL W. BINFORD is a professor of geography at the University of Florida in Gainesville. He specializes in the use of remotely sensed imagery in paleoecological studies of human interactions with aquatic ecosystems.

RANDALL B. BOONE is a wildlife scientist with the Natural Resource Ecology Laboratory at Colorado State University. He is currently working on African conservation issues and ecological modeling, and has written about the prediction of species occurrences, biogeography, and animal movement patterns.

MARIA G. BORGOGNONE currently holds a teaching and research position in the statistics department at Universidad Nacional de Rosario, Argentina.

BRIAN B. BOROSKI is a certified wildlife biologist with H. T. Harvey and Associates who specializes in quantifying, at various spatial and temporal scales, the functional relationships between species and their habitats and in using geographic information systems for vertebrate habitat assessments.

URS BREITENMOSER is the leader of KORA (a program coordinating research projects for the conservation and management of carnivores in Switzerland). He is now a senior researcher at the University of Bern, specializing in carnivore ecology and rabies epidemiology.

FRED C. BRYANT is director of the Caesar Kleberg Wildlife Research Institute at Texas A&M University in Kingsville. His research interests include livestock/wildlife interrelationships and habitat management. He has coauthored over fifty journal articles and a reference book entitled *Habitat Management of Forestlands, Rangelands, and Farmlands*.

DAVID A. BUEHLER is an associate professor of wildlife science in the Department of Forestry, Wildlife, and Fisheries at the University of Tennessee. His research interests include wildlife habitat relationships, endangered species management, and managing for biological diversity.

MARK BURGMAN is an associate professor and reader in the School of Botany at the University of Melbourne. He teaches environmental risk assessment and conservation biology. He directs a research program focusing on prediction of the spatial attributes and population dynamics of threatened plants and animals.

MARY CABLK is an assistant professor at the Desert Research Institute in Reno, Nevada, where she studies the direct and indirect effects of human impacts on biodiversity. Her research integrates field techniques with quantitative methods such as geostatistics, predictive and simulation modeling, image processing, and GIS.

JOSEPH D. CLARK is a wildlife biologist and laboratory director of the U.S. Geological Survey's Southern Appalachian Field Laboratory, located on the University of Tennessee campus. He specializes in carnivore population dynamics and habitat modeling.

CHRISTOPHER B. COGAN is a doctoral candidate in environmental studies at the University of California, Santa Cruz. Mr. Cogan has previously worked on the California and national gap analysis programs and spent many years as a field biologist working with California condors.

JOHN COLEMAN is currently a data analyst and natural resource modeler working for the Great Lakes Indian Fish and

Wildlife Commission in Wisconsin. He received his Ph.D. and M.S. degrees in wildlife ecology from the University of Wisconsin, Madison, in 1994 and from Virginia Polytechnic Institute and State University in 1983, respectively. His research has focused on analysis of the spatial distribution of species and the resources they use.

MICHAEL J. CONROY is a research scientist and assistant unit leader with the U.S. Geological Survey, Georgia Cooperative Fish and Wildlife Research Unit, although he began his career as a biometrician and biologist with the USGS Patuxent Wildlife Research Center in Maryland. He advises graduate students and teaches an advanced course in quantitative wildlife biology.

STEVE G. CUMMING is a forest ecologist and modeler with Boreal Ecosystems Research Ltd. in Edmonton, Alberta, Canada.

DONALD L. DEANGELIS is ATLSS project manager for the U.S. Geological Survey's Biological Resources Division and is a professor of biology at the University of Miami. His research has included development of dynamic models for nutrients and food webs in aquatic systems.

DIANE M. DEBINSKI is an associate professor in the Department of Animal Ecology at Iowa State University in Ames. Her areas of research include conservation biology, landscape ecology, and restoration ecology.

STEPHEN J. DEMASO is program coordinator for the Upland Wildlife Ecology Program, run by the Texas Parks and Wildlife Department's Wildlife Division.

DAVID F. DESANTE is executive director of The Institute for Bird Populations, Point Reyes Station, California. He specializes in avian demography and population dynamics, environmental correlates of productivity and survivorship in land birds, reproductive ecology of birds, avian monitoring and censusing techniques, and population and ecological implications of bird migration.

RANDY DETTMERS is a regional nongame bird biologist for the U.S. Fish and Wildlife Service, where he works on bird conservation plans and population monitoring. His research interests include avian ecology, wildlife habitat modeling, and animal behavior.

C. ANDREW DOLLOFF is Project Leader of the U.S. Forest Service Southern Research Station's Coldwater Fisheries Research Unit and associate professor of fisheries science in the Department of Fisheries and Wildlife, Virginia Tech, Blacksburg, Virginia. His research interests include the influence of natural and anthropogenic disturbance on stream communities, stream fish ecology, and ecology and management of coarse woody debris. He also is active in the development of strategies and methods for the protection and recovery of aquatic and riparian ecosystems.

TINA A. DREISBACH, a botanist with the U.S. Forest Service Pacific Northwest Research Station in Corvallis, Oregon, serves on the Forest Mycology Team, which researches fungal ecology and the biological and functional diversity of fungi and develops strategies for conservation and management of forest fungi in the Pacific Northwest.

JASON B. DUNHAM is a fisheries scientist with the U.S. Forest Service's Rocky Mountain Research Station in Boise, Idaho. He has worked with a variety of marine and freshwater fishes in tropical and temperate environments. His current research is focused on conservation and landscape ecology of threatened salmonid fishes.

JEFFREY R. DUNK is a wildlife biologist at the Redwood Sciences Laboratory in Arcata, California. He is also a lecturer at Humboldt State University in the departments of Wildlife, and Natural Resources Planning and Interpretation. His research has focused on avian ecology, primarily with raptors and ravens.

JAMES E. DUNN is chairman of the Statistics Division and head of the Statistical Laboratory at the University of Arkansas. He is a fellow of the American Statistical Association and is a long-term collaborator with the U.S. Environmental Protection Agency. His primary research interests concern development of statistical methodology applicable to problems of the environment.

JANE ELITH is a Ph.D. student at the School of Botany at the University of Melbourne. Her doctoral thesis topic concerns the prediction of plant distributions.

GEORGE L. FARNSWORTH is an assistant professor of ecology at the University of Houston-Downtown. His research interests focus on breeding ecology of forest songbirds.

ALAN H. FIELDING is a senior lecturer in ecology in the Department of Biological Sciences at Manchester Metropolitan University in England. His research interests are the conservation of birds of prey and the application of machine-learning methods to ecological problems. He is the recipient of the Leverhulme Research Fellowship, which has allowed him to pursue his research full time.

SIMON FERRIER is a principal GIS research officer in the New South Wales National Parks and Wildlife Service, Australia. His research interests include development, application, and evaluation of new approaches to biodiversity assessment, and employment of this information in regional conservation planning.

WALTER FERTIG is staff botanist with the Wyoming Natural Diversity Database, University of Wyoming, and is currently completing his doctoral dissertation on predictive modeling of vascular plant species in Wyoming.

ERICA FLEISHMAN is research associate at the Center for Conservation Biology, Stanford University. Her research focuses on integration of conservation science and land management, particularly in the arid western United States. Her current projects include analytic and predictive modeling of species distributions, conservation planning for a metapopulation of a rare butterfly, and developing and testing methods for the selection of umbrella species.

JANET FRANKLIN is a professor of geography at San Diego State University. Her research interests include biogeography, vegetation science, biophysical remote sensing, digital terrain analysis, and conservation biology.

BARRETT A. GARRISON is an environmental specialist with the California Department of Fish and Game in Rancho Cordova, California. His interests include developing and testing models of wildlife habitat relationships and distribution, and conserving California's woodlands and forests.

EDWARD O. GARTON is a professor in the Department of Fish and Wildlife Resources, College of Natural Resources, at the University of Idaho in Moscow. His work focuses on the study of population ecology.

MIKE GERTSCH has worked for the U.S. Forest Service for twenty-seven years. During this time, he spent seven years as the Region 5 representative to the California Bald Eagle Working Team, and he is currently the avian biota coordinator for the Sierra Nevada Framework Project, the Region 5 Threatened and Endangered Species (TES) Program Manager, the Forest Service Provincial Liaison to the U.S. Fish and Wildlife Service, and the Shasta-Trinity National Forest TES Program Manager.

GREG GOLDSMITH is a biologist with the U.S. Fish and Wildlife Service in Arcata, California. He is currently the geographic information systems coordinator for the Arcata Fish and Wildlife Office, which oversees federally listed species in northwest California.

CARLOS GONZALEZ-REBELES is an associate professor at the Facultad de Medicina Veterinaria y Zootecnia, Universidad Nacional Autonoma de Mexico, Mexico City, Mexico. His research interests include conservation planning and evaluation, and the wise use of wildlife; he teaches courses in wildlife management and sustainable development.

MICHAEL F. GOODCHILD is a professor of geography at the University of California, Santa Barbara. He is chair of the Executive Committee of the National Center for Geographic Information and Analysis and associate director of the Alexandria Digital Library. His research interests include GIS, digital libraries, and the accuracy of spatial databases.

STEVEN E. GRECO is an assistant professor in the Landscape Architecture Program in the Department of Environmental Design at the University of California, Davis.

LOUIS J. GROSS is a professor of ecology and evolutionary biology at the University of Tennessee and director of The Institute for Environmental Modeling. His research focuses on individual-based models and computational ecology, and on developing educational projects that enhance the quantitative training of life science students.

ANTOINE GUISAN is assistant Professor at the Institute of Ecology, University of Lausanne (Switzerland), where he teaches plant biogeography and quantitative geobotany. His main field of interest is currently the study of plant species distribution and the development of predictive species distribution models. He worked formerly at the Swiss Center for Faunal Cartography (CSCF), where he developed several predictive models of animal distribution.

FRED S. GUTHERY is professor and Bollenbach Chair in wildlife in the Department of Forestry at Oklahoma State University.

D. CALDWELL HAHN is a behavioral ecologist with the USGS Patuxent Wildlife Research Center in Laurel, Maryland. She has worked on cowbird parasitism patterns at the community scale using GIS, molecular genetics, and endocrine techniques to define fundamental aspects of brood parasitism. She is co-chair of the North American Cowbird Advisory Council.

LINNEA S. HALL was assistant professor of ornithology and wildlife ecology at California State University, Sacramento, from July 1996 to January 2000. She is currently residing and relaxing in Southern California.

HAYDEE M. HAMPTON is a spatial analyst in the Laboratory of Applied Ecology at Northern Arizona University. She holds master's degrees in geography from Northern Arizona University and in environmental engineering from Stanford University.

JoANN M. HANOWSKI is a senior research fellow at the University of Minnesota's Natural Resources Research Institute in Duluth. She is currently working on research projects involving the response of birds to various forest management practices in stream and pond riparian areas and the development of indicators of forest and water health and sustainability.

STEVEN J. HARPER, a postdoctoral associate at Miami University, conducts theoretical and empirical studies of ecological communities. He is currently investigating how omnivorous fish affect the stability of aquatic food webs in response to nutrient perturbations.

CHRISTINE S. HARTLESS is a wildlife biologist for the U.S. Environmental Protection Agency in the Office of Prevention, Pesticides, and Toxic Substances. Her interests are in ecological and environmental research and modeling.

JONATHAN B. HAUFLER is executive director of the Ecosystem Management Research Institute providing research, training, and implementation assistance to ecosystem management and biodiversity conservation initiatives. Prior to this, he was manager of wildlife and ecology at Boise Cascade Corporation and professor of wildlife ecology at Michigan State University.

AIDAN HEERDEGEN is a Ph.D. candidate at the Research School of Chemistry, Australian National University, in Canberra, Australia. His doctoral research was focused on diffuse X-ray scattering.

PATRICIA J. HEGLUND worked as a research scientist with the National Biological Survey (now USGS-BRD) in Alaska before taking a position as research faculty in the Department of Biological Sciences at the University of Idaho in Moscow. She recently had the opportunity to serve, albeit briefly, as the wildlife biologist for Potlatch Corporation before moving to her current position as regional refuge biologist for the U.S. Fish and Wildlife Service in Anchorage, Alaska. Her main interests are in wildlife habitat relationships modeling.

GEOFFREY M. HENEBRY is a research associate professor with the School of Natural Resource Sciences and an associate geoscientist with the Center for Advanced Land Management Information Technologies, Institute of Agriculture and Natural Resources, at the University of Nebraska in Lincoln. He is co-Principal Investigator of the Nebraska Gap Analysis Program. His research interests include landscape ecology, ecological remote sensing, and ecological modeling.

JEFFREY A. HEPINSTALL recently completed his doctorate in forest resources at the University of Maine. His research interests include the use of GIS and remote-sensing techniques in modeling species habitat relationships and the effects of land management on species occurrences.

KRISTINA E. HILL, associate professor at the University of Washington in Seattle, began writing about the use of fuzzy sets in GIS and habitat models while she was an assistant professor at the Massachusetts Institute of Technology. Her current research is on the impacts of urban patterns on ecological processes.

N. THOMPSON HOBBS is a research scientist at the Natural Resource Ecology Lab at Colorado State University and a life scientist for the Habitat Section, Colorado Division of Wildlife. He teaches a class on ecological modeling and conducts research sponsored by the National Science Founda-

tion, the Environmental Protection Agency, the U.S. Geological Survey, and the National Park Service.

RICHARD S. HOLTHAUSEN is national wildlife ecologist for the U.S. Forest Service and is based at the Rocky Mountain Research Station in Flagstaff, Arizona.

CAROLYN T. HUNSAKER is a certified senior ecologist with the U.S. Forest Service interested in quantification of spatial relationships for both terrestrial and aquatic ecosystems, especially with regard to uncertainty. She is the research coordinator for the Kings River Sustainable Forest Ecosystem Project on the Sierra National Forest in California.

MICHAEL A. HUSTON is a senior scientist in the Environmental Sciences Division at Oak Ridge National Laboratory and author of *Biological Diversity: The Coexistence of Species on Changing Landscapes* (Cambridge University Press 1994).

RICHARD L. HUTTO is a professor in the Division of Biological Sciences at the University of Montana in Missoula. His professional interests include studying the habitat relationships in birds and the ecology of Neotropical migrants, and interpreting ornithology for the lay public.

MARK E. JAKUBAUSKAS is a research assistant professor with the Kansas Biological Survey at the University of Kansas in Lawrence.

FRANCES C. JAMES is an ecologist at Florida State University, where she holds the Pasquale Graziadei Chair in the Department of Biological Science. Her research interests are centered on quantifying the relationships between species and their habitats and on avian population ecology.

JONATHAN A. JENKS is professor of wildlife and fisheries sciences at South Dakota State University in Brookings, South Dakota, and principle investigator of the South Dakota Gap Analysis Project.

CATHERINE M. JOHNSON is a wildlife ecologist with the Natural Resources Research Institute at the University of Minnesota in Duluth. Her primary interests are in conservation biology and habitat analysis, with recent research focusing on the incorporation of multiple spatial and temporal scales in habitat modeling for coastal and freshwater wetland-dependent species.

DOUGLAS H. JOHNSON is a statistician and Leader of the Grasslands Ecosystem Initiative at the USGS Northern Prairie Wildlife Research Center in Jamestown, North Dakota.

KRISTINE JOHNSON is a vegetation management specialist with the National Park Service in Great Smoky Mountains National Park. Her expertise is in exotic plant species control and insect and disease monitoring.

LUCINDA B. JOHNSON is the associate director of the Center for Water and the Environment at the University of Minnesota's Natural Resources Research Institute. Her primary research interests lie in quantifying the biological and physical interactions between terrestrial and aquatic ecosystems.

MALCOLM T. JONES is currently a research associate with the Natural Resources Research Institute at the University of Minnesota-Duluth. Professional areas of expertise include avian, landscape, and quantitative ecology. His current research focuses on modeling avian response to forest management and integrating model results into a forest-planning tool for natural resource managers.

JASON W. KARL is a GIS analyst with the Pacific Biodiversity Institute. His research interests are directed at better understanding the relationships between species and their habitats and the sampling protocols and sample sizes needed to test the accuracy of model outputs.

CHIRRE KECKLER is a forest wildlife biologist for the U.S. Forest Service on the Mendocino National Forest, headquartered in Willows, California. She is responsible for the overall coordination of the wildlife, plant, and fishery programs on the Forest.

TODD KEELER-WOLF is senior vegetation ecologist with the California Department of Fish and Game. He is coauthor of the *Manual of California Vegetation*, published by the California Native Plant Society.

A. ROSS KIESTER is a research biologist with the U.S. Forest Service. His current research interests include the ecology of turtles and the identification and selection of nature reserves.

KELLY KINDSCHER is a plant community ecologist at the Kansas Biological Survey and a courtesy associate professor in environmental studies at the University of Kansas where he teaches courses on environmental impact assessment and ethnobotany.

DAVID S. KLUTE is currently a postdoctoral research associate in the School of Forest Resources at Pennsylvania State University. His current research interests include the development of multiscale habitat models and assessing methods for increasing sampling efficiency of avian communities.

DANIEL A. KLUZA is a doctoral candidate in the Department of Ecology and Evolutionary Biology at the University of Kansas. He has been studying the effects of habitat disturbance and fragmentation on the tropical dry forest bird assemblages of western Mexico. His research is focused on how landscape composition and structure influence the distribution and abundance of birds.

STEVEN T. KNICK is a research ecologist and former station leader at the U.S. Geological Survey's Forest and Rangeland Ecosystem Science Center, Snake River Field Station, in Boise, Idaho. He is interested in the role of disturbance in shaping ecological systems. Most of his work has been conducted in shrubsteppe habitats of the Intermountain West.

WILLIAM B. KROHN is leader of the Maine Cooperative Fish and Wildlife Research Unit, U.S. Geological Survey's Biological Resources Division, housed at the University of Maine. His research interests include the habitat ecology of forest carnivores, small-scale species/habitat relationships, habitat assessments, and land conservation planning.

KIRK L. KRUEGER is a graduate student in the Department of Fisheries and Wildlife Sciences, Virginia Polytechnic Institute and State University, in Blacksburg, Virginia

RONALD F. LABISKY is a professor in the Department of Wildlife Ecology and Conservation at the University of Florida. He specializes in population biology.

WILLIAM A. LINK is a mathematical statistician with the USGS Patuxent Wildlife Research Center in Laurel, Maryland. His research interests include estimation of population change from count data, Bayes and empirical Bayes methods, and the development of statistical methods for the analysis of ecological problems.

MATTHEW J. LOVALLO is currently employed as a furbearer biologist with the Pennsylvania Game Commission. His current research focuses on the development of methods to assess density and distribution of mammalian predators.

THOMAS LUPO was a GIS specialist with the California Department of Fish and Game in Sacramento, California, at the time of this study. His work focused on wildlife management applications such as mapping species distributions and observations, wildlife habitat suitability modeling, and mapping hunting regulatory information. He now works as the chief of the Department's Wildlife and Habitat Data Analysis Branch in Sacramento, California.

JEFFREY J. LUSK is a doctoral student in the Department of Forestry at Oklahoma State University. He is investigating the effects of climate on bobwhites in Oklahoma and Texas.

BRIAN A. MAURER is an associate professor in the Department of Fisheries and Wildlife at Michigan State University. He specializes in vertebrate population and community ecology, with emphasis on processes occurring at different scales.

MICK A. MCCARTHY is a lecturer in environmental science at the University of Melbourne in Melbourne, Australia.

CHARLES E. MCCULLOCH is a professor and head of the Division of Biostatistics, University of California, San Francisco.

Previously he was professor and founding chair of the Department of Statistical Science at Cornell University. He conducts primary research on the development of statistical methodology, especially in the areas of generalized linear mixed models and longitudinal and repeated measures data analysis.

DANIEL W. MCKENNEY is the chief of the Landscape Analysis and Applications Section, Canadian Forest Service, in Sault Ste. Marie, Ontario, Canada.

JAMES W. MERCHANT is a professor in the School of Natural Resource Sciences and associate director of the Center for Advanced Land Management Information Technologies (CALMIT), Institute of Agriculture and Natural Resources, at the University of Nebraska in Lincoln. Dr. Merchant has been engaged in basic and applied research in remote sensing and GIS since 1971. He is principal investigator with the Nebraska Gap Analysis Program.

SHERRI L. MILLER is a wildlife biologist at the Redwood Sciences Laboratory, a unit of the U.S. Forest Service, at the Pacific Southwest Research Station. Ms. Miller has served as collaborator for research on population size and trends and habitat relationships for marine and landbird species.

RANDY MOLINA is forest mycology team leader with the U.S. Forest Service, Pacific Northwest Research Station, in Corvallis, Oregon. The Forest Mycology Team researches fungal ecology and the biological and functional diversity of fungi. This information is used to develop strategies for conservation and management of forest fungi in the Pacific Northwest.

CLINTON T. MOORE, a statistician with the U.S. Geological Survey, Patuxent Wildlife Research Center, is a doctoral candidate in forest resources at the University of Georgia. He previously served as a wildlife biometrician for the U.S. Fish and Wildlife Service and for the Florida Fish and Wildlife Conservation Commission.

MICHAEL L. MORRISON is manager of the White Mountain Research Station, University of California, Bishop. His research interests include assessment of resource use by vertebrates, especially as it relates to habitat assessment and restoration.

BARRY S. MULDER is a wildlife biologist with the U.S. Fish and Wildlife Service in Portland, Oregon. He is currently the acting state supervisor of the Service's Ecological Services program in Oregon. Since 1987, he has served as the lead coordinator for the Service's efforts to manage the northern spotted owl and related forest ecosystem activities in the Pacific Northwest.

DENNIS D. MURPHY is a research professor at the University of Nevada, Reno and a Pew Scholar in Conservation and the Environment. Dr. Murphy's research focuses on the biology of butterflies and conservation planning under the federal Endangered Species Act. He currently serves as team leader of the Lake Tahoe Watershed Assessment and as science advisor to the CALFED Bay-Delta environmental restoration program for the Sacramento and San Joaquin Rivers.

GERALD J. NIEMI is director of the Center for Water and the Environment, Natural Resources Research Institute, and professor in the Department of Biology at the University of Minnesota. His interests are birds, biostatistics, natural systems, Great Lakes issues, and the sustainability of natural resources.

BARRY NOON is a professor in the Department of Fisheries and Wildlife Biology at Colorado State University in Fort Collins. His research interests are in avian ecology, application of science to land management planning, and modeling of species occurrences in varying habitats and management scenarios.

M. PHILIP NOTT is a landscape ecologist at The Institute for Bird Populations at Point Reyes Station, California. His research interests include spatially explicit population models, species response to habitat and landscape patterns, abiotic influences on population dynamics, and monitoring, modeling, and management.

RAYMOND J. O'CONNOR is a professor of wildlife ecology at the University of Maine and was previously (1978–1987) director of the British Trust for Ornithology. His research has focused on bird population dynamics in relation to climate and habitat, particularly on farmland, and in modeling human dimensions of the environment. He teaches quantitative and statistical ecology.

JENNIE L. PEARCE is a visiting fellow in the Landscape Analysis and Applications Section of the Canadian Forest Service in Sault Ste. Marie, Ontario.

SCOTT M. PEARSON is associate professor of biology at Mars Hill College. He studies the influence of landscape-level habitat heterogeneity on community composition and the persistence of native species. Recent work includes the effect of forest fragmentation and historical land uses on the composition of forest communities in the southern Appalachian Mountains.

A. TOWNSEND PETERSON is a curator in ornithology at the Natural History Museum and Biodiversity Research Center at the University of Kansas in Lawrence. His research interests are in biodiversity of Neotropical birds; diversification, evolution, and speciation and biogeography.

JAMES T. PETERSON is currently a research biologist and assistant unit leader at the U.S. Geological Survey's Georgia Cooperative Fish and Wildlife Research Unit in Athens,

Georgia. He has conducted stream fish research with several agencies throughout North America and is coauthor of the statistical package CATDAT. His current research focuses on stream fish sampling, modeling, and developing decision support tools for natural resource management.

RICHARD E. PLANT is a professor at the University of California, Davis, in the Departments of Agronomy and Range Science, and Biological and Agricultural Engineering.

KENNETH M. PORTIER is an associate professor in the Department of Statistics at the University of Florida. He specializes in statistical research relating to environmental and ecological issues.

C. JOHN RALPH conducted early research on bird migration and orientation. He was co-founder (with L. R. Mewaldt) of the Point Reyes Bird Observatory and served as its director for a period. After teaching at Dickinson College in Pennsylvania, he moved to Hawaii with the U.S. Forest Service's research branch, investigating forest birds. Since 1981 he has been involved with landbird and seabird research, primarily in the Pacific Northwest.

MARTIN G. RAPHAEL is a chief research wildlife biologist and team leader with the U.S. Forest Service Pacific Northwest Research Station. His research interests are in wildlife habitat relationships of forest vertebrates, application of principles of landscape ecology to large-scale forest management problems, and application of science to land management planning.

PETER H. RAVEN is director of the Missouri Botanical Garden in St. Louis. As a botanist, he has many diverse research interests but focuses primarily on mapping worldwide biodiversity and on studying the natural history and conservation of plants.

RONALD R. REGAL is a professor in the Department of Mathematics and Statistics at the University of Minnesota in Duluth. His areas of expertise are experimental design, statistical modeling, and data analysis in environmental and medical sciences applications.

WILLIAM A. REINERS has pursued a career in ecosystem ecology with an emphasis on biogeochemistry at the University of Minnesota, Dartmouth College, and most recently the University of Wyoming. While at the University of Wyoming his interests have focused on the spatial distribution of species, vegetation, and ecosystem processes, and he is involved in the state gap analysis programs for Wyoming and Colorado.

CARL RICHARDS is a professor of biology and director of the Minnesota Sea Grant College Program at the University of Minnesota.

BRUCE E. RIEMAN is currently a fisheries scientist with the U.S. Forest Service Rocky Mountain Research Station in Boise, Idaho. He has worked in both research and management with several agencies throughout the Pacific Northwest. His current research interests are with conservation biology and landscape ecology of stream fishes.

LYNN M. ROBERTS is wildlife biologist with the U.S. Fish and Wildlife Service in Arcata, California. She previously worked as a wildlife biologist for the U.S. Forest Service from 1988 to 1994 and has held her present position with the U.S. Fish and Wildlife Service since 1994. Her interests include developing and implementing large-scale conservation plans for rare or federally listed species both on public and privately owned lands.

MARGARET J. ROBERTSEN is a wildlife biologist with the U.S. Forest Service on the Wrangell Ranger District, Tongass National Forest, in Alaska. Previously, she worked at the U.S. Forest Service's North Central Forest Experiment Station researching habitat relationships of migrant and breeding songbirds in Wisconsin.

GARY J. ROLOFF is a wildlife management specialist with Boise Cascade Corporation. His duties include integrating wildlife into Boise Cascade's forest planning process, assisting forestry operations with wildlife issues, and researching and modeling the effects of resource management on wildlife.

JOHN T. ROTENBERRY is a professor of biology and associate director of the Center for Conservation Biology at the University of California, Riverside. His research concerns population and community ecology of birds, particularly with respect to habitat relationships. Much of his work has been in semiarid shrubland ecosystems in the northern Great Basin and in Southern California, with a recent emphasis on landscape-level effects.

J. ANDREW ROYLE is a statistician with the U.S. Fish and Wildlife Service's Migratory Bird Management Office. His research interests include spatial statistics, Bayesian methods, model-based analysis of count survey data, and models for capture-recapture data.

STEVEN A. SADER is director of the Maine Image Analysis Laboratory and professor of forest resources and forest management at the University of Maine. His research interests include forest resource information systems, remote sensing and GIS applications in natural resources, and land use change analysis.

FRED B. SAMSON is a biologist with the U.S. Forest Service.

GLEN A. SARGEANT is a wildlife biologist and statistician at the USGS Northern Prairie Wildlife Research Center in Jamestown, North Dakota.

JOHN R. SAUER is a research wildlife biologist at the USGS Patuxent Wildlife Research Center. His interests include design and analysis of wildlife surveys, estimation of population and community change from count data, and studies of landscape associations between bird populations and their habitats.

ERIKA H. SAVERAID completed a master's degree in ecology and evolutionary biology at Iowa State University and is currently living in Brunswick, Maine.

SANDRA M. SCHAEFER is currently a GIS analyst with the James W. Sewall Co., Old Town, Maine. She is also completing her master of science degree in wildlife ecology at the University of Maine and enjoys using GIS technology to research topics related to biodiversity and land conservation planning.

SCOTT E. SCHLARBAUM is an associate professor of forest genetics in the Department of Forestry, Wildlife and Fisheries at the University of Tennessee. He is currently the project leader of the University's Tree Improvement Program through which research is conducted on various hardwood and coniferous species using traditional breeding and testing methodologies coupled with recent advances in clonal propagation.

FIONA K. A. SCHMIEGELOW is assistant professor of conservation biology with the Department of Renewable Resources at the University of Alberta in Edmonton, Alberta, Canada.

J. MICHAEL SCOTT is a senior research biologist and unit leader with the U.S. Geological Survey Cooperative Fish and Wildlife Research Unit and professor in the Department of Fish and Wildlife at the University of Idaho in Moscow. His research interests include modeling of species occurrences, conservation biology, estimation of animal numbers, and reserve identification selection and design.

ANN-MARIE SHAPIRO is a research ecologist at the U.S. Army Construction Engineering Research Laboratory in Champaign, Illinois. Her research efforts have included development of ecological community-based management recommendations for military training activities, development and testing of spatially explicit models, and development of guidelines for estimating species population goals for military installations.

SUSAN A. SHRINER is a graduate student in the Department of Zoology at North Carolina State University. Her research interests are currently focused on the ecology of forest songbirds.

THEODORE R. SIMONS is a research biologist and assistant unit leader with the U.S. Geological Survey's Cooperative Fish and Wildlife Research Unit, and associate professor of zoology at North Carolina State University. His research addresses problems in wildlife conservation and the development of multidisciplinary approaches to the conservation of protected areas and rare and endangered species. Recent studies include the breeding birds and salamanders of Great Smoky Mountains National Park and the stopover ecology of migrants birds in the southern United States.

THOMAS D. SISK is an ecologist with the Center for Environmental Sciences and Education at Northern Arizona University. He teaches courses in ecology, conservation biology, and environmental policy, and oversees a research group studying habitat fragmentation and restoration, livestock grazing, and long-term changes in land use and land cover.

PER SJÖGREN-GULVE is associate professor in conservation biology at Uppsala University in Sweden and principal research officer at the Swedish Environmental Protection Agency. He specializes in metapopulation biology with particular interest in applied population modeling, population ecology, and population genetics.

K. SHAWN SMALLWOOD is an independent consultant and researcher, a part-time faculty member of the Department of Biology at California State University, Sacramento, and a member of Consulting in the Public Interest (www.cipi. com). His research interests include animal density and habitat analysis, but his work is focused on promoting effective conservation methods and policy.

JANE E. SMITH is a research botanist with the U.S. Forest Service at the Pacific Northwest Research Station in Corvallis, Oregon. She is a member of the Forest Mycology Team, which researches fungal ecology and the biological and functional diversity of fungi and develops strategies for conservation and management of forest fungi in the Pacific Northwest.

VICKIE J. SMITH is a gap analysis specialist at South Dakota State University in Brookings, South Dakota. As state coordinator for the South Dakota Gap Analysis Project, she is responsible for vertebrate modeling.

DAVID SOLIS is a senior staff biologist for the U.S. Fish and Wildlife Service at the Arcata Fish and Wildlife Office in Arcata, California, a position he has filled for the past three years. He supervises staff overseeing implementation of various sections of the federal Endangered Species Act. Previously, David was a wildlife biologist for sixteen years with the U.S. Forest Service.

DEAN F. STAUFFER is an associate professor of wildlife science in the College of Natural Resources at Virginia Polytechnic Institute and State University (Virginia Tech). His primary interests are in analysis of wildlife habitat relationships. Current work includes modeling ruffed grouse populations and habitat use, and integrating GIS and wildlife habitat models on military lands.

HOWARD B. STAUFFER is a professor of applied statistics at Humboldt State University and statistician for the U.S. Forest Service Redwood Sciences Laboratory. His research and applied interests include sampling design and analysis, experimental design and analysis, model selection and inference, Bayesian statistics, and mixed- and random-effects models, with applications to forestry and wildlife management.

GEORGE N. STEGER is a biological technician with the U.S. Forest Service and has been involved in a variety of research studies involving radiotelemetry, aerial photograph interpretation, and demographic studies of wildlife. He has thirteen years of experience researching the California spotted owl and currently supervises two demographic studies in the southern Sierra Nevada.

DAVID R. B. STOCKWELL is an assistant research scientist at the San Diego Supercomputer Center. His research interests include the application of machine learning to modeling biodiversity and the construction of the environmental informatics infrastructure.

LEONA K. SVANCARA is a senior GIS analyst and lab manager for the Landscape Dynamics Lab, Idaho Cooperative Fish and Wildlife Research Unit, at the University of Idaho, Moscow, Idaho.

GLENN TAYLOR is a forestry technician with the National Park Service in Great Smoky Mountains National Park. His duties in the Smokies include long-term forest insect and disease monitoring. He is a graduate of Pennsylvania State University and holds a bachelor's degree in environmental resource management.

STANLEY A. TEMPLE is a Beers-Bascom Professor in Conservation in the Department of Wildlife Ecology at the University of Wisconsin in Madison. He is past president of the Society for Conservation Biology and currently chairs the University of Wisconsin's graduate program in conservation biology and sustainable development.

DAVID M. THEOBALD is a research scientist at the Natural Resource Ecology Lab at Colorado State University and a fellow with The Nature Conservancy's David H. Smith Conservation Research Program. He develops analytical techniques to improve biodiversity conservation.

KATHRYN THOMAS is a vegetation ecologist with the U.S. Geological Survey in Flagstaff, Arizona. She is currently conducting a number of projects researching vegetation distribution and dynamics throughout the arid Southwest.

BRUCE C. THOMPSON is a research biologist and unit leader with the U.S. Geological Survey's New Mexico Cooperative Fish and Wildlife Research Unit, a partnership of the U.S. Geological Survey, New Mexico Department of Game and Fish, New Mexico State University, and the Wildlife Management Institute in Las Cruces, New Mexico.

CLAUDINE TOBALSKE is a GIS specialist for the Oregon Natural Heritage Program in Portland, Oregon.

MARGARET KATHERINE TRANI (GRIEP) is the Regional Wildlife Habitat Relationships Coordinator with the USDA Forest Service, Southern Region. Dr. Trani has extensive experience with regional- and landscape-scale assessments. Her research interests include landscape pattern analysis and model development of species-habitat relationships.

WALTER M. TZILKOWSKI is an associate professor at the Pennsylvania State University where for the past twenty-two years he has had the pleasure of interacting with many excellent undergraduate and graduate students.

BEATRICE VAN HORNE is a professor in the Department of Biology at Colorado State University in Fort Collins.

FRANK T. VAN MANEN is a research ecologist with the U.S. Geological Survey's Southern Appalachian Field Laboratory located at the University of Tennessee. His primary research interest is in landscape ecology as it relates to large carnivores and rare plants and trees.

LISA A. VENIER is a research scientist working in the Landscape Analysis and Applications Section at the Great Lakes Forest Research Centre, Sault Ste. Marie, Ontario, Canada.

PIERRE R. VERNIER is a research scientist with the Centre for Applied Conservation Biology in the Faculty of Forestry at the University of British Columbia in Vancouver, British Columbia, Canada.

WILLIAM A. WALL is a wildlife ecologist and manager and currently Senior Scientist for Wildlife Conservation with Safari Club International Foundation. He directs research and management for sustainable use wildlife programs in Latin America and Asia. He was formerly leader of research and development of an integrated landscape management planning program for Potlatch Corporation in Idaho.

JAMES WESTERVELT is a Visiting Research Associate in the Department of Agricultural and Consumer Economics at the University of Illinois at Urbana-Champaign. He specializes in the design and development of geographical information systems and spatially explicit simulation modeling.

DENNIS WHITE is a geographer with the U.S. Environmental Protection Agency in Corvallis, Oregon.

JOHN A. WIENS is University Distinguished Professor at Colorado State University and a lead scientist with The Nature Conservancy.

KELLY WOLCOTT is a wildlife biologist with a background in forestry and forest habitat analysis. He currently coordinates leadership programs for the U.S. Fish and Wildlife Service at the National Conservation Training Center in Shepherdstown, West Virginia. During the development phase of the Northern Spotted Owl Baseline for the Northern California Project, he coordinated the efforts of the U.S. Fish and Wildlife Service consultation team on the Shasta-Trinity and Mendocino national forests.

BRIAN WOODBRIDGE is a wildlife biologist at the Klamath National Forest in Yreka, California, working primarily on international forestry and endangered species programs. His research focuses on avian ecology and conservation, migration studies, and tropical ecology.

AMANDA WRIGHT was formerly a senior lecturer in the Department of Environmental and Geographical Sciences at Manchester Metropolitan University in Manchester, United Kingdom. Presently, she is a senior consultant at Eunite Ltd, a media technology company.

ADRIENNE WRIGHT has been a GIS analyst for the Redwood Sciences Laboratory since 1994, working with scientists on wildlife habitat analyses. She is currently involved in northern spotted owl habitat research in Northern California.

NANCY M. WRIGHT is a GIS analyst for the Marine Region of the California Department of Fish and Game. Her work focuses on mapping nearshore subtidal habitats.

JOCK S. YOUNG is a research assistant in the Division of Biological Sciences at the University of Montana, where he manages and analyzes data for the U.S. Forest Service Northern Region Landbird Monitoring Program. His main research interests are wildlife habitat relationships and conservation biology.

CYNTHIA J. ZABEL is a research wildlife biologist with the U.S. Forest Service at the Redwood Sciences Laboratory in Arcata, California. She has conducted studies on red foxes, walrus, and spotted hyenas. More recently her research has focused on vertebrate habitat relations, with species such as spotted owls, martens, flying squirrels, and bats.

FRIDOLIN ZIMMERMANN is a research ecologist at KORA (coordinated research projects for the conservation and management of carnivores in Switzerland). He is working on his doctoral thesis at the University of Lausanne developing a spatially explicit population model for the lynx in the Alps. His research interests are home range analyses, habitat models, dispersal, and population dynamics.

Index